Foundations of MIMO Communication

Understand the fundamentals of wireless and MIMO communication with this accessible and comprehensive text. Viewing the subject through an information theory lens, but also drawing on other perspectives, it provides a sound treatment of the key concepts underpinning contemporary wireless communication and MIMO, all the way to massive MIMO. Authoritative and insightful, it includes over 330 worked examples and 450 homework problems, with solutions and MATLAB® code and data available online. Altogether, this is an excellent resource for instructors and graduate students, and a great reference for researchers and practicing engineers.

Robert W. Heath Jr. is a Cullen Trust for Higher Education Endowed Professor in the Department of Electrical and Computer Engineering at the University of Texas at Austin. He is a Fellow of the IEEE.

Angel Lozano is a Professor in the Department of Information and Communication Technologies at Universitat Pompeu Fabra, Barcelona. He is a Fellow of the IEEE.

"*Foundations of MIMO Communication* by Heath and Lozano is a much needed tour-de-force on a critically important topic for wireless system design. The book provides an in-depth comprehensive treatment of the theoretical fundamentals in single-user, multiuser, and massive MIMO communications. It covers information-theoretic performance limits as well as estimation, channel modeling, matrix theory, and optimization techniques that are key to many aspects of MIMO system design. Hundreds of worked examples drive home the value of the theory in solving practical problems, and over one thousand important references are provided for further reading. This masterfully written book will serve as an invaluable textbook for students as well as a definitive reference for researchers and practicing engineers."

Andrea Goldsmith, *Stanford University*

"This masterpiece from two preeminent scholars and innovators was worth the wait: this book took over a decade to write. Heath and Lozano are renowned information and communication theorists who have made many pioneering contributions to the theory of MIMO. However, they are unusual in also having done much to bring MIMO into practice: Heath at a pioneering Silicon Valley startup and Lozano at Bell Labs. They hold dozens of fundamental patents on MIMO and their broad perspective gives this book a unique depth and relevance. The book is well-suited for teaching or self study, and stands apart from others in the beauty and completeness of the exposition."

Jeffrey Andrews, *The University of Texas at Austin*

"Remarkably, and almost three decades after its inception, MIMO technology is holding on to its status as one of the most impactful and versatile technologies, likely to underpin the design radio access networks for years to come still. With this highly complete textbook, Heath and Lozano are sharing their deep expertise and long-term experience in an area that has seen so many successes and revolutions. This material will be invaluable to teachers, students, and practitioners alike."

David Gesbert, *Eurecom*

"MIMO and Massive MIMO are key technologies for enhancing the capacity and coverage of 3G/4G/5G and Beyond 5G wireless systems across all frequency bands. Heath and Lozano provide a comprehensive treatment of this essential technology from both a theoretical and practical point of view. An excellent and must read for graduate students and researchers in the field of wireless communication."

Amitava Ghosh, *Nokia Bell Labs*

Foundations of MIMO Communication

ROBERT W. HEATH JR.
University of Texas at Austin

ANGEL LOZANO
Universitat Pompeu Fabra

CAMBRIDGE
UNIVERSITY PRESS

CAMBRIDGE
UNIVERSITY PRESS

Shaftesbury Road, Cambridge CB2 8EA, United Kingdom

One Liberty Plaza, 20th Floor, New York, NY 10006, USA

477 Williamstown Road, Port Melbourne, VIC 3207, Australia

314–321, 3rd Floor, Plot 3, Splendor Forum, Jasola District Centre, New Delhi – 110025, India

103 Penang Road, #05–06/07, Visioncrest Commercial, Singapore 238467

Cambridge University Press is part of Cambridge University Press & Assessment,
a department of the University of Cambridge.

We share the University's mission to contribute to society through the pursuit of
education, learning and research at the highest international levels of excellence.

www.cambridge.org
Information on this title: www.cambridge.org/9780521762281

DOI: 10.1017/9781139049276

First published 2019
Reprinted 2019

A catalogue record for this publication is available from the British Library

Library of Congress Cataloging-in-Publication data
Names: Heath, Robert W., Jr, 1973– author. | Lozano, Angel, 1968– author.
Title: Foundations of MIMO communication / Robert W. Heath Jr.,
University of Texas, Austin, Angel Lozano, Universitat Pompeu Fabra.
Description: Cambridge, United Kingdom ; New York, NY, USA :
Cambridge University Press, 2019.
Identifiers: LCCN 2018035156 | ISBN 9780521762281 (hardback : alk. paper)
Subjects: LCSH: MIMO systems.
Classification: LCC TK5103.4836 .H43 2019 | DDC 621.384–dc23
LC record available at https://lccn.loc.gov/2018035156

ISBN 978-0-521-76228-1 Hardback

Additional resources for this publication at www.cambridge.org/heathlozano

To Pia and Rohan, for putting up with their busy Papa, and to Nuria.

R. W. H.

For Ester, Carla and Bernat, beams of light in my life. And for my parents and siblings.

A. L.

Contents

Preface page xv
Acknowledgments xvii
A brief historical account xix
About this book xxii

Part I Wireless Communication Theory 1

1 A primer on information theory and MMSE estimation 3
1.1 Introduction 3
1.2 Signal distributions 4
1.3 Information content 6
1.3.1 Entropy 7
1.3.2 Differential entropy 9
1.3.3 Entropy rate 10
1.4 Information dependence 11
1.4.1 Relative entropy 11
1.4.2 Mutual information 12
1.5 Reliable communication 17
1.5.1 Information-theoretic abstraction 17
1.5.2 Capacity 20
1.5.3 Coding and decoding 22
1.5.4 Bit-interleaved coded modulation 30
1.5.5 Finite-length codewords 35
1.5.6 Hybrid-ARQ 36
1.5.7 Extension to MIMO 37
1.6 MMSE estimation 39
1.6.1 The conditional-mean estimator 40
1.6.2 MMSE estimation in Gaussian noise 41
1.6.3 The I-MMSE relationship in Gaussian noise 44
1.7 LMMSE estimation 47
1.7.1 Random variables 48
1.7.2 Random processes 50
1.8 Summary 51
Problems 51

2 A signal processing perspective 57
2.1 Introduction 57
2.2 Signal, channel, and noise representations 58
2.2.1 Passband signals and complex baseband equivalents 58
2.2.2 Complex baseband channel response 61
2.2.3 Time discretization 64
2.2.4 Pulse shaping 66
2.2.5 Additive noise 71
2.2.6 Energy and power 73
2.2.7 Channel normalization 74
2.2.8 Vector representation 75
2.3 Signal, channel, and noise representations: extension to MIMO 77
2.3.1 Vector and matrix representations 77
2.3.2 Channel normalization 78
2.3.3 Stacked vector representation 79
2.3.4 Precoding 80
2.3.5 Signal constraints 83
2.4 Linear channel equalization 87
2.4.1 Linear ZF equalization 88
2.4.2 LMMSE equalization 94
2.5 Single-carrier frequency-domain equalization 98
2.5.1 Basic formulation 98
2.5.2 Extension to MIMO 102
2.6 OFDM 104
2.6.1 Basic formulation 104
2.6.2 Extension to MIMO 110
2.7 Channel estimation 110
2.7.1 Single-carrier channel estimation 111
2.7.2 OFDM channel estimation 121
2.8 Summary and outlook 122
Problems 124

3 Channel modeling 131
3.1 Introduction 131
3.2 Preliminaries 132
3.2.1 Basics of radio propagation 132
3.2.2 Modeling approaches 133
3.3 Large-scale phenomena 135
3.3.1 Pathloss and shadow fading 135
3.3.2 Free-space model 138
3.3.3 Macrocell models 139
3.3.4 Microcell models 141
3.3.5 Picocell and indoor models 142
3.4 Small-scale fading 143

3.4.1 Multipath propagation 143
3.4.2 Space selectivity 146
3.4.3 Time selectivity 151
3.4.4 Frequency selectivity 158
3.4.5 Time–frequency double selectivity 166
3.5 Interlude: essential notions of antenna arrays 170
3.5.1 Array steering vectors 170
3.5.2 Array factor and beamforming 173
3.6 Modeling of MIMO channels 175
3.6.1 Analytical models 177
3.6.2 Parametric models 188
3.7 Channel estimation revisited 192
3.7.1 Large-scale phenomena 193
3.7.2 Small-scale fading 195
3.8 MIMO channel models in standards 199
3.8.1 3GPP spatial channel model 200
3.8.2 SUI models for IEEE 802.16 201
3.8.3 IEEE 802.11 channel model 201
3.9 Summary and outlook 202
Problems 205

4 Single-user SISO 209
4.1 Introduction 209
4.2 Interplay of bit rate, power, and bandwidth 209
4.2.1 Low-SNR regime 215
4.2.2 High-SNR regime 218
4.3 AWGN channel 220
4.3.1 Capacity 220
4.3.2 Discrete constellations 224
4.3.3 Sneak preview of link adaptation 225
4.4 Frequency-selective channel 228
4.4.1 Partition into parallel subchannels 231
4.4.2 Waterfilling power allocation 233
4.4.3 Capacity 235
4.4.4 Discrete constellations 238
4.4.5 CESM, MIESM, and EESM mapping methods 241
4.5 Frequency-flat fading channel 244
4.5.1 CSIR and CSIT 245
4.5.2 No CSIT 249
4.5.3 No CSI 261
4.6 Frequency-selective fading channel 267
4.7 Which fading setting applies? 268
4.8 Pilot-assisted communication 271
4.8.1 Frequency-flat fading 271

4.8.2 Pilot power boosting 278
4.8.3 Frequency-selective fading 278
4.9 Channels with interference 279
4.10 Summary and outlook 284
Problems 287

Part II Single-user MIMO 295

5 SU-MIMO with optimum receivers 297
5.1 Introduction 297
5.2 Initial considerations 298
5.3 CSIR and CSIT 301
5.3.1 Quasi-static setting 301
5.3.2 Ergodic setting 307
5.4 No CSIT 311
5.4.1 Quasi-static setting 311
5.4.2 Ergodic setting 314
5.5 No CSI 338
5.6 Pilot-assisted communication 341
5.7 Channels with interference 345
5.8 Optimum transmitter and receiver structures 348
5.8.1 Single codeword versus multiple codewords 348
5.8.2 LMMSE-SIC receiver 349
5.8.3 The layered architecture 352
5.8.4 BICM implementations 353
5.9 Link adaptation 358
5.9.1 Single codeword 358
5.9.2 Multiple codewords 359
5.10 Reciprocity and CSI feedback 361
5.10.1 Channel reciprocity 362
5.10.2 Analog feedback 363
5.10.3 Digital feedback 364
5.11 Summary and outlook 374
Problems 378

6 SU-MIMO with linear receivers 386
6.1 Introduction 386
6.2 General characteristics of linear MIMO receivers 387
6.3 Linear ZF receiver 388
6.3.1 Receiver structure 388
6.3.2 Output SNR distribution 390
6.3.3 Ergodic spectral efficiency 392
6.4 LMMSE receiver 396

6.4.1 Receiver structure 396
6.4.2 Output SINR distribution 398
6.4.3 Ergodic spectral efficiency 403
6.5 Relationship between the LMMSE and the optimum receiver 407
6.6 Summary and outlook 408
Problems 410

Part III Multiuser MIMO 413

7 Multiuser communication prelude 415
7.1 Introduction 415
7.2 Spectral efficiency region 416
7.3 Orthogonal channel sharing 418
7.3.1 Time-division 418
7.3.2 Frequency-division 419
7.3.3 OFDMA 420
7.4 Non-orthogonal channel sharing 420
7.5 Scalar metrics 422
7.5.1 Sum of the spectral efficiencies 422
7.5.2 Weighted sum of the spectral efficiencies 422
7.5.3 Equal spectral efficiencies 424
7.5.4 Minimum of the spectral efficiencies 424
7.5.5 Proportional fairness 424
7.5.6 Generalized proportional fairness 425
7.6 User selection and resource allocation 426
7.6.1 The proportional-fair algorithm 428
7.7 Low-SNR regime 429
7.8 Summary and outlook 431
Problems 433

8 MU-MIMO with optimum transceivers 436
8.1 Introduction 436
8.2 The multiple-access channel 437
8.3 Multiple-access channel with CSIR and CSIT 440
8.3.1 Quasi-static setting 440
8.3.2 Optimum receiver structure 444
8.3.3 Precoder optimization 447
8.3.4 High-SNR regime 449
8.3.5 Ergodic setting 454
8.4 Multiple-access channel with no CSIT 457
8.4.1 Quasi-static setting 457
8.4.2 Ergodic setting 458
8.5 Multiple-access channel with no CSI 460

8.6 Pilot-assisted multiple-access channel 461
8.7 Duality between the multiple access and broadcast channels 462
8.7.1 Description and significance 462
8.7.2 Dual versus actual multiple-access channels 467
8.8 The broadcast channel 467
8.9 Broadcast channel with CSIR and CSIT 469
8.9.1 Optimum transmitter structure 469
8.9.2 Quasi-static setting 472
8.9.3 Precoder and power allocation optimization 478
8.9.4 High-SNR regime 480
8.9.5 Ergodic setting 484
8.10 Broadcast channel with no CSIT 487
8.11 Summary and outlook 490
Problems 490

9 MU-MIMO with linear transceivers 497
9.1 Introduction 497
9.2 Linear receivers for the multiple-access channel 498
9.3 Linear ZF receiver for the multiple-access channel 500
9.3.1 Receiver structure 500
9.3.2 Output SNR distribution 501
9.3.3 Ergodic spectral efficiency 501
9.3.4 High-SNR regime 505
9.4 LMMSE receiver for the multiple-access channel 507
9.4.1 Receiver structure 507
9.4.2 Output SINR distribution 509
9.4.3 Ergodic spectral efficiency 509
9.4.4 High-SNR regime 513
9.5 Duality with linear transceivers 514
9.6 Linear transmitters for the broadcast channel 516
9.7 Linear ZF transmitter for the MU-MISO broadcast channel 517
9.7.1 Transmitter structure 517
9.7.2 SNR distribution 518
9.7.3 Power allocation 520
9.7.4 Ergodic spectral efficiency 520
9.7.5 High-SNR regime 524
9.7.6 Pilot-assisted ZF transmission 528
9.8 Block-diagonalization for the broadcast channel 554
9.8.1 Transmitter structure 555
9.8.2 Power allocation 556
9.8.3 Ergodic spectral efficiency 558
9.8.4 High-SNR regime 559
9.9 Regularized ZF transmitter for the broadcast channel 562
9.9.1 Regularizing term 563

9.9.2 Power allocation and ergodic spectral efficiency 566
9.9.3 High-SNR regime 571
9.10 Summary and outlook 571
Problems 574

10 Massive MIMO 578
10.1 Introduction 578
10.2 Going massive 579
10.2.1 The massive MIMO regime 579
10.2.2 Excess antennas 581
10.3 Reverse-link channel estimation 582
10.3.1 Pilot reuse 582
10.3.2 Pilot contamination 583
10.4 Reverse-link data transmission 587
10.4.1 Channel hardening 590
10.4.2 Matched-filter receiver 592
10.4.3 LMMSE receiver 607
10.5 Forward-link data transmission 615
10.5.1 Matched-filter transmitter 616
10.5.2 Regularized ZF transmitter 620
10.6 Mitigation of pilot contamination 623
10.6.1 Subspace methods 624
10.6.2 Coordinated pilot assignment 624
10.6.3 Reception and precoding with other-cell awareness 625
10.6.4 Large-scale multicell processing 626
10.7 Practical considerations 628
10.8 Summary and outlook 631
Problems 636

11 Afterword 643
11.1 Beyond cellular 643
11.2 Beyond wireless 644

Appendices 647

Appendix A Transforms 649
A.1 Fourier transforms 649
A.2 Z-transform 651

Appendix B Matrix algebra 653
B.1 Column space, row space, null spaces 653
B.2 Special matrices 654
B.3 Matrix decompositions 656

B.4 Trace and determinant 658
B.5 Frobenius norm 658
B.6 Moore–Penrose pseudoinverse 659
B.7 Matrix inversion lemma 659
B.8 Kronecker product 659

Appendix C Random variables and processes 661
C.1 Random variables 661
C.2 Large random matrices 671
C.3 Random processes 672

Appendix D Gradient operator 674

Appendix E Special functions 676
E.1 Gamma function 676
E.2 Digamma function 677
E.3 Exponential integrals 677
E.4 Bessel functions 678
E.5 Q-function 678
E.6 Hypergeometric functions 679

Appendix F Landau symbols 680

Appendix G Convex optimization 681
G.1 Convex sets 681
G.2 Convex and concave functions 681
G.3 Convex optimization problems 681
G.4 KKT optimality conditions 682
G.5 Lagrange multipliers 683
G.6 Jensen's inequality 684

References 685
Index 752

Preface

The art of wireless communications is arguably one of the biggest technological revolutions in history, and a crowning achievement of modern engineering. Its impact on the functioning of contemporary society cannot be overstated. It seems utterly impossible to conceive today's world without the myriad devices that are wirelessly connected at all times.

In the decades since their inception, cellular systems have undergone five generational transitions. The first generation unfolded during the 1980s and offered only analog telephony. The second generation (2G), rolled out in the 1990s and whose dominant standard was the global system for mobile communications (GSM), saw the changeover to digital and the birth of text messaging. The third generation (3G), circa the 2000s, incorporated data and multimedia applications. Subsequently, the fourth generation (4G) brought about, besides faster bit rates and lower latencies, a complete adoption of packet switching as a platform for the mobile internet and the convergence of all standards worldwide into the long-term evolution (LTE) system. Honoring its name, LTE then evolved into the fifth generation (5G), augmented by another radio access interface termed new radio (NR) that allows operating on a much wider range of frequencies.

In the transitional period between 2G and 3G, as the soaring costs of spectrum collided with the pressures to increase bit rates so as to accommodate data and multimedia applications, the interest in radically improving the spectral and power efficiencies became acute. This propelled research initiatives that blossomed into major advances, chief among which stands multiple-input multiple-output (MIMO) communication, the subject of this book. In short, MIMO amounts to the transmission of concurrent signals from multiple antennas at one end of the link, with multiple antennas also at the receiving end.

Although MIMO is sometimes defined as the incorporation of the space domain to the communication process, this is not quite precise. The space domain has been at the crux of wireless systems since their onset, and in fact it is inherent to the concept of a cellular network: a region is tesselated into cells and the time/frequency signaling resources are reused repeatedly over such cells, i.e., they are spatially reused.

A more satisfying definition of MIMO might be the reusing of signaling resources, not over faraway cells, but rather across antennas—typically collocated—belonging to a common transmitter/receiver, such that new signaling dimensions are unlocked as antennas are added. The many advantages that this brings about are spelled out in detail throughout the text. By arranging these spatial dimensions, each seen as a scalar quantity, into a vector, it can be said that MIMO amounts to a vectorization of the transmission. This subsumes specific instances, such as *phased arrays* and antenna *diversity*, which predate the general formulation of MIMO.

Being in space rather than in time/frequency, the signaling dimensions harnessed with

MIMO behave differently, and this book is about how to communicate over them. These new signaling dimensions multiply the signaling opportunities in time/frequency, hence it can be said that MIMO provides additional "bandwidth" without an increase in the actual electromagnetic bandwidth [1].

Acknowledgments

There are a number of people without whose helping hand this project would not have come to fruition.

First of all, we are indebted to Ramya Bhavagatula (Google), Chanbyoung Chae (Yonsei University), Robert Daniels (PHAZR), Antonio Forenza (Artemis Networks), Kaibin Huang (University of Hong Kong), Insoo Hwang (Google), Takao Inoue (National Instruments), and Rahul Vaze (Tata Institute of Fundamental Research), all of whom contributed to preliminary chapter drafts of an earlier book that never made it to press. That work served as an inspiration for this book.

Second, we are grateful to research collaborators who have influenced the contents and their presentation. Chief among these stand Jeffrey Andrews (The University of Texas at Austin), Federico Boccardi (Ofcom), Helmut Bölcskei (ETH Zürich), Iain Collings (Macquarie University), Gerard Foschini (Bell Labs), David Gesbert (Eurecom), Martin Haenggi (University of Notre Dame), Andrea Goldsmith (Stanford University), Nuria González-Prelcic (Universidade de Vigo), Howard Huang (Bell Labs), Nihar Jindal (Google), Geert Leus (TU Delft), David Love (Purdue University), Thomas Marzetta (New York University), Matt McKay (HKUST), Constantinos Papadias (AIT), Steven Peters (Sterne Kessler Goldstein Fox), Farrokh Rashid-Farrokhi, Shilpa Talwar (Intel), Jose Tellado (HP Enterprise), Ted Rappaport (New York University), Dragan Samardzija (Bell Labs), Anthony Soong (Huawei), Antonia Tulino (Bell Labs), Reinaldo Valenzuela (Bell Labs), Sivarama Venkatesan (Bell Labs), Sergio Verdú (Princeton University), and Harish Viswanathan (Bell Labs).

We also want to thank all the students and colleagues who revised portions of the text. At The University of Texas at Austin, this includes Rebal AlJurdi, Ahmed Alkhateeb, Raquel Flores Buckley, Caleb Lo, Preeti Kumari, Khurram U. Mazher, Amine Mezghani, Nitin Jonathan Myers, Sungwoo Park, Andrew Thornburg, Yuyang Wang, Hongxiang Xie, Yi Zhang, Dalin Zhu, as well as the graduate students of EE 381V and EE 381S. At Universitat Pompeu Fabra, in turn, it includes Masoud Attarifar, Geordie George, David Morales-Jimenez, Rasoul Nikbakht, and Ratheesh Mungara, as well as Albert Guillen and Alfonso Martinez, who clarified aspects of BICM at a moment's notice. Further mention must be made of Dmitry Chizhik (Bell Labs), who kindly examined the channel modeling descriptions, of Giovanni Geraci (Nokia), who reviewed the massive MIMO material, of Luca Sanguinetti (Università di Pisa), who helped polish the description of pilot decontamination methods, and of Aldebaro Klautau (Universidade Federal do Pará), who scrutinized the signal processing coverage. Ezio Biglieri (Universitat Pompeu Fabra) provided invaluable assistance with coding aspects, and his endless wisdom on all the facets of technical writing and of the presentation of ideas in general. Markus Rupp (Technische Universität

Wien) provided references and shared his know-how on performance mapping methods. Very special thanks go to Matthew Valenti (West Virginia University), who supplied the turbo coding data from his coded modulation library, and to Jack Winters, who helped us keep the historical account honest.

Last, but not least, we appreciate the support received from the editorial staff at Cambridge University Press over the years. Phil Meyler provided the initial impetus, Elizabeth Horne carried the torch, and Julie Lancashire brought it home. Additional gratitude goes to Annie Toynbee for overseeing the production, to Gary Smith and Susan Parkinson for thoroughly proofreading the manuscript, and to the texline team for their assistance with all TeX-related matters. And, since first impressions matter, we are thankful to Carla Gironella for the picture that graces the cover.

This book would also not have been possible without support from a number of sponsors, including the National Science Foundation, the European Research Council, the Office of Naval Research, the Army Research Labs, AGAUR, DARPA, AT&T Labs, ICREA, Intel, Huawei, Freescale, MINECO, Motorola, Nokia, CommScope, Crown Castle, Mitsubishi Electronic Research Laboratories, and Samsung.

Robert W. Heath Jr.
The University of Texas at Austin
Austin, Texas

Angel Lozano
Universitat Pompeu Fabra
Barcelona

A brief historical account

We usually imagine that invention occurs in a flash, with a eureka moment that leads a lone inventor towards a startling epiphany. In truth, large leaps forward in technology rarely have a precise point of origin. At the start, forces that precede an invention merely begin to align, often imperceptibly, as a group of people and ideas converge, until over the course of months or years they gain clarity and momentum and the help of additional ideas and actors. Luck seems to matter, and so does timing, for it tends to be the case that the right answers, the right people, the right place—perhaps all three—require a serendipitous encounter with the right problem. And then, sometimes, a leap. Only in retrospect do such leaps look obvious.

Jon Gertner
The idea factory: Bell Labs and the great age of American innovation
Penguin Press, 2012.

MIMO builds on a long pedigree of the application of antenna arrays to communication, starting with Marconi's transatlantic experiment in 1901 [2]. While a comprehensive formulation and a broad understanding did arguably not materialize until the 1990s and 2000s, many of the constituent pieces were present in seemingly unconnected developments well before that. Let us briefly recount these various strands and how they came together.

A first thread relates to phased arrays, a subject of interest since the 1950s [3]. By adjusting the signal's phase at each antenna, either before transmission or upon reception, the overall radiation pattern of an array can be shaped into a beam and pointed in a desired direction, say toward a dominant propagation path. If, besides the phase, the amplitude can also be adjusted, then we have an *adaptive array*. In the 1980s, adaptive arrays evolved into smart antennas, whose array patterns can further null-out interference [4]. All of these are effectively spatial filters that rely on signal coherence across the arrays, an aspect discussed at length throughout the book and that has implications for the physical structure of those arrays and for the radio propagation. Precisely, this coherence is associated with specular propagation and with cleanly defined directions of arrival and departure.

In a second thread, arrays had long been a source of diversity. As detailed later, when multiple propagation paths exist, a signal conveyed over a wireless channel exhibits severe fluctuations, a deleterious phenomenon that can be mitigated by procuring independently fluctuating copies of that signal. An array can serve this purpose, as in such propagation conditions the antennas become uncorrelated—an opposite situation to that of adaptive arrays. Receive diversity was studied as early as 1954 [5] and implemented in early cellular generations already. Being more involved, transmit diversity is more recent [6–10].

Interestingly, diversity was shown to be compatible with the nulling-out of interference, provided the channel response was known [11].

A third thread is constituted by a variety of works that considered channels coupling multiple inputs into multiple outputs. These ranged from abstract formulations [12–14] to analyses motivated by the specific problem of communicating over coupled telephone lines [15–18]. These important precursors were the seed of MIMO, and also of multiuser detection for code-division multiple access (CDMA) [19, 20], two developments with substantial common ground. In drastic contrast with adaptive arrays and diversity, which in essence involve a single signal, these developments entail the concurrent transmission of multiple signals.

Particularly prescient was the contribution of Noach Amitay and Jack Salz, who, as early as 1984, considered a link with two orthogonally polarized antennas at each end [21]. This was extended by Jack Winters in a piece that featured multiantenna transmitters and receivers with many of the ingredients of contemporary MIMO communication: concurrent signals were transmitted from antennas collocated on a device—in time this would be known as *spatial multiplexing*—and a receiver equipped with multiple antennas recovered each of those signals by nulling-out the interference from the rest [22]. In hindsight, both of these milestone papers deserved more credit than they were given, but at the time there was no demand for enhanced wireless performance; in fact, funding for such research was being curtailed.

It was not until the 1990s that the atmosphere was primed for these ideas to coalesce. Arguably, the main catalyst was the work of Gerhard Foschini and Michael Gans, who set out to design the perfect antenna from an information-theoretic standpoint. Starting with an array and no preset conditions on how to use it, they found that, if the antennas were uncorrelated, the optimum strategy was to have each one radiate an independent signal, i.e., spatial multiplexing. This strategy was radically novel in that it sought to exploit, rather than counter or avoid, the fluctuations caused by multipath propagation. Foschini went further and proposed an architecture to effect spatial multiplexing, the so-called *layered architecture*, which was remarkable in that it could be built with off-the-shelf encoders and decoders and did not require the transmitter to know the channel's response [23]. Under the leadership of Reinaldo Valenzuela, a prototype with 12 transmit and 16 receive antennas confirmed the practicality of this proposition [24] and dispelled concerns about the feasibility of nulling-out interference from collocated antennas, i.e., concerns that in practice these signals would drown each other and be unrecoverable. Additional results by Emre Telatar consolidated the initial theoretical underpinnings of MIMO [25]. It behoves us to mention that Amitay, Salz, Winters, Foschini, Gans, Valenzuela, and Telatar were all associated with Bell Laboratories, placing this organization in a distinguished place in the history of MIMO. Angel Lozano's views on MIMO were strongly influenced by his interaction with these pioneers.

Another noteworthy advance in the 1990s was from Paulraj and Kailath, at Stanford University, who applied spatial multiplexing to TV distribution from multiple towers [26]. That idea became a core part of Iospan Wireless, an early developer of MIMO technology for fixed wireless access. Robert Heath's perspectives on MIMO were shaped by his role as an early employee of Iospan Wireless.

Also at Stanford, in yet another development in the 1990s, Raleigh and Cioffi provided a generalization to multiantenna channels of a technique (expounded in Section 4.4) that does require the transmitter to know the channel response [27, 28].

Altogether, after a slow ripening of ideas over an extensive period, a critical mass was reached. Perhaps because of the impetus provided by the soaring costs of spectrum or perhaps because, simply, the time was right, a chain reaction was sparked and spread rapidly through academia and industry. From an academic thought experiment, MIMO grew into a foundation of wireless communication standards. In the span of a few years, it was adopted by the IEEE 802.16 fixed wireless access system and by cellular standardization bodies. In particular, the 3G Partnership Project (3GPP) embraced it in a limited fashion for 3G and then as an integral part of the designs beginning with 4G LTE. In subsequent revisions of LTE, the numbers of supported antennas progressively increased and, for 5G, so-called *massive MIMO* deployments are intrinsic. However, the first commercial application of MIMO was not in the cellular arena.

In 1997, the IEEE 802.11 working group completed its original wireless local-area network (WLAN) standard. Successive revisions incorporated new features, with the earliest version to include MIMO being IEEE 802.11n, certified in 2007 and supporting four antennas (although two or three were more common at that time). IEEE 802.11n was succeeded by IEEE 802.11ac, supporting up to eight antennas, and further by IEEE 802.11ad, 802.11ax, and 802.11ay.

Cohorts of researchers have played a part in the advancement of MIMO, and it is nontrivial to apportion credit among individuals and institutions. A reasonable criterion to assign credit for any invention is the influence that each contribution ends up having [29], and this is the criterion applied in our exposition.

About this book

A wealth of texts on MIMO are available, including books with varying flavors and perspectives [30–40], edited books [41–43], monographs [44–46], and tutorial papers [47–49]. In addition, treatments of MIMO are included in general wireless communication books [50, 51]. The present volume, which builds on this diverse literature, is the result of 20 years of research and teaching. It is intended as a full-dress textbook for instructors and students at the graduate level, and a reference tool for researchers and practicing engineers. With this audience in mind, its aim is to be both accessible and comprehensive, and the conjunction of these objectives is the reason for the considerable length. We hope that readers will appreciate the organization of the contents, as well as the complementary features:

- 160 illustrations.
- 19 topical discussions.
- 339 examples. Some of these are titled examples, which encapsulate results of particular interest, making it easier for them to be identified and located. The rest of the examples are in the format of a problem accompanied by the corresponding solution.
- 463 homework problems proposed at the ends of the chapters.
- A companion website with the solutions to all the problems, and additional material.

There are various theoretical lenses under which MIMO, and in fact digital communication at large, can be seen. Here, the problem of MIMO communication is viewed through the lens of information theory, appropriately complemented with signal processing, estimation theory, channel modeling, optimization, linear algebra, and random matrix theory, and with a touch of stochastic geometry. The choice of information theory is not capricious, but rather the appreciation that, besides being digital, modern communications—certainly the forms that MIMO is relevant to—are built from the ground up with coding, operating very close to the fundamental limits that information theory delineates. Moreover, information theory yields surprisingly many design insights and the opportunity for extensive and informative analysis, greatly facilitating the exposition. As eloquently argued by James Massey [52], "information theory is the proper scientific basis for communication."

In opening his treatise on quantum theory, the Nobel laureate Steven Weinberg declared that "there are parts of this book that will bring tears to the eyes of the mathematically inclined reader." While not expecting to bring tears to the eyes of our readers, we also do not intend to sacrifice clarity or understanding at the altar of rigor. Rather, we abide by the principle that rigor should be at the service of the problem at hand, rather than the resolution of the problem being slaved to the formalisms of rigor.

Organization

The book is organized into three parts, plus a set of appendices.

- Part I, labeled *wireless communication theory*, exposes the pillars on which the edifice rests, with Chapters 1–3 successively introducing the perspectives of information theory and estimation, signal processing, and channel modeling. Along the way, the radio channel is in turn interpreted as a random transformation, as a linear impulse response, and as a stochastic process; three interpretations, each one fitting the perspective of its chapter. These perspectives converge in Chapter 4, which deals with non-MIMO communication, crisply presenting all the concepts that are then to be generalized.
- Part II is devoted to single-user MIMO communication, meaning a single transmitter and a single receiver. In this clean setting, we elaborate on how to transmit and receive with multiple antennas, first without receiver restrictions in Chapter 5, and then with a linearity restriction in Chapter 6.
- Although, with orthogonal multiplexing, single-user conditions can be created within a network, the signaling dimensions created by MIMO are best exploited when the multiuser aspects are brought into the picture. Part III deals with these aspects, first by introducing them broadly in Chapter 7, and then by delving into how to transmit to and receive from a plurality of multiantenna users. Again, this is first covered without restrictions, in Chapter 8, and then with a linearity restriction in Chapter 9. Finally, Chapter 10 broadens the scope in two respects: an entire cellular network is considered, and MIMO becomes massive MIMO.
- The appendices provide a compact tour of various mathematical results that are invoked throughout the book, conveniently couched in our notation, with the objective of rendering the text as self-contained as possible.

Based on the foregoing structure, a variety of itineraries can be defined in support of graduate courses. Some potential ones are as follows.

- A course on the information-theoretic principles of wireless communication, leading up to MIMO, can rest on Chapter 1, Chapter 2 (Sections 2.1, 2.2, and 2.6), and Chapter 4, with the necessary channel modeling taken from Sections 3.1–3.4.
- A basic course on MIMO communication can rely on Chapters 1–6 (with the possible exclusion of Sections 2.4, 2.5, and 2.7).
- An advanced course on multiuser MIMO, which could be concatenated with the one above, can rest on Chapters 7–9, with extension to massive MIMO via Chapter 10.
- A course specifically geared toward massive MIMO can be designed with Chapters 6, 9, and 10, plus whatever material is needed from Chapters 1–4 and 7.

Requisites

The background assumed on the part of the readers corresponds to senior-level or first-year graduate-level courses on signals and systems, digital communication, linear algebra, and probability. A firm grasp of random variables is particularly desirable.

No expertise on cellular networks is presumed, beyond the basic notion that such networks are organized in cells, each featuring a base station that wirelessly communicates with the population of users within. The *forward link* or *downlink* embodies the communication from the base to its users, while the *reverse link* or *uplink* embodies the return communication from the users to the base. This naturally leads to the notion of *duplexing*, which refers to how these two directions of communication are arranged. If they are simultaneous over disjoint frequency bands, we speak of frequency-division duplexing (FDD), while, if they take place on alternating time intervals over a common frequency band, we speak of time-division duplexing (TDD). Full-duplexing, deemed unfeasible in the past, is now becoming possible thanks to advanced self-interference cancelation techniques [53–55], and thus it is also considered.

Among the numerous acronyms that—MIMO aside—sprinkle the text, two stand out and deserve introductory remarks.

- CSI, which stands for channel-state information, alludes to complete knowledge of the channel's response. Since a signature attribute of wireless channels is their variablity, acquiring and employing CSI is instrumental when communicating over such channels. In fact, the availability of CSI is one of the main axes of the exposition in the book.
- OFDM, which stands for orthogonal frequency-division multiplexing, is the signaling technique that has come to dominate communication, wireline and wireless [56, 57]. LTE, NR, and the mentioned WLAN standards all feature OFDM. Although the mainstream alternatives are entertained too, we acknowledge the dominance of OFDM and rely on it extensively for our formulations.

Notation

In a long text such as this one, notation can be a minefield. While striving for consistency and intuition, a modicum of flexibility becomes necessary at points. As part of the effort to convey meaning consistently, some points are worth noting.

- Whenever possible, variables are named to directly reminisce of the quantities they represent, e.g., SNR for the signal-to-noise ratio, MMSE for the minimum mean-square error, or N_{t} and N_{r} for the numbers of transmit and receive antennas.
- Bold symbols denote vectors and matrices, while nonbold symbols correspond to scalars.
- Capitalization distinguishes matrices from vectors, and large-scale from small-scale quantities. (The meaning of these scales is to become clear throughout the text.) Variables that are in general matrices retain their capitalization even in the special cases in which they may adopt a vector form.
- Frequency-domain quantities are represented with sans serif fonts, in contrast with time-domain quantities, which are denoted with serif fonts.
- Dummy variables and counters match, whenever possible, their respective quantities, e.g., we use $\mathbf{A}$ to denote a realization of the random matrix $\boldsymbol{A}$, and we use n to run a counter over N positions.

The common notational schemes and all the relevant symbols, excluding only variables appearing in intermediate derivations, are listed in the pages that follow. The notational schemes are, as much as possible, inspired by MATLAB®. For instance, $[\boldsymbol{A}]_{:,j}$ denotes the jth column of matrix $\boldsymbol{A}$ whereas $[\boldsymbol{A}]_{:,j:j'}$ denotes the submatrix containing columns j through j'.

A number of the variables are further indexed by a user indicator u whenever they are applied in multiuser contexts. For instance, in single-user settings, $\boldsymbol{H}$ is the time-domain normalized channel matrix; then, in multiuser settings, $\boldsymbol{H}_u$ indicates the normalized channel matrix for user u. If, besides multiple users, multiple cells are present, then the indexing is further augmented to identify the cells involved: $\boldsymbol{H}_{l;\ell,u}$ denotes the normalized channel matrix linking the base station at cell l with the uth user at cell ℓth.

Further scripting, not explicitly distinguished in the listings that follow, is applied to discriminate variations in the quantities. For instance, $\boldsymbol{W}$ is a generic linear receiver while $\boldsymbol{W}^{\mathsf{MF}}$, $\boldsymbol{W}^{\mathsf{ZF}}$, $\boldsymbol{W}^{\mathsf{MMSE}}$ are specific types thereof, namely matched-filter, zero-forcing, and minimum mean-square error linear receivers.

Common notation

$\propto$	proportionality
$\approx$	approximation
$\simeq$	asymptotic equality
$\sim$	distribution
$\subseteq$	subset
$((\cdot))_K$	modulo-K
$\otimes$	Kronecker product
$\odot$	Hadamard (entry-wise) product
$*$	convolution
a, A	nonbold letters denote scalars
$\boldsymbol{a}$	bold lowercase letters denote column vectors
$\boldsymbol{A}$	bold uppercase letters denote matrices
$(\cdot)$	indexing for continuous signals
$a(t)$, $\boldsymbol{a}(t)$, $\boldsymbol{A}(t)$	time-domain continuous signals at time t
$\mathsf{a}(f)$, $\mathbf{a}(f)$, $\mathbf{A}(f)$	frequency-domain continuous signals at frequency f
$[\cdot]$	indexing for discrete signals
$a[n]$, $\boldsymbol{a}[n]$, $\boldsymbol{A}[n]$	time-domain discrete signals at time n
$\mathsf{a}[k]$, $\mathbf{a}[k]$, $\mathbf{A}[k]$	frequency-domain discrete signals at frequency k
$\{\cdot\}$	sequence
$[a]^+$	$= \max(0, a)$
$a\|_{\text{dB}}$	$= 10 \log_{10} a$
$a\|_{3\,\text{dB}}$	$= \log_2 a$
$\|a\|$	magnitude of a
$\\|\boldsymbol{a}\\|$	Euclidean norm of $\boldsymbol{a}$
$\\|\boldsymbol{A}\\|_{\text{F}}$	Frobenius norm of $\boldsymbol{A}$
$\nabla_{\boldsymbol{x}}$	gradient with respect to $\boldsymbol{x}$
$\boldsymbol{A}^{\text{T}}$	matrix transpose
$\boldsymbol{A}^{*}$	matrix conjugate transpose
$\boldsymbol{A}^{\text{c}}$	matrix conjugate
$\boldsymbol{A}^{-1}$	matrix inverse
$\boldsymbol{A}^{\dagger}$	Moore–Penrose matrix pseudoinverse
$\boldsymbol{A}^{\star}$	value of $\boldsymbol{A}$ that solves an optimization problem
$\hat{\boldsymbol{A}}$	estimate of $\boldsymbol{A}$
$\tilde{\boldsymbol{A}}$	error in the estimation of $\boldsymbol{A}$
$[\boldsymbol{a}]_j$	jth entry of $\boldsymbol{a}$
$[\boldsymbol{A}]_{i,j}$	(i, j)th entry of $\boldsymbol{A}$
$[\boldsymbol{A}]_{:,j}$	jth column of $\boldsymbol{A}$
$[\boldsymbol{A}]_{:,j:j'}$	submatrix containing columns j through j' of $\boldsymbol{A}$

$[\boldsymbol{A}]_{:,-j}$	submatrix obtained by removing column j from $\boldsymbol{A}$
$\bar{\boldsymbol{A}}_{N,M}$	$N \times M$ matrix containing $\boldsymbol{A}$ at various times, frequencies or antennas
$\dot{f}(\cdot)$	first derivative of $f(\cdot)$
$\ddot{f}(\cdot)$	second derivative of $f(\cdot)$

Symbols

$\mathbf{0}_N$	all-zero matrix (the dimension N may be omitted)
$\mathbf{1}_N$	all-one matrix (the dimension N may be omitted)
$1\{\cdot\}$	indicator function
$A(\cdot)$	array factor
α	common (unprecoded) pilot symbol overhead
α_{d}	dedicated (precoded) pilot symbol overhead
α_{fb}	feedback overhead
$\boldsymbol{a}_{\mathrm{r}}(\theta)$	receive array steering vector for an angle θ
$\boldsymbol{a}_{\mathrm{t}}(\theta)$	transmit array steering vector for an angle θ
$\arg\max$	value that maximizes a function
$\arg\min$	value that minimizes a function
$\mathsf{b}[\mathsf{n}]$	nth bit of a message
$b_\ell[n]$	ℓth bit in the nth symbol of a codeword (n may be omitted)
b	excess bandwidth due to pulse shaping
B	bandwidth
B_{c}	channel coherence bandwidth
β	ratio of transmit-to-receive antenna numbers
$\upbeta$	fudge factor in the CESM, MIESM, and EESM methods
$\mathcal{B}_1^m$	subset of coded bits mapped to the mth constellation point that equal 1
c	speed of light
$c(\tau)$	continuous-time unnormalized pseudo-baseband impulse response
$c_{\mathrm{p}}(\tau)$	continuous-time unnormalized passband impulse response
$c_{\mathrm{b}}(\tau)$	continuous-time unnormalized baseband impulse response
$c_{\mathrm{b}}[\ell]$	discrete-time unnormalized baseband impulse response
$\mathsf{c}(f)$	frequency-domain unnormalized transfer function
$C(\mathsf{SNR}, \boldsymbol{H})$	capacity of a channel $\boldsymbol{H}$ as a function of SNR
$C(\mathsf{SNR})$	ergodic capacity or sum-capacity as a function of SNR
$C_\epsilon(\mathsf{SNR})$	outage capacity at outage level ϵ as a function of SNR
$\mathsf{C}(\frac{E_{\mathrm{b}}}{N_0})$	ergodic capacity as a function of $\frac{E_{\mathrm{b}}}{N_0}$
$\mathcal{C}$	set of cells reusing the same pilots as a cell of interest
$\boldsymbol{C}$	unnormalized channel matrix
χ	shadow fading
χ^2_{2M}	chi-square distribution with $2M$ degrees of freedom
Ξ	cross-polar discrimination
$\mathsf{CM}(\cdot)$	cubic metric

d	diversity order
d_{min}	minimum distance among constellation points
d_{r}	spacing between receive antennas in a ULA array
$d_{\mathrm{r},i,i'}$	spacing between receive antennas i and i'
d_{t}	spacing between transmit antennas in a ULA array
$d_{\mathrm{t},j,j'}$	spacing between transmit antennas j and j'
$\mathsf{d}(\boldsymbol{x},\boldsymbol{y})$	subspace distance between $\boldsymbol{x}$ and $\boldsymbol{y}$
$\mathsf{d}_{\mathsf{min}}(\cdot)$	minimum subspace distance of a codebook
D	distance between transmitter and receiver
D_{c}	coherence distance
D_{ref}	pathloss reference distance
D	distortion in the average SNR
$\mathcal{D}$	set of symbols bearing payload data
$\mathcal{D}(\boldsymbol{x}\|\boldsymbol{y})$	relative entropy (information divergence) between $\boldsymbol{x}$ and $\boldsymbol{y}$
$\delta(\cdot)$	delta function
Δ	equalizer delay
Δ_D	distance shift
Δ_t	time shift
$\det(\boldsymbol{A})$	determinant of $\boldsymbol{A}$
$\mathrm{diag}(\cdot)$	diagonal matrix
$\mathrm{DFT}_N\{\cdot\}$	N-point DFT
E	energy per symbol transmitted to a user
E_{s}	total energy per symbol
$E_{\mathsf{s}}^{\mathsf{r}}$	reverse-link total energy per symbol
E_{b}	energy per bit
$\mathbb{E}[\cdot]$	expectation
$\boldsymbol{E}$	MMSE matrix
$\mathcal{E}_n(\cdot)$	exponential integral of order n
F	frequency share in FDMA
$\mathcal{F}$	digital-feedback precoding codebook
$\boldsymbol{F}$	precoder
$f_{\boldsymbol{A}}(\cdot)$	PDF of $\boldsymbol{A}$
$F_{\boldsymbol{A}}(\cdot)$	CDF of $\boldsymbol{A}$
f_{c}	carrier frequency
$g(\tau)$	delay-domain pulse shape
$g_{\mathsf{rx}}(\tau)$	delay-domain receive pulse shaping filter
$g_{\mathsf{tx}}(\tau)$	delay-domain transmit pulse shaping filter
$\mathsf{g}(f)$	frequency-domain pulse shape
$\mathsf{g}_{\mathsf{rx}}(f)$	frequency-domain receive pulse shaping filter
$\mathsf{g}_{\mathsf{tx}}(f)$	frequency-domain transmit pulse shaping filter
G	large-scale channel gain

G_{r}	receive antenna gain
G_{t}	transmit antenna gain
$\mathrm{G}(N, M)$	Grassmannian manifold of M-dimensional subspaces on the Nth-dimensional space
$\Gamma(\cdot)$	gamma function
$\Gamma(\cdot,\cdot)$	upper incomplete gamma function
$\gamma(\cdot,\cdot)$	lower incomplete gamma function
γ_{EM}	Euler–Mascheroni constant
$h(t,\tau)$	continuous-time normalized impulse response
$\hbar(t,f)$	time-frequency normalized transfer function
$\hbar(\nu,\tau)$	Doppler-delay normalized spreading function
h_{b}	base station height
h_{m}	mobile user height
$\mathfrak{h}(\boldsymbol{x})$	differential entropy of $\boldsymbol{x}$
$\mathcal{H}(\boldsymbol{x})$	entropy of $\boldsymbol{x}$
$\boldsymbol{H}[n]$	discrete-time normalized channel (n may be dropped)
$\boldsymbol{H}_{\mathsf{ind}}$	normalized channel with IND entries
$\boldsymbol{H}_{\mathrm{w}}$	normalized channel with IID complex Gaussian entries
$\boldsymbol{H}_{\mathsf{LOS}}$	normalized LOS channel component
$\boldsymbol{H}_{\mathsf{vir}}$	virtual channel
$\mathsf{H}[k]$	discrete-frequency normalized channel
$\mathsf{\eta}$	reciprocal of the water level in waterfilling
η	pathloss exponent
$i(\boldsymbol{x};\boldsymbol{x})$	information density between $\boldsymbol{x}$ and $\boldsymbol{y}$
$I(\boldsymbol{x};\boldsymbol{y})$	mutual information between $\boldsymbol{x}$ and $\boldsymbol{y}$
$I_n(\cdot)$	modified Bessel function of the first kind and order n
$\Im\{\cdot\}$	imaginary part
$\mathcal{I}(\mathsf{SNR})$	Gaussian-noise mutual information as a function of SNR
$\boldsymbol{I}_N$	identity matrix (the dimension N may be omitted)
$\mathrm{IDFT}_N\{\cdot\}$	N-point inverse DFT
j	imaginary unit
$J_n(\cdot)$	Bessel function of the first kind and order n
k	Boltzmann's constant
$\kappa(\cdot)$	kurtosis
K	number of OFDM subcarriers
$K_n(\cdot)$	modified Bessel function of the second kind
K_{ref}	pathloss intercept
K	Rice factor
L	channel order

L_{c}	length of the cyclic prefix
L_{cluster}	pilot reuse factor
L_{eq}	equalizer order
L_{netw}	number of cells in the network
L_{p}	pathloss
$\mathrm{L}_{\mathrm{A}}(b)$	a-priority L-value for bit b
$\mathrm{L}_{\mathrm{D}}(b)$	a-posteriority L-value for bit b
$\mathrm{L}_{\mathrm{E}}(b)$	log-likelihood ratio for bit b
$\mathcal{L}_{\infty}$	high-SNR power offset
λ	Lagrange multiplier
λ_{c}	carrier wavelength
$\lambda_k(\boldsymbol{A})$	kth eigenvalue of $\boldsymbol{A}$ in decreasing order
$\lambda_{\mathsf{max}}(\boldsymbol{A})$	maximum eigenvalue of $\boldsymbol{A}$
$\lambda_{\mathsf{min}}(\boldsymbol{A})$	minimum eigenvalue of $\boldsymbol{A}$
$\boldsymbol{\Lambda}_{\boldsymbol{A}}$	square diagonal matrix containing the eigenvalues of $\boldsymbol{A}$
m	Nakagami fading parameter
M	constellation cardinality
$\mu_{\boldsymbol{A}}$	mean of $\boldsymbol{A}$
$\max(\cdot)$	maximum of various quantities
$\min(\cdot)$	minimum of various quantities
MMSE	MMSE
$\overline{\mathsf{MMSE}}$	local-average MMSE
N	number of symbols per codeword
N_0	noise spectral density
N_{a}	number of antennas
N_{bits}	number of bits per message
N_{c}	fading coherence
N_{f}	number of entries in a digital-feedback codebook
N_{max}	$= \max(N_{\mathrm{t}}, N_{\mathrm{r}})$
N_{min}	$= \min(N_{\mathrm{t}}, N_{\mathrm{r}})$
N_{p}	number of pilot symbols
N_{r}	number of receive antennas
N_{s}	number of signal streams
N_{t}	number of transmit antennas
$\mathcal{N}(\boldsymbol{\mu}, \boldsymbol{R})$	real Gaussian with mean μ and covariance matrix $\boldsymbol{R}$
$\mathcal{N}_{\mathbb{C}}(\boldsymbol{\mu}, \boldsymbol{R})$	complex Gaussian with mean $\boldsymbol{\mu}$ and covariance matrix $\boldsymbol{R}$
$\boldsymbol{\Omega}$	matrix of variances
$\mathsf{P}(f)$	frequency-domain power allocation
$\mathbb{P}[\cdot]$	probability
$\boldsymbol{P}[n]$	time-domain power allocation (n may be omitted)

P_r	receive power
P_t	transmit power
$\mathcal{P}_\mathrm{r}(\theta)$	receive PAS as a function of θ
$\mathcal{P}_\mathrm{t}(\theta)$	transmit PAS as function of θ
$p_a(\cdot)$	PMF of a
p_e	error probability
p_out	outage probability
$\mathsf{PAPR}(\cdot)$	peak-to-average power ratio
ϕ	phase
$\psi(\cdot)$	digamma function
q	user weight
$Q(\cdot)$	Q-function
r	multiplexing gain
r	code rate
R	bit rate
$R_a(\cdot)$	autocorrelation of a
$\Re\{\cdot\}$	real part
$\mathcal{R}_m$	decision region for codeword m
$\boldsymbol{R}_\mathrm{r}$	receive correlation matrix
$\boldsymbol{R}_\mathrm{t}$	transmit correlation matrix
$\boldsymbol{R}_{\boldsymbol{x}}$	covariance/correlation matrix of $\boldsymbol{x}$
$\boldsymbol{R}_{\boldsymbol{xy}}$	cross-covariance/cross-correlation matrix of $\boldsymbol{x}$ and $\boldsymbol{y}$
$\mathbf{R}_{\boldsymbol{A}}$	correlation tensor for $\boldsymbol{A}$
ρ	forward–reverse power ratio
ρ_d	payload data power boosting coefficient
ρ_p	pilot power boosting coefficient
$\boldsymbol{s}[n]$	time-domain codeword symbol (n may be omitted)
$\mathbf{s}[k]$	frequency-domain codeword symbol
S_0	low-SNR slope
S_∞	number of spatial DOF
$S_a(\cdot)$	power spectrum of a
$S_h(\nu)$	Doppler spectrum
$S_\mathsf{h}(\tau)$	power delay profile
$S_\hbar(\nu,\tau)$	scattering function
$\mathcal{S}$	set of cells with pilots staggered relative to a cell of interest
$\mathcal{S}_0^\ell$	subset of constellation points whose ℓth bit is 0
$\mathcal{S}_1^\ell$	subset of constellation points whose ℓth bit is 1
SINR	local-average receive SINR
SIR	local-average receive SIR
SNR	local-average receive SNR
SNR^r	reverse-link local-average receive SNR

$\mathsf{SNR}_{\mathsf{eff}}$	effective SNR
$\mathsf{SNR}_{\mathsf{eq}}$	equivalent SNR
$\mathsf{SNR}_{\mathsf{xESM}}$	equivalent SNR in the xESM method
sir	output SIR of a signal stream
sinr	output SINR of a signal stream
snr	output SNR of a signal stream
$\overline{\mathsf{sir}}$	hardening-based output SIR of a signal stream
$\overline{\mathsf{sinr}}$	hardening-based output SINR of a signal stream
$\text{sign}(\cdot)$	sign function
$\text{sinc}(x)$	$= \frac{\sin(\pi x)}{\pi x}$
σ_a^2	variance of a
σ_{dB}	standard deviation of the shadow fading (in dB)
σ_θ	angle spread
$\sigma_k(\boldsymbol{A})$	kth singular value of $\boldsymbol{A}$ in decreasing order
$\boldsymbol{\Sigma}_{\boldsymbol{A}}$	rectangular diagonal matrix with the singular values of $\boldsymbol{A}$
t	time
τ	delay
T	symbol period in single-carrier transmission
T_{c}	coherence time
T_{d}	delay spread
T_{eff}	effective temperature
T_{OFDM}	OFDM symbol period
T	time share in TDMA
$\text{tr}(\boldsymbol{A})$	trace of $\boldsymbol{A}$
θ	angle
U	number of active users
U_{tot}	total number of users
$\mathcal{U}$	subset of users
$\boldsymbol{U}$	unitary matrix
$\boldsymbol{U}_{\boldsymbol{F}}$	matrix containing the left singular vectors of $\boldsymbol{F}$
v	velocity
V	variance of the information density
$\boldsymbol{V}$	unitary matrix
$\boldsymbol{V}_{\boldsymbol{F}}$	matrix containing the right singular vectors of $\boldsymbol{F}$
$\boldsymbol{v}[n]$	discrete-time baseband noise vector (n may be omitted)
$\boldsymbol{v}_{\text{p}}(t)$	continuous-time passband noise vector
$\mathbf{v}[k]$	discrete-frequency baseband noise vector
ν_{M}	maximum Doppler frequency
ϑ	fractional power control exponent
$\text{var}[\cdot]$	variance

$\text{vec}(\boldsymbol{A})$	vector created by stacking the columns of $\boldsymbol{A}$
$\boldsymbol{W}$	linear receiver
$\mathcal{W}_N(M, \boldsymbol{R})$	N-dimensional central Wishart with M degrees of freedom and covariance $\boldsymbol{R}$
x_{i}	in-phase transmit signal component
x_{p}	passband transmit signal
x_{q}	quadrature transmit signal component
$\boldsymbol{x}[n]$	discrete-time transmit signal (n may be omitted)
$\mathsf{x}[k]$	discrete-frequency transmit signal
y_{i}	in-phase received signal component
y_{p}	passband received signal
y_{q}	quadrature received signal component
$\boldsymbol{y}[n]$	discrete-time received signal (n may be omitted)
$\mathsf{y}[k]$	discrete-frequency received signal (k may be omitted)

Acronyms

2G	second generation
3G	third generation
3GPP	third-generation Partnership Project
4G	fourth generation
5G	fifth generation
APP	a-posteriori probability
ARQ	automatic repeat request
a.s.	almost surely
AWGN	additive white Gaussian noise
BC	broadcast channel
BCJR	Bahl–Cocke–Jelinek–Raviv
BICM	bit-interleaved coded modulation
BPSK	binary phase shift keying
CDF	cumulative distribution function
CDMA	code-division multiple access
CESM	capacity-effective SNR mapping
CM	cubic metric
COST	European Cooperation in Science and Technology
CSI	channel-state information
CSIR	channel-state information at the receiver
CSIT	channel-state information at the transmitter
DFT	discrete Fourier transform
DMT	diversity–multiplexing tradeoff
DOF	degrees of freedom
DPC	dirty-paper coding
DSL	digital subscriber line
EESM	exponential-effective SNR mapping
ESPAR	electronically steerable parasitic array radiators
FDD	frequency-division duplexing
FDMA	frequency-division multiple access
FFT	fast Fourier transform
FIR	finite impulse response
GSM	global system for mobile communications
H-ARQ	hybrid-automatic repeat request

IA	interference alignment
IEEE	Institute of Electrical and Electronics Engineers
IID	independent identically distributed
IIR	infinite impulse response
IND	independent nonidentically distributed
INR	interference-to-noise ratio
ISI	intersymbol interference
ITU	International Telecommunications Union
JSDM	joint spatial division and multiplexing
LDPC	low-density parity check
LGB	Linde–Buzo–Gray
LMMSE	linear minimum mean-square error
LOS	line-of-sight
LTE	long-term evolution
MAC	multiple-access channel
MAP	maximum a-posteriori
MCS	modulation and coding scheme
MIESM	mutual-information-effective SNR mapping
MIMO	multiple-input multiple-output
MISO	multiple-input single-output
ML	maximum likelihood
MMSE	minimum mean-square error
MU-MIMO	multiuser MIMO
MU-MISO	multiuser MISO
MU-SIMO	multiuser SIMO
MU-SISO	multiuser SISO
NLOS	non-line-of-sight
NOMA	non-orthogonal multiple access
NR	new radio
OFDM	orthogonal frequency-division multiplexing
OFDMA	orthogonal frequency-division multiple access
PAM	pulse-amplitude modulation
PAPR	peak-to-average power ratio
PARC	per-antenna rate control
PAS	power angle spectrum
PDF	probability density function
PDP	power delay profile

PHY	PHYsical layer
PMF	probability mass function
PPP	Poisson point process
PSK	phase shift keying
QAM	quadrature amplitude modulation
QPSK	quadrature phase shift keying
RMS	root mean-square
SC-FDE	single-carrier frequency-domain equalization
SCM	spatial channel model
SDMA	space-division multiple access
SIC	successive interference cancelation
SIMO	single-input multiple-output
SINR	signal-to-interference-plus-noise ratio
SIR	signal-to-interference ratio
SISO	single-input single-output
SLNR	signal-to-leakage-plus-noise ratio
SNR	signal-to-noise ratio
SUI	Stanford University Interim
SU-MIMO	single-user MIMO
SU-SISO	single-user SISO
SVD	singular value decomposition
TCM	trellis-coded modulation
TDD	time-division duplexing
TDMA	time-division multiple access
UCA	uniform circular array
ULA	uniform linear array
WLAN	wireless local area network
WSSUS	wide-sense stationary uncorrelated scattering
ZF	zero-forcing

PART I

WIRELESS COMMUNICATION THEORY

1 A primer on information theory and MMSE estimation

Theory is the first term in the Taylor series expansion of practice.

Thomas Cover

1.1 Introduction

Information theory deals broadly with the science of information, including compressibility and storage of data, as well as reliable communication. It is an exceptional discipline in that it has a precise founder, Claude E. Shannon, and a precise birthdate, 1948. The publication of Shannon's seminal treatise, "A mathematical theory of communication" [58], represents one of the scientific highlights of the twentieth century and, in many respects, marks the onset of the information age. Shannon was an engineer, yet information theory is perhaps best described as an outpost of probability theory that has extensive applicability in electrical engineering as well as substantial overlap with computer science, physics, economics, and even biology. Since its inception, information theory has been distilling practical problems into mathematical formulations whose solutions cast light on those problems. A staple of information theory is its appreciation of elegance and harmony, and indeed many of its results possess a high degree of aesthetic beauty. And, despite their highly abstract nature, they often do reveal much about the practical problems that motivated them in the first place.

Although Shannon's teachings are by now well assimilated, they represented a radical departure from time-honored axioms [52]. In particular, it was believed before Shannon that error-free communication was only possible in the absence of noise or at vanishingly small transmission rates. Shannon's channel coding theorem was nothing short of revolutionary, as it proved that every channel had a characterizing quantity (the capacity) such that, for transmission rates not exceeding it, the error probability could be made arbitrarily small. Ridding the communication of errors did not require overwhelming the noise with signal power or slowing down the transmission rate, but could be achieved in the face of noise and at positive rates—as long as the capacity was not exceeded—by embracing the concept of coding: information units should not be transmitted in isolation but rather in coded blocks, with each unit thinly spread over as many symbols as possible; redundancy and interdependency as an antidote to the confusion engendered by noise. The notion of channel capacity is thus all-important in information theory, being something akin to the speed of light in terms of reliable communication. This analogy with the speed of light, which is common and enticing, must however be viewed with perspective. While, in the early years of information theory, the capacity might have been perceived as remote (wireline modems were transmitting on the order of 300 bits/s in telephone channels whose Shannon capacity was computed as being 2–3 orders of magnitude higher), nowadays it can be closely approached in important channels. Arguably, then, to the daily lives of people the capacity is a far more relevant limitation than the speed of light.

The emergence of information theory also had an important unifying effect, proving an umbrella under which all channels and forms of communication—each with its own toolbox of methodologies theretofore—could be studied on a common footing. Before Shannon, something as obvious today as the transmission of video over a telephone line would have been inconceivable.

As anecdotal testimony of the timeless value and transcendence of Shannon's work, we note that, in 2016, almost seven decades after its publication, "A mathematical theory of communication" ranked as a top-three download in IEEE *Xplore*, the digital repository that archives over four million electrical engineering documents—countlessly many of which elaborate on aspects of the theory spawned by that one paper.

This chapter begins by describing certain types of signals that are encountered throughout the text. Then, the chapter goes on to review those concepts in information theory that are needed throughout, with readers interested in more comprehensive treatments of the matter referred to dedicated textbooks [14, 59, 60]. In addition to the relatively young discipline of information theory, the chapter also touches on the much older subject of MMSE estimation. The packaging of both topics in a single chapter is not coincidental, but rather a choice that is motivated by the relationship between the two—a relationship made of bonds that have long been known, and of others that have more recently been unveiled [61]. Again, we cover only those MMSE estimation concepts that are needed in the book, with readers interested in broader treatments referred to estimation theory texts [62].

1.2 Signal distributions

The signals described next are in general complex-valued. The interpretation of complex signals, as well as complex channels and complex noise, as baseband representations of real-valued passband counterparts is provided in Chapter 2, and readers needing background on this interpretation are invited to peruse Section 2.2 before proceeding. We advance that the real and imaginary parts of a signal are respectively termed the *in-phase* and the *quadrature* components.

Consider a complex scalar s, zero-mean and normalized to be of unit variance, which is to serve as a signal. From a theoretical vantage, a distribution that is all-important because of its optimality in many respects is the complex Gaussian, $s \sim \mathcal{N}_{\mathbb{C}}(0, 1)$, details of which are offered in Appendix C.1.9. In practice though, a scalar signal s is drawn from a discrete distribution defined by M points, say $\mathrm{s}_0, \ldots, \mathrm{s}_{M-1}$, taken with probabilities $p_0, \ldots, p_{M-1}$. These points are arranged into constellations such as the following.

Table 1.1 Constellation minimum distances

Constellation	d_{min}
M-PSK	$2\sin\left(\frac{\pi}{M}\right)$
Square M-QAM	$\sqrt{\frac{6}{M-1}}$

- M-ary phase shift keying (M-PSK), where

$$s_m = e^{j2\pi\frac{m}{M}+\phi_0} \qquad m = 0, \ldots, M-1 \tag{1.1}$$

  with ϕ_0 an arbitrary phase. Because of symmetry, the points are always equiprobable, $p_m = 1/M$ for $m = 0, \ldots, M-1$. Special mention must be made of binary phase-shift keying (BPSK), corresponding to $M = 2$, and quadrature phase-shift keying (QPSK), which corresponds to $M = 4$.
- Square M-ary quadrature amplitude modulation (M-QAM), where the in-phase and quadrature components of s independently take values in the set

$$\left\{\sqrt{\frac{3}{2\,(M-1)}}\left(2m-1-\sqrt{M}\right)\right\} \qquad m = 0, \ldots, \sqrt{M}-1 \tag{1.2}$$

  with $\sqrt{M}$ integer. (Nonsquare M-QAM constellations also exist, and they are employed regularly in wireline systems, but seldom in wireless.) Although making the points in a M-QAM constellation equiprobable is not in general optimum, it is commonplace. Note that, except for perhaps an innocuous rotation, 4-QAM coincides with QPSK.

For both M-PSK and square M-QAM, the minimum distance between constellation points is provided in Table 1.1.

Example 1.1

Depict the 8-PSK and 16-QAM constellations and indicate the distance between nearest neighbors within each.

Solution

See Fig. 1.1.

It is sometimes analytically convenient to approximate discrete constellations by means of continuous distributions over a suitable region on the complex plane. These continuous distributions can be interpreted as limits of dense M-ary constellations for $M \to \infty$. For equiprobable M-PSK and M-QAM, the appropriate unit-variance continuous distributions are:

- ∞-PSK, where $s = e^{j\phi}$ with ϕ uniform on $[0, 2\pi)$.
- ∞-QAM, where s is uniform over the square $\left[-\sqrt{3/2}, \sqrt{3/2}\right] \times \left[-\sqrt{3/2}, \sqrt{3/2}\right]$ on the complex plane.

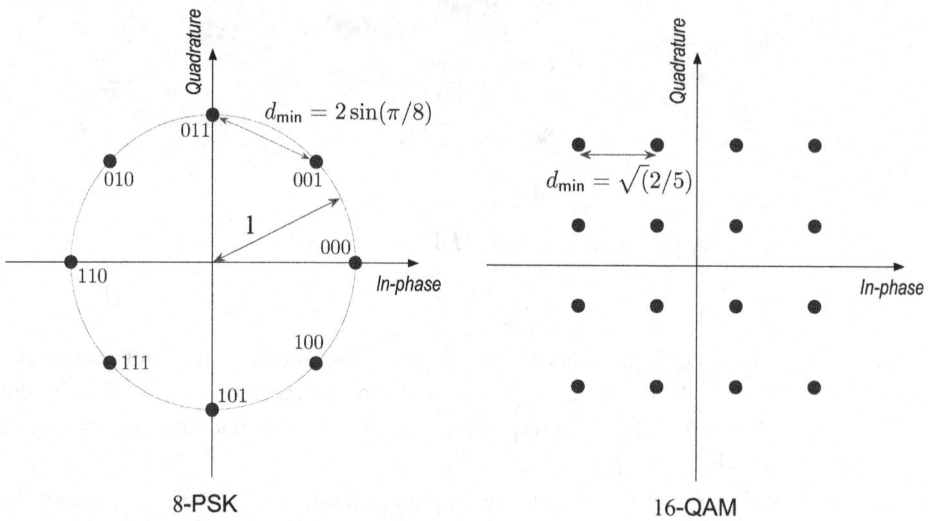

Fig. 1.1 Unit-variance 8-PSK and 16-QAM constellations.

Except for BPSK, all the foregoing distributions, both continuous and discrete, are *proper complex* in the sense of Section C.1.4.

Lastly, a distribution that is relevant for ultrawideband communication is "on–off" keying [63, 64]

$$s = \begin{cases} 0 & \text{with probability } 1-\epsilon \\ \sqrt{1/\epsilon} & \text{with probability } \epsilon \end{cases} \tag{1.3}$$

parameterized by ϵ. Practical embodiments of this distribution include pulse-position modulation [65] and impulse radio [66]. Generalizations of (1.3) to multiple "on" states are also possible.

1.3 Information content

Information equals uncertainty. If a given quantity is certain, then knowledge of it provides no information. It is therefore only natural, as Shannon recognized, to model information and data communication using probability theory. All the elements that play a role in communications (signals, channel, noise) are thereby abstracted using random variables and random processes. For the reader's convenience, reviews of the basic results on random variables and random processes that are necessary for the derivations in this chapter are respectively available in Appendices C.1 and C.3.

As the starting point of our exposition, let us see how to quantify the information content of random variables and processes. We adopt the *bit* as our information currency and, consequently, all applicable logarithms are to the base 2; other information units can be obtained by merely modifying that base, e.g., the *byte* (base 256), the *nat* (base e), and the *ban* (base 10).

All the summations and integrals that follow should be taken over the support of the corresponding random variables, i.e., the set of values on which their probabilities are nonzero.

1.3.1 Entropy

Let x be a discrete random variable with PMF $p_x(\cdot)$. Its *entropy*, denoted by $\mathcal{H}(x)$, is defined as

$$\mathcal{H}(x) = -\sum_{\mathrm{x}} p_x(\mathrm{x}) \log_2 p_x(\mathrm{x}) \tag{1.4}$$

$$= -\mathbb{E}\big[\log_2 p_x(x)\big]. \tag{1.5}$$

Although the entropy is a function of $p_x(\cdot)$ rather than of x, it is rather standard to slightly abuse notation and write it as $\mathcal{H}(x)$. The entropy is nonnegative and it quantifies the amount of uncertainty associated with x: the larger the entropy, the more unpredictable x. Not surprisingly then, the uniform PMF is the entropy-maximizing one. If the cardinality of x is M, then its entropy under a uniform PMF trivially equals $\mathcal{H}(x) = \log_2 M$ bits and thus we can affirm that, for any x with cardinality M, $\mathcal{H}(x) \leq \log_2 M$ bits. At the other extreme, variables with only one possible outcome (i.e., deterministic quantities) have an entropy of zero. The entropy $\mathcal{H}(x)$ gives the number of bits required to describe x on average. Note that the actual values taken by x are immaterial in terms of $\mathcal{H}(x)$; only the probabilities of those values matter.

Similar to Boltzmann's entropy in statistical mechanics, the entropy was introduced as a measure of information by Shannon with the rationale of being the only measure that is continuous in the probabilities, increasing in the support if $p_x(\cdot)$ is uniform, and additive when x is the result of multiple choices [67].

Example 1.2

Express the entropy of the Bernoulli random variable

$$x = \begin{cases} 0 & \text{with probability } p \\ 1 & \text{with probability } 1-p. \end{cases} \tag{1.6}$$

Solution

The entropy of x is the so-called binary entropy function,

$$\mathcal{H}(x) = -p \log_2 p - (1-p)\log_2(1-p), \tag{1.7}$$

which satisfies $\mathcal{H}(x) \leq 1$ with equality for $p = 1/2$.

Example 1.3

Express the entropy of an equiprobable M-ary constellation.

Solution

For s conforming to a discrete constellation with M equiprobable points,

$$\mathcal{H}(s) = -\sum_{m=0}^{M-1} \frac{1}{M} \log \frac{1}{M} \tag{1.8}$$

$$= \log_2 M. \tag{1.9}$$

These $\log_2 M$ bits can be mapped onto the M constellation points in various ways. Particularly relevant is the so-called *Gray mapping*, characterized by nearest-neighbor constellation points differing by a single bit. This ensures that, in the most likely error event, when a constellation point is confused with its closest neighbor, a single bit is flipped. Gray mapping is illustrated for a PSK constellation in Fig. 1.1.

Having seen how to quantify the amount of information in an individual variable, we now extend the concept to multiple ones. Indeed, because of the multiple inputs and outputs, the most convenient MIMO representation uses vectors for the signals and matrices for the channels.

Let x_0 and x_1 be discrete random variables with joint PMF $p_{x_0 x_1}(\cdot,\cdot)$ and marginals $p_{x_0}(\cdot)$ and $p_{x_1}(\cdot)$. The joint entropy of x_0 and x_1 is

$$\mathcal{H}(x_0, x_1) = -\sum_{\mathrm{x}_0}\sum_{\mathrm{x}_1} p_{x_0 x_1}(\mathrm{x}_0, \mathrm{x}_1) \log_2 p_{x_0 x_1}(\mathrm{x}_0, \mathrm{x}_1) \tag{1.10}$$

$$= -\mathbb{E}\big[\log_2 p_{x_0 x_1}(x_0, x_1)\big]. \tag{1.11}$$

If x_0 and x_1 are independent, then $\mathcal{H}(x_0, x_1) = \mathcal{H}(x_0) + \mathcal{H}(x_1)$. Furthermore, by regarding x_0 and x_1 as entries of a vector, we can claim (1.10) as the entropy of such a vector. More generally, for any discrete random vector $\boldsymbol{x}$,

$$\mathcal{H}(\boldsymbol{x}) = -\mathbb{E}\big[\log_2 p_{\boldsymbol{x}}(\boldsymbol{x})\big]. \tag{1.12}$$

Often, it is necessary to appraise the uncertainty that remains in a random variable x once a related random variable y has been observed. This is quantified by the conditional entropy of x given y,

$$\mathcal{H}(x|y) = -\sum_{\mathrm{x}}\sum_{\mathrm{y}} p_{xy}(\mathrm{x}, \mathrm{y}) \log_2 p_{x|y}(\mathrm{x}|\mathrm{y}). \tag{1.13}$$

If x and y are independent, then naturally $\mathcal{H}(x|y) = \mathcal{H}(x)$ whereas, if x is a deterministic function of y, then $\mathcal{H}(x|y) = 0$.

The joint and conditional entropies are related by the chain rule

$$\mathcal{H}(x, y) = \mathcal{H}(x) + \mathcal{H}(y|x), \tag{1.14}$$

which extends immediately to vectors. When more than two variables are involved, the chain rule generalizes as

$$\mathcal{H}(x_0, \ldots, x_{N-1}) = \sum_{n=0}^{N-1} \mathcal{H}(x_n | x_0, \ldots, x_{n-1}). \tag{1.15}$$

1.3.2 Differential entropy

A quantity seemingly analogous to the entropy, the *differential entropy*, can be defined for continuous random variables. If $f_x(\cdot)$ is the probability density function (PDF) of x, its differential entropy is

$$\mathfrak{h}(x) = -\int f_x(\mathrm{x}) \log_2 f_x(\mathrm{x})\, d\mathrm{x} \tag{1.16}$$

$$= -\mathbb{E}\left[\log_2 f_x(x)\right] \tag{1.17}$$

where the integration in (1.16) is over the complex plane. Care must be exercised when dealing with differential entropies, because they may be negative. Indeed, despite the similarity in their forms, the entropy and differential entropy do not admit the same interpretation: the former measures the information contained in a random variable whereas the latter does not. Tempting as it may be, $\mathfrak{h}(x)$ cannot be approached by discretizing $f_x(\cdot)$ into progressively smaller bins and computing the entropy of the ensuing discrete random variable. The entropy of a b-bit quantization of x is approximately $\mathfrak{h}(x)+b$, which diverges as $b \to \infty$. This merely confirms what one may have intuitively guessed, namely that the amount of information in a continuous variable, i.e., the number of bits required to describe it, is generally infinite. The physical meaning of $\mathfrak{h}(x)$ is thus not the amount of information in x. In fact, the differential entropy is devoid—from an engineering viewpoint—of operational meaning and ends up serving mostly as a stepping stone to the mutual information, which does have plenty of engineering significance.

Example 1.4

Calculate the differential entropy of a real random variable x uniformly distributed in $[0, b]$.

Solution

$$\mathfrak{h}(x) = -\int_0^b \frac{1}{b} \log_2\left(\frac{1}{b}\right) dx \tag{1.18}$$

$$= \log_2 b. \tag{1.19}$$

Note that $\mathfrak{h}(x) < 0$ for $b < 1$.

Example 1.5 (Differential entropy of a complex Gaussian scalar)

Let $x \sim \mathcal{N}_{\mathbb{C}}(\mu, \sigma^2)$. Invoking the PDF in (C.14),

$$\mathfrak{h}(x) = \mathbb{E}\left[\frac{|x-\mu|^2}{\sigma^2} \log_2 e + \log_2(\pi\sigma^2)\right] \tag{1.20}$$

$$= \log_2(\pi e \sigma^2). \tag{1.21}$$

Observe how, in Example 1.5, the mean μ is immaterial to $\mathfrak{h}(x)$. This reflects the property of differential entropy being translation-invariant, meaning that $\mathfrak{h}(x+a) = \mathfrak{h}(x)$ for any constant a; it follows from this property that we can always translate a random variable and set its mean to zero without affecting its differential entropy.

In the context of information content, the importance of the complex Gaussian distribution stems, not only from its prevalence, but further from the fact that it is the distribution that maximizes the differential entropy for a given variance [14]. Thus, for any random variable x with variance σ^2, $\mathfrak{h}(x) \leq \log_2(\pi e \sigma^2)$.

As in the discrete case, the notion of differential entropy readily extends to the multivariate realm. If $\boldsymbol{x}$ is a continuous random vector with PDF $f_{\boldsymbol{x}}(\cdot)$, then

$$\mathfrak{h}(\boldsymbol{x}) = -\mathbb{E}\big[\log_2 f_{\boldsymbol{x}}(\boldsymbol{x})\big]. \tag{1.22}$$

Example 1.6 (Differential entropy of a complex Gaussian vector)

Let $\boldsymbol{x} \sim \mathcal{N}_{\mathbb{C}}(\boldsymbol{\mu}, \boldsymbol{R})$. From (C.15) and (1.22),

$$\begin{aligned}
\mathfrak{h}(\boldsymbol{x}) &= -\mathbb{E}\big[\log_2 f_{\boldsymbol{x}}(\boldsymbol{x})\big] && (1.23)\\
&= \log_2 \det(\pi \boldsymbol{R}) + \mathbb{E}\big[(\boldsymbol{x}-\boldsymbol{\mu})^* \boldsymbol{R}^{-1} (\boldsymbol{x}-\boldsymbol{\mu})\big] \log_2 e && (1.24)\\
&= \log_2 \det(\pi \boldsymbol{R}) + \mathrm{tr}\big(\mathbb{E}\big[(\boldsymbol{x}-\boldsymbol{\mu})^* \boldsymbol{R}^{-1} (\boldsymbol{x}-\boldsymbol{\mu})\big]\big) \log_2 e && (1.25)\\
&= \log_2 \det(\pi \boldsymbol{R}) + \mathrm{tr}\big(\mathbb{E}\big[\boldsymbol{R}^{-1} (\boldsymbol{x}-\boldsymbol{\mu})(\boldsymbol{x}-\boldsymbol{\mu})^*\big]\big) \log_2 e && (1.26)\\
&= \log_2 \det(\pi \boldsymbol{R}) + \mathrm{tr}\big(\boldsymbol{R}^{-1}\mathbb{E}\big[(\boldsymbol{x}-\boldsymbol{\mu})(\boldsymbol{x}-\boldsymbol{\mu})^*\big]\big) \log_2 e && (1.27)\\
&= \log_2 \det(\pi \boldsymbol{R}) + \mathrm{tr}(\boldsymbol{I}) \log_2 e && (1.28)\\
&= \log_2 \det(\pi e \boldsymbol{R}), && (1.29)
\end{aligned}$$

where in (1.25) we used the fact that a scalar equals its trace, while in (1.26) we invoked the commutative property in (B.26).

As in the scalar case, the complex Gaussian distribution maximizes the differential entropy for a given covariance matrix. For any complex random vector $\boldsymbol{x}$ with covariance $\boldsymbol{R}$, therefore, $\mathfrak{h}(\boldsymbol{x}) \leq \log_2 \det(\pi e \boldsymbol{R})$.

The conditional differential entropy of x given y equals

$$\mathfrak{h}(x|y) = -\mathbb{E}\big[\log_2 f_{x|y}(x|y)\big] \tag{1.30}$$

with expectation over the joint distribution of x and y. The chain rule that relates joint and conditional entropies is

$$\mathfrak{h}(x_0, \ldots, x_{N-1}) = \sum_{n=0}^{N-1} \mathfrak{h}(x_n | x_0, \ldots, x_{n-1}), \tag{1.31}$$

which extends verbatim to vectors.

1.3.3 Entropy rate

To close the discussion on information content, let us turn our attention from random variables to random processes. A discrete random process $x_0, \ldots, x_{N-1}$ is a sequence of discrete random variables indexed by time. If $x_0, \ldots, x_{N-1}$ are independent identically distributed (IID), then the entropy of the process grows linearly with N at a rate $\mathcal{H}(x_0)$. More generally, the entropy grows linearly with N at the so-called *entropy rate*

$$\mathcal{H} = \lim_{N\to\infty} \frac{1}{N}\mathcal{H}(x_0,\ldots,x_{N-1}). \tag{1.32}$$

If the process is stationary, then the entropy rate can be shown to equal

$$\mathcal{H} = \lim_{N\to\infty} \mathcal{H}(x_N|x_0,\ldots,x_{N-1}). \tag{1.33}$$

When the distribution of the process is continuous rather than discrete, the same definitions apply to the differential entropy and a classification that proves useful in the context of fading channels can be introduced: a process is said to be *nonregular* if its present value is perfectly predictable from noiseless observations of the entire past, while the process is *regular* if its present value cannot be perfectly predicted from noiseless observations of the entire past [68]. In terms of the differential entropy rate $\mathfrak{h}$, the process is regular if $\mathfrak{h} > -\infty$ and nonregular otherwise.

1.4 Information dependence

Although it could be—and has been—argued that Shannon imported the concept of entropy from statistical mechanics, where it was utilized to measure the uncertainty surrounding the state of a physical system, this was but a step toward something radically original: the idea of measuring with information (e.g., with bits) the interdependence among different quantities. In the context of a communication channel, this idea opens the door to relating transmit and receive signals, a relationship from which the capacity ultimately emerges.

1.4.1 Relative entropy

Consider two PMFs, $p(\cdot)$ and $q(\cdot)$. If the latter is nonzero over the support of the former, then their *relative entropy* is defined as

$$\mathcal{D}(p||q) = \sum_{\mathrm{x}} p(\mathrm{x}) \log_2 \frac{p(\mathrm{x})}{q(\mathrm{x})} \tag{1.34}$$

$$= \mathbb{E}\left[\log_2 \frac{p(x)}{q(x)}\right] \tag{1.35}$$

where the expectation is over $p(\cdot)$. The relative entropy, also referred to as the *Kullback–Leibler divergence* or the *information divergence*, can be interpreted as a measure of the similarity of $p(\cdot)$ and $q(\cdot)$. Note, however, that it is not symmetric, i.e., $\mathcal{D}(p||q) \neq \mathcal{D}(q||p)$ in general. It is a nonnegative quantity, and it is zero if and only if $p(\mathrm{x}) = q(\mathrm{x})$ for every x.

Similarly, for two PDFs $f(\cdot)$ and $g(\cdot)$,

$$\mathcal{D}(f||g) = \int f(\mathrm{x}) \log_2 \frac{f(\mathrm{x})}{g(\mathrm{x})}\, d\mathrm{x}. \tag{1.36}$$

1.4.2 Mutual information

A quantity that lies at the heart of information theory is the *mutual information* between two or more random variables. Although present already in Shannon's original formulation [58], the mutual information did not acquire its current name until years later [67, 69]. Given two random variables s and y, the mutual information between them, denoted by $I(s;y)$, quantifies the reduction in uncertainty about the value of s that occurs when y is observed, and vice versa. The mutual information is symmetric and thus $I(s;y) = I(y;s)$. Put in the simplest terms, the mutual information measures the information that one random variable contains about another. As one would expect, $I(s;y)$ is nonnegative, equaling zero if and only if s and y are independent. At the other extreme, $I(s;y)$ cannot exceed the uncertainty contained in either s or y.

For discrete random variables, the mutual information can be computed on the basis of entropies as

$$I(s;y) = \mathcal{H}(s) - \mathcal{H}(s|y) \tag{1.37}$$

$$= \mathcal{H}(y) - \mathcal{H}(y|s), \tag{1.38}$$

or also as the information divergence between the joint PMF of s and y, on the one hand, and the product of their marginals on the other, i.e.,

$$I(s;y) = \mathcal{D}(p_{sy}||p_s p_y) \tag{1.39}$$

$$= \sum_{\mathrm{s}} \sum_{\mathrm{y}} p_{sy}(\mathrm{s},\mathrm{y}) \log_2 \frac{p_{sy}(\mathrm{s},\mathrm{y})}{p_s(\mathrm{s})\, p_y(\mathrm{y})} \tag{1.40}$$

$$= \sum_{\mathrm{s}} \sum_{\mathrm{y}} p_{sy}(\mathrm{s},\mathrm{y}) \log_2 \frac{p_{y|s}(\mathrm{y}|\mathrm{s})}{p_y(\mathrm{y})}. \tag{1.41}$$

Recalling that the information divergence measures the similarity between distributions, the intuition behind (1.39) is as follows: if the joint distribution is "similar" to the product of the marginals, it must be that s and y are essentially independent and thus one can hardly inform about the other. Conversely, if the joint and marginal distributions are "dissimilar," it must be that s and y are highly dependent and thus one can provide much information about the other.

For continuous random variables, relationships analogous to (1.37) and (1.39) apply, precisely

$$I(s;y) = \mathfrak{h}(s) - \mathfrak{h}(s|y) \tag{1.42}$$

$$= \mathfrak{h}(y) - \mathfrak{h}(y|s) \tag{1.43}$$

and

$$I(s;y) = \mathcal{D}(f_{sy}||f_s f_y) \tag{1.44}$$

$$= \iint f_{sy}(\mathrm{s},\mathrm{y}) \log_2 \frac{f_{sy}(\mathrm{s},\mathrm{y})}{f_s(\mathrm{s}) f_y(\mathrm{y})} \,\mathrm{ds}\,\mathrm{dy} \tag{1.45}$$

$$= \iint f_{sy}(\mathrm{s},\mathrm{y}) \log_2 \frac{f_{y|s}(\mathrm{y}|\mathrm{s})}{f_y(\mathrm{y})} \,\mathrm{ds}\,\mathrm{dy}. \tag{1.46}$$

In contrast with the differential entropies, which cannot be obtained as the limit of the entropy of the discretized variables, $I(s; y)$ can be perfectly computed as the limit of the mutual information between discretized versions of s and y. Albeit the entropies and conditional entropies of the discretized variables diverge, their differences remain well behaved.

Since, because of their translation invariance, the entropies and differential entropies are not influenced by the mean of the corresponding random variables, neither is the mutual information. In the derivations that follow, therefore, we can restrict ourselves to zero-mean distributions.

As shorthand notation, we introduce the informal term *Gaussian mutual information* to refer to the function $\mathcal{I}(\rho) = I(s; \sqrt{\rho}s + z)$ when z is complex Gaussian and ρ is a fixed parameter. If we interpret s as a transmit symbol and z as noise, then ρ plays the role of the signal-to-noise ratio (SNR) and the mutual information between s and the received symbol $\sqrt{\rho}s + z$ is given by $\mathcal{I}(\rho)$. Because of this interpretation, attention is paid to how $\mathcal{I}(\rho)$ behaves for small and large ρ, in anticipation of low-SNR and high-SNR analyses later on. We examine these specific behaviors by expanding $\mathcal{I}(\rho)$ and making use of the Landau symbols $\mathcal{O}(\cdot)$ and $o(\cdot)$ described in Appendix F.

Example 1.7 (Gaussian mutual information for a complex Gaussian scalar)

Let us express, as a function of ρ, the mutual information between s and $y = \sqrt{\rho}s + z$ with s and z independent standard complex Gaussians, i.e., $s \sim \mathcal{N}_{\mathbb{C}}(0, 1)$ and $z \sim \mathcal{N}_{\mathbb{C}}(0, 1)$. Noting that $y \sim \mathcal{N}_{\mathbb{C}}(0, 1 + \rho)$ and $y|s \sim \mathcal{N}_{\mathbb{C}}(\sqrt{\rho}s, 1)$, and invoking Example 1.5,

$$\mathcal{I}(\rho) = I(s; \sqrt{\rho}s + z) \tag{1.47}$$
$$= \mathfrak{h}(\sqrt{\rho}s + z) - \mathfrak{h}(\sqrt{\rho}s + z\,|s) \tag{1.48}$$
$$= \mathfrak{h}(\sqrt{\rho}s + z) - \mathfrak{h}(z) \tag{1.49}$$
$$= \log_2\big(\pi e(1 + \rho)\big) - \log_2(\pi e) \tag{1.50}$$
$$= \log_2(1 + \rho). \tag{1.51}$$

For small ρ,

$$\mathcal{I}(\rho) = \left(\rho - \frac{1}{2}\rho^2\right) \log_2 e + o(\rho^2), \tag{1.52}$$

which turns out to apply in rather wide generality: provided that s is proper complex as per the definition in Appendix C.1, its second-order expansion of $\mathcal{I}(\cdot)$ abides by (1.52) [64].

In turn, for complex Gaussian s and large ρ,

$$\mathcal{I}(\rho) = \log_2 \rho + \mathcal{O}\left(\frac{1}{\rho}\right). \tag{1.53}$$

Example 1.8 (Gaussian mutual information for ∞-PSK)

Let us reconsider Example 1.7, only with s drawn from the ∞-PSK distribution defined in Section 1.2. The corresponding mutual information cannot be expressed in closed form, but meaningful expansions can be given. For low ρ, (1.52) holds verbatim because of the properness of ∞-PSK. In turn, the high-ρ behavior is [70]

$$\mathcal{I}^{\infty\text{-PSK}}(\rho) = \frac{1}{2}\log_2 \rho + \frac{1}{2}\log_2\left(\frac{4\pi}{e}\right) + \mathcal{O}\left(\frac{1}{\rho}\right). \tag{1.54}$$

Example 1.9 (Gaussian mutual information for ∞-QAM)

Let us again reconsider Example 1.7, this time with s drawn from the ∞-QAM distribution. As with ∞-PSK, the mutual information cannot be expressed in closed form, but meaningful expansions can be found. For low ρ, and since s is proper complex, (1.52) holds whereas for high ρ [71]

$$\mathcal{I}^{\infty\text{-QAM}}(\rho) = \log_2 \rho - \log_2\left(\frac{\pi e}{6}\right) + \mathcal{O}\left(\frac{1}{\rho}\right). \tag{1.55}$$

With respect to the high-ρ mutual information in (1.53), ∞-QAM suffers a power penalty of $\frac{\pi e}{6}|_{\text{dB}} = 1.53$ dB, where we have introduced the notation $a|_{\text{dB}} = 10\log_{10} a$ that is to appear repeatedly in the sequel.

Example 1.10 (Gaussian mutual information for BPSK)

Let us reconsider Example 1.7 once more, now with s drawn from a BPSK distribution, i.e., $s = \pm 1$. The PDF of y equals

$$f_y(\mathrm{y}) = \frac{1}{2\pi}\left(e^{-|\mathrm{y}+\sqrt{\rho}|^2} + e^{-|\mathrm{y}-\sqrt{\rho}|^2}\right) \tag{1.56}$$

whereas $y|s \sim \mathcal{N}_{\mathbb{C}}(\sqrt{\rho}s, 1)$. Thus,

$$\mathcal{I}^{\text{BPSK}}(\rho) = \mathfrak{h}(y) - \mathfrak{h}(y|s) \tag{1.57}$$

$$= -\int f_y(\mathrm{y}) \log_2 f_y(\mathrm{y})\, \mathrm{dy} - \log_2(\pi e) \tag{1.58}$$

$$= 2\rho \log_2 e - \frac{1}{\sqrt{\pi}}\int_{-\infty}^{\infty} e^{-\xi^2} \log_2 \cosh\left(2\rho - 2\sqrt{\rho}\xi\right) \mathrm{d}\xi \tag{1.59}$$

where, by virtue of the real nature of s, the integration over the complex plane in (1.58) reduces, after some algebra, to the integral on the real line in (1.59). In turn, this integral can be alternatively expressed as the series [72, example 4.39]

$$\mathcal{I}^{\text{BPSK}}(\rho) = 1 + \left[(4\rho - 1)\, Q\left(\sqrt{2\rho}\right) - \sqrt{\frac{4\rho}{\pi}}\, e^{-\rho} + \sum_{\ell=1}^{\infty} \frac{(-1)^\ell}{\ell(\ell+1)}\, e^{4\ell(\ell+1)\rho}\, Q\left((2\ell+1)\sqrt{2\rho}\right)\right] \log_2 e \tag{1.60}$$

where $Q(\cdot)$ is the Gaussian Q-function (see Appendix E.5).

For small ρ, using the identity

$$\log_e \cosh\left(2\rho - 2\sqrt{\rho}\xi\right) = 2\xi^2\rho - 4\xi\rho^{3/2} + \left(2 - \frac{4\xi^4}{3}\right)\rho^2 + o(\rho^2) \tag{1.61}$$

we can reduce (1.59) to

$$\mathcal{I}^{\text{BPSK}}(\rho) = \left(\rho - \rho^2\right) \log_2 e + o(\rho^2) \tag{1.62}$$

whereas, for large ρ [71]

$$\mathcal{I}^{\text{BPSK}}(\rho) = 1 - \frac{e^{-\rho}}{\sqrt{\rho/\pi}} + \epsilon \tag{1.63}$$

with $\log \epsilon = o(\rho)$.

Example 1.11 (Gaussian mutual information for QPSK)

Since QPSK amounts to two BPSK constellations in quadrature with the power evenly divided between them,

$$\mathcal{I}^{\text{QPSK}}(\rho) = 2\,\mathcal{I}^{\text{BPSK}}\left(\frac{\rho}{2}\right). \tag{1.64}$$

Another way to see this equivalence is by considering that, given a BPSK symbol, we can add a second BPSK symbol of the same energy in quadrature without either BPSK symbol perturbing the other. The mutual information doubles while twice the energy is spent, i.e., $\mathcal{I}^{\text{QPSK}}(2\rho) = 2\,\mathcal{I}^{\text{BPSK}}(\rho)$.

Discrete constellations beyond QPSK, possibly nonequiprobable, are covered by the following example.

Example 1.12 (Gaussian mutual information for an arbitrary constellation)

Let s be a zero-mean unit-variance discrete random variable taking values in $\mathrm{s}_0, \ldots, \mathrm{s}_{M-1}$ with probabilities $p_0, \ldots, p_{M-1}$. This subsumes M-PSK, M-QAM, and any other discrete constellation. The PDF of $y = \sqrt{\rho}s + z$ equals

$$f_y(\mathrm{y}) = \frac{1}{\pi} \sum_{m=0}^{M-1} p_m \, e^{-|\mathrm{y} - \sqrt{\rho}\,\mathrm{s}_m|^2} \tag{1.65}$$

whereas $y|s \sim \mathcal{N}_{\mathbb{C}}(\sqrt{\rho}s, 1)$. Thus,

$$\mathcal{I}^{M\text{-ary}}(\rho) = I\left(s; \sqrt{\rho}s + z\right) \tag{1.66}$$

$$= -\int f_y(\mathrm{y}) \log_2 f_y(\mathrm{y}) \, d\mathrm{y} - \log_2(\pi e) \tag{1.67}$$

with integration over the complex plane.

For low ρ, an arduous expansion of $f_y(\cdot)$ and the subsequent integration leads, provided that s is proper complex, again to (1.52). For high ρ, it can be shown [71] that

$$\mathcal{I}^{M\text{-ary}}(\rho) = \log_2 M - \epsilon \tag{1.68}$$

with

$$\log \epsilon = -\frac{d_{\text{min}}^2}{4}\rho + o(\rho) \tag{1.69}$$

where, recall,

$$d_{\text{min}} = \min_{k \neq \ell} |\mathrm{s}_k - \mathrm{s}_\ell| \tag{1.70}$$

is the minimum distance between constellation points. The mutual information is capped at $\log_2 M$, as one would expect, and the speed at which this limit is approached for $\rho \to \infty$ is regulated by d_{min}.

The definition of mutual information extends also to vectors. For continuous random vectors $\boldsymbol{s}$ and $\boldsymbol{y}$, specifically,

$$I(\boldsymbol{s};\boldsymbol{y}) = \mathfrak{h}(\boldsymbol{y}) - \mathfrak{h}(\boldsymbol{y}|\boldsymbol{s}) \tag{1.71}$$

$$= \mathfrak{h}(\boldsymbol{s}) - \mathfrak{h}(\boldsymbol{s}|\boldsymbol{y}) \tag{1.72}$$

$$= \mathcal{D}(f_{\boldsymbol{s}\boldsymbol{y}} \| f_{\boldsymbol{s}} f_{\boldsymbol{y}}). \tag{1.73}$$

Example 1.13 (Gaussian mutual information for a complex Gaussian vector)

Let $\boldsymbol{y} = \sqrt{\rho}\boldsymbol{A}\boldsymbol{s} + \boldsymbol{z}$ where $\boldsymbol{s} \sim \mathcal{N}_{\mathbb{C}}(\boldsymbol{0}, \boldsymbol{R}_{\boldsymbol{s}})$ and $\boldsymbol{z} \sim \mathcal{N}_{\mathbb{C}}(\boldsymbol{0}, \boldsymbol{R}_{\boldsymbol{z}})$ while $\boldsymbol{A}$ is a deterministic matrix. With $\boldsymbol{s}$ and $\boldsymbol{z}$ mutually independent, let us express $I(\boldsymbol{s};\boldsymbol{y})$ as a function of ρ. Since $\boldsymbol{y} \sim \mathcal{N}_{\mathbb{C}}(\boldsymbol{0}, \rho\boldsymbol{A}\boldsymbol{R}_{\boldsymbol{s}}\boldsymbol{A}^* + \boldsymbol{R}_{\boldsymbol{z}})$ and $\boldsymbol{y}|\boldsymbol{s} \sim \mathcal{N}_{\mathbb{C}}(\sqrt{\rho}\boldsymbol{A}\boldsymbol{s}, \boldsymbol{R}_{\boldsymbol{z}})$, leveraging Example 1.6,

$$\mathcal{I}(\rho) = I\big(\boldsymbol{s}; \sqrt{\rho}\boldsymbol{A}\boldsymbol{s} + \boldsymbol{z}\big) \tag{1.74}$$

$$= \mathfrak{h}\big(\sqrt{\rho}\boldsymbol{A}\boldsymbol{s} + \boldsymbol{z}\big) - \mathfrak{h}\big(\sqrt{\rho}\boldsymbol{A}\boldsymbol{s} + \boldsymbol{z}\,|\boldsymbol{s}\big) \tag{1.75}$$

$$= \mathfrak{h}\big(\sqrt{\rho}\boldsymbol{A}\boldsymbol{s} + \boldsymbol{z}\big) - \mathfrak{h}(\boldsymbol{z}) \tag{1.76}$$

$$= \log_2 \det\big(\pi e\,(\rho\boldsymbol{A}\boldsymbol{R}_{\boldsymbol{s}}\boldsymbol{A}^* + \boldsymbol{R}_{\boldsymbol{z}})\big) - \log_2 \det\big(\pi e \boldsymbol{R}_{\boldsymbol{z}}\big) \tag{1.77}$$

$$= \log_2 \det\big(\boldsymbol{I} + \rho\boldsymbol{A}\boldsymbol{R}_{\boldsymbol{s}}\boldsymbol{A}^*\boldsymbol{R}_{\boldsymbol{z}}^{-1}\big). \tag{1.78}$$

For low ρ, using

$$\frac{\partial}{\partial \rho} \log_e \det(\boldsymbol{I} + \rho\boldsymbol{B})\bigg|_{\rho=0} = \text{tr}(\boldsymbol{B}) \tag{1.79}$$

$$\frac{\partial^2}{\partial \rho^2} \log_e \det(\boldsymbol{I} + \rho\boldsymbol{B})\bigg|_{\rho=0} = -\text{tr}\big(\boldsymbol{B}^2\big) \tag{1.80}$$

it is found that

$$\mathcal{I}(\rho) = \left[\text{tr}\big(\boldsymbol{A}\boldsymbol{R}_{\boldsymbol{s}}\boldsymbol{A}^*\boldsymbol{R}_{\boldsymbol{z}}^{-1}\big)\,\rho - \frac{1}{2}\,\text{tr}\Big(\big(\boldsymbol{A}\boldsymbol{R}_{\boldsymbol{s}}\boldsymbol{A}^*\boldsymbol{R}_{\boldsymbol{z}}^{-1}\big)^2\Big)\,\rho^2\right] \log_2 e + o\left(\rho^2\right) \tag{1.81}$$

whose applicability extends beyond complex Gaussian vectors to any proper complex vector $\boldsymbol{s}$. For high ρ, in turn, provided $\boldsymbol{A}\boldsymbol{R}_{\boldsymbol{s}}\boldsymbol{A}^*\boldsymbol{R}_{\boldsymbol{z}}^{-1}$ is nonsingular,

$$\mathcal{I}(\rho) = \min(N_{\boldsymbol{s}}, N_{\boldsymbol{y}}) \log_2 \rho + \log_2 \det\big(\boldsymbol{A}\boldsymbol{R}_{\boldsymbol{s}}\boldsymbol{A}^*\boldsymbol{R}_{\boldsymbol{z}}^{-1}\big) + \mathcal{O}\left(\frac{1}{\rho}\right), \tag{1.82}$$

where $N_{\boldsymbol{s}}$ and $N_{\boldsymbol{y}}$ are the dimensions of $\boldsymbol{s}$ and $\boldsymbol{y}$, respectively.

Example 1.14 (Gaussian mutual information for a discrete vector)

Reconsider Example 1.13, only with $\boldsymbol{s}$ an $N_{\boldsymbol{s}}$-dimensional discrete complex random vector and $\boldsymbol{z} \sim \mathcal{N}_{\mathbb{C}}(\boldsymbol{0}, \boldsymbol{I})$. The vector $\boldsymbol{y} = [y_0 \cdots y_{N_{\boldsymbol{y}}-1}]^{\mathrm{T}}$ is $N_{\boldsymbol{y}}$-dimensional and hence $\boldsymbol{A}$ is $N_{\boldsymbol{y}} \times N_{\boldsymbol{s}}$. Each entry of $\boldsymbol{s}$ can take one of M possible values and therefore $\boldsymbol{s}$ can take one of M^{N_s} values, $\mathbf{s}_0, \ldots, \mathbf{s}_{M^{N_s}-1}$, with probabilities $p_0, \ldots, p_{M^{N_s}-1}$. With a smattering of algebra, the PDF of $\boldsymbol{y}$ can be found to be

$$f_{\boldsymbol{y}}(\mathbf{y}) = \frac{1}{\pi^{N_y}} \sum_{m=0}^{M^{N_s}-1} p_m \, e^{-\|\mathbf{y}-\sqrt{\rho}\boldsymbol{A}\mathbf{s}_m\|^2} \tag{1.83}$$

whereas $\boldsymbol{y}|\boldsymbol{s} \sim \mathcal{N}_{\mathbb{C}}(\sqrt{\rho}\boldsymbol{A}\boldsymbol{s}, \boldsymbol{I})$. Then,

$$\mathcal{I}^{M\text{-ary}}(\rho) = \mathfrak{h}(\boldsymbol{y}) - \log_2 \det(\pi e \boldsymbol{I}) \tag{1.84}$$

$$= -\int \cdots \int f_{\boldsymbol{y}}(\mathbf{y}) \log_2 f_{\boldsymbol{y}}(\mathbf{y}) \, \mathrm{d}\mathrm{y}_0 \cdots \mathrm{d}\mathrm{y}_{N_y-1} - N_{\boldsymbol{y}} \log_2(\pi e). \tag{1.85}$$

The number of terms in the summation in (1.83) grows exponentially with $N_{\boldsymbol{s}}$, whereas the integration in (1.85) becomes unwieldy as $N_{\boldsymbol{y}}$ grows large. Except in very special cases, numerical integration techniques are called for [73]. Alternatively, it is possible to resort to approximations of the integral of a Gaussian function multiplied with an arbitrary real function [74].

For low ρ, and as long as $\boldsymbol{s}$ is proper complex, $\mathcal{I}^{M\text{-ary}}(\rho)$ expands as in (1.81) [75]. For $\rho \to \infty$, in turn, $\mathcal{I}(\rho) \to N_{\boldsymbol{s}} \log_2 M$.

Like the entropy and differential entropy, the mutual information satisfies a chain rule, specifically

$$I(x_0, \ldots, x_{N-1}; y) = \sum_{n=0}^{N-1} I(x_n; y | x_0, \ldots, x_{n-1}), \tag{1.86}$$

which applies verbatim to vectors.

1.5 Reliable communication

1.5.1 Information-theoretic abstraction

One of the enablers of Shannon's ground-breaking work was his ability to dissect a problem into simple pieces, which he could solve and subsequently put together to construct the full solution to the original problem [76]. This ability was manifest in the extremely simple abstraction of a communication link from which he derived quantities of fundamental interest, chiefly the capacity. This simple abstraction, echoed in Fig. 1.2, indeed contained all the essential ingredients.

- An encoder that parses the bits to be communicated into *messages* containing N_{bits}, meaning that there are $2^{N_{\text{bits}}}$ possible such messages, and then maps each message onto a *codeword* consisting of N unit-power complex symbols, $s[0], \ldots, s[N-1]$. The codeword is subsequently amplified, subject to the applicable constraints, into the transmit signal $x[0], \ldots, x[N-1]$.

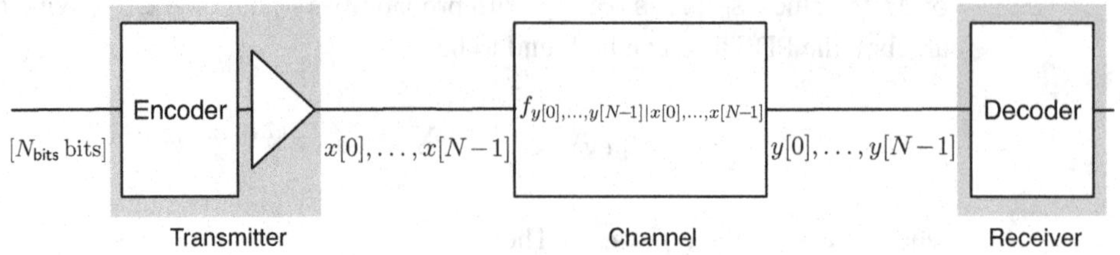

Fig. 1.2 Basic abstraction of a communication link.

- The channel, viewed as the random transformation experienced by the transmit signal and fully described, from such viewpoint, by the conditional probability of its output given every possible input, the *channel law* $f_{y[0],\ldots,y[N-1]|x[0],\ldots,x[N-1]}(\cdot)$. Accounting for the power amplification, and for any other transformation involved in converting the codeword into the transmit signal, $f_{y[0],\ldots,y[N-1]|s[0],\ldots,s[N-1]}(\cdot)$ readily derives from $f_{y[0],\ldots,y[N-1]|x[0],\ldots,x[N-1]}(\cdot)$.
- A decoder that, cognizant of the channel law, maps its observation of the channel output $y[0]\ldots,y[N-1]$ onto a guess of which codeword, and thereby which of the $2^{N_{\text{bits}}}$ possible messages, has been transmitted.

The functions used by the encoder and decoder to map messages (N_{bits} bits) onto codewords (N symbols) define the channel code. The set of all possible codewords is termed a *codebook* and the rate of information being transmitted (in bits/symbol) is N_{bits}/N.

Two observations are in order with respect to the foregoing abstraction.

(1) The abstraction is discrete in time, yet actual channels are continuous in time. As long as the channel is bandlimited, though, the sampling theorem ensures that a discrete-time equivalent can be obtained [77]. This discretization is tackled in Chapter 2 and its implications for time-varying channels are further examined in Section 3.4.5. To reconcile this discrete-time abstraction with the continuous-time nature of actual channels, the "channel" in Fig. 1.2 can be interpreted as further encompassing the transmit and receive filters, $g_{\text{tx}}(\cdot)$ and $g_{\text{rx}}(\cdot)$, plus a sampling device; this is reflected in Fig. 1.3.

(2) The abstraction is digital, i.e., the information to be transmitted is already in the form of bits. The digitization of information, regardless of its nature, underlies all modern forms of data storage and transmission, and is yet again a legacy of Shannon's work. We do not concern ourselves with the origin and meaning of the information, or with how it was digitized. Furthermore, we regard the bits to be transmitted as IID, side-stepping the source encoding process that removes data redundancies and dependencies before transmission as well as the converse process that reintroduces them after reception.

For the sake of notational compactness, we introduce vector notation for time-domain sequences (and in other chapters also for frequency-domain sequences). And, to distinguish these vectors from their space-domain counterparts, we complement the bold font types

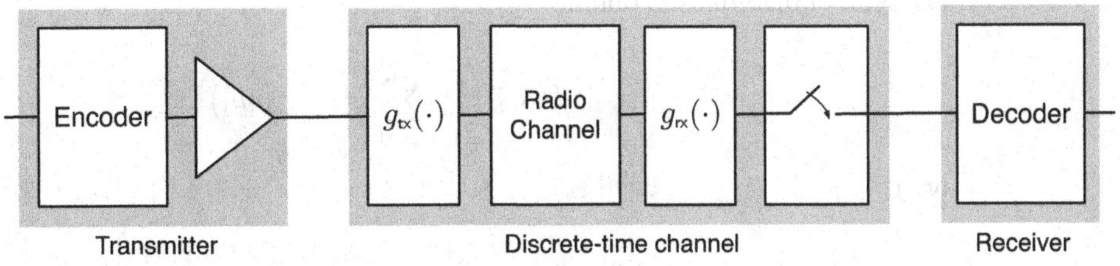

Fig. 1.3 Basic abstraction of a communication link, including the discrete-to-continuous and continuous-to-discrete interfaces.

with an overbar. The sequence $s[0], \ldots, s[N-1]$, for instance, is assembled into the vector

$$\bar{s} = \begin{bmatrix} s[0] \\ \vdots \\ s[N-1] \end{bmatrix}. \tag{1.87}$$

The channel law $f_{\bar{y}|\bar{s}}(\cdot)$ is a key element in the computation of the capacity, and the mutual information between $s[0], \ldots, s[N-1]$ and $y[0], \ldots, y[N-1]$ can be expressed as a function thereof. Recalling (1.46), we can write

$$I(\bar{s};\bar{y}) = \mathbb{E}\left[\log_2 \frac{f_{\bar{y}|\bar{s}}(\bar{y}|\bar{s})}{f_{\bar{y}}(\bar{y})}\right] \tag{1.88}$$

$$= \mathbb{E}\left[\log_2 \frac{f_{\bar{y}|\bar{s}}(\bar{y}|\bar{s})}{\frac{1}{2^{N_{\text{bits}}}}\sum_{m=0}^{2^{N_{\text{bits}}}-1} f_{\bar{y}|\bar{s}}(\bar{y}|\bar{s}_m)}\right], \tag{1.89}$$

where the expectations are over $\bar{s}$ and $\bar{y}$. In (1.89), the $2^{N_{\text{bits}}}$ codewords have been assumed equiprobable, with $\bar{s}_m$ the mth such codeword.

Example 1.15 (Channel law with Gaussian noise)

Let

$$y[n] = \sqrt{\rho}\,[\boldsymbol{A}]_{n,\mathrm{n}}\, s[\mathrm{n}] + z[n] \qquad n, \mathrm{n} = 0, \ldots, N-1 \tag{1.90}$$

or, more compactly, $\bar{y} = \sqrt{\rho}\boldsymbol{A}\bar{s} + \bar{z}$ where $\boldsymbol{A}$ is a fixed matrix whose (n, n)th entry determines how the nth transmit symbol affects the nth received one, while $\bar{z} \sim \mathcal{N}_{\mathbb{C}}(\boldsymbol{0}, \boldsymbol{I})$. For this linear channel impaired by Gaussian noise,

$$f_{\bar{y}|\bar{s}}(\bar{y}|\bar{s}) = \frac{1}{\pi^N} e^{-\|\bar{y} - \sqrt{\rho}\boldsymbol{A}\bar{s}\|^2}. \tag{1.91}$$

If the channel law factors as $f_{\bar{y}|\bar{s}}(\cdot) = \prod_{n=0}^{N-1} f_{y[n]|s[n]}(\cdot)$, meaning that its output at symbol n depends only on the input at symbol n, the channel is said to be *memoryless*. Then, there is no loss of optimality in having codewords with statistically independent entries [14] and thus $f_{\bar{s}}(\cdot)$, and subsequently $f_{\bar{y}}(\cdot)$, can also be factored as a product of per-symbol marginals to obtain

$$I(\bar{s};\bar{y}) = \sum_{n=0}^{N-1} I(s[n];y[n]) \tag{1.92}$$

with

$$I(s[n];y[n]) = \mathbb{E}\left[\log_2 \frac{f_{y[n]|s[n]}(y[n]|s[n])}{f_{y[n]}(y[n])}\right]. \tag{1.93}$$

For channels both memoryless and stationary, we can drop the index n and write

$$I(s;y) = \mathbb{E}\left[\log_2 \frac{f_{y|s}(y|s)}{f_y(y)}\right] \tag{1.94}$$

$$= \mathbb{E}\left[\log_2 \frac{f_{y|s}(y|s)}{\sum_{m=0}^{M-1} f_{y|s}(y|s_m)\, p_m}\right], \tag{1.95}$$

where (1.95) applies if the signal conforms to an M-point constellation; with those constellation points further equiprobable, $p_m = 1/M$ for $m = 0, \ldots, M-1$. This convenient formulation involving a single symbol is said to be *single-letter*. Conversely, the codeword-wise formulation that is needed for channels with memory such as the one in Example 1.15 is termed *nonsingle-letter*. Although the direct discretization of a wireless channel generally does not yield a memoryless law, with equalizing countermeasures at the receiver the effects of the memory can be reduced to a minimum (see Chapter 2). Moreover, with OFDM, the signals are structured such that their joint discretization in time and frequency ends up being basically memoryless. Altogether, most—but not all—settings in this book are memoryless.

Example 1.16 (Memoryless channel law with Gaussian noise)

Let

$$y[n] = \sqrt{\rho}\, s[n] + z[n] \qquad n = 0, \ldots, N-1 \tag{1.96}$$

where $z[0], \ldots, z[N-1]$ are IID with $z \sim \mathcal{N}_{\mathbb{C}}(0,1)$. For this memoryless channel impaired by Gaussian noise,

$$f_{y|s}(\mathrm{y}|\mathrm{s}) = \frac{1}{\pi}\, e^{-|\mathrm{y}-\sqrt{\rho}\mathrm{s}|^2}. \tag{1.97}$$

1.5.2 Capacity

In an arbitrary channel, not necessarily memoryless, the average probability of making an error when decoding a codeword equals

$$p_{\mathrm{e}} = \sum_{\mathrm{m}=0}^{2^{N_{\mathrm{bits}}}-1} \mathbb{P}[\hat{\mathrm{w}} \neq \mathrm{m}\,|\mathrm{w}=\mathrm{m}]\,\mathbb{P}[\mathrm{w}=\mathrm{m}] \tag{1.98}$$

where w is the index of the codeword actually transmitted while $\hat{\mathsf{w}}$ is the index guessed by the decoder. With the codewords equiprobable, the above reduces to

$$p_{\mathrm{e}} = \frac{1}{2^{N_{\mathsf{bits}}}} \sum_{\mathsf{m}=0}^{2^{N_{\mathsf{bits}}}-1} \mathbb{P}[\hat{\mathsf{w}} \neq \mathsf{m} \,|\mathsf{w}=\mathsf{m}]. \tag{1.99}$$

We term p_{e} the *error probability*, noting that it can be alternatively referred to as *word error probability*, *block error probability*, or *frame error probability*. A rate of information N_{bits}/N (in bits/symbol) can be communicated reliably if there exists a code of such rate for which $p_{\mathrm{e}} \to 0$ as $N \to \infty$. Note that we do not require the error probability to be zero for arbitrary N, but only that it vanishes as $N \to \infty$. In the channels of interest to this text, error-free communication at positive rates is possible only asymptotically in the codeword length.

The capacity C (in bits/symbol) is then the highest rate achievable reliably and, once exceeded, the error probability rises rapidly [60, section 10.4]. Most importantly, if the channel is *information stable* then the capacity is the maximum mutual information between the transmit and receive sequences. The concept of information stability can be explained by means of the so-called information density

$$i(\bar{\boldsymbol{s}};\bar{\boldsymbol{y}}) = \log_2 \frac{f_{\bar{\boldsymbol{s}},\bar{\boldsymbol{y}}}(\bar{\mathbf{s}},\bar{\mathbf{y}})}{f_{\bar{\boldsymbol{s}}}(\bar{\mathbf{s}})\, f_{\bar{\boldsymbol{y}}}(\bar{\mathbf{y}})}, \tag{1.100}$$

which is the quantity whose expectation, recalling (1.46), equals the mutual information. The channel is information stable if [78]

$$\lim_{N\to\infty} \frac{1}{N}\, i(\bar{\boldsymbol{s}};\bar{\boldsymbol{y}}) = \lim_{N\to\infty} \frac{1}{N}\, \mathbb{E}\Big[i(\bar{\boldsymbol{s}};\bar{\boldsymbol{y}})\Big] \tag{1.101}$$

$$= \lim_{N\to\infty} \frac{1}{N}\, I(\bar{\boldsymbol{s}};\bar{\boldsymbol{y}}), \tag{1.102}$$

which means that the information density does not deviate (asymptotically) from the mutual information. Intuitively, this indicates that the information that $y[0],\ldots,y[N-1]$ conveys about $s[0],\ldots,s[N-1]$ is invariant provided that N is large enough. This seemingly abstract concept is best understood by examining specific manifestations of stable and unstable channels, such as the ones encountered later in the context of fading. For our purposes, it is enough to point out that a sufficient condition for information stability is that the channel be stationary and ergodic, conditions that, as reasoned in Chapter 3, are satisfied within a certain time horizon by virtually all wireless channels of interest. For a more general capacity formulation that encompasses channels that are not information stable, the reader is referred to [79, 80].

If the channel is stationary and ergodic, then [81],

$$C = \max_{\mathsf{signal\ constraints}} \lim_{N\to\infty} \frac{1}{N}\, I(\bar{\boldsymbol{s}};\bar{\boldsymbol{y}}), \tag{1.103}$$

where the maximization is over the joint distribution of the unit-power codeword symbols $s[0],\ldots,s[N-1]$, with subsequent amplification subject to whichever constraints apply to the signal's power and/or magnitude (see Section 2.3.5).

Shannon originally dealt with channels not only stationary and ergodic, but also memoryless, in which case [58]

$$C = \max_{\text{signal constraints}} I(s;y), \tag{1.104}$$

with the maximum taken over the distribution of the unit-power variable s, and with the subsequent amplification subject to the applicable constraints. The capacity then entails optimizing the marginal distribution of the symbols that make up the codewords. Because of the memorylessness and stationarity of the channel, such symbols may be not only independent but IID and thus the optimization is over any one of them. In this case, the capacity admits a *single-letter* formulation.

As argued earlier, the mean of the symbols $s[0], \ldots, s[N-1]$ does not contribute to the mutual information. However, a nonzero-mean would increase the power of the transmit signal. It follows that, irrespective of the specific type of power constraint, the maximizations of mutual information invariably yield signals that are zero-mean and hence only zero-mean signals are contemplated throughout the book.

From C (in bits/symbol) and from the symbol period T, the bit rate R (in bits/s) that can be communicated reliably satisfies $R \leq C/T$. And, since the sampling theorem dictates that $1/T \leq B$ with B the (passband) bandwidth, we have that

$$\frac{R}{B} \leq C, \tag{1.105}$$

evidencing the alternative measure of C in bits/s/Hz, often preferred to bits/symbol.[1] With a capacity-achieving codebook and $1/T = B$ symbols/s, the inequality in (1.105) becomes (asymptotically) an equality. If the pulse shape induced by the transmit and receive filters $g_{\text{tx}}(\cdot)$ and $g_{\text{rx}}(\cdot)$ incurs a bandwidth larger than $1/T$, the resulting shortfall from capacity must be separately accounted for. Indeed, as discussed in Chapter 2, pulse shapes with a modicum of excess bandwidth are common to diminish the sensitivity to synchronization inaccuracies.

Throughout this text, we resist utilizing the term "capacity" to describe the performance for specific distributions of $s[0], \ldots, s[N-1]$ that may be of interest but that are not optimum in the sense of maximizing (1.103) or (1.104). Rather, we then apply the term "spectral efficiency" and the description R/B, reserving "capacity" and C for the highest value over all possible signal distributions.

1.5.3 Coding and decoding

Before proceeding, let us establish some further terminology concerning the probabilistic relationship over the channel.

- We have introduced $f_{\bar{y}|\bar{s}}(\cdot)$ as the channel law, a function of both the transmit codeword and the observation at the receiver. For a fixed codeword, this defines the distribution of

[1] Our conversion of bits/symbol to bits/s/Hz, perfectly sufficient for complex baseband symbols representing real passband signals, can be generalized to real baseband signals and to spread-spectrum signals through the notion of *Shannon bandwidth* [82].

$y[0], \ldots, y[N-1]$ given that such codeword is transmitted while, for a fixed observation, it defines the *likelihood function* of $s[0], \ldots, s[N-1]$.

- With the conditioning reversed, $f_{\bar{s}|\bar{y}}(\cdot)$ is the *posterior probability* of a codeword given the observation at the receiver.

Optimum decoding rules

To establish the decoding rule that minimizes p_e, let us rewrite (1.98) into [83]

$$p_e = \sum_{m=0}^{2^{N_{bits}}-1} \mathbb{P}[\bar{y} \notin \mathcal{R}_m \,|\mathrm{w}=\mathrm{m}]\, \mathbb{P}[\mathrm{w}=\mathrm{m}] \tag{1.106}$$

$$= \sum_{m=0}^{2^{N_{bits}}-1} \Big(1 - \mathbb{P}[\bar{y} \in \mathcal{R}_m \,|\mathrm{w}=\mathrm{m}]\Big)\, \mathbb{P}[\mathrm{w}=\mathrm{m}] \tag{1.107}$$

$$= 1 - \sum_{m=0}^{2^{N_{bits}}-1} \int_{\mathcal{R}_m} f_{\bar{s},\bar{y}}(\bar{s}_m, \bar{y})\, d\bar{y} \tag{1.108}$$

$$= 1 - \sum_{m=0}^{2^{N_{bits}}-1} \int_{\mathcal{R}_m} f_{\bar{s}|\bar{y}}(\bar{s}_m|\bar{y})\, f_{\bar{y}}(\bar{y})\, d\bar{y} \tag{1.109}$$

where $\mathcal{R}_m$ denotes the decision region associated with codeword m, that is, the set of observations $y[0], \ldots, y[N-1]$ being mapped by the receiver onto message m. The $2^{N_{bits}}$ decision regions are disjoint. To minimize p_e, each term in (1.109) can be separately maximized. By inspection, the mth term is maximized by defining $\mathcal{R}_m$ as the region that contains all observations $\bar{y}$ for which the posterior probability $f_{\bar{s}|\bar{y}}(\bar{s}_m|\bar{y})$ is maximum. The optimum decoding strategy is thus to select the most probable codeword given what has been observed, a rule that is naturally termed maximum a-posteriori (MAP).

Applying Bayes' theorem (see Appendix C.1.1),

$$f_{\bar{s}|\bar{y}}(\bar{s}_m|\bar{y}) = \frac{f_{\bar{y}|\bar{s}}(\bar{y}|\bar{s}_m)\, f_{\bar{s}}(\bar{s}_m)}{f_{\bar{y}}(\bar{y})} \tag{1.110}$$

and, when the codewords are equiprobable,

$$f_{\bar{s}|\bar{y}}(\bar{s}_m|\bar{y}) = \frac{f_{\bar{y}|\bar{s}}(\bar{y}|\bar{s}_m)}{2^{N_{bits}}\, f_{\bar{y}}(\bar{y})}, \tag{1.111}$$

where the right-hand side denominator does not depend on m and is thus irrelevant to a maximization over m. It follows that, with equiprobable codewords, maximizing the posterior probability on the left-hand side is equivalent to maximizing the likelihood function on the right-hand side numerator. MAP decoding is then equivalent to maximum-likelihood (ML) decoding, which, faced with an observation $\bar{y}$, guesses the message m that maximizes $f_{\bar{y}|\bar{s}}(\bar{y}|\bar{s}_m)$.

Example 1.17 (ML decoding rule with Gaussian noise)

Consider the channel with memory and Gaussian noise in Example 1.15. The likelihood function to maximize is

$$f_{\bar{y}|\bar{s}}(\bar{y}|\bar{s}_m) = \frac{1}{\pi^N} e^{-\|\bar{y}-\sqrt{\rho}A\bar{s}_m\|^2} \tag{1.112}$$

and, because the logarithm is a monotonic function, the ensuing maximization yields the same result as the maximization of

$$\log_e f_{\bar{y}|\bar{s}}(\bar{y}|\bar{s}_m) = -N\log_e \pi - \|\bar{y} - \sqrt{\rho}A\bar{s}_m\|^2 \tag{1.113}$$

whose first term is constant and so inconsequential to the maximization. The decision made by an ML decoder is thus the message m whose codeword $\bar{s}_m$ minimizes $\|\bar{y} - \sqrt{\rho}A\bar{s}_m\|^2$, i.e., the codeword $\bar{s}_m$ that induces the channel output $\sqrt{\rho}A\bar{s}_m$ closest in Euclidean distance to the observation $\bar{y}$. This rule is therefore termed *minimum-distance* (or *nearest-neighbor*) decoding.

Example 1.18 (ML decoding rule for a memoryless channel with Gaussian noise)

For

$$y[n] = \sqrt{\rho}\, s[n] + z[n] \qquad n = 0, \ldots, N-1 \tag{1.114}$$

with $z \sim \mathcal{N}_{\mathbb{C}}(0,1)$, the ML guess when the receiver observes $y[0], \ldots, y[N-1]$ is the codeword $s[0], \ldots, s[N-1]$ that minimizes $\sum_{n=0}^{N-1} \left|y[n] - \sqrt{\rho}\, s[n]\right|^2$.

From hard to soft decoding

In classic receivers of yore, the decoding rules were applied upfront on a symbol-by-symbol basis. From the observation of $y[n]$, a hard decision was made on the value of $s[n]$. This procedure, whereby the MAP or ML rules were applied to individual symbols, was regarded as the demodulation of the underlying constellation. Subsequently, the N hard decisions for $s[0], \ldots, s[N-1]$ were assembled and fed into a decoder, with two possible outcomes. If the block of hard decisions was a valid codeword, success was declared. Alternatively, some of the hard decisions were taken to be erroneous and an attempt was made, exploiting the algebraic structure of the code, to correct them by modifying the block into a valid codeword. In these receivers, then, the decoder was essentially a corrector for the mistakes made by the demodulator. Moreover, in making a hard decision on a given symbol, the demodulator was throwing away information that could have been valuable to the decoder when deciding on other symbols [83, 84].

The extreme instance of this approach is uncoded transmission, where the message bits are directly mapped onto a constellation at the transmitter and recovered via ML-based hard decision at the receiver. Each bit is then at the mercy of the channel experienced by the particular symbol in which it is transmitted, without the protection that being part of a long codeword can afford. Only a strong SNR or a low spectral efficiency could guarantee certain reliability in this pre-Shannon framework.

Except when simplicity is the utmost priority or no latency can be tolerated, transmissions are nowadays heavily coded and decoders operate directly on $y[0], \ldots, y[N-1]$, avoiding any preliminary discarding of information.

Near-capacity coding

As far as the codebooks are concerned, the coding theorems that establish the capacity as the maximum mutual information rely on random coding arguments—championed by Shannon—whereby the codewords are constructed by drawing symbols at random from a to-be-optimized distribution. However, because such codebooks have no structure, their optimum decoding would require an exhaustive search through the $2^{N_{\text{bits}}}$ codewords making up the codebook in order to find the one codeword that maximizes the MAP or ML criteria. This is an impossible task even for modest values of N_{bits}; with $N_{\text{bits}} = 30$, a meager value by today's standards, the number of codewords is already over 1000 million. Thus, random coding arguments, while instrumental to establishing the capacity, do not provide viable ways to design practical codes for large N_{bits}. For decades after 1948, coding theorists concentrated on the design of codebooks with algebraic structures that could be decoded optimally with a complexity that was polynomial, rather than exponential, in the codeword length [85]. Then, in the 1990s, with the serendipitous discovery of *turbo codes* [86] and the rediscovery of low-density parity check (LDPC) codes—formulated by Robert Gallager in the 1960s but computationally unfeasible at that time—the emphasis shifted to codebook constructions that could be decoded *suboptimally* in an efficient fashion. Staggering progress has been made since, and today we have powerful codes spanning hundreds to thousands of symbols and operating very close to capacity. These codes offer the random-like behavior leveraged by coding theorems with a relatively simple inner structure; in particular, turbo codes are obtained by concatenating lower-complexity codes through a large pseudo-random interleaver.

A comprehensive coverage of codebook designs and decoding techniques is beyond the scope of this book, and the interested reader is referred to dedicated texts [83, 87]. Here, we mostly regard encoders and decoders as closed boxes and discuss how these boxes ought to be arranged and/or modified to fit the MIMO *transceivers* (our term to compactly subsume both transmitters and receivers) under consideration.

Signal-space coding versus binary coding

Thus far we have implicitly considered codes constructed directly over the signal alphabet, say the points of a discrete constellation. The art of constructing such codes is referred to as *signal-space coding* (or *coded modulation*). Practical embodiments of signal-space coding exist, chiefly the trellis-coded modulation (TCM) schemes invented by Ungerboeck in the 1970s [88]. Signal-space coding conforms literally to the diagram presented in Fig. 1.3.

As an alternative to signal-space coding, it is possible to first run the message bits through a binary encoder, which converts messages onto binary codewords; subsequently, the coded bits are mapped onto symbols $s[0], \ldots, s[N-1]$ having the desired distribution, in what can be interpreted as a modulation of the constellation. This alternative is attractive because it keeps the signal distribution arbitrary while allowing the codebook to be designed over the simple and convenient binary alphabet. If the rate of the binary code is r message bits per coded bit and the spectral efficiency is R/B (in message b/s), the constellation must accommodate $\frac{1}{\mathsf{r}} R/B$ coded bits. When the M constellation points are equiprobable, the number of bits it can accommodate equals $\log_2 M$ and thus we can write

$$\frac{R}{B} = \mathsf{r} \log_2 M. \tag{1.115}$$

This expression suggests how the transmit rate can be controlled by adjusting r and M, a procedure that is explored in detail later in the book.

Fundamentally, there is no loss of optimality in implementing signal-space coding by mapping the output of a binary encoder onto constellation points as long as the receiver continues to decode as if those constellation points were the actual coding alphabet. Indeed, if we take a string of random bits, parse them onto groups, and map each such group to a constellation point, the resulting codeword is statistically equivalent to a codeword defined randomly on the constellation alphabet itself. At the transmitter end, therefore, coding and mapping can be separated with no penalty as long as the receiver performs joint demapping and decoding. Ironically, then, what defines signal-space coding is actually the signal-space *decoding*.

Rather than decoding on the signal alphabet, however, the preferred approach is to first demap the binary code from the constellation and then separately decode it by means of a binary decoder. However, to avoid the pitfalls of hard demodulation and prevent an early loss of information, what is fed to the decoder is not a hard decision on each bit but rather a soft value.

Soft-input binary decoding

Consider a bit b. We can characterize the probability that b is 0 or 1 directly via $\mathbb{P}[b=0]$ or $\mathbb{P}[b=1] = 1 - \mathbb{P}[b=0]$ but also, equivalently, through the ratio $\frac{\mathbb{P}[b=1]}{\mathbb{P}[b=0]}$ [89]. More conveniently (because products and divisions become simpler additions and subtractions), we may instead use a logarithmic version of this ratio, the so-called *L-value*

$$\mathrm{L}(b) = \log \frac{\mathbb{P}[b=1]}{\mathbb{P}[b=0]}, \tag{1.116}$$

where, as done with entropies and differential entropies, notation has been slightly abused by expressing $\mathrm{L}(\cdot)$ as a function of b when it is actually a function of its distribution. A positive L-value indicates that the bit in question is more likely to be a 1 than a 0, and vice versa for a negative value, with the magnitude indicating the confidence of the decision. An L-value close to zero indicates that the decision on the bit is unreliable.

Now denote by $b_\ell[n]$ the ℓth coded bit within $s[n]$. A soft demapper should feed to the binary decoder a value quantifying how close $b_\ell[n]$ is to being a 0 or a 1 in light of what the receiver has observed, and that information can be conveyed through the posterior L-value

$$\mathrm{L_D}\big(b_\ell[n]\,|\,\bar{\boldsymbol{y}}\big) = \log \frac{\mathbb{P}\big[b_\ell[n]=1\,|\,\bar{\boldsymbol{y}}\big]}{\mathbb{P}\big[b_\ell[n]=0\,|\,\bar{\boldsymbol{y}}\big]}. \tag{1.117}$$

Applying Bayes' theorem, we have that

$$\mathbb{P}\big[b_\ell[n]=0\,|\bar{\boldsymbol{y}}=\bar{\mathbf{y}}\big] = \frac{f_{\bar{\boldsymbol{y}}|b_\ell[n]}(\bar{\mathbf{y}}|0)}{f_{\bar{\boldsymbol{y}}}(\bar{\mathbf{y}})}\,\mathbb{P}\big[b_\ell[n]=0\big] \tag{1.118}$$

$$\mathbb{P}\big[b_\ell[n]=1\,|\bar{\boldsymbol{y}}=\bar{\mathbf{y}}\big] = \frac{f_{\bar{\boldsymbol{y}}|b_\ell[n]}(\bar{\mathbf{y}}|1)}{f_{\bar{\boldsymbol{y}}}(\bar{\mathbf{y}})}\,\mathbb{P}\big[b_\ell[n]=1\big] \tag{1.119}$$

and thus

$$\mathrm{L_D}\big(b_\ell[n]\,|\bar{\boldsymbol{y}}=\bar{\mathbf{y}}\big) = \underbrace{\log\frac{\mathbb{P}[b_\ell[n]=1]}{\mathbb{P}[b_\ell[n]=0]}}_{\mathrm{L_A}(b_\ell[n])} + \underbrace{\log\frac{f_{\bar{\boldsymbol{y}}|b_\ell[n]}(\bar{\mathbf{y}}|1)}{f_{\bar{\boldsymbol{y}}|b_\ell[n]}(\bar{\mathbf{y}}|0)}}_{\mathrm{L_E}(b_\ell[n]\,|\,\bar{\boldsymbol{y}})}, \tag{1.120}$$

whose first term is whatever a-priori information the receiver may already have about $b_\ell[n]$; in the absence of any such information, $\mathrm{L_A}\big(b_\ell[n]\big)=0$. The second term, in turn, captures whatever fresh information the demapper supplies about $b_\ell[n]$ in light of what the receiver observes. More precisely, $\mathrm{L_E}\big(b_\ell[n]\,|\,\bar{\boldsymbol{y}}\big)$ is the logarithm of the ratio of the likelihood function for $b_\ell[n]$ evaluated at its two possible values, hence it is a *log-likelihood ratio*. This convenient separation into a sum of two terms conveying old (or *intrinsic*) and new (or *extrinsic*) information is what makes L-values preferable to probabilities and sets the stage for iterative decoding schemes.

When the channel is memoryless, $s[n]$ does not influence received symbols other than $y[n]$ and thus $\mathrm{L_D}\big(b_\ell[n]\,|\,\bar{\boldsymbol{y}}\big) = \mathrm{L_D}\big(b_\ell[n]\,|\,y[n]\big)$. Dropping the symbol index, the single-letter L-value can then be written as $\mathrm{L_D}\big(b_\ell\,|\,y\big)$ and further insight can be gained. Assuming that the coded bits mapped to each codeword symbol are independent—a condition discussed in the next section—such that their probabilities can be multiplied, and that the signal conforms to the discrete constellation defined by $\mathrm{s}_0,\ldots,\mathrm{s}_{M-1}$,

$$\mathrm{L_D}\big(b_\ell|y=\mathrm{y}\big) = \log\frac{\mathbb{P}\big[b_\ell=1\,|y=\mathrm{y}\big]}{\mathbb{P}\big[b_\ell=0\,|y=\mathrm{y}\big]} \tag{1.121}$$

$$= \log\frac{\sum_{\mathrm{s}_m\in\mathcal{S}_1^\ell} p_{s|y}(\mathrm{s}_m|\mathrm{y})}{\sum_{\mathrm{s}_m\in\mathcal{S}_0^\ell} p_{s|y}(\mathrm{s}_m|\mathrm{y})} \tag{1.122}$$

$$= \log\frac{\sum_{\mathrm{s}_m\in\mathcal{S}_1^\ell} f_{y|s}(\mathrm{y}|\mathrm{s}_m)\,p_m}{\sum_{\mathrm{s}_m\in\mathcal{S}_0^\ell} f_{y|s}(\mathrm{y}|\mathrm{s}_m)\,p_m} \tag{1.123}$$

$$= \log\frac{\sum_{\mathrm{s}_m\in\mathcal{S}_1^\ell} f_{y|s}(\mathrm{y}|\mathrm{s}_m)\,\mathbb{P}[b_\ell=1]\prod_{\ell'\neq\ell}\mathbb{P}[b_{\ell'}=\ell'\text{th bit of }\mathrm{s}_m]}{\sum_{\mathrm{s}_m\in\mathcal{S}_0^\ell} f_{y|s}(\mathrm{y}|\mathrm{s}_m)\,\mathbb{P}[b_\ell=0]\prod_{\ell'\neq\ell}\mathbb{P}[b_{\ell'}=\ell'\text{th bit of }\mathrm{s}_m]} \tag{1.124}$$

$$= \log\frac{\mathbb{P}[b_\ell=1]\sum_{\mathrm{s}_m\in\mathcal{S}_1^\ell} f_{y|s}(\mathrm{y}|\mathrm{s}_m)\prod_{\ell'\neq\ell}\mathbb{P}[b_{\ell'}=\ell'\text{th bit of }\mathrm{s}_m]}{\mathbb{P}[b_\ell=0]\sum_{\mathrm{s}_m\in\mathcal{S}_0^\ell} f_{y|s}(\mathrm{y}|\mathrm{s}_m)\prod_{\ell'\neq\ell}\mathbb{P}[b_{\ell'}=\ell'\text{th bit of }\mathrm{s}_m]} \tag{1.125}$$

$$= \mathrm{L_A}(b_\ell) + \log\frac{\sum_{\mathrm{s}_m\in\mathcal{S}_1^\ell} f_{y|s}(\mathrm{y}|\mathrm{s}_m)\prod_{\ell'\neq\ell}\mathbb{P}[b_{\ell'}=\ell'\text{th bit of }\mathrm{s}_m]}{\sum_{\mathrm{s}_m\in\mathcal{S}_0^\ell} f_{y|s}(\mathrm{y}|\mathrm{s}_m)\prod_{\ell'\neq\ell}\mathbb{P}[b_{\ell'}=\ell'\text{th bit of }\mathrm{s}_m]} \tag{1.126}$$

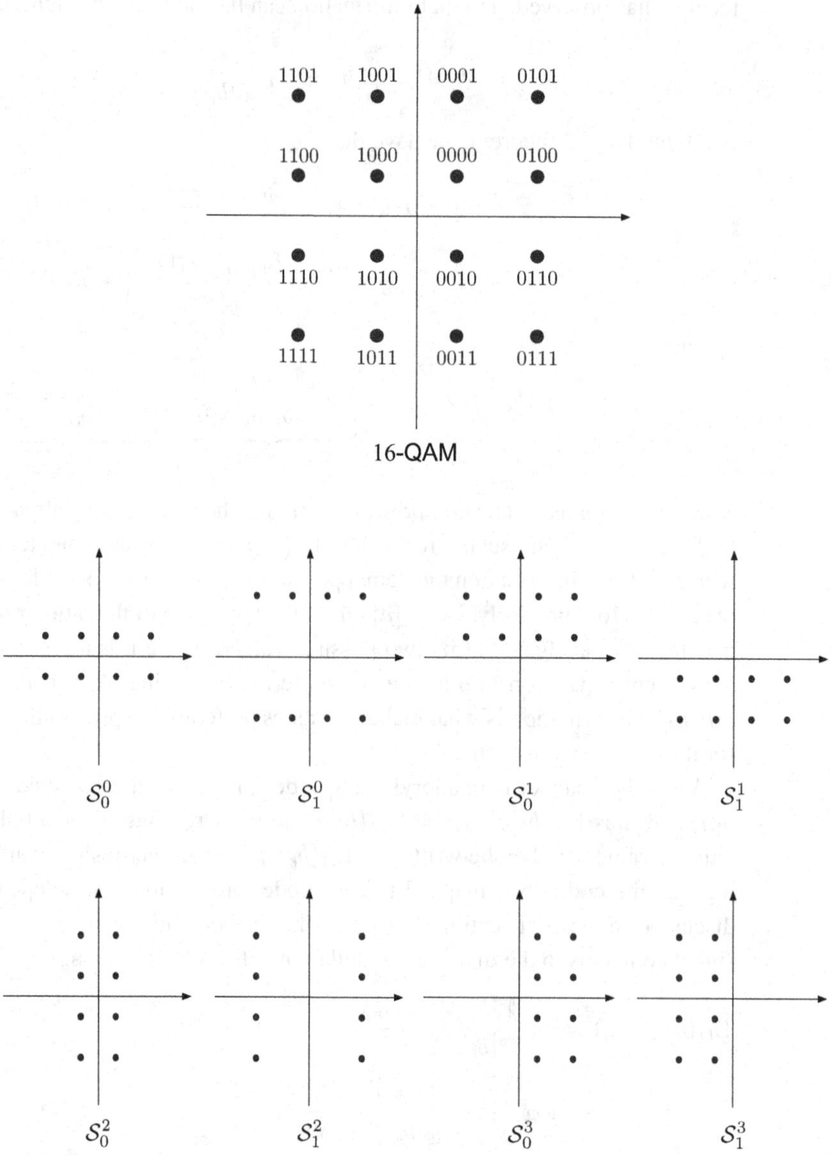

Fig. 1.4 Above, 16-QAM constellation with Gray mapping. Below, subsets $\mathcal{S}_0^\ell$ and $\mathcal{S}_1^\ell$ for $\ell = 0, 1, 2, 3$ with the bits ordered from right to left.

where $\mathcal{S}_0^\ell$ and $\mathcal{S}_1^\ell$ are the subsets of constellation points whose ℓth bit is 0 or 1, respectively (see Fig. 1.4). Dividing the second term's numerator and denominator by $\prod_{\ell' \neq \ell} \mathbb{P}[b_{\ell'} = 0]$, we further obtain

$$\mathrm{L_D}\left(b_\ell | y = \mathrm{y}\right) = \mathrm{L_A}(b_\ell) + \log \frac{\sum_{\mathrm{s}_m \in \mathcal{S}_1^\ell} f_{y|s}(\mathrm{y}|\mathrm{s}_m) \prod_{\ell' = \mathcal{B}_1^m} \frac{\mathbb{P}[b_{\ell'}=1]}{\mathbb{P}[b_{\ell'}=0]}}{\sum_{\mathrm{s}_m \in \mathcal{S}_0^\ell} f_{y|s}(\mathrm{y}|\mathrm{s}_m) \prod_{\ell' = \mathcal{B}_1^m} \frac{\mathbb{P}[b_{\ell'}=1]}{\mathbb{P}[b_{\ell'}=0]}} \tag{1.127}$$

$$= \mathrm{L_A}(b_\ell) + \log \frac{\sum_{\mathsf{s}_m \in \mathcal{S}_1^\ell} f_{y|s}(\mathsf{y}|\mathsf{s}_m) \exp\left(\sum_{\ell' \neq \ell, \ell' \in \mathcal{B}_1^m} \log \frac{\mathbb{P}[b_{\ell'}=1]}{\mathbb{P}[b_{\ell'}=0]}\right)}{\sum_{\mathsf{s}_m \in \mathcal{S}_0^\ell} f_{y|s}(\mathsf{y}|\mathsf{s}_m) \exp\left(\sum_{\ell' \neq \ell, \ell' \in \mathcal{B}_1^m} \log \frac{\mathbb{P}[b_{\ell'}=1]}{\mathbb{P}[b_{\ell'}=0]}\right)}$$

$$= \mathrm{L_A}(b_\ell) + \underbrace{\log \frac{\sum_{\mathsf{s}_m \in \mathcal{S}_1^\ell} f_{y|s}(\mathsf{y}|\mathsf{s}_m) \exp\left(\sum_{\ell' \neq \ell, \ell' \in \mathcal{B}_1^m} \mathrm{L_A}(b_{\ell'})\right)}{\sum_{\mathsf{s}_m \in \mathcal{S}_0^\ell} f_{y|s}(\mathsf{y}|\mathsf{s}_m) \exp\left(\sum_{\ell' \neq \ell, \ell' \in \mathcal{B}_1^m} \mathrm{L_A}(b_{\ell'})\right)}}_{\mathrm{L_E}(b_\ell \,|\, y)} \tag{1.128}$$

where $\mathcal{B}_1^m$ is the subset of coded bits mapped to constellation point s_m that equal 1. The crucial insight here is that the extrinsic term $\mathrm{L_E}(b_\ell|y)$ depends on $\mathrm{L_A}(b_{\ell'})$ for $\ell' \neq \ell$ but not on $\mathrm{L_A}(b_\ell)$. Hence, $\mathrm{L_E}(b_\ell|y)$ contains the information that the demapper can gather about the ℓth bit in light of what the receiver observes and of whatever a-priori information may be available about the other bits within the same symbol. (Although assumed unconditionally independent, the coded bits may exhibit dependences when conditioned on y.) This opens the door to implementing iterative receivers, as detailed in the next section. For one-shot receivers, where the soft demapping takes place only once, there is no a-priori information and thus

$$\mathrm{L_D}(b_\ell \,| y = \mathsf{y}) = \log \frac{\sum_{\mathsf{s}_m \in \mathcal{S}_1^\ell} f_{y|s}(\mathsf{y}|\mathsf{s}_m)}{\sum_{\mathsf{s}_m \in \mathcal{S}_0^\ell} f_{y|s}(\mathsf{y}|\mathsf{s}_m)}. \tag{1.129}$$

This log-likelihood ratio is what is fed into the decoder of a one-shot receiver.

Example 1.19 (Log-likelihood ratio for a memoryless channel with Gaussian noise)

For

$$y[n] = \sqrt{\rho}\, s[n] + z[n] \qquad n = 0, \ldots, N-1 \tag{1.130}$$

with $z \sim \mathcal{N}_{\mathbb{C}}(0,1)$ and equiprobable constellation points, the log-likelihood ratios fed into the decoder for each symbol are

$$\mathrm{L_D}(b_\ell \,| y = \mathsf{y}) = \log \frac{\sum_{\mathsf{s}_m \in \mathcal{S}_1^\ell} e^{-|\mathsf{y} - \sqrt{\rho}\, \mathsf{s}_m|^2}}{\sum_{\mathsf{s}_m \in \mathcal{S}_0^\ell} e^{-|\mathsf{y} - \sqrt{\rho}\, \mathsf{s}_m|^2}} \qquad \ell = 0, \ldots, \log_2 M - 1. \tag{1.131}$$

From the log-likelihood ratios based on the receiver observations and from its own knowledge of the code structure, a decoder can then compute posterior L-values for each of the message bits, namely

$$\mathrm{L_D}(\mathsf{b}[\mathsf{n}] \,| \bar{y}) = \log \frac{\mathbb{P}[\mathsf{b}[\mathsf{n}] = 1 \,| \bar{y}]}{\mathbb{P}[\mathsf{b}[\mathsf{n}] = 0 \,| \bar{y}]}, \tag{1.132}$$

where $\mathsf{b}[\mathsf{n}]$ for $\mathsf{n} = 0, \ldots, N_{\mathsf{bits}} - 1$ are the bits making up the message. A processor producing these posterior L-values, a decidedly challenging task when the codewords are long, is referred to as an a-posteriori probability (APP) decoder, or also as a soft-input soft-output decoder. The APP decoder is one of the key engines of modern receivers, with different flavors depending on the class of code, e.g., the Bahl–Cocke–Jelinek–Raviv (BCJR) algorithm for convolutional codes [90]. In the case of turbo codes, where two constituent codes are concatenated, a breakdown of $L_D(b[n] \,|\, \bar{y})$ into a-priori and extrinsic information about each message bit is the key to iterative decoding procedures whereby two APP decoders operate on the constituent codes exchanging information. Specifically, the extrinsic information generated by a first decoder is fed as a-priori information to the second decoder, allowing it to produce a better guess on the message bits as well as new extrinsic information for the first decoder, and so on. As the constituent codes are concatenated through an interleaver, the extrinsic information exchanged by the decoders must be interleaved and deinterleaved on each pass. This reduces the probability that the decoding process gets stuck in loops, and thus every iteration reduces the error probability with a handful of iterations sufficing to reach satisfactory levels. LDPC codes, although made up of a single block code, are also decoded iteratively.

Whichever the type of code, the sign of the L-values generated by an APP decoder for the message bits directly gives the MAP decisions,

$$\hat{b}[n] = \text{sign}\Big(L_D(b[n] \,|\, \bar{y})\Big) \qquad n = 0, \ldots, N_{\text{bits}} - 1. \tag{1.133}$$

Although it takes the entire codeword into account, an APP decoder maximizes the posterior probability on a bit-by-bit basis, thereby minimizing the average bit error probability rather than p_e. If the probabilities $\mathbb{P}\big[\hat{b}[n] = b[n] \,|\, \bar{y}\big]$ for $n = 0, \ldots, N_{\text{bits}} - 1$ are conditionally independent given the observations, then the minimization of the bit error probability also minimizes p_e. Otherwise it need not, yet in practice it hardly matters: although there is no guarantee that capacity can then be achieved for $N \to \infty$, APP decoders perform superbly. In simple settings with turbo or LDPC codes, operation at the brink of capacity with satisfactorily small error probabilities has been demonstrated [91, 92].

1.5.4 Bit-interleaved coded modulation

As mentioned, there is no fundamental loss of optimality in the conjunction of binary encoding and constellation mapping: a signal-space decoder can recover from such signals as much information as could have been transmitted with a nonbinary code defined directly on the constellation alphabet. Is the same true when the receiver features a combination of soft demapping and binary decoding?

To shed light on this issue at a fundamental level, let us posit a stationary and memoryless channel as well as an M-point equiprobable constellation. In this setting, codewords with IID entries are optimum and thus bits mapped to distinct symbols can be taken to be independent. However, the channel does introduce dependencies among same-symbol bits. Even with the coded bits produced by the binary encoder being statistically independent, after demapping at the receiver dependencies do exist among the soft values for bits that traveled on the same symbol. Unaware, a binary decoder designed for IID bits ignores these dependencies and regards the channel as being memoryless, not only at a symbol level but further at a bit level [93]. Let us see how much information can be recovered under this premise.

The binary-decoding counterpart to the memoryless mutual information in (1.94) and (1.95) is $\sum_{\ell=0}^{\log_2 M-1} I(b_\ell; y)$, and this binary-decoding counterpart can be put as a function of the channel law $f_{y|s}(\cdot)$ via [94]

$$\sum_{\ell=0}^{\log_2 M-1} I(b_\ell; y) = \sum_{\ell=0}^{\log_2 M-1} \mathbb{E}\left[\log_2 \frac{f_{y|b_\ell}(y|b_\ell)}{f_y(y)}\right] \tag{1.134}$$

$$= \sum_{\ell=0}^{\log_2 M-1} \frac{1}{2}\left(\mathbb{E}\left[\log_2 \frac{f_{y|b_\ell}(y|0)}{f_y(y)}\right] + \mathbb{E}\left[\log_2 \frac{f_{y|b_\ell}(y|1)}{f_y(y)}\right]\right) \tag{1.135}$$

$$= \sum_{\ell=0}^{\log_2 M-1} \frac{1}{2}\left(\mathbb{E}\left[\log_2 \frac{\sum_{\mathsf{s}_m \in \mathcal{S}_0^\ell} f_{y|s}(y|\mathsf{s}_m)\,\frac{1}{M/2}}{\sum_{m=0}^{M-1} f_{y|s}(y|\mathsf{s}_m)\,\frac{1}{M}}\right]\right.$$
$$\left. + \mathbb{E}\left[\log_2 \frac{\sum_{\mathsf{s}_m \in \mathcal{S}_1^\ell} f_{y|s}(y|\mathsf{s}_m)\,\frac{1}{M/2}}{\sum_{m=0}^{M-1} f_{y|s}(y|\mathsf{s}_m)\,\frac{1}{M}}\right]\right) \tag{1.136}$$

$$= \sum_{\ell=0}^{\log_2 M-1} \frac{1}{2}\left(\mathbb{E}\left[\log_2 \frac{\sum_{\mathsf{s}_m \in \mathcal{S}_0^\ell} f_{y|s}(y|\mathsf{s}_m)}{\frac{1}{2}\sum_{m=0}^{M-1} f_{y|s}(y|\mathsf{s}_m)}\right]\right.$$
$$\left. + \mathbb{E}\left[\log_2 \frac{\sum_{\mathsf{s}_m \in \mathcal{S}_1^\ell} f_{y|s}(y|\mathsf{s}_m)}{\frac{1}{2}\sum_{m=0}^{M-1} f_{y|s}(y|\mathsf{s}_m)}\right]\right), \tag{1.137}$$

where $\mathcal{S}_0^\ell$ and $\mathcal{S}_1^\ell$ are as defined in the previous section. In (1.136), the factors $1/(M/2)$ and $1/M$ correspond, respectively, to the probability of a constellation point s_m within the sets $\mathcal{S}_0^\ell$ and $\mathcal{S}_1^\ell$ (whose cardinality is $M/2$) and within the entire constellation (whose cardinality is M).

Whenever no dependencies among same-symbol coded bits are introduced by the channel, the binary decoder is not disregarding information and thus $\sum_\ell I(b_\ell; y) = I(s; y)$. This is the case with BPSK and QPSK, where a single coded bit is mapped to the in-phase and quadrature dimensions of the constellation. However, if each coded bit does acquire information about other ones within the same symbol, as is the case when multiple coded bits are mapped to the same dimension, then, with binary decoding not taking this information into account, $\sum_\ell I(b_\ell; y) < I(s; y)$.

Example 1.20

Consider a binary codeword mapped onto a QPSK constellation. The coded bits are parsed into pairs and the first and second bit within each pair are mapped, respectively, to the in-phase and quadrature components of the constellation, e.g., for a particular string 010010 within the binary codeword,

$$\cdots \underbrace{\underbrace{0}_{\mathrm{I}}\;\underbrace{1}_{\mathrm{Q}}}_{s[n-1]}\;\underbrace{\underbrace{0}_{\mathrm{I}}\;\underbrace{0}_{\mathrm{Q}}}_{s[n]}\;\underbrace{\underbrace{1}_{\mathrm{I}}\;\underbrace{0}_{\mathrm{Q}}}_{s[n+1]} \cdots \tag{1.138}$$

The resulting QPSK codeword $s[0], \ldots, s[N-1]$ is transmitted, contaminated by noise, and demapped back into a binary codeword at the receiver. The noise affects the bits as follows:

$$\cdots \underbrace{0}_{\Re\{z[n-1]\}} \underbrace{1}_{\Im\{z[n-1]\}} \underbrace{0}_{\Re\{z[n]\}} \underbrace{0}_{\Im\{z[n]\}} \underbrace{1}_{\Re\{z[n+1]\}} \underbrace{0}_{\Im\{z[n+1]\}} \cdots \tag{1.139}$$

Provided the noise samples are independent and the real and imaginary parts of each noise sample are also mutually independent, no dependences are introduced among the bits. Even with binary coding and decoding, it is as if the code were defined on the QPSK alphabet itself and the performance limits are dictated by $I(s; y)$.

Example 1.21

Consider a binary codeword mapped onto a 16-QAM constellation. The bits are parsed into groups of four, from which the in-phase and quadrature components must be determined. Among the various possible mappings, suppose we choose to map the first two bits of each group to the in-phase component and the final two bits to the quadrature component, e.g., for a particular string 011010110100 within the binary codeword,

$$\cdots \underbrace{\underbrace{01}_{\mathrm{I}} \underbrace{10}_{\mathrm{Q}}}_{s[n-1]} \underbrace{\underbrace{10}_{\mathrm{I}} \underbrace{11}_{\mathrm{Q}}}_{s[n]} \underbrace{\underbrace{01}_{\mathrm{I}} \underbrace{00}_{\mathrm{Q}}}_{s[n+1]} \cdots \tag{1.140}$$

The resulting 16-QAM codeword $s[0], \ldots, s[N-1]$ is transmitted, contaminated by noise, and soft-demapped back into a binary codeword at the receiver. The noise affects the bits as

$$\cdots \underbrace{1}_{\Im\{z[n-1]\}} \underbrace{0}_{\Im\{z[n-1]\}} \underbrace{1}_{\Re\{z[n]\}} \underbrace{0}_{\Re\{z[n]\}} \underbrace{1}_{\Im\{z[n]\}} \underbrace{1}_{\Im\{z[n]\}} \underbrace{0}_{\Re\{z[n+1]\}} \underbrace{1}_{\Re\{z[n+1]\}} \cdots \tag{1.141}$$

and thus pairs of coded bits are subject to the same noise. While a signal-space decoder would take these additional dependences into account and be limited by $I(s; y)$, a binary decoder ignores them and is instead limited by $I(b_0; y) + I(b_1; y) + I(b_2; y) + I(b_3; y)$ where b_ℓ is the ℓth bit within each group of four.

Remarkably, the difference between $\sum_\ell I(b_\ell; y)$ and $I(s; y)$ is tiny provided the mapping of coded bits to constellation points is chosen wisely. Gray mapping, where nearest-neighbor constellation points differ by only one bit, has been identified as a robust and attractive choice [94, 95].

Once all the ingredients that lead to (1.137) are in place, only one final functionality is needed to have a complete information-theoretic abstraction of a modern wireless transmission chain: interleaving. Although symbol-level interleaving would suffice to break off bursts of poor channel conditions, bit-level interleaving has the added advantage of shuffling also the bits contained in a given symbol; if the interleaving were deep enough to push any bit dependencies beyond the confines of each codeword, then the gap between $\sum_\ell I(b_\ell; y)$ and $I(s; y)$ would be closed. Although ineffective for $N \to \infty$, and thus not captured by mutual information calculations, bit-level interleaving does improve the performance of actual codes with finite N.

The coalition of binary coding and decoding, bit-level interleaving, and constellation

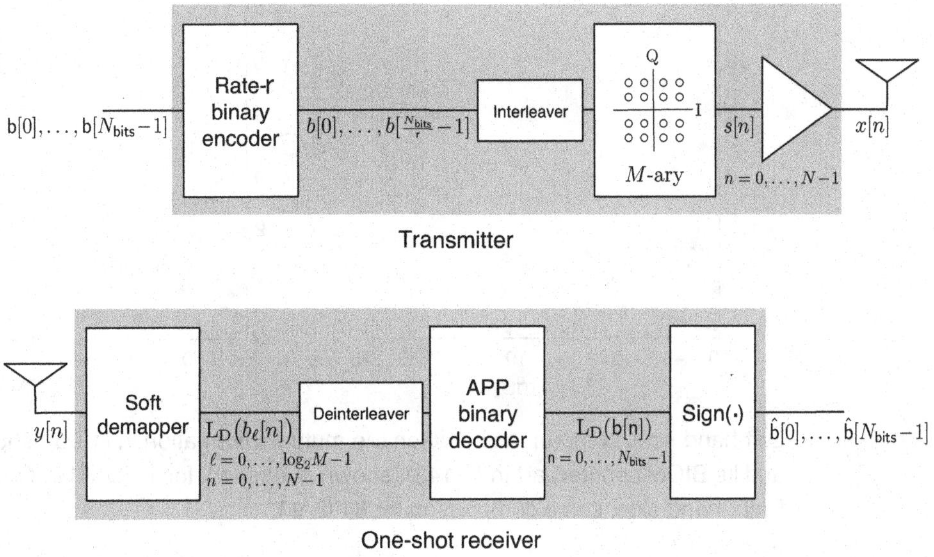

Fig. 1.5 BICM architecture with a one-shot receiver.

mappers and soft demappers constitutes the so-called *bit-interleaved coded modulation* (BICM) architecture, depicted in Fig. 1.5 and standard in wireless transceivers nowadays [96]. As mentioned, BICM is information-theoretically equivalent to signal-space coding for the cases of BPSK (single bit per symbol) and Gray-mapped QPSK (two quadrature bits per symbol). For higher-order constellations, even if the dependencies among same-symbol bits are not fully eradicated by interleaving and the receiver ignores them, it is only slightly inferior.

Example 1.22 (BICM mutual information in Gaussian noise)

Let $y = \sqrt{\rho}s + z$ with $z \sim \mathcal{N}_{\mathbb{C}}(0,1)$. Recalling from Example (1.16) the corresponding channel law, (1.137) specializes to

$$\sum_{\ell=0}^{\log_2 M-1} I(b_\ell; y) = \sum_{\ell=0}^{\log_2 M-1} \frac{1}{2}\left(\mathbb{E}\left[\log_2 \frac{\sum_{s_m \in \mathcal{S}_0^\ell} e^{-|y-\sqrt{\rho}s_m|^2}}{\frac{1}{2}\sum_{m=0}^{M-1} e^{-|y-\sqrt{\rho}s_m|^2}}\right] + \mathbb{E}\left[\log_2 \frac{\sum_{s_m \in \mathcal{S}_1^\ell} e^{-|y-\sqrt{\rho}s_m|^2}}{\frac{1}{2}\sum_{m=0}^{M-1} e^{-|y-\sqrt{\rho}s_m|^2}}\right]\right). \quad (1.142)$$

Example 1.23

For 16-QAM and 64-QAM constellations, compare the mutual information $I(s; y)$ in (1.95) against (1.142) with Gray mapping of bits to constellation points.

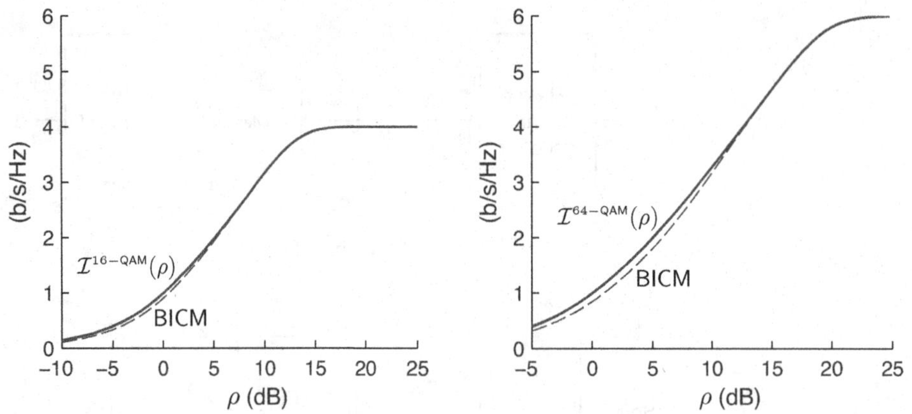

Fig. 1.6 Left-hand side, comparison between the mutual information in (1.95), shown in solid, and its BICM counterpart in (1.142), shown in dashed, for 16-QAM in Gaussian noise. Right-hand side, same comparison for 64-QAM.

Solution

The comparisons are shown in Fig. 1.6.

By incorporating iterative procedures at the receiver, most of the tiny loss incurred by a one-shot BICM receiver could be recovered at the price of decoding latency [97, 98]. In essence, an iterative BICM receiver can progressively learn the dependencies among bits given the observation of $y[0], \ldots, y[N-1]$, thereby closing the gap with signal-space coding. Although arguably not worthwhile given the tininess of this gap, the idea of iterative reception becomes more appealing in MIMO, where the gap broadens, and thus it is worthwhile that we introduce its basic structure here.

Figure 1.7 depicts an iterative BICM receiver, where the soft demapping is aided by a-priori information $\mathrm{L}_{\mathrm{A}}^{\mathsf{map}}(\cdot)$ about the coded bits. This improves the L-values $\mathrm{L}_{\mathrm{D}}^{\mathsf{map}}(\cdot)$ produced by the demapper, and the ensuing extrinsic information $\mathrm{L}_{\mathrm{E}}^{\mathsf{map}}(\cdot) = \mathrm{L}_{\mathrm{D}}^{\mathsf{map}}(\cdot) - \mathrm{L}_{\mathrm{A}}^{\mathsf{map}}(\cdot)$ is deinterleaved and fed to the APP decoder as $\mathrm{L}_{\mathrm{A}}^{\mathsf{cod}}(\cdot)$. The APP decoder then generates extrinsic information on the coded bits, $\mathrm{L}_{\mathrm{E}}^{\mathsf{cod}}(\cdot) = \mathrm{L}_{\mathrm{D}}^{\mathsf{cod}}(\cdot) - \mathrm{L}_{\mathrm{A}}^{\mathsf{cod}}(\cdot)$, which, properly interleaved, becomes the new a-priori information for the soft demapper, thereby completing an iteration. The APP decoder also generates L-values for the message bits and, once sufficient iterations have been run, the sign of these directly gives the final MAP decisions. Notice that only extrinsic L-values, representing newly distilled information, are passed around in the iterations. That prevents the demapper from receiving as a-priori information knowledge generated by itself in the previous iteration, and likewise for the decoder. Interestingly, with iterative reception, departing from Gray mapping is preferable so as to enhance the bit dependencies chased by the iterative process. Pushing things further in that direction, once could even consider multidimensional mappers operating, rather than on individual symbols, on groups of symbols [99, 100].

Altogether, the main take-away point from this section is the following: because of the

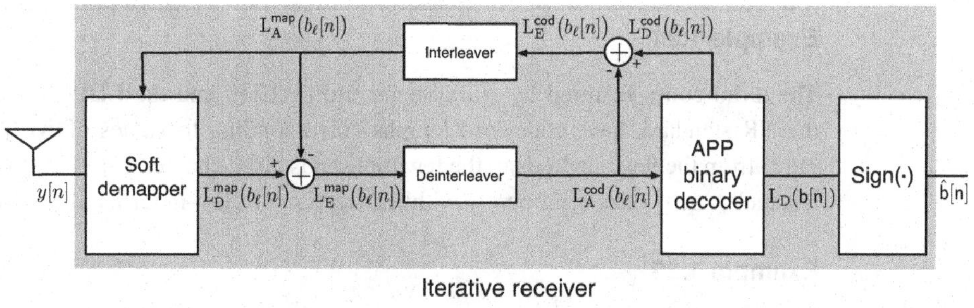

Fig. 1.7 BICM architecture with an iterative receiver.

coincidence of (1.137) and the actual mutual information for BPSK and QPSK, and their minute—and recoverable—difference for other constellations of interest, no distinction is customarily made between these quantities. This is also the principle followed in this text, where the performance limits of systems featuring BICM are investigated by means of the mutual information directly.

1.5.5 Finite-length codewords

For the most part we concern ourselves with the performance for $N \to \infty$, a stratagem that relies on this limit being representative of the performance of finite—but long—codewords. To substantiate this representativity, it is useful to briefly touch on an information-theoretic result that sheds light on the spectral efficiencies that can be fundamentally achieved when the length of the codewords is finite [101, 102]. Since error-free communication is generally not possible nonasymptotically, an acceptable error probability must be specified. If the acceptable codeword error probability is p_e, then, in many channels admitting a single-letter characterization it is possible to transmit at

$$\frac{R}{B} = C - \sqrt{\frac{V}{N}}\, Q^{-1}(p_e) + \mathcal{O}\left(\frac{\log N}{N}\right), \tag{1.143}$$

where $Q(\cdot)$ is the Gaussian Q-function while V is the variance of the information density, i.e.,

$$V = \mathrm{var}\big[i(s; y)\big], \tag{1.144}$$

with s conforming to the capacity-achieving distribution. This pleasing result says that the backoff from capacity is approximately $\sqrt{V/N}\, Q^{-1}(p_e)$, which for codeword lengths and error probabilities of interest is generally small; quantitative examples for specific channels are given in Chapter 4. Hence, the capacity indeed retains its significance for finite—but long—codewords.

Example 1.24

The turbo codes featured by 3G systems and by LTE, and the LDPC codes featured by the NR standard, have codeword lengths corresponding to values of N_{bits} that typically range from the few hundreds to the few thousands [103, chapter 12][104]. In certain simple channels, such codes can operate within a fraction of dB—in terms of SNR—of capacity.

Example 1.25

Over a bandwidth of $B = 20$ MHz, every 1000 codeword symbols incur a latency of $\frac{1000}{20 \cdot 10^6} = 0.05$ ms. If such bandwidth, typical of LTE, is shared by U users, then the latency is multiplied correspondingly. Given that LTE end-to-end latencies stand at about 10 ms, the contribution of coding to those latencies is minor.

For NR, the latency target is on the order of 1 ms [105, 106], but this reduction is to be accompanied by major increases in bandwidth and thus codeword lengths need not suffer major contractions.

The robustness of the capacity to finite codeword lengths, in conjunction with its computability for many relevant channels, renders it a quantity of capital importance. At the same time, for finite N and $p_{\text{e}} > 0$, in addition to the transmit bit rate R and the ensuing spectral efficiency R/B, a companion quantity of interest is the *throughput* that measures the rate (in b/s) within the successfully decoded codewords; this throughput is given by $(1 - p_{\text{e}})R$.

For small N, the expansion in (1.143) ceases to be precise and, in the limit of $N = 1$, the communication would become uncoded and every individual symbol would then be left at the mercy of the particular noise realization it experienced, without the protection that coding affords. The error probability would be comparatively very high and the throughput would suffer. Uncoded communication, the rule in times past, seems unnatural in the post-Shannon world and it is nowadays found only in systems priming simplicity.

1.5.6 Hybrid-ARQ

In relation to the codeword length, hybrid-ARQ has become established as an indispensable ingredient from 3G onwards. Blending channel coding with the traditional automatic repeat request (ARQ) procedure whereby erroneously received data are retransmitted, hybrid-ARQ turns the length and rate of the codewords into flexible—rather than fixed—quantities [107–109].

In a nutshell, hybrid-ARQ works as follows: when a received codeword is decoded incorrectly, rather than discarding it and receiving its retransmission anew as is the case in standard ARQ, the received codeword is stored and subsequently combined with the retransmission once it arrives at the receiver. This combination has a higher chance of successful decoding than either of the (re)transmissions individually. Moreover, the procedure may be repeated multiple times, until either decoding is indeed successful or the number of retransmissions reaches a certain value and an error is declared. Two variants of hybrid-ARQ stand out:

- *Chase combining*, where every (re)transmission contains an identical copy of the codeword. The receiver simply adds its observations thereof, increasing the SNR with each new retransmission.
- *Incremental redundancy*, where every retransmission contains additional coded symbols that extend the codeword and lower its rate. Indeed, the result of appending each retransmission to the previous one(s) is a longer codeword that represents the original N_{bits} message bits with a larger number of symbols N.

Incremental redundancy is the most powerful incarnation of hybrid-ARQ and the one we implicitly refer to unless otherwise stated.

Example 1.26

How could incremental redundancy be implemented if every (re)transmission had length N and the maximum number of retransmissions were four?

Solution

The transmitter could generate a codeword of length $4N$ and then transmit N of those symbols, say every fourth one. The receiver, privy to the codebook and hybrid-ARQ scheme, would attempt decoding. If that failed, another set of N symbols could be sent and the receiver could again attempt decoding, this time the ensuing codeword of $2N$ symbols, and so on. If the final decoding with all $4N$ symbols failed, an error would be declared.

1.5.7 Extension to MIMO

How does the formulation of the channel capacity change with MIMO? In essence, the abstraction gets vectorized. Referring back to Fig. 1.2:

- The encoder parses the source bits into messages of N_{bits} and maps those onto codewords made up of N vector symbols, $\boldsymbol{s}[0], \ldots, \boldsymbol{s}[N-1]$. Each codeword is possibly transformed (e.g., via OFDM or MIMO precoding) and amplified into $\boldsymbol{x}[0], \ldots, \boldsymbol{x}[N-1]$ as per the applicable constraints.
- The channel, which connects every input (transmit antenna) with every output (receive antenna), is described by the conditional distribution $f_{\boldsymbol{y}[0],\ldots,\boldsymbol{y}[N-1]|\boldsymbol{x}[0],\ldots,\boldsymbol{x}[N-1]}(\cdot)$. With transformations and amplification accounted for, $f_{\boldsymbol{y}[0],\ldots,\boldsymbol{y}[N-1]|\boldsymbol{s}[0],\ldots,\boldsymbol{s}[N-1]}(\cdot)$ follows from $f_{\boldsymbol{y}[0],\ldots,\boldsymbol{y}[N-1]|\boldsymbol{x}[0],\ldots,\boldsymbol{x}[N-1]}(\cdot)$.
- The decoder maps every possible channel output, $\boldsymbol{y}[0], \ldots, \boldsymbol{y}[N-1]$, onto a guess of the original block of N_{bits} bits.

The encoder can be implemented as a true vector encoder, as a bank of parallel scalar encoders, or as a combination thereof, and the tradeoffs involved as well as the structure of the corresponding receivers are examined later in the text. At this point, we do not peek inside the encoder, but only posit that it produces codewords $\boldsymbol{s}[0], \ldots, \boldsymbol{s}[N-1]$.

Under information stability, the capacity is then

$$C = \max_{\text{power constraints}} \lim_{N\to\infty} \frac{1}{N} I\big(\boldsymbol{s}[0],\ldots,\boldsymbol{s}[N-1];\boldsymbol{y}[0],\ldots,\boldsymbol{y}[N-1]\big) \tag{1.145}$$

with maximization over the distribution of $\boldsymbol{s}[0],\ldots,\boldsymbol{s}[N-1]$, with the subsequent transformation and amplification having to respect the applicable constraints. If the channel is stationary and memoryless, then the transmit vector symbols are IID and the optimization in (1.145) becomes single-letter over a single vector symbol.

MIMO BICM

Recall that the norm in modern communication systems is to rely on powerful binary codes mapped to discrete constellations at the transmitter and with soft demapping and binary decoding at the receiver. With the complement of bit-level interleaving, this comprises the BICM architecture. The dependencies that may exist among same-symbol bits are disregarded in one-shot BICM receivers and progressively learned in their iterative counterparts.

BICM extends to the MIMO realm. With parallel scalar encoders, one per transmit antenna, the remarks made for single-input single-output (SISO) BICM apply verbatim. With a vector encoder, a one-shot BICM receiver regards as mutually independent all the bits transmitted from the various antennas at each symbol. The role of $f_{y|b_\ell}$ is then played by $f_{\boldsymbol{y}|b_{\ell,j}}(\cdot)$, defined as the PDF of $\boldsymbol{y}$ conditioned on the ℓth bit from the jth transmit antenna equaling 0 or 1. With M-ary equiprobable constellations,

$$f_{\boldsymbol{y}|b_{\ell,j}}(\mathbf{y}|0) = \frac{1}{\frac{1}{2}M^{N_s}} \sum_{\mathbf{s}_m \in \mathcal{S}_0^{\ell,j}} f_{\boldsymbol{y}|\boldsymbol{s}}(\mathbf{y}|\mathbf{s}_m) \tag{1.146}$$

$$f_{\boldsymbol{y}|b_{\ell,j}}(\mathbf{y}|1) = \frac{1}{\frac{1}{2}M^{N_s}} \sum_{\mathbf{s}_m \in \mathcal{S}_1^{\ell,j}} f_{\boldsymbol{y}|\boldsymbol{s}}(\mathbf{y}|\mathbf{s}_m), \tag{1.147}$$

where N_s is the dimensionality of $\boldsymbol{s}$ while $\mathcal{S}_0^{\ell,j}$ and $\mathcal{S}_1^{\ell,j}$ are the subsets of transmit vectors whose ℓth coded bit at the jth transmit antenna is 0 and 1, respectively. From the unconditioned equiprobability of the coded bits, the cardinality of each subset is $\frac{1}{2}M^{N_s}$, hence the scaling factors in (1.146) and (1.147). Extending the SISO expression in (1.137), the information-theoretic performance of a one-shot MIMO BICM receiver is characterized by [110, 111]

$$\sum_{j=0}^{N_s-1} \sum_{\ell=0}^{\log_2 M-1} I(b_{\ell,j};\boldsymbol{y}) = \sum_{j=0}^{N_s-1} \sum_{\ell=0}^{\log_2 M-1} \frac{1}{2}\left(\mathbb{E}\left[\log_2 \frac{\sum_{\mathbf{s}_m\in\mathcal{S}_0^{\ell,j}} f_{\boldsymbol{y}|\boldsymbol{s}}(\boldsymbol{y}|\mathbf{s}_m)}{\frac{1}{2}\sum_{m=0}^{M^{N_s}-1} f_{\boldsymbol{y}|\boldsymbol{s}}(\boldsymbol{y}|\mathbf{s}_m)}\right] + \mathbb{E}\left[\log_2 \frac{\sum_{\mathbf{s}_m\in\mathcal{S}_1^{\ell,j}} f_{\boldsymbol{y}|\boldsymbol{s}}(\boldsymbol{y}|\mathbf{s}_m)}{\frac{1}{2}\sum_{m=0}^{M^{N_s}-1} f_{\boldsymbol{y}|\boldsymbol{s}}(\boldsymbol{y}|\mathbf{s}_m)}\right]\right). \tag{1.148}$$

In contrast with SISO, where BICM one-shot reception experiences no loss relative to signal-space coding for BPSK and QPSK, in MIMO there may be a nonzero loss even in those cases because of possible dependences introduced by the channel among the bits emitted from different transmit antennas. The loss is more significant than in SISO, yet still relatively modest (about 1 dB at most) if the bit mappings are wisely chosen [96]. And, as in SISO, this loss can be largely recovered through the use of iterative decoding [112, 113].

Altogether, the mutual information $I(\boldsymbol{s};\boldsymbol{y})$ continues to be satisfyingly representative of the fundamental performance achievable with binary encoding and decoding.

1.6 MMSE estimation

Estimation theory deals with the questions of how and with what accuracy one can infer the value taken by a certain quantity on the basis of related observations. Normally built upon an underlying statistical model that connects those observations with the unknown quantity, estimation theory involves devising estimators according to different fidelity criteria and analyzing their performances [62, 114, 115].

To begin with, let us again consider the basic transformation that lies at the heart of any noisy linear channel, namely

$$y = \sqrt{\rho}s + z, \tag{1.149}$$

where ρ is fixed and the noise z can be taken to be zero-mean and of unit variance, but otherwise arbitrarily distributed for now. The problem at hand is to produce the "best possible" estimate $\hat{s}(y)$ for the variable s based on the following.

- The observation of y.
- A fidelity criterion specifying the sense in which "best possible" is to be understood.
- Some knowledge of the probabilistic relationship between s and y, and in particular knowledge of the posterior probability $f_{s|y}(\cdot)$ and the likelihood function $f_{y|s}(\cdot)$. The marginal distribution $f_s(\cdot)$, termed the *prior probability*, is further available to the estimator whenever s is part of the system design (say, if s is a signal) but may or may not be available otherwise (say, if s is a channel coefficient produced by nature).

Among the fidelity criteria that could be considered, a few are, for good reasons, prevalent in information theory and communications.

- The MAP criterion gives $\hat{s}(\mathrm{y}) = \arg\max_{\mathrm{s}} f_{s|y}(\mathrm{s}|\mathrm{y})$ with maximization over all values taken by s. Just like a MAP decoder identifies the most probable codeword, a MAP estimator returns the most probable value of s given what has been observed.
- The *maximum-likelihood* (ML) criterion gives $\hat{s}(\mathrm{y}) = \arg\max_{\mathrm{s}} f_{y|s}(\mathrm{y}|\mathrm{s})$, again with maximization over all values taken by s. Just like an ML decoder selects the most likely codeword, an ML estimator returns the value of s whose likelihood is highest. As argued via Bayes' theorem in the context of optimum decoding, if the prior is uniform, i.e., if s takes equiprobable values, then the ML and the MAP criteria coincide.
- The MMSE criterion, which is the one herein entertained.

The mean-square error measures the power of the estimation error, that is, the power of $|s - \hat{s}(y)|$. A rather natural choice in Gaussian-noise contexts, given how the defining feature of such noise is its power, the mean-square error was introduced early in the nineteenth century (by Gauss himself, as well as by Legendre [116, 117]) and it is by now a ubiquitous metric. The mean-square error for a given estimate $\hat{s}(y)$ thus equals

$$\mathbb{E}\left[|s - \hat{s}(y)|^2\right], \tag{1.150}$$

with expectation over both s and y or, equivalently, over s and z. The minimization of this quantity gives the MMSE and the corresponding $\hat{s}(\cdot)$ is the MMSE estimator.

1.6.1 The conditional-mean estimator

As it turns out, and regardless of the noise distribution (refer to Problem 1.34), the MMSE estimator is

$$\hat{s}(y) = \mathbb{E}\left[s|y\right], \tag{1.151}$$

whose rather intuitive form indicates that, in the MMSE sense, the best guess for s is its expected value given whatever observations are available; if no observations are available, then the MMSE estimate is directly the mean. This *conditional-mean estimator* is unbiased in the sense that

$$\mathbb{E}\left[\hat{s}(y)\right] = \mathbb{E}\left[\mathbb{E}[s|y]\right] \tag{1.152}$$

$$= \mathbb{E}[s], \tag{1.153}$$

but it is biased in the sense that, for a realization s, it may be that $\mathbb{E}\left[\hat{s}(\sqrt{\rho}\,\mathrm{s} + z)\right] \neq \mathrm{s}$. Put differently, the estimation error over all possible values of s is always zero-mean, but achieving the MMSE may require that the estimation error for given values of s be nonzero-mean. This dichotomy may cause confusion as the estimator can be declared to be both biased and unbiased, and it is important to make the distinction precise.

Crucially, the conditional-mean estimator $\hat{s}(y) = \mathbb{E}[s|y]$ complies with the *orthogonality principle* whereby $\mathbb{E}\left[g(y^*)\left(s - \hat{s}(y)\right)\right] = 0$ for every function $g(\cdot)$. In particular,

$$\mathbb{E}\left[y^*(s - \hat{s})\right] = 0 \tag{1.154}$$

$$\mathbb{E}\left[\hat{s}^*(s - \hat{s})\right] = 0. \tag{1.155}$$

Plugged into (1.150), the conditional-mean estimator yields

$$\mathsf{MMSE}(\rho) = \mathbb{E}\left[s - \left|\mathbb{E}[s|y]\right|^2\right] \tag{1.156}$$

with outer expectation over both s and y or, equivalently, over s and z. Alternatively, we can write $\mathsf{MMSE}(\rho) = \mathbb{E}\left[\mathrm{var}[s|y]\right]$ with expectation over y and with

$$\mathrm{var}[s|y] = \mathbb{E}\left[\left|s - \mathbb{E}[s|y]\right|^2 |y\right] \tag{1.157}$$

the conditional variance of s given y. For given $f_s(\cdot)$ and $f_z(\cdot)$, i.e., for a certain signal format and some noise distribution, $\mathsf{MMSE}(\rho)$ is a decreasing function of ρ. Also, the mean of the signal being estimated does not influence the MMSE, and hence we can restrict ourselves to zero-mean signal distributions.

1.6.2 MMSE estimation in Gaussian noise

Homing in on Gaussian-noise settings, with $z \sim \mathcal{N}_{\mathbb{C}}(0,1)$, we have that $y|s \sim \mathcal{N}_{\mathbb{C}}(\sqrt{\rho}\,\mathrm{s},1)$ and thus

$$f_{y|s}(\mathrm{y}|\mathrm{s}) = \frac{1}{\pi}\, e^{-|\mathrm{y}-\sqrt{\rho}\mathrm{s}|^2}. \tag{1.158}$$

Then, the posterior probability equals, via Bayes' theorem,

$$f_{s|y}(\mathrm{s}|\mathrm{y}) = \frac{f_{y|s}(\mathrm{y}|\mathrm{s})\, f_s(\mathrm{s})}{f_y(\mathrm{y})} \tag{1.159}$$

$$= \frac{f_{y|s}(\mathrm{y}|\mathrm{s})\, f_s(\mathrm{s})}{\int f_{y|s}(\mathrm{y}|\mathrm{s})\, f_s(\mathrm{s})\, d\mathrm{s}} \tag{1.160}$$

from which the conditional-mean estimator can be expressed as

$$\hat{\mathrm{s}}(\mathrm{y}) = \mathbb{E}\big[s\,|y{=}\mathrm{y}\big] \tag{1.161}$$

$$= \int \mathrm{s}\, f_{s|y}(\mathrm{s}|\mathrm{y})\, d\mathrm{s} \tag{1.162}$$

$$= \int \frac{\mathrm{s}\, f_{y|s}(\mathrm{y}|\mathrm{s})\, f_s(\mathrm{s})}{\int f_{y|s}(\mathrm{y}|\mathrm{s})\, f_s(\mathrm{s})\, d\mathrm{s}}\, d\mathrm{s} \tag{1.163}$$

$$= \frac{\int \mathrm{s}\, f_{y|s}(\mathrm{y}|\mathrm{s})\, f_s(\mathrm{s})\, d\mathrm{s}}{\int f_{y|s}(\mathrm{y}|\mathrm{s})\, f_s(\mathrm{s})\, d\mathrm{s}} \tag{1.164}$$

$$= \frac{\int \mathrm{s}\, e^{-|\mathrm{y}-\sqrt{\rho}\mathrm{s}|^2}\, f_s(\mathrm{s})\, d\mathrm{s}}{\int e^{-|\mathrm{y}-\sqrt{\rho}\mathrm{s}|^2}\, f_s(\mathrm{s})\, d\mathrm{s}} \tag{1.165}$$

with integrations over the complex plane.

Example 1.27 (MMSE estimation of a complex Gaussian scalar)

Consider $y = \sqrt{\rho}s + z$ with $s \sim \mathcal{N}_{\mathbb{C}}(0,1)$. Then,

$$f_s(\mathrm{s}) = \frac{1}{\pi}\, e^{-|\mathrm{s}|^2} \tag{1.166}$$

and, applying (1.165),

$$\hat{\mathrm{s}}(\mathrm{y}) = \frac{\displaystyle\int \mathrm{s}\, e^{-|\mathrm{y}-\sqrt{\rho}\mathrm{s}|^2} e^{-|\mathrm{s}|^2}\, d\mathrm{s}}{\displaystyle\int e^{-|\mathrm{y}-\sqrt{\rho}\mathrm{s}|^2} e^{-|\mathrm{s}|^2}\, d\mathrm{s}} \tag{1.167}$$

$$= \frac{\displaystyle\int \mathrm{s}\, e^{-\frac{|\mathrm{y}|^2}{1+\rho}}\, e^{-\left|\sqrt{1+\rho}\,\mathrm{s}-\sqrt{\frac{\rho}{1+\rho}}\,\mathrm{y}\right|^2} d\mathrm{s}}{\displaystyle\int e^{-\frac{|\mathrm{y}|^2}{1+\rho}}\, e^{-\left|\sqrt{1+\rho}\,\mathrm{s}-\sqrt{\frac{\rho}{1+\rho}}\,\mathrm{y}\right|^2} d\mathrm{s}} \tag{1.168}$$

$$= \frac{\displaystyle\int \mathrm{s}\, e^{-\left|\mathrm{s}-\frac{\sqrt{\rho}}{1+\rho}\mathrm{y}\right|^2/\frac{1}{1+\rho}}\, d\mathrm{s}}{\displaystyle\int e^{-\left|\mathrm{s}-\frac{\sqrt{\rho}}{1+\rho}\mathrm{y}\right|^2/\frac{1}{1+\rho}}\, d\mathrm{s}} \tag{1.169}$$

$$= \frac{\displaystyle\int \mathrm{s} \frac{1}{\pi\left(\frac{1}{1+\rho}\right)} e^{-\left|\mathrm{s}-\frac{\sqrt{\rho}}{1+\rho}\mathrm{y}\right|^2 / \frac{1}{1+\rho}} d\mathrm{s}}{\displaystyle\int \frac{1}{\pi\left(\frac{1}{1+\rho}\right)} e^{-\left|\mathrm{s}-\frac{\sqrt{\rho}}{1+\rho}\mathrm{y}\right|^2 / \frac{1}{1+\rho}} d\mathrm{s}}. \tag{1.170}$$

Recognizing that

$$\frac{1}{\pi\left(\frac{1}{1+\rho}\right)} \exp\left(-\frac{\left|\mathrm{s}-\frac{\sqrt{\rho}}{1+\rho}\mathrm{y}\right|^2}{\frac{1}{1+\rho}}\right) \tag{1.171}$$

is the PDF of a complex Gaussian variable with mean $\frac{\sqrt{\rho}}{1+\rho}\mathrm{y}$, the expectation in the numerator of (1.170) equals that mean, whereas the denominator equals unity and thus

$$\hat{s}(\mathrm{y}) = \frac{\sqrt{\rho}}{1+\rho}\mathrm{y}, \tag{1.172}$$

which is a linear function of the observed value of y, hence the result that the MMSE estimator of a Gaussian quantity is *linear*. This estimator then yields

$$\mathsf{MMSE}(\rho) = \mathbb{E}\left[\left|s - \frac{\sqrt{\rho}}{1+\rho}y\right|^2\right] \tag{1.173}$$

$$= \mathbb{E}\left[|s|^2\right] - 2\frac{\sqrt{\rho}}{1+\rho}\Re(\mathbb{E}[ys^*]) + \frac{\rho}{(1+\rho)^2}\mathbb{E}\left[|y|^2\right] \tag{1.174}$$

and, using $\mathbb{E}\left[|s|^2\right] = 1$ as well as $\mathbb{E}\left[ys^*\right] = \sqrt{\rho}$ and $\mathbb{E}\left[|y|^2\right] = 1+\rho$, finally

$$\mathsf{MMSE}(\rho) = \frac{1}{1+\rho}. \tag{1.175}$$

Interestingly, the MMSE estimate of a Gaussian variable coincides with its MAP estimate (but not with the ML one). And, unsurprisingly given that the Gaussian distribution maximizes the differential entropy for a given variance, Gaussian variables are the hardest to estimate, meaning that any non-Gaussian variable of the same variance is bound to incur a lower estimation MMSE [118].

Example 1.28

Verify that the MMSE estimator in the previous example may be biased for a specific value of s but is unbiased over the distribution thereof.

Solution

For a given s,

$$\mathbb{E}\left[\hat{s}(\sqrt{\rho}s + z)\,|s{=}\mathrm{s}\right] = \mathbb{E}\left[\frac{\sqrt{\rho}}{1+\rho}(\sqrt{\rho}s + z)\,|s{=}\mathrm{s}\right] \tag{1.176}$$

$$= \mathbb{E}\left[\frac{\rho}{1+\rho}\mathrm{s} + \frac{\sqrt{\rho}}{1+\rho}z\right] \tag{1.177}$$

$$= \frac{\rho}{1+\rho}\mathrm{s} \tag{1.178}$$

$$= \mathrm{s} - \frac{1}{1+\rho}\,\mathrm{s} \tag{1.179}$$

$$\neq \mathrm{s} \tag{1.180}$$

with a bias $-\frac{1}{1+\rho}\mathrm{s}$. The expectation of this bias over the distribution of s is zero.

Example 1.29 (MMSE estimation of a BPSK scalar)

Consider $y = \sqrt{\rho}s + z$ with s drawn from a BPSK constellation. The conditional-mean estimate (refer to Problem 1.36) is

$$\hat{s}(y) = \tanh\big(2\sqrt{\rho}\,\Re\{y\}\big), \tag{1.181}$$

while the corresponding MMSE reduces to the real integral

$$\mathsf{MMSE}^{\mathsf{BPSK}}(\rho) = 1 - \frac{1}{\sqrt{\pi}}\int_{-\infty}^{\infty} \tanh\big(2\rho - 2\sqrt{\rho}\,\xi\big)\, e^{-\xi^2} \mathrm{d}\xi. \tag{1.182}$$

Example 1.30 (MMSE estimation of a QPSK scalar)

Since QPSK amounts to two BPSK constellations in quadrature, each with half the power, the conditional-mean estimators for the in-phase and quadrature components are given by (1.181) applied to the real and imaginary parts of the observation, respectively, with $\rho/2$ in place of ρ. Then, the MMSE function equals

$$\mathsf{MMSE}^{\mathsf{QPSK}}(\rho) = \mathsf{MMSE}^{\mathsf{BPSK}}\left(\frac{\rho}{2}\right). \tag{1.183}$$

The low-ρ expansion of (1.175) reveals that, for a complex Gaussian variable,

$$\mathsf{MMSE}(\rho) = 1 - \rho + o(\rho), \tag{1.184}$$

which extends to the estimation of any variable that is proper complex, i.e., that occupies both noise dimensions in a balanced manner [119]. The prime example is QPSK.

In contrast,

$$\mathsf{MMSE}^{\mathsf{BPSK}}(\rho) = 1 - 2\rho + o(\rho) \tag{1.185}$$

and this expansion applies, beyond BPSK, whenever a one-dimensional variable is being estimated in complex Gaussian noise.

In the high-ρ regime, in turn, the MMSE decays as $1/\rho$ when the variable being estimated is Gaussian and possibly faster otherwise [120]. In particular, for discrete constellations the decay is exponential and details on the exponents for certain types of constellations are given in [121, 122].

Generalization to vectors

The generalization of the preceding derivations to vector transformations is straightforward. Given

$$\boldsymbol{y} = \sqrt{\rho}\boldsymbol{A}\boldsymbol{s} + \boldsymbol{z}, \tag{1.186}$$

where $\boldsymbol{A}$ is fixed while $\boldsymbol{s}$ and $\boldsymbol{z}$ are independent with $\boldsymbol{z} \sim \mathcal{N}_{\mathbb{C}}(\boldsymbol{0}, \boldsymbol{R}_{\boldsymbol{z}})$, the conditional-mean estimator

$$\hat{\boldsymbol{s}}(\boldsymbol{y}) = \mathbb{E}\big[\boldsymbol{s}|\boldsymbol{y}\big] \tag{1.187}$$

attains the MMSE simultaneously for every entry of $\boldsymbol{s}$ and

$$\boldsymbol{E} = \mathbb{E}\Big[\big(\boldsymbol{s} - \hat{\boldsymbol{s}}(\boldsymbol{y})\big)\big(\boldsymbol{s} - \hat{\boldsymbol{s}}(\boldsymbol{y})\big)^*\Big] \tag{1.188}$$

is the MMSE matrix, which equals the covariance of the estimation error vector and fully describes the accuracy of the conditional-mean vector estimator. The jth diagonal entry of $\boldsymbol{E}$ indicates the MMSE incurred in the estimation of the jth entry of $\boldsymbol{s}$. From $\boldsymbol{E}$, scalar quantities with various significances may be derived as needed, say weighted arithmetic or geometric averages of the diagonal entries, or directly the largest diagonal entry [123].

Example 1.31 (MMSE estimation of a complex Gaussian vector)

Consider $\boldsymbol{y} = \sqrt{\rho}\boldsymbol{A}\boldsymbol{s} + \boldsymbol{z}$ with $\boldsymbol{s} \sim \mathcal{N}_{\mathbb{C}}(\boldsymbol{0}, \boldsymbol{R}_{\boldsymbol{s}})$ and $\boldsymbol{z} \sim \mathcal{N}_{\mathbb{C}}(\boldsymbol{0}, \boldsymbol{I})$. Extending to the vector realm the derivations of Example 1.27, the MMSE estimator is found to be

$$\hat{\boldsymbol{s}}(\boldsymbol{y}) = \sqrt{\rho}\, \boldsymbol{R}_{\boldsymbol{s}}\boldsymbol{A}^*(\boldsymbol{I} + \rho\boldsymbol{A}\boldsymbol{R}_{\boldsymbol{s}}\boldsymbol{A}^*)^{-1}\boldsymbol{y}, \tag{1.189}$$

which, as in the case of a Gaussian scalar, is linear in the observation. Then, from the above and (1.188),

$$\boldsymbol{E} = \mathbb{E}[\boldsymbol{s}\boldsymbol{s}^*] - \mathbb{E}[\boldsymbol{s}\hat{\boldsymbol{s}}^*] - \mathbb{E}[\hat{\boldsymbol{s}}\boldsymbol{s}^*] + \mathbb{E}[\hat{\boldsymbol{s}}\hat{\boldsymbol{s}}^*] \tag{1.190}$$

$$= \boldsymbol{R}_{\boldsymbol{s}} - 2\rho\boldsymbol{R}_{\boldsymbol{s}}\boldsymbol{A}^*(\boldsymbol{I} + \rho\boldsymbol{A}\boldsymbol{R}_{\boldsymbol{s}}\boldsymbol{A}^*)^{-1}\boldsymbol{A}\boldsymbol{R}_{\boldsymbol{s}}$$
$$+ \rho\boldsymbol{R}_{\boldsymbol{s}}\boldsymbol{A}^*(\boldsymbol{I} + \rho\boldsymbol{A}\boldsymbol{R}_{\boldsymbol{s}}\boldsymbol{A}^*)^{-1}(\boldsymbol{I} + \rho\boldsymbol{A}\boldsymbol{R}_{\boldsymbol{s}}\boldsymbol{A}^*)(\boldsymbol{I} + \rho\boldsymbol{A}\boldsymbol{R}_{\boldsymbol{s}}\boldsymbol{A}^*)^{-1}\boldsymbol{A}\boldsymbol{R}_{\boldsymbol{s}} \tag{1.191}$$

$$= \boldsymbol{R}_{\boldsymbol{s}} - \rho\boldsymbol{R}_{\boldsymbol{s}}\boldsymbol{A}^*(\boldsymbol{I} + \rho\boldsymbol{A}\boldsymbol{R}_{\boldsymbol{s}}\boldsymbol{A}^*)^{-1}\boldsymbol{A}\boldsymbol{R}_{\boldsymbol{s}}. \tag{1.192}$$

Applying the matrix inversion lemma (see Appendix B.7) in a reverse fashion to (1.192), we can also rewrite $\boldsymbol{E}$ into the alternative form

$$\boldsymbol{E} = \big(\boldsymbol{R}_{\boldsymbol{s}}^{-1} + \rho\boldsymbol{A}^*\boldsymbol{A}\big)^{-1}. \tag{1.193}$$

Expanding (1.192), the generalization of the low-ρ expansion in (1.184) to proper complex vectors comes out as

$$\boldsymbol{E} = \boldsymbol{R}_{\boldsymbol{s}} - \rho\, \boldsymbol{R}_{\boldsymbol{s}}\boldsymbol{A}^*\boldsymbol{A}\boldsymbol{R}_{\boldsymbol{s}} + \mathcal{O}(\rho^2), \tag{1.194}$$

which holds whenever $\boldsymbol{s}$ is a proper complex vector, irrespective of its distribution.

1.6.3 The I-MMSE relationship in Gaussian noise

The random transformation invoked extensively throughout this chapter, namely

$$y = \sqrt{\rho}s + z, \tag{1.195}$$

where s and z are independent and $z \sim \mathcal{N}_{\mathbb{C}}(0, 1)$, is the cornerstone of any linear scalar channel impaired by Gaussian noise and, as we have seen in the formulation of the capacity, the mutual information functions $\mathcal{I}(\rho) = I\big(s; \sqrt{\rho}s + z\big)$ for relevant distributions of s are exceedingly significant. The derivative of $\mathcal{I}(\rho)$ turns out to have significance as well. Regardless of the distribution of s, it holds that [124]

$$\frac{1}{\log_2 e} \cdot \frac{\mathrm{d}}{\mathrm{d}\rho}\mathcal{I}(\rho) = \mathsf{MMSE}(\rho), \tag{1.196}$$

where the right-hand side equals the MMSE when estimating s from its noisy observation, y. The identity in (1.196) is termed the *I-MMSE relationship*, and its integration yields an alternative form for the mutual information function, precisely

$$\frac{1}{\log_2 e}\mathcal{I}(\rho) = \int_0^{\rho} \mathsf{MMSE}(\xi)\,\mathrm{d}\xi. \tag{1.197}$$

Example 1.32 (I-MMSE relationship for a complex Gaussian scalar)

Consider $y = \sqrt{\rho}s + z$ with $s \sim \mathcal{N}_{\mathbb{C}}(0,1)$. As derived in Examples 1.7 and 1.27,

$$\mathcal{I}(\rho) = \log_2(1+\rho) \tag{1.198}$$

and

$$\mathsf{MMSE}(\rho) = \frac{1}{1+\rho}, \tag{1.199}$$

which satisfy the I-MMSE relationship in (1.196).

Example 1.33 (I-MMSE relationship for a BPSK scalar)

Let $y = \sqrt{\rho}s + z$ with s drawn from a BPSK constellation. From Examples 1.10 and 1.29,

$$\mathcal{I}(\rho) = 2\rho\log_2 e - \frac{1}{\sqrt{\pi}}\int_{-\infty}^{\infty} e^{-\xi^2}\log_2\cosh\big(2\rho - 2\sqrt{\rho}\xi\big)\,\mathrm{d}\xi \tag{1.200}$$

and

$$\mathsf{MMSE}(\rho) = 1 - \frac{1}{\sqrt{\pi}}\int_{-\infty}^{\infty}\tanh\big(2\rho - 2\sqrt{\rho}\xi\big)\,e^{-\xi^2}\,\mathrm{d}\xi, \tag{1.201}$$

which satisfy the I-MMSE relationship.

Example 1.34 (I-MMSE relationship for a QPSK scalar)

As shown in Examples 1.11 and 1.30,

$$\mathcal{I}^{\mathsf{QPSK}}(\rho) = 2\,\mathcal{I}^{\mathsf{BPSK}}\left(\frac{\rho}{2}\right). \tag{1.202}$$

and

$$\mathsf{MMSE}^{\mathsf{QPSK}}(\rho) = \mathsf{MMSE}^{\mathsf{BPSK}}\left(\frac{\rho}{2}\right), \tag{1.203}$$

which are consistent with the I-MMSE relationship.

In the low-ρ regime, the I-MMSE relationship bridges the distinctness of proper complex signals in terms of mutual information and MMSE. Indeed, recalling (1.52) and (1.185), for such signals

$$\mathcal{I}(\rho) = \left(\rho - \frac{1}{2}\rho^2\right) \log_2 e + o(\rho^2) \tag{1.204}$$

and

$$\mathsf{MMSE}(\rho) = 1 - 2\rho + o(\rho). \tag{1.205}$$

Generalization to vectors

The I-MMSE relationship also extends to the vector realm. Consider again the random transformation

$$\boldsymbol{y} = \sqrt{\rho}\boldsymbol{A}\boldsymbol{s} + \boldsymbol{z}, \tag{1.206}$$

where $\boldsymbol{A}$ is fixed while $\boldsymbol{s}$ and $\boldsymbol{z}$ are independent with $\boldsymbol{z} \sim \mathcal{N}_{\mathbb{C}}(\boldsymbol{0}, \boldsymbol{I})$. Then, regardless of the distribution of $\boldsymbol{s}$ [125]

$$\frac{1}{\log_2 e} \nabla_{\boldsymbol{A}} I(\boldsymbol{s}; \sqrt{\rho}\boldsymbol{A}\boldsymbol{s} + \boldsymbol{z}) = \rho \boldsymbol{A}\boldsymbol{E}, \tag{1.207}$$

where $\nabla_{\boldsymbol{A}}$ denotes the gradient with respect to $\boldsymbol{A}$ (see Appendix D) while $\boldsymbol{E}$ is the MMSE matrix defined in (1.188) for the estimation of $\boldsymbol{s}$, i.e., the generalization to multiple dimensions of the scalar MMSE.

Example 1.35 (I-MMSE relationship for a complex Gaussian vector)

As established in Example 1.13, when the noise is $\boldsymbol{z} \sim \mathcal{N}_{\mathbb{C}}(\boldsymbol{0}, \boldsymbol{I})$ while $\boldsymbol{s} \sim \mathcal{N}_{\mathbb{C}}(\boldsymbol{0}, \boldsymbol{R_s})$,

$$?I(\boldsymbol{s}; \sqrt{\rho}\boldsymbol{A}\boldsymbol{s} + \boldsymbol{z}) = \log_2 \det\left(\boldsymbol{I} + \rho \boldsymbol{A}\boldsymbol{R_s}\boldsymbol{A}^*\right) \tag{1.208}$$

$$= \log_2 \det\left(\boldsymbol{I} + \rho \boldsymbol{A}^*\boldsymbol{A}\boldsymbol{R_s}\right) \tag{1.209}$$

and, applying the expression for the gradient of a log-determinant function given in Appendix D, we obtain

$$\frac{1}{\log_2 e} \nabla_{\boldsymbol{A}} I(\boldsymbol{s}; \sqrt{\rho}\boldsymbol{A}\boldsymbol{s} + \boldsymbol{z}) = \rho \boldsymbol{A}\boldsymbol{R_s}\left(\boldsymbol{I} + \rho \boldsymbol{A}^*\boldsymbol{A}\boldsymbol{R_s}\right)^{-1} \tag{1.210}$$

$$= \rho \boldsymbol{A}\left(\boldsymbol{R_s}^{-1} + \rho \boldsymbol{A}^*\boldsymbol{A}\right)^{-1}, \tag{1.211}$$

which indeed equals $\rho\boldsymbol{A}\boldsymbol{E}$ with $\boldsymbol{E} = \left(\boldsymbol{R_s}^{-1} + \rho\boldsymbol{A}^*\boldsymbol{A}\right)^{-1}$ as determined in Example 1.31 for a complex Gaussian vector.

Example 1.36

Use the I-MMSE relationship to express $\frac{\partial}{\partial\rho} I(\boldsymbol{s}; \sqrt{\rho}\boldsymbol{A}\boldsymbol{s} + \boldsymbol{z})$ as a function of $\boldsymbol{A}$ and the MMSE matrix $\boldsymbol{E}$, for an arbitrarily distributed $\boldsymbol{s}$.

Solution

Let us first narrow the problem to $\boldsymbol{s} \sim \mathcal{N}_{\mathbb{C}}(\boldsymbol{0}, \boldsymbol{R_s})$. Denoting by $\lambda_j(\cdot)$ the jth eigenvalue of a matrix, we can rewrite (1.209) as

$$I(\boldsymbol{s}; \sqrt{\rho}\boldsymbol{A}\boldsymbol{s}+\boldsymbol{z}) = \log_2 \det(\boldsymbol{I}+\rho \boldsymbol{A}^*\boldsymbol{A}\boldsymbol{R}_{\boldsymbol{s}}) \tag{1.212}$$
$$= \log_2 \prod_j \lambda_j(\boldsymbol{I}+\rho \boldsymbol{A}^*\boldsymbol{A}\boldsymbol{R}_{\boldsymbol{s}}) \tag{1.213}$$
$$= \sum_j \log_2 \lambda_j(\boldsymbol{I}+\rho \boldsymbol{A}^*\boldsymbol{A}\boldsymbol{R}_{\boldsymbol{s}}) \tag{1.214}$$
$$= \sum_j \log_2\big(1+\rho\,\lambda_j(\boldsymbol{A}^*\boldsymbol{A}\boldsymbol{R}_{\boldsymbol{s}})\big). \tag{1.215}$$

Then, differentiating with respect to ρ, we obtain

$$\frac{\partial}{\partial\rho} I(\boldsymbol{s}; \sqrt{\rho}\boldsymbol{A}\boldsymbol{s}+\boldsymbol{z}) = \sum_j \frac{\lambda_j(\boldsymbol{A}^*\boldsymbol{A}\boldsymbol{R}_{\boldsymbol{s}})}{1+\rho\,\lambda_j(\boldsymbol{A}^*\boldsymbol{A}\boldsymbol{R}_{\boldsymbol{s}})} \log_2 e \tag{1.216}$$
$$= \sum_j \lambda_j\big(\boldsymbol{A}^*\boldsymbol{A}\boldsymbol{R}_{\boldsymbol{s}}(\boldsymbol{I}+\rho\boldsymbol{A}^*\boldsymbol{A}\boldsymbol{R}_{\boldsymbol{s}})^{-1}\big)\log_2 e \tag{1.217}$$
$$= \mathrm{tr}\big(\boldsymbol{A}^*\boldsymbol{A}\boldsymbol{R}_{\boldsymbol{s}}(\boldsymbol{I}+\rho\boldsymbol{A}^*\boldsymbol{A}\boldsymbol{R}_{\boldsymbol{s}})^{-1}\big)\log_2 e \tag{1.218}$$
$$= \mathrm{tr}\big(\boldsymbol{A}^*\boldsymbol{A}(\boldsymbol{R}_{\boldsymbol{s}}^{-1}+\rho\boldsymbol{A}^*\boldsymbol{A})^{-1}\big)\log_2 e \tag{1.219}$$
$$= \mathrm{tr}\big(\boldsymbol{A}\,(\boldsymbol{R}_{\boldsymbol{s}}^{-1}+\rho\boldsymbol{A}^*\boldsymbol{A})^{-1}\boldsymbol{A}^*\big)\log_2 e \tag{1.220}$$

and thus we can write

$$\frac{1}{\log_2 e}\frac{\partial}{\partial\rho} I(\boldsymbol{s}; \sqrt{\rho}\boldsymbol{A}\boldsymbol{s}+\boldsymbol{z}) = \mathrm{tr}\big(\boldsymbol{A}\boldsymbol{E}\boldsymbol{A}^*\big). \tag{1.221}$$

Although derived for a complex Gaussian vector $\boldsymbol{s}$, as a corollary of the I-MMSE relationship this identity can be claimed regardless of the distribution of $\boldsymbol{s}$. Indeed, the evaluation of $\frac{\partial}{\partial\rho} I(\boldsymbol{s}; \sqrt{\rho}\boldsymbol{A}\boldsymbol{s}+\boldsymbol{z})$ for an arbitrary $\boldsymbol{s}$ can be effected through the gradient with respect to $\sqrt{\rho}\boldsymbol{A}$, and the application of (1.207) then leads to (1.221) all the same.

Evaluated at $\rho = 0$, the identity in (1.221) gives the formula

$$\frac{1}{\log_2 e}\frac{\partial}{\partial\rho} I(\boldsymbol{s}; \sqrt{\rho}\boldsymbol{A}\boldsymbol{s}+\boldsymbol{z})\bigg|_{\rho=0} = \mathrm{tr}\left(\boldsymbol{A}\boldsymbol{R}_{\boldsymbol{s}}\boldsymbol{A}^*\right), \tag{1.222}$$

which is a generalization of (1.79).

1.7 LMMSE estimation

While the precise distribution of certain quantities (say, the signals being transmitted) is entirely within the control of the system designer, there are other quantities of interest (say, the channel gain) that are outside that control. When quantities of the latter type are to be estimated, it is often the case that we are either unable or unwilling to first obtain their distributions beyond the more accessible mean and variance. With the MMSE retained as the estimation criterion, a sensible approach is to regard the distribution as that whose MMSE estimation is the hardest, namely the Gaussian distribution. This leads to the application of the linear MMSE (LMMSE) estimators derived in Examples 1.27 and 1.31 to quantities that need not be Gaussian.

Alternatively, LMMSE estimators may be featured as a design choice, even if the relevant distribution is known, simply because of the appeal and simplicity of a linear filter.

And then, of course, LMMSE estimators may be in place simply because the quantities to be estimated are known to be Gaussian (say, capacity-achieving signals).

For all the foregoing reasons, LMMSE estimators are prevalent in wireless communications and throughout this text. Except when estimating a truly Gaussian quantity, an LMMSE estimator is bound to be inferior to a conditional-mean estimator, but also more versatile and robust.

1.7.1 Random variables

Under the constraint of a linear structure, the LMMSE estimator for a vector $\boldsymbol{s}$ based on the observation of a related vector $\boldsymbol{y}$ is to be

$$\hat{\boldsymbol{s}} = \boldsymbol{W}^{\mathsf{MMSE}*}\boldsymbol{y} + \boldsymbol{b}^{\mathsf{MMSE}}. \tag{1.223}$$

The mean $\boldsymbol{\mu}_{\boldsymbol{s}}$ can be regarded as known and the role of the constant term $\boldsymbol{b}^{\mathsf{MMSE}}$ is to ensure that $E[\hat{\boldsymbol{s}}(\boldsymbol{y})] = \boldsymbol{\mu}_{\boldsymbol{s}}$ (refer to Problem 1.45). With the unbiasedness in that sense taken care of, the LMMSE estimator is embodied by the matrix $\boldsymbol{W}^{\mathsf{MMSE}}$, which can be inferred from (1.189) to equal

$$\boldsymbol{W}^{\mathsf{MMSE}} = \boldsymbol{R}_{\boldsymbol{y}}^{-1}\boldsymbol{R}_{\boldsymbol{y}\boldsymbol{s}} \tag{1.224}$$

and that is indeed its form in broad generality. To see that, we can write the mean-square error on the estimation of the jth entry of $\boldsymbol{s}$ via a generic linear filter $\boldsymbol{W}$ as

$$\mathbb{E}\Big[|[\boldsymbol{s}-\hat{\boldsymbol{s}}]_j|^2\Big] = \mathbb{E}\Big[|[\boldsymbol{s}-\boldsymbol{W}^*\boldsymbol{y}]_j|^2\Big] \tag{1.225}$$

$$= \mathbb{E}\Big[|s_j - \boldsymbol{w}_j^*\boldsymbol{y}|^2\Big] \tag{1.226}$$

$$= \mathbb{E}\Big[|s_j|^2\Big] - \mathbb{E}\Big[\boldsymbol{w}_j^*\boldsymbol{y}\,s_j^*\Big] - \mathbb{E}\Big[s_j\,\boldsymbol{y}^*\boldsymbol{w}_j\Big] + \mathbb{E}\Big[\boldsymbol{w}_j^*\boldsymbol{y}\boldsymbol{y}^*\boldsymbol{w}_j\Big], \tag{1.227}$$

where $\boldsymbol{w}_j = [\boldsymbol{W}]_{:,j}$ is shorthand for the part of $\boldsymbol{W}$—its jth column—that is responsible for estimating that particular entry, $s_j = [\boldsymbol{s}]_j$. The gradient of (1.227) with respect to $\boldsymbol{w}_j$, obtained by applying (D.5)–(D.7), equals

$$\nabla_{\boldsymbol{w}_j}\mathbb{E}\Big[|[\boldsymbol{s}-\hat{\boldsymbol{s}}]_j|^2\Big] = -\mathbb{E}\Big[\boldsymbol{y}\,s_j^*\Big] + \mathbb{E}\Big[\boldsymbol{y}\boldsymbol{y}^*\boldsymbol{w}_j\Big] \tag{1.228}$$

$$= -\mathbb{E}\Big[\boldsymbol{y}\left(s_j - \boldsymbol{w}_j^*\boldsymbol{y}\right)^*\Big] \tag{1.229}$$

$$= -\mathbb{E}\Big[\boldsymbol{y}\,[\boldsymbol{s}-\hat{\boldsymbol{s}}]_j^*\Big], \tag{1.230}$$

which, equated to zero, is nothing but a manifestation of the orthogonality principle exposed earlier in this chapter. Assembling the expressions corresponding to (1.229) for every column of $\boldsymbol{W}$ and equating the result to zero, we find that the LMMSE filter must satisfy

$$-\mathbb{E}\Big[\boldsymbol{y}\left(\boldsymbol{s}-\boldsymbol{W}^{\text{MMSE}*}\boldsymbol{y}\right)^*\Big] = \mathbb{E}\big[\boldsymbol{y}\boldsymbol{y}^*\big]\boldsymbol{W}^{\text{MMSE}} - \mathbb{E}\big[\boldsymbol{y}\boldsymbol{s}^*\big] \tag{1.231}$$

$$= \mathbf{0} \tag{1.232}$$

and, since the mean-square error is a quadratic—and thus convex—function of the linear filter, this condition is not only necessary but sufficient (see Appendix G). Rewritten as

$$\boldsymbol{R}_{\boldsymbol{y}}\boldsymbol{W}^{\text{MMSE}} - \boldsymbol{R}_{\boldsymbol{y}\boldsymbol{s}} = \mathbf{0}, \tag{1.233}$$

its solution does give $\boldsymbol{W}^{\text{MMSE}} = \boldsymbol{R}_{\boldsymbol{y}}^{-1}\boldsymbol{R}_{\boldsymbol{y}\boldsymbol{s}}$ as anticipated in (1.224).

Moving on, the covariance of the estimate $\hat{\boldsymbol{s}}$ emerges as

$$\boldsymbol{R}_{\hat{\boldsymbol{s}}} = \mathbb{E}[\hat{\boldsymbol{s}}\hat{\boldsymbol{s}}^*] \tag{1.234}$$

$$= \boldsymbol{W}^{\text{MMSE}*}\,\mathbb{E}[\boldsymbol{y}\boldsymbol{y}^*]\boldsymbol{W}^{\text{MMSE}} \tag{1.235}$$

$$= \boldsymbol{R}_{\boldsymbol{y}\boldsymbol{s}}^*\boldsymbol{R}_{\boldsymbol{y}}^{-1}\boldsymbol{R}_{\boldsymbol{y}}\boldsymbol{R}_{\boldsymbol{y}}^{-1}\boldsymbol{R}_{\boldsymbol{y}\boldsymbol{s}} \tag{1.236}$$

$$= \boldsymbol{R}_{\boldsymbol{y}\boldsymbol{s}}^*\boldsymbol{R}_{\boldsymbol{y}}^{-1}\boldsymbol{R}_{\boldsymbol{y}\boldsymbol{s}}, \tag{1.237}$$

while

$$\boldsymbol{R}_{\hat{\boldsymbol{s}}\boldsymbol{s}} = \mathbb{E}\big[\boldsymbol{W}^{\text{MMSE}*}\boldsymbol{y}\boldsymbol{s}^*\big] \tag{1.238}$$

$$= \boldsymbol{R}_{\boldsymbol{y}\boldsymbol{s}}^*\boldsymbol{R}_{\boldsymbol{y}}^{-1}\boldsymbol{R}_{\boldsymbol{y}\boldsymbol{s}} \tag{1.239}$$

$$= \boldsymbol{R}_{\hat{\boldsymbol{s}}}. \tag{1.240}$$

It follows that the MMSE matrix is given by

$$\boldsymbol{E} = \mathbb{E}\Big[\left(\boldsymbol{s}-\hat{\boldsymbol{s}}(\boldsymbol{y})\right)\left(\boldsymbol{s}-\hat{\boldsymbol{s}}(\boldsymbol{y})\right)^*\Big] \tag{1.241}$$

$$= \boldsymbol{R}_{\boldsymbol{s}} - \boldsymbol{R}_{\hat{\boldsymbol{s}}\boldsymbol{s}} - \boldsymbol{R}_{\hat{\boldsymbol{s}}\boldsymbol{s}}^* + \boldsymbol{R}_{\hat{\boldsymbol{s}}} \tag{1.242}$$

$$= \boldsymbol{R}_{\boldsymbol{s}} - \boldsymbol{R}_{\boldsymbol{y}\boldsymbol{s}}^*\boldsymbol{R}_{\boldsymbol{y}}^{-1}\boldsymbol{R}_{\boldsymbol{y}\boldsymbol{s}}. \tag{1.243}$$

Specialized to the linear random transformation $\boldsymbol{y} = \sqrt{\rho}\boldsymbol{A}\boldsymbol{s} + \boldsymbol{z}$, the foregoing expressions for $\boldsymbol{W}^{\text{MMSE}}$ and $\boldsymbol{E}$ become

$$\boldsymbol{W}^{\text{MMSE}} = \sqrt{\rho}\,(\boldsymbol{R}_{\boldsymbol{z}} + \rho\boldsymbol{A}\boldsymbol{R}_{\boldsymbol{s}}\boldsymbol{A}^*)^{-1}\boldsymbol{A}\boldsymbol{R}_{\boldsymbol{s}} \tag{1.244}$$

and

$$\boldsymbol{E} = \boldsymbol{R}_{\boldsymbol{s}} - \rho\boldsymbol{R}_{\boldsymbol{s}}\boldsymbol{A}^*(\boldsymbol{R}_{\boldsymbol{z}} + \rho\boldsymbol{A}\boldsymbol{R}_{\boldsymbol{s}}\boldsymbol{A}^*)^{-1}\boldsymbol{A}\boldsymbol{R}_{\boldsymbol{s}}, \tag{1.245}$$

consistent with (1.189) and (1.192) if $\boldsymbol{R}_{\boldsymbol{z}} = \boldsymbol{I}$. Derived in the context of conditional-mean MMSE estimation for white Gaussian noise and Gaussian signals, within the broader confines of the LMMSE these expressions apply regardless of the distributions thereof. Only the second-order statistics of noise and signals enter the relationships, as a result of which the formulation is characterized by the presence of quadratic forms.

Applying the matrix inversion lemma (see Appendix B.7) to (1.244), we can rewrite $\boldsymbol{W}^{\text{MMSE}}$ into the alternative form

$$\boldsymbol{W}^{\text{MMSE}} = \sqrt{\rho}\left[\boldsymbol{R}_{\boldsymbol{z}}^{-1} - \rho\boldsymbol{R}_{\boldsymbol{z}}^{-1}\boldsymbol{A}\left(\boldsymbol{R}_{\boldsymbol{s}}^{-1} + \rho\boldsymbol{A}^*\boldsymbol{R}_{\boldsymbol{z}}^{-1}\boldsymbol{A}\right)^{-1}\boldsymbol{A}^*\boldsymbol{R}_{\boldsymbol{z}}^{-1}\right]\boldsymbol{A}\boldsymbol{R}_{\boldsymbol{s}} \tag{1.246}$$

$$= \sqrt{\rho}\, \boldsymbol{R}_{\boldsymbol{z}}^{-1} \left[\boldsymbol{I} - \rho \boldsymbol{A} \left(\boldsymbol{R}_{\boldsymbol{s}}^{-1} + \rho \boldsymbol{A}^* \boldsymbol{R}_{\boldsymbol{z}}^{-1} \boldsymbol{A}\right)^{-1} \boldsymbol{A}^* \boldsymbol{R}_{\boldsymbol{z}}^{-1}\right] \boldsymbol{A} \boldsymbol{R}_{\boldsymbol{s}} \tag{1.247}$$

$$= \sqrt{\rho}\, \boldsymbol{R}_{\boldsymbol{z}}^{-1} \left[\boldsymbol{A} - \rho \boldsymbol{A} \left(\boldsymbol{R}_{\boldsymbol{s}}^{-1} + \rho \boldsymbol{A}^* \boldsymbol{R}_{\boldsymbol{z}}^{-1} \boldsymbol{A}\right)^{-1} \boldsymbol{A}^* \boldsymbol{R}_{\boldsymbol{z}}^{-1} \boldsymbol{A}\right] \boldsymbol{R}_{\boldsymbol{s}} \tag{1.248}$$

$$= \sqrt{\rho}\, \boldsymbol{R}_{\boldsymbol{z}}^{-1} \boldsymbol{A} \left[\boldsymbol{I} - \rho \left(\boldsymbol{R}_{\boldsymbol{s}}^{-1} + \rho \boldsymbol{A}^* \boldsymbol{R}_{\boldsymbol{z}}^{-1} \boldsymbol{A}\right)^{-1} \boldsymbol{A}^* \boldsymbol{R}_{\boldsymbol{z}}^{-1} \boldsymbol{A}\right] \boldsymbol{R}_{\boldsymbol{s}} \tag{1.249}$$

$$= \sqrt{\rho}\, \boldsymbol{R}_{\boldsymbol{z}}^{-1} \boldsymbol{A} \left(\boldsymbol{R}_{\boldsymbol{s}}^{-1} + \rho \boldsymbol{A}^* \boldsymbol{R}_{\boldsymbol{z}}^{-1} \boldsymbol{A}\right)^{-1} \left[\left(\boldsymbol{R}_{\boldsymbol{s}}^{-1} + \rho \boldsymbol{A}^* \boldsymbol{R}_{\boldsymbol{z}}^{-1} \boldsymbol{A}\right) - \rho \boldsymbol{A}^* \boldsymbol{R}_{\boldsymbol{z}}^{-1} \boldsymbol{A}\right] \boldsymbol{R}_{\boldsymbol{s}}$$

$$= \sqrt{\rho}\, \boldsymbol{R}_{\boldsymbol{z}}^{-1} \boldsymbol{A} \left(\boldsymbol{R}_{\boldsymbol{s}}^{-1} + \rho \boldsymbol{A}^* \boldsymbol{R}_{\boldsymbol{z}}^{-1} \boldsymbol{A}\right)^{-1}, \tag{1.250}$$

while applying the matrix inversion lemma in a reverse fashion to (1.245), $\boldsymbol{E}$ can be rewritten as

$$\boldsymbol{E} = \left(\boldsymbol{R}_{\boldsymbol{s}}^{-1} + \rho \boldsymbol{A}^* \boldsymbol{R}_{\boldsymbol{z}}^{-1} \boldsymbol{A}\right)^{-1}. \tag{1.251}$$

If both noise and signal are scalars, rather than vectors, then the two expressions for $\boldsymbol{W}^{\mathsf{MMSE}}$ coincide, yielding

$$W^{\mathsf{MMSE}} = \frac{\sqrt{\rho}}{1+\rho}, \tag{1.252}$$

while the two expressions for $\boldsymbol{E}$ reduce to

$$\mathsf{MMSE} = \frac{1}{1+\rho}, \tag{1.253}$$

as derived earlier, in the context of conditional-mean MMSE estimation, for Gaussian noise and Gaussian signals. In LMMSE, these equations acquire broader generality.

1.7.2 Random processes

The LMMSE estimation problem becomes richer when formulated for random processes, as it then splits into several variants:

- *Noncausal.* The value of some signal at time n is estimated on the basis of the entire observation of another signal, a procedure also termed *smoothing*. If the observed signal is decimated relative to its estimated brethren, then the smoothing can also be regarded as *interpolation* in the MMSE sense.
- *Causal.* The value of some signal at time n is estimated on the basis of observations of another signal at times $n-1, \ldots, n-N$. This variant, for which the term *filtering* is sometimes formally reserved in the estimation literature, and which can also be regarded as *prediction* in the MMSE sense, can be further subdivided depending on whether N is finite or unbounded.

For stationary processes, the problem was first tackled by Norbert Wiener in the 1940s [126], hence the common designation of the corresponding estimator as a *Wiener filter*. (For nonstationary processes, the more general Kalman filter was developed years later.)

Without delving extensively into the matter, on which excellent textbooks exist already [62, 114, 127], we introduce herein a couple of results that are invoked throughout the text. These results pertain to the discrete-time scalar channel $y[n] = \sqrt{\rho}\, s[n] + z[n]$ where $s[n]$

is a zero-mean unit-variance stationary signal with power spectrum $S(\cdot)$ while $z[n]$ is an IID noise sequence. For such a setting, the noncausal LMMSE filter yields [68]

$$\mathsf{MMSE} = 1 - \int_{-1/2}^{1/2} \frac{\rho\, S^2(\nu)}{1 + \rho\, S(\nu)}\, d\nu, \tag{1.254}$$

while its causal counterpart gives

$$\mathsf{MMSE} = \frac{1}{\rho}\left[\exp\left(\int_{-1/2}^{1/2} \log_e\big(1 + \rho\, S(\nu)\big)\, d\nu\right) - 1\right]. \tag{1.255}$$

Letting $\rho \to \infty$ in (1.255) returns the causal MMSE when predicting $s[n]$ based on past noiseless observations of the same process,

$$\mathsf{MMSE} = \exp\left(\int_{-1/2}^{1/2} \log_e\big[S(\nu)\big]\, d\nu\right), \tag{1.256}$$

which is zero if $s[n]$ is nonregular while strictly positive if it is regular. By inspecting (1.256) it can be deduced that, in the context of stationary processes, nonregularity is tantamount to a bandlimited power spectrum—whereby the integrand diverges over part of the spectrum—while regularity amounts to a power spectrum that is not bandlimited and strictly positive.

1.8 Summary

From the coverage in this chapter, we can distill the points listed in the accompanying summary box.

Problems

1.1 Show that, for s to be proper complex, its in-phase and quadrature components must be uncorrelated and have the same variance.

1.2 Let s conform to a 3-PSK constellation defined by $\mathrm{s}_0 = \frac{1}{\sqrt{2}}(1-\mathrm{j})$, $\mathrm{s}_1 = \frac{1}{\sqrt{2}}(-1-\mathrm{j})$, and $\mathrm{s}_2 = \mathrm{j}$. Is this signal proper complex? Is it circularly symmetric?

1.3 Let s conform to a ternary constellation defined by $\mathrm{s}_0 = -1$, $\mathrm{s}_1 = 0$, and $\mathrm{s}_2 = 1$. Is this signal proper complex? Is it circularly symmetric?

1.4 Give an expression for the minimum distance between neighboring points in a one-dimensional constellation featuring M points equidistant along the real axis.

1.5 Let x be a discrete random variable and let $y = g(x)$ with $g(\cdot)$ an arbitrary function. Is $\mathcal{H}(y)$ larger or smaller than $\mathcal{H}(x)$?

Take-away points

1. The mutual information between two random variables measures the information that one of them can supply about the other.
2. The channel capacity is the highest spectral efficiency at which reliable communication is possible in the sense that the probability of erroneous codeword decoding vanishes as the codeword length N grows.
3. If the channel is information stable, meaning that the information that the received sequence $y[0], \ldots, y[N-1]$ conveys about the transmit sequence $x[0], \ldots, x[N-1]$ is invariable for large N, then the capacity equals the maximum mutual information between $x[0], \ldots, x[N-1]$ and $y[0], \ldots, y[N-1]$ for $N \to \infty$. This maximization entails finding the optimum distribution for $x[0], \ldots, x[N-1]$ subject to the applicable constraints on the transmit signal (e.g., the power).
4. The capacity is robust in that the spectral efficiencies with finite-length (but long) codewords and reasonably small error probabilities hardly depart from it. Moreover, such long codewords can be featured without incurring excessive latencies. And, through hybrid-ARQ, the codeword length can be made adaptive.
5. The use of binary codes with binary decoding incurs only a minute information-theoretic penalty with respect to coding on the constellation's alphabet. The penalty is actually nil for BPSK and QPSK, and can be largely recovered for other constellations through iterative reception. It is thus routine, in terms of performance limits, to treat binary codes mapped to arbitrary constellations as if the coding took place on that constellation's alphabet.
6. BICM is the default architecture for coding and modulation. At the transmitter, this entails binary coding, bit-level interleaving, and constellation mapping. At the receiver, it entails soft demapping, deinterleaving, and APP binary decoding.
7. In the MMSE sense, the best estimate of a quantity is the one delivered by the conditional-mean estimator. When both the quantity being estimated and the noise contaminating the observations are Gaussian, such conditional-mean estimator is a linear function of the observations, the LMMSE estimator.
8. The I-MMSE relationship establishes that the derivative of the Gaussian-noise mutual information between two quantities equals the MMSE when observing one from the other.
9. While inferior to the conditional-mean for non-Gaussian quantities, the LMMSE estimator remains attractive because of the simplicity of linear filtering and the robustness, as only second-order statistics are required.

1.6 Express the entropy of a discrete random variable x as a function of the information divergence between x and a uniformly distributed counterpart.

1.7 Express the differential entropy of a real Gaussian variable $x \sim \mathcal{N}(\mu, \sigma^2)$.

1.8 Compute the differential entropy of a random variable that takes the value 0 with probability $1/3$ and is otherwise uniformly distributed in the interval $[-1, 1]$.

1.9 Calculate the differential entropy of a random variable x that abides by the exponential distribution

$$f_x(\mathrm{x}) = \frac{1}{\mu} e^{-x/\mu}. \tag{1.257}$$

1.10 Consider a random variable s such that $\Re\{s\} \sim \mathcal{N}(0, 1/2)$ and $\Im\{s\} = q\,\Re(s)$ where $q = \pm 1$ equiprobably. Compute the differential entropy of s, which is complex and Gaussian but not proper, and compare it with that of a standard complex Gaussian.

1.11 Prove that $\mathfrak{h}(x + a) = \mathfrak{h}(x)$ for any constant a.

1.12 Prove that $\mathfrak{h}(ax) = \mathfrak{h}(x) + \log_2 |a|$ for any constant a.

1.13 Express the differential entropy of the real Gaussian vector $\boldsymbol{x} \sim \mathcal{N}(\mu, \boldsymbol{R})$.

1.14 Consider the first-order Gauss–Markov process

$$h[n] = \sqrt{1-\varepsilon}\, h[n-1] + \sqrt{\varepsilon}\, w[n] \tag{1.258}$$

where $\{w[n]\}$ is a sequence of IID random variables with $w \sim \mathcal{N}_{\mathbb{C}}(0, 1)$.
(a) Express the entropy rate as a function of ε.
(b) Quantify the entropy rate for $\varepsilon = 10^{-3}$.
Note: The Gauss–Markov process underlies a fading model presented in Chapter 3.

1.15 Verify (1.79) and (1.80).
Hint: Express det$(\cdot)$ *as the product of the eigenvalues of its argument.*

1.16 Show that $I(x_0; x_1; y) \geq I(x_0; y)$ for any random variables x_0, x_1, and y.

1.17 Let $y = \sqrt{\rho}\,(s_0 + s_1) + z$ where s_0, s_1, and z are independent standard complex Gaussian variables.
(a) Show that $I(s_0, s_1; y) = I(\boldsymbol{s}; \sqrt{\rho}\boldsymbol{A}\boldsymbol{s} + z)$ for $\boldsymbol{s} = [s_0\ s_1]^{\mathrm{T}}$ and a suitable $\boldsymbol{A}$.
(b) Characterize $I(s_0, s_1; y) - I(s_0; y)$ and approximate its limiting behaviors for $\rho \ll 1$ and $\rho \gg 1$.
(c) Repeat part (b) for the case that s_0 and s_1 are partially correlated. What do you observe?
(d) Repeat part (b) for the modified relationship $y = \sqrt{\rho/2}\,(s_0 + s_1) + z$.
Can you draw any conclusion related to MIMO from this problem?

1.18 Let s be of unit variance and uniformly distributed on a disk while $z \sim \mathcal{N}_{\mathbb{C}}(0, 1)$.
(a) What is the first-order expansion of $\mathcal{I}(\rho) = I(s; \sqrt{\rho}s + z)$ for small ρ?
(b) What is the leading term in the expansion of $\mathcal{I}(\rho)$ for large ρ?
Note: The signal distribution in this problem can be interpreted as a dense set of concentric ∞-PSK rings, conveying information in both phase and magnitude.

1.19 Repeat Problem 1.18 with s conforming to a one-dimensional discrete constellation featuring M points equidistant along a line forming an angle ϕ with the real axis.

1.20 Let s and z conform to BPSK distributions. Express $\mathcal{I}(\rho) = I(s; \sqrt{\rho}s + z)$ and obtain expansions thereof for small and large ρ. How much is $\mathcal{I}(\rho)$ for $\rho = 5$?

1.21 Compute $I(s; \sqrt{\rho}s + z)$ with $s \sim \mathcal{N}_{\mathbb{C}}(0, 1)$ and with z having a BPSK distribution.

1.22 Compute $I(s; \sqrt{\rho}s + z)$ with both s and z having BPSK distributions.

1.23 Verify that, as argued in Example 1.11,

$$\mathcal{I}^{\text{QPSK}}(\rho) = 2\,\mathcal{I}^{\text{BPSK}}\left(\frac{\rho}{2}\right). \tag{1.259}$$

1.24 Express the Gaussian mutual information of a square QAM signal as a function of the Gaussian mutual information of another signal whose points are equiprobable and uniformly spaced over the real line.

Note: This relationship substantially simplifies the computation of the Gaussian mutual information of square QAM signals, and it is exploited to perform such computations in this book.

1.25 Let $y = \sqrt{\rho}s + z$. If z were not independent of s, would that increase or decrease $I(s; y)$ relative to the usual situation where they are independent? Can you draw any communication-theoretic lesson from this?

1.26 Let $\boldsymbol{s} \sim \mathcal{N}_{\mathbb{C}}(\boldsymbol{0}, \boldsymbol{I})$ and $\boldsymbol{z} \sim \mathcal{N}_{\mathbb{C}}(\boldsymbol{0}, \boldsymbol{I})$ while

$$\boldsymbol{A} = \begin{bmatrix} 0.7 & 1+0.5\,\mathrm{j} & 1.2\,\mathrm{j} \\ 0.2+\mathrm{j} & -2.1 & 0 \end{bmatrix}. \tag{1.260}$$

(a) Plot the exact $I(\boldsymbol{s}; \sqrt{\rho}\boldsymbol{A}\boldsymbol{s} + \boldsymbol{z})$ against its low-ρ expansion for $\rho \in [0, 1]$. Up to which value of ρ is the difference below 10%?

(b) Plot the exact $I(\boldsymbol{s}; \sqrt{\rho}\boldsymbol{A}\boldsymbol{s} + \boldsymbol{z})$ against its high-ρ expansion for $\rho \in [10, 100]$. Beyond which value of ρ is the difference below 10%?

1.27 Let $\boldsymbol{s}$ have two independent unit-variance entries and let $\boldsymbol{z} \sim \mathcal{N}_{\mathbb{C}}(\boldsymbol{0}, \boldsymbol{I})$ while $\boldsymbol{A} = [0.7 \;\; 1+0.5\,\mathrm{j}]$. On a common chart, plot $\mathcal{I}(\rho) = I(\boldsymbol{s}; \sqrt{\rho}\boldsymbol{A}\boldsymbol{s} + \boldsymbol{z})$ for $\rho \in [0, 10]$ under the following distributions for the entries of $\boldsymbol{s}$:

(a) Real Gaussian.

(b) Complex Gaussian.

(c) BPSK.

(d) QPSK.

1.28 Compute and plot, as function of $\rho \in [-5, 25]$ dB, the Gaussian mutual information function for the following constellations:

(a) 8-PSK.

(b) 16-QAM.

1.29 Establish the law of the channel

$$\bar{\boldsymbol{y}} = \sqrt{\rho}\boldsymbol{A}\bar{\boldsymbol{s}} + \bar{\boldsymbol{z}}, \tag{1.261}$$

where $\boldsymbol{A}$ is a fixed matrix whose (n, n)th entry determines how the nth transmit symbol affects the nth received one, while $\bar{\boldsymbol{z}} \sim \mathcal{N}_{\mathbb{C}}(\boldsymbol{0}, \boldsymbol{R}_{\bar{\boldsymbol{z}}})$ with the (n, n)th entry of $\boldsymbol{R}_{\bar{\boldsymbol{z}}}$ determining the correlation between the noise afflicting symbols n and n.

1.30 Consider the channel

$$y[n] = \sqrt{\rho}\,h[n]s[n] + z[n] \qquad n = 0, \ldots, N-1 \tag{1.262}$$

where $z[0], \ldots, z[N-1]$ are IID with $z \sim \mathcal{N}_{\mathbb{C}}(0, 1)$.

(a) If $h[0],\ldots,h[N-1]$ are also IID with $h \sim \mathcal{N}_{\mathbb{C}}(0,1)$, what is the channel law? Is the channel memoryless?

(b) Now suppose that $h[n+1] = h[n]$ for $n = 0,2,4,\ldots,N-2$ while $h[n+1]$ and $h[n]$ are independent for $n = 1,3,5,\ldots,N-1$, meaning that every pair of symbols shares the same coefficient but then the coefficients change across symbol pairs in an IID fashion. For $h \sim \mathcal{N}_{\mathbb{C}}(0,1)$, what is the channel law? Is the channel memoryless?

1.31 Express, to first order, the number of codeword symbols N required to achieve a certain share of the capacity C as a function of V and p_e. Then, for $V/C^2 = 4$, use the found expression to gauge the following.

(a) The value of N required to achieve 90% of capacity at $p_e = 10^{-2}$.

(b) The value of N required to achieve 95% of capacity at $p_e = 10^{-3}$.

1.32 Consider a system with $B = 100$ MHz, equally divided among $U = 10$, and with a coding latency target of 1 ms. If the operating point is $p_e = 10^{-2}$ and $V/C^2 = 2$, what fraction of the capacity can each user attain?

1.33 Reproduce the BICM curve on the left-hand side of Fig. 1.6.

1.34 Consider the transformation $y = \sqrt{\rho}s + z$.

(a) Prove that, for any arbitrary function $g(\cdot)$, $\mathbb{E}\big[g(y)(s - \mathbb{E}[s|y])\big] = 0$. This is the so-called *orthogonality principle*.

(b) Taking advantage of the orthogonality principle, prove that the MMSE estimate is given by $\hat{s}(y) = \mathbb{E}[s|y]$.

1.35 Consider the transformation $y = \sqrt{\rho}s + z$ with z a standard complex Gaussian and with $s \sim \mathcal{N}_{\mathbb{C}}(\mu_s, \sigma_s^2)$.

(a) Obtain the conditional-mean estimator.

(b) Express the corresponding $\mathsf{MMSE}(\rho)$.

(c) Verify that, when $\mu_s = 0$ and $\sigma_s^2 = 1$, such estimator reduces to (1.172) while $\mathsf{MMSE}(\rho)$ reduces to (1.175).

(d) Verify that $\mathsf{MMSE}(\cdot)$ does not depend on μ_s.

1.36 Prove that, for the transformation $y = \sqrt{\rho}s + z$ with z a standard complex Gaussian and with s being BPSK-distributed, the following are true.

(a) The conditional-mean estimate equals (1.181).

(b) The MMSE as a function of ρ equals (1.182).

1.37 Given the transformation $y = \sqrt{\rho}s + z$ with z a standard complex Gaussian, derive the function $\mathsf{MMSE}(\rho)$ for s conforming to a 16-QAM distribution.

1.38 Consider the vector transformation $\boldsymbol{y} = \boldsymbol{A}\boldsymbol{s} + \boldsymbol{z}$ where $\boldsymbol{A}$ is fixed while $\boldsymbol{s}$ and $\boldsymbol{z}$ are independent with $\boldsymbol{s} \sim \mathcal{N}_{\mathbb{C}}(\boldsymbol{0}, \boldsymbol{R_s})$ and $\boldsymbol{z} \sim \mathcal{N}_{\mathbb{C}}(\boldsymbol{0}, \boldsymbol{R_z})$.

(a) Obtain the conditional-mean estimator.

(b) Express the corresponding MMSE matrix.

(c) Verify that, for $\boldsymbol{R_z} = \boldsymbol{I}$, the MMSE matrix equals (1.192).

1.39 Let s be BPSK-distributed while $z \sim \mathcal{N}_{\mathbb{C}}(0,1)$. Compute the dB-difference between the MMSEs achieved by conditional-mean and LMMSE estimates of s based on observations of $\sqrt{\rho}s + z$ for two cases:

(a) $\rho = 1$.
(b) $\rho = 10$.

1.40 Verify that the application of (1.196) to (1.200) yields (1.201).

1.41 Let s be of unit variance while $z \sim \mathcal{N}_{\mathbb{C}}(0,1)$. Provide first-order low-ρ expansions of $\mathsf{MMSE}(\rho)$ as achieved by the conditional-mean estimate of s based on observations of $\sqrt{\rho}s + z$ under the following distributions for s:
(a) Real Gaussian.
(b) Complex Gaussian.
(c) BPSK.
(d) QPSK.
(e) ∞-PSK.
(f) ∞-QAM.
What can be observed?

1.42 On a common chart, plot $\mathsf{MMSE}(\rho)$ for the estimation of s based on observing $\sqrt{\rho}s + z$ with $z \sim \mathcal{N}_{\mathbb{C}}(0,1)$ and under the following distributions for s:
(a) Real Gaussian.
(b) Complex Gaussian.
(c) BPSK.
(d) QPSK.
Further plot, on the same chart, the corresponding low-ρ expansions of $\mathsf{MMSE}(\rho)$.

1.43 Let $y = \sqrt{\rho}s + z$ with s zero-mean unit-variance and with z a standard complex Gaussian. For $\rho \in [0, 10]$, plot the dB-difference between the mean-square error achieved by a regular LMMSE estimator and by a modified version thereof in which the estimation bias for each realization of s has been removed.

1.44 Consider the vector transformation $\boldsymbol{y} = \boldsymbol{A}\boldsymbol{s} + \boldsymbol{z}$ where $\boldsymbol{s} \sim \mathcal{N}_{\mathbb{C}}(\boldsymbol{0}, \boldsymbol{R}_{\boldsymbol{s}})$ and $\boldsymbol{z} \sim \mathcal{N}_{\mathbb{C}}(\boldsymbol{0}, \boldsymbol{R}_{\boldsymbol{z}})$.
(a) Express the MMSE matrix $\boldsymbol{E}$ when estimating $\boldsymbol{s}$ based on the observation of $\boldsymbol{y}$.
(b) Based on the expression obtained for $\boldsymbol{E}$, generalize to colored Gaussian noise the I-MMSE relationship for white Gaussian noise given in (1.207).
Note: Although derived for a Gaussian signal in this problem, the generalized version of the I-MMSE relationship does hold for arbitrarily distributed $\boldsymbol{s}$.

1.45 For the LMMSE estimator $\hat{\boldsymbol{s}}(\boldsymbol{y}) = \boldsymbol{W}^{\mathsf{MMSE}*}\boldsymbol{y} + \boldsymbol{b}^{\mathsf{MMSE}}$, determine the value of $\boldsymbol{b}^{\mathsf{MMSE}}$ as a function of the known means $\boldsymbol{\mu}_{\boldsymbol{s}}$ and $\boldsymbol{\mu}_{\boldsymbol{y}}$.

1.46 Let $\boldsymbol{s}$ be a vector containing two unit-variance entries exhibiting 50% correlation and let $z \sim \mathcal{N}_{\mathbb{C}}(0,1)$ while $\boldsymbol{A} = [0.7 \quad 1 + 0.5\,\mathrm{j}]$. Plot the MMSE as a function of $\rho \in [0, 10]$ when LMMSE-estimating $\boldsymbol{s}$ from $\sqrt{\rho}\boldsymbol{A}\boldsymbol{s} + z$.

2 A signal processing perspective

Mathematics may be compared to a mill of exquisite workmanship, which grinds your stuff of any degree of fineness; but, nevertheless, what you get out depends upon what you put in.

Thomas Huxley

2.1 Introduction

Signal processing deals with topics such as analysis (e.g., decomposing a signal to look for important features), processing (e.g., designing systems to create specified input–output relationships), or estimation, and it is often oblivious to the coded nature of the signals: symbols are processed without regard for the fact that they are pieces of a codeword. In a sense, these signal processing tasks interface the information-theoretic abstractions of Figs. 1.2 and 1.3 with the physical signals.

We begin the chapter by establishing the mathematical relationships that connect the input and output of the wireless channel. The foundations for this development are the concepts of passband and baseband representations described in Section 2.2. While this might be familiar to those well versed in digital communication, a construction from first principles ensures the highest level of understanding. A key observation is that communication signals are well modeled as being narrowband in the sense that the bandwidth they occupy around some carrier frequency is small relative to that carrier. This narrowbandness makes it possible to work with a complex baseband signal, which is carrier independent, instead of the more cumbersome real passband signal. And, because this complex baseband signal is bandlimited, the transmit–receive relationship can be represented in discrete time. Once in discrete time, the signal and noise powers are formally defined.

The complex baseband representation is extended to MIMO in Section 2.3. The multiplicity of transmit antennas, in particular, gives rise to the possibility of spatially formatting signals, and to various types of signal constraints.

Based on the models, it becomes clear that channel equalization is one of the signal processing functions that may be required in a MIMO receiver. Equalization is the term for removing convolutive distortion. In Section 2.4, we derive two linear equalizers: zero-forcing and LMMSE. A surprising result is that, in the absence of noise, perfect equalization is possible provided there are more receive than transmit antennas. Equalization can be simplified by operating in the frequency domain. In Sections 2.5 and 2.6, we introduce the two most common such techniques: single-carrier frequency-domain equalization

(SC-FDE) and OFDM. Each approach uses a *cyclic prefix* to convert the linear convolution in the channel into a circular convolution that can be conveniently handled in the frequency domain. The difference between SC-FDE and OFDM is in how the time-to-frequency transformations are split between transmitter and receiver. SC-FDE and OFDM are also explained through their connection to circulant and block circulant matrices, constructed from the convolution matrix of the channel.

Channel equalization requires CSI. In Section 2.7 we introduce the concept of pilot-aided channel estimation, whereby known pilot signals are inserted periodically within the transmission so as to facilitate the estimation of the channel response at the receiver. By writing the received signal in terms of a block convolution matrix constructed from the pilots, a problem can be formulated and solved to obtain a channel estimate. We entertain two criteria for finding an estimator, ML and MMSE, further showing that the ML estimate in this case happens to be equivalent to the least-squares estimate.

2.2 Signal, channel, and noise representations

Throughout this section, we develop mathematical models for the transmit and receive signals, for the impulse response that describes the channel, and for the additive noise. To assist readers throughout this formulation, a review of the basics of Fourier transformations is provided in Appendix A.1.

2.2.1 Passband signals and complex baseband equivalents

Wireless transmitters map data onto electromagnetic waves using a combination of digital, mixed-signal, and analog components. We can characterize the transmitted waveform by the continuous-time voltage $x_{\mathrm{p}}(t)$ that is applied to the antenna. This signal is passband, meaning that, in the frequency domain, most of its energy is concentrated around a carrier frequency f_{c}. The subscript $(\cdot)_{\mathrm{p}}$ serves as a reminder that $x_{\mathrm{p}}(t)$ is passband.

Suppose that $x_{\mathrm{p}}(t)$ is an "ideal" passband signal, where "ideal" means that, in the frequency domain, it is nonzero only over a bandwidth B centered on f_{c}. (The ideality makes the mathematical exposition more precise; actual signals are not ideally bandlimited, but do have most of their energy within B.) Except in the so-called ultrawideband systems, which require separate treatment [128–131], it further holds that $B \ll f_{\mathrm{c}}$. Under this narrowband condition, $x_{\mathrm{p}}(t)$ can be written as

$$x_{\mathrm{p}}(t) = A(t)\cos\big(2\pi f_{\mathrm{c}} t + \phi(t)\big), \tag{2.1}$$

where $A(t)$ is an magnitude function and $\phi(t)$ is a phase function. Applying trigonometric identities, (2.1) can be arranged into

$$x_{\mathrm{p}}(t) = \underbrace{A(t)\cos(\phi(t))}_{\sqrt{2}\,x_{\mathrm{i}}(t)}\cos(2\pi f_{\mathrm{c}} t) - \underbrace{A(t)\sin(\phi(t))}_{\sqrt{2}\,x_{\mathrm{q}}(t)}\sin(2\pi f_{\mathrm{c}} t), \tag{2.2}$$

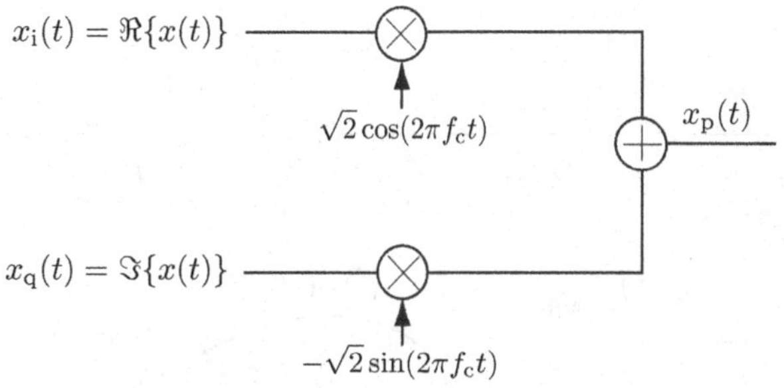

Fig. 2.1 Direct upconversion of a baseband signal to passband.

where, as advanced in the previous chapter, $x_i(t)$ is known as the *in-phase* component and $x_q(t)$ is known as the *quadrature* component; $x_i(t)$ modulates a cosine carrier while $x_q(t)$ modulates a sine carrier.

Now, let us define the *complex envelope* or *complex baseband equivalent* of $x_p(t)$ as $x(t) = x_i(t) + j\,x_q(t)$. Noting that the factors $\sqrt{2}$ in (2.2) simply ensure that $x(t)$ is defined as having the same power as $x_p(t)$, we can apply Euler's formula, $e^{j\theta} = \cos\theta + j\sin\theta$, to rewrite (2.2) as

$$x_p(t) = \sqrt{2}\left(x_i(t)\cos(2\pi f_c t) - x_q(t)\sin(2\pi f_c t)\right) \tag{2.3}$$

$$= \sqrt{2}\,\Re\left\{x(t)\,e^{j2\pi f_c t}\right\}. \tag{2.4}$$

While $x(t)$ is complex, the voltage $x_p(t)$ applied to the antenna is real because it represents a physical quantity.

In the frequency domain, (2.4) corresponds to

$$\mathsf{x}_p(f) = \frac{1}{\sqrt{2}}\left(\mathsf{x}(f - f_c) + \mathsf{x}^*(f + f_c)\right), \tag{2.5}$$

where $\mathsf{x}(f) = \mathsf{x}_i(f) + j\,\mathsf{x}_q(f)$ is the Fourier transform of $x(t)$. This means that if $\mathsf{x}_p(f)$ is ideally bandlimited with passband bandwidth B, then $\mathsf{x}(f)$ and its components $\mathsf{x}_i(f)$ and $\mathsf{x}_q(f)$ are bandlimited with baseband bandwidth $B/2$. Essentially, a passband signal of bandwidth B can convey two baseband signals of bandwidth $B/2$, which, although seemingly entangled because they overlap in both time and frequency, are actually orthogonal by virtue of the phase difference between the cosine and sine carriers.

The generation of a passband signal from a baseband counterpart at the transmitter is known as *upconversion*, while the reverse process at the receiver is *downconversion*.

Upconversion can be accomplished through a direct implementation of (2.2), as illustrated in Fig. 2.1. A local oscillator generates a carrier signal $\sqrt{2}\cos(2\pi f_c t)$ and, through a phase shifter, also $-\sqrt{2}\sin(2\pi f_c t)$. The in-phase signal $x_i(t)$ is applied to the cosine while $x_q(t)$ is applied to the sine. The frequency-domain interpretation of a direct upconversion is provided in Fig. 2.2. (As an alternative to this direct implementation, upconversion may be accomplished in multiple stages.)

The downconversion of the received passband signal is more involved than the upconversion. Referring to Figs. 2.3 and 2.4, let $y_p(t)$ denote the received passband signal, i.e.,

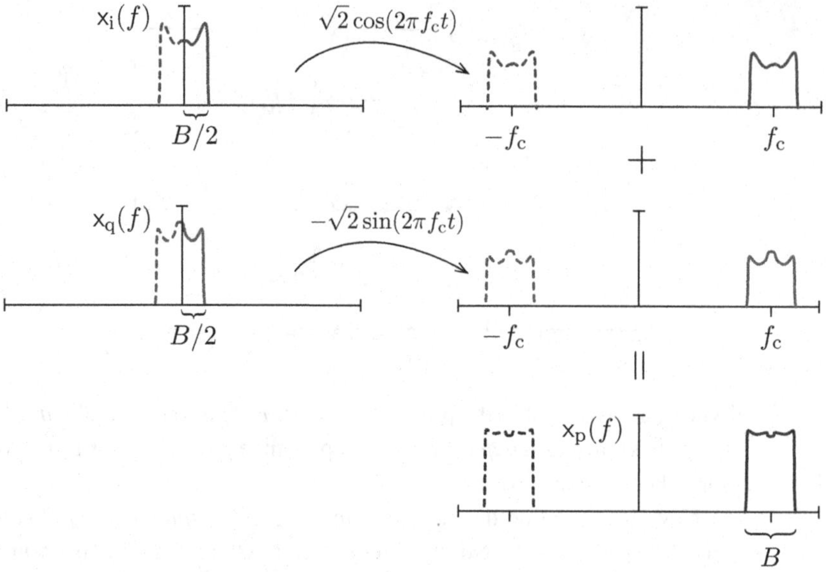

Fig. 2.2 Top, upconversion of the in-phase signal component. Middle, upconversion of its quadrature counterpart. Bottom, passband signal containing both components. (For illustration purposes, only the magnitude of the signals is depicted.)

the voltage observed at the antenna. This signal is first bandpass-filtered to reject transmissions on other frequency bands that the antenna may be picking up and that could get mixed in during the downconversion. Being passband, $y_p(t)$ has a baseband equivalent $y(t) = y_i(t) + j\, y_q(t)$ such that

$$y_p(t) = \sqrt{2}\left(y_i(t)\cos(2\pi f_c t) - y_q(t)\sin(2\pi f_c t)\right). \tag{2.6}$$

The product of $y_p(t)$ and $\sqrt{2}\cos(2\pi f_c t)$ gives, with a bit of trigonometry,

$$y_p(t)\,\sqrt{2}\cos(2\pi f_c t) = y_i(t) + y_i(t)\cos(4\pi f_c t) - y_q(t)\sin(4\pi f_c t). \tag{2.7}$$

Since $B < f_c$, there is no spectral overlap between $y_i(t)$ and the last two terms in (2.7), which are higher frequency images. Thus, a lowpass filter with cutoff at $\pm B/2$ can reproduce the in-phase component $y_i(t)$. Similarly, the product of $y_p(t)$ and $-\sqrt{2}\sin(2\pi f_c t)$ yields

$$-y_p(t)\,\sqrt{2}\sin(2\pi f_c t) = y_q(t) - y_q(t)\cos(4\pi f_c t) - y_i(t)\sin(4\pi f_c t) \tag{2.8}$$

from which, again, an appropriate lowpass filter can recover $y_q(t)$ [132].

Note that performing downconversion correctly requires that f_c be known precisely at the receiver—not an easy task! This is the subject of carrier frequency estimation, an important signal processing problem.

Signal processing and communication engineers generally prefer to work with the complex baseband equivalent of a signal instead of the passband signal. Besides the mathematical convenience of not requiring the carrier frequency f_c in the notation, this reflects the

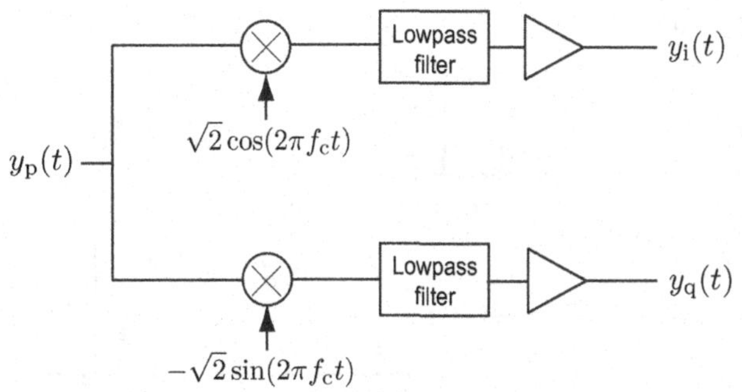

Fig. 2.3 Direct downconversion of a passband signal to baseband. The circuit includes a bandpass limiting filter right after the antenna and a low-noise amplifier with a series of gain-control circuits.

fact that most—if not all—signal processing and communication algorithms are applied in baseband, before upconversion at the transmitter and after downconversion at the receiver.

2.2.2 Complex baseband channel response

There are many kinds of impairments that distort the version of the transmitted signal that is observed at the receiver. The over-the-air propagation channel attenuates the signal and introduces (possibly time-varying) dispersion owing to the presence of multiple propagation paths. The radio-frequency circuits may exhibit nonlinear behaviors, and also introduce additive noise and phase noise. Nonideal filters may also introduce additional dispersion. Some of these impairments (e.g., nonlinearity) are minimized by careful designs; their effects on the overall system are incorporated later through simulations. Other impairments (e.g., those due to propagation and additive noise) are unavoidable and are most often modeled and analyzed explicitly because they represent inherent obstacles to communication.

In this section, a generic complex baseband model for the channel response is developed in both continuous and discrete time. In Chapter 3, we specialize this model to the wireless realm and provide a rich statistical description of those features that are relevant.

A linear time-invariant channel model suffices for the combined effects of multipath propagation and of frequency selectivity in the analog filters making up the radio front-end. Linearity follows from the linearity of electromagnetic propagation in the far-field. The time invariance is a more fragile assumption. The response of a propagation channel does vary over time due to mobility of the transmitter, receiver, or environment. Most wireless systems, though, are designed so that the channel is roughly constant over certain intervals of time and frequency. These intervals are studied in detail in Chapter 3 as part of a more general time-varying formulation, but for our purposes in this chapter a linear time-invariant impulse response is sufficient. Let us write the received signal in terms of such response, denoted by $c_p(\tau)$, which characterizes the output produced by an impulse as a function of the delay τ relative to that impulse. Although we subscript it by $(\cdot)_p$ because it

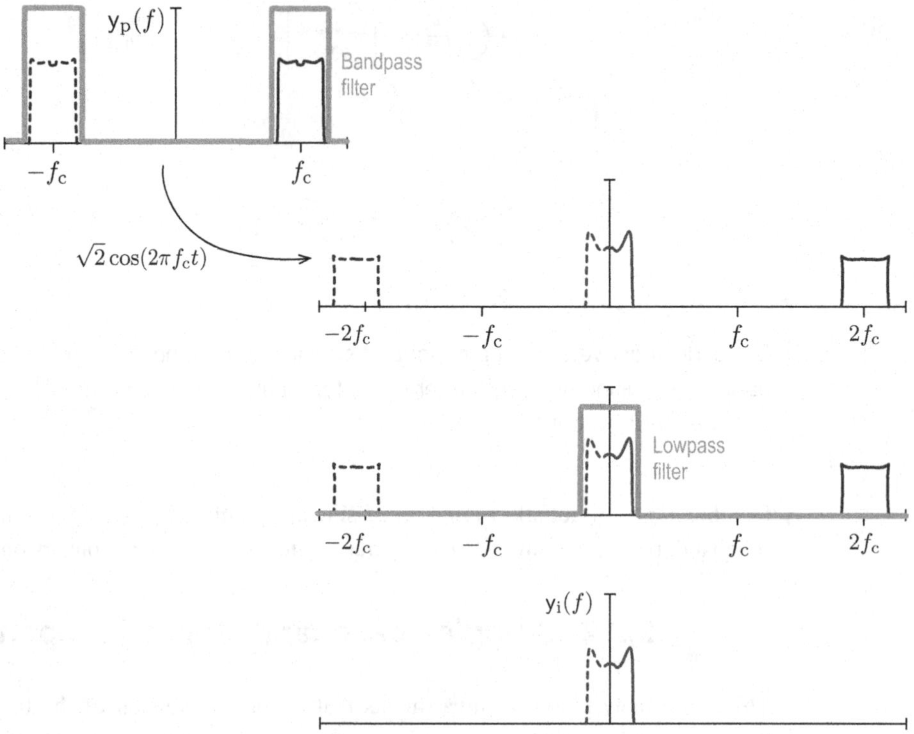

Fig. 2.4 Top, bandpass filtering and downconversion via cosine. Middle, application of a lowpass filter to remove higher frequency images. Bottom, recovered in-phase baseband signal component. (For illustration purposes, only the magnitude of the signals is depicted.)

is applied to the transmit passband signal, $c_\mathrm{p}(\tau)$ need not be itself bandlimited. Because of time invariance, $y_\mathrm{p}(t)$ relates to $x_\mathrm{p}(t)$ through the linear convolution

$$y_\mathrm{p}(t) = (c_\mathrm{p} * x_\mathrm{p})(t) \tag{2.9}$$

$$= \int_{-\infty}^{\infty} c_\mathrm{p}(\tau)\, x_\mathrm{p}(t-\tau)\, \mathrm{d}\tau, \tag{2.10}$$

where, since $x_\mathrm{p}(t)$ is passband, $y_\mathrm{p}(t)$ is also passband.

Next, we develop an expression that involves the complex baseband equivalents of the transmit and receive signals, rather than their passband counterparts. The complex baseband impulse response $c_\mathrm{b}(\tau)$ serves the same purpose as $c_\mathrm{p}(\tau)$, relating the complex baseband signals $x(t)$ and $y(t)$ through

$$y(t) = (c_\mathrm{b} * x)(t). \tag{2.11}$$

To find an expression for $c_\mathrm{b}(\tau)$ given $c_\mathrm{p}(\tau)$, let us denote by $\mathsf{g}_{B/2}(f) = \mathrm{rect}(f/B)$ an ideal lowpass filter with unit gain for $f \in [-B/2, B/2]$, with inverse Fourier transform

$$g_{B/2}(\tau) = B\,\mathrm{sinc}(B\tau), \tag{2.12}$$

where $\text{sinc}(z) = \frac{\sin(\pi z)}{\pi z}$ and the filter satisfies $g_{B/2}(0) = B$. In turn, an ideal bandpass filter with bandwidth B centered on f_c is

$$g_{\text{p},B}(\tau) = 2\, g_{B/2}(\tau) \cos(2\pi f_c \tau). \tag{2.13}$$

Recalling that $x_\text{p}(t)$ is passband, it is only necessary to model the portion of $c_\text{p}(\tau)$ that lies within the occupied spectrum of $x_\text{p}(t)$ with bandwidth B centered on f_c. Given what unfolds in the signal downconversion, $c_\text{b}(\tau)$ is a bandpass filtered version of $c_\text{p}(\tau)$, subsequently shifted down to baseband and lowpass-filtered again, i.e.,

$$c_\text{b}(\tau) = g_{B/2}(\tau) * \left[\left(c_\text{p}(\tau) * g_{\text{p},B}(\tau) \right) e^{-j2\pi f_c \tau} \right]. \tag{2.14}$$

Although, as explained, the double filtering at passband and baseband is necessary to reject adjacent signals, for our derivations here one of them suffices and thus we can directly write

$$c_\text{b}(\tau) = g_{B/2}(\tau) * \left[c_\text{p}(\tau)\, e^{-j2\pi f_c \tau} \right]. \tag{2.15}$$

Example 2.1

Consider a signal that propagates from transmitter to receiver incurring a delay τ_0 and experiencing a complex amplitude gain A_0. Determine $c_\text{p}(\tau)$ and $c_\text{b}(\tau)$.

Solution

A channel impulse response with a complex gain A_0 and a delay τ_0 corresponds to

$$c_\text{p}(\tau) = A_0\, \delta(\tau - \tau_0). \tag{2.16}$$

Applying (2.15), we obtain $c_\text{b}(t) = A_0\, g_{B/2}(\tau - \tau_0)\, e^{-j2\pi f_c \tau_0}$ which, recalling $g_{B/2}(\tau)$ from (2.12), becomes

$$c_\text{b}(\tau) = A_0 B\, \text{sinc}\big(B(\tau - \tau_0)\big)\, e^{-j2\pi f_c \tau_0}. \tag{2.17}$$

Example 2.2

Now suppose there is a second path for the signal with complex gain A_1 and delay τ_1. Determine again $c_\text{p}(\tau)$ and $c_\text{b}(\tau)$.

Solution

From linearity, $c_\text{p}(\tau) = A_0\, \delta(\tau - \tau_0) + A_1\, \delta(\tau - \tau_1)$ and

$$c_\text{b}(\tau) = A_0 B\, \text{sinc}(B(\tau - \tau_0))\, e^{-j2\pi f_c \tau_0} + A_1 B\, \text{sinc}(B(\tau - \tau_1))\, e^{-j2\pi f_c \tau_1}. \tag{2.18}$$

From Examples 2.1 and 2.2, we can infer the more general form with Q paths,

$$c_\text{p}(\tau) = \sum_{q=0}^{Q-1} A_q\, \delta(\tau - \tau_q), \tag{2.19}$$

where A_q and τ_q are the complex gain and delay of the qth signal path. Following the logic of Example 2.2,

$$c_b(\tau) = \sum_{q=0}^{Q-1} A_q \, g_{B/2}(\tau - \tau_q) \, e^{-j2\pi f_c \tau_q} \tag{2.20}$$

and, with the lowpass filter pulled out,

$$c_b(\tau) = g_{B/2}(\tau) * \sum_{q=0}^{Q-1} A_q \, \delta(\tau - \tau_q) \, e^{-j2\pi f_c \tau_q}. \tag{2.21}$$

From (2.11) then, the received baseband signal is

$$y(t) = \int_{-\infty}^{\infty} \left(g_{B/2}(\tau) * \sum_{q=0}^{Q-1} A_q \, \delta(\tau - \tau_q) \, e^{-j2\pi f_c \tau_q} \right) x(t - \tau) \, d\tau \tag{2.22}$$

which, since $x(t)$ is already bandlimited, amounts to

$$y(t) = \int_{-\infty}^{\infty} \underbrace{\left(\sum_{q=0}^{Q-1} A_q \, \delta(\tau - \tau_q) \, e^{-j2\pi f_c \tau_q} \right)}_{c(\tau)} x(t - \tau) \, d\tau \tag{2.23}$$

$$= (c * x)(t). \tag{2.24}$$

We term $c(\tau)$ the *complex pseudo-baseband channel response* since it is downshifted, but not bandlimited—hence "pseudo" baseband. The corresponding $c_b(\tau)$ can be obtained from $c(\tau)$ by lowpass filtering, i.e.,

$$c_b(\tau) = g_{B/2}(\tau) * c(\tau). \tag{2.25}$$

As an alternative to the model with Q discrete paths that leads to $c(\tau)$ containing Q impulses, a more general continuous model can be adopted for $c(\tau)$ within a certain delay interval. Also in this case, $c(\tau)$ would not be bandlimited and would have to be lowpass filtered to obtain the corresponding bandlimited $c_b(\tau)$.

2.2.3 Time discretization

The complex baseband transmit–receive relationship in (2.11) can be further simplified by exploiting the bandlimitedness of $x(\cdot)$, $y(\cdot)$, and $c_b(\cdot)$. Leveraging the sampling theorem, corresponding models entirely in discrete time can be developed. These models reflect the reality that most of the signal processing takes place digitally and in baseband, namely in programmable digital signal processors or in application-specific integrated circuits.

Digressing briefly, we can recall the essence of the sampling theorem: a bandlimited signal $z(t)$ with baseband bandwidth $B/2$ is completely determined from its samples $z[n] = z(nT)$ provided that $T \leq 1/B$; conversely, $z(t)$ can be reconstructed from $z[n]$ as

$$z(t) = \sum_n z[n] \, \text{sinc}\left(\frac{t - nT}{T}\right). \tag{2.26}$$

Thus, our $x(t)$ and $y(t)$ can be represented by their discrete-time samples with $T = 1/B$.

Now, let us find a discrete-time equivalent channel response $c[\ell]$ such that

$$y[n] = \sum_{\ell=-\infty}^{\infty} c[\ell]\, x[n-\ell]. \tag{2.27}$$

Discretizing the integral that effects the linear convolution $(c_\text{b} * x)$ with a step size $\mathrm{d}\tau = T$ and sampling the outcome, what is obtained is

$$y(nT) = \int_{-\infty}^{\infty} c_\text{b}(\tau)\, x(nT-\tau)\, \mathrm{d}\tau \tag{2.28}$$

$$\approx \sum_{\ell=-\infty}^{\infty} c_\text{b}(\ell T)\, x(nT-\ell T)\, T \tag{2.29}$$

$$= \sum_{\ell=-\infty}^{\infty} T\, c_\text{b}(\ell T)\, x[n-\ell], \tag{2.30}$$

which points to $c[\ell] \approx T c_\text{b}(\ell T)$. This intuitive result is actually exact as, for any baseband or pseudo-baseband channel, the discrete-time equivalent in (2.27) is obtained by applying a lowpass filter $T\, g_{B/2}(\tau)$ with subsequent sampling at $\tau = \ell T$ [133, ch. 4]. For an already bandlimited channel such as $c_\text{b}(\tau)$, this filtering amounts to a scaling by T and thus

$$c[\ell] = T\, c_\text{b}(\tau)|_{\tau=\ell T} \tag{2.31}$$

$$= T\, c_\text{b}(\ell T), \tag{2.32}$$

matching the intuitive result above. The scaling by T ensures the conservation of power when moving the transmit–receive relationship from continuous to discrete time, with one interpretation of this scaling being that it subsumes two factors $\sqrt{T}$ that convert the transmit and receive energies per unit time into energies per sample.

Example 2.3

Determine the discrete-time complex baseband response for Example 2.1 with $T = 1/B$.

Solution

From (2.32) and (2.17)

$$c[\ell] = A_0 \,\text{sinc}\Big(\ell - \frac{\tau_0}{T}\Big)\, e^{-\text{j}2\pi f_\text{c} \tau_0}. \tag{2.33}$$

Alternatively, if an offset τ_0 (not multiple of T) is applied to the sampler, then

$$c[\ell] = A_0 \,\text{sinc}(\ell)\, e^{-\text{j}2\pi f_\text{c} \tau_0} \tag{2.34}$$

$$= A_0\, \delta[\ell]\, e^{-\text{j}2\pi f_\text{c} \tau_0}. \tag{2.35}$$

There is a pronounced difference between (2.33) and (2.35) in that the former channel is *dispersive*, meaning that the convolution is bound to mix $x[n]$ with other symbols, while the latter merely multiplies $x[n]$ by a complex factor $A_0 e^{-\text{j}2\pi f_\text{c} \tau_0}$. Thus, the former does not allow for communication free of interference among symbols of period T, so-called *intersymbol interference* (ISI), while the latter does. Invoking the notion of memorylessness introduced in Chapter 1, (2.33) exhibits memory while (2.35) is memoryless. This distinction indicates the importance of time synchronization, again not an easy task and an important signal processing problem. Once the receiver has succeeded at synchronizing, the delay introduced by the channel can be corrected and the most convenient offset can be applied to the sampler.

Example 2.4

Determine $c[\ell]$ for Example 2.2 with $T = 1/B$.

Solution

Repeating the procedure of the previous example,

$$c[\ell] = A_0 \operatorname{sinc}\left(\ell - \frac{\tau_0}{T}\right) e^{-j2\pi f_c \tau_0} + A_1 \operatorname{sinc}\left(\ell - \frac{\tau_1}{T}\right) e^{-j2\pi f_c \tau_1}. \tag{2.36}$$

In this case, a sampling offset of τ_0 leads to

$$c[\ell] = A_0\, \delta[\ell]\, e^{-j2\pi f_c \tau_0} + A_1 \operatorname{sinc}\left(\ell - \frac{\tau_1 - \tau_0}{T}\right) e^{-j2\pi f_c \tau_1}. \tag{2.37}$$

Alternative discrete-time channels are obtained with distinct sampling offsets, but some degree of ISI is inevitable.

In light of the connection established between passband signals, bandlimited baseband signals, and discrete-time signals, most results in this book are formulated directly in discrete time. Although actual signals are not exactly bandlimited since they are time-limited, and the filters in the front-ends are never ideal, signals are close to bandlimited as they are mandated to have low adjacent channel interference. Commercial systems also feature analog-to-digital and digital-to-analog converters with finite precision. Normally, perfect conversion is assumed in the design stages and then the required tolerances are investigated after the fact via simulation. Quantization and reconstruction errors are not part of the models in this text.

2.2.4 Pulse shaping

Besides being more amenable, a discrete-time formulation contributes to bridging the transmit–receive signal relationships that we are elaborating here with the information-theoretic concepts presented in Chapter 1, and particularly with the basic ingredient of symbols. The sampling and reconstruction that we have applied to discretize the transmit–receive relationships can be interpreted as basic modulation and demodulation schemes, whereby the complex symbols that make up the codewords are embedded onto $x(t)$ and then extracted from $y(t)$. Reproducing (2.26), we can interpret the baseband transmit signal

$$x(t) = \sum_n x[n] \operatorname{sinc}\left(\frac{t - nT}{T}\right) \tag{2.40}$$

Discussion 2.1 Baseband versus pseudo-baseband channel responses

There is some confusion in the research literature between the complex baseband and the complex pseudo-baseband channel responses. Both are equivalent, by virtue of the bandlimitedness of $x(t)$, in the sense that $y(t) = c_{\mathrm{b}}(\tau) * x(t)$ and $y(t) = c(\tau) * x(t)$, and thus the simpler $c(\tau)$ is often used for convenience. However, since $c(\tau)$ is not bandlimited, it cannot be sampled and therefore care must be exercised when moving to discrete-time representations. Lowpass filtering of $c(\tau)$, which gives $c_{\mathrm{b}}(\tau)$, is necessary before discretization.

The confusion is fed by the fact that, in the important case that $c[\ell]$ exhibits a single nonzero tap, the gain of such tap can be pulled directly from $c(\tau)$. The channel merely multiplies the signal by that complex gain.

Example 2.5

If $Q = 1$ and the sampler is offset by τ_0, the discretization of $c_{\mathrm{b}}(\tau)$ in (2.20) with period $T = 1/B$ yields

$$c[\ell] = T c_{\mathrm{b}}(\ell T) \tag{2.38}$$

$$= \underbrace{A_0 \, e^{-\mathrm{j}2\pi f_{\mathrm{c}} \tau_0}}_{\text{Complex gain}} \, \delta[\ell], \tag{2.39}$$

as shown in Fig. 2.5. The complex gain of the single nonzero tap is directly that of the corresponding pseudo-baseband channel $c(\tau) = A_0 \, e^{-\mathrm{j}2\pi f_{\mathrm{c}} \tau_0} \delta(\tau - \tau_0)$ and the channel output is $y[n] = A_0 \, e^{-\mathrm{j}2\pi f_{\mathrm{c}} \tau_0} x[n]$.

Example 2.5 continues to hold approximately if $Q > 1$ with the path delays not very different relative to T; then, $c[\ell] \approx \left(\sum_q A_q \, e^{-\mathrm{j}2\pi f_{\mathrm{c}} \tau_q} \right) \delta[\ell]$. However, if $Q > 1$ and the path delays are sufficiently dissimilar relative to T, the effect of the channel is no longer merely multiplicative; rather, it is dispersive and $c[\ell]$ does not directly follow from $c(\tau)$, but only from sampling the lowpass-filtered $c_{\mathrm{b}}(\tau)$.

Example 2.6

If $Q = 2$ with $\tau_1 = \tau_0 + \frac{4}{3} T$, the discretization of $c_{\mathrm{b}}(\tau)$ with the sampler aligned with τ_0 yields $c[\ell]$ as shown in Fig. 2.6. Neither such $c[\ell]$ nor $y[n]$ can be directly derived from the pseudo-baseband $c(\tau)$.

Figure 2.7 recaps, in a graphical fashion, the relationships among the various channel representations.

as a sequence of symbols $\{x[n]\}$ linearly modulated onto the pulse shape $\mathrm{sinc}(t/T)$ and transmitted at a rate of $1/T$ symbols/s. In turn, at the receiver, we can interpret the sampling $y(nT)$ as the demodulation of those symbols from the waveform that emerges from the

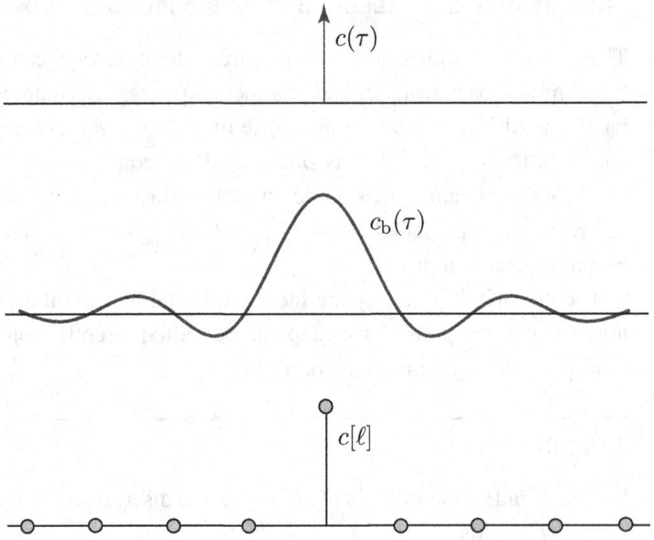

Fig. 2.5 $c(\tau)$, $c_b(\tau)$, and $c[\ell]$ for a single-path channel with properly aligned sampling.

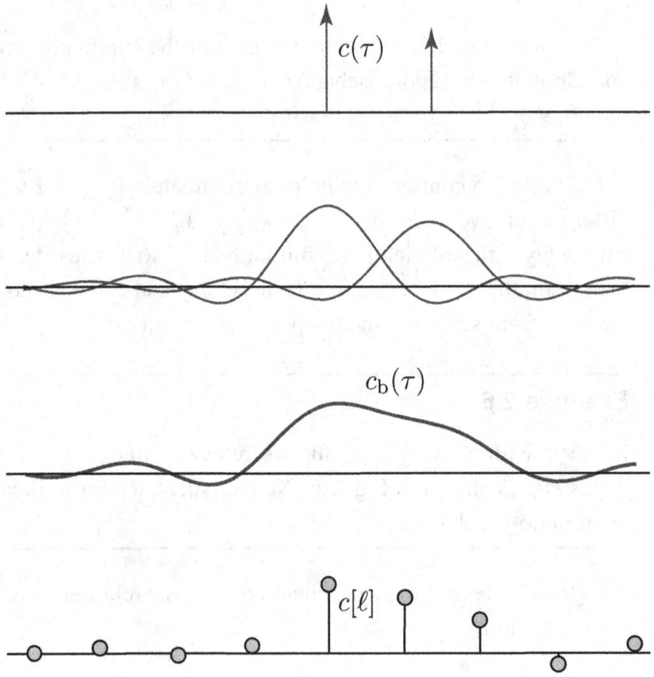

Fig. 2.6 $c(\tau)$, $c_b(\tau)$, and $c[\ell]$ for a two-path channel with $\tau_1 = \tau_0 + \frac{4}{3}T$ and $A_1 = 0.8A_0$.

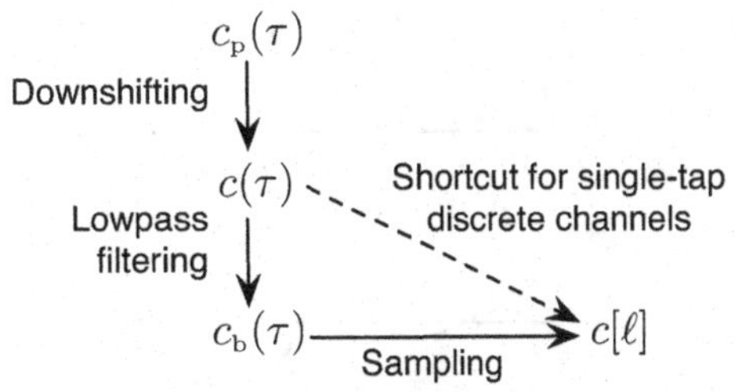

Fig. 2.7 Relationships among $c_p(\tau)$, $c_b(\tau)$, $c(\tau)$, and $c[\ell]$. When $c[\ell]$ exhibits a single nonzero tap, the gain of that tap can be gathered directly from $c(\tau)$, bypassing the need for explicit lowpass filtering.

lowpass filter and, if the channel merely effects a multiplicative gain and the lowpass filter is ideal, then the pulse shape is a scaled version of the original $\text{sinc}(\cdot)$ function.

In practice, shapes that depart from the $\text{sinc}(\cdot)$ function are preferable because this function exhibits pronounced ripples and is therefore not very robust to sampling time offsets. Thus, in lieu of (2.40), the transmit signal is built as

$$x(t) = \sum_n x[n]\, g_{\text{tx}}(t - nT), \tag{2.41}$$

where $g_{\text{tx}}(\cdot)$ is the transmit pulse shape. Effectively, symbol $x[n]$ rides the pulse $g_{\text{tx}}(t-nT)$ and the result is a complex pulse amplitude transmission. Likewise, rather than an ideal lowpass filter $g_{B/2}(\cdot)$, the receiver features a modified filter $g_{\text{rx}}(\cdot)$. Altogether, the pulse shape is the composite function

$$g(\tau) = (g_{\text{tx}} * g_{\text{rx}})(\tau). \tag{2.42}$$

A design criterion for $g(\cdot)$ is that $g(\ell T) = 1$ for $\ell = 0$ and $g(\ell T) = 0$ for $\ell \neq 0$. In particular, the condition $g(\ell T) = 0$ for $\ell \neq 0$, termed the Nyquist criterion, ensures that, although distinct pulses overlap, the on-time sampling of each is free of ISI.

A family of shapes satisfying the Nyquist criterion are the functions [134]

$$g(\tau) = \text{sinc}\left(\frac{\tau}{T}\right) \frac{\cos(\pi \mathrm{b} \tau / T)}{1 - (2\mathrm{b}\tau/T)^2} \tag{2.43}$$

parameterized by the so-called rolloff factor $\mathrm{b} \in [0, 1]$. An increase in b makes $\mathrm{g}(f)$ roll off more smoothly and attenuates the ripples in $g(\tau)$, making the sampling more forbidding to imperfect synchronization, at the expense of an increase in bandwidth. This tradeoff is better appreciated in the frequency domain, where

$$\mathrm{g}(f) = \begin{cases} T & |f| \leq \frac{1-\mathrm{b}}{2T} \\ \frac{T}{2}\left[1 + \cos\left(\frac{\pi T}{\mathrm{b}}\left[|f| - \frac{1-\mathrm{b}}{2T}\right]\right)\right] & \frac{1-\mathrm{b}}{2T} \leq |f| \leq \frac{1+\mathrm{b}}{2T} \\ 0 & \text{otherwise} \end{cases} \tag{2.44}$$

exhibits a raised-cosine shape with b the share of *excess bandwidth* over $1/T$: the passband

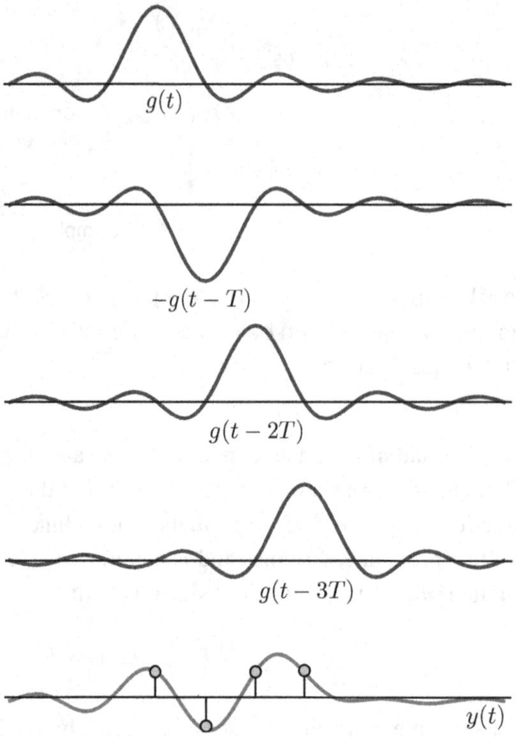

Fig. 2.8 Noiseless signal $y(t)$ carrying the BPSK symbols $+1$, -1, $+1$, and $+1$ with $g(\tau) = \text{sinc}(\tau/T)$.

bandwidth is $\frac{1+\text{b}}{T}$ and a sampling rate correspondingly faster than $1/T$ may be required. For $\text{b} = 0$, we recover $g(\tau) = \text{sinc}(\tau/T)$.

Example 2.7

For $g(\tau) = \text{sinc}(\tau/T)$, draw the noiseless received signal corresponding to the transmission of the sequence of BPSK symbols $+1$, -1, $+1$, and $+1$ over a multiplicative channel exerting a gain A_0. Verify that, despite the pulse overlaps, there is no ISI.

Solution

Depicted at the bottom of Fig. 2.8 is

$$y(t) = A_0\big(g(t) - g(t-T) + g(t-2T) + g(t-3T)\big), \tag{2.45}$$

which, sampled at multiples of T, returns a scaled version of the original BPSK symbols.

A breakdown that ensures matched transmit and receive filters with the composite pulse shape corresponding to a raised-cosine filter is

$$\text{g}_{\text{tx}}(f) = \sqrt{T\text{g}(f)} \tag{2.46}$$

$$g_{\mathrm{rx}}(f) = \sqrt{\frac{g(f)}{T}}, \tag{2.47}$$

whereby both the transmit and receive filters exhibit root raised-cosine shapes. This ensures matched filtering and ISI-free transmission provided the channel is multiplicative, as turns out to be the case with OFDM. In terms of implementation at the receiver, an analog filter $g_{\mathrm{rx}}(\cdot)$ may be followed by a sampler or else the downconverted signal may be oversampled faster than $1/T$, digitally filtered, and then downsampled to $1/T$.

If the channel is markedly dispersive, say in a non-OFDM system, a true matched-filter receiver would have to be shaped as $(g_{\mathrm{tx}} * c)$ and therefore it would be channel-dependent. Moreover, the sampled signal is then generally plagued with ISI and equalization becomes necessary. Once that is the case, the advantage of careful pulse shaping mostly evaporates. Since ISI is to arise regardless, the precise shape of $g_{\mathrm{rx}}(\cdot)$ becomes of little consequence and it can be subsumed within the subsequent equalization [135]. The downconverted signal can then be sampled (possibly faster than $1/T$) and directly equalized. In turn, the design of the transmit filter $g_{\mathrm{tx}}(\cdot)$ can concentrate on minimizing the excess bandwidth.

2.2.5 Additive noise

Besides channel distortion, communication receivers have to deal with additive noise and the most common type thereof is the thermal noise caused by the random motion of electrons in the components. Other sources of additive noise include quantization noise (mentioned in reference to analog-to-digital conversion) and interference from undesired transmissions. Phase noise, not considered here, is also possible in some instances.

Additive white Gaussian noise (AWGN), represented by $v_{\mathrm{p}}(t)$ in its passband form, is the de-facto model for thermal noise.

- It is white because its autocovariance is $R_{v_{\mathrm{p}}}(\tau) = \mathbb{E}\big[v_{\mathrm{p}}(t)v_{\mathrm{p}}^*(t+\tau)\big] = N_0\,\delta(\tau)$ and the corresponding power spectrum is $S_{v_{\mathrm{p}}}(f) = N_0$. It follows that $v_{\mathrm{p}}(t)$ is stationary and ergodic (see Appendix C.3).
- It is Gaussian because $v_{\mathrm{p}}(t) \sim \mathcal{N}_{\mathbb{C}}(0, N_0)$ at each time t. The variance N_0 is the noise spectral density (in W/Hz), computed as $N_0 = \mathrm{k}\,T_{\mathrm{eff}}$ where $\mathrm{k} = 1.38 \cdot 10^{-23} J/K$ is Boltzmann's constant and T_{eff} (in Kelvins) is the effective noise temperature that depends on the ambient temperature, the type of antennas, and the material properties of the analog front-end [132]. Sometimes, T_{eff} is conveniently expressed as the product of the ambient temperature and a noise figure subsuming all hardware-related aspects.

Although, strictly speaking, a flat noise spectral density of N_0 would indicate that the power goes to infinity with growing bandwidth, the thermal noise density eventually falls off to zero; however, this happens at frequencies that are several orders of magnitude beyond those used for communication and thus a white noise model is valid for all our purposes [136, section 4.4].

While both transmitter and receiver are noisy, only the noise in the analog portion of the receiver—where the signal is weak—is material; at the transmitter, the signal is orders of magnitude stronger than the noise. Because of the filtering in the analog front-end, it suffices to model the noise within the baseband signal bandwidth, namely $v_{\mathrm{b}}(t) =$

$(g_{\text{rx}} * v_{\text{p}})(t)$. This is a Gaussian random process with zero-mean and power spectrum

$$N_0 \, |\mathrm{g}_{\text{rx}}(f)|^2 = \frac{N_0}{T} \, \mathrm{g}(f), \tag{2.48}$$

where we have applied (2.47). The corresponding autocovariance is

$$R_{v_{\text{b}}}(\tau) = \frac{N_0}{T} \, g(\tau) \tag{2.49}$$

and it follows from $g(0) = 1$ that the noise power is $R_{v_{\text{b}}}(0) = N_0/T$, which, as intuition would have it, is proportional to the symbol rate $1/T$ and thus to the signal bandwidth. Interestingly, this noise power does not depend on the excess bandwidth. This property is sometimes expressed through the notion of *noise bandwidth*, defined as the bandwidth of an ideal filter letting in a certain noise power. In Problem 2.7 the reader is invited to verify that, with raised-cosine pulse shaping, the noise bandwidth indeed does not depend on the excess bandwidth.

With noise added to the baseband received signal, we obtain, in discrete time,

$$y[n] = \sum_{\ell=-\infty}^{\infty} c[\ell] \, x[n-\ell] + v[n], \tag{2.50}$$

where $v[n] = v_{\text{b}}(nT)$. This sampled noise is a discrete-time complex Gaussian random process with zero-mean and covariance

$$R_v[\ell] = \frac{N_0}{T} \, g(\ell T) \tag{2.51}$$

and, by virtue of the Nyquist criterion, $R_v[\ell] = N_0/T \, \delta[\ell]$ meaning that $v[n]$ is white. If the signal were oversampled, this would no longer be the case and a whitening filter could be applied to remove the effects of correlation. Although oversampling is an attractive implementation option that facilitates synchronization, it does not affect the conceptual derivations in this text and the discrete-time thermal noise is thus regarded as white.

Example 2.8

Compute the noise power within every hertz of bandwidth at the nominal ambient temperature of 290 K.

Solution

The noise power (in W) within 1 Hz of bandwidth equals

$$\mathrm{k} \, T_{\text{eff}} = 1.38 \cdot 10^{-23} \cdot 290 \tag{2.52}$$

$$= 4 \cdot 10^{-21}, \tag{2.53}$$

which is typically given as -174 dBm/Hz. Within a bandwidth B, then, the noise power (in dBm) is $-174 + B|_{\text{dB}}$. This power is further increased by the noise figure of the receiver, typically between 2 and 10 dB depending on the front-end: in the lower part of this range for base stations, higher for user devices.

2.2.6 Energy and power

Energy is a crucial resource in wireless communication. This is perhaps most evident in mobile devices, where battery life is a major concern, but it is also exceedingly important at fixed base stations; indeed, the electricity bill is often the highest operating expense of network operators. Unless otherwise stated, we concern ourselves only with the energy borne by the transmit signal and with the corresponding power, which is the rate at which such energy is radiated. Denoting the average transmit power (in watts) by P_{t}, then, if the symbols $\{x[n]\}$ are independent and zero-mean,

$$P_{\mathrm{t}} = \lim_{N\to\infty} \frac{1}{NT} \int_{NT} \mathbb{E}\big[|x(t)|^2\big]\, \mathrm{d}t \tag{2.54}$$

$$= \lim_{N\to\infty} \frac{1}{NT} \int_{NT} \mathbb{E}\left[\left|\sum_n x[n]\, g_{\mathrm{tx}}(t-nT)\right|^2\right] \mathrm{d}t \tag{2.55}$$

$$= \frac{1}{T} \int_{-\infty}^{\infty} |g_{\mathrm{tx}}(t)|^2 \,\mathrm{d}t \cdot \frac{1}{N} \lim_{N\to\infty} \sum_{n=0}^{N-1} \mathbb{E}\big[|x[n]^2|\big] \tag{2.56}$$

$$= \frac{1}{T} \int_{-\frac{1+\mathrm{b}}{2T}}^{\frac{1+\mathrm{b}}{2T}} |\mathrm{g}_{\mathrm{tx}}(f)|^2 \,\mathrm{d}f \cdot \frac{1}{N} \lim_{N\to\infty} \sum_{n=0}^{N-1} \mathbb{E}\big[|x[n]^2|\big] \tag{2.57}$$

$$= \frac{1}{T} \int_{-\frac{1+\mathrm{b}}{2T}}^{\frac{1+\mathrm{b}}{2T}} T\,\mathrm{g}(f) \,\mathrm{d}f \cdot \frac{1}{N} \lim_{N\to\infty} \sum_{n=0}^{N-1} \mathbb{E}\big[|x[n]^2|\big] \tag{2.58}$$

$$= \int_{-\frac{1+\mathrm{b}}{2T}}^{\frac{1+\mathrm{b}}{2T}} \mathrm{g}(f) \,\mathrm{d}f \cdot \frac{1}{N} \lim_{N\to\infty} \sum_{n=0}^{N-1} \mathbb{E}\big[|x[n]^2|\big] \tag{2.59}$$

$$= \frac{1}{N} \lim_{N\to\infty} \sum_{n=0}^{N-1} \mathbb{E}\big[|x[n]^2|\big], \tag{2.60}$$

where, in (2.59), the integral equals 1. With a practical view, the average power of the transmit sequence is more meaningfully written as

$$\frac{1}{N} \sum_{n=0}^{N-1} \mathbb{E}\big[|x[n]|^2\big] = P_{\mathrm{t}}, \tag{2.61}$$

with N depending on the ability of the power amplifier to sustain power peaks and be limited by its average load. In turn, as seen earlier, the power of the thermal noise contaminating the received signal is—irrespective of the excess bandwidth—given by N_0/T.

With symbols transmitted at a rate $1/T$, a convenient amount by which to parcel the energy spent at the transmitter is the amount that goes into each symbol. Denoting by E_{s} the average transmit *energy per symbol* (in joules), we have that $E_{\mathrm{s}} = P_{\mathrm{t}}T$. In turn, the thermal noise energy within a symbol period is $\frac{N_0}{T}T = N_0$. It follows that, if we scale both sequences $\{x[n]\}$ and $\{v[n]\}$ by $\sqrt{T}$, the transmit–receive relationship is preserved only with

$$\frac{1}{N} \sum_{n=0}^{N-1} \mathbb{E}\big[|x[n]|^2\big] = E_{\mathrm{s}} \tag{2.62}$$

and with $v \sim \mathcal{N}_{\mathbb{C}}(0, N_0)$. Since the powers of signal and noise only become meaningful when related to each other, simultaneous scalings of both are immaterial; moreover, any common scaling of both signal and noise can be absorbed by $g_{\text{rx}}(\cdot)$.

Both (2.61) and (2.62), respectively with $v \sim \mathcal{N}_{\mathbb{C}}(0, N_0/T)$ and $v \sim \mathcal{N}_{\mathbb{C}}(0, N_0)$, are equally valid. The former is in terms of energy per unit time, i.e., power, while the latter is in terms of energy per symbol. If the symbols had unit period, then power and energy per symbol would coincide, and the factor $\sqrt{T}$ bridging (2.61) with (2.62) can be interpreted as the stretching of the time axis that makes it so. We choose to apply (2.62) because the energy per symbol will allow us to draw parallels with another yet-to-be-defined quantity, the energy per bit. Nonetheless, the conversion of energy to power is immediate through $1/T$ and, more often than not, power is the concept we work with.

It is further useful to write the transmit symbols as

$$x[n] = \sqrt{E_s P[n]}\, s[n], \tag{2.63}$$

with E_s made explicit and with $s[n]$ a unit-variance symbol compliant with the various signal distributions in Section 1.2. If the transmit power is fixed, then $P[n] = 1$, whereas if power control is to be effected, constrained only by (2.62), then $P[0], \ldots, P[N-1]$ can register the power variations with $\frac{1}{N}\sum_{n=0}^{N-1} P[n] = 1$.

Those signal processing functionalities for which the codebook structure is immaterial can be conveniently formulated directly with $x[0], \ldots, x[N-1]$. The emphasis shifts to $s[0], \ldots, s[N-1]$ once quantities with information-theoretic underpinnings are derived, beginning in Chapter 4.

2.2.7 Channel normalization

Starting to view the channel response stochastically, it is rather customary to factor a constant $\sqrt{G}$ out of all baseband and pseudo-baseband channel representations such that

$$c(\tau) = \sqrt{G}\, h(\tau) \tag{2.64}$$

$$c_b(\tau) = \sqrt{G}\, h_b(\tau) \tag{2.65}$$

$$c[\ell] = \sqrt{G}\, h[\ell] \tag{2.66}$$

with the ensemble of realizations $h[\ell]$ satisfying $\sum_\ell \mathbb{E}\left[|h[\ell]|^2\right] = 1$. As elaborated in Chapter 3, there are sound physical justifications for this decoupling, which reflects inherent differences in the space and time scales of distinct features of the radio channel. By virtue of the normalization of $h[\ell]$, the received signal power equals $P_r = G P_t = G E_s/T$ and the average SNR at the receiver can be written as

$$\mathsf{SNR} = \frac{P_r}{N_0/T} \tag{2.67}$$

$$= \frac{G E_s}{N_0} \tag{2.68}$$

irrespective of the excess bandwidth b. Taking advantage of this, we henceforth consider the passband bandwidth to be $B = 1/T$; any excess bandwidth can be separately accounted for simply by penalizing the spectral efficiency by the appropriate factor, which in the raised-cosine case is $\frac{1}{1+\mathsf{b}}$.

Example 2.9

Let $\mathsf{b} = 0.25$ and suppose the transmit bit rate (in b/s) is R. The average SNR at the receiver is $\mathsf{SNR} = \frac{P_\mathrm{r}}{N_0 B}$ with $B = 1/T$, while the spectral efficiency (in b/s/Hz) computed as R/B must be corrected to $\frac{1}{1.25} R/B = 0.8\, R/B$.

2.2.8 Vector representation

Putting the pieces together, Fig. 2.9 provides a diagram for SISO communication that encompasses the transmit and receive front-ends (including local oscillators, upconversion and downconversion, filters and sampling devices), the channel (factored into the product of $\sqrt{G}$ with the normalized response $h(\tau)$), and the additive noise. Henceforth, all of this is abstracted into the end-to-end discrete-time complex baseband relationship

$$y[n] = \sqrt{G} \sum_{\ell=-\infty}^{\infty} h[\ell]\, x[n-\ell] + v[n], \tag{2.69}$$

where, recall, $x[n] = \sqrt{E_\mathrm{s} P[n]} s[n]$. Although, in principle, $\{h[\ell]\}$ extends indefinitely, it is reasonable to assume that wireless channels have a finite impulse response (FIR). As discussed in Chapter 3, indeed, signal paths with longer delays experience further attenuation and, beyond a certain delay, their strength is bound to be negligible relative to the noise. In turn, the pulse shapes, which also extend indefinitely in delay according to the bandlimited raised-cosine filters, are in practice truncated to a length of a few symbol periods. This truncation, in conjunction with a suitable delay, results in discrete-time impulse responses that are causal. Under the FIR and causality conditions, (2.69) becomes

$$y[n] = \sqrt{G} \sum_{\ell=0}^{L} h[\ell]\, x[n-\ell] + v[n], \tag{2.70}$$

where there are $L+1$ taps with L being the memory or *order* of the channel.

If $L = 0$, then the channel is multiplicative and (2.80) reduces to

$$y[n] = \sqrt{G}\, h\, x[n] + v[n], \tag{2.71}$$

which can be rendered time-variant by allowing $h[n]$ to be itself a function of n. In the frequency domain, an impulse response having a single tap corresponds to a flat function, and hence such channels are termed *frequency-flat*.

If $L > 0$, conversely, the channel exhibits ISI and, because its response is not flat in frequency, it is said to be *frequency-selective*. Then, (2.70) can be expressed more compactly in an alternative vector form that we derive in the remainder of this section. This vector representation hinges on zooming out and considering a block of N consecutive observations, $y[0], \ldots, y[N-1]$, rather than an individual symbol $y[n]$. Besides facilitating notational compactness, this representation becomes fitting once codewords enter the

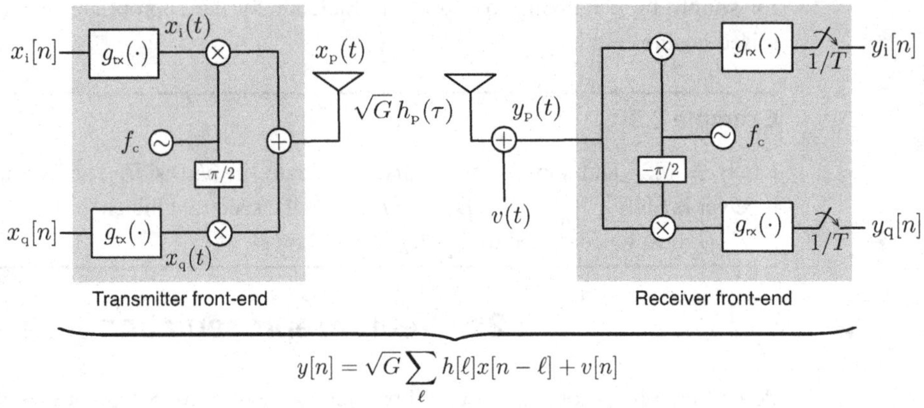

Fig. 2.9 Diagram relating the transmit and receive front-ends, the radio channel, and the additive noise.

formulation and thus it is a handy recourse for later analyses. The central ingredient in a vector representation is the $N \times (N+L)$ convolution matrix

$$\bar{\boldsymbol{H}}_{N,N+L} = \begin{bmatrix} h[L] & \cdots & h[0] & 0 & 0 & \cdots & 0 \\ 0 & h[L] & \cdots & h[0] & 0 & \cdots & 0 \\ \vdots & & & \ddots & & & \vdots \\ 0 & \cdots & 0 & 0 & h[L] & \cdots & h[0] \end{bmatrix}, \tag{2.72}$$

which displays a characteristic Toeplitz structure (see Appendix B.2.4). This structure can be exploited to develop fast algorithms [137–139].

To effect a linear convolution by means of $\bar{\boldsymbol{H}}_{N,N+L}$, we need to collect $N+L$ consecutive samples of $x[n]$ into a vector

$$\bar{\boldsymbol{x}}_{N+L} = \begin{bmatrix} x[-L] \\ \vdots \\ x[-1] \\ x[0] \\ x[1] \\ \vdots \\ x[N-1] \end{bmatrix}, \tag{2.73}$$

where past values $x[-L], \ldots, x[-1]$ are needed because of the memory in the convolution; often, these samples are part of a guard region containing zeros or else they may contain repeated data in the form of a cyclic prefix. From $\bar{\boldsymbol{H}}_{N,N+L}$ and $\bar{\boldsymbol{x}}_{N+L}$, we can now write

$$\bar{\boldsymbol{y}}_N = \sqrt{G}\bar{\boldsymbol{H}}_{N,N+L}\bar{\boldsymbol{x}}_{N+L} + \bar{\boldsymbol{v}}_N \tag{2.74}$$

where $\bar{\boldsymbol{y}}_N = \left[y[0] \cdots y[N-1]\right]^{\mathrm{T}}$ and $\bar{\boldsymbol{v}}_N = \left[v[0] \cdots v[N-1]\right]^{\mathrm{T}}$. It can be verified that the nth row in (2.74) is equivalent to (2.70).

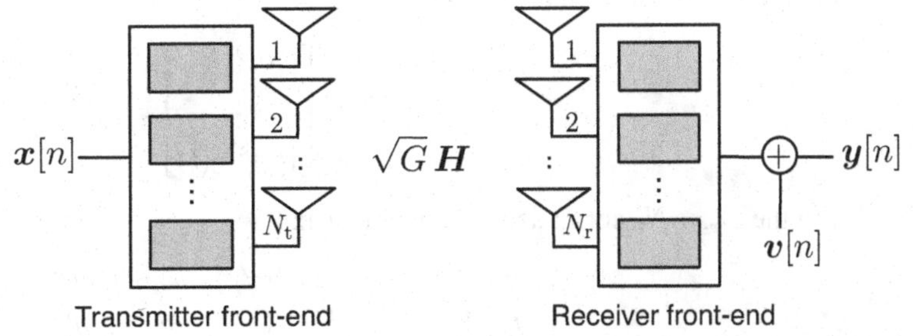

Fig. 2.10 MIMO setting with N_t transmit and N_r receive antennas.

2.3 Signal, channel, and noise representations: extension to MIMO

Let us now extend the SISO representations seen thus far to the realm of a MIMO setting such as the one sketched in Fig. 2.10, where transmitter and receiver are equipped with N_t and N_r antennas, respectively. With the understanding that each antenna is connected to a radio-frequency front-end for upconversions and downconversions, we proceed directly to the complex baseband and pseudo-baseband representations.

2.3.1 Vector and matrix representations

As a result of the linearity of the propagation environment, the ith receive antenna observes the superposition

$$y^{(i)}(t) = \sqrt{G} \sum_{j=0}^{N_t-1} \int_{-\infty}^{\infty} h^{(i,j)}(\tau)\, x^{(j)}(t-\tau)\, d\tau + v_b^{(i)}(t)$$
$$i = 0, \ldots, N_r - 1, \qquad (2.75)$$

where $x^{(j)}(t)$ is the signal emitted by the jth transmit antenna and $\sqrt{G}\, h^{(i,j)}(\tau)$ is the impulse response from the jth transmit to the ith receive antenna; the factor $\sqrt{G}$ is regarded as common to the $N_t N_r$ impulse responses, a point that is amply justified in Chapter 3. In turn, $v_b^{(i)}(t)$ is the thermal noise at the ith receiver, statistically independent from the noise at other receivers due to the separateness of the corresponding components.

For notational and analytical reasons, it seems desirable to bring together all the transmit signals, all the channel responses, and all the receive signals. As advanced in the previous chapter, vectors and matrices are again the natural means to achieve this. Let us define the $N_r \times 1$ receive signal and noise vectors

$$\boldsymbol{y}(t) = \begin{bmatrix} y^{(0)}(t) \\ \vdots \\ y^{(N_r-1)}(t) \end{bmatrix} \qquad \boldsymbol{v}(t) = \begin{bmatrix} v_b^{(0)}(t) \\ \vdots \\ v_b^{(N_r-1)}(t) \end{bmatrix}, \qquad (2.76)$$

the $N_{\mathrm{t}} \times 1$ transmit signal vector

$$\boldsymbol{x}(t) = \begin{bmatrix} x^{(0)}(t) \\ \vdots \\ x^{(N_{\mathrm{t}}-1)}(t) \end{bmatrix}, \tag{2.77}$$

and the $N_{\mathrm{r}} \times N_{\mathrm{t}}$ normalized channel matrix

$$\boldsymbol{H}(\tau) = \begin{bmatrix} h^{(0,0)}(\tau) & h^{(0,1)}(\tau) & \cdots & h^{(0,N_{\mathrm{t}}-1)}(\tau) \\ h^{(1,0)}(\tau) & h^{(1,1)}(\tau) & \cdots & h^{(1,N_{\mathrm{t}}-1)}(\tau) \\ \vdots & \vdots & \ddots & \vdots \\ h^{(N_{\mathrm{r}}-1,0)}(\tau) & h^{(N_{\mathrm{r}}-1,1)}(\tau) & \cdots & h^{(N_{\mathrm{r}}-1,N_{\mathrm{t}}-1)}(\tau) \end{bmatrix}. \tag{2.78}$$

With this notation, the MIMO transmit–receive relationship can be written compactly as

$$\boldsymbol{y}(t) = \sqrt{G} \int_{-\infty}^{\infty} \boldsymbol{H}(\tau)\, \boldsymbol{x}(t-\tau)\, \mathrm{d}\tau + \boldsymbol{v}(t). \tag{2.79}$$

The matrix function $\boldsymbol{H}(\tau)$ is a multivariate impulse response, a notion that arose historically in control applications [140] and that, herein, serves to describe a linear (for now time-invariant but generally time-variant) channel with multiple inputs and outputs.

Since signals and channels require further indexing later on, for the sake of clarity we begin using the notation $[\boldsymbol{a}]_j$ and $[\boldsymbol{A}]_{i,j}$ to refer, respectively, to the jth entry of a vector $\boldsymbol{a}$ and to the (i, j)th entry of a matrix $\boldsymbol{A}$.

To time-discretize (2.79), let $\boldsymbol{y}[n] = \boldsymbol{y}(nT)$, $\boldsymbol{x}[n] = \boldsymbol{x}(nT)$, and $\boldsymbol{v}[n] = \boldsymbol{v}(nT)$. Then, denote the discrete-time multivariate impulse response at tap ℓ as $\boldsymbol{H}[\ell]$ where $\big[\boldsymbol{H}[\ell]\big]_{i,j}$ is obtained by filtering and sampling $h^{(i,j)}(t)$. With that, the discrete-time transmit-receive MIMO relationship becomes

$$\boldsymbol{y}[n] = \sqrt{G} \sum_{\ell=0}^{L} \boldsymbol{H}[\ell]\, \boldsymbol{x}[n-\ell] + \boldsymbol{v}[n], \tag{2.80}$$

where the arguments that support a causal FIR response in SISO apply verbatim.

If $L = 0$, then the frequency-selective MIMO relationship in (2.80) reduces to the frequency-flat MIMO relationship

$$\boldsymbol{y}[n] = \sqrt{G}\, \boldsymbol{H} \boldsymbol{x}[n] + \boldsymbol{v}[n], \tag{2.81}$$

which can be rendered time-variant by allowing $\boldsymbol{H}[n]$ to be itself a function of n. Much of the theory developed for MIMO communication relies on (2.81) and, indeed, a variation of the frequency-flat channel turns out to be a satisfactory model with OFDM.

2.3.2 Channel normalization

Although a strict entry-wise extension to MIMO of the SISO channel normalization would result in $\sum_{\ell=0}^{L} \mathbb{E}\big[|[\boldsymbol{H}[\ell]]_{i,j}|^2\big] = 1$ for $i = 0, \ldots, N_{\mathrm{r}} - 1$ and $j = 0, \ldots, N_{\mathrm{t}} - 1$, a more flexible extension allows us to also embrace channel models that exhibit power asymmetries across antennas. We therefore establish the normalization of MIMO channels jointly over all their entries as

$$\sum_{\ell=0}^{L} \sum_{i=0}^{N_r-1} \sum_{j=0}^{N_t-1} \mathbb{E}\left[\left|[\boldsymbol{H}[\ell]]_{i,j}\right|^2\right] = N_t N_r. \tag{2.82}$$

This can be compactly formulated via the Frobenius norm (see Appendix B.5), giving

$$\sum_{\ell=0}^{L} \mathbb{E}\left[\|\boldsymbol{H}[\ell]\|_F^2\right] = \sum_{\ell=0}^{L} \mathbb{E}\left[\mathrm{tr}\left(\boldsymbol{H}[\ell]\boldsymbol{H}^*[\ell]\right)\right] \tag{2.83}$$

$$= \sum_{\ell=0}^{L} \mathbb{E}\left[\mathrm{tr}\left(\boldsymbol{H}^*[\ell]\boldsymbol{H}[\ell]\right)\right] \tag{2.84}$$

$$= N_t N_r. \tag{2.85}$$

In frequency-flat channels, the normalization reduces to

$$\mathbb{E}\left[\|\boldsymbol{H}\|_F^2\right] = \mathbb{E}\left[\mathrm{tr}\left(\boldsymbol{H}\boldsymbol{H}^*\right)\right] \tag{2.86}$$

$$= \mathbb{E}\left[\mathrm{tr}\left(\boldsymbol{H}^*\boldsymbol{H}\right)\right] \tag{2.87}$$

$$= N_t N_r. \tag{2.88}$$

2.3.3 Stacked vector representation

Adopting simultaneously for both the time and the antenna dimensions the block-wise representation that led to (2.74) in SISO, it is possible to express in a similarly compact manner an entire MIMO relationship. Applying (2.74) between the N_t transmit antennas and the ith receive antenna, we obtain

$$\bar{\boldsymbol{y}}_N^{(i)} = \sqrt{G} \sum_{j=0}^{N_t-1} \bar{\boldsymbol{H}}_{N,N+L}^{(i,j)} \bar{\boldsymbol{x}}_{N+L}^{(j)} + \bar{\boldsymbol{v}}_N^{(i)} \tag{2.89}$$

where $\bar{\boldsymbol{y}}_N^{(i)} = \left[y^{(i)}[0] \cdots y^{(i)}[N-1]\right]^T$ and $\bar{\boldsymbol{v}}_N^{(i)} = \left[v^{(i)}[0] \cdots v^{(i)}[N-1]\right]^T$ while

$$\bar{\boldsymbol{x}}_{N+L}^{(j)} = \begin{bmatrix} x^{(j)}[-L] \\ \vdots \\ x^{(j)}[-1] \\ x^{(j)}[0] \\ x^{(j)}[1] \\ \vdots \\ x^{(j)}[N-1] \end{bmatrix} \tag{2.90}$$

and

$$\bar{\boldsymbol{H}}_{N,N+L}^{(i,j)} = \begin{bmatrix} h^{(i,j)}[L] & \cdots & h^{(i,j)}[0] & 0 & 0 & \cdots & 0 \\ 0 & h^{(i,j)}[L] & \cdots & h^{(i,j)}[0] & 0 & \cdots & 0 \\ \vdots & & & \ddots & & & \vdots \\ 0 & \cdots & 0 & 0 & h^{(i,j)}[L] & \cdots & h^{(i,j)}[0] \end{bmatrix}. \tag{2.91}$$

Collecting the temporal block observations for all receive antennas into a larger vector, i.e., stacking $\bar{\boldsymbol{y}}_N^{(0)}$ through $\bar{\boldsymbol{y}}_N^{(N_r-1)}$, we obtain the $N_r N \times 1$ observation vector

$$\bar{\boldsymbol{y}}_{N_r N} = \begin{bmatrix} \bar{\boldsymbol{y}}_N^{(0)} \\ \vdots \\ \bar{\boldsymbol{y}}_N^{(N_r-1)} \end{bmatrix}. \tag{2.92}$$

Likewise, stacking $\bar{\boldsymbol{v}}_N^{(0)}$ through $\bar{\boldsymbol{v}}_N^{(N_r-1)}$ we obtain the $N_r N \times 1$ noise vector $\bar{\boldsymbol{v}}_{N_r N}$ and stacking $\bar{\boldsymbol{x}}_{N+L}^{(0)}$ through $\bar{\boldsymbol{x}}_{N+L}^{(N_t-1)}$ we obtain the $N_t(N+L) \times 1$ transmit vector

$$\bar{\boldsymbol{x}}_{N_t(N+L)} = \begin{bmatrix} \bar{\boldsymbol{x}}_{N+L}^{(0)} \\ \vdots \\ \bar{\boldsymbol{x}}_{N+L}^{(N_t-1)} \end{bmatrix}. \tag{2.93}$$

These vectors are connected through the block Toeplitz matrix

$$\bar{\boldsymbol{H}}_{N_r N, N_t(N+L)} = \begin{bmatrix} \bar{\boldsymbol{H}}_{N,N+L}^{(0,0)} & \bar{\boldsymbol{H}}_{N,N+L}^{(0,1)} & \cdots & \bar{\boldsymbol{H}}_{N,N+L}^{(0,N_t-1)} \\ \bar{\boldsymbol{H}}_{N,N+L}^{(1,0)} & \bar{\boldsymbol{H}}_{N,N+L}^{(1,1)} & \cdots & \bar{\boldsymbol{H}}_{N,N+L}^{(1,N_t-1)} \\ \vdots & \vdots & \ddots & \vdots \\ \bar{\boldsymbol{H}}_{N,N+L}^{(N_r-1,0)} & \bar{\boldsymbol{H}}_{N,N+L}^{(N_r-1,1)} & \cdots & \bar{\boldsymbol{H}}_{N,N+L}^{(N_r-1,N_t-1)} \end{bmatrix}. \tag{2.94}$$

With these definitions, (2.80) becomes

$$\bar{\boldsymbol{y}}_{N_r N} = \sqrt{G}\,\bar{\boldsymbol{H}}_{N_r N, N_t(N+L)} \bar{\boldsymbol{x}}_{N_t(N+L)} + \bar{\boldsymbol{v}}_{N_r N}, \tag{2.95}$$

which resembles the frequency-flat MIMO relationship in (2.81), only with augmented dimensions. The structured perspective offered by (2.95) can enable the application of frequency-flat MIMO results to frequency-selective MIMO channels.

2.3.4 Precoding

In SISO, recall, the transmit signal can be written as $x[n] = \sqrt{E_s P[n]}\, s[n]$, where E_s is the average energy per symbol and $P[n]$ allows performing power control. When one considers generalizing this transformation to MIMO, it becomes clear that the scalar $P[n]$ can be replaced by a matrix entailing more than a power gain. Specifically, with an $N_t \times N_s$ matrix we can transform an $N_s \times 1$ vector $\boldsymbol{s}[n]$ into the $N_t \times 1$ vector $\boldsymbol{x}[n]$, meaning that $N_s \leq N_t$ symbols are embedded into $\boldsymbol{x}[n]$ and thus $N_s \leq N_t$ data streams are transmitted at once.

Denoting the $N_\mathrm{t} \times N_\mathrm{s}$ transformation matrix by $\boldsymbol{F}[n]$, what ensues is

$$\boldsymbol{x}[n] = \sqrt{\frac{E_\mathrm{s}}{N_\mathrm{t}}}\boldsymbol{F}[n]\boldsymbol{s}[n], \tag{2.96}$$

where the normalization by N_t appears because we define $\boldsymbol{F}[\cdot]$ such that

$$\frac{1}{N}\sum_{n=0}^{N-1} \|\boldsymbol{F}[n]\|_\mathrm{F}^2 = \frac{1}{N}\sum_{n=0}^{N-1} \mathrm{tr}(\boldsymbol{F}[n]\boldsymbol{F}^*[n]) = N_\mathrm{t}. \tag{2.97}$$

(Alternatively, one could define $\boldsymbol{F}[\cdot]$ such that $\frac{1}{N}\sum_{n=0}^{N-1}\|\boldsymbol{F}[n]\|_\mathrm{F}^2 = 1$ and do away with the explicit normalization by N_t. However, (2.97) is the more standard normalization, and we abide by it.)

Since codeword symbols are IID, $\mathbb{E}\big[\boldsymbol{s}[n]\boldsymbol{s}^*[n]\big] = \boldsymbol{I}$ and the covariance of $\boldsymbol{x}[n]$ equals

$$\boldsymbol{R}_{\boldsymbol{x}}[n] = \mathbb{E}\big[\boldsymbol{x}[n]\boldsymbol{x}^*[n]\big] \tag{2.98}$$

$$= \frac{E_\mathrm{s}}{N_\mathrm{t}}\boldsymbol{F}[n]\boldsymbol{F}^*[n]. \tag{2.99}$$

Somewhat paradoxically given that it operates on coded symbols $\boldsymbol{s}[n]$, hence *after* the encoder, the filter $\boldsymbol{F}[\cdot]$ is termed *precoder*. The reasons for this are historical, as the term was coined in a context where uncoded or weakly coded signals were transformed and then subsequently run through an inner space-time code (providing mostly antenna diversity) prior to transmission [9, 10, 141, 142]. Nowadays, powerful outer channel codes are almost ubiquitous and diversity tends to be abundant, such that the subsequent inner code is largely unnecessary. Altogether, "postcoder" would be a more fitting name for $\boldsymbol{F}[\cdot]$, but by now the term precoder has stuck.

The precoder plays a very important role in MIMO communication, enabling a spatial formatting of the transmission on the basis of whatever CSI is available to the transmitter.

Being a matrix, $\boldsymbol{F}[\cdot]$ can be subject to a singular-value decomposition (SVD, see Appendix B.3.2) and expressed as

$$\boldsymbol{F}[n] = \boldsymbol{U}_{\boldsymbol{F}}[n]\,\boldsymbol{\Sigma}_{\boldsymbol{F}}[n]\,\boldsymbol{V}_{\boldsymbol{F}}^*[n], \tag{2.100}$$

where $\boldsymbol{U}_{\boldsymbol{F}}[\cdot]$ and $\boldsymbol{V}_{\boldsymbol{F}}[\cdot]$ are unitary, respectively $N_\mathrm{t} \times N_\mathrm{t}$ and $N_\mathrm{s} \times N_\mathrm{s}$, while $\boldsymbol{\Sigma}_{\boldsymbol{F}}[\cdot]$ is the $N_\mathrm{t} \times N_\mathrm{s}$ matrix

$$\boldsymbol{\Sigma}_{\boldsymbol{F}}[n] = \begin{bmatrix} \boldsymbol{P}^{1/2} \\ \boldsymbol{0} \end{bmatrix} \tag{2.101}$$

whose upper portion is the square-root of an $N_\mathrm{s} \times N_\mathrm{s}$ matrix $\boldsymbol{P} = \mathrm{diag}(P_0, \ldots, P_{N_\mathrm{s}-1})$ satisfying $\sum_{j=0}^{N_\mathrm{s}-1} P_j = N_\mathrm{t}$ while the lower portion is an $(N_\mathrm{t} - N_\mathrm{s}) \times N_\mathrm{s}$ all-zero matrix. This decomposition is analytically convenient and, more importantly, it invites a meaningful interpretation based on which we term the ingredients as follows: $\boldsymbol{V}_{\boldsymbol{F}}[\cdot]$ as the *mixing matrix*, $\boldsymbol{P}[\cdot]$ as the *power allocation matrix*, and $\boldsymbol{U}_{\boldsymbol{F}}[\cdot]$ as the *steering matrix*:

- The unit-variance entries of $\boldsymbol{s}[n]$, which conform to arbitrary distributions, are mixed by $\boldsymbol{V}_{\boldsymbol{F}}[n]$ into a vector $\boldsymbol{V}_{\boldsymbol{F}}[n]\boldsymbol{s}[n]$ whose entries are still of unit variance (because of the unitary nature of $\boldsymbol{V}_{\boldsymbol{F}}[n]$), but which exhibit modified distributions.

- The jth mixed signal is allocated a power $\frac{E_s}{N_t} P_j$, for $j = 0, \ldots, N_s - 1$.
- The jth amplified mixed signal is launched from all transmit antennas weighted by the coefficients on the jth column of $\boldsymbol{U_F}[n]$.

Because $\boldsymbol{U_F}[n]$ is unitary, the final step does not alter the powers but merely endows each signal with a spatial orientation, making it possible to transmit on channel directions that may be particularly favorable. Details on this are deferred to Section 3.5, where the necessary background on array processing is provided.

If $\boldsymbol{U_F}[n]$ and $\boldsymbol{V_F}[n]$ are identity matrices, and $\boldsymbol{P} = \frac{N_t}{N_s} \boldsymbol{I}$, then

$$\boldsymbol{F}[n] = \frac{N_t}{N_s} \begin{bmatrix} \boldsymbol{I}_{N_s} \\ \boldsymbol{0} \end{bmatrix}. \tag{2.102}$$

With such trivial precoder, each of N_s antennas directly radiates one of the data streams within $\boldsymbol{s}[n]$ while the remaining $N_t - N_s$ antennas are silent; the transmit signals are independent. Notwithstanding the possible directivity of the antennas, the transmission is then said to be isotropic from a precoding standpoint, or outright *unprecoded*.

Example 2.10

Let $N_t = N_s = 2$ and suppose that the two signals within $\boldsymbol{s}[n]$ are QPSK-distributed. Examine the structure of the transmit signal $\boldsymbol{x}[n]$ produced by a time-invariant precoder.

Solution

Dropping the dependence on n, let $\boldsymbol{P} = \text{diag}(P_1, P_2)$ and

$$\boldsymbol{U_F} = \begin{bmatrix} U_{00} & U_{01} \\ U_{10} & U_{11} \end{bmatrix} \qquad \boldsymbol{V_F} = \begin{bmatrix} V_{00} & V_{01} \\ V_{10} & V_{11} \end{bmatrix}. \tag{2.103}$$

The application of the mixing matrix to $\boldsymbol{s}$ yields the mixed signal vector

$$\boldsymbol{V_F^*}\, \boldsymbol{s} = \begin{bmatrix} V_{00}^*[\boldsymbol{s}]_0 + V_{10}^*[\boldsymbol{s}]_1 \\ V_{01}^*[\boldsymbol{s}]_0 + V_{11}^*[\boldsymbol{s}]_1 \end{bmatrix} \tag{2.104}$$

whose unit-variance entries conform to 16-ary distributions, in general distinct. Then, after power allocation,

$$\sqrt{\frac{E_s}{2}} \boldsymbol{P}^{1/2} \boldsymbol{V_F^*}\, \boldsymbol{s} = \begin{bmatrix} \sqrt{\frac{P_0 E_s}{2}} \left(V_{00}^*[\boldsymbol{s}]_0 + V_{10}^*[\boldsymbol{s}]_1 \right) \\ \sqrt{\frac{P_1 E_s}{2}} \left(V_{01}^*[\boldsymbol{s}]_0 + V_{11}^*[\boldsymbol{s}]_1 \right) \end{bmatrix}. \tag{2.105}$$

The top signal is launched from the two antennas, weighted by U_{00} and U_{10}, while the bottom signal is launched weighted by U_{01} and U_{11}. If $\boldsymbol{F} = \boldsymbol{I}$, then each signal is radiated from only one antenna. Conversely, if $\boldsymbol{F} \neq \boldsymbol{I}$, each signal is transmitted from both antennas and steered in a certain direction. Both possibilities are illustrated in Fig. 2.11.

Besides providing operational intuition, the breakdown of $\boldsymbol{F}[\cdot]$ into its constituent parts, $\boldsymbol{U_F}[\cdot]$, $\boldsymbol{P}[\cdot]$, and $\boldsymbol{V_F}[\cdot]$, also facilitates the optimization of the precoder. Although it is possible—in fact rather common—to optimize precoders based on estimation-theoretic

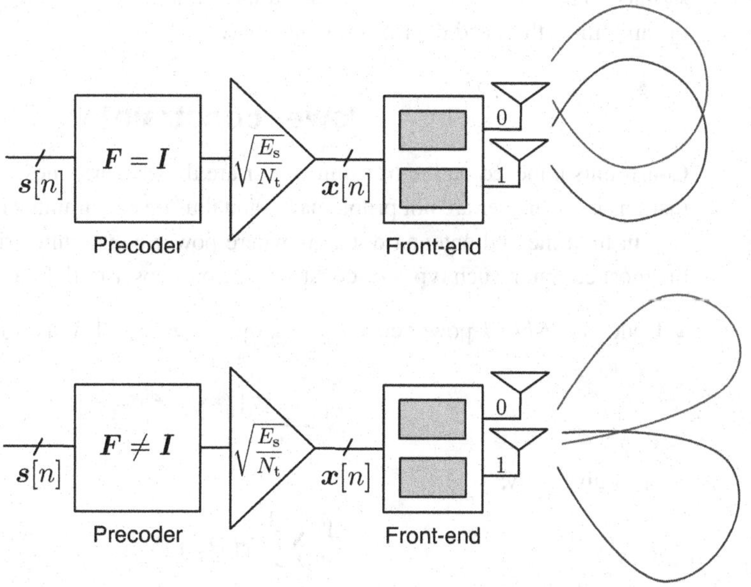

Fig. 2.11 Above, isotropic transmission with $N_t = N_s = 2$ and $\boldsymbol{F} = \boldsymbol{I}$; each signal is radiated from a single antenna according to the pattern of that antenna. Below, nonisotropic transmission; complementing the antenna patterns, each of the two signals is steered in a distinct spatial direction.

criteria or to minimize the uncoded error probability [18, 123, 143–145], what should drive this optimization in heavily coded systems is maximizing the amount of information that can be conveyed through the channel, i.e., the precoder should spatially format the signals so that as much information as possible can couple into the channel, with the suppression of errors left to encoder and decoder. Put differently, the cascade of precoder and channel should yield the widest possible information pipe. This naturally leads to the mutual-information-based precoder optimizations posed in later chapters. Although it may be tempting to anticipate that one would always want to transmit $N_s = N_t$ signal streams, in some cases the optimum strategy turns out to entail $N_s < N_t$, hence our insistence in keeping these two quantities differentiated.

A precoding case of the utmost relevance, because of the virtues of complex Gaussian codebooks, occurs when the entries of $\boldsymbol{s}[n]$ are complex Gaussian. Then, the mixing matrix $\boldsymbol{V_F}[\cdot]$ becomes immaterial because the IID complex Gaussian distribution is rotationally invariant (see Appendix C.1.9) and thus $\boldsymbol{V}_{\boldsymbol{F}}^*[n]\boldsymbol{s}[n] \sim \boldsymbol{s}[n]$. In this all-important case, then, the precoder can take the simpler form $\boldsymbol{F}[n] = \boldsymbol{U_F}[n]\boldsymbol{P}^{1/2}[n]$.

2.3.5 Signal constraints

Having extended the signal models to MIMO and introduced the notion of precoding, we are now ready to formalize the various ways in which the transmit signal may be constrained. These constraints are not arbitrary; rather, they embody physical limitations of the amplifiers that feed the transmit antennas.

Power constraints

Constraints imposed on the transmit power are the primary ones. Frequency-domain fluctuations in the power are not problematic, and thus no constraint is imposed in that domain. It is in the time and antenna domains where power constraints arise. Let us contemplate the most common such types of constraints, from looser to tighter.

- Under a *per-block* power constraint, $\boldsymbol{x}[0], \ldots, \boldsymbol{x}[N-1]$ must satisfy

$$\frac{1}{N} \sum_{n=0}^{N-1} \mathbb{E}\left[\|\boldsymbol{x}[n]\|^2\right] = E_{\mathrm{s}} \tag{2.106}$$

or, equivalently,

$$\frac{1}{N} \sum_{n=0}^{N-1} \operatorname{tr}\left(\boldsymbol{R}_{\boldsymbol{x}}[n]\right) = E_{\mathrm{s}}. \tag{2.107}$$

With this constraint, which applies if the power amplifiers tolerate signal crests and are limited only by their average loads over N symbols, it is possible to allocate power unevenly over time/frequency and across the transmit antennas as long as the average energy per symbol does not exceeded E_{s}. If we identifying the blocklength N with the extension of a codeword, the per-block power constraint becomes a *per-codeword* power constraint. Translated to the precoder, this type of constraint gives, from (2.99) and (2.107),

$$\frac{1}{N} \sum_{n=0}^{N-1} \operatorname{tr}\left(\boldsymbol{F}[n]\boldsymbol{F}^*[n]\right) = N_{\mathrm{t}}, \tag{2.108}$$

as anticipated in (2.97). Moreover, since the rotations effected by the unitary matrices $\boldsymbol{U}_{\boldsymbol{F}}[n]$ and $\boldsymbol{V}_{\boldsymbol{F}}[n]$ do not affect the trace, this is further equivalent to

$$\frac{1}{N} \sum_{n=0}^{N-1} \operatorname{tr}\left(\boldsymbol{P}[n]\right) = N_{\mathrm{t}}. \tag{2.109}$$

- Under a *per-symbol* power constraint, for every n,

$$\mathbb{E}\left[\|\boldsymbol{x}[n]\|^2\right] = E_{\mathrm{s}} \tag{2.110}$$

or, equivalently,

$$\operatorname{tr}\left(\boldsymbol{R}_{\boldsymbol{x}}[n]\right) = E_{\mathrm{s}}. \tag{2.111}$$

This constraint prohibits uneven power allocation over time, directly capturing the peak capability of the amplifiers, but does allow it across antennas. In terms of the precoder, it amounts to

$$\operatorname{tr}\left(\boldsymbol{F}[n]\boldsymbol{F}^*[n]\right) = \operatorname{tr}\left(\boldsymbol{P}[n]\right) = N_{\mathrm{t}}. \tag{2.112}$$

- Under a *per-antenna* power constraint,

$$\big[\boldsymbol{R}_{\boldsymbol{x}}[n]\big]_{j,j} = \frac{E_{\rm s}}{N_{\rm t}}, \tag{2.113}$$

  which fixes the power of the scalar symbols transmitted by each antenna. This can serve to explicitly reflect the existence of a separate amplifier behind each antenna. Translated to the precoder, it requires that every row of $\boldsymbol{F}$ have unit norm, such that $\big[\boldsymbol{F}[n]\boldsymbol{F}^*[n]\big]_{j,j} = 1$ for $j = 0, \ldots, N_{\rm t} - 1$.

All the foregoing constraints ensure that the total radiated power is preserved regardless of $N_{\rm t}$, evincing the MIMO challenge of improving the performance without increasing the transmit power, only through the addition of antennas. This conservation of the radiated power further registers limitations imposed by regulatory agencies that relate to environmental safety and other matters. From a practical vantage, the per-antenna power constraint is arguably the most pertinent, and the volume of related results has slowly grown over time, yet the per-block and per-symbol power constraints are the most prevalent ones in MIMO analysis. Although their undeniable analytical convenience is by itself hardly a sound reason to justify their use, with OFDM the argument can be made that the average transmit power under these constraints should not vary much from antenna to antenna if the number of subcarriers is large, and thus (2.113) might also be—approximately—satisfied. When this is not the case, the per-codeword and per-symbol constraints yield upper bounds to what can be achieved under the more stringent per-antenna constraint.

Sometimes the type of power constraint ends up being somewhat inconsequential. If the transmitter is not privy to the channel response, for instance, then it cannot effect time-domain power control and it ends up abiding by (2.110) or (2.113) even if a per-block constraint does apply. It is only when the transmitter has CSI that the distinctiveness of a per-block constraint becomes material.

Magnitude constraints

A secondary type of constraints are those related to the signal magnitude, whose peakedness directly impacts the nonlinear behavior of power amplifiers: the more peaky the magnitude, the higher the chances that the amplifiers will be driven into a range where their response is nonlinear and, ultimately, into saturation [146, 147].

If x is the zero-mean scalar symbol transmitted by one of the antennas, the most relevant measures of its peakedness are as follows. (As with entropies and differential entropies, we slightly abuse notation and express these measures as a function of $|x|$ when they are really a function of its distribution.)

- The peak-to-average power ratio (PAPR), which quantifies the maximum excursion of the squared magnitude over the average power, i.e.,

$$\mathsf{PAPR}\big(|x|\big) = \frac{\max\big(|x|^2\big)}{\mathbb{E}\big[|x|^2\big]} \tag{2.114}$$

  whose square root is sometimes termed *crest factor*. A finite PAPR puts a hard limit on the signal magnitude.

- The kurtosis (see Appendix C.1.8)

$$\kappa(|x|) = \frac{\mathbb{E}\left[|x|^4\right]}{\mathbb{E}\left[|x|^2\right]^2}. \tag{2.115}$$

- The cubic metric (CM), which characterizes the effects of the third-order nonlinearity of the power amplifier and has been identified as a leading indicator of nonlinear behaviors [148]. It is defined as

$$\mathsf{CM}(|x|) = \frac{\mathbb{E}\left[|x|^6\right]}{\mathbb{E}\left[|x|^2\right]^3}. \tag{2.116}$$

Example 2.11

Compute the peakedness measures for $x \sim \mathcal{N}_{\mathbb{C}}(0, \sigma^2)$.

Solution

$$\mathsf{PAPR}(|x|) = \infty \tag{2.117}$$

$$\kappa(|x|) = 2 \tag{2.118}$$

$$\mathsf{CM}(|x|) = 6. \tag{2.119}$$

Example 2.12

Compute the peakedness measures for s drawn from an M-PSK constellation.

Solution

For M-PSK, all measures of peakedness are unity; it is the ultimate nonpeaky distribution.

Example 2.13

Compute the peakedness measures for x drawn from a square M-QAM constellation.

Solution

For square M-QAM,

$$\kappa(|s|) = \frac{1}{5}\frac{7M - 13}{M - 1} \tag{2.120}$$

$$\mathsf{PAPR}(|s|) = 3\,\frac{\sqrt{M} - 1}{\sqrt{M} + 1} \tag{2.121}$$

with the cubic metric best computed numerically.

The infinite PAPR of Gaussian-distributed signals may seem problematic because such signals are theoretically optimum in numerous situations. As the next example shows, however, the peakedness of the Gaussian distribution, as measured by the more informative kurtosis and CM, is actually rather modest.

Table 2.1 PAPR, kurtosis, and CM for signal distributions of interest

Distribution	PAPR	κ	CM
$\mathcal{N}_{\mathbb{C}}$	∞	2	$9\pi/16$
$\mathcal{N}_{\mathbb{C}}$ (clipped)	10	1.9	1.74
M-PSK	1	1	1
M-QAM	$[1,3]$	$[1,1.4]$	$[1,1.33]$
on–off	$1/\delta$	$1/\delta$	$1/\delta$

Example 2.14

Consider $x \sim \mathcal{N}_{\mathbb{C}}(0, \sigma^2)$ and let z be a clipped version of x satisfying $\mathsf{PAPR}(|z|) = 10$ dB. Compute the kurtosis and CM of z.

Solution

$$\kappa(|z|) = 1.94 \tag{2.122}$$

$$\mathsf{CM}(|z|) = 5.4. \tag{2.123}$$

By contrasting these values with those in Example 2.11 we observe that this very substantive clipping reduces the kurtosis and CM by only about 3% and 10%, respectively. Gaussian-distributed signals may be clipped to relatively small PAPR values with hardly any disruption in their information-carrying ability.

Table 2.1 summarizes the peakedness measures (or range thereof given all possible cardinalities) of various signal distributions of interest, including a clipped complex Gaussian distribution. Also included is the on–off distribution in (1.3); for $\delta \to 0$, the "on" mass point diverges to infinity and, with it, all the measures of peakedness.

Besides the marginal distribution of individual symbols, in the case of OFDM the peakedness in the time domain depends on the frequency-domain signal structure (refer to Problem 2.16) [146]. And, since what is ultimately of essence is the peakedness of the transmit analog waveforms, a final—but minor—contributor to such peakedness is the pulse shape.

2.4 Linear channel equalization

A frequency-selective channel smears the signals in delay, causing ISI, and if the channel is MIMO it further introduces interference among the signals emitted by the various transmit antennas. Equalization is the removal of these channel effects and an assortment of equalization methods exist, both linear and nonlinear. This section concentrates on linear strategies, which require minimal assumptions on the signal structure.

2.4.1 Linear ZF equalization

For pedagogical purposes, equalization is first considered with noise neglected. An ideal equalizer under these circumstances completely removes the effect of the channel. Such equalizers, when they exist, are known as zero-forcing (ZF). In this section we devise linear ZF equalizers for SISO and MIMO channels, providing the foundation for their existence and then describing algorithms to compute exactly or approximately their coefficients.

Basic formulation

We start by explaining the challenge of equalizing a scalar channel with impulse response $\sqrt{G}\,h[0], \ldots, \sqrt{G}\,h[L]$. Because this exposition depends on concepts from system theory, we make use of basic notions of the Z-transform that are summarized in Appendix A.2. Let $\mathsf{h}(z)$ denote the Z-transform of $h[n]$ and let $\mathsf{w}(z)$ denote the Z-transform of an equalizer. A ZF equalizer satisfies

$$\mathsf{w}^*(z)\sqrt{G}\,\mathsf{h}(z) = z^{-\Delta} \tag{2.124}$$

for some delay Δ (in number of taps). This delay parameter provides an additional degree of freedom in the design and can be cleared through subsequent optimization.

The ideal solution to (2.124) is an equalizer with an infinite impulse response (IIR), yet such equalizers are challenging to implement. Most often, one wants to employ an FIR equalizer. Unfortunately, it is known that, except in trivial cases, an FIR impulse response does not have an FIR equalizer satisfying (2.124). When $\mathsf{w}(z)$ must be FIR of order L_{eq}, approximate ZF solutions may be obtained by truncating the IIR response, with a valid rule of thumb for satisfactory inversion being $L_{\mathsf{eq}} \approx 3L$. With an FIR equalizer of order L_{eq}, the delay is bounded in the range $0 \leq \Delta \leq (L_{\mathsf{eq}} + L)$ as determined by the length of the linear convolution of the time-domain counterparts to $\mathsf{w}(z)$ and $\mathsf{h}(z)$.

Example 2.15

For a two-tap channel where $\sqrt{G}\,h[0] = 1$ and $\sqrt{G}\,h[1] = -a$, find an IIR equalizer satisfying (2.124) with a delay of $\Delta = 1$.

Solution

The Z-transform of $\sqrt{G}\,h[\ell]$ is $\sqrt{G}(1 - az^{-1})$, giving as ideal ZF equalizer

$$\mathsf{w}^*_{\text{IIR}}(z) = \frac{1}{\sqrt{G}} \frac{z^{-1}}{1 - az^{-1}}. \tag{2.125}$$

For $a < 1$, the inverse Z-transform of (2.125) returns the causal filter

$$w^*_{\text{IIR}}[\ell] = \frac{1}{\sqrt{G}}\, a^{\ell-1}\, u[\ell - 1], \tag{2.126}$$

where $u[\ell]$ is the unit-step function.

Example 2.16

Following up on Example 2.15, identify an FIR equalizer $w[0], \ldots, w[L_{\mathsf{eq}}]$ by truncating the IIR solution.

Solution

Expanding (2.126),

$$w^*_{\mathrm{IIR}}[\ell] = \frac{\delta[\ell-1] - a\,\delta[\ell-2] + a^2\delta[\ell-3] - \cdots}{\sqrt{G}} \tag{2.127}$$

whose truncation gives

$$w^*[\ell] = \frac{\delta[\ell-1] - a\,\delta[\ell-2] + a^2\delta[\ell-3] - \cdots + (-a)^{L_{\mathsf{eq}}}\delta[\ell - L_{\mathsf{eq}} - 1]}{\sqrt{G}}. \tag{2.128}$$

In Problem 2.19, the reader is invited to check that the convolution of $w^*[\ell]$ and $\sqrt{G}h[\ell]$ yields an approximation to $\delta[\ell-1]$.

Extension to MIMO

Under certain conditions, FIR ZF equalizers turn out to exist for many types of FIR multivariate channels—an interesting oddity of multidimensional signal processing. Let the Z-transform of $\boldsymbol{H}[0], \ldots, \boldsymbol{H}[L]$ be

$$\mathsf{H}(z) = \sum_{\ell=0}^{L} \boldsymbol{H}[\ell]\, z^{-\ell} \tag{2.129}$$

whose (i,j)th entry is the Z-transform of $h^{(i,j)}[0], \ldots, h^{(i,j)}[L]$, and let $\mathsf{W}(z)$ denote the Z-transform of the matrix equalizer (see Fig. 2.12).

A ZF matrix equalizer is a left inverse of $\mathsf{H}(z)$ that satisfies

$$\mathsf{W}^*(z)\sqrt{G}\,\mathsf{H}(z) = \mathsf{D}(z), \tag{2.130}$$

where $\mathsf{D}(z)$ is a diagonal matrix with (possibly different) delay terms of the form $z^{-\Delta}$. As in the SISO case, it is in practice desirable to restrict the focus to FIR equalizers, meaning that each entry of $\mathsf{W}(z)$ corresponds to an FIR filter of order L_{eq}. Do exact left inverses of $\mathsf{H}(z)$ exist? Yes, as it turns out. More precisely, the answer is found in what is called perfect recoverability [149]. A MIMO transfer function $\mathsf{H}(z)$ has a left FIR inverse if and only if $\mathsf{H}(z)$ has full column rank (see Appendix B) for every complex $z \neq 0$.

The existence of left FIR inverses is a type of coprimeness in the polynomials that comprise $\mathsf{H}(z)$. While the result is somewhat esoteric, it turns out that exact FIR ZF MIMO equalizers exist for $N_{\mathrm{r}} > N_{\mathrm{t}}$ provided the polynomial coefficients are sufficiently random [150]. We explore this with an example for $N_{\mathrm{t}} = 1$ and $N_{\mathrm{r}} = 2$.

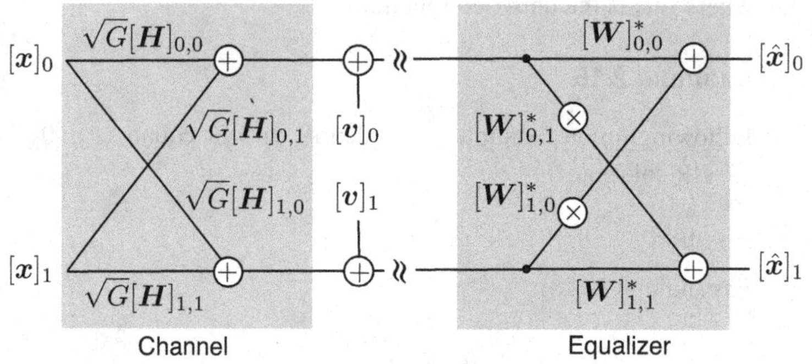

Fig. 2.12 MIMO equalizer for a single-tap channel with $N_t = N_r = 2$. Notice how the signal received on each antenna contributes to recovering each of the transmit signals.

Example 2.17

For $N_t = 1$ and $N_r = 2$, find the conditions on $\mathsf{H}(z)$ such that FIR ZF equalizers exist.

Solution

Satisfying the ZF criterion requires that

$$[\mathsf{w}(z)]_0^* \sqrt{G}\,[\mathsf{H}(z)]_0 + [\mathsf{w}(z)]_1^* \sqrt{G}\,[\mathsf{H}(z)]_1 = z^{-\Delta}. \tag{2.131}$$

To have a left inverse, the results on perfect recoverability mandate that $\mathsf{H}(z)$ have full column rank for every complex $z \neq 0$. Since $\mathsf{H}(z)$ is a column vector, the only way it cannot have full column rank is if $\mathsf{H}(z) = \mathbf{0}$ for some z. This would require $[\mathsf{H}(z)]_0 = [\mathsf{H}(z)]_1 = 0$, which can only be true if $[\mathsf{H}(z)]_0$ and $[\mathsf{H}(z)]_1$ share a common root. Therefore, as long as $[\mathsf{H}(z)]_0$ and $[\mathsf{H}(z)]_1$ are coprime (meaning they do not share a common root) it is possible to find an exact FIR inverse.

The result in Example 2.17 extends to any $N_r \times 1$ channel: the polynomials $[\mathsf{H}(z)]_i$ for $i = 0, \ldots, N_r - 1$ must be coprime. If the channel coefficients are sufficiently random, it is increasingly unlikely as N_r grows that all these polynomials share a common root.

With their existence established, the next question is how to find FIR ZF MIMO equalizers, exactly if possible or approximately otherwise. Taking the inverse Z-transform of (2.130), the objective is to find $\boldsymbol{W}[0], \ldots, \boldsymbol{W}[L_{\mathsf{eq}}]$ such that

$$\sqrt{G} \sum_{\ell=0}^{L_{\mathsf{eq}}} \boldsymbol{W}^*[\ell]\, \boldsymbol{H}[n-\ell] = \operatorname{diag}\Big(\delta[n-\Delta_0], \ldots, \delta[n-\Delta_{N_t-1}]\Big) \tag{2.132}$$

where Δ_j corresponds to the allowable reconstruction delay for the signal emanating from the jth transmit antenna. An inspection of (2.132) reveals that only the jth column of $\boldsymbol{W}[\ell]$ contributes to equalizing the signal emitted by the jth transmit antenna and thus, barring other constraints, the columns of $\boldsymbol{W}$ can be designed independently. Using the shorthand notation $\boldsymbol{w}_j[\ell] = \big[\boldsymbol{W}[\ell]\big]_{:,j}$, perfect ZF requires that

$$\sqrt{G}\sum_{\ell=0}^{L_{\text{eq}}} \boldsymbol{w}_j^*[\ell]\,\boldsymbol{H}[n-\ell] = \boldsymbol{d}_j^*[n] \qquad n = 0,1,\ldots,L_{\text{eq}}+L \tag{2.133}$$

where $\boldsymbol{d}_j^*[n]$ is a row vector with $\delta[n-\Delta_j]$ on the jth entry and zero elsewhere and where, recall, $\Delta_j \in \{0,\ldots,L_{\text{eq}}+L\}$. If perfect ZF is not possible, then we seek a solution that approximates (2.133) in some sense.

To unravel (2.133), which is in essence a convolution, it is useful to resort once more to stacked vectors and block matrices. Let the $N_{\text{r}}(L_{\text{eq}}+1)\times 1$ vector of equalizer coefficients for the jth signal be

$$\bar{\boldsymbol{w}}_j = \begin{bmatrix} \boldsymbol{w}_j[0] \\ \vdots \\ \boldsymbol{w}_j[L_{\text{eq}}] \end{bmatrix} \tag{2.134}$$

and let the $N_{\text{t}}(L_{\text{eq}}+L+1)\times 1$ vector of target delays be similarly defined as

$$\bar{\boldsymbol{d}}_j = \begin{bmatrix} \boldsymbol{d}_j[0] \\ \vdots \\ \boldsymbol{d}_j[L_{\text{eq}}+L] \end{bmatrix}. \tag{2.135}$$

Note that $\bar{\boldsymbol{d}}_j$ is identically zero except for a 1 in the position $N_{\text{t}}\Delta_j + j$. Finally, define the $N_{\text{r}}(L_{\text{eq}}+1)\times N_{\text{t}}(L_{\text{eq}}+L+1)$ block Toeplitz matrix

$$\bar{\boldsymbol{T}} = \begin{bmatrix} \boldsymbol{H}[0] & \cdots & \boldsymbol{H}[L] & \boldsymbol{0} & \boldsymbol{0} & \cdots & \boldsymbol{0} \\ \boldsymbol{0} & \boldsymbol{H}[0] & \cdots & \boldsymbol{H}[L] & \boldsymbol{0} & \cdots & \boldsymbol{0} \\ \vdots & & & \ddots & & & \vdots \\ \boldsymbol{0} & \cdots & \boldsymbol{0} & \boldsymbol{0} & \boldsymbol{H}[0] & \cdots & \boldsymbol{H}[L] \end{bmatrix} \tag{2.136}$$

such that we can rewrite (2.133) as

$$\sqrt{G}\,\bar{\boldsymbol{w}}_j^*\bar{\boldsymbol{T}} = \bar{\boldsymbol{d}}_j^*. \tag{2.137}$$

The rank of $\bar{\boldsymbol{T}}$ is an important consideration when solving for $\bar{\boldsymbol{w}}_0,\ldots,\bar{\boldsymbol{w}}_{N_{\text{t}}-1}$. If $\bar{\boldsymbol{T}}$ is not full-rank, then it is not possible to recover what was sent from at least one antenna; we thus assume that $\bar{\boldsymbol{T}}$ is full-rank. This is in fact equivalent to assuming that the conditions for perfect recoverability hold, and is also related to results on block Toeplitz matrices.

From standard linear algebra considerations, the existence and uniqueness of an exact solution depend on the dimensionality of $\bar{\boldsymbol{T}}$. Since $\bar{\boldsymbol{T}}$ is $N_{\text{r}}(L_{\text{eq}}+1)\times N_{\text{t}}(L_{\text{eq}}+L+1)$, its shape can be controlled through the equalizer order L_{eq} with three regimes of interest.

(1) $\bar{\boldsymbol{T}}$ is fat, meaning that $N_{\text{t}}(L_{\text{eq}}+L+1) > N_{\text{r}}(L_{\text{eq}}+1)$. Then, (2.137) is overdetermined and has no solution; the common approach is to minimize the squared equalization error for the chosen $\bar{\boldsymbol{d}}_j^*$, an error that is proportional to

$$\big\|\sqrt{G}\,\bar{\boldsymbol{w}}_j^*\bar{\boldsymbol{T}} - \bar{\boldsymbol{d}}_j^*\big\|^2 = \big(\sqrt{G}\,\bar{\boldsymbol{w}}_j^*\bar{\boldsymbol{T}} - \bar{\boldsymbol{d}}_j^*\big)\big(\sqrt{G}\,\bar{\boldsymbol{w}}_j^*\bar{\boldsymbol{T}} - \bar{\boldsymbol{d}}_j^*\big)^* \tag{2.138}$$

$$= G\,\bar{w}_j^*\bar{T}\bar{T}^*\bar{w}_j + \|\bar{d}_j\|^2 - \sqrt{G}\,\bar{w}_j^*\bar{T}\bar{d}_j - \sqrt{G}\,\bar{d}_j^*\bar{T}^*\bar{w}_j. \tag{2.139}$$

The equalizer $\bar{w}_j$ minimizing the above expression must satisfy

$$\nabla_{\bar{w}^{\text{ZF}}{}_j}\left\|\sqrt{G}\,\bar{w}_j^{\text{ZF}*}\bar{T} - \bar{d}_j^*\right\|^2 = \mathbf{0}. \tag{2.140}$$

Applying the gradient expressions (D.5), (D.6), and (D.7) in Appendix D, this gives the necessary condition

$$G\bar{T}\bar{T}^*\bar{w}_j^{\text{ZF}} - \sqrt{G}\,\bar{T}\bar{d}_j = \mathbf{0} \tag{2.141}$$

whose unique solution is

$$\bar{w}_j^{\text{ZF}} = \frac{1}{\sqrt{G}}(\bar{T}\bar{T}^*)^{-1}\bar{T}\bar{d}_j. \tag{2.142}$$

In the absence of delay constraints, one can further select $\bar{d}_j$ to render the ensuing squared error as small as possible.

(2) $\bar{T}$ is square, meaning that $N_{\text{t}}(L_{\text{eq}} + L + 1) = N_{\text{r}}(L_{\text{eq}} + 1)$. In this case, (2.137) has a unique solution that is readily given by

$$\bar{w}_j^{\text{ZF}} = \frac{1}{\sqrt{G}}(\bar{T}^*)^{-1}\bar{d}_j. \tag{2.143}$$

(3) $\bar{T}$ is tall, meaning that $N_{\text{t}}(L_{\text{eq}} + L + 1) < N_{\text{r}}(L_{\text{eq}} + 1)$. In this case, (2.137) is underdetermined and there are infinitely many solutions. A common criterion for selecting one among those is the minimum norm, which gives (refer to Problem 2.21)

$$\bar{w}_j^{\text{ZF}} = \frac{1}{\sqrt{G}}\bar{T}(\bar{T}^*\bar{T})^{-1}\bar{d}_j. \tag{2.144}$$

A general form for the ZF equalizer in all three cases above can be given by means of the Moore–Penrose pseudoinverse of $\bar{T}^*$, a notion that is described in Appendix B.6. Given a rectangular matrix A, its pseudoinverse $A^\dagger$ satisfies $AA^\dagger A = A$ and $A^\dagger AA^\dagger = A^\dagger$.

The ZF equalizer

$$\bar{w}_j^{\text{ZF}} = \frac{1}{\sqrt{G}}(\bar{T}^*)^\dagger\,\bar{d}_j \tag{2.145}$$

unifies the three solutions. Moreover, assembling the equalizers for the N_{t} signals into the $N_{\text{r}}(L_{\text{eq}} + 1) \times N_{\text{t}}$ matrix $\bar{W}^{\text{ZF}} = \left[\bar{w}_0^{\text{ZF}} \cdots \bar{w}_{N_{\text{t}}-1}^{\text{ZF}}\right]$ and letting $\bar{D} = \left[\bar{d}_0 \cdots \bar{d}_{N_{\text{t}}-1}\right]$, we can compactly write

$$\bar{W}^{\text{ZF}} = \frac{1}{\sqrt{G}}(\bar{T}^*)^\dagger\bar{D}. \tag{2.146}$$

Unstacking the $L_{\text{eq}} + 1$ vertical blocks within $\bar{W}^{\text{ZF}}$, we finally obtain the $(L_{\text{eq}} + 1)$-tap $N_{\text{r}} \times N_{\text{t}}$ MIMO equalizer $W^{\text{ZF}}[0], \ldots, W^{\text{ZF}}[L_{\text{eq}}]$. The factor $1/\sqrt{G}$ permeating all the terms cleanly separates what is mere compensation for the average channel attenuation, implemented via low-noise amplification, from the signal processing that disentangles the ISI and the multiantenna interference. The choice of L_{eq} impacts the dimensions of $\bar{T}^*$ and, through them, the performance of such signal processing.

- For $N_t > N_r$, $\bar{T}$ is always fat and perfect ZF is not feasible. Linear equalization is generally unappealing when $N_t > N_r$.
- For $N_t = N_r$, it becomes possible to have $N_t(L_{eq} + L + 1) = N_r(L_{eq} + 1)$ and hence a square $\bar{T}^*$. This occurs when $L = 0$, i.e., in frequency-flat channels, and it holds regardless of the value of L_{eq}. It follows that, in frequency-flat channels, a single-tap equalizer ($L_{eq} = 0$) suffices. However, in frequency-selective channels, perfect ZF equalization is in general not possible with $N_t = N_r$.
- For $N_r > N_t$, $\bar{T}$ is rendered square when it is possible to find a nonnegative integer $L_{eq} = N_t L/(N_r - N_t) - 1$, the minimum equalizer order that ensures perfect ZF. For $L_{eq} > N_t L/(N_r - N_t) - 1$ we have a fat $\bar{T}^*$ and a multiplicity of ZF solutions. Increasing the number of receive antennas N_r yields a decrease in L_{eq} for a given L and, altogether, $N_r > N_t$ is most desirable when operating with FIR equalizers.

Although, in the absence of noise, it may seem harmless to select L_{eq} above the value strictly necessary to ensure ZF feasibility, this changes once noise is considered. When noise runs through the equalizer, its power is enhanced in proportion to $\sum_{\ell=0}^{L_{eq}} \|W^{ZF}[\ell]\|_F^2$, hence the minimum-norm criterion when a multiplicity of solutions exist. And, while it does not play a role in the feasibility of ZF, the delay Δ_j can be optimized to further minimize the norm among all possible delays. If delay optimization is not possible, a rule of thumb is $\Delta_j \approx L$.

Example 2.18 (Linear ZF equalizer for frequency-flat channels)

Compute the ZF equalizer for a frequency-flat channel with $N_r \geq N_t$.

Solution

In a frequency-flat channel, $L = 0$ and therefore $\bar{T} = H$. Setting $L_{eq} = 0$ to minimize the equalizer norm, the optimum delay is also $\Delta_j = 0$ for $j = 0, \ldots, N_t - 1$. It follows that $\bar{d}_j = [I]_{:,j}$ and hence that $\bar{D} = I$, from which W^{ZF} as given by (2.146) satisfies

$$W^{ZF*} = \frac{1}{\sqrt{G}} H^{\dagger}. \tag{2.147}$$

Example 2.19

Verify that the W^{ZF} derived in Example 2.18 effects perfect ZF equalization.

Solution

The application of the equalizer to the channel gives

$$W^{ZF*}\sqrt{G}H = (HH^*)^{-1}H^*H \tag{2.148}$$

$$= I. \tag{2.149}$$

As Example 2.18 evidences, ZF equalization simplifies drastically in frequency-flat channels and it essentially amounts to solving a set of linear equations for the transmit vector. Without ISI, only multiantenna interference needs to be dealt with.

The ZF equalizer is at the heart of a suboptimum yet appealing MIMO linear receiver that is studied in detail in Chapter 6.

2.4.2 LMMSE equalization

The main drawback of ZF equalization is that noise is neglected in its formulation. To address this shortcoming, a metric is required that accounts not only for self-interference (both ISI and multiantenna interference) but further for noise, and the most natural such metric is the MMSE introduced in Chapter 1. The equalizer is then nothing but an MMSE estimator of the transmit signal, and in the context of linear equalization it is an LMMSE estimator. LMMSE equalization is but one example of the application of Wiener theory [126] within statistical signal processing [127] and an early application of MIMO channel equalization was in the ocean acoustic channel [151]. In this section, we derive the FIR MIMO equalizer, putting to work the LMMSE concepts laid down in Section 1.7. For this derivation, both $\boldsymbol{x}[n]$ and $\boldsymbol{v}[n]$ are regarded as wide-sense stationary—this assumption is supported in Chapter 2—with correlation functions $\boldsymbol{R}_{\boldsymbol{x}}[\ell] = \mathbb{E}\big[\boldsymbol{x}[n]\boldsymbol{x}^*[n+\ell]\big]$ and $\boldsymbol{R}_{\boldsymbol{v}}[\ell] = N_0\delta[\ell]\boldsymbol{I}$. It follows from the transmit–receive relationship in (2.80) that

$$\boldsymbol{R}_{\boldsymbol{y}}[\ell] = \mathbb{E}\big[\boldsymbol{y}[n]\,\boldsymbol{y}^*[n+\ell]\big] \tag{2.150}$$

$$= \mathbb{E}\left[\sqrt{G}\sum_{\ell_1=0}^{L}\boldsymbol{H}[\ell_1]\,\boldsymbol{x}[n-\ell_1]\left(\sqrt{G}\sum_{\ell_2=0}^{L}\boldsymbol{H}[\ell_2]\,\boldsymbol{x}[n+\ell-\ell_2]\right)^*\right]$$

$$+\,\mathbb{E}\big[\boldsymbol{v}[n]\boldsymbol{v}^*[n+\ell]\big] \tag{2.151}$$

$$= G\,\mathbb{E}\left[\sum_{\ell_1=0}^{L}\sum_{\ell_2=0}^{L}\boldsymbol{H}[\ell_1]\,\boldsymbol{x}[n-\ell_1]\,\boldsymbol{x}^*[n+\ell-\ell_2]\,\boldsymbol{H}^*[\ell_2]\right] + N_0\delta[\ell]\boldsymbol{I} \tag{2.152}$$

$$= G\sum_{\ell_1=0}^{L}\sum_{\ell_2=0}^{L}\boldsymbol{H}[\ell_1]\,\mathbb{E}\Big[\boldsymbol{x}[n-\ell_1]\,\boldsymbol{x}^*[n+\ell-\ell_2]\Big]\boldsymbol{H}^*[\ell_2] + N_0\delta[\ell]\boldsymbol{I} \tag{2.153}$$

$$= G\sum_{\ell_1=0}^{L}\sum_{\ell_2=0}^{L}\boldsymbol{H}[\ell_1]\,\boldsymbol{R}_{\boldsymbol{x}}[\ell+\ell_1-\ell_2]\,\boldsymbol{H}^*[\ell_2] + N_0\delta[\ell]\boldsymbol{I}, \tag{2.154}$$

while

$$\boldsymbol{R}_{\boldsymbol{yx}}[\ell] = \mathbb{E}\big[\boldsymbol{y}[n]\,\boldsymbol{x}^*[n+\ell]\big] \tag{2.155}$$

$$= \sqrt{G}\,\mathbb{E}\left[\sum_{\ell_1=0}^{L}\boldsymbol{H}[\ell_1]\,\boldsymbol{x}[n-\ell_1]\,\boldsymbol{x}^*[n+\ell]\right] + \mathbb{E}\big[\boldsymbol{v}[n]\,\boldsymbol{x}^*[n+\ell]\big] \tag{2.156}$$

$$= \sqrt{G}\sum_{\ell_1=0}^{L}\boldsymbol{H}[\ell_1]\,\mathbb{E}\big[\boldsymbol{x}[n-\ell_1]\,\boldsymbol{x}^*[n+\ell]\big] \tag{2.157}$$

$$= \sqrt{G}\sum_{\ell_1=0}^{L}\boldsymbol{H}[\ell_1]\,\boldsymbol{R}_{\boldsymbol{x}}[\ell+\ell_1]. \tag{2.158}$$

Recall that a linear FIR MIMO equalizer is $\boldsymbol{W}[0], \ldots, \boldsymbol{W}[L_{\text{eq}}]$ where $\boldsymbol{W}[\ell]$ is $N_r \times N_t$. The error between the equalizer output and the desired signal at symbol n is

$$\tilde{\boldsymbol{x}}[n] = \boldsymbol{x}[n-\Delta] - \sum_{\ell=0}^{L_{\text{eq}}} \boldsymbol{W}^*[\ell]\, \boldsymbol{y}[n-\ell], \tag{2.159}$$

with a target delay Δ taken to be common for all signals, though the formulation can be further generalized to distinct Δ_j for $j = 0, \ldots, N_t - 1$. Stacking items in the usual way, (2.159) can be rewritten as

$$\tilde{\boldsymbol{x}}[n] = \boldsymbol{x}[n-\Delta] - \bar{\boldsymbol{W}}^*\, \bar{\boldsymbol{y}}[n], \tag{2.160}$$

where

$$\bar{\boldsymbol{W}} = \begin{bmatrix} \boldsymbol{W}[0] \\ \vdots \\ \boldsymbol{W}[L_{\text{eq}}] \end{bmatrix} \qquad \bar{\boldsymbol{y}}[n] = \begin{bmatrix} \boldsymbol{y}[n] \\ \vdots \\ \boldsymbol{y}[n-L_{\text{eq}}] \end{bmatrix}. \tag{2.161}$$

with $\boldsymbol{x}[n-\Delta]$ being estimated on the basis of the observation of $\bar{\boldsymbol{y}}[n]$.

Applying directly the findings of Section 1.7, the MMSE is achieved by

$$\begin{aligned} \bar{\boldsymbol{W}}^{\text{MMSE}} &= \boldsymbol{R}_{\bar{\boldsymbol{y}}[n]}^{-1}\, \boldsymbol{R}_{\bar{\boldsymbol{y}}[n]\boldsymbol{x}[n-\Delta]} && (2.162) \\ &= \begin{bmatrix} \boldsymbol{R}_{\boldsymbol{y}}[0] & \boldsymbol{R}_{\boldsymbol{y}}[1] & \cdots & \boldsymbol{R}_{\boldsymbol{y}}[L_{\text{eq}}] \\ \boldsymbol{R}_{\boldsymbol{y}}[1] & \boldsymbol{R}_{\boldsymbol{y}}[0] & \cdots & \boldsymbol{R}_{\boldsymbol{y}}[L_{\text{eq}}-1] \\ \vdots & \vdots & \ddots & \vdots \\ \boldsymbol{R}_{\boldsymbol{y}}[L_{\text{eq}}] & \boldsymbol{R}_{\boldsymbol{y}}[L_{\text{eq}}-1] & \cdots & \boldsymbol{R}_{\boldsymbol{y}}[0] \end{bmatrix}^{-1} \begin{bmatrix} \boldsymbol{R}_{\boldsymbol{yx}}[-\Delta] \\ \boldsymbol{R}_{\boldsymbol{yx}}[1-\Delta] \\ \vdots \\ \boldsymbol{R}_{\boldsymbol{yx}}[L_{\text{eq}}-\Delta] \end{bmatrix}, && (2.163) \end{aligned}$$

with $\boldsymbol{R}_{\boldsymbol{y}}[\ell]$ and $\boldsymbol{R}_{\boldsymbol{yx}}[\ell]$ as established in (2.154) and (2.158), respectively. In turn, the MMSE matrix characterizing the performance of the equalizer is

$$\begin{aligned} \boldsymbol{E} &= \mathbb{E}\big[\tilde{\boldsymbol{x}}[n]\, \tilde{\boldsymbol{x}}^*[n]\big] && (2.164) \\ &= \boldsymbol{R}_{\boldsymbol{x}}[0] - \boldsymbol{R}_{\boldsymbol{x}[n-\Delta]\bar{\boldsymbol{y}}[n]}\, \boldsymbol{R}_{\bar{\boldsymbol{y}}[n]}^{-1}\, \boldsymbol{R}_{\bar{\boldsymbol{y}}[n]\boldsymbol{x}[n-\Delta]} && (2.165) \\ &= \boldsymbol{R}_{\boldsymbol{x}}[0] - \boldsymbol{R}_{\bar{\boldsymbol{y}}[n]\boldsymbol{x}[n-\Delta]}^*\, \boldsymbol{R}_{\bar{\boldsymbol{y}}[n]}^{-1}\, \boldsymbol{R}_{\bar{\boldsymbol{y}}[n]\boldsymbol{x}[n-\Delta]} && (2.166) \\ &= \boldsymbol{R}_{\boldsymbol{x}}[0] - \begin{bmatrix} \boldsymbol{R}_{\boldsymbol{yx}}[-\Delta] \\ \boldsymbol{R}_{\boldsymbol{yx}}[1-\Delta] \\ \vdots \\ \boldsymbol{R}_{\boldsymbol{yx}}[L_{\text{eq}}-\Delta] \end{bmatrix}^* \begin{bmatrix} \boldsymbol{R}_{\boldsymbol{y}}[0] & \cdots & \boldsymbol{R}_{\boldsymbol{y}}[L_{\text{eq}}] \\ \boldsymbol{R}_{\boldsymbol{y}}[1] & \cdots & \boldsymbol{R}_{\boldsymbol{y}}[L_{\text{eq}}-1] \\ \vdots & \ddots & \vdots \\ \boldsymbol{R}_{\boldsymbol{y}}[L_{\text{eq}}] & \cdots & \boldsymbol{R}_{\boldsymbol{y}}[0] \end{bmatrix}^{-1} \begin{bmatrix} \boldsymbol{R}_{\boldsymbol{yx}}[-\Delta] \\ \boldsymbol{R}_{\boldsymbol{yx}}[1-\Delta] \\ \vdots \\ \boldsymbol{R}_{\boldsymbol{yx}}[L_{\text{eq}}-\Delta] \end{bmatrix}. && (2.167) \end{aligned}$$

In the special but highly relevant cases that the transmit symbols are IID and/or the channel is frequency-flat, the structure of the LMMSE equalizer simplifies considerably.

Example 2.20 (LMMSE equalizer with IID signaling)

For $\boldsymbol{R}_{\boldsymbol{x}}[\ell] = \frac{E_s}{N_t}\delta[\ell]\boldsymbol{I}$, which corresponds to IID transmissions in time and across antennas compliant with the per-block, per-symbol, and per-antenna power constraints, the covariances in (2.154) and (2.158) reduce respectively to

$$\boldsymbol{R}_{\boldsymbol{y}}[\ell] = \frac{G E_{\rm s}}{N_{\rm t}} \sum_{\ell_1=0}^{L} \boldsymbol{H}[\ell_1] \, \boldsymbol{H}^*[\ell+\ell_1] + N_0 \delta[\ell] \boldsymbol{I} \tag{2.168}$$

and

$$\boldsymbol{R}_{\boldsymbol{y}\boldsymbol{x}}[\ell] = \frac{\sqrt{G} E_{\rm s}}{N_{\rm t}} \boldsymbol{H}[-\ell], \tag{2.169}$$

which, plugged into (2.163), yield $\bar{\boldsymbol{W}}^{\sf MMSE}$.

This LMMSE equalizer can be related back to the exposition of its ZF counterpart. Recovering the Toeplitz matrix $\bar{\boldsymbol{T}}$ introduced in (2.136), we can write

$$\boldsymbol{R}_{\bar{\boldsymbol{y}}[n]} = \frac{G E_{\rm s}}{N_{\rm t}} \bar{\boldsymbol{T}} \bar{\boldsymbol{T}}^* + N_0 \boldsymbol{I} \tag{2.170}$$

and

$$\boldsymbol{R}_{\bar{\boldsymbol{y}}[n] \bar{\boldsymbol{x}}[n-\Delta]} = \frac{\sqrt{G} E_{\rm s}}{N_{\rm t}} \bar{\boldsymbol{T}} \bar{\boldsymbol{D}}, \tag{2.171}$$

where the $N_{\rm t}(L + L_{\sf eq} + 1) \times N_{\rm t}$ matrix $\bar{\boldsymbol{D}}$ has zeros everywhere except for the $N_{\rm t} \times N_{\rm t}$ submatrix $[\bar{\boldsymbol{D}}]_{\Delta N_{\rm t} : \Delta N_{\rm t} + N_{\rm t} - 1, :} = \boldsymbol{I}$. Then,

$$\begin{aligned} \bar{\boldsymbol{W}}^{\sf MMSE} &= \left(\frac{G E_{\rm s}}{N_{\rm t}} \bar{\boldsymbol{T}} \bar{\boldsymbol{T}}^* + N_0 \boldsymbol{I} \right)^{-1} \frac{\sqrt{G} E_{\rm s}}{N_{\rm t}} \bar{\boldsymbol{T}} \bar{\boldsymbol{D}} & (2.172) \\ &= \frac{1}{\sqrt{G}} \left(\bar{\boldsymbol{T}} \bar{\boldsymbol{T}}^* + \frac{N_0 N_{\rm t}}{G E_{\rm s}} \boldsymbol{I} \right)^{-1} \bar{\boldsymbol{T}} \bar{\boldsymbol{D}} & (2.173) \\ &= \frac{1}{\sqrt{G}} \left(\bar{\boldsymbol{T}} \bar{\boldsymbol{T}}^* + \frac{N_{\rm t}}{\sf SNR} \boldsymbol{I} \right)^{-1} \bar{\boldsymbol{T}} \bar{\boldsymbol{D}}, & (2.174) \end{aligned}$$

where we have recalled ${\sf SNR} = \frac{G E_{\rm s}}{N_0}$. Replicating the steps in Section 1.7.1, the expression in (2.174) can further be rewritten into the alternative form

$$\bar{\boldsymbol{W}}^{\sf MMSE} = \frac{1}{\sqrt{G}} \bar{\boldsymbol{T}} \left(\bar{\boldsymbol{T}}^* \bar{\boldsymbol{T}} + \frac{N_{\rm t}}{\sf SNR} \boldsymbol{I} \right)^{-1} \bar{\boldsymbol{D}}. \tag{2.175}$$

Both forms are correct. Normally, (2.174) is applied when $\bar{\boldsymbol{T}}$ is fat while (2.175) is preferable when $\bar{\boldsymbol{T}}$ is tall, as these choices result in lower-dimensional matrix inversions. Both forms also evidence how, for growing SNR, the LMMSE equalizer converges to the ZF solution (either the least-squares or the minimum-norm solution depending on the case).

Inserting (2.170) and (2.171) into (2.166), the MMSE matrix with IID signaling becomes

$$\boldsymbol{E} = \frac{E_{\rm s}}{N_{\rm t}} \boldsymbol{I} - \frac{E_{\rm s}}{N_{\rm t}} \bar{\boldsymbol{D}}^* \bar{\boldsymbol{T}}^* \left(\bar{\boldsymbol{T}} \bar{\boldsymbol{T}}^* + \frac{N_{\rm t}}{\sf SNR} \boldsymbol{I} \right)^{-1} \bar{\boldsymbol{T}} \bar{\boldsymbol{D}}. \tag{2.176}$$

Example 2.21 (LMMSE equalizer for frequency-flat channels)

For $L = 0$, with the single-tap channel being $\sqrt{G}\boldsymbol{H}$, and with arbitrary $\boldsymbol{R_x}[\ell]$, the covariances in (2.154) and (2.158) reduce respectively to

$$\boldsymbol{R_y}[\ell] = G\boldsymbol{H}\boldsymbol{R_x}[\ell]\boldsymbol{H}^* + N_0\delta[\ell]\boldsymbol{I} \tag{2.177}$$

and

$$\boldsymbol{R_{yx}}[\ell] = \sqrt{G}\boldsymbol{H}\boldsymbol{R_x}[\ell], \tag{2.178}$$

which, plugged into (2.163) with $L_{\text{eq}} = \Delta = 0$, yield

$$\boldsymbol{W}^{\text{MMSE}} = \left(G\boldsymbol{H}\boldsymbol{R_x}[0]\boldsymbol{H}^* + N_0\boldsymbol{I}\right)^{-1}\sqrt{G}\boldsymbol{H}\boldsymbol{R_x}[0] \tag{2.179}$$

$$= \frac{1}{\sqrt{G}}\left(\boldsymbol{H}\frac{\boldsymbol{R_x}[0]}{\frac{E_s}{N_t}}\boldsymbol{H}^* + \frac{N_t}{\text{SNR}}\boldsymbol{I}\right)^{-1}\boldsymbol{H}\frac{\boldsymbol{R_x}[0]}{\frac{E_s}{N_t}}. \tag{2.180}$$

In turn, from (2.166),

$$\boldsymbol{E} = \boldsymbol{R_x}[0] - \boldsymbol{R_x}[0]\,\boldsymbol{H}^*\left(\boldsymbol{H}\frac{\boldsymbol{R_x}[0]}{\frac{E_s}{N_t}}\boldsymbol{H}^* + \frac{N_t}{\text{SNR}}\boldsymbol{I}\right)^{-1}\boldsymbol{H}\frac{\boldsymbol{R_x}[0]}{\frac{E_s}{N_t}}. \tag{2.181}$$

Recalling that $\boldsymbol{R_x}[0] = \frac{E_s}{N_t}\boldsymbol{F}\boldsymbol{F}^*$, both $\boldsymbol{W}^{\text{MMSE}}$ and $\boldsymbol{E}$ can alternatively be expressed as a function of the precoder $\boldsymbol{F}$, to wit

$$\boldsymbol{W}^{\text{MMSE}} = \frac{1}{\sqrt{G}}\left(\boldsymbol{H}\boldsymbol{F}\boldsymbol{F}^*\boldsymbol{H}^* + \frac{N_t}{\text{SNR}}\boldsymbol{I}\right)^{-1}\boldsymbol{H}\boldsymbol{F} \tag{2.182}$$

and

$$\boldsymbol{E} = \frac{E_s}{N_t}\boldsymbol{F}\boldsymbol{F}^* - \frac{E_s}{N_t}\boldsymbol{F}\boldsymbol{F}^*\boldsymbol{H}^*\left(\boldsymbol{H}\boldsymbol{F}\boldsymbol{F}^*\boldsymbol{H}^* + \frac{N_t}{\text{SNR}}\boldsymbol{I}\right)^{-1}\boldsymbol{H}\boldsymbol{F}\boldsymbol{F}^* \tag{2.183}$$

$$= \frac{E_s}{N_t}\boldsymbol{F}\left[\boldsymbol{I} - \boldsymbol{F}^*\boldsymbol{H}^*\left(\boldsymbol{H}\boldsymbol{F}\boldsymbol{F}^*\boldsymbol{H}^* + \frac{N_t}{\text{SNR}}\boldsymbol{I}\right)^{-1}\boldsymbol{H}\boldsymbol{F}\right]\boldsymbol{F}^*. \tag{2.184}$$

An important difference between the ZF and LMMSE equalizers is that the latter can deal with channels that are ill-conditioned, meaning not full-rank. The *regularization* prior to inversion ensures the stability of the LMMSE equalizer.

And, just like the ZF equalizer is at the heart of a linear MIMO receiver, the LMMSE equalizer is the basis of another linear MIMO receiver that is studied in Chapter 6. Moreover, such an LMMSE receiver turns out to be optimum within the class of linear receivers and can be an ingredient of an optimum nonlinear receiver, hence its importance is capital.

2.5 Single-carrier frequency-domain equalization

FIR equalization of a MIMO channel entails calculating the taps of a matrix equalizer and applying this filter to the observed signal by means of $N_t N_r$ convolution operations. An alternative is to equalize in the frequency domain [152]. This allows for an ideal inversion of the channel without requiring an IIR implementation. Frequency-domain equalization, however, is not straightforward. From a computational perspective, it is preferable to use the discrete Fourier transform (DFT) since this involves only a finite number of frequencies and can be implemented efficiently via the fast Fourier transform (FFT). The challenge with direct application of the DFT is that frequency-domain products correspond to *circular* convolutions in discrete time, whereas what the channel effects is a *linear* convolution. This issue can be resolved by adding a cyclic prefix to the transmit signal, which renders the linear convolution equivalent to a circular one.

This section presents SC-FDE, first for SISO, to cleanly explain the key ideas, and subsequently for MIMO. The qualifier *single-carrier* is applied to distinguish this approach from its *multicarrier* counterpart embodied by OFDM. For interested readers, a brief review of the DFT is included in Appendix A.1.4.

2.5.1 Basic formulation

Consider the product of the K-point DFTs of the transmit signal and of the channel response, namely

$$\mathsf{y}[k] = \sqrt{G}\,\mathsf{h}[k]\,\mathsf{x}[k] \qquad 0 = 1, \ldots, K-1, \tag{2.185}$$

which corresponds, in the discrete-time domain, to the circular convolution

$$y[n] = \sqrt{G} \sum_{\ell=0}^{K-1} h[\ell]\, x\big[((n-\ell))_K\big] \qquad n = 0, \ldots, K-1, \tag{2.186}$$

where $((\cdot))_K$ indicates modulo-K. From (2.185), and provided the channel response is nonzero, ZF equalization only requires computing $\frac{1}{\sqrt{G}}\mathsf{y}[k]/\mathsf{h}[k]$ for $k = 0, \ldots, K-1$. The complexity is low, requiring only divisions in the frequency domain in addition to the transformations between time and frequency.

Let us see how to take advantage of such a low-complexity equalization possibility, starting by explaining why (2.185) cannot be applied directly. Suppose that $h[n]$ is FIR of order L and that $K > L$, such that the K-point DFT of $h[n]$ is computed by zero-padding $h[0], \ldots, h[L]$ with $K - L + 1$ zeros. Then, (2.186) becomes

$$y[n] = \sqrt{G} \sum_{\ell=0}^{L} h[\ell]\, x\big[((n-\ell))_K\big] \qquad n = 0, \ldots, K-1 \tag{2.187}$$

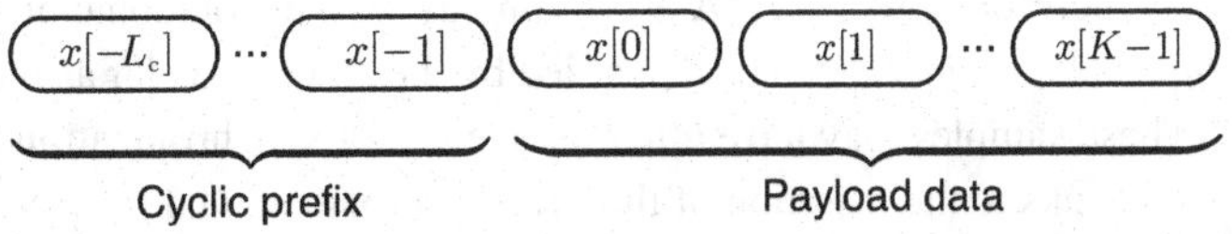

Fig. 2.13 Extended signal $\bar{x}[n]$ obtained by appending a cyclic prefix of length L_c to a block of K payload symbols.

whose inspection for different values of n gives

$$y[n] = \begin{cases} \sqrt{G} \sum_{\ell=0}^{n} h[\ell]\, x[n-\ell] + \sqrt{G} \sum_{\ell=n+1}^{L} h[\ell]\, x[n+K-\ell] & n = 0, \ldots, L-1 \\ \sqrt{G} \sum_{\ell=0}^{L} h[\ell]\, x[n-\ell] & n = L, \ldots, K-1 \end{cases}$$

where the portion corresponding to $n > L$ does look like a linear convolution, but the rest depends on values at the end of the signal block. The issue is precisely that the propagation channel (determined by nature and by the electronics) effects a linear convolution in discrete time, but SC-FDE is only possible under a circular convolution.

One solution to this problem is to insert a cyclic prefix, whereby the last few symbols are duplicated at the beginning of the block so as to create the effect of a circular convolution from a purely linear one. Denoting the length of such cyclic prefix by $L_c \geq L$, this results in the extended discrete-time signal block (see Fig. 2.13)

$$\bar{x}[n] = \begin{cases} x[n+K] & n = -L_c, \ldots, -1 \\ x[n] & n = 0, \ldots, K-1. \end{cases} \tag{2.188}$$

This cyclic prefix does make the linear convolution output look as if it came from a circular convolution. To see that, consider the output

$$\bar{y}[n] = \sqrt{G} \sum_{\ell=0}^{L} h[\ell]\, \bar{x}[n-\ell] \qquad n = 0, \ldots, K-1. \tag{2.189}$$

Evaluated for $n = 0$, the above gives

$$\bar{y}[0] = \sqrt{G} \left(h[0]\, \bar{x}[0] + h[1]\, \bar{x}[-1] + \cdots + h[L]\, \bar{x}[-L] \right) \tag{2.190}$$

which, by virtue of the structure of $\bar{x}[n]$, equals

$$\bar{y}[0] = \sqrt{G} \left(h[0] x[0] + h[1] x[K-1] + \cdots h[L] x[K-L] \right) \tag{2.191}$$

$$= \sqrt{G} \sum_{\ell=0}^{L} h[\ell]\, x\big[((-\ell))_K\big]. \tag{2.192}$$

The same relationship can be seen to hold for an arbitrary n, precisely

$$\bar{y}[n] = \sqrt{G} \sum_{\ell=0}^{L} h[\ell]\, x\big[((n-\ell))_K\big] \qquad n = 0, \ldots, K-1 \tag{2.193}$$

and therefore $\bar{y}[0], \ldots, \bar{y}[K-1]$ coincide with the outcome of a circular convolution while $\bar{y}[-L_{\rm c}], \ldots, \bar{y}[-1]$ can be discarded by the receiver as far as the equalization is concerned. (These samples may have other uses, e.g., for synchronization.)

Thanks to the insertion of the cyclic prefix then, and provided that $L_{\rm c} \geq L$, it is possible to implement SC-FDE by computing $\frac{1}{\sqrt{G}} {\rm IDFT}_K\{\mathsf{y}[k]/\mathsf{h}[k]\}$. In matrix form, the output of the equalizer is

$$\begin{bmatrix} \hat{x}[0] \\ \vdots \\ \hat{x}[K-1] \end{bmatrix} = \boldsymbol{U}^* \, {\rm diag}\left(\frac{1}{\sqrt{G}\,\mathsf{h}[0]}, \ldots, \frac{1}{\sqrt{G}\,\mathsf{h}[K-1]}\right) \boldsymbol{U} \begin{bmatrix} y[0] \\ \vdots \\ y[K-1] \end{bmatrix} \tag{2.194}$$

where $\boldsymbol{U}$ is a Fourier matrix that effects a DFT (see Appendix B.2.3) while $\boldsymbol{U}^*$ effects the IDFT. A diagram of SC-FDE is provided in Fig. 2.14. The equalizer complexity is fixed and determined by the DFT and IDFT, which with an FFT implementation entail on the order of $K \log_2 K$ operations. From a complexity perspective it is thus desirable to keep K small, which also promotes time-invariance over the block being equalized. However, the overhead associated with the cyclic prefix is $\frac{L_{\rm c}}{K+L_{\rm c}}$ and hence, from that perspective, it is desirable to increase K. The value for K emerges from resolving this tradeoff, an issue that we shall revisit in the context of OFDM. In turn, $L_{\rm c}$ should be as small as possible while respecting the condition $L_{\rm c} \geq L$; consequently, for the remainder of this section and for the ensuing formulation of OFDM we directly set

$$L_{\rm c} = L. \tag{2.195}$$

Let us now see how the application of SC-FDE affects the noise. In the presence of noise, the output of an SC-FDE ZF equalizer is

$$\begin{aligned} \begin{bmatrix} \hat{x}[0] \\ \vdots \\ \hat{x}[K-1] \end{bmatrix} &= \frac{1}{\sqrt{G}} \boldsymbol{U}^* \, {\rm diag}\left(\frac{1}{\mathsf{h}[0]}, \ldots, \frac{1}{\mathsf{h}[K-1]}\right) \boldsymbol{U} \begin{bmatrix} y[0] \\ \vdots \\ y[K-1] \end{bmatrix} \\ &= \begin{bmatrix} x[0] \\ \vdots \\ x[K-1] \end{bmatrix} + \underbrace{\frac{1}{\sqrt{G}} \boldsymbol{U}^* \, {\rm diag}\left(\frac{1}{\mathsf{h}[0]}, \ldots, \frac{1}{\mathsf{h}[K-1]}\right) \boldsymbol{U} \begin{bmatrix} v[0] \\ \vdots \\ v[K-1] \end{bmatrix}}_{\text{Output noise, } \bar{\boldsymbol{v}}}. \end{aligned} \tag{2.196}$$

The noise contaminating this output is zero-mean with covariance

$$\begin{aligned} \mathbb{E}\left[\bar{\boldsymbol{v}}\bar{\boldsymbol{v}}^*\right] &= \frac{1}{G} \boldsymbol{U}^* \, {\rm diag}\left(\frac{1}{\mathsf{h}[0]}, \ldots, \frac{1}{\mathsf{h}[K-1]}\right) \boldsymbol{U} \, \mathbb{E}\left[\begin{bmatrix} v[0] \\ \vdots \\ v[K-1] \end{bmatrix} \begin{bmatrix} v[0] \\ \vdots \\ v[K-1] \end{bmatrix}^*\right] \boldsymbol{U}^* \\ &\quad \cdot {\rm diag}\left(\frac{1}{\mathsf{h}[0]}, \ldots, \frac{1}{\mathsf{h}[K-1]}\right)^* \boldsymbol{U} \\ &= \frac{N_0}{G} \boldsymbol{U}^* \, {\rm diag}\left(\frac{1}{\mathsf{h}[0]}, \ldots, \frac{1}{\mathsf{h}[K-1]}\right) \boldsymbol{U}\boldsymbol{U}^* {\rm diag}\left(\frac{1}{\mathsf{h}[0]}, \ldots, \frac{1}{\mathsf{h}[K-1]}\right)^* \boldsymbol{U} \end{aligned}$$

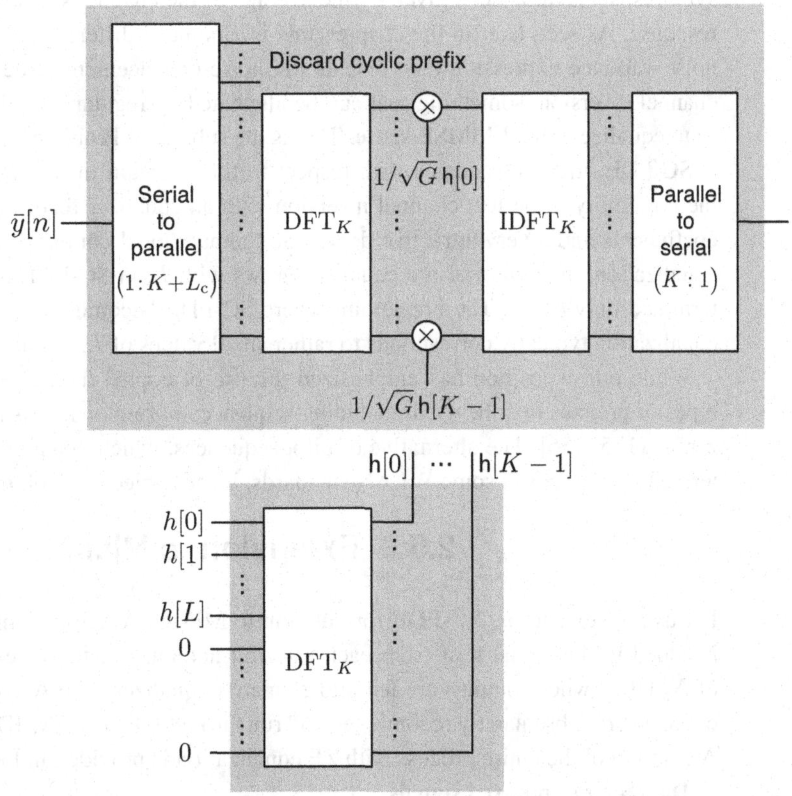

Fig. 2.14 Above, an SC-FDE equalizer. Below, K-point DFT of the channel response that is fed to the equalizer.

$$= \frac{N_0}{G} \boldsymbol{U}^* \operatorname{diag}\left(\frac{1}{\mathsf{h}[0]}, \ldots, \frac{1}{\mathsf{h}[K-1]}\right) \operatorname{diag}\left(\frac{1}{\mathsf{h}[0]}, \ldots, \frac{1}{\mathsf{h}[K-1]}\right)^* \boldsymbol{U} \quad (2.197)$$

$$= \frac{N_0}{G} \boldsymbol{U}^* \operatorname{diag}\left(\frac{1}{|\mathsf{h}[0]|^2}, \ldots, \frac{1}{|\mathsf{h}[K-1]|^2}\right) \boldsymbol{U}. \quad (2.198)$$

The above covariance matrix is, in general, not proportional to the identity, indicating that the noise is correlated. The variance of the noise on the nth equalized symbol is

$$\left[\frac{N_0}{G} \boldsymbol{U}^* \operatorname{diag}\left(\frac{1}{|\mathsf{h}[0]|^2}, \ldots, \frac{1}{|\mathsf{h}[K-1]|^2}\right) \boldsymbol{U}\right]_{n,n} \quad (2.199)$$

$$= \frac{N_0}{G} \frac{1}{K} \sum_{k=0}^{K-1} e^{-\mathrm{j}\frac{2\pi}{K}nk} \frac{1}{|\mathsf{h}[k]|^2} e^{\mathrm{j}\frac{2\pi}{K}nk} \quad (2.200)$$

$$= \frac{N_0}{G} \frac{1}{K} \sum_{k=0}^{K-1} \frac{1}{|\mathsf{h}[k]|^2}, \quad (2.201)$$

which does not depend on n, revealing that the noise variance is constant over the output symbols; such variance is given by the average of the channel's inverted power frequency response. As seen later in the chapter, this is a point of differentiation with OFDM. This noise variance expression also reveals the noise-enhancement issue that arises with ZF channel inversion, something that can be alleviated by regularizing the strict inversion so as to equalize in the LMMSE sense. This is the subject of Problem 2.27.

SC-FDE offers advantages with respect to time-domain linear equalization, including the possibility of perfect channel inversion with an equalizer featuring a finite number of coefficients and a very attractive degree of computational complexity. Whereas, in linear equalization, the length of the equalizer grows with L, in SC-FDE the complexity is determined only by K. The breakpoint where SC-FDE becomes more efficient than linear equalization typically corresponds to rather small values of L, possibly as small as $L = 4$.

While our exposition has emphasized the use of copied data as a cyclic prefix, other types of prefixes may be used, including sequences of pilot symbols [153, 154] or simply zeroes [155, 156]. The alternative of pilot sequences, which when utilized as prefixes are termed *pilot words* in some WLAN standards, is the subject of Problem 2.29.

2.5.2 Extension to MIMO

Let us now extend the SC-FDE formulation to MIMO. At the transmitter, a cyclic prefix is added to the signal sent from each transmit antenna. At the receiver, there is a bank of $N_{\rm r}$ DFTs whose outputs are fed into K matrix equalizers; the K equalized vectors thus obtained are subsequently recombined and run through a bank of $N_{\rm t}$ IDFT transformations. A diagram of the entire process with ZF equalization is provided in Fig. 2.15.

The vector of prefixed signals

$$\bar{\boldsymbol{x}}[n] = \begin{cases} \boldsymbol{x}[n+K] & n = -L, \ldots, -1 \\ \boldsymbol{x}[n] & n = 0, \ldots, K-1 \end{cases} \tag{2.202}$$

engenders at the receiver, with the noise ignored,

$$\bar{\boldsymbol{y}}[n] = \sqrt{G} \sum_{\ell=0}^{L} \boldsymbol{H}[\ell]\, \bar{\boldsymbol{x}}[n-\ell] \qquad n = 0, \ldots, K-1. \tag{2.203}$$

Applying the same logic as in (2.193), we have that

$$\boldsymbol{y}[n] = \sqrt{G} \sum_{\ell=0}^{L} \boldsymbol{H}[\ell]\, \boldsymbol{x}\left[((n-\ell))_K\right] \qquad n = 0, \ldots, K-1, \tag{2.204}$$

which, in the frequency domain, corresponds to

$$\mathsf{y}[k] = \sqrt{G}\, \mathsf{H}[k]\, \mathsf{x}[k] \qquad k = 0, \ldots, K-1, \tag{2.205}$$

where $\mathsf{y}[k] = \mathrm{DFT}_K\{\boldsymbol{y}[n]\}$, $\mathsf{x}[k] = \mathrm{DFT}_K\{\boldsymbol{x}[n]\}$, and

$$\mathsf{H}[k] = \mathrm{DFT}_K\{\boldsymbol{H}[n]\} \tag{2.206}$$

$$= \sum_{\ell=0}^{L} \boldsymbol{H}[\ell]\, e^{-\mathrm{j}\frac{2\pi}{K}k\ell} \qquad k = 0, \ldots, K-1. \tag{2.207}$$

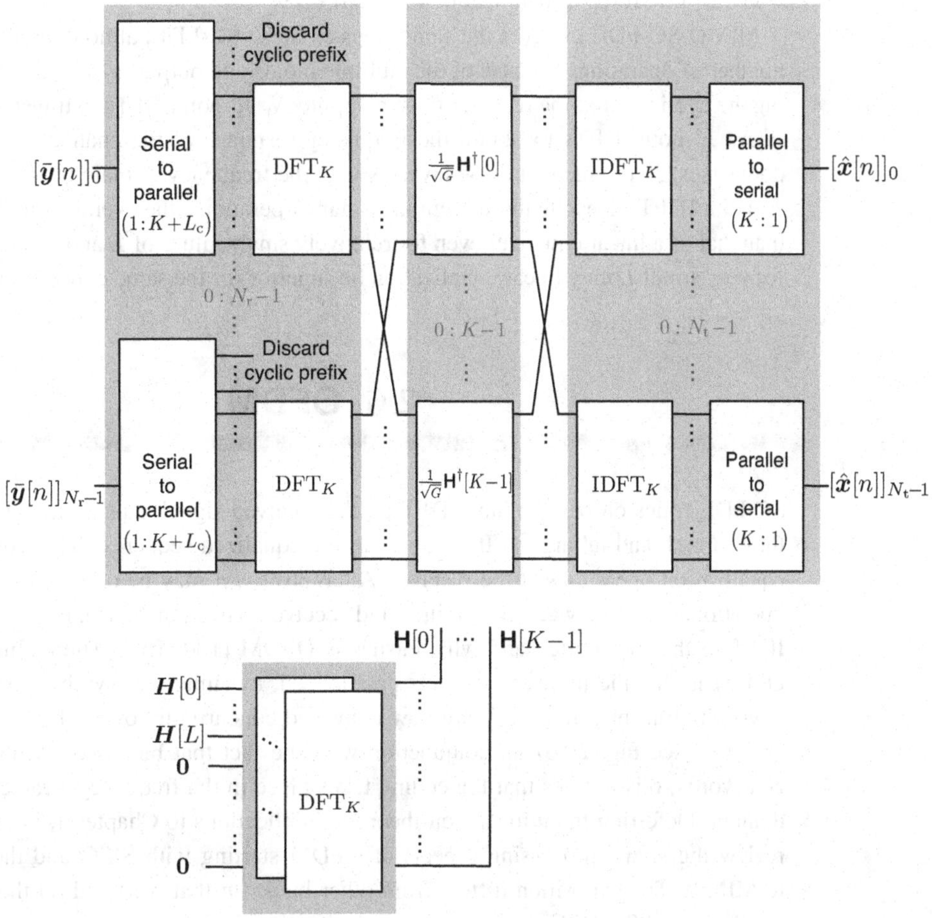

Fig. 2.15 Above, a MIMO SC-FDE equalizer. Below, bank of K-point DFTs of the channel matrix entries that are fed to the equalizer.

To perform ZF equalization, the receiver must compute

$$\hat{\boldsymbol{x}}[n] = \frac{1}{\sqrt{G}} \, \mathrm{IDFT}_K \left\{ \mathbf{H}^{\dagger}[k] \, \mathbf{y}[k] \right\} \qquad n = 0, \ldots, K-1, \tag{2.208}$$

which amounts to K channel inversions, one per discrete frequency. Alternatives to ZF are of course also possible, say LMMSE equalization on a per-frequency basis.

As far as the noise, for all but trivial channels it is spatially correlated at the equalizer's output. Its covariance depends on the specifics of the equalizer, but it is the same at all frequencies. With ZF, the output noise experienced by the signal sent from antenna j has variance

$$\frac{N_0}{G} \frac{1}{K} \sum_{k=0}^{K-1} \frac{1}{\left[\left(\mathbf{H}^*[k] \mathbf{H}[k] \right)^{-1} \right]_{j,j}} \tag{2.209}$$

which is the MIMO generalization of (2.201).

MIMO SC-FDE involves the same steps as SISO SC-FDE, although with an increased number of operations because of the multiple inputs and outputs [157, 158]. After discarding the cyclic prefix, the receiver must compute N_{r} K-point DFTs to transform the inputs, $N_{\mathrm{t}}N_{\mathrm{r}}$ K-point DFTs to obtain the frequency response of the channel, K matrix pseudoinverses, K products of those inverses with the frequency-domain observations, and N_{r} K-point IDFTs. Despite these seemingly many operations, the overall complexity is lower than that of a linear equalizer even for relatively small values of L and K [152, 159]. Only for very small L may linear equalization be superior for the same complexity.

2.6 OFDM

SC-FDE relies on performing a DFT on the received signal, equalizing with the DFT of the channel, and taking the IDFT to form the equalized sequence. This concentrates the equalization operations at the receiver. Alternatively, it may be of interest to balance the operational load between transmitter and receiver. This can be achieved by shifting the IDFT to the transmitter, and what ensues is OFDM [160, 161]. The defining feature of OFDM is that the transmission gets parallelized, meaning that symbols that would otherwise be transmitted serially are now conveyed concurrently over a bank of *subcarriers* or *tones* (see Fig. 2.16). In conjunction with the fact that blocks of symbols constitute codewords, this implies that the coding takes place in the frequency—rather than time—domain. Deferring the information-theoretic implications to Chapter 4, in this section we review the signal processing aspects of OFDM starting with SISO and then graduating to MIMO. The exposition differs from other books in that we build on the mathematics developed for SC-FDE.

2.6.1 Basic formulation

Reconsider the SC-FDE block diagram in Fig. 2.14. To convert it to OFDM, an IDFT is inserted after the serial-to-parallel converter at the transmitter while removing the IDFT from the receiver. The resulting diagram is illustrated in Fig. 2.17 and expounded next.

Transmitter

The symbols to be sent are treated as being in the frequency domain, denoted by $\mathsf{x}[k]$. The transmitter parses them into blocks of K and runs each block through an IDFT to produce

$$x[n] = \frac{1}{K} \sum_{k=0}^{K-1} \mathsf{x}[k]\, e^{\mathrm{j}\frac{2\pi}{K}nk} \qquad n = 0, \ldots, K-1, \qquad (2.210)$$

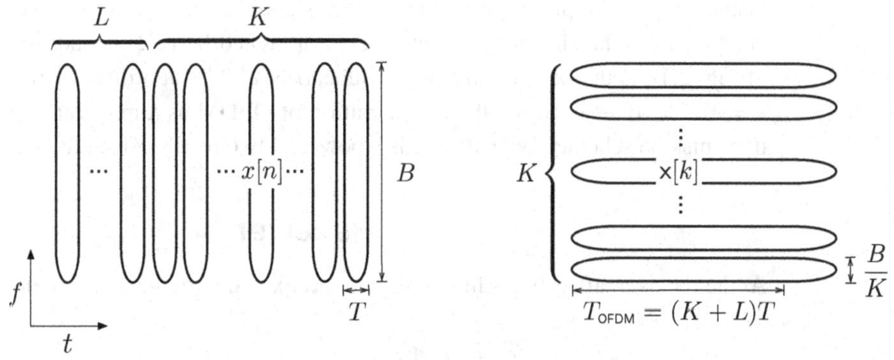

Fig. 2.16 Left, serial transmission of $K + L$ symbols in time. Right, parallel transmission of K subcarriers in frequency. (The nonoverlapping shapes are for illustration purposes only, and to emphasize the orthogonality in each domain; in actuality, symbols and subcarriers do overlap with their neighbors.) The footprint on the time–frequency plane is $(K + L)T \times B$ in both cases.

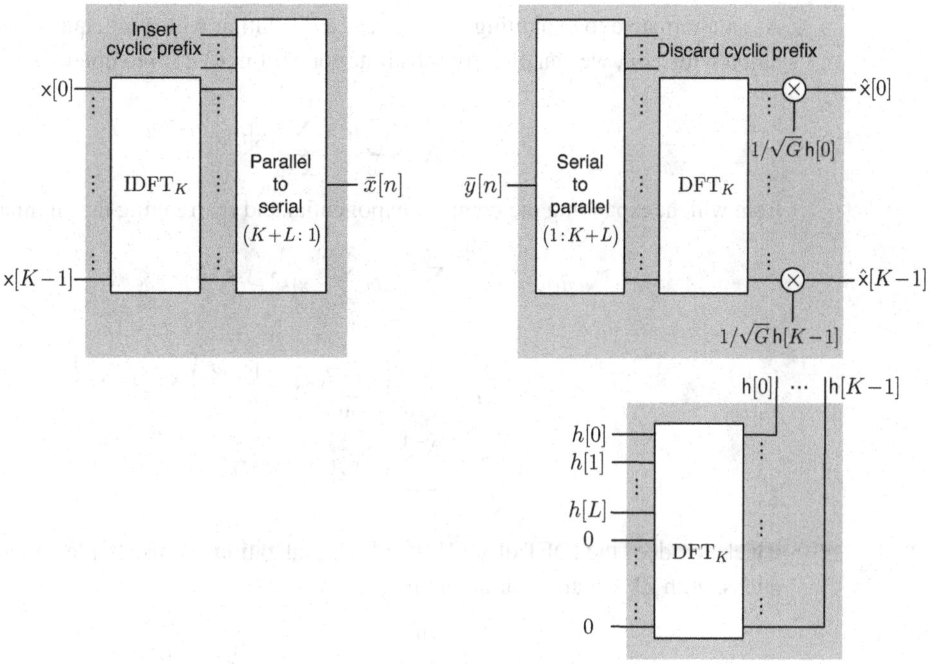

Fig. 2.17 SISO-OFDM block diagram with ZF equalization.

which, after insertion of an L-symbol cyclic prefix, becomes

$$\bar{x}[n] = \frac{1}{K} \sum_{k=0}^{K-1} \mathsf{x}[k]\, e^{\mathrm{j}\frac{2\pi}{K}nk} \qquad n = -L, ..., K-1. \tag{2.211}$$

Featuring a cyclic prefix, this time-domain signal is equipped to generate a circular convolution from the channel. The interpretation invited by (2.211) and illustrated at the top of Fig. 2.18 is that $\mathsf{x}[k]$ is sent over a subcarrier $e^{j\frac{2\pi}{K}kn}$ at a discrete frequency k/K, for $k = 0, \ldots, K-1$. Hence the consideration of OFDM as a multicarrier—or multitone—transmission scheme. Typically, K is a power of two to allow for the FFT to be employed.

Receiver

At the receiver, after discarding the cyclic prefix and with the noise neglected for now,

$$y[n] = \sqrt{G} \sum_{\ell=0}^{L} h[\ell]\, \bar{x}[n-\ell] \tag{2.212}$$

$$= \sqrt{G} \sum_{\ell=0}^{L} h[\ell]\, x[((n-\ell))_K] \qquad n = 0, \ldots, K-1, \tag{2.213}$$

which, in the frequency domain, corresponds to

$$\mathsf{y}[k] = \sqrt{G}\, \mathsf{h}[k]\, \mathsf{x}[k] \qquad k = 0, \ldots, K-1. \tag{2.214}$$

As an alternative to exploiting that a linear convolution with $\bar{x}[n]$ equals a circular convolution with $x[n]$, we can directly substitute for $\bar{x}[n]$ in (2.212) to obtain

$$y[n] = \frac{\sqrt{G}}{K} \sum_{\ell=0}^{L} h[\ell] \sum_{k=0}^{K-1} \mathsf{x}[k]\, e^{j\frac{2\pi}{K}k(n-\ell)} \tag{2.215}$$

from which, expanding the complex exponential and rearranging the summations,

$$y[n] = \frac{\sqrt{G}}{K} \sum_{\ell=0}^{L} h[\ell] \sum_{k=0}^{K-1} \mathsf{x}[k]\, e^{j\frac{2\pi}{K}kn}\, e^{-j\frac{2\pi}{K}k\ell} \tag{2.216}$$

$$= \frac{\sqrt{G}}{K} \sum_{k=0}^{K-1} \left(\sum_{\ell=0}^{L} h[\ell]\, e^{-j\frac{2\pi}{K}k\ell} \right) \mathsf{x}[k]\, e^{j\frac{2\pi}{K}kn} \tag{2.217}$$

$$= \frac{\sqrt{G}}{K} \sum_{k=0}^{K-1} \mathsf{h}[k]\, \mathsf{x}[k]\, e^{j\frac{2\pi}{K}kn}, \tag{2.218}$$

which is indeed the IDFT of $\sqrt{G}\, \mathsf{h}[k]\, \mathsf{x}[k]$. Equalization entails simple per-subcarrier operations, with ZF in particular amounting to

$$\hat{\mathsf{x}}[k] = \frac{\mathsf{y}[k]}{\sqrt{G}\, \mathsf{h}[k]} \qquad k = 0, \ldots, K-1. \tag{2.219}$$

The absence of ISI in the OFDM transmit–receive relationship indicates that each subcarrier experiences a frequency-flat channel, and this is one of the main justifications for the prevalence and claimed generality of frequency-flat analyses and algorithms.

Incorporating noise onto the formulation,

$$\mathsf{y}[k] = \sqrt{G}\, \mathsf{h}[k]\, \mathsf{x}[k] + \mathsf{v}[k], \tag{2.220}$$

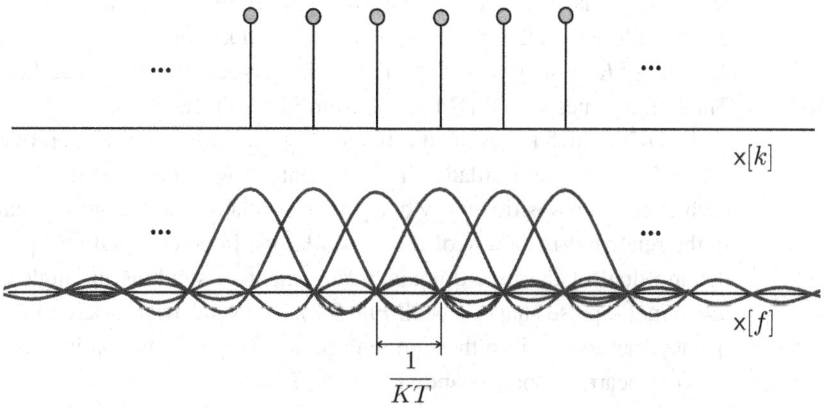

Fig. 2.18 Above, discrete-frequency signal $\mathsf{x}[k]$ where, for illustration purposes, each shown subcarrier is modulated by a $+1$ BPSK symbol. Below, corresponding continuous-frequency signal $\mathsf{x}(f)$ with a rectangular time-domain pulse shape.

where $\mathsf{v}[0], \ldots, \mathsf{v}[K-1]$ is IID with $\mathsf{v}[k] \sim \mathcal{N}_{\mathbb{C}}(0, N_0)$ because the IID complex Gaussian distribution of $v[0], \ldots, v[K-1]$ is invariant to unitary transformations such as a DFT (see Appendix C.1.6).

With ZF as in (2.219), the variance of the output noise on subcarrier k equals

$$\frac{N_0}{G} \frac{1}{|\mathsf{h}[k]|^2}, \tag{2.221}$$

which depends on k.

Pulse shaping

Before proceeding, and given that OFDM underlies many of the derivations in this text, it is appropriate that we establish some terminology:

- An IDFT sample, $x[n]$, is termed an OFDM *chip*. The interval between successive chips, the chip period, equals what the single-carrier symbol period would be in the absence of OFDM and is thus denoted by T.
- The number of subcarriers is K, the dimension of the IDFT/DFT.
- The collection of chips $\bar{x}[-L], \ldots, \bar{x}[K-1]$ forms an *OFDM symbol*.
- The *OFDM symbol period* is $T_{\text{OFDM}} = (K + L)T$.
- The *subcarrier spacing* indicates the frequency-domain displacement between adjacent subcarriers.

Prior to upconversion, the discrete-time OFDM symbols need to be turned into an analog waveform, for which purpose a pulse shape needs to be defined. Recalling the single-carrier pulse shaping discussion earlier in this chapter, the simplest starting point is to consider a rectangular shape. Excluding the cyclic prefix, which is added after the IDFT and discarded before the DFT, a rectangular pulse of duration KT endows the subcarriers with a $KT\,\mathrm{sinc}(fKT)$ shape and gives rise to the classical illustration at the bottom of Fig. 2.18. The attractiveness of OFDM stems from the tight interlacing of the subcarriers, courtesy of the DFT, which keeps them orthogonal yet closely packed. Moreover, all K subcarriers are tied to a single oscillator, in stark contrast with multicarrier schemes of yore where each subcarrier was driven by a separate oscillator and the orthogonality was vulnerable to the relative drift of any of those oscillators. In essence, OFDM produces a frequency-domain dual of the time-domain arrangement of symbols in single-carrier transmission with $\mathrm{sinc}(\cdot)$ pulse shapes (recall Fig. 2.8). The subcarrier spacing of $1/K$ in discrete frequency translates, given the sampling period T, to $\frac{1}{KT}$ in continuous frequency, and thus the K subcarriers span a bandwidth of $1/T$.

Example 2.22

Let $T_{\mathsf{OFDM}} = 3.2$ μs with $K = 256$ subcarriers and a cyclic prefix of length $L = 64$. Find the chip period T, the passband bandwidth $1/T$, the subcarrier spacing, and the guard interval spanned by the cyclic prefix.

Solution

The chip period T satisfies $(256 + 64)T = 3.2$ μs and thus $T = 10$ ns. Then, $1/T = 100$ MHz and the subcarrier spacing is $\frac{1}{KT} = 390.6$ kHz. The guard interval is $LT = 0.64$ μs and therefore channels with impulse responses of this duration can be handled.

Increasing K renders the subcarriers narrower and faster-decaying in frequency, and in that respect a larger number of subcarriers is advantageous, but nevertheless the overall signal is bound to spill outside the interval $1/T$ (refer to Problems 2.32 and 2.33), making it necessary to zero-out some edge subcarriers.

Precisely because of the possibility of turning off edge subcarriers, pulse shaping is less critical in OFDM than in single-carrier transmission. Nonetheless, to steepen the spectral decay and reduce the number of wasted subcarriers, the rectangular pulse shape in the time domain can be smoothed into a raised cosine, reinforcing the time–frequency duality between OFDM and single-carrier transmission. This can be done by multiplying the OFDM chips by coefficients chiseling every OFDM symbol into the desired shape, in a process termed *windowing*. Alternatively, the OFDM signal can be filtered before transmission, attenuating out-of-band power at the expense of compromising part of the cyclix prefix and thus the tolerance to channel dispersion [162].

Proceeding as in single-carrier transmission, we declare the passband bandwidth to be $B = 1/T$, with the understanding that there is inevitably some excess bandwidth—the idle edge subcarriers—and that any spectral efficiency expression derived in the book can be corrected by applying an appropriate penalty factor.

Example 2.23

In LTE, the subcarrier spacing is 15 kHz and every 5 MHz of bandwidth accommodates $K = 300$ subcarriers. What penalty factor should be applied to the spectral efficiency computed with $B = 1/T$?

Solution

Since $300 \cdot 15 \cdot 10^3$ Hz out of $5 \cdot 10^6$ Hz are used, the penalty factor equals $\frac{300 \cdot 15 \cdot 10^3}{5 \cdot 10^6} = 0.9$. For NR, this tightens up to 0.96.

Besides influencing the containment of the signal within the bandwidth $1/T$, the subcarrier spacing determines the sensitivity to time variations in the channel and to residual frequency offsets, which respectively distort and displace the subcarriers. It also determines the resolution of channel frequency selectivity. We provide guidelines on how to select the subcarrier spacing—which, for a given bandwidth, amounts to selecting K—in Chapter 4, once detailed models for the wireless channel are in place.

Comparisons abound between OFDM and single-carrier techniques [152, 159, 163]. OFDM is highly scalable and flexible, as each subcarrier can be loaded with a distinct power—even turned off by making that power zero—as well as modulated with a different constellation. These virtues come at the expense of a pronounced increase in peakedness, with the PAPR being essentially proportional to K, as well as enhanced sensitivity to RF impairments such as nonlinearity [164], frequency offsets [165], gain and phase imbalances between transmitter and receiver [166], and phase noise [167]. In terms of complexity, the overall volume of operations is similar in OFDM and in SC-FDE and, except for the computation of the channel frequency response, it is independent of L. This is advantageous over time-domain equalization where, as mentioned, complexity does grows with L. The OFDM complexity is balanced between transmitter and receiver, while in SC-FDE and in time-domain equalization it is heavily concentrated at the receiver.

For LTE, the balance of the foregoing tradeoffs led to the selection of SC-FDE and OFDM for the reverse and forward links, respectively. For NR, OFDM is employed in both directions.

Resource elements

As the parallelization effected by OFDM is only partial, and the signals are parceled in both time (OFDM symbols) and frequency (subcarriers), the most convenient canvas for their representation is the time–frequency plane introduced in Fig. 2.16 and that makes repeated appearances throughout the text. The basic unit that emerges from the simultaneous parceling in time and frequency is the *OFDM resource element*, which corresponds to an OFDM symbol in time and a subcarrier in frequency (see Fig. 2.19). Being the basic unit over which codewords can be constructed, the resource element is the direct generalization to OFDM of the notion of single-carrier symbol.

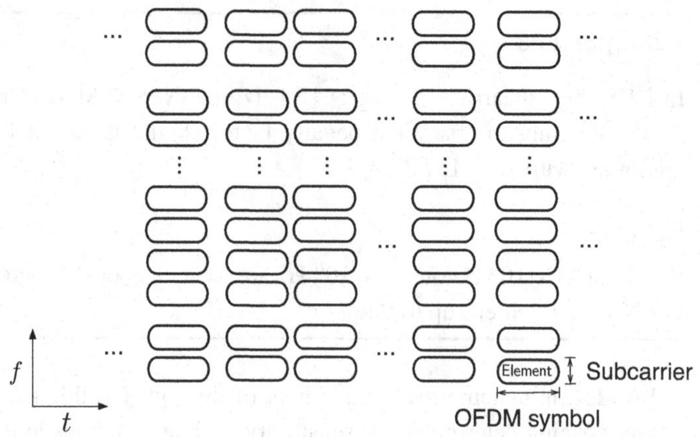

Fig. 2.19 OFDM symbols, subcarriers, and resource elements on the time–frequency plane.

2.6.2 Extension to MIMO

A block diagram of MIMO-OFDM is presented in Fig. 2.20. The signal to be transmitted from each antenna is subject to the OFDM procedures of IDFT and cyclic prefix insertion and then, at the receiver, removal of cyclic prefix followed by FFT; the difference with SISO appears when it comes to the equalization. The MIMO counterpart to (2.214) is

$$\mathbf{y}[k] = \sqrt{G}\,\mathbf{H}[k]\,\mathbf{x}[k] + \mathbf{v}[k] \qquad k = 0, \ldots, K-1, \tag{2.222}$$

which, on a subcarrier basis, does not differ from the frequency-flat MIMO channels for which equalizers are derived in this chapter, say the ZF equalizer in Example 2.18 or the LMMSE equalizer in Example 2.21. This justifies the anticipated emphasis on frequency-flat channels throughout this text and in much of the research in this area. Because of the simple equalization that it enables, also with MIMO, a majority of wireless and wireline standards have come to adopt OFDM [168–170].

2.7 Channel estimation

Equalization and coherent data detection require CSI at the receiver. While blind algorithms are available that exploit the structure of the signal to gather information about the channel (see, e.g., [171–174]), most wireless systems rely on pilot-assisted channel estimation. The defining feature of this approach is the insertion of known pilot symbols, also termed *reference* or *training* symbols, within the transmit data [175–177]. The receiver explicitly estimates the channel coefficients on the basis of the corresponding observations and then applies those estimates to tasks such as equalization and data detection; as a by-product, pilot symbols facilitate other procedures such as synchronization.

In this section, we take a first look at the issue of estimating a MIMO channel on the

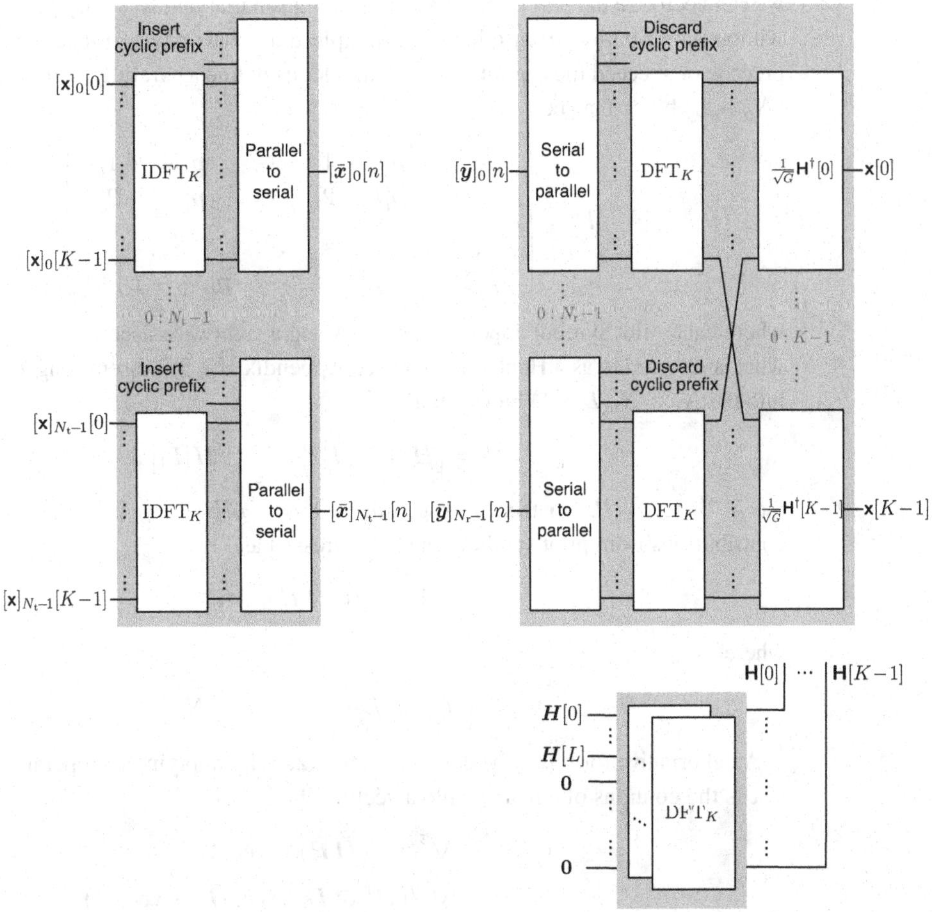

Fig. 2.20 MIMO-OFDM block diagram with ZF equalization.

basis of pilot observations for both single-carrier and OFDM transmission. Some of the derivations that follow are then retaken in Section 3.7, with the benefit of the time-varying wireless channel models presented throughout Chapter 3, so as to obtain relevant expressions for these models that then sprinkle the rest of the book.

2.7.1 Single-carrier channel estimation

Consider a sequence of known time-domain pilots $\boldsymbol{p}[0], \ldots, \boldsymbol{p}[N_{\rm p}-1]$ where $N_{\rm p} \geq 1$ denotes the number of pilots, each an $N_{\rm t} \times 1$ vector. These pilots are transmitted through an L-tap MIMO channel $\sqrt{G}\boldsymbol{H}[\ell]$, for $\ell = 0, \ldots, L$, where G is known but the matrices $\boldsymbol{H}[0], \ldots, \boldsymbol{H}[L]$ are not. From observations of the received signal and the knowledge of the pilots, plus whatever is known about the channel and noise distributions, optimization problems can be formulated whose solutions are the channel estimates $\hat{\boldsymbol{H}}[0], \ldots, \hat{\boldsymbol{H}}[L]$.

Recalling from Section 2.3.3 how convolutions can be formulated through vectorization, we can vectorize the part of the channel convolution that involves only $\boldsymbol{p}[0],\ldots,\boldsymbol{p}[N_{\mathrm{p}}-1]$ without contribution from other (in principle unknown) payload data symbols that may precede or succeed these pilots. To that end, let us define a carefully curtailed $N_{\mathrm{t}}(L+1)\times(N_{\mathrm{p}}-L)$ block matrix

$$\bar{\boldsymbol{P}}=\begin{bmatrix}\boldsymbol{p}[L] & \boldsymbol{p}[L+1] & \cdots & \boldsymbol{p}[N_{\mathrm{p}}-1]\\ \boldsymbol{p}[L+1] & \boldsymbol{p}[L+2] & \cdots & \boldsymbol{p}[N_{\mathrm{p}}-2]\\ \vdots & \cdot^{\cdot^{\cdot}} & \cdot^{\cdot^{\cdot}} & \vdots\\ \boldsymbol{p}[0] & \boldsymbol{p}[1] & \cdots & \boldsymbol{p}[N_{\mathrm{p}}-L-1]\end{bmatrix}, \tag{2.223}$$

where each pilot symbol appears repeated along a rightward-ascending diagonal, forming what is referred to as a Hankel matrix (see Appendix B.2.5). Assembling $\boldsymbol{H}[0],\ldots,\boldsymbol{H}[L]$ into the $N_{\mathrm{r}}\times N_{\mathrm{t}}(L+1)$ block matrix

$$\bar{\boldsymbol{H}}=\begin{bmatrix}\boldsymbol{H}[0] & \boldsymbol{H}[1] & \cdots & \boldsymbol{H}[L]\end{bmatrix}, \tag{2.224}$$

the $N_{\mathrm{r}}\times(N_{\mathrm{p}}-L)$ matrix of observations $\bar{\boldsymbol{Y}}=\begin{bmatrix}\boldsymbol{y}[L]\cdots\boldsymbol{y}[N_{\mathrm{p}}-1]\end{bmatrix}$ containing only contributions from pilot symbols can be expressed as

$$\bar{\boldsymbol{Y}}=\sqrt{G}\,\bar{\boldsymbol{H}}\bar{\boldsymbol{P}}+\bar{\boldsymbol{V}}, \tag{2.225}$$

where

$$\bar{\boldsymbol{V}}=\begin{bmatrix}\boldsymbol{v}[L] & \boldsymbol{v}[L+1] & \cdots & \boldsymbol{v}[N_{\mathrm{p}}-1]\end{bmatrix}. \tag{2.226}$$

An alternative form may be derived from (2.225) by applying the operator $\mathrm{vec}(\cdot)$, which stacks the columns of a matrix onto a vector. This gives

$$\begin{aligned}\mathrm{vec}(\bar{\boldsymbol{Y}}) &= \sqrt{G}\,\mathrm{vec}(\bar{\boldsymbol{H}}\bar{\boldsymbol{P}})+\mathrm{vec}(\bar{\boldsymbol{V}}) && (2.227)\\ &= \sqrt{G}(\bar{\boldsymbol{P}}^{\mathrm{T}}\otimes\boldsymbol{I}_{N_{\mathrm{r}}})\mathrm{vec}(\bar{\boldsymbol{H}})+\mathrm{vec}(\bar{\boldsymbol{V}}) && (2.228)\\ &= \sqrt{G}\bar{\boldsymbol{P}}_{\otimes}\,\mathrm{vec}(\bar{\boldsymbol{H}})+\mathrm{vec}(\bar{\boldsymbol{V}}), && (2.229)\end{aligned}$$

where $\otimes$ denotes the Kronecker product (see Appendix B.8) such that $\bar{\boldsymbol{P}}_{\otimes}=\bar{\boldsymbol{P}}^{\mathrm{T}}\otimes\boldsymbol{I}_{N_{\mathrm{r}}}$ is an $N_{\mathrm{r}}(N_{\mathrm{p}}-L)\times N_{\mathrm{t}}N_{\mathrm{r}}(L+1)$ matrix. In (2.228) we used $\mathrm{vec}(\boldsymbol{ABC})=(\boldsymbol{C}^{\mathrm{T}}\otimes\boldsymbol{A})\mathrm{vec}(\boldsymbol{B})$.

Maximum-likelihood and least-squares channel estimation

Let us begin by assuming that the conditional distribution of the received signal given the channel is known. This amounts to knowing that the channel is linear and contaminated by AWGN, i.e., that the transmit–receive relationship abides by (2.229). No other statistical assumptions are made for now, and in particular the distribution of the channel itself need not be known. The conditional distribution of the received signal given the channel suffices for ML estimation, a classical approach mentioned briefly in Chapter 1 that selects as an estimate the quantity with the highest likelihood. The ML estimate converges in probability (see Appendix C.1.10) to its true value as the number of observations grows large [115].

Consider (2.229). Since $\bar{\boldsymbol{P}}_{\otimes}$ is known, the observation elicited by a certain channel realization $\mathbf{h}$ is complex Gaussian with mean $\sqrt{G}\bar{\boldsymbol{P}}_{\otimes}\mathbf{h}$ and covariance $N_0\boldsymbol{I}$. The likelihood function we need is the corresponding PDF, to wit

$$f_{\mathrm{vec}(\bar{Y})|\mathrm{vec}(\bar{H})}(\mathbf{y}|\mathbf{h}) = \frac{1}{(\pi N_0)^{(N_\mathrm{p}-L)N_\mathrm{r}}} \exp\left(-\frac{\|\mathbf{y} - \sqrt{G}\bar{P}_\otimes \mathbf{h}\|^2}{N_0} \right). \tag{2.230}$$

The ML estimate of $\mathrm{vec}(\bar{H})$ given the observation $\mathbf{y}$ is

$$\mathrm{vec}(\hat{\bar{H}})(\mathbf{y}) = \arg\max_{\mathbf{h}} f_{\mathrm{vec}(\bar{Y})|\mathrm{vec}(\bar{H})}(\mathbf{y}|\mathbf{h}), \tag{2.231}$$

where the maximization is over all vectors $\mathbf{h}$ of dimension $N_\mathrm{t}N_\mathrm{r}(L+1)$. Since the exponential function is monotonic, the above is equivalent to minimizing $-\log f_{\mathrm{vec}(\bar{Y})|\mathrm{vec}(\bar{H})}(\cdot)$. Neglecting constants, which are immaterial to the minimization, this leads to

$$\mathrm{vec}(\hat{\bar{H}})(\mathbf{y}) = \arg\min_{\mathbf{h}} \|\mathbf{y} - \sqrt{G}\bar{P}_\otimes \mathbf{h}\|^2. \tag{2.232}$$

This least-squares minimization appears in Section 2.4.1, in the context of ZF equalization, and if $\bar{P}_\otimes$ is tall (or at least square) and full-rank, then the solution is given by

$$\mathrm{vec}(\hat{\bar{H}})(\mathbf{y}) = \underbrace{\frac{1}{\sqrt{G}}(\bar{P}_\otimes^* \bar{P}_\otimes)^{-1}\bar{P}_\otimes^*}_{W^{\mathrm{LS}*}} \mathbf{y}, \tag{2.233}$$

which amounts to applying the linear estimator $W^{\mathrm{LS}} = \frac{1}{\sqrt{G}}\bar{P}_\otimes(\bar{P}_\otimes^* \bar{P}_\otimes)^{-1}$ to the observation. Since $\bar{P}_\otimes$ is $N_\mathrm{r}(N_\mathrm{p}-L) \times N_\mathrm{t}N_\mathrm{r}(L+1)$, ensuring that it is square or tall requires

$$N_\mathrm{p} \geq (L+1)N_\mathrm{t} + L, \tag{2.234}$$

while ensuring that $\bar{P}_\otimes$ is full-rank requires that the constituent matrix $\bar{P}$ be itself full-rank, something that can be achieved through careful design of the pilot sequence. Estimators that solve problems of the form of (2.232) are known, because of the structure of those problems, as *least-squares estimators*, and they are relatively easy to compute using linear algebra [178, chapter 5]. Least-squares estimators thus happen to be, owing to the form of the complex Gaussian PDF, also ML estimators in channels contaminated by AWGN; this is a welcome occurrence.

Problem 2.36 proposes verifying that, for a channel realization $\mathbf{h}$, the least-squares estimate satisfies

$$\mathbb{E}\Big[\mathrm{vec}(\hat{\bar{H}})\,|\mathrm{vec}(\bar{H}){=}\mathbf{h}\Big] = \mathbf{h}, \tag{2.235}$$

meaning that the estimation error is zero-mean and there is no bias. Over the distributions of channel and noise, the mean-square error matrix is

$$\begin{aligned} E &= \mathbb{E}\Big[\big(\mathrm{vec}(\bar{H}) - \mathrm{vec}(\hat{\bar{H}})\big)\big(\mathrm{vec}(\bar{H}) - \mathrm{vec}(\hat{\bar{H}})\big)^*\Big] && (2.236)\\ &= \frac{1}{G}\mathbb{E}\Big[\big((\bar{P}_\otimes^* \bar{P}_\otimes)^{-1}\bar{P}_\otimes^* \mathrm{vec}(\bar{V})\big)\big((\bar{P}_\otimes^* \bar{P}_\otimes)^{-1}\bar{P}_\otimes^* \mathrm{vec}(\bar{V})\big)^*\Big] && (2.237)\\ &= \frac{1}{G}(\bar{P}_\otimes^* \bar{P}_\otimes)^{-1}\bar{P}_\otimes^*\, \mathbb{E}\big[\mathrm{vec}(\bar{V})\mathrm{vec}(\bar{V})^*\big]\bar{P}_\otimes(\bar{P}_\otimes^* \bar{P}_\otimes)^{-1} && (2.238)\\ &= \frac{N_0}{G}(\bar{P}_\otimes^* \bar{P}_\otimes)^{-1}. && (2.239) \end{aligned}$$

To unravel $\text{vec}(\hat{\bar{H}})$ and express it as a function of the original $\bar{P}$ rather than $\bar{P}_\otimes$, we can exploit certain properties of the Kronecker product (see Appendix B.8) and write

$$(\bar{P}_\otimes^* \bar{P}_\otimes)^{-1} \bar{P}_\otimes^* = \left((\bar{P}^{\mathrm{T}} \otimes I)^* (\bar{P}^{\mathrm{T}} \otimes I)\right)^{-1} (\bar{P}^{\mathrm{T}} \otimes I)^* \tag{2.240}$$
$$= \left((\bar{P}^{\mathrm{c}} \otimes I)(\bar{P}^{\mathrm{T}} \otimes I)\right)^{-1} (\bar{P}^{\mathrm{c}} \otimes I) \tag{2.241}$$
$$= \left((\bar{P}^{\mathrm{c}} \bar{P}^{\mathrm{T}} \otimes I)\right)^{-1} (\bar{P}^{\mathrm{c}} \otimes I) \tag{2.242}$$
$$= \left((\bar{P}^{\mathrm{c}} \bar{P}^{\mathrm{T}})^{-1} \otimes I\right)(\bar{P}^{\mathrm{c}} \otimes I) \tag{2.243}$$
$$= (\bar{P}^{\mathrm{c}} \bar{P}^{\mathrm{T}})^{-1} \bar{P}^{\mathrm{c}} \otimes I, \tag{2.244}$$

which, in conjunction with (2.233), gives

$$\text{vec}(\hat{\bar{H}}) = \frac{1}{\sqrt{G}} \left((\bar{P}^{\mathrm{c}} \bar{P}^{\mathrm{T}})^{-1} \bar{P}^{\mathrm{c}} \otimes I\right) \text{vec}(\bar{Y}). \tag{2.245}$$

It is also possible to unvectorize the estimate and put it in terms of the unstacked matrix $\bar{H}$ rather than $\text{vec}(\bar{H})$. Using again $\text{vec}(ABC) = (C^{\mathrm{T}} \otimes A)\text{vec}(B)$, only this time backwards, we obtain

$$\hat{\bar{H}} = \frac{1}{\sqrt{G}} \bar{Y} \left((\bar{P}^{\mathrm{c}} \bar{P}^{\mathrm{T}})^{-1} \bar{P}^{\mathrm{c}}\right)^{\mathrm{T}} \tag{2.246}$$
$$= \frac{1}{\sqrt{G}} \bar{Y} \bar{P}^* (\bar{P} \bar{P}^*)^{-1}, \tag{2.247}$$

which admits the interpretation of $\bar{Y}\bar{P}^*$ correlating the observations with the pilot sequence while $(\bar{P}^*\bar{P})^{-1}$ removes from the ensuing channel estimate the autocorrelation of that very pilot sequence. This relates to the LMMSE channel estimate discussed later in the section.

In comparison with (2.233), the form in (2.247) entails the inversion of a smaller matrix as well as fewer scalar products. Moreover, since the pilot sequence is known, $\bar{P}^*(\bar{P}\bar{P}^*)^{-1}$ can be precomputed and the evaluation of the estimate only requires a matrix multiplication for each $i = 0, \ldots, N_{\mathrm{r}} - 1$. Also notice that the ith row of $\bar{H}$ corresponds to the channel coefficients between the N_{t} transmit and the ith receive antenna, while the ith row of $\bar{Y}$ corresponds to the observation at the ith receive antenna. From (2.247) then, each row of $\bar{H}$ can be estimated simultaneously and independently of the other rows.

Not all pilot sequences are equally effective and it has been shown that, for least-squares channel estimation in AWGN, it is best if $\bar{P}\bar{P}^*$ is a scaled identity matrix with the largest scaling factor allowed by the transmit power constraint (refer to Problem 2.39). In light of the Hankel form of $\bar{P}$, the product $\bar{P}\bar{P}^*$ contains partial autocorrelations as well as cross-correlations and thus an ideal pilot sequence requires perfect partial autocorrelation and zero partial cross-correlation. This is sensible, as correlations in the pilot sequence would reduce the information gathered by the receiver while power imbalances would improve the estimation of some coefficients at the expense of others. Structures that are attractive because of their correlation properties include Gold codes and Zadoff–Chu sequences [179–181]. For the latter in particular, and with a prime value of N_{p}, so-called Zadoff–Chu root sequences can be obtained as follows. The nth symbol of the jth root sequence is

$$e^{-j\pi n(n+1)j/N_p} \qquad n = 0, \ldots, N_p - 1 \qquad j = 1, \ldots, N_p - 1. \tag{2.248}$$

Each length-N_p root sequence is of unit-magnitude and, most importantly, orthogonal to any cyclic version of itself; therefore, from each root sequence a family of orthogonal pilot sequences can be obtained through mere shifts. Moreover, the cross-correlation between distinct root sequences is well-behaved and the DFT of a Zadoff–Chu sequence is also a (scaled) Zadoff–Chu sequence. These attractive properties have led Zadoff–Chu sequences to be adopted by LTE, among other standards.

Example 2.24 (Least-squares estimator for a frequency-flat SISO channel)

Let $L = 0$ and $N_t = N_r = 1$. For this simplest of settings, a single scalar pilot suffices. With such a pilot being $\sqrt{E_s}e^{j\phi}$ where ϕ is an arbitrary phase, (2.247) reduces to

$$\hat{h} = \frac{e^{-j\phi}}{\sqrt{GE_s}}\, y \tag{2.249}$$

while $\boldsymbol{E}$ in (2.239) reduces to

$$\mathsf{MMSE} = \frac{N_0}{GE_s} \tag{2.250}$$

$$= \frac{1}{\mathsf{SNR}}, \tag{2.251}$$

which does not depend on ϕ; the pilot's phase can thus be set to zero.

For $N_p > 1$, $\bar{\boldsymbol{P}}$ becomes a row vector and the pilot sequence requires no special structure: in this setting, $\bar{\boldsymbol{P}}\bar{\boldsymbol{P}}^*$ is always a scalar, even if the same pilot symbol is simply repeated. It is easy to verify that

$$\mathsf{MMSE} = \frac{1}{N_p \mathsf{SNR}}, \tag{2.252}$$

indicating that the accrual of energy over multiple pilots has the same effect as increasing the energy of a single pilot.

Example 2.25

Let $N_t = N_r = 1$, but now with arbitrary L. Show how a sequence with perfect periodic correlation can be utilized to construct a pilot sequence of the desired length N_p.

Solution

Consider a sequence $x[0], \ldots, x[Q-1]$ with perfect periodic cross correlation, i.e., with

$$\sum_{n=0}^{Q-1} x^*[((n-\ell))_Q]\, x[n] = Q\delta[n]. \tag{2.253}$$

Now, let $N_p = Q + L$ and create a pilot sequence by adding to $x[0], \ldots, x[Q-1]$ a cyclic prefix of length L. The corresponding $\bar{\boldsymbol{P}}$ defined as per (2.223) satisfies $\bar{\boldsymbol{P}}\bar{\boldsymbol{P}}^* = Q\boldsymbol{I}$. Scaled up to the maximum permissible transmit power, such $\bar{\boldsymbol{P}}$ conforms to the desiderata.

The idea in this example extends to MIMO, and readers interested in this generalization are referred to [182, 183].

Example 2.26 (Least-squares estimator for a frequency-flat MIMO channel)

For $L = 0$ with the channel embodied by the $N_\mathrm{r} \times N_\mathrm{t}$ matrix $\boldsymbol{H}$, we have that $\bar{\boldsymbol{H}} = \boldsymbol{H}$ directly while $\bar{\boldsymbol{P}} = \left[\boldsymbol{p}[0] \cdots \boldsymbol{p}[N_\mathrm{p} - 1]\right]$ is $N_\mathrm{t} \times N_\mathrm{p}$ and $\bar{\boldsymbol{Y}} = \left[\boldsymbol{y}[0] \cdots \boldsymbol{y}[N_\mathrm{p} - 1]\right]$ is $N_\mathrm{r} \times N_\mathrm{p}$. Then, under the condition $N_\mathrm{p} \geq N_\mathrm{t}$, (2.247) reduces to

$$\hat{\boldsymbol{H}} = \frac{1}{\sqrt{G}} \frac{\sum_{n=0}^{N_\mathrm{p}-1} \boldsymbol{y}[n]\, \boldsymbol{p}^*[n]}{\sum_{n=0}^{N_\mathrm{p}-1} \|\boldsymbol{p}[n]\|^2}, \tag{2.254}$$

where the numerator correlates the observations with the pilot sequence while the denominator normalizes by the energy of the latter.

In a frequency-flat channel, the jth row of $\bar{\boldsymbol{P}}$ is the pilot sequence sent from the jth transmit antenna and the condition that $\bar{\boldsymbol{P}}\bar{\boldsymbol{P}}^*$ be proportional to the identity translates to those N_t sequences being orthogonal.

Example 2.27

Design a set of N_t orthogonal pilot sequences of length $N_\mathrm{p} \geq N_\mathrm{t}$.

Solution

Select N_t out of the N_p rows of an $N_\mathrm{p} \times N_\mathrm{p}$ unitary matrix (see Appendix B) and scale as allowed by the applicable power constraint.

The least-squares estimator can be generalized to settings where $\bar{\boldsymbol{P}}$ may not be fully known ahead of time. For instance, detected data symbols can be treated as additional pilots in what is known as data-aided estimation [184]. In this case, $\bar{\boldsymbol{P}}^*(\bar{\boldsymbol{P}}\bar{\boldsymbol{P}}^*)^{-1}$ cannot be precomputed because the data part is unknown. Adaptive algorithms such as the recursive least-squares, normalized least-squares, and least mean-squares algorithms can be employed in this case [185]. Adaptive algorithms are extensible to the case where the channel is time-varying and the estimate needs to be updated periodically.

Variations of the least-square solution may also be entertained, for instance a scaled version of the estimator may achieve a lower mean-square error at the expense of such estimator no longer being unbiased for every channel realization. Readers interested in how to tweak such fixed scaling are referred to [183].

LMMSE channel estimation

The least-squares channel estimator that we obtained by applying the ML criterion only involves prior knowledge of the pilot sequence and the observation of that sequence at the receiver. The statistics of the channel do not enter the formulation and yet, as elaborated in Chapter 3, those statistics are rather stable and thus they can be known with relative certitude. This opens the door to formulating an MMSE channel estimator, in which the estimate of each channel realization is optimally biased such that the mean-square error is minimized. The expectation of these biases over all channel realizations, however, is still zero.

Particularly accessible are the second-order channel statistics, based on which an LMMSE estimator can be implemented. To continue subsuming both MIMO and frequency selectivity, we stay with the vectorized transmit–receive relationship in (2.229), namely

$$\text{vec}(\bar{\boldsymbol{Y}}) = \sqrt{G}\bar{\boldsymbol{P}}_{\otimes}\,\text{vec}(\bar{\boldsymbol{H}}) + \text{vec}(\bar{\boldsymbol{V}}). \tag{2.255}$$

Since the mean of a quantity being estimated does not influence the MMSE, we can restrict ourselves to zero-mean channels for which $\boldsymbol{R}_{\text{vec}(\bar{\boldsymbol{H}})} = \mathbb{E}\big[\text{vec}(\bar{\boldsymbol{H}})\,\text{vec}(\bar{\boldsymbol{H}})^*\big]$ is known. Then, from $\boldsymbol{R}_{\text{vec}(\bar{\boldsymbol{V}})} = N_0\boldsymbol{I}$, also $\boldsymbol{R}_{\text{vec}(\bar{\boldsymbol{Y}})}$ and $\boldsymbol{R}_{\text{vec}(\bar{\boldsymbol{Y}})\text{vec}(\bar{\boldsymbol{H}})}$ are known.

Exercising the teachings of Section 1.7, the LMMSE channel estimate is

$$\text{vec}(\hat{\bar{\boldsymbol{H}}}) = \boldsymbol{W}^{\text{MMSE}*}\text{vec}(\bar{\boldsymbol{Y}}) \tag{2.256}$$

where, applying (1.244),

$$\boldsymbol{W}^{\text{MMSE}} = \boldsymbol{R}^{-1}_{\text{vec}(\bar{\boldsymbol{Y}})}\boldsymbol{R}_{\text{vec}(\bar{\boldsymbol{Y}})\text{vec}(\bar{\boldsymbol{H}})} \tag{2.257}$$

$$= \sqrt{G}\left(G\bar{\boldsymbol{P}}_{\otimes}\boldsymbol{R}_{\text{vec}(\bar{\boldsymbol{H}})}\bar{\boldsymbol{P}}^*_{\otimes} + N_0\boldsymbol{I}\right)^{-1}\bar{\boldsymbol{P}}_{\otimes}\boldsymbol{R}_{\text{vec}(\bar{\boldsymbol{H}})} \tag{2.258}$$

or, equivalently, applying (1.250),

$$\boldsymbol{W}^{\text{MMSE}} = \sqrt{G}\bar{\boldsymbol{P}}_{\otimes}\left(N_0\boldsymbol{R}^{-1}_{\text{vec}(\bar{\boldsymbol{H}})} + G\bar{\boldsymbol{P}}^*_{\otimes}\bar{\boldsymbol{P}}_{\otimes}\right)^{-1} \tag{2.259}$$

while, from (1.245), the MMSE matrix is

$$\boldsymbol{E} = \mathbb{E}\Big[\big(\text{vec}(\bar{\boldsymbol{H}}) - \text{vec}(\hat{\bar{\boldsymbol{H}}})\big)\big(\text{vec}(\bar{\boldsymbol{H}}) - \text{vec}(\hat{\bar{\boldsymbol{H}}})\big)^*\Big] \tag{2.260}$$

$$= \boldsymbol{R}_{\text{vec}(\bar{\boldsymbol{H}})} - \boldsymbol{R}^*_{\text{vec}(\bar{\boldsymbol{Y}})\text{vec}(\bar{\boldsymbol{H}})}\boldsymbol{R}^{-1}_{\text{vec}(\bar{\boldsymbol{Y}})}\boldsymbol{R}_{\text{vec}(\bar{\boldsymbol{Y}})\text{vec}(\bar{\boldsymbol{H}})} \tag{2.261}$$

$$= \boldsymbol{R}_{\text{vec}(\bar{\boldsymbol{H}})} - G\boldsymbol{R}^*_{\text{vec}(\bar{\boldsymbol{H}})}\bar{\boldsymbol{P}}^*_{\otimes}\left(G\bar{\boldsymbol{P}}_{\otimes}\boldsymbol{R}_{\text{vec}(\bar{\boldsymbol{H}})}\bar{\boldsymbol{P}}^*_{\otimes} + N_0\boldsymbol{I}\right)^{-1}\bar{\boldsymbol{P}}_{\otimes}\boldsymbol{R}_{\text{vec}(\bar{\boldsymbol{H}})}. \tag{2.262}$$

Example 2.28

Verify that, for a given realization of a zero-mean channel, the LMMSE estimate exhibits a bias and that the expectation of this bias over the channel distribution is zero.

Solution

For $\text{vec}(\bar{\boldsymbol{H}}) = \text{h}$, applying the LMMSE estimator form in (1.250),

$$\mathbb{E}\Big[\text{vec}(\hat{\bar{\boldsymbol{H}}})\,|\text{vec}(\bar{\boldsymbol{H}})=\text{h}\Big] = \mathbb{E}\Big[\boldsymbol{W}^{\text{MMSE}*}\big(\sqrt{G}\bar{\boldsymbol{P}}_{\otimes}\text{vec}(\bar{\boldsymbol{H}}) + \text{vec}(\bar{\boldsymbol{V}})\big)\,|\text{vec}(\bar{\boldsymbol{H}})=\text{h}\Big]$$

$$= \mathbb{E}\Bigg[\left(\frac{N_0}{G}\boldsymbol{R}^{-1}_{\text{vec}(\bar{\boldsymbol{H}})} + \bar{\boldsymbol{P}}^*_{\otimes}\bar{\boldsymbol{P}}_{\otimes}\right)^{-1}\bar{\boldsymbol{P}}^*_{\otimes}\bar{\boldsymbol{P}}_{\otimes}\,\text{vec}(\bar{\boldsymbol{H}}) \tag{2.263}$$

$$+ \sqrt{G}\left(N_0\boldsymbol{R}^{-1}_{\text{vec}(\bar{\boldsymbol{H}})} + G\bar{\boldsymbol{P}}^*_{\otimes}\bar{\boldsymbol{P}}_{\otimes}\right)^{-1}\bar{\boldsymbol{P}}^*_{\otimes}\,\text{vec}(\bar{\boldsymbol{V}})\,|\text{vec}(\bar{\boldsymbol{H}})=\text{h}\Bigg]$$

$$= \underbrace{\left(\frac{N_0}{G} \boldsymbol{R}^{-1}_{\text{vec}(\bar{\boldsymbol{H}})} + \bar{\boldsymbol{P}}^*_{\otimes} \bar{\boldsymbol{P}}_{\otimes} \right)^{-1} \bar{\boldsymbol{P}}^*_{\otimes} \bar{\boldsymbol{P}}_{\otimes}}_{\neq \boldsymbol{I}} \, \mathrm{h} \tag{2.264}$$

$$\neq \mathrm{h} \tag{2.265}$$

while $\mathbb{E}\big[\text{vec}(\hat{\bar{\boldsymbol{H}}})\big] = \mathbb{E}\big[\text{vec}(\bar{\boldsymbol{H}})\big] = \boldsymbol{0}$.

While the least-squares form for $\text{vec}(\hat{\bar{\boldsymbol{H}}})$ in (2.233) can be unvectorized—by unraveling its underlying Kronecker structure—and written in terms of $\hat{\bar{\boldsymbol{H}}}$ as in (2.247), in general the LMMSE estimator in (2.256) does not enable such unvectorization. Because it exploits potential correlations in the channel coefficients, the LMMSE estimator cannot generally be applied separately to the observations at each receive antenna. In some relevant special cases though, when correlations within the channel are absent or they abide by certain structures, simplification is forthcoming. For SISO settings also, the LMMSE formulation becomes illuminatingly simple.

Example 2.29 (LMMSE estimation for a frequency-flat SISO channel)

For $L = 0$ and $N_\text{t} = N_\text{r} = 1$, a single pilot symbol suffices. With such pilot symbol having energy E_s, the LMMSE estimator reduces to

$$\hat{h} = \frac{\sqrt{GE_\text{s}}}{N_0 + GE_\text{s}} \, y \tag{2.266}$$

$$= \frac{\mathsf{SNR}}{1+\mathsf{SNR}} \frac{1}{\sqrt{GE_\text{s}}} \, y, \tag{2.267}$$

whereas the MMSE matrix condenses down to

$$\mathsf{MMSE} = \frac{1}{1+\mathsf{SNR}}, \tag{2.268}$$

which is always lower than its least-squares cousin, $\mathsf{MMSE} = 1/\mathsf{SNR}$. By incorporating an SNR-dependent bias to the least-squares estimate of each channel realization, the LMMSE estimator reduces the MMSE down to its minimum.

As in least-squares estimation, increasing the number of pilots is tantamount to increasing the SNR and thus, for arbitrary N_p,

$$\mathsf{MMSE} = \frac{1}{1+N_\text{p}\mathsf{SNR}}. \tag{2.269}$$

The LMMSE advantage over least-squares vanishes for $\mathsf{SNR} \to \infty$ and/or $N_\text{p} \to \infty$, in the latter case because, as the number of observations grows, a least-squares estimator becomes privy to the statistics required to perform LMMSE estimation in this setting.

Example 2.30 (LMMSE estimation for an IID MIMO channel)

Let the entries of $\bar{\boldsymbol{H}}$ be IID and, by normalization, of unit variance. Then, $\boldsymbol{R}_{\text{vec}(\bar{\boldsymbol{H}})} = \boldsymbol{I}$ and

$$\boldsymbol{W}^{\mathsf{MMSE}} = \sqrt{G}\bar{\boldsymbol{P}}_{\otimes}\left(N_0 + G\bar{\boldsymbol{P}}^*_{\otimes}\bar{\boldsymbol{P}}_{\otimes}\right)^{-1}, \tag{2.270}$$

from which

$$\text{vec}(\hat{\bar{H}}) = \sqrt{G}\left(G\bar{P}_{\otimes}^{*}\bar{P}_{\otimes} + N_0 I_{(L+1)N_t N_r}\right)^{-1}\bar{P}_{\otimes}^{*}\text{vec}(\bar{Y}), \tag{2.271}$$

while the MMSE matrix in (2.262) specializes to

$$E = I - G\bar{P}_{\otimes}^{*}\left(G\bar{P}_{\otimes}\bar{P}_{\otimes}^{*} + N_0 I\right)^{-1}\bar{P}_{\otimes}. \tag{2.272}$$

In the foregoing expression for $\text{vec}(\hat{\bar{H}})$, the dimensionality of the identity matrix has been noted explicitly to set the stage for the decomposition $N_0 I_{(L+1)N_t N_r} = N_0 I_{(L+1)N_t} \otimes I_{N_r}$. Recalling that $\bar{P}_{\otimes} = \bar{P}^{\mathrm{T}} \otimes I_{N_r}$, we can then write

$$\begin{aligned}\text{vec}(\hat{\bar{H}}) &= \sqrt{G}\left(G\bar{P}^{\mathrm{c}}\bar{P}^{\mathrm{T}} \otimes I_{N_r} + N_0 I_{(L+1)N_t} \otimes I_{N_r}\right)^{-1}(\bar{P}^{\mathrm{c}} \otimes I_{N_r})\,\text{vec}(\bar{Y}) & (2.273)\\ &= \left[\sqrt{G}\left(G\bar{P}^{\mathrm{c}}\bar{P}^{\mathrm{T}} + N_0 I_{(L+1)N_t}\right)^{-1}\bar{P}^{\mathrm{c}} \otimes I_{N_r}\right]\text{vec}(\bar{Y}) & (2.274)\end{aligned}$$

and, resorting once more to the identity $\text{vec}(ABC) = (C^{\mathrm{T}} \otimes A)\,\text{vec}(B)$, what ensues is

$$\hat{\bar{H}} = \sqrt{G}\,\bar{Y}\bar{P}^{*}\left(G\bar{P}\bar{P}^{*} + N_0 I\right)^{-1}, \tag{2.275}$$

which coincides with its least-squares counterpart in (2.247), save for the argument of the inverse being regularized in proportion to the noise strength. This similarity is consistent with the fact that the channel statistics, which differentiate the least-squares and LMMSE estimators, are trivial when the entries of $\bar{H}$ are IID; an SNR-dependent regularization suffices to optimally bias the estimate of each realization. The possibility of unvectorizing the estimate in the form (2.275) is also a consequence of the receive antennas being uncorrelated, as then the LMMSE estimate of each row of $\bar{H}$ can be produced separately, based only on the corresponding row of $\bar{Y}$.

Example 2.31

Elaborating on the previous example, suppose that $\bar{P}_{\otimes}$ is constructed by taking a unitary matrix (to balance the estimation error across the IID channel coefficient) and scaling it to the maximum allowable transmit power (to keep that error to a minimum). Express the MMSE incurred in the estimation of each of the $N_t N_r(L+1)$ channel coefficient as a function of $\mathsf{SNR} = \frac{GE_s}{N_0}$ where E_s is the total transmit energy across all N_t antennas, i.e., such that $\mathbb{E}[\|p[n]\|^2] = E_s$ for every n. How does this MMSE behave for growing SNR?

Solution

Irrespective of the type of power constraint (because with $\bar{P}_{\otimes}$ unitary the power is balanced across time and antennas) we have that $\bar{P}_{\otimes}\bar{P}_{\otimes}^{*} = \bar{P}_{\otimes}^{*}\bar{P}_{\otimes} = \frac{E_s}{N_t}I$. Then,

$$\begin{aligned}E &= I - \left(1 + \frac{N_0 N_t}{GE_s}\right)^{-1} I & (2.276)\\ &= I - \left(1 + \frac{N_t}{\mathsf{SNR}}\right)^{-1} I & (2.277)\\ &= \frac{1}{1 + \frac{\mathsf{SNR}}{N_t}} I & (2.278)\end{aligned}$$

and thus, on each channel coefficient,

$$\mathsf{MMSE} = \frac{1}{1 + \frac{\mathsf{SNR}}{N_t}}. \tag{2.279}$$

Save for the dependence on N_t, this is in perfect agreement with the SISO result in (2.268), intuitively indicating that, when the channel coefficients are independent, there is no loss of optimality in estimating them separately. The dependence on N_t reflects that, for a fixed transmit power, the MMSE per channel coefficient shrinks with N_t because the power of the scalar pilot symbol transmitted by each antenna diminishes. In contrast, the MMSE per channel coefficient depends neither on N_r nor on L: if these increase, for each additional IID coefficient to estimate, additional IID observations of undiminished power are made available.

Example 2.32

Moving beyond IID channels, let $L = 0$ and $\boldsymbol{H} = \boldsymbol{R}_r^{1/2} \boldsymbol{H}_w \boldsymbol{R}_t^{1/2}$ where $\boldsymbol{H}_w$ contains IID entries while $\boldsymbol{R}_t$ and $\boldsymbol{R}_r$ are, respectively, transmit and receive antenna correlation matrices. For this particular antenna correlation model, commonly encountered in the literature and physically justified in Section 3.6.1, what form does $\boldsymbol{W}^{\mathsf{MMSE}}$ take?

Solution

For this model, with antenna correlation at both transmitter and receiver,

$$\begin{aligned}
\boldsymbol{R}_{\mathrm{vec}(\bar{\boldsymbol{H}})} &= \mathbb{E}\left[\mathrm{vec}(\boldsymbol{H})\,\mathrm{vec}(\boldsymbol{H})^*\right] && (2.280)\\
&= \mathbb{E}\left[\left(\boldsymbol{R}_t^{\mathrm{T}/2} \otimes \boldsymbol{R}_r^{1/2}\right) \mathrm{vec}(\boldsymbol{H}_w)\,\mathrm{vec}(\boldsymbol{H}_w)^* \left(\boldsymbol{R}_t^{\mathrm{T}/2} \otimes \boldsymbol{R}_r^{1/2}\right)^*\right] && (2.281)\\
&= \left(\boldsymbol{R}_t^{\mathrm{T}/2} \otimes \boldsymbol{R}_r^{1/2}\right)\left(\boldsymbol{R}_t^{\mathrm{T}/2} \otimes \boldsymbol{R}_r^{1/2}\right)^* && (2.282)\\
&= \boldsymbol{R}_t^{\mathrm{T}} \otimes \boldsymbol{R}_r && (2.283)
\end{aligned}$$

and

$$\begin{aligned}
\bar{\boldsymbol{P}}_{\otimes} \boldsymbol{R}_{\mathrm{vec}(\bar{\boldsymbol{H}})} &= (\bar{\boldsymbol{P}}^{\mathrm{T}} \otimes \boldsymbol{I}_{N_r})(\boldsymbol{R}_t^{\mathrm{T}} \otimes \boldsymbol{R}_r) && (2.284)\\
&= \bar{\boldsymbol{P}}^{\mathrm{T}} \boldsymbol{R}_t^{\mathrm{T}} \otimes \boldsymbol{R}_r && (2.285)
\end{aligned}$$

where we applied $(\boldsymbol{A} \otimes \boldsymbol{B})(\boldsymbol{C} \otimes \boldsymbol{D}) = \boldsymbol{AC} \otimes \boldsymbol{BD}$. It follows that

$$\begin{aligned}
\boldsymbol{W}^{\mathsf{MMSE}} &= \sqrt{G}\left(G(\bar{\boldsymbol{P}}^{\mathrm{T}} \boldsymbol{R}_t^{\mathrm{T}} \otimes \boldsymbol{R}_r)(\bar{\boldsymbol{P}}^{\mathrm{c}} \otimes \boldsymbol{I}_{N_r}) + N_0 \boldsymbol{I}\right)^{-1} (\bar{\boldsymbol{P}}^{\mathrm{T}} \boldsymbol{R}_t^{\mathrm{T}} \otimes \boldsymbol{R}_r) && (2.286)\\
&= \sqrt{G}\left(G\bar{\boldsymbol{P}}^{\mathrm{T}} \boldsymbol{R}_t^{\mathrm{T}} \bar{\boldsymbol{P}}^{\mathrm{c}} \otimes \boldsymbol{R}_r + N_0 \boldsymbol{I}\right)^{-1} (\bar{\boldsymbol{P}}^{\mathrm{T}} \boldsymbol{R}_t^{\mathrm{T}} \otimes \boldsymbol{R}_r), && (2.287)
\end{aligned}$$

which is not conducive to an unvectorized form similar to (2.247) whenever $\boldsymbol{R}_r$ is not diagonal. Indeed, a nondiagonal $\boldsymbol{R}_r$ entails correlation across the receive antennas, forcing the LMMSE estimator to jointly process the corresponding observations and precluding the decomposition required for an unvectorized form.

The case of frequency-flat channels whose structure is $\boldsymbol{H} = \boldsymbol{H}_w \boldsymbol{R}_t^{1/2}$, with $\boldsymbol{R}_r = \boldsymbol{I}$ such that the antennas are correlated only at the transmitter, is treated in Problem 2.41.

2.7.2 OFDM channel estimation

In OFDM, pilots are inserted in the frequency domain. Focusing on a specific OFDM symbol, the dual of the situation we considered for time-invariant single-carrier transmission would be to have a pilot sequence occupy the first N_{p} subcarriers. However, in anticipation of frequency selectivity it is more sensible to intersperse the pilot sequence as uniformly as possible over the K subcarriers [182, 186, 187].

Suppose that the subcarriers with indices $k_0, \ldots, k_{N_{\mathrm{p}}-1}$ are reserved for pilots. Recalling (2.222), the receiver observation of the pilot transmissions $\mathbf{p}[k_0], \ldots, \mathbf{p}[k_{N_{\mathrm{p}}-1}]$ is

$$\mathbf{y}[k] = \sqrt{G}\, \mathbf{H}[k]\, \mathbf{p}[k] + \mathbf{v}[k] \tag{2.288}$$

$$= \sqrt{G} \left(\sum_{\ell=0}^{L} \boldsymbol{H}[\ell]\, e^{-\mathrm{j}\frac{2\pi}{K}\ell k} \right) \mathbf{p}[k] + \mathbf{v}[k] \qquad k = k_0, \ldots, k_{N_{\mathrm{p}}-1}, \tag{2.289}$$

which, rewritten in terms of $\bar{\boldsymbol{H}} = \begin{bmatrix} \boldsymbol{H}[0] & \cdots & \boldsymbol{H}[L] \end{bmatrix}$, becomes

$$\mathbf{y}[k] = \sqrt{G}\bar{\boldsymbol{H}} \begin{bmatrix} \boldsymbol{I}_{N_{\mathrm{t}}} \\ e^{-\mathrm{j}\frac{2\pi}{K}k} \boldsymbol{I}_{N_{\mathrm{t}}} \\ \vdots \\ e^{-\mathrm{j}\frac{2\pi}{K}kL} \boldsymbol{I}_{N_{\mathrm{t}}} \end{bmatrix} \mathbf{p}[k] + \mathbf{v}[k] \tag{2.290}$$

$$= \sqrt{G}\bar{\boldsymbol{H}} \begin{bmatrix} \mathbf{p}[k] \\ e^{-\mathrm{j}\frac{2\pi}{K}k} \mathbf{p}[k] \\ \vdots \\ e^{-\mathrm{j}\frac{2\pi}{K}kL} \mathbf{p}[k] \end{bmatrix} + \mathbf{v}[k] \tag{2.291}$$

and, letting

$$\boldsymbol{u}[k] = \begin{bmatrix} 1 & e^{-\mathrm{j}\frac{2\pi}{K}k} & \cdots & e^{-\mathrm{j}\frac{2\pi}{K}kL} \end{bmatrix}^{\mathrm{T}}, \tag{2.292}$$

further

$$\mathbf{y}[k] = \sqrt{G}\bar{\boldsymbol{H}}\big(\boldsymbol{u}[k] \otimes \mathbf{p}[k]\big) + \mathbf{v}[k]. \tag{2.293}$$

Applying $\mathrm{vec}(\cdot)$ to both sides of the equality, using $\mathrm{vec}(\mathbf{y}[k]) = \mathbf{y}[k]$ and $\mathrm{vec}(\mathbf{v}[k]) = \mathbf{v}[k]$ as these are column vectors, and invoking yet again $\mathrm{vec}(\boldsymbol{ABC}) = (\boldsymbol{C}^{\mathrm{T}} \otimes \boldsymbol{A})\,\mathrm{vec}(\boldsymbol{B})$,

$$\mathbf{y}[k] = \sqrt{G}\left(\big(\boldsymbol{u}[k] \otimes \mathbf{p}[k]\big)^{\mathrm{T}} \otimes \boldsymbol{I}_{N_{\mathrm{r}}} \right) \mathrm{vec}(\bar{\boldsymbol{H}}) + \mathbf{v}[k] \tag{2.294}$$

$$= \sqrt{G}\left(\boldsymbol{u}[k]^{\mathrm{T}} \otimes \mathbf{p}[k]^{\mathrm{T}} \otimes \boldsymbol{I}_{N_{\mathrm{r}}} \right) \mathrm{vec}(\bar{\boldsymbol{H}}) + \mathbf{v}[k]. \tag{2.295}$$

Stacking all observations corresponding to pilot-bearing subcarriers, we obtain [188]

$$\underbrace{\begin{bmatrix} \mathbf{y}[k_0] \\ \mathbf{y}[k_1] \\ \vdots \\ \mathbf{y}[k_{N_{\mathrm{p}}-1}] \end{bmatrix}}_{\bar{\mathbf{y}}} = \sqrt{G} \underbrace{\begin{bmatrix} \boldsymbol{u}[k_0]^{\mathrm{T}} \otimes \mathbf{p}[k_0]^{\mathrm{T}} \otimes \boldsymbol{I}_{N_{\mathrm{r}}} \\ \boldsymbol{u}[k_1]^{\mathrm{T}} \otimes \mathbf{p}[k_1]^{\mathrm{T}} \otimes \boldsymbol{I}_{N_{\mathrm{r}}} \\ \vdots \\ \boldsymbol{u}[k_{N_{\mathrm{p}}-1}]^{\mathrm{T}} \otimes \mathbf{p}[k_{N_{\mathrm{p}}-1}]^{\mathrm{T}} \otimes \boldsymbol{I}_{N_{\mathrm{r}}} \end{bmatrix}}_{\bar{\mathbf{P}}_{\otimes}} \mathrm{vec}(\bar{\boldsymbol{H}}) + \underbrace{\begin{bmatrix} \mathbf{v}[k_0] \\ \mathbf{v}[k_1] \\ \vdots \\ \mathbf{v}[k_{N_{\mathrm{p}}-1}] \end{bmatrix}}_{\bar{\mathbf{v}}} \tag{2.296}$$

which is isomorphic with (2.229). The least-squares and LMMSE solutions derived for single-carrier transmission can thus be applied verbatim with $\text{vec}(\bar{\boldsymbol{Y}})$, $\bar{\boldsymbol{P}}_{\otimes}$, and $\text{vec}(\bar{\boldsymbol{V}})$ respectively replaced by $\bar{\mathbf{y}}$, $\bar{\mathbf{P}}_{\otimes}$, and $\bar{\mathbf{v}}$. Ensuring that $\bar{\mathbf{P}}_{\otimes}$ is square or tall requires

$$N_{\rm p} \geq (L+1)N_{\rm t} \tag{2.297}$$

and a quick comparison with (2.234) indicates that, while in frequency-flat channels the number of pilots coincides, in frequency-selective channels that number is reduced with OFDM. The reduction equals precisely L, the channel memory, which is the number of pilot observations that have to be discarded because they are influenced by payload data symbols—see the construction in (2.223).

As an alternative to estimating the delay-domain matrices $\boldsymbol{H}[0], \ldots, \boldsymbol{H}[L]$ in the foregoing formulation, the channel can be estimated directly in the frequency domain. Applying a least-squares or LMMSE estimator to (2.288), $\hat{\mathbf{H}}[k_0], \ldots, \hat{\mathbf{H}}[k_{N_{\rm p}-1}]$ are readily obtained and, from them, the channel at other subcarriers can be derived by interpolation—in the MMSE sense if the channel statistics are available, or through other methods (e.g., polynomially) otherwise [189]. Moreover, under the mild premise that the channel be identically distributed, i.e., that all subcarriers look the same statistically, $k_0, \ldots, k_{N_{\rm p}-1}$ should be regularly spaced over the K subcarriers, facilitating the interpolation.

2.8 Summary and outlook

The gist of this chapter is captured in the form of take-away points within the companion summary box.

Given the predominance of OFDM, and the attractiveness of SC-FDE for those transmissions (such as the LTE reverse link) that resist adopting OFDM on account of its peakedness, the interest in time-domain equalization has diminished. This is reflected, in the remainder of this book, with a decided inclination of the analysis toward OFDM signaling. And, whenever ergodicity applies, the analysis then often folds onto frequency-flat settings, which is welcome news from the standpoint of clarity and presentation.

As of channel estimation, given that pilots are overhead and that fresh pilots need to be transmitted as soon as the channel has experienced any substantial variation, it is important to keep them to a minimum. Since the number of required pilots grows with $N_{\rm t}$, the proposition of massive MIMO has infused new momentum into the problem of channel estimation. Specifically, there is much interest in the possibility of exploiting channel structures to improve and/or simplify the estimation process. Experimental measurements have shown that high-dimensional and/or high-carrier-frequency wireless channels may exhibit a sparse structure such that the estimator could concentrate on learning only those few coefficients that are nonnegligible. Unfortunately, the location of those coefficients tends itself to be unknown, which leads to posing a problem that also tries to find the vector that is most sparse. There is considerable research activity on designing sparsity-aware channel estimation algorithms and on revisiting classic channel estimation problems leveraging the framework of compressed sensing (see, e.g., [190–193] and references therein).

Take-away points

1. Complex baseband equivalent models allow representing passband communication more compactly and independently of f_c.
2. Because of the bandlimitedness of the transmit signal, the complex baseband representation can be further sampled and time-discretized.
3. A physical channel may not be bandlimited, in which case it cannot be sampled, but its restriction to the finite signal bandwidth can always be time-discretized.
4. Given a symbol period T and a pulse shape compliant with the Nyquist criterion, the bandwidth in terms of noise power is $B = 1/T$. Typically the signal exceeds this value, and a suitable penalty factor must be applied to the spectral efficiency.
5. The baseband discrete-time transmit–receive relationship over an $(L+1)$-tap time-invariant MIMO channel with N_t transmit and N_r receive antennas is
$$\boldsymbol{y}[n] = \sqrt{G} \sum_{\ell=0}^{L} \boldsymbol{H}[\ell]\, \boldsymbol{x}[n-\ell] + \boldsymbol{v}[n], \tag{2.298}$$
where G is such that $\mathbb{E}\big[\mathrm{tr}(\boldsymbol{H}\boldsymbol{H}^*)\big] = N_t N_r$ and $\boldsymbol{v}[n] \sim \mathcal{N}_{\mathbb{C}}(\boldsymbol{0}, N_0 \boldsymbol{I})$. For $L = 0$, this reduces to the frequency-flat relationship
$$\boldsymbol{y}[n] = \sqrt{G} \boldsymbol{H} \boldsymbol{x}[n] + \boldsymbol{v}[n]. \tag{2.299}$$
6. Stacked vectors and structured matrices can be used to convert (2.298) into a single matrix relationship, simplifying tasks such as channel estimation or equalization.
7. The transmit vector at time n can be expressed as
$$\boldsymbol{x}[n] = \sqrt{\frac{E_s}{N_t}} \boldsymbol{F}[n] \boldsymbol{s}[n], \tag{2.300}$$
where E_s is the average energy per vector symbol, $\boldsymbol{F}[n]$ is the precoder, and $\boldsymbol{s}[n]$ contains N_s independent unit-variance scalar symbols.
8. The precoder can be decomposed as
$$\boldsymbol{F}[n] = \boldsymbol{U}_{\boldsymbol{F}}[n] \begin{bmatrix} \boldsymbol{P}^{1/2}[n] \\ \boldsymbol{0} \end{bmatrix} \boldsymbol{V}_{\boldsymbol{F}}^*[n], \tag{2.301}$$
where $\boldsymbol{V}_{\boldsymbol{F}}[n]$ is a unitary mixing matrix, $\boldsymbol{P}[n]$ is a diagonal power allocation matrix, and $\boldsymbol{U}_{\boldsymbol{F}}[n]$ is a unitary steering matrix.
9. The number of transmitted data streams, $N_s \leq N_t$, determines the dimensions of $\boldsymbol{P}[n]$ as well as the rank of $\boldsymbol{F}[n]$.
10. If $\boldsymbol{F}[n] = \frac{N_t}{N_s} [\boldsymbol{I}\ \boldsymbol{0}]^{\mathrm{T}}$, then N_s antennas radiate independent signal streams while $N_t - N_s$ antennas are off. More generally, the precoding mixes the N_s streams, assigns them distinct powers, and steers them into specific spatial directions.
11. The average SNR at the receiver is
$$\mathsf{SNR} = \frac{G E_s}{N_0}. \tag{2.302}$$

12. Under a per-block or per-codeword power constraint with a time horizon of N symbols, $\frac{1}{N}\sum_{n=0}^{N-1}\text{tr}(\boldsymbol{P}[n]) = N_\text{t}$; power can be allocated unevenly over time and across antennas. Under a stricter per-symbol constraint, $\text{tr}(\boldsymbol{P}[n]) = N_\text{t}$ for every n; power can be shifted across antennas, but not in time. Finally, under a per-antenna constraint, $\left[\boldsymbol{F}[n]\boldsymbol{F}^*[n]\right]_{j,j} = 1$ for every n and j; each transmit antenna features a separate amplifier.
13. Perfect FIR ZF linear equalizers exist for almost all MIMO channels whenever $N_\text{r} > N_\text{t}$, enabling perfect reconstruction in the absence of noise.
14. The MMSE criterion yields linear equalizers that are more robust than their ZF counterparts in the presence of noise, balancing interference suppression and noise enhancement rather than stubbornly seeking to invert the channel response.
15. The parsing of the transmit sequence into blocks of K symbols with insertion of an L-symbol cyclic prefix per block allows for efficient frequency-domain equalization using the DFT, so-called SC-FDE.
16. With OFDM, the transmission takes place in parallel over K subcarriers with simple per-subcarrier equalization. In the time domain, every burst—an OFDM symbol—has a duration $T_\text{OFDM} = (K+L)T$. The subcarrier spacing is $\frac{1}{KT}$. The intersection of a specific subcarrier and a specific OFDM symbol defines an OFDM resource element, the basic unit in which signals are parceled. A larger K reduces the cyclic prefix overhead and the excess bandwidth (which here adopts the form of idle edge subcarriers) at the expense of higher complexity and increased vulnerability to time-variability in the channel.
17. In contrast with linear equalizers, the complexities of SC-FDE and OFDM do not depend on the channel memory L.
18. Through the observation of how known pilot symbols are received, the channel response can be learned. In additive Gaussian noise, the ML estimator coincides with the least-squares estimator, which does not rely on the channel distribution. If the channel covariance is known, the superior LMMSE estimator can be used. In both cases, the required number of pilot symbols grows in proportion to N_t.

Problems

2.1 Prove these results concerning the Kronecker product and the vec$(\cdot)$ operator.

(a) If $\boldsymbol{A}$ and $\boldsymbol{B}$ are square, then $\text{tr}(\boldsymbol{A}\otimes\boldsymbol{B}) = \text{tr}(\boldsymbol{A})\,\text{tr}(\boldsymbol{B})$.

(b) $\text{vec}(\boldsymbol{ABC}) = (\boldsymbol{C}^\text{T}\otimes\boldsymbol{A})\,\text{vec}(\boldsymbol{B})$.

(c) $\text{vec}(\boldsymbol{a}\boldsymbol{b}^\text{T}) = \boldsymbol{b}\otimes\boldsymbol{a}^\text{T}$.

2.2 Determine the vector $\boldsymbol{x}$ that maximizes $\boldsymbol{x}^*\boldsymbol{R}\boldsymbol{x}$ subject to $\|\boldsymbol{x}\|^2 = 1$.

Hint: You may use the method of Lagrange multipliers and matrix calculus, both topics covered in the Appendix.

2.3 Find the vector $\boldsymbol{x}$ that minimizes $\|\boldsymbol{y} - \boldsymbol{A}\boldsymbol{x}\|^2$ for $\boldsymbol{A}$ tall but low-rank. Further express the corresponding $\|\boldsymbol{y} - \boldsymbol{A}\boldsymbol{x}\|^2$.

2.4 With $f_c = 1.9$ GHz and $B = 5$ MHz, consider the channel

$$c_p(\tau) = 0.05\,\delta(\tau - \tau_0) - 0.01\,\delta(\tau - \tau_1) \tag{2.303}$$

where $\tau_0 = 0.3$ μs and $\tau_1 = 0.5$ μs.

(a) Express the bandpass-filtered channel by applying an ideal filter with center frequency f_c and passband bandwidth B.

(b) Express the complex pseudo-baseband equivalent channel $c(\tau)$.

(c) Express the complex baseband equivalent channel $c_b(\tau)$.

(d) Express the discrete-time complex baseband equivalent channel $c[\ell]$ with sampling rate $1/T$.

(e) Express the normalized discrete-time complex baseband equivalent channel $h[\ell]$.

2.5 Consider the channel $c_p = A_0\delta(\tau - \tau_0) + A_1\delta(\tau - \tau_1 + A_2\delta(\tau - \tau_2)$.

(a) Obtain the discrete-time baseband representation $c[\ell]$ with ideal lowpass filtering.

(b) Under which conditions would such representation equal $c[\ell] = A_0\, e^{-j2\pi f_c \tau_0} + A_1\, e^{-j2\pi f_c \tau_1} + A_2\, e^{-j2\pi f_c \tau_2}$?

(c) Give $c[n]$ for $\tau_0 = 0$, $\tau_1 = T/2$ and $\tau_2 = T$.

2.6 Prove that $B\,\text{sinc}(B\tau) * B\,\text{sinc}(B\tau) = B\,\text{sinc}(B\tau)$, confirming that applying a unit-gain ideal lowpass filter twice has the same effect as applying it once.

2.7 Show that, with raised-cosine pulse shaping, the baseband bandwidth W satisfying

$$N_0 W = \int_0^{\frac{1+b}{2T}} N_0\, |g_{rx}(f)|^2\, df \tag{2.304}$$

is $W = \frac{1}{2T}$. Thus, irrespective of b, the passband noise bandwidth is $B = 1/T$.

2.8 Redo Example 2.6 and the bottom drawing in Fig. 2.6, but have the sampler aligned with τ_1 rather than τ_0. Verify that a different discrete-time baseband channel arises.

2.9 Let $N_t = 1$ and $N_r = 2$ with

$$c^{(0,0)}(\tau) = A_0\,\delta(\tau - \tau_0) \tag{2.305}$$

$$c^{(1,0)}(\tau) = A_1\,\delta(\tau - \tau_1) + A_2\,\delta(\tau - \tau_2). \tag{2.306}$$

(a) Determine the complex pseudo-baseband equivalent channel.

(b) Determine the complex baseband equivalent channel.

(c) Determine the complex baseband equivalent channel in the continuous-frequency domain.

(d) Determine the discrete-time complex baseband equivalent channel.

2.10 Let $N_t = N_r = 2$ with

$$\sqrt{G}\big[\boldsymbol{H}[\ell]\big]_{0,0} = \delta[\ell] + 0.5\,\delta[\ell - 1] \tag{2.307}$$

$$\sqrt{G}\big[\boldsymbol{H}[\ell]\big]_{0,1} = -0.5\,\delta[\ell] + \mathrm{j}\,\delta[\ell - 1] \tag{2.308}$$

$$\sqrt{G}\big[\boldsymbol{H}[\ell]\big]_{1,0} = 0.75\,\delta[\ell] + (1 - 0.5\,\mathrm{j})\,\delta[\ell - 1] \tag{2.309}$$

$$\sqrt{G}\left[\boldsymbol{H}[\ell]\right]_{1,1} = -\delta[\ell]. \tag{2.310}$$

For $N = 3$, compose the block Toeplitz matrix in (2.94).

2.11 Write out the stacked-vector MIMO relationship for $L = 2$ and $N = 4$.

2.12 Plot, as a function of ξ (in dB) between 5 dB and 30 dB, the differential entropy of the random variable obtained by clipping $s \sim \mathcal{N}(0, 1)$ such that its PAPR equals ξ.

2.13 Given $x \sim \mathcal{N}(0, \sigma^2)$, let z be a clipped version of x satisfying $\mathsf{PAPR}(|z|) = 5$ dB. Compute the kurtosis and CM of z.

2.14 Calculate the PAPR and the kurtosis of a 16-QAM constellation where the four inner points have twice the probability of the 12 outer points.

2.15 Consider a signal distribution made up of two concentric ∞-PSK rings, such that symbols are drawn equiprobably from either ring and uniformly in angle.
(a) Relate the radii of the inner and outer rings, keeping in mind the unit-variance normalization of the distribution.
(b) Express the PAPR, the kurtosis, and the CM as a function of the inner radius.
(c) Plot the PAPR, the kurtosis, and the CM as a function of the inner radius. The plot should cover the range of feasible values of the inner radius.

2.16 Compute the PAPR, the kurtosis and the CM for a 10-MHz OFDM discrete-time signal with 500-kHz guards at both ends and 15-kHz subcarriers. Consider that each subcarrier carries an independent signal conforming to:
(a) QPSK.
(b) 16-QAM.

2.17 Show that, for the ∞-QAM distribution, the peakedness measures are $\mathsf{PAPR} = 3$, $\kappa = 1.4$, and $\mathsf{CM} = 1.33$.

2.18 Consider an FIR equalizer of order $L_{\mathsf{eq}} = 10$ operating on a SISO channel where $h[0] = 0.8$, $h[1] = 0.5\, e^{\mathrm{j}\pi/4}$, and $h[2] = -0.33$.
(a) Obtain the ZF equalizer coefficients by approximating the IIR solution.
(b) Obtain the LMMSE equalizer assuming that $\{x[n]\}$ and $\{v[n]\}$ are IID sequences.

2.19 For a two-tap SISO channel where $\sqrt{G}\, h[0] = 1$ and $\sqrt{G}\, h[1] = a$.
(a) Express the error at the output of an FIR ZF equalizer of order L_{eq}.
(b) For $a = 0.1$ and $L_{\mathsf{eq}} = 4$, quantify the dB-ratio between the output signal power and that of the error.
(c) Repeat part (b) for $a = 0.1$ and $L_{\mathsf{eq}} = 10$.

2.20 Let $N_{\mathrm{t}} = 1$ and $N_{\mathrm{r}} = 2$ with

$$\sqrt{G}\left[\boldsymbol{H}[\ell]\right]_{0,0} = \delta[\ell] + 0.5\, e^{\mathrm{j}\pi/4}\delta[\ell-1] - 0.25\,\delta[\ell-2] \tag{2.311}$$

$$\sqrt{G}\left[\boldsymbol{H}[\ell]\right]_{1,0} = \mathrm{j}\,\delta[\ell] - 0.5\,\mathrm{j}\,\delta[\ell-1] + 0.125\, e^{\mathrm{j}\pi/3}\delta[\ell-2]. \tag{2.312}$$

(a) Determine the lowest-order perfect FIR equalizer.
(b) Compute the ZF equalizer coefficients for $L_{\mathsf{eq}} = 10$.
(c) Compute the LMMSE equalizer for $L_{\mathsf{eq}} = 10$ assuming that $\{x[n]\}$ and $\{v[n]\}$ are IID sequences.

2.21 For $\bar{\boldsymbol{T}}$ tall, prove that the ZF equalizer with the minimum norm is

$$\bar{\boldsymbol{w}}_j^{\text{ZF}} = \frac{1}{\sqrt{G}} \bar{\boldsymbol{T}} (\bar{\boldsymbol{T}}^* \bar{\boldsymbol{T}})^{-1} \bar{\boldsymbol{d}}_j \tag{2.313}$$

and obtain the corresponding minimum squared norm.

Hint: You may use the method of Lagrange multipliers and matrix calculus.

2.22 Let $L = 1$ with

$$\boldsymbol{H}[0] = \begin{bmatrix} 1 & \mathrm{j} \\ 0 & 0 \end{bmatrix} \qquad \boldsymbol{H}[1] = \begin{bmatrix} 0 & 1 \\ 0.8 & -0.6\,\mathrm{j} \end{bmatrix}. \tag{2.314}$$

(a) Find an order-3 ZF equalizer. Show the effectiveness of the equalizer by computing $\boldsymbol{W}^{\text{ZF}*}[\ell] * \boldsymbol{H}[\ell]$ and plotting the magnitude of the result.

(b) Find an order-3 LMMSE equalizer for $\text{SNR} = 10$ dB and with $\{\boldsymbol{x}[n]\}$ and $\{\boldsymbol{v}[n]\}$ being IID sequences. Show the effectiveness of the equalizer by computing $\boldsymbol{W}^{\text{ZF}*}[\ell] * \boldsymbol{H}[\ell]$ and plotting the magnitude of the result. Further plot the MMSE for each of the two equalized signals.

2.23 For a frequency-flat channel $\boldsymbol{H}$ whose entries are drawn from IID standard complex Gaussian distributions, express the mean-square error with ZF equalization.

2.24 Reconsider Example 2.20 for signals IID in time but not necessarily across antennas, such that $\boldsymbol{R}_{\boldsymbol{x}}[\ell] = \delta[\ell] \boldsymbol{R}_{\boldsymbol{x}}$.

(a) Express $\boldsymbol{R}_{\boldsymbol{y}}[\ell]$ and $\boldsymbol{R}_{\boldsymbol{yx}}[\ell]$.

(b) Express $\bar{\boldsymbol{W}}^{\text{MMSE}}$ and $\boldsymbol{E}$.

2.25 Consider the frequency-flat channel

$$\boldsymbol{H} = \frac{1}{\sqrt{3 + \rho^2}} \begin{bmatrix} 1 & 1 \\ \rho & 1 \end{bmatrix}. \tag{2.315}$$

(a) Compute the ZF equalizer.

(b) Compute the LMMSE equalizer.

(c) For $\text{SNR} = 10$ dB, plot the two mean-square errors (in dB) for each equalizer as a function of $\rho \in [-1, 0.99]$. Explain the intuition behind the results.

2.26 Let $N_{\text{t}} = 1$ and $N_{\text{r}} = 2$. Suppose that a sequence of N_{p} known pilots is transmitted to trigger observations $\boldsymbol{y}[0], \ldots, \boldsymbol{y}[N_{\text{p}} + L - 1]$. Formulate and solve a least-squares problem for the equalizer $\boldsymbol{w}[0], \ldots, \boldsymbol{w}[L_{\text{eq}}]$ directly from the observations, without first estimating the channel explicitly.

(a) Write the output of the equalizer as a function of $\boldsymbol{y}[0], \ldots, \boldsymbol{y}[N_{\text{p}} + L - 1]$, noting that you may not need to use all these outputs.

(b) Rewrite in matrix form.

(c) Find the least-squares equalizer solution.

(d) Comment on the choices of N_{p} and L_{eq} relative to L.

(e) What are the pros and cons of direct equalization versus first estimating the channel and then computing the equalizer?

2.27 Give an expression for the output SNR in a SISO channel with SC-FDE and the following frequency-domain equalizers:

(a) ZF.

(b) LMMSE.

Assume that $\{x[n]\}$ and $\{v[n]\}$ are IID sequences.

2.28 For the MIMO channel in (2.314), plot the per-stream output SNR with SC-FDE and ZF equalization. Set the cyclic prefix length to $L_c = L$ and let $K = 4, 8, 16, 64$. Interpret the results.

2.29 In certain WLAN standards, the data are parsed into blocks of K symbols and a pilot word of L_c symbols is inserted between consecutive blocks. Show how the pilot word can be interpreted as a cyclic prefix and explain how to choose K and L_c.

2.30 Prove that, with MIMO SC-FDE and ZF, the variance of the noise contaminating the signal sent from the jth antenna is given, at the output of the equalizer, by (2.209).

2.31 Compute an expression for the per-subcarrier output SNR for SISO-OFDM with the following frequency-domain equalizers:

(a) ZF.

(b) LMMSE.

Assume that $\{x[n]\}$ and $\{v[n]\}$ are IID sequences.

2.32 Consider a SISO OFDM signal with rectangular time-domain pulse shaping.

(a) Plot, as a function of fT, the power spectrum (in dB) for $K = 16$, $K = 64$, and $K = 256$ subcarriers.

(b) For each value of K, what is the excess bandwidth measured at -30 dB? How about at -40 dB?

2.33 Reconsider Problem 2.32 for $K = 64$, but with the time-domain pulse shape modified into a raised cosine.

(a) Plot, as a function of fT, the power spectrum for raised-cosine rolloff factors 0.05 and 0.1.

(b) For each rolloff factor, recompute the excess bandwidth at -30 dB and -40 dB.

2.34 Reconsider the MIMO channel in (2.314), but now with OFDM and a cyclic prefix of length L. Plot the per-stream per-subcarrier output SNR with ZF equalization for $K = 4, 8, 16, 64$. Interpret the results.

2.35 Find the values for N_t, N_r, T, N, T_{OFDM}, L_c, the excess bandwidth and the subcarrier spacing for the following WLAN standards:

(a) IEEE 802.11ac.

(b) IEEE 802.11ax.

2.36 Verify that, for a channel realization h, the least-squares estimate $\text{vec}(\hat{\bar{\boldsymbol{H}}}) = \boldsymbol{W}^{\text{LS}*}\boldsymbol{y}$ satisfies

$$\mathbb{E}\Big[\text{vec}(\hat{\bar{\boldsymbol{H}}})\,|\,\text{vec}(\bar{\boldsymbol{H}}){=}\mathrm{h}\Big] = \mathrm{h}. \tag{2.316}$$

2.37 Let $N_t = N_r = 2$.

(a) From the first Zadoff–Chu root sequence of length 3, design the pilot sequences for the two transmit antennas. Verify that they are orthogonal.

(b) For a frequency-flat channel, assemble the Hankel matrix $\bar{\boldsymbol{P}}$ as well as $\bar{\boldsymbol{P}}_{\otimes}$ and the ensuing least-square estimator W^{LS}. Generate a realization of the receiver observation (three symbols) for the channel

$$H = \begin{bmatrix} 1 & -1 \\ \mathrm{j} & \mathrm{j} \end{bmatrix} \tag{2.317}$$

at SNR $= 5$ dB. From this observation, produce a least-square estimate of H.

(c) Repeat part (b) for a frequency-selective channel with $L = 1$, $H[0] = H$, and

$$H[1] = \begin{bmatrix} 0.3\mathrm{j} & -0.2 \\ 0.1 & -0.5\mathrm{j} \end{bmatrix}. \tag{2.318}$$

In this case, the receiver observation to generate must have four symbols.

2.38 In a frequency-flat SISO channel, consider the ratio between the power of the channel estimate $\mathbb{E}[|\hat{h}|^2]$ and the power of the estimation error $\mathbb{E}[|h - \hat{h}|^2]$.

(a) At SNR $= 5$ dB, how many pilot symbols does a least-squares estimator require to make this ratio 20 dB?

(b) Repeat part (a) for an LMMSE estimator.

(c) Taking into account that pilot symbols are overhead, at SNR $= 5$ dB, how much bandwidth does an LMMSE estimator save relative to a least-squares estimator?

(d) Beyond which SNR does the number of pilots become equal for both estimators?

2.39 Consider a least-squares MIMO channel estimator. Show that the sum of the powers of the estimation error terms,

$$\mathbb{E}\left[\|\bar{H} - \hat{\bar{H}}\|_F^2\right] \tag{2.319}$$

is minimized when the pilot sequence is such that $\bar{P}\bar{P}^*$ is proportional to the identity matrix with $\|\bar{P}\|_F^2$ as large as allowed by the applicable power constraint.

2.40 Consider the frequency-flat channel

$$H = \frac{2}{\sqrt{2.85}} \begin{bmatrix} 1 & 0.2 \\ 0.9 & \mathrm{j} \end{bmatrix} \tag{2.320}$$

and suppose that the pilot sequence $p[0], \ldots, p[N_p - 1]$, with N_p even, consists of successive columns of the identity matrix scaled to the available power. The SNR is 10 dB. Plot the channel estimation MMSE. What is the value of N_p that pushes the MMSE below 0.01?

2.41 Consider the frequency-flat MIMO channel $H = R_r^{1/2} W R_t^{1/2}$ as in Example 2.32 and let $R_r = I$. Express the LMMSE channel estimate $\hat{H}$ for an arbitrary pilot sequence.

2.42 Consider a frequency-flat MIMO channel H with $N_t = N_r = 2$ and zero-mean IID entries. Referring to the estimation of each entry of H:

(a) If the transmit antennas emit equal-power orthogonal pilot sequences of length $N_p = N_t$, what is the dB-difference between the mean-square error achieved by a least-squares and an LMMSE estimator at SNR $= 0$, 10, and 20 dB?

(b) Repeat part (a) for $N_p = 2N_t$.

2.43 Repeat Problem 2.42 for $\boldsymbol{H} = \boldsymbol{R}_\mathrm{r}^{1/2} \boldsymbol{H}_\mathrm{w}$ where the entries of $\boldsymbol{H}_\mathrm{w}$ are zero-mean IID complex Gaussian, $N_\mathrm{t} = 2$, and the receive antennas are correlated as per

$$\boldsymbol{R}_\mathrm{r} = \begin{bmatrix} 1 & 0.5 \\ 0.5 & 1 \end{bmatrix}. \tag{2.321}$$

Has the receive antenna correlation increased or decrease the estimation error?

2.44 Let $N_\mathrm{t} = N_\mathrm{r} = L = 2$. The channel coefficients are IID complex Gaussian and the pilot sequence is taken from the first two rows of a Fourier matrix of size N_p. Consider $N_\mathrm{p} = 4, 8, 16, 32, 64, 256$.

(a) Plot the channel estimation MMSE (in dB) for $\mathsf{SNR} \in [0, 10]$ dB.

(b) Plot the channel estimation mean-square error (in dB) attained by a least-squares estimator for $\mathsf{SNR} \in [0, 10]$ dB.

2.45 Repeat Problem 2.44 for the fixed channel in (2.314), assumed drawn from an IID complex Gaussian distribution.

2.46 Consider a MIMO link subject to interference from another transmitter. The intended user has N_t transmit antennas while the interferer features N_1 antennas and the receiver observes

$$\boldsymbol{y}[n] = \sqrt{G}\boldsymbol{H}\boldsymbol{x}[n] + \sqrt{G_1}\boldsymbol{H}_1\boldsymbol{x}_1[n] + \boldsymbol{v}[n], \tag{2.322}$$

where both $\boldsymbol{H}$ and $\boldsymbol{H}_1$ contain IID complex Gaussian entries while $\boldsymbol{x}_1[n]$ is an IID sequence satisfying $\mathbb{E}\big[\boldsymbol{x}_1[n]\boldsymbol{x}_1^*[n]\big] = \frac{E_\mathrm{s}}{N_\mathrm{t}}\boldsymbol{I}$. Based on a sequence of N_p pilots, derive the LMMSE estimator for $\boldsymbol{H}$ and the corresponding MMSE matrix.

2.47 Consider a SISO-OFDM signal with a cyclic prefix of length $L_\mathrm{c} = L$. The receiver conducts channel estimation over multiple OFDM symbols. The observation of symbol n over subcarrier k is

$$\mathsf{y}[k,n] = \sqrt{G}\,\mathsf{h}[k]\,\mathsf{x}[k,n] + \mathsf{v}[k,n], \tag{2.323}$$

where $\mathsf{h}[k]$ is the time-invariant channel response in the frequency domain while $\mathsf{x}[k,n]$ and $\mathsf{v}[k,n]$ are the transmit signal and noise. Let the indices of the Q pilot-carrying subcarriers be $k = \{0, K/Q, 2K/Q, \ldots, (Q-1)K/Q\}$ for some Q that evenly divides K.

(a) Suppose that N_p/Q pilot symbols are sent on each pilot subcarrier. Derive a least-squares channel estimate $\hat{\mathsf{h}}[k]$ for k corresponding to pilot subcarriers.

(b) Now assume that $\hat{\mathsf{h}}[k] = \mathsf{h}[k]$ on the pilot subcarriers and use linear interpolation to find the rest of the coefficients, precisely

$$\hat{\mathsf{h}}[k] = \sum_{q=0}^{Q-1} \varrho[k,q]\,\mathsf{h}[qK/Q+1]\,, \tag{2.324}$$

where $\varrho[k,q]$ for $q = 0, \ldots, Q-1$ are scalar interpolator coefficients for subcarrier k. Assuming that the channel taps $h[0], \ldots, h[L]$ are IID complex Gaussian, find the interpolator coefficients that minimize $\mathbb{E}\big[|\mathsf{h}[k] - \hat{\mathsf{h}}[k]|^2\big]$.

3 Channel modeling

A communication channel is the part of a communication system that we are unwilling or unable to change.

John L. Kelly

While all models are wrong, some are actually useful.

George Box

3.1 Introduction

The wireless channel, understood as everything that stands between the transmit and receive antennas, has a decisive influence on the quality and characteristics of the received signal. It impacts virtually every performance metric that we can associate with the ability of such a signal to convey information.

Channel models are mathematical constructs that intend to capture and represent the most relevant aspects of actual propagation channels. The art of channel modeling is faced with the challenge of striking a balance between realism and tractability. The world in which wireless systems operate is rife with complexity, and hence the mechanisms of radio propagation are highly involved. While it may be tempting to keep adding further layers of detail and realism to increase the generality and accuracy of a given channel model, this must be weighed against the danger of cluttering the ensuing analysis and observations. This tension is never fully resolved, and thus a wireless engineer must exercise judgment and apply, from a channel modeling toolbox as comprehensive as possible, those particular tools that capture what is essential to the problem at hand. The choice of models should promote usefulness and traction while avoiding artifacts that can misguide and confuse, always keeping in mind that the modeling assumptions largely condition the scope and validity of the end results [194, 195].

We begin this chapter, in Section 3.2, with an overview of the radio propagation mechanisms that are important for terrestrial wireless communication: transmission, reflection, diffraction, and scattering. On the basis of these mechanisms, we motivate a general channel model composed of the product of two terms: large-scale and small-scale. The large-scale (or macroscopic) term models phenomena that cause signal variations noticeable only over a scale of many wavelengths. The small-scale (or microscopic) term, in turn, models phenomena that affect the signal over scales comparable to the wavelength. Section 3.3 is devoted to the large-scale term. We present models that are inspired by measurements and actually employed in cellular and WLAN system design; for these models, the potential MIMO nature of the wireless link is largely immaterial. Next, in Section 3.4, we review models for the small-scale term; the key aspect captured by these models is multipath propagation and the fading fluctuations that it elicits, highly selective in time, frequency, and space. Then, Section 3.5 brings into the discussion some essential notions related to antenna arrays, setting the stage for the presentation in Section 3.6 of a number of MIMO channel models with emphasis on those that are to be invoked throughout the book. With the benefit of the preceding exposition, Section 3.7 revisits the issue of channel estimation, already touched on in Chapter 2, and obtains handy expressions of great analytical value. Finally, the chapter concludes in Section 3.8 with a review of models that have helped to develop actual wireless standards. The use of standardized channel models facilitates comparisons and design decisions during the development, in a manner that is unified and agreed upon by different parties. We briefly introduce some of these models and direct the reader to suitable references for a more detailed coverage.

The chapter focuses squarely on channel modeling at microwave frequencies, ranging between a few hundred megahertz and a few gigahertz, which is where existing wireless systems largely operate. At the end of the chapter, some observations are made with respect to the current interest in extending the operation to higher frequencies where vast amounts of unused bandwidth are available.

3.2 Preliminaries

3.2.1 Basics of radio propagation

Radio propagation has been studied extensively and continues to be a favorite research topic. A vast body of empirical measurements as well as numerous analytical results are available [196–198]. Distilling that accumulated knowledge, let us briefly introduce various physical mechanisms that enable the propagation of signals in terrestrial wireless channels at microwave frequencies.

A signal transmitted into a wireless medium can reach the receiver by means of a number of mechanisms. A signal component that reaches the receiver through a single path, without suffering any reflections, diffractions, or scattering, is referred to as *line-of-sight* (LOS). Necessarily, a LOS component has the shortest delay among all possible received components and it is typically the strongest one. Signal components that reach the receiver through reflection, scattering, or diffraction, are referred to as *non-line-of-sight* (NLOS).

- When a wave bounces off a smooth object that is exceedingly large relative to the wavelength, the wave is said to have undergone *reflection*. Each incoming wave is mapped onto a single reflected direction.
- *Scattering* is said to have occurred when a wave travels through a medium that contains objects whose dimensions are smaller than or comparable to the wavelength, and where

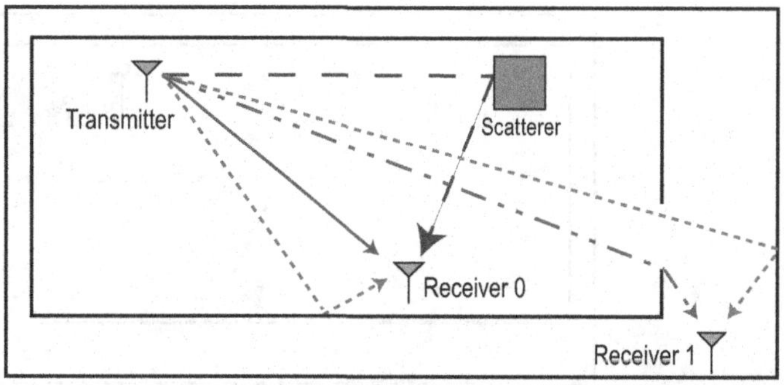

Fig. 3.1 The three types of NLOS propagation: dotted lines indicate reflection, dashed lines indicate scattering, and dash-dot lines represent diffraction. The solid line, in turn, indicates an LOS component.

the number of such objects is large. Then, each incoming wave is mapped onto many scattered ones.

- Radio waves also bend around sharp edges, a phenomenon that is referred to as *diffraction* and that plays a pivotal role in urban settings. Indeed, a major contributor to the propagation of radio signals in such settings is rooftop diffraction.

These various NLOS propagation mechanisms, illustrated in Fig. 3.1, give rise to the phenomena that we study in the sequel.

A consequential distinction for outdoor transceivers is whether they are located above or within the so-called *propagation clutter* of buildings, trees, vehicles, and suchlike. Roughly speaking, a transceiver is elevated if it is above the rooftop level, and it is within the clutter otherwise.

3.2.2 Modeling approaches

Channel models can be classified as either deterministic or stochastic. The former attempt to reproduce the propagation mechanisms (reflection, scattering, and diffraction) as they occur, at the expense of high computational complexity and limited applicability. The latter, alternatively, choose to regard the channel response as a stochastic process whose properties can be tuned on the basis of empirical measurements and/or physical considerations.

Deterministic modeling

The principal deterministic channel modeling technique is ray tracing, which aims at predicting the propagation characteristics of a specific site. This technique reproduces the propagation mechanisms in such a site, or more precisely in a computer recreation of it. This recreation must be highly accurate, not only in terms of the geometry but also of the constituent materials and their electromagnetic properties. Since the recreation of outdoor

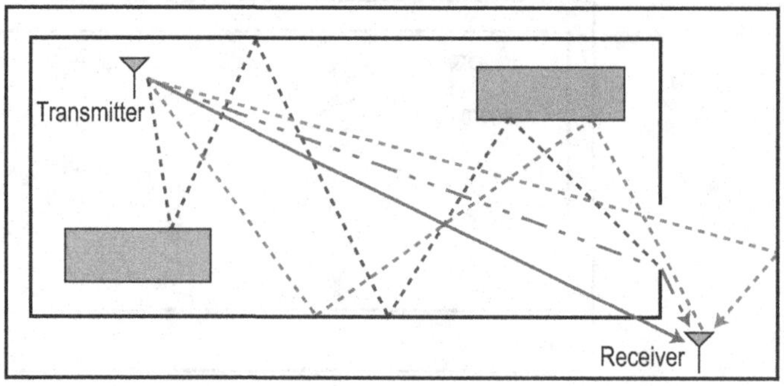

Fig. 3.2 Ray tracing in an indoor setting.

environments tends to be unwieldy, ray tracing is mostly restricted to indoor sites. As illustrated in Fig. 3.2, the simulating computer launches numerous waves from the transmitter and tracks their trajectories. The channel behavior is then reconstructed through an account of the features (e.g., delay, angle, and power) of all those waves that end up reaching the receiver. A number of software packages exist to implement ray tracing [199–201].

Ray tracing techniques offer the convenience of virtually testing the impact of modifying transmitter and receiver locations, antenna configurations, or other such parameters. The disadvantages of ray tracing include high computational requirements and the potential unavailability of fine environmental details such as interior furniture, which negatively impacts the accuracy.

As an alternative to ray tracing, it is possible to directly store channel responses measured in the field and apply them to offline performance evaluations.

Stochastic modeling

Quoting the French mathematician Henri Poincaré, "chance is only a measure of our ignorance." And yet, stochastic channel models offer so much analytical tractability, flexibility, and convenience that they deservingly are the workhorse of wireless communications and our focus henceforth. In general, a stochastic model is embodied by the conditional distribution of the channel output given its input. When the channel is linear, as is the case in the wireless medium, this is tantamount to a stochastic impulse response.

A critical modeling consideration in virtually all stochastic wireless channel models is the distinction between large-scale and small-scale phenomena. Specifically, what is assumed is that the distribution of the channel response is locally stationary within certain neighborhoods around transmitter and receiver [198]. (For the derivations in this chapter, wide-sense stationarity would suffice, but there is no reason not to directly assume strict-sense stationarity.) Although it is difficult to pinpoint the precise size of the local neighborhoods where stationarity does hold approximately, it is on the order of tens to hundreds of wavelengths. Large-scale phenomena impact the local distribution to which the small-scale phenomena conform, chiefly the local-average power. With the possible

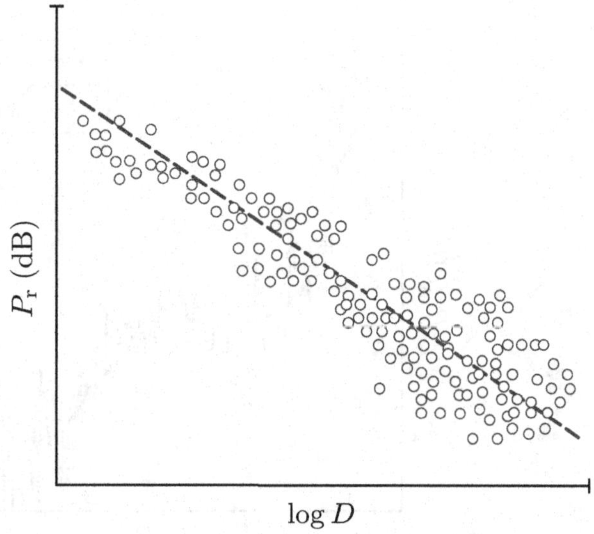

Fig. 3.3 Local-average received power (in some dB scale) as a function of the transmit–receive distance D (in log-scale). Cartoon example of a cloud of empirical values and the corresponding linear regression.

exception of massive MIMO, where antenna arrays may become exceedingly large [202], this clean separation holds fairly approximately in practice and greatly facilitates analytical characterizations that could otherwise be unwieldy.

Because of the importance of the decoupling of large- and small-scale phenomena, we use distinct variables to model them and carry those throughout the derivations in the entire book. While this takes a small toll in terms of the compactness of some expressions, it reinforces the separation between quantities that change at very different scales in time and frequency, and which have markedly different effects on the communication.

3.3 Large-scale phenomena

3.3.1 Pathloss and shadow fading

Under the separation argued in the previous paragraph, large-scale phenomena are those that determine the local distribution of the stochastic channel response. The prime feature of that distribution is, of course, the power. If one plots, for a given transmit power, the local-average received power (in log-scale) obtained from many empirical measurements in otherwise identical conditions as a function of the distance (also in log-scale) between transmitter and receiver, the result is invariably a cloud of scattered points (see Fig. 3.3). Through a linear regression, one can then find a law relating the local-average received power and the distance. Such a law represents what we call the *pathloss*, denoted by L_p,

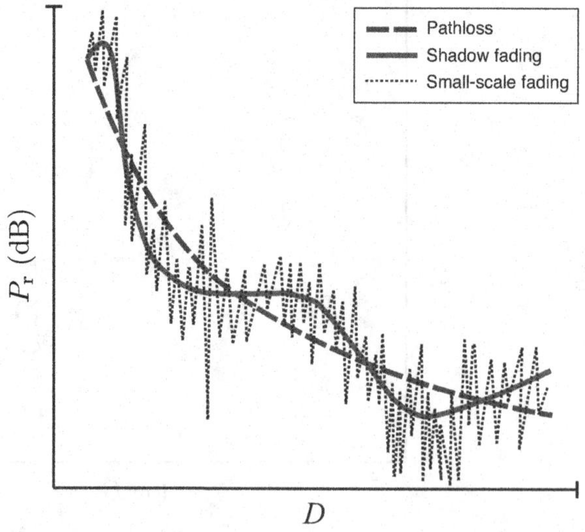

Fig. 3.4 Received signal power as a function of distance, illustrating the large-scale (pathloss and shadow fading) and the small-scale propagation phenomena.

and the spread of the scattered measurements around that law represents what we call *shadow fading*, denoted by χ. Together, the pathloss and the shadow fading conform the large-scale description of virtually every existing channel model. Although recent work indicates, on solid analytical footings based on random walks and diffusion theory, that nonlinear regressions may sometimes be superior [203, 204], the combination of pathloss and shadow fading suffices for our purposes.

The pathloss can be interpreted as the expected attenuation over an ensemble of realizations of a given propagation scenario, whereas the shadow fading individualizes, through randomization, the attenuation of each such potential realization. Another common interpretation of the large-scale phenomena is that the pathloss quantifies the local-average attenuation tied exclusively to distance whereas the shadow fading accounts for the presence of obstacles that cause the local-average attenuation to deviate from the value predicted by the pathloss. This is illustrated in Fig. 3.4, which depicts how both pathloss and shadow fading affect the evolution of the received signal power as the distance between transmitter and receiver increases over a generic trajectory; the pathloss is monotonic, whereas the shadow fading can oscillate as obstacles come in and out of the way.

It was suggested in [205] that the shadow fading attenuation, χ, could follow a log-normal distribution and empirical measurements have repeatedly confirmed the excellent fit to data offered by such distribution [206, 207]. The most widely accepted justification for this log-normal nature is that shadow fading represents the product of the losses introduced by the several obstacles traversed by the signal and, in log-scale, such a product becomes a sum of independent terms. Despite the fact that those terms can hardly be assumed to conform to the same distribution, and that the number of such terms may not be particularly large, the robustness of the central limit theorem (see Appendix C.1.10) renders the sum surprisingly well modeled by a Gaussian random variable. Under the log-normal model, $\chi|_{\text{dB}}$ is Gaussian with PDF

$$f_{\chi|_{\text{dB}}}(\xi) = \frac{1}{\sqrt{2\pi}\,\sigma_{\text{dB}}} \exp\left(-\frac{\xi^2}{2\,\sigma_{\text{dB}}^2}\right), \tag{3.1}$$

where σ_{dB} is the standard deviation, typically assumed independent of the distance for the sake of simplicity. (Dependences have been observed, but they are not easily characterized.) Outdoors, σ_{dB} tends to range between 8 and 12 dB. Indoors, it is usually smaller.

We note that, although $\chi|_{\text{dB}}$ is defined as being zero-mean in (3.1), there is no loss of generality once both pathloss and shadow fading are considered together: any nonzero-mean in $\chi|_{\text{dB}}$ can be incorporated directly to the log-scale pathloss.

Through a change of variables, the PDF of χ in linear scale can be found to be

$$f_{\chi}(\xi) = \frac{1}{\sqrt{2\pi}\,\sigma\xi} \exp\left(-\frac{(\log_e \xi)^2}{2\,\sigma^2}\right) \qquad \xi \geq 0, \tag{3.2}$$

where

$$\sigma = \frac{\sigma_{\text{dB}} \log_e(10)}{10}. \tag{3.3}$$

A relevant property confirmed through empirical observations is that, as intuition would have it, shadow fading can be broken up into distinct pieces: one that depends on the location of the transmitter and one that depends on the location of the receiver [207]. Two mutually distant receivers communicating with the same transmitter (or two mutually distant transmitters communicating with the same receiver) experience the same shadow fading for one piece but a completely different shadow fading for the other. These two pieces, whose values add up in log-scale, tend to have similar strengths. Furthermore, measurements reported in [208] point to an exponential model for the spatial autocorrelation of each piece, with a correlation distance that is on the order of tens of meters outdoors, depending on the environment, and substantially shorter indoors [209]. For more refinements on the modeling of shadow fading, the reader is referred to [210, 211] and references therein.

The local-average received power P_{r} can be expressed as a balance of gains and losses according to the so-called *link budget* equation [212]

$$P_{\text{r}}|_{\text{dB}} = P_{\text{t}}|_{\text{dB}} + G_{\text{t}}|_{\text{dB}} + G_{\text{r}}|_{\text{dB}} - (L_{\text{p}}|_{\text{dB}} + \chi|_{\text{dB}}), \tag{3.4}$$

where P_{t} is the transmit power, herein considered fixed, while G_{t} and G_{r} are, respectively, the gains of the single transmit and single receive antennas in the directions of propagation. Additional terms that can be added to the link budget equation include building penetration losses, cable losses, polarization mismatch between the transmit and receive antennas, or protection margins against impairments such as small-scale fading. In particular, additional power gains may arise with channel-dependent power control and with MIMO; since yet-to-see notions are required to express these gains, they are deferred to later in the book, respectively to Sections 4.2 and 5.2.

The link budget equation in (3.4) can be written more compactly as

$$P_{\text{r}}|_{\text{dB}} = P_{\text{t}}|_{\text{dB}} + G|_{\text{dB}}, \tag{3.5}$$

where

$$G = \frac{G_t G_r}{L_p \chi} \tag{3.6}$$

is the *large-scale channel gain*, a quantity already introduced in Chapter 2 to normalize the channel response and that now acquires its full significance.

Homing in on the pathloss, its most recurrent representation as a function of the transmission distance D is

$$L_p(D) = K_{\text{ref}} \left(\frac{D}{D_{\text{ref}}} \right)^{\eta}, \tag{3.7}$$

where η is the pathloss exponent while D_{ref} is a reference distance and K_{ref} is the pathloss at D_{ref}. In log-scale, this becomes

$$L_p(D)|_{\text{dB}} = K_{\text{ref}}|_{\text{dB}} + 10\,\eta \log_{10} \left(\frac{D}{D_{\text{ref}}} \right) \qquad D > D_{\text{ref}}. \tag{3.8}$$

The exponent η is the most consequential parameter, with typical values ranging between 3.5 and 4 if either the transmitter or the receiver are elevated. If both are within the propagation clutter, higher values can be encountered. Values below 2 are rare but possible in settings such as indoor corridors, outdoor urban canyons, or tunnels.

The parameter K_{ref} is sometimes referred to as the *intercept* at D_{ref} because it is the value at which a plot of $L_p(D)$ intercepts a vertical axis drawn at D_{ref}. Given a reference distance, K_{ref} can be measured empirically. Alternatively, if the reference distance is chosen properly, K_{ref} can be computed under the assumption that the propagation from the antenna up to that point follows the free-space law described next. In this case, the representation in (3.8) becomes the two-slope model in Fig. 3.5, namely free-space decay up to D_{ref} and decay with exponent η thereafter. This two-slope behavior can be justified analytically [51, section 2.4.1] and on the basis of experimental observations [213].

Besides the representation in (3.8), suitably complemented by log-normal shadow fading, several pathloss modeling alternatives are available. These alternatives are the subject of the remainder of this section. While the focus here is exclusively on how the large-scale propagation phenomena affect the local-average received power, models do exist for how other features of the local distribution (introduced later in this chapter) are affected, e.g., the Rice factor, the power delay profile, or the angle spread [214, 215].

3.3.2 Free-space model

The pathloss in free space, named after the radio pioneer Harald T. Friis [216], is given by

$$L_p(D)|_{\text{dB}} = 20 \log_{10} \left(\frac{4\pi D}{\lambda_c} \right), \tag{3.9}$$

where λ_c is the carrier wavelength. It holds only when D lies in the far-field region of the antennas, the rule of thumb being $D \geq 2\, d_a^2/\lambda_c$ where d_a is the largest physical dimension of either antenna. By definition, this model applies only when there are no obstructions between transmitter and receiver or in the vicinity of the trajectory linking them; this restricts

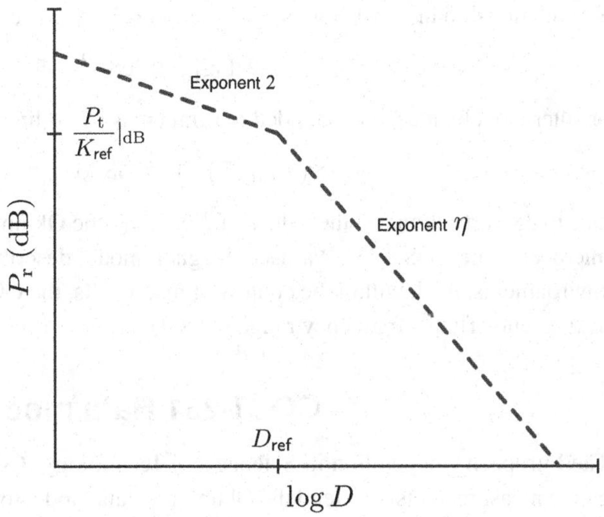

Fig. 3.5 Two-slope pathloss model.

its applicability to very specific situations, e.g., the computation of the intercept at a nearby point. There is no accompanying shadow fading.

The presence of objects of any kind, even simply the ground, alters the square-law distance decay in (3.9), yielding a more general exponent η.

3.3.3 Macrocell models

This section describes two models tailored to macrocells, defined as cells with radius exceeding roughly 1 km. These are empirical models, based on fitting parameters to match observed pathloss dependences on quantities such as frequency or antenna heights.

Okumura–Hata model

The Okumura–Hata model (or, in short, the Hata model) is a pathloss model based on empirical measurements conducted in Tokyo, Japan. In urban areas, it is given by [217]

$$\begin{aligned} L_p(D)|_{dB} &= 69.55 + 26.16 \log_{10} f_c - 13.82 \log_{10} h_b - b(h_m) \\ &\quad + (44.9 - 6.55 \log_{10} h_b) \log_{10} D, \end{aligned} \tag{3.10}$$

where f_c is the carrier frequency in megahertz ($150 \le f_c \le 1500$ MHz), h_b is the base station antenna height in meters ($30 \le h_b \le 200$ m), h_m is the mobile user antenna height also in meters ($1 \le h_m \le 10$ m), D is in kilometers ($D \ge 1$ km), and

$$b(h_m) = \begin{cases} (1.1 \log_{10} f_c - 0.7)\, h_m - 1.56 \log_{10} f_c + 0.8 & \text{small-to-medium cities} \\ 8.29\, [\log_{10}(1.54\, h_m)]^2 - 1.1 & \text{large cities, } f_c \le 300 \text{ MHz} \\ 3.2\, [\log_{10}(11.75\, h_m)]^2 - 4.97 & \text{large cities, } f_c \ge 300 \text{ MHz.} \end{cases} \tag{3.11}$$

The relationship in (3.10) can be extended to suburban areas by adding

$$-2\left[\log_{10}(f_c/28)\right]^2 - 5.4 \tag{3.12}$$

or, alternatively, it can be extended to rural areas by adding

$$-4.78\,(\log_{10} f_c)^2 + 18.33\,\log_{10} f_c - 40.94. \tag{3.13}$$

Due to the restrictions on the values of D and h_b, the Okumura–Hata model is not valid for microcells; the COST-231 Walfisch–Ikegami model described later extends it to microcell environments. And, within the context of macrocells, the COST-231 Hata model presented next extends Hata's frequency range to 2 GHz.

COST-231 Hata model

The European Cooperation in Science and Technology (COST) agency conducted propagation measurements to extend the Okumura–Hata model to frequencies between 150 MHz and 2 GHz [218]. In urban areas, the COST-231 Hata pathloss is

$$\begin{aligned} L_p(D)|_{dB} &= 46.3 + 33.9\,\log_{10} f_c - 13.82\,\log_{10} h_b - b(h_m) \\ &\quad + (44.9 - 6.55\,\log_{10} h_b)\log_{10} D \end{aligned} \tag{3.14}$$

with $b(h_m)$ as in (3.11). In dense metropolitan areas, an extra loss of 3 dB must be added. Further corrections are available for suburban and rural areas.

Example 3.1

Compute the pathlosses predicted by the Okumura–Hata and COST-231 Hata models in an urban environment (small city) for a base station antenna height of $h_b = 32$ m, a user height of $h_m = 1.5$ m, and $f_c = 1$ GHz at $D = 2$ km.

Solution

For a small city, the Okumura–Hata equations in (3.10) and (3.11) give $L_p = 137.76$ dB. The COST-231 Hata pathloss computed via (3.14) and (3.11) is $L_p = 137.73$ dB. At this frequency, as expected, the agreement is basically complete.

SUI models

The Stanford University interim (SUI) pathloss models can be applied to predict the pathloss in rural, suburban, and urban macrocells [213, 219]. The targeted application is fixed wireless access, with the user antennas rooftop-mounted and the base stations elevated 15–40 m over the clutter. The SUI models are divided depending on the type of terrain:

(A) Hilly terrains with heavy tree density.
(B) Mostly flat terrains with moderate-to-heavy tree densities or hilly terrains with light tree densities.
(C) Flat terrains with light tree densities.

Table 3.1 Values of a, b, and c in (3.16)

Parameter	Terrain A	Terrain B	Terrain C
a	4.6	4.0	3.6
b (m^{-1})	0.0075	0.0065	0.005
c (m)	12.6	17.1	20

The main equation is a refinement of (3.8), namely

$$L_p(D)|_{dB} = K_{ref}|_{dB} + 10\,\eta \log_{10}\left(\frac{D}{D_{ref}}\right) + X(f_c) + X(h_m) \qquad D > D_{ref}, \tag{3.15}$$

where $D_{ref} = 100$ m and $K_{ref}|_{dB}$ is the intercept at D_{ref} (computed assuming free-space propagation up to this reference distance). The pathloss exponent is defined as

$$\eta = \mathsf{a} - \mathsf{b}\,h_b + \frac{\mathsf{c}}{h_b}, \tag{3.16}$$

where the constants a, b, and c are given in Table 3.1 while $10 \leq h_b \leq 80$ m. In turn, $X(f_c)$ and $X(h_m)$ are correction factors associated with the carrier frequency and the user height, respectively, and defined as

$$X(f_c) = 6.0 \log_{10}\left(\frac{f_c}{2000}\right) \tag{3.17}$$

and

$$X(h_m) = \begin{cases} -10.8 \log_{10}(h_m/2000) & \text{Terrains A and B} \\ -20.0 \log_{10}(h_m/2000) & \text{Terrain C.} \end{cases} \tag{3.18}$$

3.3.4 Microcell models

Let us now shift our attention to microcells, defined as those with a radius ranging roughly between 100 m and 1 km. For such cells, we present the COST-231 Walfisch–Ikegami pathloss model.

COST-231 Walfisch–Ikegami model

This model, defined for frequencies between 800 MHz to 2 GHz, is an elaborate construction that accounts for parameters such as the distance between buildings, the street width, or the distance between the rooftop horizon and the base station. For the 3GPP parameters of base station height $h_b = 12.5$ m, building height of 12 m, building-to-building distance of 50 m, street width of 25 m, and user height $h_m = 1.5$ m, the pathloss under NLOS propagation is

$$L_p|_{dB} = -55.9 + 38 \log_{10} D + \left(24.5 + 1.5\,\frac{f_c}{925}\right) \log_{10} f_c, \tag{3.19}$$

where f_c is in megahertz while D, here in meters, must satisfy $D \geq 20$. Log-normal shadow fading should be added, with a recommended standard deviation $\sigma_{dB} = 10$ dB.

Under LOS propagation, the COST-231 Walfisch–Ikegami street canyon model with the aforementioned 3GPP parameters gives

$$L_p|_{\mathrm{dB}} = -35.4 + 26 \log_{10} D + 20 \log_{10} f_c. \qquad (3.20)$$

Here, the recommended standard deviation for the shadow fading is $\sigma_{\mathrm{dB}} = 4$ dB.

Example 3.2

Compute the pathloss at $D = 100$ m for a 2-GHz microcell under the 3GPP parameters.

Solution

Application of (3.19) gives $L_p = 111.7$ dB, which is 26 dB lower than the pathloss obtained in Example 3.1 for a 2-km macrocell. By transmitting only $P_t = 1$ W, a microcellular base station can generate an SNR that is 13 dB higher—everything else being equal—than what a macrocellular base station would generate with $P_t = 20$ W.

3.3.5 Picocell and indoor models

Like in the SUI case, the pathloss models for picocell and indoor environments tend to adopt a version of (3.8) with free-space decay up to D_{ref} and with an exponent on the order of $\eta = 3.5$ thereafter. The shadowing standard deviation in indoor picocells is markedly smaller than in outdoor large cells, typically ranging between 3 and 6 dB.

Example 3.3

Compute the pathloss at $D = 20$ m for a 5-GHz indoor WLAN with $D_{\mathrm{ref}} = 5$ m.

Solution

Applying the free-space equation in (3.9) up to $D_{\mathrm{ref}} = 5$ m we obtain $K_{\mathrm{ref}} = 30.2$ dB from which, subsequently, (3.8) gives $L_p = 51.3$ dB. Notice the several tens of dB of difference between this pathloss and the one obtained for an outdoor macrocell in Example 3.1, implying that, thanks to the much shorter range, a WLAN can attain the same SNR as a cellular system with a transmit power orders of magnitude lower and in the face of wider bandwidths (and thus higher noise powers). Even if the macrocell benefited from a typical 16.5-dBi base station antenna gain and the WLAN had ten times more bandwidth, the WLAN would still have a 60-dB advantage in link budget, meaning that despite transmitting 20 mW in lieu of 20 W, the WLAN would have a 30-dB edge in SNR.

Discussion 3.1 A note of caution

Although blurred by shadow fading, significant discrepancies may sometimes arise among the pathloss values returned by different models; this merely illustrates the difficulty of the modeling process [220]. Particular predictions are neither correct nor incorrect, but simply a better fit to specific scenarios. An Okumura–Hata prediction, for instance, is likely to be more accurate in Japan than in other parts of the world. In fact, it can be argued that stochastic channel modeling does not intend to reproduce specific environments, but rather to come up with descriptions that are representative of classes of environments of interest for the purpose of design and performance assessment.

3.4 Small-scale fading

Let us now zoom in and enter the small-scale realm. When dealing with continuous-time representations, we use the complex pseudo-baseband channel response. We also recall from Section 2.2.3 that, to discretize this representation, lowpass filtering is in general required before sampling, although in those cases when the discretization returns a single-tap response and the effect is only multiplicative, the complex gain of the pseudo-baseband channel can be borrowed directly.

3.4.1 Multipath propagation

Small-scale fading, also known as *multipath* fading because it is caused by multipath propagation, is perhaps the most defining feature of wireless channels. The propagation mechanisms of reflection, diffraction, and scattering, acting on objects surrounding the transmitter and the receiver within the respective local neighborhoods, give rise to a number of distinct propagation paths. Thus, what the receiver gets to observe is the superposition of a number of replicas of the transmitted signal; the overall channel response is the sum of the contributions over all those paths. Each such contribution is described by the attenuation and the delay that the signal traveling through that particular path is bound to experience, i.e., the complex pseudo-baseband impulse response over the qth path has the form $A_q\,\delta(\tau-\tau_q)\,e^{-j2\pi f_c \tau_q}$. Since they stem from objects within the local neighborhoods, all the relevant paths can be assumed to be subject to similar pathloss and shadow fading, suitably modified by the local reflection/diffraction/scattering, and thus the complex gains $\{A_q\}$ may not exhibit major differences. However, because at carrier frequencies of interest the phase $2\pi f_c \tau_q$ shifts radically with the slightest variation in τ_q, the differences in path delays are always sufficient to regard the path phases as independent random variables.

Suppose for now that the transmit signal $x(t)$ corresponds to a passband sinusoid at frequency f_c. Since the effect of a delay on such a signal is merely a phase shift, we can absorb each delay τ_q as a shift $2\pi f_c \tau_q$ in the phase of A_q. Then, the overall channel impulse response can be written as

$$\underbrace{\left(\sum_q A_q\, e^{-j2\pi f_c \tau_q}\right)}_{\text{Gain} = \sqrt{G}\cdot h} \delta(\tau) \tag{3.21}$$

and the signal reaching the receiver is $y(t) = \sqrt{G}h \cdot x(t)$. Thus, upon a sinusoid the channel does have only a multiplicative effect, i.e., a complex gain. Furthermore, the preceding consideration on the path phases suggests modeling this gain as a sum of IID random variables; if the number of paths is large enough and some mild conditions are satisfied [221], application of the central limit theorem leads to a complex Gaussian distribution. The result of the sum in (3.21) changes rapidly with the position of both transmitter and receiver: a mere displacement on the order of a wavelength alters the path lengths, and thus the delays thereon, by an amount that shifts the path phases drastically, causing major swings in the sum. Put differently, the paths can add constructively or destructively depending on their relative phases, which are completely modified by small changes in position. The same effect transpires if, instead of a change in position, there is a change in frequency that shifts the phases substantially.

The picture that emerges is thus as follows: the large-scale channel features (pathloss and shadow fading) determine the average channel conditions within the local neighborhood while the small-scale fading determines, subject to that local-average, the instantaneous channel conditions at each position and frequency. This invites decoupling the complex gain as in (3.21), namely $\sqrt{G} \cdot h$ where $\sqrt{G}$ determines the local-average received power, P_r, as per (3.5), whereas the small-scale fading h is normalized to have unit power. This largely achieves the desired separation between large-scale and small-scale features, yet a mild coupling remains in that certain other parameters of the local small-scale fading, e.g., the Rice factor, may depend on (or be correlated with) the large-scale gain. Ideally, the value of the large-scale features should be taken into account when setting those small-scale parameters, or else there is a danger that the hugely convenient decoupling between large- and small-scale gains entails a loss of relevant information.

The small-scale fading is locally modeled as a stationary random process, in principle zero-mean and complex Gaussian, normalized to have unit variance. At a given position and frequency, then, it is $h \sim \mathcal{N}_{\mathbb{C}}(0,1)$ with uniformly distributed phase and with a magnitude that abides by the Rayleigh distribution (see Appendix C.1.9), hence the popular designation of small-scale fading as simply *Rayleigh fading*. Precisely,

$$f_{|h|}(\xi) = \xi\, e^{-\frac{1}{2}\xi^2} \tag{3.22}$$

while the power $|h|^2$ follows the exponential distribution

$$f_{|h|^2}(\xi) = e^{-\xi}. \tag{3.23}$$

If a multipath term dominates over the rest, e.g., because it propagates through LOS and it is subject to a distinct pathloss, then the receiver can lock onto that signal component and render it unfaded. It follows that $h \sim \mathcal{N}_{\mathbb{C}}(\mu_h, \sigma_h^2)$ with mean $\mu_h \neq 0$ and with variance

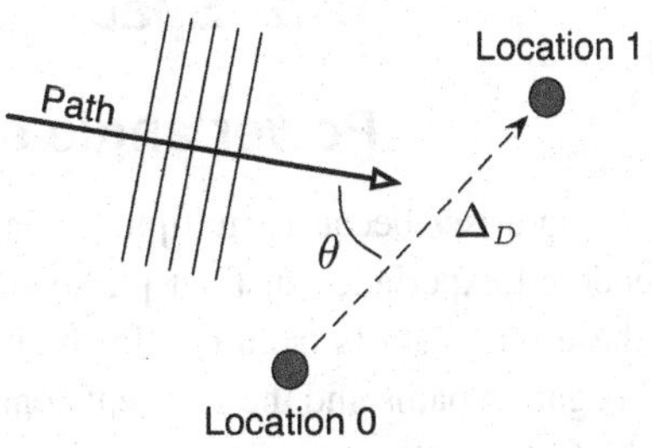

Fig. 3.6 Two nearby locations spaced by Δ_D and a propagation path at an angle θ relative to the segment connecting these locations.

$\sigma_h^2 = 1 - |\mu_h|^2$. For this so-called *Rice fading*, named after Stephen O. Rice [222],

$$f_{|h|}(\xi) = 2\,(\mathsf{K}+1)\,\xi\, e^{-(\mathsf{K}+1)\xi^2 - \mathsf{K}}\, I_0\!\left(2\sqrt{\mathsf{K}(\mathsf{K}+1)}\,\xi\right), \tag{3.24}$$

where $\mathsf{K} = |\mu_h|^2/\sigma_h^2$ is the Rice factor while $I_0(\cdot)$ is the zero-order modified Bessel function of the first kind (see Appendix E). For $\mathsf{K} = 0$, Rice fading reverts to Rayleigh; for $\mathsf{K} \to \infty$, the fading vanishes as the value of h becomes deterministic. Thus, the Rice distribution bridges these extremes and allows, through K, to regulate the severity of the fading. Analytically speaking, however, the PDF in (3.24) is not particularly friendly because of the presence of $I_0(\cdot)$. A more tractable alternative that also allows modeling varying degrees of fading severity is the Nakagami-m distribution, whereby

$$f_{|h|}(\xi) = \frac{2\,\mathsf{m}^{\mathsf{m}}\,\xi^{2\mathsf{m}-1}}{\Gamma(\mathsf{m})}\, e^{-\mathsf{m}\xi^2} \tag{3.25}$$

$$f_{|h|^2}(\xi) = \frac{\mathsf{m}^{\mathsf{m}}\,\xi^{\mathsf{m}-1}}{\Gamma(\mathsf{m})}\, e^{-\mathsf{m}\xi}, \tag{3.26}$$

with $\Gamma(\cdot)$ the Gamma function (see Appendix E) and with $\mathsf{m} \geq 1/2$ a tunable parameter. For $\mathsf{m} = 1$, the Rayleigh distribution is recovered while, for $\mathsf{m} \to \infty$, the fading vanishes. Moreover, for $\mathsf{m} = (\mathsf{K}+1)^2/(2\mathsf{K}+1)$, the Nakagami-m distribution closely approximates the Rice distribution with factor K. The flexibility of the Nakagami-m distribution allows modeling situations where the central limit theorem applies as well as others where it does not quite hold, either because the number of propagation paths is not large enough or because their gains are not sufficiently IID.

Distributions that further subsume the Nakagami-m as well as other forms of fading arising in satellite channels have been put forth, with the most general ones being the κ–μ and η–μ distributions [223]. To model terrestrial channels, though, the Rice and Nakagami-m distributions tend to suffice.

Altogether, an assortment of options that fits empirical data satisfyingly are available to model the marginal distribution of the small-scale fading. A more imposing challenge is to model its dynamics, i.e., the joint distribution of the channel response at multiple nearby positions/frequencies. This modeling is the topic of the remainder of this section.

3.4.2 Space selectivity

Power angle spectrum

Small-scale fading arises because multiple propagation paths exist and the signal replicas thereby engendered experience distinct phase shifts. Within the local neighborhood, the variation of these phase shifts from one location to another is governed by the angle between the propagation paths and the segment connecting those locations. This is illustrated in Fig. 3.6, which shows the segment connecting two locations and a propagation path at an angle θ thereabout. The change in path length from one location to the other is simply the projection of this segment along the path direction. Since this projection depends on θ, a central quantity in the characterization of small-scale fading should be a function establishing the angular distribution of paths, and indeed this is the case. This function is the *power angle spectrum* (PAS), which expresses the received signal power as a function of angle when acting as a receiver; when acting as a transmitter, equivalently, the PAS expresses the transmitted signal power *that reaches the receiver at the other end of the link* as a function of the angle of departure. The PAS is a continuous function of angle, normalized so it can be interpreted as a PDF, which is a definition consistent with the unit-power normalization of the small-scale fading. The standard deviation of the PAS, viewed as a distribution, gives the root-mean-square (RMS) *angle spread.*

Although, in principle, the definition of the PAS as a continuous function of angle models a diffuse propagation scenario, it does not preclude modeling situations where power is received only from a finite number of discrete angles; these situations can be accommodated by constructing the PAS on the basis of discrete delta functions.

Formally, the PAS is a function of both azimuth and elevation. However, empirical measurements have indicated that in most outdoor wireless systems the dependence on elevation plays only a secondary role [224]. Historically then, the focus has been on characterizing the marginal PAS on the azimuth plane—in fact, PAS can also be taken to stand for *power azimuth spectrum*—yet the elevation dimension should not be carelessly dismissed in other types of deployment, e.g., indoors or in vertical urban picocells. We return to this aspect at the end of the chapter.

Several functions have been proposed and applied to model the marginal PAS in azimuth, which in this text is denoted by $\mathcal{P}(\cdot)$.

Example 3.4 (Clarke–Jakes PAS)

For transceivers immersed in the propagation clutter, a simple model is the uniform PAS

$$\mathcal{P}(\theta) = \frac{1}{2\pi} \qquad \theta \in [-\pi, \pi). \tag{3.27}$$

Corresponding to the multipath setting in Fig. 3.7, this model was proposed in 1968 by Clarke [225] and subsequently popularized by Jakes [212].

Despite its simplicity, or perhaps because of it, the Clarke–Jakes PAS has been a powerful tool for decades. It fits surprisingly well the behavior observed at mobile users, cap-

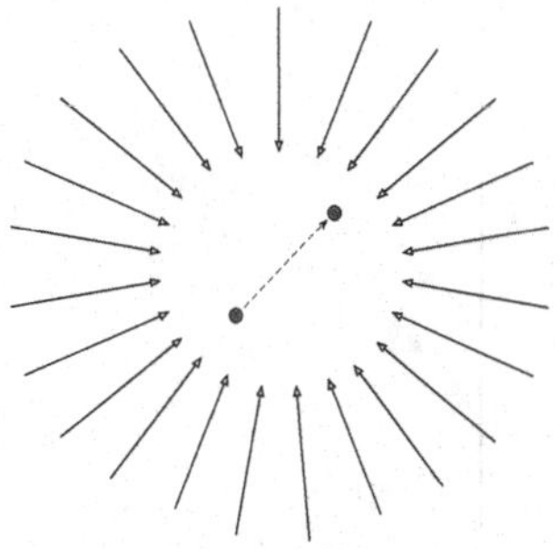

Fig. 3.7 Multipath setting modeled by a Clarke–Jakes uniform PAS.

turing the essence of multipath propagation in a manner that is friendly to analysis. Such friendliness, however, drops markedly as soon as the angular range is reduced below 2π [226, 227] and more convenient alternatives exist if directional preference (particularly important in transceivers located above the clutter, e.g., elevated base stations) is to be introduced. We next catalog several such options.

Example 3.5 (Power cosine PAS)

A directional azimuth PAS centered on an angle μ_θ can be modeled as [228]

$$\mathcal{P}(\theta) = K_c \left[\cos\left(\frac{\theta - \mu_\theta}{2} \right) \right]^{2q} \qquad \theta \in [\mu_\theta - \pi, \mu_\theta + \pi), \tag{3.28}$$

where q is an integer and

$$K_c = \frac{(q!)^2 \, 4^q}{2\pi \, (2q)!} \tag{3.29}$$

ensures that the function integrates to one.

Example 3.6 (Truncated Gaussian PAS)

An alternative to the power cosine PAS is the truncated Gaussian distribution [229, 230]

$$\mathcal{P}(\theta) = \frac{K_G}{\sqrt{2}\,\sigma_\theta} \exp\left(-\frac{(\theta - \mu_\theta)^2}{2\,\sigma_\theta^2} \right) \qquad \theta \in [\mu_\theta - \pi, \mu_\theta + \pi), \tag{3.30}$$

where

$$K_G = \frac{1}{\sqrt{\pi}\,[1 - 2\,Q(\pi/\sigma_\theta)]} \tag{3.31}$$

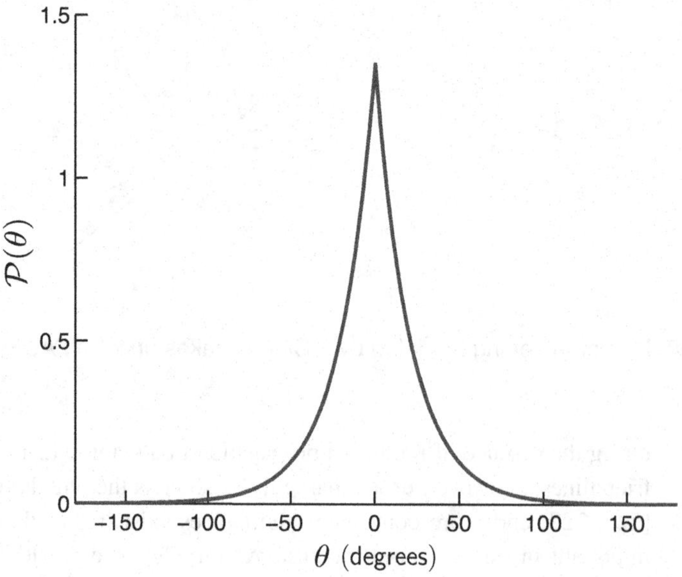

Fig. 3.8 Truncated Laplacian PAS for $\mu_\theta = 0$ and $\sigma_\theta = 0.52$ (which corresponds to 30°).

is the normalization factor needed to ensure that the function integrates to one, σ_θ is a measure of its spread (it would equal the RMS angle spread in the absence of truncation), and $Q(\cdot)$ is the Gaussian Q-function (see Appendix E).

Example 3.7 (Truncated Laplacian PAS)

Yet another alternative is a truncated Laplacian distribution (see Fig. 3.8)

$$\mathcal{P}(\theta) = \frac{K_{\mathrm{L}}}{\sqrt{2}\,\sigma_\theta} \exp\left(- \left| \frac{\sqrt{2}\,(\theta - \mu_\theta)}{\sigma_\theta} \right| \right) \qquad \theta \in [\mu_\theta - \pi, \mu_\theta + \pi), \tag{3.32}$$

where

$$K_{\mathrm{L}} = \frac{1}{1 - \exp\left(-\sqrt{2}\,\pi / \sigma_\theta\right)} \tag{3.33}$$

is, again, the normalization factor needed to ensure that the function integrates to one.

Besides the foregoing ones, yet other functions have been applied to model the PAS, e.g., the von-Mises distribution detailed in Section 3.6.2. Extensive measurement campaigns, both indoors [231–234] and outdoors [235–237], have shown a particularly satisfactory fit to the Laplacian PAS with a properly tuned σ_θ. As a result, (3.32) is a favorite choice.

It is also possible, at the expense of more cumbersome expressions, to consider a superposition of several functions of a given type to account for the existence of multiple scattering clusters [238].

Space correlation

For the sake of the exposition, let us for now posit that the transmitter remains fixed and let us study the fading dynamics as a function of the position of the receiver. The simultaneous dependence on the positions of both transmitter and receiver, essential for MIMO channel modeling, is deferred to later in the chapter.

The baseband noiseless received signal at a given location can be expressed as

$$y(t) = \sqrt{G} \int_{-\pi}^{\pi} \sqrt{\mathcal{P}(\theta)\, G_{\rm r}(\theta)}\, x(t - D(\theta)/{\rm c})\, {\rm d}\theta, \tag{3.34}$$

where $D(\theta)$ is the length of the path reaching the receiver at an angle θ and $D(\theta)/{\rm c}$ is its propagation delay given the speed of light c. Recalling that $G_{\rm r}$ denotes the receive antenna gain, i.e., the pinnacle of its pattern, we use $G_{\rm r}(\theta)$ to denote the entire antenna pattern as a function of angle. And, since the gain $G_{\rm r}$ was already incorporated within the large-scale channel gain G, the function $G_{\rm r}(\theta)$ is normalized to keep the small-scale fading unit power. For a sinusoid with wavelength $\lambda_{\rm c} = {\rm c}/f_{\rm c}$, which in baseband is a constant x, the delay of the signal replica received through angle θ maps to a phase shift of $\phi(\theta) = 2\pi D(\theta)/\lambda_{\rm c}$ and thus

$$y = \sqrt{G} \int_{-\pi}^{\pi} \sqrt{\mathcal{P}(\theta)\, G_{\rm r}(\theta)}\; x\, e^{-{\rm j}\phi(\theta)} {\rm d}\theta \tag{3.35}$$

$$= \sqrt{G}\, x \int_{-\pi}^{\pi} \sqrt{\mathcal{P}(\theta)\, G_{\rm r}(\theta)}\, e^{-{\rm j}\phi(\theta)} {\rm d}\theta. \tag{3.36}$$

It follows that $y = \sqrt{G}\, h_0\, x$ with

$$h_0 = \int_{-\pi}^{\pi} \sqrt{\mathcal{P}(\theta)\, G_{\rm r}(\theta)}\, e^{-{\rm j}\phi(\theta)}\, {\rm d}\theta, \tag{3.37}$$

which, as reasoned earlier, can be modeled as a complex Gaussian random variable under certain assumptions regarding the uniformity of the PAS and the independence of the phases viewed as random variables.

At a second location spaced Δ_D from the first one, the length of each propagation path changes by $\Delta_D \cos(\theta)$ as shown in Fig. 3.6 and thus the complex channel gain becomes

$$h_1 = \int_{-\pi}^{\pi} \sqrt{\mathcal{P}(\theta)\, G_{\rm r}(\theta)}\, e^{-{\rm j}[\phi(\theta) + 2\pi \Delta_D \cos(\theta)/\lambda_{\rm c}]}\, {\rm d}\theta. \tag{3.38}$$

The correlation between h_0 and h_1 is

$$R_h(\Delta_D) = \mathbb{E}\big[h_0\, h_1^*\big] \tag{3.39}$$

$$= \mathbb{E}\left[\int_{-\pi}^{\pi}\int_{-\pi}^{\pi} \sqrt{\mathcal{P}(\theta_0)\, G_{\rm r}(\theta_0)\, \mathcal{P}(\theta_1)\, G_{\rm r}(\theta_1)}\, e^{-{\rm j}[\phi(\theta_0) - \phi(\theta_1) - 2\pi\Delta_D \cos(\theta_1)/\lambda_{\rm c}]}\, {\rm d}\theta_0 {\rm d}\theta_1\right]$$

$$= \int_{-\pi}^{\pi}\int_{-\pi}^{\pi} \sqrt{\mathcal{P}(\theta_0)\, G_{\rm r}(\theta_0)\, \mathcal{P}(\theta_1)\, G_{\rm r}(\theta_1)}\, \mathbb{E}\Big[e^{-{\rm j}[\phi(\theta_0) - \phi(\theta_1) - 2\pi\Delta_D \cos(\theta_1)/\lambda_{\rm c}]}\Big] {\rm d}\theta_0 {\rm d}\theta_1 \tag{3.40}$$

with expectation over the phase ϕ at each angle. Under the very reasonable assumption

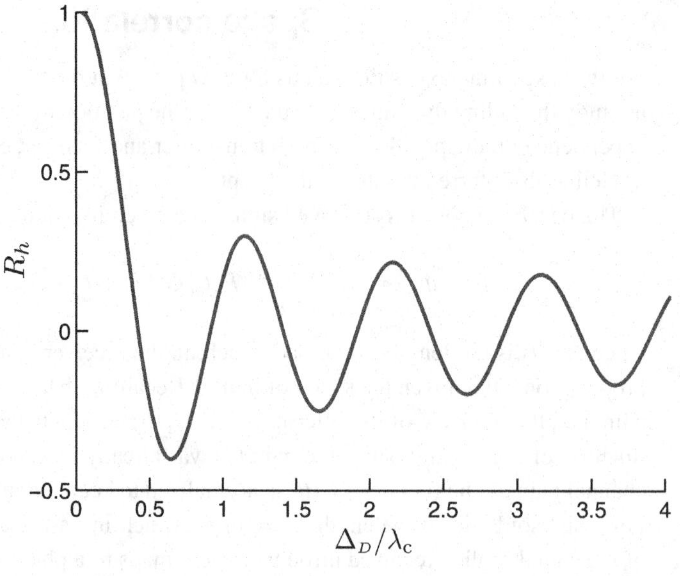

Fig. 3.9 Space correlation function corresponding to the Clarke–Jakes PAS as a function of distance (in wavelengths).

that the phase at each angle is uniformly distributed in $[-\pi, \pi)$, this expectation is nonzero only for $\theta_0 = \theta_1$; then, the dependence on ϕ vanishes, giving

$$R_h(\Delta_D) = \int_{-\pi}^{\pi} \mathcal{P}(\theta)\, G_r(\theta)\, e^{j2\pi\Delta_D \cos(\theta)/\lambda_c}\, d\theta. \tag{3.41}$$

This expression, which is a transformation of the function $\mathcal{P}(\theta)\, G_r(\theta)$, can now be particularized to specific antenna patterns and to specific PAS functions such as the ones surveyed earlier. The antenna pattern $G_r(\cdot)$ can in principle be designed to modify the environment's PAS, and thus $R_h(\cdot)$, at the expense of failing to capture all the incoming power [239, 240]. However, the habitual design principle is to apply an antenna pattern that is broader than the PAS, and approximately flat thereupon, so as to essentially capture all of the power. If the PAS is unknown or the antenna cannot be properly pointed, as is often the case, a uniform pattern is then preferable and (3.41) simplifies to

$$R_h(\Delta_D) = \int_{-\pi}^{\pi} \mathcal{P}(\theta)\, e^{j2\pi\Delta_D \cos(\theta)/\lambda_c}\, d\theta. \tag{3.42}$$

Example 3.8 (Space correlation function for the Clarke–Jakes PAS)

For $\mathcal{P}(\theta) = \frac{1}{2\pi}$, the integral in (3.42) gives

$$R_h(\Delta_D) = J_0(2\pi\Delta_D/\lambda_c), \tag{3.43}$$

where $J_0(\cdot)$ is the zero-order Bessel function of the first kind (see Appendix E). This space correlation function is depicted in Fig. 3.9.

For most other PAS functions, (3.42) does not admit a closed form but it can be computed as a series in cylindrical coordinates [241] or else numerically.

In settings where the dependence on elevation is relevant, it is natural to resort to spherical coordinates, and for the simplest case of a PAS uniform over all points of a sphere the space correlation function emerges also in closed form.

Example 3.9 (Space correlation function for a uniform spherical PAS)

Uniform integration over a sphere gives [242]

$$R_h(\Delta_D) = \mathrm{sinc}(2\pi\Delta_D/\lambda_c). \tag{3.44}$$

For certain other PAS functions involving both azimuth and elevation, series expansions in spherical coordinates have been given [241]. Henceforth, we again focus on PAS functions involving only azimuth.

Often, the space correlation function is reduced to a single characterizing quantity, the *coherence distance*.

Example 3.10 (Coherence distance for the Clarke–Jakes PAS)

For the Clarke–Jakes PAS, a natural choice for the coherence distance is the first zero of the function $J_0(\cdot)$, which occurs when its argument equals 2.4. Denoting by D_c such coherence distance, the condition $2\pi D_c/\lambda_c = 2.4$ gives

$$D_c = 0.38\,\lambda_c. \tag{3.45}$$

Example 3.11

Compute the coherence distance with a Clarke–Jakes PAS at a frequency of 2 GHz.

Solution

Applying (3.45), $D_c = 5.7$ cm. The value at this typical microwave frequency illustrates the tiny displacements over which the small-scale fading can exhibit sweeping variations.

The value $D_c = 0.38\,\lambda_c$, often rounded to $\lambda_c/2$, has become over the years a rule of thumb for the distance over which small-scale fading decorrelates in cluttered environments. Besides the first zero of $R_h(\cdot)$, other definitions of coherence distance are possible, and in fact necessary if $R_h(\cdot)$ does not exhibit zeros.

3.4.3 Time selectivity

Time correlation

The function $R_h(\Delta_D)$ studied in the previous section characterizes the correlation between the fading experienced at two nearby locations. This can serve to characterize the correlation between two nearby antennas, and also the time correlation of the time-varying complex gain $h(t)$ undergone by a single antenna in motion. At a velocity v, a distance Δ_D is covered in a time $\Delta_t = \Delta_D/\mathrm{v}$ and thus, with a simple change of variables, the expressions for space correlation map directly onto expressions for time correlation. From (3.42) then, the time correlation function is

$$R_h(\Delta_t) = \int_{-\pi}^{\pi} \mathcal{P}(\theta)\, e^{\mathrm{j}2\pi f \Delta_t \cos(\theta)\mathrm{v}/c}\, \mathrm{d}\theta, \tag{3.46}$$

where, taking advantage of the fact that their arguments are scaled versions of one another, we have slightly abused notation and reused $R_h(\cdot)$ for both space and time correlation.

Example 3.12 (Time correlation function for the Clarke–Jakes PAS)

Applying $\Delta_t = \Delta_D/\mathrm{v}$ to Example 3.8, or plugging $\mathcal{P}(\theta) = \frac{1}{2\pi}$ into (3.46),

$$R_h(\Delta_t) = J_0\Big(2\pi \frac{\mathrm{v}}{c} f_\mathrm{c}\, \Delta_t\Big). \tag{3.47}$$

Applying the same change of variables, the coherence time of the small-scale fading observed by a moving antenna is seen to be $T_\mathrm{c} = D_\mathrm{c}/\mathrm{v}$.

Example 3.13

For velocities $\mathrm{v} = 5$ km/h and $\mathrm{v} = 100$ km/h, compute the coherence time with a Clarke–Jakes PAS at a frequency of 2 GHz.

Solution

From Example (3.11), $D_\mathrm{c} = 5.7$ cm. Thus, $T_\mathrm{c} = 41$ ms at 5 km/h and $T_\mathrm{c} = 2$ ms at 100 km/h.

Notice how fading changes over time scales measured in milliseconds. Wireless systems need to react to dramatic channel fluctuations at these time scales.

Caution must be exercised when assessing the coherence time at $\mathrm{v} = 0$, which the above models would declare unbounded. Background motion of objects in the local neighborhood (e.g., vehicles, people, foliage), relatively negligible when v is positive, comes to the fore at $\mathrm{v} = 0$ curbing the coherence time at values of a few hundred milliseconds.

Doppler spectrum

Recall that, as the receiver travels a distance Δ_D, the length of a path at an angle θ changes by $\Delta_D \cos(\theta)$. This rotates the phase of a signal at frequency f_c by $2\pi\Delta_D\cos(\theta)/\lambda_\mathrm{c} = 2\pi f_\mathrm{c}\Delta_t \mathrm{v}\cos(\theta)/\mathrm{c}$. Since the time incurred to travel Δ_D is precisely Δ_t, this phase rotation corresponds to a frequency $f_\mathrm{c}\mathrm{v}\cos(\theta)/\mathrm{c}$. In other words, a signal of frequency f_c arriving over a path an angle θ relative to the direction of motion suffers a frequency increase (which can be negative, hence a decrease) equal to $\nu = f_c \mathrm{v} \cos(\theta)/\mathrm{c}$. This frequency shift, a mere manifestation of the Doppler effect, is within the range $[-\nu_M, \nu_M]$, where

$$\nu_M = \frac{\mathrm{v}}{\mathrm{c}} f_c \tag{3.48}$$

is aptly termed the *maximum Doppler shift*. Since every pair of paths at angles $\pm\theta$ map to a unique Doppler shift, (3.46) can be rewritten as an integral over Doppler shifts rather than angles. Precisely, introducing the change of variables $\nu = \nu_M \cos(\pm\theta)$ and carefully discriminating the two opposite angles that map to the same ν [212, 243]

$$R_h(\Delta_t) = \int_{-\pi}^{0} \mathcal{P}(\theta)\, e^{j2\pi f_c \Delta_t \cos(\theta)\mathrm{v}/\mathrm{c}}\, d\theta + \int_{0}^{\pi} \mathcal{P}(\theta)\, e^{j2\pi f_c \Delta_t \cos(\theta)\mathrm{v}/\mathrm{c}}\, d\theta \tag{3.49}$$

$$= \int_{-\nu_M}^{\nu_M} \frac{\mathcal{P}(-\arccos(\nu/\nu_M))}{\sqrt{\nu_M^2 - \nu^2}}\, e^{j2\pi\nu\Delta_t} d\nu + \int_{-\nu_M}^{\nu_M} \frac{\mathcal{P}(\arccos(\nu/\nu_M))}{\sqrt{\nu_M^2 - \nu^2}}\, e^{j2\pi\nu\Delta_t} d\nu \tag{3.50}$$

$$= \int_{-\nu_M}^{\nu_M} \frac{\mathcal{P}(-\arccos(\nu/\nu_M)) + \mathcal{P}(\arccos(\nu/\nu_M))}{\sqrt{\nu_M^2 - \nu^2}}\, e^{j2\pi\nu\Delta_t} d\nu \tag{3.51}$$

where we have used

$$\frac{d}{d\xi} \arccos(\xi) = -\frac{1}{\sqrt{1-\xi^2}}. \tag{3.52}$$

Eq. (3.51) is nothing but a Fourier transformation and thus the time correlation function $R_h(\Delta_t)$ is seen to be the Fourier transform of

$$S_h(\nu) = \begin{cases} \dfrac{\mathcal{P}(-\arccos(\nu/\nu_M)) + \mathcal{P}(\arccos(\nu/\nu_M))}{\sqrt{\nu_M^2 - \nu^2}} & \nu \in [-\nu_M, \nu_M] \\ 0 & \nu \notin [-\nu_M, \nu_M] \end{cases} \tag{3.53}$$

which is termed *Doppler spectrum* because it gives the density of power received at each Doppler shift (under the condition, as we recall, that the antenna pattern does not modify the PAS). For a stationary random process, indeed, the Fourier transform of the autocorrelation function gives its power spectrum (see Appendix C.3). When that random process corresponds to a multipath fading channel, the power spectrum is embodied by the Doppler spectrum and it can be interpreted as follows: a transmit sinusoid of frequency f_c gets smeared into received components at every frequency within $[f_c - \nu_M, f_c + \nu_M]$ as indicated by $S_h(\nu)$. As can be verified

$$\int_{-\pi}^{\pi} \mathcal{P}(\theta)\, d\theta = \int_{-\nu_M}^{\nu_M} S_h(\nu)\, d\nu. \tag{3.54}$$

Using (3.53), the Doppler spectrum can be readily obtained for the PAS functions introduced earlier in this section. Moreover, for all these PAS functions the Doppler spectrum exists free of delta functions and hence, as explained in Appendix C.3.2, the assumption of (local) stationarity carries with it the characteristic of (local) ergodicity. The correspondence between stationarity and ergodicity would not hold if the Doppler spectrum featured

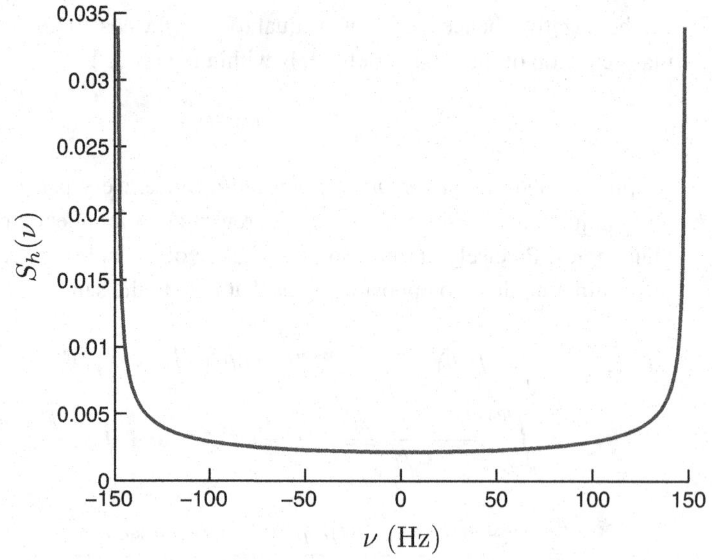

Fig. 3.10 Clarke–Jakes Doppler spectrum for $\nu_{\rm M} = 148$ Hz. This corresponds, for instance, to a velocity of $\mathrm{v} = 80$ km/h at $f_{\rm c} = 2$ GHz.

delta functions, i.e., in the face of LOS components; this is just as well because LOS components are regarded as deterministic as far as the small-scale distributions are concerned and thus they can be excluded from the spectrum.

Example 3.14 (Clarke–Jakes Doppler spectrum)

For $\mathcal{P}(\theta) = \frac{1}{2\pi}$,

$$S_h(\nu) = \begin{cases} \dfrac{1}{\pi\sqrt{\nu_{\rm M}^2 - \nu^2}} & \nu \in [-\nu_{\rm M}, \nu_{\rm M}] \\ 0 & \nu \notin [-\nu_{\rm M}, \nu_{\rm M}] \end{cases} \tag{3.55}$$

which is the Fourier transform of the time correlation function in Example 3.12, more compactly rewritten by means of $\nu_{\rm M}$ as

$$R_h(\Delta_t) = J_0(2\pi\nu_{\rm M}\Delta_t). \tag{3.56}$$

The Clarke–Jakes spectrum has the distinctive U-shape depicted in Fig. 3.10.

Besides the ones that can be obtained by transforming PAS functions given earlier in this text, additional Doppler spectra have been proposed, including the following.

Example 3.15 Bell-shape Doppler spectrum)

For $\mathrm{v} = 0$, when the dynamics are only caused by environmental motion, the fading is strongly nonzero-mean and the Doppler spectrum of the random component—which, with

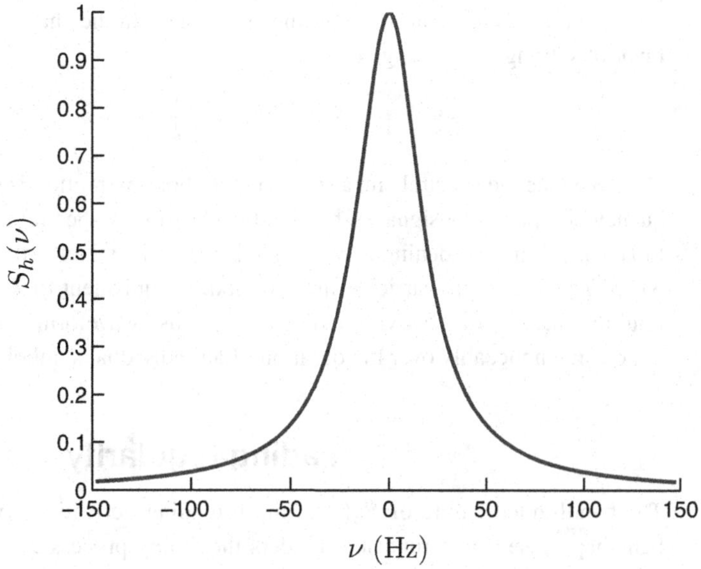

Fig. 3.11 Bell-shape Doppler spectrum for $\nu_{\mathrm{M}} = 148$ Hz. This corresponds, for instance, to a velocity of $\mathrm{v} = 80$ km/h at $f_{\mathrm{c}} = 2$ GHz.

a slight abuse of notation, we continue to denote by $S_h(\cdot)$—has been found to firmly fit the bell-shape spectral function [244]

$$S_h(\nu) = \frac{1}{1 + K_{\mathrm{B}} \left(\nu/\nu_{\mathrm{M}}\right)^2}. \tag{3.57}$$

Example 3.16

Depict the bell-shape spectrum for a velocity of $\mathrm{v} = 80$ km/h at a carrier frequency of $f_{\mathrm{c}} = 2$ GHz.

Solution

The maximum Doppler shift is $\nu_{\mathrm{M}} = f_{\mathrm{c}}\mathrm{v}/\mathrm{c} = 148.15$ Hz. Then, for the right-hand side of (3.57) to integrate to unity, we must set $K_{\mathrm{D}} = 55.23$. The resulting spectrum is shown in Fig. 3.11.

From the scaling property of the Fourier transform relating $R_h(\cdot)$ and $S_h(\cdot)$, namely that compression by a factor in one domain causes expansion by the same factor in the other domain, and vice versa, we can affirm that the coherence time satisfies

$$T_{\mathrm{c}} \propto \frac{1}{\nu_{\mathrm{M}}} \tag{3.58}$$

where the precise relationship depends on the shapes of $R_h(\cdot)$ and $S_h(\cdot)$, as well as on the degree of decorrelation in the definition of T_{c}. Although we cannot render (3.58) precise in a way that is universal in $R_h(\cdot)$ and $S_h(\cdot)$, later in the chapter we present arguments in favor of settling this scaling at

$$T_c = \frac{1}{2\,\nu_M}. \tag{3.59}$$

Also, since the spectral smearing brought about by motion is experienced by every frequency component, a signal of bandwidth B suffers a spectral broadening that expands it to $B + \nu_M$. This broadening is negligible if $\nu_M \ll B$, which, utilizing (3.58) and $B = 1/T$ where T is the single-carrier symbol period, is equivalent to $T \ll T_c$. This confirms the intuition that, for time-varying fading not to cause waveform distortion, the fading should not change noticeably over the duration of an individual symbol.

Fading regularity

The bandlimited nature of $S_h(\cdot)$ in the foregoing derivations (with the exception of the bell-shape case if not truncated) renders the fading process a *nonregular* random process in the sense of Section 1.3.3. In such a process, the present value is perfectly predictable from noiseless observations of the entire past. The bandlimited nature of $S_h(\cdot)$ descends directly from the modeling assumptions, specifically from having implicitly presumed that the signal incoming along each path bounces only once and from a stationary object. If we allowed for the possibility that the signal bounces back and forth between the receiver and the objects, or we allowed for such objects to be themselves in motion, then Doppler shifts larger than $\nu_M = f_c \mathrm{v}/c$ could be observed. In the limit, if $S_h(\cdot)$ were not bandlimited, then we would be in the presence of a *regular* fading process, for which the present value cannot be perfectly predicted from noiseless observations of the entire past.

Unless otherwise stated, we consider nonregular fading bandlimited to ν_M. While Doppler shifts exceeding this value are possible, the corresponding signal components are likely to be very weak relative to those within $[-\nu_M, \nu_M]$ and hence, barring exceedingly high SNRs, those components beyond $\pm\nu_M$ are likely to be buried well below the noise level.

Time discretization

For fading conforming to $h(t,\tau) = h(t)\delta(\tau)$ as we are considering thus far, the channel has the multiplicative effect $y(t) = \sqrt{G}h(t)x(t)$ and, as we learned in Example 2.3, the sampled pseudo-baseband gain $h[n] = h(nT)$ suffices for a discrete-time representation; we can directly write $y[n] = \sqrt{G}h[n]x[n]$. Strictly speaking, the time-discretization of $y(t)$ should entail a sampling period $\frac{1}{B+\nu_M}$ because the bandwidth B of the transmit signal increases to $B + \nu_M$ as the signal undergoes time-varying fading. However, provided that $\nu_M \ll B$, there is little loss in sampling with a period $T = 1/B$. By the same token that $h[n] = h(nT)$, the autocorrelation of $h[n]$ equals $R_h(\ell T)$ where $R_h(\tau)$ is the autocorrelation of $h(t)$. And, given $S_h(\nu)$, the Doppler spectrum of $h[n]$ equals

$$\frac{1}{T}\, S_h\!\left(\frac{\mathrm{v}}{T}\right) \qquad \mathrm{v} \in [-\nu_M T, \nu_M T], \tag{3.60}$$

with the factor $1/T$ ensuring that the scaling of the frequency axis ν to $\mathrm{v} = \nu T$ does not affect the power.

Simplified models

As mentioned at the outset of the chapter, the art of channel modeling must seek a compromise between realism and tractability, distilling what is essential for each purpose. When it comes to fading dynamics, it is sometimes desirable to have models that capture the essence of time selectivity but are otherwise stripped down to the bones. We next introduce two such models, formulated directly for the discrete-time complex channel $h[n] = h(nT)$.

Gauss–Markov model

A first possibility is to resort to a first-order Gauss–Markov process, or first-order autoregressive process, whereby the fading at symbol n satisfies

$$h[n] = \sqrt{1-\varepsilon}\, h[n-1] + \sqrt{\varepsilon}\, w[n], \tag{3.61}$$

where $\{w[n]\}$ is a sequence of IID random variables with $w \sim \mathcal{N}_{\mathbb{C}}(0,1)$. In turn, ε is the one-step prediction error, i.e., the variance of the error to which $h[n]$ can be predicted from a noiseless observation of $h[n-1]$, something that in a first-order Markov process is as informative as a noiseless observation of the entire past. By definition, a Gauss–Markov process is regular and its degree of decorrelation after ℓ symbols is determined by ε via

$$R_h[\ell] = \mathbb{E}\big[h[n]\, h^*[n+\ell]\big] \tag{3.62}$$

$$= (1-\varepsilon)^{\ell/2}. \tag{3.63}$$

Example 3.17

Let $T = 5$ μs and suppose we want a first-order Gauss–Markov fading channel that decorrelates by 50% over 10 ms. What should the value of ε be?

Solution

The number of 5-μs single-carrier symbols within a 10-ms interval equals 2000 and thus we can pose the equality

$$0.5 = (1-\varepsilon)^{1000} \tag{3.64}$$

from which $\varepsilon = 6.93 \cdot 10^{-4}$.

Block-fading model

A second possibility, illustrated in Fig. 3.12, is to resort to a block-fading structure, whereby the fading remains constant over blocks of duration T_c while changing values across blocks. It is worth noting that this model is not stationary but cyclostationary with period T_c. The fading values at consecutive blocks may be correlated, but are most often assumed IID.

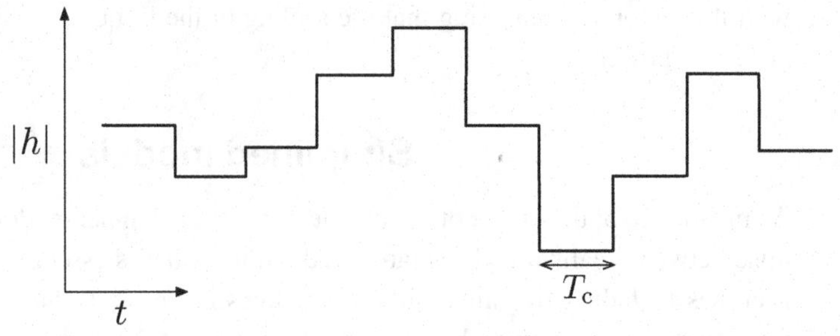

Fig. 3.12 Block-fading model.

Denoting by N_c the number of single-carrier symbols per block, we have that

$$N_c = \frac{T_c}{T} \tag{3.65}$$

$$= B\,T_c \tag{3.66}$$

and, if the blocks are IID, then it makes sense to set T_c to coincide with the coherence time of a stationary fading process. Recalling (3.59), this leads to

$$N_c = \frac{1}{2\,\nu_M T}. \tag{3.67}$$

However, care must be exercised with both (3.66) and (3.67) because they hold only if the bandwidth B is "not too large," where the precise meaning of "not too large" is to become clear in the next section.

The block-fading model can be broadened to allow for continuous variability within each block, with transitions to independent values across blocks. This more general block-fading model can sometimes serve as a convenient bridge with a true continuous fading model [245, 246].

3.4.4 Frequency selectivity

Frequency correlation

In our characterization of space and time selectivity, we have focused on the fading experienced by a sinusoid at frequency f_c. The derivations hold approximately for frequencies close enough to f_c, and thus the fading goes (roughly) unchanged over a certain frequency span. If the entire signal fits within a swath of frequencies that experience (roughly) the same fading, the fading is frequency-flat; the channel is then multiplicative, i.e., it merely applies a complex gain. The signal can also be said to be narrowband, although such narrowness must be understood only in reference to the fading.

If the signal bandwidth is large enough, however, it is bound to exceed the range of frequencies experiencing (roughly) the same fading. At a given location then, in static conditions, the noiseless received signal is $\mathsf{y}(f) = \sqrt{G}\,\mathsf{h}(f)\,\mathsf{x}(f)$ where $\mathsf{h}(f)$ is no longer constant over the bandwidth B. The signal undergoes frequency-selective fading, and it can be said to be fading-wise wideband; the channel is not merely multiplicative, but rather it effects a nontrivial convolution that modifies the shape of the symbol pulses. The degree of selectivity over a frequency range Δ_f is quantified by $R_{\mathsf{h}}(\Delta_f) = \mathbb{E}[\mathsf{h}(f)\mathsf{h}^*(f + \Delta_f)]$ whose chief measure is the *coherence bandwidth* B_{c}, the natural counterpart to the coherence time and the coherence distance. The fading is frequency-flat if $B \ll B_{\mathrm{c}}$.

Power delay profile

Frequency selectivity can be physically explained via the duality between the frequency and the delay domains. Thus far, we have attributed to the different delays experienced by the signal replicas traveling over distinct paths merely the effect of a different phase shift, and that is indeed the only effect upon a sinusoid. Then, in the absence of motion, the time-invariant impulse response corresponding to $\mathsf{h}(f)$ has the form $h(\tau) = h\,\delta(\tau)$ where h is the complex gain at the given position. With motion, this becomes $h(t,\tau) = h(t)\,\delta(\tau)$.

Sticking to motionless channels, when the signal is richer than a sinusoid, specifically a sequence of symbol pulses with period T, the dominant effect of each path's delay is still a phase shift as long as $\tau_{\mathsf{max}} - \tau_{\mathsf{min}} \ll T$ with τ_{max} and τ_{min}, respectively, the longest and shortest delays among all paths. This condition ensures that, relative to the much longer symbols, the channel still behaves approximately as $h(\tau) = h\,\delta(\tau)$. Put differently, delay differences much smaller than T are blurred by the lowpass filtering that the pulse shaping effects, and the corresponding signal replicas all fall within the main tap in the discretized $h[\ell]$. However, if the condition $\tau_{\mathsf{max}} - \tau_{\mathsf{min}} \ll T$ is not satisfied, the sequence of symbol pulses received over distinct paths no longer overlap fully and the replica of the nth symbol arriving over a given path collides with replicas of symbols other than n incoming over other paths (see Fig. 3.13). Multipath propagation then causes not only fading, but also signal distortion because of ISI. Relative to the symbol pulses, the channel response is revealed as not being simply an impulse scaled by a complex gain h, but rather as having a more general form $h(\tau)$; since it spreads the signals in delay, the channel is dispersive.

To be sure, any form of fading is the result of self-interference among multipath signal replicas of each symbol; the difference between frequency-flat and frequency-selective fading is that, in the former, every symbol essentially interferes exclusively with replicas of itself, whereas in the latter it experiences significant interference with adjacent symbols as well. The same exact channel can exhibit either frequency-flat or frequency-selective fading depending on T or, equivalently, depending on the bandwidth through which it is observed.

Ultimately, the delay-dispersive nature of fading channels, and thus the frequency selectivity, is rooted in the fact that the speed of light is finite. To appreciate this, the reader is referred to Fig. 3.14, which depicts a transmit sequence of symbols traveling over two distinct propagation paths. The finite value of c bestows upon the symbol separation a finite length in space and opens the door to distinct symbols colliding at the receiver if the difference in length between the paths is sufficiently large. In contrast, if c were infinite, then the symbol sequences received over the multiple paths would always align.

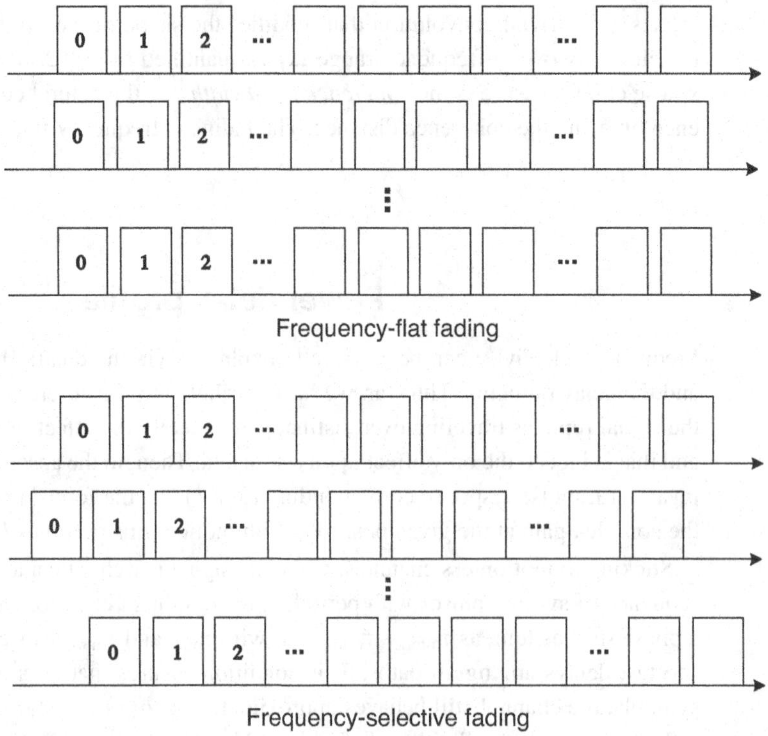

Fig. 3.13 Signal replicas received over multiple propagation paths. Above, a flat-faded channel, where each replica only experiences a different phase shift and magnitude gain. Below, a frequency-selective channel, where in addition there is ISI and distortion. (For the sake of the illustration, the symbol pulses are drawn as rectangles.)

Example 3.18

Consider Fig. 3.14 with a single-carrier symbol period $T = 3.7$ μs, which was the value employed in the 2G GSM system. Over the air, each such period maps to a length of 1.11 km. If the distance between transmitter and receiver exceeds this value, then multiple symbols are on the air at any given instant. If the difference in length between the two propagation paths shown in the figure were 500 m, then the corresponding signal replicas would have their sequences staggered by almost half a symbol, and the ensuing ISI would have to be equalized [247]. Even substantially shorter path length differences, in fact, would require equalization at the receiver.

The foregoing interpretation can be further stretched to motivate the use of OFDM: since ISI arises because of the finite length of the symbols, the antidote to ISI is to make the symbols longer.

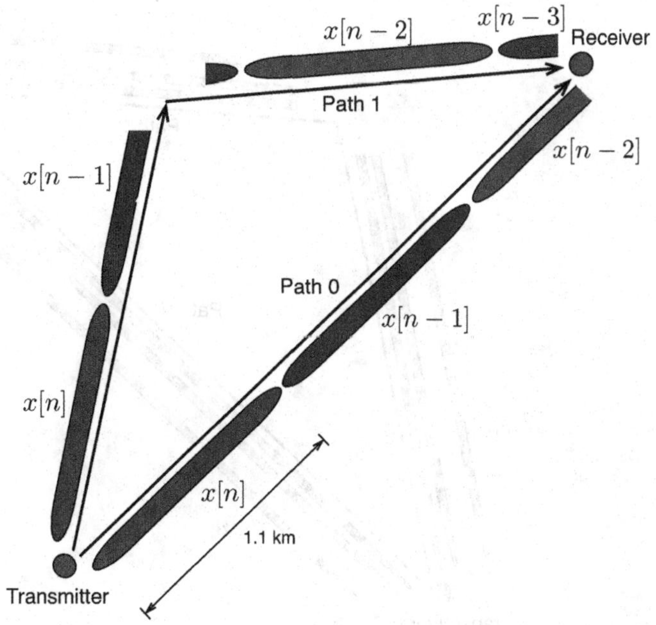

Fig. 3.14 Symbol sequence traveling over two distinct propagation paths with lengths differing by substantially more than the symbol period. The sequences received over the two paths are staggered by about half a period, which for $T = 3.7$ μs would happen if the path lengths differed by about 500 m.

Example 3.19

In LTE, OFDM is employed with $T_{\text{OFDM}} = 71$ μs, mapping to an over-the-air length of 21.3 km. For any reasonable transmission range, then, a single OFDM symbol is on the air at any given instant. Most importantly, there is a cyclic prefix of 4.7 μs between consecutive symbols, mapping to a safety length of 1.41 km. As long as the length difference between propagation paths does not exceed this considerable safety value, no ISI is experienced.

Figure 3.15 reminds us that OFDM's robustness against ISI comes at the expense of parallelizing the transmission in the frequency domain, with narrower subcarriers that are more vulnerable to the spectral smearing caused by Doppler spread.

The distribution of received power as a function of delay, normalized so it can be interpreted as a PDF, is termed power delay profile (PDP) and denoted by

$$S_{\mathsf{h}}(\tau) = \mathbb{E}\left[|h(\tau)|^2\right]. \tag{3.68}$$

Although in general a continuous function of delay, discrete PDP descriptions are also possible, and in fact fairly common. In our formulation, these are readily accommodated by means of delay-shifted delta functions, each modeling a group of paths—a *ray*—with similar delays. Because of the plurality of paths per ray, each ray is still subject to fading.

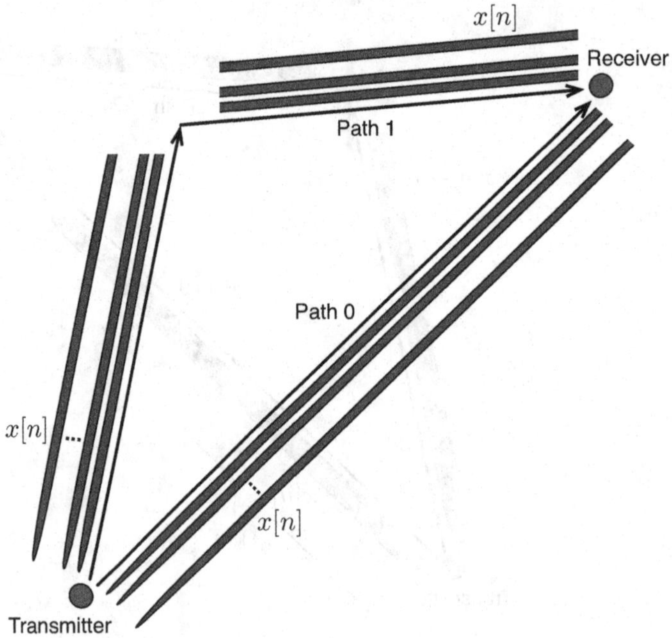

Fig. 3.15 OFDM signal traveling over two distinct propagation paths. The staggering of the two replicas does not exceed the cyclic prefix and thus there is no ISI.

Example 3.20 (Typical urban PDPs)

A family of typical urban discrete PDPs, with different numbers of rays, have been utilized to test several generations of cellular systems. Summarized in Table 3.2 are the ray delays and average power shares in a version of such PDP [248].

Example 3.21 (ITU PDPs)

Another common set of discrete PDPs are those recommended by the International Telecommunications Union (ITU) [249]. These recommendations specify different test environments: indoor, pedestrian, and vehicular. In addition, each such environment splits onto a low-dispersion version (labeled as "A") and a high-dispersion version (labeled as "B"). Altogether then, six different variations exist, two of which are reproduced in Table 3.3.

For indoor environments in particular, the PDP described in the following example has particular historical relevance.

Example 3.22 (Saleh–Valenzuela PDP)

In the 1980s, Adel Saleh and Reinaldo Valenzuela conducted wideband indoor measurements that revealed that the PDP rays tended to exhibit a clustered structure with a double-

Table 3.2 Six-ray typical urban PDP

Delay (μs)	Power share
0	0.189
0.2	0.379
0.5	0.239
1.6	0.095
2.3	0.061
5.0	0.037

Table 3.3 ITU vehicular PDP

Vehicular A		Vehicular B	
Delay (μs)	Power share	Delay (μs)	Power share
0	0.485	0	0.322
0.31	0.385	0.3	0.574
0.71	0.061	8.9	0.03
1.09	0.048	12.9	0.057
1.73	0.015	17.1	0.002
2.51	0.005	20.0	0.014

exponential shape [230]. These observations formed the foundation of the so-called Saleh–Valenzuela model, illustrated in Fig. 3.16. The power of the rays within each cluster decays exponentially at a rate γ_{SV} while the power of the set of rays leading the respective clusters also decays exponentially, at a different rate Γ_{SV}. Each ray is then Rayleigh-faded.

The delay of the lead ray of the kth cluster is τ_k, normalized by setting $\tau_0 = 0$. In turn, the delay of the ℓth ray within the kth cluster, relative to τ_k, is $\tau_{k,\ell}$. The aggregate delay of the ℓth ray within the kth cluster is thus $\tau_k + \tau_{k,\ell}$. The inter- and intra-cluster delays are mutually independent and Poisson-distributed.

Often, the PDP is succinctly reduced down to a key parameter quantifying its spread. Relative to the average delay

$$\mu_\tau = \int_{-\infty}^{\infty} \tau \, S_{\mathsf{h}}(\tau) \, \mathrm{d}\tau, \tag{3.69}$$

such RMS *delay spread* is simply the standard deviation

$$T_{\mathsf{d}} = \sqrt{\int_{-\infty}^{\infty} (\tau - \mu_\tau)^2 \, S_{\mathsf{h}}(\tau) \, \mathrm{d}\tau}. \tag{3.70}$$

Example 3.23

Compute the RMS delay spread for the PDP

$$S_{\mathsf{h}}(\tau) = \begin{cases} K_{\mathsf{e}} \exp(-4 \cdot 10^5 \tau) & \tau \in [0, 10\,\mu\mathrm{s}] \\ 0 & \tau \notin [0, 10\,\mu\mathrm{s}]. \end{cases} \tag{3.71}$$

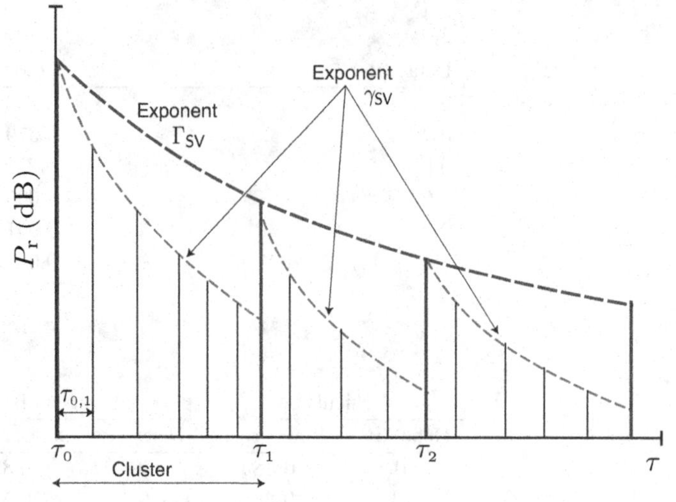

Fig. 3.16 PDP based on the Saleh–Valenzuela model.

Solution

Setting K_e such that the PDP integrates to unity and applying (3.69) and (3.70), we obtain $\mu_\mathcal{T} = 2.31$ μs and subsequently $T_d = 2.08$ μs.

If the PDP is discrete, the definition of delay spread in (3.70) becomes

$$T_d = \sqrt{\sum_{q=0}^{Q-1} (\tau_q - \mu_\mathcal{T})^2 P_q}, \tag{3.72}$$

where τ_q and P_q are the delay and power shares for the qth ray, respectively, while the average delay is $\mu_\mathcal{T} = \sum_{q=0}^{Q-1} \tau_q P_q$.

Example 3.24

Compute the delay spread for the six-ray typical urban PDP in Table 3.2.

Solution

Application of (3.72) to the ray delays and power shares in Table 3.2 yields $T_d = 1.05$ μs.

Example 3.25

Compute the delay spread for the vehicular PDPs in Table 3.3.

Solution

Application of (3.72) gives $T_d = 0.37$ μs for vehicular A and $T_d = 4$ μs for vehicular B.

As the preceding examples illustrate, the habitual values for the delay spread in outdoor macrocells range between a fraction of a microsecond and a few microseconds. In urban macrocells specifically, a very representative value is 1 μs. Indoors, alternatively, the typical values range between 50 and 100 ns.

Using the delay spread, the condition $B_c \gg B$ for frequency flatness can be equivalently written as $T_d \ll T$.

Uncorrelated scattering

Not surprisingly, the frequency-domain correlation $R_h(\cdot)$ and the PDP can be related through a Fourier transform, namely

$$R_h(\Delta_f) = \mathbb{E}\big[h(f)h^*(f+\Delta_f)\big] \tag{3.73}$$

$$= \mathbb{E}\left[\int_{-\infty}^{\infty} h(\tau_0)\, e^{-j2\pi f \tau_0}\, d\tau_0 \int_{-\infty}^{\infty} h^*(\tau_1)\, e^{j2\pi(f+\Delta_f)\tau_1}\, d\tau_1\right] \tag{3.74}$$

$$= \int_{-\infty}^{\infty}\int_{-\infty}^{\infty} \mathbb{E}\big[h(\tau_0)h^*(\tau_1)\big]\, e^{j2\pi[(f+\Delta_f)\tau_1 - f\tau_0]}\, d\tau_0\, d\tau_1. \tag{3.75}$$

Under the condition that $\mathbb{E}[h(\tau_0)h^*(\tau_1)] = 0$ for $\tau_0 \neq \tau_1$, (3.75) becomes

$$R_h(\Delta_f) = \int_{-\infty}^{\infty} \mathbb{E}\Big[|h(\tau)|^2\Big]\, e^{j2\pi\Delta_f \tau}\, d\tau \tag{3.76}$$

$$= \int_{-\infty}^{\infty} S_h(\tau)\, e^{j2\pi\Delta_f \tau}\, d\tau. \tag{3.77}$$

The pivotal condition that $\mathbb{E}[h(\tau_0)h^*(\tau_1)] = 0$ for $\tau_0 \neq \tau_1$ is referred to as the *uncorrelated scattering* condition; it is grounded in the reasonable premise that the propagation mechanisms giving rise to paths having resolvable delays are essentially independent. Together with the assumption of local stationarity, the uncorrelated scattering condition underpins the traditional modeling of small-scale fading conforming the wide-sense stationary uncorrelated scattering (WSSUS) framework introduced by Bello in 1963 [250]. Further note from (3.77) that the uncorrelated scattering condition in the delay domain ensures that $R_h(\cdot)$ does not depend on f, but only on Δ_f, thereby rendering the fading wide-sense stationary also in the frequency domain.

It follows from the Fourier relationship between $R_h(\cdot)$ and $S_h(\cdot)$ that

$$B_c \propto \frac{1}{T_d}, \tag{3.78}$$

with the exact scaling depending on the shapes of $R_h(\cdot)$ and $S_h(\cdot)$, as well as on the degree of decorrelation implied by B_c [197]. This inverse relation is, in a sense, the dual of (3.58).

Simplified models

Whatever modeling approaches one can apply to simplify the representation of fading in the time domain can be applied to represent it in the frequency domain. This includes the Gauss–Markov approach as well as the block-fading approach, in this case embodied by frequency blocks of size B_c over which the fading remains constant.

3.4.5 Time–frequency double selectivity

Time–frequency correlation

In order to foster conceptual clarity and intuition, we have thus far analyzed separately the fading selectivity in the time (or equivalently space) and in the frequency domains. To inspect the former we have posited no variations in frequency, i.e., $y(t) = \sqrt{G}\, h(t)\, x(t)$; to investigate the latter we have postulated no variations in time, i.e., $\mathsf{y}(f) = \sqrt{G}\,\mathsf{h}(f)\,\mathsf{x}(f)$, which is tantamount to $y(t) = \sqrt{G}\,(h * x)(t) = \sqrt{G} \int_\tau h(\tau)\, x(t-\tau)\, \mathrm{d}\tau$.

When both types of selectivity are brought together, as is the case in actual wireless channels, small-scale fading must be modeled as a time-varying relationship of the form

$$y(t) = \sqrt{G} \int_{-\infty}^{\infty} h(t,\tau)\, x(t-\tau)\, \mathrm{d}\tau, \tag{3.79}$$

where $h(t,\tau)$ is the channel response at time t to an impulse delayed by τ. By transforming delay into frequency or time into Doppler shift, $h(t,\tau)$ can be converted respectively into a time-varying transfer function [129]

$$\hbar(t,f) = \int_{-\infty}^{\infty} h(t,\tau)\, e^{-\mathrm{j}2\pi f \tau}\, \mathrm{d}\tau \tag{3.80}$$

or into the Doppler-delay spreading function

$$\hbar(\nu,\tau) = \int_{-\infty}^{\infty} h(t,\tau)\, e^{-\mathrm{j}2\pi \nu t}\, \mathrm{d}t. \tag{3.81}$$

Both of these functions offer alternative means of expressing the noiseless received signal as a function of the transmit signal, e.g.,

$$y(t) = \sqrt{G} \int_{-\infty}^{\infty} \int_{-\infty}^{\infty} \hbar(\nu,\tau)\, x(t-\tau)\, e^{\mathrm{j}2\pi \nu t}\, \mathrm{d}\tau \mathrm{d}\nu, \tag{3.82}$$

which conveys particular intuition: $y(t)$ is a superposition of replicas of $x(t)$, each delayed by a given τ and subject to a certain Doppler shift ν according to $\hbar(\nu,\tau)$. The formulation is thereby seen to capture multipath fading in all its generality.

From $\hbar(t,f)$ one can compute a joint time–frequency correlation, which in general would be four-dimensional but thanks to the stationarity in both time and frequency is a function of only Δ_t and Δ_f,

$$R_\hbar(\Delta_t,\Delta_f) = \mathbb{E}\Big[\hbar(t,f)\, \hbar^*(t+\Delta_t, f+\Delta_f)\Big]. \tag{3.83}$$

Scattering function

From $\hbar(\nu,\tau)$, one can compute a joint Doppler–delay spectrum

$$S_\hbar(\nu,\tau) = \mathbb{E}\Big[|\hbar(\nu,\tau)|^2\Big], \tag{3.84}$$

which is commonly called the *scattering function*, and which relates to the time–frequency correlation in (3.83) through the two-dimensional Fourier transform

$$S_{\hbar}(\nu,\tau) = \int_{-\infty}^{\infty}\int_{-\infty}^{\infty} R_{\hbar}(\Delta_t,\Delta_f)\, e^{-\mathrm{j}2\pi(\Delta_t\nu-\Delta_f\tau)}\, \mathrm{d}\Delta_t \mathrm{d}\Delta_f. \tag{3.85}$$

This general WSSUS formulation allows for an easy recovery of all the earlier special cases. Indeed, when there is no dependence on either time or frequency, $R_{\hbar}(\Delta_t,\Delta_f)$ reverts to the time correlation $R_h(\Delta_t)$ or the frequency correlation $R_{\mathrm{h}}(\Delta_f)$, respectively, while the marginals of the scattering function equal the Doppler spectrum and the PDP

$$S_h(\nu) = \int_{-\infty}^{\infty} S_{\hbar}(\nu,\tau)\,\mathrm{d}\tau \tag{3.86}$$

$$S_{\mathrm{h}}(\tau) = \int_{-\infty}^{\infty} S_{\hbar}(\nu,\tau)\,\mathrm{d}\nu. \tag{3.87}$$

Simplified models

The block-fading approach can be applied to model double selectivity in time and frequency, giving rise to rectangular tiles of size $T_{\mathrm{c}} \times B_{\mathrm{c}}$ (see Fig. 3.17). Fine-stretched to encompass an integer number of symbols and subcarriers, this structure is particularly amenable to the analysis of multicarrier signals; taking the spans of a resource element to be T_{OFDM} and $1/T_{\mathsf{OFDM}}$ in the time and frequency domains, respectively, the number of resource elements per tile is given by the integer rounding of

$$N_{\mathrm{c}} = \frac{T_{\mathrm{c}}}{T_{\mathsf{OFDM}}}\frac{B_{\mathrm{c}}}{1/T_{\mathsf{OFDM}}} \tag{3.88}$$

$$= B_{\mathrm{c}}T_{\mathrm{c}}, \tag{3.89}$$

which generalizes (3.66).

The product $B_{\mathrm{c}}T_{\mathrm{c}}$ is an important quantity that makes frequent appearances throughout the text, its significance descending from the fact that it returns the number of resource elements per time–frequency tile that render a block-fading model representative of more general fading processes. Typical values taken by the product $B_{\mathrm{c}}T_{\mathrm{c}}$ are given shortly.

Note that, in taking the spans of a resource element to be T_{OFDM} and $1/T_{\mathsf{OFDM}}$, the cyclic prefix overhead inherent to OFDM has been neglected. Accounting for it, the time and frequency spans become $T_{\mathsf{OFDM}} = (K+L)T$ and $1/KT$, and thus

$$N_{\mathrm{c}} = \frac{T_{\mathrm{c}}}{(K+L)T}\frac{B_{\mathrm{c}}}{\frac{1}{KT}} \tag{3.90}$$

$$= \frac{K}{K+L}B_{\mathrm{c}}T_{\mathrm{c}}, \tag{3.91}$$

where a good design ensures that $\frac{K}{K+L}$ is not far from unity. Throughout the book, we use $N_{\mathrm{c}} \approx B_{\mathrm{c}}T_{\mathrm{c}}$; the cyclic prefix overhead $\frac{L}{K+L}$ should be separately accounted for.

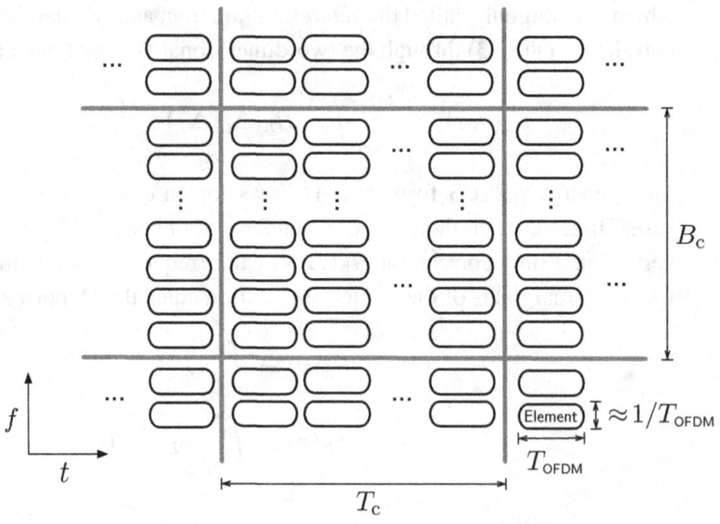

Fig. 3.17 Time–frequency tiles in a block-fading model with both time and frequency selectivity.

The underspread property

A consequential property of wireless fading is that it is highly *underspread*, meaning that the delay spread is much shorter than the coherence time ($T_d \ll T_c$) or, equivalently, that the Doppler spectrum is very narrow relative to the coherence bandwidth ($\nu_M \ll B_c$) [251]. As delay spreads are in the microsecond or sub-microsecond range while coherence times are in the millisecond range or above, wireless channels are indeed underspread. (Underwater communication channels, in contrast, are often not underspread because the speed of sound is a great deal slower than the speed of light and the delay spreads are accordingly higher.) A particularly convenient way of quantifying the underspreadness of a fading channel is through the product $B_c T_c$, which, in an underspread channel, satisfies

$$B_c T_c \gg 1. \tag{3.92}$$

Let us illustrate the range of values taken by $B_c T_c$ in the form of some test cases to be invoked throughout the text.

Example 3.26

Consider an outdoor vehicular scenario where the velocity and delay spread are, respectively, $\mathrm{v} = 100$ km/h and $T_d = 2$ µs, with carrier frequency $f_c = 2$ GHz. What is the value of $B_c T_c$?

Solution

The maximum Doppler shift is $\nu_M = f_c \mathrm{v}/c = 185$ Hz and thus $T_c \approx \frac{1}{2\,\nu_M} = 2.7$ ms. The coherence bandwidth is $B_c \approx \frac{1}{T_d} = 500$ kHz. Hence, $B_c T_c \approx 1350$, which, to allow for the possibility of higher velocities and/or carrier frequencies, or longer delay spreads, we round down to 1000 symbols.

Example 3.27

Consider an outdoor pedestrian scenario where the velocity and delay spread are, respectively, $\mathrm{v} = 3.5$ km/h and $T_{\mathrm{d}} = 2\ \mu$s, with carrier frequency $f_{\mathrm{c}} = 3.5$ GHz. What is the value of $B_{\mathrm{c}} T_{\mathrm{c}}$?

Solution

The maximum Doppler shift is $\nu_{\mathrm{M}} = f_{\mathrm{c}} \mathrm{v}/\mathrm{c} = 11.3$ Hz and thus $T_{\mathrm{c}} = 44$ ms. The coherence bandwidth is $B_{\mathrm{c}} \approx 500$ kHz. Hence, $B_{\mathrm{c}} T_{\mathrm{c}} \approx 22\,000$. Again, this is conservatively rounded down, in this case to 20 000 symbols.

Example 3.28

Consider an indoor pedestrian scenario where the velocity and delay spread are, respectively, $\mathrm{v} = 3.5$ km/h and $T_{\mathrm{d}} = 100$ ns, with the carrier frequency being $f_{\mathrm{c}} = 5$ GHz. What is the value of $B_{\mathrm{c}} T_{\mathrm{c}}$?

Solution

The maximum Doppler shift is $\nu_{\mathrm{M}} = f_{\mathrm{c}} \mathrm{v}/\mathrm{c} = 16$ Hz and thus $T_{\mathrm{c}} = 31$ ms. The coherence bandwidth is $B_{\mathrm{c}} \approx \frac{1}{T_{\mathrm{d}}} = 10$ MHz. Hence, $B_{\mathrm{c}} T_{\mathrm{c}} \approx 3.1 \cdot 10^5$.

The discretization of a time-varying channel is generally problematic because its eigenfunctions depend on the realization. Fortunately, in underspread channels a straightforward sampling of the time-varying transfer function in both time and frequency yields a discrete representation from which the original continuous channel could be reconstructed with high fidelity. This conveniently fits multicarrier schemes, where each such sample maps to a resource element. Although, in the face of both Doppler and delay spread, it is not possible to shape these elements in such a way as to ensure simultaneous orthogonality in time and frequency—i.e., zero ISI and zero intercarrier interference—in sufficiently underspread channels the residual interference among elements can be made negligible [252]. As the name suggests, the underspread property allows having resource elements that are at once "long" and "wide" relative to the delay and Doppler spreads, respectively. Ideally, the shape of these elements should be computed from the scattering function [253], but pragmatically a fixed a-priori shape is utilized. With OFDM in particular, and at the expense of the overhead represented by the cyclic prefix, the fixed shape of the subcarriers is carefully designed to enable a tight packing in the frequency domain.

This sampling of the time-varying transfer function returns, on OFDM subcarrier k, the discrete-time channel $h[n] = \hbar(nT_{\mathsf{OFDM}}, k/T_{\mathsf{OFDM}})$ with time index n and, for a given n, the discrete-frequency channel $\mathrm{h}[k] = \hbar(nT_{\mathsf{OFDM}}, k/T_{\mathsf{OFDM}})$ with frequency index k. And, because with well-designed OFDM the fading is essentially multiplicative in both time and frequency, these discrete representations are sufficient. By the same token, the time

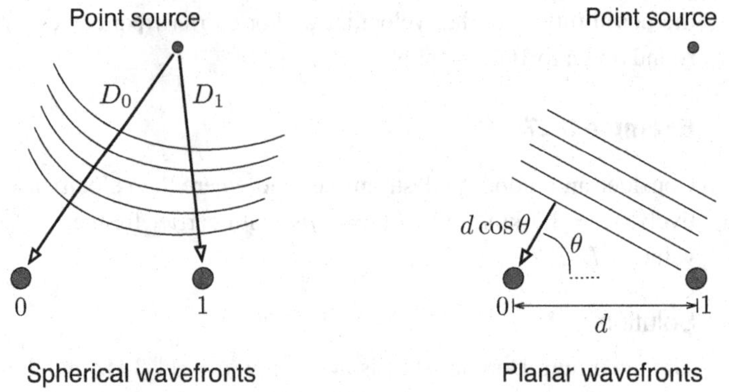

Fig. 3.18 Spherical versus planar wavefronts. While propagation is spherical, the wavefronts can be regarded as planar over an array because the constituent antennas are closely spaced relative to the distance from the source.

correlation of $h[n]$ and the frequency correlation of $\mathsf{h}[k]$ are obtained by directly sampling $R_h(\cdot)$ and $R_{\mathsf{h}}(\cdot)$, respectively.

3.5 Interlude: essential notions of antenna arrays

Before delving into MIMO channel modeling, we pause to provide some background material on antenna arrays. This material serves to describe the topology of an array from a communications standpoint. Conventional array processing is built around the mathematical description laid out in this section.

3.5.1 Array steering vectors

Array *steering vectors* are classical descriptors of the relative differences in the response between antennas within an array. As a starting point, consider an array of two receive antennas as in Fig. 3.18. A signal $x(t)$ is radiated from a point source toward this receive array, as a spherical wave. The propagation delay between the point source and the ith antenna is $\tau_i = D_i/\mathrm{c}$ and the corresponding complex pseudo-baseband LOS channel is $\sqrt{G_i}\, e^{-\mathrm{j}2\pi D_i/\lambda_\mathrm{c}}\, \delta(\tau - \tau_i)$. If the antennas are closely spaced, then $\sqrt{G_0} = \sqrt{G_1} = \sqrt{G}$ and the noiseless complex baseband received signal at the ith antenna is

$$y_i(t) = \sqrt{G}\, e^{-\mathrm{j}2\pi D_i/\lambda_\mathrm{c}}\, x(t - \tau_i). \tag{3.93}$$

Next we simplify the received signal model further by making two assumptions: frequency flatness and far-field.

First we incorporate the assumption of frequency flatness [254]. Let d_a denote the largest dimension of the array and d the spacing between the antennas. With two antennas, $d_\mathrm{a} = d$.

Under frequency flatness, the time for a wave to propagate along the longest dimension of the array satisfies $d_{\mathrm{a}}/\mathrm{c} \ll T$ and thus

$$y_i(t) \approx \sqrt{G}\, \mathrm{e}^{-\mathrm{j}2\pi D_i/\lambda_{\mathrm{c}}}\, x(t-\tau), \tag{3.94}$$

where the delay τ is now common to all antennas; at the same time, because small changes in distance can lead to major phase shifts, it is important not to simplify the phase term. The condition $d_{\mathrm{a}}/\mathrm{c} \ll T$, which extends to an array the notion of frequency flatness seen earlier for a single antenna, is reasonable in MIMO except for cases like underwater communication [255, 256], where the speed of propagation is much less than c and the bandwidths are small, or in distributed forms of MIMO [257], where the antennas may be deployed throughout a building or on different sites.

When the receive array of size d_{a} is in the far-field region, meaning that the point source is far enough (the rule of thumb being beyond $2\, d_{\mathrm{a}}^2/\lambda_{\mathrm{c}}$), the spherical wavefronts are approximately planar over the array. This condition, illustrated in Fig. 3.18, is satisfied with the only possible exception of extreme massive MIMO [258, 259].

Now, take the location of antenna 0 as a reference. From the angle of arrival θ in Fig. 3.18, the delay difference between the onset of the planar wavefront at each antenna equals $d\cos(\theta)/\mathrm{c}$. Under frequency flatness, the excess propagation delay results in only a phase shift of $2\pi d\cos(\theta)/\lambda_{\mathrm{c}}$ and thus

$$y_1(t) = e^{\mathrm{j}\frac{2\pi d}{\lambda_{\mathrm{c}}}\cos(\theta)}\, y_0(t). \tag{3.95}$$

Antenna 1 receives exactly the same signal, with a phase shift that depends only on θ. If, rather than two, there are N_{a} receive antennas, each observes a phase-shifted version of the signal. Denoting by $\phi_i(\theta)$ the shift at the ith antenna as a function of θ and of the array topology, the array steering vector (or array response) is the collection of the N_{a} phase shifts

$$\boldsymbol{a}(\theta) = \left[e^{\mathrm{j}\phi_0(\theta)} \; e^{\mathrm{j}\phi_1(\theta)} \; \cdots \; e^{\mathrm{j}\phi_{N_{\mathrm{a}}-1}(\theta)} \right]^{\mathrm{T}} \tag{3.96}$$

experienced by a signal impinging from the direction determined by θ. Equivalently, these are the phase shifts that, applied at the antennas, generate a plane wave in the direction of θ when the array acts as a transmitter. A common shift of all the phases within a steering vector, i.e., $e^{\mathrm{j}\phi}\boldsymbol{a}(\theta)$ for arbitrary ϕ, merely affects the absolute phase of the signal being received or transmitted, but not the direction. The choice of a coordinate system for the array has the same effect.

The array manifold is the set of possible array response vectors, i.e., $\{\boldsymbol{a}(\theta)\}$ for θ running from 0 to 2π. The mathematical structure of this set is exploited in array signal processing algorithms to perform tasks such as direction-of-arrival finding [4].

Example 3.29 (Uniform linear array)

For a uniform linear array (ULA), from Fig. 3.19,

$$\phi_i(\theta) = 2\pi\, \frac{i\, d\cos(\theta)}{\lambda_{\mathrm{c}}} \qquad i = 0, \ldots, N_{\mathrm{a}}-1, \tag{3.97}$$

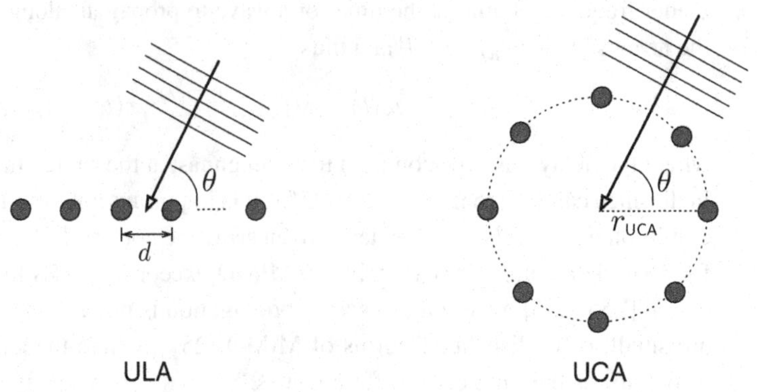

Fig. 3.19 Common antenna array topologies: the ULA and the UCA.

where θ is measured relative to the axis of the array, the so-called *endfire* direction, as indicated in the figure. With this choice of coordinate system, $\theta = \pi/2$ corresponds to the *broadside* direction. We have implicitly centered the coordinates on antenna 0, where the signal is received/transmitted with phase zero, but this coordinate center can be displaced at will by applying a common shift to $\phi_0(\theta), \ldots, \phi_{N_a-1}(\theta)$.

Example 3.30 (Nonuniform linear array)

For a nonuniform linear array, (3.97) generalizes to

$$\phi_i(\theta) = 2\pi \frac{\sum_{\iota=1}^{i} d_{\iota,\iota-1} \cos(\theta)}{\lambda_c} \qquad i = 0, \ldots, N_a - 1, \tag{3.98}$$

where $d_{\iota,\iota-1}$ is the distance between receive antennas $\iota - 1$ and ι.

Example 3.31 (Uniform circular array)

For a uniform circular array (UCA) with radius r_{UCA}, as illustrated in Fig. 3.19,

$$\phi_i(\theta) = 2\pi \frac{r_{UCA} \cos(\theta - \theta_i)}{\lambda_c} \qquad i = 0, \ldots, N_a - 1, \tag{3.99}$$

where θ_i is the angle subtended by the ith antenna [260]. All the angles are defined relative to the same direction, which is otherwise arbitrary.

The formulation of the array steering vectors for other array topologies, e.g., star or hexagonal, is straightforward. If the array and the wave are not coplanar, the formulation generally requires azimuth and elevation angles but is otherwise conceptually identical.

Note that (3.42), which returns the correlation between two positions in space, or equivalently between two closely spaced antennas, is nothing but an integral over the PAS of the array steering vector of those antennas. Thus, the steering vectors are intimately related with the antenna correlations. Precisely, they inform those correlations of the array topology and antenna spacings. In terms of diversity against fading, the use of antennas that are spaced beyond the coherence distance is naturally termed antenna diversity or *space diversity*.

The array steering vector definition in (3.96) does not include the pattern of the constituent antennas or, put differently, it presumes that every antenna has a uniform pattern in θ and that such uniform pattern is normalized by the peak gain G_r that is already included in the large-scale channel gain G. If the individual antennas exhibit a nonuniform but identical pattern, its normalized version can be incorporated as a factor multiplying $\boldsymbol{a}(\theta)$; this can be observed in (3.41).

3.5.2 Array factor and beamforming

In classic array processing over LOS channels, the antennas form a directional beam. While MIMO makes a more general use of the antennas, it is useful to review some of these classic concepts for completeness. The array factor is a function $A(\theta)$ that represents the pattern created if all the antenna outputs are linearly combined, e.g., $A(\theta) = \mathbf{1}^{\mathrm{T}}\boldsymbol{a}(\theta)$ where $\mathbf{1}$ is an all-ones vector. Used this way, the antennas work together to create a single effective antenna with a directive beam pattern. The function $G_a(\theta) = |A(\theta)|^2$ determines the power gain experienced by a signal arriving from or transmitting toward direction θ. The array factor can be interpreted as the angular response or pattern that the array would exhibit if its constituent antennas were isotropic in θ. If the constituent antennas are nonuniform in θ but identical, $A(\theta)$ is simply multiplied by the individual antenna pattern [261, chapter 20]. The generalization of $A(\theta)$ to arrays with antennas having distinct patterns is the subject of Problem 3.14.

Example 3.32 (ULA array factor)

For a ULA,

$$A(\theta) = \sum_{i=0}^{N_a-1} e^{j2\pi i \frac{d}{\lambda_c}\cos(\theta)} \tag{3.100}$$

$$= \frac{e^{j2\pi N_a \frac{d}{\lambda_c}\cos(\theta)} - 1}{e^{j2\pi \frac{d}{\lambda_c}\cos(\theta)} - 1} \tag{3.101}$$

$$= \frac{\sin\left(N_a \frac{\pi d}{\lambda_c}\cos\theta\right)}{\sin\left(\frac{\pi d}{\lambda_c}\cos\theta\right)} e^{j\pi N_a \frac{d}{\lambda_c}\cos(\theta)}, \tag{3.102}$$

where (3.101) is the sum of the geometric series in (3.100). The magnitude of the array factor is

$$|A(\theta)| = \frac{\sin\left(N_a \frac{\pi d}{\lambda_c}\cos\theta\right)}{\sin\left(\frac{\pi d}{\lambda_c}\cos\theta\right)}, \tag{3.103}$$

whose peak value is N_a. For $N_a = 4$ and various choices of d/λ_c, the function $|A(\theta)|$ is depicted in Fig. 3.20. The horizontal symmetry is due to the ULA only having an unambiguous response for $\theta \in [0, \pi)$. As d/λ_c approaches $1/2$, the main lobe becomes progressively narrower while, for $d/\lambda_c > 1/2$, spatial aliasing occurs; this is reflected in the main lobe

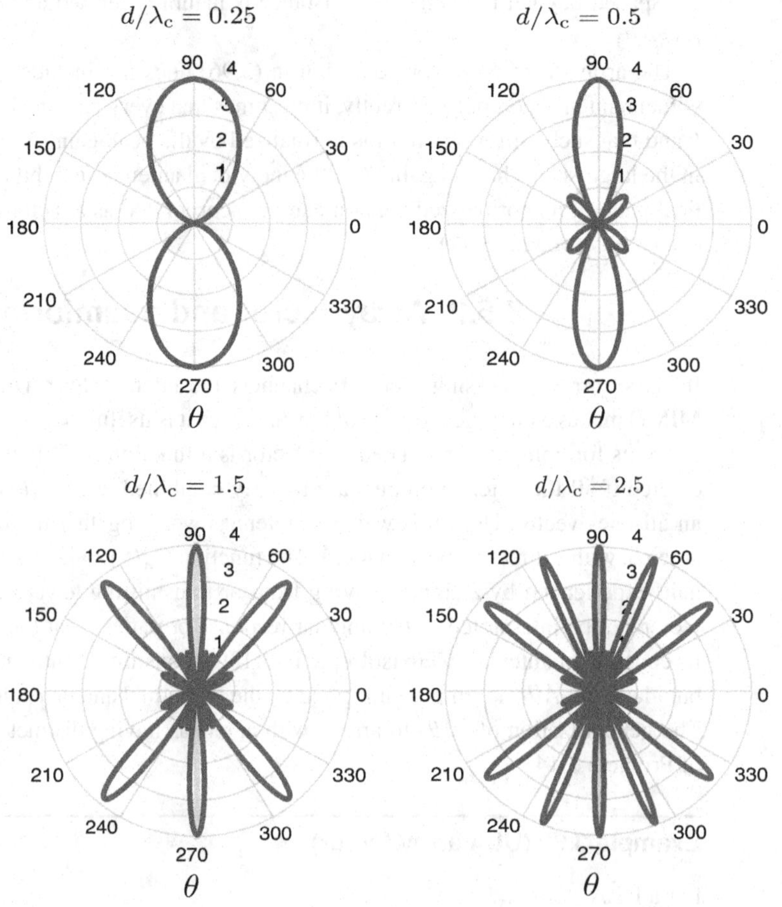

Fig. 3.20 Array factor magnitude for a ULA with $N_a = 4$ and various choices of d/λ_c.

of the antenna pointing in multiple directions. In multipath environments, the appearance of spatial aliases when d/λ_c increases beyond $1/2$ causes ripples in the spatial correlation (see Fig. 3.9), as multipath components from different directions get mixed differently.

If a common phase shift can be applied dynamically to the entire array, to change the direction in which the beams point, we speak of a phased array and of *beam steering*. If magnitude weighting can also be applied dynamically at each antenna prior to summation, to further shape the beams, we have an adaptive array and *beamforming*. Letting $\boldsymbol{w}$ contain the weight coefficients, with $\boldsymbol{w}$ having unit-magnitude entries in the case of a phased array and arbitrary magnitudes in the case of an adaptive array,

$$G_a(\theta) = |\boldsymbol{w}^* \boldsymbol{a}(\theta)|^2, \tag{3.104}$$

with $\boldsymbol{w}$ designed to achieve a specific response, for instance to accentuate signals around a certain angle and attenuate signals on other directions. In LOS propagation, beamforming

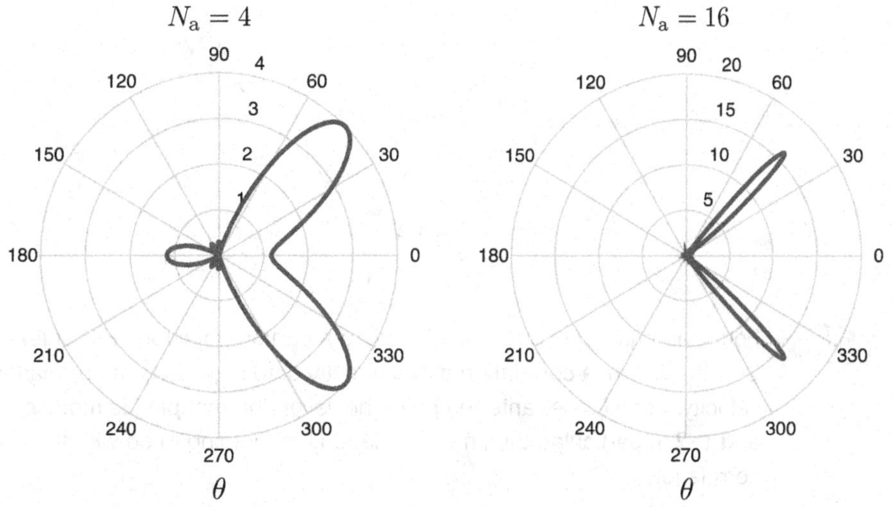

Fig. 3.21 Beam directed toward $\mu_\theta = \pi/4$ with $d = \lambda_c/2$, for matched filtering with $N_a = 4$ and $N_a = 16$. More antennas produce a sharper beam with increased gain; the relative sidelobe level also decreases.

has the literal significance of geometrically shaping a beam in θ. In Chapter 5, we provide a more general definition of beamforming that applies even when the propagation is not LOS and the geometric interpretation in the angular domain is lost. Perhaps the simplest form of adaptive beamforming is matched filtering: given a target angle μ_θ, matched filtering entails $\boldsymbol{w} = \boldsymbol{a}(\mu_\theta)$ and $G_a(\theta) = |\boldsymbol{a}^*(\mu_\theta)\boldsymbol{a}(\theta)|^2$. The design of beamforming vectors is the topic of a rich body of literature, with parallel developments in filter design given that an antenna array can be viewed as a filter in the angular domain.

Example 3.33 (ULA beamforming)

A gain pattern oriented toward $\mu_\theta = \pi/4$ is shown in Fig. 3.21 for a ULA with $d = \lambda_c/2$ under matched filtering with $N_a = 4$ and $N_a = 16$.

The introduction of adaptive arrays modifies the PAS in the same way that changing the antenna pattern modifies it in single-antenna transceivers (recall (3.41)). The space correlation function and the coherence distance are therefore altered depending on the weight coefficients [262].

3.6 Modeling of MIMO channels

Having laid a solid foundation on channel modeling for SISO channels, we are now in a position to graduate onto MIMO. The key new aspect, of course, is that with MIMO the

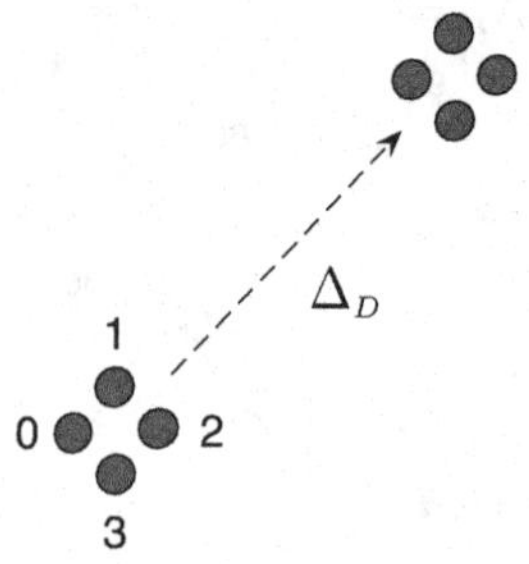

Fig. 3.22 Four-antenna transceiver moving along a certain trajectory. For antenna pairs 0–1 and 2–3, the antenna correlation maps directly onto time-domain correlation through the velocity. For all other antenna pairs, however, the mapping is modified. For pairs 0–3 and 1–2 in particular, the time correlation has nothing to do with the antenna correlation.

channel response becomes matrix-valued, and this requires that some notions be expanded and generalized.

A first consideration must refer to whether the separation of large- and small-scale phenomena is upheld, and that depends, once more, on the physical size of the antenna arrays. Unless otherwise stated, we consider that the number of antennas and their physical spacing are such that the array size does not exceed the size of the local scattering neighborhood. Under this premise, which may only need revisiting in the realm of massive MIMO, the separation is upheld—with the addition of antenna correlation as a small-scale parameter that should be made dependent, as appropriate, on the large-scale channel gain. The large-scale models presented earlier apply verbatim and, in the sequel, we focus exclusively on the small-scale domain via stochastic modeling.

Another consideration has to do with the normalization of the small-scale channel matrix, $\boldsymbol{H}$. The unit-power normalization of a scalar SISO channel can be generalized to matrices in several ways, the most obvious options being a per-entry unit-power normalization or a normalization of the Frobenius norm, $\mathbb{E}[\|\boldsymbol{H}\|_{\mathrm{F}}^2] = N_{\mathrm{t}} N_{\mathrm{r}}$. As advanced in the previous chapter, in this book we adopt the latter, which offers further flexibility, although in many of the models it is the case that every entry of $\boldsymbol{H}$ ends up being unit-power.

In contrast with SISO, where we found it convenient to bind space and time, in MIMO they need to be decoupled. The space dimension, properly sampled, becomes the antenna dimension, and it is unlocked from the time selectivity engendered by motion. More precisely, the actual trajectory of a transceiver, which induces the time selectivity, needs to be distinguished from the segments connecting every antenna pair (see Fig. 3.22).

In MIMO, the focus is evidently on how to model the joint distribution of the entries of $\boldsymbol{H}$, which is the differentiating aspect with respect to SISO [263–269]. In fact, some MIMO channel models concern themselves exclusively with this aspect, with the tacit assumption that the time/frequency dynamics of each entry can be modeled as in SISO; this disregards dependences that may actually exist, but facilitates a sharp emphasis on the MIMO nature of the link. We label these models as *analytical*, for lack of a better name. Other models, which we classify under the moniker *parametric*, do offer the possibility of capturing the space, time, and frequency dimensions jointly. All these types of models are reviewed next.

3.6.1 Analytical models

The models we package under this label share the principle of not attempting to recreate the geometry of the propagation environment, but merely the marginal statistical behavior of the channel matrix $\boldsymbol{H}$ at a given time/frequency. In most cases, the entries of $\boldsymbol{H}$ are modeled as jointly Gaussian (see Appendix C.1.9) and thus the model amounts to a description of the first- and second-order characteristics. This description can be fairly compact and analytically convenient, which is what makes these models attractive and justifies the label.

IID Rayleigh-faded model

The simplest possible analytical model, sometimes referred to as the *canonical* channel, is the one where the entries of $\boldsymbol{H}$ are IID and $[\boldsymbol{H}]_{i,j} \sim \mathcal{N}_{\mathbb{C}}(0,1)$ for $i = 0, \ldots, N_\mathrm{r} - 1$ and $j = 0, \ldots, N_\mathrm{t} - 1$. Since, besides having importance on its own, this model is the centerpiece of other analytical models, we introduce for it the specific symbol $\boldsymbol{H}_\mathrm{w}$. The PDF of $\boldsymbol{H}_\mathrm{w}$ is

$$f_{\boldsymbol{H}_\mathrm{w}}(\mathbf{H}) = \frac{1}{\pi^{N_\mathrm{t} N_\mathrm{r}}} e^{-\mathrm{tr}(\mathbf{H}^* \mathbf{H})}, \tag{3.105}$$

whereas $\boldsymbol{H}_\mathrm{w}^* \boldsymbol{H}_\mathrm{w}$ (if $N_\mathrm{t} \leq N_\mathrm{r}$) or $\boldsymbol{H}_\mathrm{w} \boldsymbol{H}_\mathrm{w}^*$ (if $N_\mathrm{t} \geq N_\mathrm{r}$) conform to the Wishart distribution detailed in Appendix C.1.9.

The IID Rayleigh-faded model is typically justified as representative of environments with rich scattering, meaning a broad PAS enabling complete decorrelation with relatively tight antenna spacings. This is a simplification of reality, since simultaneous and complete decorrelation of more than two antennas may be geometrically impossible, but the model has enormous value nevertheless; it enables exploring and demarcating what may be feasible, and in that respect it played a very important role in the early days of MIMO.

IID Rice-faded model

An obvious generalization of the IID Rayleigh-faded model is the IID Rice-faded model

$$\boldsymbol{H} = \underbrace{\sqrt{\frac{\mathsf{K}}{1+\mathsf{K}}}\, \boldsymbol{H}_\mathsf{LOS}}_{\boldsymbol{\mu}_{\boldsymbol{H}}} + \sqrt{\frac{1}{1+\mathsf{K}}}\, \boldsymbol{H}_\mathrm{w}, \tag{3.106}$$

where $\boldsymbol{H}_\mathsf{LOS}$ is deterministic and K is the Rice factor. For the normalization of $\boldsymbol{H}$ to be upheld, it is necessary that $\|\boldsymbol{H}_\mathsf{LOS}\|_\mathrm{F}^2 = N_\mathrm{t} N_\mathrm{r}$. The mean $\boldsymbol{\mu}_{\boldsymbol{H}}$ is made explicit in (3.106) while the PDF of $\boldsymbol{H}$ is

$$f_{\boldsymbol{H}}(\mathbf{H}) = \left(\frac{1+\mathsf{K}}{\pi}\right)^{N_\mathrm{t} N_\mathrm{r}} e^{-\mathrm{tr}((\mathbf{H}-\boldsymbol{\mu}_{\boldsymbol{H}})^*(\mathbf{H}-\boldsymbol{\mu}_{\boldsymbol{H}}))}, \tag{3.107}$$

which reduces to (3.105) for $\mathsf{K} = 0$. For $\mathsf{K} \to \infty$, conversely, $\boldsymbol{H}$ becomes deterministic.

Example 3.34

Establish $\boldsymbol{H}_{\text{LOS}}$ for an LOS component departing from a ULA transmitter at an angle θ_{t} and impinging on a ULA receiver at an angle θ_{r}. The antenna spacings at transmitter and receiver are d_{t} and d_{r}, respectively.

Solution

Recalling the definitions in Section 3.5.1, the transmit and receive steering vectors are

$$\boldsymbol{a}_{\text{t}}(\theta_{\text{t}}) = \begin{bmatrix} 1 & e^{\text{j}2\pi \frac{d_{\text{t}} \cos(\theta_{\text{t}})}{\lambda_{\text{c}}}} & \cdots & e^{\text{j}2\pi \frac{(N_{\text{t}}-1) d_{\text{t}} \cos(\theta_{\text{t}})}{\lambda_{\text{c}}}} \end{bmatrix}^{\text{T}} \tag{3.108}$$

$$\boldsymbol{a}_{\text{r}}(\theta_{\text{r}}) = \begin{bmatrix} 1 & e^{\text{j}2\pi \frac{d_{\text{r}} \cos(\theta_{\text{r}})}{\lambda_{\text{c}}}} & \cdots & e^{\text{j}2\pi \frac{(N_{\text{r}}-1) d_{\text{r}} \cos(\theta_{\text{r}})}{\lambda_{\text{c}}}} \end{bmatrix}^{\text{T}}. \tag{3.109}$$

The relative phase shifts between each transmit and each receive antenna are captured by the outer product $\boldsymbol{a}_{\text{r}}(\theta_{\text{r}})\boldsymbol{a}_{\text{t}}^{\text{T}}(\theta_{\text{t}})$ and, given the normalization of $\boldsymbol{H}_{\text{LOS}}$, we can directly write

$$\boldsymbol{H}_{\text{LOS}} = \boldsymbol{a}_{\text{r}}(\theta_{\text{r}})\, \boldsymbol{a}_{\text{t}}^{\text{T}}(\theta_{\text{t}}) \tag{3.110}$$

with the power of the LOS component affecting K, but not $\boldsymbol{H}_{\text{LOS}}$.

In the foregoing example, $\boldsymbol{H}_{\text{LOS}}$ is of unit rank, and this is indeed the most common situation, but it is conceivable that multiple LOS components exist and the rank of $\boldsymbol{H}_{\text{LOS}}$ be plural. The topology and physical dimension of the arrays may play some role here, and readers interested in this aspect are referred, e.g., to [270, 271].

Kronecker model

The defining feature of the two analytical models presented thus far is that the entries of $\boldsymbol{H}$ are IID. The next step toward wider generality is to allow for the dependences that almost inevitably shall arise given the close antenna proximity within a MIMO transceiver. With the channel entries jointly Gaussian, these dependences are fully captured by a second-order statistical description. For a matrix, such a description is in principle in the form of a four-dimensional tensor

$$\mathsf{R}_{\boldsymbol{H}}(i,j;i',j') = \mathbb{E}\big[[\boldsymbol{H} - \boldsymbol{\mu}_{\boldsymbol{H}}]_{i,j}\, [\boldsymbol{H} - \boldsymbol{\mu}_{\boldsymbol{H}}]^*_{i',j'}\big]. \tag{3.111}$$

Alternatively, the description can be reshuffled into a matrix by taking advantage of the vec($\cdot$) operator; the $N_{\text{r}}N_{\text{t}} \times N_{\text{r}}N_{\text{t}}$ correlation matrix

$$\boldsymbol{R}_{\text{vec}(\boldsymbol{H})} = \mathbb{E}\big[\text{vec}(\boldsymbol{H} - \boldsymbol{\mu}_{\boldsymbol{H}})\, \text{vec}(\boldsymbol{H} - \boldsymbol{\mu}_{\boldsymbol{H}})^*\big] \tag{3.112}$$

then contains all the pairwise correlation terms between any two entries (i,j) and (i',j') within $\boldsymbol{H}$.

Most often, the above fully general second-order description is skirted in favor of simpler forms. By far the most prevailing structure in the MIMO literature is the so-called

Kronecker correlation model, sometimes also referred to as the *separable* or *product* correlation model [272–275]. In this structure,

$$\mathsf{R}_{\boldsymbol{H}}(i,j;i',j') = [\boldsymbol{R}_{\rm r}]_{i,i'}\,[\boldsymbol{R}_{\rm t}]_{j,j'} \tag{3.113}$$

where $\boldsymbol{R}_{\rm r}$ and $\boldsymbol{R}_{\rm t}$ are, respectively, receive and transmit correlation matrices. Under this condition, the channel matrix can be expressed as

$$\boldsymbol{H} = \boldsymbol{\mu_H} + \boldsymbol{R}_{\rm r}^{1/2}\boldsymbol{H}_{\rm w}\boldsymbol{R}_{\rm t}^{1/2} \tag{3.114}$$

from which, given the IID and zero-mean nature of the entries of $\boldsymbol{H}_{\rm w}$, it can be verified (refer to Problem 3.16) that

$$\boldsymbol{R}_{\rm t} = \frac{1}{N_{\rm r}}\,\mathbb{E}\big[(\boldsymbol{H}-\boldsymbol{\mu_H})^*(\boldsymbol{H}-\boldsymbol{\mu_H})\big] \tag{3.115}$$

$$\boldsymbol{R}_{\rm r} = \frac{1}{N_{\rm t}}\,\mathbb{E}\big[(\boldsymbol{H}-\boldsymbol{\mu_H})(\boldsymbol{H}-\boldsymbol{\mu_H})^*\big], \tag{3.116}$$

such that $\text{tr}(\boldsymbol{R}_{\rm t}) = N_{\rm t}$ and $\text{tr}(\boldsymbol{R}_{\rm r}) = N_{\rm r}$. Recalling $\boldsymbol{R}_{\text{vec}(\boldsymbol{H})}$ as defined in (3.112),

$$\boldsymbol{R}_{\text{vec}(\boldsymbol{H})} = \mathbb{E}\Big[\text{vec}\Big({\boldsymbol{R}_{\rm r}}^{1/2}\boldsymbol{H}_{\rm w}{\boldsymbol{R}_{\rm t}}^{1/2}\Big)\,\text{vec}\Big({\boldsymbol{R}_{\rm r}}^{1/2}\boldsymbol{H}_{\rm w}{\boldsymbol{R}_{\rm t}}^{1/2}\Big)^*\Big] \tag{3.117}$$

$$= \mathbb{E}\Big[\Big({\boldsymbol{R}_{\rm t}}^{{\rm T}/2}\otimes{\boldsymbol{R}_{\rm r}}^{1/2}\Big)\,\text{vec}(\boldsymbol{H}_{\rm w})\,\text{vec}(\boldsymbol{H}_{\rm w})^*\,\Big({\boldsymbol{R}_{\rm t}}^{{\rm T}/2}\otimes{\boldsymbol{R}_{\rm r}}^{1/2}\Big)^*\Big] \tag{3.118}$$

$$= \Big({\boldsymbol{R}_{\rm t}}^{{\rm T}/2}\otimes{\boldsymbol{R}_{\rm r}}^{1/2}\Big)\,\mathbb{E}\big[\text{vec}(\boldsymbol{H}_{\rm w})\,\text{vec}(\boldsymbol{H}_{\rm w})^*\big]\,\Big({\boldsymbol{R}_{\rm t}}^{{\rm T}/2}\otimes{\boldsymbol{R}_{\rm r}}^{1/2}\Big) \tag{3.119}$$

$$= \Big({\boldsymbol{R}_{\rm t}}^{{\rm T}/2}\otimes{\boldsymbol{R}_{\rm r}}^{1/2}\Big)\Big({\boldsymbol{R}_{\rm t}}^{{\rm T}/2}\otimes{\boldsymbol{R}_{\rm r}}^{1/2}\Big) \tag{3.120}$$

$$= \boldsymbol{R}_{\rm t}^{\rm T}\otimes\boldsymbol{R}_{\rm r}, \tag{3.121}$$

where (3.118) follows from $\text{vec}(\boldsymbol{ABC}) = (\boldsymbol{C}^{\rm T}\otimes\boldsymbol{A})\,\text{vec}(\boldsymbol{B})$ [276], (3.120) follows from $\mathbb{E}[\text{vec}(\boldsymbol{H}_{\rm w})\text{vec}(\boldsymbol{H}_{\rm w})^*] = \boldsymbol{I}$, and (3.121) follows from $(\boldsymbol{A}\otimes\boldsymbol{B})(\boldsymbol{C}\otimes\boldsymbol{D}) = \boldsymbol{AC}\otimes\boldsymbol{BD}$. Thus, $\boldsymbol{R}_{\text{vec}(\boldsymbol{H})}$ emerges as the Kronecker product of $\boldsymbol{R}_{\rm t}^{\rm T}$ and $\boldsymbol{R}_{\rm r}$, justifying the name of the model.

The significance of (3.113) is the following: the correlation between the signal fading at any two receive antennas is the same irrespective of the transmit antenna from which the signal emanates, and conversely the correlation between the fading of the signals emanating from any two transmit antennas is the same at any receive antenna [277]. Intuitively, this requires that the local scattering processes around the transmitter and the receiver be decoupled. Under this condition, the entries of $\boldsymbol{R}_{\rm t}$ and $\boldsymbol{R}_{\rm r}$ can be obtained by sampling the space correlation function in (3.42) at the spacings between every transmit and every receive antenna pair, respectively, i.e.,

$$[\boldsymbol{R}_{\rm t}]_{j,j'} = \int_{-\pi}^{\pi} \mathcal{P}_{\rm t}(\theta)\, e^{-{\rm j}2\pi d_{{\rm t},j,j'}\cos(\theta)/\lambda_{\rm c}}\, {\rm d}\theta \tag{3.122}$$

$$[\boldsymbol{R}_{\rm r}]_{i,i'} = \int_{-\pi}^{\pi} \mathcal{P}_{\rm r}(\theta)\, e^{-{\rm j}2\pi d_{{\rm r},i,i'}\cos(\theta)/\lambda_{\rm c}}\, {\rm d}\theta, \tag{3.123}$$

with $d_{{\rm t},j,j'}$ and $d_{{\rm r},i,i'}$ the distances between transmit antennas j and j', and between receive antennas i and i', respectively, and with $\mathcal{P}_{\rm t}(\cdot)$ and $\mathcal{P}_{\rm r}(\cdot)$ the transmit and receive PAS.

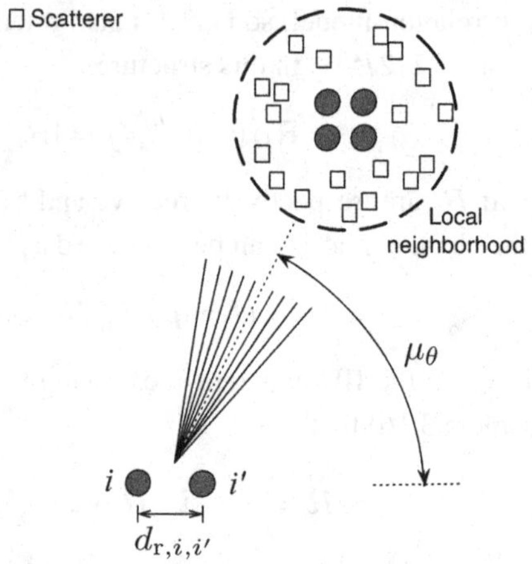

Fig. 3.23 Four-antenna user communicating with an elevated base station. The propagation paths lie within the angle width (centered at μ_θ) that the user's local scattering neighborhood subtends at the base.

Unless otherwise stated, throughout this text we consider the correlation matrices to be nonsingular, not only because that is appropriate from the perspective of modeling situations of interest, but further because dimensions in the null space of any singular correlation matrix could be taken out, leaving a reduced-dimension nonsingular model. The nonsingularity of the correlation matrices has the welcome benefit of yielding channel matrices $\boldsymbol{H}$ that are themselves full-rank with probability 1, a property that prevents certain nuisances in the formulation of subsequent chapters.

Base station correlation matrices

A common situation in cellular systems is to have a base station elevated above the clutter, deprived of local scattering, communicating with a user located within the clutter. Then, the user's PAS can be modeled as Clarke–Jakes, with the consequent applicability of the correlation function in Example 3.8. In turn, the base station's PAS can be assumed to be highly localized around its mean, μ_θ; as shown in Fig. 3.23, μ_θ is dictated by the azimuth location of the user (relative, without loss of generality, to the endfire direction) while the spread is determined by the angle subtended, from the vantage of the base station, by the local scattering neighborhood around the user. This spread depends on the elevation and the distance, among other aspects, and it can be as small as a few degrees.

Example 3.35

What would a typical base station azimuth angle spread be in a suburban environment?

Solution

In suburbia, the azimuth angle spread at a base situated 30 m above the ground has been shown to be around 1 or 2 degrees for distances between 1 and 6 km [214].

Such localized PAS functions set the stage for certain simplifications of (3.122) and (3.123) [229, 278]. Consider the reverse link, where the base station acts as a receiver. By means of the trigonometric identity $\cos(\alpha+\beta) = \cos(\alpha)\cos(\beta) - \sin(\alpha)\sin(\beta)$, we can rewrite (3.123) into

$$[\boldsymbol{R}_{\mathrm{r}}]_{i,i'} = \int_{-\pi}^{\pi} \mathcal{P}_{\mathrm{r}}(\theta)\, e^{-\mathrm{j}2\pi d_{\mathrm{r},i,i'} \cos(\mu_\theta + \theta - \mu_\theta)/\lambda_{\mathrm{c}}}\, \mathrm{d}\theta \tag{3.124}$$

$$= \int_{-\pi}^{\pi} \mathcal{P}_{\mathrm{r}}(\theta)\, e^{-\mathrm{j}2\pi d_{\mathrm{r},i,i'} \cos(\mu_\theta)\cos(\theta-\mu_\theta)/\lambda_{\mathrm{c}}}\, e^{\mathrm{j}2\pi d_{\mathrm{r},i,i'} \sin(\mu_\theta)\sin(\theta-\mu_\theta)/\lambda_{\mathrm{c}}}\, \mathrm{d}\theta$$

$$\approx e^{-\mathrm{j}2\pi d_{\mathrm{r},i,i'} \cos(\mu_\theta)/\lambda_{\mathrm{c}}} \int_{-\pi}^{\pi} \mathcal{P}_{\mathrm{r}}(\theta)\, e^{\mathrm{j}2\pi d_{\mathrm{r},i,i'} \sin(\mu_\theta)(\theta-\mu_\theta)/\lambda_{\mathrm{c}}}\, \mathrm{d}\theta, \tag{3.125}$$

where we have used $\cos(\theta - \mu_\theta) \approx 1$ and $\sin(\theta-\mu_\theta) \approx \theta - \mu_\theta$, which hold if the PAS is indeed concentrated around μ_θ. The leading term in (3.125) is a fixed phase shift, inconsequential for many purposes, while the magnitude is

$$\left|[\boldsymbol{R}_{\mathrm{r}}]_{i,i'}\right| \approx \left| \int_{-\pi}^{\pi} \mathcal{P}_{\mathrm{r}}(\theta)\, e^{\mathrm{j}2\pi d_{\mathrm{r},i,i'} \sin(\mu_\theta)(\theta-\mu_\theta)/\lambda_{\mathrm{c}}}\, \mathrm{d}\theta \right|, \tag{3.126}$$

which can serve as a stepping stone toward other convenient expressions for small angle spreads. Notice that in (3.126) we have maintained the integration from $\theta = -\pi$ to $\theta = \pi$, even though the approximation requires the integrand to be substantial only for θ in the vicinity of μ_θ; the integral could thus be compacted around μ_θ with minimal effect.

For the forward link, where the base station acts as a transmitter, starting from (3.122) and repeating the same steps,

$$\left|[\boldsymbol{R}_{\mathrm{t}}]_{j,j'}\right| \approx \left| \int_{-\pi}^{\pi} \mathcal{P}_{\mathrm{t}}(\theta)\, e^{\mathrm{j}2\pi d_{\mathrm{t},j,j'} \sin(\mu_\theta)(\theta-\mu_\theta)/\lambda_{\mathrm{c}}}\, \mathrm{d}\theta \right|. \tag{3.127}$$

Example 3.36 (Antenna correlation for a truncated Gaussian PAS)

Consider the truncated Gaussian PAS introduced in Example 3.6. Under the premise of a small σ_θ, the truncation can be neglected and the integral in (3.126) extended to obtain

$$\left|[\boldsymbol{R}_{\mathrm{r}}]_{i,i'}\right| \approx \left| \int_{-\infty}^{\infty} \frac{1}{\sqrt{2\pi}\sigma_\theta}\, e^{-\frac{\theta^2}{2\sigma_\theta^2}}\, e^{\mathrm{j}2\pi d_{\mathrm{r},i,i'} \sin(\mu_\theta)(\theta-\mu_\theta)/\lambda_{\mathrm{c}}}\, \mathrm{d}\theta \right| \tag{3.128}$$

$$= \exp\left(-2 \left[\frac{\pi\, d_{\mathrm{r},i,i'} \sin(\mu_\theta)\, \sigma_\theta}{\lambda_{\mathrm{c}}} \right]^2 \right), \tag{3.129}$$

which has been shown to offer a good fit to empirical data in elevated base stations located in suburban or rural areas [214].

Example 3.37 (Antenna correlation for a truncated Laplacian PAS)

Consider now the truncated Laplacian PDF introduced in Example 3.7. Again, under the premise of a small σ_θ, the truncation can be neglected and (3.126) becomes [279]

$$\left|[\boldsymbol{R}_{\mathrm{r}}]_{i,i'}\right| \approx \left|\int_{-\infty}^{\infty} \frac{1}{\sqrt{2}\sigma_\theta} e^{-\left|\sqrt{2}(\theta-\mu_\theta)/\sigma_\theta\right|} e^{\mathrm{j}2\pi d_{\mathrm{r},i,i'} \sin(\mu_\theta)(\theta-\mu_\theta)/\lambda_{\mathrm{c}}} \, \mathrm{d}\theta\right| \tag{3.130}$$

$$= \frac{1}{1+2\left(\pi d_{\mathrm{r},i,i'} \sin(\mu_\theta)\,\sigma_\theta/\lambda_{\mathrm{c}}\right)^2}. \tag{3.131}$$

As an alternative to the exact computation or to approximations of (3.122) on the basis of a postulated PAS, it is possible to posit an arbitrary correlation function that decays with distance. This approach is devoid of a physical interpretation, but it can suffice for certain purposes, e.g., for crude preliminary analyses.

Example 3.38 (Exponential antenna correlation model)

The magnitude of the correlation between receive antennas i and i' could be represented by [280, 281]

$$\left|[\boldsymbol{R}_{\mathrm{r}}]_{i,i'}\right| = \rho^{|i-i'|}, \tag{3.132}$$

where ρ is a parameter.

UIU model

Although the Kronecker correlation model has been shown to fit the data reasonably well in numerous instances, both indoors [272, 282–284] and outdoors [285], it also has limitations that stem from the transmit and receive PAS being fully decoupled, a condition that tends to result in optimistic assessments of capacity or of to-be-introduced metrics such as diversity [286–290].

A modeling representation that, while still short of offering the complete representation in (3.111) or (3.112), is more general than the Kronecker model is [291]

$$\boldsymbol{H} = \boldsymbol{\mu}_{\boldsymbol{H}} + \boldsymbol{U}_{\mathrm{r}} \boldsymbol{H}_{\mathsf{ind}} \boldsymbol{U}_{\mathrm{t}}^*, \tag{3.133}$$

where $\boldsymbol{U}_{\mathrm{r}}$ and $\boldsymbol{U}_{\mathrm{t}}$ are fixed unitary matrices while $\boldsymbol{H}_{\mathsf{ind}}$ has zero-mean independent non-identically distributed (IND) entries. Because of its structure, this representation is termed the *UIU model* or, in the case that the entries of $\boldsymbol{H}_{\mathsf{ind}}$ are not only IND but further complex Gaussian, also the *Weichselberger model* [292, 293]. As its main feature, the UIU model allows for the transmit and receive PAS to be coupled, something that becomes clearer once the model is further specialized in the subsequent section. Experimental data presented in [293] indicated that the UIU model mitigates the aforementioned deficiencies of its Kronecker counterpart.

An interesting interpretation of (3.133) is that it represents the Karhunen–Loève expansion of $\boldsymbol{H}$. Widely employed in data analysis, the Karhunen–Loève expansion is a transformation where the orthogonal basis functions are not fixed (e.g., complex exponentials in the Fourier case) but rather they are derived from the covariance of the process itself, in this case from the tensor $\mathsf{R}_{\boldsymbol{H}}(i,j;i',j')$; this ensures the best possible expansion for the given number of parameters. In the UIU case, it means that the columns of $\boldsymbol{U}_{\rm r}$ and $\boldsymbol{U}_{\rm t}$ contain the eigenfunctions of $\mathsf{R}_{\boldsymbol{H}}(i,j;i',j')$ while the variances of the entries of $\boldsymbol{H}_{\sf ind}$ equal the eigenvalues of $\mathsf{R}_{\boldsymbol{H}}(i,j;i',j')$, such that

$$\sum_{i'}\sum_{j'} \mathsf{R}_{\boldsymbol{H}}(i,j;i',j')\,[\boldsymbol{U}_{\rm r}]_{i,k}\,[\boldsymbol{U}_{\rm t}^*]_{j,\ell} = \lambda_{k,\ell}(\mathsf{R}_{\boldsymbol{H}})\,[\boldsymbol{U}_{\rm r}]_{i,k}\,[\boldsymbol{U}_{\rm t}^*]_{j,\ell}, \tag{3.134}$$

with the (k,ℓ)th eigenvalue of $\mathsf{R}_{\boldsymbol{H}}$ satisfying $\lambda_{k,\ell}(\mathsf{R}_{\boldsymbol{H}}) = \mathbb{E}\big[\,|[\boldsymbol{H}_{\sf ind}]_{k,\ell}|^2\,\big]$. Note from (3.133) that, indeed, the columns of $\boldsymbol{U}_{\rm t}$ and $\boldsymbol{U}_{\rm r}$ coincide with the eigenvectors of $\mathbb{E}[\boldsymbol{H}^*\boldsymbol{H}]$ and $\mathbb{E}[\boldsymbol{H}\boldsymbol{H}^*]$, respectively. The essence of the UIU model is that it maps the variances of $\boldsymbol{H}_{\sf ind}$ onto correlations within $\boldsymbol{H}$. The more uneven the variances in the former, the stronger the correlations in the latter.

To see the structure that the UIU model imposes on the complete correlation description in (3.112), it suffices to recall $\text{vec}(\boldsymbol{ABC}) = (\boldsymbol{C}^{\rm T} \otimes \boldsymbol{A})\text{vec}(\boldsymbol{B})$ in order to write $\text{vec}(\boldsymbol{H}) = (\boldsymbol{U}_{\rm t}^{\rm c} \otimes \boldsymbol{U}_{\rm r})\text{vec}(\boldsymbol{H}_{\sf ind})$, from which

$$\boldsymbol{R}_{\text{vec}(\boldsymbol{H})} = (\boldsymbol{U}_{\rm t}^{\rm c} \otimes \boldsymbol{U}_{\rm r})\,\mathbb{E}\big[\text{vec}(\boldsymbol{H}_{\sf ind})\,\text{vec}(\boldsymbol{H}_{\sf ind})^*\big](\boldsymbol{U}_{\rm t}^{\rm T} \otimes \boldsymbol{U}_{\rm r}^*), \tag{3.135}$$

where the leading and trailing matrices are unitary, because the Kronecker product of unitary matrices is unitary, while $\mathbb{E}\left[\text{vec}(\boldsymbol{H}_{\sf ind})\text{vec}(\boldsymbol{H}_{\sf ind})^*\right]$ is a diagonal matrix whose $N_{\rm t}N_{\rm r}$ diagonal entries are the variances of the entries of $\boldsymbol{H}_{\sf ind}$.

UIU versus Kronecker

The Kronecker model is but a special case of the UIU model. From (3.114), using the eigenvalue decompositions (see Appendix B) $\boldsymbol{R}_{\rm t} = \boldsymbol{U}_{\rm t}\boldsymbol{\Lambda}_{\rm t}\boldsymbol{U}_{\rm t}^*$ and $\boldsymbol{R}_{\rm r} = \boldsymbol{U}_{\rm r}\boldsymbol{\Lambda}_{\rm r}\boldsymbol{U}_{\rm r}^*$, we obtain $\boldsymbol{H} = \boldsymbol{U}_{\rm r}\boldsymbol{\Lambda}_{\rm r}^{1/2}\boldsymbol{U}_{\rm r}^*\boldsymbol{H}_{\rm w}\boldsymbol{U}_{\rm t}^*\boldsymbol{\Lambda}_{\rm t}^{1/2}\boldsymbol{U}_{\rm t}$. Because the distribution of $\boldsymbol{H}_{\rm w}$ is invariant to unitary rotations, this is statistically equivalent to $\boldsymbol{H} = \boldsymbol{U}_{\rm r}\boldsymbol{\Lambda}_{\rm r}^{1/2}\boldsymbol{H}_{\rm w}\boldsymbol{\Lambda}_{\rm t}^{1/2}\boldsymbol{U}_{\rm t}$, which in turn coincides with (3.133) if

$$\mathbb{E}\Big[\big|[\boldsymbol{H}_{\sf ind}]_{i,j}\big|^2\Big] = \lambda_i(\boldsymbol{R}_{\rm r})\,\lambda_j(\boldsymbol{R}_{\rm t}). \tag{3.136}$$

This condition is yet another way to see the constraints that the Kronecker model imposes on the structure of the correlation.

Besides the Kronecker model, another special case of (3.133) that is of interest in MIMO is the one where both $\boldsymbol{U}_{\rm t}$ and $\boldsymbol{U}_{\rm r}$ are simple identity matrices. Here, the channel matrix $\boldsymbol{H}$ has itself IND entries, which allows modeling antenna configurations where mechanisms other than physical separation are at play (see Discussion 3.2) and also situations where the antennas that are jointly transmitting or receiving are truly far apart, say distributed over a building or on distinct sites.

Discussion 3.2 Alternatives to space: pattern and polarization diversity

The antennas within an array can have different patterns, causing them to observe the same PAS through what essentially amounts to different lenses, and that tends to decorrelate them. Conceivably, two overlaid antennas can have very low correlation if their patterns are completely different and, in terms of diversity against fading, this is rightly termed *pattern diversity* [275, 294–297]. If low antenna correlation is desirable, pattern diversity has the advantage of requiring less physical space for a given number of antennas; this can be of interest at portable devices with restricted form factors. The downside is that such decorrelation comes at the expense of a loss in received power because, for the antennas to observe the same PAS differently, they must violate the design rule of having a pattern that is broader than the PAS and flat thereupon. Another consequence of utilizing different patterns is that the fading at the various antennas may be nonidentically distributed. As a special case, different patterns can be obtained by simply rotating a given (nonuniform) one; such implementation of pattern diversity is referred to as *angle diversity*.

Besides different patterns, also distinct polarizations enable more compact arrays with decreased levels of antenna correlation. The polarization of an antenna refers to the orientation of its transmit/receive electric field relative to the ground. More precisely, it refers to the projection of the electric field orientations over time onto a plane perpendicular to the direction of propagation of the radio signals. Such projection is in general an ellipse, which often is made to either specialize to a circle or to collapse onto a line. Two orthogonal polarizations are possible on a plane, namely clockwise and counterclockwise in the case of circular polarizations, and any two perpendicular lines in the case of linear polarizations. The linear orientations that naturally come to mind are horizontal, vertical, or slanted at $\pm 45°$ (see Fig. 3.24). As it turns out, man-made environments tend to favor the horizontal polarization over the vertical, possibly creating power imbalances between the two. Thus, slanted orientations are preferred because their powers are then sure to be balanced.

When a signal is transmitted on a given polarization, a receive antenna on the same polarization may collect more power than a receive antenna on the orthogonal polarization, and this difference is termed *cross-polar discrimination* [298]. At the same time, the fading at the two orthogonally polarized receive antennas tends to exhibit very low correlation [299] and hence the mixture of cross-polarized antennas allows packing twice as many of them in a given surface or volume without a significant increase in correlation. Such *polarization diversity* may result, like pattern diversity, in fading that is not identically distributed across the array antennas, and the cross-polar discrimination that determines the power differences has been found to be related to other large-scale parameters such as the large-scale channel gain or the Rice factor [300–302]. The modeling and the information-theoretic properties of MIMO channels with polarization diversity are illustrated in Examples 3.39 and 5.36. Empirical evidence of the effectiveness of polarization diversity is given, e.g., in [303–306].

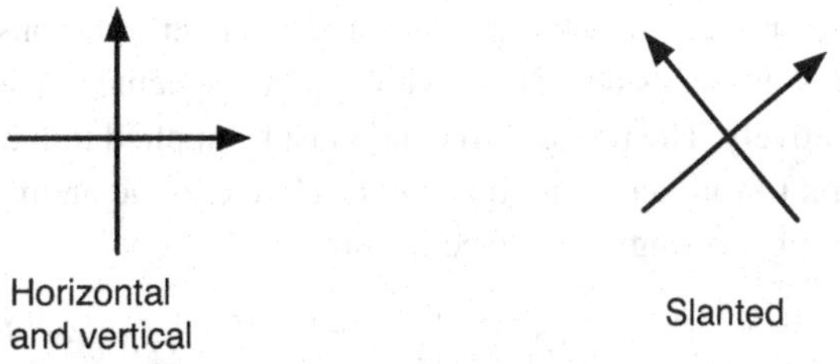

Fig. 3.24 Cross-polarization arrangements for a pair of linearly polarized antennas. Left, horizontal and vertical. Right, slanted at $\pm 45°$.

Example 3.39

Let $N_\text{t} = N_\text{r} = 2$. At both transmitter and receiver the antennas are overlaid, i.e., $d_{\text{t},0,1} = d_{\text{r},0,1} = 0$, but with the antennas having $\pm 45°$ slanted polarizations as per Fig. 3.24. There is complete fading decorrelation across polarizations while the cross-polar discrimination, i.e., the ratio of the power gain between co-polarized and cross-polarized antennas, is denoted by Ξ [298]. Express the channel matrix via the UIU model.

Solution

The channel is IND and, denoting by $\boldsymbol{\Omega}$ a matrix such that $[\boldsymbol{\Omega}]_{i,j} = \mathbb{E}\left[\left|[\boldsymbol{H}_\text{ind}]_{i,j}\right|^2\right]$,

$$\boldsymbol{\Omega} = \frac{2}{1+\Xi} \begin{bmatrix} 1 & \Xi \\ \Xi & 1 \end{bmatrix}, \tag{3.137}$$

where the leading factor guarantees the normalization. Then, $\boldsymbol{H} = \boldsymbol{\Omega}^{1/2} \odot \boldsymbol{H}_\text{w}$ where $\odot$ indicates Hadamard (entry-wise) product.

The above example illustrates the added versatility of the UIU structure. While, in the Kronecker model, zero antenna spacing entails full correlation and the entries of $\boldsymbol{H}$ must have the same variance, that need not be so on the UIU mode. This enables the analysis of antenna arrays with an assortment of polarizations and patterns, that is, arrays whose constituent antennas are not identical.

Virtual channel model

The UIU model becomes particularly meaningful when its mean is zero and $\boldsymbol{U}_\text{t}$ and $\boldsymbol{U}_\text{r}$ are Fourier matrices (see Appendix B). Then, the representation leads to the *virtual channel model* championed in [307], namely

$$\boldsymbol{H} = \boldsymbol{U}_\text{r} \boldsymbol{H}_\text{vir} \boldsymbol{U}_\text{t}^*, \tag{3.138}$$

where $\boldsymbol{H}_\text{vir}$ is, by the very nature of $\boldsymbol{U}_\text{t}$ and $\boldsymbol{U}_\text{r}$, the two-dimensional discrete Fourier transform of $\boldsymbol{H}$. In the virtual model, the entries of $\boldsymbol{H}_\text{vir}$ are zero-mean complex Gaussian, with a joint distribution detailed below. By inverting (3.138), we obtain $\boldsymbol{H}_\text{vir} = \boldsymbol{U}_\text{r}^* \boldsymbol{H} \boldsymbol{U}_\text{t}$.

This model opens the door to intuitive interpretations that borrow concepts from array processing. Consider ULAs with antenna spacings d_t and d_r at transmitter and receiver, respectively. The phase shifts that must be applied to the receive antennas in order to point a beam toward an azimuth angle θ_r relative to the endfire direction are given by the corresponding steering vector (see Section 3.5.1)

$$\boldsymbol{a}_r(\theta_r) = \left[1 \;\; e^{j2\pi \frac{d_r \cos(\theta_r)}{\lambda_c}} \;\; \cdots \;\; e^{j2\pi \frac{(N_r-1)d_r \cos(\theta_r)}{\lambda_c}}\right]^T. \tag{3.139}$$

Since the ith column of an $N_r \times N_r$ Fourier matrix $\boldsymbol{U}_r$ equals

$$[\boldsymbol{U}_r]_{:,i} = \frac{1}{\sqrt{N_r}} \left[1 \;\; e^{j2\pi \frac{i}{N_r}} \;\; \cdots \;\; e^{j2\pi \frac{(N_r-1)i}{N_r}}\right]^T, \tag{3.140}$$

the basis functions that those columns correspond to are a set of fixed beams pointing to angles

$$\theta_{r,i} = \arccos\left(\frac{\lambda_c}{d_r}\frac{i}{N_r}\right) \qquad i = 0, \ldots, N_r - 1. \tag{3.141}$$

Likewise, the columns of $\boldsymbol{U}_t$ correspond to a set of fixed transmit beams pointing to angles $\theta_{t,j} = \arccos\left(\frac{\lambda_c}{d_t}\frac{j}{N_t}\right)$ for $j = 0, \ldots, N_t - 1$. The Fourier transformation in (3.138) can be seen as a mapping from a beam domain onto the antenna domain, and the entries of $\boldsymbol{H}_{\text{vir}}$ signify the coupling between the N_t transmit and the N_r receive beams. These beams can also be interpreted as a partition of the transmit and receive PAS into N_t and N_r (overlapping) intervals, respectively, and $\boldsymbol{H}_{\text{vir}}$ determines how these intervals couple, which is something that the Kronecker model does not allow.

The resolution in the beam domain—or *virtual* domain, the term after which the model is named [307]—depends on the number of antennas in the corresponding array. As this number grows, the beams sharpen and become more orthogonal. It follows that the entries of $\boldsymbol{H}_{\text{vir}}$ characterize the coupling between the transmit and receive local neighborhoods with a resolution that depends on the number of antennas (see Fig. 3.25). Scatterers within the angular resolution of the beams are not discriminated or, put differently, their aggregate effect is captured by a single entry of $\boldsymbol{H}_{\text{vir}}$. The variance of $[\boldsymbol{H}_{\text{vir}}]_{i,j}$ indicates how intensely the jth transmit beam scatters into the ith receive beam. If there are no significant scatterers connecting those beams, then $\mathbb{E}\left[\,|[\boldsymbol{H}_{\text{vir}}]_{i,j}|^2\,\right] \approx 0$.

Example 3.40

Figure 3.25 shows a cartoon representation of the virtual channel model for $N_t = N_r = 5$. The outbound and inbound solid arrows indicate the axes of the transmit and receive beams, whose shapes are not explicitly shown. The dashed arrows indicate the coupling between beams, and thus every such arrow corresponds with one entry of $\boldsymbol{H}_{\text{vir}}$. In this case, for instance, $\mathbb{E}\left[\,|[\boldsymbol{H}_{\text{vir}}]_{1,2}|^2\,\right] \approx 0$.

Unlike in the UIU model, where the entries of $\boldsymbol{H}_{\text{ind}}$ are by definition independent, in the virtual channel model their strict definition is as the two-dimensional Fourier transform of $\boldsymbol{H}$, which need not render them independent. However, under the reasonable assumption

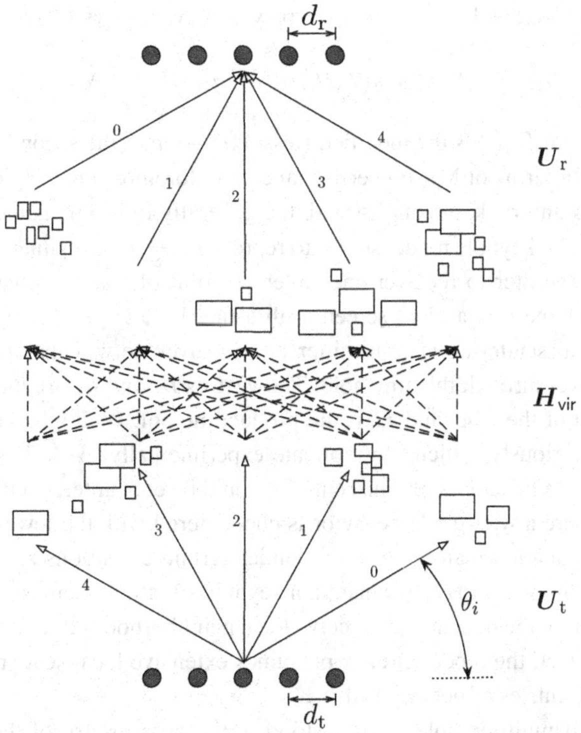

Fig. 3.25 Virtual channel model for $N_t = N_r = 5$.

that scatterers behave independently (in essence the *uncorrelated scattering* condition already invoked earlier as part of the WSSUS framework), the entries of $\boldsymbol{H}_{\mathsf{vir}}$ decorrelate as N_t and N_r grow and the beams illuminate progressively more disjoint sets of scatterers. For finite N_t and N_r, the entries of $\boldsymbol{H}_{\mathsf{vir}}$ are only approximately independent; to the extent that this approximation holds, the information about the correlation among the entries of $\boldsymbol{H}$ is reflected in the variances of those of $\boldsymbol{H}_{\mathsf{vir}}$.

When specialized to the Kronecker structure, through the imposition of condition (3.136) to $\boldsymbol{H}_{\mathsf{vir}}$, the virtual model results in transmit and receive correlation matrices $\boldsymbol{R}_t$ and $\boldsymbol{R}_r$ whose eigenvectors are the columns of a Fourier matrix.

Keyhole model

To finalize this survey of analytical models, we briefly discuss the *keyhole* or *pinhole* channel model [308–311]. In this model, the entries of $\boldsymbol{H}$ are uncorrelated yet not independent; they are therefore necessarily also non-Gaussian. Precisely,

$$\boldsymbol{H} = \boldsymbol{h}_r \, \boldsymbol{h}_t^*, \tag{3.142}$$

where $\boldsymbol{h}_t$ and $\boldsymbol{h}_r$ are, respectively, $N_t \times 1$ and $N_r \times 1$ vectors with IID zero-mean unit-variance entries. Typically the entries of $\boldsymbol{h}_t$ and $\boldsymbol{h}_r$ are further modeled as complex Gaussian, in which case each entry of $\boldsymbol{H}$ is the product of two independent complex Gaussian scalars; the distribution of the power of every entry of $\boldsymbol{H}$ can then be shown to be [312]

$$f_{|[\boldsymbol{H}]_{i,j}|^2}(\xi) = 2\,K_0\big(2\sqrt{\xi}\big) \qquad i = 0,\ldots,N_{\mathrm{r}}-1 \qquad j = 0,\ldots,N_{\mathrm{t}}-1, \tag{3.143}$$

where $K_0(\cdot)$ is the modified Bessel function of the second kind.

In terms of MIMO performance, the foremost property of the keyhole model is that $\boldsymbol{H}$ has unit rank, in contrast with the generally full-rank nature of the other models.

The keyhole model serves to represent degenerate situations where the propagation from transmitter to receiver encounters a point of spatial collapse. To visualize this, one can imagine that a giant screen with a tiny hole is placed between transmitter and receiver; local scattering at transmitter and receiver notwithstanding, the signals must inevitably travel through this tiny hole. The vectors $\boldsymbol{h}_{\mathrm{t}}$ and $\boldsymbol{h}_{\mathrm{r}}$ are then the channel responses in and out of the hole, and $\boldsymbol{H}$ is the product thereof. While this model is a theoretical construct, notoriously difficult to replicate experimentally [313, 314], there are physical situations where behaviors reminiscent of it could be encountered, chiefly tunnels or indoor hallways where a waveguide behavior is encountered with the lowest-order mode suffering far less attenuation than the rest, and under certain conditions also roof-edge diffraction [309]. It is important to realize that, in a keyhole situation, increasing the amount of local scattering within the transmit and receive local neighborhoods does not alter the nature of the channel. In fact, the model already presumes extensive local scattering, hence the IID character of the entries of both $\boldsymbol{h}_{\mathrm{t}}$ and $\boldsymbol{h}_{\mathrm{r}}$.

If multiple holes—understood in the broad sense of the above situations—exist, then $\boldsymbol{H}$ must be a superposition of several keyhole matrices, which progressively increases the rank. This richer scenario, with the further addition of local transmit and receive correlations, leads to the more elaborate model [310]

$$\boldsymbol{H} = \boldsymbol{R}_{\mathrm{r}}^{1/2}\boldsymbol{H}_{\mathrm{w}1}\boldsymbol{R}_{\mathrm{m}}^{1/2}\boldsymbol{H}_{\mathrm{w}2}\boldsymbol{R}_{\mathrm{t}}^{1/2}, \tag{3.144}$$

where $\boldsymbol{H}_{\mathrm{w}1}$ and $\boldsymbol{H}_{\mathrm{w}2}$ are mutually independent and distributed as $\boldsymbol{H}_{\mathrm{w}}$ while $\boldsymbol{R}_{\mathrm{m}}$ is an additional intermediate correlation matrix.

3.6.2 Parametric models

The analytical models reviewed thus far, and chiefly the Kronecker model, are the ones most extensively applied to explore theoretic aspects of MIMO. Other modeling approaches do exist, however, and these constitute the object of the remainder of the section. These alternatives tend to be more physically inspired, and some of them serve as building blocks for the standardized models sketched at the end of the chapter. The common trait of these alternative modeling approaches, and the reason we refer to them as parametric, is that they do not directly determine the statistical properties of the entries of $\boldsymbol{H}$ but, rather, they model parameters from which those statistical properties then derive. Let us therefore begin by resorting back to a parametric description used in Chapter 2 whereby the signals travel from transmitter to receiver through Q discrete propagation paths rather than a continuum thereof. Then, rather than an integral over a continuous PAS, as in (3.37), we encounter a sum over the paths and the (i, j)th entry of $\boldsymbol{H}$ is given by

$$[\boldsymbol{H}]_{i,j} = \frac{1}{\sqrt{Q}} \sum_{q=0}^{Q-1} A_q \, e^{\mathrm{j}\phi_i(\theta_{\mathrm{r},q})} e^{\mathrm{j}\phi_j(\theta_{\mathrm{t},q})}, \tag{3.145}$$

where $\theta_{\mathrm{t},q}$ and $\theta_{\mathrm{r},q}$ are the angles of departure and arrival of the qth path, $\phi_j(\theta_{\mathrm{t},q})$ and $\phi_i(\theta_{\mathrm{r},q})$ are the relative phase shifts accrued by such a path traveling the transmit and receive local neighborhoods (these vary from antenna to antenna, hence the indexing by j and i). In turn, A_q is a complex coefficient whose magnitude is the path gain and whose phase is random, uniform in $[-\pi, \pi)$, to represent the shift accumulated while traveling between the transmitter and receiver local neighborhoods plus whatever shift the interaction with the scatterers may introduce. The phase shift associated with A_q is common to all antennas, and the path magnitudes are normalized such that $\mathbb{E}[|[\boldsymbol{H}_{i,j}]|^2] = 1$ for $j = 0, \ldots, N_\mathrm{t} - 1$ and $i = 0, \ldots, N_\mathrm{r} - 1$. It follows that [27]

$$\boldsymbol{H} = \frac{1}{\sqrt{Q}} \sum_{q=0}^{Q-1} A_q \, \boldsymbol{a}_\mathrm{r}(\theta_{\mathrm{r},q}) \, \boldsymbol{a}_\mathrm{t}^\mathrm{T}(\theta_{\mathrm{t},q}), \tag{3.146}$$

with $\boldsymbol{a}_\mathrm{r}(\cdot)$ and $\boldsymbol{a}_\mathrm{t}(\cdot)$ the receive and transmit array steering vectors. It is generally assumed that the number of paths Q is larger than $\min(N_\mathrm{t}, N_\mathrm{r})$, such that $\boldsymbol{H}$ is full-rank with probability 1. If that were not the case, the channel would be rank-deficient and the number of spatial dimensions available for communication would be limited by the channel rather than by the number of antennas [27].

If the differences among the delays suffered by the Q paths are not insignificant relative to the symbol period, then, as we have learned, the fading becomes frequency-selective. In the delay domain, (3.146) then splits into multiple taps; every such tap corresponds to a matrix $\boldsymbol{H}[\ell]$ that has the form of (3.146) only with the summation restricted to the paths whose delays fall onto the ℓth tap, and with complex gains suitably modified by the cascade of the transmit and receive filters, $g_{\mathsf{tx}}(\cdot)$ and $g_{\mathsf{rx}}(\cdot)$, at the sampling instant of the tap.

The models presented in the remainder of this section rely either on the geometry of the transmit and receive local neighborhoods, or else in certain distributions, in order to establish concrete values for the path lengths and angles, and therefore for their magnitudes and phases, in the above parametric representation.

Before proceeding though, it is instructive to relate (3.146) with the virtual channel model. If we define $\mathcal{S}_{\mathrm{t},j}$ as the set of paths whose angle of departure falls within the jth transmit beam in that model, for $j = 0, \ldots, N_\mathrm{t} - 1$, and $\mathcal{S}_{\mathrm{r},i}$ as the set of paths whose angle of arrival falls within the ith receive beam, for $i = 0, \ldots, N_\mathrm{r} - 1$, then the variance of the (i, j)th entry of the beam–domain matrix $\boldsymbol{H}_{\mathsf{vir}}$ is

$$\mathbb{E}\left[\left|[\boldsymbol{H}_{\mathsf{vir}}]_{i,j}\right|^2\right] \approx \sum_{q \in (\mathcal{S}_{\mathrm{t},j} \cap \mathcal{S}_{\mathrm{r},i})} \mathbb{E}\left[|A_q|^2\right], \tag{3.147}$$

where the relationship is approximate because, recall, the beams become completely disjoint only as $N_\mathrm{t}, N_\mathrm{r} \to \infty$. In turn, the entries of $\boldsymbol{H}_{\mathsf{vir}}$ become complex Gaussian as the number of paths per beam grows. Then, $\boldsymbol{H}$ becomes a Rayleigh-faded MIMO channel with correlations dictated by the two-dimensional discrete Fourier transform of $\boldsymbol{H}_{\mathsf{vir}}$.

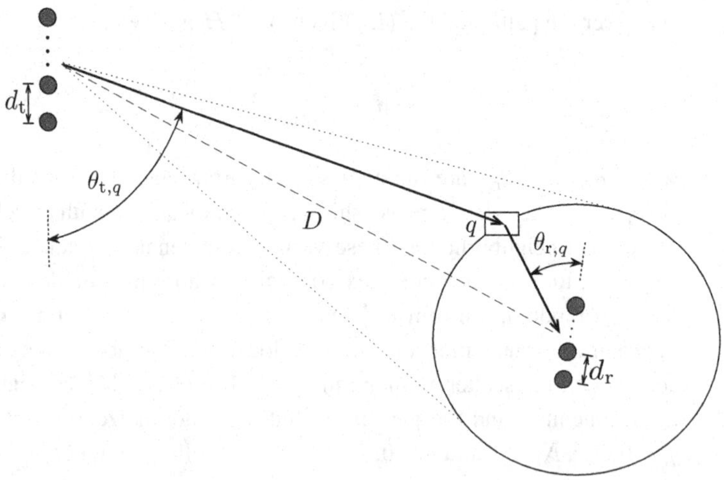

Fig. 3.26 One-ring model where, for clarity, the size of the arrays has been blown up relative to D and to the ring radius. Explicitly indicated is the path that transmits by way of scatterer q.

One-ring model

The tour of parametric models starts with a couple of geometry-based ones that address complementary situations. First, the one-ring model, which is meant to represent a situation already encountered in the context of analytical modeling: a base station elevated above the propagation clutter, deprived of local scattering, communicating with a user within the clutter. For this scenario, the correlation between any two base station antennas in the face of diffuse scattering was characterized in Examples 3.35 and 3.36. The one-ring model is merely a discrete version of the same formulation, by means of the parametric description in (3.146), which is more amenable to simulation.

This model, depicted in Fig. 3.26, considers that scatterers exist only on a ring of a certain radius centered on the user [274, 315]. Specifically, there are Q scatterers randomly distributed along the ring, each such scatterer creating a distinct path between the base station and the user's array. If the distribution of scatterers is uniform along the ring, the discrete PAS at the user is merely a sampled version of the continuous Clarke–Jakes PAS while, at the base station, the discrete PAS has a width—and thus an angle spread—that is determined by the range D and the radius of the ring.

For the sake of specificity we can focus on the forward link; the corresponding reverse-link channel would be obtained through a simple conjugate transposition. In the one-ring model, the base station local scattering neighborhood is absent and its role is played directly by the propagation between the base and the faraway ring. Then, as indicated in Fig. 3.26, $\theta_{r,q}$ and $\theta_{t,q}$ are the angles spanned by scatterer q at the user and at the base station, respectively, and $\boldsymbol{a}_r(\theta_{r,q})$ and $\boldsymbol{a}_t(\theta_{t,q})$ are the corresponding array steering vectors depending on the topology of the arrays (see Section 3.5.1). In turn, the phase of A_q equals the shift introduced by the reflection at the qth scatterer. Mimicking the Clarke–Jakes PAS at the receiver entails setting $|A_q| = 1$, for $q = 0, \ldots, Q-1$, although any set of magnitudes that respects the unit-variance normalization of the entries of $\boldsymbol{H}$ would in principle be acceptable.

As a refinement of the model, one could think about making the distribution of scatterers nonuniform over the ring, which would modify the PAS at both transmitter and receiver. A possibility here would be a distribution that, from the vantage of the transmitter, returns a sampled version of any of the continuous PAS functions given earlier, e.g., a truncated Gaussian or a truncated Laplacian. An alternative possibility is the von-Mises distribution, whereby the angles of the ring scatterers—from the vantage of the receiver—are drawn from [286]

$$f_{\theta_{\mathrm{r},q}}(\theta) = \frac{e^{\kappa \cos(\theta - \mu)}}{2\pi I_0(\kappa)} \qquad \theta \in [\mu - \pi, \mu + \pi), \tag{3.148}$$

where $I_0(\cdot)$, recall, is the zero-order modified Bessel function, $\kappa \geq 0$, and μ indicates the angle where the density of scatterers peaks. The interest of the von-Mises distribution is that it closely approximates what would be obtained by wrapping a Gaussian distribution around a circle [316], without truncations, and it does so with a PDF that is compact and tractable. The parameter κ determines the concentration of the distribution, ranging from uniform if $\kappa = 0$ to a mass point at μ if $\kappa \to \infty$.

Two-ring model

The complementary situation of that captured by the one-ring model is the one where the base station is not elevated high above the clutter, but immersed within it; that is the case, e.g., in microcells. Then, the existence of local scattering neighborhoods around both transmitter and receiver naturally calls for the two-ring model in Fig. 3.27 [317]. Focusing again on the forward link, for the sake of specificity, the rings around the base and the user contain Q_t and Q_r scatterers, respectively, such that $Q = Q_\mathrm{t} Q_\mathrm{r}$ and (3.146) can be refined as

$$\boldsymbol{H} = \frac{1}{\sqrt{Q_\mathrm{t} Q_\mathrm{r}}} \sum_{q_\mathrm{t}=0}^{Q_\mathrm{t}-1} \sum_{q_\mathrm{r}=0}^{Q_\mathrm{r}-1} A_{q_\mathrm{t},q_\mathrm{r}} \, \boldsymbol{a}_\mathrm{r}(\theta_{\mathrm{r},q_\mathrm{r}}) \, \boldsymbol{a}_\mathrm{t}^\mathrm{T}(\theta_{\mathrm{t},q_\mathrm{t}}), \tag{3.149}$$

where $\theta_{\mathrm{t},q_\mathrm{t}}$ and $\theta_{\mathrm{r},q_\mathrm{r}}$ are the angles spanned by scatterer q_t on the transmitter ring and by scatterer q_r on the receiver ring, respectively, and $\boldsymbol{a}_\mathrm{t}(\theta_{\mathrm{t},q_\mathrm{t}})$ and $\boldsymbol{a}_\mathrm{r}(\theta_{\mathrm{r},q_\mathrm{r}})$ are the corresponding array steering vectors depending on the topology of the arrays.

In principle the scatterers are uniformly spread along both rings, with path magnitudes $|A_{q_\mathrm{t},q_\mathrm{r}}| = 1$, which corresponds to sampled Clarke–Jakes PAS at either end, but again alternative distributions and any set of path magnitudes that respects the unit-variance normalization of the entries of $\boldsymbol{H}$ would be acceptable.

Clustered models

The last type of parametric models we tour are clustered models [318, 319]. Specifically, we describe an extension, suitable for MIMO, of the Saleh–Valenzuela model presented in

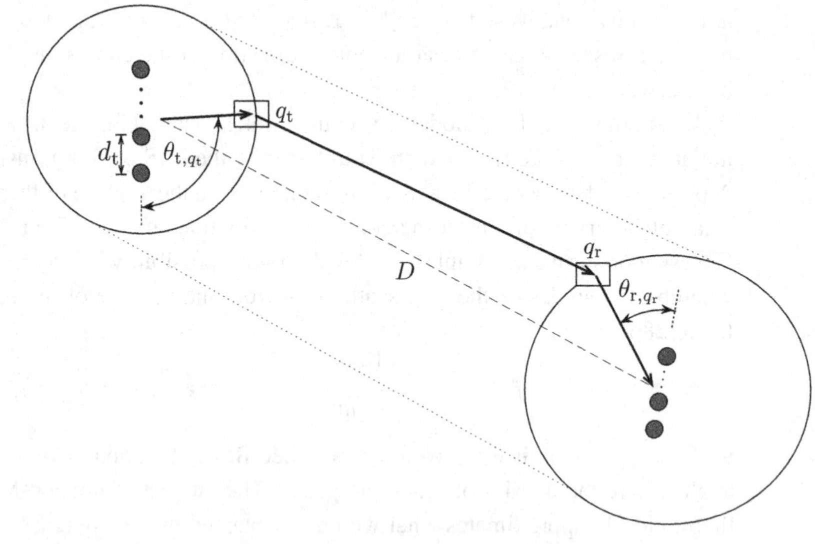

Fig. 3.27 Two-ring model where, for clarity, the size of the arrays has been blown up relative to D and to the ring radii. Explicitly indicated is the path that transmits by way of transmit scatterer q_t and receive scatterer q_r.

Example 3.22. Recall that the Saleh–Valenzuela model represents scenarios with multiple scattering clusters (whose locations are modeled statistically rather than geometrically). The key to incorporating space variability to this model is to distinguish paths on the basis, not only of their propagation delays, but also of their angles [233]. As shown in Fig. 3.28, each cluster of scatterers is then characterized by a mean delay and a mean angle around which the path delays and angles are spread. The resulting channel exhibits a clustered structure in both PDP and PAS, as illustrated in Fig. 3.29. The channel matrix at a given delay can be generated from the parametric construction in (3.146), with the angles drawn from any distribution of choice—typically the mean angle of each cluster is drawn uniformly while the distribution of angles thereabout is drawn from a truncated Laplacian—and with the magnitudes subject to fading in order to account for the fact that here each term models a multipath ray, rather than a single path. The clustered structure invites a multitap channel response with a number of taps that depends on the bandwidth.

3.7 Channel estimation revisited

One of the topics of Chapter 2 is pilot-assisted channel estimation. Indeed, knowledge of the channel response at the receiver (and sometimes at the transmitter) has a capital importance that goes beyond equalization and data detection. As the extent of such knowledge is one of the axes in our exposition, with the benefit of the material in this chapter more elaborate statements on the matter can be made at this point. Consistent with our dissocia-

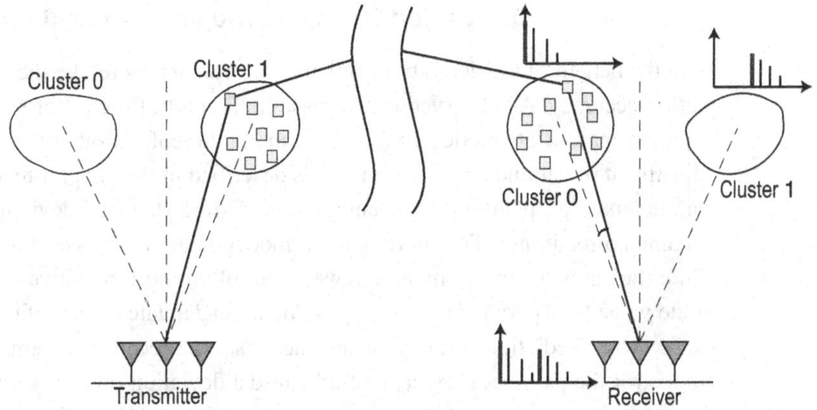

Fig. 3.28 Clustered channel model.

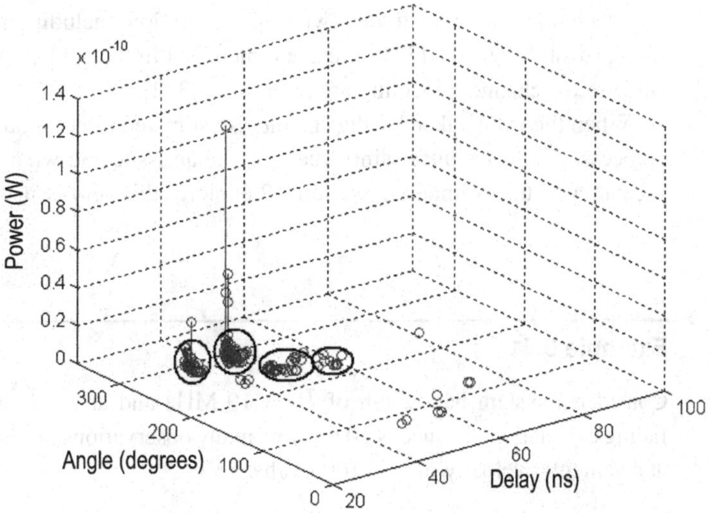

Fig. 3.29 Power density as a function of both delay and angle in a clustered model.

tion of large-scale and small-scale channel terms for modeling purposes, also for channel estimation purposes both terms are considered separately.

3.7.1 Large-scale phenomena

The large-scale channel gain G varies very slowly over space (and thus over time for a given velocity) and also very slowly over frequency, and hence lots of observations can be gathered from which to estimate it.

Discussion 3.3 A follow-up note of caution

With the benefit of the derivations hitherto, we can briefly retake the issue of model prediction accuracy. As observed in Discussion 3.1, part of the art of modeling is to select the most appropriate model from an assortment thereof. Another aspect of this art is the adjustment of parameters. All the models described in the chapter feature one or several parameters (e.g., pathloss exponent, shadow fading standard deviation, Rice factor, or antenna correlations). The more general models feature a larger number of parameters while the simpler models include fewer—but often more sensitive—ones. An inappropriate value for a parameter, not to speak of an inadequate choice of the model itself, can cause gross prediction errors. For instance, as explored in Problem 3.30, an improper choice for the pathloss exponent could cause a deviation on the predicted signal power that exceeds most of the swings of the accompanying shadow fading, It is wise to account for the uncertainty in the knowledge of these parameters, and it is best to distinguish this uncertainty from the channel variations that the model intends to reproduce. Put differently, it is wise to distinguish between "known unknowns" and "unknown unknowns," and techniques do exist to deal with this separation, including random-set theory [320] and probability boxes [321]. Readers interested in the applicability of these techniques to wireless channel modeling are referred to [322].

When the value taken by the channel is estimated directly during the communication process, parameter uncertainty ceases to be an issue and what becomes relevant is the accuracy of the estimation. Section 3.7 explores this very issue.

Example 3.41

Consider a system bandwidth of $B = 10$ MHz and an outdoor channel whose shadow fading correlation distance is 10 m. How many observations of G can (roughly) be gathered at a vehicular velocity of $\mathrm{v} = 100$ km/h?

Solution

At frequencies of interest, the shadow fading and pathloss do not vary appreciably over 10 MHz and hence all such bandwidth is available to gather observations.

On the spatial domain, at $\mathrm{v} = 100$ km/h (which equals 27.8 m/s) the shadowing correlation distance maps to a shadowing coherence time of $10/27.8 = 0.36$ s. Given $T = 1/B = 0.1$ μs, roughly the same shadow fading is observed by 3.6 million single-carrier symbols, or equivalently 3.6 million OFDM resource elements. Even if only 5% of these were devoted to channel estimation, say through the transmission of pilots, that would still amount to 180 000 observations.

The pathloss also varies over distance, but only slightly over a shadowing correlation distance and thus the effect on the number of observations is minimal.

Given the very stable nature of G, the availability of a perfect estimate thereof is almost universally taken for granted, and such is also the premise in this text. This is thoroughly justified in most conceivable scenarios, and only in extreme cases (e.g., the conjunction of very narrowband signals and utterly high velocities) may its validity warrant verification. Readers interested in how to estimate G, and other large-scale quantities such as the fading coherence and the Doppler spectrum, are directed to [323] and references therein.

3.7.2 Small-scale fading

Shifting the focus to the province of small-scale fading, the number of single-carrier symbols or OFDM resource elements over which such fading remains coherent was shown earlier to be given (roughly) by the product $B_c T_c$. This product was quantified in Examples 3.26–3.28, which essentially bracketed the range of values that are likely to be encountered: from upwards of 1000 symbols in outdoor high-mobility settings to tens of thousands in low-mobility and indoor settings. While these values tend to be sufficiently large to enable satisfactory estimation of the fading with moderate pilot overheads, such estimation can no longer be taken for granted. This is particularly true when the number of transmit antennas grows large, as the estimation of the fading coefficients associated with each such antenna essentially carves a portion of the available observations; in massive MIMO, therefore, the number of observations available to estimate each fading coefficient may shrink to a point that renders this aspect critical.

The procedures of pilot insertion and of communication supported by the ensuing estimates, as well as the possibility of communication without explicit channel estimates, are all deferred to other chapters. This section merely sets forth, for subsequent reference, expressions that arise when we apply to certain fading channels the LMMSE estimators derived for time-invariant channels in Section 2.7. Besides their applicability, the obtained expressions are relevant because they uncover an interesting relationship between fading models.

The focal interest on the LMMSE as a preferred estimator for small-scale fading has a dual motivation.

- The MMSE is intimately linked with the signal-to-interference-plus noise ratio (SINR), whose maximization turns out to be a sound criterion in the design of transmitters and receivers.
- The distribution of the fading is not controlled by the system designer and hence it is seldom available, negating the possibility of conditional-mean MMSE estimation. However, the second-order statistics required by an LMMSE estimator are indeed accessible; the variance, in particular, essentially amounts to the large-scale gain G, whose value can be precisely learned.

SISO

For the derivations that follow, we can regard the fading as frequency-flat with the understanding that the results and procedures apply on a coherence-bandwidth basis.

Let us first consider block fading. Given a scalar complex gain h, zero-mean and unit-variance but otherwise arbitrarily distributed and with IID fading blocks spanning N_c single-carrier symbols or OFDM resource elements, an LMMSE estimator would gauge the value of h in every fading block by observing a number of pilot transmissions. Over each block, this coincides exactly with the fixed-channel setting of Example 2.29. If α denotes the portion of symbols devoted to pilot transmissions, then $N_p = \alpha N_c$ observations are available and

$$\mathsf{MMSE} = \frac{1}{1+\alpha N_c \mathsf{SNR}}, \tag{3.150}$$

where SNR is the per-symbol SNR defined in the usual way.

Now, let $h(t,\tau) = h(t)\delta(\tau)$ be continuously faded with $h(t)$ having an arbitrary time correlation $R_h(\cdot)$ and a corresponding Doppler spectrum $S_h(\nu)$ confined to $\nu \in [-\nu_M, \nu_M]$. Sampled with period T, $h(t)$ yields a discrete-time fading process $h[n]$ whose Doppler spectrum is a function of the normalized Doppler shift $\mathsf{v} = \nu T$ and relates to its continuous-time counterpart via

$$\frac{1}{T} S_h\left(\frac{\mathsf{v}}{T}\right) \qquad \mathsf{v} \in [-\nu_M T, \nu_M T]. \tag{3.151}$$

If a share α of the symbols, uniformly spaced, are reserved for the transmission of pilots (see Fig. 4.22), then the discrete-time fading is further decimated and its Doppler spectrum broadens and scales into

$$S(\mathsf{v}) = \frac{\alpha}{T} S_h\left(\frac{\alpha \mathsf{v}}{T}\right) \qquad \mathsf{v} \in \left[\frac{-\nu_M T}{\alpha}, \frac{\nu_M T}{\alpha}\right]. \tag{3.152}$$

The LMMSE estimate of $h[n]$ should process all those receiver observations that have nonzero correlation with the channel at symbol n, as each of them supplies information. If *all* observations are processed, even future ones, (1.254) applies and thus

$$\mathsf{MMSE} = 1 - \int_{-\nu_M T/\alpha}^{\nu_M T/\alpha} \frac{\mathsf{SNR}\, S^2(\mathsf{v})}{1+\mathsf{SNR}\, S(\mathsf{v})} d\mathsf{v}. \tag{3.153}$$

Example 3.42 (Channel estimation MMSE for continuous fading with a Clarke–Jakes Doppler spectrum)

For the Clarke–Jakes spectrum $S_h(\cdot)$ derived in Example 3.14,

$$S(\mathsf{v}) = \frac{\alpha}{T} S_h\left(\frac{\alpha \mathsf{v}}{T}\right) \tag{3.154}$$

$$= \frac{1}{\pi\sqrt{\left(\frac{\nu_M T}{\alpha}\right)^2 - \mathsf{v}^2}} \qquad \mathsf{v} \in \left[\frac{-\nu_M T}{\alpha}, \frac{\nu_M T}{\alpha}\right], \tag{3.155}$$

whose integration in (3.153) gives [324, proposition 1]

$$\mathsf{MMSE} = 1 - \frac{\operatorname{arctanh}\sqrt{1-\left(\frac{\mathsf{SNR}}{\pi}\frac{\alpha}{\nu_M T}\right)^2}}{\frac{\pi}{2}\sqrt{\left(\frac{\pi}{\mathsf{SNR}}\frac{\nu_M T}{\alpha}\right)^2 - 1}}. \tag{3.156}$$

Example 3.43 (Channel estimation MMSE for continuous fading with a rectangular Doppler spectrum)

For a rectangular (or uniform within $\pm\nu_{\mathrm{M}}$) Doppler spectrum $S_h(\nu) = \frac{1}{2\,\nu_{\mathrm{M}}}$,

$$S(\nu) = \frac{\alpha}{2\,\nu_{\mathrm{M}}T} \qquad \nu \in \left[\frac{-\nu_{\mathrm{M}}T}{\alpha}, \frac{\nu_{\mathrm{M}}T}{\alpha}\right] \tag{3.157}$$

from which

$$\mathsf{MMSE} = 1 - \frac{\mathsf{SNR}\left(\frac{\alpha}{2\,\nu_{\mathrm{M}}T}\right)^2}{1+\mathsf{SNR}\,\frac{\alpha}{2\,\nu_{\mathrm{M}}T}} \cdot \frac{2\,\nu_{\mathrm{M}}T}{\alpha} \tag{3.158}$$

$$= \frac{1}{1+\frac{\alpha}{2\,\nu_{\mathrm{M}}T}\,\mathsf{SNR}}. \tag{3.159}$$

Notice, by contrasting (3.159) with (3.150), that the channel estimation MMSE under a rectangular Doppler spectrum coincides with its block-fading counterpart if

$$N_{\mathrm{c}} = \frac{1}{2\,\nu_{\mathrm{M}}T}, \tag{3.160}$$

which is the same relationship found in (3.67) by equating the block- and continuous-fading coherence times. This relationship is therefore rather profound as it implies that, despite its apparent coarseness, a properly dimensioned block-fading model can exactly reproduce—in terms of channel estimation MMSE—the behavior of continuous fading with a rectangular Doppler spectrum. This in turn has important ramifications because, as becomes clear in Chapter 4, the corresponding MMSE largely captures the impact of channel estimation on the performance. Certainly, the equivalence between block fading and continuous fading is only exact for a specific Doppler spectrum, but as it turns out the shape of the spectrum is irrelevant for $\mathsf{SNR} \to \infty$ (only ν_{M} matters) and relatively unimportant except possibly for $\mathsf{SNR} \to 0$. This confers a somewhat surprising relevance to the simple block-fading model, greatly facilitating performance analyses that account for channel estimation. In our SU-MIMO coverage, we consider both block fading and continuous fading, and further elaborate on the equivalence between the two, so as to fully rely on block fading in the more intricate analysis of MU-MIMO.

Recall that the foregoing derivations rely on the fading being frequency-flat, meaning $T \gg T_{\mathrm{d}}$ or, at most, $T \approx T_{\mathrm{d}}$; therefore, (3.160) holds at most up to $N_{\mathrm{c}} \approx \frac{1}{2\,\nu_{\mathrm{M}}T_{\mathrm{d}}} \approx B_{\mathrm{c}}T_{\mathrm{c}}$. This is indeed, as found, the maximum number of single-carrier symbols that fit within a coherence time without frequency selectivity; with OFDM, equivalently, $B_{\mathrm{c}}T_{\mathrm{c}}$ is the number of resource elements that fit in a time–frequency coherence tile.

In addition to unifying the block- and continuous-fading models, (3.160) also offers a way to render precise the relationship $T_{\mathrm{c}} \propto \frac{1}{\nu_{\mathrm{M}}}$ deduced in Section 3.4.3: converting the block duration from symbol units onto time units, (3.160) becomes $N_{\mathrm{c}}T = \frac{1}{2\,\nu_{\mathrm{M}}}$ and, since in block fading the coherence time is directly the block duration, we find that

$$T_{\mathrm{c}} = \frac{1}{2\,\nu_{\mathrm{M}}}, \tag{3.161}$$

as anticipated in (3.59).

Before progressing onto small-scale channel estimation for MIMO, a comment on the noncausality of the LMMSE estimator behind (3.153) and the ensuing results is in order: this noncausality is finessed through proper buffering and block processing, which as seen in Chapter 1 are underpinnings of reliable communication and therefore very much required even if we ignored the issue of channel estimation. Two possibilities arise.

- The signals are processed in blocks that exceed the coherence time of the fading, in which case (3.153) is indeed achieved with relatively high accuracy.
- The signal blocklength is shorter than the coherence time of the fading and thus the number of informative channel observations is curtailed. In this case, (3.153) is optimistic, yet the fading is so coherent that its estimation is extremely accurate and not a limiting factor in the first place.

Altogether then, the noncausal MMSE expression in (3.153) is representative when it matters. If, rather than on a buffered-block basis, the channel were to be estimated on a strict causal basis, i.e., based only on pilot observations received prior to the fading value being estimated, then (1.255) would apply and we would have

$$\mathsf{MMSE} = \frac{1}{\mathsf{SNR}} \left[\exp\left(\int_{-1/2}^{1/2} \log_e \left(1 + \mathsf{SNR}\, S(\nu)\right) d\nu \right) - 1 \right]. \tag{3.162}$$

Although less relevant to reliable data detection than the noncausal MMSE, (3.162) has the interest that its limit for $\mathsf{SNR} \to \infty$ can be used to test the regular/nonregular condition of the fading: if it is nonregular, then the Doppler spectrum is bandlimited and $\mathsf{MMSE} \to 0$, consistent with the present value being perfectly predictable from noiseless observations of the entire past; conversely, if it is regular, then the spectrum is not bandlimited and strictly positive, such that [68]

$$\mathsf{MMSE} \to \exp\left(\int_{-\nu_{\mathrm{M}} T/\alpha}^{\nu_{\mathrm{M}} T/\alpha} \log_e S(\nu)\, d\nu \right), \tag{3.163}$$

which remains strictly positive.

MIMO

In the absence of antenna correlations, the derived MMSE expressions extend to MIMO without difficulty given that, as learned in Example 2.31, when the entries of the channel matrix $\boldsymbol{H}$ are independent, there is no loss of optimality in estimating them separately. The only caveat is that the opportunities available for pilot transmissions must be divided among the N_{t} transmit antennas and hence the role played by α is now played by α/N_{t}. Conversely, the number of receive antennas is immaterial because each additional one supplies its own channel observations and these are not at the expense of other receive antennas. Thus, the noncausal estimation of each entry of $\boldsymbol{H}$ can achieve

$$\mathsf{MMSE} = \frac{1}{1 + \frac{\alpha N_{\mathrm{c}}}{N_{\mathrm{t}}} \mathsf{SNR}} \tag{3.164}$$

under block fading while, under continuous fading,

$$\text{MMSE} = 1 - \int_{-N_t\nu_M T/\alpha}^{N_t\nu_M T/\alpha} \frac{\text{SNR}\, S^2(\nu)}{1 + \text{SNR}\, S(\nu)} d\nu, \tag{3.165}$$

with the Doppler spectrum further broadened and scaled into

$$S(\nu) = \frac{\alpha}{N_t T} S_h\left(\frac{\alpha\,\nu}{N_t T}\right) \qquad \nu \in \left[\frac{-N_t\nu_M T}{\alpha}, \frac{N_t\nu_M T}{\alpha}\right]. \tag{3.166}$$

Notice how, in terms of channel estimation MMSE, an increase in the number of transmit antennas is akin to an increase in the symbol period (and thus a signal bandwidth reduction) for the same Doppler, or equivalently to an increase in ν_M for the same signal bandwidth; this merely reflects the lower pilot rate per transmit antenna. The equivalence between block and continuous fading with a rectangular Doppler spectrum is not affected.

In the face of antenna correlations, the above MMSE expressions continue to hold if the entries of $\boldsymbol{H}$ are separately estimated, but a smaller estimation error can be attained through joint estimation. For the Kronecker model in particular, the LMMSE joint estimator derived in Example 2.32 can be applied and the ensuing advantage is the subject of Problem 3.29.

Finally, broadening the view to fading selective in both time and frequency, an interpretation of the derivations in this section based on time–frequency tiles containing $N_c = B_c T_c$ single-carrier symbols or OFDM resource elements leads to

$$\text{MMSE} = \frac{1}{1 + \frac{\alpha B_c T_c}{N_t}\text{SNR}}, \tag{3.167}$$

a handy expression that, while remarkably simple, reflects the dependence on all the key quantities: coherence time, coherence bandwidth, number of transmit antennas, observation SNR, and pilot overhead.

3.8 MIMO channel models in standards

The importance of MIMO is unequivocally demonstrated by its inclusion in wireless standards such as those of 3GPP for cellular networks, IEEE 802.16 for wireless-metropolitan-area and fixed-wireless access, and IEEE 802.11 for WLANs. Channel models have been developed to suit the requirements of these standards [269], and this section briefly reviews some of these models.

Besides the standard-specific models introduced herein, and for the sake of completeness, it is worth referring to nonstandard-specific models that were also conceived for the purpose of system-level performance evaluation. Chief among these stand the COST-259 [325, 326], the COST-273 [327], and the WINNER I and WINNER II models [328, 329]. The COST-259 model, in particular, broadly includes macro-, micro-, and picocell environments and was the first model to jointly incorporate delay and angular properties. For its part, the WINNER models were developed as part of the IST-WINNER project, supported by measurement campaigns in Europe.

3.8.1 3GPP spatial channel model

The 3GPP spatial channel model (SCM) was developed jointly by 3GPP and by its North-American former counterpart, 3GPP2, for MIMO performance evaluation in outdoor deployments at frequencies of around 2 GHz. The SCM features both a calibration model and a system-level simulation model. The calibration model, highly simplified, is devised for quick testing of simulation implementations; it consists of several independently fading rays, each with a mean angle and a PAS chosen to be either Clarke–Jakes or a truncated Laplacian. The simulation model, much more detailed, is the one intended for actual performance evaluation and thus it is the one we dwell on in the remainder of this section. It distinguishes between three types of environments: urban macrocell, suburban macrocell, and urban microcell. The same parametric approach is taken irrespective of the type of environment, only with different values for the parameters.

The SCM simulation model is intended for simulations involving multiple cells and users. More precisely, the procedure that the model is designed for entails a succession of snapshots or *drops*, where a drop is characterized by a set of randomly selected user positions, orientations, and velocities. From those positions and orientations, the mean angles and the distances of each user relative to its serving base station are readily obtained, and from the distances the pathloss values derive. The rest of the relevant large-scale quantities are also set at each drop, specifically shadow fading, angle spread, and delay spread. All of these large-scale quantities are held fixed over the duration of the drop, which recreates a short displacement—within the local scattering neighborhood—over which the small-scale fading is allowed to manifest itself.

The pathloss is computed by means of the COST-231 Hata model (see Section 3.3.3) in the case of macrocells, both urban and suburban, and the COST-231 Walfisch–Ikegami model (see Section 3.3.4) in the case of microcells.

The channel response between each user and its serving base station comprises six rays subject to Rayleigh or Rice fading. The ray delays and their power shares are chosen, at each drop, from a given distribution. The angular modeling, in turn, is fairly sophisticated.

- The mean angles, from the viewpoints of the user and the base station, are determined directly by the position and orientation of the user at the given drop.
- Each ray is then considered to emanate from 20 scattering clusters, whose delays are identical and whose angles are drawn from a Gaussian distribution centered on the mean angle and with a variance that is a parameter of the model.
- Each scattering cluster within each ray gives rise to 20 propagation paths with slightly different angular offsets; these offsets are fixed and tabulated in the standard.

As the users move about during each drop at their assigned velocities, the phases shift over the propagation paths contributed by all the clusters at each ray, and the superposition of these paths yields the desired ray fading.

In addition to all the above characteristics, the SCM has a number of additional optional features related to polarization, to the existence of distant scattering clusters, or to the presence of an LOS component for the microcellular environment (which goes hand in hand with the Rice distribution for the ray fading).

Table 3.4 Doppler and delay spread for SUI channels

Doppler	Low delay spread	Moderate delay spread	High delay spread
Low	SUI-3 (low K-factor) SUI-1,2 (high K-factor)		SUI-5 (low K-factor)
High		SUI-4 (low K-factor)	SUI-6 (low K-factor)

For further details, the reader interested in implementing the SCM is referred to [330].

3.8.2 SUI models for IEEE 802.16

The models proposed by the IEEE 802.16 working group for fixed wireless access represent the outcome of the group's efforts to define models within 2–11 GHz [219]. Six SUI models were defined for the three terrain types introduced in Section 3.3.3, namely SUI-1 and SUI-2 for type A (hilly), SUI-3 and SUI-4 for type B (intermediate), and SUI-5 and SUI-6 for type C (flat). A breakdown of the Doppler and delay spread features of the six models is given in Table 3.4.

Every SUI model features three rays, the first of which is Rice-faded for models SUI-1 through SUI-4, while the rest are Rayleigh-faded. IEEE 802.16 also incorporates the angular aspects necessary for MIMO deployment [331]. Since scatterers generating propagation paths with identical delays must necessarily lie on an ellipse with the transmitter and receiver as foci [332, 333], each of the three rays can be mapped to a distinct ellipse—with the ellipse corresponding to the first ray essentially degenerating to an axis in the strong Rice cases. Scatterers can then be deployed over the ellipses in different ways.

Altogether, the SUI models are neither particularly intuitive nor easy to implement, but they are the approach of choice for IEEE 802.16.

3.8.3 IEEE 802.11 channel model

The IEEE 802.11n channel model, defined for 2-GHz and 5-GHz frequency bands, relies on the clustered approach described in Section 3.6.2. Six propagation environments (A through F) are defined in representation of small and large office spaces, residential homes, and open spaces, both with and without LOS components [334].

The pathloss, the only item in the model that depends on the frequency band, espouses the free-space model with 3-dB log-normal shadowing up to D_{ref}; thereafter, the model defaults to (3.8) with exponent $\eta = 3.5$ and a shadowing standard deviation that depends on the environment (see Table 3.5). The distance D_{ref} also varies with the environment.

The number of scattering clusters ranges between two and six, each such cluster contributing up to 18 Rayleigh-faded rays separated by at least 10 ns. From the perspectives of both transmitter and receiver, each ray is assigned a mean azimuth angle and a truncated Laplacian PAS with an angle spread ranging between 20° and 40°. Altogether, the delay spread can range between $T_{\text{d}} = 0$ (flat fading) and $T_{\text{d}} = 150$ ns.

The array topologies can be freely specified and, in conjunction with the aforesketched

Table 3.5 Large-scale parameters for the IEEE 802.11n channel model

Environment	D_{ref}	$\sigma^2_{\text{dB}}(D \geq D_{\text{ref}})$
A	5 m	4 dB
B	5 m	4 dB
C	5 m	5 dB
D	10 m	5 dB
E	20 m	6 dB
F	30 m	6 dB

cluster structure, they determine the antenna correlations. As far as the Doppler spectrum is concerned, and under the presumption that users are static and time-variability arises only because of environmental motion, the bell-shape spectrum in Example 3.15 is adopted, with an optional additional peak at a higher Doppler shift to represent passing vehicles. In addition, by means of a magnitude modulation of the rays, the model incorporates further time variations caused by fluorescent lights.

3.9 Summary and outlook

Distilling what is covered in the chapter, a number of points are enumerated in the companion summary box. In terms of the research outlook on MIMO channel modeling, we highlight some avenues of interest, beginning in the frequency axis and then proceeding onto the spatial domain.

Although the chapter deals exclusively with microwave frequencies ranging between several hundred megahertz and a few gigahertz, there is intense activity looking into wireless communication at millimeter-wave frequencies, where abundant idle spectrum awaits. (Strictly speaking, millimeter frequencies begin at 30 GHz, but the term is informally applied from 6 GHz onwards.) Historically, these high frequencies had been largely avoided because of unfavorable channel characteristics, including overly directional propagation and excessive atmosphere and rain attenuation, but these obstacles can be overcome for short-range transmissions with the aid of the dense antenna arrays that the short wavelength makes possible. NR and the IEEE 802.11ad and 802.11ay WLAN standards operate in this realm already. Thus, channel modeling at millimeter-wave frequencies is likely to be a favorite topic going forward [335].

Another line of work that has momentum is that of exploiting the elevation dimension, something that might become increasingly important as networks densify and the shrinking horizontal dimension of the links becomes more comparable to the vertical dimension [202, 336]. This requires proper three-dimensional propagation models that are no longer restricted to a plane, and the development and refinement of such models is another activity seeing increased action. Surveys of existing results can be found in [337–339], and the three-dimensional extension of the 3GPP SCM is discussed in [340]. For their part, the

Take-away points

1. The large-scale phenomena determine the distribution over local neighborhoods (tens to hundreds of wavelengths in size) of the small-scale fading, which can be regarded as stationary therewithin.
2. The large-scale phenomena are pathloss and shadow fading, respectively the expected attenuation at a given distance and the deviation with respect to it. The combination of both determines the large-scale channel gain, G.
3. The shadow fading is usually well modeled by a log-normal distribution with a standard deviation on the order of 8–12 dB outdoors and 3–6 dB indoors.
4. The most recurrent representation for the pathloss at a distance D is $L_{\mathrm{p}} = K_{\mathrm{ref}}(D/D_{\mathrm{ref}})^{\eta}$ where η is the exponent, D_{ref} is a reference distance and K_{ref} is the pathloss therein. The exponent is $\eta = 2$ in free space while typically between 3.5 and 4 if transmitter or receiver are immersed in scattering clutter.
5. The small-scale fading is caused by multipath propagation. Signal replicas traveling through different paths are subject to independent phase shifts and their superposition can be modeled as a complex Gaussian gain; its magnitude is then Rayleigh or, if one of the multipath terms dominates, Rice.
6. The small-scale fading is highly selective in space: a small displacement (order of a wavelength) drastically alters the phase shifts over the paths causing major variations in the composite gain. The PAS expresses the transmit/received signal power as a function of angle. Continuous or discrete, the PAS is normalized so it can be interpreted as a PDF and its standard deviation is the angle spread. Given a PAS $\mathcal{P}(\theta)$, the correlation between the fading at two locations spaced by Δ_D is
$$R_h(\Delta_D) = \int_{-\pi}^{\pi} \mathcal{P}(\theta)\, e^{-\mathrm{j}2\pi\Delta_D \cos(\theta)/\lambda_{\mathrm{c}}}\, \mathrm{d}\theta, \tag{3.168}$$
which is said to have decayed when Δ_D equals the coherence distance, D_{c}.
7. Given a velocity v, the space correlation maps directly onto time correlation and D_{c} maps to a coherence time $T_{\mathrm{c}} = D_{\mathrm{c}}/\mathrm{v}$. The Fourier transform of the time correlation gives the Doppler spectrum, $S_h(\nu)$, which characterizes the spectral smearing suffered by a sinusoid because of motion.
8. If the Doppler spectrum is bandlimited, typically to $\nu_{\mathrm{M}} = f_{\mathrm{c}}\mathrm{v}/\mathrm{c}$, the fading is said to be nonregular: its present value is perfectly predictable from noiseless observations of its entire past. Conversely, if the spectrum is not bandlimited, the fading is said to be regular: its present value cannot be perfectly predicted from noiseless observations of its entire past.
9. If the signal bandwidth fits within a swath of frequencies that experience (roughly) the same fading, then such fading is frequency-flat. Otherwise, it is frequency-selective. The selectivity across a frequency range Δ_f is quantified by the frequency correlation $R_{\mathrm{h}}(\Delta_f)$ whose chief measure is the coherence bandwidth, B_{c}. The fading is frequency-flat if $B \ll B_{\mathrm{c}}$. Frequency selectivity implies that the fading is not multiplicative, but rather dispersive; in fact, the PDP function expressing the received power as a function of delay is directly the Fourier transform of $R_{\mathrm{h}}(\Delta_f)$. The standard deviation of the PDP is the delay spread $T_{\mathrm{d}} \approx 1/B_{\mathrm{c}}$.

10. Even if generally selective in both time and frequency, wireless channels are highly underspread, meaning that the fading remains approximately constant over many ($B_c T_c \gg 1$) symbol periods. As a result, fixed-shape OFDM resource elements can remain approximately orthogonal in both time and frequency. It follows that, with a correct OFDM design, the fading is only multiplicative and thus one complex coefficient per resource element suffices for a discrete time–frequency representation; the pulse shaping does not affect the discretization.
11. Simplified models such as block fading can capture the essence of fading dynamics in time and/or frequency remarkably well.
12. The correlation between antennas within an array can be influenced through spacing, antenna patterns, and polarization. These correspond, respectively, to the notions of spatial, pattern, and polarization diversity.
13. In MIMO, the focus is on how to model the joint distribution of the entries of the channel matrix $\boldsymbol{H}$. In the most basic analytical models, the entries of $\boldsymbol{H}$ are IID, either Rayleigh- or Rice-distributed. Then,
$$\boldsymbol{H} = \sqrt{\frac{\mathsf{K}}{1+\mathsf{K}}}\boldsymbol{H}_{\mathsf{LOS}} + \sqrt{\frac{1}{1+\mathsf{K}}}\boldsymbol{H}_{\mathrm{w}}, \tag{3.169}$$
where $\boldsymbol{H}_{\mathsf{LOS}}$ is deterministic, $\boldsymbol{K}$ is the Rice factor, and $[\boldsymbol{H}_{\mathrm{w}}]_{i,j} \sim \mathcal{N}_{\mathbb{C}}(0,1)$.
14. Antenna correlations are incorporated, in a restricted form, by the Kronecker model
$$\boldsymbol{H} = \boldsymbol{\mu}_{\boldsymbol{H}} + \boldsymbol{R}_{\mathrm{r}}^{1/2}\boldsymbol{H}_{\mathrm{w}}\boldsymbol{R}_{\mathrm{t}}^{1/2}, \tag{3.170}$$
where $\boldsymbol{R}_{\mathrm{r}}$ and $\boldsymbol{R}_{\mathrm{t}}$ are receive and transmit correlation matrices. The restriction in this model is that the correlation between the signal fading at any two receive antennas is the same irrespective of the transmit antenna from which the signal emanates, and vice versa; this entails decoupled transmit and receive PAS. The entries of the correlation matrices can be computed by applying (3.168).
15. The restriction of decoupled transmit and receive PAS is overcome by the UIU model, whereby
$$\boldsymbol{H} = \boldsymbol{\mu}_{\boldsymbol{H}} + \boldsymbol{U}_{\mathrm{r}}\boldsymbol{H}_{\mathsf{ind}}\boldsymbol{U}_{\mathrm{t}}^{*}, \tag{3.171}$$
with $\boldsymbol{U}_{\mathrm{r}}$ and $\boldsymbol{U}_{\mathrm{t}}$ fixed unitary matrices while $\boldsymbol{H}_{\mathsf{ind}}$ has zero-mean IND entries. The Kronecker model is a special case of the UIU model.
16. If $\boldsymbol{\mu}_{\boldsymbol{H}} = \mathbf{0}$ while $\boldsymbol{U}_{\mathrm{t}}$ and $\boldsymbol{U}_{\mathrm{r}}$ are Fourier matrices, the UIU model leads to the virtual channel model $\boldsymbol{H} = \boldsymbol{U}_{\mathrm{r}}\boldsymbol{H}_{\mathsf{vir}}\boldsymbol{U}_{\mathrm{t}}^{*}$ where $\boldsymbol{H}_{\mathsf{vir}}$ is the two-dimensional discrete Fourier transform of $\boldsymbol{H}$. Here, with ULAs having antenna spacings d_{t} and d_{r}, the columns of $\boldsymbol{U}_{\mathrm{t}}$ and $\boldsymbol{U}_{\mathrm{r}}$ synthesize fixed beams pointing to angles $\theta_{\mathrm{t},j} = \arccos(\frac{\lambda_c}{d_t}\frac{j}{N_t})$ and $\theta_{\mathrm{r},i} = \arccos(\frac{\lambda_c}{d_r}\frac{i}{N_r})$ for $j = 0, \ldots, N_{\mathrm{t}} - 1$ and $i = 0, \ldots, N_{\mathrm{r}} - 1$, respectively. In turn, the variance of $[\boldsymbol{H}_{\mathsf{vir}}]_{i,j}$ indicates how intensely the jth transmit beam couples into the ith receive beam.

WINNER+ extension of the WINNER models also incorporates elevation [341]. Then, the vertical tilting of the antennas themselves becomes a relevant aspect [342, 343].

Also, and in light of the growing volume of wireless traffic that is proximal in nature, direct device-to-device transmission that bypasses base stations and the network infrastructure may become an important mode of communication [344, 345]. The elaboration of channel MIMO models for this mode is yet another research direction of interest.

Finally, site-specific reconstruction techniques that transcend ray tracing and can be applied outdoors are becoming possible thanks to the data deluge provided by the users. Based on their continual received-power reports, measurement-driven machine learning algorithms can be applied to construct maps of the large-scale channel gains across the network [346–353].

Problems

3.1 Prove that the PDF of log-normal shadow fading equals (3.2) when χ is in linear scale.

3.2 Consider a 100-m link, omnidirectional antennas, and a pathloss exponent $\eta = 3.5$. Further, let the pathloss at $D_{\text{ref}} = 1$ m be as it would be in free space, and let the shadow fading be log-normal with $\sigma_{\text{dB}} = 12$ dB.

(a) What is the probability that, because of shadow fading, the large-scale channel gain over this relatively short link exceeds its value at D_{ref}?

(b) What is the probability that the large-scale channel gain exceeds 0 dB, such that the average received power exceeds the transmit power?

(c) What is the probability that the large-scale channel gain is better than if the propagation were entirely in free space?

Note: The modeling of shadow fading by a log-normal distribution, whose upper tail extends endlessly, could raise the concern that unrealistically optimistic channels are produced by the model and, in extreme cases, that conservation of energy is violated. In this problem, we examine this issue for the situation where the effects of shadowing are more pronounced: a short link, a small pathloss exponent, and strong shadow fading.

3.3 Compute the pathloss predicted by the COST-231 Hata model at $f_c = 2$ GHz given $h_b = 30$ m and $h_m = 1.5$ m, at distances $D = 1$ km and $D = 2$ km. What is the pathloss exponent?

3.4 According to the COST-231 Hata model at $f_c = 2$ GHz, what is the percentage of increase in the power received from a user at $D = 1$ km when the base station height goes from $h_b = 30$ m to $h_b = 50$ m?

3.5 Consider a scalar Rayleigh-faded channel.

(a) What is the probability that the power received over this channel exceeds its local-average by at least 10 dB, 20 dB, or 30 dB?

(b) How about if the channel is Rice-faded with $\mathsf{K} = 5$ dB?

Note: As with shadow fading, also with multipath fading there is a concern that unrealistic channels could be produced because of the unbounded upper tail of most fading distribution models, and this has relevant implications for some multiuser techniques discussed in the book. In this problem we examine this issue and derive guidelines on how the Rayleigh and Rice distributions are disturbed if truncated at various points.

3.6 It was quantified in Example 3.11 that, for a Clarke–Jakes PAS at $f_c = 2$ GHz, the fading decorrelates completely over $D_c = 5.7$ cm. Compute the degree of decorrelation over the same distance and at the same frequency for the following cases:

(a) A power cosine PAS with $q = 1$.

(b) A truncated Gaussian PAS with $\sigma_\theta = 10°$.

(c) A truncated Laplacian PAS with $\sigma_\theta = 10°$.

In all cases, let $\mu_\theta = 60°$.

3.7 Verify (3.54).

3.8 Obtain, either exactly or approximately, the Doppler spectra for the following:

(a) A power cosine PAS with $q = 1$.

(b) A truncated Gaussian PAS.

(c) A truncated Laplacian PAS.

3.9 Consider a wireless system whose signal bandwidth is 500 times smaller than the carrier frequency. Given a user velocity of 50 km/h, what value should ε take for a Gauss–Markov frequency-flat fading model to achieve 50% decorrelation over the same number of symbols required for a Clarke–Jakes model to decorrelate by 50%?

3.10 Express in closed form the Doppler spectrum of a Gauss–Markov frequency-flat fading channel.

3.11 When servicing a high-speed train from an elevated base station, the velocity may be as large as v = 350 km/h. Argue why, in terms of fading coherence, this is more benign than what the maximum Doppler shift would indicate.

Hint: Any constant Doppler shift is automatically corrected, at the receiver, by the automatic frequency control of the local oscillator.

3.12 Compute the RMS delay spread for the PDP

$$S_h(\tau) = \begin{cases} K_0 & \tau \in [0, 2]\ \mu\text{s} \\ K_1 & \tau \in [2, 5]\ \mu\text{s} \\ 0 & \text{elsewhere.} \end{cases} \tag{3.172}$$

3.13 Consider a time–frequency doubly selective fading channel. Argue that, if we model it as continuously faded in time and block faded in frequency, the coherence in number of symbols satisfies

$$N_c \approx \frac{1}{2\,\nu_M T_d}. \tag{3.173}$$

3.14 Redo Example 3.32 and Fig. 3.20 for an array where two of the antennas are uniform in θ while the other two have a normalized pattern whose magnitude is

$$\frac{\cos\left(\frac{\pi}{2}\cos\theta\right)}{\sin\theta}. \tag{3.174}$$

3.15 Compute $\boldsymbol{H}_{\text{LOS}}$ for a Rice MIMO channel where two equal-power LOS components are present, departing from the transmitter at angles $\theta_{\text{t},0} = 30°$ and $\theta_{\text{t},1} = 45°$, respectively, and impinging on the receiver at angles $\theta_{\text{r},0} = 60°$ and $\theta_{\text{r},1} = 0°$, respectively. The transmitter features two antennas spaced by $d_{\text{t}} = \lambda_{\text{c}}/2$ while the receiver is a UCA with $r_{\text{UCA}} = \lambda_{\text{c}}$ and $N_{\text{r}} = 4$.

3.16 Prove that, if $\boldsymbol{H}$ is defined as in (3.114), then $\boldsymbol{R}_{\text{t}}$ and $\boldsymbol{R}_{\text{r}}$ satisfy (3.115) and (3.116), respectively.

3.17 Express the magnitude of the correlation between two arbitrary receive antennas, $|[\boldsymbol{R}_{\text{r}}]_{i,i'}|$, given a PAS conforming to the power cosine function in Example 3.5 with q large enough for (3.126) to apply.

3.18 Express the magnitude of the correlation between two arbitrary transmit antennas, $|[\boldsymbol{R}_{\text{t}}]_{j,j'}|$, given a PAS uniform within $[\mu_\theta - \Delta, \mu_\theta + \Delta]$ and zero elsewhere, with Δ small.

3.19 Consider the exponential antenna correlation model in Example 3.38 applied to a transceiver equipped with two antennas spaced by four wavelengths. What value should ρ take for the correlation between the antennas to coincide with the one caused by a truncated Laplacian PAS with $\sigma_\theta = 2°$?

3.20 Consider a transmitter with two antennas, 40% correlated, communicating with a receiver equipped with two antennas, 60% correlated. In a UIU channel representation, what variances should the entries of $\boldsymbol{H}_{\text{ind}}$ have?

3.21 Consider the setup of Problem 3.20. In a virtual channel representation, what correlations would remain among the entries of $\boldsymbol{H}_{\text{vir}}$?

3.22 Consider a block-fading channel with N_{c} symbols per block, of which a portion α are pilots. Treating the pilots within a block as a vector observation, prove that an LMMSE estimator attains (3.150).

3.23 Derive the counterpart to (3.150) for a unit-variance scalar channel exhibiting a Rice factor K.

Hint: The deterministic component of the channel need not be estimated.

Note: A more formal derivation of the LMMSE estimator for nonzero-mean quantities is the subject of Problem 1.35

3.24 Express the channel estimation MMSE for a continuous fading channel with a band-limited Doppler spectrum that takes a uniform value in $\left[-\frac{\nu_{\text{M}}T}{2\alpha}, \frac{\nu_{\text{M}}T}{2\alpha}\right]$ and half that value elsewhere within $\left[-\frac{\nu_{\text{M}}T}{\alpha}, \frac{\nu_{\text{M}}T}{\alpha}\right]$.

3.25 Consider a continuous fading channel with a bell-shape Doppler spectrum truncated to $\pm\nu_{\text{M}}$ where $\nu_{\text{M}} = 10$ Hz. Given a bandwidth $B = 10$ MHz and a pilot overhead $\alpha = 5\%$, compute the channel estimation MMSE.

3.26 For continuous fading with $\alpha\,\mathsf{SNR} = 1$ and $\nu_{\mathrm{M}}T = 10^{-3}$, compute the channel estimation MMSE under both rectangular and Clarke–Jakes Doppler spectra. Then, calculate the respective number of symbols N_{c} in a block-faded model that would yield the same MMSE.

Note: The difference between both values of N_{c} is the error that would be incurred if the relationship $N_{\mathrm{c}} = \frac{1}{2\,\nu_{\mathrm{M}}T}$ were applied to a Clarke–Jakes spectrum.

3.27 Expand, to first and second order in the SNR, the MMSE expression in (3.42) for high-SNR conditions. Plot the expansions and the exact MMSE expression as function of SNR for $\frac{\nu_{\mathrm{M}}T}{\alpha} = 10^{-4}$.

3.28 For a continuous fading channel with a rectangular Doppler spectrum, express the ratio between the SISO and MIMO (without antenna correlations) channel estimation MMSEs. Then, expand this ratio in the SNR for high-SNR conditions and contrast the first term in this expansion against the exact MMSE at $\mathsf{SNR} = 10$ dB, given $\frac{\nu_{\mathrm{M}}T}{\alpha} = 10^{-4}$ and $N_{\mathrm{t}} = 4$. Repeat the exercise for a Clarke–Jakes Doppler spectrum, taking advantage of the derivations in Problem 3.27.

3.29 Let the frequency-flat block-faded channel $\boldsymbol{H}$ conform to the Kronecker correlation model with

$$\boldsymbol{R}_{\mathrm{t}} = \begin{bmatrix} 1 & 0.6 \\ 0.6 & 1 \end{bmatrix} \qquad \boldsymbol{R}_{\mathrm{r}} = \begin{bmatrix} 1 & 0.4 \\ 0.4 & 1 \end{bmatrix}. \tag{3.175}$$

(a) For $\alpha N_{\mathrm{c}} = 2$ and $\mathsf{SNR} = 3$ dB, what is the dB-reduction in MMSE that a joint LMMSE estimator attains relative to an LMMSE estimator operating independently on each channel entry?

(b) Repeat part (a) for $\mathsf{SNR} = 10$ dB.

3.30 Consider a channel that is modeled with a pathloss exponent of $\eta = 3.7$ when the exponent that would have correctly matched its behavior is $\eta = 4$.

(a) Let $\sigma_{\mathrm{dB}} = 9$ dB. What is the probability that the local-average signal power predicted with the mismatched exponent deviates from the true power by more than the value of the shadow fading?

(b) What is the probability that the local-average signal power predicted with the mismatched exponent deviates from the true power by more than the small-scale variation caused by Rayleigh fading?

(c) Suppose that we want to predict the capacity of a MIMO link over this channel, with $N_{\mathrm{t}} = N_{\mathrm{r}} = 2$. Under the mismatched model, we obtain $\mathsf{SNR} = 10$ dB and compute the ensuing capacity assuming no antenna correlations. What antenna correlation would have to be introduced at the receiver to render the capacity prediction correct?

Hint: The applicable capacity expression for this problem is given in (5.96).

4 Single-user SISO

To see things in the seed, that is genius.

Lao-Tzu

4.1 Introduction

An information-theoretic approach to the study of MIMO requires a solid foundation in the application of information theory to wireless communications. Without MIMO, whose coverage starts in the next chapter, and without multiple users, which make their appearance later on, in this chapter we entertain a number of important notions involved in the communication between a single-antenna transmitter and a single-antenna receiver, the so-called single-user SISO (SU-SISO) context. Many of these notions are subsequently generalized to MIMO and to multiuser setups, but here in a SU-SISO context we can see them in their most basic and cleanest forms.

The chapter begins with a section that formulates the fundamental tradeoffs between power, bandwidth, and bit rate, and defines several quantities that play a central role in this interplay. Sections 4.3–4.9 then dwell extensively on the problem of communicating reliably over SU-SISO channels, beginning with the simplest possible setting and progressively assessing the impact of frequency selectivity, fading, CSI, and interference. In the process, distinct information-theoretic idealizations arise and are established for subsequent applicability in the study of MIMO. Finally, Section 4.10 concludes the chapter.

4.2 Interplay of bit rate, power, and bandwidth

The design of a communication system involves tradeoffs between an objective quantity, namely the bit rate R (in bits/s), and two resources, namely the transmit power P_t (in watts) and the bandwidth B (in hertz, and conveniently parceled into single-carrier symbols or OFDM resource elements). These tradeoffs are modulated by the computational complexity that can be afforded, proxies for which include the codeword length and the numbers of antennas. With MIMO put aside for now, let us focus on the interplay between R, P_t, and B for SISO and without any complexity constraints, taking into account that:

- To interpret this interplay at a fundamental level, we want to reference it to the channel capacity.

- Even though the power is constrained at the transmitter, where it is consumed, the performance depends on its value at the receiver—relative to the noise—and what bridges both ends is the channel gain.

It follows that the channel capacity and the channel gain play a central role in the formulation of the tradeoffs between R, P_t, and B. The *total channel gain* is, by definition, the ratio between the received and the transmit signal powers, i.e.,

$$\frac{\mathbb{E}\big[|\sqrt{G}hx|^2\big]}{\mathbb{E}\big[|x|^2\big]} = G\,\frac{\mathbb{E}\big[|hx|^2\big]}{E_s} \tag{4.1}$$

which, for fading-independent transmissions (x independent of the unit-power small-scale channel coefficient h) directly equals $\frac{G}{E_s}\mathbb{E}\big[|h|^2\big]\mathbb{E}\big[|x|^2\big] = G$, i.e., the total channel gain equals the large-scale channel gain introduced in Chapter 3, subsuming pathloss and shadowing. With fading-dependent power control, alternatively, the total channel gain is the complete expression on the right-hand side of (4.1), which depends on the distribution of h and on how the transmit power is controlled on the basis of h. Deferring to later in this chapter the analysis of channel-dependent power control, we take the total channel gain to equal the large-scale channel gain G unless otherwise indicated.

To illustrate the various definitions and relationships put forth in this section, we have the running example of the channel $y = \sqrt{G}hx + v$ with $h = 1$ and $v \sim \mathcal{N}_{\mathbb{C}}(0, N_0)$. This basic channel, which permeates many of the derivations in Chapter 1, is termed the *AWGN channel* because it is only impaired by AWGN; its capacity function, borrowed here and then formally derived in Section 4.3, is $C(\mathsf{SNR}) = \log_2(1 + \mathsf{SNR})$.

Delving into the tradeoffs between R, P_t, and B, we begin by observing that, rather than relate these quantities directly, it is preferable to relate the ratios R/B, P_t/B, and P_t/R, which happen to have an operational importance of their own.

- The ratio between R and B is the spectral efficiency R/B (in b/s/Hz), a quantity introduced already in Chapter 1 that quantifies how well the available bandwidth is utilized.
- The ratio between P_t and B gives the transmit energy (in joules) per single-carrier symbol or per OFDM resource element, $P_t/B = P_t T = E_s = \mathbb{E}\big[|x|^2\big]$, which, normalized by N_0, would yield a transmit-referenced SNR. Since it is more fitting to define the SNR at the receiver, applying G we can translate the transmit-referenced SNR to the receiver and write

$$\mathsf{SNR} = \frac{G P_t/B}{N_0} \tag{4.2}$$

$$= \frac{G E_s}{N_0}, \tag{4.3}$$

which is the definition given in Chapter 2 for the local-average SNR. (Alternative SNR definitions are possible and, as long as they are applied consistently, all of them can be valid.)

- The ratio between R and P_t defines the *power efficiency* (in bits/joule) that quantifies how well the available power is utilized, i.e., how many message bits are transmitted per unit energy.[1] Equivalently, the reciprocal ratio, P_t/R, measures how much transmit energy is spent per bit. Applying G, we can also translate this transmit-referenced energy per bit to the receiver to obtain

$$E_b = \frac{GP_t}{R} \tag{4.4}$$

$$= \frac{GE_s}{R/B}. \tag{4.5}$$

This quantity, referenced to N_0, yields

$$\frac{E_b}{N_0} = \frac{1}{N_0}\frac{GE_s}{R/B} \tag{4.6}$$

$$= \frac{\mathsf{SNR}}{R/B}. \tag{4.7}$$

The interplay of the ratios R/B, SNR (as proxy for P_t/B) and $\frac{E_b}{N_0}$ (as proxy for P_t/R) captures the interplay of R, P_t, and B, with the added benefit that the ratios embody very meaningful figures of merit, respectively the spectral efficiency, the local-average SNR, and the power efficiency. Since, as indicated by (1.105), R/B is bounded by the capacity, in terms of fundamental performance limits we can set $R/B = C$. With that, the quantities to interrelate become C, SNR, and $\frac{E_b}{N_0}$. As illustrated in Fig. 4.1, we can proceed to define $C(\mathsf{SNR})$ as the function that leads from SNR to the capacity while its inverse $C^{-1}(\cdot)$ returns back the SNR. Likewise, $\mathsf{C}(\frac{E_b}{N_0})$ and its inverse $\mathsf{C}^{-1}(\cdot)$ convert $\frac{E_b}{N_0}$ to capacity and vice versa. The functions $C(\cdot)$ and $\mathsf{C}(\cdot)$ map different arguments to the same value. In turn, from (4.7) and $R/B = C = \mathsf{C}$, we can connect SNR and $\frac{E_b}{N_0}$ through the direct and inverse relationships

$$\frac{E_b}{N_0} = \frac{\mathsf{SNR}}{C(\mathsf{SNR})} \tag{4.8}$$

and

$$\mathsf{SNR} = \frac{E_b}{N_0}\,\mathsf{C}\left(\frac{E_b}{N_0}\right). \tag{4.9}$$

The functions $C(\cdot)$ and $\mathsf{C}(\cdot)$, and their inverses, are not always available in explicit closed forms, but, as long as they can be computed and tabulated, the tradeoffs can be assessed.

Example 4.1 (Relationships among C, SNR, and $\frac{E_b}{N_0}$ in an AWGN channel)

In an AWGN channel, as submitted, $C(\mathsf{SNR}) = \log_2(1+\mathsf{SNR})$ with inverse $\mathsf{SNR} = 2^C - 1$, which constitute a first edge on the right-hand side triangle of Fig. 4.1. Since $C = \mathsf{C}$, we then obtain from (4.8) the inverse relationship

$$\frac{E_b}{N_0} = \frac{2^{\mathsf{C}} - 1}{\mathsf{C}}, \tag{4.10}$$

[1] The power efficiency can be generalized into the notion of *capacity per unit cost*, where an arbitrary cost function (of which the power is a special case) is defined over the signal [354]. This notion may be invoked to characterize fundamental performance limits in the face of constraints other than on the power, e.g., with constraints on the signal peakedness.

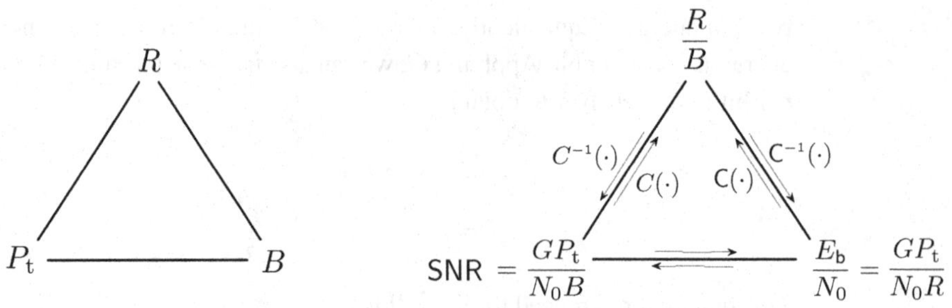

Fig. 4.1 Left, interplay between R, P_t and B. Right, interplay between C, SNR, and $\frac{E_b}{N_0}$ from which the functions $C(\mathsf{SNR})$ and $\mathsf{C}(\frac{E_b}{N_0})$, and their inverses, arise.

which, along with $\mathsf{C}(\frac{E_b}{N_0})$, is another edge in the triangle; unfortunately, $\mathsf{C}(\frac{E_b}{N_0})$ cannot be untangled into an explicit form. The final edge in the triangle is

$$\frac{E_b}{N_0} = \frac{\mathsf{SNR}}{\log_2(1+\mathsf{SNR})} \tag{4.11}$$

and its inverse, which also cannot be cleared explicitly.

Let us next see how the relationships on the right-hand-side triangle of Fig. 4.1 can be applied to the original objective of establishing the fundamental tradeoffs between bit rate, power, and bandwidth on the left-hand-side triangle. Recall that the communication is SISO with some given values for G and N_0.

Tradeoff between R and P_t for fixed B

For fixed B, the tradeoff between R and P_t is mirrored in the tradeoff between R/B and $\mathsf{SNR} = \frac{GP_t}{N_0 B}$ and thus it is captured by the function $C(\mathsf{SNR})$ via

$$R = B \cdot C\left(\frac{GP_t}{N_0 B}\right) \tag{4.12}$$

$$P_t = \frac{N_0 B}{G} C^{-1}\left(\frac{R}{B}\right). \tag{4.13}$$

By moving along the corresponding curves, we could establish how R grows with P_t and vice versa, for some given B, N_0, and G. In particular, the ratio between the bit rates achievable with powers $P_t^{(a)}$ and $P_t^{(b)}$ is given by

$$\frac{R^{(a)}}{R^{(b)}} = \frac{C\left(\frac{GP_t^{(a)}}{N_0 B}\right)}{C\left(\frac{GP_t^{(b)}}{N_0 B}\right)}, \tag{4.14}$$

while the ratio between the powers required for bit rates $R^{(a)}$ and $R^{(b)}$ is

$$\frac{P_t^{(a)}}{P_t^{(b)}} = \frac{C^{-1}\left(\frac{R^{(a)}}{B}\right)}{C^{-1}\left(\frac{R^{(b)}}{B}\right)}, \tag{4.15}$$

meaning that, to bring the bit rate from $R^{(b)}$ to $R^{(a)}$, the transmit power must change by $C^{-1}(R^{(a)}/B)|_{dB} - C^{-1}(R^{(b)}/B)|_{dB}$.

Fixing P_t and releasing N_0 or G in (4.14), we could also determine how R changes with the strength of the noise or with the large-scale channel gain.

Example 4.2 (Applicability of $C(\mathsf{SNR})$ in an AWGN channel)

Shown on the left-hand side of Fig. 4.2 is $C(\mathsf{SNR}) = \log_2(1+\mathsf{SNR})$ as corresponds to an AWGN channel. The ratio between the bit rates achievable with powers $P_t^{(a)}$ and $P_t^{(b)}$ is

$$\frac{R^{(a)}}{R^{(b)}} = \frac{\log_2\left(1+\frac{GP_t^{(a)}}{N_0 B}\right)}{\log_2\left(1+\frac{GP_t^{(b)}}{N_0 B}\right)} \tag{4.16}$$

while the ratio between the powers required for bit rates $R^{(a)}$ and $R^{(b)}$ is

$$\frac{P_t^{(a)}}{P_t^{(b)}} = \frac{2^{R^{(a)}/B}-1}{2^{R^{(b)}/B}-1}. \tag{4.17}$$

Example 4.3

On an AWGN channel with bandwidth $B = 5$ MHz, how much transmit power must be added to double the bit rate from $R = 6$ Mb/s to $R = 12$ Mb/s?

Solution

From (4.17), the change in transmit power must be

$$\frac{2^{12/5}-1}{2^{6/5}-1}\Big|_{dB} = 5.18 \text{ dB}. \tag{4.18}$$

The mapping of this tradeoff on $C(\mathsf{SNR}) = \log_2(1+\mathsf{SNR})$ is indicated in Fig. 4.2, left-hand side.

Whereas, in settings more elaborate than the AWGN channel, the function $C(\cdot)$ and its inverse may not be available in closed form, as long as these functions can be evaluated, e.g., numerically or through simulation, we can apply (4.14) and (4.15) all the same. Furthermore, while SNR is in linear scale in Fig. 4.2, it is typically represented in log-scale and the conversion between SNR and P_t, G, or N_0 is then facilitated as the factors become mere shifts along the horizontal axis.

Tradeoff between P_t and B for fixed R

This tradeoff is no longer captured by $C(\mathsf{SNR})$ as SNR depends on both P_t and B. In this case, $\mathsf{C}(\frac{E_b}{N_0})$ and its inverse are what we need: they capture the dependence between R/B and $\frac{P_t}{N_0 R}$, which for fixed R mirrors the dependence between B and P_t. By moving along the corresponding curves, we could establish how B varies as P_t is modified, and

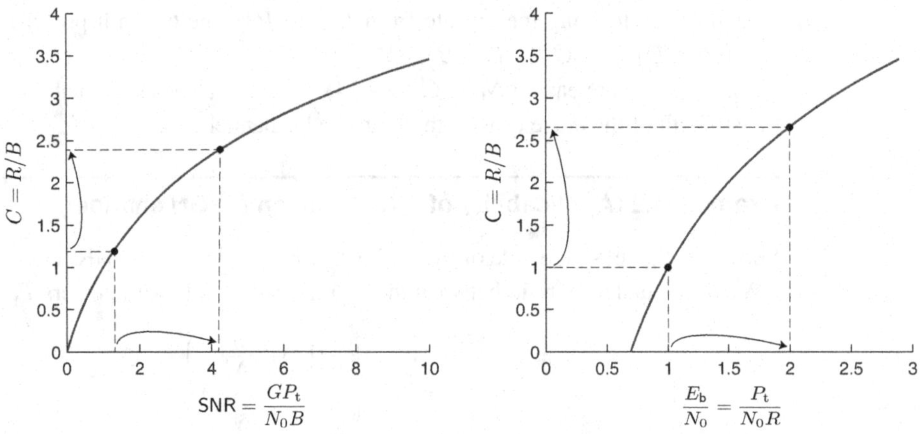

Fig. 4.2 Functions $C(\mathsf{SNR})$ and $\mathsf{C}(\frac{E_b}{N_0})$ for an AWGN channel.

vice versa, for some given R, N_0, and G. In particular, the ratio between the bandwidths required with powers $P_t^{(a)}$ and $P_t^{(b)}$ is

$$\frac{B^{(a)}}{B^{(b)}} = \frac{\mathsf{C}\left(\frac{GP_t^{(b)}}{N_0 R}\right)}{\mathsf{C}\left(\frac{GP_t^{(a)}}{N_0 R}\right)}, \tag{4.19}$$

while the ratio between the powers required with bandwidths $B^{(a)}$ and $B^{(b)}$ is

$$\frac{P_t^{(a)}}{P_t^{(b)}} = \frac{\mathsf{C}^{-1}\left(\frac{R}{B^{(a)}}\right)}{\mathsf{C}^{-1}\left(\frac{R}{B^{(b)}}\right)}. \tag{4.20}$$

Example 4.4 (Applicability of $\mathsf{C}(\frac{E_b}{N_0})$ in an AWGN channel)

Depicted on the right-hand side of Fig. 4.2 is $\mathsf{C}(\frac{E_b}{N_0})$ corresponding to $\frac{E_b}{N_0} = \frac{1}{\mathsf{C}}\left(2^{\mathsf{C}} - 1\right)$. For fixed R, G, and N_0, moving along this curve we could determine how much bandwidth can be saved with a certain increase in power and vice versa.

Example 4.5

On an AWGN channel with $G = -114$ dB, $N_0 = 4 \cdot 10^{-20}$ W/Hz, and $R = 10$ Mb/s, how much bandwidth can we save if we manage to double a transmit power of $P_t = 100$ mW?

Solution

From (4.19), the bandwidth can be reduced by the factor

$$\frac{\mathsf{C}(2\frac{GP_t}{N_0 R})}{\mathsf{C}(\frac{GP_t}{N_0 R})} = \frac{\mathsf{C}(2)}{\mathsf{C}(1)} \tag{4.21}$$

$$\approx 2.66, \tag{4.22}$$

where we have used $\mathsf{C}(1) = 1$, which can be established analytically, and $\mathsf{C}(2) \approx 2.66$, which can be read out from the right-hand side of Fig. 4.2. Thus, only about $1/2.66 \approx 38\%$ of the original bandwidth is needed once the power has been doubled.

As in the case of SNR when representing $C(\cdot)$, the argument $\frac{E_b}{N_0}$ is typically in log-scale when representing the function $\mathsf{C}(\cdot)$ and, most advantageously, this log-scale representation of $\mathsf{C}(\frac{E_b}{N_0})$ exhibits a quasi-linear behavior at low-SNR levels (see Fig. 4.4). This simplifies moving along the tradeoff between P_t and B in this regime.

Tradeoff between R and B for a fixed P_t

In this case, both $C(\mathsf{SNR})$ and $\mathsf{C}(\frac{E_b}{N_0})$ are required, each to formulate the tradeoff in a specific direction, namely

$$R = B \cdot C\left(\frac{GP_t}{N_0 B}\right) \tag{4.23}$$

and

$$B = \frac{R}{\mathsf{C}\left(\frac{GP_t}{N_0 R}\right)}. \tag{4.24}$$

All things considered, $C(\mathsf{SNR})$ is more frequently analyzed than $\mathsf{C}(\frac{E_b}{N_0})$, since more often than not the bandwidth is given and the focus is on the tradeoff between power and bit rate, yet $\mathsf{C}(\frac{E_b}{N_0})$ packs significant meaning as well. In particular, $\mathsf{C}(\frac{E_b}{N_0})$ is informative in the low-SNR regime because of the quasi-linear behavior it displays when the argument $\frac{E_b}{N_0}$ is in log-scale. Moreover, in reference to the tradeoff between transmit power and bandwidth, a general observation can be made on account of the monotonicity of the function $\mathsf{C}(\frac{E_b}{N_0})$. Power efficiency and spectral efficiency are conflicting objectives. In order to be power efficient, the transmission must take place at the lowest possible $\frac{E_b}{N_0}$ and, consequently, at a low spectral efficiency, and vice versa.

The ratio $\frac{E_b}{N_0}$ is also instrumental when assessing decoding error probabilities with non-ideal codes, as its scaling by R facilitates comparing codebooks of different rates, and $\mathsf{C}(\frac{E_b}{N_0})$ is the appropriate benchmark in that context.

Special attention is paid throughout this text to the *low-SNR* and *high-SNR* regimes, sometimes dubbed the *power-limited* and the *bandwidth-limited* regimes, respectively. The findings in these regimes not only offer superior insights, but they turn out to apply over a surprisingly wide range of SNRs. We examine these regimes by means of expansions of quantities of interest, chiefly the capacity, making use of the Landau symbols $\mathcal{O}(\cdot)$ and $o(\cdot)$ described in Appendix F.

4.2.1 Low-SNR regime

The low-SNR regime arises when the received signal power is weak relative to the noise, meaning that SNR is not far from zero. In that vicinity, $C(\mathsf{SNR})$ admits the expansion

$$C(\mathsf{SNR}) = \dot{C}(0)\,\mathsf{SNR} + \frac{1}{2}\,\ddot{C}(0)\,\mathsf{SNR}^2 + o(\mathsf{SNR}^2). \tag{4.25}$$

In terms of $\frac{E_b}{N_0}$, however, the low-SNR regime does not correspond to the vicinity of zero, as can be seen for the AWGN channel in Fig. 4.2.

Example 4.6

For $C = \log_2(1 + \mathsf{SNR})$,

$$\lim_{\mathsf{SNR}\to 0} \frac{E_b}{N_0} = \lim_{\mathsf{SNR}\to 0} \frac{\mathsf{SNR}}{\log_2(1+\mathsf{SNR})} \tag{4.26}$$

$$= \frac{1}{\log_2 e}, \tag{4.27}$$

which amounts to -1.59 dB. Below this mark, the capacity of an AWGN channel is strictly zero and reliable communication is simply unfeasible, something that was observed as early as 1949 by Shannon himself [77].

The low-SNR expansion of $\mathsf{C}(\frac{E_b}{N_0})$, therefore, must take place around a point different from zero. Moreover, as mentioned earlier, this expansion is more conveniently carried out in log-scale, as the function is then quasi-linear and only its first-order behavior needs to be captured. Denoting the minimum value of $\frac{E_b}{N_0}$ by $\frac{E_b}{N_0}_{\mathsf{min}}$, the low-SNR expansion then becomes [64]

$$\mathsf{C}\left(\frac{E_b}{N_0}\right) = S_0 \, \frac{\left.\frac{E_b}{N_0}\right|_{\mathrm{dB}} - \left.\frac{E_b}{N_0}_{\mathsf{min}}\right|_{\mathrm{dB}}}{3\,\mathrm{dB}} + \epsilon, \tag{4.28}$$

where ϵ is a second-order term and S_0 is the slope (in b/s/Hz/(3 dB)) of $C(\frac{E_b}{N_0})$ at $\frac{E_b}{N_0}_{\mathsf{min}}$.

Minimum energy per bit

Reliable communication requires $\frac{E_b}{N_0} > \frac{E_b}{N_0}_{\mathsf{min}}$. Whenever $C(\mathsf{SNR})$ is not only increasing, but further concave in SNR, this minimum is obtained for $\mathsf{SNR} \to 0$ (refer to Problem 4.1). This gives

$$\frac{E_b}{N_0}_{\mathsf{min}} = \lim_{\mathsf{SNR}\to 0} \frac{\mathsf{SNR}}{C(\mathsf{SNR})} \tag{4.29}$$

$$= \frac{1}{\dot{C}(0)}, \tag{4.30}$$

which quantifies:

- The power, relative to any chosen baseline, that is needed to communicate a nonzero bit rate reliably. Multiple examples are presented later.
- The first-order scaling of $C(\mathsf{SNR})$ whenever (4.30) holds, i.e., whenever $C(\mathsf{SNR})$ is concave, since then we can rewrite (4.25) as

$$C(\mathsf{SNR}) = \frac{1}{\frac{E_b}{N_0}_{\mathsf{min}}} \mathsf{SNR} + \frac{1}{2}\ddot{C}(0)\,\mathsf{SNR}^2 + o(\mathsf{SNR}^2). \tag{4.31}$$

The value $\frac{E_b}{N_0}_{\min} = \frac{1}{\log_2 e} = -1.59$ dB encountered in Example 4.6 for the AWGN channel turns out to be very robust, as seen throughout this chapter: as long as the transmission is SISO and fading-independent, this value is upheld. The behavior of $\frac{E_b}{N_0}_{\min}$ with channel-dependent power control and with MIMO is examined later in the chapter and in subsequent chapters, respectively.

Low-SNR slope

From the definition of $\frac{E_b}{N_0}$ and the expansion of $C(\mathsf{SNR})$,

$$\frac{E_b}{N_0} = \frac{\mathsf{SNR}}{C(\mathsf{SNR})} \tag{4.32}$$

$$= \frac{1}{\dot{C}(0) + \frac{1}{2}\ddot{C}(0)\,\mathsf{SNR} + o(\mathsf{SNR})} \tag{4.33}$$

and then, from (4.30) whenever it applies, i.e., whenever $C(\mathsf{SNR})$ is concave,

$$\frac{\frac{E_b}{N_0}_{\min}}{\frac{E_b}{N_0}} = 1 + \frac{\ddot{C}(0)}{2\,\dot{C}(0)}\,\mathsf{SNR} + o(\mathsf{SNR}). \tag{4.34}$$

Using (4.34) and $\mathsf{C}(\frac{E_b}{N_0}) = C(\mathsf{SNR})$, the definition of S_0 implied by (4.28) can be manipulated into [64]

$$S_0 = \lim_{\frac{E_b}{N_0} \to \frac{E_b}{N_0}_{\min}} \frac{\mathsf{C}\left(\frac{E_b}{N_0}\right)}{\left.\frac{E_b}{N_0}\right|_{\mathrm{dB}} - \left.\frac{E_b}{N_0}_{\min}\right|_{\mathrm{dB}}} \cdot 3\,\mathrm{dB} \tag{4.35}$$

$$= \lim_{\mathsf{SNR}\to 0} \frac{C(\mathsf{SNR})}{-10\log_{10}\left(1 + \frac{\ddot{C}(0)}{2\dot{C}(0)}\mathsf{SNR} + o(\mathsf{SNR})\right)} \cdot 10\log_{10} 2 \tag{4.36}$$

$$= \lim_{\mathsf{SNR}\to 0} \frac{C(\mathsf{SNR})}{-\log_2\left(1 + \frac{\ddot{C}(0)}{2\dot{C}(0)}\mathsf{SNR} + o(\mathsf{SNR})\right)} \tag{4.37}$$

$$= \lim_{\mathsf{SNR}\to 0} \frac{\dot{C}(0)\,\mathsf{SNR} + o(\mathsf{SNR})}{\frac{\ddot{C}(0)}{-2\dot{C}(0)}\,\mathsf{SNR}\,\log_2 e + o(\mathsf{SNR})} \tag{4.38}$$

$$= \frac{2\,[\dot{C}(0)]^2}{-\ddot{C}(0)\log_2 e}, \tag{4.39}$$

where we have further used $\log_2(1+z) = z\log_2 e + o(z)$, and relegated the negative sign to the denominator in the final expression to emphasize that $-\ddot{C}(0)$ is positive. Notice that the first-order expansion in (4.28) captures, through $\dot{C}(0)$ and $\ddot{C}(0)$, the second-order behavior of $C(\mathsf{SNR})$.

Example 4.7

Given two channels with equal $\frac{E_b}{N_0}_{\min}$, say channels (a) and (b), relate—within the confines of the low-SNR regime—the bandwidths they require to achieve the same bit rate with the same transmit power.

Solution

As seen in (4.19), the two bandwidths always satisfy

$$\frac{B^{(a)}}{B^{(b)}} = \frac{\mathsf{C}\left(\frac{GP_t^{(b)}}{N_0 R}\right)}{\mathsf{C}\left(\frac{GP_t^{(a)}}{N_0 R}\right)}. \tag{4.40}$$

Applying

$$\mathsf{C}\left(\frac{E_b}{N_0}\right) \approx S_0 \frac{\left.\frac{E_b}{N_0}\right|_{dB} - \left.\frac{E_b}{N_0}_{\min}\right|_{dB}}{3\,dB} \tag{4.41}$$

we have that, provided $\frac{E_b}{N_0}_{\min}$ is the same in channels (a) and (b),

$$\frac{B^{(a)}}{B^{(b)}} \approx \frac{S_0^{(b)}}{S_0^{(a)}} \tag{4.42}$$

where the approximation sharpens as $\frac{E_b}{N_0} \to \frac{E_b}{N_0}_{\min}$.

The foregoing example shows how low-SNR calculations can benefit from the quasi-linear behavior of $\mathsf{C}(\frac{E_b}{N_0})$ in log-scale. In many instances when this function may not be available explicitly, S_0 might be.

4.2.2 High-SNR regime

The high power regime arises when $\mathsf{SNR} \gg 1$, a condition that leads to oft-encountered expansions of the capacity.

Example 4.8

In an AWGN channel, the capacity $C(\mathsf{SNR}) = \log_2(1 + \mathsf{SNR})$ behaves as

$$C(\mathsf{SNR}) = \log_2 \mathsf{SNR} + \mathcal{O}\left(\frac{1}{\mathsf{SNR}}\right). \tag{4.43}$$

The logarithmic behavior of the high-SNR capacity in AWGN channels is not anecdotal, but rather a characteristic of the vast majority of channels of interest. For all such channels, the high-SNR behavior can be accommodated by the expansion

$$C(\mathsf{SNR}) = S_\infty \left(\log_2 \mathsf{SNR} - \mathcal{L}_\infty\right) + \mathcal{O}\left(\frac{1}{\mathsf{SNR}}\right), \tag{4.44}$$

where

$$S_\infty = \lim_{\mathsf{SNR}\to\infty} \frac{C(\mathsf{SNR})}{\log_2 \mathsf{SNR}} \tag{4.45}$$

is the pre-log factor whereas

$$\mathcal{L}_\infty = \lim_{\mathsf{SNR}\to\infty} \left(\log_2 \mathsf{SNR} - \frac{C(\mathsf{SNR})}{S_\infty}\right) \tag{4.46}$$

is the zero-order term, not dependent on SNR.

Spatial DOF and power offset

As is to be seen extensively, $\mathcal{S}_\infty$ quantifies the number of spatial DOF and it is the signature figure of merit in MIMO. One can think of $\mathcal{S}_\infty$ as the number of spatial dimensions available for communication. This dimensionality, revealed as $\mathsf{SNR} \to \infty$, is however only partially informative even in the high-SNR regime; the term $\mathcal{L}_\infty$ helps anchor the high-SNR expansion and enables discriminating channels that, despite sharing the same $\mathcal{S}_\infty$, can be rather different.

Tinkering with (4.44), we obtain

$$C(\mathsf{SNR}) = \mathcal{S}_\infty \left(\frac{10 \log_{10} \mathsf{SNR}}{10 \log_{10} 2} - \mathcal{L}_\infty \right) + \mathcal{O}\left(\frac{1}{\mathsf{SNR}}\right) \tag{4.47}$$

$$= \mathcal{S}_\infty \left(\frac{\mathsf{SNR}|_{\text{dB}}}{3\,\text{dB}} - \mathcal{L}_\infty \right) + \mathcal{O}\left(\frac{1}{\mathsf{SNR}}\right) \tag{4.48}$$

$$= \mathcal{S}_\infty \left(\mathsf{SNR}|_{3\,\text{dB}} - \mathcal{L}_\infty \right) + \mathcal{O}\left(\frac{1}{\mathsf{SNR}}\right), \tag{4.49}$$

which justifies referring to $\mathcal{L}_\infty$ as the high-SNR *power offset* in 3-dB units [355]. Put differently, $\mathcal{L}_\infty$ is the SNR shift, in units of 3 dB, with respect to the baseline

$$C(\mathsf{SNR}) = \mathcal{S}_\infty \log_2 \mathsf{SNR} + \mathcal{O}\left(\frac{1}{\mathsf{SNR}}\right). \tag{4.50}$$

As a side comment, when information is measured in bits, the quantification of power in 3-dB units arises quite naturally because, when $\log_2(\cdot)$ is applied to some quantity z,

$$\log_2 z = \frac{10 \log_{10} z}{10 \log_{10} 2} \tag{4.51}$$

$$= \frac{z|_{\text{dB}}}{3\,\text{dB}} \tag{4.52}$$

$$= z|_{3\,\text{dB}}. \tag{4.53}$$

The high-SNR expansion of $C(\mathsf{SNR})$ allows, for instance, for an easy inversion into $C^{-1}(\cdot)$, namely $\mathsf{SNR} = 2^{\frac{C}{\mathcal{S}_\infty} + \mathcal{L}_\infty}$ or, in log-scale,

$$\mathsf{SNR}|_{\text{dB}} = \left(\frac{C}{\mathcal{S}_\infty} + \mathcal{L}_\infty \right) \cdot 3\,\text{dB}. \tag{4.54}$$

Example 4.9

As seen following (4.15), the difference in transmit power required to bring the bit rate from $R^{(\text{a})}$ to $R^{(\text{b})}$ while maintaining the bandwidth at B is

$$C^{-1}\big(R^{(\text{a})}/B\big)|_{\text{dB}} - C^{-1}\big(R^{(\text{b})}/B\big)|_{\text{dB}}. \tag{4.55}$$

In the high-SNR regime, this simplifies to

$$\left(\frac{R^{(\text{a})}/B}{\mathcal{S}_\infty} + \mathcal{L}_\infty \right) 3\,\text{dB} - \left(\frac{R^{(\text{b})}/B}{\mathcal{S}_\infty} + \mathcal{L}_\infty \right) 3\,\text{dB} = \frac{R^{(\text{a})} - R^{(\text{b})}}{\mathcal{S}_\infty B}\, 3\,\text{dB} \tag{4.56}$$

which only requires the easy-to-compute $\mathcal{S}_\infty$.

When two channels share the same S_∞, their power offsets serve to establish penalties and gains that would otherwise be lost. And, if one prefers to measure such penalties or gains in terms of capacity rather than power, i.e., vertically rather than horizontally in a log-scale representation of $C(\mathsf{SNR})$, any capacity penalty or gain ΔC between two channels (a) and (b) follows from the corresponding power offset penalty or gain, $\Delta\mathcal{L}_\infty = \mathcal{L}_\infty^{(\mathrm{b})} - \mathcal{L}_\infty^{(\mathrm{a})}$, via $\Delta C = S_\infty \, \Delta\mathcal{L}_\infty$.

Example 4.10

Illustrate $\Delta\mathcal{L}_\infty$ and ΔC between the AWGN channel (for which $S_\infty = 1$ and $\mathcal{L}_\infty = 0$) and some other channel having the same S_∞.

Solution

See Fig. 4.16.

4.3 AWGN channel

With all the important quantities defined, and with their mutual relationships solidly established, we are ready to embark upon a tour of progressively more general SISO settings, beginning with the memoryless channel

$$y[n] = \sqrt{G}\, x[n] + v[n] \qquad\qquad n = 0, \ldots, N-1 \tag{4.57}$$

where the signal $x[n] = \sqrt{E_\mathrm{s}}\, s[n]$ and the noise $v[n] \sim \mathcal{N}_\mathbb{C}(0, N_0)$ are independent. The codeword symbols $s[0], \ldots, s[N-1]$, recall, are zero-mean unit-variance and it is irrelevant whether the power constraint is per-symbol or per-codeword. It is further irrelevant whether the codeword symbols are transmitted serially as single-carrier symbols or, with some parallelization, as OFDM resource elements.

This canonical channel, termed, as mentioned, the *AWGN channel* (or sometimes the *Gaussian channel*) is a key building block of most other settings and it therefore deserves extensive treatment. The single-letter channel law is

$$f_{y|s}(\mathrm{y}|\mathrm{s}) = \frac{1}{\pi N_0} e^{-\frac{|\mathrm{y} - \sqrt{G E_\mathrm{s}}\mathrm{s}|^2}{N_0}}. \tag{4.58}$$

4.3.1 Capacity

The AWGN channel is information stable and thus, from Section 1.5,

$$C = \max_{f_s : \mathbb{E}[|s|^2]=1} I\Big(s; \sqrt{G E_\mathrm{s}} s + v\Big) \tag{4.59}$$

$$= \max_{f_s : \mathbb{E}[|s|^2]=1} \Big[\mathfrak{h}\Big(\sqrt{G E_\mathrm{s}} s + v\Big) - \mathfrak{h}\Big(\sqrt{G E_\mathrm{s}} s + v \,|\, s\Big)\Big] \tag{4.60}$$

$$= \max_{f_s:\mathbb{E}[|s|^2]=1} \left[\mathfrak{h}\left(\sqrt{GE_s}s + v\right) - \mathfrak{h}(v) \right] \tag{4.61}$$

$$= \max_{f_s:\mathbb{E}[|s|^2]=1} \mathfrak{h}\left(\sqrt{GE_s}s + v\right) - \log_2(\pi e N_0), \tag{4.62}$$

where the symbol index n has been dropped, as it is unnecessary in a single-letter formulation, and where $\mathfrak{h}(v)$ has been imported from Example 1.5. The mutual information thus depends on s through the differential entropy of $\sqrt{GE_s}s + v$ and we know from Chapter 1 that, for a given variance, the differential entropy is maximized by the complex Gaussian distribution. Moreover, for $\sqrt{GE_s}s + v$ to be complex Gaussian, s must itself be complex Gaussian. Hence, and given that the channel is memoryless and stationary, the codewords should be constructed by drawing symbols independently from a standard complex Gaussian distribution. Although, from a practical viewpoint, this may seem problematic because of the infinite PAPR of the complex Gaussian distribution, it was illustrated in Problem 2.12 that such distribution can be subject to a severe truncation without much loss in its ability to pack information. Indeed, the capacity can be closely approached with signal distributions that are much more amenable to implementation.

For $s \sim \mathcal{N}_{\mathbb{C}}(0,1)$, (4.62) reduces to the mutual information in Example 1.7 and we obtain the AWGN channel capacity (in b/s/Hz) as

$$C(\mathsf{SNR}) = \log_2\left(\pi e(GE_s + N_0)\right) - \log_2(\pi e N_0) \tag{4.63}$$

$$= \log_2(1 + \mathsf{SNR}). \tag{4.64}$$

This relationship, arguably the most iconic formula in all of communications, was used as a running example earlier in this chapter. It is portrayed, in log-scale, in Fig. 4.3. As shown in Example 4.1, the companion function $\mathsf{C}(\frac{E_b}{N_0})$ can only be obtained implicitly through its inverse

$$\frac{E_b}{N_0} = \frac{2^{\mathsf{C}} - 1}{\mathsf{C}}. \tag{4.65}$$

Figure 4.4 depicts $\mathsf{C}(\frac{E_b}{N_0})$, again in log-scale.

Finite blocklength

For the AWGN channel, the variance of the information density defined in (1.144) was shown in [101] to admit the closed form

$$V = \left(1 - \frac{1}{(1+\mathsf{SNR})^2}\right) \log_2^2 e \tag{4.66}$$

and thus, recalling (1.143), with a finite blocklength N and an acceptable error probability p_e it is possible to achieve

$$\frac{R}{B} \approx \log_2(1+\mathsf{SNR}) - \sqrt{1 - \frac{1}{(1+\mathsf{SNR})^2}}\, Q^{-1}(p_e)\, \frac{\log_2 e}{\sqrt{N}} \tag{4.67}$$

where the approximation becomes tighter with growing N.

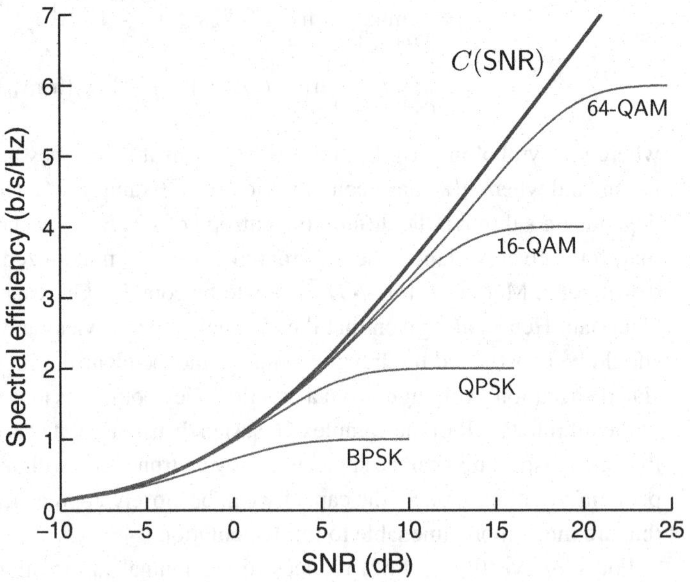

Fig. 4.3 Capacity and spectral efficiencies achieved by BPSK, QPSK, 16-QAM, and 64-QAM as a function of SNR (in dB) in an AWGN channel.

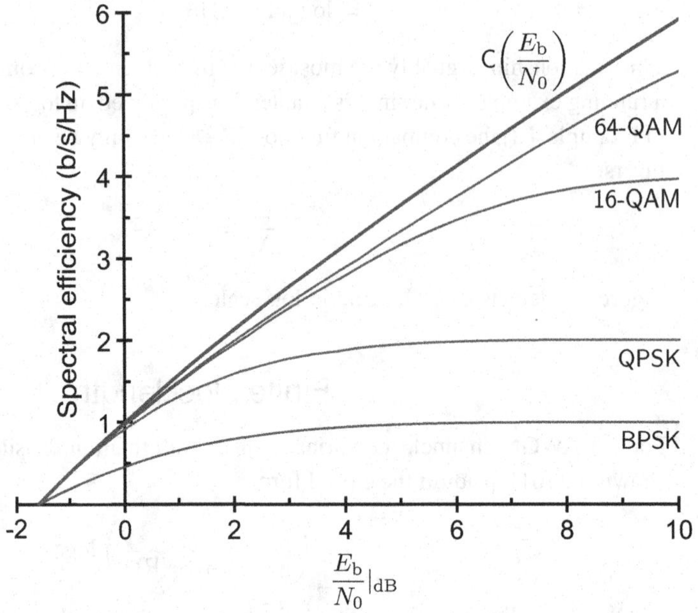

Fig. 4.4 Capacity and spectral efficiencies achieved by BPSK, QPSK, 16-QAM, and 64-QAM as a function of $\frac{E_b}{N_0}$ (in dB) in an AWGN channel.

Example 4.11

What spectral efficiency can be achieved with $N = 2000$ and $p_e = 10^{-3}$?

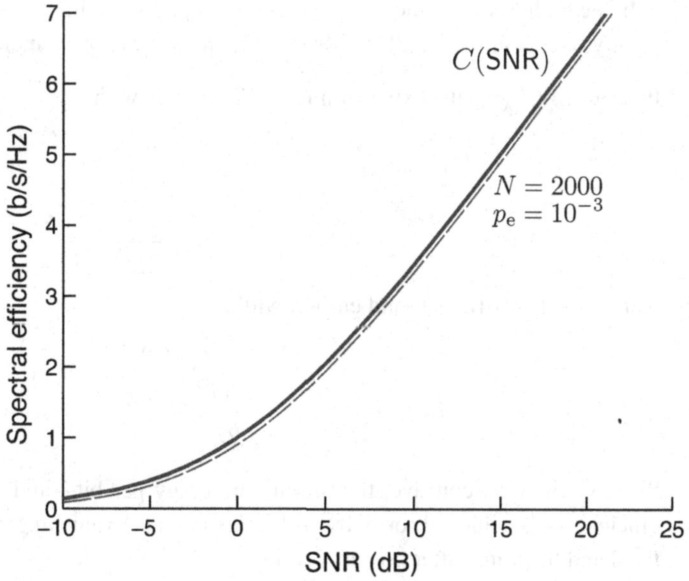

Fig. 4.5 In solid, AWGN channel capacity. In dashed, spectral efficiency achievable with $N = 2000$ and $p_{\mathrm{e}} = 10^{-3}$.

Solution

Shown in Fig. 4.5 is, alongside $C(\mathsf{SNR})$, the result of applying (4.67) with the given values of N and p_{e}. Except perhaps at very low SNR, with these conservative values the loss with respect to the capacity is anecdotal.

Low- and high-SNR regimes

In the low-SNR regime (recall Example 1.7)

$$C(\mathsf{SNR}) = \left(\mathsf{SNR} - \frac{1}{2}\,\mathsf{SNR}^2\right)\log_2 e + o(\mathsf{SNR}^2) \tag{4.68}$$

and thus $\dot{C}(0) = -\ddot{C}(0) = \log_2 e$. In turn, in the high-SNR regime, as seen earlier,

$$C(\mathsf{SNR}) = \log_2 \mathsf{SNR} + \mathcal{O}\left(\frac{1}{\mathsf{SNR}}\right) \tag{4.69}$$

$$= \frac{\mathsf{SNR}|_{\mathrm{dB}}}{3\ \mathrm{dB}} + \mathcal{O}\left(\frac{1}{\mathsf{SNR}}\right), \tag{4.70}$$

which fits the expansion in (4.48) with $\mathcal{L}_\infty = 0$; indeed, the unfaded AWGN channel serves as baseline for other channels in terms of power offset. We further observe that:

- In the low-SNR regime, $C(\mathsf{SNR})$ is—to first order—linear in the SNR. Every doubling of the received power roughly doubles the capacity. The communication is therefore power-efficient.

- In the high-SNR regime, $C(\mathsf{SNR})$ is only logarithmic in the SNR. Every doubling yields only one additional b/s/Hz. The communication is power-inefficient.

In terms of $\mathsf{C}(\frac{E_b}{N_0})$, the expansion in (4.28) applies with

$$\frac{E_b}{N_0}_{\min} = \frac{1}{\dot{C}(0)} \tag{4.71}$$

$$= \frac{1}{\log_2 e}, \tag{4.72}$$

which is -1.59 dB, as found earlier, while

$$S_0 = \frac{2\,[\dot{C}(0)]^2}{-\ddot{C}(0)\log_2 e} \tag{4.73}$$

$$= 2. \tag{4.74}$$

Since $C(\mathsf{SNR})$ is concave, the minimum energy per bit—and thus the maximum power efficiency—is achieved for $\mathsf{SNR} \to 0$. This entails a vanishing bit rate if the bandwidth is fixed and the transmitter powers down,

$$\lim_{P_t \to 0} B \log_2\left(1 + \frac{GP_t}{N_0 B}\right) = 0, \tag{4.75}$$

but not if the power is fixed and the bandwidth is increased, in which case the bit rate grows monotonically with the bandwidth and

$$\lim_{B \to \infty} B \log_2\left(1 + \frac{GP_t}{N_0 B}\right) = \frac{GP_t}{N_0}\log_2 e. \tag{4.76}$$

Although in both cases we end up operating at $\frac{E_b}{N_0} = \frac{E_b}{N_0}_{\min}$, the bit rate is very different.

Figure 4.6 depicts $\frac{E_b}{N_0}$ as a function of SNR, in log–log scale, and with that a perspective of how -1.59 dB is slowly approached as $\mathsf{SNR} \to 0$.

4.3.2 Discrete constellations

Recall that we resist utilizing the term *capacity* with suboptimum signal distributions, and instead speak only of *spectral efficiency*. (Alternatively, the terms *constellation-constrained capacity* or even simply *constrained capacity* are sometimes invoked in the literature.)

The spectral efficiency with BPSK, QPSK, and M-QAM signals is given directly by the corresponding mutual information functions, derived in Examples 1.8–1.12. The spectral efficiency of some of those signal distributions is shown in Fig. 4.3 alongside $C(\mathsf{SNR})$. Notice how, up to the vicinity of $\log_2 M$ bits, the spectral efficiency of M-QAM tightly hugs the capacity. This is a welcome result that dispels potential concerns about the practical significance of the capacity in light of the fact that the capacity-achieving signals are complex Gaussian: signals drawn from discrete constellations can perform virtually as well provided the SNR corresponding to $\log_2 M$ bits is not approached too closely. The envelope of the spectral efficiencies achieved by progressively denser M-QAM constellations coincides with the mutual information of ∞-QAM (see Example 1.9).

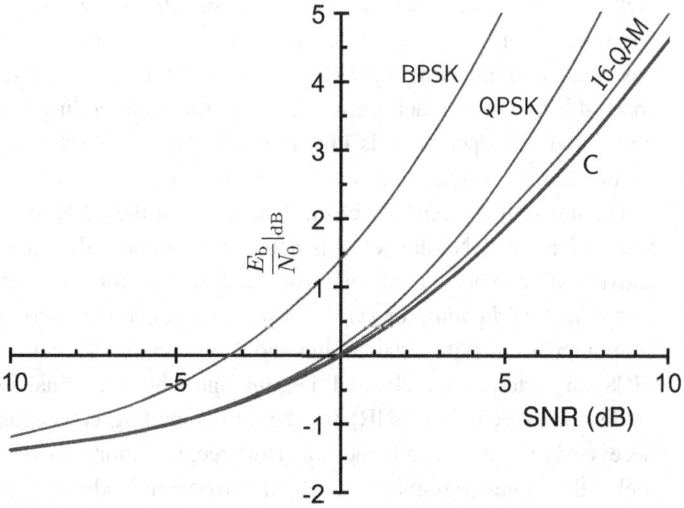

Fig. 4.6 Ratio $\frac{E_b}{N_0}$ (in dB) versus SNR (in dB) for the capacity-achieving distribution, BPSK, and 16-QAM, in an AWGN channel. For SNR $\to 0$, all curves approach -1.59 dB.

Low- and high-SNR regimes

Turning to discrete constellations in the low-SNR regime, the value of $\frac{E_b}{N_0}_{\text{min}}$ is unaffected by the signal distribution provided it is zero-mean [356]. The slope $S_0 = 2$ is also unaffected as long as the signal distribution is proper complex, and thus QPSK and M-QAM constellations exhibit the same slope. For BPSK, which fails to utilize both complex dimensions and is therefore not proper complex, (1.62) and (4.39) lead to $S_0 = 1$. Recalling Example 4.7, this indicates that, for a given low-SNR level, BPSK requires twice the bandwidth to achieve the same bit rate as a proper complex signal; this is evident in how the various curves approach the -1.59-dB mark in Fig. 4.4. For a given bandwidth, however, the bit rate achieved by BPSK in the low-SNR regime is only modestly lower than that of proper complex constellations (see Fig. 4.3 in the low-SNR range).

In the high-SNR regime, Examples 1.8–1.9 apply for ∞-PSK and ∞-QAM, respectively, up to the vicinity of $\log_2 M$, where the spectral efficiency saturates. The high-SNR power penalty suffered below that point (approximately 1.53 dB as per Example 1.9) can be partially recovered by reshaping the constellation and making the constituent points nonequiprobable in such a way that the constellation more closely resembles a complex Gaussian distribution [357–360]. This recovery, dubbed *shaping gain*, comes at the expense of a somewhat magnified peakedness.

4.3.3 Sneak preview of link adaptation

Mobile wireless transceivers feature a number of constellations; rarely BPSK (because it wastes half of each complex dimension in baseband or, equivalently, half the passband bandwidth), customarily QPSK, 16-QAM, 64-QAM, and, for evolved releases of LTE as well as for NR and for certain WLANs, even 256-QAM. Very-short-range or fixed wireless transceivers could feature even denser constellations. The combination of these various constellations with a restricted number of binary coding rates yields a set of discrete spectral efficiencies, each termed a modulation and coding scheme (MCS), at which the transmitter can operate. It is important for the set of MCSs to cover the entire range of operational interest, more or less uniformly.

The transmitter needs to select, depending on the SNR, the most appropriate MCS. In a canonical AWGN channel, this needs to be done only once. In fading channels, alternatively, it becomes a recurring procedure, termed *link adaptation* (alternatively *adaptive modulation and coding* or *rate control*), that constitutes one of the hallmarks of modern communication. With ideal coding and $N \to \infty$, the MCS to select would be the highest MCS supported by the channel, i.e., the one whose nominal spectral efficiency is largest while not exceeding $\mathcal{I}(\mathsf{SNR})$ for the corresponding constellation. With practical codes, necessarily finite in length and not error-free, it is more convenient to select the MCS that yields the largest possible throughput. Erroneous codewords are then subject to hybrid-ARQ and, if necessary, to higher-layer retransmissions to ensure that all payload bits are correctly delivered in the end. This breaks down the responsibility for the reliability of the transmission between the PHY layer and the rest of the protocol stack [361].

Example 4.12 (LTE MCSs)

Listed in Table 4.1 are the 27 MCSs available in a basic release of LTE, obtained by combining QPSK, 16-QAM, and 64-QAM with various values for the binary coding rate r.

Example 4.13 (Error probability in LTE)

Shown in Fig. 4.7 is $p_{\mathrm{e}}(\mathsf{SNR})$, the codeword error probability as a function of SNR (in dB), for MCSs 2, 5, 8, 11, 14, 17, 20, 23, and 26 with turbo coding and one-shot BICM at $N = 6480$. (Results courtesy of the *coded modulation library* maintained by Prof. Mathew Valenti [362].) Also indicated, for each MCS, is the minimum SNR at which that MCS can operate reliably, i.e., the SNR that solves

$$\mathsf{r} \log_2 M = \mathcal{I}(\mathsf{SNR}), \tag{4.77}$$

with $\mathcal{I}(\cdot)$ the Gaussian mutual information function for the corresponding MCS constellation and with M the constellation cardinality. Notice how, with turbo coding at this rather typical blocklength, the coding shortfall at the error probabilities of interest is only 1–2 dB.

Example 4.14 (Throughput and link adaptation in LTE)

Shown in Fig. 4.8 is the throughput per unit bandwidth, $\big(1 - p_{\mathrm{e}}(\mathsf{SNR})\big)\, \mathsf{r} \log_2 M$, for MCSs 2, 5, 8, 11, 14, 17, 20, 23 and 26. By selecting the MCS that provides the highest throughput at each SNR, the highlighted envelope can be achieved and, with a progressively denser set of MCSs, the dashed line could be approached. The intervals on which each MCS delivers the highest throughput are indicated right below the SNR axis, and their boundaries determine the MCS switching thresholds that maximize the throughput. (It if were of interest to

Table 4.1 MCSs in LTE

MCS	Constellation	Code rate, r	$r \log_2 M$ (b/s/Hz)
0	QPSK	0.097	0.194
1	QPSK	0.124	0.248
2	QPSK	0.156	0.312
3	QPSK	0.206	0.401
4	QPSK	0.250	0.500
5	QPSK	0.309	0.618
6	QPSK	0.368	0.737
7	QPSK	0.428	0.856
8	QPSK	0.478	0.957
9	QPSK	0.538	1.075
10	QPSK	0.617	1.233
11	16-QAM	0.353	1.411
12	16-QAM	0.397	1.589
13	16-QAM	0.442	1.767
14	16-QAM	0.500	2.000
15	16-QAM	0.522	2.089
16	16-QAM	0.567	2.267
17	16-QAM	0.633	2.533
18	16-QAM	0.678	2.711
19	16-QAM	0.736	2.944
20	16-QAM	0.795	3.181
21	64-QAM	0.589	3.537
22	64-QAM	0.630	3.780
23	64-QAM	0.655	3.928
24	64-QAM	0.704	4.225
25	64-QAM	0.729	4.373
26	64-QAM	0.845	5.070

keep the error probability below a certain level, then these thresholds could be adjusted to ensure that, at the expense of some degradation in the throughput.)

The foregoing string of examples illustrates how, with a relatively modest number of MCSs—the entire set of 27 MCSs can be indexed with five bits—and off-the-shelf turbo codes plus a one-shot BICM receiver, it is possible to operate remarkably close to capacity. At low SNR, the shortfall from capacity is tiny, all of it due to the nonideality of the coding given that QPSK is optimum in this regime. At high SNR, the shortfall increases because of the nonideality of uniform QAM constellations, but the throughput tracks the capacity closely all along. (With 256-QAM included, the high-SNR shortfall would shrink further.)

Examples 4.12–4.14 could be replicated for NR, and the coded modulation library does support the necessary LDPC codes.

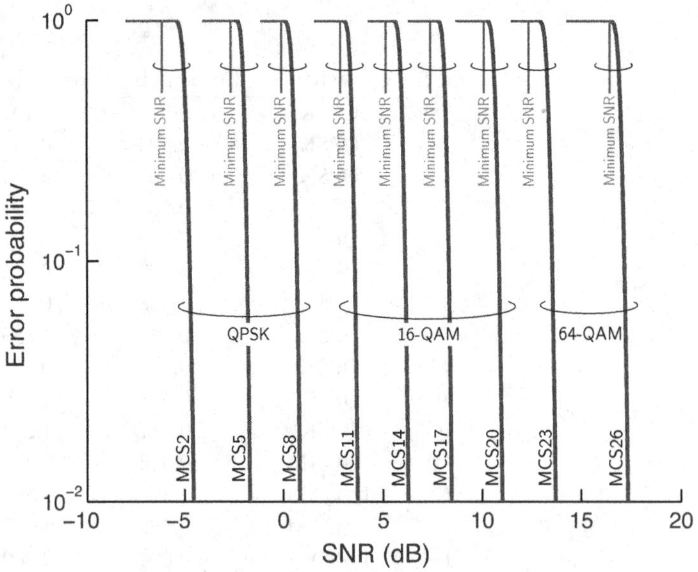

Fig. 4.7 Error probability versus SNR (in dB) for the LTE MCSs 2, 5, 8, 11, 14, 17, 20, 23, and 26 in Table 4.1 and turbo coding with $N = 6480$. Also indicated are the minimum SNRs at which reliable communication is possible for each MCS.

4.4 Frequency-selective channel

Consider now the channel

$$y[n] = \sqrt{G} \sum_{\ell=0}^{L} h[\ell]\, x[n-\ell] + v[n] \qquad n = 0, \ldots, N-1, \tag{4.78}$$

subject to a per-symbol or a per-codeword power constraint. The coefficients $h[0], \ldots, h[L]$ are fixed, i.e., not subject to fading, but normalized to satisfy $\sum_{\ell=0}^{L} |h[\ell]|^2 = 1$ such that G and SNR retain their significance. We dwell on this channel somewhat meticulously because its analysis entails a decomposition and some optimizations that are encountered, in more elaborate forms, once we deal with MIMO.

While decidedly information-stable, the channel in (4.78) is not memoryless and thus the formulation of its mutual information must necessarily be nonsingle-letter, i.e., based on blocks [363]. To better visualize that, we can recall Section 2.2.8 and re-express (4.78) block-wise by means of a convolution matrix, namely

$$\bar{\boldsymbol{y}}_N = \sqrt{G}\, \bar{\boldsymbol{H}}_{N,N+L} \bar{\boldsymbol{x}}_{N+L} + \bar{\boldsymbol{v}}_N, \tag{4.79}$$

where

$$\bar{\boldsymbol{y}}_N = \big[y[0] \;\cdots\; y[N-1]\big]^{\mathrm{T}} \tag{4.80}$$

$$\bar{\boldsymbol{x}}_{N+L} = \big[x[-L] \;\cdots\; x[-1]\, x[0]\, x[1] \;\cdots\; x[N-1]\big] \tag{4.81}$$

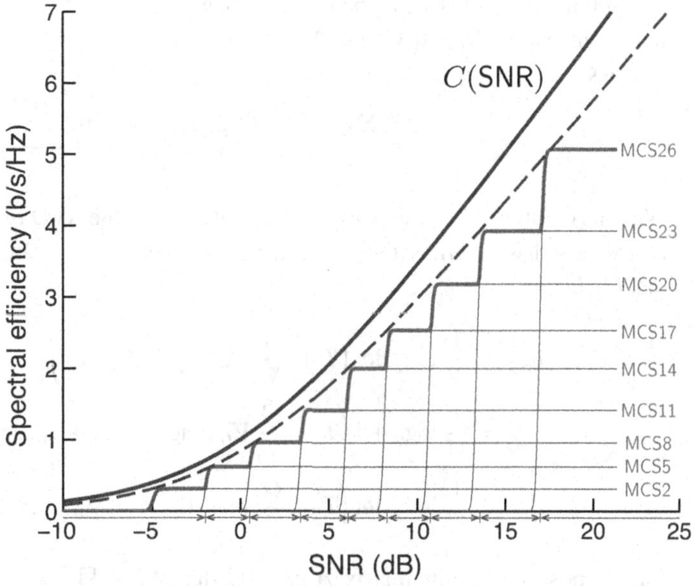

Fig. 4.8 In thin lines, throughput versus SNR (in dB) for each of the LTE MCSs 2, 5, 8, 11, 14, 17, 20, 23, and 26 in Table 4.1. The intervals on which each MCS delivers the highest throughput, delineated below the SNR axis, determine the switching thresholds. The envelope of the individual throughputs, indicated with a thicker stepped line, is the throughput achievable with link adaptation relying on this subset of MCSs and these switching thresholds while, in dashed, we have the throughput approached by this turbo code with a densifying set of MCSs. Also shown is the AWGN channel capacity.

$$\bar{\boldsymbol{v}}_N = \big[v[0] \cdots v[N-1]\big]^{\mathrm{T}} \tag{4.82}$$

whereas $\bar{\boldsymbol{H}}_{N,N+L}$ is the $N \times (N+L)$ Toeplitz matrix

$$\bar{\boldsymbol{H}}_{N,N+L} = \begin{bmatrix} h[L] & \cdots & h[0] & 0 & 0 & \cdots & 0 \\ 0 & h[L] & \cdots & h[0] & 0 & \cdots & 0 \\ \vdots & & & \ddots & & & \vdots \\ 0 & \cdots & 0 & 0 & h[L] & \cdots & h[0] \end{bmatrix}. \tag{4.83}$$

With $\bar{\boldsymbol{v}}_N \sim \mathcal{N}_{\mathbb{C}}(\boldsymbol{0}, N_0 \boldsymbol{I})$, the channel law is

$$f_{\bar{\boldsymbol{y}}_N|\bar{\boldsymbol{x}}_{N+L}}(\bar{\mathbf{y}}|\bar{\mathbf{x}}) = \frac{1}{(\pi N_0)^N} e^{-\frac{\|\bar{\mathbf{y}} - \sqrt{G}\bar{\mathbf{x}}\|^2}{N_0}} \tag{4.84}$$

and the arguments used for the AWGN channel apply to justify that the capacity-achieving transmit signal is complex Gaussian, $\bar{\boldsymbol{x}}_{N+L} \sim \mathcal{N}_{\mathbb{C}}(\boldsymbol{0}, \boldsymbol{R}_{\bar{\boldsymbol{x}}})$. Then, invoking Example 1.13,

$$\frac{1}{N} I(\bar{\boldsymbol{x}}_{N+L}; \bar{\boldsymbol{y}}_N) = \frac{1}{N} \log_2 \det\left(\boldsymbol{I} + \frac{G}{N_0} \bar{\boldsymbol{H}}^*_{N,N+L} \bar{\boldsymbol{H}}_{N,N+L} \boldsymbol{R}_{\bar{\boldsymbol{x}}} \right), \tag{4.85}$$

from which the capacity would follow by letting $N \to \infty$ while optimizing $\boldsymbol{R}_{\bar{\boldsymbol{x}}}$.

Applying the SVD (see Appendix B.3.2), we have that $\bar{\boldsymbol{H}}_{N,N+L} = \boldsymbol{U}_{\bar{H}} \boldsymbol{\Sigma}_{\bar{H}} \boldsymbol{V}_{\bar{H}}^*$ where the unitary matrices $\boldsymbol{U}_{\bar{H}}$ and $\boldsymbol{V}_{\bar{H}}$ are $N \times N$ and $(N+L) \times (N+L)$, respectively, while $\boldsymbol{\Sigma}_{\bar{H}}$ satisfies

$$\boldsymbol{\Sigma}_{\bar{H}}^* \boldsymbol{\Sigma}_{\bar{H}} = \text{diag}(\lambda_0, \ldots, \lambda_{N-1}, \underbrace{0, \ldots, 0}_{L}). \tag{4.86}$$

With the columns of $\boldsymbol{U}_{\bar{H}}$ and $\boldsymbol{V}_{\bar{H}}$ ordered differently, the SVD of $\bar{\boldsymbol{H}}_{N,N+L}$ can equivalently be expressed with $\boldsymbol{\Sigma}_{\bar{H}}$ satisfying $\boldsymbol{\Sigma}_{\bar{H}}^* \boldsymbol{\Sigma}_{\bar{H}} = \text{diag}(0, \ldots, 0, \lambda_0, \ldots, \lambda_{N-1})$. Then, given that $\boldsymbol{U}_{\bar{H}}^* \boldsymbol{U}_{\bar{H}} = \boldsymbol{I}$,

$$\frac{1}{N} I(\bar{\boldsymbol{x}}_{N+L}; \bar{\boldsymbol{y}}_N) = \frac{1}{N} \log_2 \det\left(\boldsymbol{I} + \frac{G}{N_0} \boldsymbol{V}_{\bar{H}} \boldsymbol{\Sigma}_{\bar{H}}^* \boldsymbol{\Sigma}_{\bar{H}} \boldsymbol{V}_{\bar{H}}^* \boldsymbol{R}_{\bar{x}}\right) \tag{4.87}$$

$$= \frac{1}{N} \log_2 \det\left(\boldsymbol{I} + \frac{G}{N_0} \boldsymbol{V}_{\bar{H}}\, \text{diag}(0, \ldots, 0, \lambda_0, \ldots, \lambda_{N-1})\, \boldsymbol{V}_{\bar{H}}^* \boldsymbol{R}_{\bar{x}}\right) \tag{4.88}$$

$$= \frac{1}{N} \log_2 \det\left(\boldsymbol{I} + \frac{G}{N_0} \text{diag}(0, \ldots, 0, \lambda_0, \ldots, \lambda_{N-1})\, \boldsymbol{V}_{\bar{H}}^* \boldsymbol{R}_{\bar{x}} \boldsymbol{V}_{\bar{H}}\right). \tag{4.89}$$

Since a positive-definite matrix $\boldsymbol{A}$ satisfies $\det(\boldsymbol{A}) \leq \prod_j [\boldsymbol{A}]_{j,j}$ with equality when $\boldsymbol{A}$ is diagonal, the mutual information is maximized when the argument of (4.89) is diagonal, which entails a diagonal form for $\boldsymbol{V}_{\bar{H}}^* \boldsymbol{R}_{\bar{x}} \boldsymbol{V}_{\bar{H}}$. With the IID unit-variance codeword symbols to be transmitted denoted as usual by $s[0], \ldots, s[N-1]$, this can be achieved by letting

$$\bar{\boldsymbol{x}}_{N+L} = \sqrt{E_{\mathrm{s}}}\, \boldsymbol{V}_{\bar{H}} \left[\underbrace{0 \cdots 0}_{L}\ \sqrt{P[0]} s[0]\ \cdots\ \sqrt{P[N-1]} s[N-1]\right]^{\mathrm{T}} \tag{4.90}$$

since then

$$\boldsymbol{R}_{\bar{x}} = E_{\mathrm{s}}\, \boldsymbol{V}_{\bar{H}}\, \text{diag}\Big(0, \ldots, 0, P[0], \ldots, P[N-1]\Big)\, \boldsymbol{V}_{\bar{H}}^*, \tag{4.91}$$

meaning that $\boldsymbol{V}_{\bar{H}}^* \boldsymbol{R}_{\bar{x}} \boldsymbol{V}_{\bar{H}}$ is indeed diagonal.

It is therefore optimum to transmit the codeword symbols $s[0], \ldots, s[N-1]$ by driving with them the columns of $\boldsymbol{V}_{\bar{H}}$, which can be regarded as the channel response's eigenfunctions, with power coefficients $P[0], \ldots, P[N-1]$ satisfying $\frac{1}{N} \sum_{n=0}^{N-1} P[n] = 1$ such that the power constraint is met. Unraveling the definition of $\bar{\boldsymbol{x}}_{N+L}$ in (4.90), we further obtain

$$x[n] = \begin{cases} 0 & n = -L, \ldots, -1 \\ \sqrt{E_{\mathrm{s}}} \displaystyle\sum_{\ell=0}^{N-1} \sqrt{P[\ell]}\, s[\ell]\, \big[\boldsymbol{V}_{\bar{H}}\big]_{n,\ell} & n = 0, \ldots, N-1 \end{cases} \tag{4.92}$$

indicating how the codeword symbols are to be mixed before being launched onto the channel. Indeed, since a frequency-selective channel is not memoryless, maximizing its mutual information generally requires non-IID signals, and the mixing in (4.92) creates, from the IID codeword symbols $s[0], \ldots, s[N-1]$, a complex Gaussian transmit signal with the optimum autocorrelation. Notice that symbols $x[-L], \ldots, x[-1]$ are set to zero, a guard interval that isolates consecutive codewords playing a role similar to that of the cyclic prefix in OFDM.

4.4.1 Partition into parallel subchannels

With the channel response known by both transmitter and receiver—something that is implied when $h[0], \ldots, h[L]$ are modeled as deterministic rather than random—it would be possible to implement (4.92) so as to transmit each codeword symbol using one of the columns of $\boldsymbol{V}_{\bar{\boldsymbol{H}}}$. In fact, this strategy need not be applied to the entire block of N symbols at once. Parsing the N symbols into subblocks of K, we could leverage the earlier definitions to write, for each such subblock,

$$\bar{\boldsymbol{y}}_K = \sqrt{G}\, \boldsymbol{U}_{\bar{\boldsymbol{H}}} \boldsymbol{\Sigma}_{\bar{\boldsymbol{H}}} \boldsymbol{V}^*_{\bar{\boldsymbol{H}}}\, \bar{\boldsymbol{x}}_{K+L} + \bar{\boldsymbol{v}}_K \tag{4.93}$$

and, applying at the subblock level the transmit transformation

$$\bar{\boldsymbol{x}}_{K+L} = \sqrt{E_{\rm s}}\, \boldsymbol{V}_{\bar{\boldsymbol{H}}} \Big[\underbrace{0 \cdots 0}_{L}\ \sqrt{P[0]}s[0]\ \cdots\ \sqrt{P[K-1]}s[K-1]\Big]^{\rm T} \tag{4.94}$$

while using, at the receiver, the orthonormal columns of $\boldsymbol{U}_{\bar{\boldsymbol{H}}}$ to rotate $\bar{\boldsymbol{y}}_K$ into $\bar{\boldsymbol{y}}'_K = \boldsymbol{U}^*_{\bar{\boldsymbol{H}}} \bar{\boldsymbol{y}}_K$, the output for each subblock would then be

$$\bar{\boldsymbol{y}}'_K = \boldsymbol{U}^*_{\bar{\boldsymbol{H}}} \left(\sqrt{G E_{\rm s}}\, \boldsymbol{U}_{\bar{\boldsymbol{H}}} \boldsymbol{\Sigma}_{\bar{\boldsymbol{H}}} \boldsymbol{V}^*_{\bar{\boldsymbol{H}}} \boldsymbol{V}_{\bar{\boldsymbol{H}}} \begin{bmatrix} 0 \\ \vdots \\ 0 \\ \sqrt{P[0]}s[0] \\ \vdots \\ \sqrt{P[K-1]}s[K-1] \end{bmatrix} + \bar{\boldsymbol{v}}_K \right) \tag{4.95}$$

$$= \sqrt{G E_{\rm s}}\, \boldsymbol{\Sigma}_{\bar{\boldsymbol{H}}} \begin{bmatrix} 0 \\ \vdots \\ 0 \\ \sqrt{P[0]}s[0] \\ \vdots \\ \sqrt{P[K-1]}s[K-1] \end{bmatrix} + \bar{\boldsymbol{v}}'_K, \tag{4.96}$$

where $\bar{\boldsymbol{v}}'_K = \boldsymbol{U}^*_{\bar{\boldsymbol{H}}} \bar{\boldsymbol{v}}_K$. Because IID complex Gaussian vectors are unitarily invariant (see Appendix C.1.6), we have that $\bar{\boldsymbol{v}}'_K \sim \mathcal{N}_{\mathbb{C}}(\boldsymbol{0}, N_0 \boldsymbol{I})$.

Recalling that $\boldsymbol{\Sigma}_{\bar{\boldsymbol{H}}} = \big[\boldsymbol{0}\ \text{diag}(\sqrt{\lambda_0}, \ldots, \sqrt{\lambda_{K-1}})\big]$ with λ_k the kth eigenvalue of the matrix $\bar{\boldsymbol{H}}^*_{K,K+L} \bar{\boldsymbol{H}}_{K,K+L}$, (4.96) is equivalent to

$$y'[k] = \sqrt{G E_{\rm s}} \sqrt{\lambda_k P[k]}\, s[k] + v'[k] \qquad k = 0, \ldots, K-1, \tag{4.97}$$

evidencing that the transmit and receive transformations in (4.94) and (4.95), jointly referred to as *vector coding* [364], would yield a bank of K parallel subchannels, each an AWGN channel having a distinct SNR [12].

Besides a free allocation of the available power via the coefficients $P[0], \ldots, P[K-1]$, vector coding entails channel-dependent unitary transformations at both transmitter and receiver, namely $\boldsymbol{V}_{\bar{\boldsymbol{H}}}$ and $\boldsymbol{U}^*_{\bar{\boldsymbol{H}}}$. As an alternative to vector coding, we can recall the teachings

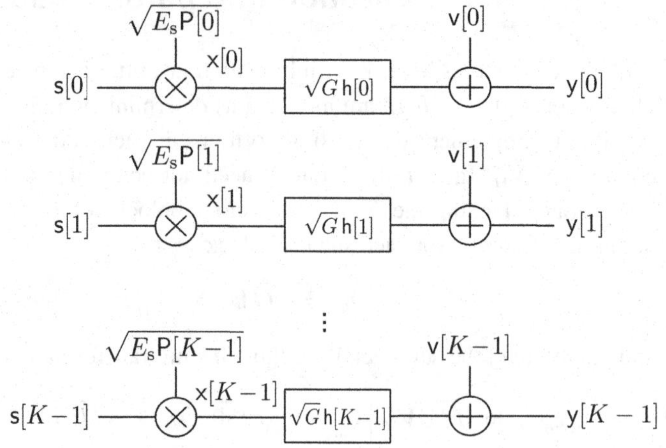

Fig. 4.9 Bank of K parallel subchannels obtained with OFDM.

of Chapter 2 and apply IDFT/DFT transformations, which are also unitary yet channel-agnostic. That allows converting each subblock—or OFDM symbol, as it is termed in that familiar context—into the frequency-domain transmit–receive relationship produced by OFDM, to wit

$$\mathsf{y}[k] = \sqrt{G}\,\mathsf{h}[k]\,\mathsf{x}[k] + \mathsf{v}[k] \tag{4.98}$$

$$= \sqrt{G E_s}\,\mathsf{h}[k]\,\sqrt{\mathsf{P}[k]}\,\mathsf{s}[k] + \mathsf{v}[k] \qquad k = 0, \ldots, K-1, \tag{4.99}$$

where $\mathsf{h}[\cdot]$ is the K-point DFT of $h[\cdot]$, with $K > L$, while $\mathsf{x}[k] = \sqrt{E_s \mathsf{P}[k]}\,\mathsf{s}[k]$ with $\frac{1}{K}\sum_{k=0}^{K-1} \mathsf{P}[k] = 1$ such that the power constraint is met (refer to Problem 4.18). The frequency-domain noise samples are IID with $\mathsf{v}[k] \sim \mathcal{N}_{\mathbb{C}}(0, N_0)$, and hence the K subcarriers thereby obtained again constitute a bank of K parallel subchannels (see Fig. 4.9).

Altogether, both vector coding and OFDM manage to partition a frequency-selective channel into a bank of parallel subchannels, and to contrast these partitioning alternatives some considerations are appropriate.

- With vector coding, the SNR on the kth subchannel is $\frac{GE_s}{N_0}\lambda_k P[k] = \mathsf{SNR}\,\lambda_k P[k]$ while, with OFDM, it is $\frac{GE_s}{N_0}|\mathsf{h}[k]|^2\mathsf{P}[k] = \mathsf{SNR}\,|\mathsf{h}[k]|^2\mathsf{P}[k]$.
- As we know, it is necessary that $N \to \infty$ for the mutual information to have the operational interpretation of a spectral efficiency that can be achieved reliably. This is compatible with a finite K as long as the codewords stretch over multiple subblocks/OFDM symbols, and the operational significance is then acquired as the number of such subblocks/OFDM symbols grows large. Both vector coding and OFDM have an overhead of L for each K codeword symbols, with such overhead taking the form of an idle guard interval in vector coding and of a cyclic prefix in OFDM. These overheads only vanish for $K \to \infty$ and thus only then can the capacity be truly achieved.
- For finite K, the spectral efficiency of vector coding always exceeds that of OFDM, given that the former transmits on the channel response's own eigenfunctions while

the latter transmits on the fixed basis functions embodied by the OFDM subcarriers. However, OFDM can exploit the processing advantage of the FFT so as to afford, for a given computational budget, a larger value of K.

4.4.2 Waterfilling power allocation

In both the vector coding and the OFDM partitionings, the transmitter is free to allocate power across the K subchannels subject to the power constraint. For the sake of specificity, let us address this power allocation problem under OFDM, with the understanding that the formulation carries over to vector coding if $|\mathsf{h}[k]|^2$ is replaced by λ_k.

With the parallel subchannels noninterfering, their individual mutual informations add up onto $\frac{1}{K}\sum_{k=0}^{K-1} I(\mathsf{s}[k];\mathsf{y}[k])$ with the scaling by K ensuring that the units are b/s/Hz (increasing K does not alter the actual bandwidth, but merely partitions it into narrower subcarriers). The optimum power allocation $\mathsf{P}[0], \ldots, \mathsf{P}[K-1]$ is the one that, subject to the power constraint, maximizes this quantity, i.e., the one that solves

$$\max_{\substack{\mathsf{P}[0],\ldots,\mathsf{P}[K-1] \\ \frac{1}{K}\sum_k \mathsf{P}[k]=1}} \frac{1}{K}\sum_{k=0}^{K-1} \log_2\left(1+\mathsf{SNR}\,|\mathsf{h}[k]|^2\,\mathsf{P}[k]\right), \tag{4.100}$$

which is a convex optimization problem with an equality constraint and can thus be solved through the method of Lagrange multipliers (see Appendix G). Forming the Lagrangian

$$\mathsf{L}(\mathsf{P}[0],\ldots,\mathsf{P}[K-1],\lambda) = \frac{1}{K}\sum_{k=0}^{K-1}\log_2\left(1+\mathsf{SNR}\,\mathsf{P}[k]\,|\mathsf{h}[k]|^2\right) + \lambda\left(\frac{1}{K}\sum_{k=0}^{K-1}\mathsf{P}[k]-1\right) \tag{4.101}$$

and setting its partial derivatives to zero, we obtain the necessary and sufficient conditions that characterize the solution $\mathsf{P}^\star[0],\ldots,\mathsf{P}^\star[K-1]$ as

$$\mathsf{P}^\star[k] = -\frac{\lambda}{\log_2 e} - \frac{1}{\mathsf{SNR}\,|\mathsf{h}[k]|^2} \qquad k=0,\ldots,K-1 \tag{4.102}$$

$$\frac{1}{K}\sum_{k=0}^{K-1}\mathsf{P}^\star[k] = 1, \tag{4.103}$$

with the added implicit condition that $\mathsf{P}^\star[k] \geq 0$. Defining a new parameter $\eta = -\frac{\log_2 e}{\lambda}$, the optimum power allocation can be compactly rewritten as

$$\mathsf{P}^\star[k] = \left[\frac{1}{\eta} - \frac{1}{\mathsf{SNR}\,|\mathsf{h}[k]|^2}\right]^+ \qquad k=0,\ldots,K-1, \tag{4.104}$$

with $[z]^+ = \max(0,z)$ and with η satisfying

$$\frac{1}{K}\sum_{k=0}^{K-1}\left[\frac{1}{\eta} - \frac{1}{\mathsf{SNR}\,|\mathsf{h}[k]|^2}\right]^+ = 1. \tag{4.105}$$

Discussion 4.1 OFDM design revisited

In the time-invariant channels considered in this section, the value of K should in principle be as large as can be afforded so as to minimize the relative overhead $\frac{L}{K+L}$. However, once mobility is taken into account this needs to be restrained to ensure that each OFDM symbol is short enough relative to the coherence time. The practice is thus to set K to the smallest possible value that renders the overhead "small."

Example 4.15

How many subcarriers should (roughly) be used in an OFDM system that is to be operated in urban and suburban outdoor environments?

Solution

Since L depends on the relationship between the delay spread and the bandwidth, the value of K is also inextricably tied to the bandwidth; what can be established in absolute terms is the subcarrier spacing.

Referring to Chapter 2, let us take the worst-case delay spread to be $T_{\mathrm{d}} \approx 8\ \mu\mathrm{s}$. For the overhead to be "small," the payload part of each OFDM symbol should be at least ten times as long, meaning $KT \gtrsim 80\ \mu\mathrm{s}$. It follows that $K \gtrsim 8 \cdot 10^{-5}/T = 8 \cdot 10^{-5} B$ and that the subcarrier spacing should be $B/K \lesssim 12.5$ kHz.

This rough calculation leads with remarkable precision to the LTE design: the subcarrier spacing is 15 kHz and every 5 MHz of bandwidth fits $K = 300$ subcarriers. In NR, subcarrier spacings that are multiples of 15 kHz (up to 960 kHz and down to 3.75 kHz) are featured to match channels with different worst-case delay spreads.

Note, by identifying their respective reciprocals, how the condition that the delay spread be "small" relative to the OFDM symbol period is mapped, in the frequency domain, onto the requirement that the subcarrier spacing be "small" relative to the coherence bandwidth. While this may agree with the intuition that each OFDM subcarrier be narrow enough to experience a (roughly) frequency-flat channel response, such frequency flatness over each subcarrier is not required for orthogonality; it suffices that $K > L$ for the subcarriers to be orthogonal at the discrete points $\mathsf{h}[0], \ldots, \mathsf{h}[K-1]$ and hence for the parallel subchannels to be noninteracting. Frequency flatness over each subcarrier is merely the frequency-domain interpretation of the condition that ensures low overhead.

At the same time, as mentioned, mobility requires the OFDM subcarriers to be wide enough to withstand the necessary Doppler spreads. It is to be celebrated that, as argued in Section 3.4.5, in underspread channels there is a broad enough gap between the Doppler spread and the coherence bandwidth for subcarriers to be wide relative to the former while narrow relative to the latter. Low overhead and robustness in the face of mobility are therefore compatible.

From (4.104) and (4.100), the maximized mutual information across the K subchannels is

$$\frac{1}{K} I(\mathsf{s}[0], \ldots, \mathsf{s}[K-1]; \mathsf{y}[0], \ldots, \mathsf{y}[K-1]) = \frac{1}{K} \sum_{k=0}^{K-1} \left[\log_2 \left(\frac{\mathsf{SNR}\, |\mathsf{h}[k]|^2}{\eta} \right) \right]^+ . \quad (4.106)$$

The power allocation policy described by (4.104) and graphically illustrated in Fig. 4.10 is termed *waterfilling*. First derived by Shannon in 1949 [77] and rigorously formalized in [59, 363, 365, 366], the waterfilling policy is not only a central result in information theory but one that is frequently encountered in other disciplines. The term *waterfilling* (or, equivalently, *waterpouring*) appears to have been coined by Robert Fano [367], and it is justified by the interpretation of pouring K units of water onto a bank of unit-base vessels solid up to a height $\frac{1}{\mathsf{SNR}\,|\mathsf{h}[k]|^2}$. The water level reaches $1/\eta$ across all subchannels and the water height within the kth vessel gives $\mathsf{P}^\star[k]$. Note that some vessels may end up with no water whatsoever, indicating that the corresponding subchannels are too weak to warrant being employed.

Example 4.16

Compute the waterfilling power allocation for $\mathsf{SNR} = 6$ dB and

$$|\mathsf{h}[0]|^2 = 0.5 \quad (4.107)$$
$$|\mathsf{h}[1]|^2 = 0.4 \quad (4.108)$$
$$|\mathsf{h}[2]|^2 = 0.1. \quad (4.109)$$

Solution

Clearing η from (4.105) under the assumption that all the terms therein are nonnegative, we obtain $\eta = 0.83$, yet that value renders the third term negative. This indicates that $\mathsf{P}[2] = 0$, something that can be verified through (4.104). Recomputing η from (4.105) with the third term set to zero, we obtain $\eta = 0.48$ and subsequently, from (4.104),

$$\mathsf{P}[0] = 1.56 \quad (4.110)$$
$$\mathsf{P}[1] = 1.44 \quad (4.111)$$
$$\mathsf{P}[2] = 0. \quad (4.112)$$

4.4.3 Capacity

From (4.100) and the overhead of L symbols for every K codeword symbols, the spectral efficiency equals

$$\frac{R}{B} = \frac{1}{K+L} \sum_{k=0}^{K-1} \log_2 \left(1 + \mathsf{SNR}\, |\mathsf{h}[k]|^2\, \mathsf{P}^\star[k] \right), \quad (4.113)$$

where the optimum power allocation $\mathsf{P}^\star[k]$ is given by waterfilling. Achieving capacity requires letting $K \to \infty$, which suffices to ensure that $N \to \infty$ while rendering the overhead

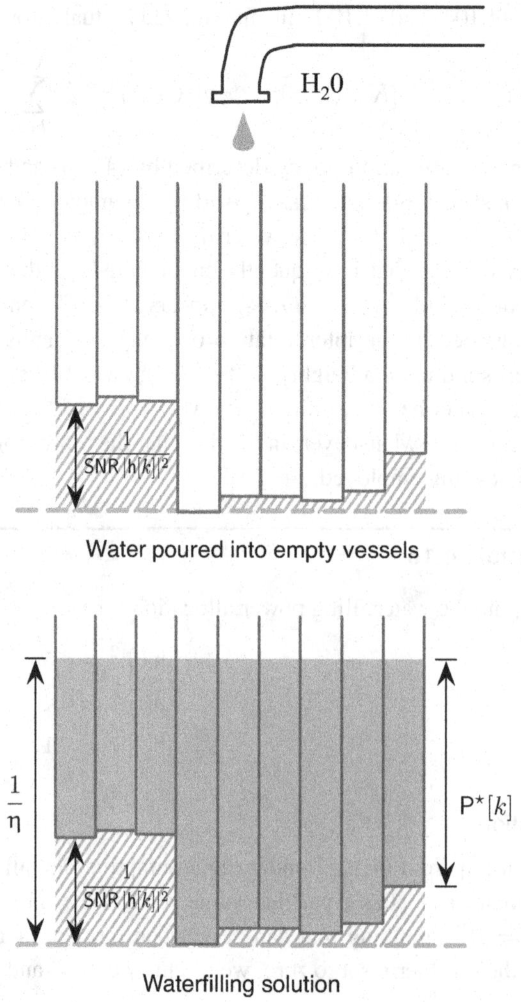

Fig. 4.10 Waterfilling power allocation.

negligible. When $K \to \infty$, the Toeplitz matrix $\bar{\boldsymbol{H}}_{K,K+L}$ becomes asymptotically equivalent to the circulant matrix obtained by augmenting $\bar{\boldsymbol{H}}_{K,K+L}$ with L additional bottom rows, namely

$$\bar{\boldsymbol{H}}_{K+L,K+L} = \begin{bmatrix} h[L] & \cdots & h[0] & 0 & 0 & \cdots & 0 \\ 0 & h[L] & \cdots & h[0] & 0 & \cdots & 0 \\ \vdots & & & \ddots & & & \vdots \\ 0 & \cdots & 0 & 0 & h[L] & \cdots & h[0] \\ h[0] & 0 & \cdots & 0 & 0 & h[L] & \cdots \\ \vdots & & & \ddots & & & \vdots \\ h[L-1] & \cdots & h[0] & 0 & \cdots & 0 & h[L] \end{bmatrix}, \tag{4.114}$$

as the edge effect associated with this augmentation vanishes with growing K. It follows from the properties of circulant matrices (see Appendix B.2.4) that, asymptotically, $\bar{\boldsymbol{H}}_{K+L,K+L}$ can be decomposed as $\boldsymbol{U}\boldsymbol{\Sigma}_{\bar{H}}\boldsymbol{U}^*$ where $\boldsymbol{U}$ is the Fourier matrix in (B.7) while the diagonal entries of $\boldsymbol{\Sigma}_{\bar{H}}$ are the DFT coefficients of an arbitrary row of $\bar{\boldsymbol{H}}_{K+L,K+L}$ (or of $\bar{\boldsymbol{H}}_{K,K+L}$), i.e., the DFT of $h[L],\ldots,h[0]$ zero-padded up to $K+L$. Since the multiplication by a Fourier matrix effects an OFDM transformation, it follows that the vector coding transmit and receive transformations converge to OFDM transformations as K grows large, and hence OFDM becomes optimum in the asymptote in which capacity is achieved.

Considering therefore OFDM with $K \to \infty$, what ensues is that the frequency resolution becomes increasingly fine and the DFTs become discrete-time Fourier transforms on the interval $[-1/2, 1/2]$ (see Section A.1). Then, (4.100) leads to the capacity

$$C(\mathsf{SNR}) = \int_{-1/2}^{1/2} \log_2\left(1 + \mathsf{SNR}\, |\mathsf{h}(\nu)|^2\, \mathsf{P}^\star(\nu)\right) d\nu, \tag{4.115}$$

with the continuous power allocation $\mathsf{P}^\star(\nu)$ given by waterfilling over $|\mathsf{h}(\nu)|^2$, where $\mathsf{h}(\nu)$ is the continuous-frequency channel response, and with the waterfilling subject to

$$\int_{-1/2}^{1/2} \mathsf{P}^\star(\nu)\, d\nu = 1. \tag{4.116}$$

If, rather than optimizing the power allocation via waterfilling, we elect to allocate power uniformly across subcarriers, then all the foregoing expressions apply only with $\mathsf{P}^\star[k]$ and $\mathsf{P}^\star(\nu)$ replaced by 1.

Single-carrier signaling

It is worthwhile to rewrite the capacity of a frequency-selective channel as that of an AWGN channel,

$$C(\mathsf{SNR}_{\mathsf{eq}}) = \log_2(1 + \mathsf{SNR}_{\mathsf{eq}}), \tag{4.117}$$

where, to preserve the equivalence with (4.115),

$$\mathsf{SNR}_{\mathsf{eq}} = 2^{\int_{-1/2}^{1/2} \log_2(1 + \mathsf{SNR}\, |\mathsf{h}(\nu)|^2 \mathsf{P}^\star(\nu))\, d\nu} - 1. \tag{4.118}$$

Given that an AWGN channel is frequency-flat, the quantity $\mathsf{SNR}_{\mathsf{eq}}$ can be interpreted as the SNR at the output of a perfect time-domain equalizer. As it happens, equalizers achieving this performance are indeed feasible, for instance in the form of a structure having an LMMSE front-end (see Section 2.4.2) plus a decision-feedback loop integrated with the decoder [368]. Thus, complexity considerations notwithstanding, directly with a single-carrier signal optimally autocorrelated, without explicit OFDM or vector coding transformations, the capacity in (4.115) could theoretically be approached arbitrarily well.

Low- and high-SNR regimes

The behavior of the optimum power allocation in the low- and high-SNR regimes can be inferred from (4.104) and from the waterfilling interpretation.

For $\mathsf{SNR} \to 0$, it is optimum to allocate power exclusively to the subcarrier(s) with the highest gain; the ensuing expansion of the spectral efficiency of frequency-selective channels in the low-SNR regime is the subject of Problem 4.30. If $\max\left(|\mathsf{h}[0]|, \ldots, |\mathsf{h}[K-1]|\right)$ is not unique, then the power should be evenly divided among the corresponding subcarriers, a fine point to which Problem 4.29 is devoted.

For $\mathsf{SNR} \to \infty$, we observe from (4.105) that $\eta \to 1 + \mathcal{O}(\frac{1}{\mathsf{SNR}})$ and, plugging this expansion into (4.104), that

$$\mathsf{P}^\star[k] = 1 + \mathcal{O}\left(\frac{1}{\mathsf{SNR}}\right) \qquad k = 0, \ldots, K-1, \tag{4.119}$$

making evident that, for $\mathsf{SNR} \to \infty$, waterfilling converges to a uniform power allocation. Combining this expansion for $\mathsf{P}^\star[0], \ldots, \mathsf{P}^\star[K-1]$ with (4.113), the spectral efficiency achievable with a given K can be specialized to the high-SNR regime; the corresponding expansion is the subject of Problem 4.32.

4.4.4 Discrete constellations

Given a bank of parallel noninteracting subchannels, say K OFDM subcarriers, what is the optimum power allocation policy if the transmit signals are independent but not Gaussian? For arbitrary signal distributions, possibly different per subcarrier, we can recast (4.100) as

$$\max_{\substack{\mathsf{P}[0],\ldots,\mathsf{P}[K-1] \\ \frac{1}{K}\sum_k \mathsf{P}[k]=1}} \frac{1}{K} \sum_{k=0}^{K-1} \mathcal{I}_k\left(\mathsf{SNR}\,|\mathsf{h}[k]|^2\,\mathsf{P}[k]\right), \tag{4.120}$$

where $\mathcal{I}_k(\cdot)$ is the Gaussian mutual information function for the signal distribution on the kth subcarrier. Since the mutual information functions are concave, the problem remains convex. By means of the I-MMSE relationship in (1.196) and the method of Lagrange multipliers, the solution can be expressed (refer to Problem 4.33) as

$$\mathsf{P}^\star[k] = \frac{1}{\mathsf{SNR}\,|\mathsf{h}[k]|^2}\,\mathsf{MMSE}_k^{-1}\left(\min\left(1, \frac{\eta}{\mathsf{SNR}\,|\mathsf{h}[k]|^2}\right)\right), \tag{4.121}$$

where η is the unique solution to the nonlinear equation

$$\frac{1}{K} \sum_{\mathsf{SNR}\,|\mathsf{h}[k]|^2 > \eta} \frac{1}{\mathsf{SNR}\,|\mathsf{h}[k]|^2}\,\mathsf{MMSE}_k^{-1}\left(\frac{\eta}{\mathsf{SNR}\,|\mathsf{h}[k]|^2}\right) = 1, \tag{4.122}$$

while $\mathsf{MMSE}_k^{-1}(\cdot)$ is the inverse of $\mathsf{MMSE}_k(\cdot)$ with respect to the composition of functions, with domain equal to $[0, 1]$. An interpretation of (4.121) that generalizes the waterfilling policy and preserves some of its intuition is put forth in [71, 369] under the name *mercury/waterfilling*. This interpretation, illustrated in Fig. 4.11, is as follows.

(a) For each subcarrier, set up a unit-base vessel solid up to $\frac{1}{\mathsf{SNR}\,|\mathsf{h}[k]|^2}$.

(b) Pour mercury onto each vessel until its height, solid included, reaches a certain level.

(c) Pour K units of water, at which point the water level reaches $1/\eta$.

(d) The water height over the mercury on the kth vessel gives $\mathsf{P}^\star[k]$.

The mercury-pouring stage regulates the water admitted by each vessel, tailoring the procedure to arbitrary signal distributions. No mercury is poured onto vessels corresponding to subcarriers fed by Gaussian signals. The mercury can be seen as playing the role of the approximate *SNR gap* invoked by certain heuristic methods [370–373], but with the advantage of being exact for any signal power and distribution.

For $\mathsf{SNR} \to 0$, mercury/waterfilling behaves exactly as regular waterfilling as long as the signals fed into all the subcarriers are proper complex; this coincidence constitutes the subject of Problem 4.36.

At high SNR, (4.119) holds with continuous distributions such as ∞-PSK and ∞-QAM but not with discrete constellations. Rather, the optimum power allocation then seeks to equalize, across the subcarriers, the product of the received power and the squared minimum distance of the constellation [71, theorem 7]. However, this solution applies at SNR levels where communication is highly power-inefficient, where the spectral efficiency of the corresponding constellations no longer hugs the capacity. Operation in this regime should be avoided if power efficiency is of minimal essence.

Single-carrier signaling

With discrete constellations, the exact equivalence between OFDM and single-carrier transmission breaks down. The equalized SNR given in (4.118) continues to be achievable, but any residual ISI is bound to be non-Gaussian if the signal itself is non-Gaussian; thus, the combined noise-plus-interference is non-Gaussian and the Gaussian mutual information functions $\mathcal{I}(\cdot)$ strictly do not apply. Nonetheless, these functions evaluated at the equalized SNR have been shown to approximate rather well the spectral efficiency achievable with single-carrier transmission [374] and therefore we can write

$$\frac{R}{B} \approx \mathcal{I}(\mathsf{SNR}_{\mathsf{eq}}), \tag{4.123}$$

where $\mathcal{I}(\cdot)$ is the Gaussian mutual information function that matches the signal distribution. While not exact, the right-hand side of (4.123) is infinitely easier to compute than

$$\lim_{N\to\infty} I\big(s[0],\ldots,s[N-1];y[0],\ldots,y[N-1]\big). \tag{4.124}$$

The values of (4.120) and (4.123) are rather analogous and thus all the OFDM-based discrete constellation derivations in this text are largely applicable to non-OFDM alternatives as well. If anything, when (4.120) and (4.123) are compared carefully, there is a slight edge for the single-carrier alternative provided the computational complexity of a joint optimum equalizer and decoder can be afforded [375].

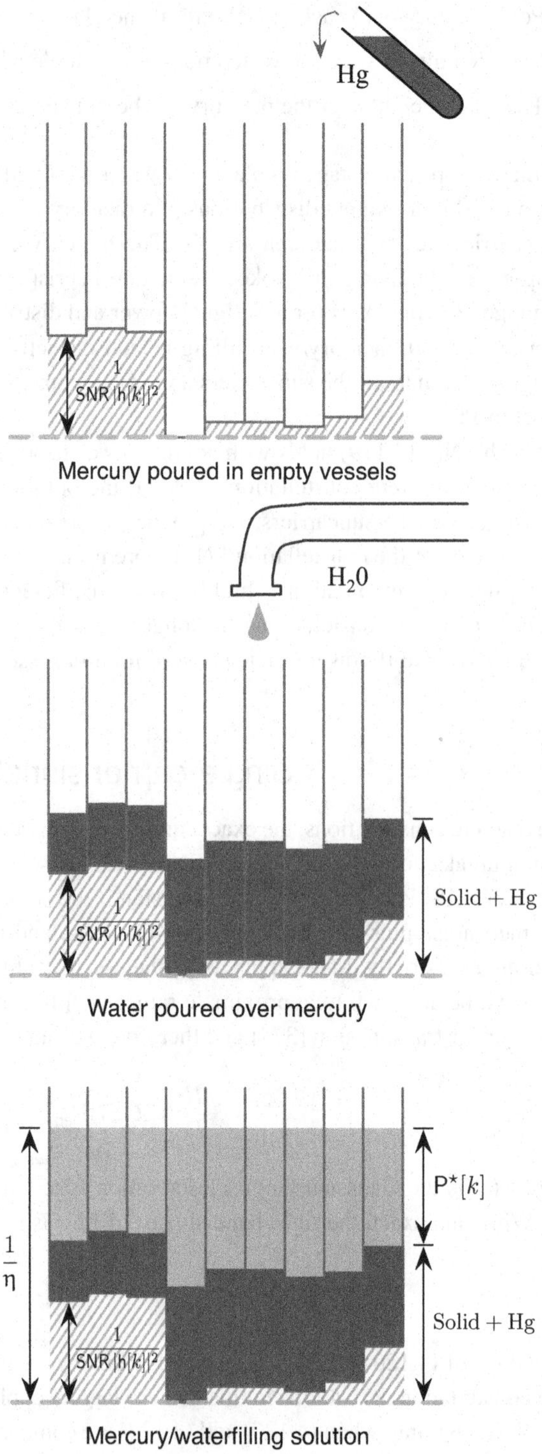

Fig. 4.11 Mercury/waterfilling power allocation.

4.4.5 CESM, MIESM, and EESM mapping methods

To wrap up our examination of fixed frequency-selective channels, it is worth examining briefly what ensues when we back off from $N \to \infty$. This setting provides a very clean context to introduce some methods that make appearances throughout the text, in the context of link adaptation. Letting N be finite, each codeword occupies K subcarriers over N/K OFDM symbols for a total of N resource elements, with an overhead of $\frac{L}{K+L}$.

The computation of the spectral efficiency achievable with a finite N and a certain codeword error probability p_e could be approached via the finite-codeword-length framework introduced in Section 1.5.5 and specialized to frequency-selective channels in [376]. The result of this approach would be in the form of an expansion equaling (4.113) with a back-off $\delta C(N, p_\mathrm{e})$, which mapped to an equivalent AWGN channel would yield

$$\mathsf{SNR}_\mathsf{eq} \approx 2^{\frac{1}{K+L}\sum_{k=0}^{K-1}\log_2(1+\mathsf{SNR}\,|\mathsf{h}[k]|^2\mathsf{P}[k])-\delta C} - 1. \tag{4.125}$$

Complementing its earlier interpretation as the SNR at the output of a perfect time-domain equalizer, SNR_eq can be seen to represent the mapping of $\mathsf{h}[0], \ldots, \mathsf{h}[K-1]$ onto an AWGN channel, and specifically the mapping in terms of the equivalent SNR at which an AWGN channel would perform as the frequency-selective channel $\mathsf{h}[0], \ldots, \mathsf{h}[K-1]$ when a single codeword spans all subcarriers.

As an alternative to the foregoing approach, which presumes finite-length but otherwise ideal coding, rather than backing off from (4.113) we can pragmatically fudge it so that its SNR mapping is

$$\mathsf{SNR}_\mathsf{CESM} \approx \beta\, 2^{\frac{1}{K+L}\sum_{k=0}^{K-1}\log_2(1+\mathsf{SNR}\,|\mathsf{h}[k]|^2\mathsf{P}[k]/\beta)} - 1, \tag{4.126}$$

where β is a fudge factor to be adjusted depending, not only on N, but further on the specific codebook being used and on other practical aspects. While not solidly grounded on information-theoretic principles, this so-called capacity-effective SNR mapping (CESM) method is highly flexible [377]. Furthermore, β does not depend on p_e. In fact, rather than adjusting β such that $\log_2(1 + \mathsf{SNR}_\mathsf{CESM})$ gives the spectral efficiency achievable at p_e, the factor β is adjusted such that $p_\mathrm{e}(\mathsf{SNR}_\mathsf{CESM})$ equals the error probability on an AWGN channel with the codebook of choice. Then, the throughput equals $(1 - p_\mathrm{e})R$.

The CESM method can be generalized as

$$\mathsf{SNR}_\mathsf{xESM} \approx \beta\, g_k^{-1}\left(\frac{1}{K+L}\sum_{k=0}^{K-1} g_k\left(\frac{\mathsf{SNR}\,|\mathsf{h}[k]|^2\mathsf{P}[k]}{\beta}\right)\right), \tag{4.127}$$

where $g_k(\cdot)$ for $k = 0, \ldots, K-1$ are concave functions and $g_k^{-1}(\cdot)$ is the inverse of $g_k(\cdot)$ with respect to the composition. Indeed, by setting $g_k(\gamma) = \log_2(1+\gamma)$, we recover the CESM method as a special case of (4.127). Two alternatives to CESM have been shown to be particularly effective.

- When $g_k(\gamma) = \mathcal{I}_k(\gamma)$, the Gaussian mutual information function for the constellation on the kth subcarrier, xESM stands for MIESM (mutual-information-effective SNR mapping) [378]. At the expense of not being in closed form, the choice of a distinct function depending on the signal distribution facilitates the task of adjusting β and renders MIESM particularly precise.

- For $g(\gamma) = e^{-\gamma}$, xESM stands for EESM (exponential-effective SNR mapping). Deriving from an uncoded error probability expression, this exponential function is slightly less efficacious and tends to feature much more dispersed values for β, yet its analytically convenience is unbeatable [379, 380].

The MIESM and EESM methods, widely employed by industry and standardization bodies, enable abstracting frequency-selective channels onto the single scalar metric $\mathsf{SNR}_{\mathsf{xESM}}$. This scalar metric is not only a succinct representation, but a convenient one, as it applies on a simple AWGN channel. Once β has been calibrated—via explicit simulation of coding and decoding in all its detail—such that the error probability as a function of SNR on some sample frequency-selective channels matches $p_{\mathsf{e}}(\mathsf{SNR}_{\mathsf{xESM}})$ on an AWGN channel, the performance of *any* frequency-selective channel of the sampled class can be read-up from $p_{\mathsf{e}}(\mathsf{SNR}_{\mathsf{xESM}})$ with the coding and decoding details forgone. Given a channel $\mathsf{h}[0], \ldots, \mathsf{h}[K-1]$ and a power allocation $\mathsf{P}[0], \ldots, \mathsf{P}[K-1]$ at a certain SNR, we can readily compute $\mathsf{SNR}_{\mathsf{xESM}}$ from (4.127) and use it to read the error probability from the AWGN curve $p_{\mathsf{e}}(\cdot)$.

Example 4.17 (LTE fudge factors)

For the LTE MCSs, the mapping has been found to be optimized by $\beta \in [0.85, 1.2]$ for MIESM and $\beta \in [0.8, 35]$ for EESM [381, appendix E]. Supporting these findings, further studies [382] have reported $\beta \in [0.8, 1.2]$ for MIESM and $\beta \in [3, 23]$ for EESM with $N = 480$, with the EESM range shrinking to $\beta \in [3, 8]$ for $N = 960$.

Example 4.18 (3G fudge factors)

For $K = 416$ subcarriers and $N = K$ such that each codeword occupies exactly one OFDM symbol, the values of β that best match the performance of 16-QAM with a turbo code of binary rate $\mathsf{r} = 1/3$ are given in [377]. The channel model applied for the calibration is the SCM (see Section 3.8.1) with its three environments included. The values obtained for β are shown in Table 4.2.

Example 4.19

Consider the setting in Example 4.18, only with a channel where in half the subcarriers $|\mathsf{h}[k]|^2 = 1.6$ while, in the other half, $|\mathsf{h}[k]|^2 = 0.4$. Knowing that $p_{\mathsf{e}}(\mathsf{SNR}_{\mathsf{EESM}}) = 10^{-2}$ at $\mathsf{SNR}_{\mathsf{EESM}} = 4.1$ dB, how far is this operating point from the fundamental limit with a complex Gaussian codebook and $N \to \infty$? Assume a uniform power allocation.

Solution

We can ignore the cyclic-prefix overhead, as it does not affect comparisons where K is fixed. On the one hand, the throughput per unit of bandwidth equals $(1 - p_{\mathsf{e}})\, \mathsf{r} \log_2 M =$

Table 4.2 Exemplary fudge factors for CESM, MIESM, and EESM with $N = K = 416$

Method	β
CESM	0.92
MIESM	1.11
EESM	3.24

$0.99 \cdot 4/3 = 1.32$ b/s/Hz while, from (4.127),

$$10^{4.1/10} = -3.24 \log_e \left(\frac{e^{-1.6\,\mathsf{SNR}/3.24} + e^{-0.4\,\mathsf{SNR}/3.24}}{2} \right) \tag{4.128}$$

and we find that $\mathsf{SNR} = 3.07$. On the other hand, the SNR at which the spectral efficiency achievable with complex Gaussian codebooks and $N \to \infty$ equals $R/B = 1.32$ b/s/Hz is the solution to

$$1.32 = \frac{\log_2(1 + 1.6\,\mathsf{SNR}) + \log_2(1 + 0.4\,\mathsf{SNR})}{2}, \tag{4.129}$$

which gives $\mathsf{SNR} = 1.7$. The ratio of the two SNRs comes to $\frac{3.07}{1.7}|_{\text{dB}} = 2.57$ dB, which approximates the loss associated with employing 16-QAM and a relatively short code rather than an unboundedly long complex Gaussian codebook. This modest loss is in line with what we found in the AWGN channel, and thus frequency selectivity need not significantly increase the shortfall of practical implementations from capacity.

Example 4.20

How would the result of the foregoing example change if CESM were applied instead of EESM?

Solution

With CESM (refer to Problem 4.40), the loss becomes 2.48 dB. The difference of 0.09 dB between both methods provides a sense of their accuracy.

As alternatives to CESM, MIESM, and EESM, all based on a single metric, methods that map $\mathsf{h}[0], \ldots, \mathsf{h}[K-1]$ onto p_{e} via multiple metrics are presented and tested in [383, 384].

Regardless of the choice of method, the mapping of a frequency-selective channel onto one or a few metrics has direct applicability in link adaptation, which is otherwise as explained in Section 4.3.3: given the channel's metric(s), the MCS that yields the largest possible throughput is selected at the transmitter.

4.5 Frequency-flat fading channel

Multipath fading is arguably the most intrinsic feature of wireless channels and a thorough study of its impact is thus imperative. Circumscribing the analysis to frequency-flat fading for now, consider

$$y[n] = \sqrt{G}\, h[n]x[n] + v[n] \qquad n = 0, \ldots, N-1, \tag{4.130}$$

where $\{h[n]\}$ is a zero-mean unit-variance random process representing the fading while the noise samples are IID with $v[n] \sim \mathcal{N}_{\mathbb{C}}(0, N_0)$. For the sake of specificity we presume single-carrier transmission throughout this long section, with the frequency flatness ensuring that the results would hold equally—cyclix prefix overheads aside—if the codeword symbols were transmitted over OFDM resource elements instead.

Frequency-flat fading is, in essence, a form of multiplicative noise. It is, however, very different from the additive noise $v[n]$ in that fading has memory, i.e., $\{h[n]\}$ has an autocorrelation function that is not an impulse, and a correspondingly nonflat Doppler spectrum. Fading can therefore be estimated, tracked, and possibly even predicted, giving rise to the notion of CSI. Throughout this text, CSI refers to knowledge of the fading realizations encountered by the codewords. The distribution of the fading, sometimes termed *statistical CSI*, is herein considered known by both transmitter and receiver on the grounds that, as submitted in Section 3.7.1, this distribution is stable over a truly large number of symbols.

The pilot symbols that are regularly inserted within the signals in the vast majority of wireless systems facilitate the task of acquiring CSI at the receiver. There is then the option of feeding some CSI back to the transmitter. Alternatively, if TDD or full duplexing is utilized and the channel is reciprocal, then every transceiver can acquire CSI while acting as receiver and subsequently exploit such CSI when acting as transmitter.

Altogether, different information-theoretic variants arise depending on the CSI. These variants are typically modeled by providing CSI as side information to the receiver and/or transmitter. By *side information* we mean that this CSI constitutes additional information granted at no cost, and to stress this point a genie is sometimes invoked as the source of the CSI.

The imprint of CSI on the capacity cannot be overstated. Not only does the value of the capacity itself vary, but especially the difficulty of its computation and the nature of the signaling strategies that achieve it can change radically. Additional CSI always increases the capacity and typically simplifies its computation and the required signaling strategies. Although, fundamentally, the only true capacity is the one without CSI, which subsumes every possible procedure whereby $\{h[n]\}$ can be inferred from the observations at the receiver, this does not render the computation of the capacity with CSI pointless. Often, while the capacity without CSI is unknown or of unwieldy computation, simple expressions can be obtained by granting CSI and these expressions can approximate very well the actual capacity and offer superior insights and guidance.

To dissect the role of CSI, and to better connect the ensuring derivations with those presented for unfaded channels, we begin by granting CSI to both transmitter and receiver. The CSI is then progressively removed, first from the transmitter and ultimately from the

Discussion 4.2 When is CSIT a reasonable assumption?

In TDD and full-duplex systems, and provided there are calibration procedures in place to compensate for potential differences between the transmit and receive circuits in the reverse and forward links (because the over-the-air portion of the channel is reciprocal, but the equipment need not be), CSIT follows from CSIR thanks to reciprocity.

In FDD systems, in contrast, the reverse- and forward-link fading coefficients are independent and thus explicit feedback is required. CSIT can then only be safely assumed when, besides being underspread (so that CSIR can be gathered with reasonable overhead), the fading exhibits a long enough coherence time (so that the feedback that converts CSIR into CSIT can also take place with reasonable overhead). This naturally leads to the notion of CSI feedback.

Reciprocity and CSI feedback are dealt with in Section 5.10, directly for MIMO.

receiver. Finally, we contemplate a specific instance where the receiver obtains channel estimates on the basis of embedded pilot symbols and then applies these estimates in lieu of ideal CSI; then, the overhead associated with the pilot symbols, as well as the nonideality of the channel estimates, come explicitly to the fore.

For notational compactness, we introduce the acronyms CSIR and CSIT to indicate, respectively, CSI at the receiver and CSI at the transmitter.

Besides the availability of CSI, the other defining aspect of fading channels in terms of reliable communication is the relationship between the codewords and the fading dynamics. In that respect, we make extensive use of two classic settings that, despite representing limiting idealizations, offer a compromise between realism and tractability and, properly applied, are invaluable sources of insight [385–387]:

- The *quasi-static setting*, where each codeword experiences essentially a single fading value (see Fig. 4.13). This setting arises whenever the codewords are compact enough relative to the fading coherences in time and frequency for the fading to remain approximately constant over each.
- The *ergodic setting*, where each codeword is exposed to sufficiently many fades (see Fig. 4.15) for the local fading distribution to be revealed thereupon and for ergodicity to apply.

4.5.1 CSIR and CSIT

Having CSIR and CSIT is tantamount to saying that $\{h[n]\}$ is a known sequence. Basically, we are then faced with a bank of parallel AWGN subchannels in the time domain and the mutual information on each is maximized when the corresponding symbol is complex Gaussian. Following the reasoning of Section 4.4, this yields

$$C(\mathsf{SNR}) = \max_{P[0],\ldots,P[N-1]} \lim_{N\to\infty} \frac{1}{N} \sum_{n=0}^{N-1} \log_2\Big(1 + \mathsf{SNR}\,|h[n]|^2\,P[n]\Big), \tag{4.131}$$

where $P[n] = \frac{\mathbb{E}\left[|x[n]|^2\right]}{E_s}$ is what the precoder reduces to in SISO, and the maximization is over all possible transmit power control policies subject to the applicable power constraint. The channel is memoryless.

Quasi-static setting

If the channel remains constant for the entire span of each codeword, $h[n] = h$ for $n = 0, \ldots, N-1$, then, regardless of the type of power constraint, the maximization in (4.131) returns $P[n] = 1$ (refer to Problem 4.37). It follows that

$$C(\mathsf{SNR}, h) = \log_2\left(1 + \mathsf{SNR}\,|h|^2\right). \tag{4.132}$$

The potential incongruity between the finite span of a fading block and an infinite codeword length ($N \to \infty$) is typically finessed by declaring each fading block—and thus each codeword—as being very long in symbol units, and indeed (recall Examples 3.26–3.28) the fading can remain roughly constant over sufficiently many symbols. The performance in this setting can be characterized by the distribution of $C(\mathsf{SNR}, h)$, viewed as a random variable induced by $|h|$ and thus subject to the vagaries of the fading, with local average

$$\mathbb{E}\left[C(\mathsf{SNR}, h)\right] = \mathbb{E}\left[\log_2\left(1 + \mathsf{SNR}\,|h|^2\right)\right]. \tag{4.133}$$

The performance over this channel is indelibly associated with link adaptation whereby, by virtue of the CSIT, the transmit bit rate is matched to the fading [388].

Example 4.21 (Link adaptation with quasi-static Rayleigh fading in LTE)

Consider a frequency-flat Rayleigh channel conforming to a block-fading structure with $N = 6480$. Available to the transmitter is the entire set of 27 LTE MCSs listed in Table 4.1. The transmit power is fixed. On each fading block, the transmitter selects, on the basis of curves such as those in Fig. 4.8, the MCS that can deliver the highest throughput given the value of $\mathsf{SNR}\,|h|^2$. The average throughput per unit bandwidth, obtained via Monte-Carlo simulation, is depicted in Fig. 4.12 as a function of SNR (in dB). Also shown, as baseline, is the capacity $C(\mathsf{SNR}, h)$ averaged over the distribution of h.

In contrast with unfaded channels, where the throughput exhibits a ladder structure, here it gets smoothed over as the fading maps each SNR to every possible value for $\mathsf{SNR}\,|h|^2$, which is what determines the MCS. At every local-average SNR, the average performance involves many—possibly all—MCSs, and their mixing changes gradually with SNR.

Despite the somewhat idealized setting of the foregoing example, sketched in Fig. 4.13, the result is actually fairly representative of what can be achieved with link adaptation assisted by only a smattering of CSIT (five bits suffice to index 27 MCSs). From Fig. 4.12, the loss caused by discrete constellations and nonideal finite-length coding with BICM in lieu of an ideal Gaussian codebook of unbounded length is within 2–3 dB. (This holds up to $\mathsf{SNR} \approx 15$ dB, beyond which the scarcity of high-spectral-efficiency MCSs is evidenced; a limitation that is not fundamental and that is addressed with the inclusion of 256-QAM.)

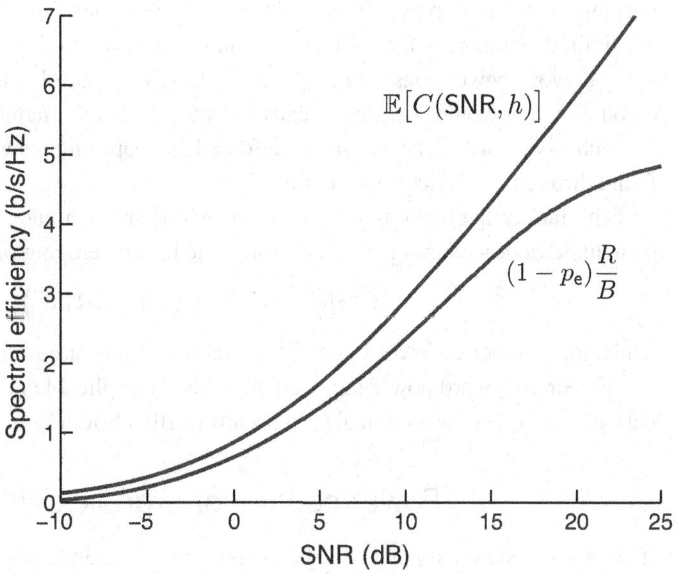

Fig. 4.12 Throughput per unit bandwidth achievable with link adaptation and the entire set of LTE MCSs in Table 4.1 as a function of SNR (in dB). Also shown, as baseline, is the local average of the channel capacity.

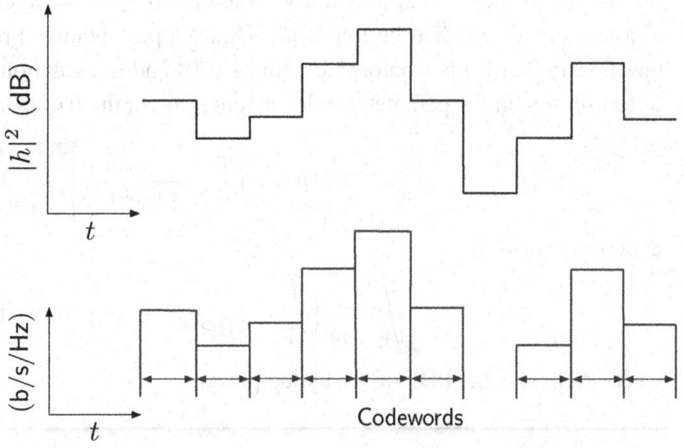

Fig. 4.13 Succession of codewords with link adaptation over a block-fading channel.

From quasi-static to ergodic settings

Let us now allow the fading to vary over each codeword, considering a block-fading structure for the sake of specificity but with the ideas applying equally to continuous fading.

Situations where each codeword spans a plurality of fading blocks, but short of what is needed for ergodicity to take hold, can be tackled through the notion of *delay-limited capacity* [389]. Under a per-symbol power constraint, there is no alternative better than keeping the transmit power fixed and the capacity is then the time-average of (4.132) over the limited number of fading blocks spanned by each codeword. Alternatively, under a per-codeword power constraint, time-domain power control could be applied yet, without ergodicity, that would require anticausal knowledge of the fading over the blocks spanned by each codeword. With causality enforced, the optimum power control policy can be found through dynamic programming [390].

If the fading fluctuations over each codeword are rich enough for ergodicity to apply, then, under a per-symbol power constraint the time-averaging of (4.132) gives

$$C(\mathsf{SNR}) = \mathbb{E}\left[\log_2(1 + \mathsf{SNR}\,|h|^2)\right] \tag{4.134}$$

while, under a per-codeword power constraint, the power control simplifies markedly. Although per-codeword power constraints are rarely applicable in frequency-flat faded channels, the ensuing power control policies are worth a brief digression.

Power control for ergodic settings

If $\{h[n]\}$ is a stationary and ergodic process, or a sequence of blocks whose fading values are ergodic, then, as $N \to \infty$, (4.131) converges almost surely (a.s., see Appendix C.1.10) to

$$C(\mathsf{SNR}) = \max_{P(|h|)} \mathbb{E}\left[\log_2\left(1 + \mathsf{SNR}\,|h|^2\, P(|h|)\right)\right], \tag{4.135}$$

where h is the fading at any arbitrary instant and $P(|h|)$ signifies the power control policy as a function of the distribution of $|h|$. Under a per-codeword power constraint, the optimization in (4.135) is isomorphic with (4.100) and the solution is therefore time-domain waterfilling with the parameter η dependent only on the fading statistics. Specifically,

$$P^\star(h) = \left[\frac{1}{\eta} - \frac{1}{\mathsf{SNR}\,|h|^2}\right]^+ \tag{4.136}$$

with η the solution to

$$\int_{\eta/\mathsf{SNR}}^{\infty} \left(\frac{1}{\eta} - \frac{1}{\mathsf{SNR}\,\xi}\right) f_{|h|^2}(\xi)\, d\xi = 1 \tag{4.137}$$

where $f_{|h|^2}(\cdot)$ is the PDF of the fading power.

Example 4.22

For the special case of Rayleigh fading, express (4.137) in closed form.

Solution

Using $f_{|h|^2}(\xi) = e^{-\xi}$, (4.137) can be written as [391]

$$\frac{e^{-\eta/\mathsf{SNR}}}{\eta} - \frac{\mathcal{E}_1(\eta/\mathsf{SNR})}{\mathsf{SNR}} = 1, \tag{4.138}$$

where $\mathcal{E}_1(\cdot)$ is an exponential integral (see Appendix E.3).

With waterfilling power control, more power is transmitted when the fading is strong and less, or even none, when it is weak. The ensuing improvement can be quantified through the reduction in $\frac{E_b}{N_0}_{\min}$ since, with channel-dependent power control, the minimum value of

$$\frac{E_b}{N_0} = \frac{\mathsf{SNR}}{C(\mathsf{SNR})} \tag{4.139}$$

still corresponds to $\mathsf{SNR} \to 0$, but that minimum is no longer -1.59 dB. Rather, $\frac{E_b}{N_0}_{\min}$ can take a (much) smaller value once x depends on h, directly reflecting an increment in total channel gain: rather than G, the total channel gain then equals the more general expression in (4.1) and the low-SNR capacity is proportionally higher. Operationally, this increase in channel gain and the corresponding reduction in $\frac{E_b}{N_0}_{\min}$ descend from the fact that fading creates peaks and nulls and, with CSIT, the transmissions can take place opportunistically only when the channel is around its peaks [392]. For Rayleigh fading, the relative improvement obtained through time-domain waterfilling is theoretically unlimited (at the expense of the equally unlimited delay that ensues from having to wait for ever higher peaks), but in actuality this is merely an artifact of the unbounded support of the Rayleigh distribution; actual fading distributions have a bounded support [194]. (The Rayleigh distribution is a legitimate and well-tested model for the fading magnitude in most respects, and in particular for the fading nulls, but not for the peaks.)

Power control policies other than waterfilling can be applied, e.g., channel inversion or truncated inversion [51, chapter 4][393]. These alternate policies, and in particular channel inversion, which is the opposite of waterfilling, always yield a lower spectral efficiency, albeit sometimes in exchange for simpler encoding and decoding.

If the codebook is not complex Gaussian, then, rather than waterfilling, the optimum power control policy is time-domain mercury/waterfilling [71, section IX].

4.5.2 No CSIT

With CSIR but no CSIT, the transmit power and bit rate must be oblivious to the fading and thus the type of power constraint becomes irrelevant. As far as the transmitter is concerned, the sequence $\{h[n]\}$ is unknown. The receiver, however, is privy to $\{h[n]\}$, a knowledge that should be made manifest by conditioning on the fading when expressing the channel law and computing the mutual information. Moreover, because of this knowledge, the channel is still memoryless: $y[n]$ does not depend on transmit symbols other than $x[n]$, because any information they could help provide about the fading is redundant to the receiver and the noise is IID. The single-letter channel law is

$$f_{y|h,s}(\mathrm{y}|\mathrm{h},\mathrm{s}) = \frac{1}{\pi N_0} e^{-\frac{|\mathrm{y}-\sqrt{GE_s}\mathrm{hs}|^2}{N_0}}. \tag{4.140}$$

Quasi-static setting

In this setting, the fading is unknown to the transmitter but constant over an entire codeword, i.e., $h[n] = h$ for $n = 0, \ldots, N-1$. The value of h does change from code-

Discussion 4.3 When is CSIR a reasonable assumption?

As seen later in the chapter, the coherence of the underspread channels of interest is more than sufficient to validate the assumption of CSIR side information in all but very extreme SNR conditions. This is especially inviting when one considers the enormous analytical simplicity that CSIR elicits. Moreover, pilot-based suboptimum schemes can approach such CSIR capacity with small overheads and penalties—which can be explicitly accounted for, to refine the results.

The above observation is made for the SU-SISO channels considered in this chapter. While it continues to hold for single-user MIMO (SU-MIMO) setups with moderate numbers of antennas, and for multiuser setups with moderate numbers of users, it may cease to hold for massive MIMO.

Taking some liberties, an analogy could be made with the notions of relativistic and classical physics: strictly speaking only the former describe nature, but in most everyday situations the latter provide excellent approximations that are far easier to obtain. Likewise, strictly speaking the capacity should be computed without side information, but in a wide range of conditions the assumption of CSIR leads to good approximations that are far easier to obtain. Just like classical physics, though, this assumption does break down beyond certain points.

word to codeword. Conditioned on h, we are faced with an AWGN channel whose SNR is $\mathsf{SNR}\,|h|^2$. The spectral efficiency that could be supported is therefore $\log_2(1 + \mathsf{SNR}\,|h|^2)$, which we can regard as a random variable with a distribution induced by that of $|h|$. As argued earlier, it is entirely reasonable to consider that the fading coherence be large enough to allow for the asymptotic behavior in N to be approached within. However, without CSIT the transmitter cannot align the codewords with the fading cycles. Consequently, the frequency-flat quasi-static setting is somewhat problematic, yet it is worth examining what unfolds when the model is taken at face value.

Unlike in the CSIT case, here the transmitter cannot match its bit rate to what the channel can support. Rather, the transmitter must blindly select a bit rate R paving the way for two distinct outcomes:

- If $\log_2(1 + \mathsf{SNR}\,|h|^2) \geq R/B$, the chosen bit rate is supported and the transmission can succeed with at most a small error probability because of the finiteness of N.
- If $\log_2(1 + \mathsf{SNR}\,|h|^2) < R/B$, the chosen bit rate is not supported and the error probability is bounded away from zero.

The contrast with the CSIT case is stark: with CSIT, the capacity of a quasi-static setting is $\log_2(1 + \mathsf{SNR}\,|h|^2)$; without CSIT, it is the minimum value of such quantity, the only amount whose reliable delivery can be guaranteed. For any fading distribution that can fade completely, e.g., Rayleigh, such amount is nil and thus the capacity in the Shannon sense is zero.

When the bit rate R is not supported, the transmission is said to be in *outage* and the outage probability is

$$p_{\text{out}}(\text{SNR}, R/B) = \mathbb{P}\Big[\log_2\big(1 + \text{SNR}\,|h|^2\big) < R/B\Big] \tag{4.141}$$

and the transmitter can make use of the fading distribution in order to select the bit rate R that ensures a desired outage probability.

For large enough N, the errors are dominated by the outage events and we can identify p_{out} with the error probability, i.e., codewords are decoded correctly whenever the transmission is not in outage but not otherwise [394]. More precisely, the error probability decays exponentially with N when not in outage, and not otherwise.

To facilitate further analysis, it is interesting to rearrange (4.141) into

$$p_{\text{out}}(\text{SNR}, R/B) = \mathbb{P}\left[|h|^2 < \frac{2^{R/B} - 1}{\text{SNR}}\right] \tag{4.142}$$

$$= F_{|h|^2}\left(\frac{2^{R/B} - 1}{\text{SNR}}\right), \tag{4.143}$$

where $F_{|h|^2}(\cdot)$ is the cumulative distribution function (CDF) of the fading power.

Example 4.23

Express the outage probability in Rayleigh fading and specialize it to the high-SNR regime.

Solution

In Rayleigh fading, $F_{|h|^2}(\xi) = 1 - e^{-\xi}$ and thus

$$p_{\text{out}}(\text{SNR}, R/B) = 1 - \exp\left(\frac{1 - 2^{R/B}}{\text{SNR}}\right), \tag{4.144}$$

which, for large SNR, behaves as

$$p_{\text{out}}(\text{SNR}, R/B) = \frac{2^{R/B} - 1}{\text{SNR}} + \mathcal{O}\left(\frac{1}{\text{SNR}^2}\right). \tag{4.145}$$

Although, being zero, the Shannon capacity is not an operationally meaningful quantity in this setting, it is often desirable to express the performance in terms of spectral efficiency rather than outage probability. The notion of *outage capacity* serves this purpose [385]. Denoted by $C_\epsilon(\text{SNR})$, it designates the highest spectral efficiency such that $p_{\text{out}} < \epsilon$, i.e.,

$$C_\epsilon(\text{SNR}) = \max_{\text{c}}(\text{c} : p_{\text{out}}(\text{SNR}, \text{c}) < \epsilon) \tag{4.146}$$

and, from (4.143),

$$C_\epsilon(\text{SNR}) = \log_2\Big(1 + \text{SNR}\, F^{-1}_{|h|^2}(\epsilon)\Big). \tag{4.147}$$

Example 4.24

Express the outage capacity in Rayleigh fading.

Solution

In Rayleigh fading, $F^{-1}_{|h|^2}(\epsilon) = -\log_e(1-\epsilon)$ and thus

$$C_\epsilon(\mathsf{SNR}) = \log_2\Big(1 - \mathsf{SNR}\log_e(1-\epsilon)\Big). \tag{4.148}$$

In the low-SNR regime, (4.147) expands as

$$C_\epsilon(\mathsf{SNR}) = F^{-1}_{|h|^2}(\epsilon)\,\mathsf{SNR}\log_2 e + \mathcal{O}\left(\frac{1}{\mathsf{SNR}}\right), \tag{4.149}$$

where the first-order term equals its counterpart without fading, given in (4.68), but scaled by $F^{-1}_{|h|^2}(\epsilon)$. Typically one wants to operate at small ϵ, which entails a small $F^{-1}_{|h|^2}(\epsilon)$ and, consequently, an outage capacity that is only a small share of the unfaded AWGN capacity.

Example 4.25

Consider a quasi-static Rayleigh-faded channel without CSIT at $\mathsf{SNR} = 0$ dB. Compute, both via (4.149) and exactly, the share of AWGN capacity achievable for $\epsilon = 10^{-2}$.

Solution

For $\epsilon = 10^{-2}$, (4.149) indicates that only 1% of the AWGN capacity can be achieved. The exact share, from Example 4.24, is 1.4%.

In the high-SNR regime, alternatively, (4.147) expands as

$$C_\epsilon(\mathsf{SNR}) = \log_2 \mathsf{SNR} - \log_2 \frac{1}{F^{-1}_{|h|^2}(\epsilon)} + \mathcal{O}\left(\frac{1}{\mathsf{SNR}}\right), \tag{4.150}$$

which exhibits, relative to an AWGN channel, an approximate SNR fading loss (in 3-dB units) of

$$\log_2 \frac{1}{F^{-1}_{|h|^2}(\epsilon)} \tag{4.151}$$

and, for small ϵ, this loss is large.

Example 4.26

Compute the SNR fading loss in Rayleigh fading at $\epsilon = 10^{-2}$.

Solution

Applying (4.151), the loss is seen to equal approximately 6.64 (in 3-dB units), just shy of 20 dB.

Albeit introduced here in the context of a quasi-static setting, the notion of outage capacity applies to any setting that is not information stable, e.g., where each codeword spans a plural but small number of fading blocks.

If the signal is not complex Gaussian, the definition of outage probability in (4.141) can be generalized to

$$p_{\text{out}}(\text{SNR}, R/B) = \mathbb{P}\Big[\mathcal{I}(\text{SNR}\,|h|^2) < R/B\Big], \tag{4.152}$$

where $\mathcal{I}(\cdot)$ is the Gaussian mutual information function of the corresponding signal distribution. The outage spectral efficiency has a form similar to (4.146).

The diversity–multiplexing tradeoff

From (4.145), if the spectral efficiency R/B is held constant then the outage probability in Rayleigh fading decays as $1/\text{SNR}$ at high SNR. Alternatively, from (4.150), if the spectral efficiency scales with $\log_2 \text{SNR}$ then the outage probability remains constant at high SNR. As it turns out, these two situations are only end-points of a more general tradeoff between spectral efficiency and outage probability, a tradeoff governed by how the spectral efficiency is controlled as a function of the SNR. (Note that this control cannot respond to the fading, since there is no CSIT, but only to the local-average SNR.) For $\text{SNR} \to \infty$, proxy quantities related to both the spectral efficiency and the outage probability can be defined and the tradeoff can be more conveniently established between them. These quantities are the *diversity* order (as proxy for the outage probability) and the *multiplexing gain* (as proxy for the spectral efficiency), and the tradeoff between them is the diversity–multiplexing tradeoff (DMT) [395]. Making explicit the dependence between R/B and SNR, the diversity order is

$$d = -\lim_{\text{SNR}\to\infty} \frac{\log p_{\text{out}}(\text{SNR}, R(\text{SNR})/B)}{\log \text{SNR}}, \tag{4.153}$$

whereas the multiplexing gain is

$$r = \lim_{\text{SNR}\to\infty} \frac{R(\text{SNR})/B}{\log_2 \text{SNR}} \tag{4.154}$$

with the justification for the term *gain* to be found in MIMO, where r can exceed 1. Altogether, the diversity order quantifies the asymptotic slope of the outage-versus-SNR relationship, in log–log scale, while the multiplexing gain quantifies the asymptotic slope of the spectral efficiency-versus-SNR relationship, in log scale.

From (4.154), we can write

$$\frac{R}{B} = r \log_2 \text{SNR} + \mathcal{L} + \mathcal{O}\left(\frac{1}{\text{SNR}}\right), \tag{4.155}$$

where $r \in [0,1]$ while $\mathcal{L}$ does not depend on SNR. As seen next, for our purposes here the value of this term is irrelevant and thus we can also drop the remainder $\mathcal{O}(1/\text{SNR})$. Applying the definition of outage probability,

$$p_{\text{out}} = \mathbb{P}\Big[\log_2(1 + \text{SNR}\,|h|^2) < r\log_2 \text{SNR} + \mathcal{L}\Big] \tag{4.156}$$

$$= \mathbb{P}\left[|h|^2 < \frac{2^{\mathcal{L}}\,\text{SNR}^r - 1}{\text{SNR}}\right] \tag{4.157}$$

$$= 1 - \exp\left(\frac{1 - 2^{\mathcal{L}}\,\mathsf{SNR}^{r}}{\mathsf{SNR}}\right) \tag{4.158}$$

$$= \frac{2^{\mathcal{L}}}{\mathsf{SNR}^{1-r}} + \mathcal{O}\left(\frac{1}{\mathsf{SNR}}\right) \tag{4.159}$$

from which (4.153) gives the DMT for the SISO Rayleigh-faded channel considered in these derivations, namely

$$d = (1 - r). \tag{4.160}$$

As long as r and d abide by the DMT, at high SNR a 3-dB power increment can simultaneously yield r additional b/s/Hz and an outage probability scaling by 2^{-d}.

The DMT for Rayleigh fading applies verbatim to Rice fading [396], can be extended to other fading types [397], and acquires further relevance with MIMO (see Section 5.4).

While it is an elegant and supremely simple formulation that allows skirting analytical hurdles associated with the exact computation of the outage capacity, the DMT also suffers from a number of weaknesses [398].

- Both d and r are asymptotic notions whose definitions entail letting $\mathsf{SNR} \to \infty$. This restricts the validity to the high-SNR regime. Nonasymptotic DMT formulations have been developed for certain transmission strategies [399], but they lack the generality and simplicity that make this framework enticing in the first place.
- d and r quantify only the asymptotic slope of outage and spectral efficiency, respectively, as a function of SNR. Arbitrarily large differences in the SNR required for a given outage probability may exist for identical diversity orders (see Fig. 4.14). These differences are simply neglected when evaluating d. Likewise, arbitrarily large disparities in the SNR required for a given spectral efficiency are lost in r (see Fig. 4.14). Only for $\mathsf{SNR} \to \infty$ can superiority be guaranteed on the basis of a better diversity/multiplexing. The DMT, in synthesis, expresses the tradeoff between the derivatives of the quantities of interest rather than between those quantities themselves.

Because of these issues, d and r provide only a coarse description of the tradeoff between spectral efficiency and outage probability [50, section 9.1.2] and hence care must be exercised when utilizing these notions to establish comparisons between schemes or to determine absolute performance standards [400].

Ergodic setting

A cartoon representation of the ergodic setting without CSIT is provided in Fig. 4.15, with each codeword spanning multiple fades in the time domain. Although, as is to become clear, this setting arises in more practically relevant situations than this one, the essence of a codeword experiencing multiple fading values is well captured by the cartoon.

It can be reasoned that, in an ergodic setting without CSIT, the capacity should be no different from the one derived with CSIT under a per-symbol power constraint, which precludes time-domain power control. Indeed, the maximization of the sequence mutual

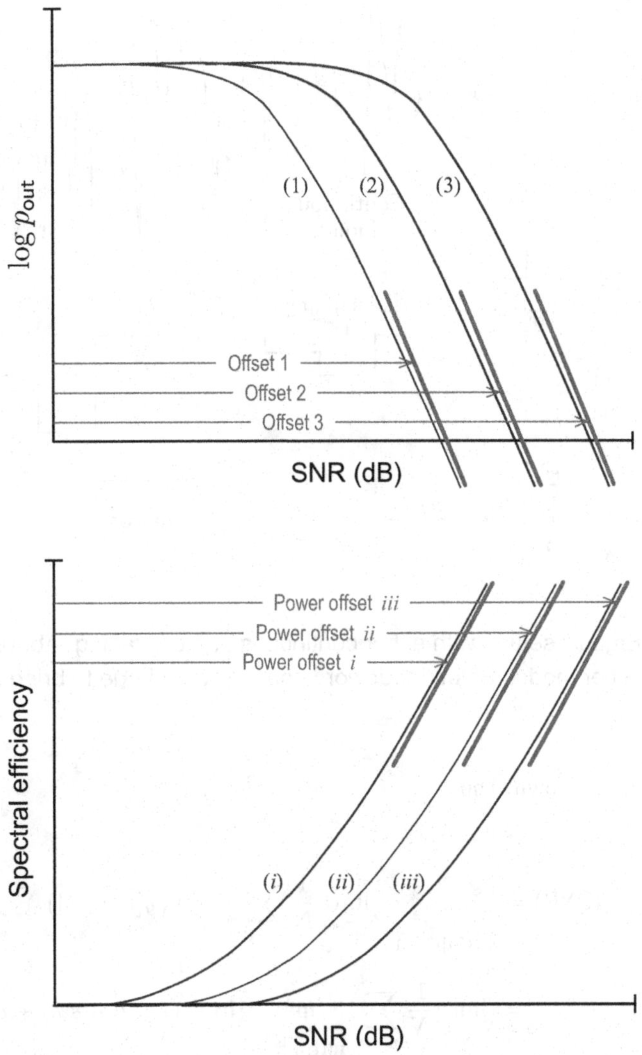

Fig. 4.14 Above, outage probability (in log scale) versus SNR (in dB) for three channels with identical diversity order. Below, spectral efficiency versus SNR (in dB) for three channels with identical multiplexing gain.

information under a per-symbol power constraint gives

$$\max_{\substack{f_{s[0]\cdots s[N-1]}:\\ \mathbb{E}[|s[n]|^2]=1}} \lim_{N\to\infty} \frac{1}{N} I\Big(s[0],\ldots,s[N-1];y[0],\ldots,y[N-1]\,|h[0],\ldots,h[N-1]\Big), \tag{4.161}$$

which becomes single-letter because, conditioned on $h[0],\ldots,h[N-1]$, the channel is rendered memoryless and the mutual information between the transmit and receive sequences

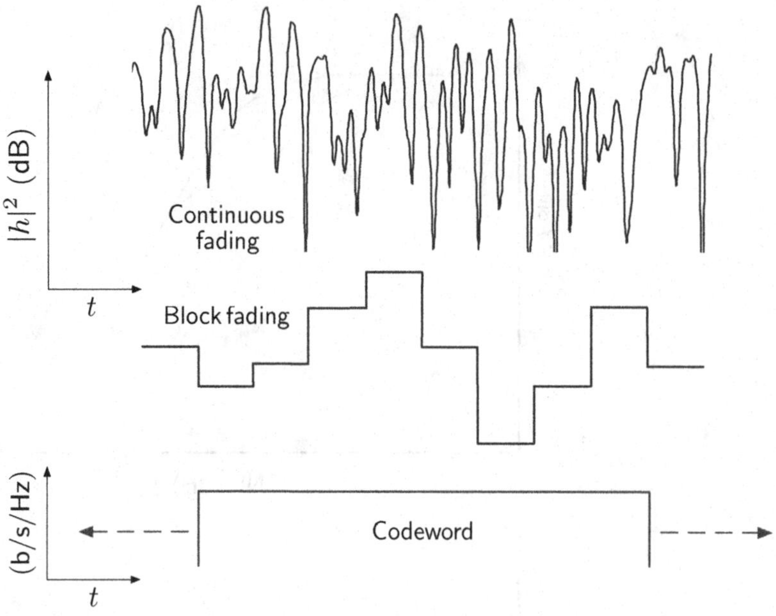

Fig. 4.15 Ergodic setting with either continuous or block fading. Above, fading realizations of either model. Below, codeword span growing large to bring about the ergodicity.

breaks down. Thus,

$$C(\mathsf{SNR}) = \max_{\substack{f_{s[n]}:\\ \mathbb{E}[|s[n]|^2]=1}} \lim_{N\to\infty} \frac{1}{N} \sum_{n=0}^{N-1} I\big(s[n]; y[n]\,|h[n]\big) \tag{4.162}$$

$$= \lim_{N\to\infty} \frac{1}{N} \sum_{n=0}^{N-1} \max_{\substack{f_{s[n]}:\\ \mathbb{E}[|s[n]|^2]=1}} \mathfrak{h}\Big(\sqrt{GE_{\mathrm{s}}}\, h[n]s[n] + v[n]\,|h[n]\Big) - \log_2(\pi e N_0)$$

$$= \lim_{N\to\infty} \frac{1}{N} \sum_{n=0}^{N-1} \log_2\Big(\pi e\big(GE_{\mathrm{s}}|h[n]|^2 + N_0\big)\Big) - \log_2(\pi e N_0) \tag{4.163}$$

$$= \lim_{N\to\infty} \frac{1}{N} \sum_{n=0}^{N-1} \log_2\Big(1 + \mathsf{SNR}\,|h[n]|^2\Big) \tag{4.164}$$

$$\overset{\text{a.s.}}{=} \mathbb{E}\Big[\log_2\Big(1 + \mathsf{SNR}\,|h|^2\Big)\Big], \tag{4.165}$$

where the maximization of the differential entropy of $\sqrt{GE_{\mathrm{s}}}h[n]s[n] + v[n]$, conditioned on $h[n]$ and with $\mathbb{E}[|s[n]|^2] = 1$, readily leads to $s[n] \sim \mathcal{N}_{\mathbb{C}}(0,1)$. Finally, the summation in (4.164) converges a.s. to the expectation in (4.165) by virtue of the strong law of large numbers.

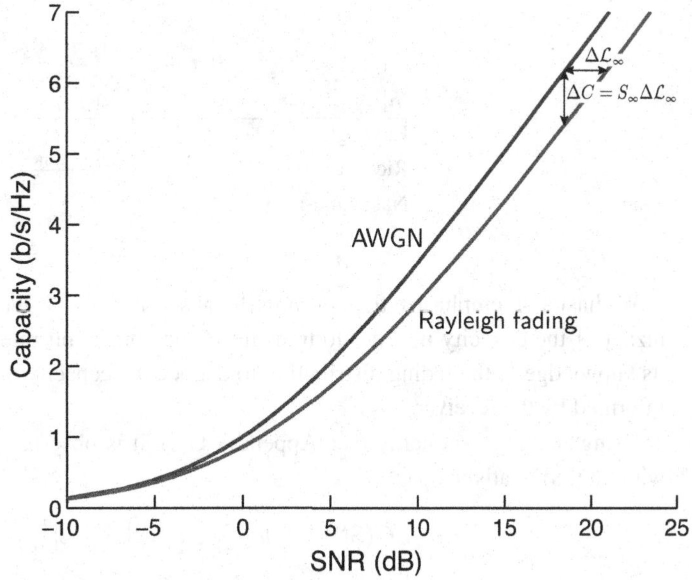

Fig. 4.16 Capacity as a function of SNR (in dB), with and without Rayleigh fading. Also indicated are the high-SNR differences in terms of both power offset and capacity.

Example 4.27 (Capacity versus SNR of a SISO Rayleigh-faded channel)

With Rayleigh fading, (4.165) yields the closed form [385, 401]

$$C(\mathsf{SNR}) = \int_0^\infty \log_2(1+\mathsf{SNR}\,\xi)\,e^{-\xi}\,d\xi \tag{4.166}$$

$$= e^{1/\mathsf{SNR}}\,\mathcal{E}_1\left(\frac{1}{\mathsf{SNR}}\right)\log_2 e, \tag{4.167}$$

where we have applied (C.40). This highly relevant $C(\mathsf{SNR})$ is depicted in Fig. 4.16.

Example 4.28 (Capacity versus SNR of a SISO Nakagami-faded channel)

With Nakagami-m fading, whose distribution is given in (3.26),

$$C(\mathsf{SNR}) = \int_0^\infty \log_2(1+\mathsf{SNR}\,\xi)\,\frac{\mathsf{m}^{\mathsf{m}}\,\xi^{\mathsf{m}-1}}{\Gamma(\mathsf{m})}\,e^{-\mathsf{m}\xi}\,d\xi \tag{4.168}$$

$$= e^{\mathsf{m}/\mathsf{SNR}}\sum_{q=0}^{\mathsf{m}-1}\left(\frac{\mathsf{m}}{\mathsf{SNR}}\right)^q\Gamma\left(-q,\frac{\mathsf{m}}{\mathsf{SNR}}\right)\log_2 e, \tag{4.169}$$

where $\Gamma(\cdot)$ and $\Gamma(\cdot,\cdot)$ are, respectively, the complete and incomplete Gamma functions (see Appendix E), and where (4.169) holds for integer m [391, 402].

For m = 1, the Nakagami fading reverts to Rayleigh fading and thus (4.169) should equal (4.167); that is indeed the case because, as indicated in (E.14), $\Gamma(0,z) = \mathcal{E}_1(z)$.

Table 4.3 Kurtosis of common fading distributions

$\lvert h \rvert$	$\kappa(\lvert h \rvert)$
Rayleigh	2
Rice (factor $\mathsf{K} \geq 0$)	$2 - \frac{4\mathsf{K}^2}{(1+2\mathsf{K})^2}$
Nakagami-m ($\mathsf{m} \geq \frac{1}{2}$)	$1 + \frac{1}{\mathsf{m}}$

We hasten to emphasize that, even in the absence of CSIT, the transmitter must be cognizant of the capacity in order to transmit at the correct bit rate. The transmitter can use its knowledge of the fading distribution to deduce the capacity on its own, or else it can be informed by the receiver.

Using Jensen's inequality (see Appendix G.6), it is observed that the ergodic capacity without CSIT satisfies

$$C(\mathsf{SNR}) \leq \log_2\left(1 + \mathsf{SNR}\,\mathbb{E}\left[|h|^2\right]\right) \tag{4.170}$$

$$= \log_2(1 + \mathsf{SNR}), \tag{4.171}$$

manifesting that fading can only diminish the CSIR-only capacity. This is in stark contrast with the CSIT case, where fading could be exploited via power control to increase the capacity.

If the signal conforms to a discrete constellation, the ergodic spectral efficiency cannot generally be expressed in closed form. For BPSK or QPSK in Rayleigh fading, an infinite series form is available [403].

Low-SNR regime

The low-SNR behavior can be characterized, for an arbitrary fading distribution, by expanding (4.165) into

$$C(\mathsf{SNR}) = \left(\mathsf{SNR} - \frac{1}{2}\,\mathbb{E}[|h|^4]\,\mathsf{SNR}^2\right)\log_2 e + o(\mathsf{SNR}^2) \tag{4.172}$$

$$= \left(\mathsf{SNR} - \frac{1}{2}\,\kappa(|h|)\,\mathsf{SNR}^2\right)\log_2 e + o(\mathsf{SNR}^2), \tag{4.173}$$

where we have taken advantage of the unit variance of h to directly express the fourth-order moment of $|h|$ as its kurtosis. Contrasting the above with its AWGN-channel counterpart in (4.68), fading is seen to be immaterial to first order and thus $\frac{E_b}{N_0}_{\min} = -1.59$ dB as in an unfaded channel. Only the second-order term reflects the impact of fading and, from (4.39) and (4.173),

$$S_0 = \frac{2}{\kappa(|h|)}, \tag{4.174}$$

with the kurtosis of the most common fading distributions listed in Table 4.3.

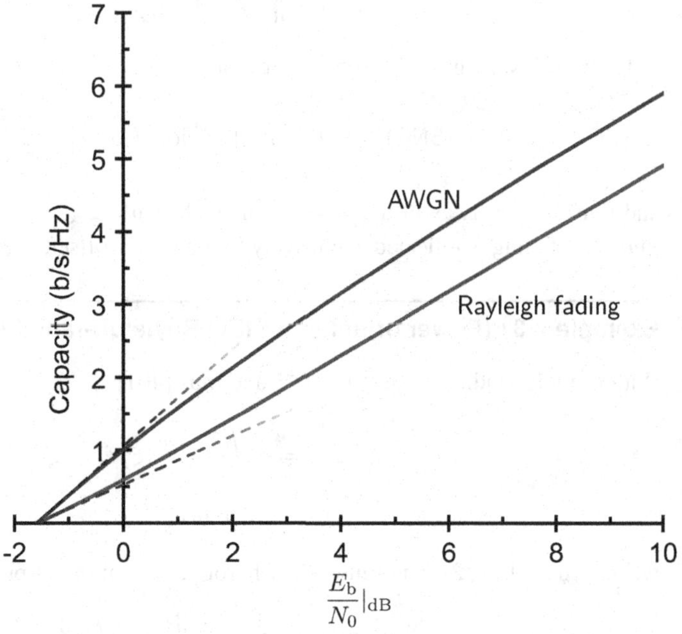

Fig. 4.17 Capacity as a function of $\frac{E_b}{N_0}$ (in dB), with and without Rayleigh fading. Also shown, in dashed, are the respective low-SNR expansions.

Example 4.29 (Capacity versus $\frac{E_b}{N_0}$ of a SISO Rayleigh-faded channel)

The function $\mathsf{C}(\frac{E_b}{N_0})$ and its low-SNR expansion in Rayleigh fading are displayed in Fig. 4.17 alongside their counterparts without fading.

Example 4.30

How much more bandwidth is required in Rayleigh fading, relative to an unfaded channel, to achieve a certain bit rate with a given power in the low-SNR regime?

Solution

In Rayleigh fading, $\kappa(|h|) = 2$ and thus $S_0 = 1$ while, without fading, $S_0 = 2$. Applying (4.42) we see that twice as much bandwidth is required in the former.

If the signal distribution is not complex Gaussian, but it is proper complex, (4.173) is upheld and thus $\frac{E_b}{N_0}_{\min} = -1.59$ dB and $S_0 = 2/\kappa(|h|)$ are preserved. In contrast, for BPSK, which is not proper complex (refer to Problem 4.53)

$$C(\mathsf{SNR}) = \Big(\mathsf{SNR} - \kappa(|h|)\, \mathsf{SNR}^2\Big) \log_2 e + o(\mathsf{SNR}^2), \tag{4.175}$$

implying that $\frac{E_b}{N_0}_{\min} = -1.59$ dB, but $S_0 = 1/\kappa(|h|)$. As in the AWGN channel, the suboptimality of BPSK in the low-SNR regime is reflected in a reduction of S_0 by half.

High-SNR regime

In the high-SNR regime, (4.165) expands as

$$C(\mathsf{SNR}) = \log_2 \mathsf{SNR} + \mathbb{E}\big[\log_2\big(|h|^2\big)\big] + \mathcal{O}\left(\frac{1}{\mathsf{SNR}}\right) \tag{4.176}$$

and thus $S_\infty = 1$, as in the AWGN channel, while $\mathcal{L}_\infty = -\mathbb{E}\big[\log_2\big(|h|^2\big)\big] > 0$. The impact of fading is reflected exclusively in the power offset, $\mathcal{L}_\infty$.

Example 4.31 (Power offset of a SISO Rayleigh-faded channel)

With Rayleigh fading, $f_{|h|^2}(\xi) = e^{-\xi}$ and thus [401]

$$\mathcal{L}_\infty = -\int_0^\infty \log_2(\xi)\, e^{-\xi}\, \mathrm{d}\xi \tag{4.177}$$

$$= \gamma_{\mathsf{EM}} \log_2 e, \tag{4.178}$$

where $\gamma_{\mathsf{EM}} \approx 0.5772$ is the Euler–Mascheroni constant (see Appendix E.2). Hence,

$$C(\mathsf{SNR}) = \log_2 \frac{\mathsf{SNR}}{e^{\gamma_{\mathsf{EM}}}} + \mathcal{O}\left(\frac{1}{\mathsf{SNR}}\right), \tag{4.179}$$

indicating that Rayleigh fading causes a power-offset penalty of $\mathcal{L}_\infty = \gamma_{\mathsf{EM}} \log_2 e$ in 3-dB units, roughly 2.5 dB.

Example 4.32 (Power offset of a SISO Rice-faded channel)

With Rice fading of factor K [404]

$$\mathcal{L}_\infty = \log_2 \frac{\mathsf{K}+1}{\mathsf{K}} - \mathcal{E}_1(\mathsf{K}) \log_2 e \tag{4.180}$$

which satisfies $\mathcal{L}_\infty \in [0, \gamma_{\mathsf{EM}} \log_2 e]$. Therefore, Rice fading causes a power-offset penalty ranging between 0 and about 2.5 dB.

Example 4.33 (Power offset of a SISO Nakagami-faded channel)

From the corresponding ergodic capacity in Example 4.28, for Nakagami-m fading with integer m the power offset equals

$$\mathcal{L}_\infty = \log_2 \mathsf{m} + \left(\gamma_{\mathsf{EM}} - \sum_{q=1}^{\mathsf{m}-1} \frac{1}{q}\right) \log_2 e. \tag{4.181}$$

It can be verified that, as in Rice fading, this power offset satisfies $\mathcal{L}_\infty \in [0, \gamma_{\mathsf{EM}} \log_2 e]$.

At high SNR, the ergodic capacity is only modestly eroded by fading, as appreciated in Fig. 4.16. Except for the power-offset penalty, the tradeoff between power and bit rate with fixed bandwidth remains largely as it is without fading.

Discrete constellations

With discrete signal constellations, the spectral efficiency of a sufficiently dense M-QAM constellation hugs the function $C(\mathsf{SNR})$, within approximately 1.53 dB, up to the vicinity of $\log_2 M$. For arbitrary SNR levels, computing the ergodic spectral efficiency achievable with a specific discrete constellation entails expecting the expressions in Examples 1.10 or 1.12 over the fading distribution, something that must generally be done numerically.

4.5.3 No CSI

In the absence of CSIR, the capacity of a fading channel is generally unknown. In some simple cases it can be computed numerically, but most of the available understanding stems from expressions obtained in different asymptotic regimes. In what follows we adhere to the ergodic setting.

A hurdle that arises in the absence of CSIR is that the channel is generally not memoryless, even when the fading is frequency-flat, because symbols other than $x[n]$ can furnish the receiver with information about the fading at time n, and hence about the distribution of $y[n]$. The capacity-achieving signal distribution is generally not IID complex Gaussian. In fact, the codewords $s[0], \ldots, s[N-1]$ should, in general, not even be composed of independent symbols. Rather, their optimum structure descends from that of the fading, which greatly complicates things. The derivation of (4.165) is not upheld once the mutual information cannot be conditioned on $h[0], \ldots, h[N-1]$ and we are therefore stuck with an optimization problem involving sequences, which does not generally break down into single-letter optimizations.

Intuition says that, in underspread channels, the capacity-achieving distribution should not differ substantially from an IID complex Gaussian sequence, and this is indeed the case. In fast-fading, however, it is radically different. This difference is best illustrated by probing the extreme case of temporally IID fading, which, despite its minimal practical relevance, is illuminating.[2] In this case, the channel does become memoryless, as any information about the fading at a certain symbol becomes useless to other symbols, and the maximization of the mutual information between the transmit and receive sequences does reduce to a single-letter optimization; the optimum signal distribution that ensues is known to be discrete with a finite number of mass points—one of them always at the origin—whose precise number and location depend on the SNR [63, 405, 406]. This signal structure lessens the uncertainty at the receiver in a gradual manner depending on the SNR. At low SNR, two mass points suffice to achieve capacity and the optimum signaling conforms to the "on–off" keying in (1.3). One of the mass points is at zero magnitude while the other one has vanishing probability and diverging magnitude as $\mathsf{SNR} \to 0$, rendering the signal increasingly peaky as the SNR shrinks. In contrast with (4.173), the low-SNR

[2] This extreme case cannot be interpreted as a very fast-fading channel since, for the sampled fading to be IID, the continuous fading would have to be so fast as to introduce considerable distortion over each symbol, i.e., the channel would not be underspread. The discretization process assumed throughout this text would then have to be modified [63]. Rather, the extreme case of IID fading must be taken as a theoretical exercise that helps understand the fundamentals.

capacity then behaves as [407]

$$C(\mathsf{SNR}) \approx \left[\mathsf{SNR} - \frac{\mathsf{SNR}}{\log_e(1/\mathsf{SNR})}\right] \log_2 e, \tag{4.182}$$

from which, applying (4.30) and (4.39), it follows that $\frac{E_b}{N_0}_{\min} = -1.59$ dB holds but with $S_0 = 0$ (see Fig. 4.18). Conversely, for growing SNR, the peakedness abates and additional mass points appear. For $\mathsf{SNR} \to \infty$, a curious behavior is observed, namely [408]

$$C(\mathsf{SNR}) = \log_2 \log \mathsf{SNR} + \mathcal{O}(1), \tag{4.183}$$

where the base of the inner logarithm is immaterial as it merely affects the term $\mathcal{O}(1)$. This peculiar expression indicates that additional transmit power becomes essentially wasted once the high-SNR regime has been entered.

Moving beyond the IID channel, consider a frequency-flat block-fading channel where codewords span multiple N_c-symbol coherence blocks. This channel is no longer memoryless because the same unknown fading affects multiple symbols and, as mentioned, information about the fading inflicting a symbol can be gathered from the observation at other symbols; thus, a nonsingle-letter formulation is required. As it turns out, the codeword distribution that achieves capacity has the following structure over each coherence block [409]:

$$s[n] = A_s \sqrt{N_c}\, u_s[n] \qquad n = 0, \ldots, N_c - 1, \tag{4.184}$$

where A_s is a positive random magnitude with $\mathbb{E}\left[A_s^2\right] = 1$ and $\bar{\boldsymbol{u}}_s = \left[u_s[0] \cdots u_s[N_c - 1]\right]$ is an isotropically distributed unit vector, i.e., a vector that is equally likely to point in any direction; such a vector can be conveniently generated as $\bar{\boldsymbol{u}}_s = \bar{\boldsymbol{z}}/\|\bar{\boldsymbol{z}}\|$ with $\bar{\boldsymbol{z}}$ having IID entries, $z[n] \sim \mathcal{N}_{\mathbb{C}}(0, 1)$. Perhaps not obviously, the entries of $\bar{\boldsymbol{u}}_s$ are *not* independent. Some relevant special cases of (4.184) are worth noting.

- For $N_c = 1$, an isotropically distributed vector reduces to a complex scalar with unit magnitude and uniform phase. If the channel is Rayleigh-faded or otherwise circularly symmetric, the phase rotation caused by $u_s[0]$ is then inconsequential and the information is conveyed exclusively by $A_s[0]$. Furthermore, as in the memoryless case, the distribution of $A_s[0]$ is discrete.
- Conversely, for $N_c \to \infty$ the distribution of A_s coalesces around a single mass at 1, i.e., it becomes deterministic, while the entries of $\bar{\boldsymbol{u}}_s$ become IID complex Gaussian because the strong law of large numbers dictates that $\|\bar{\boldsymbol{z}}\| \overset{\text{a.s.}}{\to} 1$. Altogether, the symbols become IID complex Gaussian, an expected result since a channel that fades infinitely slowly can be learned at negligible cost and thus the capacity-achieving distribution should coincide, asymptotically, with the one obtained under CSIR.

Low-SNR regime

At low SNR, as we move from $N_c = 1$ to $N_c \gg 1$, the optimum signal structure transitions from schemes based on "on–off" keying to bursty blocks of IID symbols and, ultimately, to continuous sequences of IID symbols.

As seen for the stringent case of $N_c = 1$, $\frac{E_b}{N_0}_{\min} = -1.59$ dB can be achieved without CSIR, but at the expense of signal peakedness. In Rayleigh fading specifically, the kurtosis of the transmitted sequence must grow as $\frac{1}{N_c \, \mathsf{SNR}}$ for $\mathsf{SNR} \to 0$ to achieve -1.59 dB [64]. While this is certainly problematic for small N_c, as the coherence increases the need for peakedness is swiftly alleviated and, in underspread fading, it remains significant only for exceedingly low SNRs.

In terms of the low-SNR slope, without CSIR it is $S_0 = 0$ regardless of the fading coherence, but again the value of N_c causes pronounced differences in the significance of this zero value. Figure 4.18 shows $\mathsf{C}(\frac{E_b}{N_0})$ computed numerically for IID Rayleigh fading ($N_c = 1$), with penalties of several dB relative to the CSIR capacity, alongside a cartoon representation of $\mathsf{C}(\frac{E_b}{N_0})$ for an underspread fading channel ($N_c \gg 1$), which would only exhibit a minute penalty relative to the CSIR capacity [407, 410]. Both no-CSIR curves do have $S_0 = 0$, yet they are radically different.

When the peakedness of the transmit signal is constrained, $\frac{E_b}{N_0} = -1.59$ dB cannot be reached without CSIR, but it can be approached closely provided the fading is underspread. Also, with a constrained signal peakedness the spectral efficiency is no longer a concave function of SNR; rather, it is convex up to some SNR and concave thereafter [64, 75, 411–414]. Because of the lack of concavity for low SNRs, $\frac{E_b}{N_0}_{\min}$ is no longer attained for $\mathsf{SNR} \to 0$, but rather at some $\mathsf{SNR} > 0$ that depends on N_c; SNRs below this value should be averted as they would entail both a lower spectral efficiency and a lower power efficiency (higher $\frac{E_b}{N_0}$). This observation that excessively low SNRs should be avoided is further reinforced when one takes into account practical aspects such as the radiated power being only a portion of the power consumed by the communication devices; activating the circuitries to transmit expends a fixed amount of power, rendering overly low SNRs decidedly energy-inefficient.

High-SNR regime

In the high-SNR regime, the memoryless behavior in (4.183) changes radically as soon as $N_c > 1$, conforming, under block Rayleigh fading, to (4.44) with [415, 416]

$$S_\infty = 1 - \frac{1}{N_c} \tag{4.185}$$

$$\mathcal{L}_\infty = \gamma_{\mathsf{EM}} \log_2 e + \log_2 \frac{N_c}{e} - \frac{1}{N_c - 1} \log_2 \big[(N_c - 1)! \big]. \tag{4.186}$$

For $N_c \gg 1$, the loss in DOF is minute and, for $N_c \to \infty$, Stirling's formula for the factorial of large numbers can be applied to verify that (4.186) reduces to the CSIR solution in Example 4.31.

Looking beyond block-fading, the behavior of the capacity without CSIR for $\mathsf{SNR} \to \infty$ is quite sensitive to the modeling assumptions. With continuous fading, in particular:

- If the fading coefficients are perfectly predictable from noiseless observations of their entire past (nonregular fading, see Chapter 3), a scaling akin to the one indicated by

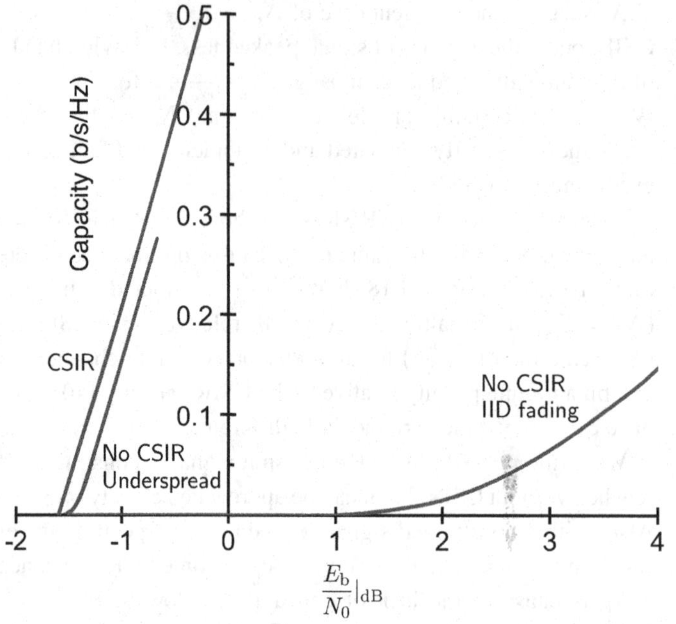

Fig. 4.18 Capacity versus $\frac{E_b}{N_0}$ (in dB) in Rayleigh fading, with and without CSIR. For the latter, exact curve for IID fading ($N_c = 1$) versus a cartoon representation of what it would be in underspread fading ($N_c \gg 1$).

(4.185) is obtained, namely [417]

$$C(\mathsf{SNR}) = S_\infty \log_2 \mathsf{SNR} + \mathcal{O}(1), \tag{4.187}$$

where $S_\infty = 1 - \mu$ with μ the share of signal bandwidth where the Doppler spectrum is null. If the Doppler spectrum is compact with ν_M its maximum frequency, then

$$S_\infty = 1 - \frac{2\,\nu_M}{B} \tag{4.188}$$

$$= 1 - 2\,\nu_M T. \tag{4.189}$$

Recalling from (3.67) the coherence equivalence between continuous and block fading derived in Chapter 3 under frequency flatness, namely $N_c = \frac{1}{2\,\nu_M T}$, it is comforting to see (4.188) agree with (4.185). In a frequency-selective channel, the equivalence would continue to hold only with $N_c = \frac{B_c}{2\,\nu_M}$ with B_c the coherence bandwidth (recall (3.173)). Altogether, and this is a recurring observation, a block-fading model offers a solid equivalence to nonregular continuous fading in terms of the number of DOF. The differences between the two arise only once finer features such as the power offset are examined.

- In contrast, if the fading coefficients cannot be perfectly predicted from noiseless observations of their entire past (regular fading), then [404]

$$C(\mathsf{SNR}) = \log_2 \log \mathsf{SNR} + \mathcal{O}(1), \tag{4.190}$$

where the leading term equals its counterpart for temporally IID fading.

The regularity of fading channels, or lack thereof, is an almost philosophical issue that becomes relevant only at SNRs beyond the operational range of interest [418]. For all practical purposes, the capacity abides by (4.187).

Impact of granting CSIR as side information

It is interesting to consider the performance, without CSIR, of the IID complex Gaussian signals that achieve the CSIR capacity. For $\mathsf{SNR} \to 0$, such performance is characterized in [412, 419]. Expressions for arbitrary SNR seem out of reach, but a semi-analytical computational procedure is put forth in [420] and bounds are given in [421].

Being suboptimal without CSIR, IID complex Gaussian signals can also be used to derive lower bounds for the no-CSIR capacity [386, 422]. Consider IID complex Gaussian signaling over a block-fading channel. On each block

$$y[n] = \sqrt{G E_{\rm s}}\, h\, s[n] + v[n] \qquad n = 0, \ldots, N_{\rm c} - 1 \tag{4.191}$$

and, applying the chain rule of mutual information twice,

$$\begin{aligned} I\big(s[0], \ldots, s[N_{\rm c}-1]; y[0], \ldots, y[N_{\rm c}-1]\big) &= I\big(h, s[0], \ldots, s[N_{\rm c}-1]; y[0], \ldots, y[N_{\rm c}-1]\big) \\ &\quad - I\big(h; y[0], \ldots, y[N_{\rm c}-1] \,|\, s[0], \ldots, s[N_{\rm c}-1]\big) \qquad (4.192) \\ &= I\big(s[0], \ldots, s[N_{\rm c}-1]; y[0], \ldots, y[N_{\rm c}-1] \,|\, h\big) \\ &\quad + I\big(h; y[0], \ldots, y[N_{\rm c}-1]\big) \\ &\quad - I\big(h; y[0], \ldots, y[N_{\rm c}-1] \,|\, s[0], \ldots, s[N_{\rm c}-1]\big). \qquad (4.193) \end{aligned}$$

Dropping $I\big(h; y[0], \ldots, y[N_{\rm c}-1]\big)$, which is expected to be small—whatever information can be inferred about h from observing its noisy product with an unknown IID complex Gaussian sequence—we obtain

$$\begin{aligned} I\big(s[0], \ldots, s[N_{\rm c}-1]; y[0], \ldots, y[N_{\rm c}-1]\big) &\\ \geq I\big(s[0], \ldots, s[N_{\rm c}-1]; y[0], \ldots, y[N_{\rm c}-1] \,|\, h\big) &\\ - I\big(h; y[0], \ldots, y[N_{\rm c}-1] \,|\, s[0], \ldots, s[N_{\rm c}-1]\big), & \end{aligned} \tag{4.194}$$

where the first term is the CSIR capacity (because, conditioned on h, IID complex Gaussian signals are optimum) and the second term bounds the penalty caused by the lack of CSIR. This second term has a curious structure, equivalent to having a bank of AWGN channels with gains $s[0], \ldots, s[N_{\rm c}-1]$ through which a single scalar h is transmitted. We know that, in such a setting, the mutual information is maximized when the transmit quantity is zero-mean complex Gaussian and thus the CSIR penalty cannot exceed its value in Rayleigh fading. For $h \sim \mathcal{N}_{\mathbb{C}}(0,1)$, invoking Example 1.13, we have that (in bits/block)

$$I\big(h; y[0], \ldots, y[N_{\rm c}-1] \,|\, s[0], \ldots, s[N_{\rm c}-1]\big) = \mathbb{E}\left[\log_2\left(1 + \mathsf{SNR} \sum_{n=0}^{N_{\rm c}-1} |s[n]|^2\right)\right] \tag{4.195}$$

$$\leq \log_2\left(1 + \mathsf{SNR} \sum_{n=0}^{N_{\rm c}-1} \mathbb{E}\left[|s[n]|^2\right]\right) \tag{4.196}$$

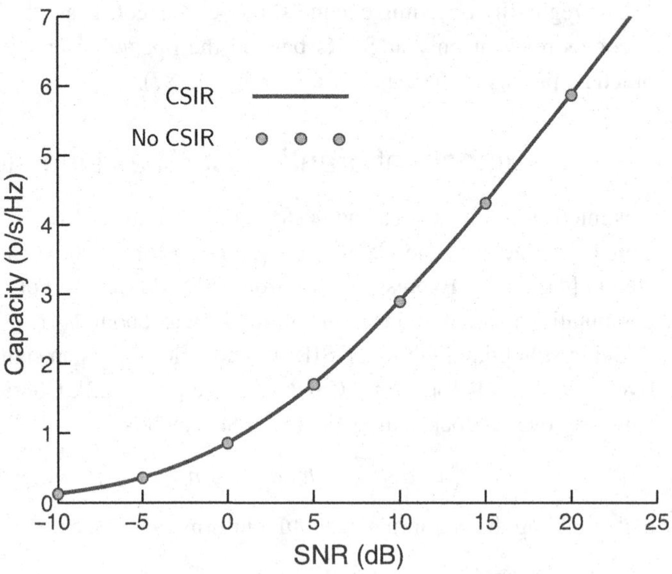

Fig. 4.19 Capacity versus SNR (in dB) in Rayleigh fading. Solid line, CSIR capacity. Circles, no-CSIR capacity lower bound for $N_c = 1000$.

$$= \log_2\Big(1 + N_c\, \mathsf{SNR}\Big), \tag{4.197}$$

where, in (4.196), Jensen's inequality has been applied. In b/s/Hz, therefore, (4.194) translates into

$$C^{\mathsf{CSIR}}(\mathsf{SNR}) - C^{\mathsf{No\,CSI}}(\mathsf{SNR}) \leq \frac{1}{N_c} \log_2\Big(1 + N_c \mathsf{SNR}\Big), \tag{4.198}$$

where we have introduced superscripting to distinguish the two capacities.

Example 4.34

Plot $C^{\mathsf{CSIR}}(\mathsf{SNR})$ alongside the lower bound for $C^{\mathsf{No\,CSI}}(\mathsf{SNR})$ in a vehicular setting.

Solution

As seen in Chapter 3, in a typical vehicular setting, $N_c = 1000$. Figure 4.19 evidences that, for such N_c, there is no appreciable difference between $C^{\mathsf{CSIR}}(\mathsf{SNR})$ and $C^{\mathsf{No\,CSI}}(\mathsf{SNR})$; the gap between the two is a meager 5% at $\mathsf{SNR} = -10$ dB and less than 1% for $\mathsf{SNR} > 0$ dB.

As the no-CSIR capacity—recall that this is the actual capacity—lies between the CSIR capacity and the no-CSIR lower bound, the closeness of these latter quantities pinpoints very precisely that actual capacity. Thus, for most SISO channels of interest, granting CSIR as side information has an essentially negligible effect at all relevant SNRs. The capacity is basically unchanged, yet the benefits in terms of analysis and design are enormous.

4.6 Frequency-selective fading channel

The generalization of the analysis of frequency-flat fading channels to the province of frequency-selective fading is relatively straightforward when the fading is underspread. The transmit–receive relationship is given by (4.78), only with the channel coefficients subject to fading. Under the underspread property, the vector coding and OFDM transformations remain approximately valid over relatively long blocks. (If the channel were not underspread, the subchannels obtained through the transformations would not be orthogonal and their signals would have to be jointly processed [423].) For vector coding, detailed CSIT—including both magnitude and phase for each fading coefficient—is required whereas, for OFDM, that is not the case.

For the remainder of the book we concentrate on OFDM, whose predominance in commercial systems has become almost absolute. Unless otherwise stated, and because it amounts to a mere penalty factor that affects all settings equally, we dismiss the cyclic prefix overhead and use the term *capacity* with a finite number of subcarriers.

We also note that:

- The fading is identically distributed across subcarriers under the reasonable premise that the time-domain fading coefficients are circularly symmetric (refer to Problem 4.58).
- The fading is highly dependent for close-by subcarriers, and essentially independent beyond the coherence bandwidth.

As has been established, the number of resource elements over which the fading remains (roughly) constant is $N_c = B_c T_c$, which the underspread property renders a large number. The availability of CSIR is therefore a reasonable premise.

With frequency selectivity, per-codeword power constraints may become applicable even if the power amplifiers do not tolerate crests: the power averaging may take place in the frequency domain, such that, with sufficiently many subcarriers, the time-domain transmit power is roughly constant.

Link adaptation is greatly facilitated by mapping methods such as CESM, MIESM, or EESM, which allow selecting on the basis of a single metric the MCS that should be employed for the transmission of each codeword. The fading variations across the resource elements spanned by the codeword, i.e., across both OFDM symbols in time and subcarriers in frequency, map onto that single metric, from which the throughput-maximizing MCS is easily identified. And only this metric, quantized down to the number of available MCSs, is required at the transmitter.

In the extreme cases of quasi-static and ergodic settings in the time domain, and depending on the availability of CSIT, the usual notions of ergodic and outage capacity apply. Without CSIT in particular, the ergodic capacity is [424]

$$C(\mathsf{SNR}) = \mathbb{E}\left[\frac{1}{K}\sum_{k=0}^{K-1} \log_2\left(1 + \mathsf{SNR}\,|\mathsf{h}[k]|^2\right)\right] \tag{4.199}$$

$$= \mathbb{E}\left[\log_2\left(1 + \mathsf{SNR}\,|\mathsf{h}|^2\right)\right], \tag{4.200}$$

where h is the fading on an arbitrary subcarrier and (4.200) follows from the identical distribution across subcarriers. In a setting already ergodic in the time domain, therefore, the capacity would be unaffected by frequency selectivity. In the opposite extreme, in contrast, frequency selectivity would have a major effect. The outage probability

$$p_{\mathsf{out}}(\mathsf{SNR}, R/B) = \mathbb{P}\left[\frac{1}{K}\sum_{k=0}^{K-1} \log_2\left(1 + \mathsf{SNR}\,|\mathsf{h}[k]|^2\right) < R/B\right] \tag{4.201}$$

would improve radically with growing K if the subcarriers involved exhibited sufficiently independent fading. An ergodic behavior would rapidly be approached, which confers additional relevance to the ergodic capacity.

4.7 Which fading setting applies?

One can legitimally feel a certain sense of confusion in light of the variety of settings that arise in fading channels, even if the CSIR is taken for granted. It is thus worthwhile to discuss the corresponding applicability.

To the generic question of whether the outage or the ergodic capacity should be invoked, the answer is that "it depends." Naïvely, one could think that it depends only on the velocity, and indeed that used to be the case in the fixed-rate narrowband systems of 1G and 2G. The classical wisdom for these systems was that the quasi-static setting and outage capacity applied in slow fading while the ergodic setting and ergodic capacity applied in fast fading. This wisdom had to be revised once adaptivity was introduced and the signal bandwidth exceeded the coherence bandwidth of the fading [195]. Since 3G, wireless systems invariably feature the following.

- Link adaptation. Rather than a single fixed MCS, a set of MCSs is available and the transmission is matched to the fading whenever a modicum of CSIT is available. In addition, the codeword length itself is made adaptive through hybrid-ARQ, which requires only one-bit notifications of decoding success/error.
- Wideband signals, whereby each codeword may span multiple coherence bandwidths.

In slow fading, adaptivity can preclude fading-induced outages and thus the outage capacity has manifestly lost relevance [195]. The ergodic setting, alternatively, has grown in relevance as time-domain ergodicity has been progessively reinforced in the frequency domain (and, with MIMO, further in the spatial domain). Altogether, the picture that emerges is as follows.

- For frequency-flat slowly fading channels, with codewords spanning OFDM resource elements within a single coherence tile, the appropriate setting is a quasi-static setting with CSIT. This is nothing but an AWGN channel with a changing SNR accorded by the fading distribution, and with the ergodic capacity measuring the local-average spectral efficiency.

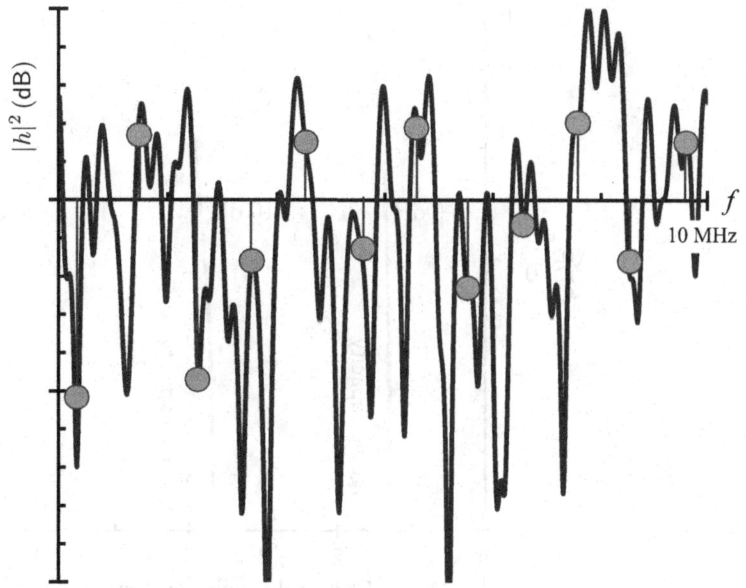

Fig. 4.20 LTE resource blocks separated by 1 MHz in the frequency domain. The resource blocks are cast on a realization of a Rayleigh-faded channel exhibiting a typical urban (recall Example 3.20) power delay profile with delay spread $T_d = 1$ ms.

- For frequency-selective fading channels, with the codeword spanning subcarriers across multiple coherence bandwidths, either with slow fading and especially with fast fading, the number of fading realizations affecting each codeword usually suffices for the ergodic setting to be a satisfactory model and for the channel to be largely information stable.

Example 4.35 (Localized transmission in LTE)

In LTE, each codeword is overlaid on a number of so-called resource blocks consisting of 12 adjacent subcarriers and seven OFDM symbols (for a total of 84 resource elements). With the LTE subcarrier spacing of 15 kHz, each such resource block thus occupies 180 kHz, narrow enough for the fading to be frequency-flat over one or even over various contiguous resource blocks. At low velocities, moreover, the fading remains constant over multiple successive resource blocks such that, altogether, each codeword experiences an AWGN channel with SNR determined by the fading.

Example 4.36 (Distributed transmission in LTE)

Rather than being contiguous, the resource blocks can also be interspersed across the available bandwidth. Provided their frequency separation is minimally large, this ensures IID fading across resource blocks bearing pieces of a given codeword. Figure 4.20 shows how

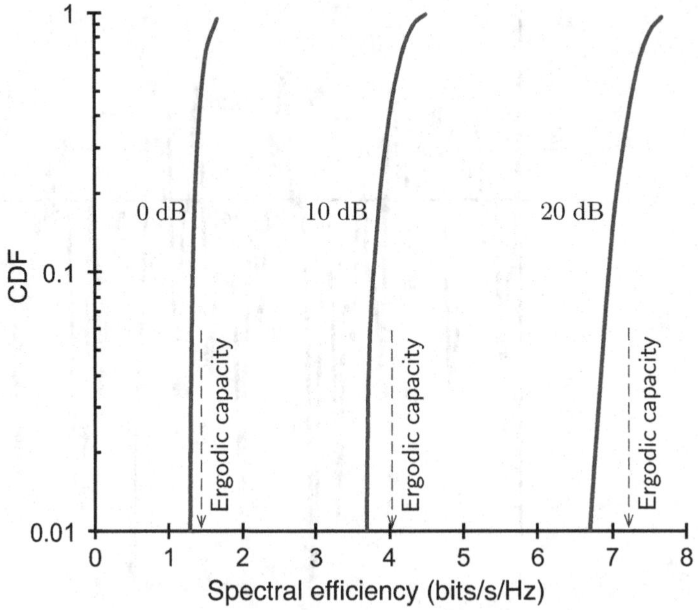

Fig. 4.21 CDF of spectral efficiency for an LTE distributed transmission at distinct local-average SNRs (0, 10, and 20 dB). Also indicated are the corresponding ergodic capacities.

resource blocks separated by 1 MHz would indeed be far apart relative to the fading coherence of a typical urban channel.

Example 4.37 (Spectral efficiency with LTE distributed transmission)

Consider an LTE transmission on a dozen interspersed resource blocks subject to IID Rayleigh fading. No additional fading variation is assumed in the time domain. Figure 4.21 shows the distribution of codeword spectral efficiency for local-average SNRs of 0, 10, and 20 dB. Also indicated are the corresponding ergodic capacities, which are seen to be very representative at each local-average SNR. The limited spread of the codeword spectral efficiency around the ergodic capacity is a measure of the information stability of the transmissions and it confirms the appropriateness of the ergodic setting. With hybrid-ARQ contributing channel variations in the time domain, and with multiple antennas contributing channel variations in the space domain, these spreads squeeze further and the spectral efficiencies tightly coalesce around the respective ergodic capacities.

The foregoing considerations, reinforced in NR with an even more flexible dimensioning of the resource blocks, give clear prominence to the AWGN and ergodic capacities, which is welcome news from an analytical standpoint. The outage capacity is somewhat relegated, although it may remain relevant to transmissions that do not abide by the operational principles of adaptivity and wideband signaling. Wireless sensor networks, for instance, may transmit at a fixed bit rate in narrowband channels, but, by the same token that these systems forgo advanced functionalities, they may not feature near-capacity codebooks; if so, the fundamental communication limits studied in this book are not directly relevant to them. The control channels would be another example that may not fully conform to our observations due to the short nature of the carried messages.

4.8 Pilot-assisted communication

Having established that the true (no CSI) capacity of underspread SISO channels is very well approximated by its CSIR value, it is instructive to explore how closely that value can be approached when the receiver explicitly estimates the fading with the aid of pilot symbols. We hasten to emphasize that this is a specific instance of communication without CSIR, and a suboptimum one in fact: the optimum signaling strategy, recall, is to transmit a signal conforming to (4.184) rather than complex Gaussian codewords with inserted pilot symbols. However, this specific instance is of enormous relevance given that virtually all existing wireless systems comply with it.

For the sake of specificity, the analysis that ensues is tailored to ergodic Rayleigh-faded settings (either continuously faded with a certain Doppler spectrum or else block-faded), yet much of it applies to other fading models as well, e.g., Gauss–Markov, and variations of the analysis for all these models can be found in [324, 407, 416, 425–434]. No attempt is made to derive the capacity-achieving signal distribution when the decoding is based on the observation of the pilots. Rather, we want to investigate how the complex Gaussian signals that are optimum under CSIR perform when the knowledge of the fading is extracted directly by the receiver rather than granted as side information.

4.8.1 Frequency-flat fading

Let us begin by considering single-carrier transmission over frequency-flat fading channels. Referring to Fig. 4.22 and denoting by α the share of single-carrier symbols that are pilots, one in every $1/\alpha$ symbols is a pilot while the rest are payload data. Moreover, $\alpha \geq \alpha_{\mathsf{min}}$ for some α_{min} to be established. Let $\mathcal{D}$ denote the set of indices corresponding to data symbols. For $n \in \mathcal{D}$, (4.130) holds with $x[0], \ldots, x[N-1]$ a sequence of IID complex Gaussian symbols. For $n \notin \mathcal{D}$, alternatively, fixed pilots $x[n] = \sqrt{E_{\mathrm{s}}}$ are transmitted and

$$y[n] = \sqrt{GE_{\mathrm{s}}}\, h[n] + v[n] \qquad n \notin \mathcal{D}. \tag{4.202}$$

Notice that pilot and data symbols have the same energy, E_{s}. Later we lift this constraint, allowing for power-boosted pilots.

Under block fading with N_{c} symbols per block, at least one pilot symbol must be inserted within each fading block and thus

$$\alpha_{\mathsf{min}} = \frac{1}{N_{\mathrm{c}}} \tag{4.203}$$

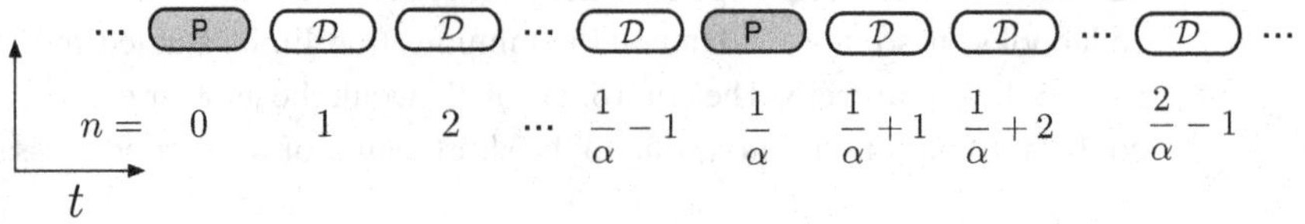

Fig. 4.22 Transmit signal with inserted pilot symbols.

while, if $N_{\rm p} \geq 1$ is the number of actual pilot symbols in every block, then

$$\alpha = \frac{N_{\rm p}}{N_{\rm c}}. \tag{4.204}$$

Under continuous frequency-flat fading with a Doppler spectrum bandlimited to $[-\nu_{\rm M}, \nu_{\rm M}]$, the Doppler spectrum of the corresponding discrete-time fading process $\{h[n]\}$ is bandlimited to $[-\nu_{\rm M}T, \nu_{\rm M}T]$. It follows from the sampling theorem that, to ensure that the decimated fading process observed through the pilots has an unaliased spectrum,

$$\alpha_{\sf min} = 2\,\nu_{\rm M}T. \tag{4.205}$$

The minimum pilot overheads dictated by the block-fading and the continuous fading models, (4.203) and (4.205) respectively, are consistent if we declare the block duration in the former to be $T_{\rm c} = \frac{1}{2\,\nu_{\rm M}}$, which, expressed in number of symbols rather than time, gives $N_{\rm c} = T_{\rm c}/T = \frac{1}{2\,\nu_{\rm M}T}$. This reinforces, and in this case irrespectively of the shape of the Doppler spectrum, the equivalence between block-fading and continuous fading unveiled for rectangular spectra in (3.160).

In pilot-assisted communication, decoding must be conducted on the basis of the observations at the receiver (both pilots and data), without genie-granted CSIR. The maximum spectral efficiency that can be achieved reliably is then the mutual information between the transmitted data symbols and the observations at the receiver (pilots and data), i.e.,

$$\lim_{N\to\infty} \frac{1}{N} I\Big(\underbrace{s[0],\ldots,s[N-1]}_{n\in\mathcal{D}}\,;\,\underbrace{y[0],\ldots,y[N-1]}_{n\in\mathcal{D}},\underbrace{y[0],\ldots,y[N-1]}_{n\notin\mathcal{D}}\Big) \tag{4.206}$$

$$= \lim_{N\to\infty} \frac{1}{N}\Bigg[I\Big(\underbrace{s[0],\ldots,s[N-1]\,;\,y[0],\ldots,y[N-1]}_{n\in\mathcal{D}}\,|\,\underbrace{y[0],\ldots,y[N-1]}_{n\notin\mathcal{D}}\Big)$$

$$+\, I\Big(\underbrace{s[0],\ldots,s[N-1]}_{n\in\mathcal{D}}\,;\,\underbrace{y[0],\ldots,y[N-1]}_{n\notin\mathcal{D}}\Big)\Bigg] \tag{4.207}$$

$$= \lim_{N\to\infty} \frac{1}{N} I\Big(\underbrace{s[0],\ldots,s[N-1]\,;\,y[0],\ldots,y[N-1]}_{n\in\mathcal{D}}\,|\,\underbrace{y[0],\ldots,y[N-1]}_{n\notin\mathcal{D}}\Big), \tag{4.208}$$

where (4.207) follows from the chain rule detailed in Section 1.4.2 and (4.208) from the independence between the transmitted data symbols and the pilot observations. Achieving (4.208), for which there is no known simplified expression, generally requires joint data decoding and channel estimation [177, 419, 435–438].

Besides being cumbersome, (4.208) is not very representative of how wireless systems actually operate because it does not capture a number of additional aspects. It is therefore common to study a different quantity, to be found in (4.216), that is both more tractable and operationally relevant. Let us briefly comment on the receiver procedures that reduce (4.208) to (4.216), step by step.

(1) Only the observed pilots, $y[n]$ for $n \notin \mathcal{D}$, assist in the processing of the data symbols. That is, the processing of a data symbol n does not benefit from information about the fading conveyed by other data symbols. This effectively breaks (4.208) into

$$\lim_{N\to\infty} \frac{1}{N} \sum_{n\in\mathcal{D}} I\Big(s[n]; y[n] \,|\, \underbrace{y[0],\ldots,y[N-1]}_{\notin\mathcal{D}}\Big). \tag{4.209}$$

The observed pilots, $y[n]$ for $n \notin \mathcal{D}$, are used to form fading estimates $\hat{h}[n]$ for $n \in \mathcal{D}$ and, provided the process of channel estimation does not destroy information, (4.209) is then equivalent to

$$\lim_{N\to\infty} \frac{1}{N} \sum_{n\in\mathcal{D}} I\Big(s[n]; y[n] \,|\hat{h}[n]\Big) = (1-\alpha)\,\mathbb{E}\Big[I\Big(s[n]; y[n] \,|\hat{h}[n]\Big)\Big] \tag{4.210}$$

where the expectation arises thanks to the law of large numbers, with the factor $(1-\alpha)$ excluding the pilot symbols from the summation. Altogether, we have a single-letter expression with expectation over the distribution of $\hat{h}[n]$. Achieving (4.210), which must be computed numerically [324, appendix A], requires that the receiver take into account the joint distribution of $h[n]$ and $\hat{h}[n]$.

(2) Further simplifying (4.210), the fading estimates are taken to be correct. Expressing each fading coefficient as $h[n] = \hat{h}[n] + \tilde{h}[n]$ where the conditional-mean estimate (recall Section 1.6)

$$\hat{h}[n] = \mathbb{E}\Big[h[n] \,|\, \underbrace{y[0],\ldots,y[N-1]}_{\notin\mathcal{D}}\Big] \tag{4.211}$$

gives the MMSE estimate of $h[n]$ on the basis of the pilot observations, the transmit–receive relationship can be rewritten as

$$\begin{aligned} y[n] &= \sqrt{G}\, h[n] x[n] + v[n] && (4.212)\\ &= \sqrt{G}\left(\hat{h}[n] + \tilde{h}[n]\right)\sqrt{E_{\mathrm{s}}}\, s[n] + v[n] && (4.213)\\ &= \sqrt{G E_{\mathrm{s}}}\, \hat{h}[n] s[n] + v'[n], && (4.214)\end{aligned}$$

where $\mathsf{MMSE} = \mathbb{E}\big[|\tilde{h}[n]|^2\big]$ and $\mathbb{E}\big[|\hat{h}[n]|^2\big] = (1 - \mathsf{MMSE})$. Now, $\hat{h}[n]$ is the fading, from the receiver's viewpoint, and the effective noise is

$$v'[n] = \sqrt{G E_{\mathrm{s}}}\, \tilde{h}[n] s[n] + v[n]. \tag{4.215}$$

Because the MMSE estimation errors $\tilde{h}[0], \ldots, \tilde{h}[N-1]$ are zero-mean and IID, the sequence $v'[0], \ldots, v'[N-1]$ is also zero-mean and IID and, by virtue of the properties of the MMSE estimator, $v'[n]$ is uncorrelated with $\hat{h}[n]x[n]$. However, $v'[n]$ is, in general, not Gaussian; rather, the distribution of $v'[n]$ and hence the optimum decoding rules become fading-dependent.

(3) We have seen that, in terms of mutual information, the best signal distribution in the face of Gaussian noise is itself Gaussian. As it turns out, this is but one side of the saddle point property of the Gaussian distribution. The other side is that, when the signal is Gaussian, the worst possible noise distribution is also Gaussian [439]. Starting from Gaussian signal and Gaussian noise, any perturbation of the signal distribution can only decrease the mutual information and any perturbation of the noise distribution can only increase it. Therefore, replacing $v'[n]$ in (4.215) with an independent complex Gaussian variable of the same variance we obtain an achievable spectral efficiency. Moreover, such spectral efficiency is attainable with minimum-distance decoding, i.e., by standard decoders designed for Gaussian noise [440]. Thereby setting $v'[n] \sim \mathcal{N}_{\mathbb{C}}(0, GE_{\mathrm{s}}\mathsf{MMSE} + N_0)$, an achievable spectral efficiency for the channel in (4.210) is

$$\frac{R}{B} = (1-\alpha)\, C(\mathsf{SNR}_{\mathsf{eff}}) \tag{4.216}$$

with $C(\cdot)$ the CSIR capacity and with

$$\mathsf{SNR}_{\mathsf{eff}} = \frac{GE_{\mathrm{s}}(1-\mathsf{MMSE})}{GE_{\mathrm{s}}\mathsf{MMSE} + N_0} \tag{4.217}$$

$$= \frac{\mathsf{SNR}\,(1-\mathsf{MMSE})}{1+\mathsf{SNR}\cdot\mathsf{MMSE}}. \tag{4.218}$$

Although not explicitly indicated, MMSE and $\mathsf{SNR}_{\mathsf{eff}}$ are functions of SNR, α, and the fading distribution. In Rayleigh fading, the conditional-mean estimator is linear and the MMSE expressions derived in Section 3.7.2 for an LMMSE estimator apply directly here. Particularly relevant is the expression that unifies block fading with continuous fading (under a rectangular Doppler spectrum), which for the reader's convenience we reproduce here as

$$\mathsf{MMSE} = \frac{1}{1+\alpha N_{\mathrm{c}}\mathsf{SNR}}. \tag{4.219}$$

In addition to being more tractable and operationally relevant, (4.216) serves, on account of the various simplifications introduced, as a lower bound to (4.208). Inspecting its expression, (4.216) evidences two differences with respect to the CSIR capacity:

- A factor $(1-\alpha)$ associated with the pilot overhead.
- A local-average *effective* SNR given by $\mathsf{SNR}_{\mathsf{eff}} \leq \mathsf{SNR}$.

These differences reflect the tension between improving the quality of the channel estimates (by increasing α) and diminishing the overhead (by reducing α). There is a unique value, $\alpha^\star$, that resolves this tradeoff and maximizes (4.216). This optimization is simple as (4.216) is a concave function of α and can be conveniently expressed by means of the CSIR capacity function $C(\cdot)$. Pleasingly, if a genie could render $\alpha = 0$ and $\mathsf{MMSE} = 0$ at once, then (4.216) would return $C(\mathsf{SNR})$.

Example 4.38

Plot, as function of SNR, the spectral efficiency in (4.216) for a frequency-flat Rayleigh-faded channel with a Clarke–Jakes Doppler spectrum corresponding to a vehicular user, with the pilot overhead optimized for each SNR; contrast it with the CSIR capacity of the same channel. Separately, plot the optimum pilot overhead.

Solution

For $N_c = 1000$, corresponding to a vehicular user, the requested results are presented in Fig. 4.23. On the left-hand side subfigure we find the ergodic spectral efficiency with the pilot overhead optimized, alongside the CSIR capacity borrowed from Example 4.27. On the right-hand side subfigure, we find the corresponding optimum pilot overhead. From (4.205), the floor on the overhead is $\alpha_{\text{min}} = 10^{-3}$, which does not take effect within the SNR range in the figure.

From the foregoing example, the following can be inferred:

- Even in this considerably fast-fading channel, with standard minimum-distance decoding treating the channel estimation errors as additional Gaussian noise, pilot-assisted communication can closely approach the CSIR capacity.
- Consistent with the minute loss with respect to the capacity, the optimum overheads are very low. Indeed, the pilot overheads in LTE and NR are, for single-user SISO transmission, on the order of 5%.

In slower-fading channels, say in pedestrian settings, these observations would be reinforced even further, lending full support to the operational meaning of the CSIR capacity: the pilot-assisted spectral efficiency can either be directly approximated by $C(\text{SNR})$, or exactly quantified as $(1-\alpha)\, C(\text{SNR}_{\text{eff}})$.

Other than for $\text{SNR} \to 0$ and $\text{SNR} \to \infty$, considered next, it is not possible to express the optimum pilot overhead in closed form. However, leveraging the underspread condition $N_c \gg 1$, it is possible to express it as a series expansion in $1/N_c$ [441].

Low-SNR regime

In the low-SNR regime under block fading, (4.216) in conjunction with (4.218) and (4.165) expands into

$$\frac{R}{B} = (1-\alpha)\alpha N_c \, \text{SNR}^2 \log_2 e + o(\text{SNR}^2). \tag{4.220}$$

With continuous fading, (4.220) holds verbatim for a rectangular Doppler spectrum and $N_c = B_c T_c$. A more general expansion, of higher order than (4.220) and valid for arbitrary Doppler spectra, is derived in [324] as

$$\frac{R}{B} = (1-\alpha)(1-\text{MMSE})\left[\text{SNR} - \text{SNR}^2 + o(\text{SNR}^2)\right] \log_2 e, \tag{4.221}$$

where MMSE depends on α and on the Doppler spectrum as per (3.153).

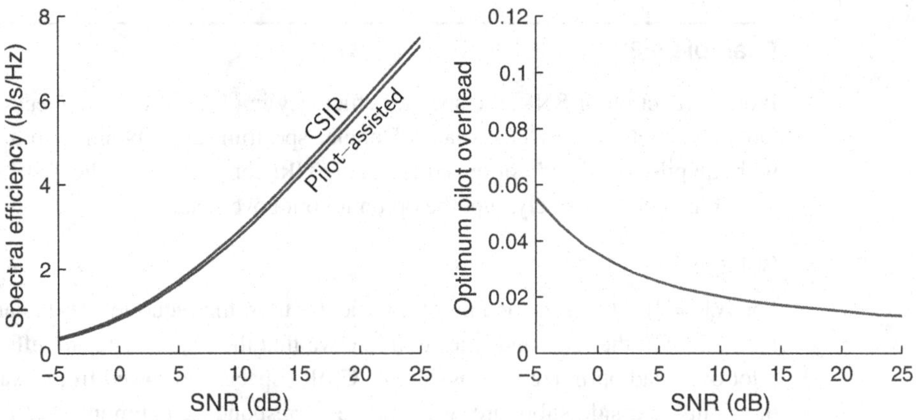

Fig. 4.23 Left, for an ergodic Rayleigh-faded channel with a Clarke–Jakes Doppler spectrum and $N_c = 1000$: optimized pilot-assisted spectral efficiency and CSIR capacity, both as a function of SNR. Right, corresponding optimum pilot overhead.

Example 4.39

Specialize (4.221) for a Clarke–Jakes spectrum.

Solution

For a Clarke–Jakes spectrum, MMSE is as given in (3.42) and thus

$$\frac{R}{B} = \frac{2\alpha(1-\alpha)}{\pi^2 \nu_M T} \log_2\left(\frac{2\pi\,\nu_M T}{\alpha\,\mathsf{SNR}}\right) \mathsf{SNR}^2 + o(\mathsf{SNR}^2). \tag{4.222}$$

The foregoing low-SNR expressions prompt a couple of remarks.

- As can be verified, for $\mathsf{SNR} \to 0$, $\alpha^\star \to 0.5$. Although, as Fig. 4.23 indicates, this limit is only relevant for *extremely low* SNR, it does confirm the intuition that the optimum overhead grows as the SNR diminishes.
- The spectral efficiency is not a concave function of SNR, corroborating that pilot-based communication is but a special case of communication without CSIR. As the signal's peakedness does not grow with diminishing SNR, the spectral efficiency becomes convex below a certain SNR and $\frac{E_b}{N_0}_{\min}$ is attained at some low but positive SNR.

Example 4.40

In a Rayleigh block-faded channel, compare the CSIR capacity with the pilot-assisted spectral efficiency for $N_c = 1000$ (vehicular user) and for $N_c = 20\,000$ (pedestrian user), all expressed as a function of $\frac{E_b}{N_0}$ (in dB). For each value of N_c, further establish $\frac{E_b}{N_0}_{\min}$ and the SNR at which it is attained.

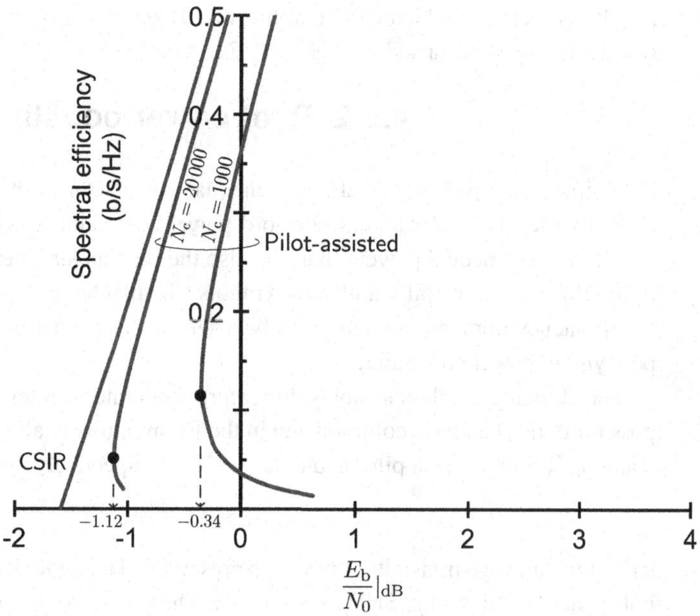

Fig. 4.24 CSIR capacity versus pilot-assisted spectral efficiency as a function of $\frac{E_b}{N_0}$ (in dB), in a Rayleigh block-faded channel. For pilot-assisted transmission, the spectral efficiency is shown for $N_c = 1000$ and $N_c = 20\,000$.

Solution

See Fig. 4.24. For $N_c = 1000$, $\frac{E_b}{N_0}_{\min} = -0.34$ dB attained at SNR $= -9$ dB. Past that point, the pilot-assisted spectral efficiency tracks the CSIR capacity to within 0.6 dB. For $N_c = 20\,000$, a substantially more underspread channel, $\frac{E_b}{N_0}_{\min} = -1.12$ dB, attained at SNR $= -14$ dB. Beyond that, the pilot-assisted efficiency falls to within only 0.15 dB of the CSIR capacity.

High-SNR regime

In the high-SNR regime, under either block fading or continuous fading with a rectangular Doppler spectrum, (4.216) in conjunction with (4.218) and (4.165) expands into

$$\frac{R}{B} = (1-\alpha)\log_2\!\left(\frac{\mathsf{SNR}}{e^{\gamma_{\mathsf{EM}}}}\frac{\alpha N_c}{1+\alpha N_c}\right) + \mathcal{O}\!\left(\frac{1}{\mathsf{SNR}}\right) \tag{4.223}$$

where γ_{EM}, recall, is the Euler–Mascheroni constant. By simple inspection, $\alpha^\star$ in (4.223) should be made as small as possible, i.e., $\alpha^\star = \alpha_{\min} = 1/N_c$ under block fading or $\alpha^\star = \alpha_{\min} = 2\,\nu_M T$ under continuous fading. Plugging such $\alpha^\star$ into (4.223) we observe, by comparison with the no-CSIR expressions in (4.185) and (4.187), that pilot-based communication suffers no penalty in terms of spatial DOF. In this regime, only a power offset penalty is incurred, which can be at most 3 dB since that is the gap to the CSIR capacity (contrast (4.223) evaluated at α_{min} with Example 4.31).

4.8.2 Pilot power boosting

Sometimes, it is possible to allocate unequal powers for pilot and data symbols. This is certainly the case under a per-codeword power constraint, since then the symbols are allowed to have unequal powers, but it is also the case under a per-symbol power constraint if OFDM is used: the pilot and data symbols can then be multiplexed (at least partially) in the frequency domain, and having subcarriers with unequal powers is compatible with a per-symbol power constraint.

Indeed, many wireless systems do feature moderate degrees of pilot power boosting, a possibility that can be accommodated in the formulation by allowing the transmit power to equal $\rho_{\mathrm{p}} P_{\mathrm{t}}$ and $\rho_{\mathrm{d}} P_{\mathrm{t}}$ on pilot and data symbols, respectively, with

$$\rho_{\mathrm{p}} \alpha + \rho_{\mathrm{d}} (1 - \alpha) = 1 \tag{4.224}$$

so that the average transmitted power is preserved. Then, the SNRs on pilot and data symbols equal $\rho_{\mathrm{p}}\mathsf{SNR}$ and $\rho_{\mathrm{d}}\mathsf{SNR}$, respectively. The spectral efficiency in (4.216) continues to hold, only with

$$\mathsf{SNR}_{\mathsf{eff}} = \frac{\mathsf{SNR}\,(1 - \mathsf{MMSE})}{1/\rho_{\mathrm{d}} + \mathsf{SNR} \cdot \mathsf{MMSE}}. \tag{4.225}$$

The MMSE expressions derived in Section 3.7.2 for an LMMSE estimator also hold, with SNR replaced with $\rho_{\mathrm{p}}\mathsf{SNR}$, and the concavity of the CSIR capacity function $C(\cdot)$ implies that the spectral efficiency is maximized by setting $\alpha = \alpha_{\mathsf{min}}$ and optimizing the pilot power boost. Moreover, with α fixed, the optimum $\rho_{\mathrm{p}}^{\star}$ is directly the one that maximizes $\mathsf{SNR}_{\mathsf{eff}}$.

For $\mathsf{SNR} \to 0$, $\rho_{\mathrm{p}} \to N_{\mathrm{c}}/2$ under block fading and $\rho_{\mathrm{p}} \to \frac{1}{4\nu_{\mathrm{M}} T}$ under continuous fading. However, pilot power boosting yields no advantage for $\mathsf{SNR} \to 0$; in the limit, power and bandwidth become interchangeable and hence optimality simply entails devoting half the transmit energy to pilots regardless of the power boosting.

For $\mathsf{SNR} \to \infty$, with block fading,

$$\rho_{\mathrm{p}} \to \frac{N_{\mathrm{c}}}{1 + \sqrt{N_{\mathrm{c}} - 1}}, \tag{4.226}$$

which does pay off: a hefty part of the 3-dB SNR loss suffered in the absence of power boosting with respect to the CSIR capacity can be recovered through pilot power boosting.

Although the peakedness implied by (4.226) and (4.224), which is $\rho_{\mathrm{p}}/\rho_{\mathrm{d}} \approx \sqrt{N_{\mathrm{c}}}$ for large N_{c}, might be excessive in practice, the corresponding analysis—together with the earlier one without power boosting—does bracket what is theoretically possible.

4.8.3 Frequency-selective fading

In frequency-selective fading channels, pilot-assisted communication is conceptually as in frequency-flat fading, with pilot symbols properly arranged on a time–frequency grid

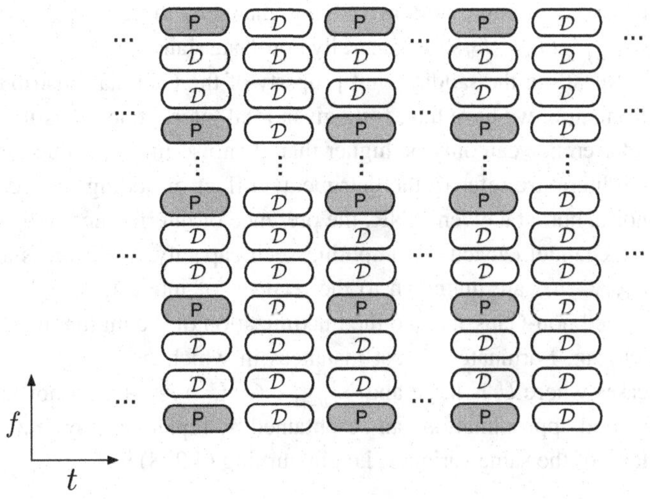

Fig. 4.25 OFDM resource elements with inserted pilot symbols.

[176, 182, 442]. Under the very mild premise that fading be identically distributed in the frequency domain, pilot symbols should be uniformly spaced thereon and thus the one-dimensional arrangement in Fig. 4.22 becomes the two-dimensional arrangement in Fig. 4.25. Most commercial systems feature variations of this arrangement (see [177, figure 1]). The optimization of the overhead now entails both the coherence time and the coherence bandwidth [443]. In block fading specifically, (4.216) and (4.225) apply verbatim with α interpreted as the share of OFDM resource elements that are reserved for pilots and with

$$\mathsf{MMSE} = \frac{1}{1 + \alpha B_{\mathrm{c}} T_{\mathrm{c}} \rho_{\mathrm{p}} \mathsf{SNR}}, \tag{4.227}$$

where, given the mixture of time and frequency multiplexing of pilots, power boosting should be feasible regardless of the type of power constraint.

4.9 Channels with interference

Although the issues surrounding multiuser communication and interference are confronted head-on in later chapters, we can take a first look at the impact of interference on a given link. To that end, let us augment our single-letter transmit–receive relationship into

$$y = \sqrt{G E_{\mathrm{s}}}\, h\, s + \sum_{u=1}^{U} \sqrt{G_u E_u}\, h_u s_u + v. \tag{4.228}$$

In addition to the noise v, the received signal is further corrupted by undesired transmissions (not be decoded by this receiver) from U other transmitters with large-scale channel gains $G_1, \ldots, G_U$, unit-variance fading coefficients $h_1, \ldots, h_U$, and per-symbol energies $E_1, \ldots, E_U$. In contrast with the noise, such interference is subject to fading. When observed through unknown fading channels, moreover, the aggregate faded interference $\sum_{u=1}^{U} \sqrt{G_u E_u} h_u s_u$ is generally non-Gaussian.

Recalling the saddle-point property of the Gaussian distribution in terms of mutual information, we have that, for a given total interference-plus-noise power, the capacity with interference can only be higher than its noise-limited value. Statements to the contrary in the literature refer to the detrimental effect of adding interference to a constant level of noise, but, at a given SINR, the presence of interference in lieu of noise can only increase the channel capacity. Computing such capacity, however, is a formidable task whenever $h_1, \ldots, h_U$ are unknown to the receiver of interest, as $\sum_{u=1}^{U} \sqrt{G_u E_u} h_u s_u + v$ is then indeed non-Gaussian and the maximization of the mutual information becomes very challenging. Fortunately, most instances in which $h_1, \ldots, h_U$ are not known correspond to cases where U is large and $\sum_{u=1}^{U} \sqrt{G_u E_u} h_u s_u + v$ is not far from Gaussian, such that a good approximation can be obtained by replacing it by a zero-mean complex Gaussian term of the same variance, i.e., by turning (4.228) into

$$y = \sqrt{G E_s}\, h\, s + v' \tag{4.229}$$

with $v' \sim \mathcal{N}_{\mathbb{C}}\big(0, \sum_{u=1}^{U} G_u E_u + N_0\big)$. This leaves us with an ordinary fading channel impaired by Gaussian noise, to which the results in this chapter apply directly and whose capacity tends to approximate very well—and in fact lower-bounds—that of the original channel in (4.228) [444]. Throughout the book, unless otherwise stated, the interference from transmissions in other cells is subsumed within the Gaussian noise, and the spectral efficiencies thereby formulated can be achieved with standard minimum-distance decoders.

Alternatively, the interference fading coefficients $h_1, \ldots, h_U$ could be considered known by the receiver, a situation that might correspond to a small value of U and that we could regard as an extended form of CSIR. The mutual information could then be conditioned on $h_1, \ldots, h_U$ and, from the vantage of the link of interest, the worst possible distribution for $s_1, \ldots, s_U$ would be complex Gaussian, in which case the conditional (on the fading) interference-plus-noise would also be complex Gaussian and the capacity-achieving signal would itself be complex Gaussian. Under these premises,

$$I(s; y \,|h, h_1, \ldots, h_U) = \log_2\left(1 + \frac{G E_s \,|h|^2}{\sum_{u=1}^{U} G_u E_u \,|h_u|^2 + N_0}\right) \tag{4.230}$$

$$= \log_2\left(1 + \frac{|h|^2}{\sum_{u=1}^{U} \frac{|h_u|^2}{\mathsf{SIR}_u} + \frac{1}{\mathsf{SNR}}}\right) \tag{4.231}$$

with individual local-average signal-to-interference (SIR) ratios $\mathsf{SIR}_u = \frac{G E_s}{G_u E_u}$ for $u = 1, \ldots, U$. With all the interference terms mutually independent, the aggregate SIR and the SINR would satisfy

$$\frac{1}{\mathsf{SIR}} = \sum_{u=1}^{U} \frac{1}{\mathsf{SIR}_u} \tag{4.232}$$

$$\frac{1}{\mathsf{SINR}} = \frac{1}{\mathsf{SIR}} + \frac{1}{\mathsf{SNR}}, \tag{4.233}$$

such that

$$\mathsf{SINR} = \frac{G E_s}{\sum_{u=1}^{U} G_u E_u + N_0} = \frac{G E_s}{N_0'} \tag{4.234}$$

where N_0' subsumes, besides the noise, also the interference.

From the mutual information in (4.231), both outage and ergodic capacities could be established depending on the setting. In quasi-static settings without CSIT, the outage capacity would conceptually be as in noise-limited channels, with the difference that outages could be triggered either by signal fades or by interference surges. In ergodic settings, the CSIR-only capacity would be

$$C(\mathsf{SNR}, \mathsf{SIR}_1, \ldots, \mathsf{SIR}_U) = \mathbb{E}\left[\log_2\left(1 + \frac{|h|^2}{\sum_{u=1}^{U} \frac{|h_u|^2}{\mathsf{SIR}_u} + \frac{1}{\mathsf{SNR}}}\right)\right], \tag{4.235}$$

with expectation over the distributions of h and $h_1, \ldots, h_U$. Since $\log(1 + 1/z)$ is convex in z, Jensen's inequality suffices to verify that having interference in lieu of noise can only increase the capacity:

$$C(\mathsf{SNR}, \mathsf{SIR}_1, \ldots, \mathsf{SIR}_U) \geq \mathbb{E}\left[\log_2\left(1 + \frac{|h|^2}{\mathbb{E}\left[\sum_{u=1}^{U} \frac{|h_u|^2}{\mathsf{SIR}_u} + \frac{1}{\mathsf{SNR}}\right]}\right)\right] \tag{4.236}$$

$$= \mathbb{E}\left[\log_2\left(1 + \frac{|h|^2}{\sum_{u=1}^{U} \frac{1}{\mathsf{SIR}_u} + \frac{1}{\mathsf{SNR}}}\right)\right] \tag{4.237}$$

$$= \mathbb{E}\left[\log_2(1 + \mathsf{SINR}\,|h|^2)\right], \tag{4.238}$$

where (4.238) is the noise-limited capacity at the same SINR. In much the same way in which fading of the desired signal is unfavorable to the CSIR-only capacity, fading of the interference would be favorable.

As intuition would have it, the inequality in (4.236) becomes an equality if the interferers are equal-power and their number grows, i.e., when $\mathsf{SIR}_u = \mathsf{SIR}$ for $u = 1, \ldots, U$ and $U \to \infty$. The superposition of many fading interference contributions of comparable power behaves essentially as noise. Conversely, the largest deviation from the noise-limited capacity occurs if there is a single dominant interferer, i.e., when $U = 1$ and $\mathsf{SNR} \to \infty$, a case for which closed forms can be found.

Example 4.41

Let $U = 1$ with h and h_1 Rayleigh-faded. Obtain the ergodic capacity as a function of SIR for $\mathsf{SNR} \to \infty$.

Solution

From (4.235),

$$C(\mathsf{SNR}, \mathsf{SIR}) \to \mathbb{E}\left[\log_2\left(1 + \frac{|h|^2}{|h_1|^2}\mathsf{SIR}\right)\right] \tag{4.239}$$

$$= \int_0^\infty \int_0^\infty \log_2\left(1 + \frac{\xi}{\zeta}\,\mathsf{SIR}\right) e^{-\xi} e^{-\zeta}\, \mathrm{d}\xi\, \mathrm{d}\zeta \tag{4.240}$$

$$= \frac{\mathsf{SIR}}{\mathsf{SIR}-1} \log_2 \mathsf{SIR}. \tag{4.241}$$

Less compact but more general closed forms allowing for multiple Rayleigh-faded interferers are given, e.g., in [445]. Another avenue for analysis would be opened by rearranging (4.235) into [446]

$$C(\mathsf{SNR}, \mathsf{SIR}_1, \ldots, \mathsf{SIR}_U) = \mathbb{E}\left[\log_2\left(1 + \mathsf{SNR}\left(|h|^2 + \sum_{u=1}^{U} \frac{|h_u|^2}{\mathsf{SIR}_u}\right)\right)\right] - \mathbb{E}\left[\log_2\left(1 + \mathsf{SNR} \sum_{u=1}^{U} \frac{|h_u|^2}{\mathsf{SIR}_u}\right)\right], \tag{4.242}$$

which is the difference between two noise-limited capacities with modified fading distributions. Unfortunately, these modified distributions are more involved than the actual ones (e.g., they are not Rayleigh even if the actual fading is Rayleigh), thereby precluding direct applicability of many results obtained for noise-limited channels.

Low-SNR regime

In the low-SNR regime, the interest would be in assessing how the capacity behaves when the signal is weak relative to the sum of noise and interference, and it therefore would not suffice to let $\mathsf{SNR} \to 0$. What would be required is to let $\mathsf{SINR} \to 0$ or, equivalently and more conveniently, to let $E_s \to 0$ for the user of interest with all other energies held fixed, in which case the definition of $\frac{E_b}{N_0}_{\min}$ in (4.30) would mutate into

$$\frac{E_b}{N_0'}_{\min} = \lim_{\mathsf{SINR}\to 0} \frac{\mathsf{SINR}}{C(\mathsf{SINR})} \tag{4.243}$$

$$= \lim_{E_s \to 0} \frac{G E_s}{N_0' \, C(\frac{G E_s}{N_0'})} \tag{4.244}$$

$$= \frac{G}{N_0' \, \frac{\mathrm{d}C}{\mathrm{d}E_s}|_{E_s=0}}. \tag{4.245}$$

From (4.235) or (4.242) expressed explicitly as a function of E_s rather than as a function of SNR and $\mathsf{SIR}_1, \ldots, \mathsf{SIR}_U$,

$$\frac{\mathrm{d}C}{\mathrm{d}E_s}|_{E_s=0} = \mathbb{E}\left[\frac{G\,|h|^2}{\sum_{u=1}^{U} G_u E_u |h_u|^2 + N_0}\right] \log_2 e \tag{4.246}$$

$$= G\,\mathbb{E}\left[\frac{1}{\sum_{u=1}^{U} G_u E_u |h_u|^2 + N_0}\right] \log_2 e, \tag{4.247}$$

where we have invoked the independence of h from the other interference fading coefficients and its unit-power nature, such that in (4.247) the outer expectation is only over

$h_1, \ldots, h_U$. Plugging this expression into (4.245),

$$\left.\frac{E_b}{N_0'}\right|_{\min} = \frac{1}{N_0' \, \mathbb{E}\left[\frac{1}{\sum_{u=1}^{U} G_u E_u |h_u|^2 + N_0}\right] \log_2 e} \tag{4.248}$$

$$= \frac{1}{\mathbb{E}\left[\frac{\sum_{u=1}^{U} G_u E_u + N_0}{\sum_{u=1}^{U} G_u E_u |h_u|^2 + N_0}\right] \log_2 e} \tag{4.249}$$

$$= \frac{1}{\mathbb{E}\left[\frac{\sum_{u=1}^{U} \mathrm{INR}_u + 1}{\sum_{u=1}^{U} \mathrm{INR}_u |h_u|^2 + 1}\right] \log_2 e}, \tag{4.250}$$

where we have invoked $N_0' = \sum_{u=1}^{U} G_u E_u + N_0$ and defined $\mathrm{INR}_u = G_u E_u / N_0$ to denote the interference-to-noise ratio for the uth interferer. Absent all interference, the above would revert to the familiar $\left.\frac{E_b}{N_0}\right|_{\min} = -1.59$ dB. With interference, this quantity could only be reduced because

$$\mathbb{E}\left[\frac{\sum_{u=1}^{U} \mathrm{INR}_u + 1}{\sum_{u=1}^{U} \mathrm{INR}_u |h_u|^2 + 1}\right] \geq 1. \tag{4.251}$$

The transmit power needed for reliable communication would thus be reduced by the left-hand side of (4.251) relative to a noise-limited channel at the same SINR; the largest departure from noise-limited conditions is for $U = 1$, which is the focus of the next example.

Example 4.42

Evaluate the left-hand side of (4.251) with a single interferer subject to Rayleigh fading.

Solution

Let $U = 1$. Recalling $\Gamma(\cdot,\cdot)$ as the incomplete Gamma function

$$\mathbb{E}\left[\frac{\mathrm{INR} + 1}{\mathrm{INR}\,|h_1|^2 + 1}\right] = \int_0^{\infty} \frac{\mathrm{INR} + 1}{\mathrm{INR}\,|h_1|^2 + 1} e^{-\xi} \, d\xi \tag{4.252}$$

$$= \left(1 + \frac{1}{\mathrm{INR}}\right) e^{1/\mathrm{INR}} \, \Gamma\left(0, \frac{1}{\mathrm{INR}}\right), \tag{4.253}$$

which can lie anywhere within $[1, \infty)$. Although indicative that reliable communication through the fades of the interference could be possible with minimal transmit power if there were no noise ($\mathrm{INR} \to \infty$), this result is fragile and should not be taken at face value. For other fading distributions, or with multiple interferers, the left-hand side of (4.251) might only be modestly above 1. With two equal-power Rayleigh-faded interferers, for instance, it lies between 1 and 2 (refer to Problem 4.68).

As far as expressions for S_0 in the face of interference are concerned, interested readers are referred to [447].

High-SNR regime

In the high-SNR regime, by simple inspection $\mathcal{S}_\infty = 1$ and $C = \log_2 \mathsf{SINR} - \mathcal{L}_\infty + o(1)$ where, from (4.46) and (4.235),

$$\mathcal{L}_\infty = -\mathbb{E}\left[\log_2\left(\frac{\sum_{u=1}^{U}\mathrm{INR}_u + 1}{\sum_{u=1}^{U}\mathrm{INR}_u|h_u|^2+1}\,|h|^2\right)\right] \tag{4.254}$$

in 3-dB units. On the one hand, the power offset could not exceed its noise-limited value (recall Examples 4.31 and 4.32). On the other hand, the smallest possible power offset must occur with a single dominant interferer and thus

$$\mathcal{L}_\infty \geq -\mathbb{E}\left[\log_2 \frac{|h|^2}{|h_1|^2}\right] \tag{4.255}$$

$$= \mathbb{E}\left[\log_2|h_1|^2\right] - \mathbb{E}\left[\log_2|h|^2\right], \tag{4.256}$$

which, if h and h_1 have the same distribution, is zero.

4.10 Summary and outlook

This chapter establishes a foundation for the rest of the book, introducing, in the simplest context of SISO communication, notions, quantities, and expressions that in the sequel are extended to MIMO and to multiple users. A listing of the main points is provided in the accompanying summary box.

The relevance of an information-theoretic approach to wireless communications is supported for SISO channels with frequency selectivity, fading, and interference. Simple constellations and off-the-shelf codes achieve throughputs within 3 dB of what information theory stipulates for complex Gaussian codebooks of unbounded length. To this shortfall, a fraction of dB must be added because of imperfect channel estimates and about 10% of the spectral efficiency must be subtracted, half of it to account for pilot overhead and half as OFDM cyclic-prefix overhead. Altogether, a gap of roughly 5 dB exists at the PHY layer whose recovery is a steep slope. Other imperfections such as feedback delays and errors (which hamper link adaptation), synchronization offsets (which cause loss of orthogonality among OFDM subcarriers and thus self-interference), and excess bandwidth (pulse shaping in single-carrier transmission or unused subcarriers in OFDM [448]) may push the gap to about 8–10 dB as reported for specific implementations [449–451]; these imperfections, however, are resolvable aspects on which steady progress is being made.

Particularly noteworthy is the role played by hybrid-ARQ, which contributes to many desirable characteristics built into modern wireless systems and alleviates some of the aforementioned imperfections. Specifically, hybrid-ARQ is beneficial in the following respects.

- It shields against feedback delays and even feedback errors, as it allows recovering from mismatched MCS selections. By diluting the coding rate over multiple hybrid-ARQ transmissions, initial MCSs selections can be corrected.

Take-away points

1. The tradeoffs between bit rate, transmit power, and bandwidth are mirrored in the functions $C(\mathsf{SNR})$ and $\mathsf{C}(\frac{E_b}{N_0})$, which express the capacity as a function of, respectively, the per-symbol/per-resource-element SNR and the per-bit SNR.
2. In the low-SNR regime, $\frac{E_b}{N_0}\big|_{\min}$ quantifies the $\frac{E_b}{N_0}$ above which reliable communication is possible, and also the first-order scaling of $C(\mathsf{SNR})$, while S_0 measures the slope of $\mathsf{C}(\frac{E_b}{N_0})$ in log-scale at $\frac{E_b}{N_0}\big|_{\min}$. Precisely

$$\mathsf{C}\left(\frac{E_b}{N_0}\right) = S_0 \frac{\frac{E_b}{N_0}\Big|_{\mathrm{dB}} - \frac{E_b}{N_0}\Big|_{\min}\Big|_{\mathrm{dB}}}{3\,\mathrm{dB}} + \epsilon. \tag{4.257}$$

3. In the high-SNR regime, S_∞ quantifies the slope of $C(\mathsf{SNR})$ in log-scale while $\mathcal{L}_\infty$ quantifies the difference (in 3-dB units) in the SNR required for a certain capacity, relative to a baseline channel where $\mathcal{L}_\infty = 0$. Precisely,

$$C(\mathsf{SNR}) = S_\infty\left(\log_2 \mathsf{SNR} - \mathcal{L}_\infty\right) + \mathcal{O}\left(\frac{1}{\mathsf{SNR}}\right). \tag{4.258}$$

4. In an unfaded SISO channel, $C(\mathsf{SNR}) = \log_2(1+\mathsf{SNR})$ while $\frac{E_b}{N_0} = (2^{\mathsf{C}}-1)/\mathsf{C}$. This channel provides the baseline values $S_\infty = 1$ and $\mathcal{L}_\infty = 0$ as well as $\frac{E_b}{N_0}\big|_{\min} = -1.59$ dB and $S_0 = 2$. The spectral efficiency with M-ary constellations hugs the capacity up to the vicinity of $\log_2 M$.
5. Frequency selectivity can be handled by means of either vector coding or OFDM. While both provide decompositions into parallel subchannels, OFDM does not rely on the specific channel response and is therefore more robust. If the transmitter is indeed privy to the channel response, then it can further allocate its power optimally across the subchannels.
6. With fading, several information-theoretic variants arise depending on whether there is CSI at the receiver and/or transmitter. Although the actual capacity is the one with no side information, simpler expressions are often obtained with CSIR granted and, in underspread fading, these expressions accurately represent the actual capacity. Under TDD/full duplex, or in channels whose coherence is long enough to accommodate feedback, CSIT can be further considered.
7. Among the various fading idealizations, strong prominence should be given to the quasi-static setting with CSIT and to the ergodic setting.
8. The quasi-static setting with CSIT amounts to an AWGN channel with SNR accorded by the fading. With fading coefficient h, $C(\mathsf{SNR}, h) = \log_2(1+\mathsf{SNR}\,|h|^2)$ with local average $\mathbb{E}\left[\log_2(1+\mathsf{SNR}\,|h|^2)\right]$.
9. In an ergodic setting with fixed transmit power, the capacity is

$$C(\mathsf{SNR}) = \mathbb{E}\left[\log_2\left(1+\mathsf{SNR}\,|h|^2\right)\right], \tag{4.259}$$

irrespective of the availability of CSIT. Its coincidence with the local-average capacity of the quasi-static setting renders this a quantity of capital relevance. In Rayleigh fading, furthermore, (4.259) admits the closed form

$$C(\mathsf{SNR}) = e^{1/\mathsf{SNR}}\, \mathcal{E}_1\left(\frac{1}{\mathsf{SNR}}\right) \log_2 e. \tag{4.260}$$

10. Frequency selectivity, when present over the codewords, contributes to the onset of ergodicity. Since the signals of current systems may span many coherence bandwidths, this reinforces the importance of the ergodic setting.
11. Link adaptation with a reduced number of MCSs and commercial codes of reasonable length can deliver throughputs within 2–3 dB of capacity. With frequency selectivity, methods such as MIESM or EESM can map the fading coefficients across the subcarriers to a single quantity from which to select the MCS.
12. To explicitly account for the pilot overhead and for the SNR penalty that reflects the imperfection of the fading estimates, the spectral efficiency can be written as $(1-\alpha)\,C(\mathsf{SNR}_{\mathsf{eff}})$ where α is the pilot overhead while

$$\mathsf{SNR}_{\mathsf{eff}} = \frac{\mathsf{SNR}\,(1-\mathsf{MMSE})}{1+\mathsf{SNR}\cdot\mathsf{MMSE}}, \tag{4.261}$$

with MMSE the variance of estimation error. This variance depends on the fading dynamics, with $\mathsf{MMSE} = \frac{1}{1+\alpha N_{\mathrm{c}}\mathsf{SNR}}$ in the case of block fading with a coherence of N_{c} single-carrier symbols or OFDM resource elements. Slight variations of $\mathsf{SNR}_{\mathsf{eff}}$ and MMSE apply if pilot power boosting is allowed.
13. With interference in lieu of additive noise, the capacity at a given SINR could only increase relative to its noise-limited value because, in contrast with noise, the interference is generally non-Gaussian and subject to fading. However, the difference is only significant if the interference is dominated by one or two terms and the corresponding fading coefficients can be tracked by the receiver.

- Likewise, it protects against interference surges that may take place between the moment of MCS selection and the decoding of the ensuing transmission.
- Finally, and most relevant to the discussions in this chapter, hybrid-ARQ may be a source of time-domain ergodicity [452, 453]. Codewords may span multiple transmissions and, because each hybrid-ARQ process is typically interlaced with other hybrid-ARQ processes intended for the same and/or other users, the constituting transmissions are non-consecutive and their fading may thus be—at least partially—independent. The only limitation to the stretching of the hybrid-ARQ process is the tolerance to latency.

In terms of power efficiency, we have seen that from an information-theoretic vantage it is maximized by operating at the lowest possible SNR (at the expense of an equally low spectral efficiency). Although the result that $\frac{E_{\mathrm{b}}}{N_0} \to \frac{E_{\mathrm{b}}}{N_0}_{\mathsf{min}}$ for $\mathsf{SNR} \to 0$ is formally obtained with CSIR, in underspread fading it essentially applies. However, this result must be toned down by noting that it refers only to radiated power. If the power consumed by the transmitter and receiver equipment is further considered, then it becomes undesirable to operate at SNRs too close to zero. Zooming out from the formal information-theoretic analysis, power-efficient communication is seen to entail operating at low, but decidedly positive, SNRs.

Problems

4.1 Show that, when $C(\mathsf{SNR})$ is increasing and concave in SNR, then $\frac{E_b}{N_0} = \frac{\mathsf{SNR}}{C(\mathsf{SNR})}$ is minimized for $\mathsf{SNR} \to 0$.

4.2 Reproduce the plot of $C(\frac{E_b}{N_0})$ given in Fig. 4.4 for an AWGN channel.

4.3 Consider the discrete constellation defined by the points

$$\left\{ -\frac{3}{\sqrt{5}}, -\frac{1}{\sqrt{5}}, \frac{1}{\sqrt{5}}, \frac{3}{\sqrt{5}} \right\}. \tag{4.262}$$

(a) Plot, as a function of SNR (in dB), the spectral efficiency achieved with this constellation on an AWGN channel.

(b) Plot, as a function of $\frac{E_b}{N_0}$ (in dB), the spectral efficiency achieved with this constellation on an AWGN channel.

4.4 Reproduce the spectral efficiency curves of BPSK, QPSK, and 16-QAM in Fig. 4.3. Further incorporate the corresponding curve for 8-PSK.

4.5 Reproduce the spectral efficiency curves of BPSK, QPSK, and 16-QAM in Fig. 4.4. Further incorporate the corresponding curve for 8-PSK.

4.6 Reproduce the curves in Fig. 4.6 corresponding to the capacity-achieving distribution as well as to BPSK.

4.7 Consider an AWGN channel operating over a finite blocklength N.

(a) Applying (1.143), show how to compute the excess energy per bit, $\Delta\frac{E_b}{N_0}$, required to achieve a certain bit rate with a certain p_e as a function of N.

(b) For $V = 10$ and $p_e = 10^{-2}$, plot $\Delta\frac{E_b}{N_0}$ as a function of $N \in [10^2, 10^4]$.

4.8 Compute and plot, as a function of SNR (in dB), the spectral efficiency achieved on an AWGN channel by an equiprobable 16-QAM constellation and by a 16-QAM constellation where the four inner points have twice the probability of the 12 outer points. Is there a positive shaping gain that recovers part of the 1.53-dB penalty experienced by QAM constellations relative to the capacity?

4.9 Consider an AWGN channel.

(a) Compute the ratio between the bandwidths required by BPSK and QPSK at any combination of P_t and R that maps to $\frac{E_b}{N_0} = 0$ dB. Repeat for $\frac{E_b}{N_0} = 3$ dB.

(b) Compute the ratio between the exact bit rates achievable by BPSK and QPSK with a bandwidth B at $\mathsf{SNR} = -3$ dB. Repeat for $\mathsf{SNR} = 0$ dB.

Note: In light of Example 4.7, and given that both signal types have the same $\frac{E_b}{N_0}\big|_{\min}$, we know that, at low SNR, BPSK requires twice the bandwidth as QPSK. In the first part of this problem we verify this result. Then, in the second part, we see how the penalty in terms of bit rate for given bandwidth and power is far less substantial.

4.10 Consider the bandwidth B required to achieve an ergodic bit rate of $R = 10$ Mb/s by a user transmitting $P_t = 20$ dBm to a base station with antenna gain $G_r = 17$ dB and $N_0 = 3.2 \cdot 10^{-20}$ W/Hz. Assume that the user's antenna is ominidirectional. The range is $D = 600$ m and the pathloss exponent is $\eta = 3.8$, while the intercept at 1 m equals $K_{\text{ref}} = 100$. Generate (via Monte-Carlo) and plot the large-scale distribution of B in the face of an 8-dB log-normal shadow fading for these small-scale settings:

(a) AWGN channel (unfaded).

(b) Rayleigh-faded channel.

(c) Rice-faded channel with $\mathsf{K} = 5$ dB.

4.11 Reconsider Problem 4.10, but with the bandwidth fixed at $B = 5$ MHz. Generate (via Monte-Carlo) and plot the large-scale distribution of the bit rate achievable in the face of a 10-dB log-normal shadow fading for the specified small-scale settings.

4.12 Reconsider Problem 4.10, with the bandwidth fixed at $B = 10$ MHz and no shadow fading.

(a) Plot the achievable R as a function of P_{t} for a range broad enough to encompass low- and high-SNR behaviors.

(b) On the same chart, plot the low- and high-SNR expansions of R versus P_{t}.

(c) Comment on the extent to which the expansions are valid.

4.13 Reconsider Problem 4.10 once more, with the transmit power fixed at $P_{\text{t}} = 20$ dBm and no shadow fading.

(a) Plot the achievable R as a function of B for a range broad enough to encompass low- and high-SNR behaviors.

(b) On the same chart, plot the low- and high-SNR expansions of R versus B.

(c) Comment on the extent to which the expansions are valid.

4.14 Reconsider the setup of Problem 4.10 for the final time, with the bit rate fixed at $R = 10$ Mb/s and no shadow fading.

(a) Plot the required B as a function of P_{t} for a range broad enough to encompass low- and high-SNR behaviors.

(b) On the same chart, plot the low- and high-SNR expansions of B versus P_{t}.

(c) Comment on the extent to which the expansions are valid.

4.15 Plot, for an AWGN channel and $\mathsf{SNR} \in [0, 20]$ dB, the percentage of capacity that can be achieved with $N = 2000$ and $p_{\text{e}} = 10^{-2}$.

4.16 Consider an AWGN channel.

(a) Find an approximate expression for the codeword length N required to achieve a certain share of the capacity with a given p_{e}.

(b) What codeword length is (approximately) required to achieve 95% of capacity at $\mathsf{SNR} = 10$ dB with $p_{\text{e}} = 10^{-3}$?

4.17 From the error probabilities as a function of SNR, $p_{\text{e}}(\mathsf{SNR})$, given in the book's companion webpage for the LTE MCSs, reproduce the individual throughput curves in Example 4.8 as well as their envelope.

4.18 Prove the following.

(a) A per-symbol or per-codeword power constraint on $x[0], \ldots, x[N-1]$ translates to $\frac{1}{N}\sum_{n=0}^{N-1} P[n] = 1$.

(b) An OFDM signal $\mathsf{x}[0], \ldots, \mathsf{x}[K-1]$ is subject to $\frac{1}{K}\sum_{k=0}^{K-1} \mathsf{P}[k] = 1$.

4.19 Repeat Example 4.15 for an indoor system, and contrast the values obtained with those of a commercial system designed for indoor operation.

4.20 Consider the time-invariant channel response described by $h[0] = 0.94$ and $h[1] = 0.34$, presumably the result of sampling some continuous impulse response $h(\tau)$, and let $\mathsf{SNR} = 5$ dB.

(a) Compute the spectral efficiency achieved by vector coding with waterfilling power allocation for $K = 2$.

(b) Recompute the spectral efficiency achieved by vector coding, but this time with a uniform power allocation.

(c) Compute the spectral efficiency achievable by OFDM, employing $K = 2$ subcarriers, under waterfilling power allocation.

(d) Recompute the spectral efficiency achievable by OFDM and $K = 2$ subcarriers, but this time with a uniform power allocation.

4.21 Consider again the time-invariant channel response of Problem 4.20 and suppose that, to reduce the cyclic-prefix overhead, we apply vector coding and OFDM with $K = 4$ rather than $K = 2$. Repeat the derivations under this increased value of K.

4.22 Consider the setting of Problem 4.20, except that the bandwidth has been quadrupled. Sampled four times faster, the impulse response $h(\tau)$ leads to $h[0] = 0.64$, $h[1] = 0.5$, $h[2] = 0.39$, $h[3] = 0.3$, $h[4] = 0.24$, and $h[5] = 0.18$. Repeat the derivations with this channel response and $K = 8$.

4.23 Consider the time-invariant channel response described by $h[0] = 0.07 + 0.12\text{j}$, $h[1] = -0.4954 + 0.86\text{j}$, and $h[2] = -0.02$.

(a) Compute the frequency-domain channel response faced by an OFDM transmission with $K = 4$ subcarriers.

(b) Establish the value SNR_0 below which a single subcarrier is allocated power when applying waterfilling.

(c) Obtain the waterfilling power allocations at $\mathsf{SNR} = 5$ dB and at $\mathsf{SNR} = 20$ dB.

(d) Calculate the spectral efficiency achievable, with waterfilling power allocations, at $\mathsf{SNR} = \mathsf{SNR}_0$ as well as at $\mathsf{SNR} = 5$ dB and 20 dB.

(e) Compute the spectral efficiency achievable with a uniform power allocation at each of those SNRs, and compare it with the respective capacities.

4.24 Formulate the counterpart to the discrete waterfilling solution in (4.104) and (4.105) for a continuous-frequency channel response $\mathsf{h}(\nu)$.

4.25 Consider the continuous-frequency channel response $\mathsf{h}(\nu) = \sqrt{2}\,\sin(\pi\nu)$ for $\nu \in [-1/2, 1/2]$ at $\mathsf{SNR} = 4$ dB.

(a) Apply waterfilling to obtain the continuous power allocation $\mathsf{P}^\star(\nu)$.

(b) Calculate the capacity.

(c) Compute the spectral efficiency achievable with a uniform power allocation.

4.26 For the time-invariant channel response described by $h[0] = 0.9$, $h[1] = 0.44$, compute the spectral efficiency achievable with $K \to \infty$ and a uniform power allocation.

4.27 Consider again the discrete-time channel response of Problem 4.20 with subblocks of size $K = 4$. Under single-carrier transmission, how should the signal within each subblock be autocorrelated in order to maximize the achievable spectral efficiency?

4.28 Consider a frequency-selective channel $h[0], \ldots, h[L]$ and a single-carrier signal $x[0], \ldots, x[N-1]$ with the constituent codeword symbols being complex Gaussian. Suppose that, rather than optimally time-correlated, these symbols are IID. What is the highest achievable spectral efficiency as a function of $h[0], \ldots, h[L]$?

4.29 Express $\frac{E_b}{N_0}_{\min}$ and S_0, accounting for the cyclic prefix overhead, for a fixed OFDM channel response $\mathsf{h}[0], \ldots, \mathsf{h}[K-1]$ having a unique $\max(|\mathsf{h}[0]|, \ldots, |\mathsf{h}[K-1]|)$.

4.30 Let $\mathsf{h}[0], \ldots, \mathsf{h}[K-1]$ be an OFDM channel response whose two strongest values are similar, which we can model by letting $\max(|\mathsf{h}[0]|, \ldots, |\mathsf{h}[K-1]|)$ have cardinality two. Show that $\frac{E_b}{N_0}_{\min}$ does not depend on how the power is split between the two while S_0 does. Further, show that S_0 is only maximized when the power split is even.

4.31 Reconsider the setting of Example 4.23, specialized to the low-SNR regime. Plot, for the range of values where the match is satisfactory, the exact functions $C(\mathsf{SNR})$ and $\mathsf{C}(\frac{E_b}{N_0})$ alongside their low-SNR expansions.

4.32 Let $\mathsf{h}[0], \ldots, \mathsf{h}[K-1]$ be a fixed OFDM channel response. Express S_∞ and $\mathcal{L}_\infty$ with the cyclic-prefix overhead accounted for.

4.33 Prove (4.121).

4.34 Consider $K = 2$ subcarriers with QPSK signals on each and let $|\mathsf{h}[0]|^2 = 0.25$ and $|\mathsf{h}[1]|^2 = 0.75$ while $\mathsf{SNR} = 9$ dB.

(a) Compute the waterfilling power allocation.

(b) Compute the mercury/waterfilling power allocation.

4.35 Consider $K = 3$ subcarriers carrying signals conforming to distinct distributions, respectively BPSK, QPSK, and complex Gaussian, with $|\mathsf{h}[0]|^2 = |\mathsf{h}[1]|^2 = |\mathsf{h}[2]|^2 = 1/\sqrt{3}$ and $\mathsf{SNR} = 8$ dB.

(a) Compute the waterfilling power allocation.

(b) Compute the mercury/waterfilling power allocation.

4.36 Consider a bank of K parallel noninteracting subchannels. Prove that, as long as the K signals fed in are proper complex, mercury/waterfilling reverts to regular waterfilling in the low-SNR regime, i.e., in terms of $\frac{E_b}{N_0}_{\min}$ and S_0.

4.37 Prove that, if $h[n] = h$ for $n = 0, \ldots, N-1$, then

$$\frac{1}{N}\sum_{n=0}^{N-1} \log_2\left(1 + \mathsf{SNR}\,|h[n]|^2\, P[n]\right) \tag{4.263}$$

is maximized by $P[n] = 1$ for $n = 0, \ldots, N-1$ regardless of whether the power constraint is per-symbol or per-codeword.

4.38 Recall Example 4.19.

(a) Recompute how far (in SNR) the performance is from that of asymptotically long complex Gaussian codebooks, with the power allocated via waterfilling.

(b) Keeping the waterfilling power allocation, recompute part (a) with respect to the performance of the best possible codebook of finite length N, capitalizing on the fact [376] that the spectral efficiency backoff due to a length-N code is

$\delta C = \sqrt{V/NQ^{-1}(p_e)} + \mathcal{O}(\log N/N)$, with

$$V = \frac{1}{K+L}\sum_{k=0}^{K-1}\left[1 - \frac{\eta^2}{|\mathsf{h}[k]|^4\,\mathsf{SNR}^2}\right]^+ \log_2^2 e \tag{4.264}$$

where $1/\eta$ is the water level.

4.39 Let $h[0] = 1$ and $h[1] = 0.5\,\mathrm{j}$. For $K = 4$, overlay plots of the squared singular values of the Toepliz matrix $\bar{\boldsymbol{H}}_{K,K+L}$ and of its circulant counterpart $\bar{\boldsymbol{H}}_{K+L,K+L}$; further overlay the sorted frequency-response coefficients $|\mathsf{h}[k]|^2$. Repeat for $K = 16$, $K = 64$ and $K = 256$. Explain what you see and why it makes sense.

4.40 Verify that the outcome of Example 4.20 is 2.48 dB.

4.41 In LTE, the OFDM symbol period is $T = 66.7$ μs. How many such symbols fit within the coherence time of a fading channel exhibiting a Clarke–Jakes Doppler spectrum, at $\mathrm{v} = 100$ km/h and $f_c = 2$ GHz? How can this result be reconciled with the value of $N_c \approx 1000$ obtained in Example 3.26?

4.42 Consider OFDM over a fixed channel $\mathsf{h}[0], \ldots, \mathsf{h}[K-1]$ with $K = 324$, which in LTE would approximately occupy $B = 5$ MHz. From the error probabilities given in the book's companion webpage for the LTE MCSs, and applying the EESM method with $\beta = 3.24$ to map the K subcarrier SNRs onto $\mathsf{SNR}_{\mathsf{EESM}}$, plot the throughput as a function of SNR for MCSs 2, 5, 8, 11, 14, 17, 20, 23, and 26 in these cases:

(a) Under frequency flatness, $\mathsf{h}[k] = 1$ for $k = 0, \ldots, K-1$, in which case we revert to Example 4.8.

(b) With $|\mathsf{h}[k]|^2 = 0.9$ for $k = 0, \ldots, \frac{K}{2}-1$ and $|\mathsf{h}[k]|^2 = 0.1$ for $k = \frac{K}{2}, \ldots, K-1$.

(c) With $\mathsf{h}[k]$ drawn once from a standard complex Gaussian distribution and normalized such that $\frac{1}{K}\sum_{k=0}^{K-1}|\mathsf{h}[k]|^2 = 1$.

4.43 Reconsider Example 4.21 with the set of MCSs available to the transmitter reduced to MCSs 2, 5, 8, 11, 14, 17, 20, 23, and 26.

(a) Plot the average throughput per unit bandwidth as a function of SNR (in dB) and verify the degradation with respect to having the entire set of MCSs available.

(b) Plot the average error probability as a function of SNR (in dB).

(c) Replot the average throughput per unit bandwidth as a function of SNR (in dB) with the MCS switching thresholds modified to ensure that the error probability does not exceed $p_e = 0.1$.

4.44 For a frequency-flat quasi-static Rayleigh-faded channel with only CSIR, plot the exact tradeoff between C_ϵ and $\epsilon \in [10^{-3}, 0.5]$ at SNR = 0 dB and SNR = 20 dB. Precisely, plot ϵ (in log-scale) as a function of C_ϵ.

Note: For $\mathsf{SNR} \to \infty$, *this relationship, scaled by* $\log_2 \mathsf{SNR}$, *gives the DMT.*

4.45 For a frequency-flat quasi-static Rayleigh-faded channel with only CSIR, where the transmitter operates at 1 b/s/Hz, plot the exact outage probability and its high-SNR expansion (from which the notion of diversity emanates) as a function of SNR, over a range broad enough for the latter to closely match the former.

4.46 Reproduce Fig. 4.16 and incorporate the function $C(\mathsf{SNR})$ for a Rice-faded channel with $\mathsf{K} = 0$ dB and $\mathsf{K} = 10$ dB.

4.47 Reproduce Fig. 4.17 and incorporate the function $C(\frac{E_b}{N_0})$ for a Rice-faded channel with $\mathsf{K} = 0$ dB and $\mathsf{K} = 10$ dB.

4.48 For a Rice-faded channel at $\mathsf{SNR} = 10$ dB, plot the ergodic capacity with CSIR as a function of the Rice factor K between -10 dB and 20 dB. Indicate on the figure the ergodic capacities of AWGN and Rayleigh-faded channels at the same SNR.

4.49 Prove (4.181) and verify that $\mathcal{L}_\infty \in [0, \gamma_{\mathsf{EM}} \log_2 e]$.

4.50 Consider an ergodic Rice-faded channel with $\mathsf{K} = 5$ dB, $\mathsf{SNR} = 0$ dB, and CSIR. Compute how much bandwidth is required relative to an unfaded channel.
(a) Approximately, based on the respective second-order low-SNR expansions.
(b) Exactly, on the basis of the exact capacities.

4.51 Consider a Rice-fading channel with $\mathsf{K} = 0$ dB and CSIR. Compute the additional power required to reach the same bit rate achievable over an AWGN channel at $\mathsf{SNR} = 10$ dB.
(a) Approximately, on the basis of the respective zero-order high-SNR expansions.
(b) Exactly, on the basis of the exact capacities.

4.52 Plot, as a function of SNR, the ergodic spectral efficiency achievable with QPSK signaling and CSIR over a Rayleigh-faded channel. On the same chart, plot the corresponding spectral efficiency over an AWGN channel.

4.53 Prove that, in the low-SNR regime, the ergodic spectral efficiency achieved with BPSK expands as (4.175).
Hint: Recall the Gaussian mutual information and its expansions in Example 1.10.

4.54 Consider a channel impaired by AWGN, possibly subject to fading, and let the information unit be the nat rather than the bit. What is the minimum energy per nat necessary per reliable communication? What observation does the result elicit?

4.55 In a block Rayleigh-faded channel, what is the minimum value of N_c required for the no-CSIR capacity to be within 1% of the CSIR capacity at $\mathsf{SNR} = 3$ dB?

4.56 Repeat Example 4.34, but plotting as a function of $\frac{E_b}{N_0}$ (in dB) and only over the low-SNR regime. What do you observe?

4.57 Consider an outdoor block-fading channel where the fading blocks have duration $T_c = 1$ ms. Which type of system would experience coherences of only 10 symbols, making the CSIR capacity no longer representative of the actual one?

4.58 Prove that the fading experienced by OFDM subcarriers is identically distributed if the time-domain fading coefficients $h[0], \ldots, h[L]$ are circularly symmetric.

4.59 Consider an ergodic frequency-flat setting where the fading is continuous and Rayleigh with a rectangular Doppler spectrum.
(a) Plot the capacity under CSIR as a function of SNR (in dB).
(b) For $\nu_M T = 10^{-3}$ plot the spectral efficiency achievable with pilot-assisted complex Gaussian signaling as a function of SNR (in dB), with the pilot overhead optimized for each SNR.
(c) Plot the optimum pilot overhead as a function of SNR (in dB).
(d) Indicate which would be the number of single-carrier symbols per fading block in the equivalent block-fading model.

Note: For parts (b) and (c), a convex optimization solver such as `fmincon` *in* MATLAB® *can be used. Alternatively, since the optimization is over a single scalar, the optimum value can be found by scanning over* $\alpha \in [0, 1]$.

4.60 Repeat Problem 4.59 for $\nu_{\mathrm{M}} T = 10^{-4}$.

4.61 Show that, in the high-SNR regime, (4.216) expands into (4.223).

4.62 Consider a Rayleigh-faded channel with a Clarke–Jakes spectrum at $\mathsf{SNR} = 0$ dB, and let $\nu_{\mathrm{M}} T = 10^{-3}$.

(a) Compute the pilot overhead $\alpha^\star$ that maximizes (4.216).

(b) Compute the approximate counterpart to $\alpha^\star$ obtained by optimizing the low-SNR expansion in (4.220).

(c) Compute the approximate counterpart to $\alpha^\star$ obtained by optimizing the low-SNR expansion in (4.222).

4.63 Reproduce Example 4.40.

4.64 Consider an ergodic frequency-flat fading channel with a Clarke–Jakes Doppler spectrum and $\nu_{\mathrm{M}} T = 10^{-3}$. Compute the SNR at which the optimum pilot overhead equals α_{min}.

4.65 Consider an ergodic frequency-flat block-fading channel with pilot-assisted transmission and the possibility of pilot power boosting.

(a) Prove that, for $\mathsf{SNR} \to 0$, $\rho_{\mathrm{p}} \to N_{\mathrm{c}}/2$.

(b) Prove that, for $\mathsf{SNR} \to \infty$, $\rho_{\mathrm{p}} \to N_{\mathrm{c}}/(1 + \sqrt{N_{\mathrm{c}} - 1})$.

(c) Express the reduction in high-SNR power loss with pilot power boosting.

4.66 Repeat Example 4.38 with the incorporation of pilot power boosting.

4.67 Consider a frequency-flat block-fading channel with $N_{\mathrm{c}} = 1000$ at $\mathsf{SNR} = 10$ dB.

(a) With no pilot power boosting, what is the maximum spectral efficiency deficit that pilot-assisted transmission with complex Gaussian signaling can incur over the no-CSIR capacity? Express this deficit in both absolute and percentual terms.

(b) Repeat part (a) with pilot power boosting.

4.68 Prove that, as per Example 4.42, with two equal-power Rayleigh-faded interferers,

$$\frac{1}{2G \log_2 e} \le \frac{E_{\mathrm{b}}}{N'_0}_{\min} \le \frac{1}{G \log_2 e}. \tag{4.265}$$

Find a general expression for the minimum possible $\frac{E_{\mathrm{b}}}{N'_0}_{\min}$ given an arbitrary number of equal-power Rayleigh-faded interferers. Verify that, for $U \to \infty$,

$$\frac{E_{\mathrm{b}}}{N'_0}_{\min} \to \frac{1}{G \log_2 e}. \tag{4.266}$$

4.69 Consider the channel $y[n] = x[n] + x_1[n]$. There is no noise, but there is one interferer whose received power equals that of the desired signal, i.e., $\mathsf{SIR} = 1$. We know that, if the signal distribution were Gaussian, the worst possible interference distribution would also be Gaussian. However, if the signal is BPSK, which is the worst possible interference distribution? Compare (*i*) the spectral efficiency given such worst-case interference distribution with (*ii*) the spectral distribution given complex Gaussian interference, in both cases with the signal being BPSK and with $\mathsf{SIR} = 1$.

PART II

SINGLE-USER MIMO

5 SU-MIMO with optimum receivers

The most exciting phrase to hear in science, the one that heralds new discoveries, is not 'Eureka!' (I found it!) but 'that's funny...'

Isaac Asimov

5.1 Introduction

This second part of the book kicks off the treatment of MIMO, bringing together the information-theoretic notions, the signal processing perspective, and the channel models presented hitherto. It expands to the MIMO realm many of the results advanced for SISO throughout Chapter 4, enriching the analysis with new aspects that arise in this realm. Specifically, the dependencies of the capacity on the numbers of antennas, and the ensuing tradeoffs, are incorporated beginning in this chapter. The coverage is circumscribed to single-user links, with the extensions to multiuser scenarios deferred to the third part of the book. Specifically, the present chapter deals with the fundamental benefits of SU-MIMO, without regard for the complexity of the receiver.

The chapter begins by generalizing to MIMO certain quantities of interest as well as the tradeoff between power, bandwidth, and bit rate. This is followed by a prolonged look at the problem of reliable communication over SU-MIMO channels, which, with the benefit of observations made in Chapter 4, prioritizes the ergodic setting. Advantageously, this reduces the need to consider flat- and frequency-selective fading separately as the ergodic capacities of the two coincide in relevant situations. The outage capacity in quasi-static settings is also treated, albeit less extensively. Within this prolonged look at SU-MIMO, more specifically, Sections 5.3–5.5 consider situations of progressively diminishing CSI, starting with the availability of both CSIR and CSIT as side information and ending with the complete absence of side information—a case that, recall, yields the true capacity. This is followed, in Section 5.6, by the consideration of the practical instance of communication without side information in which explicit channel estimates are gathered from pilot symbol observations. Subsequently, Section 5.7 brings into the picture the issue of interference from other transmissions, an aspect whose imprint becomes more pronounced with MIMO. Then, Section 5.8 overviews SU-MIMO receiver structures, including a relevant one that emanates directly from a form in which the capacity can be expressed, and Section 5.9 offers a perspective on link adaptation for SU-MIMO. Finally, Section 5.11 summarizes the contents of the chapter, briefly discusses lines of work not explicit covered, and catalogues some open problems.

As an additional analytical tool, especially for nonlimiting SNRs, the regime of large numbers of antennas is explored. This large-dimensional regime is not only convenient for the sake of tractability, but practically relevant as the asymptotic results thus obtained turn out to constitute excellent approximations even for very modest numbers of antennas [454, 455]. In fact, the large-dimensional results in this chapter do not seek to quantify the performance when many antennas are actually utilized—something that requires massive MIMO models whose study is deferred to Chapter 10—but rather to serve as simpler surrogates for settings with limited numbers of antennas. Additionally to their tractability, asymptotic results are highly robust, often invariant to the fading distribution. The applicability to MIMO of large-dimensional analysis was pioneered by Foschini while, contemporarily, it was being utilized to study multiuser detection for CDMA systems with random spreading [456–458]. The two problems are highly isomorphic and, over their parallel developments, there has been a fair amount of cross-fertilization.

5.2 Initial considerations

Recollecting the exposition in Chapter 2, the MIMO transmit–receive single-letter relationship is

$$\boldsymbol{y} = \sqrt{G}\boldsymbol{H}\boldsymbol{x} + \boldsymbol{v}, \tag{5.1}$$

where, unless otherwise stated, the noise is $\boldsymbol{v} \sim \mathcal{N}_{\mathbb{C}}(\boldsymbol{0}, N_0\boldsymbol{I})$ while the channel matrix is normalized such that

$$\begin{aligned}\mathbb{E}\big[\|\boldsymbol{H}\|_{\mathrm{F}}^2\big] &= \mathrm{tr}\big(\mathbb{E}\left[\boldsymbol{H}^*\boldsymbol{H}\right]\big) && (5.2)\\ &= N_{\mathrm{t}}N_{\mathrm{r}}. && (5.3)\end{aligned}$$

In turn,

$$\boldsymbol{x} = \sqrt{\frac{E_{\mathrm{s}}}{N_{\mathrm{t}}}}\boldsymbol{F}\boldsymbol{s}, \tag{5.4}$$

where $\boldsymbol{s}$ is an $N_{\mathrm{s}} \times 1$ vector containing independent unit-variance codeword symbols (possibly subject to distinct coding rates and/or distinct signal distributions). This makes $N_{\mathrm{s}} \leq N_{\mathrm{t}}$ the number of data streams that are spatially multiplexed via the $N_{\mathrm{t}} \times N_{\mathrm{s}}$ precoder

$$\boldsymbol{F} = \boldsymbol{U}_{\boldsymbol{F}} \begin{bmatrix} \boldsymbol{P}^{1/2} \\ \boldsymbol{0}_{(N_{\mathrm{t}}-N_{\mathrm{s}})\times N_{\mathrm{s}}} \end{bmatrix} \boldsymbol{V}_{\boldsymbol{F}}^*, \tag{5.5}$$

where $\boldsymbol{V}_{\boldsymbol{F}}$ and $\boldsymbol{U}_{\boldsymbol{F}}$ are a unitary mixing matrix and a unitary steering matrix, respectively, while $\boldsymbol{P} = \mathrm{diag}(P_1, \ldots, P_{N_{\mathrm{s}}})$ is a power allocation matrix. The precoder is of rank N_{s}, and thus $\boldsymbol{R}_{\boldsymbol{x}} = \frac{E_{\mathrm{s}}}{N_{\mathrm{t}}}\boldsymbol{F}\boldsymbol{F}^*$ is also of rank N_{s}. The transmit energy per symbol, $P_{\mathrm{t}}/B = P_{\mathrm{t}}T = E_{\mathrm{s}}$, now applies to vector symbols and thus $\mathbb{E}\big[\|\boldsymbol{x}\|^2\big] = E_{\mathrm{s}}$ irrespective of N_{t}.

Figure 5.1 pictures the transmit–receive relationship and reconciles it with the more generic abstraction in Fig. 1.2, where $\boldsymbol{F}$ is subsumed within the encoder. While in SISO the precoder can only perform power control, in MIMO it is further tasked with spatial processing aspects: the encoder produces vector symbols $\boldsymbol{s}$ with independent entries and it

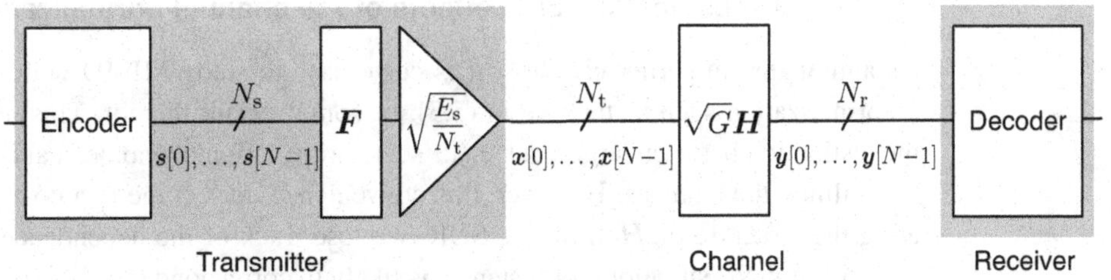

Fig. 5.1 Basic abstraction of a MIMO communication link.

is up to the precoder to mix these entries and steer them into the channel on the basis of the available CSIT. As it turns out, the restriction that the precoder be linear does not entail loss of optimality in terms of SU-MIMO capacity, which is the design metric that drives the precoder optimization throughout this text.

The covariance $\boldsymbol{R_x}$, and therefore the precoder itself, are subject to the applicable power constraint. Let us recall, from Section 2.3.5, the various types of power constraint as well as their translation onto $\boldsymbol{P}$ given our normalizations:

- Under a per-codeword power constraint, with $\boldsymbol{P}$ a function of n in this case,

$$\frac{1}{N}\sum_{n=0}^{N-1}\mathrm{tr}(\boldsymbol{P}[n]) = N_t. \tag{5.6}$$

- Under a per-symbol power constraint,

$$\mathrm{tr}(\boldsymbol{P}) = N_t. \tag{5.7}$$

- Under a per-antenna power constraint,

$$\left[\boldsymbol{F}\boldsymbol{F}^*\right]_{j,j} = 1 \qquad j = 0,\ldots,N_t-1, \tag{5.8}$$

  with the off-diagonal entries of $\left[\boldsymbol{F}\boldsymbol{F}^*\right]$ not necessarily zero. (In this case, the constraint cannot be put as function of solely $\boldsymbol{P}$.)

Let us now turn our attention to the total channel gain introduced for SISO in (4.1). With MIMO, this quantity generalizes to

$$\frac{\mathbb{E}\left[\|\sqrt{G}\boldsymbol{H}\boldsymbol{x}\|^2\right]}{\mathbb{E}\left[\|\boldsymbol{x}\|^2\right]} = G\,\frac{\mathbb{E}\left[\|\boldsymbol{H}\boldsymbol{x}\|^2\right]}{E_s}. \tag{5.9}$$

If there is no fading-dependent power control (meaning that $\boldsymbol{x}$ is independent of $\boldsymbol{H}$) and there is no precoding (i.e., $\boldsymbol{F} = \boldsymbol{I}$), then

$$G\,\frac{\mathbb{E}\left[\|\boldsymbol{H}\boldsymbol{x}\|^2\right]}{E_s} = G\,\frac{\mathbb{E}\left[\mathrm{tr}\left(\boldsymbol{H}\boldsymbol{x}\boldsymbol{x}^*\boldsymbol{H}^*\right)\right]}{E_s} \tag{5.10}$$

$$= G\,\frac{\mathbb{E}\left[\mathrm{tr}\left(\boldsymbol{H}\,\mathbb{E}[\boldsymbol{x}\boldsymbol{x}^*]\boldsymbol{H}^*\right)\right]}{E_s} \tag{5.11}$$

Discussion 5.1 Side effects of the channel normalization

As anticipated in earlier chapters, it is customary to study MIMO with $\boldsymbol{H}$ subject to a normalization such as the one in (5.3), or some variant thereof. This makes explicit the distinction between large- and small-scale propagation phenomena and also happens to facilitate the analysis. However, this convenience does come at a cost: in disassociating the structure of $\boldsymbol{H}$ from the SNR, we lose track of the dependence between the two. A low-SNR situation, for instance, is likely to correspond to a long-range transmission and therefore to certain small-scale fading distributions. In contrast, a high-SNR condition might correspond to a short-range transmission and hence to different fading distributions. While, in SISO, this can mean the distinction between Rayleigh and Rice fading, in MIMO it is more serious since it not only affects the marginal distribution of the entries of $\boldsymbol{H}$ but also their joint distribution. This should be taken into account when drawing conclusions related to the structure of $\boldsymbol{H}$, e.g., on the impact of antenna correlations.

$$= G \frac{\mathbb{E}\left[\mathrm{tr}\left(\boldsymbol{H} \frac{E_{\mathrm{s}}}{N_{\mathrm{t}}} \boldsymbol{H}^*\right)\right]}{E_{\mathrm{s}}} \tag{5.12}$$

$$= G \frac{\mathbb{E}\left[\mathrm{tr}\left(\boldsymbol{H}\boldsymbol{H}^*\right)\right]}{N_{\mathrm{t}}} \tag{5.13}$$

$$= G\, N_{\mathrm{r}}, \tag{5.14}$$

which engulfs the large-scale gain G experienced by every entry of the channel matrix as well as the aggregation of power at the N_{r} receive antennas. More generally, if the transmit power and/or the precoding take $\boldsymbol{H}$ (either its value or its distribution) into account, the total channel gain may exceed $G\, N_{\mathrm{r}}$. The additional gain that precoding can bring about relates to the notion of beamforming introduced in Section 3.5; such gain is explored in Problems 5.2–5.4 and characterized throughout the present chapter. For $\mathsf{SNR} \to 0$, not surprisingly, the precoder that achieves capacity happens to be the one that maximizes the total channel gain, as power is the limiting factor in this regime. For growing SNR, however, the capacity-achieving precoder diverges progressively from the one that maximizes the total channel gain.

The ability of each receive antenna to capture additional power, which is an intrinsic advantage of MIMO as reflected in (5.14), rests on the receive antennas not exhibiting significant electromagnetic mutual coupling. If the antennas were in excessive proximity, say a small fraction of a wavelength, then coupling would set in lessening the received power [459]. In the limit, if the antenna spacing vanished, the receive array would yield the same power as a single antenna. Although coupling is negligible if the antennas are spaced by as little as half a wavelength, arrays devoid of the power gain factor N_{r} can be invoked to study the paradigm of packing a growing number of antennas in a fixed volume of space [460–465]. This paradigm is not explicitly considered in this book, but the tools provided herein could be applied all the same, further letting the absence of the power gain factor N_{r} ripple through the analysis.

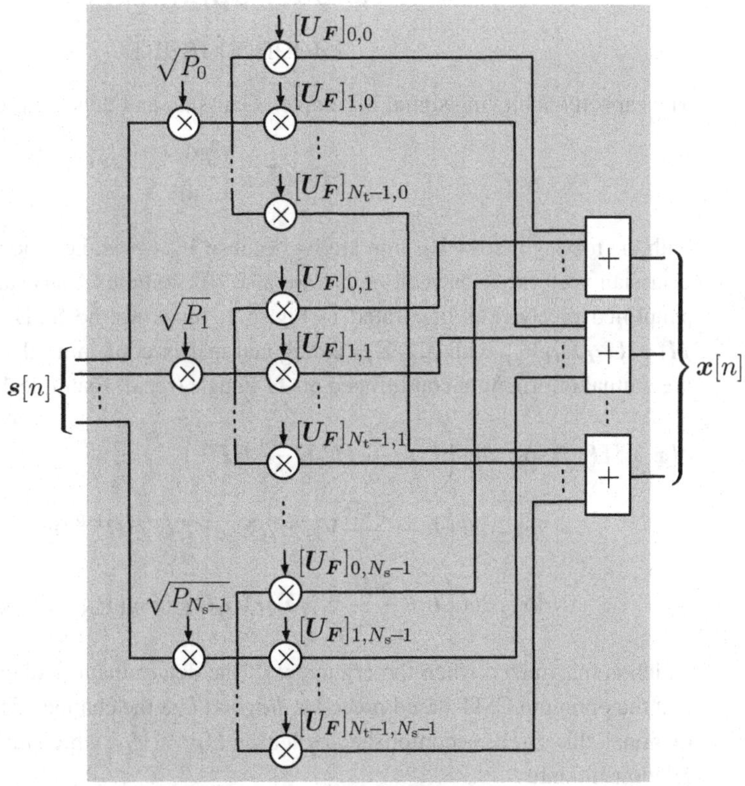

Fig. 5.2 Inner structure of a MIMO precoder $F = U_F \, [P^{1/2} \; 0]^{\mathrm{T}}$.

5.3 CSIR and CSIT

With CSIT, a per-codeword power constraint opens the door to time-domain power control. We do not dwell on this possibility beyond what is covered for SISO in Chapter 4 and rather focus on a per-symbol power constraint, which places the emphasis squarely on the space domain. This implies that $\mathrm{tr}(FF^*) = \mathrm{tr}(P) = N_t$.

5.3.1 Quasi-static setting

In this setting, with the MIMO channel fixed and known, the formulation becomes isomorphic with that of a fixed frequency-selective channel in Section 4.4. While the significance of the vector dimensions is different, and some normalizations must correspondingly be adjusted, (5.1) is indeed isomorphic with (4.79) and thus we can capitalize on the corresponding derivations.

Precoder optimization

Steering matrix

The capacity-achieving signal is complex Gaussian and thus the precoder can reduce to

$$\boldsymbol{F} = \boldsymbol{U_F} \begin{bmatrix} \boldsymbol{P}^{1/2} \\ \boldsymbol{0} \end{bmatrix}, \tag{5.15}$$

with the mixing matrix $\boldsymbol{V_F}$ immaterial because $\boldsymbol{V}_{\boldsymbol{F}}^{*}\boldsymbol{s} \sim \boldsymbol{s}$, i.e., the mixing of IID complex Gaussian vectors yields vectors that are also IID complex Gaussian. The structure of this simplified precoder is illustrated in Fig. 5.2. Invoking the SVD of the channel matrix, $\boldsymbol{H} = \boldsymbol{U_H}\boldsymbol{\Sigma_H}\boldsymbol{V}_{\boldsymbol{H}}^{*}$ with $\boldsymbol{\Sigma}_{\boldsymbol{H}}^{*}\boldsymbol{\Sigma_H}$ a diagonal matrix containing the eigenvalues of $\boldsymbol{H}^{*}\boldsymbol{H}$, the mutual information conditioned on $\boldsymbol{H}$ equals (recall Example 1.13)

$$
\begin{aligned}
I(\boldsymbol{x};\boldsymbol{y}\,|\boldsymbol{H}) &= \log_2 \det\left(\boldsymbol{I} + \frac{G}{N_0}\boldsymbol{H}^{*}\boldsymbol{H}\frac{E_{\mathrm{s}}}{N_{\mathrm{t}}}\boldsymbol{F}\boldsymbol{F}^{*}\right) &(5.16)\\
&= \log_2 \det\left(\boldsymbol{I} + \frac{\mathsf{SNR}}{N_{\mathrm{t}}}\boldsymbol{V_H}\boldsymbol{\Sigma}_{\boldsymbol{H}}^{*}\boldsymbol{\Sigma_H}\boldsymbol{V}_{\boldsymbol{H}}^{*}\boldsymbol{U_F}[\boldsymbol{P}^{1/2}\,\boldsymbol{0}]^{\mathrm{T}}[\boldsymbol{P}^{1/2}\,\boldsymbol{0}]\,\boldsymbol{U}_{\boldsymbol{F}}^{*}\right) &(5.17)\\
&= \log_2 \det\left(\boldsymbol{I} + \frac{\mathsf{SNR}}{N_{\mathrm{t}}}\boldsymbol{\Sigma}_{\boldsymbol{H}}^{*}\boldsymbol{\Sigma_H}\boldsymbol{V}_{\boldsymbol{H}}^{*}\boldsymbol{U_F}\,\mathrm{diag}(P_0,\ldots,P_{N_{\mathrm{s}}-1},0,\ldots,0)\,\boldsymbol{U}_{\boldsymbol{F}}^{*}\boldsymbol{V_H}\right)
\end{aligned}
$$

which is maximized when the argument of the determinant is diagonal, hence the saying that the optimum CSIT-based precoder *diagonalizes* the channel. Since $\boldsymbol{\Sigma}_{\boldsymbol{H}}^{*}\boldsymbol{\Sigma_H}$ is already diagonal, this can be accomplished by letting $\boldsymbol{U_F} = \boldsymbol{V_H}$, which turns the transmit–receive relationship into

$$
\begin{aligned}
\boldsymbol{y} &= \sqrt{G}\boldsymbol{H}\sqrt{\frac{E_{\mathrm{s}}}{N_{\mathrm{t}}}}\boldsymbol{F}\boldsymbol{s} + \boldsymbol{v} &(5.18)\\
&= \sqrt{G}\,\boldsymbol{U_H}\boldsymbol{\Sigma_H}\boldsymbol{V}_{\boldsymbol{H}}^{*}\sqrt{\frac{E_{\mathrm{s}}}{N_{\mathrm{t}}}}\,\boldsymbol{U_F}\begin{bmatrix} \boldsymbol{P}^{1/2} \\ \boldsymbol{0} \end{bmatrix}\boldsymbol{s} + \boldsymbol{v} &(5.19)\\
&= \sqrt{\frac{GE_{\mathrm{s}}}{N_{\mathrm{t}}}}\,\boldsymbol{U_H}\boldsymbol{\Sigma_H}\begin{bmatrix} \boldsymbol{P}^{1/2} \\ \boldsymbol{0} \end{bmatrix}\boldsymbol{s} + \boldsymbol{v}. &(5.20)
\end{aligned}
$$

Applying an innocuous unitary rotation $\boldsymbol{U}_{\boldsymbol{H}}^{*}$ at the receiver,

$$\boldsymbol{U}_{\boldsymbol{H}}^{*}\boldsymbol{y} = \sqrt{\frac{GE_{\mathrm{s}}}{N_{\mathrm{t}}}}\,\boldsymbol{\Sigma_H}\begin{bmatrix} \boldsymbol{P}^{1/2} \\ \boldsymbol{0} \end{bmatrix}\boldsymbol{s} + \boldsymbol{U}_{\boldsymbol{H}}^{*}\boldsymbol{v}, \tag{5.21}$$

where $\boldsymbol{U}_{\boldsymbol{H}}^{*}\boldsymbol{v} \sim \mathcal{N}_{\mathbb{C}}(\boldsymbol{0}, N_0\boldsymbol{I})$. Since $\left[\boldsymbol{P}^{1/2}\,\boldsymbol{0}\right]^{\mathrm{T}} = \mathrm{diag}(P_0,\ldots,P_{N_{\min}-1},0,\ldots,0)$ and

$$\boldsymbol{\Sigma_H} = \begin{bmatrix} \mathrm{diag}\left(\sqrt{\lambda_0},\ldots,\sqrt{\lambda_{N_{\min}-1}}\right) & \boldsymbol{0}_{N_{\min}\times(N_{\mathrm{t}}-N_{\min})} \\ \boldsymbol{0}_{(N_{\mathrm{r}}-N_{\min})\times N_{\min}} & \boldsymbol{0}_{(N_{\mathrm{r}}-N_{\min})\times(N_{\mathrm{t}}-N_{\min})} \end{bmatrix} \tag{5.22}$$

with $N_{\min} = \min(N_{\mathrm{t}}, N_{\mathrm{r}})$, we can finally write

$$[\boldsymbol{U}_{\boldsymbol{H}}^{*}\boldsymbol{y}]_j = \sqrt{\frac{GE_{\mathrm{s}}\,\lambda_j(\boldsymbol{H}^{*}\boldsymbol{H})\,P_j}{N_{\mathrm{t}}}}\,[\boldsymbol{s}]_j + [\boldsymbol{U}_{\boldsymbol{H}}^{*}\boldsymbol{v}]_j \qquad j = 0,\ldots,N_{\min}-1, \tag{5.23}$$

showing that the transmit–receive relationship can be cast as a bank of parallel subchannels, as many as $N_{\text{min}} = \min(N_t, N_r)$ under the default condition of $\boldsymbol{H}$ being full-rank; the number of signal streams with CSIT-based transmission is then $N_s \leq N_{\text{min}}$. The SNR on the jth subchannel is $\frac{\text{SNR}}{N_t} \lambda_j(\boldsymbol{H}^* \boldsymbol{H}) P_j$.

Power allocation

As we know from Section 4.4.2, when faced with a bank of parallel subchannels on which independent signals are conveyed, the capacity-achieving power allocation is given by the waterfilling policy. Couched for MIMO, this gives

$$P_j^\star = \left[\frac{1}{\eta} - \frac{N_t}{\text{SNR}\, \lambda_j(\boldsymbol{H}^* \boldsymbol{H})} \right]^+ \qquad j = 0, \ldots, N_{\text{min}} - 1, \tag{5.24}$$

where η must ensure that $\sum_j P_j^\star = N_t$.

Altogether, the optimum precoder features $\boldsymbol{U_F} = \boldsymbol{V_H}$ and $\boldsymbol{P} = \text{diag}(P_0^\star, \ldots, P_{N_{\text{min}}-1}^\star)$.

For future reference, the optimum $P_0^\star, \ldots, P_{N_{\text{min}}-1}^\star$ in (5.24) can be reformulated as the fixed point of the equations

$$P_j^\star = \frac{1 - \text{MMSE}_j(P_j^\star)}{\frac{1}{N_{\text{min}}} \sum_{q=0}^{N_{\text{min}}-1} (1 - \text{MMSE}_q(P_q^\star))} \qquad j = 0, \ldots, N_{\text{min}} - 1, \tag{5.25}$$

where

$$\text{MMSE}_j(P_j^\star) = \frac{1}{1 + \frac{\text{SNR}}{N_t} P_j^\star \lambda_j(\boldsymbol{H}^* \boldsymbol{H})} \tag{5.26}$$

is the MMSE incurred when estimating $[\boldsymbol{s}]_j$ by observing $\boldsymbol{y}$. It can be verified that (5.24) indeed solves (5.25).

Capacity

With the optimum precoder diagonalizing the channel and allocating power via waterfilling, the capacity of a given channel $\boldsymbol{H}$ equals

$$C(\text{SNR}, \boldsymbol{H}) = \sum_{j=0}^{N_{\text{min}}-1} \log_2 \left(1 + \frac{\text{SNR}}{N_t} \lambda_j(\boldsymbol{H}^* \boldsymbol{H}) P_j^\star \right) \tag{5.27}$$

$$= \sum_{j=0}^{N_{\text{min}}-1} \left[\log_2 \left(\frac{\text{SNR}}{N_t} \frac{\lambda_j(\boldsymbol{H}^* \boldsymbol{H})}{\eta} \right) \right]^+ . \tag{5.28}$$

Some observations are in order.

- Because the nonzero eigenvalues of $\boldsymbol{H}^* \boldsymbol{H}$ and $\boldsymbol{H} \boldsymbol{H}^*$ coincide, the capacity is unchanged if the channel matrix is transposed, i.e., if the roles of transmitter and receiver are reversed. This *duality* between the forward and reverse transmissions in SU-MIMO with CSIR and CSIT is but the tip of the iceberg of a deeper and powerful duality that is expounded later in the book, in the context of multiuser communication. (Without CSIT, this duality ceases to hold in general.)

- A further implication of the aforementioned duality is that swapping the values of N_t and N_r does not alter the capacity with CSIR and CSIT, i.e., a channel with N_r transmit and N_t receive antennas offers the same capacity as a channel with N_t transmit and N_r receive antennas. (Again, this is generally not the case without CSIT.)

Referring back to the notion of beamforming, whereby a scalar signal is simultaneously emitted from various antennas with suitable coefficients—the array response—applied, a useful interpretation of an optimum CSIT-based MIMO transmission is as a set of concurrent beams where the jth such beam conveys the jth entry of $\boldsymbol{s}$ with power determined by P_j and with an array response given by the jth column of $\boldsymbol{V_H}$, that is, by the jth eigenvector of $\boldsymbol{H}^*\boldsymbol{H}$. The rotation $\boldsymbol{U}_{\boldsymbol{H}}^*$ at the receiver can also be interpreted as a set of concurrent beams, defined by the columns of this matrix. The optimality arises because such transmit beams interlock perfectly with such receive beams along the channel's preferred directions, and these interlocked beams form orthogonal signal routes whose powers can be optimized via waterfilling. This interpretation is appealing, yet in its context the term "beam" should be used with the understanding that, while the angular distribution of power launched and received would retain the form of a beam in an LOS channel, multipath propagation distorts these angular structures—possibly beyond recognition. Hence, these beams exist in the vector space spanned by $\boldsymbol{H}$, but not necessarily in the angular domain.

The foregoing interpretation can be further stretched to construe the transmission of the jth entry of $\boldsymbol{s}$ as a scaling (by the magnitude of $[\boldsymbol{s}]_j$) and rotation (by the phase of $[\boldsymbol{s}]_j$) of the jth beam. MIMO would then amount to concurrent beams switching at the symbol rate with respect to directions dictated by the precoder and, when CSIT is available, these directions are the channel's preferred ones.

Besides optimum precoding, the other necessary ingredient in CSIT-based transmission is link adaptation, whereby the bit rate of each of the $N_{\min}$ signal streams is matched to the SNR of the corresponding subchannel. We defer the issue of MIMO link adaptation to Section 5.9, toward the end of the chapter, so as to leverage notions that are yet to be discussed and provide a more general view of this aspect.

Limiting regimes

Low-SNR regime

For $\mathsf{SNR} \to 0$, waterfilling dictates that all the power be allocated onto the strongest subchannel, which amounts to the precoder concentrating the power along the maximum-eigenvalue eigenvector of $\boldsymbol{H}^*\boldsymbol{H}$. This reduces the transmission to a single beam, which is what the term "beamforming" is typically reserved for: a transmission of unit rank, $N_s = 1$. This definition is the one espoused in this text, regardless of whether the power actually holds an angular beam shape.

If the largest eigenvalue of $\boldsymbol{H}^*\boldsymbol{H}$ has plural multiplicity, low-SNR optimality may require multiple equal-power beams determined by the corresponding eigenvectors.

High-SNR regime

For SNR $\to \infty$, the waterfilling power allocation becomes uniform and (5.27) expands as

$$C(\mathsf{SNR}, \boldsymbol{H}) = N_{\mathsf{min}} \log_2 \mathsf{SNR} + \mathcal{O}(1), \tag{5.29}$$

evidencing that the number of DOF is $\mathcal{S}_\infty = N_{\mathsf{min}}$: every 3-dB increment in SNR yields N_{min} additional b/s/Hz. This scaling is consequential, as it ignited the interest in MIMO.

The expansion in (5.29) can be further refined by means of $\mathcal{L}_\infty$, which in MIMO represents the offset (in 3-dB units) with respect to a bank of $\mathcal{S}_\infty$ parallel AWGN channels. From the definition in (4.46) and the capacity in (5.27), the power offset emerges as follows. If $N_{\mathrm{t}} \leq N_{\mathrm{r}}$, then $P_j^\star = 1$ for $j = 0, \ldots, N_{\mathrm{t}} - 1$ and

$$\mathcal{L}_\infty(\boldsymbol{H}) = \lim_{\mathsf{SNR}\to\infty} \left(\log_2 \mathsf{SNR} - \frac{C(\mathsf{SNR})}{N_{\mathrm{t}}} \right) \tag{5.30}$$

$$= \log_2 N_{\mathrm{t}} - \frac{1}{N_{\mathrm{t}}} \sum_{j=0}^{N_{\mathrm{t}}-1} \log_2 \lambda_j(\boldsymbol{H}^*\boldsymbol{H}) \tag{5.31}$$

$$= \log_2 N_{\mathrm{t}} - \frac{1}{N_{\mathrm{t}}} \log_2 \det(\boldsymbol{H}^*\boldsymbol{H}). \tag{5.32}$$

Conversely, if $N_{\mathrm{t}} \geq N_{\mathrm{r}}$, then $P_j^\star = N_{\mathrm{t}}/N_{\mathrm{r}}$ for $j = 0, \ldots, N_{\mathrm{r}} - 1$ and

$$\mathcal{L}_\infty(\boldsymbol{H}) = \log_2 N_{\mathrm{r}} - \frac{1}{N_{\mathrm{r}}} \log_2 \det(\boldsymbol{H}\boldsymbol{H}^*). \tag{5.33}$$

In both cases, the premise of a full-rank channel ensures that the log-determinants are bounded, and the refinement of (5.29) materializes as

$$C(\mathsf{SNR}, \boldsymbol{H}) = N_{\mathsf{min}} \left(\log_2 \mathsf{SNR} - \mathcal{L}_\infty(\boldsymbol{H}) \right) + \mathcal{O}\!\left(\frac{1}{\mathsf{SNR}} \right). \tag{5.34}$$

Discrete constellations

It is important to stress that the optimality of a precoder of the form $\boldsymbol{F} = \boldsymbol{U}_{\boldsymbol{F}} \left[\boldsymbol{P}^{1/2} \; \boldsymbol{0} \right]^{\mathrm{T}}$, and the ensuing diagonalization of the channel into parallel subchannels, rest on the signals being Gaussian. If the signals are drawn from other distributions, then complete precoders $\boldsymbol{F} = \boldsymbol{U}_{\boldsymbol{F}} \left[\boldsymbol{P}^{1/2} \; \boldsymbol{0} \right]^{\mathrm{T}} \boldsymbol{V}_{\boldsymbol{F}}^*$ may perform better. The incorporation of the mixing matrix $\boldsymbol{V}_{\boldsymbol{F}}$ has direct consequences for discrete constellations. On the one hand, it allows forming richer signals, which is relevant beyond some SNR; on the other hand, since the mixed signals launched into the channel are then not independent, it makes joint processing preferable and consequently it makes channel diagonalization generally undesirable. The receiver then becomes more involved than a mere bank of scalar decoders [466].

Example 5.1

Let $N_{\mathrm{t}} = 2$ with $\boldsymbol{s}$ containing two QPSK signals. When is a mixing matrix desirable?

Solution

Without a mixing matrix, at most 2 b/s/Hz can be dispatched through each of the columns of $\boldsymbol{U_F}$. If either of the two subchannels that would be obtained by diagonalizing $\boldsymbol{H}$ could accommodate a higher spectral efficiency, then a mixing matrix becomes desirable so as to produce a vector $\boldsymbol{V_F^* s}$ containing two 16-ary signals.

However, because these two 16-ary signals are not independent, the steering matrix $\boldsymbol{U_F}$ that maximizes the mutual information then no longer equals $\boldsymbol{V_H}$ and the channel is not diagonalized, meaning that joint processing of the two signals based on the observation of $\boldsymbol{y}$ becomes necessary.

The intuition provided by the foregoing example applies in broad generality: the mixing matrix becomes relevant whenever the spectral efficiency of the densest available constellation is insufficient if the power concentrates on the strongest subchannel, i.e., when

$$\log_2 M < \log_2\big(1 + \mathsf{SNR}\,\lambda_{\mathsf{max}}(\boldsymbol{H}^*\boldsymbol{H})\big), \tag{5.35}$$

with $\lambda_{\mathsf{max}}(\cdot)$ denoting the largest eigenvalue. Referring back to Fig. 4.3, the mixing matrix is relevant if we want to operate past the point where the spectral efficiency of the constellation ceases to hug the capacity. As long as this can be avoided, for instance with link adaptation that switches to denser constellations when the SNR is high, the optimum precoder devised for Gaussian signals remains effective. In well-designed systems, this should be the case.

If the constellations are constrained and insufficiently rich in a significant share of channel realizations, then the mixing matrix does help. And, just as the I-MMSE relationship for scalar channels spawns the mercury/waterfilling solution for parallel noninteracting channels, the I-MMSE relationship for vector channels can be capitalized on to optimize the complete precoder in MIMO. Unfortunately, and in contrast with the problem that arises under channel diagonalization, this more general optimization is not always concave in $\boldsymbol{F}\boldsymbol{F}^*$ and thus the conditions [467] that can be derived through the vector I-MMSE relationship are necessary but not sufficient for optimality; they are satisfied by any local maximum, minimum, and saddle point. Gradient search algorithms can be constructed to find these critical points [125], but due to their slow convergence these algorithms do not seem too well suited for wireless channels. Later developments have shown that the mutual information with arbitrary signal distributions is a concave function of the quadratic form $\boldsymbol{F}^*\boldsymbol{H}^*\boldsymbol{H}\boldsymbol{F}$, thereby opening the door to globally convergent algorithms [74, 468–470]; again, though, the slow convergence may restrict the applicability. As a simpler alternative, and despite being suboptimum, diagonalizing precoders of the form $\boldsymbol{F} = \boldsymbol{V_H}\big[\boldsymbol{P}^{1/2}\ \boldsymbol{0}\big]^{\mathrm{T}}$ can be used and then the best $\boldsymbol{P}$ for such precoders is given by mercury/waterfilling. Intermediate solutions offering a compromise between optimum precoding and pragmatic diagonalization-plus-mercury/waterfilling have also been formulated. These include the X-codes [471], which first diagonalize the channel and then further precode by mixing selected pairs of subchannels, and also the flexible reduced-mixing approach propounded in [472, 473].

For $\mathsf{SNR} \to 0$, the mixing matrix becomes unnecessary and, regardless of the signal distribution, the optimum precoder is the one derived for Gaussian signals, that is, beamforming on the maximum-eigenvalue eigenvector(s) of $\boldsymbol{H}^*\boldsymbol{H}$.

For $\mathsf{SNR} \to \infty$ with specific constellations, the mixing matrix is always eventually required and the optimum precoder turns out to be the one that maximizes the minimum distance between the constellation vectors [467]. Recall, however, that communication becomes highly power-inefficient once the spectral efficiency ceases to hug the capacity function and that operation in this regime should be avoided. In a well-designed system this is the case, as mentioned, also because receiver imperfections and other-user interference prevent the SNR from truly growing without bound [474, 475]. By and large then, the limit $\mathsf{SNR} \to \infty$ can be safely understood as indicating that the SNR becomes high but within the range where the mutual information hugs the capacity.

5.3.2 Ergodic setting

The ergodic capacity with CSIT is given by the expectation of the right-hand side of (5.28) over the distribution of $\lambda_j(\boldsymbol{H}^*\boldsymbol{H})$ for $j = 0, \ldots, N_{\mathsf{min}} - 1$. Since, unordered, these eigenvalues have the same marginal distribution,

$$C(\mathsf{SNR}) = \mathbb{E}\left[\sum_{j=0}^{N_{\mathsf{min}}-1}\left[\log_2\left(\frac{\mathsf{SNR}}{N_{\mathrm{t}}}\frac{\lambda_j(\boldsymbol{H}^*\boldsymbol{H})}{\eta}\right)\right]^+\right] \tag{5.36}$$

$$= N_{\mathsf{min}}\,\mathbb{E}\left[\left[\log_2\left(\frac{\mathsf{SNR}}{N_{\mathrm{t}}}\frac{\lambda(\boldsymbol{H}^*\boldsymbol{H})}{\eta}\right)\right]^+\right], \tag{5.37}$$

where $\lambda(\cdot)$ is an arbitrary eigenvalue. For the canonical channel with IID Rayleigh-faded entries, $\boldsymbol{H}^*\boldsymbol{H}$ is a Wishart matrix and the PDF of $\lambda(\boldsymbol{H}^*\boldsymbol{H})$ is given in Appendix C.1.9. The expectation in (5.37) can then be computed in closed form for a given η, but, unfortunately, η must still be optimized numerically [476, 477].

Large-dimensional expressions are available for the ergodic capacity of channels with IID entries [478, 479] and even with a Kronecker correlation structure [480]. These expressions, however, apply only for SNRs high enough that $P_j^\star > 0$ for $j = 0, \ldots, N_{\mathsf{min}} - 1$.

Limiting regimes

Low-SNR regime

At low SNR, recall, the waterfilling policy dictates that precoder beamform along the maximum-eigenvalue eigenvector of $\boldsymbol{H}^*\boldsymbol{H}$ for each realization of $\boldsymbol{H}$. The SNR experienced by the scalar signal sent on that eigenvector equals $\lambda_{\mathsf{max}}(\boldsymbol{H}^*\boldsymbol{H})\,\mathsf{SNR}$ and, recalling Example 1.7,

$$C(\mathsf{SNR}) = \left(\mathbb{E}[\lambda_{\mathsf{max}}(\boldsymbol{H}^*\boldsymbol{H})]\,\mathsf{SNR} - \frac{1}{2}\mathbb{E}[\lambda^2_{\mathsf{max}}(\boldsymbol{H}^*\boldsymbol{H})]\,\mathsf{SNR}^2\right)\log_2 e + o(\mathsf{SNR}^2). \tag{5.38}$$

It follows, applying (4.30) and (4.39), that

$$\frac{E_{\rm b}}{N_0}_{\rm min} = \frac{1}{\mathbb{E}\left[\lambda_{\rm max}(\boldsymbol{H}^*\boldsymbol{H})\right]\log_2 e} \tag{5.39}$$

$$\mathcal{S}_0 = \frac{2}{\kappa\left(\sqrt{\lambda_{\rm max}(\boldsymbol{H}^*\boldsymbol{H})}\right)}, \tag{5.40}$$

where, as usual, $\kappa(\cdot)$ denotes kurtosis. If $\lambda_{\rm max}(\boldsymbol{H}^*\boldsymbol{H})$ has plural multiplicity, (5.39) holds regardless of how the power is allocated among the corresponding eigenvectors (refer to Problem 5.12). To maximize $\mathcal{S}_0$, however, the transmit power should be evenly divided among those eigenvectors and then (5.38) and (5.40), which correspond to power allocation on only one eigenvector, should be modified accordingly. We also note that (5.38)–(5.40) hold for non-Gaussian signals as long as these are proper complex.

The contrast of (5.39) with its SISO counterpart

$$\frac{E_{\rm b}}{N_0}_{\rm min} = \frac{1}{\log_2 e} = -1.59 \text{ dB} \tag{5.41}$$

indicates that, thanks to the beamforming, the transmit power required to achieve $\frac{E_{\rm b}}{N_0}_{\rm min}$ can be reduced by $\mathbb{E}[\lambda_{\rm max}(\boldsymbol{H}^*\boldsymbol{H})]|_{\rm dB}$ relative to SISO. Alternatively, from (5.38), beamforming multiplies the capacity at a given (low) SNR by $\mathbb{E}[\lambda_{\rm max}(\boldsymbol{H}^*\boldsymbol{H})]$.

The expectation of $\lambda_{\rm max}(\boldsymbol{H}^*\boldsymbol{H})$ is far from simple, precluding direct general insights. In special cases though, insight is forthcoming.

Example 5.2 (Beamforming in a SIMO channel)

For $N_{\rm t} = 1$, we are faced with a single-input multiple-output (SIMO) channel. Then, $\boldsymbol{H}$ is a column vector and $\mathbb{E}[\lambda_{\rm max}(\boldsymbol{H}^*\boldsymbol{H})] = N_{\rm r}$. Receive beamforming amounts to maximum-ratio combining via the linear filter $\boldsymbol{U}_{\boldsymbol{H}} = \frac{1}{\|\boldsymbol{H}\|}\boldsymbol{H}$ at the receiver.

Example 5.3 (CSIT-based beamforming in a MISO channel)

For $N_{\rm r} = 1$, we have a multiple-input single-output (MISO) channel. Then, $\boldsymbol{H}$ is a row vector and beamforming reduces to maximum-ratio transmission whereby the precoder equals $\boldsymbol{F} = \frac{\sqrt{N_{\rm t}}}{\|\boldsymbol{H}\|}\boldsymbol{H}^*$ while $\mathbb{E}[\lambda_{\rm max}(\boldsymbol{H}^*\boldsymbol{H})] = N_{\rm t}$.

Example 5.4 (CSIT-based beamforming in a fully correlated MIMO channel)

If the antennas are fully correlated at both the transmitter and the receiver, then we have that $\mathbb{E}\left[\lambda_{\rm max}(\boldsymbol{H}^*\boldsymbol{H})\right] = N_{\rm t}N_{\rm r}$.

To unleash further intuition, we can let $N_{\rm t}$ and $N_{\rm r}$ grow large with ratio $\beta = N_{\rm t}/N_{\rm r}$ [481]. If the entries of $\boldsymbol{H}$ are IID and zero-mean, then the multiplicity of $\lambda_{\rm max}(\boldsymbol{H}^*\boldsymbol{H})$ becomes one while its value converges a.s. to a nonrandom limit. Precisely [482]

$$\frac{\lambda_{\rm max}(\boldsymbol{H}^*\boldsymbol{H})}{N_{\rm r}} \overset{\rm a.s.}{\to} \left(1+\sqrt{\beta}\right)^2. \tag{5.42}$$

Thus, in the absence of antenna correlations and for sufficiently large N_t and N_r,

$$\mathbb{E}\big[\lambda_{\mathsf{max}}(\boldsymbol{H}^*\boldsymbol{H})\big] \approx \left(\sqrt{N_t}+\sqrt{N_r}\right)^2, \tag{5.43}$$

which is substantially smaller than the $N_t N_r$ gain obtained in Example 5.4 for full correlation. Indeed, correlation is beneficial at low SNR, or whenever one chooses to beamform, because it creates preferred directions on which the channel is stronger and the beamforming more effective.

Example 5.5

Suppose that $N_t = N_r = 8$ with the entries of $\boldsymbol{H}$ being IID. By how much can the transmit power be reduced, relative to SISO, while maintaining the same low-SNR capacity?

Solution

Applying the large-dimensional approximation, we obtain

$$\mathbb{E}\big[\lambda_{\mathsf{max}}(\boldsymbol{H}^*\boldsymbol{H})\big]|_{\mathrm{dB}} \approx 20\log_{10}\left(\sqrt{N_t}+\sqrt{N_r}\right) = 15 \text{ dB}, \tag{5.44}$$

while the exact dB value, computed via Monte-Carlo, is $\mathbb{E}\big[\lambda_{\mathsf{max}}(\boldsymbol{H}^*\boldsymbol{H})\big]|_{\mathrm{dB}} = 13.8$ dB. The transmit power can be reduced by this dB amount, relative to SISO, and still achieve the same capacity at a receiver operating at low SNR. Alternatively, keeping the same transmit power, the capacity can be scaled (to first order) by the corresponding linear factor, $\mathbb{E}\big[\lambda_{\mathsf{max}}(\boldsymbol{H}^*\boldsymbol{H})\big] = 23.7$.

The foregoing results on beamforming have added importance because, in addition to being the optimum scheme at low SNR, beamforming is a popular and relatively simple technique that could be exercised irrespective of the SNR whenever CSIT is available.

High-SNR regime

The high-SNR behavior in (5.29), which applies to every full-rank channel realization, translates to any ergodic setting where such condition holds with probability 1. For all channels of interest then, $\mathcal{S}_\infty = N_{\mathsf{min}}$. As far as the power offset goes, by expecting over (5.32) and (5.33) we obtain, for $N_t \leq N_r$,

$$\mathcal{L}_\infty = \log_2 N_t - \frac{1}{N_t}\mathbb{E}\big[\log_2 \det(\boldsymbol{H}^*\boldsymbol{H})\big], \tag{5.45}$$

whereas, for $N_t \geq N_r$,

$$\mathcal{L}_\infty = \log_2 N_r - \frac{1}{N_r}\mathbb{E}\big[\log_2 \det(\boldsymbol{H}\boldsymbol{H}^*)\big]. \tag{5.46}$$

Although general closed forms for $\mathcal{L}_\infty$ seem out of reach, solutions can be found when the entries of $\boldsymbol{H}$ are Rayleigh-faded.

Example 5.6 (Power offset for an IID Rayleigh-faded MIMO channel with CSIT)

If $N_t \le N_r$, then $\boldsymbol{H}^*\boldsymbol{H}$ is a Wishart matrix. Invoking (C.28) in Appendix C.1.9, the expression in (5.45) specializes to

$$\mathcal{L}_\infty = \log_2 N_t - \frac{\log_2 e}{N_t} \sum_{j=0}^{N_t-1} \psi(N_t - j), \tag{5.47}$$

where $\psi(\cdot)$ is the digamma function (see Appendix E.2). Exploiting the recursive property in (E.9) we further obtain [479, 483–485]

$$\mathcal{L}_\infty = \log_2 N_t + \left(\gamma_{\mathsf{EM}} - \sum_{q=1}^{N_r-N_t} \frac{1}{q} - \frac{N_r}{N_t} \sum_{q=N_r-N_t+1}^{N_r} \frac{1}{q} + 1 \right) \log_2 e. \tag{5.48}$$

If $N_t \ge N_r$, then, replicating the preceding derivation with the Wishart matrix $\boldsymbol{H}\boldsymbol{H}^*$, or merely interchanging N_t and N_r, we obtain

$$\mathcal{L}_\infty = \log_2 N_r + \left(\gamma_{\mathsf{EM}} - \sum_{q=1}^{N_t-N_r} \frac{1}{q} - \frac{N_t}{N_r} \sum_{q=N_t-N_r+1}^{N_t} \frac{1}{q} + 1 \right) \log_2 e. \tag{5.49}$$

For $N_t = N_r = N_a$, both expressions reduce to the common form

$$\mathcal{L}_\infty = \log_2 N_a + \left(\gamma_{\mathsf{EM}} - \sum_{q=2}^{N_a} \frac{1}{q} \right) \log_2 e. \tag{5.50}$$

Example 5.7 (Power offset for a MIMO channel with Kronecker correlations, $N_t = N_r$ and CSIT)

For $\boldsymbol{H} = \boldsymbol{R}_r^{1/2} \boldsymbol{H}_w \boldsymbol{R}_t^{1/2}$ and $N_t = N_r = N_a$,

$$\mathcal{L}_\infty = \log_2 N_a - \frac{1}{N_a} \mathbb{E}\left[\log_2 \det \left(\boldsymbol{H}_w^* \boldsymbol{R}_r \boldsymbol{H}_w \boldsymbol{R}_t \right) \right] \tag{5.51}$$

$$= \log_2 N_a - \frac{1}{N_a} \mathbb{E}\left[\log_2 \det \left(\boldsymbol{H}_w^* \boldsymbol{R}_r \boldsymbol{H}_w \right) \right] - \frac{1}{N_a} \log_2 \det (\boldsymbol{R}_t) \tag{5.52}$$

$$= \log_2 N_a - \frac{1}{N_a} \mathbb{E}\left[\log_2 \det \left(\boldsymbol{H}_w \boldsymbol{H}_w^* \right) \right] - \frac{1}{N_a} \log_2 \det (\boldsymbol{R}_t) - \frac{1}{N_a} \log_2 \det (\boldsymbol{R}_r) \tag{5.53}$$

$$= \log_2 N_a + \left(\gamma_{\mathsf{EM}} - \sum_{q=2}^{N_a} \frac{1}{q} \right) \log_2 e - \frac{1}{N_a} \sum_{j=0}^{N_a-1} \log_2 \lambda_j(\boldsymbol{R}_t) - \frac{1}{N_a} \sum_{i=0}^{N_a-1} \log_2 \lambda_i(\boldsymbol{R}_r), \tag{5.54}$$

which equals the power offset of a channel without correlations, given in (5.50), plus two corrective terms associated with the transmit and receive correlations, respectively. These corrections are positive, e.g., for $\boldsymbol{R}_{\mathrm{t}}$, applying Jensen's inequality,

$$-\frac{1}{N_{\mathrm{a}}}\sum_{j=0}^{N_{\mathrm{a}}-1}\log_2\lambda_j(\boldsymbol{R}_{\mathrm{t}}) \geq -\log_2\left(\frac{1}{N_{\mathrm{a}}}\sum_{j=0}^{N_{\mathrm{a}}-1}\lambda_j(\boldsymbol{R}_{\mathrm{t}})\right) = 0 \tag{5.55}$$

and therefore, as they increase the power offset, both types of correlation are detrimental for $N_{\mathrm{t}} = N_{\mathrm{r}} = N_{\mathrm{a}}$.

Derivations similar to the one above can be conducted for $N_{\mathrm{t}} < N_{\mathrm{r}}$ with correlation only at the transmitter, and for $N_{\mathrm{t}} > N_{\mathrm{r}}$ with correlation only at the receiver. The general case $N_{\mathrm{t}} \neq N_{\mathrm{r}}$ with correlation at both ends requires additional mathematical machinery, and the interested reader is referred to [355, proposition 4].

5.4 No CSIT

Let us now turn our attention to the case that the transmitter is not privy to $\boldsymbol{H}$, but only to its distribution. Without CSIT, no time-domain power control is possible at the small-scale level and thus we need not distinguish between per-codeword and per-symbol power constraints. The fundamental quantity here is the mutual information between $\boldsymbol{s}$ and $\boldsymbol{y}$, conditioned on $\boldsymbol{H}$ to register the CSIR. The arguments that prove the optimality of complex Gaussian signals in SISO extend readily to MIMO and hence, recalling Example 1.13,

$$I(\boldsymbol{s};\boldsymbol{y}|\boldsymbol{H}) = \log_2\det\left(\boldsymbol{I} + \frac{\mathsf{SNR}}{N_{\mathrm{t}}}\boldsymbol{H}\boldsymbol{F}\boldsymbol{F}^*\boldsymbol{H}^*\right) \tag{5.56}$$

where the precoder satisfies $\mathrm{tr}(\boldsymbol{F}\boldsymbol{F}^*) = \mathrm{tr}(\boldsymbol{P}) = N_{\mathrm{t}}$.

5.4.1 Quasi-static setting

In a flat-faded quasi-static setting, we must resort to the notion of outage capacity introduced for SISO, only with the mutual information of the scalar channel replaced by (5.56). This gives

$$C_\epsilon(\mathsf{SNR}) = \max_c(c : p_{\mathsf{out}}(\mathsf{SNR}, c) < \epsilon), \tag{5.57}$$

with the outage probability being

$$p_{\mathsf{out}}(\mathsf{SNR}, R/B) = \mathbb{P}\left[\log_2\det\left(\boldsymbol{I} + \frac{\mathsf{SNR}}{N_{\mathrm{t}}}\boldsymbol{H}\boldsymbol{F}\boldsymbol{F}^*\boldsymbol{H}^*\right) < R/B\right]. \tag{5.58}$$

The case $N_{\mathrm{t}} = 1$, while not strictly MIMO but rather SIMO, has historical significance as it embodies receive diversity setups that preceded MIMO when the idea of multiantenna devices was still far-fetched.

Example 5.8

Express the outage probability in a Rayleigh-faded SIMO channel.

Solution

Let $N_\mathrm{t} = 1$ and let the vector $\boldsymbol{H}$ have IID Rayleigh-faded entries. Then, $\|\boldsymbol{H}\|^2 \sim \chi^2_{2N_\mathrm{r}}$, meaning that $\|\boldsymbol{H}\|^2$ follows a chi-square distribution with $2N_\mathrm{r}$ degrees of freedom (see Appendix C.1.9) and

$$p_{\mathsf{out}}(\mathsf{SNR}, R/B) = \mathbb{P}\big[\log_2\big(1 + \mathsf{SNR}\,\|\boldsymbol{H}\|^2\big) < R/B\big] \tag{5.59}$$

$$= \mathbb{P}\left[\|\boldsymbol{H}\|^2 < \frac{2^{R/B} - 1}{\mathsf{SNR}}\right] \tag{5.60}$$

$$= \int_0^{(2^{R/B}-1)/\mathsf{SNR}} f_{\|\boldsymbol{H}\|^2}(\xi)\, \mathrm{d}\xi \tag{5.61}$$

$$= \int_0^{(2^{R/B}-1)/\mathsf{SNR}} \frac{\xi^{N_\mathrm{r}-1} e^{-\xi}}{(N_\mathrm{r}-1)!}\, \mathrm{d}\xi \tag{5.62}$$

$$= \frac{\gamma\big(N_\mathrm{r}, (2^{R/B} - 1)/\mathsf{SNR}\big)}{(N_\mathrm{r} - 1)!}, \tag{5.63}$$

where the chi-square PDF appears in (5.62) while $\gamma(\cdot,\cdot)$ is the lower incomplete Gamma function (see Appendix E.1).

For $N_\mathrm{t} > 1$, the transmitter has the added freedom of selecting the precoder $\boldsymbol{F}$. Strictly speaking, the outage capacity then entails a maximization over $\boldsymbol{F}$ for each choice of p_{out}. This maximization is in general unwieldy and the term "outage capacity" is typically invoked even if $\boldsymbol{F}$ is fixed or outright absent. Nonetheless, as the following examples illustrate, in some special cases it is indeed possible to, at least partially, characterize the optimum $\boldsymbol{F}$ as a function of p_{out}.

Example 5.9

Determine the outage-optimum precoder for a channel with IID Rayleigh-faded entries.

Solution

If $\boldsymbol{H}$ has IID complex Gaussian entries, or more generally it is unitarily invariant, then the steering matrix $\boldsymbol{U}_{\boldsymbol{F}}$ is immaterial because $\boldsymbol{H}\boldsymbol{U}_{\boldsymbol{F}} \sim \boldsymbol{H}$ and we can confine the optimization to the diagonal power allocation matrix $\boldsymbol{P}$. Based on the symmetry of the problem, Telatar conjectured that the optimum precoder might have the form [25]

$$\boldsymbol{F} = \frac{N_\mathrm{t}}{Q}\,\mathrm{diag}\big(\underbrace{1,\ldots,1}_{Q\text{ ones}}, \underbrace{0,\ldots,0}_{N_\mathrm{t}-Q\text{ zeros}}\big), \tag{5.64}$$

with Q dependent on p_{out}. This conjecture was eventually proved for MISO channels [486].

Example 5.10

If $\boldsymbol{H}$ has correlations only at the transmitter, then the optimum $\boldsymbol{U}_{\boldsymbol{F}}$ always coincides with the eigenvector matrix of $\boldsymbol{R}_\mathrm{t}$ [487]. The power allocation $\boldsymbol{P}$, however, must be optimized numerically depending on p_{out} and SNR.

In the high-SNR regime, recall, the tradeoff between spectral efficiency and outage probability is very often studied through their proxies, the multiplexing gain r and the diversity order d respectively, with the DMT as the benchmark. With MIMO, the term *multiplexing gain* acquires its full significance as it becomes possible that $r > 1$. More precisely, $0 \leq r \leq N_{\mathsf{min}}$. The full DMT for Rayleigh fading is defined by the piecewise linear curve connecting the points (r, d) where $r = 0, \ldots, N_{\mathsf{min}}$ and [395]

$$d = (N_{\mathrm{t}} - r)(N_{\mathrm{r}} - r). \tag{5.65}$$

At one end of the DMT, $r = 0$ and $d = N_{\mathrm{t}} N_{\mathrm{r}}$: the spectral efficiency does not increase for $\mathsf{SNR} \to \infty$ (at least not with $\log_2 \mathsf{SNR}$) and the outage probability then decays as $1/\mathsf{SNR}^{N_{\mathrm{t}} N_{\mathrm{r}}}$, accelerated by MIMO. At the other end, $r = N_{\mathsf{min}}$ and $d = 0$: the spectral efficiency does increase as $R/B = N_{\mathsf{min}} \log_2 \mathsf{SNR} + \mathcal{O}(1)$ with the full force of MIMO directed toward a higher spectral efficiency, but the outage probability does not decay with SNR (at least not polynomially). We note the following.

- Noninteger values for r and d are possible, simply by scaling the spectral efficiency fractionally with $\log_2 \mathsf{SNR}$.
- For suboptimum transmit and receive architectures, the tradeoff between r and d may fall short of the DMT in (5.65).
- N_{s} and r should not be confused. The former is the number of signal streams that are simultaneously transmitted whereas the latter informs of how the amount of information encoded in these streams scales with $\log_2 \mathsf{SNR}$. For sure, $r \leq N_{\mathrm{s}}$.
- If there is a precoder, only its rank N_{s} affects the DMT. This does not imply that the precoder choice is irrelevant at high SNR, but that the coarse description provided by the DMT is unable to distinguish between different precoders of the same rank.
- The DMT is likewise insensitive to nonsingular antenna correlations, because they do not modify the number of spatial DOF.

As explicated in Section 4.5.2, care must be exercised not to extrapolate the meaning of the DMT beyond what it actually signifies.

Example 5.11

Draw the DMT for $N_{\mathrm{t}} = 3$ and $N_{\mathrm{r}} = 4$ in Rayleigh fading.

Solution

See Fig. 5.3.

As the number of antennas grows large, the fluctuations of the mutual information around its expected value tend, in fairly broad generality, to a normal behavior [488–497]. This is because, even though each entry of $\boldsymbol{H}$ is stuck at a given value in a quasi-static setting, with an increasing number of entries the fading distribution is revealed and outage-free communication becomes possible [498, 499]. Put differently, the quasi-static setting becomes progressively ergodic in the space domain and thus the ergodic capacity, studied next, is directly the metric of interest.

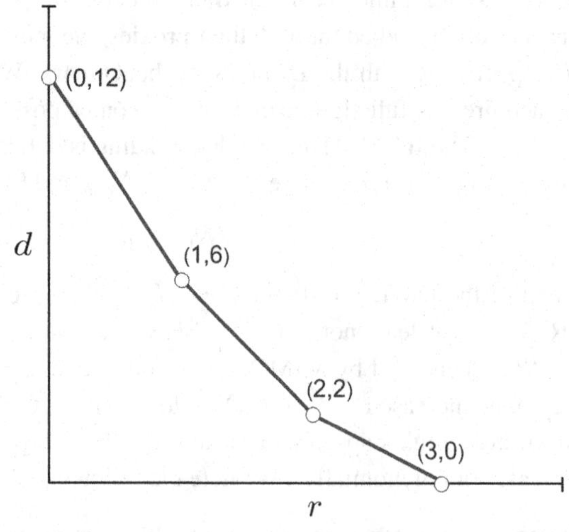

Fig. 5.3 DMT for $N_\mathrm{t} = 3$ and $N_\mathrm{r} = 4$ in Rayleigh fading.

5.4.2 Ergodic setting

From (5.56), the ergodic capacity equals

$$C(\mathsf{SNR}) = \max_{\boldsymbol{F}:\mathrm{tr}(\boldsymbol{F}\boldsymbol{F}^*)=N_\mathrm{t}} \mathbb{E}\left[\log_2 \det\left(\boldsymbol{I} + \frac{\mathsf{SNR}}{N_\mathrm{t}}\boldsymbol{H}\boldsymbol{F}\boldsymbol{F}^*\boldsymbol{H}^*\right)\right], \tag{5.66}$$

which is the vector brethren of (4.165), only involving a convex optimization over $\boldsymbol{F}\boldsymbol{F}^*$ (or equivalently over the transmit covariance $\boldsymbol{R}_{\boldsymbol{x}}$) [500]. Indeed, the expectation in (5.66) is concave in $\boldsymbol{F}\boldsymbol{F}^*$ and the set of admissible matrices $\boldsymbol{F}\boldsymbol{F}^*$, i.e., positive-semidefinite and satisfying $\mathrm{tr}(\boldsymbol{F}\boldsymbol{F}^*) = N_\mathrm{t}$, is convex.

Although one can concoct channels where the capacity-achieving precoder is not unique, in the vast majority of cases it is [501]. The precoder structure in Fig. 5.2 continues to apply, only with $\boldsymbol{U}_{\boldsymbol{F}}$ and $\boldsymbol{P}$ computed differently and kept steady throughout the fading. Completely general expressions for $\boldsymbol{U}_{\boldsymbol{F}}$ and $\boldsymbol{P}$ are not forthcoming, but a fairly precise characterization can be provided that encompasses most of the MIMO channel models described in Chapter 3.

Precoder optimization

Steering matrix

Let us first examine $\boldsymbol{U}_{\boldsymbol{F}}$. For a broad class of channels (including the Kronecker and UIU models, the IID Rice model, and the keyhole model), the optimum precoder diagonalizes $\mathbb{E}[\boldsymbol{H}^*\boldsymbol{H}]$ [291, 502–506]; this is a gratifying counterpart to the CSIT result where $\boldsymbol{H}^*\boldsymbol{H}$ was diagonalized. Without CSIT, the strategy then amounts to diagonalizing $\boldsymbol{H}^*\boldsymbol{H}$ *on average*. Although this steering strategy is not universally optimum, requiring a modicum of symmetry and structure in the distribution of $\boldsymbol{H}$ [507][508, myth 1], for most channels of interest we can indeed affirm that the optimum U_F is the eigenvector matrix of $\mathbb{E}[H^* H]$. Next, we specialize this result to specific models.

Example 5.12 (Steering matrix for a MIMO channel with Kronecker correlations and no CSIT)

Consider the Kronecker correlation model, $H = R_{\mathrm{r}}^{1/2} H_{\mathrm{w}} R_{\mathrm{t}}^{1/2}$. Defining the eigenvalue decompositions $R_{\mathrm{t}} = U_{\mathrm{t}} \Lambda_{\mathrm{t}} U_{\mathrm{t}}^*$ and $R_{\mathrm{r}} = U_{\mathrm{r}} \Lambda_{\mathrm{r}} U_{\mathrm{r}}^*$,

$$\mathbb{E}[H^* H] = \mathbb{E}\left[R_{\mathrm{t}}^{1/2} H_{\mathrm{w}}^* R_{\mathrm{r}} H_{\mathrm{w}} R_{\mathrm{t}}^{1/2}\right] \tag{5.67}$$

$$= \mathbb{E}\left[U_{\mathrm{t}} \Lambda_{\mathrm{t}}^{1/2} H_{\mathrm{w}}^* \Lambda_{\mathrm{r}} H_{\mathrm{w}} \Lambda_{\mathrm{t}}^{1/2} U_{\mathrm{t}}^*\right] \tag{5.68}$$

$$= U_{\mathrm{t}} \Lambda_{\mathrm{t}}^{1/2}\, \mathbb{E}\left[H_{\mathrm{w}}^* \Lambda_{\mathrm{r}} H_{\mathrm{w}}\right] \Lambda_{\mathrm{t}}^{1/2} U_{\mathrm{t}}^* \tag{5.69}$$

$$= N_{\mathrm{r}}\, U_{\mathrm{t}} \Lambda_{\mathrm{t}} U_{\mathrm{t}}^*, \tag{5.70}$$

where, in (5.68), the unitary matrices right and left of H_{w}^* and H_{w} are absorbed therein owing to unitary invariance whereas (5.70) follows from $\mathbb{E}[H_{\mathrm{w}}^* \Lambda_{\mathrm{r}} H_{\mathrm{w}}] = N_{\mathrm{r}} I$. The steering matrix is thus $U_F = U_{\mathrm{t}}$.

Example 5.13 (Steering matrix for a UIU MIMO channel with no CSIT)

Let $H = U_{\mathrm{r}} H_{\mathsf{ind}} U_{\mathrm{t}}^*$ where H_{ind} has zero-mean IND entries. Then,

$$\mathbb{E}[H^* H] = \mathbb{E}[U_{\mathrm{t}} H_{\mathsf{ind}}^* H_{\mathsf{ind}} U_{\mathrm{t}}^*] \tag{5.71}$$

$$= U_{\mathrm{t}}\, \mathbb{E}\left[H_{\mathsf{ind}}^* H_{\mathsf{ind}}\right] U_{\mathrm{t}}^*, \tag{5.72}$$

where $\mathbb{E}[H_{\mathsf{ind}}^* H_{\mathsf{ind}}]$ is a diagonal matrix. Thus, $U_F = U_{\mathrm{t}}$.

Example 5.14 (Steering matrix for a Rice MIMO channel with no CSIT)

Consider the uncorrelated Rice channel

$$H = \sqrt{\frac{\mathsf{K}}{\mathsf{K}+1}} H_{\mathsf{LOS}} + \sqrt{\frac{1}{\mathsf{K}+1}} H_{\mathrm{w}}. \tag{5.73}$$

Defining the eigenvalue decomposition $H_{\mathsf{LOS}}^* H_{\mathsf{LOS}} = U_{\mathsf{LOS}} \Lambda_{\mathsf{LOS}} U_{\mathsf{LOS}}^*$,

$$\mathbb{E}[H^* H] = \frac{\mathsf{K}}{\mathsf{K}+1} H_{\mathsf{LOS}}^* H_{\mathsf{LOS}} + \frac{1}{\mathsf{K}+1} \mathbb{E}[H_{\mathrm{w}}^* H_{\mathrm{w}}] \tag{5.74}$$

$$= \frac{\mathsf{K}}{\mathsf{K}+1} U_{\mathsf{LOS}} \Lambda_{\mathsf{LOS}} U_{\mathsf{LOS}}^* + \frac{N_{\mathrm{r}}}{\mathsf{K}+1} I \tag{5.75}$$

$$= \frac{1}{\mathsf{K}+1} U_{\mathsf{LOS}} (\mathsf{K} \Lambda_{\mathsf{LOS}} + N_{\mathrm{r}} I) U_{\mathsf{LOS}}^* \tag{5.76}$$

and thus $U_F = U_{\mathsf{LOS}}$.

Power allocation

Recall how, with both CSIR and CSIT, the precoder's rank was found to be $N_s \leq N_{\min}$, with equality at high SNR and then diminishing toward $N_s = 1$ (beamforming) at low SNR. This emanates directly from the formulation of the capacity.

With only CSIR, the precoder's rank, and the optimum power allocation over the corresponding data streams, should again emerge from the capacity. We equate the steering matrix, $\boldsymbol{U_F}$, to the eigenvector matrix of $\mathbb{E}[\boldsymbol{H}^*\boldsymbol{H}]$, as per the considerations in the foregoing subsection, and proceed to optimize over $\boldsymbol{P} = \text{diag}(P_0, \ldots, P_{N_s-1})$ with $\sum_j P_j = N_t$. Then, (5.66) becomes

$$C(\mathsf{SNR}) = \max_{\boldsymbol{P}:\text{tr}(\boldsymbol{P})=N_t} \mathbb{E}\left[\log_2 \det\left(\boldsymbol{I} + \frac{\mathsf{SNR}}{N_t}\boldsymbol{H}\boldsymbol{U_F}\boldsymbol{P}\boldsymbol{U_F^*}\boldsymbol{H}^*\right)\right], \tag{5.77}$$

which is a convex problem. Thus, the necessary and sufficient conditions satisfied by the optimum $\boldsymbol{P}^\star$ can be obtained by applying the techniques in Appendix G. Alternatively, these conditions can be derived from first principles [291, appendix B][509, 510] by imposing that the derivative of the spectral efficiency in the direction from $\boldsymbol{P}^\star$ to any other $\boldsymbol{P}$ be negative. Plugging $\boldsymbol{P}_\xi = (1-\xi)\boldsymbol{P}^\star + \xi\boldsymbol{P}$ into the argument of (5.77) we obtain, for $0 \leq \xi \leq 1$, the spectral efficiency over the line that connects $\boldsymbol{P}^\star$ with $\boldsymbol{P}$. Enforcing that its one-side derivative with respect to ξ at $\xi = 0^+$ be negative, we obtain

$$\mathbb{E}\left[\text{tr}\left(\left(\boldsymbol{I} + \frac{\mathsf{SNR}}{N_t}\boldsymbol{H}\boldsymbol{U_F}\boldsymbol{P}\boldsymbol{U_F^*}\boldsymbol{H}^*\right)\left(\boldsymbol{I} + \frac{\mathsf{SNR}}{N_t}\boldsymbol{H}\boldsymbol{U_F}\boldsymbol{P}^\star\boldsymbol{U_F^*}\boldsymbol{H}^*\right)^{-1} - \boldsymbol{I}\right)\right] \leq 0. \tag{5.78}$$

As it turns out, because the above condition is affine on $\boldsymbol{P}$, imposing it on the N_t extreme points of the set of admissible matrices $\boldsymbol{P}$ suffices to enforce it for the entire set. At the jth extreme point ($P_j = N_t$, $P_{j'} = 0$ for $j' \neq j$), the condition specializes to

$$\mathbb{E}\left[\text{tr}\left((\boldsymbol{I} + \mathsf{SNR}\,\boldsymbol{H}\boldsymbol{u}_j\boldsymbol{u}_j^*\boldsymbol{H}^*)\left(\boldsymbol{I} + \frac{\mathsf{SNR}}{N_t}\boldsymbol{H}\boldsymbol{U_F}\boldsymbol{P}^\star\boldsymbol{U_F^*}\boldsymbol{H}^*\right)^{-1} - \boldsymbol{I}\right)\right] \leq 0, \tag{5.79}$$

where $\boldsymbol{u}_j = [\boldsymbol{U_F}]_{:,j}$. The line connecting the jth extreme point with $\boldsymbol{P}^\star$ extends beyond $\boldsymbol{P}^\star$ if and only if $P_j^\star > 0$, i.e., if and only if $\boldsymbol{P}^\star$ is in the interior of the set (all powers strictly positive); the derivative at $\boldsymbol{P}^\star$ then vanishes and (5.79) becomes a strict equality. Otherwise, $P_j^\star = 0$, meaning that $\boldsymbol{P}^\star$ is on the boundary, and (5.79) remains an inequality. From these considerations, cartooned in Fig. 5.4, we can write

$$\mathbb{E}\left[\text{tr}\left((\boldsymbol{I} + \mathsf{SNR}\,\boldsymbol{H}\boldsymbol{u}_j\boldsymbol{u}_j^*\boldsymbol{H}^*)\left(\boldsymbol{I} + \frac{\mathsf{SNR}}{N_t}\boldsymbol{H}\boldsymbol{U_F}\boldsymbol{P}^\star\boldsymbol{U_F^*}\boldsymbol{H}^*\right)^{-1}\right)\right] \begin{array}{ll} = N_r & \text{if} \;\; P_j^\star > 0 \\ \leq N_r & \text{if} \;\; P_j^\star = 0. \end{array} \tag{5.80}$$

Interestingly, these N_t equations can be rewritten to involve the MMSE that would be incurred if we were to estimate $\boldsymbol{s}$ at the receiver, adding to the body of results that connect information theory with estimation. Couching (1.245) for the model at hand, the MMSE matrix if the receiver were to estimate $\boldsymbol{s} \sim \mathcal{N}_{\mathbb{C}}(\boldsymbol{0}, \boldsymbol{I})$ on the basis of $\boldsymbol{y}$ would be

$$\boldsymbol{E} = \boldsymbol{I} - \frac{\mathsf{SNR}}{N_t}\boldsymbol{P}^{1/2}\boldsymbol{U_F^*}\boldsymbol{H}^*\left(\boldsymbol{I} + \frac{\mathsf{SNR}}{N_t}\boldsymbol{H}\boldsymbol{U_F}\boldsymbol{P}\boldsymbol{U_F^*}\boldsymbol{H}^*\right)^{-1}\boldsymbol{H}\boldsymbol{U_F}\boldsymbol{P}^{1/2}. \tag{5.81}$$

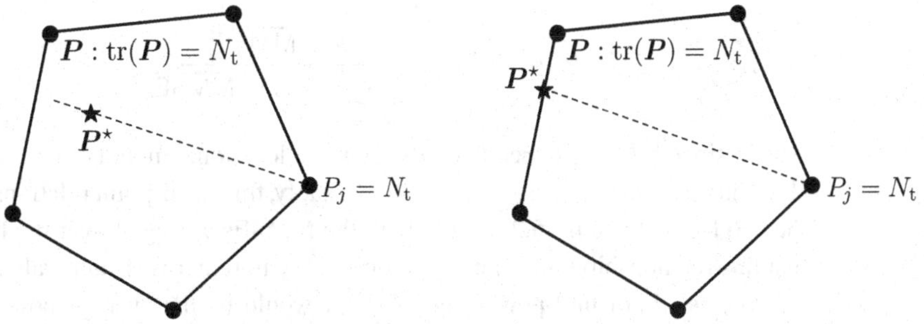

Fig. 5.4 Convex set of matrices $\boldsymbol{P}$ satisfying $\mathrm{tr}(\boldsymbol{P}) = N_\mathrm{t}$ and $P_j \geq 0$, with the boundary defined by those for which one or several powers are zero. At the extreme points, indicated by solid circles, only one power is nonzero. Left, a case where $\boldsymbol{P}^\star$ is in the interior. Right, a case where $\boldsymbol{P}^\star$ is on the boundary.

It follows that the estimation of $[\boldsymbol{s}]_j$ would incur an error with variance

$$\mathsf{MMSE}_j = [\boldsymbol{E}]_{j,j} \tag{5.82}$$

$$= 1 - \frac{\mathsf{SNR}}{N_\mathrm{t}} P_j \boldsymbol{u}_j^* \boldsymbol{H}^* \left(\boldsymbol{I} + \frac{\mathsf{SNR}}{N_\mathrm{t}} \boldsymbol{H} \boldsymbol{U_F} \boldsymbol{P} \boldsymbol{U_F^*} \boldsymbol{H}^* \right)^{-1} \boldsymbol{H} \boldsymbol{u}_j \tag{5.83}$$

while, with a bit of algebra (proposed in Problem 5.19), it can be verified that

$$\mathrm{tr}\left(\left(\boldsymbol{I} + \frac{\mathsf{SNR}}{N_\mathrm{t}} \boldsymbol{H} \boldsymbol{U_F} \boldsymbol{P} \boldsymbol{U_F^*} \boldsymbol{H}^* \right)^{-1} \right) = \sum_{j=0}^{N_\mathrm{t}-1} \mathsf{MMSE}_j + N_\mathrm{r} - N_\mathrm{t}. \tag{5.84}$$

For $P_j^\star > 0$, we can develop the left-hand side of (5.80) into

$$\mathbb{E}\left[\mathrm{tr}\left(\left(\boldsymbol{I} + \frac{\mathsf{SNR}}{N_\mathrm{t}} \boldsymbol{H} \boldsymbol{U_F} \boldsymbol{P}^\star \boldsymbol{U_F^*} \boldsymbol{H}^* \right)^{-1} \right.\right.$$

$$\left.\left. + \mathsf{SNR}\, \boldsymbol{H} \boldsymbol{u}_j \boldsymbol{u}_j^* \boldsymbol{H}^* \left(\boldsymbol{I} + \frac{\mathsf{SNR}}{N_\mathrm{t}} \boldsymbol{H} \boldsymbol{U_F} \boldsymbol{P}^\star \boldsymbol{U_F^*} \boldsymbol{H}^* \right)^{-1} \right) \right] \tag{5.85}$$

and, subsequently, into

$$\sum_{q=0}^{N_\mathrm{t}-1} \overline{\mathsf{MMSE}}_q + N_\mathrm{r} - N_\mathrm{t} + \mathbb{E}\left[\mathrm{tr}\left(\mathsf{SNR}\, \boldsymbol{u}_j^* \boldsymbol{H}^* \left(\boldsymbol{I} + \frac{\mathsf{SNR}}{N_\mathrm{t}} \boldsymbol{H} \boldsymbol{U_F} \boldsymbol{P}^\star \boldsymbol{U_F^*} \boldsymbol{H}^* \right)^{-1} \boldsymbol{H} \boldsymbol{u}_j \right) \right]$$

$$= \sum_{q=0}^{N_\mathrm{t}-1} \overline{\mathsf{MMSE}}_q + N_\mathrm{r} - N_\mathrm{t} + \mathbb{E}\left[\mathsf{SNR}\, \boldsymbol{u}_j^* \boldsymbol{H}^* \left(\boldsymbol{I} + \frac{\mathsf{SNR}}{N_\mathrm{t}} \boldsymbol{H} \boldsymbol{U_F} \boldsymbol{P}^\star \boldsymbol{U_F^*} \boldsymbol{H}^* \right)^{-1} \boldsymbol{H} \boldsymbol{u}_j \right]$$

$$= \sum_{q=0}^{N_\mathrm{t}-1} \overline{\mathsf{MMSE}}_q + N_\mathrm{r} - N_\mathrm{t} + \frac{N_\mathrm{t}}{P_j^\star} (1 - \overline{\mathsf{MMSE}}_j), \tag{5.86}$$

where we have used (5.83) and (5.84) while defining $\overline{\mathsf{MMSE}}_j = \mathbb{E}[\mathsf{MMSE}_j]$. Equated to

N_r, as per (5.80), the above gives

$$P_j^\star = \frac{1 - \overline{\mathsf{MMSE}}_j}{\frac{1}{N_t}\sum_{q=0}^{N_t-1}(1 - \overline{\mathsf{MMSE}}_q)}, \tag{5.87}$$

which, since $\overline{\mathsf{MMSE}}_j$ depends on the power allocation, amounts to a fixed-point relationship characterizing $P_0^\star, \ldots, P_{N_t-1}^\star$. Remarkably, this fixed-point relationship is identical to the CSIT-based one in (5.25), only with the MMSEs averaged over the fading, suggesting that this relationship holds the key to optimality in rather wide generality.

Since $[\boldsymbol{s}]_j$ is of unit-power and $\overline{\mathsf{MMSE}}_j$ would be the average power of the error estimating it, $1 - \overline{\mathsf{MMSE}}_j$ would be the average power of the signal estimate. Examining (5.87) then, $P_j^\star$, which is the share of the total transmit power allocated to the jth signaling eigenvector, should equal the share of average signal power that would be recovered from that eigenvector by an LMMSE estimator. If the channel is poor on average along the jth signaling eigenvector, then $P_j^\star = 0$ and $\overline{\mathsf{MMSE}}_j = 1$, and that dimension can be taken out of the optimization. Specifically, developing (5.80) with $P_j^\star = 0$, the precise condition for the jth dimension not to be active emerges as

$$\frac{1}{P_j}\mathbb{E}\left[\frac{1}{\mathsf{MMSE}_j} - 1\right] \leq \frac{1}{N_t}\sum_{q=0}^{N_t-1}(1 - \overline{\mathsf{MMSE}}_q) \tag{5.88}$$

for $P_j \to 0$.

Despite the parallelism between the CSIT-based expression in (5.25) and its no-CSIT counterpart in (5.87), two differences do exist.

- With CSIT, $N_s \leq N_{\mathsf{min}}$ while, without CSIT, the optimization yields a precoder whose rank is possibly as large as N_t. This difference, only material for $N_t > N_r$, is commented on in Discussions 5.2–5.4.
- Whereas the CSIT-based solution reduces to a waterfilling, the no-CSIT solution does not. Because the signals sent through the eigenvectors of $\mathbb{E}[\boldsymbol{H}^*\boldsymbol{H}]$ are mutually interfering at the receiver, the power allocation that solves (5.87) is in general *not* a waterfilling; in particular, a waterfilling—sometimes termed *statistical waterfilling*—on the eigenvalues of $\mathbb{E}[\boldsymbol{H}^*\boldsymbol{H}]$ does not generally lead to the capacity-achieving power allocation $\boldsymbol{P}^\star$, although it can provide solutions that are quite satisfactory [511].

Rather than the hallway to a waterfilling, (5.87) and (5.88) constitute a set of coupled equations that invites an iterative approach, and indeed simply feeding $P_0, \ldots, P_{N_t-1}$ repeatedly into these relationships tends to lead to a solution [512]. An obvious hurdle, however, is the need to average $\mathsf{MMSE}_0, \ldots, \mathsf{MMSE}_{N_t-1}$ over the fading at every iteration to find $\overline{\mathsf{MMSE}}_0, \ldots, \overline{\mathsf{MMSE}}_{N_t-1}$. This obstacle can be dodged in several ways.

- Using closed forms for $\overline{\mathsf{MMSE}}_j$ in those cases in which they are known [513–515].
- Borrowing solutions obtained in the large-dimensional regime, where the fading-related randomness disappears and the averaging is no longer needed.
- Leveraging the simpler solutions that emerge in the low- and high-SNR regimes.

Discussion 5.2 Dimensional overloading

With CSIR, but no CSIT, the optimum precoder can have a rank as large as N_t. For $N_t > N_r$, this entails overloading the channel with more signal streams than receiver dimensions. The benefit of overloading tends to be modest, and it takes a toll in receiver complexity, but it is necessary to reach capacity when the eigendirections of $\boldsymbol{H}$ are not known to the transmitter. Take the all-important channel with IID entries: the optimum precoder with CSIR but no CSIT, $\boldsymbol{F} = \boldsymbol{I}$, does entail transmission from all antennas, even if $N_t > N_r$.

Enter pilot overhead, which, as shown later in the chapter, scales with the precoder's rank. One could think that, given the modest advantage of overloading, a precoder of lower rank requiring less overhead may actually be preferable; in channels with IID entries, this would amount to activating less than N_t antennas.

In underspread channels, as it turns out, the additional overhead associated with an overloading precoder only erases the corresponding benefits at extreme SNRs (either very small or very large). Elsewhere these benefits largely stand, and altogether the CSIR results obtained throughout this chapter—possible refined by the appropriate overhead and channel-estimation-related SNR penalty—are representative of pilot-assisted transmission. The issue of dimensional overloading is revisited in Discussions 5.3 and 5.4, and it is the subject of Problems 5.41–5.44.

- Resorting to suboptimum but easier-to-compute power allocations. This is a pragmatic yet often effective option, with the most popular power allocations being the uniform one and the one derived from statistical waterfilling [516]. An additional alternative is to allocate power uniformly over a subset of the precoder's steering directions [517].

Some of these options are discussed in the balance of this section, and the last option is elaborated, in the context of CSI feedback, at the end of the chapter. And, for the important class of channels in the next example, the precoder becomes altogether unnecessary.

Example 5.15 (Precoder for MIMO channels with no transmit correlations and no CSIT)

Let $\boldsymbol{H} = \boldsymbol{R}_r^{1/2}\boldsymbol{H}_w$, whose columns are independent and have the same marginal distribution. From Example 5.12, $\boldsymbol{U_F} = \boldsymbol{I}$. Let $\boldsymbol{P}$ be any power allocation satisfying $\text{tr}(\boldsymbol{P}) = N_t$ and let $\boldsymbol{P}^{(n)}$ be its n-position cyclic shift, i.e., such that $P_j^{(n)} = P_{j'}$ with $j' = ((j-n))_{N_t}$. Clearly,

$$\frac{1}{N_t}\sum_{n=0}^{N_t-1} \boldsymbol{P}^{(n)} = \boldsymbol{I} \tag{5.89}$$

and, applying Jensen's inequality,

$$\mathbb{E}\left[\log_2 \det\left(\boldsymbol{I} + \frac{\mathsf{SNR}}{N_t}\boldsymbol{H}\boldsymbol{H}^*\right)\right] = \mathbb{E}\left[\log_2 \det\left(\boldsymbol{I} + \frac{\mathsf{SNR}}{N_t}\boldsymbol{H}\left(\frac{1}{N_t}\sum_{n=0}^{N_t-1}\boldsymbol{P}^{(n)}\right)\boldsymbol{H}^*\right)\right]$$

$$\geq \frac{1}{N_\mathrm{t}} \sum_{n=0}^{N_\mathrm{t}-1} \mathbb{E}\left[\log_2 \det\left(\boldsymbol{I} + \frac{\mathsf{SNR}}{N_\mathrm{t}} \boldsymbol{H}\boldsymbol{P}^{(n)}\boldsymbol{H}^*\right)\right] \tag{5.90}$$

$$= \mathbb{E}\left[\log_2 \det\left(\boldsymbol{I} + \frac{\mathsf{SNR}}{N_\mathrm{t}} \boldsymbol{H}\boldsymbol{P}\boldsymbol{H}^*\right)\right], \tag{5.91}$$

where (5.91) holds because the distribution of $\boldsymbol{H} = \boldsymbol{R}_\mathrm{r}^{1/2}\boldsymbol{H}_\mathrm{w}$ is invariant to any cyclic shift of its columns, meaning that the distribution of $\boldsymbol{H}\boldsymbol{P}^{(n)}\boldsymbol{H}^*$ does not depend on n. Altogether, a uniform power allocation ($\boldsymbol{P} = \boldsymbol{I}$) outperforms any other choice of $\boldsymbol{P}$ and, in conjunction with $\boldsymbol{U}_{\boldsymbol{F}} = \boldsymbol{I}$, leads to an unprecoded transmission [518, 519]. A special case of this is the canonical channel with IID Rayleigh-faded entries, for which an unprecoded transmission indeed achieves capacity [25].

Capacity

Although the lack of general explicit expressions for the optimum precoder hampers the provision of closed forms for the no-CSIT ergodic capacity, the abundance of channels for which an unprecoded transmission is optimal facilitates a number of important instances. Before graduating to full MIMO, let us consider the special case of SIMO, where the precoder optimization is immaterial.

Example 5.16 (Capacity of an IID Rayleigh-faded SIMO channel)

For $N_\mathrm{t} = 1$, (5.66) specializes to

$$C(\mathsf{SNR}) = \mathbb{E}\left[\log_2\left(1 + \mathsf{SNR}\,\|\boldsymbol{H}\|^2\right)\right], \tag{5.92}$$

which equals the capacity of a SISO channel (recall (4.165)), only with $\|\boldsymbol{H}\|^2$ in lieu of $|H|^2$. This reflects the fact that, in a SIMO channel, capacity can be achieved by simply combining the signals at the N_r receive antennas and applying a scalar decoder.

As seen in the quasi-static SIMO setting of Example 5.8, when the entries of $\boldsymbol{H}$ are IID complex Gaussian, the distribution of $\|\boldsymbol{H}\|^2$ is chi-square (see Appendix C.1.9). Then, the ergodic capacity equals

$$C(\mathsf{SNR}) = \int_0^\infty \log_2(1 + \mathsf{SNR}\,\xi)\, f_{\|\boldsymbol{H}\|^2}(\xi)\, \mathrm{d}\xi \tag{5.93}$$

$$= \int_0^\infty \log_2(1 + \mathsf{SNR}\,\xi)\, \frac{\xi^{N_\mathrm{r}-1} e^{-\xi}}{(N_\mathrm{r}-1)!}\, \mathrm{d}\xi \tag{5.94}$$

$$= e^{1/\mathsf{SNR}} \sum_{q=1}^{N_\mathrm{r}} \mathcal{E}_q\left(\frac{1}{\mathsf{SNR}}\right) \log_2 e, \tag{5.95}$$

where the integration is solved by invoking (C.37) and where $\mathcal{E}_q(\cdot)$, recall, is an exponential integral.

As it turn out, the SIMO setting appears as a building block in the analysis of certain

MIMO receiver structures and hence the expressions in Example 5.16 are to be encountered repeatedly throughout the text.

Moving onto MIMO now, we next generalize Example 5.16, giving a closed form for the capacity of the all-important canonical channel.

Example 5.17 (Capacity of an IID Rayleigh-faded MIMO channel with no CSIT)

If $\boldsymbol{H}$ has IID Rayleigh-faded entries, no precoder is needed and

$$C(\mathsf{SNR}) = \mathbb{E}\left[\log_2 \det\left(\boldsymbol{I} + \frac{\mathsf{SNR}}{N_\mathrm{t}}\boldsymbol{H}\boldsymbol{H}^*\right)\right] \tag{5.96}$$

$$= \sum_{j=0}^{N_{\mathsf{min}}-1} \mathbb{E}\left[\log_2\left(1 + \frac{\mathsf{SNR}}{N_\mathrm{t}}\lambda_j(\boldsymbol{H}\boldsymbol{H}^*)\right)\right] \tag{5.97}$$

$$= N_{\mathsf{min}}\, \mathbb{E}\left[\log_2\left(1 + \frac{\mathsf{SNR}}{N_\mathrm{t}}\lambda\right)\right] \tag{5.98}$$

$$= N_{\mathsf{min}} \int_0^\infty \log_2\left(1 + \frac{\mathsf{SNR}}{N_\mathrm{t}}\xi\right) f_\lambda(\xi)\, d\xi \tag{5.99}$$

$$= e^{N_\mathrm{t}/\mathsf{SNR}} \sum_{i=0}^{N_{\mathsf{min}}-1} \sum_{j=0}^{i} \sum_{\ell=0}^{2j} \left[\binom{2i-2j}{i-j} \binom{2j + 2N_{\mathsf{max}} - 2N_{\mathsf{min}}}{2j-\ell} \right. \\ \left. \cdot \frac{(-1)^\ell\, (2j)!\, (N_{\mathsf{max}} - N_{\mathsf{min}} + \ell)!}{2^{2i-\ell}\, j!\, \ell!\, (N_{\mathsf{max}} - N_{\mathsf{min}} + j)!} \sum_{q=1}^{N_{\mathsf{max}}-N_{\mathsf{min}}+\ell+1} \mathcal{E}_q\left(\frac{N_\mathrm{t}}{\mathsf{SNR}}\right)\right] \log_2 e, \tag{5.100}$$

where (5.98) holds because the N_{min} unordered nonzero eigenvalues of the Wishart matrix $\boldsymbol{H}\boldsymbol{H}^*$ have the same marginal distribution, and λ is any of those eigenvalues. The distribution $f_\lambda(\cdot)$ is given in Appendix C.1.9 and the result of the integration in (5.99), solved in [520], directly generalizes the SISO formula in Example 4.27. To express the final result, we introduced $N_{\mathsf{max}} = \max(N_\mathrm{t}, N_\mathrm{r})$.

An alternative closed form for $C(\mathsf{SNR})$, with a recursive structure, is [521]

$$C(\mathsf{SNR}) = \sum_{i=0}^{N_{\mathsf{min}}-1} \sum_{j=0}^{i} \sum_{\ell=0}^{i} \binom{i}{j} \frac{(i + N_{\mathsf{max}} - N_{\mathsf{min}})!\, (-1)^{j+\ell}}{(i-\ell)!\, (N_{\mathsf{max}} - N_{\mathsf{min}} + j)!\, (N_{\mathsf{max}} - N_{\mathsf{min}} + \ell)!\, \ell!} \\ \cdot I_{j+\ell+N_{\mathsf{max}}-N_{\mathsf{min}}}\left(\frac{\mathsf{SNR}}{N_\mathrm{t}}\right) \log_2 e, \tag{5.101}$$

where $I_0(\rho) = e^{1/\rho}\, \mathcal{E}_1(1/\rho)$ and

$$I_k(\rho) = k\, I_{k-1}(\rho) + (-\rho)^{-k}\left[I_0(\rho) + \sum_{q=1}^{k} (q-1)!\, (-\rho)^q\right]. \tag{5.102}$$

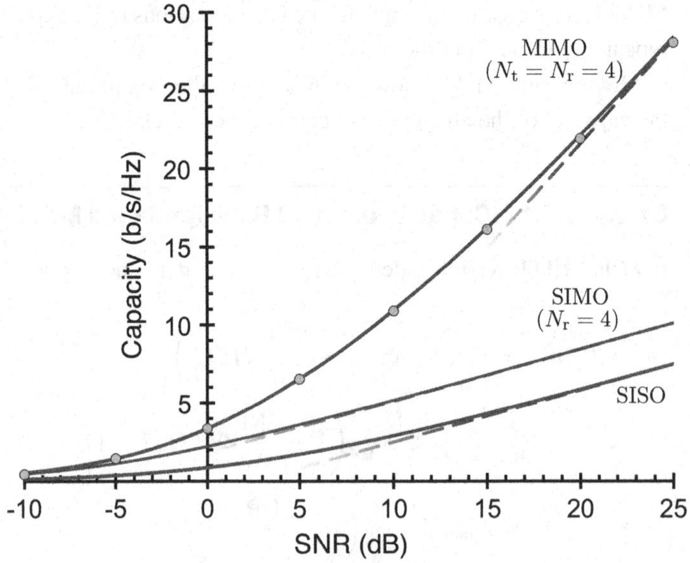

Fig. 5.5 Solid lines, exact $C(\mathsf{SNR})$ in IID Rayleigh fading for SISO, for SIMO with $N_r = 4$, and for MIMO with $N_t = N_r = 4$. Dashed, corresponding high-SNR expansions. Circles, large-dimensional approximation to the MIMO case.

Yet another form for $C(\mathsf{SNR})$, this one involving an infinite series but no exponential integral functions, is given in [522].

Example 5.18

Recover the capacity of a SIMO channel with IID Rayleigh-faded entries as a special case of MIMO.

Solution

Plugging $N_t = 1$, $N_{\mathsf{min}} = 1$, and $N_{\mathsf{max}} = N_r$ into either (5.100) or (5.101), the SIMO capacity expression in Example 5.16 is readily recovered.

Example 5.19

Under IID Rayleigh fading, plot $C(\mathsf{SNR})$ for a SISO channel, for a SIMO channel with $N_r = 4$, and for a MIMO channel with $N_t = N_r = 4$.

Solution

The plots are shown in Fig. 5.5, and the benefits of unlocking new signaling dimensions with MIMO are manifest.

The exact expressions for the capacity become rapidly unwieldy once features such as antenna correlations or Rice terms are incorporated. With only receive antenna correlations, exact closed forms for $C(\mathsf{SNR})$ can still be found [523–526]. With transmit antenna correlations and/or Rice terms, only capacity bounds—in some cases corresponding to the exact spectral efficiency without precoding—are available [274, 484, 527–536]. Although these exact expressions and bounds are analytical conquests that allow circumventing Monte-Carlo evaluations, they offer little in the way of insight. Further intuition is often sought by examining various asymptotic regimes, a task we next embark on.

Prior to that, though, a few observations can already be advanced concerning the impact of antenna correlation.

- Without precoding, both transmit and receive correlations are detrimental to the spectral efficiency at every SNR [537]. The eigenvalues of $\mathbb{E}[\boldsymbol{H}^*\boldsymbol{H}]$ become dissimilar with correlation, and the uniform power allocation in an unprecoded transmission is then inadequate.
- With optimum precoding, conversely, transmit correlations enable focusing power. For MISO specifically, this has been shown to increase the capacity at every SNR [513]. The extension of this wisdom to MIMO is to be found in the asymptotic realm.

Discrete constellations

Recalling Example 1.14, we can cast the spectral efficiency with precoder $\boldsymbol{F}$ and an equiprobable M-ary constellation per stream, such that $\boldsymbol{s}$ takes values s_m for $m = 0, \ldots, M^{N_\mathrm{s}}-1$, as

$$\frac{R}{B} = -\mathbb{E}\left[\log_2\left(\frac{1}{\pi^{N_\mathrm{r}}}\frac{1}{M^{N_\mathrm{s}}}\sum_{m=0}^{M^{N_\mathrm{s}}-1}\exp\left(\frac{-\left\|\sqrt{\frac{GE_\mathrm{s}}{N_\mathrm{t}}}\boldsymbol{H}\boldsymbol{F}(\boldsymbol{s}-\mathsf{s}_m)+\boldsymbol{v}\right\|^2}{N_0}\right)\right)\right] - N_\mathrm{r}\log_2(\pi e) \tag{5.103}$$

$$= -\frac{1}{M^{N_\mathrm{s}}}\sum_{m'=0}^{M^{N_\mathrm{s}}-1}\mathbb{E}\left[\log_2\sum_{m=0}^{M^{N_\mathrm{s}}-1}\exp\left(-\left\|\sqrt{\frac{\mathsf{SNR}}{N_\mathrm{t}}}\boldsymbol{H}\boldsymbol{F}(\mathsf{s}_{m'}-\mathsf{s}_m)+\boldsymbol{v}'\right\|^2\right)\right] + N_\mathrm{s}\log_2 M - N_\mathrm{r}\log_2 e, \tag{5.104}$$

where the expectation over the equiprobable signal vector $\boldsymbol{s}$ has been effected in (5.104), by means of the outer average over the auxiliary index m', leaving an expectation over the scaled noise $\boldsymbol{v}' \sim \mathcal{N}_\mathbb{C}(\boldsymbol{0}, \boldsymbol{I})$ and the fading $\boldsymbol{H}$. The optimization of this expression with respect to the precoder $\boldsymbol{F} = \boldsymbol{U_F}\boldsymbol{P}^{1/2}\boldsymbol{V_F^*}$ seems in general unwieldy. In order to tame the problem, [538] derives a tractable bound on (5.104) and optimizes $\boldsymbol{F}$ approximately on the basis of that bound. For M-PSK signaling and IID Rayleigh fading, a simplified version of (5.104) is obtained in [539].

In any event, as long as the constellation is dense enough for its mutual information to hug the capacity in almost all channel realizations at the SNRs of interest, the precoder that maximizes (5.104) is hardly different from the capacity-achieving precoder derived earlier and $R/B \approx C$. In a well-designed system, this should be the case.

Limiting regimes

Low-SNR regime

At low SNR, taking advantage of the mutual information expansion in Example 1.13,

$$C(\mathsf{SNR}) = \max_{\boldsymbol{F}:\mathrm{tr}(\boldsymbol{F}\boldsymbol{F}^*)=N_\mathrm{t}} \left[\frac{1}{N_\mathrm{t}} \mathrm{tr}\big(\mathbb{E}[\boldsymbol{H}^*\boldsymbol{H}]\,\boldsymbol{F}\boldsymbol{F}^*\big)\,\mathsf{SNR} - \frac{1}{2N_\mathrm{t}^2}\mathrm{tr}\Big(\mathbb{E}\Big[(\boldsymbol{H}^*\boldsymbol{H}\boldsymbol{F}\boldsymbol{F}^*)^2\Big]\Big)\,\mathsf{SNR}^2\right] \log_2 e + o(\mathsf{SNR}^2), \tag{5.105}$$

whose leading term, with a precoder diagonalizing $\mathbb{E}\big[\boldsymbol{H}^*\boldsymbol{H}\big]$, equals

$$\frac{\mathsf{SNR}}{N_\mathrm{t}} \sum_{j=0}^{N_{\mathsf{min}}-1} P_j\,\lambda_j\big(\mathbb{E}[\boldsymbol{H}^*\boldsymbol{H}]\big). \tag{5.106}$$

This value is maximized by pouring all the power on the eigenvector of $\mathbb{E}\big[\boldsymbol{H}^*\boldsymbol{H}\big]$ corresponding to the largest eigenvalue, as any allocation diverting power onto weaker-gain directions would yield a smaller sum. Likewise, with a precoder not diagonalizing $\mathbb{E}\big[\boldsymbol{H}^*\boldsymbol{H}\big]$, some power would be deflected onto eigenvectors other than the maximum-gain one. Altogether then, the optimum strategy for $\mathsf{SNR} \to 0$ is always to concentrate all the power on the maximum-eigenvalue eigenvector(s) of $\mathbb{E}[\boldsymbol{H}^*\boldsymbol{H}]$, a scheme dubbed *statistical beamforming* to distinguish it from CSIT-based beamforming. If $\lambda_{\mathsf{max}}\big(\mathbb{E}[\boldsymbol{H}^*\boldsymbol{H}]\big)$ is unique and $\boldsymbol{u}$ is the corresponding eigenvector, then $\boldsymbol{F} = \sqrt{N_\mathrm{t}}\boldsymbol{u}$ and, applying (4.30),

$$\frac{E_\mathrm{b}}{N_0}_{\mathsf{min}} = \frac{1}{\frac{1}{N_\mathrm{t}}\mathrm{tr}\big(\mathbb{E}[\boldsymbol{H}^*\boldsymbol{H}]\boldsymbol{F}\boldsymbol{F}^*\big)}\frac{1}{\log_2 e} \tag{5.107}$$

$$= \frac{1}{\mathrm{tr}\big(\mathbb{E}[\boldsymbol{H}^*\boldsymbol{H}]\boldsymbol{u}\boldsymbol{u}^*\big)}\frac{1}{\log_2 e} \tag{5.108}$$

$$= \frac{1}{\boldsymbol{u}^*\mathbb{E}[\boldsymbol{H}^*\boldsymbol{H}]\boldsymbol{u}}\frac{1}{\log_2 e} \tag{5.109}$$

$$= \frac{1}{\lambda_{\mathsf{max}}\big(\mathbb{E}[\boldsymbol{H}^*\boldsymbol{H}]\big)}\frac{1}{\log_2 e}, \tag{5.110}$$

indicating that the minimum energy per bit is reduced by $\lambda_{\mathsf{max}}\big(\mathbb{E}[\boldsymbol{H}^*\boldsymbol{H}]\big)|_{\mathrm{dB}}$ relative to SISO and, correspondingly, that $C(\mathsf{SNR})$ is scaled (to first-order) by $\lambda_{\mathsf{max}}\big(\mathbb{E}[\boldsymbol{H}^*\boldsymbol{H}]\big)$.

As far as $\mathcal{S}_0$ is concerned, plugging the asymptotically optimal ($\mathsf{SNR} \to 0$) precoder $\boldsymbol{F} = \sqrt{N_\mathrm{t}}\boldsymbol{u}$ into (5.105),

$$C(\mathsf{SNR}) = \left[\lambda_{\mathsf{max}}\big(\mathbb{E}[\boldsymbol{H}^*\boldsymbol{H}]\big)\,\mathsf{SNR} - \frac{1}{2}\mathrm{tr}\Big(\mathbb{E}\Big[(\boldsymbol{u}^*\boldsymbol{H}^*\boldsymbol{H}\boldsymbol{u})^2\Big]\,\mathsf{SNR}^2\Big)\right]\log_2 e + o\left(\mathsf{SNR}^2\right) \tag{5.111}$$

$$= \left[\lambda_{\mathsf{max}}\big(\mathbb{E}[\boldsymbol{H}^*\boldsymbol{H}]\big)\,\mathsf{SNR} - \frac{1}{2}\mathbb{E}\big[\|\boldsymbol{H}\boldsymbol{u}\|^4\big]\,\mathsf{SNR}^2\right]\log_2 e + o\left(\mathsf{SNR}^2\right), \tag{5.112}$$

which, applying (4.39), gives

$$\mathcal{S}_0 = \frac{2}{\kappa\big(\|\boldsymbol{H}\boldsymbol{u}\|\big)}. \tag{5.113}$$

Discussion 5.3 Precoder's rank at low SNR

With only CSIR, the optimum precoder's rank at low SNR equals the multiplicity of $\lambda_{\max}(\mathbb{E}[\boldsymbol{H}^* \boldsymbol{H}])$, with uniform power allocation over the applicable streams if the multiplicity is plural. In a channel with IID entries, this amounts to an equal-power transmission from all antennas, $\boldsymbol{F} = \boldsymbol{I}$, which involves dimensional overloading if $N_t > N_r$.

The pilot overhead, however, is proportional to the number of streams and it increases as the SNR shrinks. At sufficiently low SNR, in the vicinity of $\frac{E_b}{N_0}_{\min}$, transmitting a single stream is preferable once we account for the overhead, and thus a beamforming-based computation of $\frac{E_b}{N_0}_{\min}$ and S_0 would lead to a more exact asymptote for pilot-assisted communication. In underspread fading though, as the SNR inches up, additional streams are quickly activated and the expressions obtained with CSIR-based precoders immediately become appropriate for pilot-assisted communication as well. For channels with IID entries, this means $\boldsymbol{F} = \boldsymbol{I}$.

If $\lambda_{\max}(\mathbb{E}[\boldsymbol{H}^* \boldsymbol{H}])$ has plural multiplicity, then multiple signal streams can be transmitted on the corresponding eigenvectors and $\frac{E_b}{N_0}_{\min}$ does not depend on how many of those eigenvectors are activated or on how the power is divided among them (refer to Problem 5.26). However, S_0 is maximized if all those eigenvectors are activated and the power is evenly divided among them [64, 540]. The most relevant precoder in which the power is divided among multiple signal streams at low SNR—in fact at every SNR—is $\boldsymbol{F} = \boldsymbol{I}$ and, for such unprecoded transmission, (5.105) specializes to

$$C(\mathsf{SNR}) = \left[N_r \, \mathsf{SNR} - \frac{1}{2N_t^2} \operatorname{tr}\left(\mathbb{E}\left[(\boldsymbol{H}^* \boldsymbol{H})^2 \right] \right) \mathsf{SNR}^2 \right] \log_2 e + o(\mathsf{SNR}^2) \tag{5.114}$$

from which, applying (4.30) and (4.39),

$$\frac{E_b}{N_0}_{\min} = \frac{1}{N_r \log_2 e} \tag{5.115}$$

$$S_0 = \frac{2N_t^2 N_r^2}{\operatorname{tr}\left(\mathbb{E}\left[(\boldsymbol{H}\boldsymbol{H}^*)^2 \right]\right)}. \tag{5.116}$$

We hasten to emphasize that (5.115) and (5.116) hold any time that $\boldsymbol{F} = \boldsymbol{I}$, regardless of whether this is optimal or not.

Let us exemplify the applicability of the foregoing derivations of $\frac{E_b}{N_0}_{\min}$ and S_0, first on channels free of antenna correlation.

Example 5.20 ($\frac{E_b}{N_0}_{\min}$ and S_0 for a MIMO channel with IID entries and no CSIT)

When the entries of $\boldsymbol{H}$ are IID, (5.115) applies. Thus, $\frac{E_b}{N_0}_{\min}$ depends only on the number of receive antennas, irrespective of N_t and of the fading distribution. At the same time, if h denotes an arbitrary entry of $\boldsymbol{H}$, then

$$\operatorname{tr}\left(\mathbb{E}\left[(\boldsymbol{H}^* \boldsymbol{H})^2 \right] \right) = \sum_{j=0}^{N_t-1} \sum_{j'=0}^{N_t-1} \mathbb{E}\left[\left| [\boldsymbol{H}^* \boldsymbol{H}]_{j,j'} \right|^2 \right] \tag{5.117}$$

$$= N_t N_r \Big(N_t + N_r + \kappa(|h|) - 2 \Big), \tag{5.118}$$

with the reader invited, in Problem 5.28, to verify (5.118). Plugged into (5.116), this gives

$$\mathcal{S}_0 = \frac{2 N_t N_r}{N_t + N_r + \kappa(|h|) - 2}. \tag{5.119}$$

If $|h|$ is further Rayleigh distributed, then $\kappa(|h|) = 2$ and

$$\mathcal{S}_0 = \frac{2 N_t N_r}{N_t + N_r}. \tag{5.120}$$

Example 5.21

In the absence of antenna correlations, how much does MIMO improve the low-SNR capacity relative to SISO?

Solution

For MIMO with $N_t = N_r = N_a$, $\frac{E_b}{N_0}_{\min} = \frac{1}{N_a \log_2 e}$ and $\mathcal{S}_0 = N_a$, both improved by a factor of N_a relative to their SISO counterparts. To first order, $C(\mathsf{SNR})$ grows linearly with N_a or, put differently, the capacity per antenna equals the SISO capacity irrespective of the number of antennas. This is graphically illustrated later in the chapter, in Fig. 5.16, which compares the SISO capacity with the per-antenna MIMO capacity in IID Rayleigh fading. Indeed, the multiantenna interference is drowned by the noise in this regime, and it is as if we had N_a noninterfering subchannels despite the lack of CSIT.

Example 5.22

In the absence of antenna correlations, how much does MIMO improve the low-SNR capacity relative to SIMO?

Solution

There is no change in $\frac{E_b}{N_0}_{\min}$ and thus no first-order improvement in $C(\mathsf{SNR})$, as can be appreciated in Fig. 5.5. Following (5.120), the slope $\mathcal{S}_0$ does increase when additional transmit antennas are added, and the increase is highest while $N_t \ll N_r$.

Example 5.23

Under IID Rayleigh fading, plot $\mathsf{C}(\frac{E_b}{N_0})$ for a SISO channel, for a SIMO channel with $N_r = 4$, and for a MIMO channel with $N_t = N_r = 4$.

Solution

The plots, shown in Fig. 5.6, are obtained via the usual relationship

$$\frac{E_b}{N_0} = \frac{\mathsf{SNR}}{C(\mathsf{SNR})}, \tag{5.121}$$

with $C = \mathsf{C}$. Also shown in the figure are the corresponding low-SNR expansions, with both SIMO and MIMO exhibiting the anticipated 6-dB reduction in $\frac{E_b}{N_0}_{\min}$ relative to SISO.

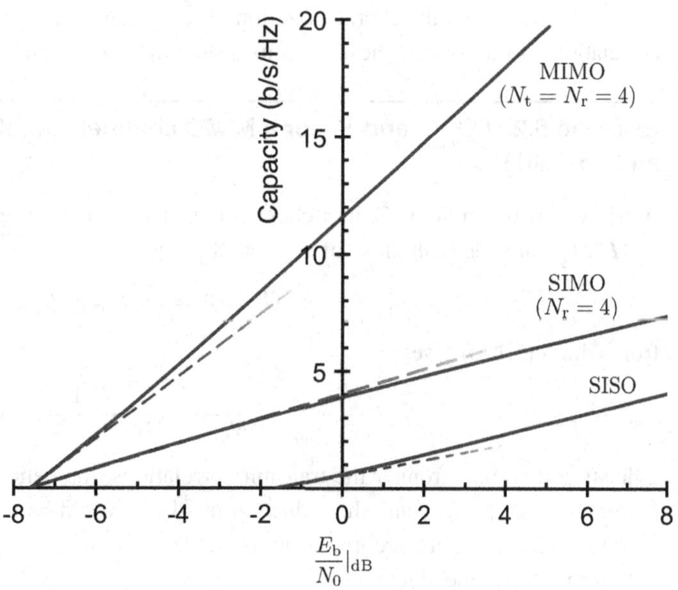

Fig. 5.6 Solid lines, $C(\frac{E_b}{N_0})$ for SISO, for SIMO with $N_r = 4$, and for MIMO with $N_t = N_r = 4$. Dashed lines, corresponding low-SNR expansions.

In addition, $S_0 = 1$ for SISO, $S_0 = 1.6$ for SIMO, and $S_0 = 4$ for MIMO. Notice the broad range of values over which the low-SNR expansions accurately track the actual function $C(\frac{E_b}{N_0})$.

Example 5.24

Let $(GP_t/N_0)|_{dB} = 60$ dB and $B = 1$ MHz, such that $\text{SNR} = 0$ dB. The fading is IID across antennas and Rayleigh-distributed. With SISO, $\frac{E_b}{N_0}_{\min} = \frac{1}{\log_2 e} = -1.59$ dB and $S_0 = 1$ b/s/Hz/(3 dB) and, applying the ergodic capacity expression in Example 4.27, the bit rate is found to be $R = B \cdot C(\text{SNR}) = 863$ Kb/s. Relative to this SISO transmission, by how much could MIMO with $N_t = 2$ and $N_r = 4$ reduce the power and the bandwidth while still achieving the same bit rate?

Solution

For $N_t = 2$ and $N_r = 4$, $\frac{E_b}{N_0}_{\min} = \frac{1}{4 \log_2 e}$ and $S_0 = 2.67$ b/s/Hz/(3 dB). Therefore, the transmitter can power down by 6 dB and simultaneously reduce its bandwidth by a factor of 2.67 while still achieving the same bit rate as with SISO. An exact calculation using the MIMO capacity expression in Example 5.17 validates this figure: the exact bit rate with $(GP_t/N_0)|_{dB} = 54$ dB and $B = 375$ kHz is 832 Kb/s, very close to the original 863 Kb/s.

Alternatively, if the original power and bandwidth are conserved, then MIMO can scale up the SISO bit rate (to first order) by a factor of 4, giving $R \approx 4 \cdot 863 = 3.45$ Mb/s. An exact calculation yields a MIMO bit rate of 2.86 Mb/s.

Let us now extend the characterization of $\frac{E_b}{N_0}_{\min}$ and S_0 to channels exhibiting antenna correlations, to show how these quantities shed light on the impact of such correlations.

Example 5.25 ($\frac{E_b}{N_0}_{\min}$ and S_0 for a MIMO channel with Kronecker correlations and no CSIT)

As shown in Example 5.12, in a channel with Kronecker correlations the eigenvalues of $\mathbb{E}[\boldsymbol{H}^*\boldsymbol{H}]$ coincide with those of $\boldsymbol{R}_t$ scaled by N_r. Thus,

$$\lambda_{\max}\big(\mathbb{E}[\boldsymbol{H}^*\boldsymbol{H}]\big) = N_r\, \lambda_{\max}(\boldsymbol{R}_t), \tag{5.122}$$

from which (5.110) gives

$$\frac{E_b}{N_0}_{\min} = \frac{1}{N_r\, \lambda_{\max}(\boldsymbol{R}_t)} \frac{1}{\log_2 e}, \tag{5.123}$$

indicating that, the stronger the transmit correlations, the better; with full transmit correlation, $\lambda_{\max}(\boldsymbol{R}_t) = N_t$ and the reduction in $\frac{E_b}{N_0}_{\min}$ would be maximum. Receive antenna correlations, conversely, are irrelevant to (5.123).

In turn, after some algebra

$$S_0 = \frac{2N_r^2}{N_r^2 + \mathrm{tr}\big(\boldsymbol{R}_r^2\big)}, \tag{5.124}$$

which is not affected by transmit correlations. Since $\mathrm{tr}\big(\boldsymbol{R}_r^2\big)$ is minimized when $\boldsymbol{R}_r = \boldsymbol{I}$, receive correlations can only diminish S_0 and are thus detrimental.

From the foregoing example, the impact of Kronecker antenna correlations on the low-SNR capacity is established as follows.

- Transmit correlations decrease $\frac{E_b}{N_0}_{\min}$ without affecting S_0 and they are therefore always beneficial. This is because they enhance the effectiveness of statistical beamforming.
- Receive correlations diminish S_0 without affecting $\frac{E_b}{N_0}_{\min}$. Thus, they are invariably detrimental.

Things are different without precoding, as then $\frac{E_b}{N_0}_{\min}$ is invariant to correlations but S_0 in (5.116) diminishes with both transmit and receive correlations. It follows that all correlations are detrimental when the transmission does not take them into account [447].

Since every proper complex signal exhibits the same second-order mutual information expansion (recall Examples 1.13 and 1.14), the foregoing low-SNR observations extend to non-Gaussian signals. For BPSK, which is not proper complex, the expressions involving S_0 have to be reworked.

In closing this examination of low-SNR behaviors we note that, as was the case with CSIT, the simplicity of beamforming (in this case statistical) makes it an attractive strategy even when the SNR is moderately high [541]. The SNR breakpoint above which beamforming is no longer optimal can be determined as a function of the two largest eigenvalues of $\mathbb{E}[\boldsymbol{H}^*\boldsymbol{H}]$ [487, 542–544]. The more disparate these eigenvalues, the higher the breakpoint. For Kronecker correlations, this translates to the rule that, the stronger the transmit correlations, the higher the breakpoint and the more the appeal of statistical beamforming.

High-SNR regime

Let us now turn to the high-SNR regime. For $N_t \le N_r$, the ergodic spectral efficiency with a precoder $\boldsymbol{F}$ can be written as

$$\frac{R}{B} = \mathbb{E}\left[\log_2 \det\left(\boldsymbol{I} + \frac{\mathsf{SNR}}{N_t}\boldsymbol{H}^*\boldsymbol{H}\boldsymbol{F}\boldsymbol{F}^*\right)\right] \tag{5.125}$$

$$= \log_2 \det(\mathsf{SNR} \cdot \boldsymbol{I}_{N_t}) + \mathbb{E}\left[\log_2 \det\left(\frac{1}{\mathsf{SNR}}\boldsymbol{I} + \frac{1}{N_t}\boldsymbol{H}^*\boldsymbol{H}\boldsymbol{F}\boldsymbol{F}^*\right)\right] \tag{5.126}$$

$$= N_t \log \mathsf{SNR} + \mathbb{E}\left[\log_2 \det\left(\frac{1}{\mathsf{SNR}}\boldsymbol{I} + \frac{1}{N_t}\boldsymbol{H}^*\boldsymbol{H}\boldsymbol{F}\boldsymbol{F}^*\right)\right]. \tag{5.127}$$

Similarly, for $N_t > N_r$,

$$\frac{R}{B} = \mathbb{E}\left[\log_2 \det\left(\boldsymbol{I}_{N_r} + \frac{\mathsf{SNR}}{N_t}\boldsymbol{H}\boldsymbol{F}\boldsymbol{F}^*\boldsymbol{H}^*\right)\right] \tag{5.128}$$

$$= N_r \log \mathsf{SNR} + \mathbb{E}\left[\log_2 \det\left(\frac{1}{\mathsf{SNR}}\boldsymbol{I} + \frac{1}{N_t}\boldsymbol{H}\boldsymbol{F}\boldsymbol{F}^*\boldsymbol{H}^*\right)\right]. \tag{5.129}$$

Altogether,

$$\mathcal{S}_\infty = N_{\min} = \min(N_t, N_r) \tag{5.130}$$

as long as $\boldsymbol{H}^*\boldsymbol{H}\boldsymbol{F}\boldsymbol{F}^*$ and $\boldsymbol{H}\boldsymbol{F}\boldsymbol{F}^*\boldsymbol{H}^*$ are of full rank, $N_{\min}$, such that the log-determinants in (5.127) and (5.129) are bounded for $\mathsf{SNR} \to \infty$. With $\boldsymbol{H}$ being full-rank with probability 1, the rank of the precoder must itself be no smaller than $N_{\min}$.

In terms of the power offset, applying to (5.66) the definition of $\mathcal{L}_\infty$ given in (4.46) we obtain, for $N_t \le N_r$,

$$\mathcal{L}_\infty = -\frac{1}{N_t} \max_{\boldsymbol{F}:\mathrm{tr}(\boldsymbol{F}\boldsymbol{F}^*)=N_t} \mathbb{E}\left[\log_2 \det\left(\frac{1}{N_t}\boldsymbol{H}^*\boldsymbol{H}\boldsymbol{F}\boldsymbol{F}^*\right)\right] \tag{5.131}$$

$$= -\frac{1}{N_t}\mathbb{E}\left[\log_2 \det(\boldsymbol{H}^*\boldsymbol{H})\right] - \frac{1}{N_t} \max_{\boldsymbol{F}:\mathrm{tr}(\boldsymbol{F}\boldsymbol{F}^*)=N_t} \log_2 \det\left(\frac{1}{N_t}\boldsymbol{F}\boldsymbol{F}^*\right), \tag{5.132}$$

where the argument of the last term satisfies, applying Jensen's inequality,

$$\log_2 \det\left(\frac{1}{N_t}\boldsymbol{F}\boldsymbol{F}^*\right) = \sum_{j=0}^{N_t-1} \log_2 \frac{\lambda_j(\boldsymbol{F}\boldsymbol{F}^*)}{N_t} \tag{5.133}$$

$$= N_t \left(\frac{1}{N_t}\sum_{j=0}^{N_t-1} \log_2 \frac{\lambda_j(\boldsymbol{F}\boldsymbol{F}^*)}{N_t}\right) \tag{5.134}$$

$$\le N_t \log_2\left(\frac{1}{N_t}\sum_{j=0}^{N_t-1} \frac{\lambda_j(\boldsymbol{F}\boldsymbol{F}^*)}{N_t}\right) \tag{5.135}$$

$$= N_t \log_2 \frac{\mathrm{tr}(\boldsymbol{F}\boldsymbol{F}^*)}{N_t^2} \tag{5.136}$$

$$= N_t \log_2 \frac{1}{N_t}. \tag{5.137}$$

The above inequality becomes an equality for $\boldsymbol{F} = \boldsymbol{I}$, and thus an unprecoded transmission is optimal in terms of both DOF and power offset whenever $N_t \leq N_r$. Precisely, $\mathcal{S}_\infty = N_t$ and

$$\mathcal{L}_\infty = \log_2 N_t - \frac{1}{N_t}\mathbb{E}\big[\log_2 \det(\boldsymbol{H}^*\boldsymbol{H})\big]. \tag{5.138}$$

Interestingly, and perhaps somewhat surprisingly, these are the same $\mathcal{S}_\infty$ and $\mathcal{L}_\infty$ encountered without CSIT in Section 5.3.2. For $N_t \leq N_r$, therefore, the absence of CSIT does not affect the high-SNR capacity. (The receiver complexity of the receiver does increase, since the signal streams are mutually interfering once they are not transmitted through the channel's eigendirections.)

In turn, for $N_t > N_r$,

$$\mathcal{L}_\infty = -\frac{1}{N_r} \max_{\boldsymbol{F}:\mathrm{tr}(\boldsymbol{F}\boldsymbol{F}^*)=N_t} \mathbb{E}\left[\log_2 \det\left(\frac{1}{N_t}\boldsymbol{H}\boldsymbol{F}\boldsymbol{F}^*\boldsymbol{H}^*\right)\right] \tag{5.139}$$

$$= \log_2 N_t - \frac{1}{N_r} \max_{\boldsymbol{F}:\mathrm{tr}(\boldsymbol{F}\boldsymbol{F}^*)=N_t} \mathbb{E}\big[\log_2 \det(\boldsymbol{H}\boldsymbol{F}\boldsymbol{F}^*\boldsymbol{H}^*)\big], \tag{5.140}$$

which need no longer be optimized by $\boldsymbol{F} = \boldsymbol{I}$, and which no longer equals its no-CSIT counterpart. Thus, the absence of CSIT does have an effect on the high-SNR capacity for $N_t > N_r$, and such is the subject of Problem 5.14.

Example 5.26 (Power offset for an IID Rayleigh-faded MIMO channel with no CSIT)

If $N_t \leq N_r$, then, borrowing the expression derived with CSIT in Example 5.6,

$$\mathcal{L}_\infty = \log_2 N_t + \left(\gamma_{\mathsf{EM}} - \sum_{q=1}^{N_r-N_t} \frac{1}{q} - \frac{N_r}{N_t} \sum_{q=N_r-N_t+1}^{N_r} \frac{1}{q} + 1\right) \log_2 e, \tag{5.141}$$

which, for the special case $N_t = N_r = N_a$, simplifies to

$$\mathcal{L}_\infty = \log_2 N_a + \left(\gamma_{\mathsf{EM}} - \sum_{q=2}^{N_a} \frac{1}{q}\right) \log_2 e. \tag{5.142}$$

If $N_t > N_r$, and capitalizing on the optimality at every SNR of $\boldsymbol{F} = \boldsymbol{I}$ for channels with IID entries, we have that

$$\mathcal{L}_\infty = \log_2 N_t - \frac{1}{N_r}\mathbb{E}\big[\log_2 \det(\boldsymbol{H}\boldsymbol{H}^*)\big], \tag{5.143}$$

where $\boldsymbol{H}\boldsymbol{H}^*$ is a Wishart matrix. This allows (refer to Problem 5.34) elaborating the power offset further into

$$\mathcal{L}_\infty = \log_2 N_t + \left(\gamma_{\mathsf{EM}} - \sum_{q=1}^{N_t-N_r} \frac{1}{q} - \frac{N_t}{N_r} \sum_{q=N_t-N_r+1}^{N_t} \frac{1}{q} + 1\right) \log_2 e, \tag{5.144}$$

Discussion 5.4 Precoder's rank at high SNR

Consider an IID Rayleigh-faded channel with CSIR, but no CSIT. By increasing N_t indefinitely and transmitting $N_s = N_t > N_r$ signal streams, sustained improvements in high-SNR capacity are attained even with N_r held fixed. Indeed, for $N_t \to \infty$,

$$\sum_{q=1}^{N_t} \frac{1}{q} - \log_e N_t \to \gamma_{\text{EM}}, \tag{5.146}$$

which, plugged into (5.144), returns $\mathcal{L}_\infty \to 0$. In contrast, the power offset given for $N_t = N_r$ in (5.142) is invariably positive, meaning that with CSIR, but no CSIT, there are benefits to be had by overloading.

At the same time, the additional pilot symbols required by the overloading antennas erode the DOF, meaning that the improvement in $\mathcal{L}_\infty$ comes at the expense of some reduction in $\mathcal{S}_\infty$. For $\mathsf{SNR} \to \infty$, it is thus best not to overload once the overhead is accounted for and, indeed, if we remove the CSIR, the optimum number of active transmit antennas does equal N_{min} at asymptotically high SNRs. Again though, in underspread fading, as soon as we back off from infinity and consider SNRs of interest, CSIR-based expressions become appropriate for pilot-assisted transmission as well. For $N_t > N_r$, this means that the overloading-based power offset in (5.144) is meaningful—possibly in conjunction with the suitable overhead discount and channel estimation penalty.

in reference to which the reader is referred to Discussion 5.4.

Example 5.27 (Power offset for an IID Rayleigh-faded SIMO channel)

For SIMO, (5.141) specializes to

$$\mathcal{L}_\infty = \left(\gamma_{\text{EM}} - \sum_{q=1}^{N_r - 1} \frac{1}{q} \right) \log_2 e, \tag{5.145}$$

which evidences the power gain of incorporating additional receive antennas to a SISO setting whose power offset is $\mathcal{L}_\infty = \gamma_{\text{EM}} \log_2 e$.

Building on the foregoing examples, we can attest to the accuracy of the high-SNR analysis based on $\mathcal{S}_\infty$ and $\mathcal{L}_\infty$.

Example 5.28

Let $\boldsymbol{H}$ have IID Rayleigh-faded entries. Shown in Fig. 5.5 is the exact function $C(\mathsf{SNR})$ for SISO and for MIMO with $N_t = N_r = 4$, along with the corresponding high-SNR expansions. Notice how these become accurate at rather modest SNRs.

Example 5.29

For $N_{\mathrm{t}} = N_{\mathrm{r}} = 4$, the power offset in 3-dB units comes at $\mathcal{L}_{\infty} = 1.27$, while its SISO counterpart is $\mathcal{L}_{\infty} = 0.83$. The difference is $\Delta\mathcal{L}_{\infty} = 0.44$, meaning $0.44{\cdot}3$ dB $= 1.31$ dB. This is precisely the high-SNR gap that can be appreciated in Fig. 5.16 between the SISO capacity and the per-antenna MIMO capacity in IID Rayleigh fading. With the spatial DOF normalized to be the same, the capacity is displaced by this amount at high SNR.

In the presence of Kronecker antenna correlations, the result for $N_{\mathrm{t}} = N_{\mathrm{r}} = N_{\mathrm{a}}$ derived with CSIT in Example 5.7 continues to apply without CSIT, namely

$$\mathcal{L}_{\infty} = \log_2 N_{\mathrm{a}} + \left(\gamma_{\mathsf{EM}} - \sum_{q=2}^{N_{\mathrm{a}}} \frac{1}{q}\right) \log_2 e - \frac{1}{N_{\mathrm{a}}} \sum_{j=0}^{N_{\mathrm{a}}-1} \log_2 \lambda_j(\boldsymbol{R}_{\mathrm{t}}) - \frac{1}{N_{\mathrm{a}}} \sum_{i=0}^{N_{\mathrm{a}}-1} \log_2 \lambda_i(\boldsymbol{R}_{\mathrm{r}}). \tag{5.147}$$

Example 5.30

Consider a channel (a) with $N_{\mathrm{t}} = N_{\mathrm{r}} = 2$ and with the Kronecker correlation matrices

$$\boldsymbol{R}_{\mathrm{t}} = \begin{bmatrix} 1 & 0.8 \\ 0.8 & 1 \end{bmatrix} \qquad \boldsymbol{R}_{\mathrm{r}} = \begin{bmatrix} 1 & 0.6 \\ 0.6 & 1 \end{bmatrix} \tag{5.148}$$

and further consider a second channel (b) with $N_{\mathrm{t}} = 2$, $N_{\mathrm{r}} = 4$, and no correlations. Establish the difference in the SNRs required to achieve a certain high-SNR capacity.

Solution

Applying (5.147),

$$\mathcal{L}_{\infty}^{(\mathrm{a})} = 2.17, \tag{5.149}$$

while, applying (5.141),

$$\mathcal{L}_{\infty}^{(\mathrm{b})} = -0.57. \tag{5.150}$$

Although an analysis based only on the fact that $\mathcal{S}_{\infty}^{(\mathrm{a})} = \mathcal{S}_{\infty}^{(\mathrm{b})} = 2$ would declare both of these channels equivalent at high SNR, the discrepancy in power offsets points to an asymptotic difference of $\mathcal{L}_{\infty}^{(\mathrm{a})} - \mathcal{L}_{\infty}^{(\mathrm{b})} = (2.17 + 0.57) \times 3 = 8.22$ dB in the SNR required to achieve a certain capacity.

A Monte-Carlo computation indicates that the capacity of channel (a) at $\mathsf{SNR} = 25$ dB is 12.5 b/s/Hz, whereas channel (b) attains that same capacity at $\mathsf{SNR} = 17.08$ dB. The difference is 7.92 dB, very close to the asymptotic prediction of 8.22 dB.

By further exercising the power offset expressions, the impact of antenna correlations on the high-SNR capacity can be established as follows.

- For $N_{\mathrm{t}} \leq N_{\mathrm{r}}$, transmit correlations can only increase $\mathcal{L}_{\infty}$ and are thus deleterious to the capacity. For $N_{\mathrm{t}} > N_{\mathrm{r}}$, however, they may be favorable. (Recall that, for MISO, transmit correlations are beneficial at every SNR [513].)

- Receive correlations can only increase $\mathcal{L}_\infty$ and they are therefore always detrimental in terms of SU-MIMO capacity.

Closed forms can be derived for the power offset of more general channels, e.g., Rice, but the expressions become highly involved [355, 545].

Large-dimensional regime

As mentioned at the beginning of the chapter, the large-dimensional results that follow do not seek to quantity the performance when many antennas are actually utilized, but rather to offer an alternative to the exact analysis for specific N_t and N_r. The idea is to let $N_t, N_r \to \infty$ with the aspect ratio of $\boldsymbol{H}$ held at

$$\beta = \frac{N_t}{N_r} \tag{5.151}$$

and to characterize the ensuing capacity as a function of β. Since, for $N_t, N_r \to \infty$, the capacity typically grows without bound, it is commonplace to study a normalized version thereof, namely the capacity per receive antenna, $\frac{1}{N_r} C(\mathsf{SNR})$ (in b/s/Hz/antenna). The large-dimensional expressions thus found then serve as approximations (sometimes termed *deterministic equivalents*) for finite N_t and N_r.

To begin with, the ergodic spectral efficiency with a certain precoder $\boldsymbol{F}$ and CSIR can be posed as a function of the eigenvalues of $\frac{1}{N_t}\boldsymbol{H}\boldsymbol{F}\boldsymbol{F}^*\boldsymbol{H}^*$, to wit

$$\frac{R}{B} = \mathbb{E}\left[\sum_{i=0}^{N_r-1} \log_2\left(1 + \mathsf{SNR}\,\lambda_i\left(\frac{1}{N_t}\boldsymbol{H}\boldsymbol{F}\boldsymbol{F}^*\boldsymbol{H}^*\right)\right)\right]. \tag{5.152}$$

Let us denote the empirical cumulative distribution of the eigenvalues of $\frac{1}{N_t}\boldsymbol{H}\boldsymbol{F}\boldsymbol{F}^*\boldsymbol{H}^*$ by

$$F^{N_r}(\xi) = \frac{1}{N_r}\sum_{i=0}^{N_r-1} 1\left\{\lambda_i\left(\frac{1}{N_t}\boldsymbol{H}\boldsymbol{F}\boldsymbol{F}^*\boldsymbol{H}^*\right) \le \xi\right\}, \tag{5.153}$$

where $1\{\cdot\}$ is the indicator function, returning 1 if its argument is true and 0 otherwise. Thus, $F^{N_r}(\xi)$ gives the fraction of the eigenvalues $\lambda_i(\frac{1}{N_t}\boldsymbol{H}\boldsymbol{F}\boldsymbol{F}^*\boldsymbol{H}^*)$, $i = 0, \ldots, N_r - 1$, that fall below ξ while the corresponding density $f^{N_r}(\xi)$ can be seen as a collection of delta functions at those eigenvalues. With that, the spectral efficiency per receive antenna can be expressed as

$$\frac{1}{N_r}\frac{R}{B} = \mathbb{E}\left[\frac{1}{N_r}\sum_{i=0}^{N_r-1} \log_2\left(1 + \mathsf{SNR}\,\lambda_i\left(\frac{1}{N_t}\boldsymbol{H}\boldsymbol{F}\boldsymbol{F}^*\boldsymbol{H}^*\right)\right)\right] \tag{5.154}$$

$$= \mathbb{E}\left[\int_0^\infty \log_2(1 + \mathsf{SNR}\,\xi)\, f^{N_r}(\xi)\, d\xi\right]. \tag{5.155}$$

For most MIMO channels (with rare exceptions being channels with rank-limited correlations and keyhole channels), $F^{N_r}(\cdot)$ converges a.s. to a nonrandom limit as $N_t, N_r \to \infty$.

Then, as the randomness recedes, the eigenvalue distribution, and by association the channel itself, are said to *harden*. Denoting by $F(\cdot)$ the nonrandom asymptotic eigenvalue distribution,

$$\frac{1}{N_\mathrm{r}}\frac{R}{B} \overset{\text{a.s.}}{\to} \int \log_2(1+\mathsf{SNR}\,\xi)\,\mathrm{d}F(\xi) \tag{5.156}$$

$$= \int \log_2(1+\mathsf{SNR}\,\xi)\,f(\xi)\,\mathrm{d}\xi, \tag{5.157}$$

where the outer expectation in (5.155) becomes immaterial, and $f(\cdot)$ denotes the density corresponding to $F(\cdot)$. Since (5.157) depends on the channel only through $f(\cdot)$, the analysis profits directly from advances in the characterization of the asymptotic eigenvalue distribution of random matrices. A primer on this topic is featured in Appendix C.2. For more extensive tutorials, the interested reader is referred to [546, 547].

If only one of the dimensions of $\boldsymbol{H}$ is allowed to grow, the other one being kept fixed, then the hardening transpires directly from the law of large numbers (see Appendix C.1.10). Specifically, if $\boldsymbol{h} = [h_0 \;\cdots\; h_{N-1}]^\mathrm{T}$ is a vector with IID zero-mean unit-variance entries, then, for growing N,

$$\frac{1}{N}\boldsymbol{h}^*\boldsymbol{h} = \frac{1}{N}\sum_{i=0}^{N-1}|h_i|^2 \tag{5.158}$$

$$\overset{\text{a.s.}}{\to} \mathbb{E}\big[|h_i|^2\big] \tag{5.159}$$

$$= 1 \tag{5.160}$$

while, if $\boldsymbol{h}$ and $\boldsymbol{g}$ are two independent such vectors,

$$\frac{1}{N}\boldsymbol{h}^*\boldsymbol{g} = \frac{1}{N}\sum_{i=0}^{N-1}h_i^* g_i \tag{5.161}$$

$$\overset{\text{a.s.}}{\to} \mathbb{E}[h_i^* g_i] \tag{5.162}$$

$$= 0. \tag{5.163}$$

Applied to the rows and columns of $\boldsymbol{H}$, these principles ensure that, if the entries are IID,

$$\frac{1}{N_\mathrm{r}}\boldsymbol{H}^*\boldsymbol{H} \overset{\text{a.s.}}{\to} \boldsymbol{I}_{N_\mathrm{t}} \qquad (\text{fixed } N_\mathrm{t} \text{ and } N_\mathrm{r}\to\infty) \tag{5.164}$$

$$\frac{1}{N_\mathrm{t}}\boldsymbol{H}\boldsymbol{H}^* \overset{\text{a.s.}}{\to} \boldsymbol{I}_{N_\mathrm{r}} \qquad (\text{fixed } N_\mathrm{r} \text{ and } N_\mathrm{t}\to\infty). \tag{5.165}$$

The above find immediate application in MIMO, whenever $\boldsymbol{H}$ has IID entries and even when correlations are present at the end of the link whose dimension is fixed.

Example 5.31

Consider a channel with $\boldsymbol{R}_\mathrm{t} = \boldsymbol{I}$, for which capacity is achieved by $\boldsymbol{F} = \boldsymbol{I}$. Determine the capacity for fixed N_r and $N_\mathrm{t}\to\infty$, and then use this result to predict the capacity of an IID channel having $N_\mathrm{r} = 2$ and $N_\mathrm{t} = 8$ at $\mathsf{SNR} = 3$ dB.

Solution

From

$$\frac{1}{N_t} H H^* \xrightarrow{\text{a.s.}} R_r, \tag{5.166}$$

the capacity per receive antenna satisfies

$$\frac{1}{N_r} C(\mathsf{SNR}) \xrightarrow{\text{a.s.}} \frac{1}{N_r} \log_2 \det\big(I + \mathsf{SNR}\, R_r\big), \tag{5.167}$$

which, if $R_r = I$, further becomes $\frac{1}{N_r} C(\mathsf{SNR}) = \log_2(1 + \mathsf{SNR})$. For $N_r = 2$ and $\mathsf{SNR} = 3$ dB, this gives 1.58 b/s/Hz/antenna for a total of 3.16 b/s/Hz. The exact ergodic capacity for $N_r = 2$ and $N_t = 8$ is 3.01 b/s/Hz, already close to the large-dimensional ($N_t \to \infty$) value.

Example 5.32

Determine the capacity with fixed N_t and growing N_r if the channel has IID zero-mean entries.

Solution

With no precoding and $\mathsf{SNR} > 0$,

$$C(\mathsf{SNR}) = N_t \log_2\left(\frac{N_r}{N_t}\mathsf{SNR}\right) + \mathcal{O}\left(\frac{1}{N_r}\right). \tag{5.168}$$

Care must be exercised whenever multiple limits are taken simultaneously, and Example 5.32 serves to illustrate this issue. The expansion in (5.168) hinges on $\frac{N_r}{N_t}\mathsf{SNR} \gg 1$, which is assured for any fixed $\mathsf{SNR} > 0$ and sufficiently large N_r. Suppose, however, that $\mathsf{SNR} = 1/N_r^2$; then, $\frac{N_r}{N_t}\mathsf{SNR} \ll 1$ and $C(\mathsf{SNR}) = \frac{1}{N_r}\log_2 e + \mathcal{O}\big(\frac{1}{N_r^2}\big)$. Thus, the low-SNR large-dimensional capacity depends entirely on the relative speed with which $N_r \to \infty$ and $\mathsf{SNR} \to 0$. In contrast, the high-SNR large-dimensional capacity abides by (5.168) regardless of the relative speeds with which $N_r \to \infty$ and $\mathsf{SNR} \to \infty$. Hence, complications do not always arise, but one must be alert when multiple asymptotes are explored.

If H has IID zero-mean entries and both N_t and N_r grow large simultaneously, with $\beta = N_t/N_r$, then something remarkable occurs: the diagonal entries of both HH^* and H^*H converge to 1 while their off-diagonal entries converge to 0, but the empirical eigenvalue distributions of these matrices *do not converge* to that of the identity, namely a step function at 1. The number of entries of HH^* and H^*H grows faster than the pace at which those entries converge to 1 (diagonal) and 0 (off-diagonal), preventing the eigenvalues from clustering around 1. Nonetheless, although not to a step function at 1, the empirical eigenvalue distribution does converge to a nonrandom limit, by means of which the large-dimensional capacity of MIMO channels with IID entries can be established in its full generality.

Example 5.33 (Large-dimensional capacity of a MIMO channel with IID zero-mean entries and no CSIT)

For this channel, capacity is achieved without precoding and the asymptotic eigenvalue density of $\frac{1}{N_t}\boldsymbol{H}\boldsymbol{H}^*$ is the Marčenko–Pastur law (see Appendix C.2)

$$f(\xi) = [1-\beta]^+ \, \delta(\xi) + \beta \, \frac{\sqrt{(\xi-a)(b-\xi)}}{2\pi\xi} \qquad \xi \in [a,b], \tag{5.169}$$

where $[1-\beta]^+ \, \delta(\xi)$ corresponds to the zero-value eigenvalues that $\boldsymbol{H}\boldsymbol{H}^*$ has if $\beta < 1$ and

$$a = \left(1 - \frac{1}{\sqrt{\beta}}\right)^2 \qquad b = \left(1 + \frac{1}{\sqrt{\beta}}\right)^2. \tag{5.170}$$

From (5.156) and (5.169), bearing in mind that the zero-value eigenvalues do not contribute to the capacity,

$$\frac{1}{N_r} C(\mathsf{SNR}) \overset{\text{a.s.}}{\to} \int_a^b \log_2(1+\mathsf{SNR}\,\xi)\, f(\xi)\, d\xi \tag{5.171}$$

$$= \frac{\beta}{2\pi} \int_a^b \log_2(1+\mathsf{SNR}\,\xi)\, \frac{\sqrt{(\xi-a)(b-\xi)}}{\xi}\, d\xi \tag{5.172}$$

$$= \log_2\big(1+\mathsf{SNR} - \mathcal{F}(\mathsf{SNR},\beta)\big) + \beta \log_2\left(1 + \frac{\mathsf{SNR}}{\beta} - \mathcal{F}(\mathsf{SNR},\beta)\right) - \beta \frac{\log_2 e}{\mathsf{SNR}} \mathcal{F}(\mathsf{SNR},\beta), \tag{5.173}$$

where

$$\mathcal{F}(\mathsf{SNR},\beta) = \frac{1}{4}\left(\sqrt{1+\mathsf{SNR}\left(1+\frac{1}{\sqrt{\beta}}\right)^2} - \sqrt{1+\mathsf{SNR}\left(1-\frac{1}{\sqrt{\beta}}\right)^2}\right)^2, \tag{5.174}$$

with the integral in (5.172) having been independently solved in [457] and [458]. Specializing the result to $\beta = 1$, i.e., for $N_t = N_r = N_a$,

$$\frac{1}{N_a} C(\mathsf{SNR}) \overset{\text{a.s.}}{\to} 2\log_2\left(\frac{1+\sqrt{1+4\,\mathsf{SNR}}}{2}\right) - \frac{\log_2 e}{4\,\mathsf{SNR}}\left(\sqrt{1+4\,\mathsf{SNR}}-1\right)^2 \tag{5.175}$$

which confirms that, at a fixed SNR, the capacity grows linearly with the number of antennas. At high SNR, the right-hand side of (5.175) expands as $\log_2 \mathsf{SNR} - \log_2 e + \mathcal{O}(\frac{1}{\mathsf{SNR}})$, an expression first derived in [23].

Example 5.34

Let $\boldsymbol{H}$ have IID Rayleigh-faded entries. From (5.173), the large-dimensional approximation to the capacity for specific N_t and N_r is

$$C(\mathsf{SNR}) \approx N_r \log_2\left(1+\mathsf{SNR} - \mathcal{F}(\mathsf{SNR}, \tfrac{N_t}{N_r})\right) + N_t \log_2\left(1 + N_r \frac{\mathsf{SNR}}{N_t} - \mathcal{F}(\mathsf{SNR}, \tfrac{N_t}{N_r})\right) - N_t \frac{\log_2 e}{\mathsf{SNR}} \mathcal{F}(\mathsf{SNR}, \tfrac{N_t}{N_r}). \tag{5.176}$$

Table 5.1 Exact ergodic capacity and large-dimensional approximation (in b/s/Hz) at SNR = 10 dB

N_r	$N_\mathrm{t}=1$	$N_\mathrm{t}=2$	$N_\mathrm{t}=3$	$N_\mathrm{t}=4$
1	2.9 (2.72)	3.17 (3.12)	3.26 (3.24)	3.32 (3.3)
2	4.06 (4.01)	5.56 (5.45)	6.05 (6.0)	6.29 (6.25)
3	4.73 (4.7)	7.05 (6.98)	8.22 (8.17)	8.84 (8.79)
4	5.2 (5.17)	8.05 (8.02)	9.87 (9.8)	10.93 (10.89)

Within brackets in Table 5.1 are these large-dimensional approximations for $1 \leq N_\mathrm{t} \leq 4$ and $1 \leq N_\mathrm{r} \leq 4$ at $\mathsf{SNR} = 10$ dB, each next to the exact ergodic capacity.

Example 5.35

Let $\boldsymbol{H}$ have IID Rayleigh-faded entries. Shown in Fig. 5.5 is the exact function $C(\mathsf{SNR})$ for $N_\mathrm{t} = N_\mathrm{r} = 4$ along with the corresponding large-dimensional approximations. Notice the accuracy despite the modest numbers of antennas.

Interestingly, the capacity expressions in Example 5.33 have been shown to apply to a class of channels broader than that of matrices with zero-mean IID entries. Specifically, consider an instance of the UIU model seen in Chapter 3 where $\boldsymbol{H}$ has zero-mean IND entries whose variances are assembled into the matrix $\boldsymbol{\Omega}$, i.e., such that

$$[\boldsymbol{\Omega}]_{i,j} = \mathbb{E}\Big[|[\boldsymbol{H}]_{i,j}|^2\Big]. \tag{5.177}$$

If $\boldsymbol{\Omega}$ is *mean doubly regular*, meaning that the average of the entries along every row and column coincides, then, under the very mild technical condition that the entries of $\boldsymbol{\Omega}$ remain bounded as $N_\mathrm{t}, N_\mathrm{r} \to \infty$, the large-dimensional spectral efficiency without precoding is equivalent to that on a channel with IID entries at the same SNR [548]. If an unprecoded transmission achieves the capacity of $\boldsymbol{H}$, then the foregoing spectral efficiency equivalence directly translates to a capacity equivalence.

Example 5.36

Suppose that antennas with alternating orthogonal polarizations are utilized and that there is no correlation, in which case [549]

$$\boldsymbol{\Omega} = \frac{2}{1+\Xi}\begin{bmatrix} 1 & \Xi & 1 & \Xi & \cdots \\ \Xi & 1 & \Xi & 1 & \cdots \\ 1 & \Xi & 1 & \Xi & \cdots \\ \Xi & 1 & \Xi & 1 & \cdots \\ \vdots & \vdots & \vdots & \vdots & \ddots \end{bmatrix}, \tag{5.178}$$

with Ξ the cross-polar discrimination introduced in Chapter 3 [298]. The large-dimensional capacity of this channel is given by (5.173).

In channels whose entries exhibit correlations, the asymptotic analysis becomes substantially more involved, but progress can still be made well beyond what is possible nonasymptotically [478, 489, 548, 550–552]. The most relevant finding is that the capacity still scales linearly with N_{min}, although the scaling varies with respect to (5.173).

Example 5.37 (Large-dimensional spectral efficiency of a MIMO channel with Kronecker correlations and no CSIT)

For the Kronecker correlation model, a large-dimensional solution is available [548]. Expressed for finite N_t and N_r,

$$\frac{R}{B} \approx \sum_{j=0}^{N_t-1} \log_2\left(\frac{1+\mathsf{SNR}\, P_j\, \lambda_j(\boldsymbol{R}_t)\Upsilon_r}{e^{\mathsf{SNR}\,\Upsilon_t\Upsilon_r}}\right) + \sum_{i=0}^{N_r-1} \log_2\big(1+\mathsf{SNR}\,\lambda_i(\boldsymbol{R}_r)\Upsilon_t\big) \quad (5.179)$$

where $P_0, \ldots, P_{N_t-1}$ is the power allocation whereas Υ_t and Υ_r must be obtained by solving the fixed-point equations

$$\Upsilon_t = \frac{1}{N_t} \sum_{j=0}^{N_t-1} \frac{P_j\, \lambda_j(\boldsymbol{R}_t)}{1+\mathsf{SNR}\, P_j\, \lambda_j(\boldsymbol{R}_t)\Upsilon_r} \quad (5.180)$$

$$\Upsilon_r = \frac{1}{N_t} \sum_{i=0}^{N_r-1} \frac{\lambda_i(\boldsymbol{R}_r)}{1+\mathsf{SNR}\,\lambda_i(\boldsymbol{R}_r)\Upsilon_t}. \quad (5.181)$$

The spectral efficiency without precoding is obtained simply by setting $P_j = 1$ for $j = 0, \ldots, N_t - 1$. If $\boldsymbol{R}_t$ and $\boldsymbol{R}_r$ are further replaced by identity matrices, the right-hand side of (5.179) then reverts to that of (5.173).

The asymptotic characterization of the capacity-achieving powers, $P_0^\star, \ldots, P_{N_t-1}^\star$, is tackled in [489, 548, 552].

A noteworthy exception to the usually rapid convergence of the ergodic capacity to its large-dimensional value occurs in Rice channels with unit-rank $\boldsymbol{H}_{\text{LOS}}$. The Rice term then perturbs a single eigenvalue of $\boldsymbol{H}\boldsymbol{F}\boldsymbol{F}^*\boldsymbol{H}^*$, not affecting the asymptotic eigenvalue distribution of this matrix [553]. It follows that the large-dimensional capacity is as if the Rice term were not there, yet the Rice term does influence the actual capacity with small N_t and N_r, and this influence disappears very slowly with the number of antennas.

5.5 No CSI

It was argued in the analysis of SISO that, in underspread settings, the assumption of CSIR is always reasonable. With MIMO, this rationale slowly weakens as N_t grows, to the point of becoming a potential limitation in massive MIMO. Thus, although CSIR is still fully justified for moderate N_t, because of its relevance to massive MIMO we briefly examine the MIMO capacity without CSIR.

Consider a frequency-flat Rayleigh-faded channel with coherence blocks spanning N_c

symbols and no antenna correlation. Since, without CSIR, this channel does have memory, a nonsingle-letter formulation is required; moreover, the codeword symbols need not be independent. Assembling the portion of those symbols transmitted over one coherence block, $\boldsymbol{s}[0], \ldots, \boldsymbol{s}[N_c - 1]$, into an $N_t \times N_c$ matrix $\boldsymbol{S}$, the capacity-achieving signal (with the precoder subsumed) has been shown to have the form

$$\boldsymbol{S} = \boldsymbol{A_S} \cdot \sqrt{N_c} \boldsymbol{U_S}, \tag{5.182}$$

where $\boldsymbol{A_S} = \text{diag}(A_0, \ldots, A_{N_t-1})$ with A_j the magnitude of the signal emitted by the jth antenna, satisfying $\mathbb{E}[A_j^2] = 1$ [409]. In turn, $\boldsymbol{U_S}$ is a $N_t \times N_c$ matrix whose rows are orthonormal isotropic vectors, equally likely to point in any direction. For $N_t < N_c$, such $\boldsymbol{U_S}$ amounts to N_t rows of an $N_c \times N_c$ unitary matrix and, since $\boldsymbol{U_S} \boldsymbol{U_S}^* = \boldsymbol{I}$ but $\boldsymbol{U_S}^* \boldsymbol{U_S} \neq \boldsymbol{I}$, the term *semiunitary* is sometimes applied. The factor $\sqrt{N_c}$ in (5.182) scales the rows of $\boldsymbol{S}$ so as to comply with the normalization in this text.

On a given coherence block, the jth antenna transmits the jth row of $\sqrt{N_c}\boldsymbol{U_S}$ scaled by A_j, i.e., the antennas transmit easy-to-discriminate orthogonal sequences with magnitudes $A_0, \ldots, A_{N_t-1}$ drawn from a suitable distribution that must be optimized numerically. Note how, for $N_t = 1$, we recover the optimum signal structure unveiled for SISO.

The transmit–receive relationship over each fading block is

$$\boldsymbol{Y} = \sqrt{\frac{GE_s}{N_t}} \boldsymbol{H} \boldsymbol{S} + \mathbf{V}, \tag{5.183}$$

where $\mathbf{V} = \left[\boldsymbol{v}[0] \cdots \boldsymbol{v}[N_c - 1]\right]$ is $N_r \times N_c$ with $[\mathbf{V}]_{i,n} \sim \mathcal{N}_\mathbb{C}(0, N_0)$. The conditional distribution of $\boldsymbol{Y}$ for a given $\boldsymbol{S}$ is (see Appendix C.1.9)

$$f_{\boldsymbol{Y}|\boldsymbol{S}}(\mathbf{Y}|\mathbf{S}) = \frac{1}{\pi^{N_r N_c} \det^{N_r}\left(\boldsymbol{I} + \frac{\mathsf{SNR}}{N_t}\mathbf{S}^*\mathbf{S}\right)} \exp\left(-\text{tr}\left(\left(\boldsymbol{I} + \frac{\mathsf{SNR}}{N_t}\mathbf{S}^*\mathbf{S}\right)^{-1} \mathbf{Y}^*\mathbf{Y}\right)\right), \tag{5.184}$$

which depends on $\boldsymbol{S}$ only through $\boldsymbol{S}^*\boldsymbol{S}$. Since $\boldsymbol{S}^*\boldsymbol{S}$ is $N_c \times N_c$, one would expect that any desired distribution for it could be realized with at most N_c transmit antennas, and that is indeed the case [409, theorem 1]. In other words: whatever capacity could be achieved with $N_t > N_c$ antennas can also be achieved with at most $N_t = N_c$ antennas. Intuitively, trying to transmit too many concurrent signals with unknown fading coefficients is counterproductive beyond a point, with the added fading uncertainty offsetting the addition of new signals. In fact, once $N_t > N_c - N_r$, signals of the form $\boldsymbol{S} = \boldsymbol{A_S} \cdot \sqrt{N_c}\boldsymbol{U_S}$ cease to be optimum altogether [554]. In any event though, other than possibly in massive MIMO, it is the case that $N_t \ll N_c$ and the sustained capacity scaling with N_t that we encounter in the face of CSIR essentially holds without CSIR.

Low-SNR regime

As seen throughout our derivations of $\frac{E_b}{N_0}_{\min}$, its value in Gaussian noise equals $\frac{1}{\log_2 e} = -1.59$ dB, reduced by the total channel gain attainable beyond G. Let us apply this principle over a block of N_c symbols, wherein for the transmit sequence $\boldsymbol{S}$ the total channel gain is given by

$$\frac{\mathbb{E}\left[\|\sqrt{\frac{GE_\mathrm{s}}{N_\mathrm{t}}}\boldsymbol{H}\boldsymbol{S}\|_\mathrm{F}^2\right]}{\mathbb{E}\left[\|\sqrt{\frac{E_\mathrm{s}}{N_\mathrm{t}}}\boldsymbol{S}\|_\mathrm{F}^2\right]} = \frac{G}{\mathbb{E}[\|\boldsymbol{S}\|_\mathrm{F}^2]}\,\mathbb{E}\left[\|\boldsymbol{H}\boldsymbol{S}\|_\mathrm{F}^2\right] \tag{5.185}$$

$$= \frac{G}{N_\mathrm{t}N_\mathrm{c}}\,\mathbb{E}\left[\|\boldsymbol{H}\boldsymbol{S}\|_\mathrm{F}^2\right], \tag{5.186}$$

from which

$$\frac{E_\mathrm{b}}{N_0}_{\mathsf{min}} = \frac{\frac{1}{\log_2 e}}{\frac{1}{N_\mathrm{t}N_\mathrm{c}}\,\mathbb{E}\left[\|\boldsymbol{H}\boldsymbol{S}\|_\mathrm{F}^2\right]} \tag{5.187}$$

$$= \frac{N_\mathrm{t}N_\mathrm{c}}{\mathrm{tr}\big(\mathbb{E}[\boldsymbol{H}\boldsymbol{S}\boldsymbol{S}^*\boldsymbol{H}^*]\big)\log_2 e} \tag{5.188}$$

$$= \frac{N_\mathrm{t}}{\mathrm{tr}\big(\mathbb{E}[\boldsymbol{H}\boldsymbol{A}_{\boldsymbol{S}}\boldsymbol{U}_{\boldsymbol{S}}\boldsymbol{U}_{\boldsymbol{S}}^*\boldsymbol{A}_{\boldsymbol{S}}^*\boldsymbol{H}^*]\big)\log_2 e} \tag{5.189}$$

$$= \frac{N_\mathrm{t}}{\mathrm{tr}\big(\mathbb{E}[\boldsymbol{H}\boldsymbol{H}^*]\big)\log_2 e} \tag{5.190}$$

$$= \frac{1}{N_\mathrm{r}\log_2 e}, \tag{5.191}$$

where (5.190) holds because $\boldsymbol{U}_{\boldsymbol{S}}\boldsymbol{U}_{\boldsymbol{S}}^* = \boldsymbol{I}$ and $\mathbb{E}[\boldsymbol{A}_{\boldsymbol{S}}\boldsymbol{A}_{\boldsymbol{S}}^*] = \boldsymbol{I}$, with $\boldsymbol{A}_{\boldsymbol{S}}$ independent of $\boldsymbol{H}$, while (5.191) holds because $\mathrm{tr}\big(\mathbb{E}[\boldsymbol{H}\boldsymbol{H}^*]\big) = N_\mathrm{t}N_\mathrm{r}$. Since (5.191) coincides with the value found for IID channels with CSIR, we confirm that, as in SISO, the lack of CSIR does not alter $\frac{E_\mathrm{b}}{N_0}_{\mathsf{min}}$.

In low-SNR no-CSIR conditions, the information is conveyed mostly by the magnitudes within $\boldsymbol{A}_{\boldsymbol{S}}$ and, again as in SISO, achieving $\frac{E_\mathrm{b}}{N_0}_{\mathsf{min}}$ exactly would entail unbounded peakedness in those magnitudes. However, in underspread settings, it can be approached very closely with standard IID complex Gaussian signaling. Indeed, the low-SNR no-CSIR slope is $\mathcal{S}_0 = 0$, but—recall Fig. 4.18—this must be properly interpreted: for $N_\mathrm{c} \gg 1$, the no-CSIR capacity tightly tracks its CSIR counterpart, despite the two having very different slopes at $\frac{E_\mathrm{b}}{N_0}_{\mathsf{min}}$ [555].

High-SNR regime

In high-SNR no-CSIR conditions, the magnitudes within $\boldsymbol{A}_{\boldsymbol{S}}$ cluster around 1 and the information is conveyed mostly by $\boldsymbol{U}_{\boldsymbol{S}}$ [415, 556, 557]. It can be established that, to maximize $\mathcal{S}_\infty$, the number of streams should equal $N_\mathrm{s} = \min(N_\mathrm{t}, N_\mathrm{r}, \lfloor N_\mathrm{c}/2 \rfloor)$ [416]. Other than possibly in extreme versions of massive MIMO, $N_{\mathsf{min}} = \min(N_\mathrm{t}, N_\mathrm{r}) < \lfloor N_\mathrm{c}/2 \rfloor$ and therefore $N_\mathrm{s} = N_{\mathsf{min}}$, a result that should be construed in light of Discussion 5.4. The maximum number of spatial DOF is then

$$\mathcal{S}_\infty = \left(1 - \frac{N_{\mathsf{min}}}{N_\mathrm{c}}\right) N_{\mathsf{min}}, \tag{5.192}$$

which, for $N_\mathrm{t} = N_\mathrm{r} = 1$, reduces to the SISO value in (4.185). Contrasting (5.192) with its CSIR counterpart, $\mathcal{S}_\infty = N_{\mathsf{min}}$, the penalty associated with the lack of side information at the receiver is seen to be the factor $(1 - N_{\mathsf{min}}/N_{\mathrm{c}})$, which, in underspread nonmassive settings, is very small.

Expressions for the power offset without CSIR are derived in [416, 554].

If the fading is continuous rather than in blocks, then, as in SISO, the capacity without CSIR exhibits extreme sensitivity to the modeling assumptions for $\mathsf{SNR} \to \infty$. With nonregular fading, the analogy between block fading and continuous fading established in Chapter 3, $N_{\mathrm{c}} = B_{\mathrm{c}} T_{\mathrm{c}}$, would mean that

$$C(\mathsf{SNR}) = \left(1 - \frac{N_{\mathsf{min}}}{B_{\mathrm{c}} T_{\mathrm{c}}}\right) N_{\mathsf{min}} \log_2 \mathsf{SNR} + \mathcal{O}(1), \tag{5.193}$$

where B_{c} and T_{c}, recall, are the coherence bandwidth and the coherence time of the fading process. With regular fading, however [404, 558]

$$C(\mathsf{SNR}) = \log_2 \log \mathsf{SNR} + \mathcal{O}(1) \tag{5.194}$$

and a single DOF is available. For SNRs of practical relevance, the regularity in the fading is immaterial and the MIMO capacity abides by (5.193).

For the IID complex Gaussian signals that are optimum with CSIR, the spectral efficiency without CSIR can be computed by means of the semi-analytical procedure in [420].

5.6 Pilot-assisted communication

Let us now extend the SISO pilot-assisted analysis presented in Section 4.8 to the province of MIMO [183, 187, 559–563]. To avoid distracting elements that might clutter the analysis, we focus on the canonical channel with IID Rayleigh-faded entries.

We denote by $\alpha \geq \alpha_{\mathsf{min}}$ the total share of symbols that are pilots, with α/N_{t} the fraction corresponding to each transmit antenna. The pilot sequences emitted by distinct antennas are taken to be mutually orthogonal, which, as argued in Chapter 2, is optimum in time-invariant channels and, by extension, quasi-optimum in underspread fading [564, 565]. A preferred embodiment are the Zadoff–Chu sequences also described in Chapter 2. Arguably the simplest manner in which the pilot symbols constituting those sequences can be mixed with the data is by insertion in time and/or frequency (see Fig. 5.7), although other approaches such as superposition or code-division multiplexing are possible.

Under block fading, at least N_{t} pilot symbols must be inserted within each coherence block, one per antenna, and thus $\alpha_{\mathsf{min}} = N_{\mathrm{t}}/N_{\mathrm{c}}$; with $N_{\mathrm{p}} \geq N_{\mathrm{t}}$ pilot symbols per block, $\alpha = N_{\mathrm{p}}/N_{\mathrm{c}}$. Under continuous frequency-flat fading, in turn, to ensure that the decimated fading process observed through the pilot sequence of each antenna has an unaliased spectrum, it is necessary that $\alpha_{\mathsf{min}} = N_{\mathrm{t}} \cdot 2\nu_{\mathrm{M}} T$.

As in our SISO analysis, rather than study the mutual information between the transmitted data symbols and the received symbols (pilots and data), we focus on the spectral efficiency achieved by receivers that form explicit channel estimates from the pilot observations and subsequently decode the data regarding these channel estimates as correct while treating the term containing the estimation error as additional Gaussian noise. With

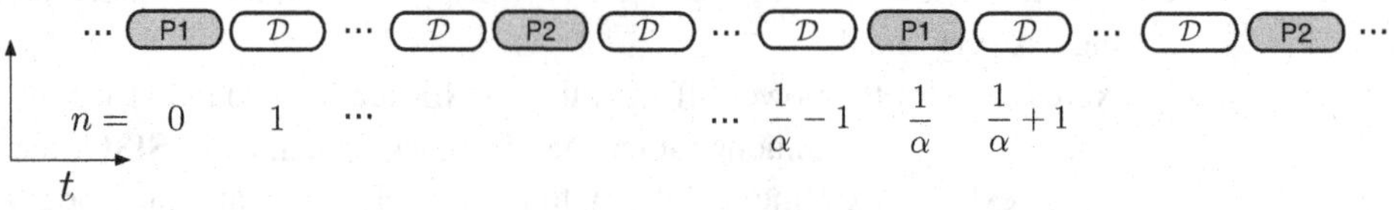

Fig. 5.7 MIMO transmit signal with $N_t = 2$ and time-multiplexed pilot symbols. On the time indices labeled as P1 and P2, only the corresponding antenna transmits a pilot. Elsewhere, both antennas transmit data symbols.

the entries of $\boldsymbol{H}$ being IID, they can be separately estimated without loss of optimality and a derivation similar to the one for SISO leads to an achievable spectral efficiency of

$$\frac{R}{B} = (1 - \alpha)\, C(\mathsf{SNR}_{\mathsf{eff}}), \tag{5.195}$$

where $C(\cdot)$ is the MIMO capacity with CSIR, closed forms for which are provided in Example 5.17, while

$$\mathsf{SNR}_{\mathsf{eff}} = \frac{\mathsf{SNR}\,(1 - \mathsf{MMSE})}{1 + \mathsf{SNR} \cdot \mathsf{MMSE}}, \tag{5.196}$$

where MMSE is the variance of the estimation error for each entry of $\boldsymbol{H}$. In Rayleigh fading, the MMSE channel estimator is linear and thus the MMSE expressions derived in Section 3.7.2 for an LMMSE estimator apply. In particular, recalling from Chapter 3 the form that unifies block fading and continuous fading,

$$\mathsf{MMSE} = \frac{1}{1 + \frac{\alpha N_c}{N_t}\mathsf{SNR}}. \tag{5.197}$$

The right-hand side of (5.195) is concave in α and thus the optimum overhead $\alpha^\star$ can be found efficiently.

Example 5.38

Let $N_t = N_r = 4$ with the antennas uncorrelated and plot, as a function of SNR, the spectral efficiency in (5.195) for a Rayleigh-faded channel with a Clarke–Jakes Doppler spectrum corresponding to vehicular and to outdoor pedestrian users, with the pilot overhead optimized for each SNR. Contrast these spectral efficiencies with the CSIR capacity. Separately, plot the respective optimum pilot overheads.

Solution

As computed in Examples 3.26 and 3.27, typical vehicular and outdoor pedestrian settings may correspond to $N_c = 1000$ and $N_c = 20\,000$, respectively. For these fading coherences, the requested results are presented in Figs. 5.8 and 5.9. In the left-hand subfigures we find the ergodic spectral efficiencies with the pilot overheads optimized, alongside the CSIR capacity borrowed from Example 5.17. In the right-hand subfigures, we find the corresponding optimum pilot overheads.

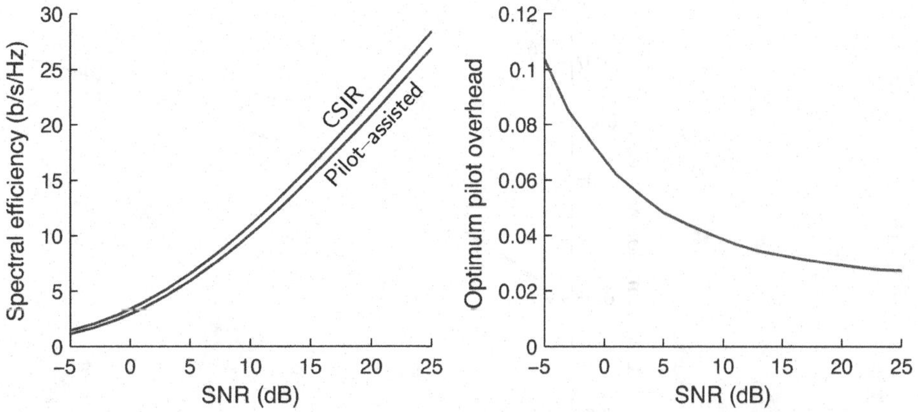

Fig. 5.8 Left, for an ergodic Rayleigh-faded channel with $N_t = N_r = 4$ uncorrelated antennas and $N_c = 1000$: optimized pilot-assisted spectral efficiency and CSIR capacity, both as function of SNR (in dB). Right: corresponding optimum pilot overhead.

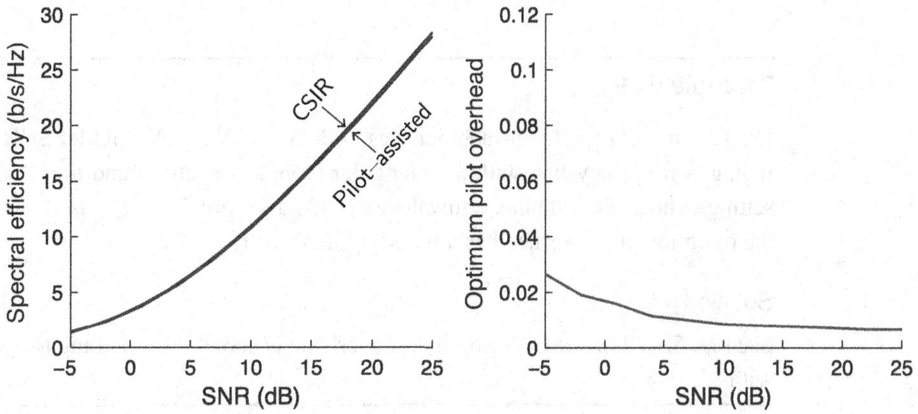

Fig. 5.9 Left, for an ergodic Rayleigh-faded channel with $N_t = N_r = 4$ uncorrelated antennas and $N_c = 20\,000$: optimized pilot-assisted spectral efficiency and CSIR capacity, both as function of SNR (in dB). Right: corresponding optimum pilot overhead.

Even in vehicular conditions, and decidedly in pedestrian situations, pilot-assisted communication can perform satisfactorily close to capacity with $N_t = N_r = 4$. With respect to SISO, here the tension between more accurate channel estimates at the expense of higher overhead is exacerbated as the number of transmit antennas increases. Interestingly, the estimation error depends on N_t only through the product $N_t \cdot \nu_M T$ (with continuous frequency-flat fading) or through the radio N_t/N_c (with block fading) and thus any scaling in the number of transmit antennas is equivalent to the inverse scaling in fading coherence.

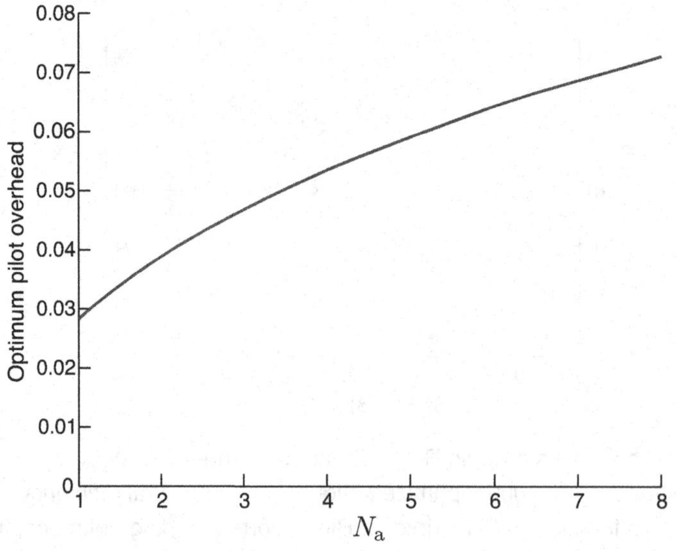

Fig. 5.10 Optimum pilot overhead for SNR $= 10$ dB and $\nu_M T = 10^{-3}$ (equivalent to $N_c = 500$ symbols) as a function of the number of antennas, $N_t = N_r = N_a$.

Example 5.39

Let $\boldsymbol{H}$ have IID Rayleigh-faded entries with $N_t = N_r = N_a$ and let SNR $= 10$ dB. The fading is frequency-flat, with a rectangular Doppler spectrum and $\nu_M T = 10^{-3}$. For this setting, whose block-fading equivalent would have only $N_c = \frac{1}{2\,\nu_M T} = 500$ symbols, plot the optimum pilot overhead as a function of $N_a = 1, \ldots, 8$.

Solution

See Fig. 5.10. For SISO, the optimum overhead is about 3%, climbing to a still modest 7% with $N_a = 8$.

From (5.195) and (5.196), low- and high-SNR expansions can be derived and $\alpha^\star$ can be expressed explicitly [426]. For SNR $\to \infty$ in particular, $\alpha^\star \to \alpha_{\text{min}}$ and hence the number of spatial DOF is

$$S_\infty = (1 - \alpha_{\text{min}}) N_{\text{min}} \tag{5.198}$$

$$= \left(1 - \frac{N_t}{N_c}\right) N_{\text{min}}, \tag{5.199}$$

which, for $N_t \leq N_r$, equals the no-CSIR value in (5.192). For $N_t > N_r$, (5.199) is smaller than (5.192), but this shortfall in DOF can be corrected by reducing the rank of the precoder to $N_s = N_{\text{min}}$, and thus the pilot overhead down to $\alpha_{\text{min}} = N_{\text{min}}/N_c$; then, (5.199) becomes equal to (5.192). The maximization of S_∞ by $N_s = N_{\text{min}}$ is consistent with our no-CSIR analysis, of which pilot-assisted communication is indeed a special case, but it is also equally modulated by Discussion 5.4.

Returning to arbitrary SNRs, the spectral efficiency can be expanded as a series in the fading coherence, leading to explicit forms for $\alpha^\star$ [441].

If pilot-power boosting is allowed, then, at every SNR, the spectral efficiency is optimized by $\alpha^\star = \alpha_{\text{min}}$ in conjunction with the power boost that maximizes SNR_{eff} [426].

Antenna correlation is bound to influence not only the channel estimation MMSE, and thus the effective SNR, but also the function $C(\text{SNR})$. The optimization becomes considerably richer, since then the separate estimation of each channel entry is no longer optimum. Rather, joint estimation of the entries of $\boldsymbol{H}$ is superior, and SNR_{eff} generalizes into a matrix of effective SNRs [566]. At the same time, in the face of correlation the payload data should ideally be nontrivially precoded. Moreover, the pilot sequences can themselves be precoded, meaning not being dispatched from the transmit antennas directly but rather from the columns of the steering matrix $\boldsymbol{U_F}$, possibly with different powers. Then, the problem of estimating $\boldsymbol{H}$ morphs into the problem of estimating $\boldsymbol{HF}$. This modified problem takes center stage in the third part of the book, once we turn to CSIT-based precoding for MU-MIMO. In the present context of SU-MIMO, the interplay of antenna correlation, precoding, and pilot-assisted channel estimation is explored in Problems 5.45–5.47, and readers interested in probing further are pointed to [567–569]; these references treat the spectral efficiency assuming that, through reciprocity or feedback, both transmitter and receiver have access to the same channel estimates.

5.7 Channels with interference

In the presence of interference from U other users, the single-letter transmit–receive relationship for a certain user of interest is

$$\boldsymbol{y} = \sqrt{G}\boldsymbol{H}\boldsymbol{x} + \sum_{u=1}^{U} \sqrt{G_u}\boldsymbol{H}_u\boldsymbol{x}_u + \boldsymbol{v}, \tag{5.200}$$

where the per-symbol energies of the interfering signals are $\mathbb{E}\big[\|\boldsymbol{x}_u\|^2\big] = E_u$ and the fading channels satisfy, as usual, $\mathbb{E}\big[\|\boldsymbol{H}_u\|_{\text{F}}^2\big] = N_u N_{\text{r}}$ where N_u is the number of transmit antennas at the uth interferer.

If $\boldsymbol{H}_1\boldsymbol{F}_1, \ldots, \boldsymbol{H}_U\boldsymbol{F}_U$ are not known by the receiver of interest, then, as argued in the SISO analysis of Section 4.9, the non-Gaussianity of the interference subject to fading can be circumvented by converting (5.200) into

$$\boldsymbol{y} = \sqrt{G}\boldsymbol{H}\boldsymbol{x} + \boldsymbol{v}', \tag{5.201}$$

where

$$\boldsymbol{v}' \sim \mathcal{N}_{\mathbb{C}}\left(\boldsymbol{0}, \sum_{u=1}^{U} G_u \mathbb{E}\big[\boldsymbol{H}_u \boldsymbol{R}_{\boldsymbol{x}_u} \boldsymbol{H}_u^*\big] + N_0\, \boldsymbol{I}\right). \tag{5.202}$$

The capacity of this modified channel lower-bounds, usually very tightly [444], that of (5.200). Therefore, the entire body of results in this chapter applies under the premise that $\boldsymbol{H}_1\boldsymbol{F}_1, \ldots, \boldsymbol{H}_U\boldsymbol{F}_U$ are unknown to the receiver of interest.

Alternatively, we can posit a receiver that does know $\boldsymbol{H}_1\boldsymbol{F}_1, \ldots, \boldsymbol{H}_U\boldsymbol{F}_U$ in addition to $\boldsymbol{H}$ and $\boldsymbol{F}$. The conditional covariance of the received interference-plus-noise then equals

$$\begin{aligned} \boldsymbol{R}(\boldsymbol{H}_1, \ldots, \boldsymbol{H}_U) &= \mathbb{E}\left[\left(\sum_{u=1}^{U} \sqrt{G_u}\boldsymbol{H}_u\boldsymbol{x}_u + \boldsymbol{v}\right)\left(\sum_{u=1}^{U} \sqrt{G_u}\boldsymbol{H}_u\boldsymbol{x}_u + \boldsymbol{v}\right)^{*} \Big| \boldsymbol{H}_1, \ldots, \boldsymbol{H}_U\right] \\ &= \sum_{u=1}^{U} G_u \boldsymbol{H}_u \boldsymbol{R}_{\boldsymbol{x}_u} \boldsymbol{H}_u^{*} + N_0 \boldsymbol{I}. \end{aligned} \tag{5.203}$$

The mutual information when the desired and interfering signals are complex Gaussian is

$$I(\boldsymbol{x}; \boldsymbol{y} \,|\boldsymbol{H}, \boldsymbol{H}_1, \ldots, \boldsymbol{H}_U) = \log_2 \det\big(\boldsymbol{I} + G\boldsymbol{H}\boldsymbol{R}_{\boldsymbol{x}}\boldsymbol{H}^{*}\boldsymbol{R}^{-1}(\boldsymbol{H}_1, \ldots, \boldsymbol{H}_U)\big), \tag{5.204}$$

which generalizes the SISO expression in (4.231). From the mutual information, the capacity, with and without CSIT, can be evaluated.

With CSIT, the capacity that derives from (5.204) is studied in [570–575], with the main complication being the optimization of the precoder. In fact, if not only the desired transmitter but also the interferers selfishly optimize their precoders, we have a game-theoretic scenario where every transmitter is a player and the payoff is the mutual information [576, 577]. When confronted with this scenario, one naturally gravitates towards the idea of having the transmitters cooperate and jointly optimize their precoders for the common good, an idea that is surveyed in works such as [577–579].

Without CSIT, alternatively,

$$C = \max_{\boldsymbol{R}_{\boldsymbol{x}}:\mathrm{tr}(\boldsymbol{R}_{\boldsymbol{x}})=E_{\mathrm{s}}} \mathbb{E}\Big[\log_2 \det\big(\boldsymbol{I} + G\boldsymbol{H}\boldsymbol{R}_{\boldsymbol{x}}\boldsymbol{H}^{*}\boldsymbol{R}^{-1}(\boldsymbol{H}_1, \ldots, \boldsymbol{H}_U)\big)\Big] \tag{5.205}$$

$$= \max_{\boldsymbol{F}:\mathrm{tr}(\boldsymbol{F}\boldsymbol{F}^{*})=N_{\mathrm{t}}} \mathbb{E}\left[\log_2 \det\left(\boldsymbol{I} + \frac{\mathsf{SNR}}{N_{\mathrm{t}}}\boldsymbol{H}\boldsymbol{F}\boldsymbol{F}^{*}\boldsymbol{H}^{*}\left(\boldsymbol{I} + \sum_{u=1}^{U}\frac{\mathsf{INR}_u}{N_u}\boldsymbol{H}_u\boldsymbol{F}_u\boldsymbol{F}_u^{*}\boldsymbol{H}_u^{*}\right)^{-1}\right)\right] \tag{5.206}$$

where we have used $\boldsymbol{R}_{\boldsymbol{x}_u} = \mathbb{E}\big[\boldsymbol{x}_u\boldsymbol{x}_u^{*}|\boldsymbol{F}_u\big] = \sqrt{E_u/N_u}\boldsymbol{F}_u\boldsymbol{F}_u^{*}$ and recalled the definition of INR, namely $\mathsf{INR}_u = G_u E_u/N_0$. The expectation in (5.206) is over $\boldsymbol{H}$ and $\boldsymbol{H}_1, \ldots, \boldsymbol{H}_U$, and the precoder optimization is again complicated relative to a noise-limited channel. From the viewpoint of the receiver of interest, it would be desirable if the interferers applied precoders with the lowest possible ranks: as shown in Chapter 6, interference with dimensionality inferior to N_{r} can be completely suppressed through mere linear processing [11]. At the same time, from a selfish standpoint, each interferer might want to apply a precoder with higher rank. Therefore, if all the transmitters could optimize their precoders, we would again be faced with a game-theoretic situation [580].

Let us examine, in the remainder of this section, what unfolds if the interferers are unprecoded. As in SISO, having interference in lieu of noise is beneficial—at a given SINR—because the interference undergoes fading. With MIMO, furthermore, the interference exhibits spatial color that, thanks to its knowledge of $\boldsymbol{H}_1\boldsymbol{F}_1, \ldots, \boldsymbol{H}_U\boldsymbol{F}_U$, the receiver can exploit. For a given power, the more structured the interference, the less harmful, with unstructured noise being the worst possible impairment. The ability to discern spatial structure is determined by N_{r} whereas the degree of structure is determined by $\sum_{u=1}^{U} N_u$. If

$N_r > \sum_{u=1}^{U} N_u$, the receiver can theoretically nullify the interference completely. At the other extreme, if N_r is fixed and $\sum_{u=1}^{U} N_u \to \infty$, the structure becomes hopelessly fine and the aggregate interference looks like noise to the receiver.

The analysis at arbitrary SNR of the no-CSIT capacity with interference can be tackled directly [581], or else by rewriting (5.206) as the difference of two noise-limited capacities, namely [446]

$$C = \max_{F:\mathrm{tr}(FF^*)=N_t} \mathbb{E}\left[\log_2 \det\left(I + \frac{\mathsf{SNR}}{N_t} HFF^*H^* + \sum_{u=1}^{U} \frac{\mathrm{INR}_u}{N_u} H_u F_u F_u^* H_u^*\right)\right]$$
$$- \mathbb{E}\left[\log_2 \det\left(I + \sum_{u=1}^{U} \frac{\mathrm{INR}_u}{N_u} H_u F_u F_u^* H_u^*\right)\right]. \tag{5.207}$$

Then, as usual, the large-dimensional regime offers an alternative arena for the analysis [489, 582, 583].

Low-SNR regime

Let us now turn our attention to the low-SNR regime, recalling the SISO definition of $N_0' = \sum_{u=1}^{U} G_u E_u + N_0$ to subsume both noise and interference. Following a similar derivation, the formula for $\frac{E_b}{N_0}_{\min}$ found in (4.250) generalizes to

$$\frac{E_b}{N_0'}_{\min} = \frac{1}{\left(1 + \sum_{u=1}^{U} \mathrm{INR}_u\right) \log_2 e} \tag{5.208}$$
$$\cdot \min_{F:\mathrm{tr}(FF^*)=N_t} \frac{1}{\mathrm{tr}\left(\mathbb{E}\left[\frac{1}{N_t} HFF^*H^*\right] \mathbb{E}\left[\left(I + \sum_{u=1}^{U} \frac{\mathrm{INR}_u}{N_u} H_u F_u F_u^* H_u^*\right)^{-1}\right]\right)}$$

where the minimization can be trivial, in canonical examples like the one that follows, but is in general rather involved.

Example 5.40

Consider a transmission impaired by a single unprecoded interferer, i.e., $U = 1$, $F_1 = I$, and $\mathrm{INR} \to \infty$. Evaluate $\frac{E_b}{N_0'}_{\min}$.

Solution

Applying (5.208),

$$\frac{E_b}{N_0'}_{\min} = \min_{F:\mathrm{tr}(FF^*)=N_t} \frac{1}{\mathrm{tr}\left(\mathbb{E}\left[\frac{1}{N_t} HFF^*H^*\right] \mathbb{E}\left[\left(\frac{1}{N_1} H_1 H_1^*\right)^{-1}\right]\right)} \cdot \frac{1}{\log_2 e}. \tag{5.209}$$

If both H and H_1 have IID Rayleigh-faded entries, then the same arguments invoked for noise serve to claim that capacity is achieved with $F = I$. Thus, $\mathbb{E}\left[\frac{1}{N_t} HFF^*H^*\right] = I$

such that

$$\frac{E_b}{N'_0}_{\min} = \frac{1}{N_1 \operatorname{tr}\left(\mathbb{E}\left[(\boldsymbol{H}_1 \boldsymbol{H}_1^*)^{-1}\right]\right)} \cdot \frac{1}{\log_2 e}, \tag{5.210}$$

where $\boldsymbol{H}_1 \boldsymbol{H}_1^*$ is a Wishart matrix to which we can apply (C.26). If $N_1 \leq N_r$, then $\operatorname{tr}(\mathbb{E}[(\boldsymbol{H}_1 \boldsymbol{H}_1^*)^{-1}])$ is infinite whereas, if $N_1 > N_r$, $\operatorname{tr}(\mathbb{E}[(\boldsymbol{H}_1 \boldsymbol{H}_1^*)^{-1}]) = N_r/(N_1 - N_r)$. Altogether,

$$\frac{E_b}{N'_0}_{\min} = \frac{1}{\log_2 e}\left[\frac{1}{N_r} - \frac{1}{N_1}\right]^+, \tag{5.211}$$

consistent with the fact that, if the interference has dimensionality inferior to N_r, some dimensions at the receiver of interest remain free of interference and, in the absence of noise, communication over those dimensions is possible with negligible power. Of course, in actuality there is always residual noise beneath the interference, and reliable communication necessitates a received E_b that is -1.59 dB above that residual noise.

Alternatively, if the interference dimensionality is substantially larger than N_r, then $\frac{E_b}{N'_0}_{\min} \approx \frac{E_b}{N_0}_{\min}$, meaning that the interference looks approximately like Gaussian noise to the receiver. Then, we revert to the same situation encountered when the receiver is not privy to $\boldsymbol{H}_1 \boldsymbol{F}_1, \ldots, \boldsymbol{H}_U \boldsymbol{F}_U$ and the results derived throughout the chapter fully apply.

High-SNR regime

In the high-SNR regime, provided that $\boldsymbol{H}^* \boldsymbol{R}^{-1}(\boldsymbol{H}_1, \ldots, \boldsymbol{H}_U)\boldsymbol{H}$ is nonsingular with probability 1, the number of spatial DOF is as in noise-limited conditions, namely $\mathcal{S}_\infty = N_{\min}$. For analyses of the power offset in the face of interference, the reader is referred to [355].

5.8 Optimum transmitter and receiver structures

5.8.1 Single codeword versus multiple codewords

Henceforth in this chapter, no constraints have been imposed on the structure of the MIMO transmitter and receiver. At this point, we peek inside these units to discuss how, depending on the number of codewords being transmitted, distinct capacity-achieving structures are possible. Indeed, the N_s data streams in the vector sequence $\boldsymbol{s}[0], \ldots, \boldsymbol{s}[N-1]$ can embed one or various codewords [584]. Although intermediate cases are possible, conceptually it suffices to consider the two limiting cases where the number of codewords is either 1 or N_s (see Fig. 5.11). While the capacity does not depend on the number of codewords, the structure of the receiver may. Moreover, for finite N, a single codeword has an N_s-fold advantage in length, which can help close whatever performance deficit there may be relative to $N \to \infty$.

If a single codeword occupies the N_s data streams, then to achieve capacity the receiver

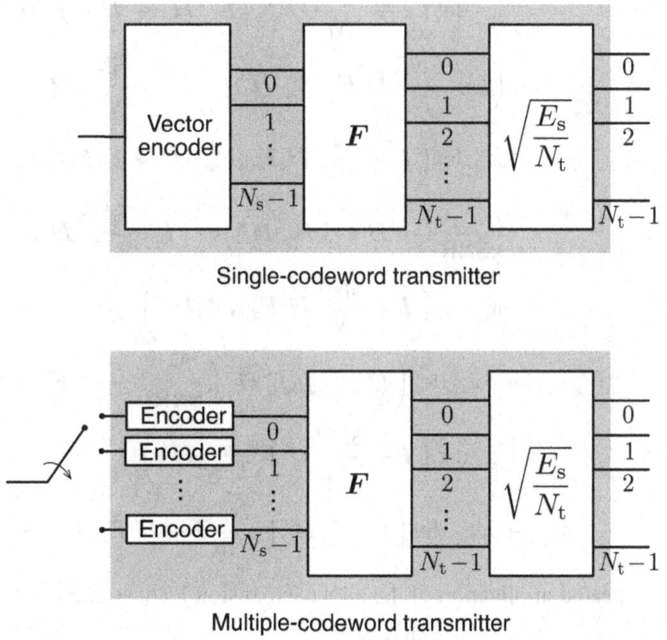

Fig. 5.11 MIMO transmission of a single codeword (above) or of N_s codewords (below).

must decode these streams jointly with a complexity that, barring possible simplifications, grows exponentially in N_s.

5.8.2 LMMSE-SIC receiver

If separate encoders feed the N_s data streams, then joint decoding continues to be optimum but other possibilities arise. Transmission of multiple codewords is particularly alluring with CSIT, as in this case parallel noninteracting subchannels are created and a bank of scalar decoders suffices to achieve capacity with a complexity that scales only polynomially with N_s. Without CSIT, the spatially multiplexed signals are in general mutually interfering but, remarkably, separate codeword decoding can still achieve capacity under certain conditions. To see this, first observe that

$$\boldsymbol{F}\boldsymbol{F}^* = \boldsymbol{f}_0\boldsymbol{f}_0^* + \boldsymbol{F}_{>0}\boldsymbol{F}_{>0}^*, \tag{5.212}$$

where we used the shorthand notation $\boldsymbol{f}_j = [\boldsymbol{F}]_{:,j}$ and $\boldsymbol{F}_{>j} = [\boldsymbol{F}]_{:,j+1:N_t-1}$. Exploiting this decomposition, we can elaborate the mutual information with a given precoder $\boldsymbol{F}$ into

$$\begin{aligned}\log_2 \det\left(\boldsymbol{I} + \frac{\mathsf{SNR}}{N_t}\boldsymbol{H}\boldsymbol{F}\boldsymbol{F}^*\boldsymbol{H}^*\right) \\ = \log_2 \det\left(\boldsymbol{I} + \frac{\mathsf{SNR}}{N_t}\left(\boldsymbol{H}\,\boldsymbol{F}_{>0}\boldsymbol{F}_{>0}^*\boldsymbol{H}^* + \boldsymbol{H}\boldsymbol{f}_0\boldsymbol{f}_0^*\boldsymbol{H}^*\right)\right)\end{aligned} \tag{5.213}$$

$$= \log_2 \det\left(\boldsymbol{I} + \frac{\mathsf{SNR}}{N_t} \boldsymbol{H} \boldsymbol{F}_{>0} \boldsymbol{F}^*_{>0} \boldsymbol{H}^* + \boldsymbol{H} \boldsymbol{f}_0 \boldsymbol{f}_0^* \boldsymbol{H}^* \frac{\mathsf{SNR}}{N_t} \right.$$
$$\left. \cdot \left(\boldsymbol{I} + \frac{\mathsf{SNR}}{N_t} \boldsymbol{H} \boldsymbol{F}_{>0} \boldsymbol{F}^*_{>0} \boldsymbol{H}^* \right)^{-1} \left(\boldsymbol{I} + \frac{\mathsf{SNR}}{N_t} \boldsymbol{H} \boldsymbol{F}_{>0} \boldsymbol{F}^*_{>0} \boldsymbol{H}^* \right) \right) \tag{5.214}$$

$$= \log_2 \det\left(\boldsymbol{I} + \frac{\mathsf{SNR}}{N_t} \boldsymbol{H} \boldsymbol{F}_{>0} \boldsymbol{F}^*_{>0} \boldsymbol{H}^* + \boldsymbol{H} \boldsymbol{f}_0 \boldsymbol{f}_0^* \boldsymbol{H}^* \right.$$
$$\left. \cdot \left(\frac{N_t}{\mathsf{SNR}} \boldsymbol{I} + \boldsymbol{H} \boldsymbol{F}_{>0} \boldsymbol{F}^*_{>0} \boldsymbol{H}^* \right)^{-1} \left(\boldsymbol{I} + \frac{\mathsf{SNR}}{N_t} \boldsymbol{H} \boldsymbol{F}_{>0} \boldsymbol{F}^*_{>0} \boldsymbol{H}^* \right) \right) \tag{5.215}$$

$$= \log_2 \det\left(\boldsymbol{I} + \frac{\mathsf{SNR}}{N_t} \boldsymbol{H} \boldsymbol{F}_{>0} \boldsymbol{F}^*_{>0} \boldsymbol{H}^* \right)$$
$$+ \log_2 \det\left(\boldsymbol{I} + \boldsymbol{H} \boldsymbol{f}_0 \boldsymbol{f}_0^* \boldsymbol{H}^* \left(\frac{N_t}{\mathsf{SNR}} \boldsymbol{I} + \boldsymbol{H} \boldsymbol{F}_{>0} \boldsymbol{F}^*_{>0} \boldsymbol{H}^* \right)^{-1} \right) \tag{5.216}$$

$$= \log_2 \det\left(\boldsymbol{I} + \frac{\mathsf{SNR}}{N_t} \boldsymbol{H} \boldsymbol{F}_{>0} \boldsymbol{F}^*_{>0} \boldsymbol{H}^* \right)$$
$$+ \log_2 \det\left(1 + \boldsymbol{f}_0^* \boldsymbol{H}^* \left(\frac{N_t}{\mathsf{SNR}} \boldsymbol{I} + \boldsymbol{H} \boldsymbol{F}_{>0} \boldsymbol{F}^*_{>0} \boldsymbol{H}^* \right)^{-1} \boldsymbol{H} \boldsymbol{f}_0 \right). \tag{5.217}$$

Repeated application of the above expansion leads to the telescopic expression

$$\log_2 \det\left(\boldsymbol{I} + \frac{\mathsf{SNR}}{N_t} \boldsymbol{H} \boldsymbol{F} \boldsymbol{F}^* \boldsymbol{H}^* \right)$$
$$= \sum_{j=0}^{N_s - 1} \log_2 \left(1 + \boldsymbol{f}_j^* \boldsymbol{H} \left(\frac{N_t}{\mathsf{SNR}} \boldsymbol{I} + \boldsymbol{H} \boldsymbol{F}_{>j} \boldsymbol{F}^*_{>j} \boldsymbol{H}^* \right)^{-1} \boldsymbol{H} \boldsymbol{f}_j \right), \tag{5.218}$$

which deserves careful inspection. As made clear in the next chapter,

$$\boldsymbol{f}_j^* \boldsymbol{H} \left(\frac{N_t}{\mathsf{SNR}} \boldsymbol{I} + \boldsymbol{H} \boldsymbol{F}_{>j} \boldsymbol{F}^*_{>j} \boldsymbol{H}^* \right)^{-1} \boldsymbol{H} \boldsymbol{f}_j \tag{5.219}$$

is the SINR experienced by stream j at the output of an LMMSE receiver in the absence of interference from streams $0, \ldots, j-1$, interfered only by streams $j+1, \ldots, N_s - 1$. Hence, the jth term in (5.218) is the spectral efficiency achievable by stream j at the output of an LMMSE receiver with interference from streams $0, \ldots, j-1$ previously canceled out. This naturally invites the multistage receiver structure in Fig. 5.12, where, at stage j, the following unfolds.

(1) The sequence of observations $\boldsymbol{y}[0], \ldots, \boldsymbol{y}[N-1]$ passes through an LMMSE filter that targets stream j.
(2) Stream j is decoded.
(3) The interference caused by stream j onto streams $j+1, \ldots, N_s - 1$ is reconstructed and canceled from $\boldsymbol{y}[0], \ldots, \boldsymbol{y}[N-1]$.

The optimality of this strategy, aptly dubbed *successive interference cancelation* (SIC), hinges critically on each stream corresponding to a separate codeword, for only in that case can the cancelation be based on highly reliable decoded data. As importantly, each codeword must be encoded at the correct rate such that, after SIC and LMMSE filtering, the

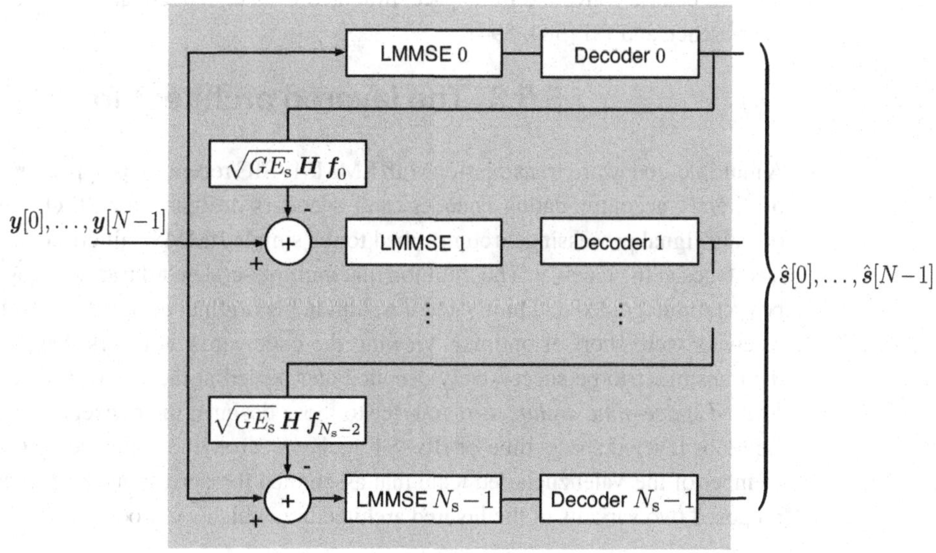

Fig. 5.12 LMMSE-SIC receiver.

spectral efficiency of the corresponding stream equals the jth term in (5.218). These codeword rates depend on $\boldsymbol{H}$ and thus, even when there is no CSIT and the precoder is fixed, the transmitter must be furnished the limited information of these rates for (5.218) to be achieved. Put differently, link adaptation must be conducted separately for each codeword, which is sometimes referred to as *per-antenna rate control* (PARC) [585–587].

Encoding all codewords at the same rate, i.e., enforcing a common link adaptation process, would incur a significant penalty [588, 589]. Moreover, while PARC allows achieving capacity regardless of the decoding order because (5.218) holds with the columns of $\boldsymbol{HF}$ arbitrarily reordered, if all codewords were encoded at some common rate, then the spectral efficiency would depend on the decoding order [23]. Thus, at each stage, the receiver would have to compute the LMMSE filter for every remaining codeword and select the highest-SINR one, a procedure that can be simplified by exploiting structure commonalities in the filters [590] but that is nevertheless tedious.

The information-theoretic optimality of the LMMSE-SIC receiver was first observed by Varanasi and Guess in the context of the multiple-access channel (MAC) [591]. Indeed, when each stream corresponds to a separate codeword, a MIMO channel can be viewed as a MAC where each such stream is a separate user, and then direct analogies arise with multiuser detection [20, 592, 593]. This MAC result was subsequently rederived for MIMO [584]. Alternative interpretations of the optimality of the LMMSE-SIC receiver can be devised, for instance, by means of the chain rule of mutual information [50, section 8.3.4].

In non-OFDM systems facing frequency selectivity, the filters within an LMMSE-SIC receiver can be converted into space–time LMMSE equalizers able to handle both spatial interference and ISI [584, 594].

5.8.3 The layered architecture

A multiple-codeword transmission with LMMSE-SIC reception is appealing on the grounds of directly accommodating encoders and decoders designed for SISO, with the spatial-domain signal processing circumscribed to the simple LMMSE filters connecting the various stages at the receiver. The intuition that multiple-codeword transmissions could indeed be preferable guided and motivated Foschini in his original designs, even if the considered receivers were short of optimal. Viewing the codewords as layers that were piled on at the transmitter to be successively decoded and peeled at the receiver, he coined the term *layered space–time architecture* to refer to those designs; the concept was popularized as Bell-labs layered space–time or BLAST, a clever acronym concocted by Glenn Golden, a member of the Valenzuela-led team that assembled the corresponding prototype. Foschini proposed two variants of the layered architecture, which we proceed to describe.

Vertical and Horizontal BLAST

The first variant consisted of an unprecoded multiple-codeword transmitter and a SIC receiver where the various stages, rather than by LMMSE filters, were connected by ZF filters. A prototype devoid of coding was built to demonstrate the practical feasibility of the concept [24, 595]. Although, without coding ensuring low error probabilities, SIC receivers are prone to error propagation [584, 596–598] and a hefty back-off in bit rate is required, the prototype did succeed at proving the viability of MIMO. The term vertical BLAST (V-BLAST) caught on because of the vertical piling of the layers, which in this case were not codewords but merely raw uncoded symbols.

With coding incorporated and the layers thereby stretching from individual symbols to long codewords, the more fitting name of horizontal BLAST (H-BLAST) was put forth [599]. Fundamentally, this version of the layered architecture was not capacity-achieving in the following respects.

- The transmitter lacked a precoder. (This was not an issue for the IID channels initially considered, but would have been in wider generality.)
- The linear filters connecting the receiver stages were ZF, rather than LMMSE. (Not an issue for the high SNRs initially considered, but again it would have been in general.)
- All codewords were encoded at the same rate.

With the top two of these issues being easily correctable, it is actually the latter point that fundamentally limited the ability to achieve capacity. As argued, this limitation can be sidestepped by releasing the restriction that all codeword rates be equal and adjusting each one individually as per (5.218), yet Foschini was focused on quasi-static settings without link adaptation. In such conditions, indeed, V-BLAST is decidedly deficient in terms of outage capacity as each codeword is at the mercy of the fading at a specific transmit antenna [395]. This led Foschini to formulate a second type of layered architecture.

Diagonal BLAST

In the second variant of Foschini's layered architecture, aptly nicknamed diagonal BLAST (D-BLAST), each codeword was not associated with a specific transmit antenna. Rather, the codewords were cycled around the N_t antennas to ensure equal exposition to all fading coefficients. This closed the gap in terms of outage capacity, ensuring optimality in quasi-static settings without link adaptation. The rotating association between codewords and transmit antennas can be interpreted as a time-varying precoder.

Example 5.41

For $N_t = 2$, alternating the precoders

$$\boldsymbol{F}_0 = \begin{bmatrix} 1 & 0 \\ 0 & 1 \end{bmatrix} \qquad \boldsymbol{F}_1 = \begin{bmatrix} 0 & 1 \\ 1 & 0 \end{bmatrix} \tag{5.220}$$

ensures that each codeword is equally exposed to the two columns of $\boldsymbol{H}$, rather than to only one of them. The receiver must be aware of which precoder is applied at each symbol so as to map its outputs correspondingly.

Refinements of D-BLAST by other authors went on to propose more intricate ways of permuting the codewords across antennas and time, including the possibility of blending the time-varying precoder and the interleaver into a space–time interleaver [600, 601]. Despite their optimality, these architectures, and D-BLAST itself as a matter of fact, have been largely relegated for the same reasons that the outage capacity has been relegated: in ergodic settings, the additional diversity brought about by associating each codeword with all transmit antennas is immaterial while, in quasi-static settings, link adaptation is universally employed and easily upgradeable to PARC.

5.8.4 BICM implementations

As developed in earlier chapters, the de-facto implementation of SISO transceivers entails binary codebooks and BICM. Let us now see how to extend those principles to MIMO, depending on whether one or multiple codewords are transmitted. There are various flavors of MIMO BICM, all bound—within the tiny loss that BICM incurs if the constellations are higher-order and the receiver is one-shot, and with the caveat of any nonidealities—by the fundamental limits established throughout this chapter.

Single-codeword BICM

For a single codeword, the MIMO BICM transmitter and receiver are depicted in Fig. 5.13. The coded and interleaved bits are parsed into groups of $N_s \log_2 M$; of the bits within each group, $\log_2 M$ are mapped to each of N_s M-ary constellations, giving the $N_s \times 1$ vector $\boldsymbol{s}$. Finally, $\boldsymbol{s}$ is precoded and amplified into the transmit signal $\boldsymbol{x}$. As far as the receiver is concerned, Fig. 5.13 illustrates its iterative form [112, 113, 602]. From the observation $\boldsymbol{y}$, and a-priori information on the coded bits sent back from the decoder, a soft

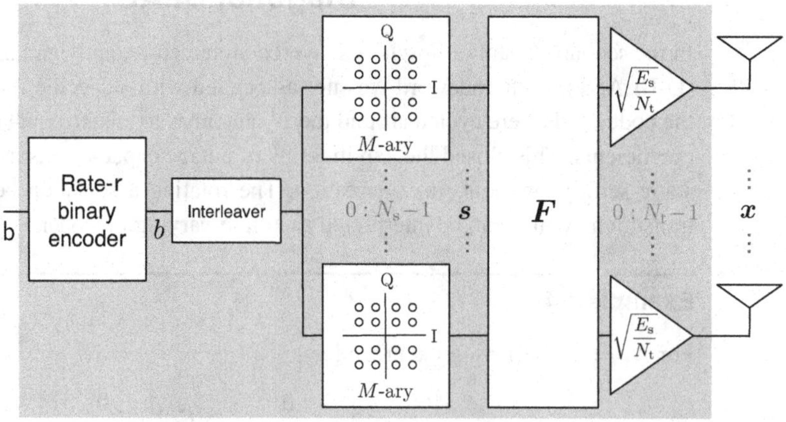

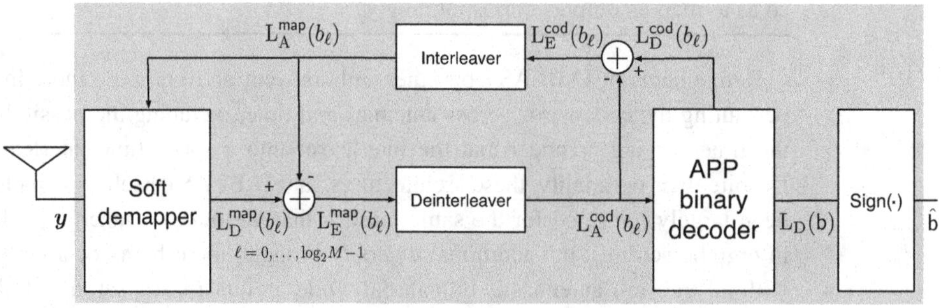

Fig. 5.13 MIMO BICM architecture with a single codeword and iterative reception.

demapper computes L-values $\mathrm{L}_{\mathrm{D}}^{\mathsf{map}}(\cdot)$ and subsequently extrinsic information $\mathrm{L}_{\mathrm{E}}^{\mathsf{map}}(\cdot) = \mathrm{L}_{\mathrm{D}}^{\mathsf{map}}(\cdot) - \mathrm{L}_{\mathrm{A}}^{\mathsf{map}}(\cdot)$ on those coded bits. The extrinsic information is deinterleaved and fed into the APP decoder as $\mathrm{L}_{\mathrm{A}}^{\mathsf{cod}}(\cdot)$. In turn, the decoder outputs its own extrinsic information on the coded bits, $\mathrm{L}_{\mathrm{E}}^{\mathsf{cod}}(\cdot) = \mathrm{L}_{\mathrm{D}}^{\mathsf{cod}}(\cdot) - \mathrm{L}_{\mathrm{A}}^{\mathsf{cod}}(\cdot)$, which, properly interleaved, becomes the new a-priori information for the soft demapper, completing an iteration. The APP decoder also generates L-values for the message bits and, once sufficient iterations have been run, the sign of these directly gives the final MAP decisions. As mentioned for SISO, when the receiver is iterative, departing from Gray mapping is preferable.

A one-shot receiver is a special case of the structure in Fig. 5.13 where latency is reduced at the expense of the soft demapper not benefitting from a-priori information. Then, Gray mapping becomes advisable.

To zoom in on the computation of the L-values produced by the soft demapper, the SISO expression in (1.128) readily generalizes to MIMO CSIR settings as

$$\mathrm{L_D}(b_\ell \,|\, \boldsymbol{y}{=}\mathbf{y}, \boldsymbol{HF}{=}\mathbf{HF}) \tag{5.221}$$

$$= \mathrm{L_A}(b_\ell) + \underbrace{\log \frac{\sum_{\mathbf{s}_m \in \mathcal{S}_1^\ell} f_{\boldsymbol{y}|\boldsymbol{s},\boldsymbol{HF}}(\mathbf{y}|\mathbf{s}_m, \mathbf{HF}) \exp\left(\sum_{\ell' \neq \ell, \ell' \in \mathcal{B}_1^m} \mathrm{L_A}(b_{\ell'})\right)}{\sum_{\mathbf{s}_m \in \mathcal{S}_0^\ell} f_{\boldsymbol{y}|\boldsymbol{s},\boldsymbol{HF}}(\mathbf{y}|\mathbf{s}_m, \mathbf{HF}) \exp\left(\sum_{\ell' \neq \ell, \ell' \in \mathcal{B}_1^m} \mathrm{L_A}(b_{\ell'})\right)}}_{\mathrm{L_E}(b_\ell \,|\, \boldsymbol{y}, \boldsymbol{HF})},$$

where $\mathcal{S}_0^\ell$ and $\mathcal{S}_1^\ell$ are the subsets of M^{N_s}-ary vectors $\mathbf{s}_m$ obtained by mapping every group of $N_\mathrm{s} \log_2 M$ coded bits whose ℓth element is 0 or 1, respectively. In turn, $\mathcal{B}_1^m$ is the subset of coded bits that equal 1 within the group that maps to $\mathbf{s}_m$. The channel law $f_{\boldsymbol{y}|\boldsymbol{s},\boldsymbol{HF}}(\cdot)$ is given by

$$f_{\boldsymbol{y}|\boldsymbol{s},\boldsymbol{HF}}(\mathbf{y}|\mathbf{s}_m, \mathbf{HF}) = \frac{1}{(\pi N_0)^{N_\mathrm{r}}} \exp\left(-\frac{1}{N_0}\left\| \boldsymbol{y} - \sqrt{\frac{GE_\mathrm{s}}{N_\mathrm{t}}} \mathbf{HF}\mathbf{s}_m \right\|^2\right) \tag{5.222}$$

and hence

$$\mathrm{L_E}(b_\ell \,|\, \boldsymbol{y}, \boldsymbol{HF}) = \log \frac{\sum_{\mathbf{s}_m \in \mathcal{S}_1^\ell} \exp\left(-\frac{1}{N_0}\left\| \boldsymbol{y} - \sqrt{\frac{GE_\mathrm{s}}{N_\mathrm{t}}} \mathbf{HF}\mathbf{s}_m \right\|^2 + \sum_{\ell' \neq \ell, \ell' \in \mathcal{B}_1^m} \mathrm{L_A}(b_{\ell'})\right)}{\sum_{\mathbf{s}_m \in \mathcal{S}_0^\ell} \exp\left(-\frac{1}{N_0}\left\| \boldsymbol{y} - \sqrt{\frac{GE_\mathrm{s}}{N_\mathrm{t}}} \mathbf{HF}\mathbf{s}_m \right\|^2 + \sum_{\ell' \neq \ell, \ell' \in \mathcal{B}_1^m} \mathrm{L_A}(b_{\ell'})\right)} \tag{5.223}$$

where the summations in numerator and denominator feature $\frac{1}{2} M^{N_\mathrm{s}}$ terms; this is half the number of possible M^{N_s} vectors, a potentially very large value. The numerical evaluation of (5.223) is often simplified by means of the so-called max-log approximation [603]

$$\log_e\left(e^{a_0} + \cdots + e^{a_{Q-1}}\right) \approx \max(a_0, \ldots, a_{Q-1}), \tag{5.224}$$

whereby the aforementioned summations revert to finding the vector $\mathbf{s}_m$ that minimizes

$$\left\| \boldsymbol{y} - \sqrt{\frac{GE_\mathrm{s}}{N_\mathrm{t}}} \mathbf{HF}\mathbf{s}_m \right\|^2 \tag{5.225}$$

over each of the subsets, $\mathcal{S}_0^\ell$ and $\mathcal{S}_1^\ell$. Even so, the number of candidate vectors continues to be $\frac{1}{2} M^{N_\mathrm{s}}$.

Example 5.42

How many vectors have to be considered in the numerator and denominator of (5.223) if $N_\mathrm{s} = 4$ and the constellations are 64-QAM? How about for $N_\mathrm{s} = 8$ with the constellations being 16-QAM?

Solution

For $M = 64$ and $N_\mathrm{s} = 4$, the number of vectors equals $\frac{1}{2}\, 64^4 = 8.4$ million. In turn, for $M = 16$ and $N_\mathrm{s} = 8$, the number is $\frac{1}{2}\, 16^8 = 2150$ million.

Several approaches have been set forth to further approximate these computations, drastically shrinking the number of vectors to consider. A favorite such approach is the *list sphere decoding* algorithm, which excludes all those vectors s_m for which (5.225) exceeds a certain value [604–609]. Viewing the set of M^{N_s} vectors as a lattice, this amounts to considering only vectors that lie within a sphere centered on

$$\sqrt{\frac{N_\mathrm{t}}{GE_\mathrm{s}}}(\boldsymbol{F}^*\boldsymbol{H}^*\boldsymbol{H}\boldsymbol{F})^{-1}\boldsymbol{H}^*\boldsymbol{F}^*\boldsymbol{y}, \tag{5.226}$$

which (see Section 2.7.1) is the unconstrained ML estimate of the vector minimizing (5.225). Evidently, the efficacy of the algorithm is closely tied to the choice of the sphere radius, which should be large enough to contain—with high probability—the sought vector yet small enough to bring about a sufficient reduction in complexity [610]. Another approach to lessen the complexity of the soft demapping is the tree decoder [611], which also reduces the search space by exploiting the structure of the space of M^{N_s}-ary vectors.

Multiple-codeword BICM

A multiple-codeword BICM transmitter is illustrated in Fig. 5.14. After being separately interleaved, each codeword is parsed into groups of $\log_2 M$ coded bits and mapped to respective M-ary constellations. The vector $\boldsymbol{s}$ thus obtained is precoded and amplified into the transmit signal $\boldsymbol{x}$.

At the receiver, it is certainly possible to disregard the fact that separate codewords are built into the signal and apply the same structure described for single-codeword transmissions. However, if joint decoding is to take place, then it is arguably preferable to transmit a single (and therefore longer) codeword in the first place. The main point of using multiple codewords is to take advantage of that at the receiver, with an LMMSE-SIC structure. The BICM version of this structure is depicted in the bottom part of Fig. 5.14. An LMMSE filter isolates an initial codeword, for which a soft demapper then computes $\mathrm{L}_\mathrm{D}^\mathrm{map}(\cdot)$. Without a-priori information, this is directly the extrinsic information on the coded bits, and its deinterleaved version is fed into the corresponding APP decoder. The decoder goes on to generate L-values for the message bits, whose sign gives the final MAP decisions, as well as L-values for the coded bits, whose sign is a reconstruction of the binary codeword itself. Reinterleaved and remapped again, this yields a reconstruction of the corresponding entry of $\boldsymbol{s}$ and subsequently—via precoder, channel estimate, and amplification—of its interference contribution on the observed vector $\boldsymbol{y}$. This interference is subtracted and the process is repeated for the next codeword, and so on, until all the N_s codewords have been decoded.

The one-shot demapping and decoding of each codeword could also be rendered iterative if high-order constellations are to be used, an idea inspired by turbo multiuser receivers [612] and that could be imported to LMMSE-SIC MIMO receivers [584, 613].

Also, and despite hard decisions at each stage sufficing for the information-theoretic optimality of LMMSE-SIC reception, to render the implementation robust against imperfections the SIC process could be made soft, i.e., rather than using hard bit decisions to reconstruct and cancel interference, L-values could be used instead [614].

Hybrid receiver structures having aspects of joint decoding and aspects of SIC are also possible, for instance by jointly decoding groups of two or more codewords and then

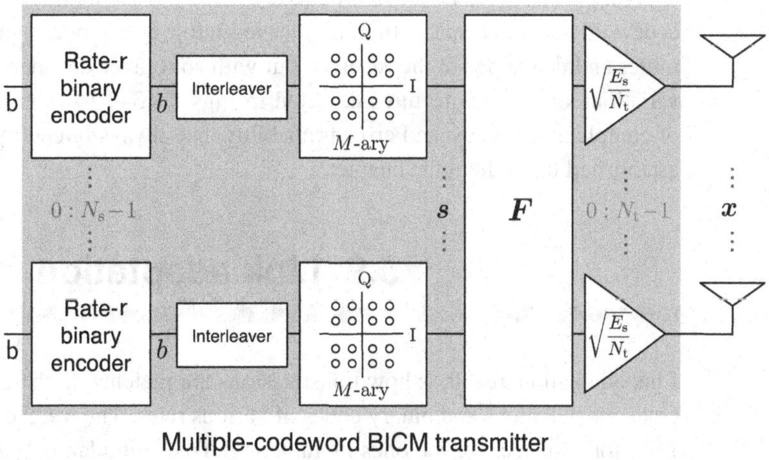

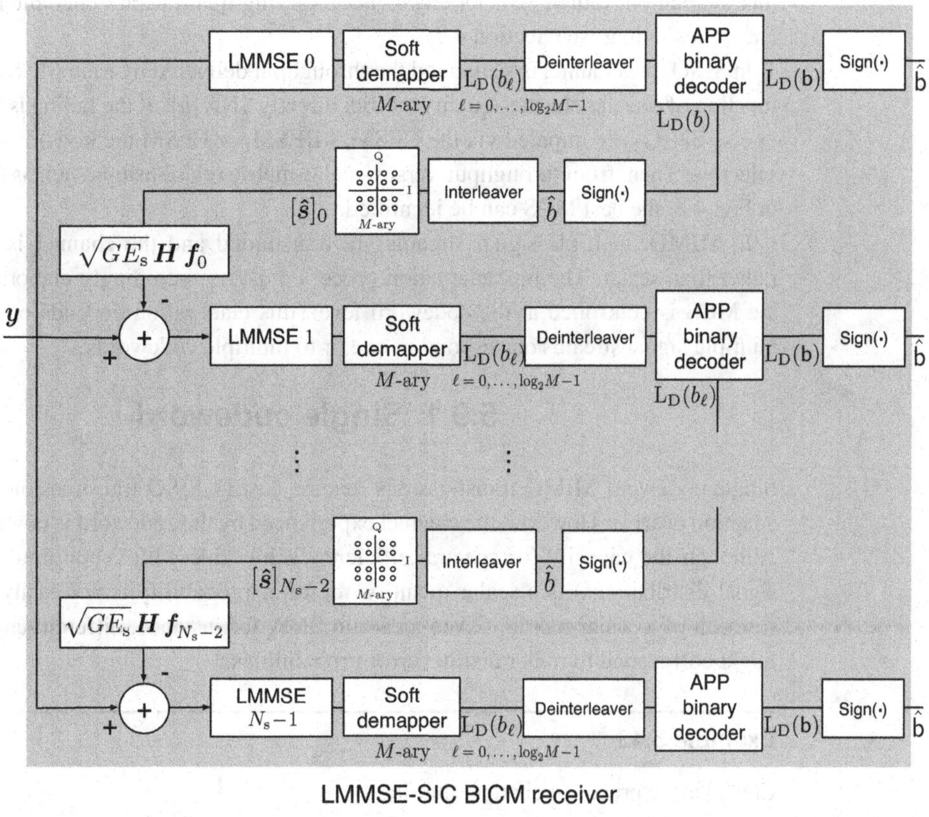

Fig. 5.14 MIMO BICM architecture with multiple codewords and LMMSE-SIC reception.

applying group-wise LMMSE-SIC [615]. Or, as in the so-called *turbo BLAST*, multiple-codewords could be space–time interleaved into a single one, jointly mapped at the transmitter and demapped at the receiver, but with split outer decoders [601]. These, and other variations of the architectures depicted in Figs. 5.13–5.14 offer distinct options in terms of complexity, latency, and error probability, but always bound by the fundamental limits established throughout the chapter.

5.9 Link adaptation

Link adaptation, recall, is how transmissions are matched to the channel by means of discrete constellations and binary codes of various rates. The MCS combinations available in LTE, for instance, are the ones in Table 4.1. Transmit adaptivity is particularly important in quasi-static settings, so as to avoid being trapped in the outage framework. With proper link adaptation, rather, AWGN channels of varying qualities are encountered as the link undergoes fading (see Section 4.7).

In SISO, the channel is scalar and the throughput delivered by each MCS can be put as a function of a scalar channel-quality metric: directly $\mathsf{SNR}\,|h|^2$ if the fading is frequency-flat, or else $\mathsf{SNR}_{\times\mathrm{ESM}}$ (computed via the CESM, MIESM, or EESM methods) if it is frequency-selective. Then, from throughput-versus-scalar-metric relationships such as the LTE curves in Fig. 4.8, the best MCS can be identified.

In MIMO, multiple signal streams are transmitted and the channel is matrix-valued rather than scalar. The link adaptation process must be accordingly elaborated and, since the MCS is controlled at the codeword level, this elaboration depends on whether these multiple signal streams correspond to one or to multiple codewords.

5.9.1 Single codeword

Single-codeword MIMO transmissions feature, just as SISO transmissions, a single link adaption process. However, the channel experienced by the codeword is now matrix-valued. Although the capacity (or spectral efficiency achievable with suboptimum precoding or signal distributions) is a scalar quantity, the error probability is not easily captured as a function of a scalar metric. Given a certain SNR, for instance, different channel matrices might correspond to rather distinct error probabilities.

Example 5.43

Consider the precoded channel matrices

$$\boldsymbol{H}_0\boldsymbol{F}_0 = \begin{bmatrix} 1 & 1 \\ 1 & 1 \end{bmatrix} \qquad \boldsymbol{H}_1\boldsymbol{F}_1 = \begin{bmatrix} \sqrt{2} & 0 \\ 0 & \sqrt{2} \end{bmatrix}. \tag{5.227}$$

Although both of these precoded matrices yield the same value for the MIMO counterpart to $\mathsf{SNR}\,|h|^2$, namely for $\mathsf{SNR}\,\|\boldsymbol{H}\boldsymbol{F}\|_{\mathrm{F}}^2$, the former suffers from strong spatial interference whereas the latter is free of it. The error probability of a certain MCS is likely to be significantly different in either case.

The natural approaches to obtain a scalar metric more representative than $\mathsf{SNR}\,\|\boldsymbol{H}\boldsymbol{F}\|_{\mathrm{F}}^2$ are again the CESM, MIESM, and EESM methods introduced in Chapter 4, applied here in the spatial domain. Specifically, these methods could be applied to the nonzero eigenvalues of $\boldsymbol{H}\boldsymbol{F}\boldsymbol{F}^*\boldsymbol{H}^*$. And, in MIMO OFDM, by applying these methods to the nonzero eigenvalues of $\boldsymbol{H}[k]\boldsymbol{F}[k]\boldsymbol{F}[k]^*\boldsymbol{H}[k]^*$ for subcarriers $k = 0, \ldots, K-1$, the mapping onto a scalar metric in the frequency and in the space domains would be conveniently blended. However, the calibration of the fudge factor for these methods would then be much more challenging, particularly if the precoded channel is not diagonal and the various signal streams are mutually interfering. Beyond the marginal distribution of each entry of $\boldsymbol{H}$, meaning the type of fading, the fudge factor would go on to depend on the joint distribution, meaning the antenna correlations, and on the numbers of antennas themselves. A large stock of fudge factors, and the ability to know which one to apply based on the distribution of $\boldsymbol{H}$, might be needed to cover most likely scenarios. Besides a colossal amount of offline simulation work, this would likely lead to significant performance variability.

As an alternative to the CESM, MIESM, and EESM methods, and with a view to BICM implementations, scalar metrics based on the L-values at the output of the decoder can be entertained. Specifically, the average (over the entire codeword) of the mutual information between the distribution of the coded bits and that of the posterior L-values has been identified as a robust indicator of the error probability [616]. Besides monitoring how the error probability decays as iterations unfold, therefore, this average mutual information could serve to gauge the throughput attainable by each MCS and to effect link adaptation [617].

Regardless of the metric, as mentioned, there is likely to be a higher performance variability for each value of the metric than there is in SISO. This makes hybrid-ARQ even more invaluable in MIMO, as its ability to recover from mismatched MCS selections avoids the introduction of hefty safety margins and the ensuing loss in spectral efficiency.

5.9.2 Multiple codewords

When multiple codewords are transmitted, scalar channel-quality metrics are more forthcoming. In particular, with CSIT-based precoding diagonalizing the channel, each codeword is conveyed over a noninterfering subchannel and the corresponding SNR can directly be such a metric. If frequency selectivity is present, then CESM, MIESM, or EESM can be applied to each subchannel separately, exactly as in SISO.

With precoders that do not diagonalize the channel, the signal streams bearing the codewords are mutually interfering. If the receiver features a joint decoder, then the remarks made in relation to single-codeword link adaptation apply. Alternatively, if an LMMSE-SIC receiver is featured, then scalar metrics can be formulated based on the effective scalar channel experienced by each codeword. As anticipated in the derivation of the LMMSE-SIC and detailed in the next chapter, the SINR experienced by the jth signal stream is given by (5.219), and this quantity can perfectly serve as a scalar metric for the link adaptation of the jth codeword. This enables PARC, whereby each codeword is subject to a separate

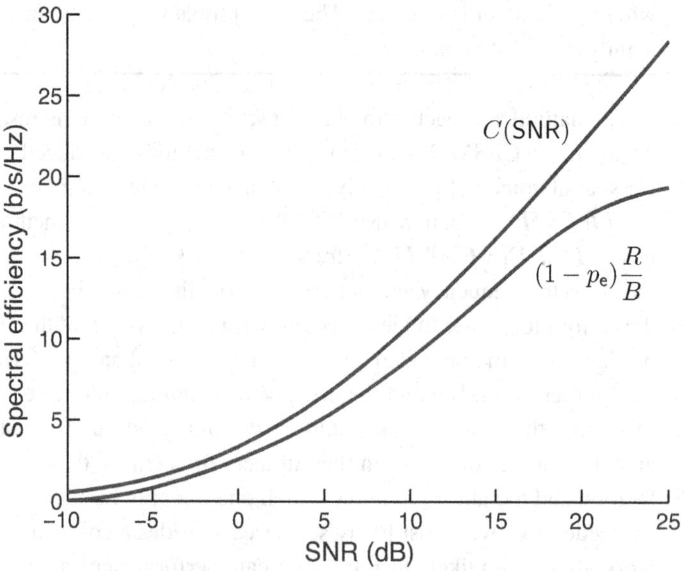

Fig. 5.15 For an IID Rayleigh-faded MIMO channel with $N_\mathrm{t} = N_\mathrm{r} = 4$, BICM, PARC, and no precoding, throughput per unit bandwidth achievable with LMMSE-SIC reception and the entire set of 27 LTE MCSs. Also shown is the no-CSIT ergodic capacity.

adaptation. Again, if frequency selectivity is present, then CESM, MIESM, or EESM can be applied across the SINR values on the OFDM subcarriers.

Example 5.44 (Link adaptation with MIMO IID Rayleigh fading and LMMSE-SIC reception in LTE)

Let $N_\mathrm{t} = N_\mathrm{r} = 4$ with the channel being IID Rayleigh-faded and frequency-flat. The transmission is BICM and unprecoded, with a separate codeword emitted from each antenna and with PARC. The receiver is LMMSE-SIC and the channel is quasi-static over each codeword.

Presented in Fig. 5.15 is the throughput achievable with the set of 27 LTE MCSs. Also shown is the ergodic spectral efficiency achievable by an unprecoded transmission, which, in an IID channel, directly equals the no-CSIT ergodic capacity. With only the modicum of information at the transmitter represented by the MCS selections, the throughput can closely track such capacity. A set of 27 MCSs over $N_\mathrm{t} = 4$ antennas can be indexed with only 19 bits, which could be compressed into fewer by removing those MCS combinations that occur with very low probability, given the fact that the first decoded codeword suffers from interference from all the rest, whereas the last decoded codeword is interference-free.

Example 5.45

By how much is the gap between the throughput and the capacity exacerbated by MIMO, relative to a SISO transmission utilizing the same set of MCSs?

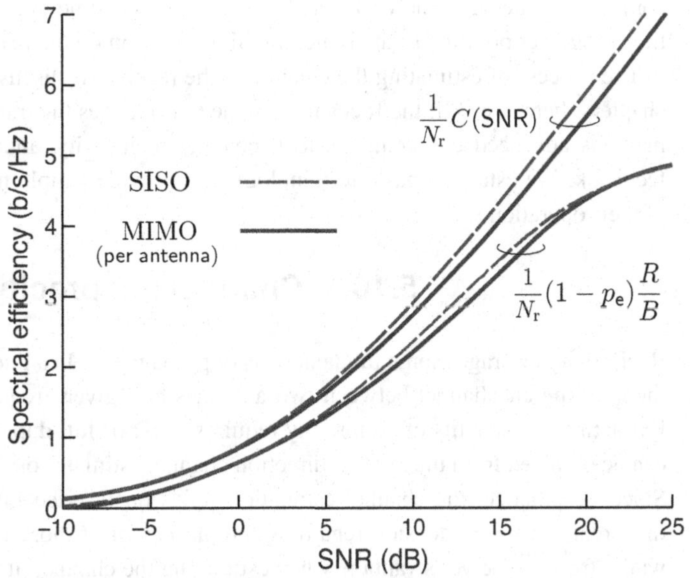

Fig. 5.16 Solid lines, per-antenna throughput per unit bandwidth and per-antenna capacity as a function of SNR for a MIMO channel with $N_t = N_r = 4$, IID Rayleigh fading, unprecoded BICM with PARC and LMMSE-SIC reception. Dashed lines, same quantities for a SISO channel.

Solution

This question can be addressed by bringing Figs. 4.12 and 5.15, respectively corresponding to SISO and MIMO, together on an equal footing. Fig. 5.16 shows per-antenna throughputs and capacities for both settings. The per-antenna capacities coincide at low SNR, with MIMO having only a small loss of 1.31 dB at high SNR (see Example 5.29). Remarkably, the per-antenna throughout is also preserved at low SNR and exhibits a similarly small loss at high SNR—before eventually saturating at the value dictated by the highest-rate MCS. Thus, MIMO does not worsen the gap between throughput and capacity in this specific but very relevant example. Other cases are proposed as problems at the end of the chapter.

As far as hybrid-ARQ is concerned, the transmission of multiple codewords enables flexibility. Rather than trigger a complete retransmission of all N_s codewords if any of them is erroneously decoded, it becomes possible to retransmit only those specific codewords that are in error, and this is indeed advantageous [618, 619]

5.10 Reciprocity and CSI feedback

Link adaptation requires an MCS indicator per codeword, and CSIT-based precoding—desirable not only from a performance standpoint, but also on the grounds of transceiver complexity—needs somewhat detailed CSI for each time–frequency channel coherence tile, altogether pointing to the issue of CSI acquisition on the part of the transmitter.

The process of estimating the channel at the receiver is discussed in detail in preceding chapters; here, we shift the focus to how the CSI reaches the transmitter and discuss three methods advanced earlier in the text: channel reciprocity, analog feedback, and digital feedback. We establish basic terminology and provide simple mathematical descriptions of their operation.

5.10.1 Channel reciprocity

Reciprocity leverages some fundamentals of propagation. In a linear and isotropic medium, the over-the-air channel between two antennas at a given frequency is reciprocal [620]. Extracting the benefits of reciprocity requires TDD or full duplexing, such that the CSIT can be obtained from the reverse-direction channel estimates on the same frequency range. Since, in practice, the channel estimation relies on inserted pilots, in TDD specifically this creates a certain tension regarding the placement of those pilots in the time domain: while, from a receiver's vantage when estimating the channel, it is desirable to have pilots regularly sprinkled, from the reverse-direction transmitter's vantage it is preferable to have the pilots placed toward the end of the duplexing duty cycle so as to have fresh estimates.

To adjust them to a reciprocity situation, the CSIT-based results in this chapter, derived under the premise of perfect CSIT, need to be tinkered with so as to incorporate the reverse-direction estimation errors. And it must be taken into account that both communication directions benefit from a pilot overhead that is nominally incurred in the reverse direction. (Note, for instance, that the maximization of the forward spectral efficiency, disregarding the costs to the reverse link, would entail devoting this link exclusively to pilots [621].) We do not delve here into the inclusion of reverse-direction channel estimation errors and into the optimization of the overhead for reciprocal SU-MIMO settings, but rather defer these issues directly to Chapter 9, in the context of multiuser MIMO. The single-user setup is a special case of the results therein.

The primary challenge associated with exploiting reciprocity is that the transmit and receive circuits are themselves not reciprocal, i.e., the tandem of base station transmitter and user receiver may differ from the tandem of user transmitter and base station receiver. The main sources of mismatch are due to differences between the power amplifier output impedance at the transmitter and the input impedance of the low-noise amplifier at the receiver [622]. Mathematically, these can be treated as an additional per-antenna complex gain such that the effective channel becomes $\boldsymbol{D}_{\mathrm{A}}\boldsymbol{H}\boldsymbol{D}_{\mathrm{B}}$ where $\boldsymbol{D}_{\mathrm{A}}$ and $\boldsymbol{D}_{\mathrm{B}}$ are diagonal matrices that differ at transmitter and receiver [623]. These matrices drift over time, for instance due to variations in temperature, but slowly relative to $\boldsymbol{H}$, meaning that their calibration and compensation entail a negligible additional overhead [624–626].

There are two methods for calibration.

- Self-calibration, where an extra transceiver is used [623, 627]. This extra transceiver determines differences between the transmit and receive channels successively for each antenna during a special calibration stage.

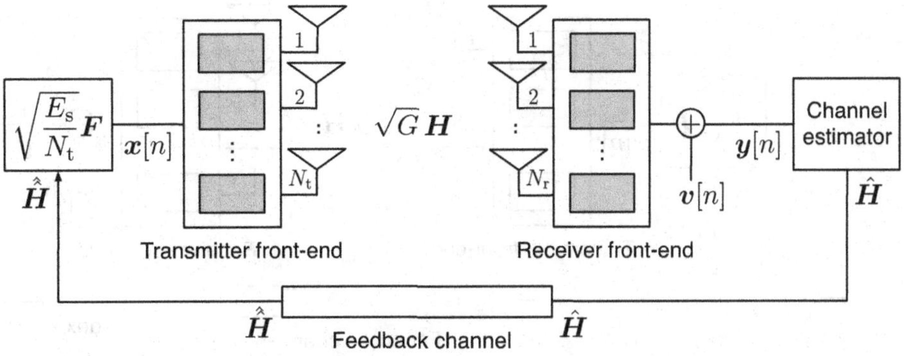

Fig. 5.17 SU-MIMO transmission based on analog feedback.

- Over-the-air calibration, where the channel estimate gathered at the receiver is fed back to the transmitter so that the required calibration parameters can be determined [628].

Calibration mismatches are typically neglected in analyses, and this is also the approach in this text.

5.10.2 Analog feedback

In FDD systems there is no reciprocity and thus explicit feedback is required, as well as a coherence time much longer than the feedback delay. The quantity being fed back can be the observation, $\boldsymbol{y}$, the ensuing channel estimate, $\hat{\boldsymbol{H}}$, or even the precoder obtained from that estimate. If the quantity of choice, possibly subject to a linear or nonlinear transformation (if nothing else, scaled to match the feedback power), is sent back without quantization, the feedback is said to be analog [629–633]. Referring to Fig. 5.17, where the quantity being fed back is $\hat{\boldsymbol{H}}$, the analysis of analog feedback should incorporate three sources of error.

- Channel estimation errors, which alter the actual channel matrix $\boldsymbol{H}$ into $\hat{\boldsymbol{H}}$. The relevance of these errors is determined by the quantity and power of the forward pilots.
- Errors in the estimation of the feedback channel over which $\hat{\boldsymbol{H}}$ is subsequently conveyed. These errors depend on the quantity and power of whatever pilots accompany the feedback.
- Errors in the estimation of $\hat{\boldsymbol{H}}$ at the output of the inaccurately known feedback channel, which further alter $\hat{\boldsymbol{H}}$ into $\hat{\hat{\boldsymbol{H}}}$. These errors depend on the size of the feedback interval, which limits how many repetitions are possible, and on the feedback power.

These various errors compound to deviate the final CSIT, $\hat{\hat{\boldsymbol{H}}}$, from its true value, $\boldsymbol{H}$, and the overall error minimization entails a joint optimization of the various overheads and powers [634]. A detailed exercise of all of this is undertaken in Chapter 9.

An advantage of analog feedback over its digital counterpart is that, for growing SNRs (in both directions), all three types of estimation error vanish and $\hat{\hat{\boldsymbol{H}}} \to \boldsymbol{H}$.

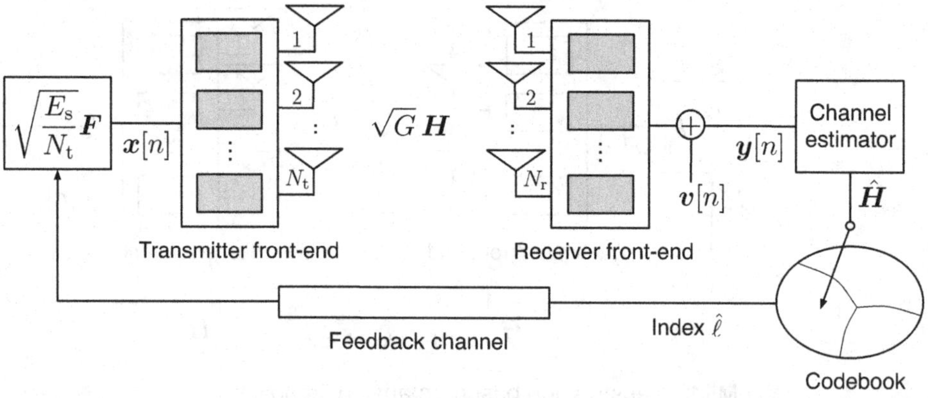

Fig. 5.18 SU-MIMO transmission based on codebook-based digital feedback.

5.10.3 Digital feedback

In FDD systems, the alternative to analog feedback is to relay back a digital representation of the CSI. Since, as any overhead, the number of feedback bits should be limited, this approach is sometimes known as *limited feedback* [635].

To effect digital feedback, the quantity of choice needs first of all to be quantized. This entails selecting a representative value from a finite set, i.e., from a codebook shared by transmitter and receiver. Suppose for instance that the quantity to be fed back is the channel estimate itself, as in Fig. 5.18, and let the codebook be $\{\mathbf{H}_0, \ldots, \mathbf{H}_{N_f-1}\}$ where N_f is the number of entries. From the channel estimate $\hat{\boldsymbol{H}}$ at the receiver, the index $\hat{\ell}$ of a selected codebook entry is fed back to the transmitter and, by virtue of the codebook being shared, the transmitter can then generate $\mathbf{H}_{\hat{\ell}}$ as its CSIT.

The distortion introduced by this process is a function of the codebook and the channel distribution. Relative to its analog counterpart, the main drawback of digital feedback is that, in the former, the distortion vanishes as the SNRs grow, whereas in the latter it does not. With digital feedback, rather, the distortion only vanishes for $N_f \to \infty$. This is not catastrophic in single-user channels, but it acquires great relevance in multiuser setups [147] and it requires N_f to grow with the SNRs (or, in practice, to be dimensioned for the highest envisioned SNRs).

Feedback need not take place for every symbol and subcarrier, but only once per coherence tile, and many works have explored how to tailor the quantization and feedback process to changing fading coherences in both frequency [636, 637] and time [638–640]. Antenna correlations can be leveraged to reduce the amount of feedback [641, 642].

Beamforming

Before opening the discussion to precoders of arbitrary rank, it is instructive to examine digital feedback for beamforming, a transmission option whose precoder, recall, is ideally given by the maximum-eigenvalue eigenvector of $\boldsymbol{H}^*\boldsymbol{H}$. A beamforming situation allows illustrating why quantizing the channel itself may be inefficient.

The main motivation comes from an argument about dimensionality. An $N_\mathrm{r} \times N_\mathrm{t}$ channel matrix is parameterized by $2N_\mathrm{r}N_\mathrm{t}$ real values. With direct quantization, the beamforming vector would be computed from the quantized channel after feedback, a computation whose accuracy generally requires much resolution in the quantized feedback. Alternatively, the receiver could quantize and feed back the beamforming vector itself, which is parameterized by only $2N_\mathrm{t}$ real values. Furthermore, since the beamforming vector has norm N_t, there is one free quantity and only $2N_\mathrm{t} - 1$ real values are needed.

Instead of a codebook of channel matrices then, a codebook $\mathcal{F} = \{\mathbf{F}_0, \ldots, \mathbf{F}_{N_\mathrm{f}-1}\}$ of beamforming vectors can advantageously be featured. Selecting an entry from this codebook on the basis of minimum distance, $\hat{\ell} = \arg\min \|\boldsymbol{v}_{\mathsf{max}} - \mathbf{F}_\ell\|^2$ where $\boldsymbol{v}_{\mathsf{max}}$ is the maximum eigenvalue eigenvector of $\hat{\boldsymbol{H}}^*\hat{\boldsymbol{H}}$, would lead to a low distortion in the quantization of the precoder, but not necessarily to the best communication performance. Rather, it is better to maximize the ergodic spectral efficiency by selecting [643–645]

$$\hat{\ell} = \arg\max_{\ell} \mathbb{E}\left[\log_2\left(1 + \frac{\mathsf{SNR}}{N_\mathrm{t}} \left\|\hat{\boldsymbol{H}}\mathbf{F}_\ell\right\|^2\right)\right] \tag{5.228}$$

or, since beamforming is mostly of interest at low SNR, to directly maximize the average SNR at the receiver by selecting

$$\hat{\ell} = \arg\max_{\ell} \mathbb{E}\left[\left\|\hat{\boldsymbol{H}}\mathbf{F}_\ell\right\|^2\right]. \tag{5.229}$$

Note that, for any phase ϕ,

$$\|\hat{\boldsymbol{H}}\,(e^{\mathrm{j}\phi}\mathbf{F}_\ell)\|^2 = (e^{\mathrm{j}\phi}\mathbf{F}_\ell)^*\hat{\boldsymbol{H}}^*\hat{\boldsymbol{H}}\,(e^{\mathrm{j}\phi}\mathbf{F}_\ell) \tag{5.230}$$

$$= e^{-\mathrm{j}\phi}e^{\mathrm{j}\phi}\mathbf{F}_\ell^*\hat{\boldsymbol{H}}^*\hat{\boldsymbol{H}}\mathbf{F}_\ell \tag{5.231}$$

$$= \|\hat{\boldsymbol{H}}\mathbf{F}_\ell\|^2, \tag{5.232}$$

indicating that the optimum beamforming vector is not unique: any phase-rotated version performs equally. This phase invariance amounts to another free quantity, leaving $2N_\mathrm{t} - 2$ real values to quantify. A good design for $\mathcal{F}$ should incorporate the norm constraint and the phase invariance.

Let us see a codebook design based on (5.229), i.e., a codebook $\mathcal{F}$ that minimizes the distortion

$$\mathsf{D}(\mathcal{F}) = \frac{\mathsf{SNR}}{N_\mathrm{t}} \mathbb{E}\left[\|\hat{\boldsymbol{H}}\boldsymbol{v}_{\mathsf{max}}\|^2 - \max_{\mathbf{F}_\ell \in \mathcal{F}} \|\hat{\boldsymbol{H}}\mathbf{F}_\ell\|^2\right]. \tag{5.233}$$

We can rewrite $\mathsf{D}(\mathcal{F})$ in terms of the maximizer $\mathbf{F}_{\hat{\ell}}$ (which depends on the codebook) as

$$\mathsf{D}(\mathcal{F}) = \frac{\mathsf{SNR}}{N_\mathrm{t}} \mathbb{E}\left[\|\hat{\boldsymbol{H}}\boldsymbol{v}_{\mathsf{max}}\|^2 - \|\hat{\boldsymbol{H}}\mathbf{F}_{\hat{\ell}}\|^2\right] \tag{5.234}$$

$$= \mathsf{SNR}\,\mathbb{E}\left[\lambda_{\mathsf{max}}(\hat{\boldsymbol{H}}^*\hat{\boldsymbol{H}}) - \sum_{j=0}^{N_{\mathsf{min}}-1} \lambda_j(\hat{\boldsymbol{H}}^*\hat{\boldsymbol{H}}) \frac{|\boldsymbol{v}_j^*\mathbf{F}_{\hat{\ell}}|^2}{N_\mathrm{t}}\right]. \tag{5.235}$$

Discarding the terms corresponding to smaller eigenvalues gives an upper bound on the average distortion, to wit

$$\mathsf{D}(\mathcal{F}) \leq \mathsf{SNR}\, \mathbb{E}\left[\lambda_{\mathsf{max}}(\hat{\boldsymbol{H}}^* \hat{\boldsymbol{H}}) \left(1 - \frac{\left|\boldsymbol{v}^*_{\mathsf{max}} \mathbf{F}_{\hat{\ell}}\right|^2}{N_{\mathrm{t}}}\right)\right] \tag{5.236}$$

$$= \mathsf{SNR}\, \mathbb{E}\left[\lambda_{\mathsf{max}}(\hat{\boldsymbol{H}}^* \hat{\boldsymbol{H}})\right] \mathbb{E}\left[1 - \frac{\left|\boldsymbol{v}^*_{\mathsf{max}} \mathbf{F}_{\hat{\ell}}\right|^2}{N_{\mathrm{t}}}\right] \tag{5.237}$$

$$= \mathsf{SNR}\, \mathbb{E}\left[\lambda_{\mathsf{max}}(\hat{\boldsymbol{H}}^* \hat{\boldsymbol{H}})\right] \mathbb{E}\left[\mathsf{d}^2\left(\boldsymbol{v}_{\mathsf{max}}, \frac{\mathbf{F}_{\hat{\ell}}}{\sqrt{N_{\mathrm{t}}}}\right)\right], \tag{5.238}$$

where (5.237) follows from the independence of the eigenvalues and eigenvectors of complex Wishart matrices [646, 647], while in (5.238) we have introduced the function

$$\mathsf{d}(\boldsymbol{x}, \boldsymbol{y}) = \sqrt{1 - |\boldsymbol{x}^* \boldsymbol{y}|^2}. \tag{5.239}$$

Applied to unit-norm vectors, whose inner product equals the cosine of the angle they subtend, $\mathsf{d}(\boldsymbol{x}, \boldsymbol{y}) = \sin(\theta)$ with $\theta \in [0, \pi/2]$ being the angle between $\boldsymbol{x}$ and $\boldsymbol{y}$. As it turns out, $\mathsf{d}(\cdot, \cdot)$ is a measure of distance on the set of one-dimensional subspaces (i.e., lines) within the N_{t}-dimensional space, a set known as the *Grassmann manifold* $\mathrm{G}(N_{\mathrm{t}}, 1)$. It is important to distinguish the *subspace distance* $\mathsf{d}(\cdot, \cdot)$ from the Euclidean distance employed elsewhere in the book. Among other properties, the subspace distance function is phase invariant, i.e., $\mathsf{d}(e^{\mathrm{j}\phi} \boldsymbol{x}, \boldsymbol{y}) = \mathsf{d}(\boldsymbol{x}, \boldsymbol{y})$, as desired in a good codebook.

Returning to (5.238), we have that the distortion upper bound depends on the subspace distance between the ideal beamforming vector, $\boldsymbol{v}_{\mathsf{max}}$, and its best codebook entry, $\mathbf{F}_{\hat{\ell}}$. If the entries of $\hat{\boldsymbol{H}}$ are IID Rayleigh-faded, then $\boldsymbol{v}_{\mathsf{max}}$ is isotropically distributed and $\mathbb{E}\left[\mathsf{d}^2(\mathbf{F}_{\hat{\ell}}, \boldsymbol{v}_{\mathsf{max}})\right]$ in (5.238) can itself be upper-bounded. For a given codebook $\mathcal{F}$, let

$$\mathsf{d}_{\mathsf{min}}(\mathcal{F}) = \min_{0 \leq \ell < \ell' \leq N_{\mathrm{f}}-1} \sqrt{1 - \frac{|\mathbf{F}^*_{\ell} \mathbf{F}_{\ell'}|^2}{N_{\mathrm{t}}^2}} \tag{5.240}$$

be the smallest subspace distance between any pair of vectors in the codebook. Then [648]

$$\mathsf{D}(\mathcal{F}) \leq \mathsf{SNR}\, \mathbb{E}\left[\lambda_{\mathsf{max}}(\hat{\boldsymbol{H}}^* \hat{\boldsymbol{H}})\right] \left[1 + N_{\mathrm{f}} \left(\frac{\mathsf{d}_{\mathsf{min}}(\mathcal{F})}{2}\right)^{2(N_{\mathrm{t}}-1)} \left(\frac{\mathsf{d}^2_{\mathsf{min}}(\mathcal{F})}{4} - 1\right)\right] \tag{5.241}$$

where $\mathsf{SNR}\, \mathbb{E}\left[\lambda_{\mathsf{max}}(\hat{\boldsymbol{H}}^* \hat{\boldsymbol{H}})\right]$ is the average receive SNR with unquantized CSIT, and thus the remaining term reflects the penalty of a quantized CSIT. This penalty is minimized by the codebook $\mathcal{F}$ with the largest $\mathsf{d}_{\mathsf{min}}(\mathcal{F})$. Designs of maximally spaced lines are known as *Grassmannian line packings* [649, 650].

The bound in (5.241) provides a fundamental connection between digital feedback beamforming for IID channels, the Grassmann manifold, and Grassmannian line packings. For this reason, the term Grassmannian beamforming is sometimes employed, and the corresponding codebooks (which exactly or nearly maximize the minimum distance) are known as Grassmannian codebooks [651–656]. The minimum distances in Grassmannian codebooks are often the benchmark by which other beamforming codebooks for IID Rayleigh-faded channels are measured.

The upper bound in (5.241) can be complemented by lower bounds, and any specific codebook yields one such bound. In particular [654],

$$\mathsf{D}(\mathcal{F}) \gtrsim \mathsf{SNR}\Big(\mathbb{E}[\lambda_{\max}(\hat{\boldsymbol{H}}^*\hat{\boldsymbol{H}})] - N_\mathrm{r}\Big) N_\mathrm{f}^{-\frac{1}{N_\mathrm{t}-1}}, \tag{5.242}$$

which is a rigorous inequality for MISO ($N_\mathrm{r} = 1$) and an approximate one for MIMO ($N_\mathrm{r} > 1$). This expression shows how the distortion decreases with the size of the codebook, raised to a power that depends on N_t. For a certain codebook size, additional transmit antennas increase the distortion.

As an alternative to characterizing the performance for specific codebooks, the performance can be averaged over a randomly drawn codebook [644, 645, 657, 658]. In the case of beamforming, this corresponds to vectors distributed on the unit sphere and, for MISO IID Rayleigh-faded channels specifically [645],

$$\mathbb{E}[\mathsf{D}(\mathcal{F})] = \mathsf{SNR}\, \mathbb{E}[\lambda_{\max}(\hat{\boldsymbol{H}}^*\hat{\boldsymbol{H}})]\, \frac{N_\mathrm{f}(N_\mathrm{f}-1)!\,\Gamma\big(\frac{N_\mathrm{t}}{N_\mathrm{t}-1}\big)}{\Gamma\big(N_\mathrm{f} + \frac{N_\mathrm{t}}{N_\mathrm{t}-1}\big)}. \tag{5.243}$$

As the codebook grows large, $\mathbb{E}[\mathsf{D}(\mathcal{F})]$ decays approximately with $N_\mathrm{f}^{-\frac{1}{N_\mathrm{t}-1}}$, as in (5.242).

Arbitrary-rank precoders

An arbitrary precoder for complex Gaussian signals comprises a steering matrix $\boldsymbol{U_F}$ and a power allocation matrix $\boldsymbol{P}$. For feedback purposes, the power allocation coefficients are sometimes rendered binary, meaning that N_s is set to the number of eigenvectors of $\hat{\boldsymbol{H}}^*\hat{\boldsymbol{H}}$ on which waterfilling would allocate nonzero power, and then power is allocated uniformly thereon [659]. With that, only a properly dimensioned $\boldsymbol{U_F}$ must be fed back, a semiunitary precoder satisfying $\boldsymbol{U_F}\boldsymbol{U_F}^* = \boldsymbol{I}_{N_\mathrm{s}}$ (but $\boldsymbol{U_F}^*\boldsymbol{U_F} \neq \boldsymbol{I}_{N_\mathrm{t}}$ for $N_\mathrm{s} < N_\mathrm{t}$). Note that, for both $\mathsf{SNR} \to 0$ and $\mathsf{SNR} \to \infty$, this entails no loss in optimality, as the optimum precoders in these limiting regimes do allocate power uniformly, respectively over the maximum-eigenvalue eigenvector(s) of $\boldsymbol{H}^*\boldsymbol{H}$ and over all the eigenvectors. At intermediate SNRs there is a loss, and Problem 5.58 is devoted to its quantification.

As in beamforming, there are invariances in arbitrary-rank precoding that reduce the dimensionality. Because of the unitary invariance of IID complex Gaussian signals, for any $N_\mathrm{s} \times N_\mathrm{s}$ unitary matrix $\boldsymbol{V}$, the precoder $\boldsymbol{U_F}\boldsymbol{V}$ achieves the same mutual information as $\boldsymbol{U_F}$, meaning that the spectral efficiency with an optimum receiver is unaffected. The orthogonality of the columns of $\boldsymbol{U_F}$ and the unitary invariance reduces the number of free parameters, and digital-feedback semiunitary precoding with relatively small codebooks can have surprisingly satisfactory performance.

The equivalence of $\boldsymbol{U_F}\boldsymbol{V}$ for any unitary $\boldsymbol{V}$ implies that what matters is the subspace spanned by the columns of $\boldsymbol{U_F}$. With the Grassmann manifold $\mathrm{G}(N_\mathrm{t}, N_\mathrm{s})$ being the set of all possible N_s-dimensional subspaces on the N_t-dimensional space, each precoding subspace corresponds to a point in the Grassmann manifold and the optimization of the codebook entails selecting N_f such points depending on the channel distribution. For IID Rayleigh-faded channels specifically, a good codebook should feature precoders that are regularly far in terms of subspace distance; again, a tight connection emerges with Grassmannian packings [660].

To allow for N_s itself to be a parameter that is fed back—this complement of link adaptation is termed *rank adaptation*—and that can vary between 1 and N_{min}, the codebook needs to consist of N_{min} subcodebooks, one for each possible rank [659, 661–665].

Codebook designs

Although a detailed analysis of digital-feedback codebooks is beyond the scope of this book, we do present a brief survey of the main families of codebooks and provide references for readers interested in further exploring the topic.

Vector-quantization-based codebooks

Vector quantization extends the classical notion of scalar quantization to vector-valued sources [666]. The source is quantized by selecting one vector from a codebook. This tends to rely on squared-error distortion functions, which makes its direct application to MIMO precoding difficult in light of the different distortion functions that arise in this context. With some generalizations though, vector quantization can be applied [654, 665, 667, 668].

Most approaches are based on modifications of the Linde–Buzo–Gray (LGB) algorithm [669], which is itself a variation of the Lloyd or the k-means clustering algorithms in signal processing [670]. The operation of these algorithms can be sketched as follows. The codebook is initialized with N_f entries drawn at random. Then, a number of source realizations are produced, in our case precoders. Each precoder is assigned to its closest—in some desired sense—codebook entry, forming N_f clusters, and the centroids of these clusters become the entries of a new codebook. The process is repeated until convergence. While such convergence may be to a local optimum, the technique is enticing because it can find a codebook for any channel distribution, possibly in real time. Dynamically optimized codebooks would keep the distortion at a minimum, at the expense of frequent updates.

Grassmannian codebooks

As seen earlier in the section, codebooks maximizing the minimum subspace distance are desirable for IID Rayleigh-faded channels. The entries of a Grassmannian codebook correspond to subspaces, e.g., lines for $N_s = 1$ or planes for $N_s = 2$. Maximizing the minimum distance ensures that these subspaces are as far as possible. Essentially, Grassmannian codebooks yield a near-uniform sampling of the precoding space, and their design is a packing problem.

Example 5.46

Shown in Fig. 5.19 is the toy example of packing points in a square with the largest possible minimum distance [671, 672]. Some packings are more efficient than others, leaving smaller gaps.

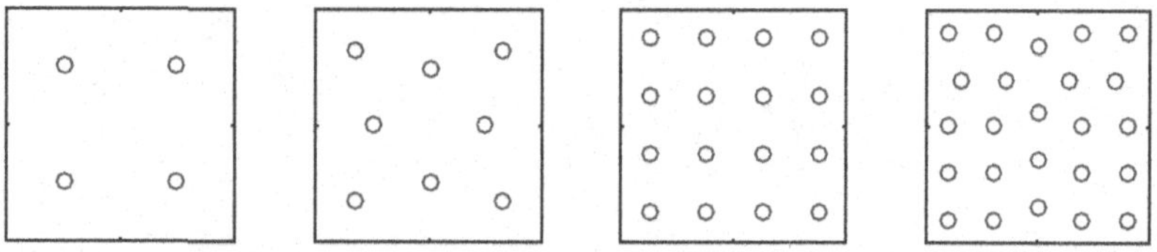

Fig. 5.19 Packing points in a box [672].

Example 5.47

Shown in Fig. 5.20 are packings in $\mathrm{G}(3,1)$, i.e., packings of lines in a three-dimensional space [673]. For $N_\mathrm{f} = 3$, the lines are orthogonal. In the other cases they are not orthogonal, but they are apart in terms of subspace distance

Example 5.48

Shown in Fig. 5.21 are packings in $\mathrm{G}(3,2)$, i.e., packings of planes in a three-dimensional space. Exploiting the fact that a packing for $\mathrm{G}(N_\mathrm{f},1)$ can serve to fabricate a packing for $\mathrm{G}(N_\mathrm{f}, N_\mathrm{f}-1)$, each plane in this example is determined by a line.

One of the challenges of Grassmannian codebook design is that the problem of finding the best subspace packings has received attention in the mathematics community only in recent years [649]. Finding optimum packings is notoriously difficult and closed forms are known only in special cases; most solutions have to be derived numerically. In [649] specifically, a nonlinear optimization software package is employed; resulting codebooks can be downloaded online [673]. Possibly the most relevant practical approach to design new codebooks is the alternating projection technique described in [650]. A few closed form solutions are also available for optimum packings, for instance a construction with $N_\mathrm{f} = 2N_\mathrm{t}$ entries for $N_\mathrm{s} = 1$ and power-of-two values of N_t [674] and optimum designs for $N_\mathrm{s} = 1$ and several values of N_t [651]. Yet another source of potential codebooks for digital feedback is the literature on communication over no-CSIR MIMO channels, whose capacity-achieving signals, recall from Section 5.5, are $N_\mathrm{t} \times N_\mathrm{c}$ signal matrices of the form $\boldsymbol{S} = \boldsymbol{A}_{\boldsymbol{S}} \cdot \sqrt{N_\mathrm{c}}\, \boldsymbol{U}_{\boldsymbol{S}}$, where $\boldsymbol{U}_{\boldsymbol{S}}$ is semiunitary; the design of signal codebooks from which to draw $\boldsymbol{U}_{\boldsymbol{S}}$ relates to optimum packings on the Grassmann manifold [647, 675].

For beamforming on IID Rayleigh-faded channels specifically, the Grassmannian codebooks are packings in $\mathrm{G}(N_\mathrm{t},1)$ and the minimum subspace distance in (5.240) satisfies

$$\mathsf{d}_{\min}(\mathcal{F}) \geq \sqrt{1 - \frac{N_\mathrm{f} - N_\mathrm{t}}{N_\mathrm{t}\,(N_\mathrm{f}-1)}}. \tag{5.244}$$

Equality can be achieved in only a few special cases [651, 674], a necessary condition being $N_\mathrm{f} \leq N_\mathrm{t}^2$. When the equality holds, the constructions are equiangular, meaning that, for every $\boldsymbol{F}_\ell, \boldsymbol{F}_{\ell'} \in \mathcal{F}$, $\ell \neq \ell'$, the value of $|\boldsymbol{F}_\ell^* \boldsymbol{F}_{\ell'}|$ is constant.

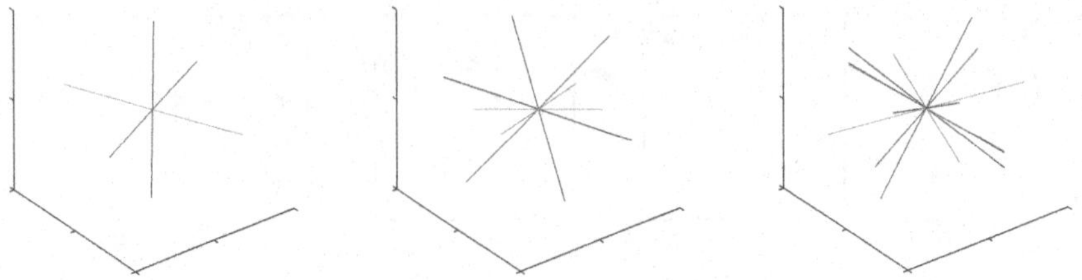

Fig. 5.20 Packings in $G(3,1)$ for $N_f = 3$, 5, and 8.

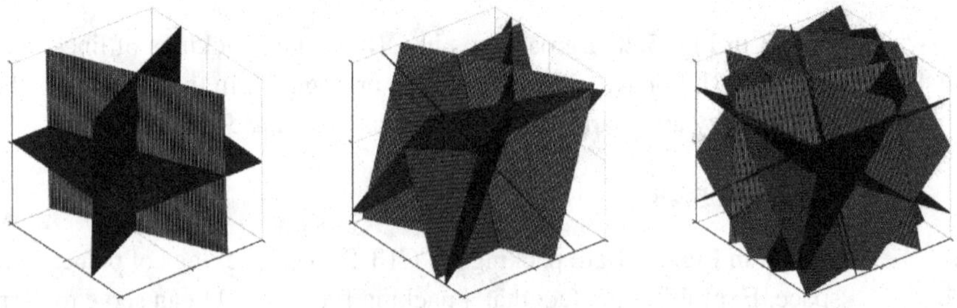

Fig. 5.21 Packings in $G(3,2)$ for $N_f = 3$, 5, and 8.

Example 5.49

The Grassmannian beamforming codebook for $N_t = 2$ and $N_f = 4$, which satisfies (5.244) with equality, is

$$\mathcal{F} = \left\{ \begin{array}{c} \sqrt{2}\begin{bmatrix} -0.16125 - \mathrm{j}\,0.73479 \\ -0.51347 - \mathrm{j}\,0.41283 \end{bmatrix} \\ \sqrt{2}\begin{bmatrix} -0.07869 - \mathrm{j}\,0.31920 \\ -0.25059 + \mathrm{j}\,0.91056 \end{bmatrix} \\ \sqrt{2}\begin{bmatrix} -0.23994 + \mathrm{j}\,0.59849 \\ -0.76406 - \mathrm{j}\,0.02120 \end{bmatrix} \\ \sqrt{2}\begin{bmatrix} -0.95406 \\ 0.29961 \end{bmatrix} \end{array} \right\}. \tag{5.245}$$

Example 5.50

Consider an IID Rayleigh-faded MISO channel with $N_t = 2$. The receiver has CSIR and at every coherence tile it feeds back two bits indexing the entry of the codebook in (5.245)

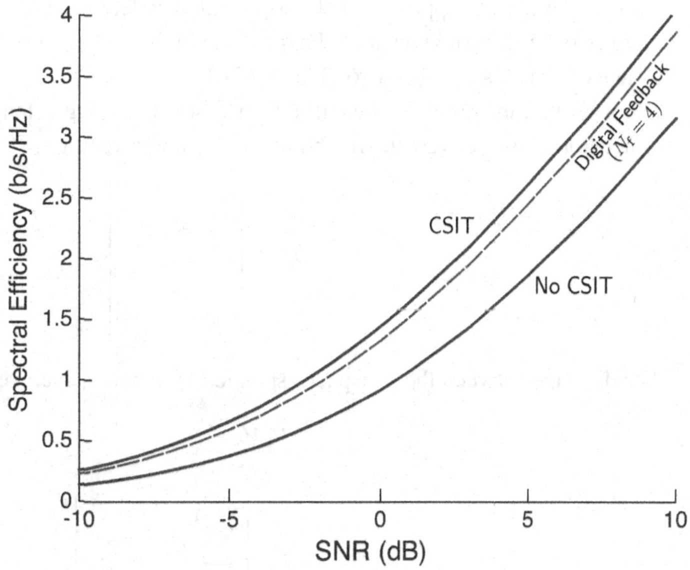

Fig. 5.22 In solid, CSIT and no-CSIT capacities of a MISO IID Rayleigh-faded channel with $N_t = 2$. In dashed, spectral efficiency with two bits of feedback and a Grassmannian beamforming codebook.

that maximizes the receive SNR. The feedback delay is much shorter than the coherence time, such that the selected precoder can be assumed to apply over the entire tile. Plot the achievable spectral efficiency alongside the capacities with and without CSIT.

Solution

See Fig. 5.22. With only two bits of feedback per coherence tile, most of the gap between the CSIT and no-CSIT capacities is bridged. With additional bits and a richer codebook, it could be sealed completely.

As in any Grassmannian codebook, the vectors in (5.245) can be subject to unitary rotations: for $\boldsymbol{F}_\ell$ representing a point in $\mathrm{G}(N_t, N_s)$, $\boldsymbol{F}_\ell \boldsymbol{V}$ represents the same point for any unitary $\boldsymbol{V}$ because the columns of $\boldsymbol{F}_\ell$ and $\boldsymbol{F}_\ell \boldsymbol{V}$ span the same subspace. Indeed, subjecting the lines or planes in Figs. 5.20 and 5.21 to a common rotation is inconsequential to their packing and to their effectiveness as precoders for IID Rayleigh-faded channels.

Fourier codebooks

Grassmannian codebooks maximize the minimum subspace distance between entries, minimize bounds on several average distortion metrics for IID Rayleigh-faded channels, are ideally equiangular, and have intuitively appealing properties. However, as most packings are found through numerical search, they do not exhibit much structure, complicating the implementation both in terms of search and of storage. Consequently, there has been interest in developing approximate Grassmannian codebooks that possibly sacrifice the strict optimality of the packing in exchange for additional structure. The Fourier design is an instance of such a codebook [660, 675, 676].

Let us first examine the beamforming case. By taking a length-N_t portion of an N_f-dimensional Fourier vector, we obtain the harmonic vector

$$\boldsymbol{\nu}_\ell = \begin{bmatrix} 1 \\ e^{j2\pi \frac{\ell}{N_f}} \\ \vdots \\ e^{j2\pi \frac{(N_t-1)\ell}{N_f}} \end{bmatrix}. \tag{5.246}$$

The distance between the subspaces spanned by any two such vectors satisfies

$$\mathrm{d}^2(\boldsymbol{\nu}_\ell, \boldsymbol{\nu}_{\ell'}) = 1 - \frac{|\boldsymbol{\nu}_\ell^* \boldsymbol{\nu}_{\ell'}|^2}{N_t^2} \tag{5.247}$$

$$= 1 - \frac{1}{N_t^2} \left| \sum_{j=0}^{N_t-1} e^{j2\pi \frac{(\ell-\ell')j}{N_f}} \right|^2 \tag{5.248}$$

$$= \begin{cases} 1 & \ell = \ell' \\ 1 - \frac{1}{N_t^2} \left| \frac{\sin\left(\frac{\pi N_t(\ell-\ell')}{N_f}\right)}{\sin\left(\frac{\pi(\ell-\ell')}{N_f}\right)} \right|^2 & \ell \neq \ell', \end{cases} \tag{5.249}$$

which depends on ℓ and ℓ' only through $((\ell - \ell'))_{N_f}$, i.e., it has a circulant structure. Thus, it suffices to consider $\mathrm{d}(\boldsymbol{\nu}_0, \boldsymbol{\nu}_\ell)$ for $\ell = 1, \ldots, N_t - 1$. The key idea behind Fourier codebooks is to reorder the entries of (5.246) into beamforming vectors with the best possible such distances for the channel distribution of interest. To that end, consider the diagonal matrix

$$\boldsymbol{\Theta} = \begin{bmatrix} e^{j\frac{2\pi}{N_f}u_0} & 0 & \cdots & 0 \\ 0 & e^{j\frac{2\pi}{N_f}u_1} & 0 & 0 \\ \vdots & \ddots & \ddots & \vdots \\ 0 & 0 & \cdots & e^{j\frac{2\pi}{N_f}u_{N_t-1}} \end{bmatrix}, \tag{5.250}$$

where $0 \leq u_0, \ldots, u_{N_t-1} \leq N_f - 1$. From $\boldsymbol{\Theta}$ and a generating Fourier vector $\boldsymbol{\nu}_0$, a Fourier beamforming codebook is obtained as

$$\mathcal{F} = \left\{ \boldsymbol{\nu}_0, \boldsymbol{\Theta}\boldsymbol{\nu}_0, \boldsymbol{\Theta}^2\boldsymbol{\nu}_0, \ldots, \boldsymbol{\Theta}^{N_f-1}\boldsymbol{\nu}_0 \right\}, \tag{5.251}$$

corresponding to N_f points arranged on an N_t-dimensional complex circle. The choice of $\boldsymbol{\nu}_0$ and $\{u_0, \ldots, u_{N_t-1}\}$ should optimize this arrangement depending on the channel distribution, something that can be done through exhaustive search, for small codebooks, and more generally through random search.

If the channel is IID Rayleigh-faded, then the N_f points should be regularly spaced over the N_t-dimensional complex circle, so as to maximize the minimum distance. There are Fourier such codebooks that are also Grassmannian codebooks [651]. Based on this connection, Fourier constructions that meet (5.244) with equality can be found.

Looking beyond beamforming, for semiunitary precoders of rank $N_s > 1$, the same ideas apply only with the generator containing N_s columns of an N_t-dimensional Fourier matrix.

Fourier codebooks possess much structure and small storage requirements, as only the generator and $\boldsymbol{\Theta}$ need to be stored and the codebook entries can then be generated in real time. Because every codebook entry is of unit magnitude, the calculation of $\boldsymbol{\Theta}^{\ell}\boldsymbol{\nu}_0$ only entails phase rotations. Moreover, subcodebooks for different precoder ranks may admit nested constructions whereby all the subcodeword items derive from those of the highest rank. Because of all of these features, Fourier codebooks are featured in LTE.

Kerdock codebooks

Kerdock codebooks are based on the concept of mutually unbiased bases, which are collections of unitary matrices with constant cross-correlation [677, 678]. Precisely, the set $\mathcal{U} = \{\boldsymbol{U}_0, \ldots, \boldsymbol{U}_{L-1}\}$ with $\boldsymbol{U}_\ell$ an $N_t \times N_t$ unitary matrix is a mutually unbiased collection if any two columns $\boldsymbol{u}$ and $\boldsymbol{v}$ of different matrices therein satisfy

$$|\boldsymbol{u}^*\boldsymbol{v}|^2 = \frac{1}{N_t}. \tag{5.252}$$

In turn, any two columns of the same matrix are orthogonal because of the unitary nature of these matrices. It is not possible to find collections satisfying (5.252) for every N_t and L, and in fact a necessary condition is $L \le N_t + 1$, which upper-bounds the size of the feasible codebooks at $N_f = N_t\,(N_t + 1)$. There are known constructions for important values of N_t, and specifically for $N_t = 1, 2, 3, 4$, and 8 [679–683].

Kerdock codebooks for digital feedback have been proposed for $N_t = 2, 4$, and 8 [680, 684, 685]. For beamforming specifically, the codebooks are assembled from columns in one or various bases. These designs have a systematic construction and a quaternary alphabet, reducing storage and search complexity. Other values of N_t may entail larger alphabets. Links to Grassmannian packings further justify their potential as digital feedback codebooks [686].

Householder codebooks

A Householder linear transformation describes a reflection about a hyperplane containing the origin. This hyperplane can be determined by an $N_t \times 1$ vector $\boldsymbol{g}$ orthogonal to it, and the reflection of a vector with respect to the hyperplane can be effected by the so-called Householder matrix

$$\boldsymbol{U}(\boldsymbol{g}) = \boldsymbol{I} - \frac{2}{\|\boldsymbol{g}\|^2}\boldsymbol{g}\boldsymbol{g}^*, \tag{5.253}$$

which satisfies

$$\boldsymbol{U}(\boldsymbol{g})\boldsymbol{U}(\boldsymbol{g}) = \boldsymbol{I} \tag{5.254}$$

$$\boldsymbol{U}(\boldsymbol{g})\boldsymbol{U}(\boldsymbol{g})^* = \boldsymbol{I} \tag{5.255}$$

$$\boldsymbol{U}(\boldsymbol{g})\,\boldsymbol{g} = -\boldsymbol{g}. \tag{5.256}$$

Together, these properties indicate that $\boldsymbol{U}(\boldsymbol{g})$ is full-rank and unitary, that it has $N_{\mathrm{t}} - 1$ eigenvalues that equal 1 and one eigenvalue that equals -1, and that $\boldsymbol{g}/\|\boldsymbol{g}\|$ is the eigenvector corresponding to $\lambda(\boldsymbol{U}(\boldsymbol{g})) = -1$. A Householder matrix derives from one of its eigenvectors and, if $\boldsymbol{g}$ is of unit norm, then no divisions are required to compute it.

There are several ways to build a digital-feedback codebook from a Householder matrix. All constructions start from a generating codebook of vectors having the same norm. This could be a Grassmannian codebook or some other construction described in this section. The generators are then used to create a set of Householder matrices, whose columns can subsequently be employed for either beamforming or arbitrary-rank precoding. Householder codebooks are also featured in LTE [687].

5.11 Summary and outlook

The chief results derived throughout this chapter are restated in the companion summary box. In consistency with the prioritization of the ergodic setting, only those results are included in the summary; readers interested in the outage capacity and the DMT are referred back to Section 5.4.1.

All in all, the problem of communicating over SU-MIMO channels impaired by underspread fading and Gaussian noise can be regarded as a relatively mature topic, yet interesting research avenues do remain. One such avenue is the capacity-achieving precoding, with and without CSIT, under a per-antenna power constraint. A pragmatic approach with such type of constraint is to discard precoding altogether, yet the ability to precode is not lost—it is merely diminished. This problem has been solved for certain channels [688–690].

- In MISO channels with CSIT, the optimum precoder transmits equal-power signals that add coherently at the receiver, i.e., $[\boldsymbol{F}]_j = [\boldsymbol{H}]_j/\|[\boldsymbol{H}]_j\|$.
- In ergodic MIMO channels without CSIT and with no transmit antenna correlations, the optimum precoding strategy is the same one that applies under a per-symbol power constraint, namely $\boldsymbol{F} = \boldsymbol{I}$.

Further results can be found in [691, 692], but complete solutions for the precoding and the ensuing capacity under a per-antenna power constraint are still lacking. Likewise, the inclusion of magnitude constraints into the design of the precoders is a relevant extension of the results presented in this chapter, and interested readers are referred to [693–695].

Another interesting direction relates to the impact of delay constraints, both from a pure physical-layer perspective (finite-length codewords in MIMO [696]) as well as from the viewpoint of a protocol stack that has to meet delay guarantees that are possibly different for different data flows. At the higher level of abstraction required for the latter, a modified metric termed *effective capacity* has been proposed [697].

Yet other research avenues open up if complexity and hardware constraints are allowed priority over the raw performance. The reduction in baseband complexity by means of linear receivers is deferred to Chapter 6, which is devoted entirely to such receivers. If, alternatively, the limitations are placed at the radio-frequency level, then one can formulate the

Take-away points

1. For moderate numbers of antennas, results obtained under CSIR (and possibly CSIT, if the coherence is long enough to accommodate feedback) are fully meaningful. To represent pilot-assisted transmission, these results can be refined via the appropriate overhead discount and the replacement of SNR by $\mathsf{SNR}_{\mathsf{eff}}$.
2. With CSIT, the capacity-achieving precoder is $\boldsymbol{F} = \boldsymbol{U_F}\left[\boldsymbol{P}^{1/2}\ \boldsymbol{0}\right]^{\mathrm{T}}$ where $\boldsymbol{U_F}$ diagonalizes the channel and creates $N_{\mathsf{min}} = \min(N_{\mathrm{t}}, N_{\mathrm{r}})$ parallel subchannels over which $\boldsymbol{P}$ effects a waterfilling. This can be interpreted as the steering of N_{min} beams with optimized powers toward the nonzero-eigenvalue eigenvectors of $\boldsymbol{H}^*\boldsymbol{H}$, giving
$$C(\mathsf{SNR}) = N_{\mathsf{min}}\,\mathbb{E}\left[\left[\log_2\left(\frac{\mathsf{SNR}}{N_{\mathrm{t}}}\frac{\lambda(\boldsymbol{H}^*\boldsymbol{H})}{\eta}\right)\right]^+\right], \tag{5.257}$$
which is unchanged if transmitter and receiver swap roles.
3. At low SNR, the CSIT-based precoder reduces to beamforming on the maximum-eigenvalue eigenvector(s) of $\boldsymbol{H}^*\boldsymbol{H}$. In an ergodic setting, this gives
$$\frac{E_{\mathrm{b}}}{N_0}_{\mathsf{min}} = \frac{1}{\mathbb{E}[\lambda_{\mathsf{max}}(\boldsymbol{H}^*\boldsymbol{H})]\log_2 e} \tag{5.258}$$
$$S_0 = \frac{2}{\kappa\left(\sqrt{\lambda_{\mathsf{max}}(\boldsymbol{H}^*\boldsymbol{H})}\right)}. \tag{5.259}$$
4. At high SNR, $S_\infty = N_{\mathsf{min}}$. Transmission of N_{min} equal-power streams delivers such DOF and the optimum $\mathcal{L}_\infty$.
5. With signals drawn from discrete constellations rather than complex Gaussian, the CSIT-based capacity-achieving precoder remains valid as long as the constellation cardinalities do not become a limiting factor. Otherwise, the precoder must incorporate a mixing matrix and channel diagonalization becomes suboptimum.
6. Without CSIT, transmission on the eigenvectors of $\mathbb{E}[\boldsymbol{H}\boldsymbol{H}^*]$ is optimal for a broad class of channels. This on-average diagonalization does not create parallel subchannels, and hence spatial interference does arise. The optimum power allocation is no longer a waterfilling; rather, the share of power allocated to each of the concurrent beams should equal the share of useful signal that an LMMSE estimator would recover from that beam. If the transmit antennas are uncorrelated, this reduces to a uniform power allocation.
7. With $\boldsymbol{F}$ set to the optimum no-CSIT precoder,
$$C(\mathsf{SNR}) = \mathbb{E}\left[\log_2\det\left(\boldsymbol{I} + \frac{\mathsf{SNR}}{N_{\mathrm{t}}}\boldsymbol{H}\boldsymbol{F}\boldsymbol{F}^*\boldsymbol{H}^*\right)\right], \tag{5.260}$$
which in general does change if transmitter and receiver swap roles. For $\boldsymbol{H}$ having IID Rayleigh-faded entries, $C(\mathsf{SNR})$ admits a closed form involving only exponential integral functions.

8. Without CSIT, beamforming on the maximum-eigenvalue eigenvector(s) of $\mathbb{E}[\boldsymbol{H}^*\boldsymbol{H}]$ is low-SNR-optimal in all channels. This yields
$$\frac{E_{\rm b}}{N_0}_{\rm min} = \frac{1}{\lambda_{\rm max}\big(\mathbb{E}\left[\boldsymbol{H}^*\boldsymbol{H}\right]\big)\log_2 e}. \tag{5.261}$$
Denoting by $\boldsymbol{u}$ the maximum-eigenvalue eigenvector of $\mathbb{E}[\boldsymbol{H}^*\boldsymbol{H}]$, $\mathcal{S}_0 = \frac{2}{\kappa(\|\boldsymbol{H}\boldsymbol{u}\|^2)}$ which reverts, in IID Rayleigh fading, to
$$\mathcal{S}_0 = \frac{2N_{\rm t}N_{\rm r}}{N_{\rm t}+N_{\rm r}}. \tag{5.262}$$
9. Without CSIT and $N_{\rm t} \leq N_{\rm r}$, transmission of $N_{\rm s} = N_{\rm min}$ equal-power streams ensures $\mathcal{S}_\infty = N_{\rm min}$ and the optimum $\mathcal{L}_\infty$. For $N_{\rm t} > N_{\rm r}$, $N_{\rm s} \geq N_{\rm min}$ suffices to ensure $\mathcal{S}_\infty$, but overloading with $N_{\rm s} = N_{\rm t}$ is required to optimize $\mathcal{L}_\infty$.
10. Transmit antenna correlations are beneficial at low SNR, as they enable a more focused beam and higher received power. At high SNR, transmit antenna correlations may also be beneficial if $N_{\rm t} > N_{\rm r}$ but they are detrimental if $N_{\rm t} \leq N_{\rm r}$. Receive antenna correlations reduce the capacity at every SNR.
11. If the transmission is unprecoded, all correlations are detrimental at every SNR.
12. As $N_{\rm t}$ and/or $N_{\rm r}$ grow,
$$\frac{1}{N_{\rm r}}\frac{R}{B} \overset{\text{a.s.}}{\to} \int \log_2(1+\mathsf{SNR}\,\xi)\, f(\xi)\, d\xi, \tag{5.263}$$
with $f(\cdot)$ the asymptotic PDF of the eigenvalues of $\frac{1}{N_{\rm t}}\boldsymbol{H}\boldsymbol{F}\boldsymbol{F}^*\boldsymbol{H}^*$. This nonrandom limit is more amenable to analysis while serving as an excellent approximation to the ergodic spectral efficiency with very modest numbers of antennas.
13. Incorporating pilot overhead and the impact of imperfect channel estimation, the spectral efficiency in an unprecoded IID Rayleigh-faded channel can be written as $(1-\alpha)\,C(\mathsf{SNR}_{\rm eff})$ where α is the total pilot overhead whereas
$$\mathsf{SNR}_{\rm eff} = \frac{\mathsf{SNR}\,(1-\mathsf{MMSE})}{1+\mathsf{SNR}\cdot\mathsf{MMSE}} \tag{5.264}$$
with $\mathsf{MMSE} = 1/(1+\frac{\alpha N_{\rm c}}{N_{\rm t}}\mathsf{SNR})$ the variance of the estimation error on each entry of $\boldsymbol{H}$. As the overhead must be divided among the transmit antennas, any scaling in $N_{\rm t}$ is equivalent to the inverse scaling in fading coherence.
14. With interference in lieu of additive noise, the capacity at a given SINR can only increase relative to its noise-limited value because, in contrast with noise, the interference has structure. However, the difference is only significant if the dimensionality of the interference is not large relative to $N_{\rm r}$ and its fading can be tracked.
15. Capacity can be achieved by threading a single codeword across all signal streams, but also by transmitting a separate codeword per stream and sequentially decoding with LMMSE filters at each stage and SIC. Both approaches admit BICM embodiments, either one-shot or with iterations between soft demappers and decoders.

16. With TDD or full duplexing, the CSIT can be obtained through reciprocity. With FDD, it requires feedback, either analog or digital, and sufficient time coherence for that feedback to be effective. In the digital case, a codebook of N_f precoders can be featured, and the design of such codebooks relates to the problem of packing N_f subspaces of dimension N_s on an N_t-dimensional space.

problem of communicating with a restricted number of radio-frequency chains—possibly a single one—even if the number of antennas and the baseband processing are unrestricted. A family of schemes, emerged from this formulation, are briefly reviewed in the sequel.

Spatial modulation

Spatial modulation is a technique devised for transmitters equipped with multiple antennas, but a single radio-frequency chain [698–705]. At each symbol, the signal is emitted from only one antenna, and the selection of one from among the N_t antennas conveys $\log_2 N_t$ bits. Additional information can be embedded in the magnitude and phase of the signal emitted from the chosen antenna such that, with an M-ary constellation, $\log_2 N_t + \log_2 M$ coded bits/symbol can be sent; depending on the SNR, then, $R/B \leq \log_2 N_t + \log_2 M$ is achievable. The reception can take place with either one or multiple radio chains.

To bridge the gap between pure spatial modulation (which transmits from a single antenna) and standard MIMO (which transmits from all N_t antennas), the former can be extended so as to transmit from a number of antennas ranging between 1 and N_t [706, 707].

Parasitic antennas

Parasitic arrays also rely on a single radio-frequency chain, but in this case the active antenna is fixed and all other antennas are parasitic, deprived of any power feed [708–711]. The arrangement of the array is such that mutual coupling is strong and the parasitic antennas capture and reradiate significant amounts of the power emitted by the active one; then, by controlling the coupling through tunable reactive elements, information can be embedded in this reradiated power and by default in the overall radiation pattern. Indeed, by controlling the coupling at the symbol rate, fast beam-switching is possible, which—recall—is an interpretation of MIMO and justifies the common designation of these transmitters as electrically-steerable parasitic array radiators (ESPARs). However, in contrast with a standard MIMO transceiver, in ESPAR far less power is effectively transmitted and the amount of control that can be effected through the coupling is limited.

Media-based communication

This scheme is built around the idea of controlling, not the radiation pattern, but rather the propagation channel itself. Put differently, it relies on meddling with the transmission medium, hence the term *media*-based communication [712]. Information is imprinted onto the radio signals *after* they have left the transmit antenna(s), rather than before. In principle, any technique that allows for intentional alterations of the electromagnetic properties in the vicinity of the transmitter at a sufficiently fast rate can be utilized to realize this scheme. One such possibility is to install a ring of radio-frequency mirrors around each transmit antenna, with each mirror being individually reconfigurable as transparent or reflective at the frequencies of interest [713]. The channel can then be blocked in specific directions, thereby modifying the multipath propagation characteristics of the channel, and the amount of information that can be embedded on the signals depends on the number of mirrors and the multipath richness. And, as in the case of spatial modulation, additional information can be modulated onto the magnitude and phase of the signals themselves.

Hybrid precoding

As carrier frequencies shift into the millimeter-wave range, the pathloss and atmospheric attenuation increase, and the interference becomes angularly sparse, such that the communication tends to be mostly noise-limited; thus, beamforming becomes progressively attractive. At the same time, with the shrinking wavelength, a growing number of antennas can be packed on any given area. Altogether then, the situation becomes one of many antennas but few signal streams, possibly a single stream. This is just as well because, at such high frequencies, the analog-to-digital and digital-to-analog converters consume enormous amounts of power and it is desirable to keep their number down.

Hybrid precoders and hybrid receive filters reconcile the sufficiency of a small number of streams, the desire for a small number of converters, and the need for a large number of antennas. The idea is to have an N_{s}-dimensional digital stage cascaded with an analog stage that brings the over dimensionality up to the number of antennas. Readers interested in the design of such precoders are referred to [191–193, 714, 715] and references therein.

Problems

5.1 Show that, for $N_{\mathrm{s}} = N_{\mathrm{t}}$, a per-antenna power constraint requires $\boldsymbol{F}\boldsymbol{F}^* = \boldsymbol{I}$.

5.2 Consider the channel gain within (5.9) that is in excess of the gain $G N_{\mathrm{r}}$ achieved without precoding. This additional gain equals

$$\frac{\mathbb{E}\big[\|\boldsymbol{H}\boldsymbol{x}\|^2\big]}{E_{\mathrm{s}} N_{\mathrm{r}}} = \frac{\mathbb{E}\big[\mathrm{tr}\big(\boldsymbol{H}\boldsymbol{F}\boldsymbol{F}^*\boldsymbol{H}^*\big)\big]}{N_{\mathrm{t}} N_{\mathrm{r}}}. \tag{5.265}$$

For $\boldsymbol{H}$ having IID Rayleigh-faded entries, plot (5.265) expressed in dB as a function of $N_{\mathrm{t}} = N_{\mathrm{r}}$ between 1 and 10 antennas for the following precoders.

(a) CSIT-based beamforming on the maximum-eigenvalue eigenvector of $\boldsymbol{H}^*\boldsymbol{H}$.

(b) CSIT-based beamforming on the minimum-eigenvalue eigenvector of $\boldsymbol{H}^*\boldsymbol{H}$.

(c) Equal-power allocation on all the eigenvectors of $\boldsymbol{H}^*\boldsymbol{H}$.

What do you observe?

5.3 Let $N_t = 2$ with the transmit correlation matrix

$$\boldsymbol{R}_t = \begin{bmatrix} 1 & \rho \\ \rho & 1 \end{bmatrix} \tag{5.266}$$

while $N_r = 2$ with $\boldsymbol{R}_r = \boldsymbol{I}$. Suppose the precoder beamforms on the maximum-eigenvalue eigenvector of $\boldsymbol{R}_t$.

(a) Plot (5.265) in dB as a function of $\rho \in [0, 1]$.

(b) Repeat part (a) with $\boldsymbol{R}_r = \boldsymbol{R}_t$.

5.4 Let $\boldsymbol{H}$ have IID Rice-faded entries with $N_t = N_r = 3$ and with $\boldsymbol{H}_{\text{LOS}} = \mathbf{1}$. Suppose that the precoder beamforms on the maximum-eigenvalue eigenvector of $\mathbb{E}[\boldsymbol{H}^*\boldsymbol{H}]$. Plot (5.265) in dB as a function of the Rice factor $\mathsf{K} \in [-5, 10]$ dB.

5.5 Consider the quasi-static channel

$$\boldsymbol{H} = \begin{bmatrix} 1 + 0.5\,\mathrm{j} & -0.6 \\ 0.2 + \mathrm{j} & 0.4 + 0.1\,\mathrm{j} \end{bmatrix} \tag{5.267}$$

with CSIT. Establish the SNR below which beamforming is optimal and give the precoder that effects such beamforming.

5.6 Consider again the channel in Problem 5.5 and let the antenna spacings at transmitter and receiver be $d_t = \lambda_c$ and $d_r = \lambda_c/2$, respectively. Compute the optimum CSIT-based precoder at SNR = 10 dB and, from it, the angles of the two beams launched by the transmitter and those of the two beams formed by the receiver.

5.7 As established, the mixing matrix within a CSIT-based precoder becomes relevant whenever

$$\log_2 M < \log_2\left(1 + \mathsf{SNR}\,\lambda_{\max}(\boldsymbol{H}^*\boldsymbol{H})\right). \tag{5.268}$$

Let us examine this condition for an IID Rayleigh-faded channel with $N_t = 2$ and $N_r = 3$.

(a) Plot the CDF of the right-hand side of (5.268) for SNR = 5 dB. What is the probability that (5.268) is satisfied for QPSK? How about for 16-QAM? How about for 64-QAM?

(b) Repeat part (a) for SNR = 15 dB.

5.8 Consider an IID Rayleigh-faded channel with $N_t = 4$ and $N_r = 2$ and with CSIT.

(a) Plot the ergodic capacity as a function of $\mathsf{SNR} \in [-5, 20]$ dB.

(b) On the same chart as part (a), plot the ergodic spectral efficiency without precoding.

(c) Plot the ergodic capacity as a function of $\frac{E_b}{N_0}|_{\text{dB}} \in [\frac{E_b}{N_0}_{\min}|_{\text{dB}}, \frac{E_b}{N_0}_{\min}|_{\text{dB}} + 6\text{ dB}]$.

(d) On the same chart as part (c), plot the ergodic spectral efficiency without precoding.

5.9 Consider an ergodic channel with IID Rice-faded entries and $\boldsymbol{H}_{\text{LOS}} = \mathbf{1}$, and with CSIT.

(a) For $\mathsf{K} = 0$ dB and $\mathsf{K} = 10$ dB, plot the power gain achieved by beamforming as function of $N_t = N_r = 1, \ldots, 8$. Alongside, plot the largest possible gain, $N_t N_r$.

(b) On the same chart as part (a), plot the power gain achieved by beamforming for an IID Rayleigh-faded channel as well as its large-dimensional approximation, $(\sqrt{N_t} + \sqrt{N_r})$.

5.10 Let $N_t = N_r = 2$ with $\boldsymbol{H}$ having IID Rayleigh-faded entries and with CSIT. Plot the low-SNR expansion of $\mathsf{C}(\frac{E_b}{N_0})$ based on $\frac{E_b}{N_0}_{\mathsf{min}}$ and S_0, alongside the actual $\mathsf{C}(\frac{E_b}{N_0})$. All plots in log-scale.

5.11 Suppose that an antenna array is to be used in a certain low-SNR environment, with the benefit of CSIT. What can the system designer do to ensure its best performance?

5.12 Consider a MIMO channel with CSIT and with $\lambda_{\mathsf{max}}(\boldsymbol{H}^*\boldsymbol{H})$ having multiplicity two. Prove that $\frac{E_b}{N_0}_{\mathsf{min}}$ has the same value regardless of how the transmit power is allocated between the two corresponding eigenvectors of $\boldsymbol{H}^*\boldsymbol{H}$.

5.13 At low SNR and with CSIT, antenna correlation is beneficial. Is the same true at high SNR?

5.14 Consider an ergodic MIMO channel with $N_t > N_r$ and CSIR.

(a) By means of the power offset, express the power advantage brought about by CSIT at high SNR.

(b) For $\boldsymbol{H}$ having IID Rayleigh-faded entries, further express part (a) in closed form.

(c) Quantify the power advantage in part (b) for $N_t = 4$ and $N_r = 2$.

5.15 Express the outage capacity of a Rayleigh-faded channel having $N_r = 1$ and $N_t > 1$ with those N_t transmit antennas fully correlated.

5.16 Consider a SIMO channel with $N_r = 3$ and IID Rayleigh fading.

(a) Plot the DMT.

(b) Separately plot, in log-scale, p_{out} versus R/B for $\mathsf{SNR} = 10$ dB, $\mathsf{SNR} = 20$ dB, and $\mathsf{SNR} = 30$ dB.

(c) On the same chart as part (a), plot $\frac{\log p_{\mathsf{out}}}{\log \mathsf{SNR}}$ versus $\frac{R/B}{\log_2 \mathsf{SNR}}$ for $\mathsf{SNR} = 10$ dB, $\mathsf{SNR} = 20$ dB, and $\mathsf{SNR} = 30$ dB.

Note: This problem allows gauging the efficacy of the DMT as a representation of the tradeoff between p_{out} and R/B.

5.17 Repeat Problem 5.16 for a MIMO channel with $N_t = N_r = 2$ and the following precoders.

(a) $\boldsymbol{F} = \boldsymbol{I}$.

(b) $\boldsymbol{F} = \mathrm{diag}(2, 0)$.

What do you observe?

5.18 Show that, by enforcing that the one-side derivative of $\boldsymbol{P}_\xi = (1 - \xi)\boldsymbol{P}^\star + \xi \boldsymbol{P}$ with respect to ξ be negative at $\xi = 0^+$, the condition (5.78) follows from (5.77).

5.19 Prove (5.84).

5.20 Consider a Rice-faded channel subject to transmit and receive correlations $\boldsymbol{R}_t$ and $\boldsymbol{R}_r$, with no CSIT.

(a) Express the optimum $\boldsymbol{U}_{\boldsymbol{F}}$.

(b) Indicate three cases in which $\boldsymbol{U_F} = \boldsymbol{U}_{\text{LOS}}$.

(c) Is there any situation in which $\boldsymbol{U_F} = \boldsymbol{I}$ is optimal?

5.21 Let $N_t = 3$ and $N_r = 4$, with no CSIT. Suppose that the transmitter is elevated above the propagation clutter, such that Example 3.36 applies and, specifically, $[\boldsymbol{R}_t]_{i,j} = e^{-0.05(i-j)^2}$. The receiver is located within the clutter, such that $\boldsymbol{R}_r = \boldsymbol{I}$.

(a) Obtain the steering matrix, $\boldsymbol{U_F}$.

(b) For $\mathsf{SNR} = -3$ dB, obtain the optimum power allocation $\boldsymbol{P}^\star$ after ten iterations of (5.87) and (5.88). Compute the corresponding capacity.

(c) Repeat part (b) for $\mathsf{SNR} = 5$ dB.

5.22 Let $N_t = N_r = 2$, with no CSIT. Suppose the channel is Rice-faded with no antenna correlations, $\mathsf{K} = 0$ dB, and

$$\boldsymbol{H}_{\text{LOS}} = \begin{bmatrix} 1 & -1 \\ \mathrm{j} & \mathrm{j} \end{bmatrix}. \tag{5.269}$$

(a) Obtain the steering matrix, $\boldsymbol{U_F}$.

(b) For $\mathsf{SNR} = 5$ dB, obtain the optimum power allocation $\boldsymbol{P}^\star$ after ten iterations of (5.87) and (5.88). Compute the corresponding capacity.

5.23 Reproduce Fig. 5.5.

5.24 Let $N_t = N_r = 4$ with no CSIT.

(a) Letting the fading be IID, plot the capacity as a function of $\mathsf{SNR} \in [-10, 25]$ dB.

(b) On the same chart as part (a), plot the capacity with receive antenna correlations $[\boldsymbol{R}_r]_{i,j} = e^{-0.05(i-j)^2}$.

5.25 Consider a channel with IID Rayleigh-faded entries and $N_t = N_r = N_a$.

(a) Plot the capacity without CSIT as a function of $N_a = 1, \ldots, 8$, for $\mathsf{SNR} = 0$ dB, $\mathsf{SNR} = 10$ dB, and $\mathsf{SNR} = 20$ dB.

(b) On the same chart, repeat part (a) with CSIT.

5.26 Show that, if the largest eigenvalue of $\mathbb{E}[\boldsymbol{H}^*\boldsymbol{H}]$ has plural multiplicity, the expression for $\frac{E_b}{N_0}_{\min}$ in (5.110) holds as long as the precoder concentrates its power on the corresponding eigenspace and regardless of how that power is divided among the constituent eigenvectors.

5.27 Generalize the expression for $\mathcal{S}_0$ in (5.113) for the case that the largest eigenvalue of $\mathbb{E}[\boldsymbol{H}^*\boldsymbol{H}]$ has plural multiplicity and the precoder divides its power evenly among the corresponding eigenvectors.

5.28 Verify that (5.118) follows from (5.117).

5.29 Consider the setting of Example 5.24, only with the two transmit antennas in MIMO exhibiting a 60% correlation. The receive antennas are uncorrelated. Calculate the reduction in power and bandwidth with MIMO such that the original SISO bit rate of 863 Kb/s can still be achieved.

5.30 In ergodic channels without CSIT, the key quantity at low SNR is $\lambda_{\max}(\mathbb{E}[\boldsymbol{H}^*\boldsymbol{H}])$. Show that this quantity satisfies

$$N_r \le \lambda_{\max}(\mathbb{E}[\boldsymbol{H}^*\boldsymbol{H}]) \le N_t N_r. \tag{5.270}$$

5.31 Consider the setting of Example 5.24, only with the receive antennas in MIMO subject to the correlations

$$\boldsymbol{R}_{\mathrm{r}} = \begin{bmatrix} 1 & \rho & \rho^4 & \rho^9 \\ \rho & 1 & \rho & \rho^4 \\ \rho^4 & \rho & 1 & \rho \\ \rho^9 & \rho^4 & \rho & 1 \end{bmatrix}, \tag{5.271}$$

with $\rho = 0.5$. The transmit antennas are uncorrelated. Using $\frac{E_{\mathrm{b}}}{N_0}_{\mathrm{min}}$ and $\mathcal{S}_0$, calculate the reduction in power and bandwidth with MIMO such that the original SISO bit rate of 863 Kb/s can still be achieved.

5.32 Consider a channel with CSIR but no CSIT, subject to Kronecker antenna correlations. Verify that the general expression for $\mathcal{S}_0$ in (5.113) specializes to (5.124).

5.33 Expand the high-SNR spectral efficiency of an ergodic MIMO channel where we choose to transmit $N_{\mathrm{s}} < N_{\mathrm{min}}$ signal streams.

5.34 Derive the expression in (5.144).

5.35 Consider a channel with $N_{\mathrm{t}} = N_{\mathrm{r}} = 2$ and no CSIT. The transmit antennas are uncorrelated while the receive antennas are 50% correlated. Plot the capacity for $\mathsf{SNR} \in [0, 30]$ dB, as well as the corresponding high-SNR expansion.

5.36 Let $N_{\mathrm{t}} = N_{\mathrm{r}}$ in the absence of CSIT. On a given chart, and for $\mathsf{SNR} \in [0, 20]$ dB, plot the following.

(a) The large-dimensional capacity per receive antenna of an IID channel.

(b) The ergodic capacity of an IID Rayleigh-faded channel with $N_{\mathrm{t}} = N_{\mathrm{r}} = 4$.

(c) The ergodic capacity of an IND Rayleigh-faded channel, the variance of whose entries conforms to (5.178) with $\Xi = 0.4$.

5.37 Consider a Rice-faded channel with $\boldsymbol{H}_{\mathrm{LOS}} = 1$, $\mathsf{K} = 0$ dB, and $\mathsf{SNR} = 10$ dB. Plot, as a function of $N_{\mathrm{t}} = N_{\mathrm{r}} = 4, \ldots, 10$, the ergodic spectral efficiency without precoding and its large-dimensional approximation.

5.38 Verify that, in a MIMO channel with IID entries, pilot-assisted transmission with complex Gaussian signaling, and LMMSE channel estimation at the receiver, the spectral efficiency achievable with a pilot overhead α is given by (5.195).

5.39 Repeat Problem 4.59 for a Clarke–Jakes Doppler spectrum, $N_{\mathrm{t}} = 2$, and $N_{\mathrm{r}} = 3$, with the antennas uncorrelated.

Hint: A convex optimization solver such as `fmincon` *in MATLAB® can be used. Alternatively, since each optimization is over a single scalar, the optimum value can be found by scanning over* $\alpha \in [0, 1]$.

5.40 Repeat Problem 5.39 for $N_{\mathrm{t}} = N_{\mathrm{r}} = 4$.

5.41 Consider an ergodic channel that is continuous, frequency-flat, and Rayleigh-faded with a rectangular spectrum, and where there is no antenna correlation. Let $N_{\mathrm{r}} = 1$ and $\mathsf{SNR} = 0$ dB.

(a) Beyond which value of $\nu_{\mathrm{M}} T$ does going from $N_{\mathrm{t}} = 1$ to $N_{\mathrm{t}} = 2$ become detrimental, rather than beneficial, in pilot-assisted communication without pilot power boosting?

(b) For a reasonable outdoors coherence bandwidth (or, equivalently, a reasonable delay spread), to how many symbols or OFDM resource elements of coherence in time–frequency does that correspond?

5.42 Consider an ergodic frequency-flat Rayleigh-faded channel with $N_c = 1000$, and with no antenna correlation. Let $N_r = 1$. Plot the pilot-assisted spectral efficiency for $\mathsf{SNR} \in [-10, 10]$ dB, for both $N_t = 1$ and $N_t = 2$, without pilot power boosting. What do you observe?

5.43 Consider a channel with IID Rayleigh-faded entries.

(a) Assuming CSIR, reproduce the plots of $\mathsf{C}(\frac{E_b}{N_0})$ in Fig. 5.6.

(b) Repeat the exercise with pilot-assisted transmission in a block-faded setting with $N_c = 1000$, corresponding to a vehicular channel. What do you observe?

5.44 Repeat Problem 5.43 for the function $C(\mathsf{SNR})$ and $\mathsf{SNR} \in [-10, 0]$ dB.

5.45 Consider a vehicular frequency-flat block-fading channel that conforms to the Kronecker model with

$$\boldsymbol{R}_t = \begin{bmatrix} 1 & 0.6 \\ 0.6 & 1 \end{bmatrix} \qquad \boldsymbol{R}_r = \begin{bmatrix} 1 & 0.4 \\ 0.4 & 1 \end{bmatrix}. \tag{5.272}$$

Neither pilots nor data are precoded, with pilot power boosting allowed and thus with $N_p = N_t$. Plot, for $\mathsf{SNR} \in [-5, 25]$ dB, the ergodic spectral efficiency and the corresponding pilot overhead achieved in the following cases.

(a) Independent LMMSE estimation of the channel entries.

(b) Joint LMMSE estimation of the channel entries.

Further plot, over the same SNR range, the relative gain of the latter over the former.

Hint: The joint LMMSE channel estimator for channels with Kronecker correlations is derived in Example 2.32.

5.46 Repeat Problem 5.45 with the pilots unprecoded, but with a payload data precoder optimized for $\boldsymbol{R}_t$.

5.47 Repeat Problem 5.45 with both the pilots and the payload data sent through a precoder optimized for $\boldsymbol{R}_t$.

Hint: The receiver should estimate $\boldsymbol{HF}$; furthermore, only those spatial dimensions on which data are to be sent need to be estimated [566].

5.48 Consider a vehicular frequency-flat block-fading Rayleigh channel with no precoding and $N_t = N_r = 2$.

(a) For $\mathsf{SNR} = 0$ dB, and with independent LMMSE estimation of each channel entry, compare the spectral efficiency without antenna correlation and with the correlations in (5.272).

(b) Repeat part (a) with joint LMMSE estimation of the channel entries and quantify the reduction in the spectral efficiency gap.

(c) Repeat parts (a) and (b) at $\mathsf{SNR} = 10$ dB.

5.49 Repeat Problem 5.48 with a precoder optimized for $\boldsymbol{R}_t$.

5.50 Derive the expression in (5.208) for an ergodic MIMO channel impaired by noise and by U interferers, with CSIR (including the fading of the interferers) but no CSIT.

5.51 We know that, with a separate codeword per stream and optimized rate per codeword, an LMMSE-SIC receiver achieves capacity. Suppose, however, that all the codewords must be encoded at the same rate. Then, what share of the ergodic capacity of an IID Rayleigh-faded channel could be achieved at $\mathsf{SNR} = 10$ dB?

5.52 Consider a layered architecture where each antenna transmits a separate codeword and the receiver is LMMSE-SIC, but all codewords must be encoded at the same rate. Focus on IID Rayleigh-faded channels.

(a) Derive expressions for $\frac{E_b}{N_0}_{\min}$ and S_0.

(b) Derive expressions for S_∞ and $\mathcal{L}_\infty$.

(c) Using the derived expressions, characterize how the tradeoff between power, bandwidth, and bit rate worsens relative to a capacity-achieving architecture.

5.53 From the error probabilities as a function of SNR given in the book's companion webpage for the LTE MCSs, reproduce the throughput per unit bandwidth in Example 5.44. Also reproduce the ergodic spectral efficiency in the same example.

5.54 Repeat Problem 5.53 with the receive antennas correlated according to $\boldsymbol{R}_r$ as in (5.271) and $\rho = 0.6$. Plot the throughput per unit bandwidth and the spectral efficiency on a per-antenna basis and contrast them with their counterparts in Fig. 5.16.

5.55 Repeat Problem 5.53 with $N_t = 2$ and $N_r = 3$. The transmit antennas are correlated according to $\boldsymbol{R}_t$ as in (5.272) while the receive antennas are uncorrelated. Suppose that, while the channel is quasi-static over each codeword, the precoder is held fixed at the optimum value dictated by $\boldsymbol{R}_t$.

5.56 Verify that (5.235) follows from (5.234).

5.57 Consider the average distortion with a random codebook as given in (5.243).

(a) Determine the leading term in the expansion of this expression for growing N_f.

(b) For $N_t = 4$, contrast the decay in average distortion when going from $N_f = 32$ to $N_f = 64$ as predicted by this term versus the exact value.

5.58 Consider an IID Rayleigh-faded channel with $N_t = N_r = 4$.

(a) Plot, for $\mathsf{SNR} \in [0, 20]$ dB, the ergodic capacity with CSIT.

(b) On the same chart, plot the ergodic spectral efficiency with a precoder allocating its power uniformly over the N_s eigenvectors of $\boldsymbol{H}^*\boldsymbol{H}$ on which waterfilling would allocate nonzero power.

Note: This problem allows gauging the spectral efficiency deficit of a semiunitary precoder, an interesting alternative in FDD systems requiring CSIT feedback.

5.59 Applying (5.240), quantify the worst-case dB-loss in average receive SNR when beamforming on an IID Rayleigh-faded channel and with digital feedback via the codebook in Example 5.49.

5.60 For $N_t = 3$, design a digital feedback codebook made up of three subcodebooks, one for each possible precoder rank N_s. The subcodebook corresponding to a certain value of N_s should allow for the simultaneous cophased transmission from N_s antennas. How many feedback bits are required to index the overall codebook?

5.61 Reproduce Example 5.50.

5.62 Repeat Example 5.50, but with the precoder selection at the receiver based on an LMMSE channel estimate rather than on CSIR side information. Assume that a single pilot with optimized power boosting is inserted within each coherence tile, with the underspreadness ensuring that $(1 - 1/N_{\mathrm{c}}) \approx 1$.

5.63 Repeat Example 5.50 with the same beamforming codebook, but over a MIMO channel with $N_{\mathrm{t}} = N_{\mathrm{r}} = 2$.

5.64 Repeat Example 5.50, concentrating on the low-SNR regime and with the spectral efficiencies depicted as a function of $\frac{E_{\mathrm{b}}}{N_0}$ rather than SNR.

5.65 Verify (5.249).

5.66 Design an optimum Fourier beamforming codebook for $N_{\mathrm{t}} = 2$ and $N_{\mathrm{f}} = 4$ in the following situations.

(a) An IID Rayleigh-faded channel.

(b) The LOS channel $\boldsymbol{H}_{\mathrm{LOS}} = \boldsymbol{a}_{\mathrm{r}}(\theta_{\mathrm{r}})\boldsymbol{a}_{\mathrm{t}}^{\mathrm{T}}(\theta_{\mathrm{t}})$.

(c) A channel with a uniform PAS over $\theta \in [\mu_\theta - \Delta_\theta, \mu_\theta + \Delta_\theta]$.

5.67 Express the capacity of a MISO channel with CSIT and a per-antenna power constraint.

5.68 Express $\frac{E_{\mathrm{b}}}{N_0}_{\min}$ for a MISO channel with CSIT and a per-antenna power constraint.

6 SU-MIMO with linear receivers

The most profound technologies are those that disappear. They weave themselves into the fabric of everyday life until they are indistinguishable from it.

Mark Weiser

6.1 Introduction

While, in SISO, the degree of decoding complexity at the receiver is largely dominated by the codeword length, in MIMO it becomes further associated with the number of spatially multiplexed signal streams, which itself depends on the numbers of antennas. As illustrated in Example 5.42, the mere computation of L-values at the soft demapper is exponential in the number of streams and may become prohibitively complex even for small numbers thereof if constellations of considerable cardinality are employed—and the trend is precisely to move to such rich constellations. The transmission of multiple codewords alleviates this issue, pointing to a receiver structure that separately decodes such codewords and whose complexity scales more gracefully with the number of signal streams, while still achieving capacity: the LMMSE-SIC receiver. Relaxing the need for optimality, this follow-up chapter on SU-MIMO delves into even simpler receivers within the class that separately decodes multiple codewords. Specifically, the chapter is devoted to strictly linear receivers, devoid of nonlinear SIC components. Without backing off from the asymptote of long codewords, and building upon the linear ZF and LMMSE equalizers derived in Chapter 2, we examine the information-theoretic potential of the linear receivers that emanate from those equalizers, and relate their performance with that of the optimum receivers considered heretofore.

The disposition of the chapter is as follows. Section 6.2 presents a general overview of linear receivers. This is followed, respectively in Sections 6.3 and 6.4, by detailed analyses of the linear ZF and the LMMSE receivers. These analyses entail characterizations of the linear filters themselves, of the SNR and SINR distributions at the output of those filters, and of the resulting ergodic spectral efficiencies. Finally, Section 6.5 revisits and expands the relationship, advanced in the previous chapter, between the LMMSE and the optimum receiver, while Section 6.6 concludes the chapter.

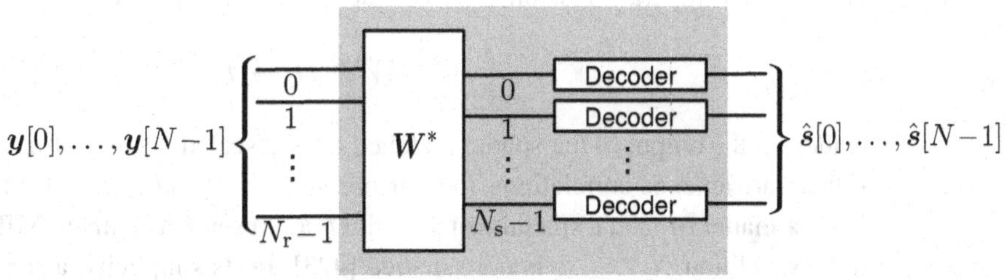

Fig. 6.1 Linear MIMO receiver.

6.2 General characteristics of linear MIMO receivers

A generic linear MIMO receiver consists of a linear filter W followed by a bank of decoders operating separately on each signal stream (see Fig. 6.1). This separate decoding, which rests on each stream bearing a distinct codeword, is responsible for both the reduction in complexity and for the suboptimality of linear reception, two sides of the same coin. Note that, in an LMMSE-SIC receiver, the decoding of the codewords is sequential but each benefits from the decoding of all other codewords in earlier stages; thanks to this, such a receiver can achieve capacity. In a pure linear receiver, in contrast, each codeword is decoded oblivious to the rest; only during the linear filtering do the respective signal streams interact. We concentrate on frequency-flat fading channels, with the understanding that frequency selectivity can be readily accommodated by having the linear filter be $\mathsf{W}[0],\ldots,\mathsf{W}[K-1]$, computed on the basis of $\mathsf{H}[0],\ldots,\mathsf{H}[K-1]$ over K subcarriers. Thus, the formulations in this chapter can be interpreted as pertaining to a specific subcarrier, with the ensuing ergodic spectral efficiencies subsuming any frequency selectivity.

Taking advantage of the wisdom expounded in preceding chapters concerning the applicability of side information and the efficacy of pilot-assisted channel estimation, CSIR is considered throughout the analysis that follows. As far as CSIT, when it is available, linear SU-MIMO receivers emerging from the SVD have been seen to achieve capacity. With CSIT, therefore, linear SU-MIMO receivers are directly the ones studied in the previous chapter, and we can refer to the treatment of their structure and performance therein. It is when CSIT is unavailable that linear SU-MIMO receivers materialize as a suboptimum alternative to their more involved capacity-achieving counterparts, and it is hence in the absence of CSIT that the analysis is conducted.

There is one limiting regime in which linear receivers do achieve capacity in the absence of CSIT: for $N_r/N_t \to \infty$. The columns of any channel matrix with independent entries then become naturally orthogonal and separate decoding of the corresponding signals suffices for optimality. In this situation, which may arise in the reverse link of massive MIMO systems, the linear receiver needs only be a simple spatial matched filter, namely

$W^{\mathsf{MF}} \propto H$ if the transmission is unprecoded such that, by virtue of (5.164),

$$\frac{1}{N_r} W^{\mathsf{MF}*} H \xrightarrow{\text{a.s.}} I \tag{6.1}$$

rendering the output of the spatial matched-filter a scaled version of s, plus filtered noise. If there are antenna correlations, the transmission can then be precoded and $W^{\mathsf{MF}} \propto HF$.

As a matter of fact, a spatial matched filter could serve as a linear MIMO receiver even if the condition $N_r \gg N_t$ is not satisfied [423]. In its simplicity, a spatial matched filter merely maximizes the received signal power, disregarding the interference among streams and, for arbitrary N_t and N_r, this maximization only pays off as $\mathsf{SNR} \to 0$, once the noise overwhelms everything else. Other than for $N_r \gg N_t$ or $\mathsf{SNR} \to 0$, the performance of a spatial matched-filter tends to be poor (refer to Problems 6.1 and 6.3), and thus in this chapter we dwell exclusively on receivers featuring the superior ZF and LMMSE filters. The analysis of matched-filter receivers is deferred to Chapter 10, in the context of massive MIMO.

The ZF and LMMSE filters are derived in Section 2.4 in the framework of channel equalization, with both ISI and spatial interference present. In the absence of ISI, either because the fading is itself frequency-flat or by virtue of OFDM, the ZF and LMMSE equalizers revert to the spatial filters (see Examples 2.18 and 2.21) that equip the receivers studied in this chapter.

As far as the precoder is concerned, when dealing with Gaussian signals it reduces as usual to $F = U_F \left[P^{1/2} \; 0\right]^{\mathrm{T}}$. While the optimum choice for U_F appears to be the same one that achieves capacity [716], not as much progress has been made on characterizing the optimum P. In the sequel, no attempt is made to optimize the precoder; rather, the formulation unfolds for a generic precoder F and the results are left as functions thereof.

6.3 Linear ZF receiver

6.3.1 Receiver structure

Consider a frequency-flat MIMO setting with CSIR. A ZF filter, also termed a *decorrelator*, projects—in the spatial domain—each of the signal streams onto a subspace orthogonal to the one spanned by all other streams. This completely rids each signal stream of interference from the rest, at the expense of some loss in SNR. The jth signal stream is observed by the receiver through the vector corresponding to the jth column of HF, which is what connects the jth entry of s with the N_r receive antennas, and thus a ZF receiver needs to project this vector onto a subspace orthogonal to the one defined by columns $0, \ldots, j-1, j+1, \ldots, N_s - 1$ of the same matrix, HF. Geometrically, a necessary condition for this projection to succeed is to have at least as many receiver dimensions as data streams, i.e., $N_s \leq N_r$, which in IID channels with trivial precoding is tantamount to $N_t \leq N_r$. This is the same conclusion reached in Section 2.4.1, on the basis of linear algebra considerations.

As intuition would have it, and Example 2.18 confirms, a ZF filter is essentially the inverse of the channel experienced by the target signal. To recover $\boldsymbol{s}$, the ZF filter must essentially invert $\boldsymbol{HF}$ with proper scaling. Invoking (see Appendix B.6) the notion of Moore–Penrose pseudoinverse $\boldsymbol{A}^\dagger$ for a matrix $\boldsymbol{A}$, whereby $\boldsymbol{A}^\dagger \boldsymbol{A} = \boldsymbol{I}$, the ZF filter $\boldsymbol{W}^{\text{ZF}}$ satisfies

$$\boldsymbol{W}^{\text{ZF}*} = \left(\sqrt{\frac{G E_\text{s}}{N_\text{t}}} \boldsymbol{HF} \right)^\dagger \tag{6.2}$$

$$= \sqrt{\frac{N_\text{t}}{G E_\text{s}}} \left(\boldsymbol{F}^* \boldsymbol{H}^* \boldsymbol{HF} \right)^{-1} \boldsymbol{F}^* \boldsymbol{H}^*. \tag{6.3}$$

The ouput of this filter is, as desired,

$$\boldsymbol{W}^{\text{ZF}*} \boldsymbol{y} = \boldsymbol{W}^{\text{ZF}*} \left(\sqrt{\frac{G E_\text{s}}{N_\text{t}}} \boldsymbol{HFs} + \boldsymbol{v} \right) \tag{6.4}$$

$$= \boldsymbol{s} + \breve{\boldsymbol{v}} \tag{6.5}$$

where the filtered noise $\breve{\boldsymbol{v}} = \boldsymbol{W}^{\text{ZF}*} \boldsymbol{v}$ has, for a given channel realization, the conditional covariance

$$\mathbb{E}\left[\breve{\boldsymbol{v}} \breve{\boldsymbol{v}}^* | \boldsymbol{H} \right] = \mathbb{E}\left[\boldsymbol{W}^{\text{ZF}*} \boldsymbol{v} \boldsymbol{v}^* \boldsymbol{W}^{\text{ZF}} | \boldsymbol{H} \right] \tag{6.6}$$

$$= N_0 \, \boldsymbol{W}^{\text{ZF}*} \boldsymbol{W}^{\text{ZF}} \tag{6.7}$$

$$= \frac{N_\text{t}}{\mathsf{SNR}} \left(\boldsymbol{F}^* \boldsymbol{H}^* \boldsymbol{HF} \right)^{-1}. \tag{6.8}$$

Example 6.1

Express $\boldsymbol{W}^{\text{ZF}}$ for an invertible channel $\boldsymbol{H}$ with $N_\text{t} = N_\text{r} = 2$ and $\boldsymbol{F} = \boldsymbol{I}$.

Solution

In this case, $\boldsymbol{W}^{\text{ZF}*} = \sqrt{\frac{N_\text{t}}{G E_\text{s}}} \boldsymbol{H}^{-1}$. Letting

$$\boldsymbol{H} = \begin{bmatrix} h_{00} & h_{01} \\ h_{10} & h_{11} \end{bmatrix} \tag{6.9}$$

and applying the 2×2 matrix inverse formula

$$\boldsymbol{H}^{-1} = \frac{1}{h_{00} h_{11} - h_{01} h_{10}} \begin{bmatrix} h_{11} & -h_{01} \\ -h_{10} & h_{00} \end{bmatrix}, \tag{6.10}$$

we obtain

$$\boldsymbol{W}^{\text{ZF}} = \sqrt{\frac{N_\text{t}}{G E_\text{s}}} \frac{h_{00} h_{11} - h_{01} h_{10}}{|h_{00} h_{11} - h_{01} h_{10}|^2} \begin{bmatrix} h_{11}^* & -h_{10}^* \\ -h_{01}^* & h_{00}^* \end{bmatrix}. \tag{6.11}$$

6.3.2 Output SNR distribution

It follows from (6.5) and (6.8) that, for a given $\boldsymbol{H}$, the output (or post-processing) SNR for stream j is

$$\mathsf{snr}_j^{\mathsf{ZF}} = \frac{\left[\mathbb{E}[\boldsymbol{s}\boldsymbol{s}^*]\right]_{j,j}}{\left[\mathbb{E}[\breve{\boldsymbol{v}}\breve{\boldsymbol{v}}^* | \boldsymbol{H}]\right]_{j,j}} \tag{6.12}$$

$$= \frac{1}{\left[\mathbb{E}[\breve{\boldsymbol{v}}\breve{\boldsymbol{v}}^* | \boldsymbol{H}]\right]_{j,j}} \tag{6.13}$$

$$= \frac{\mathsf{SNR}}{N_{\mathrm{t}}} \frac{1}{\left[(\boldsymbol{F}^*\boldsymbol{H}^*\boldsymbol{H}\boldsymbol{F})^{-1}\right]_{j,j}}, \tag{6.14}$$

which we can regard as a random variable whose distribution is induced by that of $\boldsymbol{H}$. If desired, $\mathsf{snr}_j^{\mathsf{ZF}}$ can be put as a function of the constituent elements of the precoder, namely

$$\mathsf{snr}_j^{\mathsf{ZF}} = \frac{\mathsf{SNR}}{N_{\mathrm{t}}} \frac{P_j}{\left[(\boldsymbol{U}_{\boldsymbol{F}}^*\boldsymbol{H}^*\boldsymbol{H}\boldsymbol{U}_{\boldsymbol{F}})^{-1}\right]_{j,j}}. \tag{6.15}$$

Using the cofactor expression for the (j, j)th entry of a matrix inverse $(\boldsymbol{A}^*\boldsymbol{A})^{-1}$, which is given by

$$\left[(\boldsymbol{A}^*\boldsymbol{A})^{-1}\right]_{j,j} = \frac{\det(\boldsymbol{A}_{-j}^*\boldsymbol{A}_{-j})}{\det(\boldsymbol{A}^*\boldsymbol{A})}, \tag{6.16}$$

where $\boldsymbol{A}_{-j}$ is the submatrix obtained by removing the jth column from $\boldsymbol{A}$, the expression in (6.14) can equivalently be written as

$$\mathsf{snr}_j^{\mathsf{ZF}} = \frac{\mathsf{SNR}}{N_{\mathrm{t}}} \frac{\det(\boldsymbol{F}^*\boldsymbol{H}^*\boldsymbol{H}\boldsymbol{F})}{\det(\boldsymbol{F}_{-j}^*\boldsymbol{H}^*\boldsymbol{H}\boldsymbol{F}_{-j})}. \tag{6.17}$$

Note that the removal of the jth column of $\boldsymbol{H}\boldsymbol{F}$ has been carried out by removing the jth column of $\boldsymbol{F}$. Yet alternatively, (6.14) can be expressed (refer to Problem 6.2) as [717]

$$\mathsf{snr}_j^{\mathsf{ZF}} = \frac{\mathsf{SNR}}{N_{\mathrm{t}}} \boldsymbol{f}_j^*\boldsymbol{H}^*\left(\boldsymbol{I} - \boldsymbol{H}\boldsymbol{F}_{-j}\left(\boldsymbol{F}_{-j}^*\boldsymbol{H}^*\boldsymbol{H}\boldsymbol{F}_{-j}\right)^{-1}\boldsymbol{F}_{-j}^*\boldsymbol{H}^*\right)\boldsymbol{H}\boldsymbol{f}_j, \tag{6.18}$$

where, as in the previous chapter, $\boldsymbol{f}_j = [\boldsymbol{F}]_{:,j}$.

Playing a trick with the columns of the identity matrix, we can garner that

$$\max_j \left[\boldsymbol{F}^*\boldsymbol{H}^*\boldsymbol{H}\boldsymbol{F}\right]_{j,j}^{-1} = \max_j [\boldsymbol{I}]_{:,j}^*\left[\boldsymbol{F}^*\boldsymbol{H}^*\boldsymbol{H}\boldsymbol{F}\right]^{-1}[\boldsymbol{I}]_{:,j} \tag{6.19}$$

$$\leq \max_{\boldsymbol{a}:||\boldsymbol{a}||^2=1} \boldsymbol{a}^*\left[\boldsymbol{F}^*\boldsymbol{H}^*\boldsymbol{H}\boldsymbol{F}\right]^{-1}\boldsymbol{a} \tag{6.20}$$

$$= \lambda_{\mathsf{max}}\left(\left[\boldsymbol{F}^*\boldsymbol{H}^*\boldsymbol{H}\boldsymbol{F}\right]^{-1}\right) \tag{6.21}$$

$$= \frac{1}{\lambda_{\mathsf{min}}(\boldsymbol{F}^*\boldsymbol{H}^*\boldsymbol{H}\boldsymbol{F})}, \tag{6.22}$$

where (6.20) holds because extending the maximization from the unit-norm columns of $\boldsymbol{I}$ to the entire set of unit-norm vectors cannot reduce the maximum while (6.21) results from making $\boldsymbol{a}$ the maximum-eigenvalue eigenvector of $\boldsymbol{F}^*\boldsymbol{H}^*\boldsymbol{H}\boldsymbol{F}$. From (6.14) and (6.22), we have that

$$\mathsf{snr}_j^{\mathsf{ZF}} \geq \frac{\mathsf{SNR}}{N_\mathrm{t}} \lambda_{\mathsf{min}}(\boldsymbol{F}^*\boldsymbol{H}^*\boldsymbol{H}\boldsymbol{F}), \tag{6.23}$$

which points to the importance of the smallest eigenvalue of $\boldsymbol{F}^*\boldsymbol{H}^*\boldsymbol{H}\boldsymbol{F}$. Channel realizations where this eigenvalue is close to zero are indeed ill-conditioned for ZF purposes.

Letting $\boldsymbol{H}$ now conform to MIMO channel models of interest, let us proceed to examine the distribution of $\mathsf{snr}_j^{\mathsf{ZF}}$ for $j = 0, \ldots, N_\mathrm{s} - 1$. With N_r receive antennas having to null-out $N_\mathrm{s} - 1$ competing signal streams, the distribution of $\mathsf{snr}_j^{\mathsf{ZF}}$ amounts to the SNR distribution in a SIMO channel with $N_\mathrm{r} - N_\mathrm{s} + 1$ effective antennas at the receiver and we can therefore leverage SIMO results—where linear reception suffices to achieve capacity—obtained in Section 5.4.2.

Example 6.2 (ZF output SNR distribution in an IID Rayleigh-faded MIMO channel)

Let $\boldsymbol{H}$ have IID Rayleigh-faded entries, in which case the precoder can be set to $\boldsymbol{F} = \boldsymbol{I}$ with $N_\mathrm{s} = N_\mathrm{t}$.

As shown in Example 5.16, the SNR distribution in a SIMO channel with IID Rayleigh-faded entries is chi-square with a number of degrees of freedom equal to twice the number of receive antennas. The SIMO interpretation of how each signal stream is processed under ZF, with $N_\mathrm{r} - N_\mathrm{s} + 1$ effective antennas at the receiver, gives

$$\mathsf{snr}_j^{\mathsf{ZF}} \sim \chi^2_{2(N_\mathrm{r}-N_\mathrm{t}+1)} \qquad j = 0, \ldots, N_\mathrm{t} - 1. \tag{6.24}$$

Furthermore, since this SIMO channel is part of a MIMO channel with N_t transmit antennas, the role of the local-average SNR is played by $\frac{\mathsf{SNR}}{N_\mathrm{t}}$. Altogether, from the chi-square PDF in (5.62), we have that, for $\xi \geq 0$,

$$f_{\mathsf{snr}_j^{\mathsf{ZF}}}(\xi) = \frac{N_\mathrm{t}}{\mathsf{SNR}\,(N_\mathrm{r} - N_\mathrm{t})!} \exp\left(-\frac{N_\mathrm{t}}{\mathsf{SNR}}\xi\right) \left(\frac{N_\mathrm{t}}{\mathsf{SNR}}\xi\right)^{N_\mathrm{r}-N_\mathrm{t}} \tag{6.25}$$

which, for $N_\mathrm{t} = N_\mathrm{r}$, reduces to an exponential distribution.

From $f_{\mathsf{snr}_j^{\mathsf{ZF}}}(\cdot)$, the average SNR for stream j at the output of the ZF filter is

$$\mathbb{E}\left[\mathsf{snr}_j^{\mathsf{ZF}}\right] = \int_0^\infty \xi\, f_{\mathsf{snr}_j^{\mathsf{ZF}}}(\xi)\, \mathrm{d}\xi \tag{6.26}$$

$$= \frac{N_\mathrm{r} - N_\mathrm{t} + 1}{N_\mathrm{t}}\, \mathsf{SNR}. \tag{6.27}$$

The generalization of $f_{\mathsf{snr}_j^{\mathsf{ZF}}}(\cdot)$ to Rayleigh-faded channels with transmit antenna correlation is addressed in [717–719] and distilled in the next two examples.

Example 6.3 (ZF output SNR distribution in a correlated Rayleigh-faded MIMO channel with optimum precoding)

Let $\boldsymbol{H}$ be Rayleigh-faded with transmit correlation $\boldsymbol{R}_{\mathrm{t}}$ and no receive antenna correlations. Further, let the precoder transmit $N_{\mathrm{s}} = N_{\mathrm{t}}$ signal streams aligned with the eigenvectors of $\boldsymbol{R}_{\mathrm{t}}$. The generalization of (6.25) is, for $\xi \geq 0$,

$$f_{\mathsf{snr}_j^{\mathsf{ZF}}}(\xi) = \frac{N_{\mathrm{t}}}{\mathsf{SNR}\, P_j \lambda_j(\boldsymbol{R}_{\mathrm{t}})(N_{\mathrm{r}} - N_{\mathrm{t}})!} \exp\left(-\frac{N_{\mathrm{t}}}{\mathsf{SNR}\, P_j \lambda_j(\boldsymbol{R}_{\mathrm{t}})} \xi\right) \left(\frac{N_{\mathrm{t}}}{\mathsf{SNR}\, P_j \lambda_j(\boldsymbol{R}_{\mathrm{t}})} \xi\right)^{N_{\mathrm{r}} - N_{\mathrm{t}}} \tag{6.28}$$

where, as usual, $\lambda_j(\cdot)$ denotes the jth eigenvalue of a matrix. Correspondingly,

$$\mathbb{E}\big[\mathsf{snr}_j^{\mathsf{ZF}}\big] = \frac{N_{\mathrm{r}} - N_{\mathrm{t}} + 1}{N_{\mathrm{t}}} P_j \, \lambda_j(\boldsymbol{R}_{\mathrm{t}}) \, \mathsf{SNR}. \tag{6.29}$$

Example 6.4 (ZF output SNR distribution in a correlated Rayleigh-faded MIMO channel without precoding)

With $\boldsymbol{H}$ being Rayleigh-faded with transmit correlation $\boldsymbol{R}_{\mathrm{t}}$ and no correlations at the receiver, now let $\boldsymbol{F} = \boldsymbol{I}$. In this case,

$$f_{\mathsf{snr}_j^{\mathsf{ZF}}}(\xi) = \frac{N_{\mathrm{t}} \, [\boldsymbol{R}_{\mathrm{t}}^{-1}]_{j,j}}{\mathsf{SNR}\,(N_{\mathrm{r}} - N_{\mathrm{t}})!} \exp\left(-\frac{N_{\mathrm{t}} \, [\boldsymbol{R}_{\mathrm{t}}^{-1}]_{j,j}}{\mathsf{SNR}} \xi\right) \left(\frac{N_{\mathrm{t}} \, [\boldsymbol{R}_{\mathrm{t}}^{-1}]_{j,j}}{\mathsf{SNR}} \xi\right)^{N_{\mathrm{r}} - N_{\mathrm{t}}}, \tag{6.30}$$

whose average

$$\mathbb{E}\big[\mathsf{snr}_j^{\mathsf{ZF}}\big] = \frac{N_{\mathrm{r}} - N_{\mathrm{t}} + 1}{N_{\mathrm{t}} \, [\boldsymbol{R}_{\mathrm{t}}^{-1}]_{j,j}} \mathsf{SNR} \tag{6.31}$$

equals that of a channel with IID entries, divided by the (j, j)th entry of $\boldsymbol{R}_{\mathrm{t}}^{-1}$.

For more general channel models, say simultaneous correlations at both transmitter and receiver, the analytical characterization of the distribution of $\mathsf{snr}_j^{\mathsf{ZF}}$ complicates very considerably. For the case of Rice-faded channels, the interested reader is referred to [720, 721].

6.3.3 Ergodic spectral efficiency

Effectively, the jth signal stream experiences a scalar fading channel known by the receiver and with SNR given by $\mathsf{snr}_j^{\mathsf{ZF}}$. The spectral efficiency is therefore maximized when the corresponding codeword is drawn from a complex Gaussian distribution, and the overall ergodic spectral efficiency is then

$$C^{\mathsf{ZF}}(\mathsf{SNR}) = \sum_{j=0}^{N_{\mathrm{s}}-1} \mathbb{E}\Big[\log_2\big(1 + \mathsf{snr}_j^{\mathsf{ZF}}\big)\Big] \tag{6.32}$$

$$= \sum_{j=0}^{N_{\mathrm{s}}-1} \mathbb{E}\left[\log_2\left(1 + \frac{\mathsf{SNR}}{N_{\mathrm{t}}} \frac{1}{\big[(\boldsymbol{F}^* \boldsymbol{H}^* \boldsymbol{H} \boldsymbol{F})^{-1}\big]_{j,j}}\right)\right]. \tag{6.33}$$

Example 6.5 (ZF spectral efficiency in an IID Rayleigh-faded MIMO channel)

Let $\boldsymbol{H}$ have IID Rayleigh-faded entries, in which case the precoder can be set to $\boldsymbol{F} = \boldsymbol{I}$ with $N_s = N_t$. In this case, the N_t streams exhibit the same SNR distribution and thus

$$C^{\text{ZF}}(\text{SNR}) = N_t \, \mathbb{E}\left[\log_2\left(1 + \text{snr}_j^{\text{ZF}}\right)\right] \tag{6.34}$$

$$= N_t \int_0^{\infty} \log_2(1+\xi) \frac{N_t}{\text{SNR}\,(N_r - N_t)!} \exp\!\left(-\frac{N_t}{\text{SNR}}\xi\right) \left(\frac{N_t}{\text{SNR}}\xi\right)^{N_r - N_t} d\xi, \tag{6.35}$$

where j is the index of an arbitrary antenna and $f_{\text{snr}_j^{\text{ZF}}}(\cdot)$ is as derived in Example 6.2. Applying (C.37),

$$C^{\text{ZF}}(\text{SNR}) = N_t \, e^{N_t/\text{SNR}} \sum_{q=1}^{N_r - N_t + 1} \mathcal{E}_q\!\left(\frac{N_t}{\text{SNR}}\right) \log_2 e, \tag{6.36}$$

which amounts to N_t times the ergodic capacity of a SIMO channel (see Example 5.16) with $N_r - N_t + 1$ receive antennas and local-average SNR equal to $\frac{\text{SNR}}{N_t}$. For $N_t = N_r$, the ZF spectral efficiency reduces to

$$C^{\text{ZF}}(\text{SNR}) = N_t \, e^{N_t/\text{SNR}} \, \mathcal{E}_1\!\left(\frac{N_t}{\text{SNR}}\right) \log_2 e, \tag{6.37}$$

which is N_t times the ergodic capacity of a SISO channel (see Example 4.22) with local-average SNR equal to $\frac{\text{SNR}}{N_t}$.

Example 6.6

For a channel with IID Rayleigh-faded entries and $N_t = N_r = 4$, plot, as a function of SNR (in dB), the ZF ergodic spectral efficiency in Example 6.5 alongside the ergodic capacity of an optimum receiver given in Example 5.17.

Solution

See Fig. 6.2. (Also included in the figure is the ergodic spectral efficiency of an LMMSE receiver, derived in the next section.)

The characterization of $C^{\text{ZF}}(\text{SNR})$ for channels with transmit antenna correlations is the subject of Problems 6.6 and 6.7. For even more general channel models, the characterization of the spectral efficiency may not be within reach analytically, but Monte-Carlo evaluations are always possible for any of the models described in Chapter 3.

The $\frac{E_b}{N_0}$ perspective on the fundamental performance of linear ZF receivers is obtained the usual way, namely through the relationship

$$\frac{E_b}{N_0} = \frac{\text{SNR}}{C^{\text{ZF}}(\text{SNR})}, \tag{6.38}$$

which yields $C^{\text{ZF}}\!\left(\frac{E_b}{N_0}\right)$ in an implicit form.

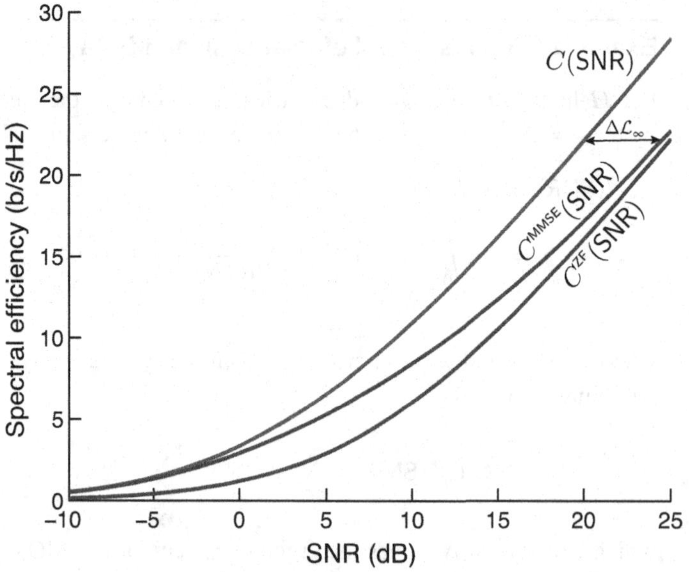

Fig. 6.2 Ergodic capacity, ZF spectral efficiency, and LMMSE spectral efficiency as a function of SNR (in dB) for $N_{\mathrm{t}} = N_{\mathrm{r}} = 4$. The difference between the capacity and the spectral efficiency of the linear receivers approaches $\Delta\mathcal{L}_{\infty} = 4.7$ dB already within the range of the plot.

Example 6.7

For a channel with IID Rayleigh-faded entries and $N_{\mathrm{t}} = N_{\mathrm{r}} = 4$, plot, as a function of $\frac{E_{\mathrm{b}}}{N_0}$ (in dB), the ergodic spectral efficiency of a linear ZF receiver alongside the ergodic capacity with an optimum receiver.

Solution

See Fig. 6.3. (Also included in the figure is the ergodic spectral efficiency of an LMMSE receiver, derived in the next section.)

In terms of the performance with signal distributions other than complex Gaussian, if stream j conforms to a discrete constellation with mutual information function $\mathcal{I}_j(\cdot)$, then the right-hand side of (6.33) morphs into $\sum_{j=0}^{N_{\mathrm{s}}-1} \mathbb{E}\left[\mathcal{I}_j(\mathsf{snr}_j^{\mathsf{ZF}})\right]$.

Low- and high-SNR regimes

The low- and high-SNR behaviors of ZF receivers can be examined using the tools derived in previous chapters.

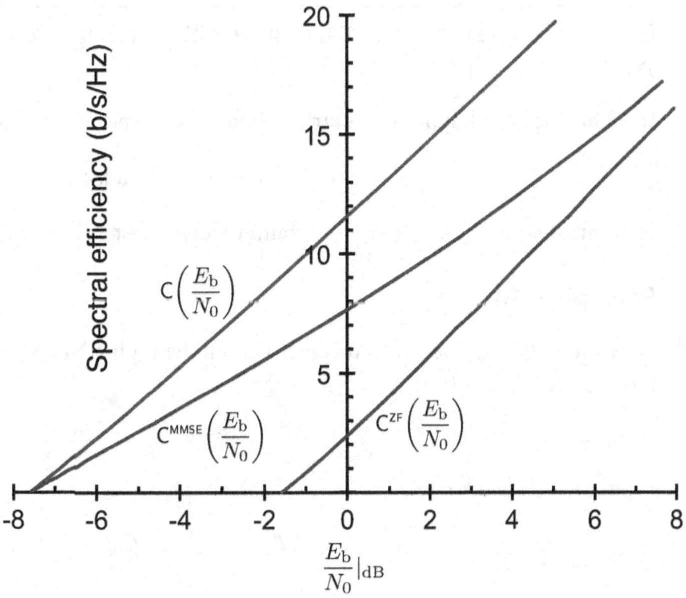

Fig. 6.3 Ergodic capacity, ZF spectral efficiency, and LMMSE spectral efficiency as a function of $\frac{E_b}{N_0}$ (in dB) for $N_t = N_r = 4$.

Example 6.8 ($\frac{E_b}{N_0}{}^{\text{ZF}}_{\text{min}}$ in an IID Rayleigh-faded MIMO channel)

For a channel with IID Rayleigh-faded entries

$$\frac{E_b}{N_0}{}^{\text{ZF}}_{\text{min}} = \frac{1}{N_r - N_t + 1} \cdot \frac{1}{\log_2 e}, \tag{6.39}$$

in contrast with $\frac{E_b}{N_0}{}_{\text{min}} = \frac{1}{N_r \log_2 e}$ for an optimum MIMO receiver. The ZF strategy of stubbornly suppressing interference while ignoring the noise is decidedly flawed in the regime where it is the noise that dominates the performance, and the first-order scaling of the spectral efficiency with SNR is heavily penalized. For $N_r = N_r$, in fact, $\frac{E_b}{N_0}{}^{\text{ZF}}_{\text{min}}$ reverts back to -1.59 dB as in SISO, regardless of how many antennas are present, and thus—to first order—all MIMO gains are forsaken.

From the previous example it is evident that, as one might have anticipated, ZF receivers are ill-advised in the low-SNR regime and we thus move, without further ado, to the high-SNR regime. There, interference is the dominant impairment and the ZF strategy does appear to be fitting. The maximum number of signal streams by definition satisfies $N_s \leq N_t$, and for ZF also $N_s \leq N_r$ so that the inversion is feasible, altogether meaning that the number of spatial DOF is $S_\infty^{\text{ZF}} = N_{\text{min}} = \min(N_t, N_r)$; there is no loss in DOF with respect to the optimum receiver. However, the power offset does reflect a shortfall with respect to capacity, which vanishes only as $N_r/N_t \to \infty$ and linear receivers become optimum.

Example 6.9 (ZF power offset in an IID Rayleigh-faded MIMO channel with $N_t = N_r = N_a$)

If $\boldsymbol{H}$ has IID Rayleigh-faded entries, then $\boldsymbol{F} = \boldsymbol{I}$ and, from (4.46) and (6.36),

$$\mathcal{L}_\infty^{\mathsf{ZF}} = \log_2 N_a + \gamma_{\mathsf{EM}} \log_2 e \tag{6.40}$$

in contrast with (5.142) for an optimum receiver. For $N_a > 1$, ZF incurs a power loss.

Example 6.10

How much power does a ZF receiver lose in the high-SNR regime, relative to an optimum receiver, if $N_a = 4$?

Solution

From (5.142) and (6.40),

$$\Delta\mathcal{L}_\infty = \mathcal{L}_\infty^{\mathsf{ZF}} - \mathcal{L}_\infty \tag{6.41}$$

$$= \sum_{q=2}^{N_a} \frac{1}{q} \log_2 e, \tag{6.42}$$

which, evaluated for $N_a = 4$, returns 1.56 3-dB units. Therefore, the power loss for $N_a = 4$ amounts to 4.7 dB, as visualized in Fig. 6.2.

To finalize the discussion of ZF receivers, let us briefly comment on their performance in quasi-static settings without CSIT. These receivers can operate at any multiplexing gain $r \leq N_t$. The diversity available to each codeword, however, is capped at $N_r - N_t + 1$ and hence the corresponding DMT falls short of the optimum one in (5.65). Precisely [722],

$$d^{\mathsf{ZF}} = (N_r - N_t + 1)\left[1 - \frac{r}{N_t}\right]^+. \tag{6.43}$$

6.4 LMMSE receiver

6.4.1 Receiver structure

In contrast with the ZF filter, which completely eliminates the interference among streams disregarding its strength relative to the noise, the LMMSE filter strikes a balance between interference reduction and noise enhancement in the sense of minimizing the mean-square error between the filter output and the actual signal. This in principle allows relaxing the constraint that $N_s \leq N_r$, but, as seen later in the chapter, pushing the number of streams beyond N_{min} is ill-advised even disregarding the additional pilot overheads.

Derived in Section 1.7.1, the filter $\boldsymbol{W}$ that simultaneously minimizes every diagonal entry of $\mathbb{E}\left[(\boldsymbol{s}-\boldsymbol{W}^*\boldsymbol{y})(\boldsymbol{s}-\boldsymbol{W}^*\boldsymbol{y})^*\right]$ for a given $\boldsymbol{H}$ is

$$\boldsymbol{W}^{\mathsf{MMSE}} = \boldsymbol{R}_{\boldsymbol{y}}^{-1}\boldsymbol{R}_{\boldsymbol{ys}} \tag{6.44}$$

$$= \sqrt{\frac{N_{\mathrm{t}}}{GE_{\mathrm{s}}}}\left(\boldsymbol{HFF}^*\boldsymbol{H}^* + \frac{N_{\mathrm{t}}}{\mathsf{SNR}}\boldsymbol{I}\right)^{-1}\boldsymbol{HF} \tag{6.45}$$

and, as intuition would have it, $\boldsymbol{W}^{\mathsf{MMSE}}$ acquires ZF behavior for $\mathsf{SNR}\to\infty$ and matched-filter behavior for $\mathsf{SNR}\to 0$.

Replicating the steps in Section 1.7.1, by means of the matrix inversion lemma (given in Appendix B.7) we can rewrite (6.45) in the alternative form

$$\boldsymbol{W}^{\mathsf{MMSE}} = \sqrt{\frac{N_{\mathrm{t}}}{GE_{\mathrm{s}}}}\left[\frac{\mathsf{SNR}}{N_{\mathrm{t}}}\boldsymbol{I} - \left(\frac{\mathsf{SNR}}{N_{\mathrm{t}}}\right)^2\boldsymbol{HF}\left(\boldsymbol{I}+\frac{\mathsf{SNR}}{N_{\mathrm{t}}}\boldsymbol{F}^*\boldsymbol{H}^*\boldsymbol{HF}\right)^{-1}\boldsymbol{F}^*\boldsymbol{H}^*\right]\boldsymbol{HF}$$

$$= \sqrt{\frac{N_{\mathrm{t}}}{GE_{\mathrm{s}}}}\frac{\mathsf{SNR}}{N_{\mathrm{t}}}\left[\boldsymbol{HF} - \boldsymbol{HF}\left(\frac{N_{\mathrm{t}}}{\mathsf{SNR}}\boldsymbol{I}+\boldsymbol{F}^*\boldsymbol{H}^*\boldsymbol{HF}\right)^{-1}\boldsymbol{F}^*\boldsymbol{H}^*\boldsymbol{HF}\right] \tag{6.46}$$

$$= \sqrt{\frac{N_{\mathrm{t}}}{GE_{\mathrm{s}}}}\frac{\mathsf{SNR}}{N_{\mathrm{t}}}\boldsymbol{HF}\left[\boldsymbol{I} - \left(\frac{N_{\mathrm{t}}}{\mathsf{SNR}}\boldsymbol{I}+\boldsymbol{F}^*\boldsymbol{H}^*\boldsymbol{HF}\right)^{-1}\boldsymbol{F}^*\boldsymbol{H}^*\boldsymbol{HF}\right] \tag{6.47}$$

$$= \sqrt{\frac{N_{\mathrm{t}}}{GE_{\mathrm{s}}}}\frac{\mathsf{SNR}}{N_{\mathrm{t}}}\boldsymbol{HF}\left(\frac{N_{\mathrm{t}}}{\mathsf{SNR}}\boldsymbol{I}+\boldsymbol{F}^*\boldsymbol{H}^*\boldsymbol{HF}\right)^{-1} \cdot\left[\left(\frac{N_{\mathrm{t}}}{\mathsf{SNR}}\boldsymbol{I}+\boldsymbol{F}^*\boldsymbol{H}^*\boldsymbol{HF}\right)-\boldsymbol{F}^*\boldsymbol{H}^*\boldsymbol{HF}\right] \tag{6.48}$$

$$= \sqrt{\frac{N_{\mathrm{t}}}{GE_{\mathrm{s}}}}\boldsymbol{HF}\left(\frac{N_{\mathrm{t}}}{\mathsf{SNR}}\boldsymbol{I}+\boldsymbol{F}^*\boldsymbol{H}^*\boldsymbol{HF}\right)^{-1}. \tag{6.49}$$

Since $\boldsymbol{y} = \sqrt{GE_{\mathrm{s}}/N_{\mathrm{t}}}\,\boldsymbol{HFs}+\boldsymbol{v}$, the MMSE matrix for a given $\boldsymbol{H}$, i.e., the conditional covariance of the estimation error, is

$$\boldsymbol{E}(\boldsymbol{H}) = \mathbb{E}\left[(\boldsymbol{s}-\boldsymbol{W}^{\mathsf{MMSE}*}\boldsymbol{y})(\boldsymbol{s}-\boldsymbol{W}^{\mathsf{MMSE}*}\boldsymbol{y})^*|\boldsymbol{H}\right] \tag{6.50}$$

$$= \boldsymbol{I} - \sqrt{\frac{GE_{\mathrm{s}}}{N_{\mathrm{t}}}}\boldsymbol{W}^{\mathsf{MMSE}*}\boldsymbol{HF} - \sqrt{\frac{GE_{\mathrm{s}}}{N_{\mathrm{t}}}}\boldsymbol{F}^*\boldsymbol{H}^*\boldsymbol{W}^{\mathsf{MMSE}} + \boldsymbol{W}^{\mathsf{MMSE}*}\left(\frac{GE_{\mathrm{s}}}{N_{\mathrm{t}}}\boldsymbol{HFF}^*\boldsymbol{H}^*+N_0\boldsymbol{I}\right)\boldsymbol{W}^{\mathsf{MMSE}}, \tag{6.51}$$

which, invoking (6.45) and the fact that

$$\boldsymbol{W}^{\mathsf{MMSE}*}\left(\frac{GE_{\mathrm{s}}}{N_{\mathrm{t}}}\boldsymbol{HFF}^*\boldsymbol{H}^*+N_0\boldsymbol{I}\right)\boldsymbol{W}^{\mathsf{MMSE}} = \boldsymbol{F}^*\boldsymbol{H}^*\left(\boldsymbol{HFF}^*\boldsymbol{H}^*+\frac{N_{\mathrm{t}}}{\mathsf{SNR}}\boldsymbol{I}\right)^{-1}\boldsymbol{HF}$$

simplifies, after a bit of algebra, into

$$\boldsymbol{E}(\boldsymbol{H}) = \boldsymbol{I} - \boldsymbol{F}^*\boldsymbol{H}^*\left(\boldsymbol{HFF}^*\boldsymbol{H}^*+\frac{N_{\mathrm{t}}}{\mathsf{SNR}}\boldsymbol{I}\right)^{-1}\boldsymbol{HF}. \tag{6.52}$$

Applying the matrix inversion lemma in reverse, the above can alternatively be put as

$$\boldsymbol{E}(\boldsymbol{H}) = \left(\boldsymbol{I}+\frac{\mathsf{SNR}}{N_{\mathrm{t}}}\boldsymbol{F}^*\boldsymbol{H}^*\boldsymbol{HF}\right)^{-1}. \tag{6.53}$$

It follows from (6.52) and (6.53) that, given a channel realization $\boldsymbol{H}$, the MMSE for the jth stream is

$$\begin{aligned}\mathsf{MMSE}_j &= \mathbb{E}\left[\left|[\boldsymbol{s}]_j - [\boldsymbol{W}^{\mathsf{MMSE}}]^*_{:,j}\,\boldsymbol{y}\right|^2\right] && (6.54)\\ &= \left[\boldsymbol{E}(\boldsymbol{H})\right]_{j,j} && (6.55)\\ &= 1 - \boldsymbol{f}_j^*\boldsymbol{H}^*\left(\boldsymbol{H}\boldsymbol{F}\boldsymbol{F}^*\boldsymbol{H}^* + \frac{N_{\mathrm{t}}}{\mathsf{SNR}}\boldsymbol{I}\right)^{-1}\boldsymbol{H}\boldsymbol{f}_j && (6.56)\\ &= \left[\left(\boldsymbol{I} + \frac{\mathsf{SNR}}{N_{\mathrm{t}}}\boldsymbol{F}^*\boldsymbol{H}^*\boldsymbol{H}\boldsymbol{F}\right)^{-1}\right]_{j,j} && (6.57)\end{aligned}$$

where we once more used $\boldsymbol{f}_j = [\boldsymbol{F}]_{:,j}$.

6.4.2 Output SINR distribution

Let us now evaluate the SINR at the output of the LMMSE filter. For a generic receive filter $\boldsymbol{W}$, and with the shorthand notation $\boldsymbol{w}_j = [\boldsymbol{W}]_{:,j}$, the output corresponding to the jth signal stream is

$$\begin{aligned}[\boldsymbol{W}^*\boldsymbol{y}]_j &= \boldsymbol{w}_j^*\,\boldsymbol{y} && (6.58)\\ &= \sqrt{\frac{GE_{\mathrm{s}}}{N_{\mathrm{t}}}}\,\boldsymbol{w}_j^*\boldsymbol{H}\boldsymbol{F}\boldsymbol{s} + \boldsymbol{w}_j^*\boldsymbol{v} && (6.59)\\ &= \underbrace{\sqrt{\frac{GE_{\mathrm{s}}}{N_{\mathrm{t}}}}\,\boldsymbol{w}_j^*\boldsymbol{H}\boldsymbol{f}_j\,[\boldsymbol{s}]_j}_{\text{Desired signal}} + \underbrace{\sqrt{\frac{GE_{\mathrm{s}}}{N_{\mathrm{t}}}}\sum_{\ell\neq j}\boldsymbol{w}_j^*\boldsymbol{H}\boldsymbol{f}_\ell\,[\boldsymbol{s}]_\ell + \boldsymbol{w}_j^*\boldsymbol{v}}_{\text{Interference + filtered noise}} && (6.60)\end{aligned}$$

with SINR given by

$$\mathsf{sinr}_j = \frac{\frac{GE_{\mathrm{s}}}{N_{\mathrm{t}}}\left|\boldsymbol{w}_j^*\boldsymbol{H}\boldsymbol{f}_j\right|^2}{\frac{GE_{\mathrm{s}}}{N_{\mathrm{t}}}\sum_{\ell\neq j}\left|\boldsymbol{w}_j^*\boldsymbol{H}\boldsymbol{f}_\ell\right|^2 + N_0\left\|\boldsymbol{w}_j\right\|^2}. \qquad (6.61)$$

While a ZF receiver nullifies the interference and leaves only filtered noise, with other receivers both are present at the output. For an LMMSE receiver, in particular, recalling the expression in (6.45) we have that

$$\begin{aligned}\boldsymbol{w}_j^{\mathsf{MMSE}} &= [\boldsymbol{W}^{\mathsf{MMSE}}]_{:,j} && (6.62)\\ &= \sqrt{\frac{N_{\mathrm{t}}}{GE_{\mathrm{s}}}}\left(\boldsymbol{H}\boldsymbol{F}\boldsymbol{F}^*\boldsymbol{H}^* + \frac{N_{\mathrm{t}}}{\mathsf{SNR}}\boldsymbol{I}\right)^{-1}\boldsymbol{H}\boldsymbol{f}_j && (6.63)\end{aligned}$$

which, plugged into (6.61), yields for $\mathsf{sinr}_j^{\mathsf{MMSE}}$ a ratio with numerator

$$\begin{aligned}\mathrm{Num} &= \boldsymbol{f}_j^*\boldsymbol{H}^*\left(\boldsymbol{H}\boldsymbol{F}\boldsymbol{F}^*\boldsymbol{H}^* + \frac{N_{\mathrm{t}}}{\mathsf{SNR}}\boldsymbol{I}\right)^{-1}\boldsymbol{H}\boldsymbol{f}_j\boldsymbol{f}_j^*\boldsymbol{H}^*\left(\boldsymbol{H}\boldsymbol{F}\boldsymbol{F}^*\boldsymbol{H}^* + \frac{N_{\mathrm{t}}}{\mathsf{SNR}}\boldsymbol{I}\right)^{-1}\boldsymbol{H}\boldsymbol{f}_j\\ &= \left[\boldsymbol{f}_j^*\boldsymbol{H}^*\left(\boldsymbol{H}\boldsymbol{F}\boldsymbol{F}^*\boldsymbol{H}^* + \frac{N_{\mathrm{t}}}{\mathsf{SNR}}\boldsymbol{I}\right)^{-1}\boldsymbol{H}\boldsymbol{f}_j\right]^2 && (6.64)\end{aligned}$$

and denominator

$$\text{Den} = \sum_{\ell \neq j} \boldsymbol{f}_j^* \boldsymbol{H}^* \left(\boldsymbol{H}\boldsymbol{F}\boldsymbol{F}^*\boldsymbol{H}^* + \frac{N_t}{\mathsf{SNR}} \boldsymbol{I} \right)^{-1} \boldsymbol{H}\boldsymbol{f}_\ell \boldsymbol{f}_\ell^* \boldsymbol{H}^* \left(\boldsymbol{H}\boldsymbol{F}\boldsymbol{F}^*\boldsymbol{H}^* + \frac{N_t}{\mathsf{SNR}} \boldsymbol{I} \right)^{-1} \boldsymbol{H}\boldsymbol{f}_j$$

$$+ \frac{N_t}{\mathsf{SNR}} \boldsymbol{f}_j^* \boldsymbol{H}^* \left(\boldsymbol{H}\boldsymbol{F}\boldsymbol{F}^*\boldsymbol{H}^* + \frac{N_t}{\mathsf{SNR}} \boldsymbol{I} \right)^{-1} \left(\boldsymbol{H}\boldsymbol{F}\boldsymbol{F}^*\boldsymbol{H}^* + \frac{N_t}{\mathsf{SNR}} \boldsymbol{I} \right)^{-1} \boldsymbol{H}\boldsymbol{f}_j$$

$$= \boldsymbol{f}_j^* \boldsymbol{H}^* \left(\boldsymbol{H}\boldsymbol{F}\boldsymbol{F}^*\boldsymbol{H}^* + \frac{N_t}{\mathsf{SNR}} \boldsymbol{I} \right)^{-1} \left[\sum_{\ell \neq j} \boldsymbol{H}\boldsymbol{f}_\ell \boldsymbol{f}_\ell^* \boldsymbol{H}^* \right]$$

$$\cdot \left(\boldsymbol{H}\boldsymbol{F}\boldsymbol{F}^*\boldsymbol{H}^* + \frac{N_t}{\mathsf{SNR}} \boldsymbol{I} \right)^{-1} \boldsymbol{H}\boldsymbol{f}_j \tag{6.65}$$

$$+ \frac{N_t}{\mathsf{SNR}} \boldsymbol{f}_j^* \boldsymbol{H}^* \left(\boldsymbol{H}\boldsymbol{F}\boldsymbol{F}^*\boldsymbol{H}^* + \frac{N_t}{\mathsf{SNR}} \boldsymbol{I} \right)^{-1} \left(\boldsymbol{H}\boldsymbol{F}\boldsymbol{F}^*\boldsymbol{H}^* + \frac{N_t}{\mathsf{SNR}} \boldsymbol{I} \right)^{-1} \boldsymbol{H}\boldsymbol{f}_j$$

$$= \boldsymbol{f}_j^* \boldsymbol{H}^* \left(\boldsymbol{H}\boldsymbol{F}\boldsymbol{F}^*\boldsymbol{H}^* + \frac{N_t}{\mathsf{SNR}} \boldsymbol{I} \right)^{-1} \left[\boldsymbol{H}\boldsymbol{F}\boldsymbol{F}^*\boldsymbol{H}^* - \boldsymbol{H}\boldsymbol{f}_j \boldsymbol{f}_j^* \boldsymbol{H}^* \right]$$

$$\cdot \left(\boldsymbol{H}\boldsymbol{F}\boldsymbol{F}^*\boldsymbol{H}^* + \frac{N_t}{\mathsf{SNR}} \boldsymbol{I} \right)^{-1} \boldsymbol{H}\boldsymbol{f}_j \tag{6.66}$$

$$+ \boldsymbol{f}_j^* \boldsymbol{H}^* \left(\boldsymbol{H}\boldsymbol{F}\boldsymbol{F}^*\boldsymbol{H}^* + \frac{N_t}{\mathsf{SNR}} \boldsymbol{I} \right)^{-1} \cdot \frac{N_t}{\mathsf{SNR}} \boldsymbol{I} \cdot \left(\boldsymbol{H}\boldsymbol{F}\boldsymbol{F}^*\boldsymbol{H}^* + \frac{N_t}{\mathsf{SNR}} \boldsymbol{I} \right)^{-1} \boldsymbol{H}\boldsymbol{f}_j$$

$$= \boldsymbol{f}_j^* \boldsymbol{H}^* \left(\boldsymbol{H}\boldsymbol{F}\boldsymbol{F}^*\boldsymbol{H}^* + \frac{N_t}{\mathsf{SNR}} \boldsymbol{I} \right)^{-1} \left[\boldsymbol{H}\boldsymbol{F}\boldsymbol{F}^*\boldsymbol{H}^* + \frac{N_t}{\mathsf{SNR}} \boldsymbol{I} - \boldsymbol{H}\boldsymbol{f}_j \boldsymbol{f}_j^* \boldsymbol{H}^* \right]$$

$$\cdot \left(\boldsymbol{H}\boldsymbol{F}\boldsymbol{F}^*\boldsymbol{H}^* + \frac{N_t}{\mathsf{SNR}} \boldsymbol{I} \right)^{-1} \boldsymbol{H}\boldsymbol{f}_j \tag{6.67}$$

$$= \boldsymbol{f}_j^* \boldsymbol{H}^* \left(\boldsymbol{H}\boldsymbol{F}\boldsymbol{F}^*\boldsymbol{H}^* + \frac{N_t}{\mathsf{SNR}} \boldsymbol{I} \right)^{-1} \boldsymbol{H}\boldsymbol{f}_j$$

$$- \left[\boldsymbol{f}_j^* \boldsymbol{H}^* \left(\boldsymbol{H}\boldsymbol{F}\boldsymbol{F}^*\boldsymbol{H}^* + \frac{N_t}{\mathsf{SNR}} \boldsymbol{I} \right)^{-1} \boldsymbol{H}\boldsymbol{f}_j \right]^2, \tag{6.68}$$

altogether giving

$$\mathsf{sinr}_j^{\mathsf{MMSE}} = \frac{\text{Num}}{\text{Den}} \tag{6.69}$$

$$= \frac{\boldsymbol{f}_j^* \boldsymbol{H}^* \left(\boldsymbol{H}\boldsymbol{F}\boldsymbol{F}^*\boldsymbol{H}^* + \frac{N_t}{\mathsf{SNR}} \boldsymbol{I} \right)^{-1} \boldsymbol{H}\boldsymbol{f}_j}{1 - \boldsymbol{f}_j^* \boldsymbol{H}^* \left(\boldsymbol{H}\boldsymbol{F}\boldsymbol{F}^*\boldsymbol{H}^* + \frac{N_t}{\mathsf{SNR}} \boldsymbol{I} \right)^{-1} \boldsymbol{H}\boldsymbol{f}_j} \tag{6.70}$$

$$= \frac{1 - \mathsf{MMSE}_j}{\mathsf{MMSE}_j} \tag{6.71}$$

where in (6.71) we have invoked (6.56). This is indeed the expression that one would have expected in light of the unit power of the quantity being estimated, $[\boldsymbol{s}]_j$. Equivalently,

$$\mathsf{sinr}_j^{\mathsf{MMSE}} = \frac{1}{\mathsf{MMSE}_j} - 1 \tag{6.72}$$

$$= \frac{1}{\left[\left(I + \frac{\mathsf{SNR}}{N_t} F^* H^* H F\right)^{-1}\right]_{j,j}} - 1, \tag{6.73}$$

where we have applied (6.57).

Yet an alternative form can be derived for $\mathsf{sinr}_j^{\mathsf{MMSE}}$ by further manipulating (6.70) into

$$\mathsf{sinr}_j^{\mathsf{MMSE}} = f_j^* H^* \left(H F_{-j} F_{-j}^* H^* + \frac{N_t}{\mathsf{SNR}} I \right)^{-1} H f_j \tag{6.74}$$

where we used F_{-j} as the shorthand for $[F]_{:,-j}$, the submatrix obtained by removing the jth column from F. Verifying the coincidence of (6.70) and (6.74) entails showing that

$$f_j^* H^* \left(H F F^* H^* + \frac{N_t}{\mathsf{SNR}} I \right)^{-1} H f_j = \left[1 - f_j^* H^* \left(H F F^* H^* + \frac{N_t}{\mathsf{SNR}} I \right)^{-1} H f_j \right] \cdot f_j^* H^* \left(H F_{-j} F_{-j}^* H^* + \frac{N_t}{\mathsf{SNR}} I \right)^{-1} H f_j$$

which, with the right-hand side expanded, amounts to the equality

$$f_j^* H^* \left(H F F^* H^* + \frac{N_t}{\mathsf{SNR}} I \right)^{-1} H f_j = f_j^* H^* \left[\left(H F_{-j} F_{-j}^* H^* + \frac{N_t}{\mathsf{SNR}} I \right)^{-1} - \left(H F F^* H^* + \frac{N_t}{\mathsf{SNR}} I \right)^{-1} H f_j f_j^* H^* \cdot \left(H F_{-j} F_{-j}^* H^* + \frac{N_t}{\mathsf{SNR}} I \right)^{-1} \right] H f_j$$

and therefore to

$$\left(H F F^* H^* + \frac{N_t}{\mathsf{SNR}} I \right)^{-1} = \left[I - \left(H F F^* H^* + \frac{N_t}{\mathsf{SNR}} I \right)^{-1} H f_j f_j^* H^* \right] \cdot \left(H F_{-j} F_{-j}^* H^* + \frac{N_t}{\mathsf{SNR}} I \right)^{-1}, \tag{6.75}$$

which can be seen to indeed hold by multiplying both sides by $\left(H F F^* H^* + \frac{N_t}{\mathsf{SNR}} I \right)$; the left-hand side then becomes an identity matrix while the right-hand side becomes

$$\left[\left(H F F^* H^* + \frac{N_t}{\mathsf{SNR}} I \right) - H f_j f_j^* H^* \right] \left(H F_{-j} F_{-j}^* H^* + \frac{N_t}{\mathsf{SNR}} I \right)^{-1}, \tag{6.76}$$

which is also an identity matrix because $F F^* - f_j f_j^* = F_{-j} F_{-j}^*$.

As intuition would have it, in minimizing the mean-square error for each of the streams, the LMMSE receiver maximizes each of the SINRs. To verify that, let us rewrite the SINR of the jth stream for a generic receiver, given in (6.61), as

$$\mathsf{sinr}_j = \frac{w_j^* H f_j f_j^* H^* w_j}{w_j^* \left(\sum_{\ell \neq j} H f_\ell f_\ell^* H^* + \frac{N_t}{\mathsf{SNR}} I \right) w_j} \tag{6.77}$$

$$= \frac{\boldsymbol{w}_j^* \boldsymbol{H} \boldsymbol{f}_j \boldsymbol{f}_j^* \boldsymbol{H}^* \boldsymbol{w}_j}{\boldsymbol{w}_j^* \left(\boldsymbol{H} \boldsymbol{F}_{\text{-}j} \boldsymbol{F}_{\text{-}j}^* \boldsymbol{H}^* + \frac{N_t}{\mathsf{SNR}} \boldsymbol{I} \right) \boldsymbol{w}_j} \tag{6.78}$$

$$= \frac{\boldsymbol{d}_j^* \left(\boldsymbol{H} \boldsymbol{F}_{\text{-}j} \boldsymbol{F}_{\text{-}j}^* \boldsymbol{H}^* + \frac{N_t}{\mathsf{SNR}} \boldsymbol{I} \right)^{-1/2} \boldsymbol{H} \boldsymbol{f}_j \boldsymbol{f}_j^* \boldsymbol{H}^* \left(\boldsymbol{H} \boldsymbol{F}_{\text{-}j} \boldsymbol{F}_{\text{-}j}^* \boldsymbol{H}^* + \frac{N_t}{\mathsf{SNR}} \boldsymbol{I} \right)^{-1/2} \boldsymbol{d}_j}{\boldsymbol{d}_j^* \boldsymbol{d}_j}$$

$$= \frac{\boldsymbol{d}_j^*}{\|\boldsymbol{d}_j\|} \left(\boldsymbol{H} \boldsymbol{F}_{\text{-}j} \boldsymbol{F}_{\text{-}j}^* \boldsymbol{H}^* + \frac{N_t}{\mathsf{SNR}} \boldsymbol{I} \right)^{-1/2} \boldsymbol{H} \boldsymbol{f}_j \boldsymbol{f}_j^* \boldsymbol{H}^* \left(\boldsymbol{H} \boldsymbol{F}_{\text{-}j} \boldsymbol{F}_{\text{-}j}^* \boldsymbol{H}^* + \frac{N_t}{\mathsf{SNR}} \boldsymbol{I} \right)^{-1/2} \frac{\boldsymbol{d}_j}{\|\boldsymbol{d}_j\|}$$

where we have defined the column vector

$$\boldsymbol{d}_j = \left(\boldsymbol{H} \boldsymbol{F}_{\text{-}j} \boldsymbol{F}_{\text{-}j}^* \boldsymbol{H}^* + \frac{N_t}{\mathsf{SNR}} \boldsymbol{I} \right)^{1/2} \boldsymbol{w}_j. \tag{6.79}$$

Further defining $\boldsymbol{u}_j = \boldsymbol{d}_j / \|\boldsymbol{d}_j\|$ and

$$\boldsymbol{B}_j = \left(\boldsymbol{H} \boldsymbol{F}_{\text{-}j} \boldsymbol{F}_{\text{-}j}^* \boldsymbol{H}^* + \frac{N_t}{\mathsf{SNR}} \boldsymbol{I} \right)^{-1/2} \boldsymbol{H} \boldsymbol{f}_j \boldsymbol{f}_j^* \boldsymbol{H}^* \left(\boldsymbol{H} \boldsymbol{F}_{\text{-}j} \boldsymbol{F}_{\text{-}j}^* \boldsymbol{H}^* + \frac{N_t}{\mathsf{SNR}} \boldsymbol{I} \right)^{-1/2} \tag{6.80}$$

we obtain $\mathsf{sinr}_j = \boldsymbol{u}_j^* \boldsymbol{B}_j \boldsymbol{u}_j$, which is maximized when $\boldsymbol{u}_j$ is the maximum-eigenvalue eigenvector of $\boldsymbol{B}_j$ because any other choice would partially project on directions on which $\boldsymbol{B}_j$ has less gain. In the problem at hand, moreover, $\boldsymbol{B}_j$ is a rank-1 matrix with a single nonzero eigenvalue; any choice for $\boldsymbol{u}_j$ other than the corresponding eigenvector would partially project on zero-gain directions. Such nonzero-eigenvalue eigenvector is

$$\boldsymbol{u}_j = \frac{\left(\boldsymbol{H} \boldsymbol{F}_{\text{-}j} \boldsymbol{F}_{\text{-}j}^* \boldsymbol{H}^* + \frac{N_t}{\mathsf{SNR}} \boldsymbol{I} \right)^{-1/2} \boldsymbol{H} \boldsymbol{f}_j}{\left\| \left(\boldsymbol{H} \boldsymbol{F}_{\text{-}j} \boldsymbol{F}_{\text{-}j}^* \boldsymbol{H}^* + \frac{N_t}{\mathsf{SNR}} \boldsymbol{I} \right)^{-1/2} \boldsymbol{H} \boldsymbol{f}_j \right\|}, \tag{6.81}$$

giving the maximum value

$$\mathsf{sinr}_j = \boldsymbol{u}_j^* \boldsymbol{B}_j \boldsymbol{u}_j \tag{6.82}$$

$$= \frac{\left[\boldsymbol{f}_j^* \boldsymbol{H}^* \left(\boldsymbol{H} \boldsymbol{F}_{\text{-}j} \boldsymbol{F}_{\text{-}j}^* \boldsymbol{H}^* + \frac{N_t}{\mathsf{SNR}} \boldsymbol{I} \right)^{-1} \boldsymbol{H} \boldsymbol{f}_j \right]^2}{\left\| \left(\boldsymbol{H} \boldsymbol{F}_{\text{-}j} \boldsymbol{F}_{\text{-}j}^* \boldsymbol{H}^* + \frac{N_t}{\mathsf{SNR}} \boldsymbol{I} \right)^{-1/2} \boldsymbol{H} \boldsymbol{f}_j \right\|^2} \tag{6.83}$$

$$= \frac{\left[\boldsymbol{f}_j^* \boldsymbol{H}^* \left(\boldsymbol{H} \boldsymbol{F}_{\text{-}j} \boldsymbol{F}_{\text{-}j}^* \boldsymbol{H}^* + \frac{N_t}{\mathsf{SNR}} \boldsymbol{I} \right)^{-1} \boldsymbol{H} \boldsymbol{f}_j \right]^2}{\boldsymbol{f}_j^* \boldsymbol{H}^* \left(\boldsymbol{H} \boldsymbol{F}_{\text{-}j} \boldsymbol{F}_{\text{-}j}^* \boldsymbol{H}^* + \frac{N_t}{\mathsf{SNR}} \boldsymbol{I} \right)^{-1} \boldsymbol{H} \boldsymbol{f}_j} \tag{6.84}$$

$$= \boldsymbol{f}_j^* \boldsymbol{H}^* \left(\boldsymbol{H} \boldsymbol{F}_{\text{-}j} \boldsymbol{F}_{\text{-}j}^* \boldsymbol{H}^* + \frac{N_t}{\mathsf{SNR}} \boldsymbol{I} \right)^{-1} \boldsymbol{H} \boldsymbol{f}_j, \tag{6.85}$$

which coincides with SINR at the output of an LMMSE receiver as given in (6.74). This confirms that an LMMSE receiver simultaneously maximizes the output SINRs of all streams and, given that the mutual information over a scalar channel is monotonic in the SINR, the LMMSE is the optimum linear receiver in the sense of maximizing the spectral efficiency—which is the truly relevant sense of optimality as far as reliable communication is concerned. Indeed, it should not be forgotten that, while for signal processing purposes the minimization of the mean-square error can be an end in itself, in a communication design this minimization is but a conduit to the maximization of the spectral efficiency.

As a matter of fact, any scaled version of $\boldsymbol{W}^{\mathsf{MMSE}}$ achieves the same maximum SINR because the scaling affects equally the desired signal, the noise, and the interference. (When the LMMSE filter is scaled, it continues to minimize the mean-square error only with respect to a scaled version of $\boldsymbol{s}$ rather than $\boldsymbol{s}$ itself.) At the same time, from the expression for $\boldsymbol{u}_j$ in (6.81), obtained via $\boldsymbol{u}_j = \boldsymbol{d}_j / \|\boldsymbol{d}_j\|^2$, we can go back and clear from (6.79) that

$$\boldsymbol{w}_j = \left(\boldsymbol{H} \boldsymbol{F}_{\text{-}j} \boldsymbol{F}^*_{\text{-}j} \boldsymbol{H}^* + \frac{N_\mathrm{t}}{\mathsf{SNR}} \boldsymbol{I} \right)^{-1} \boldsymbol{H} \boldsymbol{f}_j \tag{6.86}$$

or any scaled version thereof is an alternative linear filter that maximizes the output SINR for signal stream j. It follows that $\boldsymbol{w}_j^{\mathsf{MMSE}}$ in (6.63) and $\boldsymbol{w}_j$ in (6.86) must be colinear vectors, and that is indeed the case: the removal of the jth column of $\boldsymbol{H}\boldsymbol{F}$ from within the inverse in (6.86) merely scales the projection of that very column vector, $\boldsymbol{H}\boldsymbol{f}_j$, onto the inverse. This is an instance of the result

$$\left(\boldsymbol{A}_{\text{-}j} \boldsymbol{A}^*_{\text{-}j} + \boldsymbol{a}_j \boldsymbol{a}_j^* + \boldsymbol{I} \right)^{-1} \boldsymbol{a}_j \propto \left(\boldsymbol{A}_{\text{-}j} \boldsymbol{A}^*_{\text{-}j} + \boldsymbol{I} \right)^{-1} \boldsymbol{a}_j, \tag{6.87}$$

which the reader is invited to prove in Problem 6.19.

With $\mathsf{sinr}_j^{\mathsf{MMSE}}$ expressed, let us now turn our attention to its distribution. Such distribution is substantially more involved than that of $\mathsf{snr}_j^{\mathsf{ZF}}$, but an expression for the canonical channel with IID Rayleigh-faded entries does exist.

Example 6.11 (LMMSE output SINR distribution in an IID Rayleigh-faded MIMO channel)

Let $\boldsymbol{H}$ have IID Rayleigh-faded entries while $\boldsymbol{F} = \boldsymbol{I}$ with $N_\mathrm{s} = N_\mathrm{t} \le N_\mathrm{r}$. Some lengthy derivations in [723, 724], not reproduced here, yield

$$\begin{aligned} f_{\mathsf{sinr}_j^{\mathsf{MMSE}}}(\xi) = e^{-\xi N_\mathrm{t}/\mathsf{SNR}} \Bigg[& \sum_{i=N_\mathrm{r}-N_\mathrm{t}+2}^{N_\mathrm{r}} \left(\frac{N_\mathrm{t}}{\mathsf{SNR}} - \frac{i-1}{\xi} + \frac{N_\mathrm{t}-1}{1+\xi} \right) t_0(i,\xi) - t_1(i,\xi) \\ & - \frac{N_\mathrm{t}}{\mathsf{SNR}} \sum_{i=1}^{N_\mathrm{r}-N_\mathrm{t}+1} \frac{i-1-\xi N_\mathrm{t}/\mathsf{SNR}}{(i-1)!} \left(\frac{\xi N_\mathrm{t}}{\mathsf{SNR}} \right)^{i-2} \Bigg], \end{aligned} \tag{6.88}$$

where

$$t_k(i,\xi) = \frac{(\xi N_\mathrm{t}/\mathsf{SNR})^{i-1}}{(i-1)!} \sum_{j=k}^{N_\mathrm{r}-i} \binom{N_\mathrm{t}-1}{j} \frac{j^k \xi^{j-k}}{(1+\xi)^{N_\mathrm{t}-1}} \tag{6.89}$$

with $j^k = 1$ whenever $j = k = 0$. For $\mathsf{SNR} \to \infty$, the above PDF approaches its ZF counterpart in Example 6.2.

The CDF of $\mathsf{sinr}_j^{\mathsf{MMSE}}$ can also be found in [723, 724].

The derivations in [723, 724] can accommodate transmit antenna correlations with suitable precoding. Beyond that, for more general channel structures, closed forms for the distribution of $\mathsf{sinr}_j^{\mathsf{MMSE}}$ do not seem forthcoming, but it can be shown [719, 725] that

$$\mathsf{sinr}_j^{\mathsf{MMSE}} = \mathsf{snr}^{\mathsf{ZF}} + \mathrm{T}_j, \tag{6.90}$$

where T_j is independent of $\mathsf{snr}_j^{\mathsf{ZF}}$. The distribution of $\mathsf{snr}_j^{\mathsf{ZF}}$ was discussed in Section 6.3 whereas that of T_j can be well approximated, for channels with certain correlations, by means of suitably tuned Gamma or generalized Gamma distributions [719]. In turn, the exceedingly more involved analysis of the SINR distribution for Rice-faded channels is tackled in [726, 727].

6.4.3 Ergodic spectral efficiency

It must be mentioned that, in contrast with the ZF receiver, where the filter output is impaired strictly by noise, in the LMMSE case the impairment contains a mixture of noise and interference from other streams. When the signaling is complex Gaussian, this mixture is, conditioned on $\boldsymbol{H}$, also complex Gaussian and the spectral efficiency is directly

$$C^{\mathsf{MMSE}}(\mathsf{SNR}) = \sum_{j=0}^{N_s-1} \mathbb{E}\left[\log_2\left(1 + \mathsf{sinr}_j^{\mathsf{MMSE}}\right)\right] \tag{6.91}$$

$$= \sum_{j=0}^{N_s-1} \mathbb{E}\left[\log_2 \frac{1}{\mathsf{MMSE}_j}\right]. \tag{6.92}$$

There is no guarantee that (6.91) and (6.92) represent the highest possible spectral efficiency because of the separate decoding of each stream: with optimum decoders, non-Gaussian codewords could constitute worse signals for the corresponding streams, but also more benign interference to the rest. Nevertheless, it has been shown that, regardless of the signal distribution, the noise-plus-interference mixture becomes rapidly Gaussian as the number of antennas grows [728]. Hence, not only is complex Gaussian signaling close-to-optimum with LMMSE reception, but we can further approximate the spectral efficiency with other types of signals as $\sum_{j=0}^{N_t-1} \mathbb{E}[\mathcal{I}_j(\mathsf{sinr}_j^{\mathsf{MMSE}})]$.

Whatever the signal distribution, the LMMSE receiver is the best within the class of linear ones because, recall, an LMMSE filter maximizes the SINRs and the mutual information conditioned on the fading is monotonic in the SINR.

Example 6.12

For a channel with $N_t = N_r = 4$ and IID Rayleigh-faded entries, for which $\boldsymbol{F} = \boldsymbol{I}$ is optimum, plot $C^{\mathsf{MMSE}}(\mathsf{SNR})$ and $\mathsf{C}^{\mathsf{MMSE}}(\frac{E_b}{N_0})$.

Solution

See Figs. 6.2 and 6.3, where the LMMSE spectral efficiency has been computed via Monte-Carlo.

An important remark is forthcoming in reference to linear receivers, and to the LMMSE receiver in particular. In contrast with the spectral efficiency of an optimum receiver, which is invariant to unitary precoding (for any unitary $\boldsymbol{F}$, the distribution of the quadratic form $\boldsymbol{H}\boldsymbol{F}\boldsymbol{F}^*\boldsymbol{H}^*$ is unaltered), the spectral efficiency of the LMMSE receiver does change with unitary precoding: the distribution of $\mathsf{sinr}_j^{\mathsf{MMSE}}$ varies and, with that, the spectral efficiency. This is because unitary precoding amounts to a spatial rotation that modifies the signal and interference mix for each of the streams at the output of the filters attempting to isolate them. In contrast, an optimum receiver processing the signal vector as a whole is unaffected by mere signal rotations; what is not captured in one dimension is captured in another.

Low-SNR regime

Applying (6.91) and the results in earlier sections, the low-SNR performance measures for an LMMSE receiver can be obtained.

Example 6.13 ($\frac{E_b}{N_0}{}_{\min}^{\mathsf{MMSE}}$ and S_0^{MMSE} in an IID Rayleigh-faded MIMO channel)

Let $\boldsymbol{H}$ have IID Rayleigh-faded entries and let $\boldsymbol{F} = \boldsymbol{I}$. Applying to (6.91) the corresponding definitions [729, 730],

$$\frac{E_b}{N_0}{}_{\min}^{\mathsf{MMSE}} = \frac{1}{N_r \log_2 e} \tag{6.93}$$

$$S_0^{\mathsf{MMSE}} = \frac{2N_t N_r}{2N_t + N_r - 1}, \tag{6.94}$$

where $\frac{E_b}{N_0}{}_{\min}^{\mathsf{MMSE}}$ coincides with its counterpart for an optimum receiver, revealing that the spatial matched-filter behavior of the LMMSE receiver renders it first-order optimum for $\mathsf{SNR} \to 0$. However, and except for $N_t = 1$, $S_0^{\mathsf{MMSE}} < S_0$ and thus the LMMSE MIMO receiver is not optimum to second order.

High-SNR regime

It is easily verified from (6.91) that, for $N_t \leq N_r$, the number of spatial DOF with an LMMSE receiver is $S_\infty^{\mathsf{MMSE}} = N_t$, exactly as with an optimum receiver. For $N_t > N_r$, however, $S_\infty^{\mathsf{MMSE}} = 0$; this evinces that, as $\mathsf{SNR} \to \infty$, the signals at the LMMSE output would get clogged with interference from other streams that the filter would be unable to reject. It is thus ill-advised to insist on transmitting more than $N_{\min}$ signal streams with an LMMSE receiver in the high-SNR regime.

For $N_s \leq N_{\min}$, the power offset with an LMMSE receiver is larger than with an optimum receiver (see Fig. 6.2 and Problem 6.21), indicating that there is always a gap between the spectral efficiency of an LMMSE receiver and the actual capacity.

As one would expect, for $\mathsf{SNR} \to \infty$ the LMMSE receiver converges to ZF; indeed, letting $\mathsf{SNR} \to \infty$ in $\boldsymbol{W}^{\mathsf{MMSE}}$ immediately yields $\boldsymbol{W}^{\mathsf{ZF}}$. The difference $\mathrm{T}_j = \mathsf{sinr}_j^{\mathsf{MMSE}} - \mathsf{snr}_j^{\mathsf{ZF}}$, in turn, converges in distribution to a fixed-variance random variable for every stream j [725]. It follows that

$$\frac{\mathsf{sinr}_j^{\mathsf{MMSE}}}{\mathsf{snr}_j^{\mathsf{ZF}}} \to 1. \tag{6.95}$$

The difference $\mathrm{T}_j = \mathsf{sinr}_j^{\mathsf{MMSE}} - \mathsf{snr}_j^{\mathsf{ZF}}$ does not vanish, but it is inconsequential to C^{MMSE} in this regime because, for growing SNR,

$$\log_2\left(\mathsf{snr}_j^{\mathsf{ZF}} + \mathrm{T}_j\right) = \log_2 \mathsf{snr}_j^{\mathsf{ZF}} + \mathcal{O}\left(\frac{1}{\mathsf{snr}^{\mathsf{ZF}}}\right), \tag{6.96}$$

irrespective of the bounded value T_j. (The nonvanishing difference between $\mathsf{sinr}_j^{\mathsf{MMSE}}$ and $\mathsf{snr}_j^{\mathsf{ZF}}$ does have an impact on the high-SNR *uncoded* performance [725].)

In quasi-static settings, the DMT is also unaffected by T_j and hence it equals that of a ZF receiver, given in (6.43) [722].

Large-dimensional regime

Let us finally consider the regime where $N_{\mathrm{t}}, N_{\mathrm{r}} \to \infty$ with $\beta = N_{\mathrm{t}}/N_{\mathrm{r}}$.

Example 6.14 (Large-dimensional LMMSE spectral efficiency of a channel with IID zero-mean entries)

Consider a channel whose entries are IID and zero-mean, but otherwise arbitrarily distributed, and let $\boldsymbol{F} = \boldsymbol{I}$. For every j, as $N_{\mathrm{t}}, N_{\mathrm{r}} \to \infty$ [457]

$$\mathbb{E}\left[\mathsf{sinr}_j^{\mathsf{MMSE}}\right] = \mathbb{E}\left[\boldsymbol{h}_j^* \left(\boldsymbol{H}_{\text{-}j}\boldsymbol{H}_{\text{-}j}^* + \frac{N_{\mathrm{t}}}{\mathsf{SNR}}\boldsymbol{I}\right)^{-1} \boldsymbol{h}_j\right] \tag{6.97}$$

$$= \mathbb{E}\left[\sum_{i=0}^{N_{\mathrm{r}}-1} \left[\left(\boldsymbol{H}_{\text{-}j}\boldsymbol{H}_{\text{-}j}^* + \frac{N_{\mathrm{t}}}{\mathsf{SNR}}\boldsymbol{I}\right)^{-1}\right]_{i,i}\right] \tag{6.98}$$

$$= \mathbb{E}\left[\mathrm{tr}\left(\left(\boldsymbol{H}_j\boldsymbol{H}_j^* + \frac{N_{\mathrm{t}}}{\mathsf{SNR}}\boldsymbol{I}\right)^{-1}\right)\right] \tag{6.99}$$

$$= \mathbb{E}\left[\sum_{i=0}^{N_{\mathrm{r}}-1} \lambda_i\left(\left(\boldsymbol{H}_{\text{-}j}\boldsymbol{H}_{\text{-}j}^* + \frac{N_{\mathrm{t}}}{\mathsf{SNR}}\boldsymbol{I}\right)^{-1}\right)\right] \tag{6.100}$$

$$= \mathbb{E}\left[\sum_{i=0}^{N_{\mathrm{r}}-1} \frac{1}{\lambda_i\left(\boldsymbol{H}_{\text{-}j}\boldsymbol{H}_{\text{-}j}^* + \frac{N_{\mathrm{t}}}{\mathsf{SNR}}\boldsymbol{I}\right)}\right] \tag{6.101}$$

$$= N_{\mathrm{r}}\, \mathbb{E}\left[\frac{1}{\lambda\left(\boldsymbol{H}_{\text{-}j}\boldsymbol{H}_{\text{-}j}^* + \frac{N_{\mathrm{t}}}{\mathsf{SNR}}\boldsymbol{I}\right)}\right] \tag{6.102}$$

$$= \frac{N_{\mathrm{r}}}{N_{\mathrm{t}}}\, \mathbb{E}\left[\frac{1}{\lambda\left(\frac{1}{N_{\mathrm{t}}}\boldsymbol{H}_{\text{-}j}\boldsymbol{H}_{\text{-}j}^*\right) + \frac{1}{\mathsf{SNR}}}\right] \tag{6.103}$$

$$\to \frac{1}{\beta}\int_0^\infty \frac{1}{\xi + \frac{1}{\mathsf{SNR}}} f(\xi)\, \mathrm{d}\xi \tag{6.104}$$

$$= \frac{\mathsf{SNR}}{\beta} - \mathcal{F}(\mathsf{SNR}, \beta), \tag{6.105}$$

where in (6.98) we have exploited the independence between $\boldsymbol{h}_j$ and $\boldsymbol{H}_{-j}$ (refer to Problem 6.22) whereas, in (6.102) and (6.103), $\lambda(\cdot)$ is any of the $N_{\rm r}$ identically distributed eigenvalues of its argument. In (6.104), $f(\cdot)$ is the Marčenko–Pastur asymptotic empirical eigenvalue density of $\frac{1}{N_{\rm t}}\boldsymbol{H}\boldsymbol{H}^*$ (see Appendix C.2), which coincides with its brethren for $\frac{1}{N_{\rm t}}\boldsymbol{H}_{-j}\boldsymbol{H}_{-j}^*$ because the deletion of a single column of $\boldsymbol{H}$ is asymptotically immaterial. The integration over $f(\cdot)$ yields the final result where, as defined in (5.174),

$$\mathcal{F}(\mathsf{SNR},\beta) = \frac{1}{4}\left(\sqrt{1+\mathsf{SNR}\left(1+\frac{1}{\sqrt{\beta}}\right)^2} - \sqrt{1+\mathsf{SNR}\left(1-\frac{1}{\sqrt{\beta}}\right)^2}\right)^2. \qquad (6.106)$$

As it turns out, not only does the expectation of $\mathsf{sinr}_j^{\mathsf{MMSE}}$ converge to (6.105), but actually every realization of $\mathsf{sinr}_j^{\mathsf{MMSE}}$ converges a.s. to (6.105) and hence the expectation in the foregoing derivation is unnecessary [546]. The spectral efficiency of an LMMSE receiver follows suit, giving

$$\frac{1}{N_{\rm r}} C^{\mathsf{MMSE}}(\mathsf{SNR}) \overset{\text{a.s.}}{\to} \beta \log_2\left(1+\frac{\mathsf{SNR}}{\beta} - \mathcal{F}(\mathsf{SNR},\beta)\right), \qquad (6.107)$$

which is the counterpart to the large-dimensional capacity expression in (5.173).

With antenna correlations, each stream generally experiences a distinct SINR. While the asymptotic empirical distribution of such SINRs still converges to a nonrandom limit, it generally cannot be expressed in closed form but only as the solution of a fixed-point equation.

Example 6.15 (Large-dimensional LMMSE spectral efficiency of a channel with Kronecker correlations)

Let $\boldsymbol{H} = \boldsymbol{R}_{\rm r}^{1/2}\boldsymbol{H}_{\rm w}\boldsymbol{R}_{\rm t}^{1/2}$ with $\boldsymbol{R}_{\rm r} = \boldsymbol{U}_{\rm r}\boldsymbol{\Lambda}_{\rm r}\boldsymbol{U}_{\rm r}^*$ and $\boldsymbol{R}_{\rm t} = \boldsymbol{U}_{\rm t}\boldsymbol{\Lambda}_{\rm t}\boldsymbol{U}_{\rm t}^*$, and let the precoder be $\boldsymbol{F} = \boldsymbol{U}_{\boldsymbol{F}}\boldsymbol{P}^{1/2}$ with the steering matrix set to $\boldsymbol{U}_{\boldsymbol{F}} = \boldsymbol{U}_{\rm t}$. The precoded channel is then $\boldsymbol{H}\boldsymbol{F} = \boldsymbol{U}_{\rm r}\boldsymbol{\Lambda}_{\rm r}\boldsymbol{U}_{\rm r}^*\boldsymbol{H}_{\rm w}\boldsymbol{U}_{\rm t}\boldsymbol{\Lambda}_{\rm t}\boldsymbol{P}^{1/2}$ whose distribution, because of the unitary invariance of $\boldsymbol{H}_{\rm w}$, coincides with that of $\boldsymbol{U}_{\rm r}\boldsymbol{\Lambda}_{\rm r}\boldsymbol{H}_{\rm w}\boldsymbol{\Lambda}_{\rm t}\boldsymbol{P}^{1/2}$. Furthermore, since the application at the receiver of a unitary rotation $\boldsymbol{U}_{\rm r}^*$ does not alter the noise distribution, we can equivalently consider the channel $\boldsymbol{\Lambda}_{\rm r}\boldsymbol{H}_{\rm w}\boldsymbol{\Lambda}_{\rm t}\boldsymbol{P}^{1/2}$ whose entries are IND. Then [548]

$$\mathsf{sinr}_j^{\mathsf{MMSE}} \approx P_j\,\lambda_j(\boldsymbol{R}_{\rm t})\,\Upsilon_{\rm r}\,\mathsf{SNR}, \qquad (6.108)$$

where $\Upsilon_{\rm r}$ solves the fixed-point equations in (5.180) and where the approximation sharpens as $N_{\rm t}$ and $N_{\rm r}$ grow large. From the large-dimensional distribution of output SINRs, the corresponding spectral efficiency is readily obtained by applying (6.91) with the expectations rendered unnecessary.

With correlation only at the transmitter, the expression in Example 6.15 maps to a well-known large-dimensional solution derived for multiuser detection [456].

6.5 Relationship between the LMMSE and the optimum receiver

As established in Section 5.8.2, an optimum receiver can be constructed as a cascade of LMMSE filters coupled by SIC elements, such that

$$C(\mathsf{SNR}) = \sum_{j=0}^{N_s-1} \log_2\left(1+\mathsf{sinr}_j^{\mathsf{MMSE\text{-}SIC}}\right), \tag{6.109}$$

whose jth term is the spectral efficiency achievable by signal stream j at the output of an LMMSE receiver with interference from streams $0,\ldots,j-1$ canceled out. This allows expressing the capacity as a function of the LMMSE spectral efficiency.

A converse relationship can also be derived, so as to express the LMMSE spectral efficiency as a function of the capacity of an optimum receiver. To get to this relationship, let us replicate the derivation in (5.213)–(5.217), but for an arbitrary column of $\boldsymbol{HF}$ rather than for the column $j=0$. This gives

$$\log_2\det\left(\boldsymbol{I}+\frac{\mathsf{SNR}}{N_t}\boldsymbol{HFF}^*\boldsymbol{H}^*\right) = \log_2\det\left(\boldsymbol{I}+\frac{\mathsf{SNR}}{N_t}\boldsymbol{HF}_{\text{-}j}\boldsymbol{F}^*_{\text{-}j}\boldsymbol{H}^*\right) + \log_2\left(1+\boldsymbol{f}_j^*\boldsymbol{H}^*\left(\frac{N_t}{\mathsf{SNR}}+\boldsymbol{HF}_{\text{-}j}\boldsymbol{F}^*_{\text{-}j}\boldsymbol{H}^*\right)^{-1}\boldsymbol{Hf}_j\right)$$

or, more compactly,

$$\log_2\det\left(\boldsymbol{I}+\frac{\mathsf{SNR}}{N_t}\boldsymbol{HFF}^*\boldsymbol{H}^*\right) = \log_2\det\left(\boldsymbol{I}+\frac{\mathsf{SNR}}{N_t}\boldsymbol{HF}_{\text{-}j}\boldsymbol{F}^*_{\text{-}j}\boldsymbol{H}^*\right) + \log_2\left(1+\mathsf{sinr}_j^{\mathsf{MMSE}}\right). \tag{6.110}$$

Taking expectations over the distribution of $\boldsymbol{H}$, this leads to

$$C(\mathsf{SNR}) = C_{\text{-}j}\left(\frac{N_t-P_j}{N_t}\mathsf{SNR}\right) + \mathbb{E}\left[\log_2\left(1+\mathsf{sinr}_j^{\mathsf{MMSE}}\right)\right], \tag{6.111}$$

where $C_{\text{-}j}(\cdot)$ is the capacity with the jth signal stream removed and its power not relocated to other streams, i.e., $C_{\text{-}j}(\cdot)$ is the capacity of the $N_r\times(N_s-1)$ precoded channel $\boldsymbol{HF}_{\text{-}j}$ with per-symbol power constraint $\|\boldsymbol{F}_{\text{-}j}\|_F^2 = N_t - P_j$. Note that we have implicitly assumed that $\boldsymbol{F}$ is the capacity-achieving precoder; if that is not the case, the term "capacity" should formally not be employed, but the derivation holds all the same. Then, summing both sides of (6.111) over $j=0,\ldots,N_s-1$, we obtain

$$N_s\, C(\mathsf{SNR}) = \sum_{j=0}^{N_s-1} C_{\text{-}j}\left(\frac{N_t-P_j}{N_t}\mathsf{SNR}\right) + C^{\mathsf{MMSE}} \tag{6.112}$$

from which, finally,

$$C^{\mathsf{MMSE}} = N_s\, C(\mathsf{SNR}) - \sum_{j=0}^{N_s-1} C_{\text{-}j}\left(\frac{N_t-P_j}{N_t}\mathsf{SNR}\right). \tag{6.113}$$

This relationship enables direct application to the LMMSE realm of all the results available for the optimum receiver (refer to Problems 6.25 and 6.26). For IID Rayleigh-faded channels, for instance, a closed form for C^{MMSE} emerges from (6.113) and Example 5.17.

6.6 Summary and outlook

The contents of the chapter, reliant on the premise of CSIR, are condensed into the take-away points within the accompanying summary box.

The chapter has not delved into link adaptation, which was covered extensively in the context of optimum receivers and whose applicability to linear receivers entails no conceptual novelty. A problem is proposed, at the end of the chapter, to exercise this aspect; readers interested in the specificity of link adaptation for linear receivers are further referred to [731, 732].

Likewise, the chapter has not probed the issue of pilot-assisted transmission because there is no conceptual difference in how pilot symbols are inserted and channel estimates gathered when linear receivers are in place. From the estimate $\hat{\boldsymbol{H}}$, the linear filter of choice can be readily computed using the expressions given in the chapter. The LMMSE filter in particular can alternatively be computed via

$$\boldsymbol{W}^{\text{MMSE}} = \boldsymbol{R}_{\boldsymbol{y}}^{-1} \boldsymbol{R}_{\boldsymbol{y}\boldsymbol{s}} \tag{6.122}$$

with

$$\boldsymbol{R}_{\boldsymbol{y}\boldsymbol{s}} = \sqrt{\frac{G E_{\mathrm{s}}}{N_{\mathrm{t}}}}\, \hat{\boldsymbol{H}} \boldsymbol{F}, \tag{6.123}$$

but with $\boldsymbol{R}_{\boldsymbol{y}}$ estimated directly from the observed data rather than computed from $\hat{\boldsymbol{H}}$. From observations $\boldsymbol{y}[0], \ldots, \boldsymbol{y}[N-1]$ at the receiver, the sample average

$$\hat{\boldsymbol{R}}_{\boldsymbol{y}} = \frac{1}{N} \sum_{n=0}^{N-1} \boldsymbol{y}[n] \boldsymbol{y}^*[n] \tag{6.124}$$

can be plugged directly into (6.122). This approach does not presume a specific covariance for the noise but rather incorporating it automatically, which is advantageous in the face of spatially colored interference.

The matter of pilot-assisted transmission with linear receivers is addressed in detail later in the book, in multiuser settings of which SU-MIMO is a particular case, and we therefore do not dwell on it further at this point.

In terms of open research problems in the context of linear receivers for SU-MIMO, one that transpires from the treatment in this chapter is the information-theoretic optimization of the precoder in the absence of CSIT.

Take-away points

1. Within the class of linear receivers, meaning a linear filter followed by a bank of scalar decoders operating separately on each stream, the LMMSE receiver is the one that maximizes the spectral efficiency. With precoder $\boldsymbol{F}$, such a filter is

$$\boldsymbol{W}^{\mathsf{MMSE}} = \sqrt{\frac{N_t}{G E_s}} \left(\boldsymbol{H F F^* H^*} + \frac{N_t}{\mathsf{SNR}} \boldsymbol{I} \right)^{-1} \boldsymbol{H F} \tag{6.114}$$

$$= \sqrt{\frac{N_t}{G E_s}} \boldsymbol{H F} \left(\frac{N_t}{\mathsf{SNR}} \boldsymbol{I} + \boldsymbol{F^* H^* H F} \right)^{-1}. \tag{6.115}$$

2. For $\mathsf{SNR} \to 0$, the LMMSE filter reverts to a spatial matched filter. For $\mathsf{SNR} \to \infty$, the LMMSE filter reverts to the ZF filter

$$\boldsymbol{W}^{\mathsf{ZF}} = \sqrt{\frac{N_t}{G E_s}} \boldsymbol{H F} \left(\boldsymbol{F^* H^* H F} \right)^{-1}. \tag{6.116}$$

3. The ergodic LMMSE spectral efficiency with precoder $\boldsymbol{F}$ equals

$$C^{\mathsf{MMSE}} = \sum_{j=0}^{N_s-1} \mathbb{E}\left[\log_2\left(1 + \mathsf{sinr}_j^{\mathsf{MMSE}}\right)\right] \tag{6.117}$$

where

$$\mathsf{sinr}_j^{\mathsf{MMSE}} = \frac{\boldsymbol{f}_j^* \boldsymbol{H}^* \left(\boldsymbol{H F F^* H^*} + \frac{N_t}{\mathsf{SNR}} \boldsymbol{I} \right)^{-1} \boldsymbol{H} \boldsymbol{f}_j}{1 - \boldsymbol{f}_j^* \boldsymbol{H}^* \left(\boldsymbol{H F F^* H^*} + \frac{N_t}{\mathsf{SNR}} \boldsymbol{I} \right)^{-1} \boldsymbol{H} \boldsymbol{f}_j} \tag{6.118}$$

$$= \frac{1}{\left[\left(\boldsymbol{I} + \frac{\mathsf{SNR}}{N_t} \boldsymbol{F^* H^* H F} \right)^{-1} \right]_{j,j}} - 1 \tag{6.119}$$

$$= \boldsymbol{f}_j^* \boldsymbol{H}^* \left(\boldsymbol{H F}_{-j} \boldsymbol{F}_{-j}^* \boldsymbol{H}^* + \frac{N_t}{\mathsf{SNR}} \boldsymbol{I} \right)^{-1} \boldsymbol{H} \boldsymbol{f}_j. \tag{6.120}$$

4. The ergodic ZF spectral efficiency of an IID Rayleigh-faded channel amounts to N_t times the capacity of a SIMO channel having $N_r - N_t + 1$ receive antennas and local-average SNR equal to $\frac{\mathsf{SNR}}{N_t}$, namely

$$C^{\mathsf{ZF}}(\mathsf{SNR}) = N_t \, e^{N_t/\mathsf{SNR}} \sum_{q=1}^{N_r - N_t + 1} \mathcal{E}_q\left(\frac{N_t}{\mathsf{SNR}} \right) \log_2 e. \tag{6.121}$$

5. The LMMSE and spatial matched filter receivers achieve the same $\frac{E_b}{N_0}_{\min}$ as an optimum receiver, but a reduced $\mathcal{S}_0$.
6. The LMMSE and ZF receivers achieve the same number of spatial DOF as an optimum receiver, but incur a loss in terms of power offset.
7. C^{MMSE} can be expressed as a function of the difference between the capacities with N_s and $N_s - 1$ signal streams and, through this relationship, any related quantities can be obtained by harnessing the results derived for optimum receivers.

Problems

6.1 Express $\frac{E_b}{N_0}\big|_{\text{min}}$ for a MIMO channel with a spatial matched-filter receiver.

6.2 Applying the identity $[\boldsymbol{A}^{-1}]_{00} = (\boldsymbol{A}_{00} - \boldsymbol{A}_{01}\boldsymbol{A}_{11}^{-1}\boldsymbol{A}_{10})^{-1}$ to a matrix with block partitionings

$$\boldsymbol{A} = \begin{bmatrix} \boldsymbol{A}_{00} & \boldsymbol{A}_{01} \\ \boldsymbol{A}_{10} & \boldsymbol{A}_{11} \end{bmatrix} \qquad \boldsymbol{A}^{-1} = \begin{bmatrix} [\boldsymbol{A}^{-1}]_{00} & [\boldsymbol{A}^{-1}]_{01} \\ [\boldsymbol{A}^{-1}]_{10} & [\boldsymbol{A}^{-1}]_{11} \end{bmatrix}, \tag{6.125}$$

verify that (6.14) and (6.18) coincide.

6.3 Reconsider Example 6.6.

(a) Reproduce the plot of $C^{\text{ZF}}(\text{SNR})$ in Fig. 6.2.

(b) Reproduce the plot of $C^{\text{MMSE}}(\text{SNR})$ in Fig. 6.2.

(c) Plot the corresponding ergodic spectral efficiency with a spatial matched-filter receiver.

6.4 Plot, as a function of $\text{SNR} \in [-10, 25]$ dB, $C(\text{SNR})$ for SISO and for MIMO with $N_t = N_r = 4$. In the same chart, plot $C^{\text{ZF}}(\text{SNR})$ and $C^{\text{MMSE}}(\text{SNR})$ for $N_t = N_r = 4$.

6.5 Let $\boldsymbol{H}$ have IID Rayleigh-faded entries.

(a) Plot, as a function of $\text{SNR} \in [-10, 25]$ dB, the ZF spectral efficiency per antenna, $\frac{1}{N_r} C^{\text{ZF}}(\text{SNR})$, for $N_t = N_r = 1$, for $N_t = N_r = 2$, for $N_t = N_r = 4$, and for $N_t = N_r = 8$.

(b) Repeat part (a) for an LMMSE receiver.

What do you observe?

6.6 Derive $C^{\text{ZF}}(\text{SNR})$ for a Rayleigh-faded MIMO channel with transmit correlation $\boldsymbol{R}_t$ and with a precoder aligned with the eigenvectors of $\boldsymbol{R}_t$.

6.7 Consider a Rayleigh-faded MIMO channel with transmit correlation $\boldsymbol{R}_t$ and no precoding.

(a) Derive $C^{\text{ZF}}(\text{SNR})$.

(b) For $N_t = N_r = 2$ and

$$\boldsymbol{R}_t = \begin{bmatrix} 1 & 0.7 \\ 0.7 & 1 \end{bmatrix}, \tag{6.126}$$

what is the optimum power allocation (in terms of maximizing the ergodic spectral efficiency with ZF reception) at $\text{SNR} = 7$ dB?

6.8 Let $N_t = N_r = 2$ with $\boldsymbol{H}$ having Rayleigh-faded entries and with $\boldsymbol{R}_t$ as in (6.126). Plot, as a function of $\text{SNR} \in [-10, 25]$ dB, the ergodic ZF spectral efficiency without precoding and with the optimum precoding. What do you observe?

6.9 Repeat Problem 6.8 for an LMMSE receiver, without precoding.

6.10 Reconsider Example 6.7.

(a) Reproduce the plot of $\mathsf{C}^{\text{ZF}}(\frac{E_b}{N_0})$ in Fig. 6.3.

(b) Reproduce the plot of $\mathsf{C}^{\text{MMSE}}(\frac{E_b}{N_0})$ in Fig. 6.3.

(c) Plot the corresponding ergodic spectral efficiency with a spatial matched-filter receiver.

6.11 Prove (6.39) and further derive $\mathcal{S}_0^{\text{ZF}}$ for a ZF receiver in a MIMO channel with IID Rayleigh-faded entries.

6.12 Using $\frac{E_b}{N_0}{}^{\text{ZF}}_{\text{min}}$ and $\mathcal{S}_0^{\text{ZF}}$, compute the power and bandwidth reduction that a MIMO transmission with $N_t = N_r = 4$ and ZF reception enjoys, relative to a SISO transmission at the same bit rate. Assume low-SNR conditions.

6.13 For a Rayleigh-faded MIMO channel with transmit correlation $\boldsymbol{R}_t$ and no receive correlations, express $\frac{E_b}{N_0}{}^{\text{ZF}}_{\text{min}}$ in the following conditions.

(a) With an optimum precoder.

(b) Without precoding.

6.14 Generalize the power offset expression in Example 6.9 to arbitrary N_t and N_r.

6.15 Verify (6.52).

6.16 Verify (6.53).

6.17 Produce a table similar to Table 5.1 contrasting the exact spectral efficiency of an LMMSE receiver and its large-dimensional approximation, for $N_t, N_r = 1, \ldots, 4$ and $\text{SNR} = 10$ dB.

6.18 Provide an alternative proof that $\text{sinr}_j^{\text{MMSE}}$ is given by (6.74), by formulating a linear receiver for stream j that whitens the interference from all other streams and then performs matched filtering. Since a matched filter is the optimum receiver with white noise, this strategy does ensure the maximization of sinr_j.

6.19 Prove (6.87).

Hint: Start from $\boldsymbol{A}\boldsymbol{A}^* + \boldsymbol{I} = \boldsymbol{A}_{-j}\boldsymbol{A}_{-j}^* + \boldsymbol{a}_j\boldsymbol{a}_j + \boldsymbol{I}$.

6.20 Repeat Example 6.6 with $N_t = 2$ and $N_r = 4$. What are the key differences in behavior when N_r is substantially larger than N_t?

6.21 Given a channel with IID Rayleigh-faded entries and $N_t \leq N_r$, write in closed form the difference between the high-SNR power offsets of an LMMSE receiver and an optimum receiver.

6.22 Verify (6.98).

6.23 Specialize to $\beta = 1$ the large-dimensional LMMSE spectral efficiency expression for a channel with IID entries.

6.24 For an unprecoded channel with IID entries and LMMSE reception in the large-dimensional regime, verify that, for $\beta \to \infty$, the following holds.

(a) The average output SINR of every signal stream drops to zero.

(b) The number of spatial DOF also drops to zero.

Note: The spectral efficiency would seem not to vanish, but rather to approach a nonzero constant. However, the overhead required to sustain it would also grow unboundedly with β. Again, for $N_s > N_{\text{min}}$, we find that results that rely on side information are modeling artifacts.

6.25 Use (6.113) to rederive, for an unprecoded transmission, the following.

(a) $\frac{E_b}{N_0}{}^{\text{MMSE}}_{\text{min}}$ for an arbitrary MIMO channel.

(b) $\mathcal{L}_{\infty}^{\mathsf{MMSE}}$ for a MIMO channel with IID Rayleigh-faded entries and $N_{\mathrm{t}} = N_{\mathrm{r}}$.

6.26 Use (6.113) to express C^{MMSE} for a channel with $N_{\mathrm{t}} = N_{\mathrm{r}} = 2$ and IID Rayleigh-faded entries.

6.27 Let $N_{\mathrm{t}} = N_{\mathrm{r}} = 4$, with the channel being IID Rayleigh-faded and frequency-flat. The transmission is unprecoded, with a separate codeword emitted from each antenna and with PARC. The receiver is LMMSE and the channel can be regarded as quasi-static over each codeword. From the error probabilities as a function of SNR given in the book's companion webpage for the 27 LTE MCSs, plot the throughput per unit bandwidth. Further plot, on the same chart, C^{MMSE}.

6.28 Consider a vehicular block-fading channel ($N_{\mathrm{c}} = 1000$) with IID Rayleigh-faded entries and $N_{\mathrm{t}} = N_{\mathrm{r}} = 2$. Pilot power boosting is not allowed. Plot, for $\mathsf{SNR} \in [-5, 25]$ dB, the ergodic LMMSE spectral efficiency of a pilot-based transmission alongside C^{MMSE} with CSIR. Further plot the overhead of the pilot-based transmission.

Hint: A convex optimization solver such as `fmincon` *in* MATLAB® *can be used. Alternatively, since each optimization is over a single scalar, the optimum value can be found by scanning over* $\alpha \in [0, 1]$.

6.29 Suppose that, rather than a complete precoder, an $N_{\mathrm{t}} \times N_{\mathrm{s}}$ semiunitary precoder $\boldsymbol{U_F}$ is applied.

(a) Prove that $\sum_{j=0}^{N_{\mathrm{s}}} \mathsf{MMSE}_j$ is invariant to right unitary rotations of the precoder.

(b) Verify that, for an IID Rayleigh-faded channel with $N_{\mathrm{t}} = N_{\mathrm{r}} = 4$ and with $\mathsf{SNR} = 10$ dB, $\mathbb{E}[\mathsf{MMSE}_j]$ for $j = 0, \ldots, N_{\mathrm{s}}$ are invariant to unitary rotations of the precoder.

(c) Does the invariance in part (b) extend to $\mathbb{E}[\mathsf{sinr}_j^{\mathsf{MMSE}}]$ for $j = 0, \ldots, N_{\mathrm{s}}$? How about to the per-stream average spectral efficiencies with LMMSE reception?

(d) Now let the transmit antennas be correlated via

$$\boldsymbol{R}_{\mathrm{t}} = \begin{bmatrix} 1 & 0.8 & 0.6 & 0.4 \\ 0.8 & 1 & 0.8 & 0.6 \\ 0.6 & 0.8 & 1 & 0.8 \\ 0.4 & 0.6 & 0.8 & 1 \end{bmatrix}. \tag{6.127}$$

Which of the foregoing invariances apply?

Note: As elaborated in Chapter 5, the unitary invariance of the precoder is relevant to the CSIT quantization required for digital feedback. With an optimum receiver, the spectral efficiency is always invariant.

PART III

MULTIUSER MIMO

7 Multiuser communication prelude

He that will not apply new remedies must expect new evils; for time is the greatest innovator.

Francis Bacon

7.1 Introduction

Throughout the leading parts of the book, we have laid down a foundation for SU-MIMO communication, that is, with a single transmitter and a single receiver. It is now time to broaden the scope and consider setups with multiple users. From an information-theoretic viewpoint, the understanding of multiuser channels is decidedly less mature than that of their single-user counterparts, and the capacity is not fully determined even for seemingly simple cases. Fortuitously, the two setups that are arguably most relevant to wireless networks are relatively well characterized. These are the so-called multiple-access channel (MAC) and broadcast channel (BC), respectively abstracting the reverse link (or uplink) and the forward link (or downlink) within a given system cell. The MIMO incarnations of these two setups, and some intriguing duality relationships between them, are what we set out to cover next, as a stepping stone to the study of massive MIMO.

The present chapter kicks off this coverage, generically introducing those concepts that are necessary to move beyond single-user communication. Some of these concepts have appeared tangentially in our analyses of channels with interference and are now fully embraced, while others appear for the first time. For the sake of crispness, they are presented and exemplified for single-antenna transceivers in the present chapter. With these concepts established, and with multiple antennas incorporated, subsequent chapters then delve into the analysis of the MU-MIMO MAC and BC, with the treatment of optimum and linear transceivers deferred to separate respective chapters as done in the exposition of SU-MIMO; this provides a clean distinction between the study of the capacity, along with the transmitter and receiver structures that achieve it, and the examination of linear transmitters and receivers that need not be capacity-achieving but are more attractive to implement.

Since the necessary information-theoretic notions are already well defined by now, the starting points for the derivations are directly, on the one hand, the power and bandwidth constraints, and, on the other hand, the bit rate, the spectral and power efficiencies, and the capacity, all suitably generalized to the multiuser realm.

- Although we focus directly on error-free communication for $N \to \infty$, the considerations made in Section 1.5 concerning communication at nonzero error probabilities over a finite blocklength apply.

- We consider multiuser communication in fading channels impaired by Gaussian noise, with the wisdom expounded in Sections 4.9 and 5.7 in relation to any additional interference (say, from neighboring cells) very present: at a given SINR, having interference in lieu of noise is beneficial, but the difference is significant only if the overall interference dimensionality is small relative to the number of receive antennas. Since this is seldom the case, the noise-limited performance, with the interference regarded as additional noise, is wholly meaningful.
- Likewise, we consider coded modulation signals keeping present that in most instances the actual implementation is in the form of BICM (see Sections 1.5.4 and 5.8.4).
- We emphasize the dependence on the number of users, which is the distinguishing aspect in the multiuser realm, and de-emphasize (except in those instances where it runs counter to what it was in SU-MIMO) the impact of channels features such as antenna correlation or Rice factors.

The chapter begins in Section 7.2 with the generalization to multiuser contexts of the concept of capacity, followed in Sections 7.3 and 7.4 by a dissection of the various ways in which a channel can be shared by multiple users; orthogonal and non-orthogonal sharing schemes are inspected in these respective sections. This sets the stage for Section 7.5, which introduces several scalar metrics that are commonly employed to succinctly represent the communication performance on multiuser channels. Then, Section 7.6 discusses how to dynamically implement the channel-sharing process, introducing notions such as user selection, scheduling, and resource allocation. Finally, Section 7.7 justifies the high-SNR emphasis of non-orthogonal multiuser sharing schemes and Section 7.8 wraps up the chapter.

7.2 Spectral efficiency region

We are interested in these two multiuser setups:

- The MAC, where multiple transmitting users communicate with a single receiver.
- The BC, where a single transmitter communicates with multiple receiving users.

With respect to single-user setups, a chief variable in the multiuser realm is the number of users U that transmit to a common receiver (in the MAC) or that receive from a common transmitter (in the BC). These U active users are generally selected from a larger pool of U_{tot} users, through criteria that are important and considered later on, but which are of no concern at this point. What is immediately relevant is that, rather than an individual spectral efficiency, we now need to refer to a plurality of U efficiencies. Whereas the spectral efficiency of a single user lies on a segment (ranging between zero and a scalar capacity), the combination of spectral efficiencies simultaneously achievable by U users determine

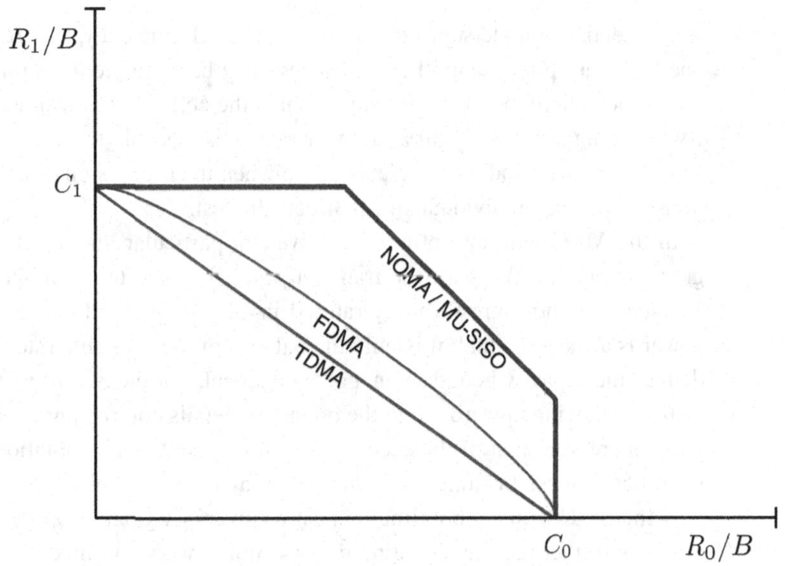

Fig. 7.1 Example of MAC regions of achievable spectral efficiencies for $U = 2$.

a region in a U-dimensional space. This multidimensional nature of the performance is a key point from which the rest of the multiuser concepts then emanate.

Although the capacity is really only the boundary of the largest region of achievable spectral efficiencies, it is not uncommon in the literature to have the entire region referred to as the capacity region. (It is also not uncommon to have the region of achievable spectral efficiencies referred to as the rate region, with the bandwidth B implicitly normalized.) For $U = 1$, the region of achievable spectral efficiencies collapses to a segment and the boundary particularizes to the familiar scalar capacity of a single-user channel.

For illustration purposes, we often resort to the case $U = 2$, which captures the multiuser essence with a simple-to-visualize capacity boundary.

Example 7.1

Draw the two-dimensional region of achievable spectral efficiency pairs, $(R_0/B, R_1/B)$, for a single-antenna MAC with $U = 2$.

Solution

See Fig. 7.1, which portrays the region for the various channel-sharing schemes described throughout the chapter.

For $U > 2$, the achievable spectral efficiencies are no longer pairs but rather U-tuples, $(R_0/B, \ldots, R_{U-1}/B)$, and the region they form as well as its boundary become difficult to visualize, yet they are conceptually identical.

Discussion 7.1 Interpretation of a spectral efficiency region

It is essential to understand that a user's spectral efficiency, say R_u/B, changes from one boundary point to another, not necessarily because the transmit power of that user varies, but chiefly because the rate at which the corresponding signal is encoded varies. By encoding a user's signal at a lower rate, it is possible to accommodate higher rates from other users, and vice versa, because when the users' signals are jointly received or transmitted, each individual signal affects the rest.

In the MAC with an optimum receiver, in particular, the spectral efficiency region grows monotonically with the transmit power of each user—this point is explored in Problem 7.4 and more formally ratified in Chapter 8—and hence transmission at full power is always optimal; it is only the balance of the encoding rates by the U users that defines the capacity boundary in a given channel. For the two-user setup in Fig. 7.1, for instance, moving upward along the boundary entails contracting the rate of the first user while increasing that of the second. Needless to say, link adaptation is instrumental in the process of establishing a specific operating point of choice over the boundary.

In the MAC with a suboptimum linear receiver (see Chapter 9), delineating the boundary does further require adjusting the transmit powers, but link adaptation continues to be instrumental.

7.3 Orthogonal channel sharing

The channel signaling dimensions (time, frequency, and space) must be shared by the U active users, with the most straightforward approach being a division of the time and/or frequency axes into U disjoint intervals, each assigned to one of the users.

7.3.1 Time-division

In time-division, the time axis gets divided into slots. Then, if slots are assigned to users statically and $\mathsf{T}_u \in [0,1]$ is the time share of user u, such that $\sum_{u=0}^{U-1} \mathsf{T}_u = 1$, the spectral efficiencies of the users are

$$\frac{R_u}{B} = \mathsf{T}_u\, C_u(\mathsf{SNR}_u) \qquad u = 0, \ldots, U-1 \tag{7.1}$$

with $C_u(\cdot)$ the individual capacity of the corresponding user in absence of the rest, and with SNR_u its local-average SNR. This relationship applies to both the MAC and the BC. Strictly speaking, time-division should be termed time-division multiple access (TDMA) in the MAC and time-division multiplexing in the BC; however, the acronym TDMA is often abused to denote time-division for both the MAC and the BC.

Example 7.2

Consider a MAC where $U = 2$ users communicate with a common receiver. Which spectral efficiency pairs $(R_0/B, R_1/B)$ are achievable with TDMA?

Solution

If $\mathsf{T}_0 = 1$ and $\mathsf{T}_1 = 0$, then $R_0/B \leq C_0$ and $R_1/B = 0$ whereas, if $\mathsf{T}_0 = 0$ and $\mathsf{T}_1 = 1$, then $R_0/B = 0$ and $R_1/B \leq C_1$. The straight segment connecting $(C_0, 0)$ and $(0, C_1)$ can be achieved by varying T_0 and $\mathsf{T}_1 = 1 - \mathsf{T}_0$, and all the spectral efficiency pairs within the inner triangular region are therefore achievable (see Fig. 7.1).

For $U = 3$, the TDMA region boundary becomes a plane while the region itself is the volume under that plane. For $U > 3$, the boundary becomes a hyperplane.

7.3.2 Frequency-division

In frequency-division, it is the frequency axis that gets partitioned into bands. As with time-division, a strict denomination would call for the term frequency-division multiple access (FDMA) in the MAC and frequency-division multiplexing in the BC, yet the acronym FDMA often refers to both.

Let us suppose that the frequency bands are assigned statically and that F_u is the share of the bandwidth B assigned to user u, such that $\sum_{u=0}^{U-1} \mathsf{F}_u = 1$.

Multiple-access channel

In the MAC, user u concentrates its power on a bandwidth $\mathsf{F}_u B$ and thus faces a noise power $\mathsf{F}_u N_0 B$. With SNR_u denoting the local-average SNR that this user would experience signaling over the entire bandwidth B, the local-average SNR experienced over $\mathsf{F}_u B$ is $\mathsf{SNR}_u/\mathsf{F}_u$ and therefore

$$\frac{R_u}{B} = \mathsf{F}_u\, C_u\left(\frac{\mathsf{SNR}_u}{\mathsf{F}_u}\right) \qquad u = 0, \ldots, U-1. \tag{7.2}$$

Example 7.3

Consider the same setup of Example 7.2. Which spectral efficiency pairs $(R_0/B, R_1/B)$ are achievable with FDMA?

Solution

From (7.2) with $\mathsf{F}_0 \in [0, 1]$ and $\mathsf{F}_1 = 1 - \mathsf{F}_0$, we obtain the FDMA boundary in Fig. 7.1, whose contained region is somewhat larger than its TDMA counterpart.

The reason why the MAC spectral efficiency in (7.2) does not coincide with its TDMA brethren in (7.1) is insinuated in the definition of FDMA: each user concentrates its power on only a fraction of the bandwidth, hence subject to only a fraction of the noise. If, in TDMA, through power control subject to only an average constraint, the power not transmitted during idle slots could be reclaimed for active slots, then the signal power on the active slots would increase by $1/\mathsf{T}_u$ and the resulting spectral efficiencies would coincide with their FDMA counterparts.

Broadcast channel

In the BC, the transmitter has to divide its power among the U frequency bands assigned to the U active users. If the power is divided among those users in direct proportion to how the bandwidth is allocated to them, then on every band there are identical shares of signal power and of noise, meaning that the local-average SNR experienced by user u is precisely SNR_u and hence

$$\frac{R_u}{B} = \mathsf{F}_u\, C_u(\mathsf{SNR}_u) \qquad u = 0, \ldots, U-1, \tag{7.3}$$

as in time-division. However, the power need not be divided proportionally to the bandwidth; rather, its allocation to users can be separately controlled. Denoting by $\frac{E_u}{E_s} \in [0,1]$ the share of transmit power allocated to user u, we then have

$$\frac{R_u}{B} = \mathsf{F}_u\, C_u\left(\frac{\frac{E_u}{E_s}\mathsf{SNR}_u}{\mathsf{F}_u}\right) \qquad u = 0, \ldots, U-1. \tag{7.4}$$

By separately optimizing $\frac{E_0}{E_s}, \ldots, \frac{E_{U-1}}{E_s}$, subject to $\sum_{u=0}^{U-1} \frac{E_u}{E_s} = 1$, it is possible to enlarge the frequency-division spectral efficiency region [733]. The advantage of frequency-division with an optimized power allocation with respect to time-division increases with the disparity in user SNRs.

7.3.3 OFDMA

In systems featuring OFDM signaling, frequency-division can be conveniently implemented by assigning blocks of subcarriers to different users. Referred to as orthogonal frequency division multiple access (OFDMA), this modern form of frequency-division is often implemented in conjunction with time-division and with a resource allocation engine making decisions on subcarrier and time slot assignments. This approach represents an enticing alternative that underpins most modern communication standards.

Besides time- and frequency-division, possibly via OFDMA, other orthogonal sharing strategies are possible, chiefly code-division.

7.4 Non-orthogonal channel sharing

While time- and frequency-division are channel-sharing approaches that readily come to mind, their orthogonal structure is imposed and comes at a cost in terms of information-theoretic optimality. The principles of MIMO clearly suggest that higher efficiencies can be attained if all the transmissions are allowed to take place concurrently, in a non-orthogonal fashion, and that is indeed the case. With single-antenna transceivers, this is sometimes termed superposition coding, or simply *superposition*, an approach long known to be information-theoretically optimum for such setups [14] and that is now getting practical traction under the moniker of non-orthogonal multiple access (NOMA). With NOMA, the combination of time slotting and OFDMA that forms the basis of most of today's wireless systems is augmented by a certain degree of superposition, meaning the possibility of $U > 1$ overlapping transmissions on each time slot and subcarrier [734–737]. With single-antenna transceivers, NOMA entails dimensional overloading.

Once multiple antennas enter the picture, NOMA reverts to the broader scheme of MU-MIMO and, to avoid an excessive proliferation of acronyms, in this book we directly refer to all non-orthogonal multiuser communication types as MU-MIMO, regardless of whether there is dimensional overloading. If transmitters or/and receivers feature a single-antenna, then we have the special cases of MU-SIMO, MU-MISO, and MU-SISO. Note that, in all these MU-xxxx setups, the channel features multiple inputs and multiple outputs, jointly processed at the base station and separately processed at each user, with the qualifiers "SIMO," "MISO," and "SISO" indicating the multiantenna nature of the individual links. In this chapter specifically, all examples correspond to MU-SISO: a single-antenna base station communicating with various single-antenna users.

With properly optimized transmit signals, the MU-MIMO region of spectral efficiencies cannot be improved upon. In certain cases, e.g., in an MU-SISO MAC, the FDMA and MU-MIMO boundaries touch at one point and thus FDMA is also optimum thereupon, but the latter boundary always contains the former.

Example 7.4

Draw the MU-SISO capacity boundary for a MAC with $U = 2$.

Solution

See Fig. 7.1.

The unique local-average SNR at the receiver, which for single-user setups is defined as $\mathsf{SNR} = \frac{G P_{\rm t}}{N_0 B} = \frac{G E_{\rm s}}{N_0}$, ramifies for MU-MIMO into

$$\mathsf{SNR}_u = \frac{G_u E_{\rm s}}{N_0} \qquad u = 0, \ldots, U-1, \tag{7.5}$$

where G_u is, in general, distinct for each user.

While, strictly speaking, optimality entails having *all* users transmit concurrently, the more pragmatic approach given a large population of $U_{\rm tot}$ users is to have subsets of U users operating in an MU-MIMO fashion, with orthogonal multiplexing of the subsets [738]. This approach, which is far simpler and hardly suboptimum if U is chosen wisely depending on the numbers of antennas, is the one presumed throughout this text.

7.5 Scalar metrics

Consider a multiuser setup with U users. Having the capacity be a region boundary rather a scalar quantity complicates its role as a performance benchmark and, often, it is desirable to select a few points on that boundary that are representative in some respect and whose dependence on parameters of interest (e.g., the numbers of antennas) can be more conveniently evaluated [34]. Indeed, while in general the entire boundary is of interest, certain points thereon command special attention because of their operational relevance.

7.5.1 Sum of the spectral efficiencies

Perhaps the most obvious scalar metric is the point corresponding to the largest sum of the spectral efficiencies of the U users. This quantity, for which we reserve the function $C(\cdot)$ in our multiuser analysis, is intimately related with the SU-MIMO spectral efficiency. In fact, in SU-MIMO one could characterize a boundary involving the spectral efficiencies of the various transmit antennas. However, since all these antennas belong to the same user, only their sum is relevant. In multiuser setups, such is no longer the case and the sum of the spectral efficiencies does not suffice as a characterization, yet it continues to have considerable relevance.

With MU-MIMO and properly optimized transmit signals, the sum of the spectral efficiencies becomes the *sum-capacity*.

7.5.2 Weighted sum of the spectral efficiencies

More generally, we may be interested in characterizing the points on the boundary that correspond, not to the highest possible value for $\sum_{u=0}^{U-1} R_u/B$, but rather for $\sum_{u=0}^{U-1} q_u R_u/B$ given some nonnegative weights $q_0, \ldots, q_{U-1}$ establishing a relative priority for each of the users. While maximizing the sum of the spectral efficiencies may lead to operating points where the performance is considerably skewed for the various users, a weighted sum can characterize the performance under any degree of fairness imposed through the weights. And, if $q_0 = \cdots = q_{U-1}$, the weighted sum yields the sum itself.

Example 7.5

Consider the same setup of Example 7.2. For given q_0 and q_1, how can we determine the boundary point where $q_0 R_0/B + q_1 R_1/B$ is maximized?

Solution

For any q_0 and q_1, the contours of constant weighted sum spectral efficiency, i.e., the pairs $(R_0/B, R_1/B)$ such that $q_0 R_0/B + q_1 R_1/B$ is constant, form lines having a slope $-q_0/q_1$. Figure 7.2 shows the MU-SISO, TDMA, and FDMA two-dimensional regions with such contours explicitly depicted. For each region, the sought boundary point (indicated with a solid circle) is the one touching the farthest contour.

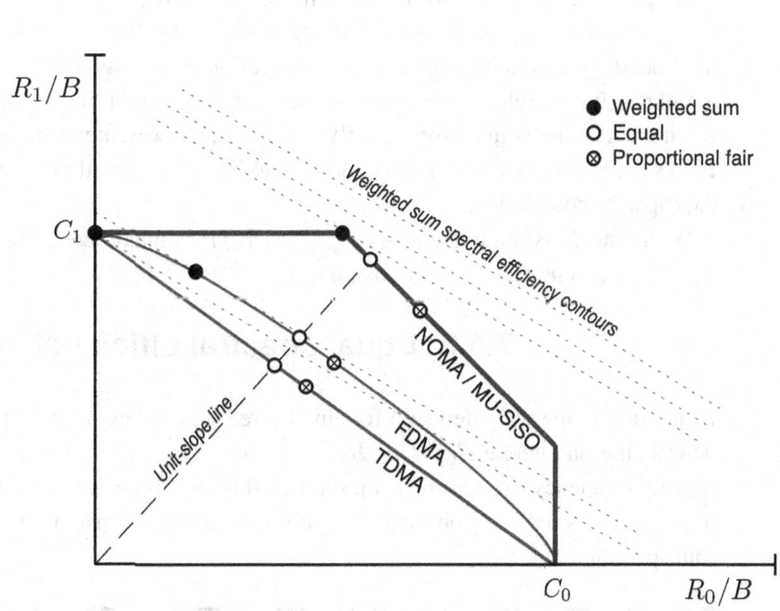

Fig. 7.2 Example of region boundaries for $U = 2$ with TDMA, FDMA, and MU-SISO. Shown with dotted lines are the weighted sum spectral efficiency contours for some specific q_0 and q_1 and, marked with solid circles, the ensuing weighted sum points. Shown with dashes is a unit-slope line going through the origin and, marked with clear circles, are the equal spectral efficiency points. Marked with crossed circles are the proportional-fair points.

For TDMA in particular, the point of maximum weighted sum corresponds to $R_0/B = 0$ if $q_0 C_0 < q_1 C_1$ (as in the figure) and to $R_1/B = 0$ if $q_0 C_0 > q_1 C_1$.

The weights need not be normalized in any specific way, but it must be understood that scaled versions thereof give the same boundary point, and thus the same spectral efficiency U-tuple, yet different weighted sums. Comparisons and evaluations are perfectly meaningful as long as the weights are utilized consistently.

Example 7.6

Reconsider Example 7.5. If we double both q_0 and q_1, the contours are unaffected and thus the pairs $(R_0/B, R_1/B)$ that maximize the weighted sum for each of the regions (TDMA, FDMA, and MU-SISO) do not change. However, $q_0 R_0/B + q_1 R_1/B$ doubles in each case.

The ideas of Examples 7.5 and 7.6 generalize straightforwardly to $U > 2$. The MU-MIMO weighted sum is maximized by the spectral efficiency U-tuple at the boundary point touching the farthest contour hyperplane for the given $q_0, \ldots, q_{U-1}$.

As graphical intuition would have it, by computing the U-tuples that maximize the weighted sum $\sum_{u=0}^{U-1} q_u R_u/B$ for every possible combination of $q_0, \ldots, q_{U-1}$, the entire boundary is obtained; thus, the maximization of the weighted sum spectral efficiency for all possible weight combinations is itself a way of computing the region boundary. And, when the transmit signals are properly optimized, the weighted sum spectral efficiency becomes the *weighted sum-capacity* and its maximization for all possible weights delineates the capacity boundary.

With time-division, in contrast, $\sum_{u=0}^{U-1} q_u R_u/B$ is maximized when only the user u for which $q_u C_u$ is largest gets to transmit.

7.5.3 Equal spectral efficiencies

In addition to the weighted sum for any desired set of weights, another relevant boundary point is the one where $R_0/B = R_1/B = \cdots = R_{U-1}/B$, which indicates the highest spectral efficiency that can be achieved by all users simultaneously. This enforces a strict performance equality among the users, possibly at the expense of exacting a steep price in sum spectral efficiency.

Example 7.7

Consider, yet again, the same setup of Example 7.2. How is the equal spectral efficiency point determined?

Solution

As illustrated in Fig. 7.2, this boundary point is the one intersecting the unit-slope line that crosses the origin.

7.5.4 Minimum of the spectral efficiencies

As yet another alternative to the weighted sum, one can focus on maximizing the worst among the spectral efficiencies of the U users [739]. While this falls short of enforcing a strict performance equality, it does make the sum spectral efficiency take a back seat relative to fairness. And, indeed, a substantial price is often paid also in this case in terms of sum spectral efficiency.

7.5.5 Proportional fairness

As mentioned, a shortcoming of the straight sum as a scalar metric is that it completely disregards fairness, yielding skewed spectral efficiencies if the user SNRs are highly dissimilar. This can be corrected by resorting to a weighted sum, at the cost of involving further parameters (the weights) whose choice is itself not devoid of difficulty. The equal spectral efficiency point suffers from the opposite limitation: an excess of fairness that penalizes users with favorable channel conditions, forcing them to operate at the low spectral efficiencies imposed by users having worse channels. Altogether, it would be desirable to identify a point offering a balance between aggregate performance and fairness, and an attractive such point is the one corresponding to the largest product of the spectral efficiencies of the U users [740]. Since the logarithm is a monotonic function, this is equivalently the spectral efficiency U-tuple that maximizes

$$\log\left(\prod_{u=0}^{U-1}\frac{R_u}{B}\right)=\sum_{u=0}^{U-1}\log\frac{R_u}{B}. \tag{7.6}$$

This is the so-called *proportional-fair* point, which provides a satisfying balance between fairness and sum spectral efficiency. Specifically, multiplying any of the user's spectral efficiency by some factor has the same effect on (7.6), regardless of the user. Moreover, denoting by $R_0^\star/B,\ldots,R_{U-1}^\star/B$ the proportional-fair U-tuple, it can be shown that [740]

$$\sum_{u=0}^{U-1}\frac{R_u/B-R_u^\star/B}{R_u^\star/B}\leq 0, \tag{7.7}$$

meaning that, if we move the operating point from $R_0^\star/B,\ldots,R_{U-1}^\star/B$ to any other boundary point $R_0/B,\ldots,R_{U-1}/B$, the sum of the fractional increases in user spectral efficiency cannot be positive. In this specific sense, the proportional-fair point is optimal. Also, this point is implementationally relevant because a time-multiplexed system can be made to operate on it through the simple and popular proportional-fair scheduling algorithm described later in the chapter.

Example 7.8

For the setup of Example 7.2, indicate the points maximizing (7.6).

Solution

See Fig. 7.2. For TDMA in particular, at the proportional-fair point each of the two users transmits half the time.

7.5.6 Generalized proportional fairness

The range of operating points that stretches from maximizing the sum of spectral efficiencies down to maximizing exclusively the minimum thereof, crossing through the proportional-fair point, can be bridged through the utility function [741]

$$\sum_{u=0}^{U-1} g_\zeta\left(\frac{R_u}{B}\right) \tag{7.8}$$

with

$$g_\zeta(z)=\begin{cases}\log_e z & \zeta=1\\ \frac{z^{1-\zeta}}{1-\zeta} & \zeta\neq 1\end{cases} \tag{7.9}$$

where ζ is a single parameter that regulates the fairness. For $\zeta = 0$, the maximization of (7.8) yields the sum spectral efficiency while for $\zeta = 1$ it gives the proportional-fair solution, and for $\zeta \to \infty$ it leads to maximizing the minimum of the U spectral efficiencies.

The ability to tune the degree of fairness with a single parameter, rather than a set of U weights, makes this generalized proportional fairness an attractive alternative to determine the operating point in a multiuser channel.

7.6 User selection and resource allocation

As mentioned, the pervading approach to channel sharing is, given a population of U_{tot} users, to have subsets of U users operate in an MU-MIMO fashion, with orthogonal multiplexing of the U_{tot}/U subsets. This gives rise to two intertwined challenges:

- *User selection*, meaning the decision of which users constitute each subset, i.e., which U users are served concurrently.
- *Resource allocation*, meaning the policy that determines the subset of U users to which each time–frequency resource is assigned.

Altogether, this amounts to a tortuous problem that could warrant an entire book. We touch on it tangentially, in this chapter and in subsequent ones, only to the extent necessary for our coverage of MIMO.

The starting point for resource allocation is the decision of how to partition the available time and frequency resources. In older systems, this was rather rudimentary: time slots and fixed frequency bands. With OFDM, much finer granularity is possible in the frequency domain. Typically, a basic resource block is defined containing a specified number of resource elements, i.e., a certain number of OFDM symbols in time and a certain number of subcarriers.

Example 7.9

In both LTE and NR, a basic resource block spans 7 OFDM symbols and 12 subcarriers, for a total of 84 resource elements. Whereas, in LTE, this always corresponds to a time–frequency tile of 0.5 ms by 180 kHz, in NR the aspect ratio of this tile depends on the subcarrier spacing: every time the subcarrier spacing is doubled, the tile contracts in time while stretching in frequency.

By concatenating resource blocks, a variety of different time and frequency shares (with the ensuing variety of bit rates) can be made available.

The second issue, once the resources have been partitioned into blocks on the time–frequency plane, is to allocate those signaling resources to users. In older systems with telephony as the reigning application, this was done statically, even simply round-robin; most importantly, the allocations were blind to channel conditions. In modern systems, conversely, the allocations are determined dynamically on the basis of the users' demands

Discussion 7.2 Multiuser diversity

Besides accounting for large-scale channel gains, resource allocation schemes opportunistically exploit small-scale fading swings, attempting to constantly reallocate each resource block to users enjoying favorable channel conditions thereon. This brings about a phenomenon commonly referred to as *multiuser diversity*, and asymptotic laws have been put forth that quantify the sustained improvements that arise as $U_{\text{tot}} \to \infty$ if the users having the most favorable fading are assigned to each resource block [742–745]. As is often the case, the asymptotic expressions are simple and insightful [746, 747]. However, and letting alone the latency and the ceiling posed by the cardinality of the available constellations, one must be cautious about these asymptotic laws for two reasons.

- The multiuser diversity gains are significant for small numbers of users, quickly diminishing thereafter. In this range of interest, the gains cannot be quantified through asymptotic expressions.
- For $U_{\text{tot}} \to \infty$, modeling artifacts take hold if an unbounded support is assumed for the user fading distributions. Beyond a point, the upper tails of such distributions are no longer representative of actual fading and results that rely on operating ever further along such tails are suspect. Put differently, fading distributions such as Rayleigh, Rice, or Nakagami are excellent models—except precisely for their upper tail, which is unbounded. Because the transmit power is finite, the support of actual fading distributions is necessarily bounded and thus multiuser diversity inevitably saturates. This is explored in Problems 7.14–7.16.

Altogether, resource allocation schemes are effective at avoiding downfades, but one must be prudent not to assume excessively favorable upfades on the allocated resources [748]. Values moderately above the local-average is what a fine resource allocation engine may be able to achieve.

and their channel conditions over each resource block. The relative latency tolerance and bursty nature of many non-telephony applications grant a high degree of flexibility in the sense that, to a substantial extent, each resource can be allocated to the U users that stand to make the most efficient utilization thereof. However, as reasoned in reference to the operating points on a multiuser capacity boundary, the push toward the most efficient utilization of resources—allocating each block to the U users reaping the highest sum spectral efficiency—needs to be balanced with the need for some degree of fairness and, also, with the latency constraints of the overlaying applications. Altogether, the resource allocation exercise is extraordinarily complicated, and the algorithms addressing it represent one of the proprietary elements in the wireless ecosystem; not being subject to standardization, resource allocation engines enable vendor differentiation.

7.6.1 The proportional-fair algorithm

The idea behind this algorithm is to allocate each resource block to the U users that can collectively achieve the highest spectral efficiency thereon relative to their own running average [749] (see also [34, section 5.4.3]).

To begin with, suppose that the resource blocks are defined only in time, spanning the entire bandwidth B, and that they are of equal duration. Further, let such time slots be indexed by n. The resource allocation engine then reduces to a time scheduler. Denoting by $C_u[\mathrm{n}]$ the spectral efficiency that could be achieved—with whatever transmission scheme is in place—by the uth user on the nth block, such a block should be assigned to the U users that can achieve the highest weighted sum spectral efficiency, $\sum_{u=0}^{U-1} q_u C_u[\mathrm{n}]$, with weights given by

$$q_u = \frac{1}{\bar{C}_u[\mathrm{n}]}, \tag{7.10}$$

where $\bar{C}_u[\mathrm{n}]$ is the average spectral efficiency achieved by the uth user hitherto. The more deficit that a user has accrued in spectral efficiency, the higher that user's weight at block n. Then, once the nth resource block has been allocated, the running average for all users is updated via

$$\bar{C}_u[\mathrm{n}+1] = \left(1 - \frac{1}{\varepsilon}\right) \bar{C}_u[\mathrm{n}] + \frac{1}{\varepsilon} C_u[\mathrm{n}] \qquad u = 0, \ldots, U_{\mathrm{tot}} - 1, \tag{7.11}$$

with $C_u[\mathrm{n}] = 0$ for those users that were not allocated to the nth block.

With this algorithm, the U_{tot} users dynamically share the channel in such a way that, on average, each one occupies a fraction U/U_{tot} of the time–frequency resources. The parameter $\varepsilon \geq 1$ determines the memory of the running average and, with that, the number of consecutive blocks that a user can go without service. At most, a user with poor channel conditions will have to wait (roughly) ε/U blocks to be served, and thus (approximate) latency guarantees can be established. A small value for ε ensures shorter service latencies, sacrificing sum spectral efficiency; for $\varepsilon = 1$ in particular, the algorithm reverts to simple round-robin, regardless of channel conditions. Conversely, a larger ε gives the scheduler permission to wait longer for each user to experience relatively strong channel conditions and, for $\varepsilon \to \infty$, the performance converges to proportional fairness [750]. As per (7.7), any other scheduling algorithm that increased the spectral efficiency of a specific user by some percentage would cause an aggregate reduction across all other users at least as large as that percentage.

Rather than on the basis of the spectral efficiencies that could have been achieved in the absence of errors, the running averages that underlie the proportional-fair algorithm can alternatively be based on actually achieved throughput, so as to discard from the decision metric those previous transmissions that were unsuccessful [751, 752].

For applications with strict requirements, the algorithm can also be modified to require that a minimum average spectral efficiency be granted to every user [753].

If the resource blocks are of unequal size, then (7.11) should be refined by applying to $C_u[\mathrm{n}]$ an appropriate coefficient. Moreover, if the resource blocks are defined as tiles on the time–frequency plane, then in effect there are multiple parallel channels (in the frequency domain) over which to schedule [748, 754].

7.7 Low-SNR regime

In the low-SNR regime, the spectral efficiency regions of frequency-division and of MU-MIMO become progressively similar because other-user interference takes a back seat relative to the noise. Recalling (4.25),

$$C_u(\mathsf{SNR}_u) = \dot{C}_u(0)\,\mathsf{SNR}_u + \mathcal{O}(\mathsf{SNR}_u^2), \tag{7.12}$$

which is a valid expansion for the single-user capacity with CSIR; the value of $\dot{C}_u(0)$ depends on the numbers of antennas, the presence or absence of CSIT, and possibly some channel features. Let us now see what transpires in multiuser setups.

Multiple-access channel

In the case of the MAC, from (7.2), the FDMA spectral efficiency for any user with $\mathsf{F}_u > 0$ behaves as

$$\frac{R_u}{B} = \mathsf{F}_u \left[\dot{C}_u(0) \frac{\mathsf{SNR}_u}{\mathsf{F}_u} + \mathcal{O}(\mathsf{SNR}_u^2) \right] \tag{7.13}$$

$$= \dot{C}_u(0)\,\mathsf{SNR}_u + \mathcal{O}(\mathsf{SNR}_u^2) \qquad u = 0, \ldots, U-1 \tag{7.14}$$

and thus, regardless of the choice of $\mathsf{F}_0, \ldots, \mathsf{F}_{U-1}$, all users having strictly positive F_u simultaneously enjoy (to first order) their single-user capacity. Put differently, each user can enjoy (to first order) its own single-user capacity irrespective of what the other users are doing. We note that, if any of the F_u is very small, for that user the range of validity of the first-order expansion squeezes down.

With MU-MIMO, there is other-user interference, but its effect vanishes as the SNRs drop and the behavior also conforms to (7.14). This can be verified (refer to Problems 8.10 and 8.23) by expanding specific MU-MIMO MAC spectral efficiency expressions derived in Chapter 8. Altogether, the FDMA and the MU-MIMO regions become similar in the sense that their ratio goes to unity. This implies a first-order equivalence between FDMA and MU-MIMO and therefore, by virtue of (4.30), equality in terms of $\frac{E_b}{N_0}_{\min}$.

Example 7.10

Replot the two-dimensional spectral efficiency regions of Fig. 7.1, which corresponded to intermediate SNRs, but this time for $\mathsf{SNR}_0 \ll 1$ and $\mathsf{SNR}_1 \ll 1$.

Solution

See Fig. 7.3. Notice how the FDMA and MU-SISO capacity boundaries become progres-

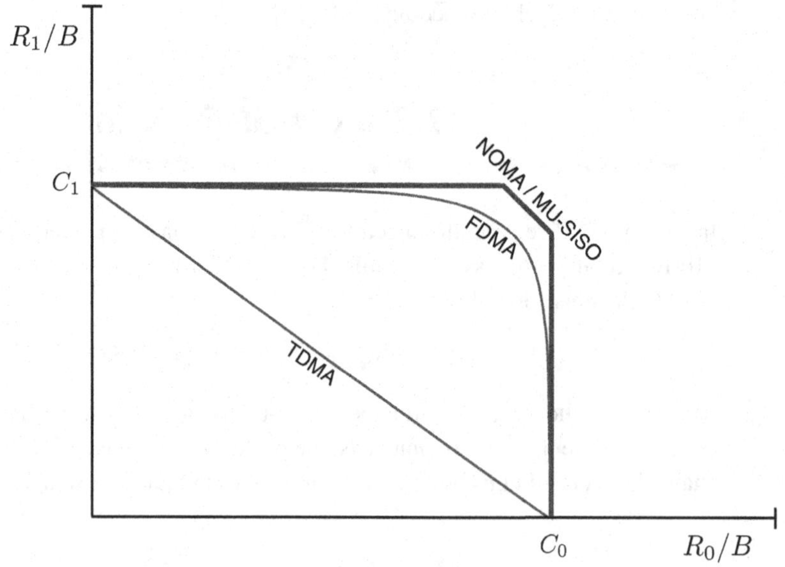

Fig. 7.3 Spectral efficiency regions for $U = 2$ when $\text{SNR}_0 \ll 1$ and $\text{SNR}_1 \ll 1$.

sively rectangular, indicating that the spectral efficiency of one user is not influenced by that of the other.

In contrast, there is no such equivalence between TDMA and MU-MIMO because of the more inefficient utilization of power in the former, a matter that is of the essence in the low-SNR regime. Indeed, from (7.1),

$$\frac{R_u}{B} = \mathsf{T}_u \, \dot{C}_u(0) \, \text{SNR}_u + \mathcal{O}(\text{SNR}_u^2) \qquad u = 0, \ldots, U-1 \tag{7.15}$$

no longer conforming to (7.14).

The equivalence between the FDMA and the MU-MIMO regions need not extend to the second-order behavior and thus the low-SNR slopes do not necessarily coincide [755]. And, just as the corresponding spectral efficiencies, the low-SNR slopes achievable by the U users form a *slope region*. For $U = 2$ and $N_{\mathrm{t}} = N_{\mathrm{r}} = 1$, this slope region is characterized in [755].

Broadcast channel

In the BC, we saw that time-division leads to (7.1) while frequency-division leads to (7.4). The respective spectral efficiency regions coincide, for arbitrary SNRs, only in the special case that the power allocation in the frequency-division approach is proportional to the bandwidth allocation. Let us now see how these regions behave at the low SNR. From

(7.12), in the time-division case we have that

$$\frac{R_u}{B} = \mathsf{T}_u \, \dot{C}_u(0) \, \mathsf{SNR}_u + \mathcal{O}(\mathsf{SNR}_u^2) \qquad u = 0, \ldots, U-1 \tag{7.16}$$

while, in the frequency-division case,

$$\frac{R_u}{B} = \mathsf{F}_u \, \dot{C}_u(0) \, \frac{\frac{E_u}{E_\mathrm{s}} \mathsf{SNR}_u}{\mathsf{F}_u} + \mathcal{O}(\mathsf{SNR}_u^2) \tag{7.17}$$

$$= \frac{E_u}{E_\mathrm{s}} \, \dot{C}_u(0) \, \mathsf{SNR}_u + \mathcal{O}(\mathsf{SNR}_u^2) \qquad u = 0, \ldots, U-1. \tag{7.18}$$

By identifying T_u in (7.16) with $\frac{E_u}{E_\mathrm{s}}$ in (7.18), the corresponding expressions for R_u/B become identical to first order; the spectral efficiency region achievable with all possible time-divisions in the former coincides with the region achievable with all possible power allocations in the latter, in the sense that their ratios go to unity. Even more importantly for our purposes, the ratio of either of these regions to the region achievable with MU-MIMO also goes to unity as the SNRs shrink. Again, this can be verified (refer to Problem 8.34) by expanding specific BC spectral efficiency expressions derived in Chapter 8.

7.8 Summary and outlook

The main lessons from this chapter are summarized in the box that accompanies this section.

As a result of the first-order optimality of frequency-division (and, in the case of the BC, also time-division) at low SNR, this regime is downplayed throughout the MU-MIMO analysis in the chapters that follow. The values of $\frac{E_\mathrm{b}}{N_0}_{\mathsf{min}}$ for MU-MIMO can be directly obtained by borrowing the corresponding single-user expressions and the signaling strategies that yield those values are the applicable forms of beamforming or statistical beamforming, with other-user interference disregarded. The slope regions, in turn, could be characterized by extending the derivations of [755].

The above point is reinforced even further in subsequent chapters, once we apply linear transceivers, since then orthogonal single-user transmissions may not only be first-order equivalent to MU-MIMO, but, depending on the channel and the type of transceiver, possibly superior. This phenomenon is exemplified in Chapter 9.

In our MU-MIMO analysis, therefore, the spotlight is on the high-SNR regime, which is where the advantages over orthogonal SU-MIMO become more pronounced and crisply evident. In particular, MU-MIMO increases the number of spatial dimensions and, in the high-SNR regime, this has a direct reflection in the number of spatial DOF. Moreover, the understanding that emerges from a high-SNR analysis—whose range of validity is determined by the lowest among the U SNRs—is decidedly illuminating.

A further consequence of focusing the MU-MIMO analysis on the high-SNR regime is that the importance of optimizing the power allocations weakens. As in SU-MIMO, spreading the transmit power uniformly over the signaling dimensions available to each

Take-away points

1. The two most relevant multiuser setups are the MAC and the BC, respectively featuring U transmitters for a single receiver and a single transmitter for U receivers.
2. Rather than a scalar $R/B \in [0, C]$, the spectral efficiency in a multiuser setup is a U-dimensional region and the capacity is the boundary of this region. Each point on the capacity boundary corresponds to a U-tuple of user spectral efficiencies.
3. The U users can share the channel via time- and/or frequency-division. This orthogonal partition is straightforward, yet in general not optimum.
4. Alternatively, the U transmissions can take place concurrently. This approach, termed NOMA or directly MU-MIMO, entails no a-priori loss of optimality.
5. Given a set of nonnegative weights $q_0, \ldots, q_{U-1}$ establishing relative user priorities, a boundary point can be determined that maximizes the corresponding weighted sum spectral efficiency. If all weights are equal, then the sum spectral efficiency is obtained. Other points of interest include the one where all spectral efficiencies are equal, and the one where the minimum thereof is largest.
6. Particularly relevant in terms of the tradeoff between aggregate performance and fairness is the proportional-fair point, where the product (or sum of the logarithms) of the user spectral efficiencies is maximized.
7. Given a population of U_{tot} users, it is customary to orthogonally multiplex subsets of U users, with each such subset operating in MU-MIMO mode. The selection and allocation of user subsets to orthogonal resource blocks is done dynamically on the basis of the users' channel conditions thereon. In the time domain specifically, the proportional-fair scheduling algorithm can be applied.
8. At low SNR, the spectral efficiency regions of orthogonal frequency-division and of MU-MIMO are equivalent to first order. For the BC, this equivalence extends also to time-division.

user is often satisfactory and frequently optimal in this regime [756]. And, when it comes to allocating the power of the BC transmitter among users, spreading such power in direct proportion to the weights $q_0 \ldots, q_{U-1}$ turns out to be the generalization of the uniform power allocation whenever the objective is not the sum of the spectral efficiencies, but rather some weighted sum. Despite this weakened importance of the power allocation, for the sake of completeness, and because it sometimes contributes to the overall understanding, we entertain the general optimization of the transmit powers in each of the multiuser settings considered in the sequel. Also entertained, and crucial for the BC at all SNRs, are those precoding aspects related to the spatial orientation of the signals, i.e, the steering matrices.

The proportional-fair scheduling algorithm—suitably extended to time–frequency resource partitions and tweaked to accommodate strict latency and bit rate constraints for some applications—is a satisfactory approach to determine which U users should be allocated to each resource block, with $U = 1$ for SU-MIMO and $U > 1$ for MU-MIMO. These U users are then the ones maximizing the weighted sum spectral efficiency with user weights $q_0, \ldots, q_{U-1}$ given by the reciprocals of the respective average spectral efficiencies. This delivers the arguably most relevant point on the capacity boundary. At the same time, obtaining these specific weights requires dynamic simulations instantiating the large-scale as well as the time–frequency small-scale channel coefficients and actually implementing the scheduling algorithm. For the purpose of our exposition of MIMO, it is preferable to resort to operating points corresponding to some fixed weights and thus, in the chapters that follow, the MU-MIMO coverage touches on several of the scalar metrics presented earlier but prioritizes the sum, especially when it comes to the choice of examples and illustrations. This preference is further reinforced by the fact that in the MAC, as is to be seen, often the choice of weights alters the spectral efficiencies achieved by the users but not their sum, which remains constant.

Problems

7.1 Consider an unfaded single-antenna MAC with $U = 2$ and with $\mathsf{SNR}_0 = 15$ dB and $\mathsf{SNR}_1 = 10$ dB. Plot the TDMA/FDMA spectral efficiency region.

7.2 Consider an unfaded single-antenna BC with $U = 2$ and with $\mathsf{SNR}_0 = 15$ dB and $\mathsf{SNR}_1 = 10$ dB.
(a) Plot the time-division spectral efficiency region.
(b) Plot the frequency-division spectral efficiency region.
Hint: Remember that SNR_u *is the SNR that the uth user would experience signaling over the entire bandwidth B.*

7.3 Consider an unfaded single-antenna MAC. Interpret the aggregation of the U single-antenna users as a U-antenna transmitter with precoder $\boldsymbol{F} = \boldsymbol{I}$, such that no power sharing or collaboration can take place between those users.
(a) Provide an expression for the capacity boundary.
(b) For $U = 2$ and $\mathsf{SNR}_0 = 15$ dB, $\mathsf{SNR}_1 = 10$ dB, draw the capacity boundary.

7.4 Prove that, in a single-antenna MAC, reducing the power from any of the users can only shrink the capacity boundary.

7.5 For an unfaded single-antenna BC operated in a time-division fashion, with $U = 2$, establish the boundary pairs $(R_0/B, R_1/B)$ for the following operating points.
(a) Equal spectral efficiency for both users.
(b) Maximization of the minimum spectral efficiency.

7.6 Repeat Problem 7.5 for frequency-division.

7.7 Let a base station transmit independent QPSK signals to two users via superposition, with 90% of the power allocated to user 0 and the remaining 10% to user 1.
(a) Plot the composite transmit constellation.
(b) Is Gray mapping feasible?

7.8 For an unfaded MU-SISO MAC with $U = 2$, give the boundary pairs $(R_0/B, R_1/B)$ for the following operating points.

(a) Equal spectral efficiency for both users.

(b) Maximization of the minimum spectral efficiency.

7.9 Consider an unfaded single-antenna BC operated in a time-division fashion and let $U = 3$ with $\mathsf{SNR}_0 = 10$ dB, $\mathsf{SNR}_1 = 7$ dB, and $\mathsf{SNR}_2 = 3$ dB. Compute the triplet $(R_0/B, R_1/B, R_2/B)$ whose weighted sum is maximized for the following weight combinations.

(a) $q_0 = 0.3$, $q_1 = 0.3$ and $q_2 = 0.4$.

(b) $q_0 = 0.2$, $q_1 = 0.3$ and $q_2 = 0.5$.

(c) $q_0 = 0.3$, $q_1 = 0.4$ and $q_2 = 0.3$.

7.10 Consider an unfaded MU-SISO MAC and let $U = 2$ with $\mathsf{SNR}_0 = \mathsf{SNR}_1 = 10$ dB. Compute the pair $(R_0/B, R_1/B)$ whose weighted sum is maximized for $q_0 = 0.8$ and $q_1 = 0.2$.

7.11 Consider an unfaded single-antenna BC operated in a time-division fashion with $\mathsf{SNR}_0 = 10$ dB and $\mathsf{SNR}_1 = 0$ dB. Compute the boundary pair $(R_0/B, R_1/B)$ corresponding to the proportional-fair point.

7.12 Prove that, in an unfaded single-antenna MAC or BC operated in a time-division fashion, the point corresponding to $\mathsf{T}_0 = \mathsf{T}_1 = 0.5$ satisfies (7.7) and it is thus the proportional-fair point.

Note: Without fading, simple round-robin transmission suffices. With fading and a proportional-fair scheduler, the generalization of this result is that every user is allocated the same share of time, but, rather than round-robin, the user selected at each slot is the one whose fading is the most favorable relative to its own average.

7.13 Consider an unfaded MU-SISO MAC. Let $U = 2$ with $\mathsf{SNR}_0 = 10$ dB. What is the range of SNR_1 such that operating at the equal-spectral-efficiency point entails a sacrifice in sum spectral efficiency?

7.14 Consider a population of U_{tot} users with $\mathsf{SNR}_u = 10$ dB for $u = 0, \ldots, U_{\text{tot}} - 1$ and independent Rayleigh block fading per user. The channel is shared via time-division. Suppose that a scheduler selects, at each time slot, the single user having the highest SNR.

(a) Plot the ergodic spectral efficiency attained over the channel as a function of $U_{\text{tot}} \in [1, 1000]$, with U_{tot} in log-scale.

(b) On the same chart, repeat part (a) with the fading at each user clipped to have a PAPR of 6 dB.

What do you observe?

7.15 Consider a population of U_{tot} users with each SNR_u drawn independently and uniformly between 0 and 20 dB. Each user is also subject to unclipped Rayleigh block fading. The channel is shared via time-division. Suppose that a scheduler selects, at each time slot, the single user having the highest SNR.

(a) Plot the ergodic spectral efficiency, further averaged over the distributions of SNR_u for $u = 0, \ldots, U_{\text{tot}} - 1$, as a function of $U_{\text{tot}} \in [1, 1000]$ in log-scale.

(b) On the same chart, repeat part (a) with the fading at each user clipped to have a PAPR of 6 dB.

What do you observe?

7.16 Consider a population of U_{tot} users with $\text{SNR}_u = 10$ dB for $u = 0, \dots, U_{\text{tot}} - 1$ and independent Rayleigh block fading per user. The channel is shared via time-division. Suppose that a scheduler selects, at each time slot, the single user having the highest SNR.

(a) Plot the ergodic spectral efficiency attained over the channel as a function of $U_{\text{tot}} \in [1, 1000]$, with U_{tot} in log-scale.

(b) On the same chart, repeat part (a) with the constellation cardinality restricted to 16-QAM.

What do you observe?

Hint: For (b), it may be useful to precompute the 16*-QAM mutual information function into a look-up table from which values for arbitrary SNRs can be interpolated.*

7.17 Contemplate a single-antenna BC with single-user transmission ($U = 1$) via time-division among U_{tot} users. All time slots are of equal duration. Let $U_{\text{tot}} = 1000$ and $\text{SNR}_u = 5$ dB for $u = 0, \dots, U_{\text{tot}} - 1$, with the users further subject to unclipped Rayleigh fading.

(a) Suppose that a scheduler selects, at each slot, the user having the highest SNR. Plot the CDF of the spectral efficiency over the channel. Further compute the ergodic spectral efficiency and the average latency (in number of slots) between successive transmissions to the same user.

(b) Repeat part (a) with a proportional-fair scheduler for $\varepsilon = 10$, $\varepsilon = 100$, and $\varepsilon = 1000$ slots.

(c) Repeat part (a) for a scheduler that allocates each slot to the user with the lowest spectral efficiency running average.

(d) Repeat part (a) for a round-robin scheduler.

7.18 Contemplate a single-antenna BC with single-user transmission ($U = 1$) via time-division among U_{tot} users. All time slots are of equal duration. Let $\text{SNR}_u = 10$ dB for $u = 0, \dots, U_{\text{tot}} - 1$, with the users further subject to unclipped Rayleigh fading.

(a) Suppose that a scheduler selects, at each slot, the user having the highest SNR. Plot the ergodic spectral efficiency as a function of $U_{\text{tot}} \in [1, 1000]$ in log-scale.

(b) Repeat part (a) with a proportional-fair scheduler for $\varepsilon = 10$, $\varepsilon = 100$, and $\varepsilon = 1000$ slots.

(c) Repeat part (a) for a scheduler that allocates each slot to the user with the lowest spectral efficiency running average.

(d) Repeat part (a) for a round-robin scheduler.

8 MU-MIMO with optimum transceivers

Every great movement must experience three stages: ridicule, discussion, adoption.

John Stuart Mill

8.1 Introduction

As argued in the previous chapter, MU-MIMO is hardly better than frequency-division SU-MIMO in the low-SNR regime. At high SNR, conversely, MU-MIMO can be decidedly superior. Given the lesser complexity of orthogonal channel sharing, this invites applying SU-MIMO and MU-MIMO for low- and high-SNR users, respectively. Good system designs should thus feature both SU-MIMO ($U = 1$) and MU-MIMO ($U > 1$), and indeed they do. At the same time, as seen in this chapter, it is ineffective to have U exceed a certain value that depends on the numbers of antennas; thus, even MU-MIMO needs to be combined with time- and frequency-division because of the need to accommodate a potentially large population of U_{tot} users.

The information-theoretic principles of SU-MIMO were covered in detail in the second part of the book, and their applicability in concert with scheduled time- and frequency-division is immediate. The present part is centered on MU-MIMO, with SU-MIMO as an occasional baseline and with the low-SNR regime de-emphasized in favor of high-SNR conditions.

This specific chapter initiates the treatment of MU-MIMO, with the objective of establishing the MAC and BC capacity boundaries as well as transmitter and receiver architectures that can attain them. Optimality is the driving force, rather than complexity. Perhaps counterintuitively, the MAC is notoriously less intricate than the BC, and thus we begin by analyzing the former and then, taking advantage of a duality relationship between the two, move on to the latter. Within both the MAC and BC portions of this chapter, the organization of the discussion mirrors that of the SU-MIMO exposition, with the availability of CSI as a guiding axis. And, on account of the observations made in Section 4.7, we prioritize the ergodic setting.

The chapter is organized as follows. Sections 8.2 through 8.6 address the MAC, with progressively diminishing degrees of CSI, while Sections 8.8 through 8.10 cover the BC, again with diminishing levels of CSI. Between these two blocks, Section 8.7 plays the pivotal role of presenting the duality that connects the MAC and the BC, facilitating the analysis of the latter based on the former. Finally, Section 8.11 wraps up the chapter.

Discussion 8.1 Fading distribution in multiuser channels

In a multiuser setup, the resource allocation policy modifies the fading distribution. Specifically, in assigning users to selected time–frequency resource blocks, the resource allocation procedure tends to render the fading distribution more benign. Adverse situations such as pronounced downfades tend to be avoided. This multiuser diversity is disregarded if the fading distribution is taken to be a standard one, e.g., Rayleigh, Rice, or their MIMO generalizations.

The impact of multiuser diversity is major on outage metrics. The impact is much more moderate on ergodic quantities, where expectations—such as the ones computed in this and subsequent chapters—over standard fading distributions should be construed as conservative assessments of the achievable performance. If multiuser diversity were taken into account, then, besides the fading distributions themselves, the actual performance would go on to depend on U_{tot} and on the resource allocation policy.

8.2 The multiple-access channel

The MAC, illustrated in Fig. 8.1, consists of U transmitters and a single receiver; it abstracts a reverse link with the users and the base station playing the roles of transmitters and receiver, respectively. The capacity of the MAC was first established, for SISO links, in the 1970s [757, 758]. In what follows, we proceed to formulate it directly for MIMO.

With N_u antennas at the uth transmitter and N_{r} antennas at the common receiver, the single-letter MAC transmit–receive relationship under frequency-flat fading is

$$\boldsymbol{y} = \sum_{u=0}^{U-1} \sqrt{G_u} \boldsymbol{H}_u \boldsymbol{x}_u + \boldsymbol{v}, \tag{8.1}$$

where $\boldsymbol{H}_u$ is the $N_{\text{r}} \times N_u$ channel matrix linking the uth user with the receiver, normalized as usual such that $\mathbb{E}\big[\|\boldsymbol{H}_u\|_{\text{F}}^2\big] = N_u N_{\text{r}}$, while G_u is the corresponding large-scale channel gain and

$$\boldsymbol{x}_u = \sqrt{\frac{E_u}{N_u}} \boldsymbol{F}_u \boldsymbol{s}_u. \tag{8.2}$$

The precoder $\boldsymbol{F}_u$, and thus the covariance $\boldsymbol{R}_{\boldsymbol{x}_u} = \frac{E_u}{N_u} \boldsymbol{F}_u \boldsymbol{F}_u^*$, are subject to the type of power constraint (per-codeword, per-symbol, or per-antenna) that applies to user u, whose energy per symbol is $E_u \leq E_{\text{s}}$. The signal vectors $\boldsymbol{s}_0, \ldots, \boldsymbol{s}_{U-1}$ have IID entries whereas the noise is $\boldsymbol{v} \sim \mathcal{N}_{\mathbb{C}}(\boldsymbol{0}, N_0 \boldsymbol{I})$. Recalling, from the previous chapter, the definition

$$\mathsf{SNR}_u = \frac{G_u E_{\text{s}}}{N_0} \tag{8.3}$$

we have that the SNR experienced by user u is determined by the product $\frac{E_u}{E_{\text{s}}} \mathsf{SNR}_u$, where SNR_u is a local-average inherent to the channel (pathloss, shadow fading, bandwidth,

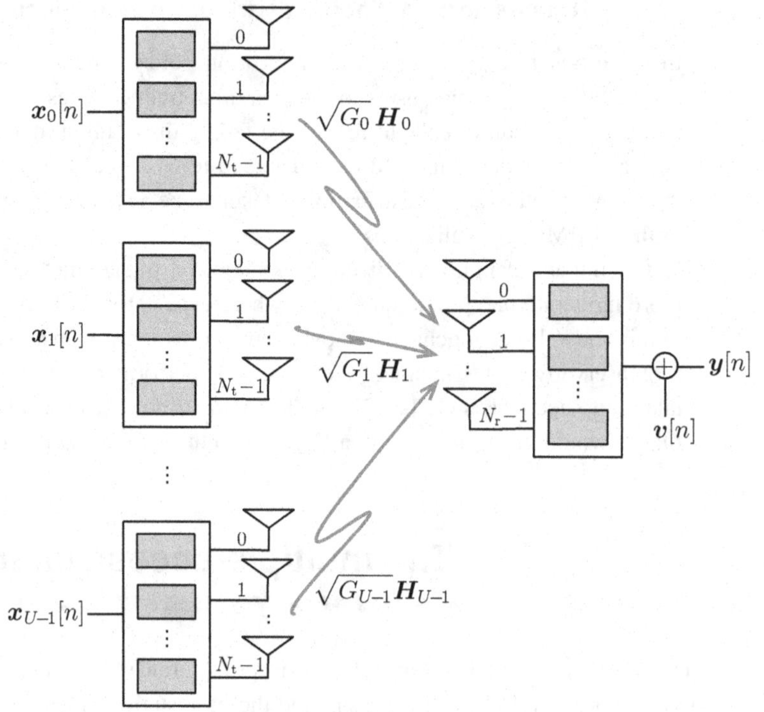

Fig. 8.1 MAC with U N_t-antenna transmitters and one N_r-antenna receiver.

power budget) whereas $\frac{E_u}{E_s} \in [0, 1]$ is a power control coefficient that can be adjusted on the basis of SNR_u, or of other factors.

As a consequence of the focus on the high-SNR regime, in our multiuser analysis we let—unless otherwise stated—the number of transmit signal streams be as large as possible, not contemplating beamforming, but rather full spatial multiplexing.

It is sometimes useful to interpret the MAC in Fig. 8.1 as an SU-MIMO channel where the U transmitters are dislodged pieces—unable to jointly precode or pool their transmit powers—of a larger aggregate transmitter. This forces the precoder and spatial covariance of this aggregate transmitter to have a block-diagonal structure, but otherwise the isomorphism does hold. To reflect this interpretation, we can rewrite (8.1) as

$$\boldsymbol{y} = \underbrace{\left[\sqrt{G_0}\boldsymbol{H}_0 \; \sqrt{G_1}\boldsymbol{H}_1 \; \cdots \; \sqrt{G_{U-1}}\boldsymbol{H}_{U-1}\right]}_{\boldsymbol{C}} \underbrace{\begin{bmatrix} \boldsymbol{x}_0 \\ \vdots \\ \boldsymbol{x}_{U-1} \end{bmatrix}}_{\boldsymbol{x}} + \boldsymbol{v} \tag{8.4}$$

$$= \boldsymbol{C}\boldsymbol{x} + \boldsymbol{v}, \tag{8.5}$$

which is the familiar SU-MIMO relationship, only with a more structured $N_r \times \sum_u N_u$

channel matrix $\boldsymbol{C}$ and with

$$\boldsymbol{R}_{\boldsymbol{x}} = \begin{bmatrix} \frac{E_0}{N_0}\boldsymbol{F}_0\boldsymbol{F}_0^* & 0 & \cdots & 0 \\ 0 & \frac{E_1}{N_1}\boldsymbol{F}_1\boldsymbol{F}_1^* & 0 & \cdots \\ \vdots & & \ddots & \vdots \\ 0 & \cdots & 0 & \frac{E_{U-1}}{N_{U-1}}\boldsymbol{F}_{U-1}\boldsymbol{F}_{U-1}^* \end{bmatrix} \tag{8.6}$$

having more restrictive constraints. In the special case that all users feature a single antenna (an MU-SIMO MAC), we have that $\boldsymbol{R}_{\boldsymbol{x}} = \text{diag}(E_0, \ldots, E_{U-1})$.

Alternatively to (8.4), we can write

$$\boldsymbol{y} = \left[\sqrt{\frac{G_0 E_0}{N_0}}\boldsymbol{H}_0 \quad \cdots \quad \sqrt{\frac{G_{U-1}E_{U-1}}{N_{U-1}}}\boldsymbol{H}_{U-1} \right] \underbrace{\begin{bmatrix} \boldsymbol{F}_0 & 0 & \cdots & 0 \\ 0 & \boldsymbol{F}_1 & 0 & \cdots \\ \vdots & & \ddots & \vdots \\ 0 & \cdots & 0 & \boldsymbol{F}_{U-1} \end{bmatrix}}_{\boldsymbol{F}} \boldsymbol{s} + \boldsymbol{v}, \tag{8.7}$$

where $\boldsymbol{F}$ is an aggregate precoder and $\boldsymbol{s} = \left[\boldsymbol{s}_0^{\mathrm{T}} \cdots \boldsymbol{s}_{U-1}^{\mathrm{T}}\right]^{\mathrm{T}}$. In an MU-SIMO MAC, $\boldsymbol{F} = \boldsymbol{I}$ and $\boldsymbol{s} = \left[s_0 \cdots s_{U-1}\right]^{\mathrm{T}}$.

For any aggregate precoder complying with the above structure, expressions derived for SU-MIMO can be directly imported to compute the sum spectral efficiency. Thus, even before we formally begin our analysis of the MU-MIMO MAC we can already make a powerful observation: except for further restrictions on the precoding, an MU-MIMO MAC looks like an SU-MIMO channel participated by all U users at once. And, because the SU-MIMO capacity is monotonic in the transmit power, we can infer that users should transmit at full power, meaning that $\frac{E_u}{E_s} = 1$ for $u = 0, \ldots, U-1$ such that (8.3) is directly the local-average SNR of user u. (The optimality of full-power transmission holds as long as the receiver is optimum, but not necessarily once we consider linear receivers in the next chapter.)

Also note that, in the foregoing SU-MIMO interpretation of the MAC, the transmissions from the various users have been assumed to be symbol-synchronous. This is not unreasonable in the context of OFDM; coarse synchronization is ensured by advancing or delaying each user's signals and any remaining time offset is automatically handled as part of the corresponding delay spread. Readers interested in how to explicitly incorporate asynchronicities into the MAC analysis are referred to [759, 760].

To avoid an explosion in the number of parameters that would hardly add further intuition, the exposition in the sequel lets $N_u = N_{\mathrm{t}}$ for $u = 0, \ldots, U-1$, with qualifying comments wherever appropriate.

Provided there is CSIR, complex Gaussian codebooks continue to maximize the mutual information, in this case between the U transmitters and the receiver. The argument used to justify this optimality in Chapter 5 applies verbatim: for a given power, the differential entropy of $\boldsymbol{y}$ is maximized when its distribution (conditioned on $\boldsymbol{H}_0, \ldots, \boldsymbol{H}_{U-1}$) is complex Gaussian and, in the face of Gaussian noise, this occurs when $\boldsymbol{x}_0, \ldots, \boldsymbol{x}_{U-1}$ are themselves complex Gaussian. Then, the only quantities that remain free in the derivation of the MU-MIMO MAC capacity are the user precoders, depending on the availability of CSIT and the type of power constraint; as seen in the sequel though, the optimization of these precoders is considerably more involved than in SU-MIMO.

8.3 Multiple-access channel with CSIR and CSIT

With CSIT, a per-codeword power constraint would open the door to time-domain power control. The optimum power control in a MAC shares some of the features of time-domain waterfilling, namely that users are allocated more power in favorable fading conditions and less in poor fading conditions, but with the differentiating aspect that multiple users compete for the channel; thus, the transmit powers become in general coupled and the optimum policy allows for only a limited number of simultaneous transmissions, depending on $N_{\rm r}$. Rather than elaborate on this further, we henceforth focus directly on the more practically relevant per-symbol power constraint, meaning $\|\boldsymbol{F}_u\|_{\rm F}^2 = N_{\rm t}$ for $u = 0, \ldots, U-1$. Readers interested in power control policies for the MAC under a per-codeword power constraint are referred to [761–765]

8.3.1 Quasi-static setting

In this setting, as usual, $\boldsymbol{H}_0, \ldots, \boldsymbol{H}_{U-1}$ are fixed over the codeword span. It is instructive to begin with a basic setup where $U = 2$ and $N_{\rm t} = 1$.

Example 8.1

Determine the capacity boundary for a MAC with $U = 2$ and with $\boldsymbol{H}_0$ and $\boldsymbol{H}_1$ being $N_{\rm r} \times 1$ vectors.

Solution

The spectral efficiency of each user cannot exceed the respective individual capacity and thus $R_0/B \leq C_0$ and $R_1/B \leq C_1$ with

$$C_0(\mathsf{SNR}_0, \boldsymbol{H}_0) = \log_2\big(1 + \mathsf{SNR}_0 \, \|\boldsymbol{H}_0\|^2\big) \tag{8.8}$$

$$C_1(\mathsf{SNR}_1, \boldsymbol{H}_1) = \log_2\big(1 + \mathsf{SNR}_1 \, \|\boldsymbol{H}_1\|^2\big). \tag{8.9}$$

Invoking the analogy with an SU-MIMO setup whose channel is

$$\boldsymbol{C} = \left[\sqrt{G_0}\boldsymbol{H}_0 \;\; \sqrt{G_1}\boldsymbol{H}_1\right] \tag{8.10}$$

and where the aggregate precoder is the identity matrix, we see that $R_0/B + R_1/B \leq C$, with the sum-capacity being (see Example 1.13)

$$C(\mathsf{SNR}_0, \mathsf{SNR}_1, \boldsymbol{H}_0, \boldsymbol{H}_1) = \log_2 \det\left(\boldsymbol{I} + \frac{E_{\rm s}}{N_0}\boldsymbol{C}\boldsymbol{C}^*\right) \tag{8.11}$$

$$= \log_2 \det\left(\boldsymbol{I} + \mathsf{SNR}_0 \boldsymbol{H}_0 \boldsymbol{H}_0^* + \mathsf{SNR}_1 \boldsymbol{H}_1 \boldsymbol{H}_1^*\right). \tag{8.12}$$

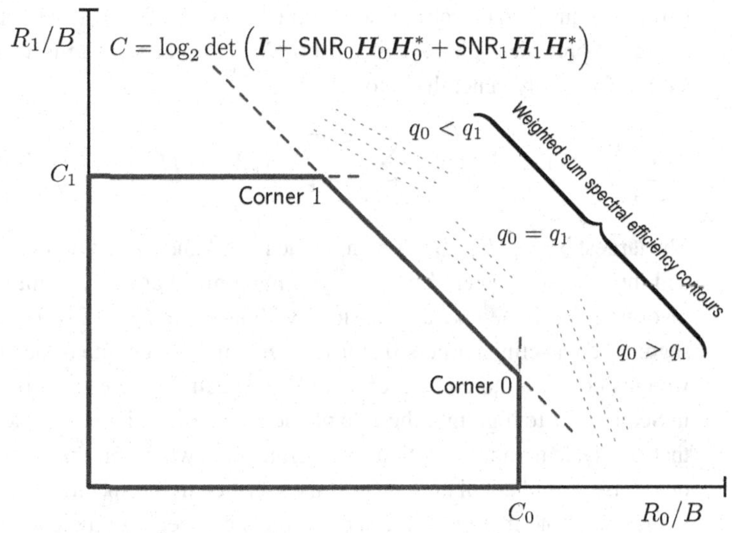

Fig. 8.2 MAC with $U = 2$ and $N_t = 1$. Dashed, conditions $R_0/B = C_0$, $R_1/B = C_1$, and $R_0/B + R_1/B = C$. Solid, MAC capacity boundary. Also shown with dotted lines are the contours of constant weighted sum spectral efficiency for $q_0 < q_1$, $q_0 = q_1$, and $q_0 > q_1$.

The conditions $R_0/B = C_0$, $R_1/B = C_1$ and $R_0/B + R_1/B = C$ map, on a two-dimensional representation, to lines that intersect at two corner points and define the capacity boundary depicted in Fig. 8.2. In combination with the axes, which can be interpreted as incorporating the trivial conditions $R_0/B \geq 0$ and $R_1/B \geq 0$, this boundary encloses a pentagonal region of spectral efficiency pairs $(R_0/B, R_1/B)$.

A compact way of describing the capacity boundary in Example 8.1 is

$$\sum_{u \in \mathcal{U}} \frac{R_u}{B} \leq \log_2 \det\left(\boldsymbol{I} + \sum_{u \in \mathcal{U}} \mathsf{SNR}_u \boldsymbol{H}_u \boldsymbol{H}_u^*\right) \tag{8.13}$$

applied with $\mathcal{U} = \{0\}$, $\mathcal{U} = \{1\}$, and $\mathcal{U} = \{0, 1\}$. This description integrates the three conditions defining the segments that form the capacity boundary in Fig. 8.2. Extending this description to an arbitrary number of users, the achievable spectral efficiency U-tuples satisfy (8.13) with the summations expanded to encompass all possible subsets $\mathcal{U} \subseteq \{0, \ldots, U-1\}$. This gives $2^U - 1$ conditions, which for $U = 3$ correspond to planes instead of segments, and for $U > 3$ to hyperplanes. Intersecting at $U!$ corner points, these hyperplanes define a capacity boundary that generalizes the two-dimensional pentagon in Fig. 8.2. For $U = 3$, this region acquires the form of a polyhedron and, beyond that, of a U-dimensional polytope.

Note that, since every user transmits a signal stream, dimensional overloading takes place if $U > N_r$.

As the final step in the characterization of the MAC capacity boundary with CSIR and CSIT in a quasi-static setting, we need to allow for $N_t > 1$, which brings into the formulation the precoders $\boldsymbol{F}_0, \ldots, \boldsymbol{F}_{U-1}$. For $N_t > 1$ and any admissible $\boldsymbol{F}_0, \ldots, \boldsymbol{F}_{U-1}$, (8.13) straightforwardly generalizes to

$$\sum_{u \in \mathcal{U}} \frac{R_u}{B} \leq \log_2 \det\left(\boldsymbol{I} + \sum_{u \in \mathcal{U}} \frac{\mathsf{SNR}_u}{N_t} \boldsymbol{H}_u \boldsymbol{F}_u \boldsymbol{F}_u^* \boldsymbol{H}_u^* \right) \qquad \mathcal{U} \subseteq \{0, \ldots, U-1\}. \quad (8.14)$$

The largest $\sum_{u=0}^{U-1} R_u/B$ is given by the right-hand side of (8.14) with $\mathcal{U} = \{0, \ldots, U-1\}$ and thus the sum spectral efficiency is maximized at every point on the main hyperplane involving simultaneous transmission by all users; in Fig. 8.2, this means every point on the segment connecting corners 0 and 1. Interestingly then, the imposition of fairness through weights comes at no cost in terms of sum spectral efficiency, a point that was anticipated in Section 7.7 to highlight the role of such sum as a chief scalar metric. This also indicates that the system should preferably operate somewhere on that main hyperplane, as doing otherwise would entail an unnecessary sacrifice in the spectral efficiency of at least one of the users. (Note that the invariance of the sum spectral efficiency to the fairness imposed through weights may not hold if it were mandated that $R_0/B = \ldots = R_{U-1}/B$, as that could push the system outside the referred hyperplane. This is the subject of Problem 7.13.)

The weighted sum, $\sum_{u=0}^{U-1} q_u R_u/B$, is maximized by the spectral efficiency U-tuple at a corner point of that main hyperplane, depending on $q_0, \ldots, q_{U-1}$. This is evident in Fig. 8.2, where corner 0 is optimum for $q_0 > q_1$ while corner 1 is optimum for $q_0 < q_1$.

Example 8.2

Let $U = 2$ with $\mathsf{SNR}_0 = 10$ dB and $\mathsf{SNR}_1 = 3$ dB, and with

$$\boldsymbol{H}_0 = \begin{bmatrix} 1 + 0.5\mathrm{j} & 2 \\ 0.3 - \mathrm{j} & 0.7 + 1.2\mathrm{j} \end{bmatrix} \qquad \boldsymbol{H}_1 = \begin{bmatrix} 0.4\mathrm{j} & 2 + 0.3\mathrm{j} \\ 1.1 + \mathrm{j} & 1.2 \end{bmatrix}. \quad (8.15)$$

Further let the precoders be $\boldsymbol{F}_0 = \boldsymbol{F}_1 = \boldsymbol{I}$. For $q_0 = 1/4$ and $q_1 = 3/4$, compute the pair $(R_0/B, R_1/B)$ that maximizes $q_0 R_0/B + q_1 R_1/B$.

Solution

The SU-MIMO capacities (in b/s/Hz) are

$$C_0 = \log_2 \det\left(\boldsymbol{I} + \frac{\mathsf{SNR}_0}{2} \boldsymbol{H}_0 \boldsymbol{H}_0^* \right) = 8.5 \quad (8.16)$$

$$C_1 = \log_2 \det\left(\boldsymbol{I} + \frac{\mathsf{SNR}_1}{2} \boldsymbol{H}_1 \boldsymbol{H}_1^* \right) = 4, \quad (8.17)$$

whereas (8.14) gives $R_0/B + R_1/B \leq 9$ b/s/Hz. The pair $(R_0/B, R_1/B)$ that maximizes $q_0 R_0/B + q_1 R_1/B$ is found at corner 1 under the labeling in Fig. 8.2, where $R_1/B = C_1 = 4$ b/s/Hz and $R_0/B = 9 - C_1 = 5$ b/s/Hz.

Let us now release the precoders from being fixed. The MU-MIMO MAC capacity is, by

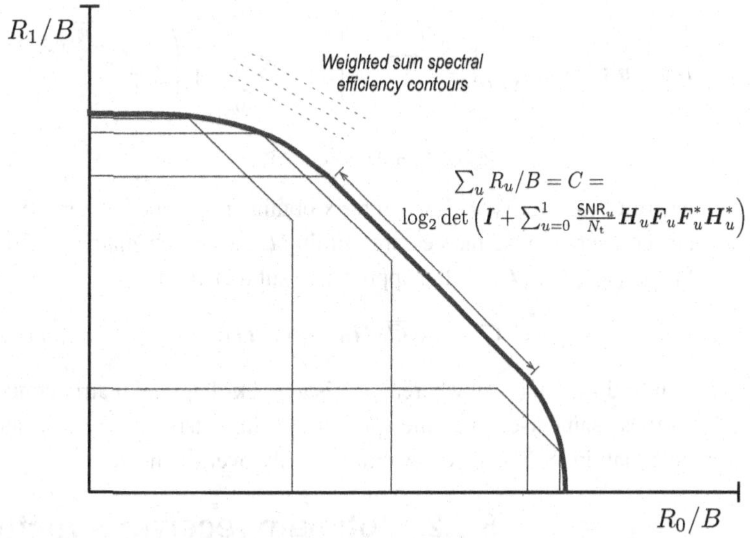

Fig. 8.3 MAC capacity boundary with $U = 2$. With thin lines, boundaries of the pentagonal regions achievable with some specific precoders. With a thicker line, boundary of the union of all possible such pentagonal regions. Also shown with dotted lines are the weighted sum spectral efficiency contours for some q_0 and q_1.

definition, the boundary of the union of the regions spawned by all admissible precoders, i.e., the boundary of

$$\bigcup_{\|F_u\|_F^2 = N_t} \left[\sum_{u \in \mathcal{U}} \frac{R_u}{B} \leq \log_2 \det\left(I + \sum_{u \in \mathcal{U}} \frac{\mathsf{SNR}_u}{N_t} H_u F_u F_u^* H_u^* \right) \qquad \mathcal{U} \subseteq \{0, \ldots, U-1\} \right] \tag{8.18}$$

which is exemplified, for $U = 2$, in Fig. 8.3. The precoders that correspond to the points on the boundary cannot in general be expressed explicitly, but they can be obtained numerically as explained later in the section.

Discrete constellations

With signals drawn from equiprobable M-ary constellations, rather than the complex Gaussian distribution, (8.14) naturally becomes

$$\sum_{u \in \mathcal{U}} \frac{R_u}{B} \leq I(s^{\mathcal{U}}; y \,|C^{\mathcal{U}}) \qquad \mathcal{U} \subseteq \{0, \ldots, U-1\} \tag{8.19}$$

where $s^{\mathcal{U}}$ is the $|\mathcal{U}|N_s \times 1$ vector obtained by stacking the signal vectors of all users whose indices are within the set $\mathcal{U}$. To express $I(s^{\mathcal{U}}; y\,|C^{\mathcal{U}})$, we can again resort to the SU-MIMO analogy and rewrite (5.104) into

$$I(s^{\mathcal{U}};y|C^{\mathcal{U}}) = \frac{-1}{M^{|\mathcal{U}|N_s}} \sum_{m'=0}^{M^{|\mathcal{U}|N_s}-1} \mathbb{E}\left[\log_2 \sum_{m=0}^{M^{|\mathcal{U}|N_s}-1} \exp\left(\frac{-\left\|\sqrt{\frac{E_s}{N_t}}C^{\mathcal{U}}F^{\mathcal{U}}(\mathbf{s}_{m'}^{\mathcal{U}} - \mathbf{s}_m^{\mathcal{U}}) + v\right\|^2}{N_0}\right)\right]$$
$$+ |\mathcal{U}|N_s \log_2 M - N_r \log_2 e, \tag{8.20}$$

where $C^{\mathcal{U}}$ is the $N_r \times |\mathcal{U}|N_t$ matrix obtained by concatenating the channel matrices from all the users whose indices are within $\mathcal{U}$, each such matrix scaled by the corresponding factor $\sqrt{G_u}$, i.e., $C^{\mathcal{U}}$ is the appropriate submatrix of

$$C = \begin{bmatrix} \sqrt{G_0}H_0 & \sqrt{G_1}H_1 & \cdots & \sqrt{G_{U-1}}H_{U-1} \end{bmatrix} \tag{8.21}$$

as introduced in (8.4), whereas $F^{\mathcal{U}}$ is a block-diagonal matrix containing the precoders for all those same users, i.e., the appropriate submatrix of the aggregate precoder in (8.7). We note that, in (8.20), the expectation is only over the noise v.

8.3.2 Optimum receiver structure

As seen for SU-MIMO, information theory not only serves to quantify the limits of reliable communication, but it can further offer design guidelines. Specifically, an LMMSE-SIC receiver can achieve capacity, and it can do so regardless of the order in which the codewords are decoded. In an MU-MIMO MAC, the optimality of LMMSE-SIC not only holds, but it acquires further significance. (Recall from Chapter 5 that, in fact, the information-theoretic optimality of the LMMSE-SIC receiver was first observed in the context of the MAC [591] and only in due course recognized for SU-MIMO.)

In MU-MIMO, the codewords are transmitted by distinct users and while, as in SU-MIMO, the sum-capacity is invariant to their decoding order, the individual spectral efficiencies of the U users are not. Hence, different boundary points can be attained with different decoding orders.

Example 8.3

Consider again Example 8.1 and Fig. 8.2. Which pairs $(R_0/B, R_1/B)$ can be achieved by an LMMSE-SIC receiver?

Solution

Referring back to Section 5.7, if user 0 is decoded with interference from user 1,

$$\frac{R_0}{B} = \log_2\left(1 + G_0 E_s\, H_0^*\left(N_0 I + G_1 E_s H_1 H_1^*\right)^{-1} H_0\right) \tag{8.22}$$
$$= \log_2\left(1 + \mathsf{SNR}_0 H_0^*\left(I + \mathsf{SNR}_1 H_1 H_1^*\right)^{-1} H_0\right) \tag{8.23}$$

while subsequently, with $\sqrt{G_0}H_0 x_0$ canceled from the received signal, $R_1/B = C_1$. With a modicum of algebra, the sum of (8.23) and C_1 can be seen to equal the sum-capacity in (8.12) and thus we are operating at corner 1 of the capacity boundary under the labeling in Fig. 8.2. Conversely, if the decoding order were reversed, we would operate at corner 0. And, by time-dividing both decoding orders, we could operate at any point on the sum-capacity segment connecting the corners.

Since corner 0 befits $q_0 > q_1$ while corner 1 befits $q_0 < q_1$, the highest weight user should be decoded last so it does not experience interference.

When $U > 2$, the capacity boundary has $U!$ corners and each can be achieved with a distinct decoding order. To operate at the correct corner for some $q_0, \ldots, q_{U-1}$, i.e., to maximize the weighted sum spectral efficiency, users should be decoded in order of increasing weights. The sum-capacity, though, is achieved regardless of the order because it is the same at every corner point, and in fact all over the main hyperplane [766, 767].

The LMMSE-SIC optimality and significance also apply with $N_\text{t} > 1$, as the next example illustrates.

Example 8.4

Reconsider Example 8.3, only with $N_\text{t} > 1$. Which spectral efficiency pairs can be achieved by an LMMSE-SIC receiver?

Solution

If the signal of user 0 is decoded with interference from user 1, then

$$\frac{R_0}{B} = \log_2 \det\left(\boldsymbol{I} + \frac{\mathsf{SNR}_0}{N_\text{t}} \boldsymbol{H}_0 \boldsymbol{F}_0 \boldsymbol{F}_0^* \boldsymbol{H}_0^* \left(\boldsymbol{I} + \frac{\mathsf{SNR}_1}{N_\text{t}} \boldsymbol{H}_1 \boldsymbol{F}_1 \boldsymbol{F}_1^* \boldsymbol{H}_1^* \right)^{-1} \right) \tag{8.24}$$

while subsequently, with $\sqrt{G_0} \boldsymbol{H}_0 \boldsymbol{x}_0$ canceled from the received signal,

$$\frac{R_1}{B} = C_1(\mathsf{SNR}_1, \boldsymbol{H}_1) = \log_2 \det\left(\boldsymbol{I} + \frac{\mathsf{SNR}_1}{N_\text{t}} \boldsymbol{H}_1 \boldsymbol{F}_1 \boldsymbol{F}_1^* \boldsymbol{H}_1^* \right). \tag{8.25}$$

The sum of (8.24) and (8.25) gives the sum-capacity and thus this pair $(R_0/B, R_1/B)$ continues to lie at corner 1 under the labeling in Fig. 8.2. The reverse decoding order would give the pair at corner 0.

The above is irrespective of whether each user transmits one or multiple codewords. In the former case, joint decoding of the signals transmitted by the N_t user antennas is required whereas, in the latter, an inner LMMSE-SIC process could be applied instead (with the decoding order therein immaterial).

Example 8.4 stretches rather effortlessly to $U > 2$. With the users indexed and decoded in order of increasing weight, user u suffers interference from users $u+1, \ldots, U-1$ but not from users $0, \ldots, u-1$ and thus

$$\frac{R_u}{B} = \log_2 \det\left(\boldsymbol{I} + \frac{\mathsf{SNR}_u}{N_\text{t}} \boldsymbol{H}_u \boldsymbol{F}_u \boldsymbol{F}_u^* \boldsymbol{H}_u^* \left(\boldsymbol{I} + \sum_{\mathrm{u}=u+1}^{U-1} \frac{\mathsf{SNR}_\mathrm{u}}{N_\text{t}} \boldsymbol{H}_\mathrm{u} \boldsymbol{F}_\mathrm{u} \boldsymbol{F}_\mathrm{u}^* \boldsymbol{H}_\mathrm{u}^* \right)^{-1} \right)$$

$$u = 0, \ldots, U-1, \tag{8.26}$$

which, denoting the inverted matrix by $\boldsymbol{A}$ and decomposing the leading identity matrix as $\boldsymbol{I} = \boldsymbol{A}\boldsymbol{A}^{-1}$, can be rewritten as

$$\frac{R_u}{B} = \log_2 \det\left(\boldsymbol{I} + \sum_{\mathrm{u}=u}^{U-1} \frac{\mathsf{SNR}_{\mathrm{u}}}{N_{\mathrm{t}}} \boldsymbol{H}_{\mathrm{u}} \boldsymbol{F}_{\mathrm{u}} \boldsymbol{F}_{\mathrm{u}}^{*} \boldsymbol{H}_{\mathrm{u}}^{*}\right) - \log_2 \det\left(\boldsymbol{I} + \sum_{\mathrm{u}=u+1}^{U-1} \frac{\mathsf{SNR}_{\mathrm{u}}}{N_{\mathrm{t}}} \boldsymbol{H}_{\mathrm{u}} \boldsymbol{F}_{\mathrm{u}} \boldsymbol{F}_{\mathrm{u}}^{*} \boldsymbol{H}_{\mathrm{u}}^{*}\right) \qquad u = 0, \ldots, U-1. \tag{8.27}$$

Such $(R_0/B, \ldots, R_{U-1}/B)$ corresponds to one of the $U!$ corners of the region spawned by $\boldsymbol{F}_0, \ldots, \boldsymbol{F}_{U-1}$, precisely the corner that befits $q_0 \leq \cdots \leq q_{U-1}$. All that remains is to optimize the precoders, a task on which we embark shortly.

As mentioned, the sum-capacity is achieved at any of the corners, or at any point on the hyperplane that connects those corners, and therefore it is not affected by the decoding order. The decoding order affects the spectral efficiencies of the individual users, but not their sum.

We hasten to emphasize that LMMSE-SIC reception is not the only way to operate at any given corner, and that a joint decoder can achieve the same spectral efficiency U-tuples, but an LMMSE-SIC receiver does facilitate deriving and interpreting (8.27). In fact, caution must be exercised when actually employing an LMMSE-SIC receiver in an MU-MIMO MAC because of the potentially large power differences between users, given their different locations. (This is in contrast with SU-MIMO, where all the codewords emanate from the same location.) Imperfections in the CSIR, even if small, are amplified by these power differences and can give rise to highly restrictive levels of interference. This is sometimes referred to as the *near–far effect*.

Example 8.5

Let $U = 2$ and $\mathsf{SNR}_0|_{\mathrm{dB}} - \mathsf{SNR}_1|_{\mathrm{dB}} = 20$ dB. Suppose that user 0 is decoded first, with its signal subsequently reconstructed and cancelled from the observation, $\boldsymbol{y}$. If the CSIR imperfection led to a 1% error in that reconstruction and cancelation, user 1 would encounter a residual interference as strong as its own signal. How small must the error be to ensure that the residual interference is at least 10 dB below?

Solution

An additional 10 dB means that the error cannot exceed 0.1%.

The near–far effect can be mitigated by power control—which makes additional sense when one considers discrete constellations, rather than Gaussian signals, as then it is pointless to transmit at power levels beyond those needed to support the available constellation cardinalities.

8.3.3 Precoder optimization

Recall that the users are indexed such that $q_0 \leq \cdots \leq q_{U-1}$. From (8.27), the weighted sum spectral efficiency equals

$$\sum_{u=0}^{U-1} q_u \frac{R_u}{B} = \sum_{u=0}^{U-1} q_u \left[\log_2 \det\left(\boldsymbol{I} + \sum_{\mathrm{u}=u}^{U-1} \frac{\mathsf{SNR}_\mathrm{u}}{N_\mathrm{t}} \boldsymbol{H}_\mathrm{u} \boldsymbol{F}_\mathrm{u} \boldsymbol{F}_\mathrm{u}^* \boldsymbol{H}_\mathrm{u}^* \right) - \log_2 \det\left(\boldsymbol{I} + \sum_{\mathrm{u}=u+1}^{U-1} \frac{\mathsf{SNR}_\mathrm{u}}{N_\mathrm{t}} \boldsymbol{H}_\mathrm{u} \boldsymbol{F}_\mathrm{u} \boldsymbol{F}_\mathrm{u}^* \boldsymbol{H}_\mathrm{u}^* \right) \right]. \tag{8.28}$$

Although it may not be immediately obvious whether the optimization of (8.28) over the precoders is convex, because we have the difference of two concave functions, which itself need not be concave, the concavity of (8.28) can be made evident by regrouping it into

$$\sum_{u=0}^{U-1} q_u \frac{R_u}{B} = q_0 \log_2 \det\left(\boldsymbol{I} + \sum_{\mathrm{u}=0}^{U-1} \frac{\mathsf{SNR}_\mathrm{u}}{N_\mathrm{t}} \boldsymbol{H}_\mathrm{u} \boldsymbol{F}_\mathrm{u} \boldsymbol{F}_\mathrm{u}^* \boldsymbol{H}_\mathrm{u}^* \right) + \sum_{u=1}^{U-1} (q_u - q_{u-1}) \log_2 \det\left(\boldsymbol{I} + \sum_{\mathrm{u}=u}^{U-1} \frac{\mathsf{SNR}_\mathrm{u}}{N_\mathrm{t}} \boldsymbol{H}_\mathrm{u} \boldsymbol{F}_\mathrm{u} \boldsymbol{F}_\mathrm{u}^* \boldsymbol{H}_\mathrm{u}^* \right), \tag{8.29}$$

where we now have a sum of positive terms, each concave in a subset of the covariance matrices $\boldsymbol{F}_0 \boldsymbol{F}_0^*, \ldots, \boldsymbol{F}_{U-1} \boldsymbol{F}_{U-1}^*$. Moreover, the feasible $\boldsymbol{F}_0 \boldsymbol{F}_0^*, \ldots, \boldsymbol{F}_{U-1} \boldsymbol{F}_{U-1}^*$ take values on a convex set. These covariances, and thus the precoders yielding the weighted sum-capacity for any $q_0, \ldots, q_{U-1}$, can be found with general-purpose convex optimization tools (see Appendix G). For a tailored algorithm, readers are referred to [34, section 3.5.1].

Referring back to the two-user example in Fig. 8.3, the optimization of the precoders is tantamount to identifying the specific pentagon whose appropriate corner (the one determined by $q_0 \leq q_1$) is at the boundary wherever the contours of constant weighted sum spectral efficiency touch it for the given q_0 and q_1.

For the sum-capacity in particular, the optimum precoders can alternatively be found through the elegant *iterative waterfilling* algorithm [768], which exploits the structure of the problem. This algorithm rests on the intuitive idea that the optimum precoder for the uth user must solve the corresponding SU-MIMO optimization (see Section 5.3) with the received signals from all other users acting as interference, i.e., with a Gaussian noise-plus-interference vector having covariance

$$\boldsymbol{\Sigma}_u = N_0 \boldsymbol{I} + \sum_{\mathrm{u} \neq u} \frac{G_\mathrm{u} E_\mathrm{s}}{N_\mathrm{t}} \boldsymbol{H}_\mathrm{u} \boldsymbol{F}_\mathrm{u} \boldsymbol{F}_\mathrm{u}^* \boldsymbol{H}_\mathrm{u}^*, \tag{8.30}$$

and this must hold for all users simultaneously. The SU-MIMO solution for the uth user is then to diagonalize $\boldsymbol{\Sigma}_u^{-1/2} \boldsymbol{H}_u$ via SVD and allocate its power via waterfilling thereupon. The fact that this needs to be true for all users invites an iterative procedure whereby users take turns at optimizing their respective precoders, each updated while the others are held fixed. Striking a compromise between the benefits for the channel of interest and the impairment to the rest, each precoder evolves and the algorithm rapidly converges. Multiple solutions mapping to the same sum-capacity may exist for $\boldsymbol{F}_0, \ldots, \boldsymbol{F}_{U-1}$ and the algorithm always converges to one of those, depending on the initialization. Iterative waterfilling is attractive when the goal is to seek the sum-capacity, without regard for how that capacity breaks down among users.

Discrete constellations

Consider now discrete constellations. From (8.19)–(8.20), the problem of finding the precoders $\boldsymbol{F}_0, \ldots, \boldsymbol{F}_{U-1}$ that maximize the region of achievable spectral efficiencies—or, equivalently, the problem of maximizing the weighted sum spectral efficiency for some arbitrary weights—is no longer a concave function of those precoders. Thus, as in SU-MIMO with discrete constellations, the conditions that can be derived through the vector I-MMSE relationship are necessary but not sufficient for optimality; they are satisfied by any local maximum, minimum, and saddle point. An iterative algorithm designed to find maxima has been proposed [769], but global optimality cannot be guaranteed.

A relevant observation made in [769, 770] is that, when the constellations are fixed, the performance with the capacity-achieving precoders (designed for complex Gaussian signals) can be substantially worse than with the constellation-optimized precoders. More surprisingly, the performance with the capacity-achieving precoders can be substantially worse than *without* precoding. This is a sobering reminder that the results derived for Gaussian signaling are representative up to some SNR that depends on the constellation richness. Beyond that SNR, there is little point in going through the optimization (e.g., iterative waterfilling) necessary to compute the capacity-achieving precoders.

Example 8.6

Let $U = 2$ with $N_\text{t} = N_\text{r} = 2$ and with $\mathsf{SNR}_0 = \mathsf{SNR}_1 = 10$ dB. Further, let

$$\boldsymbol{H}_0 = \begin{bmatrix} 1.39 & 0.11\,\text{j} \\ -0.11\,\text{j} & 0.21 \end{bmatrix} \qquad \boldsymbol{H}_1 = \begin{bmatrix} 1.22 & 0 \\ 0 & 0.7 \end{bmatrix}. \tag{8.31}$$

Compute the achievable spectral efficiency regions with QPSK signaling, both with the capacity-achieving precoders and without precoding.

Solution

The regions, obtained by evaluating (8.19) and (8.20), on the one hand with the precoders obtained from a convex optimization of the weighted sum-capacity and on the other hand with $\boldsymbol{F}_0 = \boldsymbol{F}_1 = \boldsymbol{I}$, are shown in Fig. 8.4.

As the mutual information of QPSK signals departs markedly from its complex Gaussian counterpart at this SNR, the capacity-achieving precoders are actually detrimental. In particular, the combined strength of noise and other-user interference drives the capacity-achieving precoders to effect beamforming: each user transmits a single signal stream, and this is ill-advised when the signal is QPSK because it limits R_0/B and R_1/B to 2 b/s/Hz each. Simple unprecoded signaling from all users ends up yielding a substantially larger region that exhibits the familiar pentagonal shape associated with fixed precoders. With precoders tailored to QPSK, an even larger region would be attained (see [769, figure 7]).

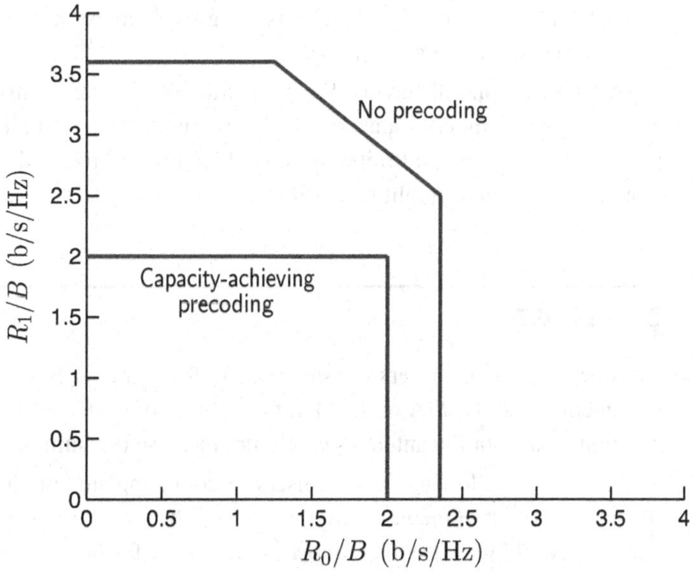

Fig. 8.4 MU-MIMO MAC spectral efficiency regions with $U = 2$, $N_t = N_r = 2$, $\text{SNR}_0 = \text{SNR}_1 = 10$ dB, and QPSK signaling. Capacity-achieving precoders versus unprecoded transmission.

8.3.4 High-SNR regime

Spatial DOF

Recall that, for SNR $\to \infty$, the waterfilling power allocation in SU-MIMO becomes uniform. Rewriting the expansion in (5.29) as a function of E_s/N_0 rather than SNR, the ensuing SU-MIMO capacity behaves as

$$C = S_\infty \log_2 \frac{E_s}{N_0} + \mathcal{O}(1), \tag{8.32}$$

where G has been lumped into the $\mathcal{O}(1)$ term and where the number of spatial DOF is the customary $S_\infty = \min(N_t, N_r)$. In a multiuser channel where SU-MIMO is applied in conjunction with FDMA or TDMA, the number of DOF does not increase further.

Enter MU-MIMO. The aggregate channel of the U users is described by the $N_r \times UN_t$ matrix defined in (8.4) and (8.21) and, with the usual caveat of full-rank channel matrices, the sum-capacity behaves as

$$C = \min(UN_t, N_r) \log_2 \frac{E_s}{N_0} + \mathcal{O}(1), \tag{8.33}$$

where the advantage of expanding with respect to E_s/N_0 becomes clear, as a single term is obtained despite the plurality of SNRs. It can be verified, from the rank of the argument of the determinant in (8.14), that (8.33) is indeed the expansion thereof with $G_0, \ldots, G_{U-1}$ and $\boldsymbol{F}_0, \ldots, \boldsymbol{F}_{U-1}$ relegated to the $\mathcal{O}(1)$ term. This expansion illuminates the key advantage of MU-MIMO over SU-MIMO: the number of spatial DOF is $S_\infty = \min(UN_t, N_r)$

rather than $S_\infty = \min(N_t, N_r)$, registering the increase in the number of spatial dimensions available for communication.

The implications of having $S_\infty = \min(UN_t, N_r)$ are immediate. Rather than being limited by the number of antennas at the user end, often a small number, by having multiple concurrent users the limitation in DOF can be shifted to the base station, where large numbers of antennas might be feasible.

Example 8.7

Consider two-antenna users transmitting to a four-antenna base station. With SU-MIMO in conjunction with FDMA or TDMA, the number of spatial DOF is $S_\infty = 2$. Furthermore, additional base station antennas would not increase the number of DOF.

However, by allowing $U = 2$ users to communicate simultaneously via MU-MIMO, the number of DOF increases to $S_\infty = 4$. With $U > 2$, moreover, additional base station antennas would increase the number of DOF even further.

The increase in spatial DOF that it may bring about crisply justifies the space-division multiple access (SDMA) interpretation that is sometimes made of MU-MIMO, i.e., the interpretation of MU-MIMO as a way to organize channel access in the space domain, in contrast with (or, more appropriately, in addition to) TDMA and FDMA. Much in the same way that TDMA and FDMA orthogonalize the transmissions in their respective domains, in an orthodox conception of SDMA the transmissions would be as orthogonal as possible in the space domain [771]. If the angle spreads are narrow enough to render the antennas strongly correlated, and there is sufficient angular separation among users, the SDMA receiver could form U beams pointing to the respective users and achieve a certain degree of isolation [772]. This orthodox conception is actually transcended in MU-MIMO, where the users are jointly processed and, by virtue of fading and CSIR, can be discriminated even if not angularly separated [773].

Note that, in terms of spatial DOF, there is no point in making U larger than the value that makes $UN_t = N_r$. If $UN_t > N_r$, then we have dimensional overloading and there are not enough spatial dimensions at the receiver to simultaneously accommodate N_t disentangled signals from each of U users. Suppose, though, that we do have $UN_t > N_r$ and we insist on each user transmitting N_t signal streams. At whatever corner point of the capacity boundary is optimal for the given $q_0, \ldots, q_{U-1}$, the user decoded last communicates in interference-free conditions thereby enjoying $\min(N_t, N_r)$ spatial DOF. The user decoded next-to-last communicates in the face of an N_t-dimensional interferer whose power does not vanish with growing E_s/N_0; as learned in earlier chapters, linearly removing such interference requires N_t receiver dimensions and thus that next-to-last user enjoys $\min(N_t, N_r - N_t)$ spatial DOF, and so on. Finally, the user decoded first enjoys $\min(N_t, N_r - (U-1)N_t)$. The foregoing DOF should be taken as zero whenever negative and hence, for $UN_t > N_r$, some of the early-decoded users are bound to have zero DOF.

Example 8.8

Consider a three-user MAC with $N_{\text{t}} = 2$ and $N_{\text{r}} = 4$. At the corner point that maps to LMMSE-SIC processing in some order, the user decoded first enjoys zero DOF while the other two users enjoy two DOF each. The sum-capacity scales with $S_{\infty} = 4$ b/s/Hz/(3 dB).

If $UN_{\text{t}} > N_{\text{r}}$, to have all users enjoy some spatial DOF it is necessary that $N_{\text{s}} < N_{\text{t}}$ signal streams be transmitted (e.g., beamforming with $N_{\text{s}} = 1$) such that the total number of streams does not exceed S_{∞}.

Power offset

The legitimate excitement elicited by the prospects of an increased number of spatial DOF thanks to MU-MIMO must be slightly tempered, as argued for SU-MIMO in Section 4.2, by the fact that the number of DOF is only a partially informative metric. Conscious of this, we proceed to refine the coarse expansion in (8.33). Precisely, we characterize the power offset for $UN_{\text{t}} \leq N_{\text{r}}$. Under this condition, each user can find N_{t} usable dimensions and therefore the signals from the U users become progressively disentangled as the SNRs grow large. It can be shown [774, lemma 1] that, if we denote by $\boldsymbol{H}_{\perp,u}$ the projection of $\boldsymbol{H}_u$ onto the null space of $\boldsymbol{H}_{u+1}, \ldots, \boldsymbol{H}_{U-1}$, then

$$\lim_{E_{\text{s}}/N_0 \to \infty} \left[\boldsymbol{H}_{\perp,u}^* \boldsymbol{H}_{\perp,u} - \boldsymbol{H}_u^* \left(\boldsymbol{I} + \sum_{\text{u}=u+1}^{U-1} \frac{\mathsf{SNR}_{\text{u}}}{N_{\text{t}}} \boldsymbol{H}_{\text{u}} \boldsymbol{F}_{\text{u}} \boldsymbol{F}_{\text{u}}^* \boldsymbol{H}_{\text{u}}^* \right)^{-1} \boldsymbol{H}_u \right] = \boldsymbol{0} \tag{8.34}$$

and thus, for the purpose of what takes place for $E_{\text{s}}/N_0 \to \infty$, we can replace (8.26) by

$$\frac{R_u}{B} = \log_2 \det \left(\boldsymbol{I} + \frac{\mathsf{SNR}_u}{N_{\text{t}}} \boldsymbol{H}_{\perp,u} \boldsymbol{F}_u \boldsymbol{F}_u^* \boldsymbol{H}_{\perp,u}^* \right) \qquad u = 0, \ldots, U-1. \tag{8.35}$$

This implies that, in the high-SNR regime, user u transmits on the null space of the channels of users $u+1, \ldots, U-1$; user u could thus experience interference only from users $0, \ldots, u-1$, but these have already been decoded and removed in an MMSE-SIC procedure by the time the signals of user u come up for decoding. Asymptotically released from all interference, user u then undergoes the SU-MIMO channel embodied by $\boldsymbol{H}_{\perp,u}$ and impaired only by Gaussian noise, as reflected by (8.35); the ensuing waterfilling is known to converge to a uniform power allocation over all dimensions with nonzero gain. Since the condition $UN_{\text{t}} \leq N_{\text{r}}$ ensures that the projections $\boldsymbol{H}_{\perp,0}, \ldots, \boldsymbol{H}_{\perp,U-1}$ are full-rank, it follows that $\boldsymbol{F}_u \to \boldsymbol{I}$ for $u = 0, \ldots, U-1$.

Plugging $\boldsymbol{F}_u = \boldsymbol{I}$ into (8.35) and expanding, we obtain

$$\frac{R_u}{B} = \log_2 \det \left(\frac{\mathsf{SNR}_u}{N_{\text{t}}} \boldsymbol{H}_{\perp,u}^* \boldsymbol{H}_{\perp,u} \right) + \mathcal{O}\left(\frac{1}{\mathsf{SNR}_u} \right) \tag{8.36}$$

$$= N_{\text{t}} \log_2 \mathsf{SNR}_u - N_{\text{t}} \log_2 N_{\text{t}} + \log_2 \det(\boldsymbol{H}_{\perp,u}^* \boldsymbol{H}_{\perp,u}) + \mathcal{O}\left(\frac{1}{\mathsf{SNR}_u} \right). \tag{8.37}$$

Discussion 8.2 Collaborative upper bound for the MU-MIMO MAC

Imagine the U users in a MU-MIMO MAC truly functioning as a single user, pooling their powers and jointly precoding. Since cost-free collaboration can only be beneficial, the ensuing single-user capacity upper bounds the sum-capacity of the actual MU-MIMO MAC. While not generally tight (refer to Problem 8.15), this collaborative upper bound does becomes tight for $UN_\mathrm{t} \leq N_\mathrm{r}$ and once all users have entered the high-SNR regime. Indeed, the collaborative user's precoder would satisfy $\boldsymbol{F} \to \boldsymbol{I}$ with uniform power allocation across all UN_t antennas while, in the actual MU-MIMO MAC, $\boldsymbol{F}_u \to \boldsymbol{I}$ for $u = 0, \ldots, U-1$. Remarkably, even if admittedly only in the asymptote of all SNRs, the MU-MIMO capacity is exactly as if the U users could fully collaborate and pool their powers, even though such collaboration is not taking place.

Each user enjoys N_t spatial DOF and, adapting the SU-MIMO definition in Section 4.2, the uth user exhibits a power offset

$$\mathcal{L}_{\infty,u}(\boldsymbol{H}_u, \ldots, \boldsymbol{H}_{U-1}) = \lim_{\mathsf{SNR}_u \to \infty} \left(\log_2 \mathsf{SNR}_u - \frac{R_u/B}{N_\mathrm{t}} \right) \tag{8.38}$$

$$= \log_2 N_\mathrm{t} - \frac{1}{N_\mathrm{t}} \log_2 \det\left(\boldsymbol{H}^*_{\perp,u} \boldsymbol{H}_{\perp,u}\right), \tag{8.39}$$

where we have emphasized that, in a quasi-static setting, the power offset is a function of the channel realizations; moreover, it is not only a function of $\boldsymbol{H}_u$ but, through the projection on their null space, also of $\boldsymbol{H}_{u+1}, \ldots, \boldsymbol{H}_{U-1}$. Utilizing the power offset, we can write the high-SNR expansion of R_u/B as

$$\frac{R_u}{B} = N_\mathrm{t} \left(\log_2 \mathsf{SNR}_u - \mathcal{L}_{\infty,u}(\boldsymbol{H}_u, \ldots, \boldsymbol{H}_{U-1}) \right) + \mathcal{O}\left(\frac{1}{\mathsf{SNR}_u} \right). \tag{8.40}$$

For given channel realizations, the power offset of each user depends on the decoding order and it is through that order that the weights $q_0, \ldots, q_{U-1}$ are accounted for. What does not depend on the decoding order is the sum-capacity, which always satisfies

$$C = \sum_{u=0}^{U-1} N_\mathrm{t} \left(\log_2 \mathsf{SNR}_u - \mathcal{L}_{\infty,u}(\boldsymbol{H}_u, \ldots, \boldsymbol{H}_{U-1}) \right) + \mathcal{O}\left(\frac{1}{E_\mathrm{s}/N_0} \right) \tag{8.41}$$

$$= N_\mathrm{t} \left(\sum_{u=0}^{U-1} \log_2 \mathsf{SNR}_u - U\mathcal{L}_{\infty}(\boldsymbol{H}_u, \ldots, \boldsymbol{H}_{U-1}) \right) + \mathcal{O}\left(\frac{1}{E_\mathrm{s}/N_0} \right) \tag{8.42}$$

$$= N_\mathrm{t} \sum_{u=0}^{U-1} \left(\log_2 \mathsf{SNR}_u - \mathcal{L}_{\infty}(\boldsymbol{H}_u, \ldots, \boldsymbol{H}_{U-1}) \right) + \mathcal{O}\left(\frac{1}{E_\mathrm{s}/N_0} \right), \tag{8.43}$$

and which represents the sought refinement of (8.33) for $UN_\mathrm{t} \leq N_\mathrm{r}$. The quantity

$$\mathcal{L}_{\infty}(\boldsymbol{H}_0, \ldots, \boldsymbol{H}_{U-1}) = \frac{1}{U} \sum_{u=0}^{U-1} \mathcal{L}_{\infty,u}(\boldsymbol{H}_u, \ldots, \boldsymbol{H}_{U-1}) \tag{8.44}$$

$$= \log_2 N_t - \frac{1}{UN_t} \sum_{u=0}^{U-1} \log_2 \det\left(\boldsymbol{H}_{\perp,u}^* \boldsymbol{H}_{\perp,u}\right) \tag{8.45}$$

naturally emerges as the sum-capacity version of the single-user power offset. The expression in (8.45) has the inconvenience of involving the matrices $\boldsymbol{H}_{\perp,0}, \ldots, \boldsymbol{H}_{\perp,U-1}$. An analytically more convenient form can be found by plugging $\boldsymbol{F}_u = \boldsymbol{I}$ into the right-hand side of (8.14) and expand the ensuing sum-capacity expression

$$C = \log_2 \det\left(\boldsymbol{I} + \sum_{u=0}^{U-1} \frac{\mathsf{SNR}_u}{N_t} \boldsymbol{H}_u \boldsymbol{H}_u^*\right) \tag{8.46}$$

$$= \log_2 \det\left(\boldsymbol{I} + \frac{E_s/N_0}{N_t} \sum_{u=0}^{U-1} \boldsymbol{H}_u G_u \boldsymbol{H}_u^*\right) \tag{8.47}$$

$$= \log_2 \det\left(\boldsymbol{I} + \frac{E_s}{N_t N_0} \boldsymbol{H} \operatorname{diag}\big(\underbrace{G_0, \ldots, G_0}_{N_t}, \ldots, \underbrace{G_{U-1}, \ldots, G_{U-1}}_{N_t}\big) \boldsymbol{H}^*\right) \tag{8.48}$$

$$= \log_2 \det\left(\boldsymbol{I} + \frac{E_s}{N_t N_0} \operatorname{diag}\big(\underbrace{G_0, \ldots, G_0}_{N_t}, \ldots, \underbrace{G_{U-1}, \ldots, G_{U-1}}_{N_t}\big) \boldsymbol{H}^* \boldsymbol{H}\right), \tag{8.49}$$

where we have introduced the $N_r \times UN_t$ matrix $\boldsymbol{H} = \left[\boldsymbol{H}_0 \cdots \boldsymbol{H}_{U-1}\right]$. The expansion of (8.49) gives

$$C = \log_2 \det\left(\frac{E_s}{N_t N_0} \operatorname{diag}\big(\underbrace{G_0, \ldots, G_0}_{N_t}, \ldots, \underbrace{G_{U-1}, \ldots, G_{U-1}}_{N_t}\big) \boldsymbol{H}^* \boldsymbol{H}\right) + \mathcal{O}\left(\frac{1}{E_s/N_0}\right) \tag{8.50}$$

$$= UN_t \log_2 \frac{E_s}{N_t N_0} + \log_2\left(\prod_{u=0}^{U-1} G_u^{N_t}\right) + \log_2 \det\left(\boldsymbol{H}^* \boldsymbol{H}\right) + \mathcal{O}\left(\frac{1}{E_s/N_0}\right) \tag{8.51}$$

$$= UN_t \log_2 \frac{E_s}{N_0} + \sum_{u=0}^{U-1} N_t \log_2 G_u - UN_t \log_2 N_t + \log_2 \det\left(\boldsymbol{H}^* \boldsymbol{H}\right) + \mathcal{O}\left(\frac{1}{E_s/N_0}\right) \tag{8.52}$$

$$= \sum_{u=0}^{U-1} N_t \log_2 \mathsf{SNR}_u - UN_t \log_2 N_t + \log_2 \det(\boldsymbol{H}^* \boldsymbol{H}) + \mathcal{O}\left(\frac{1}{E_s/N_0}\right) \tag{8.53}$$

$$= N_t \sum_{u=0}^{U-1} \Big(\log_2 \mathsf{SNR}_u - \mathcal{L}_\infty(\boldsymbol{H}_0, \ldots, \boldsymbol{H}_{U-1})\Big) + \mathcal{O}\left(\frac{1}{E_s/N_0}\right), \tag{8.54}$$

from which we can identify for the sum-capacity power offset the alternative form

$$\mathcal{L}_\infty(\boldsymbol{H}_0, \ldots, \boldsymbol{H}_{U-1}) = \log_2 N_t - \frac{1}{UN_t} \log_2 \det\left(\boldsymbol{H}^* \boldsymbol{H}\right), \tag{8.55}$$

which depends only on $\boldsymbol{H}$, far simpler to manipulate than $\boldsymbol{H}_{\perp,0},\ldots,\boldsymbol{H}_{\perp,U-1}$. We return to this power offset (which, recall, is for $UN_{\rm t} \leq N_{\rm r}$) in the ergodic analysis that follows.

8.3.5 Ergodic setting

Expecting (8.14) over the fading distribution, the boundary of the ergodic spectral efficiency region achievable with CSIT-based precoders $\boldsymbol{F}_0,\ldots,\boldsymbol{F}_{U-1}$ is characterized by

$$\sum_{u\in\mathcal{U}} \frac{R_u}{B} = \mathbb{E}\left[\log_2 \det\left(\boldsymbol{I} + \sum_{u\in\mathcal{U}} \frac{\mathsf{SNR}_u}{N_{\rm t}} \boldsymbol{H}_u \boldsymbol{F}_u \boldsymbol{F}_u^* \boldsymbol{H}_u^*\right)\right] \qquad \mathcal{U} \subseteq \{0,\ldots,U-1\}, \tag{8.56}$$

where, though not indicated explicitly, the precoders (subject to the applicable power constraints) are a function of $\boldsymbol{H}_0,\ldots,\boldsymbol{H}_{U-1}$ and thus the expectation is over them as well.

Let us now focus on one of the $U!$ corners of the region enclosed by (8.56). Invoking an LMMSE-SIC receiver and assuming as usual that the users are indexed and decoded in order of increasing weight, the expectation of (8.27) over $\boldsymbol{H}_0,\ldots,\boldsymbol{H}_{U-1}$ gives

$$\begin{aligned}\frac{R_u}{B} = \mathbb{E}\Bigg[&\log_2 \det\left(\boldsymbol{I} + \sum_{\mathrm{u}=u}^{U-1} \frac{\mathsf{SNR}_{\mathrm{u}}}{N_{\rm t}} \boldsymbol{H}_{\mathrm{u}} \boldsymbol{F}_{\mathrm{u}} \boldsymbol{F}_{\mathrm{u}}^* \boldsymbol{H}_{\mathrm{u}}^*\right) \\ &- \log_2 \det\left(\boldsymbol{I} + \sum_{\mathrm{u}=u+1}^{U-1} \frac{\mathsf{SNR}_{\mathrm{u}}}{N_{\rm t}} \boldsymbol{H}_{\mathrm{u}} \boldsymbol{F}_{\mathrm{u}} \boldsymbol{F}_{\mathrm{u}}^* \boldsymbol{H}_{\mathrm{u}}^*\right)\Bigg] \qquad u = 0,\ldots,U-1,\end{aligned} \tag{8.57}$$

where, again, the precoders are functions of $\boldsymbol{H}_0,\ldots,\boldsymbol{H}_{U-1}$ and hence the expectation is over them as well. The ergodic weighted sum-capacity for some $q_0,\ldots,q_{U-1}$ is obtained by expecting $\sum_u q_u R_u/B$ over $\boldsymbol{H}_0,\ldots,\boldsymbol{H}_{U-1}$ with the precoders optimized at each realization.

Example 8.9

Consider a three-user MAC with $N_{\rm t} = 2$ and $N_{\rm r} = 6$. The respective local-average channel strengths are referenced to a common SNR such that

$$\mathsf{SNR}_0|_{\rm dB} = \mathsf{SNR}|_{\rm dB} \tag{8.58}$$

$$\mathsf{SNR}_1|_{\rm dB} = \mathsf{SNR}|_{\rm dB} + 5\,\text{dB} \tag{8.59}$$

$$\mathsf{SNR}_2|_{\rm dB} = \mathsf{SNR}|_{\rm dB} + 8\,\text{dB}, \tag{8.60}$$

with $\boldsymbol{H}_0$, $\boldsymbol{H}_1$ and $\boldsymbol{H}_2$ having IID Rayleigh-faded entries. Obtain the ergodic sum-capacity and compare it with its unprecoded counterpart. Further compare it with the SU-MIMO ergodic capacity for each of the users.

Solution

Shown in Fig. 8.5 are the ergodic sum-capacity, obtained through iterative waterfilling by all users on each channel realization, and the ergodic sum spectral efficiency with

in Appendix C.1.9 as well as (E.9) in Appendix E,

$$\mathcal{L}_\infty = \log_2 N_t + \left(\gamma_{\mathsf{EM}} - \sum_{q=1}^{N_r - UN_t} \frac{1}{q} - \frac{N_r}{UN_t} \sum_{q=N_r-UN_t+1}^{N_r} \frac{1}{q} + 1 \right) \log_2 e, \tag{8.63}$$

which, for $UN_t = N_r$, becomes

$$\mathcal{L}_\infty = \log_2 N_r + \left(\gamma_{\mathsf{EM}} - \sum_{q=2}^{N_r} \frac{1}{q} \right) \log_2 e. \tag{8.64}$$

Example 8.12

Consider $U = 3$ users, each equipped with $N_t = 2$ antennas, transmitting to a base station having $N_r = 6$ antennas. Under IID Rayleigh fading, compute $\mathcal{L}_\infty$ and utilize it to approximate the ergodic sum-capacity at high SNR.

Solution

Applying (8.64), we obtain $\mathcal{L}_\infty = -0.26$ (in 3-dB units). Along with (8.61), this gives

$$C(\mathsf{SNR}_0, \mathsf{SNR}_1, \mathsf{SNR}_2) \approx 2 \sum_{u=0}^{2} \big(\log_2 \mathsf{SNR}_u + 0.26 \big). \tag{8.65}$$

In terms of individual user spectral efficiencies, the breakdown of DOF discussed earlier for given channel realizations holds: the user decoded first enjoys $\min(N_t, N_r - (U-1)N_t)$ DOF, the user decoded second enjoys $\min(N_t, N_r - (U-2)N_t)$ DOF, and so on till the user decoded last, which enjoys $\min(N_t, N_r)$ DOF. As long as $UN_t \leq N_r$, each user enjoys N_t spatial DOF.

8.4 Multiple-access channel with no CSIT

Let us now suppose that each transmitter is not privy to the realization of its channel matrix, but only to its distribution. No time-domain power control is then possible and thus the per-symbol power constraint we consider effectively subsumes also a per-codeword constraint.

8.4.1 Quasi-static setting

As we know, the ruling notions in a flat-faded quasi-static setting without CSIT are those of outage probability and outage capacity. Extending these notions to the MAC,

$$\begin{aligned} p_{\mathsf{out}}(\mathsf{SNR}_0, \ldots, \mathsf{SNR}_{U-1}, R_0/B, \ldots, R_{U-1}/B) \\ = \mathbb{P}\left[\log_2 \det\left(\boldsymbol{I} + \sum_{u \in \mathcal{U}} \frac{\mathsf{SNR}_u}{N_t} \boldsymbol{H}_u \boldsymbol{F}_u \boldsymbol{F}_u^* \boldsymbol{H}_u^* \right) < \sum_{u \in \mathcal{U}} \frac{R_u}{B} \right. \\ \text{for any } \mathcal{U} \subseteq \{0, \ldots, U-1\} \end{aligned} \tag{8.66}$$

with the outage capacity region at outage ϵ being the largest spectral efficiency U-tuple $(R_0/B,\ldots,R_{U-1}/B)$ for which $p_{\text{out}} \leq \epsilon$ at some certain SNRs.

As in SU-MIMO, the term *capacity* is often employed in reference to this spectral efficiency U-tuple even when the precoders are not optimized for the given ϵ but simply fixed, typically at $\boldsymbol{F}_u = \boldsymbol{I}$ for $u = 0,\ldots,U-1$. Even for such unprecoded transmission, the distribution of the mutual informations within (8.66) is not easily characterized analytically. This again motivates the interest in assessing the tradeoff between spectral efficiency and outage through the respective proxies, multiplexing gain and diversity, and a MAC extension of the DMT can be formulated between the two [775].

8.4.2 Ergodic setting

The ergodic spectral efficiency region without CSIT differs from (8.56) only in that the precoders $\boldsymbol{F}_0,\ldots,\boldsymbol{F}_{U-1}$ are not allowed to depend on $\boldsymbol{H}_0,\ldots,\boldsymbol{H}_{U-1}$, but only on their distribution. The capacity is the boundary of [776]

$$\bigcup_{\|\boldsymbol{F}_u\|_F^2=N_t} \left\{ \sum_{u\in\mathcal{U}} \frac{R_u}{B} \leq \mathbb{E}\left[\log_2 \det\left(\boldsymbol{I} + \sum_{u\in\mathcal{U}} \frac{\mathsf{SNR}_u}{N_t} \boldsymbol{H}_u \boldsymbol{F}_u \boldsymbol{F}_u^* \boldsymbol{H}_u^*\right)\right] \quad \mathcal{U} \subseteq \{0,\ldots,U-1\} \right\} \tag{8.67}$$

and the transmit precoders that achieve a given point on the boundary can be obtained again through a convex optimization.

Example 8.13 (Precoders for a MU-MIMO MAC with IID fading and no CSIT)

Although less evident than in SU-MIMO, it is still the case [777, 778] that, with IID fading at all users, the capacity-achieving precoders are $\boldsymbol{F}_u^\star = \boldsymbol{I}$ for $u = 0,\ldots,U-1$.

Example 8.14 (Capacity of a MU-MIMO MAC with IID fading and no CSIT)

With $\boldsymbol{F}_u^\star = \boldsymbol{I}$ for $u = 0,\ldots,U-1$, the ergodic capacity is the boundary of

$$\sum_{u\in\mathcal{U}} \frac{R_u}{B} \leq \mathbb{E}\left[\log_2 \det\left(\boldsymbol{I} + \sum_{u\in\mathcal{U}} \frac{\mathsf{SNR}_u}{N_t} \boldsymbol{H}_u \boldsymbol{H}_u^*\right)\right] \qquad \mathcal{U} \subseteq \{0,\ldots,U-1\}, \tag{8.68}$$

with the sum-capacity corresponding to $\mathcal{U} = \{0,\ldots,U-1\}$. For a plot of the ergodic sum-capacity with no CSIT, the reader is referred to Fig. 8.5; the unprecoded transmission curve therein equals the sum-capacity in the absence of CSIT. As far as the user spectral efficiencies are concerned, letting $q_0 \leq \cdots \leq q_{U-1}$ and plugging $\boldsymbol{F}_u^\star = \boldsymbol{I}$ into (8.57),

$$\frac{R_u}{B} = \mathbb{E}\left[\log_2 \det\left(\boldsymbol{I} + \sum_{\mathrm{u}=u}^{U-1} \frac{\mathsf{SNR}_{\mathrm{u}}}{N_t} \boldsymbol{H}_{\mathrm{u}} \boldsymbol{H}_{\mathrm{u}}^*\right) - \log_2 \det\left(\boldsymbol{I} + \sum_{\mathrm{u}=u+1}^{U-1} \frac{\mathsf{SNR}_{\mathrm{u}}}{N_t} \boldsymbol{H}_{\mathrm{u}} \boldsymbol{H}_{\mathrm{u}}^*\right)\right] \quad u = 0,\ldots,U-1. \tag{8.69}$$

From the weighted sum-capacity, $\sum_{q_u=0}^{U-1} q_u R_u/B$, swept over all weight combinations, we could reconstruct the boundary in (8.68).

For a number of more general channel structures of interest, precoder characterizations that hold regardless of the user weights and SNRs can be provided.

Example 8.15 (Precoders for a transmit-correlated MU-MIMO MAC with no CSIT)

Let the uth user have a transmit correlation matrix $\boldsymbol{R}_u = \boldsymbol{U}_u \boldsymbol{\Lambda}_u \boldsymbol{U}_u^*$ while $\boldsymbol{R}_\mathrm{r} = \boldsymbol{I}$. Then, $\boldsymbol{F}_u^\star = \boldsymbol{U}_u \boldsymbol{P}_u^{1/2}$, meaning that the uth user's steering matrix should equal $\boldsymbol{U}_u$. The diagonal power allocation matrix $\boldsymbol{P}_u$ remains to be optimized.

Example 8.16 (Precoders for a transmit-uncorrelated MU-MIMO MAC with no CSIT)

If the transmit correlation matrix of user u is $\boldsymbol{R}_u = \boldsymbol{I}$, then $\boldsymbol{F}_u^\star = \boldsymbol{I}$ regardless of the transmit correlations at the other users and of the receive correlation matrix [777, 778]. As a corollary, if there are no transmit or receive correlations, unprecoded signaling by all users is optimum, as claimed earlier.

Gratifyingly, the foregoing precoder characterizations are consistent with the SU-MIMO intuition of having the uth user diagonalize $\mathbb{E}[\boldsymbol{H}_u^* \boldsymbol{H}_u]$ whenever there is a modicum of structure in the correlations. This consistency carries over to uncorrelated Rice channels.

In terms of the power allocation matrices, $\boldsymbol{P}_0, \ldots, \boldsymbol{P}_{U-1}$, an iterative algorithm for their optimization is provided in [512]. And it is known that, as in SU-MIMO, if the transmit correlations are strong enough—in relation to the user's SNR—then statistical beamforming becomes the optimum strategy [778]. Interestingly, statistical beamforming is always optimum (regardless of the SNRs and correlations) for $U \to \infty$ with fixed N_t and N_r [778]; by minimizing the rank of each individual transmission, the system can better accommodate a growing number of users within the fixed receiver dimensionality.

The impact of correlation in an MU-MIMO MAC is no different than in SU-MIMO: receive correlations are detrimental, as they squeeze the dimensionality spanned by the receiver, whereas transmit correlations are beneficial at low SNR and may be either beneficial or detrimental at high SNR depending on the balances of U, N_t, and N_r.

If one does not want to bother optimizing the precoders, pragmatically adopting $\boldsymbol{F}_u = \boldsymbol{I}$ for $u = 0, \ldots, U-1$ regardless of the distribution of the channel matrices, the spectral efficiency expressions in Example 8.14 apply and all correlations become damaging.

Discrete constellations

With discrete constellations in lieu of Gaussian signals, the ergodic spectral efficiency region for some fixed precoders is defined by (8.19) and (8.20) with the expectation in the latter further involving the fading coefficients within $C^{\mathcal{U}}$. The precoder optimization continues to be nonconvex [779].

High-SNR regime

Without CSIT, the number of spatial DOF continues to be $\mathcal{S}_\infty = \min(UN_\mathrm{t}, N_\mathrm{r})$. As far as $\mathcal{L}_\infty$ goes, the expressions derived with CSIT and $UN_\mathrm{t} \leq N_\mathrm{r}$ also apply here because they correspond to a uniform power allocation, which does not depend on the channel realization, meaning that the CSIT is actually immaterial to those expressions. Hence, $\mathcal{L}_\infty$ for $UN_\mathrm{t} \leq N_\mathrm{r}$ is given by (8.62) all the same.

Large-dimensional regime

A large-dimensional MU-MIMO MAC framework is presented in [780, 781] and, as in SU-MIMO, it enables extending the reach of the finite-dimensional analysis. In particular, an iterative algorithm to compute the optimum transmit power allocations is put forth and, advantageously with respect to the finite-dimensional algorithm given in [512], no expectations over the fading are required because the corresponding randomness has disappeared asymptotically. This power allocation algorithm is a welcome addition to the MU-MIMO MAC understanding, with the necessary qualification that MU-MIMO is of interest mostly at medium and high SNRs, when the precoder and power optimizations are less vital. This is illustrated in [781, figures 2 and 3], which (as observed under CSIT in Fig. 8.5) shows limited improvements over unprecoded signaling when the SNRs are high.

Additional MU-MIMO large-dimensional results can be found, e.g., in [782, 783].

8.5 Multiple-access channel with no CSI

As we learned earlier in the book, in SU-MIMO the assumption of CSIR is sound provided that the fading coherence in symbols, N_c, satisfies $N_\mathrm{t} \ll N_\mathrm{c}$. Resorting to its interpretation as a dislodged SU-MIMO setup, in the MU-MIMO MAC the same rationale is bound to apply with this condition generalized to $UN_\mathrm{t} \ll N_\mathrm{c}$. This condition is met in many MU-MIMO scenarios, making the results obtained thus far in the chapter very relevant. As N_t and/or U grow, care must be exercised and appropriate corrections must be introduced; this becomes acutely important once the realm of massive MIMO is entered.

Much of what is laid down in Section 5.5 concerning SU-MIMO without CSI extends to the MU-MIMO MAC. In particular, the translation of SU-MIMO to MU-MIMO readily indicates that the number of streams that maximizes $\mathcal{S}_\infty$ is $N_\mathrm{s} = \min\big(UN_\mathrm{t}, N_\mathrm{r}, \lfloor N_\mathrm{c}/2 \rfloor\big)$. With massive MIMO deferred to Chapter 10, it holds that $\min(UN_\mathrm{t}, N_\mathrm{r}) < \lfloor N_\mathrm{c}/2 \rfloor$; therefore, $N_\mathrm{s} = \min(UN_\mathrm{t}, N_\mathrm{r})$ and

$$\mathcal{S}_\infty = \left(1 - \frac{\min(UN_\mathrm{t}, N_\mathrm{r})}{N_\mathrm{c}}\right) \min(UN_\mathrm{t}, N_\mathrm{r}) \tag{8.70}$$

meaning that, with respect to the DOF obtained under CSIR, there is a penalty factor of

$$1 - \frac{\min(UN_\mathrm{t}, N_\mathrm{r})}{N_\mathrm{c}}. \tag{8.71}$$

As the next example confirms, even under adverse combinations of the various parameters this penalty is very minor.

Example 8.17

Let $UN_t = N_r = 8$, with the users of a vehicular nature. What percentage of the eight spatial DOF stipulated by a CSIR analysis is actually achievable?

Solution

As calculated in Example 3.26, we can take vehicular users to correspond to a coherence of $N_c = 1000$. Applying (8.71), we see that 99.2% of the DOF stipulated by a CSIR analysis is achievable.

A substantial difference between SU-MIMO and MU-MIMO, immaterial in terms of $\mathcal{S}_\infty$ but which does affect $\mathcal{L}_\infty$, is the power constraint: in SU-MIMO, additional transmit antennas do not alter the power budget while, in MU-MIMO, additional users mean additional power. With this proviso, and the necessary independence of the signals transmitted by distinct users, the computation of the power offset without CSI may borrow from the SU-MIMO derivations in [416, 554].

Once we pull back from the high-SNR regime, the MU-MIMO MAC capacity without CSI becomes analytically elusive. Interested readers are referred to the bounding techniques in [784].

8.6 Pilot-assisted multiple-access channel

In multiuser contexts, every user may have its own distinct fading coherence; strictly speaking then, user-specific values should be utilized for N_c (or for the Doppler spread if the fading is modeled as continuous). At the same time, having a multitude of fading coherences complicates the formulation without much conceptual enrichment, hence it is customary to retain a single fading coherence for all users. We abide by this practice and, invoking a worst-case design guideline, note that the value of N_c can be regarded as that of the least-underspread active user.

When pilot symbols are explicitly transmitted, the corrections required by the CSIR spectral efficiency of user u are the usual ones [785]:

- A factor $(1 - \alpha_u)$, where α_u is that user's pilot overhead.
- In place of SNR_u, a lower effective SNR that accounts for channel estimation errors.

The optimum overheads $\alpha_0^\star, \ldots, \alpha_{U-1}^\star$ emerge from balancing these two corrections in an optimization process. To pose such optimization for the MAC, we can resort once more to the analogy with an SU-MIMO channel; the main difference in the formulation is that, in a MAC, multiple SNRs and pilot overheads are now involved, one per user [784]. Alternatively, the overheads can be set to their minimum values, $\alpha_u = N_t/N_c$ for

$u = 0, \ldots, U-1$, with the optimization transplanted to the pilot power boosting at each user [786].

As SNR_u grows, shrinking the linear penalty $(1-\alpha_u)$ takes precedence over improving the effective SNR within the logarithm and the number of pilots per transmit antenna and per coherence block is sure to approach one; as that happens, $\alpha_u \to N_t/N_c$ such that the total overhead satisfies

$$\sum_{u=0}^{U-1} \alpha_u \to \frac{UN_t}{N_c}. \tag{8.72}$$

The number of spatial DOF is thus

$$S_\infty = \left(1 - \frac{UN_t}{N_c}\right) \min(UN_t, N_r), \tag{8.73}$$

which, for $UN_t \leq N_r$, equals (8.70), the highest possible number of DOF. The condition $UN_t \ll N_c$ ensures that each antenna at each user can transmit at least one pilot symbol within each coherence interval without the total overhead becoming significant.

8.7 Duality between the multiple access and broadcast channels

It is rather convenient, before delving into the analysis of the BC, to introduce a duality relationship that connects it with the by-now-familiar MAC. This duality is instrumental in the computation of the BC capacity because, as it turns out, the corresponding precoder optimization is nonconvex whereas, recall, it is convex for the MAC. Through the duality relationship between the two, the nonconvex BC optimization can be mapped onto a computationally friendlier convex dual-MAC problem. In addition to this important simplification in the computation of the capacity boundary, the MAC–BC duality reveals an optimum transmitter structure for the BC that mirrors the LMMSE-SIC receiver structure in the dual MAC.

8.7.1 Description and significance

Some early indications that there might be a certain duality between the MAC and the BC appeared in works such as [787, 788], and also in the reciprocal nature of the encoding/decoding schemes that achieve the capacity of the SISO MAC and BC [14]. Confirmation that this duality was profound in an information-theoretic sense was eventually provided [789–791]. We purposely avoid plunging into the proofs, for which the reader is referred to the foregoing references, and cut straight to the essence of the result and its implications.

The MAC, illustrated in Fig. 8.1, consists of U transmitters and a single receiver; it abstracts a reverse link. The BC, in Fig. 8.7, is obtained by reversing the direction of transmission; correspondingly, it abstracts a forward link where a single base station transmits

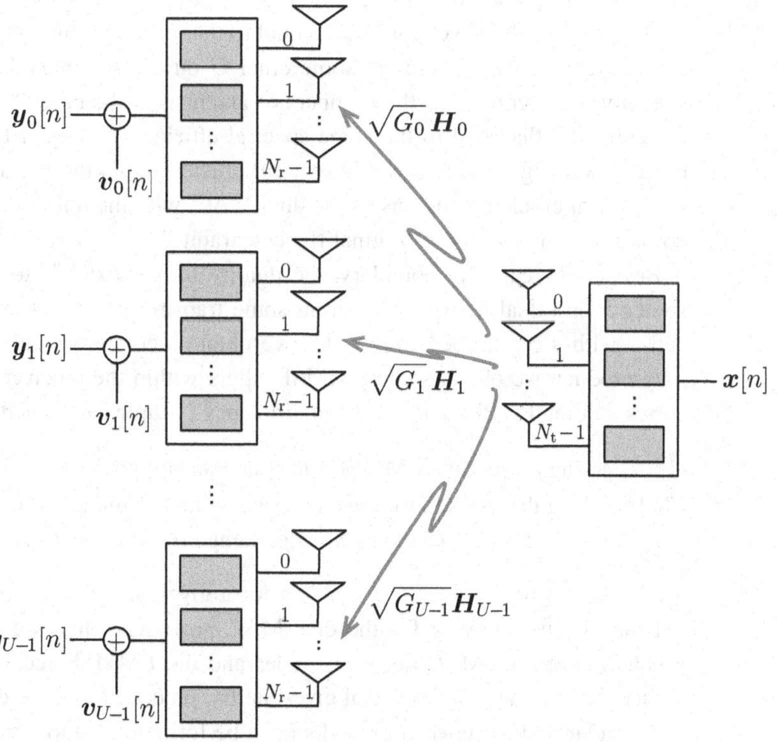

Fig. 8.7 BC with an N_t-antenna transmitter and U multiantenna receivers.

to U users. In a nutshell, what the MAC–BC duality states is that the capacity boundaries of the two setups are identical under some conditions.

- CSIR and CSIT.
- Reciprocal channel matrices, such that the MAC gain from transmit antenna j at user u to receive antenna i coincides with the BC gain from transmit antenna i to receive antenna j at user u. Applying superscripts to distinguish between dual MAC and BC, what this means is that the channels between the base station and the uth user satisfy $G_u^{\mathsf{MAC}} = G_u^{\mathsf{BC}}$ and $\boldsymbol{H}_u^{\mathsf{MAC}} = \boldsymbol{H}_u^{\mathsf{BC}*}$; identical magnitudes and opposite phases on account of the reverse wave travel directions.
- Complex Gaussian noise with the same power per antenna in both setups.
- Power constraint in the BC transmitter equal to the sum of the power constraints at the U MAC transmitters.

The last point serves to bridge a defining difference between a BC and a MAC, namely that, in the latter, the signals transmitted by the various users are subject to separate power constraints whereas, in the former, the signals transmitted to the various users can be subject to a joint power constraint at the base station.

Duality therefore implies that any spectral efficiency U-tuple achievable in the BC can also be achieved in a dual MAC where the U users are allowed to share their powers according to a single power constraint equal to that of the original BC. This implies that, for any U BC precoders, there are some other U dual-MAC precoders (of different dimensionality in general, since the numbers of antennas at the base station and the users need not coincide) that lead to the same spectral efficiency U-tuple. Putting all the U-tuples together, the region of achievable spectral efficiencies in the BC amounts to the union of the spectral efficiency regions of the dual MAC, with the union taken over all user power constraints that sum to the original BC constraint.

Besides the capacity boundary, the duality further extends to the optimum decoding strategy. In a dual MAC, by positing some transmit precoders and an LMMSE-SIC receiver with a certain decoding order, we obtain a certain spectral efficiency U-tuple; as a by-product, we obtain some LMMSE filters within the receiver structure. Then, in the corresponding BC, the same spectral efficiency U-tuple can be achieved by:

(1) Applying those same LMMSE filters as transmit precoders.
(2) Encoding the users in the reverse order. (The optimum successive encoding structure that serves as the dual of the SIC decoding procedure is described in Section 8.9.1.)

Thanks to duality, the BC precoders for a given setup and given user weights can be obtained by first solving for the dual-MAC precoders; this is a straightforward convex problem. The dual-MAC decoding order and the LMMSE receive filters follow as per earlier sections, and their reversal gives the BC precoders and the encoding order.

The relationship between precoders can be formalized into a transformation that gives the BC precoders directly as a function of the dual-MAC precoders obtained from the convex optimization. Precisely, with $\boldsymbol{F}_0^{\text{dMAC}}, \ldots, \boldsymbol{F}_{U-1}^{\text{dMAC}}$ denoting the dual-MAC precoders and with users decoded in their indexing order in the MAC, the BC precoders with reverse encoding order satisfy [789]

$$\boldsymbol{F}_u^{\text{BC}} = \sqrt{\frac{N_\text{t}}{N_\text{r}}} \boldsymbol{B}_u^{-1/2} \boldsymbol{U}_u \boldsymbol{V}_u^* \boldsymbol{A}_u^{1/2} \cdot \boldsymbol{F}_u^{\text{dMAC}} \qquad u = 0, \ldots, U-1, \tag{8.74}$$

where N_t and N_r are the transmit and per-user receive antennas in the BC, whereas

$$\boldsymbol{A}_u = \boldsymbol{I} + \frac{\mathsf{SNR}_u}{N_\text{t}} \boldsymbol{H}_u \left(\sum_{\text{u}=0}^{u-1} \frac{E_\text{u}}{E_\text{s}} \boldsymbol{F}_\text{u}^{\text{BC}} \boldsymbol{F}_\text{u}^{\text{BC}*} \right) \boldsymbol{H}_u^* \tag{8.75}$$

$$\boldsymbol{B}_u = \boldsymbol{I} + \frac{\mathsf{SNR}_u}{N_\text{r}} \sum_{\text{u}=u+1}^{U-1} \frac{E_\text{u}}{E_\text{s}} \boldsymbol{H}_\text{u}^* \boldsymbol{F}_\text{u}^{\text{dMAC}} \boldsymbol{F}_\text{u}^{\text{dMAC}*} \boldsymbol{H}_\text{u}, \tag{8.76}$$

with $\boldsymbol{U}_u$ and $\boldsymbol{V}_u^*$ arising from the SVD of $\boldsymbol{B}_u^{-1/2} \boldsymbol{H}_u^* \boldsymbol{A}_u^{-1/2}$, i.e., such that

$$\boldsymbol{B}_u^{-1/2} \boldsymbol{H}_u^* \boldsymbol{A}_u^{-1/2} = \boldsymbol{U}_u \boldsymbol{\Sigma}_u \boldsymbol{V}_u^*. \tag{8.77}$$

Importantly, this SVD must be zero-padded such that $\boldsymbol{\Sigma}_u$ is not only diagonal but also square, implying the following.[1]

[1] Standard software packages such as MATLAB® by default return the most compact form of the SVD, without unnecessary zeroes, and hence corrections may be required or else the matrix dimensionalities will not match.

- If $N_t \leq N_r$, then Σ_u is $N_r \times N_r$. Thus, U_u is $N_t \times N_r$ (the first N_t columns contain the left singular vectors of $B_u^{-1/2} H_u^* A_u^{-1/2}$ and the last $N_r - N_t$ columns are zero) whereas V_u is $N_r \times N_r$ (it contains the right singular vectors). Correspondingly, $V_u V_u^* = V_u^* V_u = I_{N_r}$ and $U_u U_u^* = I_{N_t}$, while

$$U_u^* U_u = \begin{bmatrix} I_{N_t} & 0 \\ 0 & 0 \end{bmatrix}. \tag{8.78}$$

- If $N_t \geq N_r$, then Σ_u is $N_t \times N_t$. Thus, U_u is $N_t \times N_t$ (it contains the left singular vectors) whereas V_u is $N_r \times N_t$ (the first N_r columns contain the right singular vectors and the last $N_t - N_r$ columns are zero). Correspondingly, $U_u U_u^* = U_u^* U_u = I_{N_t}$ and $V_u V_u^* = I_{N_r}$, while

$$V_u^* V_u = \begin{bmatrix} I_{N_r} & 0 \\ 0 & 0 \end{bmatrix}. \tag{8.79}$$

From the normalization $\|F_u^{\mathsf{BC}}\|_F^2 = N_t$ and the transformation in (8.74), it can be verified that the corresponding dual-MAC precoder satisfies $\|F_u^{\mathsf{dMAC}}\|_F^2 = N_r$. This verification is the subject of Problem 8.27.

Although, as mentioned, we do not delve into the technical proofs of the information-theoretic equivalence of a BC and its dual MAC, we do illustrate it for the specific case of $U = 1$; this connects with the SU-MIMO analysis in Chapter 5, where the capacity with CSIR and CSIT is seen not to change if the roles of transmitter and receiver are reversed. Indeed, for $U = 1$, the dual-MAC coincides with the actual MAC.

Example 8.18

Show that, for $U = 1$, the BC and dual-MAC capacities at a given SNR are equal.

Solution

For $U = 1$, $A_0 = I_{N_r}$, and $B_0 = I_{N_t}$, and thus $B_0^{-1/2} H_0^* A_0^{-1/2} = H_0^*$. Dropping, for notational compactness, the single-user index, we can express the SVD of the $N_t \times N_r$ matrix H^* as $H^* = U \Sigma V^*$, where Σ is square with dimensions given by $\max(N_t, N_r)$.

The precoders can be decomposed as $F^{\mathsf{BC}} = U_{\mathsf{BC}} P_{\mathsf{BC}}^{1/2}$ and $F^{\mathsf{dMAC}} = U_{\mathsf{dMAC}} P_{\mathsf{dMAC}}^{1/2}$, where U_{BC} and P_{BC} are $N_t \times N_t$ while U_{dMAC} and P_{dMAC} are $N_r \times N_r$.

For $N_t \leq N_r$, the BC capacity is

$$C = \log_2 \det\left(I_{N_r} + \frac{\mathsf{SNR}}{N_t} H F^{\mathsf{BC}} F^{\mathsf{BC}*} H^* \right) \tag{8.80}$$

$$= \log_2 \det\left(I_{N_r} + \frac{\mathsf{SNR}}{N_t} V \Sigma U^* U_{\mathsf{BC}} P_{\mathsf{BC}} U_{\mathsf{BC}}^* U \Sigma V^* \right) \tag{8.81}$$

$$= \log_2 \det\left(I_{N_r} + \frac{\mathsf{SNR}}{N_t} U^* U_{\mathsf{BC}} P_{\mathsf{BC}} U_{\mathsf{BC}}^* U \Sigma^2 \right), \tag{8.82}$$

where (8.82) follows from $V^* V = I$. The capacity-achieving strategy is to transmit along the channel's singular vectors, for which the columns of U_{BC} must equal the first

N_t columns within the $N_t \times N_r$ matrix $\boldsymbol{U}$. Then, $\boldsymbol{U}_{\mathsf{BC}}^* \boldsymbol{U} = \begin{bmatrix} \boldsymbol{I} & \boldsymbol{0} \end{bmatrix}$ and, consequently,

$$\boldsymbol{U}^* \boldsymbol{U}_{\mathsf{BC}} \boldsymbol{P}_{\mathsf{BC}} \boldsymbol{U}_{\mathsf{BC}}^* \boldsymbol{U} = \begin{bmatrix} \boldsymbol{I} \\ \boldsymbol{0} \end{bmatrix} \boldsymbol{P}_{\mathsf{BC}} \begin{bmatrix} \boldsymbol{I} & \boldsymbol{0} \end{bmatrix} \tag{8.83}$$

$$= \begin{bmatrix} \boldsymbol{P}_{\mathsf{BC}} & \boldsymbol{0} \\ \boldsymbol{0} & \boldsymbol{0} \end{bmatrix} \tag{8.84}$$

while

$$\boldsymbol{\Sigma}^2 = \begin{bmatrix} \boldsymbol{\Lambda} & \boldsymbol{0} \\ \boldsymbol{0} & \boldsymbol{0} \end{bmatrix}, \tag{8.85}$$

with $\boldsymbol{\Lambda}$ an $N_t \times N_t$ diagonal matrix containing the nonzero eigenvalues of $\boldsymbol{H}^* \boldsymbol{H}$ or, equivalently, of $\boldsymbol{H}\boldsymbol{H}^*$. Altogether,

$$C = \log_2 \det \left(\boldsymbol{I}_{N_r} + \frac{\mathsf{SNR}}{N_t} \begin{bmatrix} \boldsymbol{P}_{\mathsf{BC}} & \boldsymbol{0} \\ \boldsymbol{0} & \boldsymbol{0} \end{bmatrix} \begin{bmatrix} \boldsymbol{\Lambda} & \boldsymbol{0} \\ \boldsymbol{0} & \boldsymbol{0} \end{bmatrix} \right) \tag{8.86}$$

$$= \log_2 \det \left(\boldsymbol{I}_{N_r} + \frac{\mathsf{SNR}}{N_t} \boldsymbol{P}_{\mathsf{BC}} \boldsymbol{\Lambda} \right), \tag{8.87}$$

with the powers within $\boldsymbol{P}_{\mathsf{BC}}$ optimized via waterfilling.

Applying duality, the above capacity should coincide with

$$C = \log_2 \det \left(\boldsymbol{I}_{N_t} + \frac{\mathsf{SNR}}{N_r} \boldsymbol{H}^* \boldsymbol{F}^{\mathsf{dMAC}} \boldsymbol{F}^{\mathsf{dMAC}*} \boldsymbol{H} \right)$$

$$= \log_2 \det \left(\boldsymbol{I}_{N_t} + \frac{\mathsf{SNR}}{N_r} \boldsymbol{U} \boldsymbol{\Sigma} \boldsymbol{V}^* \boldsymbol{U}_{\mathsf{dMAC}} \boldsymbol{P}_{\mathsf{dMAC}} \boldsymbol{U}_{\mathsf{dMAC}}^* \boldsymbol{V} \boldsymbol{\Sigma} \boldsymbol{U}^* \right) \tag{8.88}$$

$$= \log_2 \det \left(\boldsymbol{I}_{N_t} + \frac{\mathsf{SNR}}{N_r} \boldsymbol{V}^* \boldsymbol{U}_{\mathsf{dMAC}} \boldsymbol{P}_{\mathsf{dMAC}} \boldsymbol{U}_{\mathsf{dMAC}}^* \boldsymbol{V} \boldsymbol{\Sigma} \boldsymbol{U}^* \boldsymbol{U} \boldsymbol{\Sigma} \right), \tag{8.89}$$

where $\boldsymbol{U}^* \boldsymbol{U}$ is given by (8.78), such that

$$\boldsymbol{\Sigma} \boldsymbol{U}^* \boldsymbol{U} \boldsymbol{\Sigma} = \begin{bmatrix} \boldsymbol{\Lambda} & \boldsymbol{0} \\ \boldsymbol{0} & \boldsymbol{0} \end{bmatrix}. \tag{8.90}$$

In turn, the capacity-achieving precoder features $\boldsymbol{U}_{\mathsf{dMAC}} = \boldsymbol{V}$ and, all in all,

$$C = \log_2 \det \left(\boldsymbol{I}_{N_t} + \frac{\mathsf{SNR}}{N_r} \boldsymbol{P}_{\mathsf{dMAC}} \begin{bmatrix} \boldsymbol{\Lambda} & \boldsymbol{0} \\ \boldsymbol{0} & \boldsymbol{0} \end{bmatrix} \right), \tag{8.91}$$

which does equal its BC counterpart provided that

$$\boldsymbol{P}_{\mathsf{dMAC}} = \begin{bmatrix} \frac{N_r}{N_t} \boldsymbol{P}_{\mathsf{BC}} & \boldsymbol{0} \\ \boldsymbol{0} & \boldsymbol{0} \end{bmatrix}. \tag{8.92}$$

This is sensible: the optimum allocation over the $\min(N_t, N_r)$ parallel subchannels is the same in both directions, only scaled to comply with the respective precoder normalizations.

For $N_t \geq N_r$, a similar derivation, relegated to Problem 8.28, yields

$$\boldsymbol{P}_{\mathsf{BC}} = \begin{bmatrix} \frac{N_t}{N_r} \boldsymbol{P}_{\mathsf{dMAC}} & \boldsymbol{0} \\ \boldsymbol{0} & \boldsymbol{0} \end{bmatrix}. \tag{8.93}$$

An alternative interpretation of duality is that, when a BC is recast in a form that is concave on the precoders, it emerges in a disposition that resembles a MAC, only with a joint power constraint.

In the case of a BC with a per-antenna (rather than per-symbol) power constraint, a different duality relationship exists [792].

Some final considerations on duality are that the relationship does not in general hold without CSIT and that, in ergodic settings with CSIT, duality holds for the ergodic spectral efficiency regions irrespective of whether the dual-MAC and BC fading realizations coincide at all times—as long as their distributions do coincide.

8.7.2 Dual versus actual multiple-access channels

For $U > 1$, it is important to distinguish between the dual MAC and the actual MAC that embodies the reverse link. On the one hand, the dual MAC is a fictitious construction in the following sense.

- The U users can pool their powers, subject to a single power constraint.
- Irrespective of U, such single power constraint equals the power transmitted by the base station in the BC.

On the other hand, the actual MAC is a factual channel.

- Each user has a separate power constraint.
- The sum of the U separate power constraints may be very different from the power transmitted by the base station in the BC, and this sum increases with U as each extra user contributes additional power.

Note that, in the dual MAC, the transmit power can be shared among users but, other than that, each user has a separate precoder that spans only that user's signals. Differently from the collaborative upper bound entertained in Discussion 8.2, in both the dual and the actual MAC there is no joint precoding across users.

8.8 The broadcast channel

The BC, already pictured in Fig. 8.7, consists of a single transmitter communicating with U receivers. The establishment of the BC capacity was a far less forthcoming exercise than that of the MAC, and the efforts that began in the 1970s [793–795] did not culminate until the 2000s [789, 790, 796, 797]. The attention in this text is on the transmission of independent information to each of the users, yet the BC can be further generalized to include mixtures of independent and common information [14, section 14.6].

With $N_{\rm t}$ transmit antennas and N_u antennas at the uth receiver, the BC transmit–receive single-letter relationship under frequency-flat fading is

$$\boldsymbol{y}_u = \sqrt{G_u}\boldsymbol{H}_u\boldsymbol{x} + \boldsymbol{v}_u \qquad u = 0, \dots, U-1, \tag{8.94}$$

where $\boldsymbol{H}_u$ is the $N_u \times N_t$ channel matrix linking the transmitter with the uth user, satisfying $\mathbb{E}\left[\|\boldsymbol{H}_u\|_F^2\right] = N_t N_u$, while

$$\boldsymbol{x} = \sum_{u=0}^{U-1} \sqrt{\frac{E_u}{N_t}} \boldsymbol{F}_u \boldsymbol{s}_u. \tag{8.95}$$

The covariance

$$\boldsymbol{R_x} = \sum_{u=0}^{U-1} \frac{E_u}{N_t} \boldsymbol{F}_u \boldsymbol{F}_u^* \tag{8.96}$$

is subject to the applicable power constraint (per-codeword, per-symbol, or per-antenna). Consistent with the normalizations in this text, we can translate (8.96) into

$$\sum_{u=0}^{U-1} \frac{E_u}{E_s} = 1, \tag{8.97}$$

with each of the precoders, $\boldsymbol{F}_0, \ldots, \boldsymbol{F}_{U-1}$, satisfying the applicable type of constraint as given in (5.6), (5.7), and (5.8) for SU-MIMO. (Any other combination of power and precoder constraints that preserves the overall constraint on $\boldsymbol{R_x}$ would also be valid, but it would modify the operational significance of $E_0, \ldots, E_{U-1}$ as the total energy-per-symbol transmitted to each of the users.)

As in the single-user and MAC setups, we avail ourselves of the local-average quantities

$$\mathsf{SNR}_u = \frac{G_u E_s}{N_0} \qquad u = 0, \ldots, U-1, \tag{8.98}$$

which do not depend on the allocation of power among users. Then, the SNR experienced by user u is determined by the product $\frac{E_u}{E_s}\mathsf{SNR}_u$ as in the MAC, with the difference that the MAC power control coefficients are individually constrained whereas the BC (and the dual MAC) power allocation coefficients are joint constrained by (8.97).

Example 8.19

Verify the constraint on the precoders $\boldsymbol{F}_0, \ldots, \boldsymbol{F}_{U-1}$ when $\boldsymbol{x}$ is subject to a per-symbol power constraint.

Solution

Under a per-symbol power constraint,

$$\mathrm{tr}(\boldsymbol{R_x}) = E_s \tag{8.99}$$

and, upholding $\sum_{u=0}^{U-1} \frac{E_u}{E_s} = 1$, the precoders must satisfy, as in the MAC,

$$\|\boldsymbol{F}_u\|_F^2 = N_t \qquad u = 0, \ldots, U-1. \tag{8.100}$$

Example 8.20

Verify the constraint on the precoders $\boldsymbol{F}_0, \ldots, \boldsymbol{F}_{U-1}$ when $\boldsymbol{x}$ is subject to a per-antenna power constraint?

Solution

Under a per-antenna power constraint,

$$[\boldsymbol{R_x}]_{j,j} = \frac{E_{\rm s}}{N_{\rm t}} \qquad j = 0, \ldots, N_{\rm t} - 1. \tag{8.101}$$

Upholding $\sum_{u=0}^{U-1} \frac{E_u}{E_{\rm s}} = 1$, each precoder must satisfy $[\boldsymbol{F}_u \boldsymbol{F}_u^*]_{j,j} = 1$ for $j = 0, \ldots, N_{\rm t}$–1. Equivalently, we need every row of every precoder to be of unit norm.

Although one could be inclined to dismiss the possibility of interpreting the BC as an aggregate SU-MIMO channel, because the BC users cannot jointly process their received signals, duality makes this interpretation possible—as long as there is both CSIR and CSIT. Indeed, duality recognizes that the MU-MIMO BC is equivalent to a certain MU-MIMO MAC, which *does* admit an SU-MIMO interpretation. This is a first indication that, in the BC, CSIT is instrumental.

The assumption of symbol-synchronous transmission is well justified in the BC, given that all the signals emanate from the same transmitter.

As for the MAC, and in order to avoid an unnecessary explosion in the number of parameters, the exposition in the sequel considers that $N_u = N_{\rm r}$ for $u = 0, \ldots, U-1$, with qualifying comments wherever appropriate.

In the face of CSIR, the optimality of Gaussian codebooks is once again upheld, although it must be noted that for the BC the proof of this optimality is much more intricate that in the SU-MIMO and the MAC [797].

8.9 Broadcast channel with CSIR and CSIT

Once again, because of its higher practical relevance, we focus on the per-symbol power constraint setting $\sum_{u=0}^{U-1} \frac{E_u}{E_{\rm s}} = 1$ and $\|\boldsymbol{F}_u\|_{\rm F}^2 = N_{\rm t}$ for $u = 0, \ldots, U-1$.

8.9.1 Optimum transmitter structure

Recall how, in the MAC with CSIR, the LMMSE-SIC structure emerges as an important element, not only as one possible embodiment of the optimum receiver, but further as an interpretation of the spectral efficiency breakdown among users at each corner of the capacity boundary. In the BC, by virtue of duality, the role of the MAC receiver is played by the transmitter and the role of the CSIR is played by the CSIT. This suggests the existence of a transmit structure that, under CSIT, serves as one embodiment of the capacity-achieving transmitter and further as an interpretation of how the spectral efficiency breaks down among users. This is indeed the case and, just as the MAC the LMMSE-SIC structure involves U linear filters and U single-user decoders, properly chained together, in the BC the dual structure involves U linear precoders and U single-user encoders, properly chained together.

Without coding, such a transmit structure could operate on a symbol-by-symbol basis.

From the knowledge of $\boldsymbol{H}_0, \ldots, \boldsymbol{H}_{U-1}$ and of the information symbols $s_0[n], \ldots, s_{U-1}[n]$, a signal $x[n]$ can be produced such that the uth user observes $s_u[n]$ with interference from only a subset of the symbols meant for other users. To achieve this, one could exploit the parallelisms between a single-user ISI channel (recall Section 4.4) and a BC: in the former, the information symbol at a given time epoch is received with interference from information symbols corresponding to other epochs while, in the latter, the information symbol of a given user is received with interference from information symbols meant for other users. Ideas originally developed to communicate over single-user ISI channels can be applied, e.g., the Tomlinson–Harashima transmit–receive structure, which yields a performance similar to that which would be achieved by a symbol-by-symbol SIC receiver [798, 799].

Example 8.21

Consider a two-user BC where, for conceptual simplicity, $N_{\mathrm{t}} = N_{\mathrm{r}} = 1$. It follows that the precoders are $F_0 = F_1 = 1$, and we further split the transmit power evenly between the two users. Thus

$$y_u = \sqrt{\frac{G_u E_{\mathrm{s}}}{2}} h_u \left(s_0 + s_1 \right) + v_u \qquad u = 0, 1. \tag{8.102}$$

Suppose that s_u is drawn from the unit-variance M-ary pulse amplitude modulation (M-PAM) constellation

$$\left\{ (2m - 1 - M) \frac{d_{\mathsf{min}}}{2} \right\} \qquad m = 0, \ldots, M-1, \tag{8.103}$$

where the distance between neighboring points is

$$d_{\mathsf{min}} = 2 \sqrt{\frac{3}{M^2 - 1}}. \tag{8.104}$$

How can the interference that user 0 would normally inflict upon user 1 be eliminated?

Solution

As illustrated in Fig. 8.8, let us repeat the M-PAM constellation along the entire real axis so as to obtain an endlessly extended constellation [50, section 10.3.3]. Each of the M original constellation points now maps to a class of equivalent points. To convey a certain constellation point to user 1, say the point indexed by 0, the transmitter examines all the equivalent points within that class and selects the one closest to s_0. If that closest point (see Fig. 8.9) is the one with magnitude q, the symbol for user 1 becomes $s_1 = \mathsf{q} - s_0$. Upon reception, user 1 observes

$$y_1 = \sqrt{\frac{G_1 E_{\mathrm{s}}}{2}} h_1 \Big((\mathsf{q} - s_0) + s_0 \Big) + v_1 \tag{8.105}$$

or, with adequate scaling, equivalently

$$y_1' = \mathsf{q} + v_1', \tag{8.106}$$

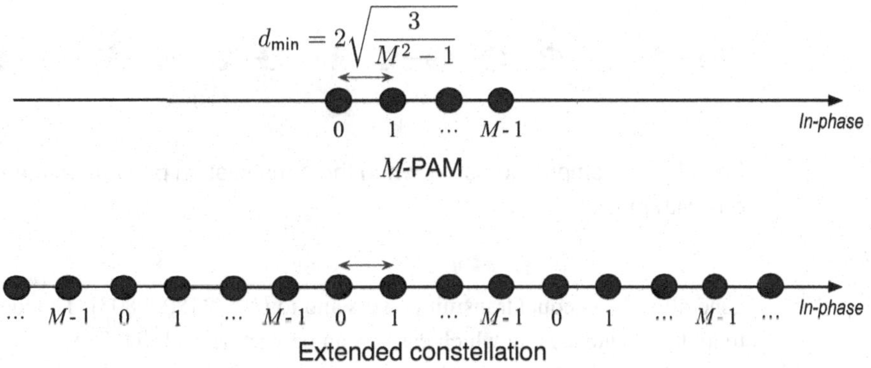

Fig. 8.8 M-PAM constellation (above) and its endlessly extended version (below). Each of the M original points gives rise to a class of equivalent points that share the same label.

where v_1' has a modified variance. User 1 then searches the endlessly extended constellation, identifies the point closest to y_1, and maps it to the original constellation point that gives rise to that class of equivalent points; if the noise is not strong enough to cause an error, this will be the point indexed by 0 in our case. From (8.106), it is clear that the uncoded error probability at user 1 equals what it would be in the absence of s_0, only for an endlessly extended constellation as opposed to the actual M-PAM constellation: the impact of the extension is rather minor, mostly noticeable only in the two edge points—which have a single nearest-neighbor in the original constellation but two in the extended constellation.

It remains to ensure that $\mathbb{E}[s_1^2] = 1$ and, if not, to scale the transmit signal appropriately. It is easily verified that $s_1 \in [-M d_{min}/2, M d_{min}/2]$. To get a feel for how much $\mathbb{E}[s_1^2]$ may depart from unity, let us assume that s_1 is uniformly distributed over the identified interval $[-M d_{min}/2, M d_{min}/2]$. Then,

$$\mathbb{E}[s_1^2] = \frac{M^2 d_{min}^2}{12} \tag{8.107}$$

$$= \frac{M^2}{M^2-1}, \tag{8.108}$$

and hence the extended constellation indeed has an increased power, although this increase is minor even for small M. This calls for an overall signal scaling that causes a correspondingly minor decrease in SNR.

What about user 0? Its normalized observation is $y_0' = \mathsf{q} + v_0'$ and, since $\mathsf{q} = s_1 + s_0$, what it observes is its intended signal s_0 subject to interference from s_1. Given an arbitrary user ordering, then, only the interference from users earlier in the order can be suppressed (approximately, because of the aforementioned edge effects).

Basic as the above toy example may be, it does convey the intuition that underpins optimum symbol-by-symbol transmission in the BC. With actual Tomlinson–Harashima, this

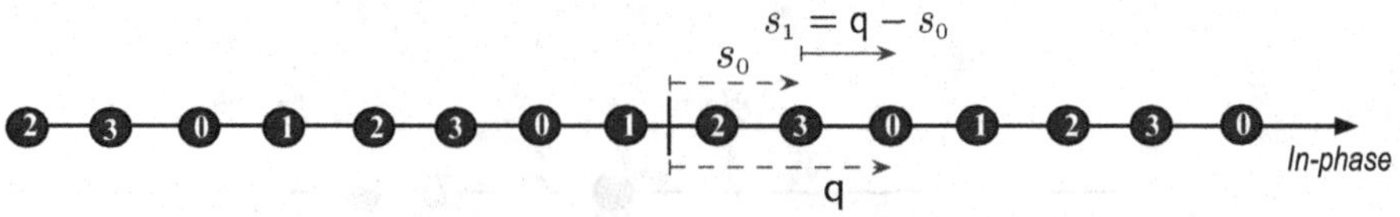

Fig. 8.9 For $M = 4$, computation of s_1 when the constellation point to be communicated is the one with index 0.

generalizes to unequal transmit powers and to $U > 2$ [800–803]. The idea also generalizes to multiple antennas, in which case it can be vectorized [804, 805].

The drawback of Tomlinson–Harashima is, precisely, that it operates on a symbol-by-symbol basis and, as such, it is essentially the dual of symbol-by-symbol SIC at the receiver. Recall, however, that achieving capacity requires codeword-wise SIC whereby the interference cancelation occurs after each entire codeword has been decoded and reconstructed. This naturally prompts the question of whether a transmit dual of codeword-wise SIC exists, and this question was answered in the affirmative by Costa in [806]. This dual, referred to as either *Costa's precoding* or—on account of the paper's title, "Writing on dirty paper"—as *dirty-paper coding* (DPC), successively encodes the users' data into codewords in such a way that each codeword does not experience interference from the codewords that preceded it. Tempting as it may be to think that this entails a form of presubtraction, DPC actually goes in the opposite direction: much like in the symbol-by-symbol procedure of Example 8.21, DPC adapts to the interference from the preceding codewords, rather than trying to counter it; the encoder looks for codewords that are compatible with that interference in terms of the power constraint yet distinguishable given the noise level. Fittingly, the ensuing U codewords turn out to be IID complex Gaussian and statistically independent, and the edge effects experienced with Tomlinson–Harashima disappear as $N \to \infty$. The precoders $\boldsymbol{F}_0, \ldots, \boldsymbol{F}_{U-1}$ and the power allocation $\frac{E_0}{E_s}, \ldots, \frac{E_{U-1}}{E_s}$ also play a role in rotating and scaling the successive codewords as the composite signal $\boldsymbol{x}$ is produced (see Fig. 8.10).

Although the description of specific DPC implementations is beyond the scope of this book (see, e.g., [807, 808]), it is worth mentioning that, just as in Example 8.21, the transmit operation can be interpreted as a scalar quantization of one user's symbol onto the other, a full DPC transmitter entails large-dimensional vector quantizers that are not only elaborate but very sensitive to CSIT inaccuracies [809]. (This motivates the interest in linear transmission strategies, which are deferred to Chapter 9.)

8.9.2 Quasi-static setting

In this setting, recall, $\boldsymbol{H}_0, \ldots, \boldsymbol{H}_{U-1}$ are fixed over the codeword span. As usual, it is instructive to begin with the most elemental case: $U = 2$ and $N_t = 1$. For this setup, DPC is not the only capacity-achieving approach, and simpler superposition suffices.

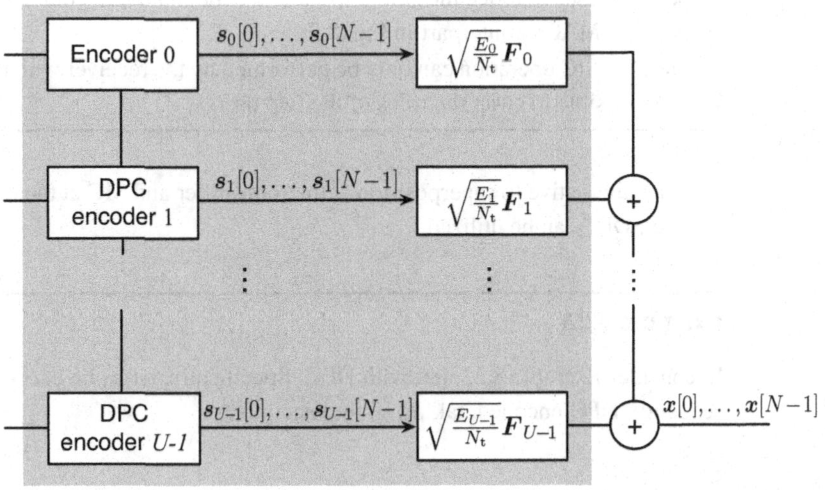

Fig. 8.10 DPC transmitter structure.

Example 8.22

Let $U = 2$ with $\boldsymbol{H}_0$ and $\boldsymbol{H}_1$ being $N_r \times 1$ vectors. Suppose that $\sqrt{G_0}\|\boldsymbol{H}_0\| < \sqrt{G_1}\|\boldsymbol{H}_1\|$. Determine the BC capacity boundary with s_0 and s_1 separately encoded utilizing IID complex Gaussian codebooks, and with subsequent superposition.

Solution

For $N_t = 1$, the precoders are $F_0 = F_1 = 1$ and the transmit signal is thus $x = E_0 s_0 + E_1 s_1$. If $\sqrt{G_0}\|\boldsymbol{H}_0\| < \sqrt{G_1}\|\boldsymbol{H}_1\|$, then, regardless of E_0 and E_1, when s_1 is encoded at the maximum rate decodable at user 1, such signal is not decodable at user 0 because the SNR at that receiver is insufficient. In terms of the capacity boundary then, user 0 must decode its signal in the face of noise plus the interference $\sqrt{G_0}\boldsymbol{H}_0 E_1 s_1$. Conditioned on $\boldsymbol{H}_0$, such interference is complex Gaussian and hence

$$\frac{R_0}{B} = \log_2\left(1 + \frac{G_0 E_0 \|\boldsymbol{H}_0\|^2}{N_0 + G_0 E_1 \|\boldsymbol{H}_0\|^2}\right) \tag{8.109}$$

$$= \log_2\left(1 + \frac{\frac{E_0}{E_s}\mathsf{SNR}_0 \|\boldsymbol{H}_0\|^2}{1 + \frac{E_1}{E_s}\mathsf{SNR}_0 \|\boldsymbol{H}_0\|^2}\right). \tag{8.110}$$

Conversely, user 1 *can* decode s_0 because such codeword is rated for a weaker channel and, after SIC, the decoding of s_1 is impaired by only noise. Thus,

$$\frac{R_1}{B} = \log_2\left(1 + \frac{E_1}{E_s}\mathsf{SNR}_1 \|\boldsymbol{H}_1\|^2\right). \tag{8.111}$$

The points on the ensuing capacity boundary (see Fig. 8.11) can be achieved by varying the power split between $\frac{E_0}{E_s}$ and $\frac{E_1}{E_s}$, subject to $\frac{E_0}{E_s} + \frac{E_1}{E_s} = 1$, with the transmit rates properly adjusted. Notice the smoothness of this boundary, in contrast with the pentagonal shape of its MAC counterpart in Fig. 8.2.

Since the SIC operation can only be performed by the receiver with the stronger channel, it is not possible to swap the roles of the two users.

As an alternative to superposition at the transmitter and SIC at the user with the stronger channel, DPC can be utilized.

Example 8.23

Reconsider Example 8.22, but with DPC. Specifically, let s_0 be encoded first with s_1 subsequently DPC-encoded, taking s_0 into account.

Solution

The capacity boundary determined by (8.110) and (8.111) continues to be achievable as the DPC process ensures that user 1 is not afflicted by interference.

Example 8.24

Still for $\sqrt{G_0}\|\boldsymbol{H}_0\| < \sqrt{G_1}\|\boldsymbol{H}_1\|$, what happens if s_1 is encoded first and s_0 is subsequently DPC-encoded taking s_1 into account?

Solution

The resulting region, also depicted in Fig. 8.11, is not only contained within the capacity boundary but contained even within the time-division region. Hence, DPC makes sense in this setup only if the channel strengths are taken into account.

A compact way of describing the capacity boundary defined by (8.110) and (8.111) is

$$\frac{R_u}{B} = \log_2\left(1 + \frac{\frac{E_u}{E_s}\mathsf{SNR}_u\|\boldsymbol{H}_u\|^2}{1+\sum_{u=u+1}^{U-1}\frac{E_u}{E_s}\mathsf{SNR}_u\|\boldsymbol{H}_u\|^2}\right) \qquad u = 0, 1. \tag{8.112}$$

Indexing the users in such a way that $\sqrt{G_0}\|\boldsymbol{H}_0\| \leq \sqrt{G_1}\|\boldsymbol{H}_1\| \leq \cdots \leq \sqrt{G_{U-1}}\|\boldsymbol{H}_{U-1}\|$, (8.112) readily extends to arbitrary U. The corresponding capacity boundary can be achieved either by superposition with SIC at the receivers, or by DPC encoding in order of channel strengths. Irrespective of $q_0, \ldots, q_{U-1}$, the weighted sum-capacity is achieved with this encoding order.

The fact that, when $N_t = 1$, the users can be absolutely ranked by their channel strength is the key to the equivalence between superposition with SIC, on the one hand, and DPC, on the other. The setups where this is possible are said to be *degraded* (for a precise definition of degradedness, see [14, chapter 15]). This condition generally ceases to hold when $N_t > 1$ as, in that case, one cannot always rank users absolutely.

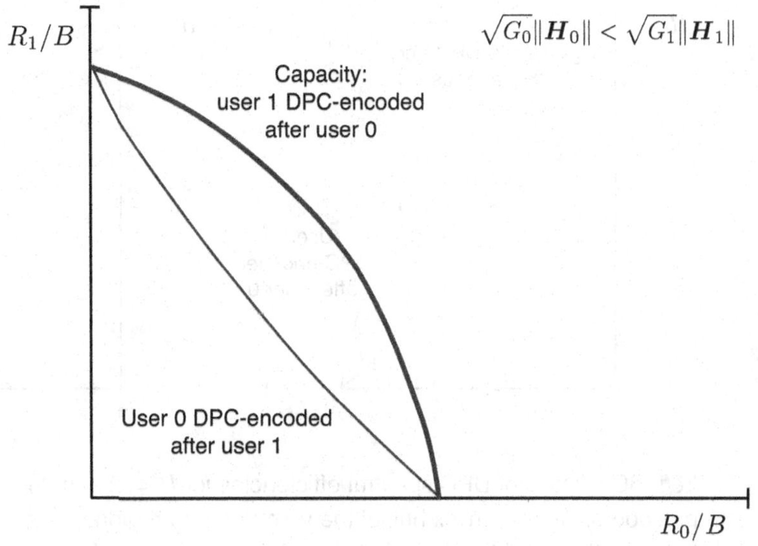

Fig. 8.11 BC regions of achievable spectral efficiencies for $U = 2$ and $N_t = 1$ with $\sqrt{G_0}\|\boldsymbol{H}_0\| < \sqrt{G_1}\|\boldsymbol{H}_1\|$.

Example 8.25

Let $U = 2$ with $N_t > 1$. For $N_r > 1$, multiple spatial dimensions exist and the ranking in one need not apply to another. However, for $N_r = 1$, is an absolute ranking not possible?

Solution

It is not. Even if $\sqrt{G_0}\|\boldsymbol{H}_0\| < \sqrt{G_1}\|\boldsymbol{H}_1\|$, there may be an admissible precoder $\boldsymbol{F}_0$ such that $\sqrt{G_0}|\boldsymbol{H}_0\boldsymbol{F}_0| > \sqrt{G_1}|\boldsymbol{H}_1\boldsymbol{F}_0|$. Consequently, it cannot be guaranteed that, when $\boldsymbol{s}_0$ is encoded at the maximum rate decodable by user 0, it is also decodable by user 1, and vice versa.

Once users cannot be absolutely ranked, there is no degradedness. DPC remains capacity-achieving, while superposition does not. Unlike in the degraded case, where the DPC spectral efficiency region with a certain user ordering contains all other ones, in a nondegraded setup that need not be the case.

Example 8.26

Determine the region of DPC spectral efficiencies for a nondegraded BC with $U = 2$ and $N_t > 1$, and with fixed $\sqrt{E_0/E_s}\boldsymbol{F}_0$ and $\sqrt{E_1/E_s}\boldsymbol{F}_1$.

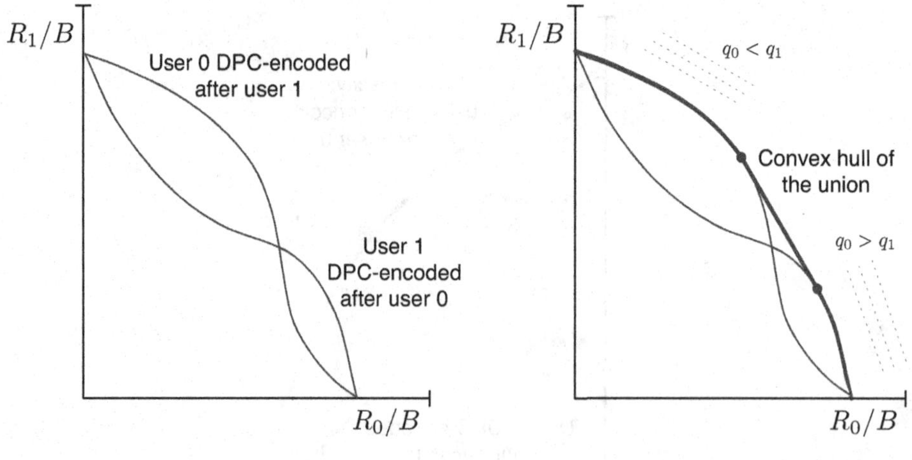

Fig. 8.12 Left, BC regions of DPC spectral efficiencies for $U = 2$ and $N_\mathrm{t} > 1$ with fixed precoders. Right, convex hull of the union of both regions; also shown with dotted lines are the weighted sum spectral efficiency contours for $q_0 < q_1$ and $q_0 > q_1$.

Solution

The regions achievable with either DPC encoding order are illustrated on the left-hand side of Fig. 8.12. As neither is contained within the other one, the union of both is achievable. Furthermore, spectral efficiency pairs anywhere within the minimum convex set of that union (termed the *convex hull* and depicted on the right-hand side of Fig. 8.12) are also achievable by time-dividing both DPC orderings with the appropriate duty cycle.

Notice from the weighted sum spectral efficiency contours included in Fig. 8.12 how, for $q_0 < q_1$, user 1 should be encoded first while, for $q_0 > q_1$, user 0 should be encoded first. In this nondegraded BC example, therefore, users should be DPC-encoded in order of decreasing weight. As seen later in the section, this is not an anecdotal observation.

For arbitrary U and N_t, as well as specific $\sqrt{E_0/E_\mathrm{s}}\boldsymbol{F}_0, \ldots, \sqrt{E_{U-1}/E_\mathrm{s}}\boldsymbol{F}_{U-1}$, and with the users DPC-encoded in the order of their indices, (8.112) generalizes to

$$\frac{R_u}{B} = \log_2 \det\left(\boldsymbol{I} + \frac{\frac{E_u}{E_\mathrm{s}}\mathsf{SNR}_u}{N_\mathrm{t}}\boldsymbol{H}_u\boldsymbol{F}_u\boldsymbol{F}_u^*\boldsymbol{H}_u^*\left(\boldsymbol{I} + \boldsymbol{H}_u\sum_{\mathrm{u}=u+1}^{U-1}\frac{\frac{E_\mathrm{u}}{E_\mathrm{s}}\mathsf{SNR}_u}{N_\mathrm{t}}\boldsymbol{F}_\mathrm{u}\boldsymbol{F}_\mathrm{u}^*\cdot\boldsymbol{H}_u^*\right)^{-1}\right)$$

$$u = 0, \ldots, U-1, \tag{8.113}$$

which can be rewritten as

$$\frac{R_u}{B} = \log_2 \det\left(\boldsymbol{I} + \boldsymbol{H}_u\sum_{\mathrm{u}=u}^{U-1}\frac{\frac{E_\mathrm{u}}{E_\mathrm{s}}\mathsf{SNR}_u}{N_\mathrm{t}}\boldsymbol{F}_\mathrm{u}\boldsymbol{F}_\mathrm{u}^*\cdot\boldsymbol{H}_u^*\right) \tag{8.114}$$

$$- \log_2 \det\left(\boldsymbol{I} + \boldsymbol{H}_u\sum_{\mathrm{u}=u+1}^{U-1}\frac{\frac{E_\mathrm{u}}{E_\mathrm{s}}\mathsf{SNR}_u}{N_\mathrm{t}}\boldsymbol{F}_\mathrm{u}\boldsymbol{F}_\mathrm{u}^*\cdot\boldsymbol{H}_u^*\right) \qquad u = 0, \ldots, U-1.$$

What can be attained in a nondegraded BC is the convex hull of the union of the regions achievable with every possible DPC ordering. The necessity of the convex hull operation, to render convex a union of regions that is otherwise not, has important consequences that become clear once we release $\sqrt{E_0/E_s}F_0, \ldots, \sqrt{E_{U-1}/E_s}F_{U-1}$ from being fixed so as to take a further union over all their admissible values.

Specifically, and in contrast with the MAC, the BC precoder optimization is not convex because $\sum_{u=0}^{U-1} q_u R_u/B$ with R_u/B as given by (8.114) is not a concave function of $\sqrt{E_0/E_s}F_0, \ldots, \sqrt{E_{U-1}/E_s}F_{U-1}$. While (8.114) may, at a first glance, appear similar to the MAC expression for R_u/B in (8.27), the subtle difference between the two suffices to break the convexity. See, for instance, how (8.114) cannot be put in the form of (8.29), which substantiates the convexity of the MAC precoder optimization.

It is precisely in the face of the nonconvexity of the BC optimization that the BC-MAC duality comes to the rescue: the BC weighted sum-capacity for any arbitrary $q_0, \ldots, q_{U-1}$ can be obtained by computing the same weighted sum-capacity for the dual MAC, which is convex. Thanks to duality, then, by merely replacing every H_u with H_u^* in (8.18) and modifying the power constraint to turn the MAC into a dual MAC, we obtain the weighted sum-capacity of the BC as the boundary of

$$\bigcup_{\substack{\|F_u^{\mathrm{dMAC}}\|_F^2 = N_r \\ \sum_{u=0}^{U-1} \frac{E_u}{E_s}=1}} \left[\sum_{u\in\mathcal{U}} \frac{R_u}{B} \leq \log_2 \det \left(I + \sum_{u\in\mathcal{U}} \frac{\frac{E_u}{E_s}\mathsf{SNR}_u}{N_r} H_u^* F_u^{\mathrm{dMAC}} F_u^{\mathrm{dMAC}*} H_u \right) \quad \mathcal{U} \subseteq \{0, \ldots, U-1\} \right]. \tag{8.115}$$

Notice that, as a by-product of the channel transpositions, the precoder normalization has changed to $1/N_r$; such is indeed the number of transmit antennas per user in the dual MAC, and the value to which the corresponding precoder is normalized.

In terms of the spectral efficiency U-tuples, the modification of the MAC expression in (8.27) gives, for the dual MAC with the users decoded in their indexing order,

$$\begin{aligned} \frac{R_u}{B} &= \log_2 \det \left(I + \sum_{\mathrm{u}=u}^{U-1} \frac{\frac{E_{\mathrm{u}}}{E_s}\mathsf{SNR}_{\mathrm{u}}}{N_r} H_{\mathrm{u}}^* F_{\mathrm{u}}^{\mathrm{dMAC}} F_{\mathrm{u}}^{\mathrm{dMAC}*} H_{\mathrm{u}} \right) \\ &\quad - \log_2 \det \left(I + \sum_{\mathrm{u}=u+1}^{U-1} \frac{\frac{E_{\mathrm{u}}}{E_s}\mathsf{SNR}_{\mathrm{u}}}{N_r} H_{\mathrm{u}}^* F_{\mathrm{u}}^{\mathrm{dMAC}} F_{\mathrm{u}}^{\mathrm{dMAC}*} H_{\mathrm{u}} \right) \qquad u = 0, \ldots, U-1. \end{aligned} \tag{8.116}$$

By applying the transformation in (8.74), it can be verified (refer to Problem 8.30) that the foregoing expression indeed equals the spectral efficiency U-tuple for the BC with the users DPC-encoded in the reverse indexing order, namely

$$\begin{aligned} \frac{R_u}{B} &= \log_2 \det \left(I + H_u \sum_{\mathrm{u}=0}^{u} \frac{\frac{E_{\mathrm{u}}}{E_s}\mathsf{SNR}_u}{N_t} F_{\mathrm{u}} F_{\mathrm{u}}^* \cdot H_u^* \right) \\ &\quad - \log_2 \det \left(I + H_u \sum_{\mathrm{u}=0}^{u-1} \frac{\frac{E_{\mathrm{u}}}{E_s}\mathsf{SNR}_u}{N_t} F_{\mathrm{u}} F_{\mathrm{u}}^* \cdot H_u^* \right) \qquad u = 0, \ldots, U-1. \end{aligned} \tag{8.117}$$

It follows that, since in the MAC it is optimum to decode the users in order of increasing weight, in a nondegraded BC it is optimum to encode the users in order of decreasing weight. Hence, $(R_0/B, \ldots, R_{U-1}/B)$ as given by (8.117) befits $q_0 \leq \cdots \leq q_{U-1}$, something that would hardly be obvious without duality. Although this encoding order seems to expose the highest-weight user to the most interference, the precoders and the power allocation also play a role and, altogether, this is indeed the correct order.

In terms of sum-capacity, that is, for $q_0 = \cdots = q_{U-1}$, the DPC encoding order becomes immaterial.

8.9.3 Precoder and power allocation optimization

With R_u/B in the dual-MAC form provided in (8.116), $\sum_{u=0}^{U-1} q_u R_u/B$ is a concave function of $\sqrt{E_0/E_s}\boldsymbol{F}_0^{\mathsf{dMAC}}, \ldots, \sqrt{E_{U-1}/E_s}\boldsymbol{F}_{U-1}^{\mathsf{dMAC}}$. Any optimization algorithm tailored to the MAC can be applied, suitably modified to solve the dual MAC where there is a single aggregate power constraint rather than individual per-user constraints; interested readers are referred, e.g., to the convex optimization procedure in [34, section 3.5.2] or, for the sum-capacity specifically, to the modified version of the iterative waterfilling algorithm in [810, 811]. It must be kept in mind that the precoders obtained from this optimization are *not* the sought BC precoders; the obtained precoders are $N_r \times N_r$, as befits the dual MAC, while the BC precoders are $N_t \times N_t$. The transformation in (8.74) must be applied to recover the BC precoders $\boldsymbol{F}_0, \ldots, \boldsymbol{F}_{U-1}$ from $\boldsymbol{F}_0^{\mathsf{dMAC}}, \ldots, \boldsymbol{F}_{U-1}^{\mathsf{dMAC}}$. However, because of the difficulties in implementing DPC, we are interested in the BC capacity for benchmarking purposes only, and by virtue of duality such capacity boundary can be determined without explicitly obtaining the BC precoders. We thus do not dwell on the optimum BC precoders any further at this point. (In the next chapter, we return with force to the issue of BC precoding under linear transmission schemes.)

In the MU-MISO BC specifically, the dual-MAC precoders become trivial and the determination of the capacity adopts a particularly simple form.

Example 8.27 (Capacity boundary for a MU-MISO BC with CSIT)

In the relevant special case that the U users are equipped with a single antenna, the dual MAC features $F_u^{\mathsf{dMAC}} = 1$ for $u = 0, \ldots, U-1$ and (8.115) simplifies into

$$\bigcup_{\sum_{u=0}^{U-1} \frac{E_u}{E_s}=1} \left[\sum_{u \in \mathcal{U}} \frac{R_u}{B} \leq \log_2\left(1 + \sum_{u \in \mathcal{U}} \frac{\frac{E_u}{E_s}\mathsf{SNR}_u}{N_r} \|\boldsymbol{H}_u\|^2\right) \qquad \mathcal{U} \subseteq \{0, \ldots, U-1\} \right]. \tag{8.118}$$

Example 8.28

Consider a base station with $N_t = 2$ antennas serving $U = 2$ single-antenna users with $\mathsf{SNR}_0 = 6$ dB and $\mathsf{SNR}_1 = 10$ dB. Let the respective channel matrices be

$$\boldsymbol{H}_0 = \begin{bmatrix} 0.3 - \mathrm{j} & 0.7 + 1.2\mathrm{j} \end{bmatrix} \tag{8.119}$$

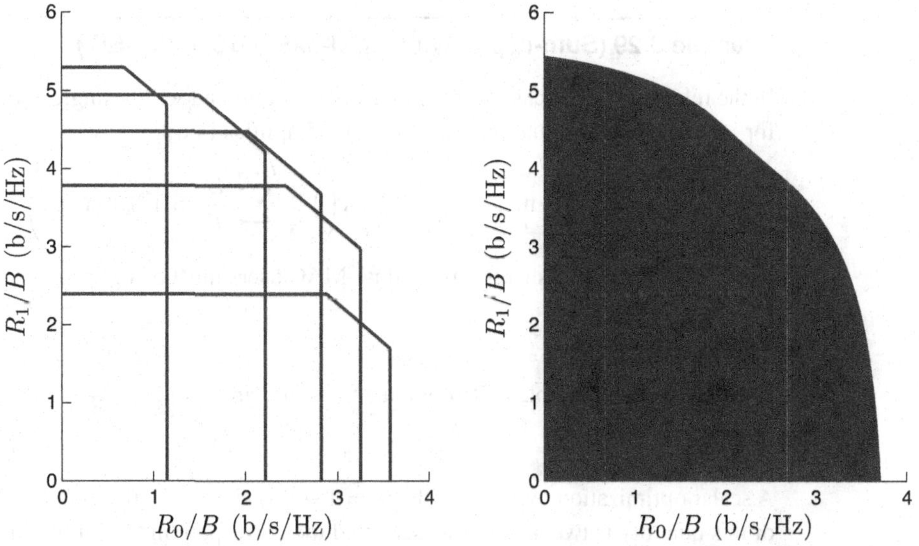

Fig. 8.13 Left, spectral efficiency regions corresponding to a few specific power shares in the dual MAC. Right, union over all possible such shares delineating the capacity boundary.

and

$$\boldsymbol{H}_1 = \left[\ 0.4\mathrm{j} \quad 2 - 0.3\mathrm{j}\ \right]. \tag{8.120}$$

For some fixed $\boldsymbol{F}_0$ and $\boldsymbol{F}_1$, the region of achievable spectral efficiencies would resemble the right-hand side of Fig. 8.12. Establishing the capacity boundary would entail a union of all such regions over all admissible two-dimensional vector precoders. Instead, establish the capacity boundary by applying duality.

Solution

From (8.118), and taking a union over only a scalar power share, the BC capacity boundary can be obtained. Shown on the left-hand side of Fig. 8.13 are the regions corresponding to a few specific power shares, with the right-hand side depicting the union over all possible power shares.

Concentrating now on the BC sum-capacity, we can obtain it from (8.115) as

$$C = \max_{\substack{\boldsymbol{F}_0^{\mathsf{dMAC}},\ldots,\boldsymbol{F}_{U-1}^{\mathsf{dMAC}}:\|\boldsymbol{F}_u^{\mathsf{dMAC}}\|_{\mathrm{F}}^2 = N_{\mathrm{r}} \\ \frac{E_0}{E_{\mathrm{s}}},\ldots,\frac{E_{U-1}}{E_{\mathrm{s}}}:\sum_{u=0}^{U-1}\frac{E_u}{E_{\mathrm{s}}}=1}} \log_2 \det\left(\boldsymbol{I} + \sum_{u=0}^{U-1} \frac{\frac{E_u}{E_{\mathrm{s}}}\mathsf{SNR}_u}{N_{\mathrm{r}}} \boldsymbol{H}_u^* \boldsymbol{F}_u^{\mathsf{dMAC}} \boldsymbol{F}_u^{\mathsf{dMAC}*} \boldsymbol{H}_u\right). \tag{8.121}$$

Example 8.29 (Sum-capacity of a MU-MISO BC with CSIT)

In the relevant special case that the U users are equipped with a single antenna, $F_u^{\text{dMAC}} = 1$ for $u = 0, \ldots, U-1$ and the sum-capacity simplifies into

$$C = \max_{\frac{E_0}{E_s}, \ldots, \frac{E_{U-1}}{E_s} : \sum_{u=0}^{U-1} \frac{E_u}{E_s} = 1} \log_2 \left(1 + \sum_{u=0}^{U-1} \frac{E_u}{E_s} \mathsf{SNR}_u \, \| \boldsymbol{H}_u \|^2 \right), \tag{8.122}$$

where only the power shares among dual-MAC users must be optimized.

Example 8.30

For the setup of Example 8.28, compute the sum-capacity.

Solution

A scalar optimization over $\frac{E_0}{E_s}$, with $\frac{E_1}{E_s} = 1 - \frac{E_0}{E_s}$, gives $C = 6.52$ b/s/Hz. The breakdown of this quantity between the two users, and the corresponding spectral efficiency pairs for proportional fairness and for other operating points, are the subject of Problem 8.36.

In terms of how multiple codewords can be conveyed to each of the users in a MU-MIMO BC, the answer is to embed a standard MIMO transmission within the DPC process. At user u, a local LMMSE-SIC receiver can be applied to extract the various codewords.

Readers interested in the optimization of the BC precoders for discrete constellations are referred to [812].

8.9.4 High-SNR regime

The argument that supports that, at high SNR, the sum-capacity of a MU-MIMO MAC having $UN_t \leq N_r$ is achieved by a uniform power allocation at each user (see Section 8.3.4) can be applied verbatim to the dual MAC of a MU-MIMO BC. It is important to realize that the corresponding BC precoders, once the duality transformations are applied, do not generally embody uniform power allocations and actually depend on the channel realizations through those transformations. Remarkably, thanks to duality, we can compute the high-SNR BC capacity boundary for $N_t \geq UN_r$ without having to identify those optimum BC precoders, dealing only with the asymptotically trivial $\boldsymbol{F}_0^{\text{dMAC}}, \ldots, \boldsymbol{F}_{U-1}^{\text{dMAC}}$.

What remains to be optimized is the asymptotic power allocation, $\frac{E_0}{E_s}, \ldots, \frac{E_{U-1}}{E_s}$, which applies to both the dual MAC and the BC.

Power allocation

For $N_t \geq UN_r$, the asymptotically uniform power allocations in the dual-MAC imply that $\boldsymbol{F}_u^{\text{dMAC}} \to \boldsymbol{I}$ for $u = 0, \ldots, U-1$. Plugged into (8.116), these give

$$\frac{R_u}{B} = \log_2 \det \left(\boldsymbol{I} + \frac{\frac{E_u}{E_s} \mathsf{SNR}_u}{N_r} \boldsymbol{H}_u^* \boldsymbol{H}_u \left(\boldsymbol{I} + \sum_{u=u+1}^{U-1} \frac{\frac{E_u}{E_s} \mathsf{SNR}_u}{N_r} \boldsymbol{H}_u^* \boldsymbol{H}_u \right)^{-1} \right) \tag{8.123}$$

and, invoking as done for the actual MAC the result in [774, lemma 1], if we denote by $\boldsymbol{H}_{\perp,u}$ the projection of $\boldsymbol{H}_u$ onto the null space of $\boldsymbol{H}_{u+1}, \ldots, \boldsymbol{H}_{U-1}$, then

$$\lim_{E_s/N_0 \to \infty} \left[\boldsymbol{H}_{\perp,u} \boldsymbol{H}^*_{\perp,u} - \boldsymbol{H}_u \left(\boldsymbol{I} + \sum_{\mathrm{u}=u+1}^{U-1} \frac{\frac{E_{\mathrm{u}}}{E_s} \mathsf{SNR}_{\mathrm{u}}}{N_r} \boldsymbol{H}^*_{\mathrm{u}} \boldsymbol{H}_{\mathrm{u}} \right)^{-1} \boldsymbol{H}^*_u \right] = \boldsymbol{0}. \tag{8.124}$$

Hence, for the purposes of what transpires for $E_s/N_0 \to \infty$, we can replace (8.123) by

$$\frac{R_u}{B} = \log_2 \det \left(\boldsymbol{I} + \frac{\frac{E_u}{E_s} \mathsf{SNR}_u}{N_r} \boldsymbol{H}_{\perp,u} \boldsymbol{H}^*_{\perp,u} \right) \tag{8.125}$$

and, to optimize the power allocation for some given weights $q_0, \ldots, q_{U-1}$, we can build the Lagrangian function (see Appendix G)

$$\begin{aligned} \mathsf{L}\left(\frac{E_0}{E_s}, \ldots, \frac{E_{U-1}}{E_s}, \lambda \right) &= \sum_{u=0}^{U-1} q_u \log_2 \det \left(\boldsymbol{I} + \frac{\frac{E_u}{E_s} \mathsf{SNR}_u}{N_r} \boldsymbol{H}_{\perp,u} \boldsymbol{H}^*_{\perp,u} \right) \\ &\quad + \lambda \cdot \left(\sum_{u=0}^{U-1} \frac{E_u}{E_s} - 1 \right) \end{aligned} \tag{8.126}$$

whose partial derivative with respect to $\frac{E_u}{E_s}$ yields the necessary and sufficient condition

$$q_u \operatorname{tr} \left(\left(\boldsymbol{I} + \frac{\frac{E_u}{E_s} \mathsf{SNR}_u}{N_r} \boldsymbol{H}_{\perp,u} \boldsymbol{H}^*_{\perp,u} \right)^{-1} \frac{\mathsf{SNR}_u}{N_r} \boldsymbol{H}_{\perp,u} \boldsymbol{H}^*_{\perp,u} \right) \log_2 e + \lambda = 0 \tag{8.127}$$

where we have applied

$$\frac{\partial}{\partial z} \log_e \det(\boldsymbol{A}(z)) = \operatorname{tr}\left(\boldsymbol{A}^{-1}(z) \frac{\partial \boldsymbol{A}(z)}{\partial z} \right). \tag{8.128}$$

For $N_t \geq U N_r$, the projection $\boldsymbol{H}_{\perp,u}$ is full-rank and, for growing SNR_u and nonzero $\frac{E_u}{E_s}$, the condition in (8.127) expands as

$$\frac{q_u}{\frac{E_u}{E_s}} \log_2 e + \lambda + \mathcal{O}\left(\frac{1}{\mathsf{SNR}_u} \right) = 0. \tag{8.129}$$

This leads to

$$\frac{E_u}{E_s} = -\frac{\log_2 e}{\lambda} q_u + \mathcal{O}\left(\frac{1}{\mathsf{SNR}_u} \right), \tag{8.130}$$

where the Lagrange multiplier λ can be cleared by enforcing $\sum_{u=0}^{U-1} \frac{E_u}{E_s} = 1$, yielding

$$\lambda = -\log_2 e \cdot \sum_{u=0}^{U-1} q_u + \mathcal{O}\left(\frac{1}{\mathsf{SNR}_u} \right). \tag{8.131}$$

Plugged into (8.130), such λ finally gives

$$\frac{E_u}{E_s} = \frac{q_u}{\sum_{\mathrm{u}=0}^{U-1} q_{\mathrm{u}}} + \mathcal{O}\left(\frac{1}{\mathsf{SNR}_u} \right), \tag{8.132}$$

Discussion 8.3 Collaborative upper bound for the MU-MIMO BC

The collaborative upper bound on the sum-capacity, introduced for the MAC in Discussion 8.2, can also be applied to the BC. In this case, it is obtained by allowing the U users to function as a single one, jointly receiving and decoding all the transmissions. As in the MAC, this bound is not generally tight (refer to Problem 8.40), but it does become so for $UN_t \geq N_r$ and once all users are at high SNR. Observed already in the early works on the MU-MIMO BC [796], this follows directly from the BC-MAC duality and the similar high-SNR tightness observed in the MAC, and in that sense it should not be surprising. However, from an operational perspective, it is rather striking that the sum-capacity is asymptotically as if all users collaborated, even though no collaboration is taking place.

which is the optimum power allocation policy at high SNR: each user should be allocated power in direct proportion to its weight. As seen in the next chapter, this policy turns out to have broad validity well beyond the context of optimum DPC transmission, holding also for other types of transmitters.

Homing in on the sum-capacity, (8.132) gives $\frac{E_u}{E_s} \to 1/U$.

Spatial DOF

Let us see how the sum-capacity behaves at high SNR. Inserting $\frac{E_u}{E_s} = 1/U$ and $\boldsymbol{F}_u^{\text{dMAC}} = \boldsymbol{I}$ into (8.121), we obtain

$$C = \log_2 \det\left(\boldsymbol{I} + \sum_{u=0}^{U-1} \frac{\mathsf{SNR}_u}{UN_r} \boldsymbol{H}_u^* \boldsymbol{H}_u\right), \tag{8.133}$$

a coarse expansion of which gives

$$C = \mathcal{S}_\infty \log_2 \frac{E_s}{N_0} + \mathcal{O}\left(\frac{1}{E_s/N_0}\right), \tag{8.134}$$

where, provided the channels are as usual full-rank,

$$\mathcal{S}_\infty = \min(N_t, UN_r). \tag{8.135}$$

Thanks to the CSIT, it holds also for the BC that MU-MIMO circumvents DOF bottlenecks resulting from limitations in the number of antennas at user devices (see Example 8.7, which applies verbatim to the BC). The number of active users U should be such that $N_t \geq UN_r$, with N_r signal streams sent to each user for a total of UN_r streams. If we insisted on transmitting more than UN_r streams, then the number of DOF per user would be, applying duality, as in the MAC but with the order reversed: the user encoded first would enjoy $\min(N_t, N_r)$ spatial DOF, the user encoded second would enjoy $\min(N_t - N_r, N_r)$ spatial DOF, and so on, with these quantities taken as zero whenever negative. In order to have all users enjoy some spatial DOF, it is necessary that the total number of streams does not exceed $\mathcal{S}_\infty$.

Power offset

For $N_t \geq UN_r$, we can rewrite (8.133) as

$$C = \log_2 \det\left(\boldsymbol{I} + \frac{E_s/N_0}{UN_r} \sum_{u=0}^{U-1} \boldsymbol{H}_u^* G_u \boldsymbol{H}_u \right) \tag{8.136}$$

$$= \log_2 \det\left(\boldsymbol{I} + \frac{E_s}{UN_rN_0} \boldsymbol{H}^* \operatorname{diag}(\underbrace{G_0, \ldots, G_0}_{N_r}, \ldots, \underbrace{G_{U-1}, \ldots, G_{U-1}}_{N_r}) \boldsymbol{H} \right)$$

$$= \log_2 \det\left(\boldsymbol{I} + \frac{E_s}{UN_rN_0} \operatorname{diag}(\underbrace{G_0, \ldots, G_0}_{N_r}, \ldots, \underbrace{G_{U-1}, \ldots, G_{U-1}}_{N_r}) \boldsymbol{H}\boldsymbol{H}^* \right),$$

where $\boldsymbol{H}^* = \left[\boldsymbol{H}_0^* \cdots \boldsymbol{H}_{U-1}^*\right]$. Expanding this expression, we obtain

$$C = \log_2 \det\left(\frac{E_s}{UN_rN_0} \operatorname{diag}(\underbrace{G_0, \ldots, G_0}_{N_r}, \ldots, \underbrace{G_{U-1}, \ldots, G_{U-1}}_{N_r}) \boldsymbol{H}\boldsymbol{H}^* \right) + \mathcal{O}\left(\frac{1}{E_s/N_0}\right) \tag{8.137}$$

$$= UN_r \log_2 \frac{E_s}{N_0} - UN_r \log_2(UN_r) + \log_2\left(\prod_{u=0}^{U-1} G_u^{N_r} \right) + \log_2 \det(\boldsymbol{H}\boldsymbol{H}^*) + \mathcal{O}\left(\frac{1}{E_s/N_0}\right) \tag{8.138}$$

$$= N_r U \log_2 \frac{E_s}{N_0} + N_r \sum_{u=0}^{U-1} \log_2 G_u - UN_r \log_2(UN_r) + \log_2 \det(\boldsymbol{H}\boldsymbol{H}^*) + \mathcal{O}\left(\frac{1}{E_s/N_0}\right) \tag{8.139}$$

$$= N_r \sum_{u=0}^{U-1} \log_2 \mathsf{SNR}_u - UN_r \log_2(UN_r) + \log_2 \det(\boldsymbol{H}\boldsymbol{H}^*) + \mathcal{O}\left(\frac{1}{E_s/N_0}\right)$$

$$= \sum_{u=0}^{U-1} N_r \left(\log_2 \mathsf{SNR}_u - \mathcal{L}_\infty(\boldsymbol{H}_0, \ldots, \boldsymbol{H}_{U-1}) \right) + \mathcal{O}\left(\frac{1}{E_s/N_0}\right), \tag{8.140}$$

where the $\mathcal{S}_\infty = UN_r$ spatial DOF, precisely N_r DOF per user, are evident while the power offset given the channel realizations $\boldsymbol{H}_0, \ldots, \boldsymbol{H}_{U-1}$ equals [774]

$$\mathcal{L}_\infty(\boldsymbol{H}_0, \ldots, \boldsymbol{H}_{U-1}) = \log_2(UN_r) - \frac{1}{UN_r} \log_2 \det(\boldsymbol{H}\boldsymbol{H}^*). \tag{8.141}$$

8.9.5 Ergodic setting

Expecting (8.114) over the fading,

$$\frac{R_u}{B} = \mathbb{E}\left[\log_2 \det\left(I + H_u \sum_{\mathrm{u}=u}^{U-1} \frac{\frac{E_{\mathrm{u}}}{E_{\mathrm{s}}}\mathsf{SNR}_u}{N_{\mathrm{t}}} F_{\mathrm{u}} F_{\mathrm{u}}^* \cdot H_u^*\right) - \log_2 \det\left(I + H_u \sum_{\mathrm{u}=u+1}^{U-1} \frac{\frac{E_{\mathrm{u}}}{E_{\mathrm{s}}}\mathsf{SNR}_u}{N_{\mathrm{t}}} F_{\mathrm{u}} F_{\mathrm{u}}^* \cdot H_u^*\right)\right] \quad u = 0, \ldots, U-1, \tag{8.142}$$

where, recall, the precoders are a function of $H_0, \ldots, H_{U-1}$ (subject to the applicable power constraints) and thus the expectation is over them as well.

The ergodic weighted sum-capacity for some $q_0, \ldots, q_{U-1}$ equals $\sum_{u=0}^{U-1} q_u R_u/B$ with the precoders optimized, at each fading realization, for $H_0, \ldots, H_{U-1}$. And, by sweeping over all possible weight combinations, the entire capacity boundary could be delimited. Unfortunately, as observed earlier, the precoder optimization for each fading realization is a nonconvex and rather inconvenient problem. Alternatively, we can invoke duality and more conveniently obtain the ergodic capacity boundary by expecting (8.115) over the fading distribution. Likewise, the ergodic sum-capacity $C(\mathsf{SNR}_0, \ldots, \mathsf{SNR}_{U-1})$ can be obtained from (8.121) as

$$C = \mathbb{E}\left[\max_{\substack{F_0^{\mathsf{dMAC}}, \ldots, F_{U-1}^{\mathsf{dMAC}} : \|F_u^{\mathsf{dMAC}}\|_F^2 = N_r \\ \frac{E_0}{E_s}, \ldots, \frac{E_{U-1}}{E_s} : \sum_{u=0}^{U-1} \frac{E_u}{E_s} = 1}} \log_2 \det\left(I + \sum_{u=0}^{U-1} \frac{\frac{E_u}{E_s}\mathsf{SNR}_u}{N_r} H_u^* F_u^{\mathsf{dMAC}} F_u^{\mathsf{dMAC}*} H_u\right)\right]. \tag{8.143}$$

Example 8.31

Consider a base station with $N_{\mathrm{t}} = 6$ antennas serving $U = 3$ users, each equipped with $N_{\mathrm{r}} = 2$ antennas. The channel matrices have IID Rayleigh-faded entries and

$$\mathsf{SNR}_0|_{\mathrm{dB}} = \mathsf{SNR}|_{\mathrm{dB}} \tag{8.144}$$

$$\mathsf{SNR}_1|_{\mathrm{dB}} = \mathsf{SNR}|_{\mathrm{dB}} + 5\ \mathrm{dB} \tag{8.145}$$

$$\mathsf{SNR}_2|_{\mathrm{dB}} = \mathsf{SNR}|_{\mathrm{dB}} + 8\ \mathrm{dB}. \tag{8.146}$$

Compute the ergodic sum-capacity with CSIR and CSIT and plot it as a function of the reference SNR.

Solution

Applying a standard convex optimization solver to (8.143), with the explicit additional constraints that $\left(F_u^{\mathsf{dMAC}} F_u^{\mathsf{dMAC}*}\right)$ be positive-semidefinite and $\frac{E_u}{E_s} \geq 0$ for $u = 0, \ldots, U-1$,

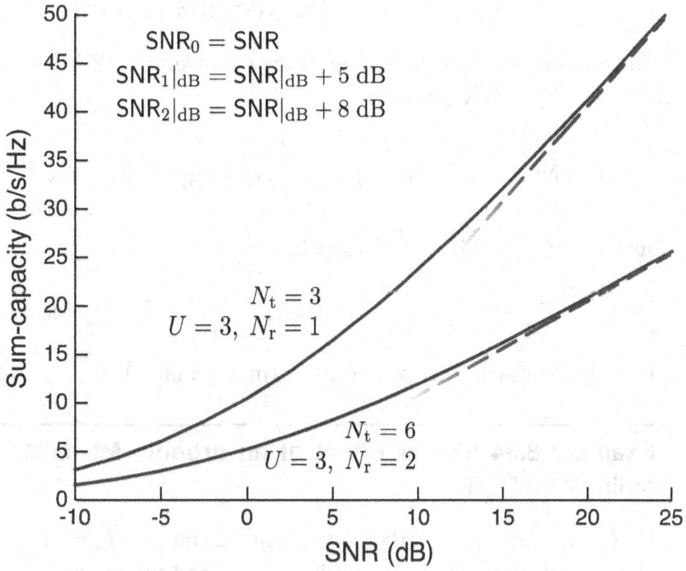

Fig. 8.14 Solid, BC ergodic sum-capacity with $N_{\mathrm{t}} = 6$, $U = 3$, and $N_{\mathrm{r}} = 2$, as well as with $U = N_{\mathrm{t}} = 3$ and $N_{\mathrm{r}} = 1$, both as functions of the reference SNR (in dB). Dashed, corresponding high-SNR expansions.

the result presented in Fig. 8.14 is readily obtained. We are to recall this result later in the book, to gauge the performance of linear transmission schemes.

Example 8.32 (Ergodic sum-capacity of a MISO BC with CSIT)

If the users are equipped with a single antenna, $F_u^{\mathrm{dMAC}} = 1$ for $u = 0, \ldots, U-1$ and the sum-capacity simplifies into

$$C = \mathbb{E}\left[\max_{\frac{E_0}{E_s},\ldots,\frac{E_{U-1}}{E_s}:\sum_{u=0}^{U-1}\frac{E_u}{E_s}=1} \log_2\left(1 + \sum_{u=0}^{U-1}\frac{E_u}{E_s}\,\mathsf{SNR}_u\,\|\boldsymbol{H}_u\|^2\right)\right], \tag{8.147}$$

where, for each fading realization, only the power shares among dual-MAC users must be optimized.

Example 8.33

Repeat Example 8.31, only with $N_{\mathrm{t}} = U = 3$ and $N_{\mathrm{r}} = 1$.

Solution

The result, obtained by applying a standard convex optimization solver to (8.147), with the explicit additional constraint that $\frac{E_u}{E_s} \geq 0$ for $u = 0, \ldots, U-1$, is shown in Fig. 8.14. We are also to recall this result later, to gauge the performance of linear transmission schemes.

High-SNR regime

The expectation over the fading of the expansion in (8.140) and (8.141), which recall applies for $N_{\text{t}} \geq U N_{\text{r}}$, returns,

$$C(\mathsf{SNR}_0, \ldots, \mathsf{SNR}_{U-1}) = \sum_{u=0}^{U-1} N_{\text{r}} \left(\log_2 \mathsf{SNR}_u - \mathcal{L}_\infty \right) + \mathcal{O}\left(\frac{1}{E_{\text{s}}/N_0} \right) \tag{8.148}$$

such that $S_\infty = \min(N_{\text{t}}, U N_{\text{r}})$ while

$$\mathcal{L}_\infty = \log_2(U N_{\text{r}}) - \frac{1}{U N_{\text{r}}} \mathbb{E}\left[\log_2 \det(\boldsymbol{H}\boldsymbol{H}^*) \right] \tag{8.149}$$

is the BC sum-capacity power offset (in 3-dB units).

Example 8.34 (Power offset of an ergodic MU-MIMO BC with IID Rayleigh fading and CSIT)

If $\boldsymbol{H}_0, \ldots, \boldsymbol{H}_{U-1}$ have IID Rayleigh-faded entries, $\boldsymbol{H}^* = \left[\boldsymbol{H}_0^* \cdots \boldsymbol{H}_{U-1}^* \right]$ is an $N_{\text{t}} \times U N_{\text{r}}$ matrix with IID complex Gaussian entries and the argument of the $\log \det(\cdot)$ function in (8.149) is a Wishart matrix. Invoking (C.28) and (E.9),

$$\mathcal{L}_\infty = \log_2(U N_{\text{r}}) + \left(\gamma_{\text{EM}} - \sum_{q=1}^{N_{\text{t}} - U N_{\text{r}}} \frac{1}{q} - \frac{N_{\text{t}}}{U N_{\text{r}}} \sum_{q = N_{\text{t}} - U N_{\text{r}} + 1}^{N_{\text{t}}} \frac{1}{q} + 1 \right) \log_2 e \tag{8.150}$$

which, for $N_{\text{t}} = U N_{\text{r}}$, becomes

$$\mathcal{L}_\infty = \log_2(U N_{\text{r}}) + \left(\gamma_{\text{EM}} - \sum_{q=2}^{U N_{\text{r}}} \frac{1}{q} \right) \log_2 e. \tag{8.151}$$

Example 8.35

Consider a base station featuring $N_{\text{t}} = 6$ antennas, communicating with $U = 3$ users each equipped with $N_{\text{r}} = 2$ antennas. Given IID Rayleigh fading, compute $\mathcal{L}_\infty$ and utilize it to approximate the sum-capacity in the high-SNR regime. Then, plot this approximation next to the actual sum-capacity for

$$\mathsf{SNR}_0|_{\text{dB}} = \mathsf{SNR}|_{\text{dB}} \tag{8.152}$$

$$\mathsf{SNR}_1|_{\text{dB}} = \mathsf{SNR}|_{\text{dB}} + 5 \text{ dB} \tag{8.153}$$

$$\mathsf{SNR}_2|_{\text{dB}} = \mathsf{SNR}|_{\text{dB}} + 8 \text{ dB}. \tag{8.154}$$

Solution

Applying (8.151), we obtain $\mathcal{L}_\infty = 1.32$ (in 3-dB units). In conjunction with (8.148), this gives

$$C(\mathsf{SNR}_0, \mathsf{SNR}_1, \mathsf{SNR}_2) \approx 2 \sum_{u=0}^{2} \left(\log_2 \mathsf{SNR}_u - 1.32 \right) \tag{8.155}$$

which is plotted in Fig. 8.14 alongside the actual sum-capacity.

Example 8.36

Repeat the previous example for $U = N_t = 3$ and $N_r = 1$.

Solution

Applying again (8.151), $\mathcal{L}_\infty = 1.22$ (in 3-dB units). The ensuing high-SNR approximation is also depicted in Fig. 8.14, next to the actual sum-capacity.

The number of DOF for individual users is, in the ergodic sense, as argued for specific channel realizations: the user encoded first enjoys $\min(N_t, N_r)$ spatial DOF, the user encoded second enjoys $\min(N_t - N_r, N_r)$ spatial DOF, and so on, with these quantities taken as zero whenever negative.

8.10 Broadcast channel with no CSIT

When the transmitter is not privy to the realization of the channel matrices, but only to their distribution, no time-domain power control is possible and thus a per-symbol power constraint effectively engulfs also a per-codeword constraint.

Rather than separately study what happens with no CSIT or with no CSI whatsoever, we can consider both cases in concert. Indeed, we have learned throughout the text that the availability of CSIR in the form of side information changes but little the performance in nonmassive MIMO conditions. Consequently, what drives our interest here is the role of CSIT. In SU-MIMO and in the MU-MIMO MAC, this role was far from critical, and altogether inconsequential in most high-SNR situations. In the MU-MIMO BC, things turns out to be radically different and not having CSIT has severe consequences:

- The MAC–BC duality no longer holds.
- DPC cannot be applied.

As a result, much of what is learned under CSIT ceases to apply. Since, without CSIT, the precoders can only depend on the fading distributions and chiefly the transmit antenna correlations, it is then of interest to have these be as pronounced as possible. Without antenna correlations, in fact, the transmitter is unable to take advantage of the availability of multiple antennas to communicate concurrently with multiple users as it does not know in which spatial directions the signals intended for each user should be launched [796]. Intuitively, then, the capacity of a MU-MIMO BC should be no better than if the transmitter multiplexed its signals orthogonally, and this is indeed the case. Specifically, it is shown in [813] that the capacity boundary of a MU-MISO BC (single-antenna users and $N_t > 1$ antennas at the base station) is contained within that of the corresponding MU-SISO BC (single-antenna users but $N_t = 1$); without CSIT, therefore, the space dimension cannot be exploited for multiplexing purposes and the spatial DOF collapse. More generally, the situation in terms of the number of DOF is as in Table 8.1.

Table 8.1 Number of spatial DOF

CSIT (side information)	No CSIT
$S_\infty = \min(N_\mathrm{t}, UN_\mathrm{r})$	$S_\infty = \min(N_\mathrm{t}, N_\mathrm{r})$

In order to transcend what can be achieved by a combination of SU-MIMO and orthogonal multiplexing, avoiding the collapse of the spatial DOF, it is necessary to acquire CSIT. Fortunately, the underspreadness of the fading makes this possible, just as it makes it possible to acquire CSIR. As discussed in Section 5.10, with TDD/full duplexing the CSIR acquired upon reception can double as CSIT while, with FDD, feedback is often effective. In either case, the analysis of how pilot symbols enable the procurement of channel estimates and, from those, the adjustment of transmitters and receivers, is certainly more involved in the BC than in SU-MIMO or in the MAC. An extensive analysis of how pilot-assisted transmission can bridge the gap between the no-CSI and the CSI extremes is presented in the next chapter, in the context of linear transmitters for the MU-MIMO BC.

In the present chapter's context of optimum transmitters, the elucidation of how exploiting the fading coherence can enable a transmitter deprived of side information to approach what is possible with CSIT is forestalled by the fact that the capacity boundary of a MU-MIMO BC without CSIT is unknown and appears to be rather unwieldy. Notwithstanding that, illuminating results in this respect have been put forth in terms of the number of DOF; despite the coarse nature of the DOF metric, informative only of the high-SNR asymptotic behavior, these results shed light on the problem and merit attention. The gist of them is encased in the example that follows, and the implications are discussed in due course.

Example 8.37

Consider a MU-MISO BC with $U \leq N_\mathrm{t}$. Antenna correlations may exist at the base station. Suppose that $\boldsymbol{H}_0$ is known by the transmitter while $\boldsymbol{H}_1, \ldots, \boldsymbol{H}_{U-1}$ are not. Then, as indicated, the capacity is no better than with single-user operation: $S_\infty = 1$. We are interested in the conditions leading to $S_\infty > 1$ and, in particular, to $S_\infty = U$. Do $\boldsymbol{H}_1, \ldots, \boldsymbol{H}_{U-1}$ have to become perfectly known as their SNRs grow? If so, how fast should the uncertainty in the transmitter knowledge of $\boldsymbol{H}_1, \ldots, \boldsymbol{H}_{U-1}$ decay as those SNRs grow?

Solution

It is proved in [814] (see also [815] and [816]) that, if the uncertainty in the transmitter knowledge of $\boldsymbol{H}_u$ for $u = 1, \ldots, U-1$ behaves as

$$\mathcal{O}\left(\frac{1}{\mathsf{SNR}_u^{\zeta_u}}\right) \qquad \zeta_u \in [0, 1], \tag{8.156}$$

then the sum-capacity exhibits

$$S_\infty = 1 + \sum_{u=1}^{U-1} \zeta_u. \tag{8.157}$$

The fact that the quality of the CSIT needs to improve with the SNR and become perfect in the limit for each DOF not to be zero is hardly surprising. Intuitively, as the noise vanishes, the interference that a signal stream experiences from others should vanish as well—lest it become the limiting impairment—and such interference relates directly to the CSIT. But (8.157) is a stronger result, as it quantifies the rate at which the CSIT needs to improve for a given number of DOF between 1 and U to be achievable. In particular:

- For $\zeta_u = 0$, the DOF of user u collapse. In fact, such a collapse occurs not only in the complete absence of CSIT, but anytime the CSIT does not improve with the SNR as per (8.156). Beware that the coarseness of the DOF metric is on display here; having more CSIT is certainly better than having less, even if the improvement does not conform to (8.156), and any difference is sure to reflect on a higher capacity even if the number of DOF is unchanged.
- For $\zeta_1 = \cdots = \zeta_{U-1} = 1$ we obtain $S_\infty = U$. An even faster rate of improvement in the CSIT quality may bring about further gains in capacity, necessarily sublogarithmic because the number of spatial DOF cannot exceed $S_\infty = U$. Again, the DOF metric fails to capture these finer aspects.

Elaborating on Example 8.37 and taking the implications to their logical conclusion we could affirm that, without side information, the number of spatial DOF in a MU-MIMO BC equals $\min(N_\text{t}, UN_\text{r})$ times a loss factor capturing the signaling resources that have to be consumed in order to satisfy (8.156). For an FDD system, the loss factor needs to further account for whatever feedback resources are necessary to ensure (8.156). In this regard, it is also argued in [814] that quantized digital feedback at the proper rate suffices.

Transmit antenna correlations, allowed in Example 8.37, as well as receive antenna correlations in the case of multiantenna users, are immaterial when it comes to the spatial DOF but would likely affect the capacity boundary. On the one hand, correlations facilitate the acquisition of CSI while, on the other hand, they may modify (either enlarge or shrink) the CSI capacity boundary. The balance of these effects is in general not obvious.

If one wanted to operate a BC without gathering and exploiting CSIT, that would favor closely spaced transmit antennas and narrow angle spreads, suggesting base stations elevated over the scattering clutter (see Section 3.6.1). With strong antenna correlations, the precoders could synthesize beams directed to each user and, if those users were sufficiently separated in angle—this could be aided by an adequate selection of the U active users from the population U_tot—the beams would interfere minimally even without nonlinear processing at the transmitter. What this points to is the orthodox notion of SDMA illustrated in the bottom part of Fig. 2.11. Since the exact capacity-achieving statistical beamformers are not easy to compute explicitly [516, 817–819], a pragmatic approach to such SDMA would be to simply apply SU-MIMO statistical beamforming to each user's signal. As learned in Section 5.4, this amounts to setting $\boldsymbol{F}_u = \sqrt{N_\text{t}}\boldsymbol{u}_u$ with $\boldsymbol{u}_u$ the maximum-eigenvalue eigenvector of $\mathbb{E}\big[\boldsymbol{H}_u^*\boldsymbol{H}_u\big]$. As mentioned, interference among beams could be kept to a minimum with a proper user selection and, if necessary, by perturbing the SU-MIMO precoding solutions. In fact, sectorization—a mainstay of traditional cellular design—can be viewed as a static form of SDMA, where the "beams" are fixed sectors that can be shaped by a single directive antenna but also by an array of closely spaced antennas.

8.11 Summary and outlook

A key insight in MU-MIMO is that, with CSIR and CSIT as well as with optimum transmitters and receivers, the high-SNR sum-capacity is essentially as if U users could collaborate as a single user. This is a powerful result that allows skirting restrictions in the number of antennas at the user devices, where form factors may be an issue, by simply activating multiple users. This advantage does not show at low SNRs, where performance is driven by the captured power, something on which MU-MIMO is not superior to SU-MIMO: in the MAC, the number of receive antennas is the same for both SU-MIMO and MU-MIMO while, in the BC, the transmit power must be split among MU-MIMO users. It is thus at high SNR that the superiority of MU-MIMO becomes pronounced, once the performance is driven by the number of communication dimensions; this is sharply registered by the increase in spatial DOF when switching from SU-MIMO to MU-MIMO.

In the MAC, the foregoing insight holds despite the U users being unable to pool their powers or precode their signals jointly. For $E_s/N_0 \to \infty$ and $UN_t \leq N_r$, the power allocation becomes uniform and this restriction is rendered irrelevant, with or without CSIT.

In the BC, alternatively, the transmit powers are inherently pooled and the precoding is localized, but the availability of CSIT is critical. It is the CSIT that enables the BC to perform as if the U receivers could collaborate, but, by the same token, if the CSIT is absent—or does not improve sufficiently fast with the SNRs—the MU-MIMO BC founders.

The more specific findings of the chapter, with the low-SNR regime de-emphasized given its reduced interest for MU-MIMO, are recapped in the form of take-away points within the companion summary box.

Because of its complexity and sensitivity to CSIT inaccuracies, DPC is used mostly for benchmarking and bounding, rather than actually implemented; we resort to this benchmark in the next chapter. And, on the matter of CSI accuracy, the all-important issue of pilot-based transmission has been touched on briefly; the bulk of it is also deferred to the next chapter, where it is covered extensively in the context of linear transceivers.

The reader may also have noticed that, in contrast with the SU-MIMO part of the book, in this part we are not resorting as intensively to the large-dimensional regime. Indeed, except for very special cases such as equals-SNR users, the matrices involved have more intricate structures and the insight advantages are largely lost. Despite this reality, some results are pointed out in Section 8.4.2, with more to come in the massive MIMO analysis.

Problems

8.1 Draw the MU-SIMO MAC capacity boundary for $U = 2$ with $\mathsf{SNR}_0 = 10$ dB, $\mathsf{SNR}_1 = 5$ dB, and

$$\boldsymbol{H}_0 = \begin{bmatrix} 1+0.5\,\mathrm{j} \\ 0.3-\mathrm{j} \end{bmatrix} \qquad \boldsymbol{H}_1 = \begin{bmatrix} 2+0.3\,\mathrm{j} \\ 1.2 \end{bmatrix}. \tag{8.163}$$

Take-away points

1. In a MAC with an optimum receiver, users should transmit at full power (or, with discrete constellations, at the highest power convenient for those constellations).
2. With CSIR and CSIT, the achievable $(R_0/B, \ldots, R_{U-1}/B)$ for channel realizations $\boldsymbol{H}_0, \ldots, \boldsymbol{H}_{U-1}$ with precoders $\boldsymbol{F}_0, \ldots, \boldsymbol{F}_{U-1}$ is defined by
$$\sum_{u \in \mathcal{U}} \frac{R_u}{B} \leq \log_2 \det \left(\boldsymbol{I} + \sum_{u \in \mathcal{U}} \frac{\mathsf{SNR}_u}{N_\mathrm{t}} \boldsymbol{H}_u \boldsymbol{F}_u \boldsymbol{F}_u^* \boldsymbol{H}_u^* \right) \qquad \mathcal{U} \subseteq \{0, \ldots, U-1\}$$
and the MAC capacity is the boundary of the union of all regions corresponding to admissible precoders.
3. The precoders that yield the boundary point corresponding to some weights $q_0, \ldots, q_{U-1}$ can be obtained via convex optimization. For the sum-capacity specifically, the precoders can also be obtained through iterative waterfilling.
4. With LMMSE-SIC reception, users should be decoded in order of increasing weight and the uth decoded user achieves
$$\frac{R_u}{B} = \log_2 \det \left(\boldsymbol{I} + \sum_{\mathrm{u}=u}^{U-1} \frac{\mathsf{SNR}_\mathrm{u}}{N_\mathrm{t}} \boldsymbol{H}_\mathrm{u} \boldsymbol{F}_\mathrm{u} \boldsymbol{F}_\mathrm{u}^* \boldsymbol{H}_\mathrm{u}^* \right) - \log_2 \det \left(\boldsymbol{I} + \sum_{\mathrm{u}=u+1}^{U-1} \frac{\mathsf{SNR}_\mathrm{u}}{N_\mathrm{t}} \boldsymbol{H}_\mathrm{u} \boldsymbol{F}_\mathrm{u} \boldsymbol{F}_\mathrm{u}^* \boldsymbol{H}_\mathrm{u}^* \right) \qquad u = 0, \ldots, U-1, \tag{8.158}$$
from which $\sum_{q=0}^{U-1} q_u R_u / B$ can be optimized over the precoders. This optimization, swept over all possible weights, yields again the capacity boundary.
5. In the MAC, the number of spatial DOF is $S_\infty = \min(UN_\mathrm{t}, N_\mathrm{r})$ whereas the sum-capacity power offset is
$$\mathcal{L}_\infty(\boldsymbol{H}_0, \ldots, \boldsymbol{H}_{U-1}) = \log_2 N_\mathrm{t} - \frac{1}{UN_\mathrm{t}} \log_2 \det(\boldsymbol{H}^* \boldsymbol{H}), \tag{8.159}$$
where $\boldsymbol{H} = [\boldsymbol{H}_0 \cdots \boldsymbol{H}_{U-1}]$.
6. Without CSIT, the MAC precoders depend on the distribution of $\boldsymbol{H}_0, \ldots, \boldsymbol{H}_{U-1}$ and can again be obtained via convex optimization. In correlated channels specifically, each user should diagonalize its transmit own correlation matrix.
7. To bypass the assumption of CSIR side information, a penalty factor $(1 - S_\infty/N_\mathrm{c})$ should be applied to $S_\infty = \min(UN_\mathrm{t}, N_\mathrm{r})$. The ensuing number of DOF can be achieved with pilot-assisted transmission.
8. The BC capacity boundary with CSIT and CSIR coincides with the capacity boundary of a fictitious dual MAC where users can pool their powers (but not jointly precode).
9. In a nondegraded BC with CSIT, DPC transmission with users encoded in order of decreasing weight is optimum. For a degraded BC, both DPC and superposition are optimal.

10. With DPC in order of decreasing weights, the BC counterpart to (8.158) is

$$\frac{R_u}{B} = \log_2 \det\left(\boldsymbol{I} + \boldsymbol{H}_u \sum_{\mathrm{u}=0}^{u} \frac{\frac{E_{\mathrm{u}}}{E_{\mathrm{s}}}\mathsf{SNR}_u}{N_{\mathrm{t}}} \boldsymbol{F}_{\mathrm{u}} \boldsymbol{F}_{\mathrm{u}}^* \cdot \boldsymbol{H}_u^* \right) \tag{8.160}$$

$$- \log_2 \det\left(\boldsymbol{I} + \boldsymbol{H}_u \sum_{\mathrm{u}=0}^{u-1} \frac{\frac{E_{\mathrm{u}}}{E_{\mathrm{s}}}\mathsf{SNR}_u}{N_{\mathrm{t}}} \boldsymbol{F}_{\mathrm{u}} \boldsymbol{F}_{\mathrm{u}}^* \cdot \boldsymbol{H}_u^* \right) \qquad u = 0, \ldots, U-1.$$

In contrast with the MAC, here the optimization of $\sum_{q=0}^{U-1} q_u R_u/B$ is not convex.

11. Applying the BC-MAC duality, (8.160) can be converted into

$$\frac{R_u}{B} = \log_2 \det\left(\boldsymbol{I} + \sum_{\mathrm{u}=u}^{U-1} \frac{\frac{E_{\mathrm{u}}}{E_{\mathrm{s}}}\mathsf{SNR}_{\mathrm{u}}}{N_{\mathrm{r}}} \boldsymbol{H}_{\mathrm{u}}^* \boldsymbol{F}_{\mathrm{u}}^{\mathsf{dMAC}} \boldsymbol{F}_{\mathrm{u}}^{\mathsf{dMAC}*} \boldsymbol{H}_{\mathrm{u}} \right) \tag{8.161}$$

$$- \log_2 \det\left(\boldsymbol{I} + \sum_{\mathrm{u}=u+1}^{U-1} \frac{\frac{E_{\mathrm{u}}}{E_{\mathrm{s}}}\mathsf{SNR}_{\mathrm{u}}}{N_{\mathrm{r}}} \boldsymbol{H}_{\mathrm{u}}^* \boldsymbol{F}_{\mathrm{u}}^{\mathsf{dMAC}} \boldsymbol{F}_{\mathrm{u}}^{\mathsf{dMAC}*} \boldsymbol{H}_{\mathrm{u}} \right) \qquad u = 0, \ldots, U-1$$

whose optimization over $\sqrt{E_0/E_{\mathrm{s}}}\boldsymbol{F}_0^{\mathsf{dMAC}}, \ldots, \sqrt{E_{U-1}/E_{\mathrm{s}}}\boldsymbol{F}_{U-1}^{\mathsf{dMAC}}$ is convex. Moreover, the BC precoders are given by the filters in the LMMSE-SIC receiver for this dual MAC.

12. In the BC with CSIT, the number of spatial DOF is $S_\infty = \min(N_{\mathrm{t}}, UN_{\mathrm{r}})$ whereas the sum-capacity power offset is

$$\mathcal{L}_\infty(\boldsymbol{H}_0, \ldots, \boldsymbol{H}_{U-1}) = \log_2(UN_{\mathrm{r}}) - \frac{1}{UN_{\mathrm{r}}} \log_2 \det(\boldsymbol{H}\boldsymbol{H}^*), \tag{8.162}$$

where $\boldsymbol{H}^* = [\boldsymbol{H}_0^* \cdots \boldsymbol{H}_{U-1}^*]$.

13. In ergodic settings, the spectral efficiencies and power offsets in this summary should be further expected over the distribution of $\boldsymbol{H}_0, \ldots, \boldsymbol{H}_{U-1}$ and of any quantities that are functions thereof.

14. In the BC, in the absence of CSIT, $S_\infty = 1 + \sum_{u=1}^{U-1} \zeta_u$ provided the uncertainty in the transmitter knowledge of $\boldsymbol{H}_u$ scales as $\mathcal{O}(\mathsf{SNR}_u^{-\zeta_u})$ for $\zeta_u \in [0,1]$ and $u = 1, \ldots, U-1$.

8.2 Write down the conditions determining the MU-SIMO MAC capacity boundary for $U = 3$ with $\mathsf{SNR}_0 = 10$ dB, $\mathsf{SNR}_1 = 3$ dB, and $\mathsf{SNR}_2 = 5$ dB, and with

$$\boldsymbol{H}_0 = \begin{bmatrix} 1+0.5\mathrm{j} \\ 0.3-\mathrm{j} \end{bmatrix} \qquad \boldsymbol{H}_1 = \begin{bmatrix} 2+0.3\mathrm{j} \\ 1.2 \end{bmatrix} \qquad \boldsymbol{H}_2 = \begin{bmatrix} 2 \\ 0.7+1.2\mathrm{j} \end{bmatrix}.$$

8.3 Repeat Example 8.2 in the following cases.

(a) $\mathsf{SNR}_0 = \mathsf{SNR}_1 = 0$ dB.

(b) $\mathsf{SNR}_0 = 0$ dB and $\mathsf{SNR}_1 = 10$ dB.

8.4 Repeat Example 8.2, but with the precoders optimized.

Hint: A convex solver is required, for instance `fmincon` *in* MATLAB®.

8.5 In a quasi-static MAC with LMMSE-SIC reception, if the users are decoded in an order other than that of increasing weights, do the sum-capacity and weighted sum-capacity diminish or stay unchanged? What is the significance of this?

8.6 Prove that, when the noise is $\boldsymbol{v} \sim \mathcal{N}_{\mathbb{C}}(\mathbf{0}, \boldsymbol{\Sigma})$, the optimum SU-MIMO precoder is the one that diagonalizes $\boldsymbol{\Sigma}^{-1/2}\boldsymbol{H}$. This sets the stage for iterative waterfilling.

8.7 For the setup of Example 8.2, apply iterative waterfilling to obtain the precoders that deliver the sum-capacity. Then apply those precoders to compute such sum-capacity.

8.8 For the setup in Example 8.2, compute $q_0 R_0/B + q_1 R_1/B$ under TDMA SU-MIMO. What is the advantage for each user of employing MU-MIMO rather than TDMA SU-MIMO at the operating point defined by such q_0 and q_1?

8.9 Prove that (8.23) plus C_1 equal the sum-capacity C in Example 8.3.

8.10 Expand (8.57) and show that it behaves as

$$\frac{R_u}{B} = \dot{C}_u(0)\,\mathsf{SNR}_u + \mathcal{O}(\mathsf{SNR}_u^2) \tag{8.164}$$

as claimed in Section 7.7. Further show that, in this expansion,

$$\dot{C}_u(0) = \mathbb{E}\big[\lambda_{\max}(\boldsymbol{H}_u^*\boldsymbol{H}_u)\big] \log_2 e \tag{8.165}$$

as in the corresponding expansion for the SU-MIMO ergodic capacity with CSIT, given in (5.38).

Note: This proves that, with CSIT, MU-MIMO and frequency-division SU-MIMO are equivalent to first order in a low-SNR MAC.

8.11 Consider a vehicular block-faded MU-SIMO MAC with $U = 2$, $\mathsf{SNR}_0 = 20$ dB, $\mathsf{SNR}_1 = 0$ dB, and $N_r = 2$. The receiver is LMMSE-SIC with user 0 decoded first. What is the minimum pilot overhead to ensure that the effective SNR experienced by user 0 is at least 10 dB above SNR_1?

8.12 Let $U = 3$ with $\mathsf{SNR}_u = \mathsf{SNR}$ for $u = 0, 1, 2$ and with

$$\boldsymbol{H}_0 = \begin{bmatrix} 1+0.5\,\mathrm{j} & 2 \\ 0.3-\mathrm{j} & 0.7+1.2\,\mathrm{j} \end{bmatrix} \qquad \boldsymbol{H}_1 = \begin{bmatrix} 0.4\,\mathrm{j} & 2+0.3\,\mathrm{j} \\ 1.1+\mathrm{j} & 1.2 \end{bmatrix}$$

$$\boldsymbol{H}_2 = \begin{bmatrix} 0.7 & -1+0.9\,\mathrm{j} \\ 1.3\,\mathrm{j} & 0 \end{bmatrix}.$$

(a) Calculate the multiuser power offset, $\mathcal{L}_\infty(\boldsymbol{H}_0, \boldsymbol{H}_1, \boldsymbol{H}_2)$.

(b) Plot the unprecoded sum spectral efficiency as a function of $\mathsf{SNR} \in [0, 30]$ dB. On the same chart, plot the high-SNR expansion of the sum-capacity.

8.13 Generalize to the MU-MIMO MAC the power offset definition in (4.46).

8.14 Derive the counterpart for $UN_t > N_r$ to the high-SNR expansion of the MU-MIMO MAC sum-capacity in (8.53).

8.15 Reproduce Example 8.9. Further plot, on the same chart, the collaborative upper bound: the ergodic sum-capacity of an SU-MIMO channel where the transmitter aggregates the U users, with their antennas and powers pooled, and joint precoding.

8.16 Repeat Example 8.9 with $N_r = 4$.

8.17 Repeat Example 8.9 with every user's two transmit antennas being fully correlated. What do you observe?

8.18 Reproduce Example 8.10.

8.19 Repeat Example 8.10 with $N_r = 4$ and corroborate that now one of the users has zero DOF.

8.20 Plot (8.65) alongside the corresponding ergodic sum spectral efficiency without precoding for $\mathsf{SNR}_u = \mathsf{SNR}$, $u = 0, 1, 2$, over the range $\mathsf{SNR} \in [0, 30]$ dB.

8.21 For $U = 3$ and $N_t = 2$, plot $\mathcal{L}_\infty$ for an ergodic MU-MIMO MAC with IID Rayleigh fading as a function of $N_r = 6, \ldots, 10$.

8.22 Consider an ergodic MU-MIMO MAC with $U = 4$, $N_t = 2$, and $N_r = 8$. Further let $\mathsf{SNR}_0|_{\text{dB}} = \mathsf{SNR}_1|_{\text{dB}} = \mathsf{SNR}|_{\text{dB}}$ and $\mathsf{SNR}_2|_{\text{dB}} = \mathsf{SNR}_3|_{\text{dB}} = \mathsf{SNR}|_{\text{dB}} + 5$ dB, with $\boldsymbol{H}_0$ and $\boldsymbol{H}_1$ having IID Rayleigh-faded entries while $\boldsymbol{H}_2$ and $\boldsymbol{H}_3$ have IID Rice-faded entries ($\mathsf{K} = 3$ dB).

(a) Plot the unprecoded ergodic sum-spectral efficiency for $\mathsf{SNR} \in [0, 30]$ dB.

(b) On the same chart, plot the corresponding high-SNR expansion.

8.23 Prove the low-SNR first-order equivalence between MU-MIMO and frequency-division SU-MIMO in a MAC, as in Problem 8.10, but this time with CSIR only (no CSIT).

8.24 Consider a MU-MIMO MAC with $U = N_r = 3$ and $N_t = 2$. The channels are Rayleigh-faded with transmit correlation matrix

$$\boldsymbol{R}_u = \begin{bmatrix} 1 & 0.7 \\ 0.7 & 1 \end{bmatrix} \qquad u = 0, 1, 2 \tag{8.166}$$

and the uth user beamforms along the maximum-eigenvalue eigenvector of $\boldsymbol{R}_u$. For $\mathsf{SNR}_0|_{\text{dB}} = \mathsf{SNR}_1|_{\text{dB}} = \mathsf{SNR}|_{\text{dB}}$ and $\mathsf{SNR}_2|_{\text{dB}} = \mathsf{SNR}|_{\text{dB}} + 5$ dB, plot the ergodic sum spectral efficiency as a function of $\mathsf{SNR} \in [0, 20]$ dB and confirm that $\mathcal{S}_\infty = 3$.

8.25 Consider a MU-SIMO MAC with $\mathsf{SNR}_u = \mathsf{SNR}$ for $u = 0, \ldots, U-1$ and with $\mathsf{SNR} = 10$ dB. Further, $\boldsymbol{H}_0, \ldots, \boldsymbol{H}_{U-1}$ have IID Rayleigh-faded entries. For this setup and fixed N_r, compute the leading term in the expansion of the ergodic sum-capacity as a function of U.

8.26 Consider a vehicular Rayleigh-faded MU-MIMO MAC with $U = N_t = 2$ and $N_r = 4$. The channel matrices have IID entries and the transmissions are unprecoded with $\mathsf{SNR}_0|_{\text{dB}} = \mathsf{SNR}|_{\text{dB}}$ and $\mathsf{SNR}_1|_{\text{dB}} = \mathsf{SNR}|_{\text{dB}} + 6$ dB.

(a) Plot the ergodic sum spectral efficiency with CSIR over $\mathsf{SNR} \in [0, 20]$ dB.

(b) On the same chart, plot the pilot-assisted sum spectral efficiency without pilot power boosting. Further plot the optimum pilot overheads, $\alpha_0^\star$ and $\alpha_1^\star$.

Hint: A convex solver is required, for instance `fmincon` *in* MATLAB®.

8.27 Show that, given $\|\boldsymbol{F}_u^{\mathsf{BC}}\|_{\text{F}}^2 = N_t$, the dual-MAC precoder satisfies $\|\boldsymbol{F}_u^{\mathsf{dMAC}}\|_{\text{F}}^2 = N_r$.

8.28 Show how, for $U = 1$ and $N_t \geq N_r$, the capacity of the BC and MAC channels coincide at any given SNR.

8.29 Verify that (8.114) equals (8.113).

8.30 Verify that (8.74) equates the BC expression in (8.117) with its dual-MAC counterpart in (8.116).

8.31 Consider an MU-SISO BC with $U = 2$ and $N_\text{t} = N_\text{r} = 1$. Let $\mathsf{SNR}_0\,|h_0|^2 = 5$ dB and $\mathsf{SNR}_1\,|h_1|^2 = 10$ dB, with CSIR and CSIT.
(a) Draw the capacity boundary directly.
(b) Draw the capacity boundary as the union of the spectral efficiency regions achievable under all possible power allocations in the dual MAC. Verify that this boundary coincides with the one in part (a).

8.32 Reconsider Problem 8.31, but this time suppose that user 1 is DPC-encoded before user 0.
(a) Draw the boundary of the region of achievable spectral efficiencies.
(b) Draw the line connecting corner 1 of the spectral efficiency pentagons achievable under all possible power allocations in the dual MAC. Verify that this line coincides with the boundary in part (a).
Note: This problem shows how an incorrect encoding order in a degraded BC (i.e., not in order of channel strengths) corresponds to a decoding order (i.e., a corner) in the dual MAC that is not the one delineating the capacity boundary when the union is taken over all power allocations.

8.33 Show that (8.127) expands as (8.129).

8.34 Expand (8.142) and show that it behaves as

$$\frac{R_u}{B} = \frac{E_u}{E_\text{s}}\,\dot{C}_u(0)\,\mathsf{SNR}_u + \mathcal{O}(\mathsf{SNR}_u^2), \tag{8.167}$$

with $\dot{C}_u(0) = \mathbb{E}\big[\lambda_{\max}(\boldsymbol{H}_u^*\boldsymbol{H}_u)\big]\log_2 e$. This expansion coincides with (7.18), proving that MU-MIMO and frequency-division SU-MIMO are equivalent to first order in a low-SNR BC.

8.35 Reproduce Example 8.28.

8.36 For the BC in Example 8.28, compute the spectral efficiency pairs corresponding to the following operating points.
(a) Sum-capacity.
(b) Proportional fairness.
(c) Equal spectral efficiencies.
Hint: A convex solver is required, for instance `fmincon` *in* MATLAB®.

8.37 Compute the sum-capacity for the BC in Example 8.28, modified such that $\mathsf{SNR}_0 = 10$ dB and $\mathsf{SNR}_1 = 3$ dB.

8.38 Reconsider the BC in Example 8.28, except with $\mathsf{SNR}_0 = \mathsf{SNR}_1 = \mathsf{SNR}$.
(a) Plot the sum spectral efficiency with an equal power allocation as a function of $\mathsf{SNR} \in [10, 20]$ dB.
(b) Compute $\mathcal{S}_\infty$ and $\mathcal{L}_\infty(\boldsymbol{H}_0, \boldsymbol{H}_1)$, and plot, on the same chart as part (a), the high-power expansion of the sum-capacity.

8.39 Reconsider Example 8.31.
(a) Reproduce the sum-capacity in Fig. 8.14.
(b) On the same chart, plot the CSIT-aware SU-MIMO ergodic capacity for each of the three users.

8.40 Plot the collaborative upper bound for Example 8.31, i.e., the ergodic sum-capacity of an SU-MIMO channel where the receiver is the aggregation of the three receivers therein.

8.41 The collaborative upper bound equals the sum-capacity of both MAC and BC, asymptotically in the SNRs. Does that equality extend to the entire high-SNR capacity boundary?

8.42 Consider a BC where $U = N_t = 2$ and $N_r = 1$ with

$$\mathsf{SNR}_0|_{\text{dB}} = \mathsf{SNR}|_{\text{dB}} \tag{8.168}$$

$$\mathsf{SNR}_1|_{\text{dB}} = \mathsf{SNR}|_{\text{dB}} + 3\ \text{dB} \tag{8.169}$$

and with IID Rayleigh fading. Applying duality, compute the ergodic sum-capacity with CSIR and CSIT and plot it as a function of SNR.

Hint: Although a convex solver may be applied, it is not essential since the dual-MAC optimization entails a single scalar parameter.

8.43 Repeat Problem 8.42 with the number of base station antennas increased to $N_t = 4$.

8.44 Repeat Problem 8.42 with $U = N_t = 4$ and with

$$\mathsf{SNR}_2|_{\text{dB}} = \mathsf{SNR}|_{\text{dB}} \tag{8.170}$$

$$\mathsf{SNR}_3|_{\text{dB}} = \mathsf{SNR}|_{\text{dB}} + 3\ \text{dB}. \tag{8.171}$$

8.45 Repeat Problem 8.42, but with $N_t = 4$ and $U = N_r = 2$. The fading is Rayleigh-distributed and each user's antennas are 70% correlated. The transmit antennas remain uncorrelated.

8.46 Repeat Problem 8.42 with $N_t = 3$ and $U = 2$; one of the users has a single antenna while the other has two antennas.

8.47 For an ergodic BC with CSIT, confirm utilizing the power offset expressions for IID Rayleigh fading that, once all users have entered the high-SNR regime, the sum-capacity is exactly as if those users could collaborate as a single receiver.

8.48 Reproduce Example 8.35.

8.49 Reconsider the setup of Example 8.35. Compute the change in high-SNR sum-capacity if, rather than IID Rayleigh-faded, the channels are as follows.

(a) IID Rice-faded with $\mathsf{K} = 0$ dB.

(b) IID Rice-faded with $\mathsf{K} = 10$ dB.

(c) Rayleigh-faded with 70% correlation between user antennas and no correlation at the base station.

8.50 Repeat Problem 8.49, but this time express the changes in terms of transmit power.

8.51 Consider a two-antenna base station communicating with two single-antenna users. The channels are Rayleigh-faded and uncorrelated, with

$$\mathsf{SNR}_0|_{\text{dB}} = \mathsf{SNR}|_{\text{dB}} \tag{8.172}$$

$$\mathsf{SNR}_1|_{\text{dB}} = \mathsf{SNR}|_{\text{dB}} + 5\ \text{dB}. \tag{8.173}$$

On a common chart, plot the MAC and the BC sum-capacities as a function of $\mathsf{SNR} \in [0, 20]$ dB with CSIR and CSIT. What do you observe?

9 MU-MIMO with linear transceivers

Leave the beaten tracks occasionally, and dive into the woods. You will be certain to find something that you have never seen before.

Alexander Graham Bell

9.1 Introduction

As in SU-MIMO, linear receivers are a lesser complexity alternative also in MU-MIMO. Moreover, for the MU-MIMO BC, linear transmitters are almost a necessity given the difficulties of implementing DPC. Altogether, the importance of linear transceivers for MU-MIMO is far-reaching and deserving of the comprehensive treatment dispensed in this chapter.

In comparison with SU-MIMO, where the performance deficit of linear receivers depends chiefly on the balance between the number of transmit and receive antennas, in MU-MIMO some additional elements influence the shortfall of linear transceivers relative to capacity.

- User selection, whereby each resource block is allocated to a selected subset of U users from the population of U_{tot} users.
- The need to furnish the BC transmitter with CSIT.

These elements permeate the exposition in the chapter, injecting into the derivations aspects such as the type of duplexing or the feedback.

With linear in place of optimum transceivers, the structure of this chapter mirrors that of the previous one, namely a block of sections dealing with the MAC and another block dealing with the BC, with both blocks connected by a pivotal section examining the MAC–BC duality. Precisely, Section 9.2 lays the ground for the study of the MAC with linear receivers while Sections 9.3 and 9.4 analyze the ZF and LMMSE receivers, respectively. Then, Section 9.5 addresses the duality with linear transceivers and leads into the BC part of the chapter. Within that part, Section 9.6 kicks things off by introducing the key ingredients of a BC with linear transmission and Section 9.7 extensively covers the ZF performance in MU-MISO channels, including the issues that arise in pilot-assisted transmission with FDD, TDD, or full duplexing. Subsequently, Section 9.8 generalizes the ZF MU-MISO transmission into block-diagonalization for MU-MIMO, and Section 9.9 further extends the ZF approach by means of regularization. Finally, Section 9.10 wraps up the chapter.

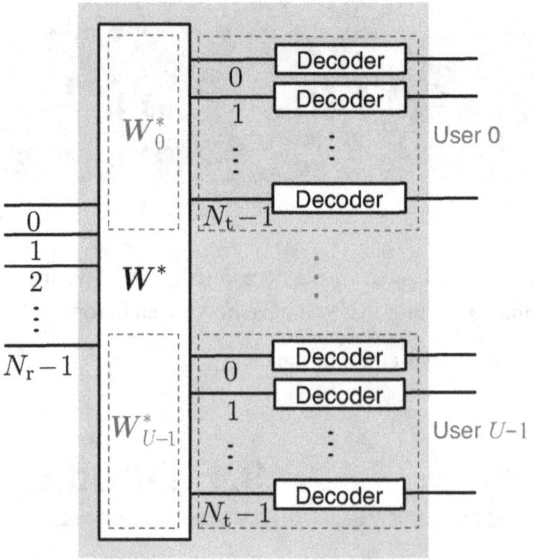

Fig. 9.1 Linear MU-MIMO MAC receiver. The linear filter $\boldsymbol{W}$ is made up of U filters $\boldsymbol{W}_0, \ldots, \boldsymbol{W}_{U-1}$, each targeting the streams of the pertinent user. If $N_s < N_t$, then some of the outputs for each user are zero.

9.2 Linear receivers for the multiple-access channel

In the study of linear receivers, the interpretation of the MAC as a SU-MIMO channel with block-diagonal precoding proves once again useful, allowing for a rather economic presentation. As in Chapter 8, we consider that every user has the same number of transmit antennas, N_t; this assumption reduces the number of variables with no appreciable loss of content in the exposition. We also consider, unless otherwise indicated, that $N_s = N_t$. However, we do retain distinct power control coefficients, $\frac{E_0}{E_s}, \ldots, \frac{E_{U-1}}{E_s}$, because, with a linear receiver, having every user transmit at full power need not be optimal. (With an optimum receiver it is optimal, which is why in the corresponding MAC treatment $\frac{E_u}{E_s} = 1$ for $u = 0, \ldots, U-1$.)

For the sake of linear reception, separate codewords should be transmitted, not only across users—always the case—but further across same-user data streams. The linear receiver, in turn, can be viewed as either a single filter estimating all data streams at once or, alternatively, as U filters, one per user (see Fig. 9.1). This is followed by a bank of scalar decoders operating separately on each stream.

With $\boldsymbol{W}_u$ the filter in charge of recovering the signals from user u, the output corresponding to the jth stream of such user is

$$[\boldsymbol{W}_u^* \boldsymbol{y}]_j = [\boldsymbol{W}_u]_{:,j}^* \, \boldsymbol{y} \tag{9.1}$$

$$= [\boldsymbol{W}_u]^*_{:,j} \left(\sum_{\mathrm{u}=0}^{U-1} \sqrt{\frac{G_{\mathrm{u}} E_{\mathrm{u}}}{N_{\mathrm{t}}}} \boldsymbol{H}_{\mathrm{u}} \boldsymbol{F}_{\mathrm{u}} \boldsymbol{s}_{\mathrm{u}} + \boldsymbol{v} \right) \tag{9.2}$$

$$= \underbrace{\sqrt{\frac{G_u E_u}{N_{\mathrm{t}}}} [\boldsymbol{W}_u]^*_{:,j} \, \boldsymbol{H}_u \, [\boldsymbol{F}_u]_{:,j} \, [\boldsymbol{s}_u]_j}_{\text{Desired signal}}$$

$$+ \underbrace{\sqrt{\frac{G_u E_u}{N_{\mathrm{t}}}} \sum_{j' \neq j} [\boldsymbol{W}_u]^*_{:,j} \, \boldsymbol{H}_u \, [\boldsymbol{F}_u]_{:,j'} \, [\boldsymbol{s}_u]_{j'}}_{\text{Interference from same-user streams}} \tag{9.3}$$

$$+ \underbrace{\sum_{\mathrm{u} \neq u} \sqrt{\frac{G_{\mathrm{u}} E_{\mathrm{u}}}{N_{\mathrm{t}}}} \sum_{j'=0}^{N_{\mathrm{t}}-1} [\boldsymbol{W}_u]^*_{:,j} \, \boldsymbol{H}_{\mathrm{u}} \, [\boldsymbol{F}_{\mathrm{u}}]_{:,j'} \, [\boldsymbol{s}_{\mathrm{u}}]_{j'}}_{\text{Interference from other users}} + \underbrace{[\boldsymbol{W}_u]^*_{:,j} \, \boldsymbol{v}}_{\text{Filtered noise}}$$

from which, for given fading realizations, we can express this stream's SINR as

$$\mathsf{sinr}_{u,j} = \frac{\frac{E_u}{E_s} \frac{\mathsf{SNR}_u}{N_{\mathrm{t}}} \left| [\boldsymbol{W}_u]^*_{:,j} \, \boldsymbol{H}_u \, [\boldsymbol{F}_u]_{:,j} \right|^2}{\mathrm{Den}} \tag{9.4}$$

where the denominator, packaging the interference-plus-noise power, is

$$\mathrm{Den} = \frac{\frac{E_u}{E_s} \mathsf{SNR}_u}{N_{\mathrm{t}}} \sum_{j' \neq j} \left| [\boldsymbol{W}_u]^*_{:,j} \, \boldsymbol{H}_u \, [\boldsymbol{F}_u]_{:,j'} \right|^2$$
$$+ \sum_{\mathrm{u} \neq u} \frac{\frac{E_{\mathrm{u}}}{E_s} \mathsf{SNR}_{\mathrm{u}}}{N_{\mathrm{t}}} \sum_{j'=0}^{N_{\mathrm{t}}-1} \left| [\boldsymbol{W}_u]^*_{:,j} \, \boldsymbol{H}_{\mathrm{u}} \, [\boldsymbol{F}_{\mathrm{u}}]_{:,j'} \right|^2 + \left\| [\boldsymbol{W}_u]_{:,j} \right\|^2. \tag{9.5}$$

In MU-SIMO MAC specifically, the receive filter for user u is a column vector and the precoders become immaterial, such that

$$\mathsf{sinr}_u = \frac{\frac{E_u}{E_s} \mathsf{SNR}_u \left| \boldsymbol{W}_u^* \boldsymbol{H}_u \right|^2}{\sum_{\mathrm{u} \neq u} \frac{E_{\mathrm{u}}}{E_s} \mathsf{SNR}_{\mathrm{u}} \left| \boldsymbol{W}_u^* \boldsymbol{H}_{\mathrm{u}} \right|^2 + \left\| \boldsymbol{W}_u \right\|^2}. \tag{9.6}$$

The simplest possible linear receiver consists of a bank of matched filters, $\boldsymbol{W}_u \propto \boldsymbol{H}_u \boldsymbol{F}_u$ for $u = 0, \ldots, U-1$, which performs well if $N_{\mathrm{r}} \gg U N_{\mathrm{t}}$ as then the columns of $\boldsymbol{H}_0, \ldots, \boldsymbol{H}_{U-1}$ are close to orthogonal even in the absence of any smart user selection; this situation may arise in massive MIMO. Otherwise, a bank of matched filters performs poorly and the more involved ZF and LMMSE receivers are called for.

User selection, which as mentioned influences the gap between linear receivers and capacity, also affects the relative performance of the ZF and LMMSE structures. When there is a large pool of U_{tot} users from which to favorably select the U active ones, with the value of U itself subject to choice, the performance of both linear structures improves and the gap between them shrinks. However, such sum-spectral-efficiency-driven user selection must be balanced with the need for long-term fairness, and the chapter includes examples that illustrate this point.

9.3 Linear ZF receiver for the multiple-access channel

9.3.1 Receiver structure

As in SU-MIMO, the first idea that comes to mind to outdo a matched filter is that of a ZF filter completely ridding each signal from the interference from the rest. Advantageously, this renders full-power transmission by each user optimal, as the intended signals become stronger without the downside of increased interference. Therefore, in this section we can let $\frac{E_u}{E_s} = 1$ for $u = 0, \ldots, U-1$. Under the premise, necessary for ZF feasibility, that $N_r \geq U N_t$, the ZF receiver for the MU-MIMO MAC could be obtained by resorting to our recurring SU-MIMO interpretation and applying the corresponding solution. The resulting ZF filter would have to invert (in the Moore–Penrose pseudoinverse sense, see Appendix B.6) the equivalent SU-MIMO channel connecting $\boldsymbol{s} = [\boldsymbol{s}_0^{\mathrm{T}} \cdots \boldsymbol{s}_{U-1}^{\mathrm{T}}]^{\mathrm{T}}$ with $\boldsymbol{y}$, i.e., the channel

$$\left[\sqrt{\tfrac{G_0 E_s}{N_t}} \boldsymbol{H}_0 \cdots \sqrt{\tfrac{G_{U-1} E_s}{N_t}} \boldsymbol{H}_{U-1}\right] \begin{bmatrix} \boldsymbol{F}_0 & \boldsymbol{0} & \boldsymbol{0} \\ \boldsymbol{0} & \ddots & \boldsymbol{0} \\ \boldsymbol{0} & \boldsymbol{0} & \boldsymbol{F}_{U-1} \end{bmatrix}$$
$$= \left[\sqrt{\tfrac{G_0 E_s}{N_t}} \boldsymbol{H}_0 \boldsymbol{F}_0 \cdots \sqrt{\tfrac{G_{U-1} E_s}{N_t}} \boldsymbol{H}_{U-1} \boldsymbol{F}_{U-1}\right]. \tag{9.7}$$

The single ZF filter $\boldsymbol{W}^{\mathrm{ZF}}$ inverting such channel satisfies

$$\boldsymbol{W}^{\mathrm{ZF}*} = \left[\sqrt{\tfrac{G_0 E_s}{N_t}} \boldsymbol{H}_0 \boldsymbol{F}_0 \cdots \sqrt{\tfrac{G_{U-1} E_s}{N_t}} \boldsymbol{H}_{U-1} \boldsymbol{F}_{U-1}\right]^{\dagger} \tag{9.8}$$

$$= \left(\begin{bmatrix} \sqrt{\frac{G_0 E_s}{N_t}} \boldsymbol{F}_0^* \boldsymbol{H}_0^* \\ \vdots \\ \sqrt{\frac{G_{U-1} E_s}{N_t}} \boldsymbol{F}_{U-1}^* \boldsymbol{H}_{U-1}^* \end{bmatrix} \left[\sqrt{\tfrac{G_0 E_s}{N_t}} \boldsymbol{H}_0 \boldsymbol{F}_0 \cdots \sqrt{\tfrac{G_{U-1} E_s}{N_t}} \boldsymbol{H}_{U-1} \boldsymbol{F}_{U-1}\right] \right)^{-1} \cdot \begin{bmatrix} \sqrt{\frac{G_0 E_s}{N_t}} \boldsymbol{F}_0^* \boldsymbol{H}_0^* \\ \vdots \\ \sqrt{\frac{G_{U-1} E_s}{N_t}} \boldsymbol{F}_{U-1}^* \boldsymbol{H}_{U-1}^* \end{bmatrix}, \tag{9.9}$$

which, for $U = 1$, reduces to the SU-MIMO solution in (6.3). Parceling out $\boldsymbol{W}^{\mathrm{ZF}}$ into U blocks of N_t columns such that

$$\boldsymbol{W}^{\mathrm{ZF}} = \left[\boldsymbol{W}_0^{\mathrm{ZF}} \cdots \boldsymbol{W}_{U-1}^{\mathrm{ZF}}\right], \tag{9.10}$$

we obtain separate ZF filters $\boldsymbol{W}_0^{\mathrm{ZF}}, \ldots, \boldsymbol{W}_{U-1}^{\mathrm{ZF}}$ that recover the individual user signals, i.e., that satisfy

$$\boldsymbol{W}_u^{\mathrm{ZF}*} \boldsymbol{y} = \boldsymbol{s}_u + \check{\boldsymbol{v}}_u \tag{9.11}$$

where the filtered noise $\breve{\boldsymbol{v}}_u = \boldsymbol{W}_u^{\text{ZF}*}\boldsymbol{v}_u$ has a conditional covariance

$$\mathbb{E}\left[\breve{\boldsymbol{v}}_u\breve{\boldsymbol{v}}_u^* | \boldsymbol{H}_0, \ldots, \boldsymbol{H}_{U-1}\right] = N_0\, \boldsymbol{W}_u^{\text{ZF}*}\boldsymbol{W}_u^{\text{ZF}}. \tag{9.12}$$

9.3.2 Output SNR distribution

From (9.11) and (9.12), the output SNR for the jth stream of the uth user equals, for given fading realizations,

$$\mathsf{snr}_{u,j}^{\text{ZF}} = \frac{1}{N_0\left[\boldsymbol{W}_u^{\text{ZF}*}\boldsymbol{W}_u^{\text{ZF}}\right]_{j,j}}. \tag{9.13}$$

The distribution of this quantity can be characterized by recognizing that, from the vantage of the jth stream of the uth user, the N_{r} receive antennas null out $UN_{\text{t}} - 1$ interfering streams, leaving the equivalent of $N_{\text{r}} - UN_{\text{t}} + 1$ receive antennas to effectively process the desired signal. The distribution of $\mathsf{snr}_{u,j}^{\text{ZF}}$ is thus identical to that of a SIMO channel with $N_{\text{r}} - UN_{\text{t}} + 1$ receive antennas.

Example 9.1 (ZF output SNR distribution in an IID Rayleigh-faded MU-MIMO MAC)

Consider unprecoded transmissions in IID Rayleigh fading. As shown in Examples 5.16 and 6.2, the SNR distribution in the corresponding MU-SIMO channel is chi-square. It follows from those examples that, with $N_{\text{r}} - UN_{\text{t}} + 1$ effective receive antennas,

$$\mathsf{snr}_{u,j}^{\text{ZF}} \sim \chi^2_{2(N_{\text{r}}-UN_{\text{t}}+1)} \qquad j = 0, \ldots, N_{\text{t}} - 1 \quad u = 0, \ldots, U - 1. \tag{9.14}$$

Precisely, for $\xi \geq 0$,

$$f_{\mathsf{snr}_{u,j}^{\text{ZF}}}(\xi) = \frac{N_{\text{t}}}{\mathsf{SNR}_u (N_{\text{r}} - UN_{\text{t}})!} \exp\left(-\frac{N_{\text{t}}}{\mathsf{SNR}_u}\xi\right)\left(\frac{N_{\text{t}}}{\mathsf{SNR}_u}\xi\right)^{N_{\text{r}}-UN_{\text{t}}}, \tag{9.15}$$

which, if $UN_{\text{t}} = N_{\text{r}}$, reduces to an exponential distribution. The average SNR output for the jth stream of the uth user is

$$\mathbb{E}\left[\mathsf{snr}_{u,j}^{\text{ZF}}\right] = (N_{\text{r}} - UN_{\text{t}} + 1)\,\frac{\mathsf{SNR}_u}{N_{\text{t}}}, \tag{9.16}$$

which, if $UN_{\text{t}} = N_{\text{r}}$, reduces to $\mathbb{E}\left[\mathsf{snr}_{u,j}^{\text{ZF}}\right] = \frac{\mathsf{SNR}_u}{N_{\text{t}}}$.

9.3.3 Ergodic spectral efficiency

The jth stream of the uth user experiences a scalar Gaussian-noise channel with fading SNR given by $\mathsf{snr}_{u,j}^{\text{ZF}}$, hence its spectral efficiency is maximized when the codewords are drawn from a complex Gaussian distribution. Adding over the corresponding streams, the ergodic spectral efficiency of the uth user then equals

$$C_u^{\text{ZF}} = \sum_{j=0}^{N_{\text{t}}-1} \mathbb{E}\left[\log_2\left(1 + \mathsf{snr}_{u,j}^{\text{ZF}}\right)\right] \tag{9.17}$$

and the weighted sum spectral efficiency is $\sum_{u=0}^{U-1} q_u C_u^{\text{ZF}}$.

Example 9.2 (ZF spectral efficiency in an IID Rayleigh-faded MU-MIMO MAC)

Consider unprecoded transmissions in IID Rayleigh fading. The N_t streams of user u exhibit the same SNR distribution and hence

$$C_u^{\text{ZF}} = N_t \, \mathbb{E}\big[\log_2\big(1 + \mathsf{snr}_{u,j}^{\text{ZF}}\big)\big] \tag{9.18}$$

$$= N_t \int_0^\infty \log_2(1+\xi) \frac{N_t}{\mathsf{SNR}_u \,(N_r - U N_t)!} \exp\!\left(-\frac{N_t}{\mathsf{SNR}_u}\xi\right) \left(\frac{N_t}{\mathsf{SNR}_u}\xi\right)^{N_r - U N_t} d\xi \tag{9.19}$$

with j the index of an arbitrary stream and with $f_{\mathsf{snr}_{u,j}^{\text{ZF}}}(\cdot)$ borrowed from Example 9.1. Then, applying (C.37),

$$C_u^{\text{ZF}} = N_t \, e^{N_t/\mathsf{SNR}_u} \sum_{q=1}^{N_r - U N_t + 1} \mathcal{E}_q\!\left(\frac{N_t}{\mathsf{SNR}_u}\right) \log_2 e. \tag{9.20}$$

Example 9.3 (ZF spectral efficiency in an IID Rayleigh-faded MU-SIMO MAC)

For $N_t = 1$ and $U = N_r$, Example 9.2 specializes to

$$C_u^{\text{ZF}} = e^{1/\mathsf{SNR}_u} \, \mathcal{E}_1\!\left(\frac{1}{\mathsf{SNR}_u}\right) \log_2 e, \tag{9.21}$$

which coincides with the capacity of a Rayleigh-faded SISO channel with the same SNR (recall Example 4.27). A linear ZF receiver therefore enables as many single-antenna users as receive antennas to communicate concurrently as if each transmission was picked up by a single receive antenna, free of interference.

Example 9.4

Consider a three-user MAC with $N_t = 2$ and $N_r = 6$. The respective local-average SNRs are referenced to a common variable SNR such that

$$\mathsf{SNR}_0|_{\text{dB}} = \mathsf{SNR}|_{\text{dB}} \tag{9.22}$$

$$\mathsf{SNR}_1|_{\text{dB}} = \mathsf{SNR}|_{\text{dB}} + 5\,\text{dB} \tag{9.23}$$

$$\mathsf{SNR}_2|_{\text{dB}} = \mathsf{SNR}|_{\text{dB}} + 8\,\text{dB}. \tag{9.24}$$

Letting $\boldsymbol{H}_0$, $\boldsymbol{H}_1$, and $\boldsymbol{H}_2$ be IID Rayleigh-faded while $\boldsymbol{F}_u = \boldsymbol{I}$ for $u = 0, 1, 2$, evaluate the sum spectral efficiency of ZF and compare it with the sum-capacity. Include in the comparison also the SU-MIMO spectral efficiency obtained when only each of the users transmits without precoding and individual ZF reception is applied.

Solution

See Fig. 9.2. At low SNR, having only the strongest user transmit in SU-MIMO mode

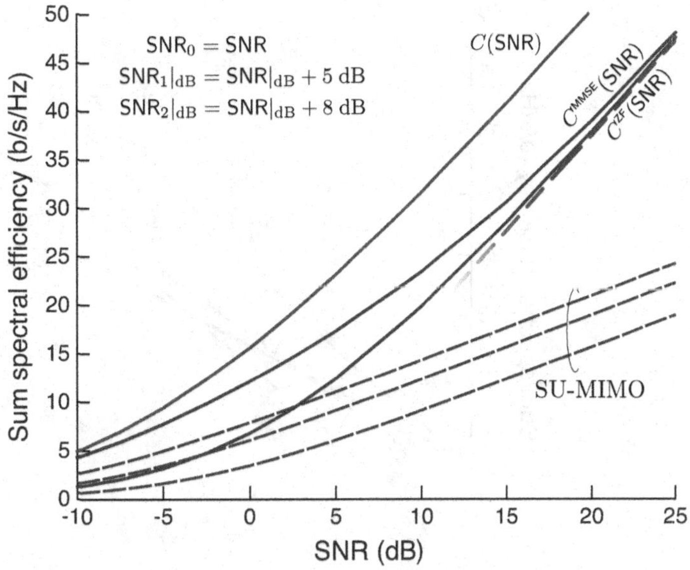

Fig. 9.2 MAC ergodic sum spectral efficiency for MU-MIMO with $U = 3$, $N_{\rm t} = 2$, and $N_{\rm r} = 6$, as a function of the reference SNR (in dB). In solid lines, linear ZF and LMMSE receivers compared against the optimum receiver. In dashed lines, high-SNR expansions and also individual spectral efficiencies achieved by each user in SU-MIMO mode with ZF reception. All transmissions are unprecoded.

is preferable with reception. Once the high-SNR regime is entered, however, MU-MIMO overpowers SU-MIMO.

While it was argued in previous chapters that the low-SNR sum-capacity of MU-MIMO roughly equals that of SU-MIMO, in Example 9.4 we observe that, with ZF reception, the low-SNR sum spectral efficiency of MU-MIMO is decidedly lower than that of SU-MIMO with a properly selected user. This points to the advantage of selecting *how many* and *which* users are actively served under ZF reception, two aspects that are worth differentiating.

- The smallest singular value of a square or quasi-square matrix with complex Gaussian random entries is known to behave poorly; such singular value tends to be very small, such that its inversion greatly enhances the noise. Thus, depending on the SNRs, having $UN_{\rm t} < N_{\rm r}$ (rather than $UN_{\rm t} = N_{\rm r}$) so as to invert a more rectangular matrix may actually be advantageous in terms of ZF performance.
- With a pool of $U_{\rm tot}$ candidate users available from which to select the U active users, additional performance improvements can be attained.

The next example illustrates the first of these aspects in a setup that, by having all users at the same SNR, conveniently seeks to decouple it as much as possible from the second.

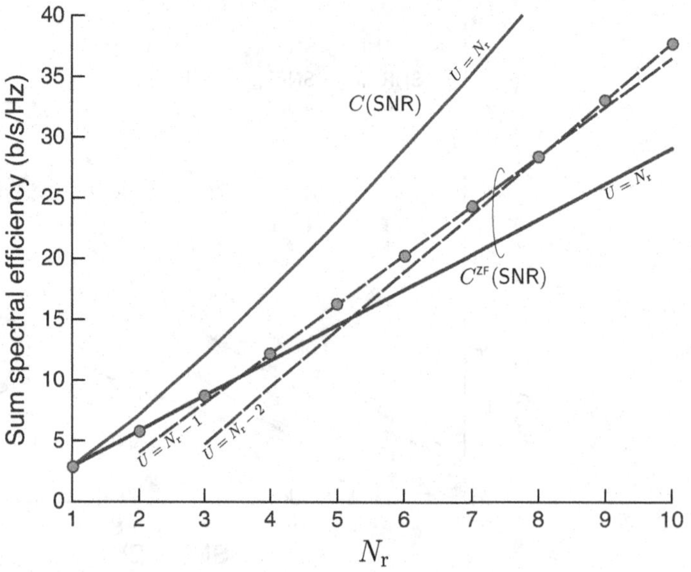

Fig. 9.3 MAC ergodic sum spectral efficiency with $N_t = 1$ as a function of N_r. The solid curves correspond to $U = N_r$, for both the optimum receiver (the sum-capacity) and for the linear ZF receiver, while the dashed lines correspond also to the linear ZF receiver but with $U = N_r - 1$ and $U = N_r - 2$. In all cases, there is no precoding and $\mathsf{SNR}_u = 10$ dB for $u = 0, \ldots, U-1$. The circles trace the envelope of the ZF curves.

Example 9.5

Consider the particular setup $\mathsf{SNR}_u = \mathsf{SNR}$ for $u = 0, \ldots, U-1$, and let $\mathsf{SNR} = 10$ dB. Further let $\boldsymbol{H}_0, \ldots, \boldsymbol{H}_{U-1}$ have IID Rayleigh-faded entries. For $N_t = 1$, compute as a function of N_r, the ergodic sum spectral efficiency with linear ZF reception with the value of U optimized for each value of N_r.

Solution

Figure 9.3 depicts, in solid, the sum spectral efficiency with ZF reception and $U = N_r$. As the dashed lines show, for some values of N_r this can be improved upon by having $U < N_r$, and the envelope of circles is the sought optimized ZF sum spectral efficiency. Also shown in the figure, as a baseline, is the sum-capacity with $U = N_r$.

The foregoing example offers a glimpse of what unfolds as the dimensionality increases, namely that the optimum ratio UN_t/N_r under ZF approaches a value that depends on the SNRs. The lower the SNRs, the further that this optimum ratio is from 1. Conversely, at sufficiently higher SNRs, it is optimum to have $UN_t = N_r$, a behavior that is explored in Problem 9.3 and formalized in the upcoming high-SNR analysis.

With unequal SNRs and a dynamic user selection, the advantage of having a progressively smaller ratio UN_t/N_r as we pull back from the high-SNR regime can only become more pronounced. However, the fairness and latency limitations discussed in Section 7.6 must be very present whenever this additional benefit is quantified.

To conclude the analysis of ZF reception at arbitrary SNRs we note that, throughout this section, we have made no attempt to optimize the precoders, placing the burden of eliminating the interference exclusively at the receiver. If the precoders could be designed to eliminate interference among the streams of each given user, then the structure of the receiver could be relaxed to zero-force only across users, but not across same-user streams. This so-called *block diagonalization* approach is explored in Section 9.6, in the context of the BC, where it acquires increased relevance.

9.3.4 High-SNR regime

At high SNR, each user enjoys N_t spatial DOF and thus the sum spectral efficiency with ZF reception exhibits $S_\infty^{\text{ZF}} = UN_t$ spatial DOF. There is no DOF penalty with respect to an optimum receiver; the suboptimality of ZF reception is registered only in the power offset.

The high-SNR expansion of the uth user's spectral efficiency is as in SU-MIMO with N_t spatial DOF, i.e.,

$$C_u^{\text{ZF}}(\text{SNR}_u) = N_t \left(\log_2 \text{SNR}_u - \mathcal{L}_{\infty,u}^{\text{ZF}}\right) + \mathcal{O}\left(\frac{1}{\text{SNR}_u}\right). \tag{9.25}$$

Applied to (9.17), this gives the power offset of the uth user as

$$\mathcal{L}_{\infty,u}^{\text{ZF}} = \lim_{\text{SNR}_u \to \infty} \left(\log_2 \text{SNR}_u - \frac{1}{N_t} \sum_{j=0}^{N_t-1} \mathbb{E}\left[\log_2 \text{snr}_{u,j}^{\text{ZF}}\right] \right) \tag{9.26}$$

$$= \lim_{\text{SNR}_u \to \infty} \frac{1}{N_t} \sum_{j=0}^{N_t-1} \mathbb{E}\left[\log_2 \frac{\text{SNR}_u}{\text{snr}_{u,j}^{\text{ZF}}}\right] \tag{9.27}$$

$$= \frac{1}{N_t} \sum_{j=0}^{N_t-1} \mathbb{E}\left[\log_2 \frac{\text{SNR}_u}{\text{snr}_{u,j}^{\text{ZF}}}\right], \tag{9.28}$$

where the limit becomes immaterial because, from (9.9), (9.10), and (9.13), the ratio $\text{SNR}_u/\text{snr}_{u,j}^{\text{ZF}}$ can be seen to be independent of SNR_u.

From (9.25), summing over the U users, the sum spectral efficiency expands as

$$C^{\text{ZF}}(\text{SNR}_0, \ldots, \text{SNR}_{U-1}) = N_t \sum_{u=0}^{U-1} \left(\log_2 \text{SNR}_u - \mathcal{L}_{\infty,u}^{\text{ZF}}\right) + \mathcal{O}\left(\frac{1}{E_s/N_0}\right) \tag{9.29}$$

$$= N_t \sum_{u=0}^{U-1} \left(\log_2 \text{SNR}_u - \mathcal{L}_{\infty}^{\text{ZF}}\right) + \mathcal{O}\left(\frac{1}{E_s/N_0}\right), \tag{9.30}$$

with the sum spectral efficiency power offset being $\mathcal{L}_\infty = \frac{1}{U}\sum_{u=0}^{U-1} \mathcal{L}_{\infty,u}^{\text{ZF}}$. For IID Rayleigh fading, as we show next, closed forms can be obtained for $\mathcal{L}_{\infty,u}^{\text{ZF}}$ and thus for $\mathcal{L}_{\infty}^{\text{ZF}}$.

Example 9.6 (ZF power offset in an IID Rayleigh-faded MU-MIMO MAC)

Let the channel be IID Rayleigh-faded and let $\boldsymbol{F}_u = \boldsymbol{I}$ for $u = 0, \ldots, U-1$. Although the power offset could be derived by applying the corresponding definition to the spectral efficiency in Example 9.2, it is more expeditious to recognize once again that such expression equals its SU-MIMO brethren only with $N_\text{r} - UN_\text{t} + 1$ in lieu of $N_\text{r} - N_\text{t} + 1$. Translating this change into (6.9), the power offset of each user in a MU-MIMO MAC with ZF reception is, in 3-dB units,

$$\mathcal{L}_{\infty,u}^{\text{ZF}} = \log_2 N_\text{t} + \left(\gamma_{\text{EM}} - \sum_{q=1}^{N_\text{r}-UN_\text{t}} \frac{1}{q} \right) \log_2 e \qquad u = 0, \ldots, U-1, \tag{9.31}$$

which, for $UN_\text{t} = N_\text{r}$, reduces to

$$\mathcal{L}_{\infty,u}^{\text{ZF}} = \log_2 N_\text{t} + \gamma_{\text{EM}} \log_2 e \qquad u = 0, \ldots, U-1. \tag{9.32}$$

Example 9.7

Compare the expansion in (9.30) with the exact sum spectral efficiency for the setup in Example 9.4.

Solution

For $U = 3$, $N_\text{t} = 2$, and $N_\text{r} = 6$, we have that $\mathcal{L}_{\infty}^{\text{ZF}} = 1.83$ in 3-dB units and thus the ZF power offset equals 5.52 dB. From (9.30) then,

$$C^{\text{ZF}}(\mathsf{SNR}_0, \mathsf{SNR}_1, \mathsf{SNR}_2) \approx \sum_{u=0}^{2} 2 \left(\log_2 \mathsf{SNR}_u - 1.83 \right) \tag{9.33}$$

which is depicted in Fig. 9.2 alongside the actual sum spectral efficiency.

Armed with the power offset expressions derived here and in Section 8.4.2, we can quantify the high-SNR performance gap between a ZF and an optimum receiver. Indeed, since the number of spatial DOF is the same in both cases, it is only through the power offset that these receivers can be discriminated at high SNR. The difference between the respective power offsets represents the power loss of ZF, relative to the sum-capacity, at high SNR.

Example 9.8

Express the high-SNR power loss of ZF for IID Rayleigh fading.

Solution

From (8.63) and (9.31), the power loss in 3-dB units is

$$\Delta \mathcal{L}_\infty = \mathcal{L}_{\infty}^{\text{ZF}} - \mathcal{L}_\infty \tag{9.34}$$

$$= \left(\frac{N_{\mathrm{r}}}{U N_{\mathrm{t}}} \sum_{q=N_{\mathrm{r}}-U N_{\mathrm{t}}+1}^{N_{\mathrm{r}}} \frac{1}{q} - 1 \right) \log_2 e, \tag{9.35}$$

which, for $UN_{\mathrm{t}} = N_{\mathrm{r}}$, simplifies into

$$\Delta\mathcal{L}_{\infty} = \left(\sum_{q=2}^{N_{\mathrm{r}}} \frac{1}{q} \right) \log_2 e. \tag{9.36}$$

Example 9.9

Evaluate $\Delta\mathcal{L}_{\infty}$ for $U = 3$, $N_{\mathrm{t}} = 2$, and $N_{\mathrm{r}} = 6$.

Solution

Applying (9.36), we obtain $\Delta\mathcal{L}_{\infty} = 2.09$ in 3-dB units and thus a power loss of 6.29 dB. This is indeed the difference between the power offsets found in Examples 8.12 and 9.7, respectively.

Although (9.35) is specific to IID Rayleigh fading, observations with broad validity can be made therefrom.

- The power loss experienced by ZF is minimized for $UN_{\mathrm{t}} \ll N_{\mathrm{r}}$ and, in fact, as $\frac{UN_{\mathrm{t}}}{N_{\mathrm{r}}}$ shrinks, the power loss vanishes. However, potential spatial DOF then go unused.
- Conversely, the power loss is maximum when $UN_{\mathrm{t}} = N_{\mathrm{r}}$, when all the potential DOF are activated.

Altogether: while, as seen earlier, at certain SNRs having $UN_{\mathrm{t}} < N_{\mathrm{r}}$ might be advantageous in terms of ZF performance, at sufficiently high SNRs it is always best to have $UN_{\mathrm{t}} = N_{\mathrm{r}}$. (This conclusion holds for nonmassive-MIMO setups, where the acquisition of CSIR is not a major issue.)

9.4 LMMSE receiver for the multiple-access channel

9.4.1 Receiver structure

Before delving into the derivation of the LMMSE receiver for the MU-MIMO MAC, a preliminary comment on the issue of estimation in multiuser setups is warranted: from the vantage of the mean-square error when estimating a given user's signal, say $\boldsymbol{s}_u$, it would be best that all other users are simply powered off, but of course that is in general undesirable from a communication perspective. The receiver structure derived in this section simultaneously minimizes the mean-square error in the linear estimation of all the data streams within $\boldsymbol{s}_0, \ldots, \boldsymbol{s}_{U-1}$ for some given transmit precoders and powers.

To derive such LMMSE receiver we could, as done for ZF in the previous section, apply the SU-MIMO solution to the corresponding interpretation of the MU-MIMO MAC, thereby obtaining a single filter that estimates $\boldsymbol{s} = [\boldsymbol{s}_0^{\mathrm{T}} \cdots \boldsymbol{s}_{U-1}^{\mathrm{T}}]^{\mathrm{T}}$ at once; this approach is the subject of Problem 9.4. Alternatively, we can proceed with a separate formulation for each user and apply the teachings of Section 1.7.1 to write down the filter that minimizes the mean-square error incurred in the estimation of $\boldsymbol{s}_u$ with the signals from all other users regarded as interference. This gives

$$\boldsymbol{W}_u^{\mathsf{MMSE}} = \boldsymbol{R}_{\boldsymbol{y}}^{-1} \boldsymbol{R}_{\boldsymbol{y}\boldsymbol{s}_u} \tag{9.37}$$

$$= \sqrt{\frac{N_{\mathrm{t}}}{G_u E_u}} \left(\sum_{\mathrm{u}=0}^{U-1} \frac{\frac{E_{\mathrm{u}}}{E_{\mathrm{s}}}\mathsf{SNR}_{\mathrm{u}}}{\frac{E_u}{E_{\mathrm{s}}}\mathsf{SNR}_u} \boldsymbol{H}_{\mathrm{u}} \boldsymbol{F}_{\mathrm{u}} \boldsymbol{F}_{\mathrm{u}}^* \boldsymbol{H}_{\mathrm{u}}^* + \frac{N_{\mathrm{t}}}{\frac{E_u}{E_{\mathrm{s}}}\mathsf{SNR}_u} \boldsymbol{I} \right)^{-1} \boldsymbol{H}_u \boldsymbol{F}_u, \tag{9.38}$$

which, for $E_{\mathrm{s}}/N_0 \to \infty$, reverts to $\boldsymbol{W}_u^{\mathsf{ZF}}$ whereas, for $E_{\mathrm{s}}/N_0 \to 0$, converges to a matched filter for user u.

Conditioned on the fading realizations, the MMSE matrix of user u is

$$\begin{aligned} \boldsymbol{E}_u(\boldsymbol{H}_0,\ldots,\boldsymbol{H}_{U-1}) &= \mathbb{E}\Big[\big(\boldsymbol{s}_u - \boldsymbol{W}_u^{\mathsf{MMSE}*}\boldsymbol{y}\big) \big(\boldsymbol{s}_u - \boldsymbol{W}_u^{\mathsf{MMSE}*}\boldsymbol{y}\big)^* \big| \boldsymbol{H}_0,\ldots,\boldsymbol{H}_{U-1} \Big] \\ &= \boldsymbol{I} - \sqrt{\frac{G_u E_u}{N_{\mathrm{t}}}} \boldsymbol{W}_u^{\mathsf{MMSE}*} \boldsymbol{H}_u \boldsymbol{F}_u - \sqrt{\frac{G_u E_u}{N_{\mathrm{t}}}} \boldsymbol{F}_u^* \boldsymbol{H}_u^* \boldsymbol{W}_u^{\mathsf{MMSE}} \\ &\quad + \boldsymbol{W}_u^{\mathsf{MMSE}*} \left(\sum_{\mathrm{u}=0}^{U-1} \frac{G_{\mathrm{u}} E_{\mathrm{u}}}{N_{\mathrm{t}}} \boldsymbol{H}_{\mathrm{u}} \boldsymbol{F}_{\mathrm{u}} \boldsymbol{F}_{\mathrm{u}}^* \boldsymbol{H}_{\mathrm{u}}^* + N_0 \boldsymbol{I} \right) \boldsymbol{W}_u^{\mathsf{MMSE}}, \end{aligned} \tag{9.39}$$

where all crossed terms involving different users have disappeared because their signals are independent and zero-mean. Combining (9.38) and (9.39), and after a bit of algebra,

$$\boldsymbol{E}_u(\boldsymbol{H}_0,\ldots,\boldsymbol{H}_{U-1}) = \boldsymbol{I} - \boldsymbol{F}_u^* \boldsymbol{H}_u^* \left(\sum_{\mathrm{u}=0}^{U-1} \frac{\frac{E_{\mathrm{u}}}{E_{\mathrm{s}}}\mathsf{SNR}_{\mathrm{u}}}{\frac{E_u}{E_{\mathrm{s}}}\mathsf{SNR}_u} \boldsymbol{H}_{\mathrm{u}} \boldsymbol{F}_{\mathrm{u}} \boldsymbol{F}_{\mathrm{u}}^* \boldsymbol{H}_{\mathrm{u}}^* + \frac{N_{\mathrm{t}}}{\frac{E_u}{E_{\mathrm{s}}}\mathsf{SNR}_u} \boldsymbol{I} \right)^{-1} \boldsymbol{H}_u \boldsymbol{F}_u \tag{9.40}$$

which is the MAC generalization of (6.52), and which can alternatively be put as

$$\begin{aligned} &\boldsymbol{E}_u(\boldsymbol{H}_0,\ldots,\boldsymbol{H}_{U-1}) \\ &= \boldsymbol{I} - \boldsymbol{F}_u^* \boldsymbol{H}_u^* \left(\boldsymbol{H}_u \boldsymbol{F}_u \boldsymbol{F}_u^* \boldsymbol{H}_u^* + \sum_{\mathrm{u} \neq u} \frac{\frac{E_{\mathrm{u}}}{E_{\mathrm{s}}}\mathsf{SNR}_{\mathrm{u}}}{\frac{E_u}{E_{\mathrm{s}}}\mathsf{SNR}_u} \boldsymbol{H}_{\mathrm{u}} \boldsymbol{F}_{\mathrm{u}} \boldsymbol{F}_{\mathrm{u}}^* \boldsymbol{H}_{\mathrm{u}}^* + \frac{N_{\mathrm{t}}}{\frac{E_u}{E_{\mathrm{s}}}\mathsf{SNR}_u} \boldsymbol{I} \right)^{-1} \boldsymbol{H}_u \boldsymbol{F}_u \\ &= \left[\boldsymbol{I} + \frac{\frac{E_u}{E_{\mathrm{s}}}\mathsf{SNR}_u}{N_{\mathrm{t}}} \boldsymbol{F}_u^* \boldsymbol{H}_u^* \left(\boldsymbol{I} + \sum_{\mathrm{u} \neq u} \frac{\frac{E_{\mathrm{u}}}{E_{\mathrm{s}}}\mathsf{SNR}_{\mathrm{u}}}{N_{\mathrm{t}}} \boldsymbol{H}_{\mathrm{u}} \boldsymbol{F}_{\mathrm{u}} \boldsymbol{F}_{\mathrm{u}}^* \boldsymbol{H}_{\mathrm{u}}^* \right)^{-1} \boldsymbol{H}_u \boldsymbol{F}_u \right]^{-1}, \end{aligned} \tag{9.41}$$

where (9.41) follows from the matrix inversion lemma; the reader is invited to verify this step in Problem 9.6. The MMSE corresponding to jth stream of user u is

$$\mathsf{MMSE}_{u,j} = \big[\boldsymbol{E}_u(\boldsymbol{H}_0,\ldots,\boldsymbol{H}_{U-1}) \big]_{j,j} \tag{9.42}$$

$$= 1 - [\boldsymbol{F}_u]^*_{:,j} \boldsymbol{H}_u^* \left(\sum_{\mathrm{u}=0}^{U-1} \frac{\frac{E_\mathrm{u}}{E_\mathrm{s}} \mathsf{SNR}_\mathrm{u}}{\frac{E_u}{E_\mathrm{s}} \mathsf{SNR}_u} \boldsymbol{H}_\mathrm{u} \boldsymbol{F}_\mathrm{u} \boldsymbol{F}_\mathrm{u}^* \boldsymbol{H}_\mathrm{u}^* + \frac{N_\mathrm{t}}{\frac{E_u}{E_\mathrm{s}} \mathsf{SNR}_u} \boldsymbol{I} \right)^{-1} \boldsymbol{H}_u [\boldsymbol{F}_u]_{:,j} \tag{9.43}$$

$$= \left[\boldsymbol{I} + \frac{\frac{E_u}{E_\mathrm{s}} \mathsf{SNR}_u}{N_\mathrm{t}} \boldsymbol{F}_u^* \boldsymbol{H}_u^* \left(\boldsymbol{I} + \sum_{\mathrm{u} \neq u} \frac{\frac{E_\mathrm{u}}{E_\mathrm{s}} \mathsf{SNR}_\mathrm{u}}{N_\mathrm{t}} \boldsymbol{H}_\mathrm{u} \boldsymbol{F}_\mathrm{u} \boldsymbol{F}_\mathrm{u}^* \boldsymbol{H}_\mathrm{u}^* \right)^{-1} \boldsymbol{H}_u \boldsymbol{F}_u \right]^{-1}_{j,j}, \tag{9.44}$$

where, here and henceforth, $[\boldsymbol{A}]^{-1}_{j,j}$ compactly denotes the (j,j)th entry of $\boldsymbol{A}^{-1}$.

9.4.2 Output SINR distribution

Let us now evaluate the SINR enjoyed by the jth stream of user u at the output of the LMMSE receiver. Plugging the expression for $\boldsymbol{W}_u^{\mathsf{MMSE}}$ given in (9.38) into (9.4) and (9.5), and after some algebra (refer to Problem 9.7), we obtain

$$\mathsf{sinr}_{u,j}^{\mathsf{MMSE}} = \frac{\frac{E_u}{E_\mathrm{s}} \mathsf{SNR}_u [\boldsymbol{F}_u]^*_{:,j} \boldsymbol{H}_u^* \left(\sum_{\mathrm{u}=0}^{U-1} \frac{E_\mathrm{u}}{E_\mathrm{s}} \mathsf{SNR}_\mathrm{u} \boldsymbol{H}_\mathrm{u} \boldsymbol{F}_\mathrm{u} \boldsymbol{F}_\mathrm{u}^* \boldsymbol{H}_\mathrm{u}^* + N_\mathrm{t} \boldsymbol{I} \right)^{-1} \boldsymbol{H}_u [\boldsymbol{F}_u]_{:,j}}{1 - \frac{E_u}{E_\mathrm{s}} \mathsf{SNR}_u [\boldsymbol{F}_u]^*_{:,j} \boldsymbol{H}_u^* \left(\sum_{\mathrm{u}=0}^{U-1} \frac{E_\mathrm{u}}{E_\mathrm{s}} \mathsf{SNR}_\mathrm{u} \boldsymbol{H}_\mathrm{u} \boldsymbol{F}_\mathrm{u} \boldsymbol{F}_\mathrm{u}^* \boldsymbol{H}_\mathrm{u}^* + N_\mathrm{t} \boldsymbol{I} \right)^{-1} \boldsymbol{H}_u [\boldsymbol{F}_u]_{:,j}} \tag{9.45}$$

$$= \frac{1 - \mathsf{MMSE}_{u,j}}{\mathsf{MMSE}_{u,j}} \tag{9.46}$$

$$= \frac{1}{\mathsf{MMSE}_{u,j}} - 1 \tag{9.47}$$

$$= \frac{1}{\left[\boldsymbol{I} + \frac{E_u}{E_\mathrm{s}} \frac{\mathsf{SNR}_u}{N_\mathrm{t}} \boldsymbol{F}_u^* \boldsymbol{H}_u^* \left(\boldsymbol{I} + \sum_{\mathrm{u} \neq u} \frac{E_\mathrm{u}}{E_\mathrm{s}} \frac{\mathsf{SNR}_\mathrm{u}}{N_\mathrm{t}} \boldsymbol{H}_\mathrm{u} \boldsymbol{F}_\mathrm{u} \boldsymbol{F}_\mathrm{u}^* \boldsymbol{H}_\mathrm{u}^* \right)^{-1} \boldsymbol{H}_u \boldsymbol{F}_u \right]^{-1}_{j,j}} - 1, \tag{9.48}$$

whose convergence to $\mathsf{snr}_{u,j}^{\mathsf{ZF}}$ as $N_0 \to 0$ abides by the observation made in the context of SU-MIMO: the ratio of both quantities approaches unity but their difference does not vanish.

For $U = 1$, the above expressions for $\mathsf{sinr}_{u,j}^{\mathsf{MMSE}}$ revert to their SU-MIMO counterparts in Section 6.4 and, resorting to the same arguments and proving technique applied therein, it can be confirmed that the LMMSE receiver is optimum in the sense of maximizing the SINR of all streams and hence the region of achievable spectral efficiencies (for given precoders and transmit powers).

9.4.3 Ergodic spectral efficiency

Repeating from Chapter 6 the disclaimer that complex Gaussian codewords need not be strictly capacity-achieving with an LMMSE receiver (because, by improving the interference distribution at the separate decoders, non-Gaussian codewords could prove slightly superior), the spectral efficiency of user u with complex Gaussian signaling is, conditioned on the fading realizations,

$$C_u^{\mathsf{MMSE}} = \sum_{j=0}^{N_t-1} \mathbb{E}\left[\log_2\left(1+\mathsf{sinr}_{u,j}^{\mathsf{MMSE}}\right)\right] \tag{9.49}$$

$$= \sum_{j=0}^{N_t-1} \mathbb{E}\left[\log_2 \frac{1}{\mathsf{MMSE}_{u,j}}\right] \tag{9.50}$$

and the weighted sum spectral efficiency is $\sum_{u=0}^{U-1} q_u C_u^{\mathsf{MMSE}}$.

Example 9.10

Considering again the setup of Example 9.4, with full-power transmission by every user ($\frac{E_u}{E_s} = 1$ for $u = 0, 1, 2$), evaluate the sum spectral efficiency achievable with an LMMSE receiver and complex Gaussian signaling.

Solution

The result is presented in Fig. 9.2, alongside its counterparts for both the ZF and the optimum receiver. Notice how LMMSE reception approaches optimality at low SNR and ZF performance at high SNR.

Let us now see how adjusting U, i.e., the number of active users, can impact the performance gap between ZF and LMMSE.

Example 9.11

Considering again the setup of Example 9.5, with full-power transmission by every user, compute, as a function of N_r, the ergodic sum spectral efficiency with LMMSE reception with the value of U optimized for each value of N_r. Show the resulting curve next to its ZF counterpart.

Solution

Figure 9.4 depicts, in solid, the sum spectral efficiency for $U = N_r$ with ZF and with LMMSE reception and, in dotted lines, the sought sum spectral efficiencies with the values of U optimized. With U properly adjusted, as opposed to fixed, the performance gap between the ZF and LMMSE receivers is largely closed.

Also shown in the figure, as a baseline, is the sum-capacity with $U = N_r$.

Before concluding this section with some remarks on the high-SNR behavior, a note on the optimization of the precoders under LMMSE reception is in order. For any choice of precoders, setting the receive filters for all users to (9.38) ensures simultaneous minimization of the mean-square error for every stream and, with that, maximization—conditioned on that choice of precoders—of the corresponding SINRs and of the spectral efficiency boundary.

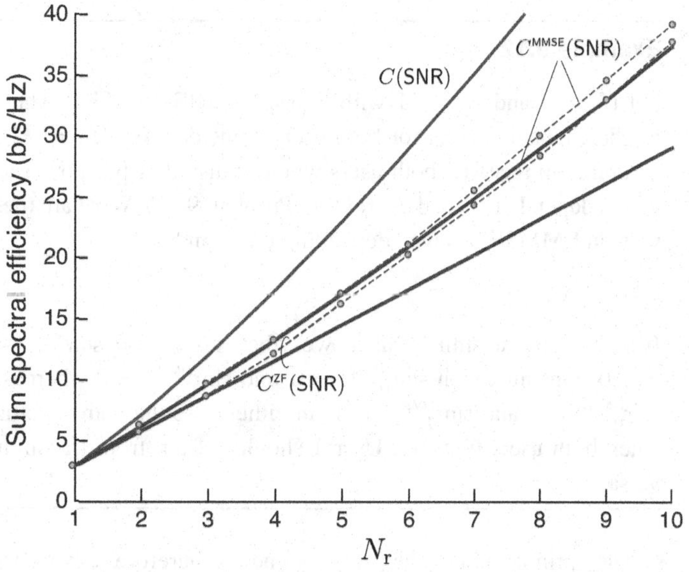

Fig. 9.4 MU-SIMO MAC sum spectral efficiency as a function of N_r. The solid curves correspond to $U = N_r$ for the optimum receiver (the sum-capacity), for the linear ZF receiver, and for the LMMSE receiver. The dotted lines correspond to optimizing U for every N_r. In all cases, $\text{SNR}_u = 10$ dB for $u = 0, \ldots, U-1$.

Should there be CSIT in the system, this information could be put toward optimizing the precoders; however, this feat is fraught with difficulties.

- The precoders that maximize the SINR for a given stream do not necessarily maximize it for the rest.
- The optimum precoder for user u is not solely a function of $\boldsymbol{W}_u^{\text{MMSE}}$, but rather a function of $\boldsymbol{W}_0^{\text{MMSE}}, \ldots, \boldsymbol{W}_{U-1}^{\text{MMSE}}$ [820].

Formally, the problem that must be solved is

$$\max_{\boldsymbol{F}_0,\ldots,\boldsymbol{F}_{U-1}} \sum_{u=0}^{U-1} q_u \sum_{j=0}^{N_t-1} \mathbb{E}\left[\log_2 \frac{1}{\text{MMSE}_{u,j}(\boldsymbol{F}_0,\ldots,\boldsymbol{F}_{U-1})}\right], \tag{9.51}$$

subject to $\text{tr}(\boldsymbol{F}_u\boldsymbol{F}_u^*) = N_t$ for $u = 0, \ldots, U-1$ and with $\text{MMSE}_{u,j}(\boldsymbol{F}_0, \ldots, \boldsymbol{F}_{U-1})$, whose dependence on the precoders has been made explicit, given by either (9.43) or (9.44). This optimization is generally nonconvex in $\boldsymbol{F}_0, \ldots, \boldsymbol{F}_{U-1}$.

The foregoing difficulties are compounded by the fact that, in contrast with an optimum MAC receiver, for which it is always optimal to have every user transmit at full power, with an LMMSE receiver this need not be the case.

Example 9.12

Let $U = 2$ and $N_t = 1$ with $\mathsf{SNR}_0 = \mathsf{SNR}_1 = \mathsf{SNR}$ where $\mathsf{SNR} \gg 1$, and further let the channel vectors for both users coincide, i.e., $\boldsymbol{H}_0 = \boldsymbol{H}_1$ [34, section 4.1.2]. With an optimum receiver, both users would transmit at full power and only their coding rates would depend on q_0 and q_1 (refer to Problem 9.10). What are the optimum transmit powers with an LMMSE receiver, depending on q_0 and q_1?

Solution

If both users transmit at full power, then $\mathsf{sinr}_0^{\mathsf{MMSE}} = \mathsf{sinr}_1^{\mathsf{MMSE}} \approx 1$. Alternatively, if only user 0 transmits, then $\mathsf{sinr}_0^{\mathsf{MMSE}} \gg 1$ and $\mathsf{sinr}_1^{\mathsf{MMSE}} = 0$ whereas, if only user 1 transmits, $\mathsf{sinr}_0^{\mathsf{MMSE}} = 0$ and $\mathsf{sinr}_1^{\mathsf{MMSE}} \gg 1$; in either case the sum spectral efficiency is superior to when both users transmit. User 1 should refrain from transmitting if $q_0 > q_1$, and vice versa.

The optimization of the precoders needs therefore to be broadened to include the optimization of the power control coefficients $\frac{E_0}{E_s}, \ldots, \frac{E_{U-1}}{E_s}$ converting (9.51) into

$$\max_{\boldsymbol{F}_0,\ldots,\boldsymbol{F}_{U-1},\frac{E_0}{E_s},\ldots,\frac{E_{U-1}}{E_s}} \sum_{u=0}^{U-1} q_u \sum_{j=0}^{N_t-1} \mathbb{E}\left[\log_2 \frac{1}{\mathsf{MMSE}_{u,j}(\boldsymbol{F}_0,\ldots,\boldsymbol{F}_{U-1},\frac{E_0}{E_s},\ldots,\frac{E_{U-1}}{E_s})}\right] \tag{9.52}$$

subject to $\text{tr}(\boldsymbol{F}_u \boldsymbol{F}_u^*) = N_t$ and $\frac{E_u}{E_s} \leq 1$ for $u = 0, \ldots, U-1$. This is an even harder, generally nonconvex problem.

Example 9.13

For $N_t = 1$, with the precoders rendered immaterial, verify that the weighted sum spectral efficiency with an LMMSE receiver is not a concave function of the power control coefficients.

Solution

Applying (9.43) and (9.46),

$$\mathsf{sinr}_u^{\mathsf{MMSE}} = \frac{\frac{E_u}{E_s}\mathsf{SNR}_u\,|\boldsymbol{H}_u|^2}{1+\sum_{\mathrm{u}\neq u}\frac{E_\mathrm{u}}{E_s}\mathsf{SNR}_\mathrm{u}\,|\boldsymbol{H}_\mathrm{u}|^2}, \tag{9.53}$$

from which the weighted sum spectral efficiency with an LMMSE receiver is

$$\sum_{u=0}^{U-1} q_u \log_2(1+\mathsf{sinr}_u) = \sum_{u=0}^{U-1} q_u \log_2\left(\frac{1+\sum_{\mathrm{u}=0}^{U-1}\frac{E_\mathrm{u}}{E_s}\mathsf{SNR}_\mathrm{u}\,|\boldsymbol{H}_\mathrm{u}|^2}{1+\sum_{\mathrm{u}\neq u}\frac{E_\mathrm{u}}{E_s}\mathsf{SNR}_\mathrm{u}\,|\boldsymbol{H}_\mathrm{u}|^2}\right) \tag{9.54}$$

$$= \log_2\left(1+\sum_{\mathrm{u}=0}^{U-1}\frac{E_\mathrm{u}}{E_s}\mathsf{SNR}_\mathrm{u}\,|\boldsymbol{H}_\mathrm{u}|^2\right)\sum_{u=0}^{U-1} q_u$$

$$- \sum_{u=0}^{U-1} q_u \log_2 \left(1 + \sum_{\mathrm{u} \neq u} \frac{E_{\mathrm{u}}}{E_{\mathrm{s}}} \mathsf{SNR}_{\mathrm{u}} \left| \boldsymbol{H}_{\mathrm{u}} \right|^2 \right), \tag{9.55}$$

which is generally not concave in $\frac{E_0}{E_s}, \ldots, \frac{E_{U-1}}{E_s}$. The reader is invited to verify this lack of concavity in Problems 9.12 and 9.13.

As argued earlier, from the vantage of a given user it is best that all other users are simply powered off, but in terms of the weighted sum spectral efficiency, in principle all should transmit at varying power levels depending on their channels and on $q_0, \ldots, q_{U-1}$. (In some extreme cases such as Example 9.12, the optimum value for some of the transmit powers may be strictly zero.) The corresponding optimization of the precoders and transmit powers can be tackled iteratively, through alternating minimization procedures whereby, in turns, the precoders and powers are re-optimized with the receivers fixed, and vice versa, in a process that progressively lowers the MMSEs. Algorithmic embodiments of this procedure have been presented for single-stream (beamforming) transmission per user [787, 788, 821] and for arbitrary-rank MU-MIMO [820]. However, although these algorithms are sure to converge to some solution, there is no guarantee that such solution is globally optimum because of the lack of concavity. Moreover, since this CSIT-based iterative optimization would have to converge well within the coherence time of the fading, its scope of applicability is limited.

If the optimization of precoders and transmit powers is to entail an iterative process, then it would be more practically relevant to have it based on channel distributions; then, the speed of convergence and the overheads would be less problematic given the radically different time scale over which distributions are stable. A pragmatic but very effective such power control algorithm, reliant only on local-average channel gains, is presented and applied in Chapter 10.

9.4.4 High-SNR regime

In terms of the high-SNR behavior of the LMMSE receiver, the comments made for SU-MIMO apply—with the dimensionalities properly adjusted—quite verbatim to the MAC.

- For $UN_{\mathrm{t}} \leq N_{\mathrm{r}}$, we have $\mathcal{S}_{\infty}^{\mathsf{MMSE}} = UN_{\mathrm{t}}$ exactly as with an optimum or a ZF receiver.
- For $UN_{\mathrm{t}} > N_{\mathrm{r}}$, we need to back off on the number of transmit streams in order not to overload the dimensionality of the receiver. Otherwise, $\mathcal{S}_{\infty}^{\mathsf{MMSE}} = 0$, indicating that it is not advisable to transmit more than N_{r} streams when an LMMSE receiver is utilized in this regime. (With a ZF receiver, it is directly not possible to separate more than N_{r} streams.)

Altogether then, we find for the sum spectral efficiency the familiar value

$$\mathcal{S}_{\infty}^{\mathsf{MMSE}} = \min(UN_{\mathrm{t}}, N_{\mathrm{r}}). \tag{9.56}$$

As far as the power offset is concerned, for $UN_{\mathrm{t}} \leq N_{\mathrm{r}}$ the expressions derived for the MU-MIMO MAC with ZF reception apply given that the spectral efficiencies of both receivers converge for $E_{\mathrm{s}}/N_0 \to \infty$.

Also noteworthy is that the optimization of the transmit powers becomes more tractable in the high-SNR regime, where it adopts the form of a geometric programming problem that can be recast as a convex problem [822, 823]. Although not consequential in terms of the number of spatial DOF, which is insensitive to the value of the transmit powers, this optimization would register on the power offset.

9.5 Duality with linear transceivers

The MU-MIMO MAC with a linear receiver and CSIR has its dual in the MU-MIMO BC with a linear transmitter and CSIT. In fact, such duality was the first to be observed [787, 788] and its clues eventually led to the discovery of the MAC–BC capacity duality presented in Chapter 8. As in our exposition of that duality, superscripts are applied in this section to distinguish between MAC and BC quantities. Furthermore, for notational clarity we introduce the variable $N_{\rm a}$ to denote both the number of receive antennas in the MAC and the coinciding number of transmit antennas in the BC.

Let us begin by considering an MU-SIMO MAC with an $N_{\rm a}$-antenna receiver, arbitrary $N_{\rm a} \times 1$ channels $\boldsymbol{H}_0, \ldots, \boldsymbol{H}_{U-1}$, and arbitrary receive filters $\boldsymbol{W}_0, \ldots, \boldsymbol{W}_{U-1}$ normalized such that $\|\boldsymbol{W}_u\|^2 = N_{\rm a}$ for $u = 0, \ldots, U-1$. Filter $\boldsymbol{W}_u$ targets the signal of user u, giving at its output

$$\boldsymbol{W}_u^* \left(\sum_{\rm u=0}^{U-1} \sqrt{G_{\rm u}} \boldsymbol{H}_{\rm u} \sqrt{E_{\rm u}^{\sf MAC}} s_{\rm u} + \boldsymbol{v} \right) = \sqrt{G_u E_u^{\sf MAC}}\, \boldsymbol{W}_u^* \boldsymbol{H}_u s_u + \sum_{{\rm u} \neq u} \sqrt{G_{\rm u} E_{\rm u}^{\sf MAC}}\, \boldsymbol{W}_u^* \boldsymbol{H}_{\rm u} s_{\rm u} + \boldsymbol{W}_u^* \boldsymbol{v}, \tag{9.57}$$

with the SINR given in (9.6) and reproduced here with the additional superscripting:

$$\mathsf{sinr}_u^{\sf MAC} = \frac{\frac{E_u^{\sf MAC}}{E_{\rm s}} \mathsf{SNR}_u \, |\boldsymbol{W}_u^* \boldsymbol{H}_u|^2}{\sum_{{\rm u} \neq u} \frac{E_{\rm u}^{\sf MAC}}{E_{\rm s}} \mathsf{SNR}_{\rm u} \, |\boldsymbol{W}_u^* \boldsymbol{H}_{\rm u}|^2 + N_{\rm a}}. \tag{9.58}$$

Consider now the dual MU-MISO BC with channels $\boldsymbol{H}_0^*, \ldots, \boldsymbol{H}_{U-1}^*$ and an $N_{\rm a}$-antenna transmitter with precoders $\boldsymbol{W}_0, \ldots, \boldsymbol{W}_{U-1}$. We release the BC transmit energies per symbol from coinciding with those in the MAC, superscripting the former by $(\cdot)^{\sf BC}$ just as we have superscripted the latter by $(\cdot)^{\sf MAC}$. User u then receives

$$y_u = \sqrt{G_u} \boldsymbol{H}_u^* \left(\sum_{\rm u=0}^{U-1} \sqrt{\frac{E_{\rm u}^{\sf BC}}{N_{\rm a}}} \boldsymbol{W}_{\rm u} s_{\rm u} \right) + v_u \tag{9.59}$$

with SINR

$$\mathsf{sinr}_u^{\sf BC} = \frac{\mathbb{E}\left[\left| \boldsymbol{H}_u^* \sqrt{\frac{G_u E_u^{\sf BC}}{N_{\rm a}}} \boldsymbol{W}_u s_u \right|^2 \right]}{\mathbb{E}\left[\left| \boldsymbol{H}_u^* \sum_{{\rm u} \neq u} \sqrt{\frac{G_u E_{\rm u}^{\sf BC}}{N_{\rm a}}} \boldsymbol{W}_{\rm u} s_{\rm u} + v_u \right|^2 \right]} \tag{9.60}$$

$$= \frac{\frac{G_u E_u^{\mathsf{BC}}}{N_a} |\boldsymbol{H}_u^* \boldsymbol{W}_u|^2}{\sum_{\mathrm{u} \neq u} \frac{G_{\mathrm{u}} E_{\mathrm{u}}^{\mathsf{BC}}}{N_a} |\boldsymbol{H}_u^* \boldsymbol{W}_{\mathrm{u}}|^2 + N_0} \tag{9.61}$$

$$= \frac{\frac{E_u^{\mathsf{BC}}}{E_s} \mathsf{SNR}_u |\boldsymbol{H}_u^* \boldsymbol{W}_u|^2}{\sum_{\mathrm{u} \neq u} \frac{E_{\mathrm{u}}^{\mathsf{BC}}}{E_s} \mathsf{SNR}_u |\boldsymbol{H}_u^* \boldsymbol{W}_{\mathrm{u}}|^2 + N_a}. \tag{9.62}$$

The duality for linear transceivers states that any feasible combination of SINRs can be achieved in both the MAC and the BC with the same $\boldsymbol{W}_0, \ldots, \boldsymbol{W}_{U-1}$ (acting as receive filters in the former and as precoders in the latter) and with the power constraint in the BC transmitter equal to the sum of the power constraints at the U MAC transmitters. Put differently, for any feasible $\mathsf{sinr}_0, \ldots, \mathsf{sinr}_{U-1}$ there exist $\boldsymbol{W}_0, \ldots, \boldsymbol{W}_{U-1}$ as well as $E_0^{\mathsf{MAC}}, \ldots, E_{U-1}^{\mathsf{MAC}}$ and $E_0^{\mathsf{BC}}, \ldots, E_{U-1}^{\mathsf{BC}}$ such that, for $u = 0, \ldots, U-1$,

$$\frac{\frac{E_u^{\mathsf{MAC}}}{E_s} \mathsf{SNR}_u |\boldsymbol{W}_u^* \boldsymbol{H}_u|^2}{\sum_{\mathrm{u} \neq u} \frac{E_{\mathrm{u}}^{\mathsf{MAC}}}{E_s} \mathsf{SNR}_{\mathrm{u}} |\boldsymbol{W}_u^* \boldsymbol{H}_{\mathrm{u}}|^2 + N_a} = \frac{\frac{E_u^{\mathsf{BC}}}{E_s} \mathsf{SNR}_u |\boldsymbol{H}_u^* \boldsymbol{W}_u|^2}{\sum_{\mathrm{u} \neq u} \frac{E_{\mathrm{u}}^{\mathsf{BC}}}{E_s} \mathsf{SNR}_{\mathrm{u}} |\boldsymbol{H}_u^* \boldsymbol{W}_{\mathrm{u}}|^2 + N_a}, \tag{9.63}$$

with

$$\sum_{u=0}^{U-1} \frac{E_u^{\mathsf{MAC}}}{E_s} = \sum_{u=0}^{U-1} \frac{E_u^{\mathsf{BC}}}{E_s} \tag{9.64}$$

although, in general, with $\frac{E_u^{\mathsf{MAC}}}{E_s} \neq \frac{E_u^{\mathsf{BC}}}{E_s}$.

Corroborating the above relationship requires solving for the transmit energies per symbol that yield the same given SINRs with some given $\boldsymbol{W}_0, \ldots, \boldsymbol{W}_{U-1}$ in both (9.58) and (9.62). This entails solving two systems of U linear equations [50, section 10.3.2]. It can be verified that the solutions to (9.58) are

$$\frac{E_u^{\mathsf{MAC}}}{E_s} = \sum_{\mathrm{u}=0}^{U-1} \left[\boldsymbol{A}_{\mathsf{MAC}}^{-1}\right]_{u,\mathrm{u}} \qquad u = 0, \ldots, U-1, \tag{9.65}$$

where $[\boldsymbol{A}_{\mathsf{MAC}}]_{u,\mathrm{u}} = -\frac{\mathsf{SNR}_{\mathrm{u}}}{N_a} |\boldsymbol{W}_u^* \boldsymbol{H}_{\mathrm{u}}|^2$ for $u \neq \mathrm{u}$ and

$$[\boldsymbol{A}_{\mathsf{MAC}}]_{u,u} = \frac{\mathsf{SNR}_u}{N_a} \frac{|\boldsymbol{W}_u^* \boldsymbol{H}_u|^2}{\mathsf{sinr}_u^{\mathsf{MAC}}}. \tag{9.66}$$

In turn, the values that solve (9.62) are

$$\frac{E_u^{\mathsf{BC}}}{E_s} = \sum_{\mathrm{u}=0}^{U-1} [\boldsymbol{A}_{\mathsf{BC}}^{-1}]_{u,\mathrm{u}} \qquad u = 0, \ldots, U-1, \tag{9.67}$$

where $[\boldsymbol{A}_{\mathsf{BC}}]_{u,\mathrm{u}} = -\frac{\mathsf{SNR}_u}{N_a} |\boldsymbol{W}_{\mathrm{u}}^* \boldsymbol{H}_u|^2$ for $u \neq \mathrm{u}$ and

$$[\boldsymbol{A}_{\mathsf{BC}}]_{u,u} = \frac{\mathsf{SNR}_u}{N_a} \frac{|\boldsymbol{W}_u^* \boldsymbol{H}_u|^2}{\mathsf{sinr}_u^{\mathsf{BC}}}. \tag{9.68}$$

If $\mathsf{sinr}_u^{\mathsf{MAC}} = \mathsf{sinr}_u^{\mathsf{BC}}$ for $u = 0, \ldots, U-1$, then from the foregoing relationships we have that $\boldsymbol{A}_{\text{BC}} = \boldsymbol{A}_{\text{MAC}}^{\text{T}}$ and hence

$$\sum_{u=0}^{U-1} \frac{E_u^{\text{MAC}}}{E_{\text{s}}} = \sum_{u=0}^{U-1}\sum_{\text{u}=0}^{U-1} \left[\boldsymbol{A}_{\text{MAC}}^{-1}\right]_{u,\text{u}} \tag{9.69}$$

$$= \sum_{u=0}^{U-1}\sum_{\text{u}=0}^{U-1} \left[\left(\boldsymbol{A}_{\text{BC}}^{\text{T}}\right)^{-1}\right]_{u,\text{u}} \tag{9.70}$$

$$= \sum_{u=0}^{U-1}\sum_{\text{u}=0}^{U-1} \left[\boldsymbol{A}_{\text{BC}}^{-1}\right]_{\text{u},u} \tag{9.71}$$

$$= \sum_{u=0}^{U-1} \frac{E_u^{\text{BC}}}{E_{\text{s}}} \tag{9.72}$$

confirming the duality.

Attempts to generalize the foregoing duality to multiantenna users would require introducing linear filters at those users as well; for each data stream, a separate filter acting as precoder in the MAC and as receiver in the BC.

Given the direct relationship between $\mathsf{sinr}_0, \ldots, \mathsf{sinr}_{U-1}$ and the corresponding spectral efficiencies, we find that with linear transceivers—just as found in Section 8.7 for the optimum transceivers—the MAC and the BC exhibit the same spectral efficiency regions if the power constraint at the BC transmitter equals the sum of the power constraints at the U MAC transmitters. And, given that the number of spatial DOF is insensitive to the power constraint, the MAC and BC exhibit the same number thereof.

In the case of a BC with a per-antenna (rather than per-symbol) power constraint, a different duality relationship can be formulated and interested readers are referred to [792].

9.6 Linear transmitters for the broadcast channel

For the BC, the appeal of a linear structure is very high, given the difficulties posed by the optimum DPC transmitter presented in the previous chapter. With a linear structure, the BC transmit signal continues to be

$$\boldsymbol{x} = \sum_{u=0}^{U-1} \sqrt{\frac{E_u}{N_{\text{t}}}} \boldsymbol{F}_u \boldsymbol{s}_u \tag{9.73}$$

only with $\boldsymbol{s}_0$ through $\boldsymbol{s}_{U-1}$ separately encoded. This rules out DPC, Tomlinson–Harashima, and any other nonlinear interaction among the codewords. Then, as far as the transmitter is concerned, the task of managing interference rests exclusively on the precoders.

In terms of the U receivers, and unless otherwise stated, we consider them to be linear as well, each conforming to Fig. 6.1. This makes our BC the exact reciprocal of the MAC considered earlier in the chapter.

It is useful, for the formulation that follows, to stack into a single matrix the channels (including both large- and small-scale components) that connect the transmitter with the U

receivers. This gives the $UN_r \times N_t$ aggregate channel matrix

$$C = \begin{bmatrix} \sqrt{G_0} H_0 \\ \vdots \\ \sqrt{G_{U-1}} H_{U-1} \end{bmatrix}. \tag{9.74}$$

9.7 Linear ZF transmitter for the MU-MISO broadcast channel

The ZF transmitter is best introduced for single-antenna users and, for that reason, this is the setup considered in the thorough analysis that we undertake in this section. The extension to $N_r > 1$, and its natural progression into block-diagonalization, follow thereafter.

9.7.1 Transmitter structure

When it comes to applying the ZF principles to the BC, the dimensionality constraint that applies is $N_t \geq UN_r$; in the MU-MISO case, this reduces to $N_t \geq U$. The aggregate channel matrix C defined in (9.74) is then $U \times N_t$ and the first idea that comes to mind, the one pursued in most of the related literature, is to transmit $C^\dagger [s_0 \cdots s_{U-1}]^T$ with the appropriate scaling to ensure that the power constraint is satisfied. This yields, at the channel outputs, scaled and noisy versions of $s_0, \ldots, s_{U-1}$ with a common SNR and thus a common spectral efficiency per user.

While indeed achieving ZF, the foregoing transmission strategy is unnecessarily restrictive. Subsuming precoding and power allocation, it enforces, in addition to ZF, equality in the spectral efficiencies of the U users, not allowing the flexibility of operating on other points of the spectral efficiency boundary. Put differently, this strategy turns the cascade of transmitter and channel experienced by $[s_0 \cdots s_{U-1}]^T$ into a scaled identity matrix, when in reality all that is required for ZF is that it be a diagonal matrix.

To relax the restriction of equal spectral efficiencies, precoding and power allocation must be decoupled as per the formulation utilized throughout this book. Vectorizing the transmit–receive relationship for the U users, we obtain

$$\begin{bmatrix} y_0 \\ \vdots \\ y_{U-1} \end{bmatrix} = \underbrace{\begin{bmatrix} \sqrt{G_0} H_0 \\ \vdots \\ \sqrt{G_{U-1}} H_{U-1} \end{bmatrix}}_{C} \left[F_0^{\text{ZF}} \cdots F_{U-1}^{\text{ZF}}\right] \begin{bmatrix} \sqrt{\frac{E_0}{N_t}} & 0 & 0 \\ 0 & \ddots & 0 \\ 0 & 0 & \sqrt{\frac{E_{U-1}}{N_t}} \end{bmatrix} \begin{bmatrix} s_0 \\ \vdots \\ s_{U-1} \end{bmatrix} + \begin{bmatrix} v_0 \\ \vdots \\ v_{U-1} \end{bmatrix}, \tag{9.75}$$

where the aggregate $N_t \times U$ precoder $\left[F_0^{\text{ZF}} \cdots F_{U-1}^{\text{ZF}}\right]$ should equal the pseudoinverse of $\boldsymbol{C}$, with each column of such aggregate precoder properly normalized to satisfy the constraint that applies to the corresponding individual precoder; with a per-symbol power constraint at each precoder, this means

$$\boldsymbol{F}_u^{\mathsf{ZF}} = \sqrt{N_{\mathrm{t}}}\frac{[\boldsymbol{C}^\dagger]_{:,u}}{\|[\boldsymbol{C}^\dagger]_{:,u}\|}, \tag{9.76}$$

satisfying $\|\boldsymbol{F}_u^{\mathsf{ZF}}\|^2 = N_{\mathrm{t}}$. The resulting transmitter diagonalizes the channel completely, conveying one data stream to each single-antenna user without interference because, normalizations notwithstanding, $\boldsymbol{H}_u \boldsymbol{F}_{\mathrm{u}}^{\mathsf{ZF}} = 0$ for $u \neq \mathrm{u}$. Thus

$$y_u = \sqrt{\frac{G_u E_u}{N_{\mathrm{t}}}}\boldsymbol{H}_u \boldsymbol{F}_u^{\mathsf{ZF}} s_u + v_u \qquad u = 0, \ldots, U-1, \tag{9.77}$$

with power allocation $\frac{E_0}{E_{\mathrm{s}}}, \ldots, \frac{E_{U-1}}{E_{\mathrm{s}}}$.

If the power constraint is per-antenna, rather than per-symbol, then the formulation requires generalizing the notion of pseudoinverse; interested readers are referred to [824].

9.7.2 SNR distribution

From (9.77), the SNR at user u equals, for given $\boldsymbol{H}_u$,

$$\mathsf{snr}_u^{\mathsf{ZF}} = \frac{G_u E_u}{N_0}\frac{|\boldsymbol{H}_u \boldsymbol{F}_u^{\mathsf{ZF}}|^2}{N_{\mathrm{t}}} \tag{9.78}$$

$$= \frac{\frac{E_u}{E_{\mathrm{s}}}\mathsf{SNR}_u |\boldsymbol{H}_u \boldsymbol{F}_u^{\mathsf{ZF}}|^2}{N_{\mathrm{t}}}. \tag{9.79}$$

We have U parallel noninterfering subchannels with generally different $\mathsf{snr}_0^{\mathsf{ZF}}, \ldots, \mathsf{snr}_{U-1}^{\mathsf{ZF}}$, which can be controlled by adjusting $\frac{E_0}{E_{\mathrm{s}}}, \ldots, \frac{E_{U-1}}{E_{\mathrm{s}}}$ subject to $\sum_{u=0}^{U-1}\frac{E_u}{E_{\mathrm{s}}} = 1$. This allows operating at any desired point on the boundary of the spectral efficiency region.

For any fixed $\frac{E_0}{E_{\mathrm{s}}}, \ldots, \frac{E_{U-1}}{E_{\mathrm{s}}}$, the distribution of $\mathsf{snr}_u^{\mathsf{ZF}}$ can be characterized by recognizing that $\boldsymbol{F}_u^{\mathsf{ZF}}$ in (9.76) is an N_{t}-dimensional fixed-norm vector that lies orthogonal to the $(U-1)$-dimensional subspace spanned by $\boldsymbol{H}_0, \ldots, \boldsymbol{H}_{u-1}, \boldsymbol{H}_{u+1}, \ldots, \boldsymbol{H}_{U-1}$. Intuitively then, $U-1$ of the N_{t} dimensions of $\boldsymbol{F}_u^{\mathsf{ZF}}$ are tied up to enforce the orthogonality and $N_{\mathrm{t}} - (U-1)$ dimensions remain free to focus power toward the intended single-antenna user. This is tantamount to $N_{\mathrm{t}} - U + 1$ transmit and one receive antenna, which, with CSIT and by virtue of duality, is equivalent to one transmit and $N_{\mathrm{t}} - U + 1$ receive antennas with CSIR. Thus, precoder normalization aside, the distribution of the SNRs with ZF transmission turns out to be similar to what is encountered with ZF reception (see Section 9.3).

Example 9.14 (ZF SNR distribution in an IID Rayleigh-faded MU-MISO BC)

Let $\frac{E_u}{E_{\mathrm{s}}}$ be fixed while $\boldsymbol{H}_u$ has IID Rayleigh-faded entries. The distribution of $|\boldsymbol{H}_u \boldsymbol{F}_u^{\mathsf{ZF}}|^2$, as per the reasoning above, must equal that of a CSIR-equipped MU-SIMO channel with one transmit and $N_{\mathrm{t}} - U + 1$ receive antennas. As argued in earlier examples, this corresponds to a chi-square distribution $\chi^2_{2(N_{\mathrm{t}}-U+1)}$, in this case with $\mathbb{E}\big[|\boldsymbol{H}_u \boldsymbol{F}_u^{\mathsf{ZF}}|^2\big] = N_{\mathrm{t}}(N_{\mathrm{t}}-U+1)$.

This confers to (9.79) the distribution (for $\xi \geq 0$),

$$f_{\mathsf{snr}_u^{\mathsf{ZF}}}(\xi) = \frac{1}{\frac{E_u}{E_s}\mathsf{SNR}_u (N_t - U)!} \exp\left(-\frac{\xi}{\frac{E_u}{E_s}\mathsf{SNR}_u}\right)\left(\frac{\xi}{\frac{E_u}{E_s}\mathsf{SNR}_u}\right)^{N_t - U}, \tag{9.80}$$

with

$$\mathbb{E}\left[\mathsf{snr}_u^{\mathsf{ZF}}\right] = (N_t - U + 1)\frac{E_u}{E_s}\mathsf{SNR}_u. \tag{9.81}$$

If the power allocation is uniform, then $\frac{E_u}{E_s} = 1/U$ and

$$f_{\mathsf{snr}_u^{\mathsf{ZF}}}(\xi) = \frac{U}{\mathsf{SNR}_u (N_t - U)!} \exp\left(-\frac{U}{\mathsf{SNR}_u}\xi\right)\left(\frac{U}{\mathsf{SNR}_u}\xi\right)^{N_t - U}, \tag{9.82}$$

with

$$\mathbb{E}\left[\mathsf{snr}_u^{\mathsf{ZF}}\right] = \frac{N_t - U + 1}{U}\mathsf{SNR}_u. \tag{9.83}$$

Example 9.15

How does Example 9.14 specialize to the case $U = N_t$?

Solution

For $U = N_t$, $\boldsymbol{F}_u^{\mathsf{ZF}}$ is completely determined by the requirement that it be orthogonal to the subspace spanned by the channels of all other users and by the constraint that its norm be fixed. Hence, $\boldsymbol{F}_u^{\mathsf{ZF}}$ is actually independent of $\boldsymbol{H}_u$. Since the entries of $\boldsymbol{H}_u$ are IID standard complex Gaussian while $\|\boldsymbol{F}_u^{\mathsf{ZF}}\|^2 = N_t$, for every feasible realization of $\boldsymbol{F}_u^{\mathsf{ZF}}$ the scalar $\boldsymbol{H}_u \boldsymbol{F}_u^{\mathsf{ZF}}$ is some linear combination of such complex Gaussian entries satisfying $\mathbb{E}[\boldsymbol{H}_u \boldsymbol{F}_u^{\mathsf{ZF}}] = 0$ and $\mathbb{E}\left[|\boldsymbol{H}_u \boldsymbol{F}_u^{\mathsf{ZF}}|^2\right] = N_t$. Altogether then,

$$\boldsymbol{H}_u \boldsymbol{F}_u^{\mathsf{ZF}} \sim \mathcal{N}_{\mathbb{C}}(0, N_t). \tag{9.84}$$

It follows that $|\boldsymbol{H}_u \boldsymbol{F}_u^{\mathsf{ZF}}|^2$, and therefore $\mathsf{snr}_u^{\mathsf{ZF}}$, are exponentially distributed and indeed this is the type of distribution that we recover by setting $U = N_t$ in (9.80), namely

$$f_{\mathsf{snr}_u^{\mathsf{ZF}}}(\xi) = \frac{1}{\frac{E_u}{E_s}\mathsf{SNR}_u} \exp\left(-\frac{\xi}{\frac{E_u}{E_s}\mathsf{SNR}_u}\right), \tag{9.85}$$

with

$$\mathbb{E}\left[\mathsf{snr}_u^{\mathsf{ZF}}\right] = \frac{E_u}{E_s}\mathsf{SNR}_u. \tag{9.86}$$

If the power allocation is further uniform, then

$$f_{\mathsf{snr}_u^{\mathsf{ZF}}}(\xi) = \frac{U}{\mathsf{SNR}_u} \exp\left(-\frac{U}{\mathsf{SNR}_u}\xi\right), \tag{9.87}$$

with

$$\mathbb{E}\left[\mathsf{snr}_u^{\mathsf{ZF}}\right] = \frac{\mathsf{SNR}_u}{U}. \tag{9.88}$$

9.7.3 Power allocation

Given the availability of CSIT, the power allocation $\frac{E_0}{E_s}, \ldots, \frac{E_{U-1}}{E_s}$ need not be fixed but may be adjusted on the basis of the fading. This optimization is a convex problem, and familiar forms emerge as solutions.

Example 9.16

What is the optimum power allocation in terms of sum spectral efficiency given a ZF transmitter and single-antenna users?

Solution

As learned in Section 4.4, the power allocation $\frac{E_0^\star}{E_s}, \ldots, \frac{E_{U-1}^\star}{E_s}$ that maximizes the sum of the mutual informations over a bank of parallel subchannels is waterfilling. Applying it,

$$\frac{E_u^\star}{E_s} = \left[\frac{1}{\eta} - \frac{N_t}{\mathsf{SNR}_u \, |\boldsymbol{H}_u \boldsymbol{F}_u^{\mathsf{ZF}}|^2}\right]^+ \qquad u = 0, \ldots, U-1, \tag{9.89}$$

with η such that $\sum_{u=0}^{U-1} \frac{E_u^\star}{E_s} = 1$. The advantage of such power allocation, as opposed to the one that equalizes the user spectral efficiencies, is the subject of Problem 9.15.

Readers interested in generalizations of this policy to arbitrary signal constellations, in essence a multiuser form of mercury/waterfilling, are referred to [825, 826].

To maximize the weighted sum spectral efficiency rather than simply the sum spectral efficiency, variations of the waterfilling policy must be applied. We defer the formulation of these variations to Problem 9.16, and address the considerably simpler form they adopt in high-SNR conditions later in this section.

9.7.4 Ergodic spectral efficiency

Turning now to the ergodic spectral efficiency, the data stream intended for the uth user experiences a scalar fading channel and hence the corresponding codeword should be drawn from a complex Gaussian distribution. Then, the spectral efficiency achieved by the uth user is

$$C_u^{\mathsf{ZF}} = \mathbb{E}\left[\log_2\left(1 + \mathsf{snr}_u^{\mathsf{ZF}}\right)\right] \tag{9.90}$$

$$= \mathbb{E}\left[\log_2\left(1 + \frac{\frac{E_u}{E_s}\mathsf{SNR}_u \, |\boldsymbol{H}_u \boldsymbol{F}_u^{\mathsf{ZF}}|^2}{N_t}\right)\right] \tag{9.91}$$

and the weighted sum spectral efficiency is $\sum_{u=0}^{U-1} q_u C_u^{\mathsf{ZF}}$, which is a concave function of $\frac{E_0}{E_s}, \ldots, \frac{E_{U-1}}{E_s}$ from which the aforementioned variations of waterfilling can be formulated.

For a power allocation not dependent on the fading, explicit expressions can be obtained for C_u^{ZF}.

Example 9.17 (ZF spectral efficiency in an IID Rayleigh-faded MU-MISO BC)

Let $\boldsymbol{H}_0, \ldots, \boldsymbol{H}_{U-1}$ have IID Rayleigh-faded entries while $\frac{E_0}{E_s}, \ldots, \frac{E_{U-1}}{E_s}$ are fixed. The distributions of $\mathsf{snr}_0^{\mathsf{ZF}}, \ldots, \mathsf{snr}_{U-1}^{\mathsf{ZF}}$ then abide by Example 9.14 and thus

$$C_u^{\mathsf{ZF}} = \int_0^\infty \log_2(1+\xi) \frac{1}{\frac{E_u}{E_s}\mathsf{SNR}_u (N_t - U)!} \exp\left(-\frac{\xi}{\frac{E_u}{E_s}\mathsf{SNR}_u}\right) \left(\frac{\xi}{\frac{E_u}{E_s}\mathsf{SNR}_u}\right)^{N_t - U} d\xi, \tag{9.92}$$

which amounts to the ergodic spectral efficiency of a scalar channel with a chi-square fading distribution, a computation we have encountered throughout the text. Applying (C.37),

$$C_u^{\mathsf{ZF}} = \exp\left(\frac{1}{\frac{E_u}{E_s}\mathsf{SNR}_u}\right) \sum_{q=1}^{N_t-U+1} \mathcal{E}_q\left(\frac{1}{\frac{E_u}{E_s}\mathsf{SNR}_u}\right) \log_2 e. \tag{9.93}$$

Example 9.18

How does Example 9.17 specialize to the case $U = N_t$?

Solution

For $U = N_t$,

$$C_u^{\mathsf{ZF}} = \exp\left(\frac{1}{\frac{E_u}{E_s}\mathsf{SNR}_u}\right) \mathcal{E}_1\left(\frac{1}{\frac{E_u}{E_s}\mathsf{SNR}_u}\right) \log_2 e, \tag{9.94}$$

which equals the ergodic capacity of a Rayleigh-faded SISO channel with an average SNR of $\frac{E_u}{E_s}\mathsf{SNR}_u$. If the power allocation is uniform, then

$$C_u^{\mathsf{ZF}} = e^{U/\mathsf{SNR}_u} \mathcal{E}_1\left(\frac{U}{\mathsf{SNR}_u}\right) \log_2 e. \tag{9.95}$$

Example 9.19

Compare the MU-MISO BC sum spectral efficiency achieved by ZF transmission under waterfilling and under a uniform power allocation if $U = N_t = 3$ with

$$\mathsf{SNR}_0|_{\text{dB}} = \mathsf{SNR}|_{\text{dB}} \tag{9.96}$$

$$\mathsf{SNR}_1|_{\text{dB}} = \mathsf{SNR}|_{\text{dB}} + 5\,\text{dB} \tag{9.97}$$

$$\mathsf{SNR}_2|_{\text{dB}} = \mathsf{SNR}|_{\text{dB}} + 8\,\text{dB} \tag{9.98}$$

and with Rayleigh fading. Further compare both ZF solutions with the ergodic sum-capacity.

Solution

See Fig. 9.5, where the result with waterfilling was obtained numerically while its uniform-power counterpart corresponds to Example 9.18. The sum-capacity, in turn, is borrowed from Example 8.33.

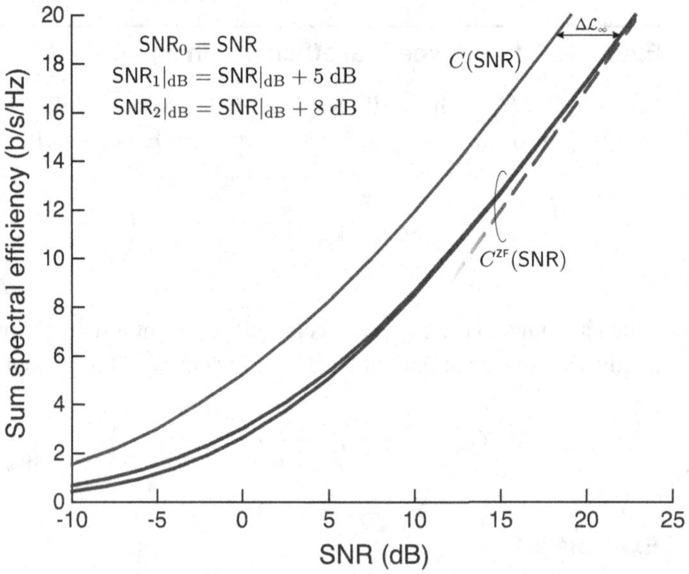

Fig. 9.5 MU-MISO BC sum spectral efficiency as a function of SNR (in dB) with $U = N_{\mathrm{t}} = 3$. The two bottom solid lines correspond to ZF transmission under both waterfilling and a uniform power allocation. The dashed line is the high-SNR expansion for both ZF solutions. The top solid line indicates the sum-capacity, borrowed from Example 8.33.

Example 9.19 suggests that the optimization of the power allocation in ZF transmission may be relatively unimportant. Its benefit appears only at SNR values low enough to warrant other types of MU-MIMO transmission or directly single-user transmission with orthogonal sharing.

As mentioned in the coverage of ZF receivers for SU-MIMO and for the MU-MIMO MAC, the inversion of a channel with small singular values suffers from strong noise enhancement. ZF transmission is afflicted by the converse phenomenon: the inversion of a channel with small singular values drives the precoder power consumption into overdrive. The consequence is the same, namely a disappointingly sublinear increase in the sum spectral efficiency with the number of antennas for $U = N_{\mathrm{t}}$ without user selection, and this consequence is further aggravated by another issue that is specific to the BC. In contrast with the MAC, where each user contributes its own power budget, in the BC a fixed transmit power must be divided among the U users. As U grows, the power per signal stream is preserved in the MAC while, in the BC, it shrinks sustainedly. Indeed, a contrast of (9.16) and (9.88) confirms that the local-average SNR of any given signal stream is not inversely proportional to U in the MAC, while it is in the BC.

The extent to which the performance of a ZF transmitter suffers if $U \approx N_{\mathrm{t}}$ and both U and N_{t} grow large is best illustrated by the example that follows, which is inspired by [827, section III].

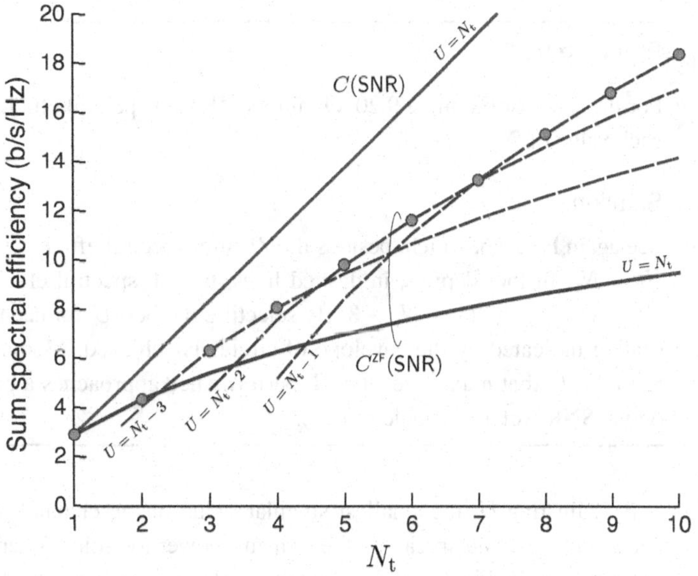

Fig. 9.6 MU-MISO BC sum spectral efficiency with ZF transmission as a function of N_t. The top solid line is the BC sum-capacity for $U = N_t$. The bottom solid curve corresponds to ZF with $U = N_t$, while the dashed lines correspond to ZF with $U = N_t - 1$, $U = N_t - 2$, and $U = N_t - 3$. The envelope of circles indicates the best possible value achievable with ZF for each N_t. In all cases, $\text{SNR}_u = 10$ dB for $u = 0, \ldots, U - 1$.

Example 9.20

Let $\text{SNR}_u = \text{SNR}$ for $u = 0, \ldots, U - 1$ with $\text{SNR} = 10$ dB and with $\boldsymbol{H}_0, \ldots, \boldsymbol{H}_{U-1}$ having IID Rayleigh-faded entries. Obtain the MU-MISO BC sum spectral efficiency with linear ZF transmission and waterfilling power allocation for $U = N_t$ and depict it as function of $N_t = 1, \ldots, 10$. Further compare such sum spectral efficiency with the corresponding BC sum-capacity.

Solution

The results are the two solid curves in Fig. 9.6, with the sum-capacity computed by applying duality and a convex optimization solver.

Although admittedly very particular because of the equality in SNRs, the above example does illustrate the fact that, at a reasonable SNR, the scaling of the ZF spectral efficiency with N_t is markedly sublinear when $U = N_t$. This is in stark contrast with the BC sum-capacity, which does exhibit a linear increase. However, things do improve for ZF if U is carefully adjusted.

Example 9.21

For the setup of Example 9.20, obtain the ZF sum spectral efficiency with U optimized for each value of N_t.

Solution

See again Fig. 9.6, which, besides the ZF sum spectral efficiency and the sum-capacity for $U = N_t$, further depicts, in dashed lines, the ZF spectral efficiencies for $U = N_t - 1$, $U = N_t - 2$, and $U = N_t - 3$. By selecting the best U for each N_t, the essentially linear scaling indicated by the envelope of circles is achieved. Moreover, for $U, N_t \to \infty$, the ratio N_t/U that maximizes the ZF performance approaches a constant ratio that depends on the SNR (refer to Problem 9.18).

By adjusting U, the smallest singular value of the channel inverse is made to behave better and we avoid spreading the transmit power too thinly over too many signal streams. As suggested by Fig. 9.6 and proved in [827], with the proper ratio N_t/U, in the equal-SNR setup of Examples 9.20 and 9.21 a ZF transmitter can achieve a hefty share of the channel capacity. Both the sum-capacity and the sum spectral efficiency of ZF with a properly adjusted U scale linearly with N_t and thus their ratio is bound to approach the ratio of the corresponding slopes. (For general unequal SNRs, the optimum number of active users and the achievable share of the capacity would depart from those in the equal-SNR case, but the qualitative observations would hold.)

If, moreover, the U active users are opportunistically selected on the basis of their channel conditions, then the slope versus N_t of the ZF sum spectral efficiency with a properly adjusted U converges, as $U_{\text{tot}} \to \infty$, to the slope of the sum-capacity versus N_t [828]; this convergence holds under either a per-antenna or a per-symbol power constraint [829]. While these results are indicative of the simple ZF transmit structure being a more enticing option than its performance for some fixed U may indicate, it must be borne in mind that an offset remains with respect to the sum-capacity and that the opportunistic selection of users is subject to the caveats discussed in Chapter 8.

As a final remark, and to reconcile the desideratum for $U < N_t$ with the upcoming high-SNR analysis, it is worth anticipating that, as the SNR grows large, the optimum ratio N_t/U approaches unity. At sufficiently high SNRs, it is optimum to have $U = N_t$.

9.7.5 High-SNR regime

For given fading realizations $\boldsymbol{H}_0, \ldots, \boldsymbol{H}_{U-1}$, the weighted sum spectral efficiency is

$$\sum_{u=0}^{U-1} q_u \log_2\left(1 + \frac{\frac{E_u}{E_s}\mathsf{SNR}_u \, |\boldsymbol{H}_u \boldsymbol{F}_u^{\mathsf{ZF}}|^2}{N_t}\right), \tag{9.99}$$

from which the power allocation $\frac{E_0}{E_s}, \ldots, \frac{E_{U-1}}{E_s}$ can be optimized by building the Lagrangian (see Appendix G)

$$\mathsf{L}\left(\frac{E_0}{E_s}, \ldots, \frac{E_{U-1}}{E_s}, \lambda\right) = \sum_{u=0}^{U-1} q_u \log_2\left(1 + \frac{\frac{E_u}{E_s}\mathsf{SNR}_u |\boldsymbol{H}_u \boldsymbol{F}_u^{\mathsf{ZF}}|^2}{N_t}\right) + \lambda\left(\sum_{u=0}^{U-1} \frac{E_u}{E_s} - 1\right) \tag{9.100}$$

and differentiating; this yields for $\frac{E_u}{E_s}$ the necessary and sufficient condition

$$q_u \frac{\frac{\mathsf{SNR}_u}{N_t}|\boldsymbol{H}_u \boldsymbol{F}_u^{\mathsf{ZF}}|^2}{1 + \frac{E_u}{E_s}\frac{\mathsf{SNR}_u}{N_t}|\boldsymbol{H}_u \boldsymbol{F}_u^{\mathsf{ZF}}|^2} \log_2 e + \lambda = 0, \tag{9.101}$$

from which the anticipated variations of waterfilling can be derived. For growing SNR_u and nonzero $\frac{E_u}{E_s}$, this condition expands as

$$\frac{q_u}{\frac{E_u}{E_s}} \log_2 e + \lambda + \mathcal{O}\left(\frac{1}{\mathsf{SNR}_u}\right) = 0, \tag{9.102}$$

from which

$$\frac{E_u}{E_s} = -\frac{\log_2 e}{\lambda} q_u + \mathcal{O}\left(\frac{1}{\mathsf{SNR}_u}\right), \tag{9.103}$$

where λ can be cleared by enforcing the power constraint

$$\sum_{u=0}^{U-1} \frac{E_u}{E_s} = -\frac{\log_2 e}{\lambda} \sum_{u=0}^{U-1} q_u + \mathcal{O}\left(\frac{1}{\mathsf{SNR}_u}\right) \tag{9.104}$$

$$= 1, \tag{9.105}$$

leading to

$$\lambda = -\log_2(e) \sum_{u=0}^{U-1} q_u + \mathcal{O}\left(\frac{1}{\mathsf{SNR}_u}\right). \tag{9.106}$$

Plugged into (9.103), this value of λ finally gives

$$\frac{E_u}{E_s} = \frac{q_u}{\sum_{u=0}^{U-1} q_u} + \mathcal{O}\left(\frac{1}{\mathsf{SNR}_u}\right), \tag{9.107}$$

which is the same high-SNR allocation policy found in Section 8.9.4 for the BC capacity: each user should be allocated power in direct proportion to its weight, a sort of "weighted uniform power allocation." For the sum spectral efficiency in particular, (9.107) reduces to the truly uniform power allocation

$$\frac{E_u}{E_s} = \frac{1}{U} + \mathcal{O}\left(\frac{1}{\mathsf{SNR}_u}\right). \tag{9.108}$$

With a single receive antenna, each user enjoys a single spatial DOF and thus the sum spectral efficiency exhibits $S_\infty^{\mathsf{ZF}} = U$ (which generalizes to $S_\infty^{\mathsf{ZF}} = UN_r$ for $N_r > 1$). The suboptimality of ZF transmission is registered in the power offset, whose computation benefits from the simplicity of the asymptotic power allocation in (9.107). Applying the single-user definition of power offset to (9.90), user u is seen to experience

$$\mathcal{L}^{\mathsf{ZF}}_{\infty,u} = \lim_{\mathsf{SNR}_u \to \infty} \left(\log_2 \mathsf{SNR}_u - C^{\mathsf{ZF}}_u \right) \tag{9.109}$$

$$= \lim_{\mathsf{SNR}_u \to \infty} \left(\log_2 \mathsf{SNR}_u - \mathbb{E}[\log_2(1 + \mathsf{snr}^{\mathsf{ZF}}_u)] \right) \tag{9.110}$$

$$= \lim_{\mathsf{SNR}_u \to \infty} \mathbb{E}\left[\log_2 \frac{\mathsf{SNR}_u}{1 + \mathsf{snr}^{\mathsf{ZF}}_u} \right] \tag{9.111}$$

$$= \log_2 N_\mathrm{t} - \mathbb{E}\left[\log_2 \left(\frac{E_u}{E_\mathrm{s}} |\boldsymbol{H}_u \boldsymbol{F}^{\mathsf{ZF}}_u|^2 \right) \right], \tag{9.112}$$

where we have recalled (9.79).

From $\mathcal{L}^{\mathsf{ZF}}_{\infty,u}$, we can write the spectral efficiency of user u as

$$C^{\mathsf{ZF}}_u(\mathsf{SNR}_u) = \log_2 \mathsf{SNR}_u - \mathcal{L}^{\mathsf{ZF}}_{\infty,u} + \mathcal{O}\left(\frac{1}{\mathsf{SNR}_u} \right) \tag{9.113}$$

and, summing over the U users, the sum spectral efficiency as

$$C^{\mathsf{ZF}}(\mathsf{SNR}_0, \ldots, \mathsf{SNR}_{U-1}) = \sum_{u=0}^{U-1} \left(\log_2 \mathsf{SNR}_u - \mathcal{L}^{\mathsf{ZF}}_{\infty,u} \right) + \mathcal{O}\left(\frac{1}{E_\mathrm{s}/N_0} \right), \tag{9.114}$$

where $\mathcal{L}^{\mathsf{ZF}}_{\infty,u}$ should be computed with $\frac{E_u}{E_\mathrm{s}} = 1/U$; then, $\mathcal{L}^{\mathsf{ZF}}_{\infty} = \frac{1}{U} \sum_{u=0}^{U-1} \mathcal{L}^{\mathsf{ZF}}_{\infty,u}$ provides the sum spectral efficiency power offset.

Under IID Rayleigh fading in particular, $\mathcal{L}^{\mathsf{ZF}}_{\infty,u}$ and $\mathcal{L}_{\infty}$ can be expressed in closed form.

Example 9.22 (ZF power offset for an IID Rayleigh-faded MU-MISO BC)

Consider the sum spectral efficiency. Setting $\frac{E_u}{E_\mathrm{s}} = 1/U$, we can apply duality between the MAC with ZF reception in Example 9.2 and the BC with ZF transmission in Example 9.17. This allows translating the MAC power offset expression in Example 9.6 into

$$\mathcal{L}^{\mathsf{ZF}}_{\infty,u} = \log_2 U + \left(\gamma_{\mathsf{EM}} - \sum_{q=1}^{N_\mathrm{t}-U} \frac{1}{q} \right) \log_2 e, \tag{9.115}$$

which, for $U = N_\mathrm{t}$, reduces to

$$\mathcal{L}^{\mathsf{ZF}}_{\infty,u} = \log_2 U + \gamma_{\mathsf{EM}} \log_2 e. \tag{9.116}$$

Example 9.23

Compare (9.114) with the exact sum spectral efficiency for $U = N_\mathrm{t} = 3$, with

$$\mathsf{SNR}_0|_{\mathrm{dB}} = \mathsf{SNR}|_{\mathrm{dB}} \tag{9.117}$$

$$\mathsf{SNR}_1|_{\mathrm{dB}} = \mathsf{SNR}|_{\mathrm{dB}} + 5\,\mathrm{dB} \tag{9.118}$$

$$\mathsf{SNR}_2|_{\mathrm{dB}} = \mathsf{SNR}|_{\mathrm{dB}} + 8\,\mathrm{dB} \tag{9.119}$$

and with Rayleigh fading.

Solution

See Fig. 9.5, where the expansion is seen to be valid over a wide range of SNRs.

From the power offset expressions in this section and in Section 8.9.4, we can quantify the performance gap between ZF and optimum DPC transmission at high SNR. Since the number of DOF is the same in both cases, it is only through the power offset that these transmitters can be discriminated.

Example 9.24

For IID Rayleigh fading, express the shortfall of the ZF sum spectral efficiency relative to the sum-capacity in terms of a high-SNR power loss.

Solution

From (8.150), particularized to $N_r = 1$, and (9.115), the power loss in 3-dB units equals

$$\Delta\mathcal{L}_\infty = \mathcal{L}_\infty^{\text{ZF}} - \mathcal{L}_\infty \tag{9.120}$$

$$= \left(\frac{N_t}{U} \sum_{q=N_t-U+1}^{N_t} \frac{1}{q} - 1 \right) \log_2 e, \tag{9.121}$$

which, for $U = N_t$, simplifies into

$$\Delta\mathcal{L}_\infty = \left(\sum_{q=2}^{N_t} \frac{1}{q} \right) \log_2 e. \tag{9.122}$$

Example 9.25

What is the high-SNR power loss of ZF for IID Rayleigh fading when $U = N_t = 3$?

Solution

In this case, $\Delta\mathcal{L}_\infty = 1.2$, indicating a power loss of 3.6 dB. As shown in Fig. 9.5, this value approximates very well the gap between the ZF sum spectral efficiency and the sum-capacity even at moderate SNRs.

The power loss $\Delta\mathcal{L}_\infty$ with single-antenna receivers and ZF transmission dualizes the one for single-antenna transmitters and ZF reception, and the corresponding observations can thus be directly imported.

- The power loss experienced by ZF is minimized for $U \ll N_t$ and, in fact, it vanishes for $N_t/U \to \infty$. However, most spatial DOF then go unused.
- Conversely, the power loss is maximum when $U = N_t$, when all the potential DOF are activated.

Altogether, while at certain SNRs having $U < N_t$ is advantageous in terms of ZF performance, at sufficiently high SNRs it is always best to have $U = N_t$. However, this conclusion is contingent on CSIT side information, and that is a natural lead into the subject of the next section.

9.7.6 Pilot-assisted ZF transmission

The preceding formulation of MU-MISO ZF transmission relies critically on the knowledge of $\boldsymbol{H}_0, \ldots, \boldsymbol{H}_{U-1}$ at the transmitter. It is imperative to assess whether the performance established under such CSIT is robust and can be actually approached when the CSI must be extracted from pilot-symbol observations, rather than being granted as side information. In fact, the issue of CSI acquisition is even more delicate here than it was in SU-MIMO and in the MU-MIMO MAC because, in contrast with those setups, in the MU-MIMO BC the CSI is required at the transmitter. It is not obvious whether the reassuring conclusions drawn for other setups apropos the CSIR continue to hold here with respect to the CSIT. Consequently, the matter deserves careful attention.

With the aim of quantifying the degree to which the performance of signaling schemes developed under CSI can be approached when these same schemes are applied with CSI acquired in operationally relevant conditions, the analysis that follows has this specificity.

- IID Rayleigh fading—purposely chosen as an adverse distribution as far as the CSI is concerned, since antenna correlations would facilitate the channel estimation and reduce the amount of feedback—with a block-fading structure having coherence N_{c}. The fading need not be frequency-flat, but rather the fading blocks can be time–frequency tiles containing N_{c} resource elements.
- Complex Gaussian signaling, optimum under CSIR.
- ZF precoding with a uniform power allocation, which as seen earlier is only slightly suboptimal in the range of operational interest to MU-MIMO while being more robust than the optimum power allocation in the face of CSI uncertainty.
- A fully loaded system with $U = N_{\mathrm{t}}$, the most vulnerable configuration. (Despite the equality between U and N_{t}, we carry both variables through the analysis in order to shed light on the role played by each and also to set the stage for potential generalizations to $U < N_{\mathrm{t}}$.)

General multistage procedure

The type of duplexing, which is immaterial when CSI is presumed, acquires considerable importance here. We therefore begin with a broad formulation that accommodates all types of duplexing, and in due course specialize the analysis. This initial broad formulation calls for a multistage procedure tailored to the fading coherence.

(1) *Unprecoded pilot transmission.* In a first stage, the transmitter emits at least one pilot symbol orthogonally in time (and/or in frequency) from each of the transmit antennas for a total of $N_{\mathrm{p}} \geq N_{\mathrm{t}}$ pilot symbols and an overhead of $\alpha = N_{\mathrm{p}}/N_{\mathrm{c}}$. These pilots are unprecoded and, since all users profit from them, they are also termed *common.* During each unprecoded pilot transmission from the jth transmit antenna, the uth user observes $\sqrt{G_u E_{\mathrm{s}}}\, [\boldsymbol{H}_u]_j + v_u$.

(2) *CSI acquisition at each user.* Accumulating its observations over the $N_\mathrm{p}/N_\mathrm{t}$ pilots emitted from the jth antenna, user u collects, for that antenna,

$$y_{u,j} = \sqrt{\frac{N_\mathrm{p}}{N_\mathrm{t}} G_u E_\mathrm{s}}\, [\boldsymbol{H}_u]_j + v_u \tag{9.123}$$

$$= \sqrt{\frac{\alpha N_\mathrm{c}}{N_\mathrm{t}} G_u E_\mathrm{s}}\, [\boldsymbol{H}_u]_j + v_u \tag{9.124}$$

and, assembling its cumulative observations for all the transmit antennas into a column vector, user u obtains

$$\boldsymbol{y}_u = \sqrt{\frac{\alpha N_\mathrm{c}}{N_\mathrm{t}} G_u E_\mathrm{s}}\, \boldsymbol{H}_u^\mathrm{T} + \boldsymbol{v}_u, \tag{9.125}$$

where $\boldsymbol{v}_u \sim \mathcal{N}_\mathbb{C}(\boldsymbol{0}, N_0 \boldsymbol{I})$.

From $\boldsymbol{y}_u$, the uth user may derive the MMSE channel estimate $\hat{\boldsymbol{H}}_u$, satisfying

$$\hat{\boldsymbol{H}}_u^\mathrm{T} = \mathbb{E}\big[\boldsymbol{H}_u^\mathrm{T} | \boldsymbol{y}_u\big] \tag{9.126}$$

$$= \big(\boldsymbol{R}_{\boldsymbol{y}_u}^{-1} \boldsymbol{R}_{\boldsymbol{y}_u \boldsymbol{H}_u^\mathrm{T}}\big)^* \boldsymbol{y}_u \tag{9.127}$$

$$= \frac{\sqrt{\frac{\alpha N_\mathrm{c}}{N_\mathrm{t}} G_u E_\mathrm{s}}}{N_0 + \frac{\alpha N_\mathrm{c}}{N_\mathrm{t}} G_u E_\mathrm{s}}\, \boldsymbol{y}_u, \tag{9.128}$$

where (9.127) holds because, as $\boldsymbol{H}_u$ is complex Gaussian, the MMSE and the LMMSE estimates coincide. Since the entries of $\boldsymbol{H}_u$ are independent, each one is estimated on the basis of only the corresponding entry of $\boldsymbol{y}_u$. We can write $\boldsymbol{H}_u = \hat{\boldsymbol{H}}_u + \tilde{\boldsymbol{H}}_u$ where $\tilde{\boldsymbol{H}}_u$ is independent of $\hat{\boldsymbol{H}}_u$ and it satisfies $\tilde{\boldsymbol{H}}_u^\mathrm{T} \sim \mathcal{N}_\mathbb{C}(\boldsymbol{0}, \mathsf{MMSE}_u \boldsymbol{I})$ with

$$\mathsf{MMSE}_u = \frac{1}{1 + \alpha \frac{N_\mathrm{c}}{N_\mathrm{t}} \mathsf{SNR}_u}. \tag{9.129}$$

(3) *CSI feedback.* The averaged observations $\boldsymbol{y}_0, \ldots, \boldsymbol{y}_{U-1}$, or else the ensuing channel estimates $\hat{\boldsymbol{H}}_0, \ldots, \hat{\boldsymbol{H}}_{U-1}$, are conveyed from the users back to the transmitter, which ends up with generally different estimates $\hat{\hat{\boldsymbol{H}}}_0, \ldots, \hat{\hat{\boldsymbol{H}}}_{U-1}$. Subsuming both reciprocity as well as actual feedback, this process can be modeled as a mapping, in general probabilistic to allow for noise and errors, from $\hat{\boldsymbol{H}}_u$ to $\hat{\hat{\boldsymbol{H}}}_u$. We denote by α_fb the associated overhead and, even though it is inflicted on the reverse channel, we carry it through our computations in recognition of the fact that it exists to enable the BC precoding.

(4) *Computation of the precoders.* From $\hat{\hat{\boldsymbol{H}}}_0, \ldots, \hat{\hat{\boldsymbol{H}}}_{U-1}$, the transmitter can now compute the ZF precoders by constructing the $U \times N_\mathrm{t}$ matrix

$$\hat{\hat{\boldsymbol{C}}} = \begin{bmatrix} \sqrt{G_0}\, \hat{\hat{\boldsymbol{H}}}_0 \\ \vdots \\ \sqrt{G_{U-1}}\, \hat{\hat{\boldsymbol{H}}}_{U-1} \end{bmatrix} \tag{9.130}$$

and applying (9.76) to $\hat{\hat{\boldsymbol{C}}}$ to obtain, for each user, a column vector precoder $\hat{\hat{\boldsymbol{F}}}_u^\mathsf{ZF}$ that is orthogonal to the subspace spanned by $\hat{\hat{\boldsymbol{H}}}_\mathrm{u}$ for u $\neq u$.

(5) *Precoded pilot transmission.* At this point, additional pilot symbols need to be transmitted such that each user can estimate its precoded channel and be ready to process data coherently. Indeed, the users are not privy to $\hat{\hat{\boldsymbol{F}}}_0^{\text{ZF}}, \ldots, \hat{\hat{\boldsymbol{F}}}_{U-1}^{\text{ZF}}$ nor can they compute these precoders on the basis of the channel estimates they gather from the unprecoded pilots, as the uth user only gets to learn $\hat{\boldsymbol{H}}_u$. The difference between unprecoded and precoded pilot transmissions is that the former are emitted orthogonally from each of the antennas whereas the latter are emitted orthogonally through each of the precoders. And, since each precoded pilot is transmitted for the benefit of a specific user, these pilots are also termed *dedicated.* We denote by α_{d} the precoded pilot overhead and thus the number of precoded pilots is $\alpha_{\text{d}} N_{\text{c}}$, evenly divided among the U precoders. The averaged (over its $\alpha_{\text{d}} N_{\text{c}}/U$ share of pilots) observation from the uth precoder at the uth user is

$$y_{u,\mathrm{u}} = \sqrt{\frac{\alpha_{\text{d}} N_{\text{c}}}{U N_{\text{t}}} G_u E_{\text{s}}} \, \underbrace{\boldsymbol{H}_u \hat{\hat{\boldsymbol{F}}}_{\mathrm{u}}^{\text{ZF}}}_{a_{u,\mathrm{u}}} + v_u, \tag{9.131}$$

where we have introduced $a_{u,\mathrm{u}} = \boldsymbol{H}_u \hat{\hat{\boldsymbol{F}}}_{\mathrm{u}}^{\text{ZF}}$ to denote the coupling coefficient between the uth precoder and the u user. Common factors aside, $a_{u,u}$ represents the intended channel coefficient for the uth stream whereas $a_{u,\mathrm{u}}$ for $\mathrm{u} \neq u$ quantifies the amount of interference that leaks from the uth stream onto the uth user; this leakage, which would be zero with CSIT, is caused by precoder misalignments induced by channel estimation and feedback inaccuracies. For $U = N_{\text{t}}$, $\hat{\hat{\boldsymbol{F}}}_{\mathrm{u}}^{\text{ZF}}$ depends on $\hat{\hat{\boldsymbol{H}}}_{\mathrm{u}}$ for $\mathrm{u} \neq u$ but not on $\hat{\hat{\boldsymbol{H}}}_u$ (see Example 9.15) and, since $\hat{\hat{\boldsymbol{H}}}_{\mathrm{u}}$ for $\mathrm{u} \neq u$ are independent of $\boldsymbol{H}_u$, it follows that $\hat{\hat{\boldsymbol{F}}}_u^{\text{ZF}}$ is independent of $\boldsymbol{H}_u$. Hence,

$$a_{u,u} = \boldsymbol{H}_u \hat{\hat{\boldsymbol{F}}}_u^{\text{ZF}} \tag{9.132}$$

is the product of a standard complex Gaussian vector with an independent random vector having norm N_{t} and uniform phase, giving $a_{u,u} \sim \mathcal{N}_{\mathbb{C}}(0, N_{\text{t}})$.

(6) *Payload data transmission.* Finally, MU-MIMO transmission can take place for the remainder of the fading coherence block, i.e., for $(1 - \alpha - \alpha_{\text{fb}} - \alpha_{\text{d}}) N_{\text{c}}$ symbols. The uth user receives

$$y_u = \sqrt{G_u} \boldsymbol{H}_u \sum_{\mathrm{u}=0}^{U-1} \sqrt{\frac{E_{\mathrm{u}}}{N_{\text{t}}}} \hat{\hat{\boldsymbol{F}}}_{\mathrm{u}}^{\text{ZF}} s_{\mathrm{u}} + v_u \tag{9.133}$$

$$= \sqrt{G_u \frac{E_{\text{s}}}{U N_{\text{t}}}} \, a_{u,u} s_u + \sum_{\mathrm{u} \neq u} \sqrt{G_u \frac{E_{\text{s}}}{U N_{\text{t}}}} \, a_{u,\mathrm{u}} s_{\mathrm{u}} + v_u, \tag{9.134}$$

where, because of the uniform power allocation, $E_u = E_{\text{s}}/U$ for $u = 0, \ldots, U-1$. To process (9.134), the uth user has access to the precoded-pilot observations in (9.131).

Spectral efficiency

To determine the maximum spectral efficiency that can be achieved reliably by the uth user, we need to evaluate the mutual information between s_u and the outputs: y_u in (9.134) and $y_{u,0}, \ldots, y_{u,U-1}$ in (9.131). That is, we need to evaluate

$$I(s_u; y_u, y_{u,0}, \ldots, y_{u,U-1}) = I(s_u; y_u | y_{u,0}, \ldots, y_{u,U-1}) + I(s_u; y_{u,0}, \ldots, y_{u,U-1}) \tag{9.135}$$

$$= I(s_u; y_u | y_{u,0}, \ldots, y_{u,U-1}), \tag{9.136}$$

where (9.135) follows from the chain rule of mutual information and (9.136) from the independence between the transmitted data and the precoded-pilot observations. As intuition would have it, then, what we need to evaluate is the mutual information between s_u and y_u with the precoded-pilot observations as (the only) side information, and indeed (9.136) is the MU-MIMO operational counterpart to the single-user expression in (4.208).

Without the possibility of conditioning on anything other than the precoded-pilot observations, and in particular without the possibility of conditioning on the fading, the computation of (9.136) is challenging. Fortunately, as in single-user communication, we can reduce this quantity to another that is both more tractable and more representative of how wireless systems operate, and that further serves as a lower bound to (9.136). This entails having the uth user form estimates of its intended channel coefficient $a_{u,u}$ on the basis of only the precoded-pilot observation $y_{u,u}$ in (9.131), in particular the MMSE estimate

$$\hat{a}_{u,u} = \mathbb{E}\big[a_{u,u} \,|y_{u,u}\big] \tag{9.137}$$

$$= \frac{\mathbb{E}\big[a_{u,u}\, y^*_{u,u}\big]}{\mathbb{E}\big[|y_{u,u}|^2\big]} y_{u,u} \tag{9.138}$$

$$= \frac{\sqrt{\alpha_{\mathrm{d}} \frac{N_{\mathrm{c}}}{U} N_{\mathrm{t}} G_u E_{\mathrm{s}}}}{N_0 + \alpha_{\mathrm{d}} \frac{N_{\mathrm{c}}}{U} G_u E_{\mathrm{s}}} y_{u,u}, \tag{9.139}$$

where (9.138) holds because $a_{u,u}$ is complex Gaussian. Further to this Gaussian nature, $\hat{a}_{u,u} \sim \mathcal{N}_{\mathbb{C}}(0, N_{\mathrm{t}}(1 - \mathsf{MMSE}_{u,u}))$ and $\tilde{a}_{u,u} \sim \mathcal{N}_{\mathbb{C}}(0, N_{\mathrm{t}}\, \mathsf{MMSE}_{u,u})$ with

$$\mathsf{MMSE}_{u,u} = \frac{1}{1 + \alpha_{\mathrm{d}} \frac{N_{\mathrm{c}}}{U} \mathsf{SNR}_u}, \tag{9.140}$$

and we can write $a_{u,u} = \hat{a}_{u,u} + \tilde{a}_{u,u}$ where $\hat{a}_{u,u}$ and $\tilde{a}_{u,u}$ are independent.

The uth user then regards $\hat{a}_{u,u}$ as its true channel coefficient, allowing for (9.134) to be rewritten as

$$y_u = \sqrt{\frac{G_u E_{\mathrm{s}}}{U N_{\mathrm{t}}}}\, \hat{a}_{u,u} s_u + \sqrt{\frac{G_u E_{\mathrm{s}}}{U N_{\mathrm{t}}}}\, \tilde{a}_{u,u} s_u + \sum_{\mathrm{u} \neq u} \sqrt{\frac{G_u E_{\mathrm{s}}}{U N_{\mathrm{t}}}}\, a_{u,\mathrm{u}} s_{\mathrm{u}} + v_u \tag{9.141}$$

$$= \sqrt{\frac{G_u E_{\mathrm{s}}}{U N_{\mathrm{t}}}}\, \hat{a}_{u,u} s_u + v'_u, \tag{9.142}$$

where

$$v'_u = \sqrt{\frac{G_u E_{\mathrm{s}}}{U N_{\mathrm{t}}}}\, \tilde{a}_{u,u} s_u + \sum_{\mathrm{u} \neq u} \sqrt{\frac{G_u E_{\mathrm{s}}}{U N_{\mathrm{t}}}}\, a_{u,\mathrm{u}} s_{\mathrm{u}} + v_u \tag{9.143}$$

incorporates the own-channel estimation errors, the other-stream interference, and the noise. While v'_u is uncorrelated with the term of interest to the uth user, $\hat{a}_{u,u} s_u$, the distribution of v'_u is not Gaussian. Nonetheless, invoking the saddle-point property of the Gaussian distribution in terms of the mutual information, we can obtain a spectral efficiency achievable with minimum-distance decoding by replacing v'_u with Gaussian noise of the same variance, i.e., of variance

$$\mathbb{E}\left[|v'_u|^2 \,|\hat{a}_{u,u}\right] = \frac{G_u E_s}{U} \,\mathsf{MMSE}_{u,u} + \sum_{u \neq u} \frac{G_u E_s}{U N_t} \mathbb{E}\left[|a_{u,u}|^2 \,|\hat{a}_{u,u}\right] + N_0 \tag{9.144}$$

where the conditioning on $\hat{a}_{u,u}$ reflects the fact that the receiver knows it and regards it as the true channel coefficient. With that, (9.142) leads to

$$I(s_u; y_u|\hat{a}_{u,u}) = \log_2\left(1 + \frac{|\hat{a}_{u,u}|^2}{N_t\,(1 - \mathsf{MMSE}_{u,u})}\,\mathsf{sinr}_{\mathsf{eff},u}(\hat{a}_{u,u})\right), \tag{9.145}$$

with $\frac{|\hat{a}_{u,u}|^2}{N_t(1-\mathsf{MMSE}_{u,u})}$ having unit power, such that the effective conditional average SINR is given by

$$\mathsf{sinr}_{\mathsf{eff},u}(\hat{a}_{u,u}) = \frac{\frac{G_u E_s}{U}\,(1 - \mathsf{MMSE}_{u,u})}{\mathbb{E}\left[|v'_u|^2 \,|\hat{a}_{u,u}\right]} \tag{9.146}$$

$$= \frac{\frac{G_u E_s}{U}\,(1 - \mathsf{MMSE}_{u,u})}{\frac{G_u E_s}{U}\,\mathsf{MMSE}_{u,u} + \sum_{u \neq u} \frac{G_u E_s}{U N_t} \mathbb{E}\left[|a_{u,u}|^2 \,|\hat{a}_{u,u}\right] + N_0} \tag{9.147}$$

$$= \frac{\frac{1}{U}\mathsf{SNR}_u\,(1 - \mathsf{MMSE}_{u,u})}{1 + \frac{1}{U}\mathsf{SNR}_u \mathsf{MMSE}_{u,u} + \sum_{u \neq u} \frac{\mathsf{SNR}_u}{N_t U} \mathbb{E}\left[|a_{u,u}|^2 \,|\hat{a}_{u,u}\right]}. \tag{9.148}$$

From (9.145), taking an outer expectation over the distribution of $\hat{a}_{u,u}$ and accounting for the fact that a share $(1 - \alpha - \alpha_{\mathsf{fb}} - \alpha_{\mathsf{d}})$ of the symbols are available for payload data transmission—the rest being consumed by unprecoded and precoded pilot transmissions and by feedback—the ergodic spectral efficiency achievable by the uth user is

$$\frac{R_u}{B} = (1 - \alpha - \alpha_{\mathsf{fb}} - \alpha_{\mathsf{d}})\,\mathbb{E}\left[\log_2\left(1 + \frac{\mathsf{sinr}_{\mathsf{eff},u}(\hat{a}_{u,u})}{N_t\,(1 - \mathsf{MMSE}_{u,u})}\,|\hat{a}_{u,u}|^2\right)\right], \tag{9.149}$$

with $\mathsf{sinr}_{\mathsf{eff},u}(\hat{a}_{u,u})$ and $\mathsf{MMSE}_{u,u}$ as given in (9.148) and (9.140), respectively. The factor $(1 - \alpha - \alpha_{\mathsf{fb}} - \alpha_{\mathsf{d}})$ is common to every user and the weighted sum spectral efficiency is given by $\sum_{u=0}^{U-1} q_u R_u / B$.

Example 9.26

How does the spectral efficiency in (9.149) behave as the fading coherence grows?

Solution

Let $N_c \to \infty$. Since α, α_{fb}, and α_{d} depend on the number of users and number of antennas, but not on N_c, for $N_c \to \infty$ the overhead vanishes. Also, from (9.140), $\mathsf{MMSE}_{u,u} \to 0$ and consequently $\hat{a}_{u,u} \to a_{u,u} \sim \mathcal{N}_{\mathbb{C}}(0, N_t)$. Altogether,

$$\frac{R_u}{B} \to \mathbb{E}\left[\log_2\left(1 + \frac{\mathsf{sinr}_{\mathsf{eff},u}(a_{u,u})}{N_t}\,|a_{u,u}|^2\right)\right] \tag{9.150}$$

$$= \mathbb{E}\left[\log_2\left(1 + \frac{\frac{\mathsf{SNR}_u}{UN_t}}{1+\sum_{\mathrm{u}\neq u}\frac{\mathsf{SNR}_u}{UN_t}\mathbb{E}\left[|a_{u,\mathrm{u}}|^2\,|a_{u,u}\right]}|a_{u,u}|^2\right)\right]. \tag{9.151}$$

From (9.129), $\mathsf{MMSE}_u \to 0$ and therefore $\hat{\boldsymbol{H}}_u \to \boldsymbol{H}_u$ meaning that the channel estimates obtained by the users from the unprecoded pilot transmissions become perfect. The feedback process then maps $\boldsymbol{H}_u$ to $\hat{\hat{\boldsymbol{H}}}_u$ and, assuming this mapping can itself become perfect for $N_c \to \infty$, then the true ZF precoders are implemented. It follows that the interference leakage vanishes ($\mathbb{E}\left[|a_{u,\mathrm{u}}|^2\right] \to 0$), giving

$$\frac{R_u}{B} = \mathbb{E}\left[\log_2\left(1+\frac{\mathsf{SNR}_u}{UN_t}|a_{u,u}|^2\right)\right] \tag{9.152}$$

$$= \mathbb{E}\left[\log_2\left(1+\frac{\mathsf{SNR}_u}{UN_t}|\boldsymbol{H}_u\boldsymbol{F}_u^{\mathsf{ZF}}|^2\right)\right], \tag{9.153}$$

which coincides with the CSI-based spectral efficiency in (9.91), specialized to a uniform power allocation.

The pilot-assisted multistage procedure that we are considering is, as shown by the foregoing example, asymptotically optimal in the fading coherence under the reasonable assumption that the feedback process is itself asymptotically optimal. This pilot-assisted multistage procedure is therefore a reasonable basis to establish the robustness of ZF transmission in the MU-MISO BC. For finite N_c, degradation arises in the following respects:

- Strictly positive overheads, $\alpha, \alpha_{\mathsf{fb}}, \alpha_{\mathsf{d}} > 0$.
- Strictly positive $\mathsf{MMSE}_{u,u}$, which diminishes both $\mathsf{sinr}_{\mathsf{eff},u}$ and the power of the intended channel coefficient, $\mathbb{E}\left[|\hat{a}_{u,u}|^2\right]$.
- Strictly positive conditional interference power, $\mathbb{E}\left[|a_{u,\mathrm{u}}|^2\,|\hat{a}_{u,u}\right]$.

While the first two effects are tractable directly from (9.149), the third one is a serious obstacle owing to the intricate dependences between $a_{u,\mathrm{u}}$ and $\hat{a}_{u,u}$. In order to circumvent this obstacle and drive home the analysis, we follow the footsteps of [830, section III] and proceed to lower-bound (9.149).

Spectral efficiency lower bound

Plugging into (9.149) the expression for $\mathsf{sinr}_{\mathsf{eff},u}(\cdot)$ obtained in (9.148), we can write

$$\frac{R_u}{B} = (1-\alpha-\alpha_{\mathsf{fb}}-\alpha_{\mathsf{d}})\Bigg(\mathbb{E}\Bigg[\log_2\Bigg(1+\frac{\mathsf{SNR}_u}{U}\mathsf{MMSE}_{u,u} + \sum_{\mathrm{u}\neq u}\frac{\mathsf{SNR}_u}{UN_t}\mathbb{E}\left[|a_{u,\mathrm{u}}|^2\,|\hat{a}_{u,u}\right]+\frac{\mathsf{SNR}_u}{UN_t}|\hat{a}_{u,u}|^2\Bigg)\Bigg]$$

$$-\mathbb{E}\left[\log_2\left(1+\frac{\mathsf{SNR}_u}{U}\mathsf{MMSE}_{u,u}+\sum_{\mathrm{u}\neq u}\frac{\mathsf{SNR}_u}{UN_t}\mathbb{E}\left[|a_{u,\mathrm{u}}|^2\,|\hat{a}_{u,u}\right]\right)\right]\Bigg) \tag{9.154}$$

$$\geq (1-\alpha-\alpha_{\mathsf{fb}}-\alpha_{\mathsf{d}})\Bigg(\mathbb{E}\bigg[\log_2\bigg(1+\frac{\mathsf{SNR}_u}{U}\mathsf{MMSE}_{u,u}+\frac{\mathsf{SNR}_u}{UN_\mathrm{t}}|\hat{a}_{u,u}|^2\bigg)\bigg]$$

$$-\mathbb{E}\bigg[\log_2\bigg(1+\frac{\mathsf{SNR}_u}{U}\mathsf{MMSE}_{u,u}+\sum_{u\neq u}\frac{\mathsf{SNR}_u}{UN_\mathrm{t}}\mathbb{E}\Big[|a_{u,u}|^2\,|\hat{a}_{u,u}\Big]\bigg)\bigg]\Bigg) \quad (9.155)$$

$$=(1-\alpha-\alpha_{\mathsf{fb}}-\alpha_{\mathsf{d}})\Bigg(\mathbb{E}\bigg[\log_2\bigg(1+\frac{\mathsf{SNR}_u}{UN_\mathrm{t}}\Big(N_\mathrm{t}\,\mathsf{MMSE}_{u,u}+|\hat{a}_{u,u}|^2\Big)\bigg)\bigg]$$

$$-\mathbb{E}\bigg[\log_2\bigg(1+\frac{\mathsf{SNR}_u}{U}\mathsf{MMSE}_{u,u}+\sum_{u\neq u}\frac{\mathsf{SNR}_u}{UN_\mathrm{t}}\mathbb{E}\Big[|a_{u,u}|^2\,|\hat{a}_{u,u}\Big]\bigg)\bigg]\Bigg) \quad (9.156)$$

$$\geq(1-\alpha-\alpha_{\mathsf{fb}}-\alpha_{\mathsf{d}})\Bigg(\mathbb{E}\bigg[\log_2\bigg(1+\frac{\mathsf{SNR}_u}{UN_\mathrm{t}}\Big(N_\mathrm{t}\,\mathsf{MMSE}_{u,u}\,|z|^2+|\hat{a}_{u,u}|^2\Big)\bigg)\bigg]$$

$$-\mathbb{E}\bigg[\log_2\bigg(1+\frac{\mathsf{SNR}_u}{U}\mathsf{MMSE}_{u,u}+\sum_{u\neq u}\frac{\mathsf{SNR}_u}{UN_\mathrm{t}}\mathbb{E}\Big[|a_{u,u}|^2\,|\hat{a}_{u,u}\Big]\bigg)\bigg]\Bigg), \quad (9.157)$$

where in (9.155) a positive term has been dropped and in (9.157) we have introduced an independent variable $z \sim \mathcal{N}_{\mathbb{C}}(0,1)$ that reduces the expectation of the corresponding logarithm. Indeed, from Jensen's inequality,

$$\mathbb{E}\bigg[\log_2\bigg(1+\frac{\mathsf{SNR}_u}{UN_\mathrm{t}}\Big(N_\mathrm{t}\,\mathsf{MMSE}_{u,u}\,|z|^2+|\hat{a}_{u,u}|^2\Big)\bigg)\bigg]$$

$$\leq\mathbb{E}\bigg[\log_2\bigg(1+\frac{\mathsf{SNR}_u}{UN_\mathrm{t}}\Big(N_\mathrm{t}\,\mathsf{MMSE}_{u,u}\,\underbrace{\mathbb{E}\big[|z|^2\big]}_{1}+|\hat{a}_{u,u}|^2\Big)\bigg)\bigg], \quad (9.158)$$

as applied in (9.157). And, since $\hat{a}_{u,u} \sim \mathcal{N}_{\mathbb{C}}(0, N_\mathrm{t}(1-\mathsf{MMSE}_{u,u}))$ as derived earlier, we have that

$$\sqrt{N_\mathrm{t}\,\mathsf{MMSE}_{u,u}}\,z+\hat{a}_{u,u} \sim \mathcal{N}_{\mathbb{C}}(0,N_\mathrm{t}) \quad (9.159)$$

and consequently

$$\mathbb{E}\bigg[\log_2\bigg(1+\frac{\mathsf{SNR}_u}{UN_\mathrm{t}}\Big(N_\mathrm{t}\,\mathsf{MMSE}_{u,u}\,|z|^2+|\hat{a}_{u,u}|^2\Big)\bigg)\bigg]$$

$$=\mathbb{E}\bigg[\log_2\bigg(1+\frac{\mathsf{SNR}_u}{UN_\mathrm{t}}\Big|\sqrt{N_\mathrm{t}\,\mathsf{MMSE}_{u,u}}\,z+\hat{a}_{u,u}\Big|^2\bigg)\bigg] \quad (9.160)$$

$$=C_u^{\mathsf{ZF}}(\mathsf{SNR}_u), \quad (9.161)$$

where $C_u^{\mathsf{ZF}}(\cdot)$, the CSI-based spectral efficiency function for ZF transmission with uniform power allocation, is obtained because the argument of (9.160) has the same distribution as that of (9.153); for IID Rayleigh fading, this function admits the closed form in (9.95). Altogether, (9.157) can be rewritten as

$$\frac{R_u}{B}\geq(1-\alpha-\alpha_{\mathsf{fb}}-\alpha_{\mathsf{d}})\Bigg(C_u^{\mathsf{ZF}}(\mathsf{SNR}_u)$$

$$- \mathbb{E}\left[\log_2\left(1 + \frac{\mathsf{SNR}_u}{U}\mathsf{MMSE}_{u,u} + \sum_{\mathrm{u}\neq u} \frac{\mathsf{SNR}_u}{U N_\mathrm{t}} \mathbb{E}\left[|a_{u,\mathrm{u}}|^2 \, |\hat{a}_{u,u}\right]\right)\right] \tag{9.162}$$

and, applying Jensen's inequality again,

$$\frac{R_u}{B} \geq (1-\alpha-\alpha_\mathrm{fb}-\alpha_\mathrm{d})\Bigg(C_u^{\mathsf{ZF}}(\mathsf{SNR}_u) \tag{9.163}$$

$$- \log_2\left(\mathbb{E}\left[1 + \frac{\mathsf{SNR}_u}{U}\mathsf{MMSE}_{u,u} + \sum_{\mathrm{u}\neq u} \frac{\mathsf{SNR}_u}{U N_\mathrm{t}} \mathbb{E}\left[|a_{u,\mathrm{u}}|^2 \, |\hat{a}_{u,u}\right]\right]\right)\Bigg)$$

$$- (1-\alpha-\alpha_\mathrm{fb}-\alpha_\mathrm{d})\Bigg(C_u^{\mathsf{ZF}}(\mathsf{SNR}_u)$$

$$- \log_2\left(1 + \frac{\mathsf{SNR}_u}{U}\mathsf{MMSE}_{u,u} + \sum_{\mathrm{u}\neq u} \frac{\mathsf{SNR}_u}{U N_\mathrm{t}} \mathbb{E}\left[|a_{u,\mathrm{u}}|^2\right]\right)\Bigg), \tag{9.164}$$

where $\mathbb{E}\left[|a_{u,\mathrm{u}}|^2\right]$ has become disentangled from $\hat{a}_{u,u}$, thereby facilitating its computation. Recalling the definitions of $\mathsf{MMSE}_{u,u}$ and $a_{u,\mathrm{u}}$, finally,

$$\frac{R_u}{B} \geq (1-\alpha-\alpha_\mathrm{fb}-\alpha_\mathrm{d})\Bigg[C_u^{\mathsf{ZF}}(\mathsf{SNR}_u)$$

$$- \log_2\left(1 + \frac{\mathsf{SNR}_u}{U + \alpha_\mathrm{d} N_\mathrm{c}\,\mathsf{SNR}_u} + \sum_{\mathrm{u}\neq u} \frac{\mathsf{SNR}_u}{U N_\mathrm{t}} \mathbb{E}\left[|\boldsymbol{H}_u \hat{\boldsymbol{F}}_\mathrm{u}^{\mathsf{ZF}}|^2\right]\right)\Bigg], \tag{9.165}$$

which lower bounds the spectral efficiency in complete generality as far as the duplexing is concerned. It can be verified that the bound is tight for $N_\mathrm{c} \to \infty$, returning $C_u^{\mathsf{ZF}}(\mathsf{SNR}_u)$.

Even in complete generality as far as the duplexing is concerned, some preliminary observations can be made from (9.165).

- There is an immediate decrease factor of $(1-\alpha-\alpha_\mathrm{fb}-\alpha_\mathrm{d})$ in the number of spatial DOF per user, and hence in the total number of spatial DOF.
- Decrease in DOF aside, the second term in (9.165) directly bounds the losses in spectral efficiency.

This second observation raises the interest in assessing how the term $\mathbb{E}\left[|\boldsymbol{H}_u \hat{\boldsymbol{F}}_\mathrm{u}^{\mathsf{ZF}}|^2\right]$ behaves and, particularly, how it behaves as a function of SNR_u. From a cursory inspection of (9.165) it is seen—in concord with the basic result in Section 8.10—that, if this term decays at least as fast as $1/\mathsf{SNR}_u$, then the power loss is sure to be in the form of only a shift in the power offset. Otherwise, the power loss swells without bound and there could be an additional decrease in the number of spatial DOF. To establish the behavior of this term, though, it becomes necessary to postulate a specific duplexing scheme.

FDD

We begin with FDD, decidedly the most adverse type of duplexing for the matter at hand, and, for the purpose that occupies us here, we consider analog feedback. A digital implementation of the feedback would entail extending to the multiuser realm the SU-MIMO concepts in Section 5.10, and readers interested in this extension are directed to [831–849].

With analog feedback, user u feeds back transformed versions of either the cumulative observation $\boldsymbol{y}_u$ in (9.125) or of the ensuing channel estimate $\hat{\boldsymbol{H}}_u$. For the sake of specificity, let us posit that user u sends back $\boldsymbol{y}_u$, possibly repeated and scaled according to the available feedback power. Let us begin by considering that the feedback transmissions from the U users are orthogonally multiplexed, such that each one extends over $\alpha_{\mathsf{fb}} \frac{N_\mathrm{c}}{U}$ symbols per block, leaving $\alpha_{\mathsf{fb}} \frac{N_\mathrm{c}}{N_\mathrm{t} U}$ feedback symbols to repeat each of the N_t entries of $\boldsymbol{y}_u$. Moreover, let us momentarily consider that the feedback transmissions are subject to noise, but not to fading. Introducing additional $(\cdot)^\mathsf{r}$ superscripting to distinguish reverse-link quantities from their BC counterparts, the feedback transmission of $\boldsymbol{y}_u$ on the part of user u is observed back at the base station, after accumulation over $\alpha_{\mathsf{fb}} \frac{N_\mathrm{c}}{U N_\mathrm{t}}$ repetitions and assembly of the N_t entries into a vector, as

$$\boldsymbol{y}_u^\mathsf{r} = \frac{\sqrt{\alpha_{\mathsf{fb}} \frac{N_\mathrm{c}}{U N_\mathrm{t}} N_\mathrm{t} G_u E_\mathrm{s}^\mathsf{r}}}{\sqrt{\frac{1}{N_\mathrm{t}} \mathbb{E}\left[\|\boldsymbol{y}_u\|^2\right]}} \boldsymbol{y}_u + \boldsymbol{v}_u^\mathsf{r} \tag{9.166}$$

where the denominator scales down to unity the variance of each entry of $\boldsymbol{y}_u$ while:

- E_s^r scales it back up to the available reverse-link power.
- G_u is the large-scale power gain of the reverse link of user u, identical to that of the corresponding forward link, with an accompanying factor N_t because each feedback symbol is picked up by N_t antennas back at the base station.
- $\alpha_{\mathsf{fb}} \frac{N_\mathrm{c}}{U N_\mathrm{t}}$ amasses power over the repetitions of each entry.

Recalling (9.125), the above unfolds into

$$\boldsymbol{y}_u^\mathsf{r} = \frac{\sqrt{\alpha_{\mathsf{fb}} \frac{N_\mathrm{c}}{U} G_u E_\mathrm{s}^\mathsf{r}}}{\sqrt{\alpha \frac{N_\mathrm{c}}{N_\mathrm{t}} G_u E_\mathrm{s} + N_0}} \boldsymbol{y}_u + \boldsymbol{v}_u^\mathsf{r} \tag{9.167}$$

$$= \frac{G_u \sqrt{\alpha_{\mathsf{fb}} \alpha \frac{N_\mathrm{c}^2}{U N_\mathrm{t}} E_\mathrm{s} E_\mathrm{s}^\mathsf{r}}}{\sqrt{\alpha \frac{N_\mathrm{c}}{N_\mathrm{t}} G_u E_\mathrm{s} + N_0}} \boldsymbol{H}_u^\mathrm{T} + \underbrace{\frac{\sqrt{\alpha_{\mathsf{fb}} \frac{N_\mathrm{c}}{U} G_u E_\mathrm{s}^\mathsf{r}}}{\sqrt{\alpha \frac{N_\mathrm{c}}{N_\mathrm{t}} G_u E_\mathrm{s} + N_0}} \boldsymbol{v}_u + \boldsymbol{v}_u^\mathsf{r}}_{\text{Total noise, } \boldsymbol{v}_u'}, \tag{9.168}$$

where, given the independence and IID complex Gaussian nature of $\boldsymbol{v}_u$ and $\boldsymbol{v}_u^\mathsf{r}$, the total noise is $\boldsymbol{v}_u' \sim \mathcal{N}_\mathbb{C}(\boldsymbol{0}, \sigma^2 \boldsymbol{I})$ with

$$\sigma^2 = \left(1 + \frac{\alpha_{\mathsf{fb}} \frac{N_\mathrm{c}}{U} \mathsf{SNR}_u^\mathsf{r}}{1 + \alpha \frac{N_\mathrm{c}}{N_\mathrm{t}} \mathsf{SNR}_u}\right) N_0, \tag{9.169}$$

where we have introduced the reverse-link local-average SNR

$$\mathsf{SNR}_u^\mathsf{r} = \frac{G_u E_\mathrm{s}^\mathsf{r}}{N_0}. \tag{9.170}$$

From $\boldsymbol{y}_u^{\mathrm{r}}$, the sought estimate $\hat{\hat{\boldsymbol{H}}}_u$ can be derived via

$$\hat{\hat{\boldsymbol{H}}}_u^{\mathrm{T}} = \mathbb{E}\left[\boldsymbol{H}_u^{\mathrm{T}} | \boldsymbol{y}_u^{\mathrm{r}}\right] \tag{9.171}$$

$$= \mathbb{E}\left[\boldsymbol{H}_u^{\mathrm{T}} \boldsymbol{y}_u^{\mathrm{r}*}\right] \left(\mathbb{E}\left[\boldsymbol{y}_u^{\mathrm{r}} \boldsymbol{y}_u^{\mathrm{r}*}\right]\right)^{-1} \boldsymbol{y}_u^{\mathrm{r}} \tag{9.172}$$

$$= \frac{\frac{N_{\mathrm{c}}}{N_{\mathrm{t}}}}{\sqrt{N_0}} \frac{\sqrt{\alpha_{\mathrm{fb}} \alpha \frac{N_{\mathrm{t}}}{U} \mathsf{SNR}_u \mathsf{SNR}_u^{\mathrm{r}}}}{\left(1 + \alpha_{\mathrm{fb}} \frac{N_{\mathrm{c}}}{U} \mathsf{SNR}_u^{\mathrm{r}}\right) \sqrt{1 + \alpha \frac{N_{\mathrm{c}}}{N_{\mathrm{t}}} \mathsf{SNR}_u}} \boldsymbol{y}_u^{\mathrm{r}}, \tag{9.173}$$

where (9.172) holds because $\boldsymbol{H}_u$ is complex Gaussian while (9.173) emerges after a bit of algebra. The reader is invited to verify this step in Problem 9.25.

As in earlier occasions, we can write $\boldsymbol{H}_u = \hat{\hat{\boldsymbol{H}}}_u + \tilde{\tilde{\boldsymbol{H}}}_u$ where $\tilde{\tilde{\boldsymbol{H}}}_u$ is independent of $\hat{\hat{\boldsymbol{H}}}_u$ and $\tilde{\tilde{\boldsymbol{H}}}_u \sim \mathcal{N}_{\mathbb{C}}(\boldsymbol{0}, \sigma_{\mathrm{e}}^2 \boldsymbol{I})$ with

$$\sigma_{\mathrm{e}}^2 = \frac{1}{1 + \alpha_{\mathrm{fb}} \frac{N_{\mathrm{c}}}{U} \mathsf{SNR}_u^{\mathrm{r}}} \left(1 + \frac{\alpha_{\mathrm{fb}} \frac{N_{\mathrm{c}}}{U} \mathsf{SNR}_u^{\mathrm{r}}}{1 + \alpha \frac{N_{\mathrm{c}}}{N_{\mathrm{t}}} \mathsf{SNR}_u}\right). \tag{9.174}$$

Armed with the foregoing expressions, we are finally in a position to establish the behavior of $\mathbb{E}\left[|\boldsymbol{H}_u \hat{\hat{\boldsymbol{F}}}_{\mathrm{u}}^{\mathrm{ZF}}|^2\right]$ in (9.165). Precisely,

$$\mathbb{E}\left[\left|\boldsymbol{H}_u \hat{\hat{\boldsymbol{F}}}_{\mathrm{u}}^{\mathrm{ZF}}\right|^2\right] = \mathbb{E}\left[\left|\left(\hat{\hat{\boldsymbol{H}}}_u + \tilde{\tilde{\boldsymbol{H}}}_u\right) \hat{\hat{\boldsymbol{F}}}_{\mathrm{u}}^{\mathrm{ZF}}\right|^2\right] \tag{9.175}$$

$$= \mathbb{E}\left[\left|\tilde{\tilde{\boldsymbol{H}}}_u \hat{\hat{\boldsymbol{F}}}_{\mathrm{u}}^{\mathrm{ZF}}\right|^2\right] \tag{9.176}$$

$$= \mathbb{E}\left[\hat{\hat{\boldsymbol{F}}}_{\mathrm{u}}^{\mathrm{ZF}*} \tilde{\tilde{\boldsymbol{H}}}_u^* \tilde{\tilde{\boldsymbol{H}}}_u \hat{\hat{\boldsymbol{F}}}_{\mathrm{u}}^{\mathrm{ZF}}\right] \tag{9.177}$$

$$= \mathbb{E}\left[\hat{\hat{\boldsymbol{F}}}_{\mathrm{u}}^{\mathrm{ZF}*} \, \mathbb{E}\left[\tilde{\tilde{\boldsymbol{H}}}_u^* \tilde{\tilde{\boldsymbol{H}}}_u\right] \hat{\hat{\boldsymbol{F}}}_{\mathrm{u}}^{\mathrm{ZF}}\right] \tag{9.178}$$

$$= \sigma_{\mathrm{e}}^2 \, \mathbb{E}\left[\|\hat{\hat{\boldsymbol{F}}}_{\mathrm{u}}^{\mathrm{ZF}}\|^2\right] \tag{9.179}$$

$$= N_{\mathrm{t}} \, \sigma_{\mathrm{e}}^2, \tag{9.180}$$

where we have exploited that, for $U = N_{\mathrm{t}}$, $\hat{\hat{\boldsymbol{F}}}_{\mathrm{u}}^{\mathrm{ZF}}$ is both orthogonal to $\hat{\hat{\boldsymbol{H}}}_u$ and independent of $\tilde{\tilde{\boldsymbol{H}}}_u$, and in (9.180) we have further invoked the precoder normalization.

By means of (9.180) and (9.174), the general lower bound in (9.165) can now be particularized to FDD with analog feedback, giving

$$\frac{R_u}{B} \geq (1 - \alpha - \alpha_{\mathrm{fb}} - \alpha_{\mathrm{d}}) \Bigg[C_u^{\mathrm{ZF}}(\mathsf{SNR}_u) - \log_2\left(1 + \frac{\frac{1}{U}\mathsf{SNR}_u}{1 + \alpha_{\mathrm{d}} \frac{N_{\mathrm{c}}}{U} \mathsf{SNR}_u} + \frac{\left(1 - \frac{1}{U}\right) \mathsf{SNR}_u}{1 + \alpha_{\mathrm{fb}} \frac{N_{\mathrm{c}}}{U} \mathsf{SNR}_u^{\mathrm{r}}} \left(1 + \frac{\alpha_{\mathrm{fb}} \frac{N_{\mathrm{c}}}{U} \mathsf{SNR}_u^{\mathrm{r}}}{1 + \alpha \frac{N_{\mathrm{c}}}{N_{\mathrm{t}}} \mathsf{SNR}_u}\right)\right)\Bigg]. \tag{9.181}$$

The final step in the characterization of the performance of pilot-assisted transmission is to optimize the overheads α, α_{fb}, and α_{d}, yielding a maximized bound that depends only on SNR_u, $\mathsf{SNR}_u^{\mathrm{r}}$, $U = N_{\mathrm{t}}$, and N_{c}. Specifically, the optimization of the weighted sum spectral efficiency is formulated as

$$\max_{\alpha,\alpha_{\mathrm{fb}},\alpha_{\mathrm{d}}} (1-\alpha-\alpha_{\mathrm{fb}}-\alpha_{\mathrm{d}}) \sum_{u=0}^{U-1} q_u \Bigg[C_u^{\mathsf{ZF}}(\mathsf{SNR}_u) \tag{9.182}$$

$$- \log_2 \left(1 + \frac{\frac{1}{U}\mathsf{SNR}_u}{1+\alpha_{\mathrm{d}}\frac{N_{\mathrm{c}}}{U}\mathsf{SNR}_u} + \frac{(1-\frac{1}{U})\,\mathsf{SNR}_u}{1+\alpha_{\mathrm{fb}}\frac{N_{\mathrm{c}}}{U}\mathsf{SNR}^{\mathsf{r}}_u} \left(1+\frac{\alpha_{\mathrm{fb}}\frac{N_{\mathrm{c}}}{U}\mathsf{SNR}^{\mathsf{r}}_u}{1+\alpha\frac{N_{\mathrm{c}}}{N_{\mathrm{t}}}\mathsf{SNR}_u}\right)\right)\Bigg]$$

with the constraints that $\alpha > 0$, $\alpha_{\mathrm{fb}} > 0$ and $\alpha_{\mathrm{d}} > 0$, as well as $\alpha + \alpha_{\mathrm{fb}} + \alpha_{\mathrm{d}} < 1$. This problem is convex and very affordable, but the function is involved and it is not easy to glean much analytical insight from it beyond some asymptotic behaviors that are explored in [850] and entertained later in this section. Furthermore, the optimization couples the performance of the various users through the common factor $(1-\alpha-\alpha_{\mathrm{fb}}-\alpha_{\mathrm{d}})$.

If all user SNRs are equal, then the optimization does decouple into per-user problems of the form

$$\max_{\alpha,\alpha_{\mathrm{fb}},\alpha_{\mathrm{d}}} (1-\alpha-\alpha_{\mathrm{fb}}-\alpha_{\mathrm{d}}) \Bigg[C_u^{\mathsf{ZF}}(\mathsf{SNR}_u) \tag{9.183}$$

$$- \log_2 \left(1 + \frac{\frac{1}{U}\mathsf{SNR}_u}{1+\alpha_{\mathrm{d}}\frac{N_{\mathrm{c}}}{U}\mathsf{SNR}_u} + \frac{(1-\frac{1}{U})\,\mathsf{SNR}_u}{1+\alpha_{\mathrm{fb}}\frac{N_{\mathrm{c}}}{U}\mathsf{SNR}^{\mathsf{r}}_u} \left(1+\frac{\alpha_{\mathrm{fb}}\frac{N_{\mathrm{c}}}{U}\mathsf{SNR}^{\mathsf{r}}_u}{1+\alpha\frac{N_{\mathrm{c}}}{N_{\mathrm{t}}}\mathsf{SNR}_u}\right)\right)\Bigg],$$

whose solution, while not general, is highly revealing. We thus choose to tackle this problem at this point, bracketing the range of values taken by the various overheads and by the ensuing bound by numerically solving (9.183) for extreme values of $U = N_{\mathrm{t}}$ and N_{c}.

Let us first consider the case of balanced forward and reverse transmit powers, such that $\mathsf{SNR}^{\mathsf{r}}_u = \mathsf{SNR}_u$, and revisit this condition a bit later.

Example 9.27

Let $U = N_{\mathrm{t}} = 2$ and $\mathsf{SNR}^{\mathsf{r}}_u = \mathsf{SNR}_u$. For fading coherences corresponding to vehicular and pedestrian users, solve (9.183). Plot the optimized lower bound alongside $C_u^{\mathsf{ZF}}(\mathsf{SNR}_u)$, which represents an upper bound to the spectral efficiency achievable with pilot-assisted ZF transmission; these two bounds demarcate the range where the actual ZF spectral efficiency lies. Further plot the various overheads, as well as their sum, as a function of SNR_u.

Solution

As calculated in Example 3.27, vehicular and pedestrian settings may correspond, respectively, to $N_{\mathrm{c}} = 1000$ and $N_{\mathrm{c}} = 20\,000$. For these fading coherences, the spectral efficiencies and overheads are presented in Figs. 9.7 and 9.8. On the left-hand side subfigures, the shaded regions indicate the range of spectral efficiencies comprised between the lower and upper bounds. On the right-hand side subfigures, the values of α, α_{d}, and α_{fb} that optimize the lower bound, as well as their sum, are depicted.

Example 9.28

Repeat Example 9.27 for $U = N_{\mathrm{t}} = 8$.

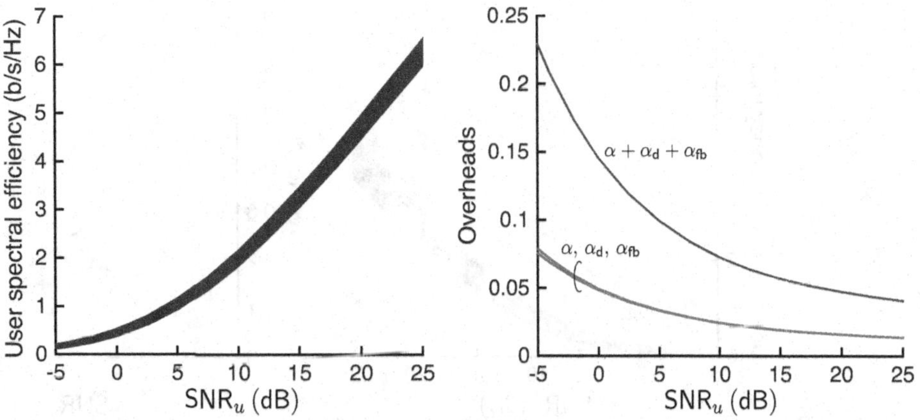

Fig. 9.7 Left, shaded region containing the spectral efficiency of a user as a function of its SNR (in dB) for pilot-assisted ZF transmission with FDD and with $U = N_t = 2$, $N_c = 1000$, and $\mathsf{SNR}_u^r = \mathsf{SNR}_u$. Right, corresponding overheads (α, α_d, and α_{fb}, as well as their sum) optimized for the spectral efficiency lower bound.

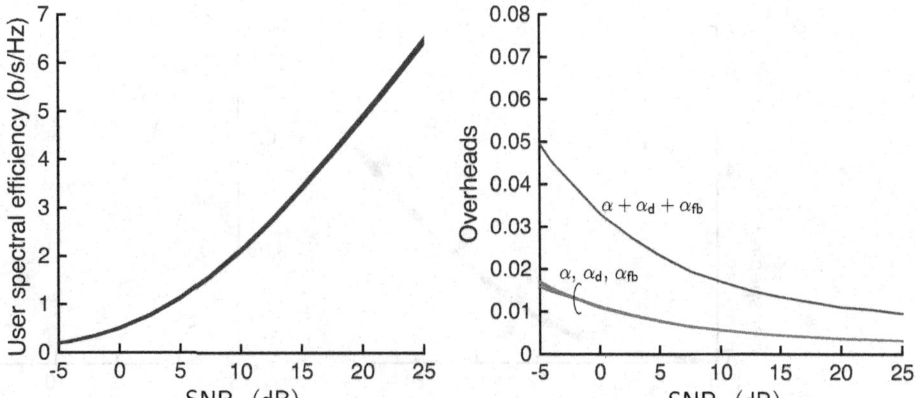

Fig. 9.8 Left, shaded region containing the spectral efficiency of a user as a function of its SNR (in dB) for pilot-assisted ZF transmission with FDD and with $U = N_t = 2$, $N_c = 20\,000$, and $\mathsf{SNR}_u^r = \mathsf{SNR}_u$. Right, corresponding overheads (α, α_d, and α_{fb}, as well as their sum) optimized for the spectral efficiency lower bound.

Solution

See Figs. 9.9 and 9.10.

From the foregoing examples, we observe the following.

- For pedestrian users, the total overhead is small (well below 10% for $U \leq 8$ in the medium-to-high-SNR range) and the reduction in spectral efficiency with respect to its CSI-based value is minute.

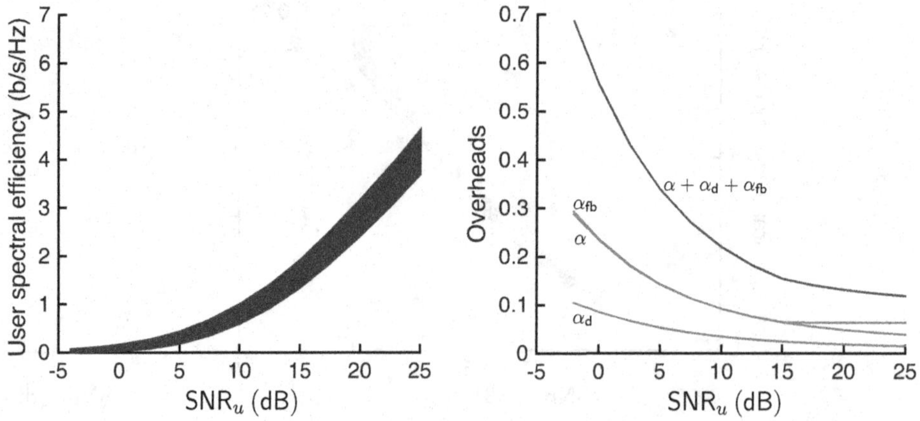

Fig. 9.9 Left, shaded region containing the spectral efficiency of a user as a function of its SNR (in dB) for pilot-assisted ZF transmission with FDD and with $U = N_t = 8$, $N_c = 1000$, and $\mathsf{SNR}_u^r = \mathsf{SNR}_u$. Right, corresponding overheads (α, α_d, and α_{fb}, as well as their sum) optimized for the spectral efficiency lower bound.

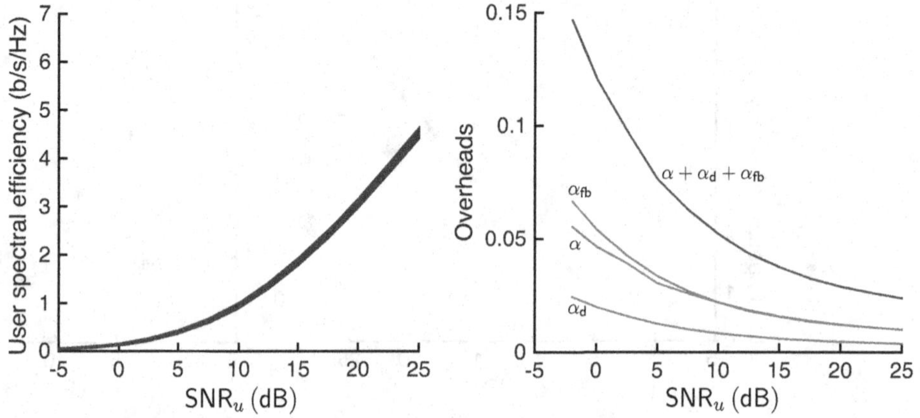

Fig. 9.10 Left, shaded region containing the spectral efficiency of a user as a function of its SNR (in dB) for pilot-assisted ZF transmission with FDD and with $U = N_t = 8$, $N_c = 20\,000$, and $\mathsf{SNR}_u^r = \mathsf{SNR}_u$. Right, corresponding overheads (α, α_d, and α_{fb}, as well as their sum) optimized for the spectral efficiency lower bound.

- For a reduced number of vehicular users, the total overhead is still small (again below 10% for $U = 2$) and the spectral efficiency continues to be largely preserved.
- Only under the combination of both vehicular conditions and a substantial number of users does the total overhead escalate and the spectral efficiency exhibit a noticeable decrease. Even here though, the degradation is far from catastrophic.
- The optimum overheads decrease monotonically with the SNR, a behavior already observed in pilot-assisted SU-MIMO and that is rooted in the fact that the cost of adding

further pilot and feedback symbols is in the form of a linear factor while the benefits from more precise channel and precoder estimates are sublinear. At sufficiently high SNR, fewer pilot and feedback symbols are sure to be preferable.

Following these examples, we are left with the reassurance that the performance of ZF transmission with FDD is robust and that the CSI analysis conducted earlier in the chapter is meaningful, with corrections that depend on the number of users and antennas and on the fading coherence. As argued earlier, ZF with large U might require—even with CSI and let alone without—that $N_{\mathrm{t}} > U$.

The repercussions of having $\mathsf{SNR}^{\mathrm{r}}_u \neq \mathsf{SNR}_u$ are considered next, as we seek to more sharply determine the corrections that the CSI-based spectral efficiency necessitates to properly represent the performance of pilot-assisted ZF transmission.

Reverse-link power asymmetry

It can be seen by inspecting (9.181) that, if $\mathsf{SNR}^{\mathrm{r}}_u$ does not grow with SNR_u, i.e., if for $\mathsf{SNR}_u \to \infty$ it holds that

$$\frac{\mathsf{SNR}^{\mathrm{r}}_u}{\mathsf{SNR}_u} \to 0 \tag{9.184}$$

then the spectral efficiency lower bound behaves poorly (refer to Problem 9.28). It being a lower bound, this does not necessarily spell doom for the actual spectral efficiency, but it supports the intuition developed in Section 8.10 that the feedback needs to become cleaner so as to pack richer CSI content as SNR_u grows and the forward estimates themselves become cleaner. Building on this intuition, let us allow SNR_u and $\mathsf{SNR}^{\mathrm{r}}_u$ to keep some positive ratio

$$\rho = \frac{E_{\mathrm{s}}}{E^{\mathrm{r}}_{\mathrm{s}}} \tag{9.185}$$

$$= \frac{\mathsf{SNR}_u}{\mathsf{SNR}^{\mathrm{r}}_u}. \tag{9.186}$$

This allows rewriting (9.181) as a function of solely SNR_u, namely

$$\frac{R_u}{B} \geq (1 - \alpha - \alpha_{\mathrm{fb}} - \alpha_{\mathrm{d}}) \Bigg[C^{\mathrm{ZF}}_u(\mathsf{SNR}_u) \tag{9.187}$$

$$- \log_2 \left(1 + \frac{\frac{1}{U}\mathsf{SNR}_u}{1 + \alpha_{\mathrm{d}} \frac{N_{\mathrm{c}}}{U} \mathsf{SNR}_u} + \frac{(1 - \frac{1}{U})\,\mathsf{SNR}_u}{1 + \alpha_{\mathrm{fb}} \frac{N_{\mathrm{c}}}{\rho U} \mathsf{SNR}_u} \left(1 + \frac{\alpha_{\mathrm{fb}} \frac{N_{\mathrm{c}}}{\rho U} \mathsf{SNR}_u}{1 + \alpha \frac{N_{\mathrm{c}}}{N_{\mathrm{t}}} \mathsf{SNR}_u} \right) \right) \Bigg],$$

which expression is an opportune point to transition the analysis to the high-SNR regime.

High-SNR regime

Recalling (9.113) and (9.116),

$$C^{\mathrm{ZF}}_u(\mathsf{SNR}_u) = \log_2 \mathsf{SNR}_u - \mathcal{L}^{\mathrm{ZF}}_{\infty,u} + \mathcal{O}\left(\frac{1}{\mathsf{SNR}_u} \right) \tag{9.188}$$

$$= \log_2 \frac{\mathsf{SNR}_u}{U} - \gamma_{\mathsf{EM}} \log_2 e + \mathcal{O}\left(\frac{1}{\mathsf{SNR}_u}\right), \tag{9.189}$$

which, plugged into (9.187) and after a further expansion with respect to SNR_u, gives

$$\frac{R_u}{B} \geq (1 - \alpha - \alpha_{\mathsf{fb}} - \alpha_{\mathsf{d}}) \left[\log_2 \frac{\mathsf{SNR}_u}{U} - \gamma_{\mathsf{EM}} \log_2 e - \log_2 \left(1 + \frac{1}{\alpha_{\mathsf{d}} N_{\mathsf{c}}} + \frac{1 - \frac{1}{U}}{N_{\mathsf{c}}} \left(\frac{\rho U}{\alpha_{\mathsf{fb}}} + \frac{N_{\mathsf{t}}}{\alpha} \right) \right) \right] + \mathcal{O}\left(\frac{1}{\mathsf{SNR}_u}\right). \tag{9.190}$$

For $\mathsf{SNR}_u \to \infty$, as argued, the optimum overheads converge to their minimum values:

- One unprecoded pilot symbol per antenna, such that $\alpha = N_{\mathsf{t}}/N_{\mathsf{c}}$.
- One precoded pilot symbol per precoder, hence $\alpha_{\mathsf{d}} = U/N_{\mathsf{c}}$.
- One feedback symbol from each user for each of the precoders, giving $\alpha_{\mathsf{fb}} = U N_{\mathsf{t}}/N_{\mathsf{c}}$.

Inserting these quantities into (9.190), we obtain

$$\frac{R_u}{B} \geq \left(1 - \frac{N_{\mathsf{t}} + U + U N_{\mathsf{t}}}{N_{\mathsf{c}}} \right) \left[\log_2 \frac{\mathsf{SNR}_u}{U} - \gamma_{\mathsf{EM}} \log_2 e - \log_2 \left(2 + \rho \frac{1 - \frac{1}{U}}{N_{\mathsf{t}}} \right) \right] + \mathcal{O}\left(\frac{1}{\mathsf{SNR}_u}\right) \tag{9.191}$$

and therefore

$$\Delta\mathcal{L}_{\infty}^{\mathsf{ZF}} = \log_2 \left(2 + \rho \frac{1 - \frac{1}{U}}{N_{\mathsf{t}}} \right) \tag{9.192}$$

gives the power offset penalty, in 3-dB units, of pilot-assisted ZF transmission relative to its CSI counterpart. And, although derived under a uniform power allocation, this power offset penalty pertains to the optimum power allocation as well because of the asymptotic uniformity of the waterfilling solution.

Altogether then, the degradation with respect to the CSI spectral efficiency in the high-SNR regime is in the form of a decrease factor

$$\left(1 - \frac{N_{\mathsf{t}} + U + U N_{\mathsf{t}}}{N_{\mathsf{c}}} \right) \tag{9.193}$$

in spatial DOF plus a power offset penalty equal to $\Delta\mathcal{L}_{\infty}^{\mathsf{ZF}}$.

Example 9.29

Reconsider Example 9.28, where $U = N_{\mathsf{t}} = 8$ and $\rho = 1$ for the vehicular setting represented by $N_{\mathsf{c}} = 1000$. Plot again the range of values between the optimized lower bound and the upper bound $C_u^{\mathsf{ZF}}(\mathsf{SNR}_u)$, as on the left-hand side of Fig. 9.9, but homing in on the high-SNR regime. Then, add to the plot the expansions in (9.189) and (9.191) and verify that they hug this range for growing SNR_u.

Solution

See Fig. 9.11.

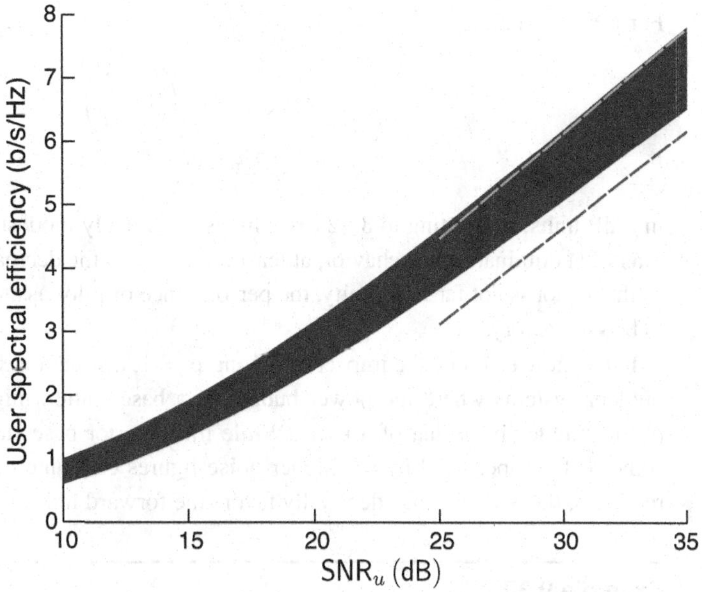

Fig. 9.11 Shaded region containing the spectral efficiency of a given user as a function of its SNR (in dB) for pilot-assisted ZF transmission with $U = N_t = 8$, $N_c = 1000$, and $\rho = 1$. The dashed lines are the high-SNR expansions in (9.189) and (9.191).

By means of the DOF decrease factor in (9.193) and of the power offset penalty $\Delta\mathcal{L}_{\infty}^{\text{ZF}}$ in (9.192), the degradation with respect to the CSI spectral efficiency can be more crisply gauged.

In terms of DOF, the decrease factor is entirely caused by overhead. Beyond that inevitable degradation, pilot-based ZF transmission is able to preserve the integrity of the $U = N_t$ spatial DOF. This can be reconciled with the basic result laid down in Section 8.10, according to which the preservation of the spatial DOF in a BC requires that the uncertainty in the transmitter knowledge of the channel coefficients decay at least inversely with the SNR. In the FDD flavor of pilot-assisted ZF that occupies us here, the rate of decay of this uncertainty can be gauged from the variance of the error between each true channel realization $\boldsymbol{H}_u$ and the corresponding estimate available to the transmitter, $\hat{\hat{\boldsymbol{H}}}_u$. This variance, derived in (9.174), expands as

$$\sigma_e^2 = \frac{N_t}{\alpha N_c \,\mathsf{SNR}_u} + \mathcal{O}\left(\frac{1}{\mathsf{SNR}_u^2}\right), \tag{9.194}$$

which indeed decays inversely with SNR_u.

In terms of power offset penalty, it can be verified (refer to Problem 9.29) that, regardless of the value of ρ, such penalty is maximum for $U = N_t = 2$. From (9.192) then,

$$\Delta\mathcal{L}_{\infty}^{\text{ZF}} \leq \log_2\left(2 + \rho\frac{1-\frac{1}{2}}{2}\right) \tag{9.195}$$

$$= \log_2\left(2 + \frac{\rho}{4}\right). \tag{9.196}$$

For $\rho = 1$ then,

$$\Delta\mathcal{L}_{\infty}^{\text{ZF}} \leq \log_2\left(2 + \frac{1}{4}\right) \tag{9.197}$$

$$= 1.17, \tag{9.198}$$

in 3-dB units, amounting to 3.52 dB. This is a relatively modest penalty at high SNR and thus what dominates the behavior, at least for $\rho = 1$, is the decrease factor in DOF: as long as this factor is not far from unity, the performance of pilot-assisted transmission is bound to be satisfactory.

Let us now examine the impact of having $\rho > 1$, a situation that is descriptive of many outdoor systems where the power budget of a base station might be one or two orders of magnitude above that of a user. (While the superior base station transmit powers are somewhat compensated by the higher noise figures encountered in consumer-grade user receivers, the SNR balance decidedly favors the forward link.)

Example 9.30

Everything else being the same, quantify the additional power offset penalty associated with having $\rho = 10$ rather than $\rho = 1$.

Solution

For $U = N_{\text{t}} = 2$, the additional power offset penalty in 3-dB units is

$$\log_2\left(2 + 10\,\frac{1 - \frac{1}{2}}{2}\right) - \log_2\left(2 + \frac{1 - \frac{1}{2}}{2}\right) = 1 \tag{9.199}$$

amounting to a 3-dB loss. For $U = N_{\text{t}} > 2$, this additional penalty is seen to decrease monotonically. Altogether then, the high-SNR cost of a 10-dB imbalance between the forward and reverse links does not exceed 3 dB relative to the balanced case.

For $\rho/N_{\text{t}} \gg 1$, we have from (9.192) that $\Delta\mathcal{L}_{\infty}^{\text{ZF}}$ grows logarithmically with ρ and thus every dB of reduction in reverse transmit power, relative to the forward power, maps directly to a dB of additional penalty in the power offset. This bodes poorly for systems with major asymmetries, although with an important mitigating factor: as can be noticed by examining (9.190), the spectral efficiency lower bound depends on ρ only through its ratio with $\alpha_{\text{fb}} N_{\text{c}}$, the number of feedback symbols, which sensibly indicates that what matters is the total energy applied to the feedback signals. Consequently, any increase in ρ can be made up by a corresponding increase in the number of feedback symbols and, at not-so-high SNRs, asymmetries indeed are bound to translate partly (depending on the fading coherence) onto additional feedback symbols. Although not necessarily optimal, we could translate the entire effect of the asymmetry onto additional feedback symbols. Precisely, increasing the number of feedback symbols from UN_{t} to $\rho\, UN_{\text{t}}$ (integer-rounded if necessary), we can keep the power offset penalty to within the modest 3.52 dB calculated in

(9.198) while turning the DOF decrease factor in (9.193) into

$$\left(1 - \frac{N_t + U + \rho U N_t}{N_c}\right). \tag{9.200}$$

This allows us to safely state that, with $U = N_t$, pilot-assisted ZF with FDD performs satisfactorily if

$$\frac{2U + \rho U^2}{N_c} \ll 1. \tag{9.201}$$

From this compact condition, involving the number of users (and transmit antennas), the fading coherence, and the forward–reverse SNR asymmetry, we can delineate the combinations of these parameters that are well tolerated.

Example 9.31

For the decrease factor on the number of spatial DOF not to exceed 10%, meaning

$$\frac{2U + \rho U^2}{N_c} < 0.1, \tag{9.202}$$

what is the largest number of users and antennas that can be accommodated with symmetric powers in pedestrian and vehicular conditions?

Solution

For $\rho = 1$, the condition becomes

$$2U + U^2 < 0.1\, N_c. \tag{9.203}$$

Setting $N_c = 1000$ to represent vehicular coherence, we obtain $U < 10$, which is consistent with earlier findings. With pedestrian users, the condition becomes much laxer.

Example 9.32

For the decrease factor in the number of spatial DOF not to exceed 10%, what is the largest number of users and antennas that can be accommodated with a 10-dB power asymmetry?

Solution

For $\rho = 10$, (9.203) morphs into

$$2U + 10U^2 < 0.1\, N_c. \tag{9.204}$$

Setting $N_c = 1000$ we obtain $U < 4$ and thus vehicular users can still be supported, albeit in limited number. Pedestrian users can be amply supported.

Example 9.33

Finally, for the decrease factor in the number of DOF not to exceed 10%, what is the largest number of users and antennas that can be accommodated with a 20-dB power asymmetry?

Solution

For $\rho = 100$,

$$2U + 100U^2 < 0.1 N_c. \tag{9.205}$$

Setting $N_c = 1000$, we obtain $U < 1$, which warns that under such strong power asymmetry vehicular users might not be supportable at all. Turning to pedestrians though, $N_c = 20\,000$ still yields a comforting value of $U < 5$.

SNR asymmetries therefore constrain and eventually compromise the ability of the system to support fast-moving users, but, even in highly asymmetric situations, pilot-assisted ZF transmission with FDD can perform rather satisfactorily for pedestrian users.

Reverse-link fading

The one weak point of the analysis up to this point is the consideration of an AWGN feedback channel. With that, we have captured the effect of noise and, most importantly, the scaling of the reverse SNR with its forward counterpart and the possible asymmetries between the two, but we have omitted reverse-link fading. While, per se, reverse-link fading diminishes the spectral efficiency only slightly (refer to Problem 9.30), it does have two notable consequences.

- To be coherently detected, the feedback transmissions must be preceded by additional reverse pilots, at least one from each user, transmitted orthogonally. This means additional overhead.
- Recognizing that the reverse channel is a MAC, the feedback transmissions themselves can be concurrent rather than orthogonal [630]. It is shown in [830] that, with LMMSE reception at the N_t-antenna base station, the spectral efficiency lower bound is maximized by having two orthogonal feedback transmissions, each involving $U/2$ users transmitting simultaneously. (It is also argued in [830] that the exact spectral efficiency might be maximized by having a single feedback transmission by all U users at once, but, since such exact spectral efficiency is unwieldy and we have instead chosen to utilize the lower bound, we abide by the two orthogonal transmissions.)

In terms of the number of spatial DOF, the balance of these two consequences is actually favorable. For $\mathsf{SNR}_u \to \infty$:

- One reverse pilot is required orthogonally from each user for a total of U additional reverse pilots.
- In exchange, rather than UN_t feedback symbols, $2N_t$ suffice with non-orthogonal feedback.

When all this is considered, (9.201) changes into

$$\frac{3 + 2\rho}{N_c} U \ll 1. \tag{9.206}$$

Example 9.34

By means of (9.201) and for the decrease factor in number of DOF not to exceed 10%, it is determined in Example 9.33 that, with a 10-dB power asymmetry, $U < 4$ vehicular users are supportable. Redo this calculation with (9.206) in place of (9.201).

Solution

For $\rho = 10$, (9.206) yields

$$23\,U < 0.1\,N_{\rm c}. \tag{9.207}$$

Setting $N_{\rm c} = 1000$, we obtain $U < 5$.

As an additional improvement, if either the base station or the user power constraints allow it, it may be possible to allocate unequal transmit powers to the various pilot symbols relative to the data and feedback symbols. As seen in the context of SU-MIMO, such pilot power boosting does not affect the number of spatial DOF but it can bring about a reduction (expected to be modest) in the power offset penalty.

TDD and full duplexing

Leaving FDD behind, let us now move on to TDD and full duplexing. These are more favorable duplexing alternatives because they open the door to the exploitation of reciprocity, and with that to a reduction in the number of pilots [851]. The performance under these alternatives can be established from the FDD expressions, without the need to start the analysis anew. To see that, let us revisit the FDD procedure, which entails these stages:

(1) Unprecoded pilot transmissions from each antenna.
(2) CSI acquisition at each user.
(3) Reverse-link pilot transmissions from each user, reverse-link channel estimation at the base station, and CSI feedback.
(4) Computation of the precoders.
(5) Precoded pilot transmission through each precoder.
(6) Payload data transmission.

Channel reciprocity enables the removal of the two initial stages as well as of the CSI feedback, leaving the procedure as follows:

(1) Reverse-link pilot transmissions from each user, and channel estimation at the base station.
(2) Computation of the precoders applying the reverse-link channel estimates to the forward direction.
(3) Precoded pilot transmission through each precoder.
(4) Payload data transmission.

Tempting as it may be to think that perhaps the precoded pilot transmissions could also be done away with, that is not the case in nonmassive MIMO, as users are unable to independently produce the precoders generated at the base station. The subsequent precoded pilot transmissions make it possible for the users to learn the precoded channels through which the data are to be ultimately sent.

As far as our analysis is concerned, the general lower bound obtained in (9.165) continues to apply by setting $\alpha_{\mathsf{fb}} = 0$. This aside, the differences between FDD and TDD emerge once the term

$$\mathbb{E}\left[\left|\boldsymbol{H}_u \hat{\hat{\boldsymbol{F}}}{}_{\mathsf{u}}^{\mathsf{ZF}}\right|^2\right] \tag{9.208}$$

is developed. Recall that $\hat{\hat{\boldsymbol{F}}}{}_0^{\mathsf{ZF}}, \ldots, \hat{\hat{\boldsymbol{F}}}{}_{U-1}^{\mathsf{ZF}}$ are generated by applying ZF to $\hat{\hat{\boldsymbol{H}}}_u = \boldsymbol{H}_u - \tilde{\tilde{\boldsymbol{H}}}_u$ for $u = 0, \ldots, U-1$. Now, rather than being given by (9.174), the variance of the entries of $\tilde{\tilde{\boldsymbol{H}}}_u$ is

$$\sigma_{\mathsf{e}}^2 = \frac{1}{1 + \frac{\alpha N_{\mathsf{c}}}{U} \mathsf{SNR}_u^{\mathsf{r}}}, \tag{9.209}$$

reflecting the fact that each such estimate is obtained directly at the base station by observing reverse pilot transmissions from the users. We have retained the variable α, previously utilized to denote the unprecoded pilot overhead in the forward direction, to now denote the unprecoded pilot overhead that takes its place in the reverse direction; each of the U users gets to transmit $\alpha \frac{N_{\mathsf{c}}}{U}$ pilot symbols per block and every pilot symbol enables simultaneous estimation of that user's N_{t} fading coefficients—one per antenna—at the base station.

From (9.180) and (9.209),

$$\mathbb{E}\left[\left|\boldsymbol{H}_u \hat{\hat{\boldsymbol{F}}}{}_{\mathsf{u}}^{\mathsf{ZF}}\right|^2\right] = \frac{N_{\mathsf{t}}}{1 + \alpha \frac{N_{\mathsf{c}}}{U} \mathsf{SNR}_u^{\mathsf{r}}}, \tag{9.210}$$

which, plugged into the general lower bound in (9.165), yields

$$\frac{R_u}{B} \geq (1 - \alpha - \alpha_{\mathsf{d}}) \left[C_u^{\mathsf{ZF}}(\mathsf{SNR}_u) - \log_2\left(1 + \frac{\frac{1}{U}\mathsf{SNR}_u}{1 + \alpha_{\mathsf{d}} \frac{N_{\mathsf{c}}}{U} \mathsf{SNR}_u} + \frac{\left(1 - \frac{1}{U}\right)\mathsf{SNR}_u}{1 + \alpha \frac{N_{\mathsf{c}}}{U} \mathsf{SNR}_u^{\mathsf{r}}}\right)\right]. \tag{9.211}$$

Invoking the forward–reverse SNR ratio

$$\rho = \frac{\mathsf{SNR}_u}{\mathsf{SNR}_u^{\mathsf{r}}}, \tag{9.212}$$

we finally obtain

$$\frac{R_u}{B} \geq (1 - \alpha - \alpha_{\mathsf{d}}) \left[C_u^{\mathsf{ZF}}(\mathsf{SNR}_u) - \log_2\left(1 + \frac{\frac{1}{U}\mathsf{SNR}_u}{1 + \alpha_{\mathsf{d}} \frac{N_{\mathsf{c}}}{U} \mathsf{SNR}_u} + \frac{\left(1 - \frac{1}{U}\right)\mathsf{SNR}_u}{1 + \alpha \frac{N_{\mathsf{c}}}{\rho U} \mathsf{SNR}_u}\right)\right], \tag{9.213}$$

which is a TDD/full duplex counterpart to the FDD expression in (9.187); to get a sense of how much the two differ, we can repeat for the former one of the examples of the latter.

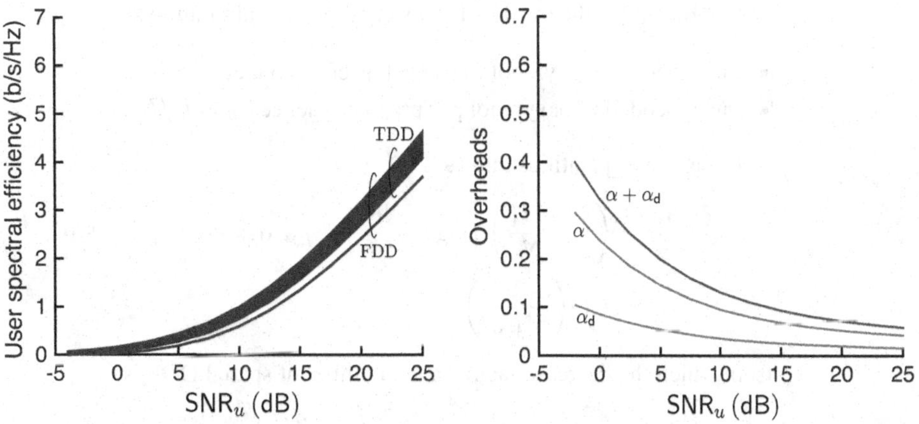

Fig. 9.12 Left, shaded region containing the spectral efficiency of a user as a function of its SNR (in dB) for pilot-assisted ZF transmission with TDD/full duplex, $U = N_t = 8$, $N_c = 1000$, and $\text{SNR}^r_u = \text{SNR}_u$. In a solid line below the shaded region, the FDD lower bound. Right, TDD/full duplex overheads (α and α_d, as well as their sum) optimized for the spectral efficiency lower bound.

Example 9.35

Repeat Example 9.28, where $U = N_t = 8$ and $\rho = 1$, for TDD/full duplex in a vehicular setting ($N_c = 1000$). Contrast the TDD/full duplex spectral efficiency lower bound with its FDD brethren.

Solution

See Fig. 9.12. On the left-hand side subfigure, the shaded region indicates the range between the CSI spectral efficiency and the TDD/full duplex lower bound. Below, a solid line indicates the FDD lower bound. On the right-hand side subfigure, α and α_d that optimize the TDD lower bound, as well as their sum.

With TDD/full duplex, the total overheads are reduced substantially and the worst-case loss in spectral efficiency is cut almost in half.

High-SNR regime

Recalling the expansion of $C_u^{\text{ZF}}(\text{SNR}_u)$ in (9.189), the lower bound in (9.213) can be seen to behave as

$$\frac{R_u}{B} \geq (1 - \alpha - \alpha_d) \left[\log_2 \frac{\text{SNR}_u}{U} - \gamma_{\text{EM}} \log_2 e - \log_2 \left(1 + \frac{1}{\alpha_d N_c} + \rho \frac{U-1}{\alpha N_c} \right) \right] + \mathcal{O}\left(\frac{1}{\text{SNR}_u} \right). \tag{9.214}$$

For $\mathsf{SNR}_u \to \infty$, the overheads converge to their minimum values:

- One reverse pilot symbol from each user, such that $\alpha = U/N_\mathrm{c}$.
- One precoded pilot symbol per precoder, hence $\alpha_\mathrm{d} = U/N_\mathrm{c}$.

Inserting these quantities into (9.214),

$$\frac{R_u}{B} \geq \left(1 - \frac{2U}{N_\mathrm{c}}\right)\left[\log_2 \frac{\mathsf{SNR}_u}{U} - \gamma_\mathsf{EM} \log_2 e - \log_2\left(1 + \rho - \frac{\rho - 1}{U}\right)\right] + \mathcal{O}\left(\frac{1}{\mathsf{SNR}_u}\right), \tag{9.215}$$

from which the decrease factor in the number of spatial DOF is

$$\left(1 - \frac{2U}{N_\mathrm{c}}\right) \tag{9.216}$$

while the power offset penalty, in 3-dB units, is

$$\Delta\mathcal{L}_\infty^\mathsf{ZF} = \log_2\left(1 + \rho - \frac{\rho - 1}{U}\right). \tag{9.217}$$

Both the decrease factor in DOF and the power offset penalty reflect a milder degradation, with respect to the CSI spectral efficiency, than we had found with FDD. As far as the DOF is concerned, this is confirmed by contrasting (9.216) with (9.193). And, as far as the power offset is concerned, since $\rho - 1 \geq 0$, the TDD penalty $\Delta\mathcal{L}_\infty^\mathsf{ZF}$ in (9.217) achieves its higher value for $U \to \infty$. Thus,

$$\Delta\mathcal{L}_\infty^\mathsf{ZF} \leq \log_2(1 + \rho) \tag{9.218}$$

which, for $\rho = 1$, gives a penalty not exceeding 3 dB. This is less than the 3.52 dB encountered when this same exercise was conducted for FDD, and thus the conclusion is reinforced: it is a relatively modest penalty at high SNR and what dominates the behavior, at least for $\rho = 1$, is the decrease factor in DOF.

Any increase in ρ above unity, that is, any asymmetry $\mathsf{SNR}_u^\mathsf{r} < \mathsf{SNR}_u$, can be compensated by a proportional increase in the number of reverse pilot symbols emitted by the users and, depending on the SNRs and fading coherence, power asymmetries are sure to translate partly onto additional reverse pilots. Translating the entire effect of any asymmetry onto additional reverse symbols, and thereby keeping the power offset penalty to within the modest 3 dB calculated above, we can safely state that pilot-assisted ZF with TDD/full duplex performs satisfactorily if

$$\frac{1 + \rho}{N_\mathrm{c}} U \ll 1. \tag{9.219}$$

Recalling the equivalent expression derived in (9.206) for FDD, in Table 9.1 we conveniently contrast the sufficient conditions in terms of number of users and antennas, forward–reverse SNR asymmetry, and fading coherence, for pilot-assisted ZF transmission to operate "close" to its CSI level with either FDD or TDD/full duplex; the meaning of "close" is made precise by quantifying the degree of inequality in the conditions.

Table 9.1 Sufficient condition in terms of $U = N_t$, power asymmetry, and fading coherence, for pilot-assisted ZF transmission to approach CSI-based ZF

FDD	TDD/full duplex
$(3+2\rho)\,U \ll N_c$	$(1+\rho)\,U \ll N_c$

TDD/full duplex is uniformly superior, meaning that for any given ρ more users can be accommodated and, for any given $U = N_t$, a steeper power asymmetry can be tolerated.

Example 9.36

Quantify how TDD or full duplex increase the tolerance to asymmetries for some given U.

Solution

Introducing additional scripting to distinguish between the power asymmetry tolerated with FDD or with TDD/full duplex, for some given $U = N_t$ the equivalence of the respective conditions in Table 9.1 gives

$$\rho_{\mathsf{TDD}} \approx 2\left(1 + \rho_{\mathsf{FDD}}\right). \tag{9.220}$$

For $\rho_{\mathsf{FDD}} = 1$ for instance, $\rho_{\mathsf{TDD}} = 4$, meaning a 6-dB enhancement in tolerance. In turn, for sufficiently pronounced asymmetries,

$$\rho_{\mathsf{TDD}} \approx 2\,\rho_{\mathsf{FDD}}, \tag{9.221}$$

meaning a 3-dB enhancement in tolerance. All in all, with TDD/full duplex an additional power asymmetry of 3–6 dB can be tolerated, everything else being equal.

In a similar vein, we could quantify how the number of users and antennas that can be accommodated for a given ρ increases with TDD/full duplex relative to FDD, an aspect that is explored in Problems 9.32 and 9.33.

Let us now present a final example where we contrast the performance of pilot-assisted ZF transmission against the ultimate performance limit, namely the BC capacity with CSIT side information. This serves the double purpose of gauging the performance of a specific practical scheme, and of appraising the role of the BC capacity as a benchmark. We conduct the example in a familiar setup for which we have already computed the BC sum-capacity.

Example 9.37

Let us revisit Example 8.33, where $U = N_t = 3$ with

$$\mathsf{SNR}_0|_{\mathrm{dB}} = \mathsf{SNR}|_{\mathrm{dB}} \tag{9.222}$$

$$\mathsf{SNR}_1|_{\mathrm{dB}} = \mathsf{SNR}|_{\mathrm{dB}} + 5\,\mathrm{dB} \tag{9.223}$$

$$\mathsf{SNR}_2|_{\mathrm{dB}} = \mathsf{SNR}|_{\mathrm{dB}} + 8\,\mathrm{dB} \tag{9.224}$$

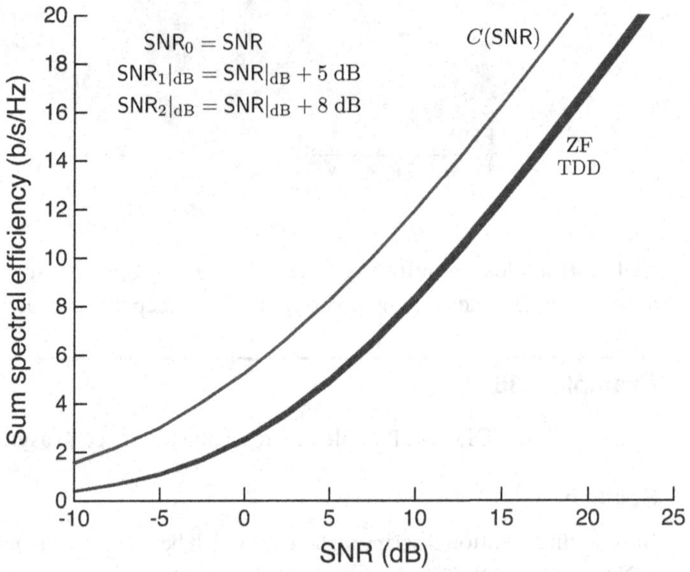

Fig. 9.13 The upper curve indicates the BC sum-capacity with $U = N_t = 3$ and CSIT while the shaded region below contains the sum spectral efficiency for pilot-assisted ZF transmission with TDD, $N_c = 20\,000$, and $\rho = 10$.

and with Rayleigh fading. Figure 8.14 presents the sum-capacity while Fig. 9.5 presents both the sum-capacity and the sum spectral efficiency achieved by ZF with CSI.

Quantify now the range of sum spectral efficiencies for pilot-based ZF transmission with TDD in pedestrian conditions ($N_c = 20\,000$) and with a 10-dB power asymmetry ($\rho = 10$).

Solution

The lower bound to the sum spectral efficiency with ZF transmission and TDD is obtained by solving

$$\max_{\alpha,\alpha_d} \; (1 - \alpha - \alpha_d) \sum_{u=0}^{2} \left[C_u^{\text{ZF}}(\mathsf{SNR}_u) - \log_2\left(1 + \frac{\frac{1}{U}\mathsf{SNR}_u}{1 + \alpha_d \frac{N_c}{U}\mathsf{SNR}_u} + \frac{(1-\frac{1}{U})\,\mathsf{SNR}_u}{1 + \alpha \frac{N_c}{\rho U}\mathsf{SNR}_u}\right)\right] \tag{9.225}$$

subject to $\alpha > 0$, $\alpha_d > 0$, and $\alpha + \alpha_d < 1$. The shaded region in Fig. 9.13 contains the range of values between the result of this convex optimization and the CSI-based ZF sum spectral efficiency while the curve above it is the BC sum-capacity.

Example 9.38

Repeat the previous example for vehicular users ($N_c = 1000$).

Solution

See Fig. 9.14.

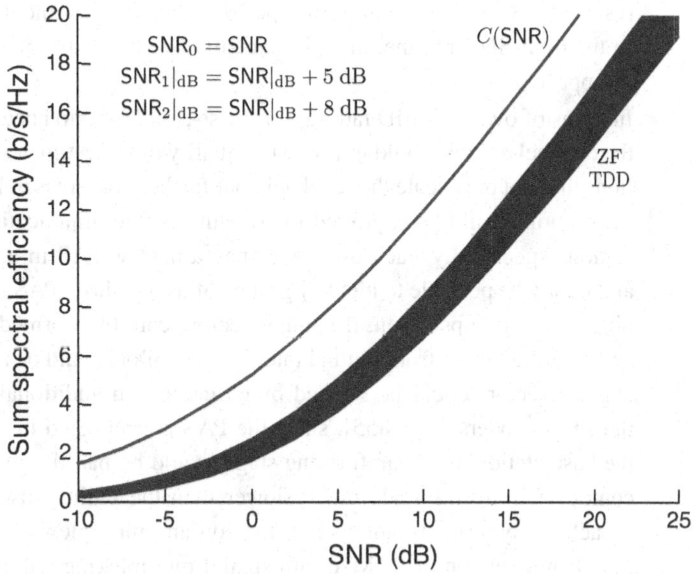

Fig. 9.14 The solid line indicates the BC sum-capacity with $U = N_t = 3$ and CSIT while the shaded region below contains the sum spectral efficiency for pilot-assisted ZF transmission with TDD, $N_c = 1000$, and $\rho = 10$.

To wrap up the coverage of pilot-assisted ZF transmission, some final remarks.

- Our analysis relies on a block-fading channel model, perfectly representative in terms of the computation of the number of DOF while possibly less-than-perfect on finer measures such as the power offset. Therefore, the conditions derived in Table 9.1 should hold up also under continuous-fading whereas the values returned by the expressions for $\Delta\mathcal{L}_{\infty}^{\text{ZF}}$ may depart slightly from the power penalties in a continuous fading channel. Specifically, it is reasonable to suspect that there may be a slightly bigger edge for TDD/full duplex over FDD than what our analysis reports, and that there may be a small differentiation, too fine for our models to capture, favoring full duplex over TDD.
- Our analysis is for $U = N_t$, the fully loaded configuration where the user SNRs are pushed down the most. For $U < N_t$, the decrease factors in the number of spatial DOF, and hence the conditions summarized in Table 9.1, continue to apply given that they are based on the number of dimensions devoted to channel estimation and feedback, and that we have taken the care of carrying both variables, U and N_t, separately throughout the foregoing analysis; the number of dimensions hence remain properly counted. In turn, the power offset penalty can only improve if $U < N_t$ and therefore our expressions for $\Delta\mathcal{L}_{\infty}^{\text{ZF}}$, evaluated for $U = N_t$, can be regarded as worst-case values.
- We have evenly divided the precoded-pilot overhead among the U precoders, and the feedback overhead among the U users, optimizing only the totals α_{d} and α_{db}. This even split is not strictly optimal except when all user SNRs are equal, and a separate optimization whereby each precoder's and each user's shares of the overheads are tailored to the

respective SNRs may yield minor performance improvements at medium power levels. In the high-SNR regime, though, once the overheads hit their minima, any differences disappear.

- In terms of overhead, IID fading is a worst-case scenario because any correlation among the channel entries would improve the quality of the estimates and allow for transformations that not only scale the feedback, but further compress it. Moreover, certain transmit correlations could be exploited to structure the communication itself. In elevated base stations specifically, each user may span a narrow PAS in azimuth (see Section 3.4.2) and it may be possible to identify groups of users whose PAS are essentially nonoverlapping across groups. Then, if angular sectors could be formed segregating those groups and shielding them from mutual interference, pilots could be reused across groups. Such angular sectors could be formed by cascading an additional beamforming stage onto the ZF precoders [852–855]; since the PAS are reflected in the antenna correlations at the base station, the beamforming stage would be based on those correlations and thus computed at a time scale much slower than the fading-based ZF precoders. This approach, aptly termed joint spatial division and multiplexing (JSDM), may be regarded as a group version of SDMA with spatial multiplexing within each group and it could allow reducing the pilot overhead by the number of groups provided the beam isolation is satisfactory and pilots can be reused [856]; in that, both the angular segregation and the antenna spacing play a role. And, just as fixed sectorization is the static version of SDMA, fixed sectorization with spatial multiplexing within each sector would be the static version of JSDM.

9.8 Block-diagonalization for the broadcast channel

After the long detour on pilot-assisted ZF transmission, let us bring back the CSI. The MU-MISO analysis of ZF could be extended to multiantenna users in a straightforward fashion by simply regarding each receive antenna as a separate user [857]; this approach is demonstrated in Example 9.40, a bit later in the text. The resulting aggregate precoder, of dimension $N_\mathrm{t} \times UN_\mathrm{r}$, would be—with suitable normalizations such that the U constituent user precoders respect their respective constraints—the pseudoinverse of the $UN_\mathrm{r} \times N_\mathrm{t}$ aggregate channel matrix $\boldsymbol{C}$ defined in (9.74) and reproduced here for convenience:

$$\boldsymbol{C} = \begin{bmatrix} \sqrt{G_0}\boldsymbol{H}_0 \\ \vdots \\ \sqrt{G_{U-1}}\boldsymbol{H}_{U-1} \end{bmatrix}. \tag{9.226}$$

This extension to multiantenna users, however, would place the burden of eliminating interference solely at the transmitter, ignoring that each multiantenna user can further apply a receive filter. Specifically, each multiantenna user can apply a local filter to help remove the interference among its streams, and thus the transmit structure can be relaxed so as to concentrate on eliminating only other-user interference. This amounts to block-diagonalizing (rather than strictly diagonalizing) the channel, hence the term *block-diagonalization* already brought up in the context of the MAC and on which we dwell next.

9.8.1 Transmitter structure

The derivations that follow rely on the concepts of column and null spaces of a matrix, for details on which the reader is referred to Appendix B.

With $\boldsymbol{F}_u^{\text{BD}}$ the $N_t \times N_r$ precoder for the uth user—each user is sent as many signal streams as antennas it has—we seek an aggregate $N_t \times UN_r$ precoder $[\boldsymbol{F}_0^{\text{BD}} \cdots \boldsymbol{F}_{U-1}^{\text{BD}}]$ that, applied to $\boldsymbol{C}$, yields a block-diagonal matrix: U blocks of dimension $N_r \times N_r$ along the diagonal and zeros elsewhere. This is tantamount to enforcing $\boldsymbol{H}_u \boldsymbol{F}_{\text{u}}^{\text{BD}} = \boldsymbol{0}$ for $u \neq \text{u}$.

From the vantage of user u, the channel to all other (unintended) users is embodied by the $(U-1)N_r \times N_t$ matrix

$$\boldsymbol{C}_{\text{-}u} = \begin{bmatrix} \sqrt{G_0}\boldsymbol{H}_0 \\ \vdots \\ \sqrt{G_{u-1}}\boldsymbol{H}_{u-1} \\ \sqrt{G_{u+1}}\boldsymbol{H}_{u+1} \\ \vdots \\ \sqrt{G_{U-1}}\boldsymbol{H}_{U-1} \end{bmatrix}, \tag{9.227}$$

which, with probability 1 under all fading distributions of interest and given the condition $N_t \geq UN_r$, has rank $(U-1)N_r$. To avoid spilling signals onto users other than u, $\boldsymbol{F}_u^{\text{BD}}$ needs to lie in the null space of $\boldsymbol{C}_{\text{-}u}$. We can write the SVD of $\boldsymbol{C}_{\text{-}u}$ as

$$\boldsymbol{C}_{\text{-}u} = \boldsymbol{U}_{\text{-}u} \boldsymbol{\Sigma}_{\text{-}u} \left[\boldsymbol{V}_{\text{-}u}^{\overline{\text{null}}} \boldsymbol{V}_{\text{-}u}^{\text{null}}\right]^*, \tag{9.228}$$

where we have grouped the initial $(U-1)N_r$ right singular vectors into $\boldsymbol{V}_{\text{-}u}^{\overline{\text{null}}}$ and the other $N_t - (U-1)N_r$ ones into $\boldsymbol{V}_{\text{-}u}^{\text{null}}$; these latter ones correspond to zero singular values and thus they span the desired null space [858–860]. Applying the $N_t \times (N_t - (U-1)N_r)$ matrix $\boldsymbol{V}_{\text{-}u}^{\text{null}}$ as a precoder for user u, we obtain for such user the $N_r \times (N_t - (U-1)N_r)$ effective channel $\boldsymbol{H}_u \boldsymbol{V}_{\text{-}u}^{\text{null}}$. Let the SVD of this effective channel be

$$\left(\boldsymbol{H}_u \boldsymbol{V}_{\text{-}u}^{\text{null}}\right) = \boldsymbol{U}_u \boldsymbol{\Sigma}_u \left[\boldsymbol{V}_u^{\overline{\text{null}}} \boldsymbol{V}_u^{\text{null}}\right]^*, \tag{9.229}$$

where again we have grouped the right singular vectors, in this case the initial N_r corresponding to nonzero singular values, into $\boldsymbol{V}_u^{\overline{\text{null}}}$ and the remaining $N_t - UN_r$ into $\boldsymbol{V}_u^{\text{null}}$. Faced with the effective channel $\left(\boldsymbol{H}_u \boldsymbol{V}_{\text{-}u}^{\text{null}}\right)$ and being in possession of CSIT, the optimum SU-MIMO signaling strategy for the uth user (see Section 5.3) is to further precode with the $(N_t - (U-1)N_r) \times N_r$ matrix $\boldsymbol{V}_u^{\overline{\text{null}}}$ and to receive with the $N_r \times N_r$ matrix $\boldsymbol{U}_u$ so as to diagonalize this effective channel matrix. Transmitting through the cascaded precoders $\boldsymbol{V}_{\text{-}u}^{\text{null}}$ and $\boldsymbol{V}_u^{\overline{\text{null}}}$, and receiving with the filter $\boldsymbol{U}_u$, what emerges at the output is

$$\breve{\boldsymbol{y}}_u = \boldsymbol{U}_u^* \boldsymbol{y}_u \tag{9.230}$$

$$= \boldsymbol{U}_u^* \sqrt{\frac{G_u E_u}{N_t}} \left(\boldsymbol{H}_u \boldsymbol{V}_{\text{-}u}^{\text{null}}\right) \boldsymbol{V}_u^{\overline{\text{null}}} \boldsymbol{s}_u + \boldsymbol{U}_u^* \boldsymbol{v}_u \tag{9.231}$$

$$= \sqrt{\frac{G_u E_u}{N_t}} U_u^* \left(U_u \Sigma_u \left[V_u^{\overline{\text{null}}} V_u^{\text{null}} \right]^* \right) V_u^{\overline{\text{null}}} s_u + \breve{v}_u \tag{9.232}$$

$$= \sqrt{\frac{G_u E_u}{N_t}} \left(U_u^* U_u \right) \Sigma_u \left(\begin{bmatrix} V_u^{\overline{\text{null}}*} \\ V_u^{\text{null}*} \end{bmatrix} V_u^{\overline{\text{null}}} \right) s_u + \breve{v}_u \tag{9.233}$$

$$= \sqrt{\frac{G_u E_u}{N_t}} \Sigma_u \begin{bmatrix} I \\ 0 \end{bmatrix} s_u + \breve{v}_u \tag{9.234}$$

$$= \sqrt{\frac{G_u E_u}{N_t}} \operatorname{diag}\left(\lambda_{u,0}^{1/2}, \ldots, \lambda_{u,N_r-1}^{1/2} \right) s_u + \breve{v}_u, \tag{9.235}$$

where in (9.232) we have invoked (9.229) while $\lambda_{u,j}$ denotes the jth nonzero eigenvalue of $\left(H_u V_{-u}^{\text{null}}\right)\left(H_u V_{-u}^{\text{null}}\right)^*$. The preceding derivation holds for $u = 0, \ldots, U-1$, with each filtered noise vector $\breve{v}_u = U_u^* v_u$ having the same IID complex Gaussian distribution as the corresponding original noise v_u because of unitary invariance.

With the channel thus diagonalized, we can equivalently write for user u the scalar transmit–receive relationships

$$[\breve{y}_u]_j = \sqrt{\frac{G_u E_u}{N_t}} \lambda_{u,j}^{1/2} [s_u]_j + [\breve{v}_u]_j \qquad j = 0, \ldots, N_r - 1. \tag{9.236}$$

Hence, with precoders $\left(V_{-u}^{\text{null}} V_u^{\overline{\text{null}}}\right)$ and receivers U_u^* for $u = 0, \ldots, U-1$ we have obtained a bank of U SU-MIMO channels, each in turn consisting of N_r parallel subchannels. Since each $\left(V_{-u}^{\text{null}} V_u^{\overline{\text{null}}}\right)$ is semiunitary, these precoders orient the signals in space but do not effect power allocation. Put differently, the power allocation is uniform, which need not be optimal with CSIT. This invites the further incorporation to each precoder of a diagonal matrix $P_u = \operatorname{diag}(P_{u,0}, \ldots, P_{u,N_r-1})$, with $\sum_{j=0}^{N_r-1} P_{u,j} = N_t$, yielding the familiar form

$$F_u^{\text{BD}} = \left(V_{-u}^{\text{null}} V_u^{\overline{\text{null}}}\right) P_u^{1/2}, \tag{9.237}$$

where $\left(V_{-u}^{\text{null}} V_u^{\overline{\text{null}}}\right)$ is the steering matrix. With that, the transmit–receive relationship for user u ends up being

$$[\breve{y}_u]_j = \sqrt{\frac{G_u E_u}{N_t} \lambda_{u,j} P_{u,j}}\, [s_u]_j + [\breve{v}_u]_j \qquad j = 0, \ldots, N_r - 1. \tag{9.238}$$

9.8.2 Power allocation

It is observed for ZF transmission that the optimization of the power allocation has a modest effect on the performance, and hence a legitimate choice—and a robust one in the face of CSI uncertainly—would be to forgo this optimization for block-diagonalization, keeping the power uniform across both signal streams and users. Nevertheless, for the sake of completeness, we next develop this optimization.

The problem of allocating power across the signal streams of each given user reduces to the problem of allocating power in an SU-MIMO setting with CSIT and hence the optimization of $P_{u,0}, \ldots, P_{u,N_r-1}$ must return a waterfilling policy. Specifically, by identifying terms in the waterfilling solution of Section 5.3 we find that

$$P^{\star}_{u,j} = \left[\frac{1}{\eta_u} - \frac{N_t N_0}{G_u E_u \lambda_{u,j}}\right]^+ \tag{9.239}$$

$$= \left[\frac{1}{\eta_u} - \frac{N_t}{\frac{E_u}{E_s}\mathsf{SNR}_u \lambda_{u,j}}\right]^+ \qquad j = 0, \ldots, N_r - 1, \tag{9.240}$$

where η_u ensures that $\sum_{j=0}^{N_r-1} P_{u,j} = N_t$. Then, the ergodic spectral efficiency achieved by the uth user is

$$C_u^{\mathsf{BD}} = \sum_{j=0}^{N_r-1} \mathbb{E}\left[\log_2\left(1 + \frac{\frac{E_u}{E_s}\mathsf{SNR}_u}{N_t} P_{u,j}\lambda_{u,j}\right)\right] \tag{9.241}$$

$$= \sum_{j=0}^{N_r-1} \mathbb{E}\left[\log_2\left(\frac{\frac{E_u}{E_s}\mathsf{SNR}_u}{N_t}\frac{\lambda_{u,j}}{\eta_u}\right)\right]^+ \tag{9.242}$$

and the weighted sum spectral efficiency is $\sum_{u=0}^{U-1} q_u C_u^{\mathsf{BD}}$. The allocation of power across users, i.e., the optimization of $\frac{E_0}{E_s}, \ldots, \frac{E_{U-1}}{E_s}$ subject to $\sum_{u=0}^{U-1} \frac{E_u}{E_s} = 1$, depends on the weights $q_0, \ldots, q_{U-1}$.

Example 9.39

What are the values $\frac{E_0}{E_s}, \ldots, \frac{E_{U-1}}{E_s}$ that maximize the sum spectral efficiency?

Solution

For the sum spectral efficiency, the optimization of $\frac{E_0}{E_s}, \ldots, \frac{E_{U-1}}{E_s}$ also adopts the form of a waterfilling and thus we can formulate a single waterfilling spanning the UN_r parallel subchannels with a single power constraint, namely

$$\left(P_{u,j}\frac{E_u}{E_s}\right)^{\star} = \left[\frac{1}{\eta} - \frac{N_t}{\mathsf{SNR}_u \lambda_{u,j}}\right]^+ \qquad u = 0, \ldots, U-1 \qquad j = 0, \ldots, N_r - 1, \tag{9.243}$$

with η such that

$$\sum_{u=0}^{U-1}\sum_{j=0}^{N_r-1} P_{u,j}\frac{E_u}{E_s} = N_t. \tag{9.244}$$

The sum spectral efficiency is

$$C^{\mathsf{BD}}(\mathsf{SNR}_u) = \sum_{u=0}^{U-1}\sum_{j=0}^{N_r-1} \mathbb{E}\left[\log_2\left(\frac{\mathsf{SNR}_u}{N_t}\frac{\lambda_{u,j}}{\eta}\right)\right]^+. \tag{9.245}$$

9.8.3 Ergodic spectral efficiency

To gauge the advantage that block-diagonalization can offer over regular ZF transmission at arbitrary SNRs, as well as its closeness to the BC capacity, let us exemplify the application of the spectral efficiency expression in (9.241).

Example 9.40

Consider a three-user BC with $N_t = 6$ and $N_r = 2$ where

$$\mathsf{SNR}_0|_{\text{dB}} = \mathsf{SNR}|_{\text{dB}} \tag{9.246}$$

$$\mathsf{SNR}_1|_{\text{dB}} = \mathsf{SNR}|_{\text{dB}} + 5\,\text{dB} \tag{9.247}$$

$$\mathsf{SNR}_2|_{\text{dB}} = \mathsf{SNR}|_{\text{dB}} + 8\,\text{dB}, \tag{9.248}$$

with $\boldsymbol{H}_0$, $\boldsymbol{H}_1$, and $\boldsymbol{H}_2$ having IID Rayleigh-faded entries. Compare, as a function of SNR, the MU-MIMO sum spectral efficiencies of strict ZF and block-diagonalization. Further compare both of them with the BC sum-capacity.

Solution

Shown in Fig. 9.15 are the sum spectral efficiencies of MU-MIMO with strict ZF and with block-diagonalization, in both cases with waterfilling power allocation at each channel realization. Also shown is the BC sum-capacity, obtained via duality and convex optimization. Observe how block-diagonalization manages to partially close the gap between the performance of a ZF transmitter and the capacity.

Example 9.41

Further compare the results generated in Example 9.40 with the individual SU-MIMO capacity for each of the users, in otherwise equal conditions (CSIR and CSIT).

Solution

Also shown in Fig. 9.15 are the individual SU-MIMO sum-capacities with single-user waterfilling at each channel realization.

Notice how, at low SNR, serving only the strongest user is preferable to MU-MIMO with either linear scheme (ZF or block-diagonalization), and serving even only the second strongest user is preferable to ZF. With linear transmission schemes, as observed, it is not only that MU-MIMO is no better than SU-MIMO at low SNR, but that it may be decidedly worse.

Once the power increases, though, all MU-MIMO alternatives prevail over SU-MIMO given the difference in DOF ($S_\infty = 6$ versus $S_\infty = 2$).

The advantage of block-diagonalization over strict ZF may grow in the face of antenna correlations (refer to Problem 9.39). Furthermore, the performance of block-diagonalization can sometimes be further improved upon by a slightly more general technique termed *multiuser eigenmode transmission*. The premise of this idea is that, with block-diagonalization,

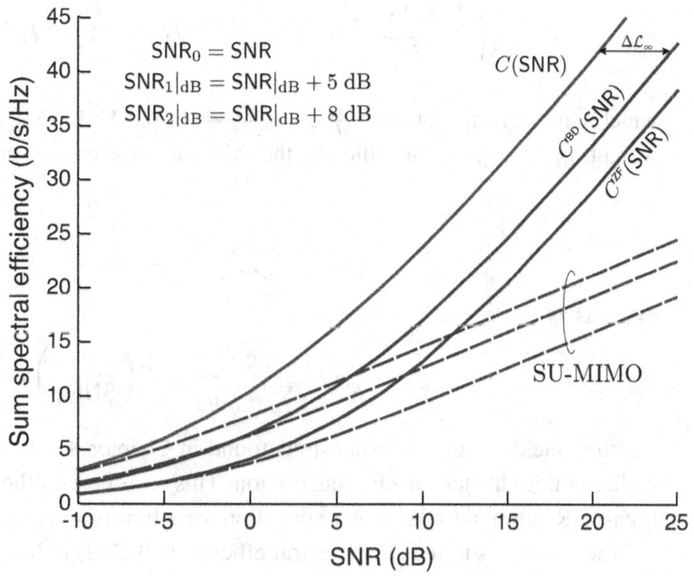

Fig. 9.15 MU-MIMO BC sum spectral efficiency with $U = 3$, $N_{\mathrm{t}} = 6$, and $N_{\mathrm{r}} = 2$, as a function of SNR (in dB); the top solid line is the sum-capacity while the two other solid lines correspond to block-diagonalization and strict ZF. Also shown, with dashed lines, are the individual SU-MIMO capacities for each of the three users.

some degrees of freedom may be consumed to enforce ZF conditions that protect from interference certain channels to which then the subsequent waterfilling process allocates no power. By avoiding such situations, further spectral efficiency improvements can sometimes be attained, particularly when antenna correlations do exist. For details on the implementation of this technique, the reader is referred to [861] and [34, section 4.2.1].

The formulation of block-diagonalization under a per-antenna power constraint is tackled in [862], while the combination of block diagonalization and user selection is addressed in [863, 864].

9.8.4 High-SNR regime

In the high-SNR regime, waterfilling is known to return a uniform power allocation and thus $P^\star_{u,j} = 1$ for $j = 0, \ldots, N_{\mathrm{r}}-1$ and $u = 0, \ldots, U-1$. Plugged into (9.241), this gives

$$C_u^{\mathrm{BD}} = \sum_{j=0}^{N_{\mathrm{r}}-1} \mathbb{E}\left[\log_2\left(1 + \frac{\frac{E_u}{E_{\mathrm{s}}}\mathrm{SNR}_u}{N_{\mathrm{t}}} \lambda_{u,j}(\boldsymbol{H}_u \boldsymbol{V}_{-u}^{\mathrm{null}} \boldsymbol{V}_{-u}^{\mathrm{null}*} \boldsymbol{H}_u^*)\right)\right]. \tag{9.249}$$

For nonzero $\frac{E_u}{E_{\mathrm{s}}}$ and growing SNR_u,

$$C_u^{\mathrm{BD}} = N_{\mathrm{r}} \log_2\left(\frac{\frac{E_u}{E_{\mathrm{s}}}\mathrm{SNR}_u}{N_{\mathrm{t}}}\right) + \mathbb{E}\left[\sum_{j=0}^{N_{\mathrm{r}}-1} \log_2 \lambda_{u,j}(\boldsymbol{H}_u \boldsymbol{V}_{-u}^{\mathrm{null}} \boldsymbol{V}_{-u}^{\mathrm{null}*} \boldsymbol{H}_u^*)\right] + \mathcal{O}\left(\frac{1}{\mathrm{SNR}_u}\right)$$

$$= N_\mathrm{r} \log_2\!\left(\frac{\frac{E_u}{E_\mathrm{s}}\mathsf{SNR}_u}{N_\mathrm{t}}\right) + \mathbb{E}\big[\log_2 \det\!\big(\boldsymbol{H}_u \boldsymbol{V}^{\mathsf{null}}_{\text{-}u} \boldsymbol{V}^{\mathsf{null}*}_{\text{-}u} \boldsymbol{H}^*_u\big)\big] + \mathcal{O}\!\left(\frac{1}{\mathsf{SNR}_u}\right) \tag{9.250}$$

which can serve to optimize $\frac{E_0}{E_\mathrm{s}}, \ldots, \frac{E_{U-1}}{E_\mathrm{s}}$ in the high-SNR regime, depending on the weights $q_0, \ldots, q_{U-1}$. Specifically, the solution to the convex problem

$$\max_{\frac{E_0}{E_\mathrm{s}},\ldots,\frac{E_{U-1}}{E_\mathrm{s}} : \sum_{u=0}^{U-1}\frac{E_u}{E_\mathrm{s}}=1} \; \sum_{u=0}^{U-1} q_u \, C^{\mathsf{BD}}_u \tag{9.251}$$

expands as

$$\frac{E_u}{E_\mathrm{s}} = \frac{q_u}{\sum_{\mathrm{u}=0}^{U-1} q_\mathrm{u}} + \mathcal{O}\!\left(\frac{1}{\mathsf{SNR}_u}\right), \tag{9.252}$$

which coincides with the expansions found in Chapter 8 with optimum transmission and earlier in this chapter for ZF transmission. This corroborates the broad applicability of this high-SNR solution to the power allocation for arbitrary weights $q_0, \ldots, q_{U-1}$.

When it comes to the sum spectral efficiency, (9.252) reduces to a uniform power split and, summing over the U users,

$$C^{\mathsf{BD}} = \sum_{u=0}^{U-1} N_\mathrm{r}\left(\log_2 \mathsf{SNR}_u - \mathcal{L}^{\mathsf{BD}}_{\infty,u}\right) + \mathcal{O}\!\left(\frac{1}{\mathsf{SNR}_u}\right), \tag{9.253}$$

with $\mathcal{L}^{\mathsf{BD}}_{\infty} = \frac{1}{U}\sum_{u=0}^{U-1}\mathcal{L}^{\mathsf{BD}}_{\infty,u}$ providing the sum spectral efficiency power offset, given

$$\mathcal{L}^{\mathsf{BD}}_{\infty,u} = \log_2(U N_\mathrm{t}) - \frac{1}{N_\mathrm{r}}\mathbb{E}\big[\log_2 \det\!\big(\boldsymbol{H}_u \boldsymbol{V}^{\mathsf{null}}_{\text{-}u} \boldsymbol{V}^{\mathsf{null}*}_{\text{-}u} \boldsymbol{H}^*_u\big)\big]. \tag{9.254}$$

For IID Rayleigh fading, as it turns out, the power offset can be expressed in a closed form that sheds light on how block diagonalization behaves at high SNR, relative to both the optimum and the ZF transmitters.

Example 9.42 (Block-diagonalization power offset in an IID Rayleigh-faded MU-MIMO BC)

In Rayleigh fading, $\boldsymbol{H}_u$ has IID complex Gaussian entries. Its product with the semiunitary precoder $\boldsymbol{V}^{\mathsf{null}}_{\text{-}u}$ yields a matrix that has IID complex Gaussian entries and dimensionality $N_\mathrm{r} \times (N_\mathrm{t} - (U-1)N_\mathrm{r})$. It follows that

$$\big(\boldsymbol{H}_u \boldsymbol{V}^{\mathsf{null}}_{\text{-}u} \boldsymbol{V}^{\mathsf{null}*}_{\text{-}u} \boldsymbol{H}^*_u\big) \sim \mathcal{W}_{N_\mathrm{r}}\!\left(N_\mathrm{t} - (U-1)N_\mathrm{r}, \boldsymbol{I}\right) \tag{9.255}$$

is an $N_\mathrm{r} \times N_\mathrm{r}$ Wishart matrix with $N_\mathrm{t} - (U-1)N_\mathrm{r}$ degrees of freedom. Invoking (C.28) in Appendix C.1.9 as well as (E.9) in Appendix E,

$$\mathcal{L}^{\mathsf{BD}}_{\infty,u} = \log_2(U N_\mathrm{t}) + \left(\gamma_{\mathsf{EM}} - \sum_{q=1}^{N_\mathrm{t}-UN_\mathrm{r}} \frac{1}{q} - \frac{N_\mathrm{t} - (U-1)N_\mathrm{r}}{N_\mathrm{r}} \sum_{q=N_\mathrm{t}-UN_\mathrm{r}+1}^{N_\mathrm{t}-(U-1)N_\mathrm{r}} \frac{1}{q} + 1\right)\log_2 e, \tag{9.256}$$

which, for $N_t = UN_r$, becomes

$$\mathcal{L}^{\text{BD}}_{\infty,u} = \log_2(UN_t) + \left(\gamma_{\text{EM}} - \sum_{q=2}^{N_r} \frac{1}{q}\right) \log_2 e. \tag{9.257}$$

From the foregoing expressions for the power offset, as well as those derived for the sum-capacity in Chapter 8, we can quantify the power loss of block-diagonalization relative to the sum-capacity at high SNR.

Example 9.43

For IID Rayleigh fading, express the high-SNR power loss of block-diagonalization relative to the sum-capacity.

Solution

From (8.150) and (9.256), the power loss in 3-dB units is

$$\Delta\mathcal{L}_\infty = \mathcal{L}^{\text{BD}}_\infty - \mathcal{L}_\infty \tag{9.258}$$

$$= \left(\frac{N_t}{UN_r} \sum_{q=N_t-UN_r+1}^{N_t} \frac{1}{q} - \frac{N_t-(U-1)N_r}{N_r} \sum_{q=N_t-UN_r+1}^{N_t-(U-1)N_r} \frac{1}{q}\right) \log_2 e, \tag{9.259}$$

which, for $N_t = UN_r$, simplifies into

$$\Delta\mathcal{L}_\infty = \left(\sum_{q=N_r+1}^{N_t} \frac{1}{q}\right) \log_2 e. \tag{9.260}$$

An alternative expression for (9.259), derived in [774], is

$$\Delta\mathcal{L}_\infty = \frac{\log_2 e}{UN_r} \sum_{q=0}^{U-1} \sum_{i=0}^{N_r-1} \sum_{\ell=qN_r+1}^{(U-1)N_r} \frac{1}{N_t - i - \ell}. \tag{9.261}$$

Example 9.44

What is the high-SNR power loss of block-diagonalization for IID Rayleigh fading when $N_t = 6$, $U = 3$, and $N_r = 2$?

Solution

In this case, $\Delta\mathcal{L}_\infty = 1.37$, which puts the power loss at 4.13 dB. As can be appreciated in Fig. 9.15, this value approximates very well the gap between the block-diagonalization sum spectral efficiency and the sum-capacity even at rather moderate SNRs.

In addition to establishing the high-SNR performance deficit with respect to the sum-capacity, $\Delta\mathcal{L}_\infty$ can further serve to quantify how much of the high-SNR losses experienced by ZF transmission can be recovered, whenever $N_r > 1$, by applying instead block-diagonalization. This calculation is exercised in Problem 9.35.

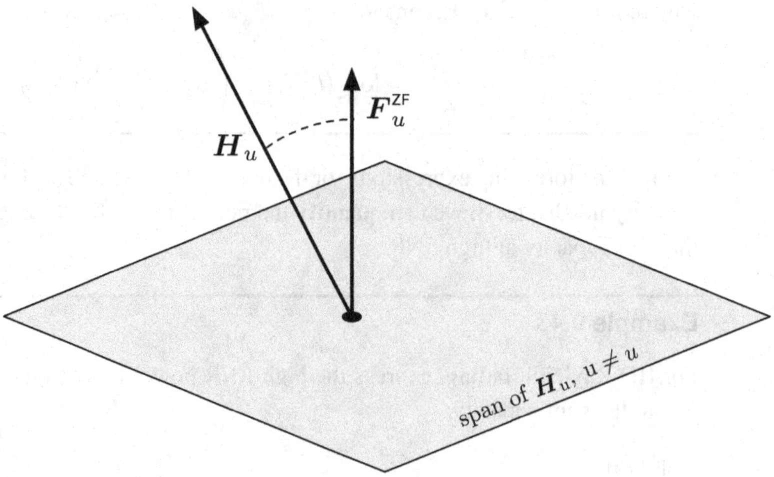

Fig. 9.16 ZF precoder orthogonal to the span of $\boldsymbol{H}_0, \ldots, \boldsymbol{H}_{u-1}, \boldsymbol{H}_{u+1}, \ldots, \boldsymbol{H}_{U-1}$, and a matched filter that aligns with $\boldsymbol{H}_u$. In between live a range of other precoders.

9.9 Regularized ZF transmitter for the broadcast channel

To complete our coverage of linear transmitters for the MU-MIMO BC, it is necessary to address how the gap between the matched-filter and ZF precoders can be bridged. This is visualized in Fig. 9.16, from the vantage of user u. The subspace spanned by $\boldsymbol{H}_0, \ldots, \boldsymbol{H}_{u-1}, \boldsymbol{H}_{u+1}, \ldots, \boldsymbol{H}_{U-1}$, which for simplicity is depicted as a plane but is in general of dimension $(U-1)N_{\text{r}}$, contains the interfering signals and thus the ZF precoder for user u is orthogonal to it. Conversely, the matched-filter precoder aligns with $\boldsymbol{H}_u$. Between these two extremes live a range of other linear structures, often termed *regularized* ZF transmitters [865].

A subtle issue that arises once we depart from the ZF solution is that the argument invoked to support the optimality of complex Gaussian codebooks, namely the Gaussian nature of the noise, ceases to hold. Now, besides noise, each receiver may be afflicted by interference from signals intended for other users. We are thus faced by the saddle point property of the Gaussian distribution encountered earlier in the text: it is both the best signal distribution if the interference is Gaussian, and the worst interference distribution if the signal is Gaussian. It is therefore possible that codebook distributions other than Gaussian be optimal, yet, as it turns out, provided we do not depart excessively from ZF such that the interference power is low or moderate, complex Gaussian codebooks do remain approximately optimal [866].

9.9.1 Regularizing term

On the one hand, the ZF structure is based on the pseudoinverse $\boldsymbol{C}^*(\boldsymbol{C}\boldsymbol{C}^*)^{-1}$ with $\boldsymbol{C}$ the $UN_\mathrm{r} \times N_\mathrm{t}$ aggregate channel matrix in (9.74) whereas, on the other hand, a matched filter is embodied by $\boldsymbol{C}^*$. A structure that naturally bridges these extremes is the familiar form

$$\boldsymbol{C}^*(\boldsymbol{C}\boldsymbol{C}^* + \varrho \boldsymbol{I})^{-1}, \tag{9.262}$$

where, given the respective asymptotic optimality of ZF and matched filter precoding at high and low SNRs, the desideratum for the regularizing term ϱ includes having $\varrho \to 0$ and $\varrho \to \infty$ in those respective regimes. However, finding the proper amount of regularization is far less straightforward here than in the MAC: while in the MAC the SINR experienced by a given data stream depends only on the corresponding receive filter (recall (9.58)), in the BC the SINR experienced by a given data stream in general depends, not only on the precoder for that stream, but on every precoder for every user.

Example 9.45

In a MU-MISO BC, we can express the SINR experienced by the one stream sent to the uth user by couching (9.62) into our regular BC notation, giving

$$\mathsf{sinr}_u = \frac{\frac{G_u E_u}{N_\mathrm{t}} |\boldsymbol{H}_u \boldsymbol{F}_u|^2}{\sum_{\mathrm{u} \neq u} \frac{G_u E_\mathrm{u}}{N_\mathrm{t}} |\boldsymbol{H}_u \boldsymbol{F}_\mathrm{u}|^2 + N_0} \tag{9.263}$$

$$= \frac{\frac{E_u}{E_\mathrm{s}} |\boldsymbol{H}_u \boldsymbol{F}_u|^2}{\sum_{\mathrm{u} \neq u} \frac{E_\mathrm{u}}{E_\mathrm{s}} |\boldsymbol{H}_u \boldsymbol{F}_\mathrm{u}|^2 + \frac{N_\mathrm{t}}{\mathsf{SNR}_u}}, \tag{9.264}$$

which indeed depends, not only on $\boldsymbol{F}_u$, but also on $\boldsymbol{F}_\mathrm{u}$ for $\mathrm{u} \neq u$. Thus, while in the MAC we can optimize (in the SINR sense) each receive filter separately, and the result is the LMMSE receiver, in the BC we cannot optimize (in the SINR sense) each precoder separately. The exception is the low-SNR regime, where

$$\mathsf{sinr}_u \approx \frac{\frac{E_u}{E_\mathrm{s}} \mathsf{SNR}_u}{N_\mathrm{t}} |\boldsymbol{H}_u \boldsymbol{F}_u|^2, \tag{9.265}$$

which depends only on $\boldsymbol{F}_u$, pointing to a matched-filter precoder. Conversely, at high SNR,

$$\mathsf{sinr}_u \approx \frac{\frac{E_u}{E_\mathrm{s}} |\boldsymbol{H}_u \boldsymbol{F}_u|^2}{\sum_{\mathrm{u} \neq u} \frac{E_\mathrm{u}}{E_\mathrm{s}} |\boldsymbol{H}_u \boldsymbol{F}_\mathrm{u}|^2}, \tag{9.266}$$

which is maximized by disregarding the numerator and having the rest of the precoders nullify the denominator, i.e., the ZF solution.

Clearly then, ϱ ought to depend on the SNR and, since in general each user enjoys a different SNR, ϱ should be individualized, meaning that we should have distinct $\varrho_0, \ldots, \varrho_{U-1}$.

Regularization in MU-MISO

Continuing for now with the MU-MISO case, the foregoing argumentation points to the uth user's precoder being the uth column of $C^*(CC^* + \varrho_u I)^{-1}$ properly normalized; with a per-symbol power constraint, this gives

$$F_u^{\text{Reg}} = \sqrt{N_t} \frac{\left[C^*(CC^* + \varrho_u I)^{-1}\right]_{:,u}}{\| \left[C^*(CC^* + \varrho_u I)^{-1}\right]_{:,u} \|}. \tag{9.267}$$

For $\varrho_u > 0$, the matrix inversion lemma indicates that

$$C^*(CC^* + \varrho_u I)^{-1} = \frac{1}{\varrho_u}\left[C^* - C^*C\left(C^*C + \varrho_u I\right)^{-1} C^*\right] \tag{9.268}$$

$$= \frac{1}{\varrho_u}\left[(C^*C + \varrho_u I) - C^*C\right]\left(C^*C + \varrho_u I\right)^{-1} C^* \tag{9.269}$$

$$= (C^*C + \varrho_u I)^{-1} C^*, \tag{9.270}$$

based on which we can rewrite (9.267) as

$$F_u^{\text{Reg}} = \sqrt{N_t} \frac{\left[(C^*C + \varrho_u I)^{-1} C^*\right]_{:,u}}{\| \left[(C^*C + \varrho_u I)^{-1} C^*\right]_{:,u} \|} \tag{9.271}$$

$$= \sqrt{N_t} \frac{(C^*C + \varrho_u I)^{-1} [C^*]_{:,u}}{\|(C^*C + \varrho_u I)^{-1} [C^*]_{:,u}\|} \tag{9.272}$$

$$= \sqrt{N_t} \frac{\left(\sum_{\mathrm{u}=0}^{U-1} G_{\mathrm{u}} H_{\mathrm{u}}^* H_{\mathrm{u}} + \varrho_u I\right)^{-1} \sqrt{G_u} H_u^*}{\left\|\left(\sum_{\mathrm{u}=0}^{U-1} G_{\mathrm{u}} H_{\mathrm{u}}^* H_{\mathrm{u}} + \varrho_u I\right)^{-1} \sqrt{G_u} H_u^*\right\|} \tag{9.273}$$

$$= \sqrt{N_t} \frac{\left(\sum_{\mathrm{u}=0}^{U-1} \frac{G_{\mathrm{u}}}{G_u} H_{\mathrm{u}}^* H_{\mathrm{u}} + \varrho_u' I\right)^{-1} H_u^*}{\left\|\left(\sum_{\mathrm{u}=0}^{U-1} \frac{G_{\mathrm{u}}}{G_u} H_{\mathrm{u}}^* H_{\mathrm{u}} + \varrho_u' I\right)^{-1} H_u^*\right\|} \tag{9.274}$$

$$= \sqrt{N_t} \frac{\left(\sum_{\mathrm{u}=0}^{U-1} \frac{\mathsf{SNR}_{\mathrm{u}}}{\mathsf{SNR}_u} H_{\mathrm{u}}^* H_{\mathrm{u}} + \varrho_u' I\right)^{-1} H_u^*}{\left\|\left(\sum_{\mathrm{u}=0}^{U-1} \frac{\mathsf{SNR}_{\mathrm{u}}}{\mathsf{SNR}_u} H_{\mathrm{u}}^* H_{\mathrm{u}} + \varrho_u' I\right)^{-1} H_u^*\right\|}, \tag{9.275}$$

where we have recalled the structure of C from (9.74) and introduced $\varrho_u' = \varrho_u / G_u$. It can be proved [867] that the precoder structure in (9.275) can maximize the SINR, yet the determination of the values of $\varrho_0', \ldots, \varrho_{U-1}'$ that renders it optimal in that sense turns out to be a hard problem [868, 869]. In light of this, we proceed down a different path.

The MAC–BC duality relationship for linear transceivers (see Section 9.6) indicates that, with the important caveat of having an aggregate power constraint in the dual MAC, the same SINRs can be achieved in both directions when the BC precoders equal the dual MAC receive filters. And we know that, for the MAC, the receive filters that maximize the SINRs for some given transmissions are the LMMSE ones in (9.38). This suggests that a good choice for the BC precoders could be the dual of (9.38) and indeed (9.275) has, normalization aside, already that form. In the MAC, and with additional superscripting to distinguish its SNR from that of the BC, the role of ϱ'_u is played by

$$\frac{N_{\mathrm{t}}}{\mathsf{SNR}^{\mathrm{r}}_{u}} = \frac{N_0 N_{\mathrm{t}}}{E_u G_u}, \tag{9.276}$$

whose application to the BC gives

$$\varrho'_u = \frac{N_0 N_{\mathrm{t}}}{E_u G_u} \tag{9.277}$$

$$= \frac{N_{\mathrm{t}}}{\frac{E_u}{E_{\mathrm{s}}}\mathsf{SNR}_u}. \tag{9.278}$$

This choice for ϱ'_u in (9.278) has the inconvenience of depending on the power allocation coefficient $\frac{E_u}{E_{\mathrm{s}}}$, which, given the availability of CSIT, should be reoptimized for each coherence tile. To decouple the structure of the precoders from the power allocation, one possibility is to fix ϱ'_u to the value it would have under a specific power allocation, with the most robust choice being the uniform one. This gives

$$\varrho'_u = \frac{U N_{\mathrm{t}}}{\mathsf{SNR}_u}, \tag{9.279}$$

which is a pleasing expression, yet a function of only SNR_u and not of $\mathsf{SNR}_{\mathrm{u}}$ for $\mathrm{u} \neq u$. Since the BC SINR for user u depends on all the SNRs, this suggests lack of optimality, and indeed the regularization in (9.279) need not maximize the BC SINRs. Rather, as one would expect given how we derived (9.279), such regularization maximizes the dual-MAC SINRs; recalling the expression for the MAC SINR in (9.58), this means that, with a uniform power allocation, the quantities being maximized are [869]

$$\mathsf{slnr}_u = \frac{\frac{1}{U}|\boldsymbol{H}_u \boldsymbol{F}_u|^2}{\sum_{\mathrm{u}\neq u} \frac{1}{U}\frac{\mathsf{SNR}_{\mathrm{u}}}{\mathsf{SNR}_u}|\boldsymbol{H}_{\mathrm{u}}\boldsymbol{F}_u|^2 + \frac{N_{\mathrm{t}}}{\mathsf{SNR}_u}} \qquad u = 0, \ldots, U-1, \tag{9.280}$$

which are the so-called *signal-to-leakage-plus-noise ratios* (SLNRs), hence the variable introduced to denote them. In Problem 9.41, readers are invited to verify that (9.280) is indeed maximized by the regularized ZF precoder

$$\boldsymbol{F}_u^{\mathrm{Reg}} = \sqrt{N_{\mathrm{t}}} \frac{\left(\sum_{\mathrm{u}=0}^{U-1} \frac{\mathsf{SNR}_{\mathrm{u}}}{\mathsf{SNR}_u}\boldsymbol{H}_{\mathrm{u}}^*\boldsymbol{H}_{\mathrm{u}} + \frac{UN_{\mathrm{t}}}{\mathsf{SNR}_u}\boldsymbol{I}\right)^{-1}\boldsymbol{H}_u^*}{\left\|\left(\sum_{\mathrm{u}=0}^{U-1} \frac{\mathsf{SNR}_{\mathrm{u}}}{\mathsf{SNR}_u}\boldsymbol{H}_{\mathrm{u}}^*\boldsymbol{H}_{\mathrm{u}} + \frac{UN_{\mathrm{t}}}{\mathsf{SNR}_u}\boldsymbol{I}\right)^{-1}\boldsymbol{H}_u^*\right\|} \tag{9.281}$$

$$= \sqrt{N_{\mathrm{t}}} \frac{\left(\sum_{\mathrm{u}=0}^{U-1} \mathsf{SNR}_{\mathrm{u}}\boldsymbol{H}_{\mathrm{u}}^*\boldsymbol{H}_{\mathrm{u}} + UN_{\mathrm{t}}\boldsymbol{I}\right)^{-1}\boldsymbol{H}_u^*}{\left\|\left(\sum_{\mathrm{u}=0}^{U-1} \mathsf{SNR}_{\mathrm{u}}\boldsymbol{H}_{\mathrm{u}}^*\boldsymbol{H}_{\mathrm{u}} + UN_{\mathrm{t}}\boldsymbol{I}\right)^{-1}\boldsymbol{H}_u^*\right\|}. \tag{9.282}$$

A close inspection of the SINR and SLNR expressions in (9.264) and (9.280) reveals that they differ in the indices being swapped within the denominator summation. While in the SINR such summation involves the total interference inflicted upon user u, in the SLNR it involves the total interference caused by the transmission of user u. Precisely:

- The SINR relates (*i*) the power meant for user u that is actually conveyed to that user, with (*ii*) the power meant for other users that leaks onto user u, plus noise.

Discussion 9.1 SLNR versus SINR

The regularizing term that maximizes the SLNR, $\varrho'_u = \frac{U N_t}{\mathsf{SNR}_u}$, behaves sensibly in that, when a user's SNR is low, it pushes that user's precoder toward selfish matched-filter beamforming while, when the SNR is high, it pushes the precoder toward selfless ZF. The lack of optimality stems from this behavior not being tempered by the SNRs of the other users.

In the very special case that all SNRs are equal, the ensuing symmetry makes the SLNRs and SINRs analogous—to the point that the regularizing term that maximizes the SLNRs also maximizes the SINRs for $U \to \infty$ [870, 871]. This strongly suggests that SLNR-maximizing precoders are close to optimal when the SNRs are not very dissimilar. It is only in the face of major SNR differences that SLNR-maximizing precoders may significantly depart from optimality in the SINR and spectral efficiency senses.

- The SLNR relates (*i*) the power meant for user u that is actually conveyed to that user, with (*ii*) the power meant for user u that leaks onto other users, plus noise.

By appropriately relaxing the ZF constraint of having zero leakage onto other users, it becomes possible to better focus onto the intended user's channel, and when the regularization is as in (9.280) the result is the maximization of the SLNR.

Despite not being SINR-maximizing in general, and therefore not optimal in a spectral efficiency sense, there is ample evidence that supports (9.282) as a satisfactory precoder for the MU-MISO BC [827, 872, 873]. In conjunction with the utterly simple form of its regularizing term, this represents a tempting shortcut when designing regularized ZF transmitters, and indeed SLNR-maximizing precoders (alternatively termed *transmit Wiener filters* or *MMSE beamformers*, among other monikers) are favorite choices.

Generalization to MU-MIMO

In MU-MIMO regularized ZF with N_r signal streams transmitted to every N_r-antenna user, the N_r column-vector precoders for user u can be brought together into that user's precoding matrix. Thus, $\boldsymbol{F}_u^{\mathsf{Reg}}$ is still essentially given by (9.282), only with $\boldsymbol{H}_0, \ldots, \boldsymbol{H}_{U-1}$ being matrices rather than vectors and with the normalization being effected column-wise to ensure that $\|[\boldsymbol{F}_u^{\mathsf{Reg}}]_{:,i}\|^2 = N_\mathrm{t}$ for $i = 0, \ldots, N_\mathrm{r} - 1$.

9.9.2 Power allocation and ergodic spectral efficiency

The MU-MISO case

To tackle the optimization of $\frac{E_0}{E_\mathrm{s}}, \ldots, \frac{E_{U-1}}{E_\mathrm{s}}$, let us again begin with the MU-MISO case. From the SINR expression in (9.264), the ergodic spectral efficiency achieved by user u is

$$C_u^{\mathsf{Reg}} = \mathbb{E}\left[\log_2\left(1 + \frac{\frac{E_u}{E_\mathrm{s}}\,|\boldsymbol{H}_u \boldsymbol{F}_u^{\mathsf{Reg}}|^2}{\sum_{\mathrm{u} \neq u} \frac{E_\mathrm{u}}{E_\mathrm{s}}\,|\boldsymbol{H}_u \boldsymbol{F}_\mathrm{u}^{\mathsf{Reg}}|^2 + \frac{N_\mathrm{t}}{\mathsf{SNR}_u}}\right)\right], \tag{9.283}$$

with $\boldsymbol{F}_0^{\text{Reg}}, \ldots, \boldsymbol{F}_{U-1}^{\text{Reg}}$ as in (9.282). The weighted sum spectral efficiency is $\sum_{u=0}^{U-1} q_u C_u^{\text{Reg}}$, with the power allocation $\frac{E_0}{E_s}, \ldots, \frac{E_{U-1}}{E_s}$ the only aspect that remains to be optimized.

From the vantage of each user, it is best that all power be simply allocated to the corresponding signal stream, with none left for the streams of other users. The tensions that this selfish tendency creates must be balanced, depending on $q_0, \ldots, q_{U-1}$, by the power allocation algorithm. As pointed out in [874], there is in general no $\frac{E_0}{E_s}, \ldots, \frac{E_{U-1}}{E_s}$ that simultaneously maximizes all SINRs, meaning that the optimization of the power allocation must be effected directly over the weighted sum spectral efficiency. With CSIT available, the power allocation should aim at maximizing

$$\sum_{u=0}^{U-1} q_u \log_2 \left(1 + \frac{\frac{E_u}{E_s} |\boldsymbol{H}_u \boldsymbol{F}_u^{\text{Reg}}|^2}{\sum_{\text{u} \neq u} \frac{E_{\text{u}}}{E_s} |\boldsymbol{H}_u \boldsymbol{F}_{\text{u}}^{\text{Reg}}|^2 + \frac{N_t}{\mathsf{SNR}_u}} \right) \tag{9.284}$$

prior to any outer expectation over the fading. However, this optimization is not convex. Building the Lagrangian function (see Appendix G)

$$\begin{aligned} \mathsf{L}\left(\frac{E_0}{E_s}, \ldots, \frac{E_{U-1}}{E_s}, \lambda \right) = & \sum_{u=0}^{U-1} q_u \log_2 \left(1 + \frac{\frac{E_u}{E_s} |\boldsymbol{H}_u \boldsymbol{F}_u^{\text{Reg}}|^2}{\sum_{\text{u} \neq u} \frac{E_{\text{u}}}{E_s} |\boldsymbol{H}_u \boldsymbol{F}_{\text{u}}^{\text{Reg}}|^2 + \frac{N_t}{\mathsf{SNR}_u}} \right) \\ & + \lambda \left(\sum_{u=0}^{U-1} \frac{E_u}{E_s} - 1 \right) \end{aligned} \tag{9.285}$$

and taking partial derivatives with respect to $\frac{E_0}{E_s}, \ldots, \frac{E_{U-1}}{E_s}$, we obtain the following necessary (but not sufficient because of the nonconvexity) conditions for the optimum power allocation: for $u = 0, \ldots, U-1$,

$$\begin{aligned} \lambda + \frac{q_u |\boldsymbol{H}_u \boldsymbol{F}_u^{\text{Reg}}|^2}{\sum_{\text{u}=0}^{U-1} \frac{E_{\text{u}}}{E_s} |\boldsymbol{H}_u \boldsymbol{F}_{\text{u}}^{\text{Reg}}|^2 + \frac{N_t}{\mathsf{SNR}_u}} = & \sum_{u' \neq u} \frac{q_{u'} |\boldsymbol{H}_{u'} \boldsymbol{F}_u^{\text{Reg}}|^2}{\sum_{\text{u}=0}^{U-1} \frac{E_{\text{u}}}{E_s} |\boldsymbol{H}_{u'} \boldsymbol{F}_{\text{u}}^{\text{Reg}}|^2 + \frac{N_t}{\mathsf{SNR}_{u'}}} \\ & \cdot \frac{\frac{E_{u'}}{E_s} |\boldsymbol{H}_{u'} \boldsymbol{F}_{u'}^{\text{BD}}|^2}{\sum_{\text{u} \neq u'} \frac{E_{\text{u}}}{E_s} |\boldsymbol{H}_{u'} \boldsymbol{F}_{\text{u}}^{\text{Reg}}|^2 + \frac{N_t}{\mathsf{SNR}_{u'}}}, \end{aligned} \tag{9.286}$$

with λ such that $\sum_{u=0}^{U-1} \frac{E_u}{E_s} = 1$. These conditions lead to various candidate power allocations and, as the next example shows, when U is small these various candidates can be simply contrasted and the optimum power allocation determined by inspection. Since the nonnegativity of the powers is not explicitly incorporated into the above conditions, the feasibility of the candidates must be verified and solutions with negative powers must be discarded.

Example 9.46

Let $U = 2$ and $q_0 = q_1 = 1$, such that the quantity being targeted is the sum spectral efficiency. The combination of (9.286) for $u = 0$ and $u = 1$ gives

$$\frac{|\boldsymbol{H}_0 \boldsymbol{F}_0^{\text{Reg}}|^2}{\sum_{\text{u}=0}^{1} \frac{E_{\text{u}}}{E_s} |\boldsymbol{H}_0 \boldsymbol{F}_{\text{u}}^{\text{Reg}}|^2 + \frac{N_t}{\mathsf{SNR}_0}} - \frac{|\boldsymbol{H}_1 \boldsymbol{F}_1^{\text{Reg}}|^2}{\sum_{\text{u}=0}^{1} \frac{E_{\text{u}}}{E_s} |\boldsymbol{H}_1 \boldsymbol{F}_{\text{u}}^{\text{Reg}}|^2 + \frac{N_t}{\mathsf{SNR}_1}} \tag{9.287}$$

$$= \frac{|\boldsymbol{H}_1 \boldsymbol{F}_0^{\text{Reg}}|^2}{\sum_{\text{u}=0}^{1} \frac{E_{\text{u}}}{E_{\text{s}}} |\boldsymbol{H}_1 \boldsymbol{F}_{\text{u}}^{\text{Reg}}|^2 + \frac{N_{\text{t}}}{\mathsf{SNR}_1}} \cdot \frac{\frac{E_1}{E_{\text{s}}} |\boldsymbol{H}_1 \boldsymbol{F}_1^{\text{Reg}}|^2}{\frac{E_0}{E_{\text{s}}} |\boldsymbol{H}_1 \boldsymbol{F}_0^{\text{Reg}}|^2 + \frac{N_{\text{t}}}{\mathsf{SNR}_1}}$$
$$- \frac{|\boldsymbol{H}_0 \boldsymbol{F}_1^{\text{Reg}}|^2}{\sum_{\text{u}=0}^{1} \frac{E_{\text{u}}}{E_{\text{s}}} |\boldsymbol{H}_0 \boldsymbol{F}_{\text{u}}^{\text{Reg}}|^2 + \frac{N_{\text{t}}}{\mathsf{SNR}_0}} \cdot \frac{\frac{E_0}{E_{\text{s}}} |\boldsymbol{H}_0 \boldsymbol{F}_0^{\text{Reg}}|^2}{\frac{E_1}{E_{\text{s}}} |\boldsymbol{H}_0 \boldsymbol{F}_1^{\text{Reg}}|^2 + \frac{N_{\text{t}}}{\mathsf{SNR}_0}}$$

with $\frac{E_0}{E_s} + \frac{E_1}{E_s} = 1$. For every fading realization, this yields two potential solutions for the power allocation. Discarding any solutions not in the admissible set defined by

$$0 \leq \frac{E_u}{E_s} \leq 1 \qquad u = 0, 1 \tag{9.288}$$

and contrasting any remaining solutions against the extremes of the admissible set, namely $(\frac{E_0}{E_s} = 0, \frac{E_1}{E_s} = 1)$ and $(\frac{E_0}{E_s} = 1, \frac{E_1}{E_s} = 0)$, the optimum power allocation can be identified. Note that the extremes of the admissible set must be examined in case the spectral efficiency is monotonic thereon.

Example 9.47

Consider the setup of Example 9.46, further with

$$\mathsf{SNR}_0|_{\text{dB}} = \mathsf{SNR}|_{\text{dB}} \tag{9.289}$$
$$\mathsf{SNR}_1|_{\text{dB}} = \mathsf{SNR}|_{\text{dB}} + 5\,\text{dB} \tag{9.290}$$

and with $\boldsymbol{H}_0$ and $\boldsymbol{H}_1$ having IID Rayleigh-faded entries. Compute, as a function of SNR, the sum spectral efficiency with regularized ZF and optimized power allocation, and compare it against the sum spectral efficiency with ZF and waterfilling. Further contrast both linear transmission schemes against the BC sum-capacity.

Solution

The requested sum spectral efficiencies and the sum-capacity are depicted in Fig. 9.17. For the regularized ZF solution, the optimum power allocation is obtained by numerically solving the conditions in Example 9.46. In turn, the sum-capacity is obtained by applying a convex solver to (8.147). Notice how the SLNR-maximizing regularization enables a successful bridging of the gap between the ZF and the optimum DPC transmitters at low and medium SNRs.

As U grows, the computation and inspection of the solutions becomes cumbersome and alternative methods become desirable. Although approaches based on game theory, convex relaxation, or branch-and-bound have been suggested [875, 876], there is still a need for power allocation schemes that could be implemented in real time. Short of that, a uniform power allocation is a robust and uncomplicated recourse.

Generalization to MU-MIMO

The generalization of the foregoing derivations to multiantenna users requires, first of all, that we generalize the SINR expression in (9.264). With arbitrary precoders $\boldsymbol{F}_0, \ldots, \boldsymbol{F}_{U-1}$

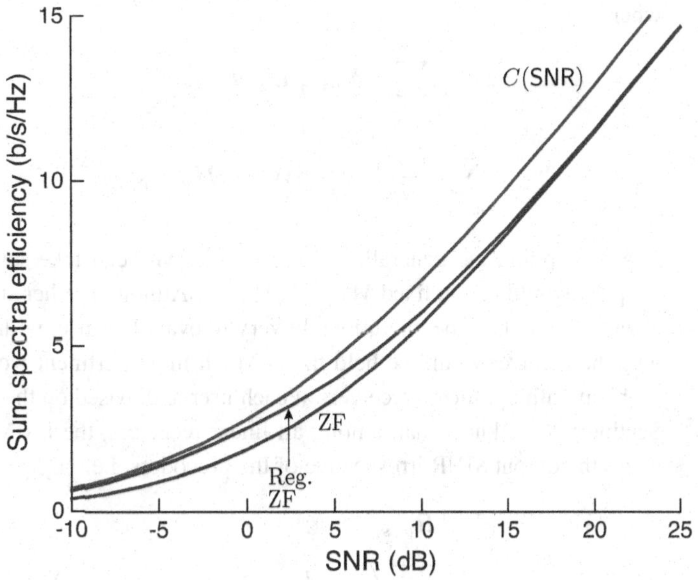

Fig. 9.17 MU-MISO BC sum spectral efficiency for $U = N_{\mathrm{t}} = 2$ as a function of SNR. The top curve is the BC sum-capacity. The lower two curves correspond to regularized (under SLNR maximization) ZF transmission with optimum power allocation, and to ZF with waterfilling.

and linear receivers $\boldsymbol{W}_0, \ldots, \boldsymbol{W}_{U-1}$, the output corresponding to the jth data stream at the uth user is

$$[\boldsymbol{W}_u^* \boldsymbol{y}_u]_j = [\boldsymbol{W}_u]_{:,j}^* \, \boldsymbol{y}_u \tag{9.291}$$

$$= [\boldsymbol{W}_u]_{:,j}^* \left(\sum_{\mathrm{u}=0}^{U-1} \sqrt{\frac{G_u E_{\mathrm{u}}}{N_{\mathrm{t}}}} \boldsymbol{H}_u \boldsymbol{F}_{\mathrm{u}} \boldsymbol{s}_{\mathrm{u}} + \boldsymbol{v}_u \right) \tag{9.292}$$

$$= \underbrace{\sqrt{\frac{G_u E_u}{N_{\mathrm{t}}}} [\boldsymbol{W}_u]_{:,j}^* \, \boldsymbol{H}_u \, [\boldsymbol{F}_u]_{:,j} \, [\boldsymbol{s}_u]_j}_{\text{Desired signal}}$$
$$+ \underbrace{\sqrt{\frac{G_u E_u}{N_{\mathrm{t}}}} \sum_{j' \neq j} [\boldsymbol{W}_u]_{:,j}^* \, \boldsymbol{H}_u \, [\boldsymbol{F}_u]_{:,j'} \, [\boldsymbol{s}_u]_{j'}}_{\text{Interference from same-user streams}}$$
$$+ \underbrace{\sum_{\mathrm{u} \neq u} \sqrt{\frac{G_u E_{\mathrm{u}}}{N_{\mathrm{t}}}} \sum_{j'=0}^{N_{\mathrm{r}}-1} [\boldsymbol{W}_u]_{:,j}^* \, \boldsymbol{H}_u \, [\boldsymbol{F}_{\mathrm{u}}]_{:,j'} \, [\boldsymbol{s}_{\mathrm{u}}]_{j'}}_{\text{Interference from other users}} + \underbrace{[\boldsymbol{W}_u]_{:,j}^* \, \boldsymbol{v}_u}_{\text{filtered noise}}, \tag{9.293}$$

from which, for given fading realizations,

$$\mathsf{sinr}_{u,j}^{\mathrm{Reg}} = \frac{\frac{G_u E_u}{N_{\mathrm{t}}} \left| [\boldsymbol{W}_u]_{:,j}^* \, \boldsymbol{H}_u \, [\boldsymbol{F}_u]_{:,j} \right|^2}{\mathrm{Den}}, \tag{9.294}$$

where

$$\text{Den} = \frac{G_u E_u}{N_\text{t}} \sum_{j' \neq j} \left| [\boldsymbol{W}_u]^*_{:,j} \boldsymbol{H}_u \, [\boldsymbol{F}_u]_{:,j'} \right|^2$$
$$+ \sum_{\text{u} \neq u} \frac{G_u E_\text{u}}{N_\text{t}} \sum_{j'=0}^{N_\text{r}-1} \left| [\boldsymbol{W}_u]^*_{:,j} \boldsymbol{H}_u \, [\boldsymbol{F}_\text{u}]_{:,j'} \right|^2 + N_0 \left\| [\boldsymbol{W}_u]_{j,:} \right\|^2. \tag{9.295}$$

At this point, the generalization to MU-MIMO can take several forms. The first and simplest would be via fixed $\boldsymbol{W}_0, \ldots, \boldsymbol{W}_{U-1}$. Although rather straightforward, this generalization would not be operationally very relevant because, with CSIR, there is no reason why the receiver would be held fixed. Much more pertinent would be to reformulate the problem with optimized receivers at each user and, based on the SU-MIMO derivations of Section 6.4 we know that, among all linear receivers, the LMMSE is the one that maximizes the output SINR irrespective of the precoders, i.e.,

$$\boldsymbol{W}_u^{\mathsf{MMSE}} = \boldsymbol{R}_{\boldsymbol{y}_u}^{-1} \boldsymbol{R}_{\boldsymbol{y}_u \boldsymbol{s}_u} \tag{9.296}$$
$$= \sqrt{\frac{N_\text{t}}{G_u E_u}} \left(\sum_{\text{u}=0}^{U-1} \frac{E_\text{u}}{E_u} \boldsymbol{H}_u \boldsymbol{F}_\text{u} \boldsymbol{F}_\text{u}^* \boldsymbol{H}_u^* + \frac{N_\text{t}}{\frac{E_u}{E_\text{s}} \mathsf{SNR}_u} \right)^{-1} \boldsymbol{H}_u \boldsymbol{F}_u. \tag{9.297}$$

Combining (9.294) and (9.295) with the above $\boldsymbol{W}_u^{\mathsf{MMSE}}$ plugged in, and after some algebra, we obtain an SINR expression that is reminiscent of—but not identical to—the ones obtained for SU-MIMO and for the MU-MIMO MAC with LMMSE reception, precisely

$$\mathsf{sinr}_{u,j}^{\mathsf{Reg}} = \frac{[\boldsymbol{F}_u]^*_{:,j} \boldsymbol{H}_u^* \left(\sum_{\text{u}=0}^{U-1} \frac{E_\text{u}}{E_u} \boldsymbol{H}_u \boldsymbol{F}_\text{u} \boldsymbol{F}_\text{u}^* \boldsymbol{H}_u^* + \frac{N_\text{t}}{\frac{E_u}{E_\text{s}} \mathsf{SNR}_u} \right)^{-1} \boldsymbol{H}_u \, [\boldsymbol{F}_u]_{:,j}}{1 - [\boldsymbol{F}_u]^*_{:,j} \boldsymbol{H}_u^* \left(\sum_{\text{u}=0}^{U-1} \frac{E_\text{u}}{E_u} \boldsymbol{H}_u \boldsymbol{F}_\text{u} \boldsymbol{F}_\text{u}^* \boldsymbol{H}_u^* + \frac{N_\text{t}}{\frac{E_u}{E_\text{s}} \mathsf{SNR}_u} \right)^{-1} \boldsymbol{H}_u \, [\boldsymbol{F}_u]_{:,j}}, \tag{9.298}$$

which is the basis from which the weighted sum spectral efficiency

$$\sum_{u=0}^{U-1} q_u \sum_{j=0}^{N_\text{r}-1} \mathbb{E}\left[\log_2\left(1 + \mathsf{sinr}_{u,j}^{\mathsf{MMSE}}\right) \right] \tag{9.299}$$

could then be maximized by adjusting $\boldsymbol{F}_0, \ldots, \boldsymbol{F}_{U-1}$ and $\frac{E_0}{E_\text{s}}, \ldots, \frac{E_{U-1}}{E_\text{s}}$. This optimization, decidedly nonconvex, is studied in [877], where a procedure that searches for maxima is derived. Unfortunately, besides the lack of guarantees that the global maximum be found, this procedure is iterative and based on CSIT, meaning that it must converge within the coherence of the small-scale fading. Hence, operationally practical algorithms to adjust the precoders and powers would be a welcome development.

As an alternative form of generalization to MU-MIMO, the transmission could adopt a regularized version of block-diagonalization rather than a regularized ZF, relaxing the requirement of controlling the interference among same-user streams [878].

Finally, the generalization to MU-MIMO could transcend the limitation of linearity at the corresponding multiantenna receivers and consider optimum receive structures, maintaining the requirement of linearity only at the transmitter. Then, rather than by (9.299), the weighted sum spectral efficiency would be given by

$$\sum_{u=0}^{U-1} q_u \, \mathbb{E}\left[\log_2 \det\left(I + \frac{\frac{E_u}{E_s}\mathsf{SNR}_u}{N_t} H_u F_u F_u^* H_u^* \right.\right.$$
$$\left.\left. \cdot \left(I + \sum_{\mathrm{u} \neq u} \frac{\frac{E_{\mathrm{u}}}{E_s}\mathsf{SNR}_{\mathrm{u}}}{N_t} H_{\mathrm{u}} F_{\mathrm{u}} F_{\mathrm{u}}^* H_{\mathrm{u}}^*\right)^{-1}\right)\right], \quad (9.300)$$

whose maximization over $F_0, \ldots, F_{U-1}$ and $\frac{E_0}{E_s}, \ldots, \frac{E_{U-1}}{E_s}$ is again a nonconvex problem. An iterative algorithm that seeks maxima of this function is put forth in [879, 880], and the same comments made in reference to [877] apply: there are no optimality guarantees and convergence should take place within the coherence of the small-scale fading.

Things simplify somewhat, as usual, in the large-dimensional regime, and readers interested in the corresponding formulation are referred to [881].

9.9.3 High-SNR regime

The spectral efficiency of a regularized ZF transmitter falls between that of a pure ZF transmitter and the capacity. Since the number of spatial DOF with ZF transmission coincides with that of the BC capacity, it follows that a regularized ZF transmitter enjoys the same number of spatial DOF: $S_\infty = \min(N_t, UN_r)$.

As in the MU-MIMO MAC, in the high-SNR regime the optimization of the MU-MIMO BC transmit powers becomes more tractable: it adopts the form of a geometric programming problem that can be recast as a convex problem [822]. Although this optimization is not consequential in terms of the number of spatial DOF, which is insensitive to the value of the transmit powers, it would register on the power offset.

9.10 Summary and outlook

In the MAC, the linear receivers reviewed in this chapter are an alternative to their nonlinear counterparts in Chapter 8. As in SU-MIMO, this offers flexibility in terms of the performance versus complexity tradeoff.

In the BC, linear transmitters play a much more central role. Because of the implementational difficulties of DPC, the structures reviewed in this chapter are not merely complementary, but arguably the prime choices. Moreover, owing to the necessity of CSIT, a realistic characterization of their performance entails delving into the duplexing, the insertion of pilots, and possibly the feedback, and a comprehensive such analysis for ZF transmitters is included in the chapter.

The main findings are summarized in the list of take-away points, while several open problems have been identified throughout the chapter.

- For the MAC, this includes the optimization of the precoders and of the power control, with and without CSIT, under LMMSE reception.

Take-away points

1. In a MAC with ZF reception, the ergodic spectral efficiency of user u is given by $C_u^{\text{ZF}} = \sum_{j=0}^{N_t-1} \mathbb{E}\left[\log_2\left(1 + \text{snr}_{u,j}^{\text{ZF}}\right)\right]$, where
$$\text{snr}_{u,j}^{\text{ZF}} = \frac{1}{N_0 \left[\boldsymbol{W}_u^{\text{ZF}*} \boldsymbol{W}_u^{\text{ZF}}\right]_{j,j}} \tag{9.301}$$
with $\boldsymbol{W}_u^{\text{ZF}}$ the uth block of N_t columns within $\boldsymbol{W}^{\text{ZF}}$, which in turn satisfies
$$\boldsymbol{W}^{\text{ZF}*} = \left[\sqrt{\frac{G_0 E_s}{N_t}} \boldsymbol{H}_0 \boldsymbol{F}_0 \cdots \sqrt{\frac{G_{U-1} E_s}{N_t}} \boldsymbol{H}_{U-1} \boldsymbol{F}_{U-1}\right]^{\dagger}. \tag{9.302}$$
2. In IID Rayleigh fading,
$$C_u^{\text{ZF}} = N_t \, e^{N_t/\text{SNR}_u} \sum_{q=1}^{N_r - U N_t + 1} \mathcal{E}_q\left(\frac{N_t}{\text{SNR}_u}\right) \log_2 e. \tag{9.303}$$
3. With ZF reception at high SNRs, it is optimum to have $U N_t = N_r$. As the SNRs diminish, the ratio $\frac{N_r}{U N_t}$ should rise above unity.
4. In a MAC with LMMSE reception, $C_u^{\text{MMSE}} = \sum_{j=0}^{N_t-1} \mathbb{E}\left[\log_2\left(1 + \text{sinr}_{u,j}^{\text{MMSE}}\right)\right]$, where $\text{sinr}_{u,j}^{\text{MMSE}} = \frac{1}{\text{MMSE}_{u,j}} - 1$ and $\text{MMSE}_{u,j}$ is given by
$$1 - [\boldsymbol{F}_u]_{:,j}^* \boldsymbol{H}_u^* \left(\sum_{u=0}^{U-1} \frac{\frac{E_u}{E_s}\text{SNR}_u}{\frac{E_u}{E_s}\text{SNR}_u} \boldsymbol{H}_u \boldsymbol{F}_u \boldsymbol{F}_u^* \boldsymbol{H}_u^* + \frac{N_t}{\frac{E_u}{E_s}\text{SNR}_u} \boldsymbol{I}\right)^{-1} \boldsymbol{H}_u \, [\boldsymbol{F}_u]_{:,j}.$$
5. The LMMSE MAC receiver for user u is
$$\boldsymbol{W}_u^{\text{MMSE}} = \sqrt{\frac{N_t}{G_u E_u}} \left(\sum_{u=0}^{U-1} \frac{\frac{E_u}{E_s}\text{SNR}_u}{\frac{E_u}{E_s}\text{SNR}_u} \boldsymbol{H}_u \boldsymbol{F}_u \boldsymbol{F}_u^* \boldsymbol{H}_u^* + \frac{N_t}{\frac{E_u}{E_s}\text{SNR}_u} \boldsymbol{I}\right)^{-1} \boldsymbol{H}_u \boldsymbol{F}_u.$$
6. Provided that $U N_t \leq N_r$, the high-SNR MAC behavior with ZF or LMMSE reception is governed by $\mathcal{S}_\infty = \min(U N_t, N_r)$.
7. With single-antenna users, duality indicates that any feasible combination of MAC SINRs achievable with receivers $\boldsymbol{W}_0, \ldots, \boldsymbol{W}_{U-1}$ can also be achieved in the BC with those same filters acting as precoders and with the BC power constraint equal to the sum of the power U MAC power constraints.
8. In the BC, the combination of ZF across users and joint processing of same-user streams yields block-diagonalization. User u's precoder is $\boldsymbol{F}_u^{\text{BD}} = \left(\boldsymbol{V}_{-u}^{\text{null}} \boldsymbol{V}_u^{\text{null}}\right) \boldsymbol{P}_u^{1/2}$ where $\boldsymbol{V}_{-u}^{\text{null}}$ contains the right singular vectors of
$$\boldsymbol{C}_{-u} = \left[\sqrt{G_0} \boldsymbol{H}_0^{\text{T}} \cdots \sqrt{G_{u-1}} \boldsymbol{H}_{u-1}^{\text{T}} \cdots \sqrt{G_{u+1}} \boldsymbol{H}_{u+1}^{\text{T}} \cdots \sqrt{G_{U-1}} \boldsymbol{H}_{U-1}^{\text{T}}\right]^{\text{T}}$$
corresponding to zero singular values while $\boldsymbol{V}_u^{\text{null}}$ contains the right singular vectors of $(\boldsymbol{H}_u \boldsymbol{V}_{-u}^{\text{null}})$ corresponding to nonzero singular values, and $\boldsymbol{P}_u$ is a power allocation matrix optimizable via waterfilling.

9. With block-diagonalization,

$$C_u^{\mathsf{BD}} = \sum_{j=0}^{N_r-1} \mathbb{E}\left[\log_2\left(1 + \frac{\frac{E_u}{E_s}\mathsf{SNR}_u}{N_t} P_{u,j}, \lambda_{u,j}\right)\right], \tag{9.304}$$

where $\lambda_{u,j}$ is the jth nonzero eigenvalue of $\left(\boldsymbol{H}_u \boldsymbol{V}_{-u}^{\mathsf{null}}\right)\left(\boldsymbol{H}_u \boldsymbol{V}_{-u}^{\mathsf{null}}\right)^*$ and the power allocation $\frac{E_0}{E_s}, \ldots, \frac{E_{U-1}}{E_s}$ can be further optimized for given weights $q_0, \ldots, q_{U-1}$. If the weights are equal, then this optimization is again a waterfilling.

10. At high SNRs, $\frac{E_u}{E_s} = \frac{q_u}{\sum_{u=0}^{U-1} q_u} + \mathcal{O}\left(\frac{1}{\mathsf{SNR}_u}\right)$ while $S_\infty = \min(N_t, UN_r)$.

11. Block-diagonalization can be simplified into ZF and, with IID Rayleigh fading specifically,

$$C_u^{\mathsf{ZF}} = \exp\left(\frac{N_r}{\frac{E_u}{E_s}\mathsf{SNR}_u}\right) \sum_{q=1}^{N_t-UN_r+1} \mathcal{E}_q\left(\frac{N_r}{\frac{E_u}{E_s}\mathsf{SNR}_u}\right) \log_2 e. \tag{9.305}$$

12. At high SNRs, it is optimum to have $UN_r = N_t$. As the SNRs diminish, the ratio $\frac{N_t}{UN_r}$ should progressively escalate.

13. Block-diagonalization and ZF can be regularized, improving the performance at low and medium SNRs. Since it is hard to establish the SINR-maximizing regularization, a suboptimum yet simple and robust alternative is to maximize the SLNRs.

14. With FDD, the following is required before payload data can be communicated on each BC coherence tile: transmitting unprecoded pilots from the N_t antennas, gathering CSI at the U users, reporting back the CSI, computing the precoders, transmitting pilots through those precoders, and obtaining precoded channel estimates at the users. The total overhead is small for pedestrian users, and modest for vehicular users provided U is not large. At high SNR, relative to the CSI-based performance, there is a decrease factor in DOF and a power penalty; altogether, the performance is satisfactory if $(3 + 2\rho)\,U \ll N_c$, with ρ the forward–reverse SNR imbalance.

15. With TDD or full duplexing, the procedure simplifies into: transmitting reverse-link pilots from the U users, effecting channel estimation, computing the precoders, transmitting pilots through those precoders, and establishing CSIR at the users. The condition ensuring satisfactory performance relaxes to $(1 + \rho)\,U \ll N_c$.

- For the BC, it includes the SINR-maximizing regularization, known to be difficult, and the optimum power allocation in a regularized transmission, which is nonconvex. For this latter problem, a possible way forward would be to perturb the ZF solution, which does result from a convex optimization.

Open problems notwithstanding, the contents of this chapter set the stage for massive MIMO, which is the subject of the next and final chapter of the book.

Problems

9.1 Reproduce the ZF curve and the SU-MIMO curves in Example 9.4.

9.2 Reproduce the ZF curves in Example 9.5.

9.3 Repeat Example 9.5 for $\mathsf{SNR} = 20$ dB and verify that, with ZF reception, as the SNR grows large the optimum number of active users and/or transmit antennas increases for each given N_r.

9.4 Formulate the MU-MIMO MAC LMMSE receiver by applying the SU-MIMO LMMSE solution to the SU-MIMO interpretation of the MAC, and verify that it is equivalent to (9.38).

9.5 Show that the expression for $\boldsymbol{W}_u^{\text{MMSE}}$ in a MAC converges to the corresponding $\boldsymbol{W}_u^{\text{ZF}}$ for $E_s/N_0 \to \infty$.

9.6 Show, by means of the matrix inversion lemma, that (9.41) is a valid expression for $\boldsymbol{E}_u(\boldsymbol{H}_0, \ldots, \boldsymbol{H}_{U-1})$.

9.7 Verify that (9.4) and (9.5), with the definition of $\boldsymbol{W}_u^{\text{MMSE}}$ given in (9.7), yield (9.45).

9.8 Reproduce the LMMSE curve in Example 9.4.

9.9 Reproduce the ZF and LMMSE dotted lines in Fig. 9.4 and provide a table containing the optimum value of U for each N_r and each type of receiver.

9.10 Compute the capacity region for the MAC in Example 9.12 and contrast it with the spectral efficiency region for an LMMSE receiver in the example.

9.11 Consider an MU-SIMO MAC with $U = N_r = 4$ operating at high SNR.

(a) Let the fading be IID Rayleigh. How much additional power is required with a ZF or LMMSE receiver to achieve the sum-capacity that an optimum receiver would attain?

(b) Concentrate now on the ZF or LMMSE receiver. Relative to the IID Rayleigh fading setting, how much more power is required to achieve the same sum spectral efficiency if the receiver conforms to the exponential correlation model (recall Example 3.38) with $\rho = 0.8$?

9.12 Specialize Example 9.13 to $U = 2$ and show analytically that the sum spectral efficiency is not a concave function of $\frac{E_0}{E_s}$ (for given $\frac{E_1}{E_s}$) and vice versa.

9.13 Specialize Example 9.13 to $U = 2$ with $|\boldsymbol{H}_0|^2 = |\boldsymbol{H}_1|^2 = 1$ and suppose that $\frac{E_0}{E_s}\mathsf{SNR}_0$ cannot exceed $\frac{E_1}{E_s}\mathsf{SNR}_1$. If $\frac{E_1}{E_s}\mathsf{SNR}_1 = 3$ dB, what value of $\frac{E_0}{E_s}\mathsf{SNR}_0$ maximizes the sum spectral efficiency? How about if $\frac{E_1}{E_s}\mathsf{SNR}_1 = 7$ dB?

9.14 Verify, for $U = 2$, that (9.65) and (9.67) yield $\mathsf{sinr}_u^{\text{MAC}}$ and $\mathsf{sinr}_u^{\text{BC}}$, respectively.

9.15 Let $N_t = 4$, $N_r = 1$, and $U = 4$, with $\mathsf{SNR}_u = \mathsf{SNR}$ for $u = 0, \ldots, 3$. Plot, as a function of SNR (in dB), the sum spectral efficiency of a ZF transmitter with waterfilling power allocation and with a uniform power allocation. Further plot the sum spectral efficiency when all users are constrained to having the same efficiency, i.e., when the transmit signal is $\boldsymbol{C}^{\dagger}[s_0 \; \cdots \; s_3]^{\mathrm{T}}$ with $\boldsymbol{C}$ as per (9.74).

9.16 Consider a linear ZF transmitter with N_t antennas and $U \leq N_t$ single-antenna users. Formulate the power allocation policy that maximizes the weighted sum spectral efficiency for arbitrary weights $q_0, \ldots, q_{U-1}$. Show that, for equal weights, it reduces to waterfilling.

9.17 Reproduce Example 9.19

9.18 Consider the setup of Examples 9.20 and 9.21. For $U, N_t \to \infty$ with $N_t/U = \beta$ the sum spectral efficiency normalized by N_t satisfies [827]

$$\frac{\sum_{u=0}^{U-1} C_u^{\text{ZF}}}{N_t} \to \frac{1}{\beta} \log_2(1 + (\beta - 1)\,\mathsf{SNR}). \tag{9.306}$$

Compute the optimum ratio β for $\mathsf{SNR} = 10$ dB and for $\mathsf{SNR} \to \infty$.

9.19 Repeat Example 9.21 for $\mathsf{SNR} = 20$ dB and verify that, with ZF transmission, as the SNR grows the optimum number of active users increases for each given N_t.

9.20 Show that, in a MU-MIMO BC, the high-SNR power loss of a linear ZF transmitter relative to an optimum transmitter admits, in addition to (9.121), the alternative expression

$$\Delta \mathcal{L}_\infty = \frac{1}{U} \sum_{q=0}^{U-1} \sum_{\ell=q+1}^{U-1} \frac{\log_2 e}{N_t - \ell}. \tag{9.307}$$

Hint: Exploit the high-SNR analysis of block-diagonalization.

9.21 Consider a six-antenna base station communicating with $U = 4$ single-antenna users having $\mathsf{SNR}_0 = 10$ dB, $\mathsf{SNR}_1 = 12$ dB, $\mathsf{SNR}_2 = 14$ dB, and $\mathsf{SNR}_3 = 16$ dB. Assume CSIT and IID Rayleigh fading. Relative to the BC sum-capacity with the foregoing SNRs, how much additional transmit power would a linear ZF transmitter require to perform equally? Contrast the exact calculation with the approximate value obtained from power offset expressions.

9.22 For the setup of Example 9.25, compute the high-SNR spectral efficiency shortfall of a ZF transmitter relative to the sum-capacity, i.e., convert the high-SNR power loss into a spectral efficiency difference.

9.23 In SU-MIMO with digital CSI feedback, reporting back a selected precoder (via codebook indexing) has some advantages with respect to reporting back the pilot observation or the ensuing channel estimate. Is this still the case in the MU-MIMO BC?

9.24 Show that, with a suitably defined sinr_u, (9.148) can be rewritten as

$$\mathsf{sinr}_{\mathsf{eff},u}(\hat{a}_{u,u}) = \frac{\frac{1}{U}\mathsf{sinr}_u \cdot (1 - \mathsf{MMSE}_{u,u})}{1 + \frac{1}{U}\mathsf{sinr}_u \cdot \mathsf{MMSE}_{u,u}}, \tag{9.308}$$

which can be recognized as a counterpart to, and in fact a generalization of, the single-user effective SNR in (4.218).

9.25 Verify (9.173) and (9.174).

9.26 Repeat Example 9.27 for $U = N_t = 3$.
Note: A convex solver is required, for instance `fmincon` *in* MATLAB®.

9.27 Reproduce Example 9.28.

9.28 Suppose that the reverse-link SNR is held fixed or simply does not improve at the same pace as the forward SNR. We can model this behavior by letting

$$\frac{\mathsf{SNR}^{\mathsf{r}}_u}{\mathsf{SNR}_u} \to 0 \tag{9.309}$$

for $\mathsf{SNR}_u \to \infty$. Show that, under this condition, the right-hand side of (9.181) does not exhibit a sustained growth with SNR_u and thus the spectral efficiency lower bound exhibits zero spatial DOF.

9.29 Prove that $\Delta\mathcal{L}^{\mathsf{ZF}}_{\infty}$ as expressed in (9.192), which is valid for $U = N_{\mathrm{t}}$, achieves its highest value for $U = N_{\mathrm{t}} = 2$.

9.30 Incorporating fading onto the reverse-link analog feedback, (9.166) becomes

$$\boldsymbol{y}^{\mathsf{r}}_u = \frac{\sqrt{\alpha_{\mathsf{fb}} \frac{N_{\mathrm{c}}}{N_{\mathrm{t}} U} G_u N_{\mathrm{t}} E^{\mathsf{r}}_{\mathsf{s}}}}{\sqrt{\frac{1}{N_{\mathrm{t}}} \mathbb{E}[\|\boldsymbol{y}_u\|^2]}} \boldsymbol{H}^{\mathsf{r}}_u \boldsymbol{y}_u + \boldsymbol{v}^{\mathsf{r}}_u, \tag{9.310}$$

where $\boldsymbol{H}^{\mathsf{r}}_u \sim \mathcal{N}_{\mathbb{C}}(\mathbf{0}, \boldsymbol{I})$ is the $N_{\mathrm{t}} \times 1$ vector connecting the uth user with the N_{t} base station antennas. Assuming $\boldsymbol{H}^{\mathsf{r}}_0, \ldots, \boldsymbol{H}^{\mathsf{r}}_{U-1}$ are known by the base station, rederive the spectral efficiency lower bound in (9.181) conditioned thereon. Although the reverse channel is hardly ergodic over a given feedback transmission—it takes a single value per block—its fading merely alters the variance of the interference leakage that impairs the forward communication and the codewords being transmitted forwardly do experience ergodicity. It is thus legitimate to average the spectral efficiency lower bound for user u over the distribution of $\boldsymbol{H}^{\mathsf{r}}_u$. Taking such expectation, and expanding with respect to SNR_u, verify that the number of spatial DOF is unaltered while the power offset penalty in (9.192) becomes

$$\Delta\mathcal{L}^{\mathsf{ZF}}_{\infty} = \mathbb{E}\left[\log_2\left(2 + \rho \frac{1 - \frac{1}{U}}{\|\boldsymbol{H}_u\|^2}\right)\right]. \tag{9.311}$$

Finally, confirm that, for $U = N_{\mathrm{t}} = 2$, reverse-link fading known by the base station increases the power offset penalty only slightly, from 3.52 dB to 3.86 dB.

9.31 Reproduce the TDD spectral efficiency lower bound and the various overheads in Example 9.35.
Note: A convex solver is required, for instance `fmincon` *in* MATLAB®.

9.32 It is found in Example 9.34 that $U < 5$ vehicular users can be supported, with FDD and a 10-dB forward–reverse SNR asymmetry, without exceeding a 10% decrease in the number of spatial DOF of pilot-assisted ZF transmission relative to CSI-based ZF transmission. How many more users could be supported with TDD, everything else being the same?

9.33 Consider an MU-MISO ZF transmission with $U = N_{\mathrm{t}} = 2$ and a 10-dB power asymmetry between the forward and reverse links. What percentage of the two spatial DOF achievable with CSI can be achieved with pilot-assisted ZF transmission? Answer the question for both FDD and TDD by means of Table 9.1.

9.34 Repeat Example 9.40 but with $U = 2$, keeping only the two strongest users in the original example.

9.35 Consider a BC with $N_r \geq 1$ and $N_t \geq U N_r$ with all users operating at high SNR. How much of the power loss experienced by linear ZF transmission is recovered by applying block-diagonalization?

9.36 Consider a four-antenna base station communicating with $U = 2$ two-antenna users having $\mathsf{SNR}_0 = 15$ dB and $\mathsf{SNR}_1 = 20$ dB. Assume CSIT and IID Rayleigh fading. Relative to the BC sum-capacity at these SNRs, how much additional transmit power would a linear ZF transmitter require to perform equally? Contrast the exact calculation with the approximate value obtained from power offset expressions.

9.37 Repeat Problem 9.36 for block-diagonalization rather than ZF.
Note: A convex solver is required to compute the BC capacity, for instance `fmincon` *in* MATLAB®.

9.38 Reconsider Example 9.40.
(a) Compute the ZF sum spectral efficiency with a uniform power allocation, and compare it with the waterfilling curve in the original example.
(b) Repeat part (a) for block-diagonalization.

9.39 Repeat Example 9.40 with correlation 0.75 between the antennas at each receiver.

9.40 Consider an MU-MIMO BC. Having CSIT, the base station chooses to apply a block-diagonalization precoding. Show that, in the high-SNR regime, the optimum power allocation among users expands as in (9.252).

9.41 Verify that (9.280) is maximized by $\boldsymbol{F}_u^{\text{Reg}}$ in (9.282).
Hint: Recall the derivation, in Chapter 6, of the linear filter that maximizes the SINR at the receiver.

9.42 Consider a MU-MISO BC with all users having the same SNR and regularized ZF transmission. Derive the regularizing term that maximizes the SINR at every user for $U \to \infty$ and verify that it equals the value that maximizes the SLNRs.

9.43 Reproduce the regularized ZF curve in Fig. 9.17.

9.44 For the setup of Examples 9.46 and 9.47, compute and plot the ergodic sum spectral efficiency as a function of SNR with regularized ZF precoding and a uniform power allocation. Contrasting this curve with its optimum-power-allocation counterpart in Fig. 9.17, what do you observe?

10 Massive MIMO

Originality consists of returning to the origin.

Antoni Gaudí

10.1 Introduction

Distilling the essence of the book up to this point, we can state that the addition of antennas at transmitters and receivers opens up new spatial signaling dimensions, and that these dimensions are indeed usable under reasonable assumptions—more precisely, under assumptions that hold for small and moderate numbers of antennas, when CSI can be taken for granted or the effects of its acquisition can be cleanly discounted. The result is a sustained increase in the spectral and/or power efficiency within the confines of such numbers of antennas. The reader may have noticed how, in the examples thus far, we have purposely avoided venturing beyond those confines, and we hasten to restate that the large-dimensional formulations scattered throughout the text are mere stratagems to simplify the analysis of settings with limited numbers of antennas.

The natural question to pose at this point is: how far can this go? Such is the question that drives this final chapter and, as the contents unfold, it is to become clear that a multicell formulation is required to address it. This entails mutually interfering MACs and BCs, one of each per cell. In this multicell context, we broaden the terminology and speak of reverse link and forward link, respectively, to refer to the transmissions by users and by base stations.

With such terminological broadening, and with some related notational adjustments, the chapter is organized in the following manner. Section 10.2 shifts things into gear and motivates key aspects, such as the payoffs of reciprocity and excess antennas, that to some extent have come to define massive MIMO itself. Then, Section 10.3 recollects earlier results on MMSE channel estimation and adapts them to a multicell environment; in the process, the issues of pilot reuse and pilot contamination—in a single-cell context these issues are immaterial—become manifest. Mirroring once more the MAC–BC coverage of previous chapters, Sections 10.4 and 10.5 tackle the transmission of payload data in the reverse and forward directions. Matched-filter structures, not considered for nonmassive settings, take center stage here, not only because of their satisfactory performance, but further owing to their instrumental value on the analytical front. Channel hardening, a phenomenon briefly encountered earlier in the text, emerges with force. Besides matched filters, the LMMSE receiver and the regularized ZF precoder, superior yet less analytically friendly, are also considered. Ramping things down, Section 10.6 surveys techniques to mitigate pilot contamination, Section 10.7 briefly addresses practical concerns that naturally arise in the formulation of massive MIMO, and Section 10.8 concludes the chapter.

10.2 Going massive

10.2.1 The massive MIMO regime

Venturing beyond the comfortable realm of moderate numbers of antennas and exploring what unfolds when these numbers are in the tens or even hundreds is sure to require a formulation devoid of side information, and to have a first taste we can recall the analysis of pilot-assisted ZF transmission in Section 9.7.6. In particular, and with the caveats associated with the high-SNR regime and a single-cell BC, we can recall from Table 9.1 that, with TDD/full duplex and $U = N_{\rm t}$, satisfactory forward-link operation is guaranteed if

$$(1+\rho)U \ll N_{\rm c}, \tag{10.1}$$

where ρ is the forward–reverse SNR ratio.

Example 10.1

Apply (10.1) to gauge the limits of massification in outdoor macrocellular deployments, which is where one can most easily imagine base stations with a multitude of antennas.

Solution

For outdoor macrocellular deployments, power asymmetries on the order of 20 dB are reasonable. Setting $\rho = 100$ in (10.1), we obtain

$$U \ll \frac{N_{\rm c}}{101} \tag{10.2}$$

and, if we interpret the "$\ll$" sign as $\frac{1}{10}$, then

$$U < \frac{N_{\rm c}}{1010}. \tag{10.3}$$

This condition indicates that, in a vehicular setting with $N_{\rm c} = 1000$, efficient MU-MIMO communication may not be possible while, in a pedestrian setting with $N_{\rm c} = 20\,000$, less than $U = 20$ users could be actively served from less than $N_{\rm t} = 20$ antennas. Hardly a huge number, which would shrink even further with FDD.

The foregoing example, based on a sufficient condition, does not imply that truly large numbers of antennas are out of reach, but rather that pushing into those numbers might require relaxing some premises. To that end, a couple of observations are in order.

- With TDD/full duplex, advantageously, the CSI-related overhead does not scale with N_t, but only with U as pilots are transmitted by users over the reverse link. This suggests decoupling N_t from U, such that the former, which does not affect the overhead, can grow larger. Taken to the limit, this leads to $N_t \gg U$.
- With hefty numbers of antennas and a correspondingly sized spectral efficiency, substantial overheads can be tolerated for a net performance that might still be very attractive. (As little as 50% of a very large figure may represent much more than 90% of a comparatively small figure.) This motivates having $U \gg 1$, such that the number of spatial DOF is large, even if the overhead is then high.

These considerations point to the regime where $N_t \gg U \gg 1$ or, bringing both link directions into the same umbrella,

$$N_a \gg U \gg 1, \tag{10.4}$$

where N_a is the number of antennas at the base station. These conditions frame the massive MIMO regime as originally envisioned by Marzetta, who spearheaded its analysis for matched-filter transceivers [882, 883]. With more sophisticated transceivers and/or the incorporation of power control, the first inequality softens into

$$N_a \geq U \gg 1, \tag{10.5}$$

which would be a more general framing of massive MIMO. The challenges and opportunities that arise in this regime quickly captured the imagination of researchers, and a string of follow-up analyses subsequently expanded its understanding [884–897].

In this chapter we concentrate on TDD/full duplex systems featuring linear transceivers at the base stations and a single antenna at each user. For such systems, we explore what is possible when the only fundamental limitation is either N_a or the real estate available at the base stations. Comments on FDD massive MIMO and on massive MIMO with multi-antenna users are provided in the closing section.

For the forward link of TDD/full duplex systems, a multistage procedure is laid down in Chapter 9. Incorporating to it a stage for reverse-link data transmission, we obtain the following scheme:

(1) Pilots are transmitted from the users.
(2) The base station effects channel estimation and computes the receive filters and forward precoders.
(3) Reverse-link data transmission takes place.
(4) Precoded pilots are transmitted through each precoder to enable precoded channel estimates at the users,
(5) Finally, forward data transmission occurs.

On each coherence tile, N_p symbols or OFDM resource elements are reserved for pilots, representing an overhead of $\alpha = N_p/N_c$. The remaining $N_c - N_p$ symbols are available for data. These can be apportioned between the forward and reverse directions according to the needs of the system. Moreover, the reverse and forward data transmission stages could be swapped.

The user channels are embodied by the column vectors $\boldsymbol{H}_0, \ldots, \boldsymbol{H}_{U-1}$ in the reverse link and by the row vectors $\boldsymbol{H}_0^*, \ldots, \boldsymbol{H}_{U-1}^*$ in the forward link. In the analysis that follows, we consider block fading with no antenna correlations at the base station; the impact of correlations is discussed at the end of the chapter. Each user is assumed to experience the same local-average SNR across the N_a antennas, a point that is also qualified at the end of the chapter.

Recognizing that the power budgets at base stations and users, E_s and E_s^r respectively, may be rather different indeed, we retain the forward–reverse SNR ratio

$$\rho = \frac{E_\mathrm{s}}{E_\mathrm{s}^\mathrm{r}} \tag{10.6}$$

such that, for the forward link, $\mathsf{SNR}_u = \frac{G_u E_\mathrm{s}}{N_0}$ with power allocation coefficient $\frac{E_u}{E_\mathrm{s}}$, whereas for the reverse link

$$\mathsf{SNR}_u^\mathrm{r} = \frac{G_u E_\mathrm{s}^\mathrm{r}}{N_0} \tag{10.7}$$

$$= \frac{\mathsf{SNR}_u}{\rho}, \tag{10.8}$$

with power control coefficient $\frac{E_u}{E_\mathrm{s}^\mathrm{r}}$.

10.2.2 Excess antennas

As argued, one of the postulates of massive MIMO is a potentially large ratio N_a/U—this ratio is the reciprocal of the system load in terms of users per spatial dimension—and thus it is worth dwelling on what this entails. Although it may seem that enlarging N_a much beyond U condemns us to a slow logarithmic improvement, two opportunities open up as N_a/U grows large.

(1) As advanced in the large-dimensional analysis of Chapter 5, the law of large numbers dictates that, if $\boldsymbol{W}_u$ and $\boldsymbol{H}_\mathrm{u}$ are N_a-dimensional vectors with IID entries, then $\frac{1}{N_\mathrm{a}}\boldsymbol{W}_u^*\boldsymbol{H}_\mathrm{u} \to \frac{1}{N_\mathrm{a}}\mathbb{E}[\boldsymbol{W}_u^*\boldsymbol{H}_\mathrm{u}]$ for $N_\mathrm{a}/U \to \infty$. Setting $\boldsymbol{W}_u = \boldsymbol{H}_u$, this leads to the channel hardening

$$\frac{1}{N_\mathrm{a}}\boldsymbol{H}_u^*\boldsymbol{H}_\mathrm{u} \overset{\text{a.s.}}{\to} 1 \qquad u,\mathrm{u} = 0,\ldots,U-1 \qquad u = \mathrm{u} \tag{10.9}$$

as well as to the asymptotic orthogonality

$$\frac{1}{N_\mathrm{a}}\boldsymbol{H}_u^*\boldsymbol{H}_\mathrm{u} \overset{\text{a.s.}}{\to} 0 \qquad u,\mathrm{u} = 0,\ldots,U-1 \qquad u \neq \mathrm{u}, \tag{10.10}$$

which is sometimes termed *favorable propagation*. Altogether, for $N_\mathrm{a} \gg U$, a receive filter $\boldsymbol{W}_u$ aligned with $\boldsymbol{H}_u$ shall reject most of the interference from users other than u; similarly, a precoder aligned with $\boldsymbol{H}_u$ shall inflict little interference onto users other than u. Hence, a simple matched filter for each user might suffice at the base station, both to receive and to transmit, drastically simplifying the tasks of interference avoidance. In fact, a condition weaker than (10.9) and (10.10) suffices: as long as $\boldsymbol{H}_u^*\boldsymbol{H}_u$

grows faster than $\boldsymbol{H}_u^* \boldsymbol{H}_{\text{u}}$ (u $\neq u$) for $N_{\text{a}}/U \to \infty$, a matched filter is ultimately effective to both receive and transmit.

(2) Because of channel hardening, forward-link precoded pilot transmissions become dispensable. Put differently, hardened channels acquire operational significance in massive MIMO because, in contrast with the large-dimensional analyses of earlier chapters, here the large number of antennas is factual rather than a mathematical artifice.

With simple matched filters then, as $N_{\text{a}}/U \to \infty$, the intended signals surge above the interference, the noise, and even the channel estimation errors, while the small-scale fading is averaged out. All of this materializes into sustainedly high SINRs for many simultaneous users and enormous spectral efficiencies, which is the promise of massive MIMO. At the same time, forward precoded pilots may be dispensable, a point whose implications are dissected later in the chapter.

10.3 Reverse-link channel estimation

Let us begin by examining the acquisition of CSI at a single base station. Suppose that there is no pilot power boosting; the same power control applies to data and pilot symbols. On every coherence tile, the uth user transmits a single pilot of energy E_u. With the U pilot symbols being mutually orthogonal, the base station observes

$$\boldsymbol{y}_u = \sqrt{G_u E_u}\, \boldsymbol{H}_u + \boldsymbol{v}_u \qquad u = 0, \ldots, U-1 \tag{10.11}$$

and, with $\boldsymbol{H}_u$ having independent entries, each such entry can be separately estimated without loss of optimality. Reproducing a basic result that appears throughout the book, the LMMSE channel estimates $\hat{\boldsymbol{H}}_0, \ldots, \hat{\boldsymbol{H}}_{U-1}$ obtained by the base satisfy

$$\boldsymbol{H}_u = \hat{\boldsymbol{H}}_u + \tilde{\boldsymbol{H}}_u \qquad u = 0, \ldots, U-1 \tag{10.12}$$

where the errors $\tilde{\boldsymbol{H}}_0, \ldots, \tilde{\boldsymbol{H}}_{U-1}$ are uncorrelated with the respective estimates and have variances

$$\mathsf{MMSE}_u = \frac{1}{1 + \frac{E_u}{E_{\text{s}}^{\text{r}}} \mathsf{SNR}_u^{\text{r}}} \qquad u = 0, \ldots, U-1. \tag{10.13}$$

10.3.1 Pilot reuse

At this point, a new ingredient must be incorporated into the analysis. Because the fading coherence N_{c} is finite, the number of orthogonal pilot dimensions $N_{\text{p}} < N_{\text{c}}$ is also finite. While it is certainly not a problem to assign, on each coherence tile, orthogonal pilots to the U active users within a certain cell, it is impossible to maintain strict pilot orthogonality over an entire network having many cells. Eventually, the system runs out of pilot dimensions and has to reuse them. Precisely, if there are L_{netw} cells, each having U users, then at least UL_{netw} pilots are required. Once $UL_{\mathsf{netw}} \geq N_{\text{c}}$, pilots are to be reused in different parts of the network. This reuse causes interference during the pilot stage and, to distinguish its effects from those of regular interference during the data transmission stages, the distinct term *pilot contamination* (or *pilot pollution*) has conveniently been coined.

To be sure, pilot reuse and contamination occur regardless of the number of antennas and thus the issue is not specific to massive MIMO. However, it is in this regime that it acquires relevance because U becomes large and because, unlike interference, noise, and fading, the contamination does not vanish for $N_{\mathrm{a}}/U \to \infty$, but rather it stubbornly persists.

10.3.2 Pilot contamination

Explicitly capturing pilot contamination requires zooming out to encompass multiple cells, and in turn an additional level of indexing. Referring to Fig. 10.1, where for illustration purposes the cells are hexagonal and arranged into a regular lattice, let us denote by $D_{l;\ell,u}$ the distance between the uth user at the ℓth cell and the base station at cell l. The pathloss associated with this distance, and the companion shadow fading, combine into a large-scale gain $G_{l;\ell,u}$, from which the channel connecting the uth user at the ℓth cell with the base station at cell l can be written as $\sqrt{G_{l;\ell,u}}\,\boldsymbol{H}_{l;\ell,u}$. In turn, the energy per symbol transmitted by user u at cell ℓ is $E_{\ell,u}$.

Since, in a large and homogeneous network, all cells are statistically equivalent, without loss of generality we declare cell 0 to be the cell of interest, meaning the cell where we assess performance. Letting $\mathcal{C}$ denote the subset of other cells reusing the same pilots as the cell of interest, the observations of the pilot transmissions at the base station of interest now become

$$\boldsymbol{y}_{0,u} = \sqrt{G_{0;0,u}E_{0,u}}\,\boldsymbol{H}_{0;0,u} + \sum_{\ell\in\mathcal{C}} \sqrt{G_{0;\ell,u}E_{\ell,u}}\,\boldsymbol{H}_{0;\ell,u} + \boldsymbol{v}_{0,u} \qquad u = 0,\ldots,U-1, \tag{10.14}$$

which reduce to (10.11) if there is no pilot reuse and the subset $\mathcal{C}$ is empty. The index u identifies the users at the cell of interest and at the cells within $\mathcal{C}$ that are sharing the same pilot dimension. Since the index of the cell of interest is uninformative, it can be dropped to avoid carrying it throughout the derivations; this compacts (10.14) into

$$\boldsymbol{y}_u = \sqrt{G_u E_u}\,\boldsymbol{H}_u + \sum_{\ell\in\mathcal{C}} \sqrt{G_{\ell,u}E_{\ell,u}}\,\boldsymbol{H}_{\ell,u} + \boldsymbol{v}_u \qquad u = 0,\ldots,U-1. \tag{10.15}$$

Jointly considering the large- and small-scale channel components and the power control, which is what the pilot symbols undergo, the base station of interest would like to estimate, for its U users,

$$\sqrt{\frac{E_u}{E_{\mathrm{s}}^{\mathrm{r}}} G_u}\,\boldsymbol{H}_u \qquad u = 0,\ldots,U-1. \tag{10.16}$$

However, it actually ends up estimating

$$\sqrt{\frac{E_u}{E_{\mathrm{s}}^{\mathrm{r}}} G_u}\,\boldsymbol{H}_u + \sum_{\ell\in\mathcal{C}} \sqrt{\frac{E_{\ell,u}}{E_{\mathrm{s}}^{\mathrm{r}}} G_{\ell,u}}\,\boldsymbol{H}_{\ell,u} \qquad u = 0,\ldots,U-1, \tag{10.17}$$

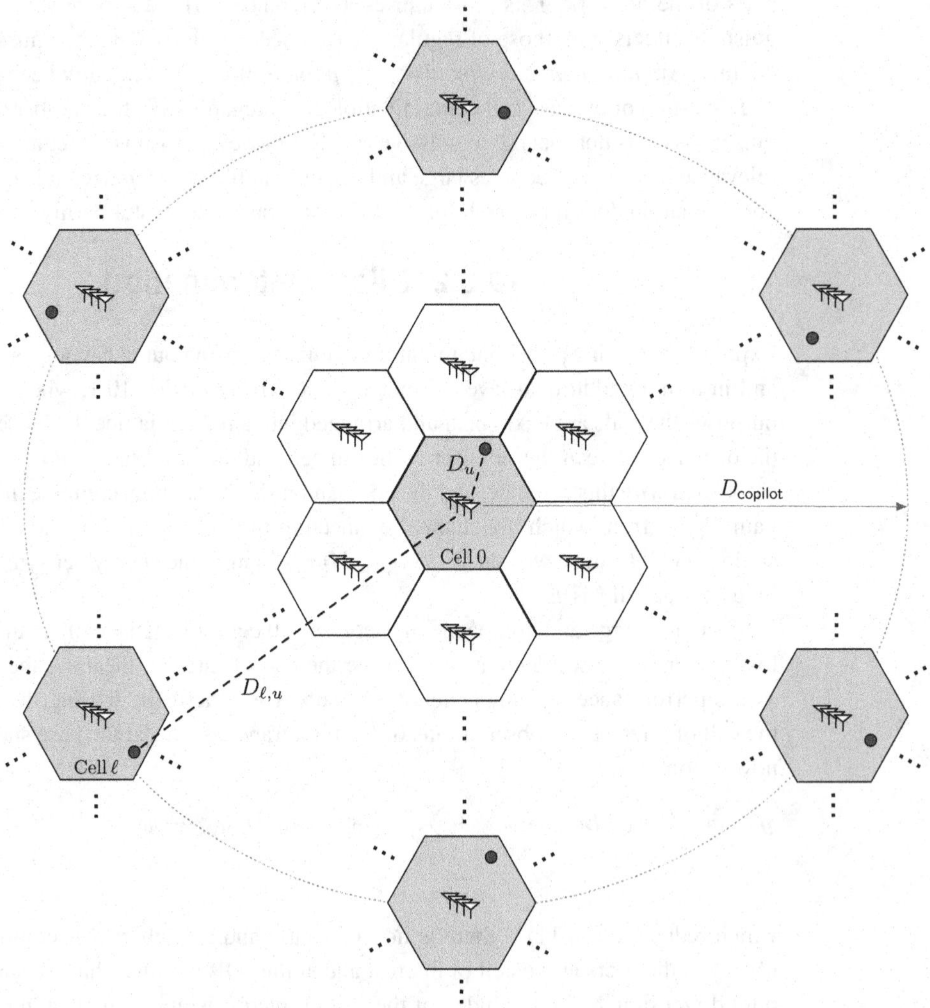

Fig. 10.1 Hexagonal lattice network. The cell of interest and the cells belonging to subset $\mathcal{C}$, i.e., all the ones reusing the same pilot dimensions, are shaded and a copilot user in each such cell is indicated by a circle. Also indicated is the distance D_{copilot} between the cell of interest and its first tier of copilot cells.

which are the composite channels between each set of copilot transmissions and that base. Scaled by the known large-scale components and power control coefficients, the small-scale channel estimates satisfy

$$\hat{\boldsymbol{H}}_u \propto \boldsymbol{H}_u + \sum_{\ell \in \mathcal{C}} \frac{\sqrt{\frac{E_{\ell,u}}{E_s^r} G_{\ell,u}}}{\sqrt{\frac{E_u}{E_s^r} G_u}} \boldsymbol{H}_{\ell,u} + \text{noise} \qquad u = 0, \ldots, U-1 \qquad (10.18)$$

or, equivalently,

$$\hat{\boldsymbol{H}}_u \propto \boldsymbol{H}_u + \underbrace{\sum_{\ell \in \mathcal{C}} \sqrt{\frac{\frac{E_{\ell,u}}{E_s^r}\mathsf{SNR}^r_{\ell,u}}{\frac{E_u}{E_s^r}\mathsf{SNR}^r_u}}\, \boldsymbol{H}_{\ell,u}}_{\text{pilot contamination}} + \text{noise} \qquad u = 0, \ldots, U-1. \tag{10.19}$$

Indeed, as the reader is invited to verify in Problem 10.4, the LMMSE channel estimate is

$$\hat{\boldsymbol{H}}_u = \frac{\frac{E_u}{E_s^r}\mathsf{SNR}^r_u}{1 + \frac{E_u}{E_s^r}\mathsf{SNR}^r_u + \sum_{\ell \in \mathcal{C}} \frac{E_{\ell,u}}{E_s^r}\mathsf{SNR}^r_{\ell,u}} \left(\boldsymbol{H}_u + \sum_{\ell \in \mathcal{C}} \sqrt{\frac{\frac{E_{\ell,u}}{E_s^r}\mathsf{SNR}^r_{\ell,u}}{\frac{E_u}{E_s^r}\mathsf{SNR}^r_u}}\, \boldsymbol{H}_{\ell,u} + \boldsymbol{v}'_u \right), \tag{10.20}$$

which does conform with (10.19) and makes precise the scaling of the intended term and of the pilot contamination, as well as the strength of the noise. The variance of the IID entries of $\boldsymbol{v}'_u$ is

$$\frac{1}{N_a}\mathbb{E}\left[\|\boldsymbol{v}'_u\|^2\right] = \frac{1}{\frac{E_u}{E_s^r}\mathsf{SNR}^r_u}. \tag{10.21}$$

Rearranging (10.20) into $\boldsymbol{H}_u = \hat{\boldsymbol{H}}_u + \tilde{\boldsymbol{H}}_u$, where $\hat{\boldsymbol{H}}_u$ and $\tilde{\boldsymbol{H}}_u$ are uncorrelated, the estimation error variance, i.e., the variance of the entries of $\tilde{\boldsymbol{H}}_u$, comes out as

$$\mathsf{MMSE}_u = \frac{1 + \sum_{\ell \in \mathcal{C}} \frac{E_{\ell,u}}{E_s^r}\mathsf{SNR}^r_{\ell,u}}{1 + \frac{E_u}{E_s^r}\mathsf{SNR}^r_u + \sum_{\ell \in \mathcal{C}} \frac{E_{\ell,u}}{E_s^r}\mathsf{SNR}^r_{\ell,u}}, \tag{10.22}$$

which reduces to (10.13) whenever there is no pilot reuse and $\mathcal{C}$ is empty. It can be verified from (10.20) that the variance of the entries of $\hat{\boldsymbol{H}}_u$ equals $(1 - \mathsf{MMSE}_u)$.

Examining (10.20), we observe that the channel estimate for user u at the base station of interest contains, besides an intended term, also an undesired term—whose power is weighted down by the ratio of local-average SNRs and power control coefficients—belonging to the corresponding user u in every copilot cell. As a result, when such channel estimate is subsequently applied to receive data, the base station of interest inadvertently welcomes small amounts of contamination-induced interference from the users in copilot cells; likewise, when the channel estimate is applied to transmit data, the base of interest inadvertently launches small amounts of power onto users in copilot cells, creating contamination-induced interference.

Typically, pilots are reused in cells that are not adjacent, but rather sufficiently apart to ensure that

$$\sum_{\ell \in \mathcal{C}} \frac{E_{\ell,u}}{E_s^r}\mathsf{SNR}^r_{\ell,u} \ll 1, \tag{10.23}$$

signifying that the total power received from copilot transmissions in other cells is well below the noise floor. Then, (10.20) and (10.22) can be seen to approximate their respective noise-limited forms, namely

$$\hat{\boldsymbol{H}}_u \approx \frac{\frac{E_u}{E_s^r}\mathsf{SNR}^r_u}{1 + \frac{E_u}{E_s^r}\mathsf{SNR}^r_u} \left(\boldsymbol{H}_u + \boldsymbol{v}'_u\right) \tag{10.24}$$

Discussion 10.1 Regular versus irregular pilot sequences

The pilot disposition considered in our analysis, consisting of a set of N_p symbols or OFDM resource elements declared as pilots of which each user is assigned a subset, can be seen as a special embodiment of a more general arrangement where there is a set of N_p orthogonal sequences of length N_p (e.g., the Zadoff–Chu sequences mentioned in Chapter 2) and each user is assigned one of those sequences. Our disposition, moreover, is *regular* in that the same set of sequences is reused over and over throughout the network.

In an alternative disposition of this more general arrangement, one could generate at random a set of N_p orthogonal sequences, for instance the columns of an $N_\mathrm{p} \times N_\mathrm{p}$ unitary isotropic random matrix; the eigenvectors of an IID complex Gaussian matrix are, recall, an example of such matrix. Then, rather than reuse the same set of orthogonal sequences once these have been exhausted, a new set could be drawn at random for every additional cluster of cells, in what we could call an *irregular* disposition.

In a regular disposition, the pilot symbols transmitted by user u in the cell of interest collide only with those transmitted by users u in cells $\ell \in \mathcal{C}$, but the collision is complete. In an irregular disposition, in contrast, the pilot sequences transmitted by user u in the cell of interest collide with those from every user in the network save for the $N_\mathrm{p} - 1$ in the same set; however, the collision is now only partial, dictated by the orientation of two random isotropic vectors. This partial overlap with a larger population of users has the effect of a spatial average of the pilot contamination [883, section VII-F]. Although an accurate comparison of how the regular and irregular dispositions perform would require detailed system-level simulations, the former appears preferable if the pilots are allocated with care whereas the latter seems more robust to careless allocations. Preferring to presume a carefully orchestrated system, in our analysis we consider a regular disposition, yet we do not expect any of the conclusions drawn in the chapter to change drastically with an irregular disposition.

and

$$\mathsf{MMSE}_u \approx \frac{1}{1 + \frac{E_u}{E_\mathrm{s}^\mathrm{r}} \mathsf{SNR}_u^\mathrm{r}}, \tag{10.25}$$

with the pilot contamination being negligible. This proviso is implicit in all pilot-assisted formulations in earlier chapters, and indeed it is amply satisfied.

However, because excess antennas effectively lower the noise floor, in terms of effective SINRs (with the processing at the receive filters or transmit precoders accounted for) the condition in (10.23) needs to be modified for massive MIMO. Moreover, this modification depends on the specific type of filters and precoders, which affect the degree of noise reduction and of other-user and other-cell interference adding to the noise. For that reason, we defer quantifying the impact of pilot contamination to later in the chapter, in the context of specific receivers and transmitters, meanwhile keeping it present in the general formulation.

Discussion 10.2 Aligned versus staggered pilot sequences

Another relevant aspect of the pilot disposition considered in our analysis is that, besides being regular, the N_p available pilot sequences are aligned. Put differently, pilot sequences are transmitted at once throughout the network, as depicted in the upper part of Fig. 10.2. Consequently, the transmission of payload data also takes place at once. With this disposition, the estimation of $\boldsymbol{H}_u$ is impaired by noise and contaminated by the pilots transmitted by users u in cells $\ell \in \mathcal{C}$. While this is perhaps the most natural disposition, it is not the only one [898–900].

Consider the alternative disposition in the lower part of Fig. 10.2, where the sets of N_p pilot sequences reused throughout the network are staggered and do not overlap. Besides noise, the estimation of $\boldsymbol{H}_u$ is now interfered by payload data from every user in every other-cluster cell. The LMMSE channel estimate in (10.20) then morphs (refer to Problem 10.5) into

$$\hat{\boldsymbol{H}}_u = \frac{\frac{E_u}{E_\mathrm{s}^\mathrm{r}}\mathsf{SNR}_u^\mathrm{r}}{1+\frac{E_u}{E_\mathrm{s}^\mathrm{r}}\mathsf{SNR}_u^\mathrm{r}+\sum_{\ell\in\mathcal{S}}\sum_{\mathrm{u}=0}^{U-1}\frac{E_{\ell,\mathrm{u}}}{E_\mathrm{s}^\mathrm{r}}\mathsf{SNR}_{\ell,\mathrm{u}}^\mathrm{r}} \cdot \left(\boldsymbol{H}_u + \sum_{\ell\in\mathcal{S}}\sum_{\mathrm{u}=0}^{U-1}\sqrt{\frac{\frac{E_{\ell,\mathrm{u}}}{E_\mathrm{s}^\mathrm{r}}\mathsf{SNR}_{\ell,\mathrm{u}}^\mathrm{r}}{\frac{E_u}{E_\mathrm{s}^\mathrm{r}}\mathsf{SNR}_u^\mathrm{r}}}\,\boldsymbol{H}_{\ell,\mathrm{u}}\, s_{\ell,\mathrm{u}} + \boldsymbol{v}_u'\right), \tag{10.26}$$

where $s_{\ell,\mathrm{u}}$ is a data symbol transmitted by user u in cell ℓ and the set $\mathcal{S}$ contains all cells whose transmissions are staggered relative to the cell of interest. The entries of $\boldsymbol{v}_u'$ are IID complex Gaussian with power

$$\frac{1}{N_\mathrm{a}}\mathbb{E}\left[\|\boldsymbol{v}_u'\|^2\right] = \frac{1}{\frac{E_u}{E_\mathrm{s}^\mathrm{r}}\mathsf{SNR}_u^\mathrm{r}}, \tag{10.27}$$

whereas the interference that the payload data inflict during the estimation stage is IID, but non-Gaussian on account of its unknown fading. Altogether, the entries of $\tilde{\boldsymbol{H}}_u$ are IID with variance

$$\mathsf{MMSE}_u = \frac{1+\sum_{\ell\in\mathcal{S}}\sum_{\mathrm{u}=0}^{U-1}\frac{E_{\ell,\mathrm{u}}}{E_\mathrm{s}^\mathrm{r}}\mathsf{SNR}_{\ell,\mathrm{u}}^\mathrm{r}}{1+\frac{E_u}{E_\mathrm{s}^\mathrm{r}}\mathsf{SNR}_u^\mathrm{r}+\sum_{\ell\in\mathcal{S}}\sum_{\mathrm{u}=0}^{U-1}\frac{E_{\ell,\mathrm{u}}}{E_\mathrm{s}^\mathrm{r}}\mathsf{SNR}_{\ell,\mathrm{u}}^\mathrm{r}}, \tag{10.28}$$

which is worse than its aligned-pilot counterpart in (10.22). Comparisons between the two dispositions are proposed in the problems at the end of the chapter.

10.4 Reverse-link data transmission

Upon data transmission from the users, the observation at the base station of interest is

$$\boldsymbol{y} = \sum_{\ell}\sum_{u=0}^{U-1}\sqrt{G_{\ell,u}E_{\ell,u}}\,\boldsymbol{H}_{\ell,u}s_{\ell,u} + \boldsymbol{v} \tag{10.29}$$

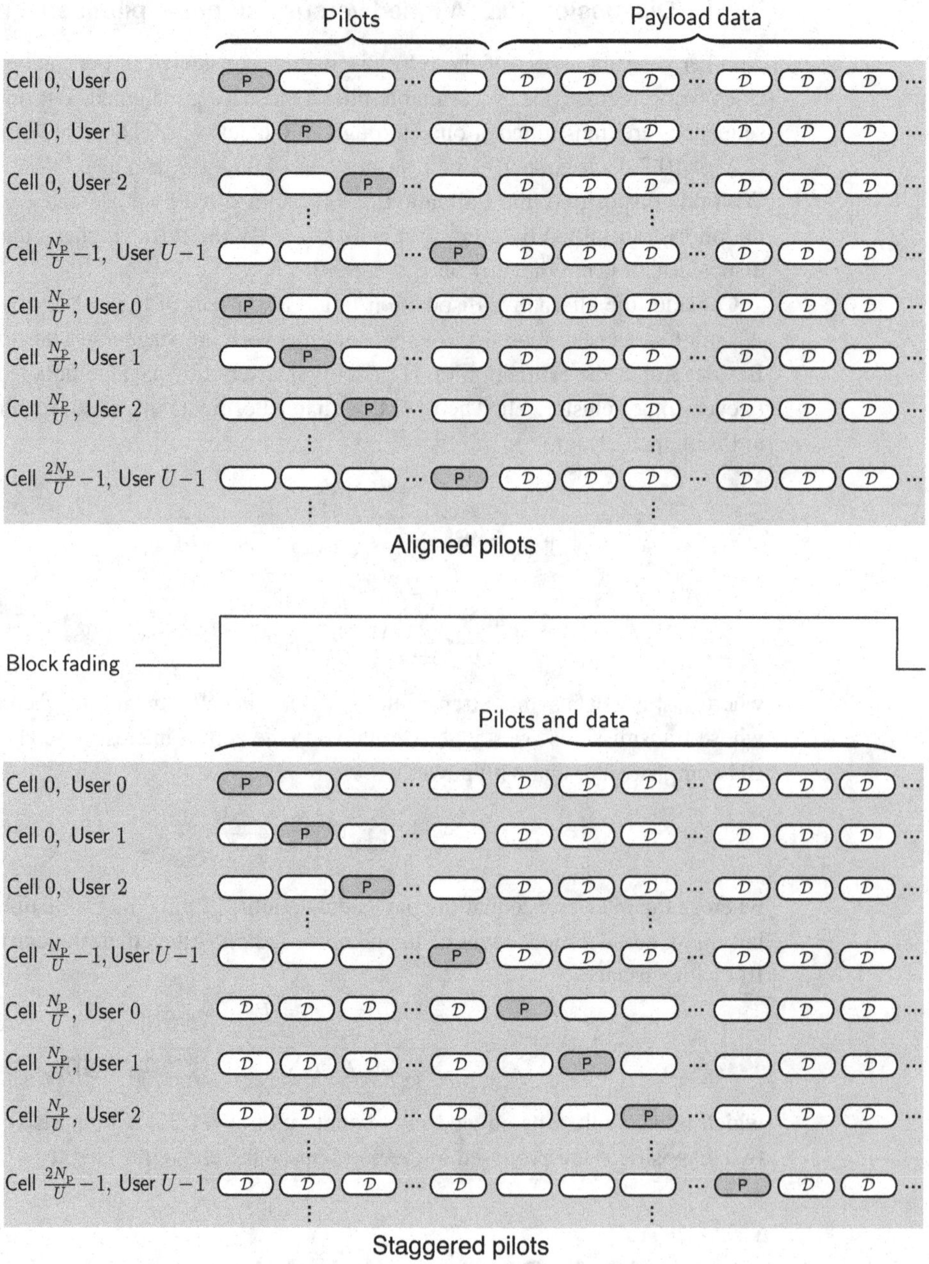

Fig. 10.2 Pilot dispositions in a network where each pilot sequence consists of a single active symbol (shown as a shaded shape) and $N_{\rm p}-1$ idle symbols (shown as clear shapes). Above, aligned pilots, with user u in cell ℓ experiencing contamination from user u in cells $\ell+N_{\rm p}/U$, $\ell+2N_{\rm p}/U$, and so on. Below, staggered pilots, with the contamination replaced by interference from payload data transmissions by all users in cells $\ell+N_{\rm p}/U$, $\ell+2N_{\rm p}/U$, and so on.

$$= \sum_{u=0}^{U-1} \sqrt{G_u E_u}\,(\hat{\boldsymbol{H}}_u + \tilde{\boldsymbol{H}}_u) s_u + \sum_{\ell \neq 0} \sum_{u=0}^{U-1} \sqrt{G_{\ell,u} E_{\ell,u}}\, \boldsymbol{H}_{\ell,u} s_{\ell,u} + \boldsymbol{v}, \tag{10.30}$$

where, in (10.30), we have segregated the same-cell and the other-cell transmissions and we have decomposed the fading for the former—the base station has estimates of these channels—as $\boldsymbol{H}_u = \hat{\boldsymbol{H}}_u + \tilde{\boldsymbol{H}}_u$ for $u = 0, \ldots, U-1$. Applying a generic linear receiver $\boldsymbol{W}_u$, the filtered observation for user u is

$$\begin{aligned} \boldsymbol{W}_u^* \boldsymbol{y} = & \underbrace{\sqrt{G_u E_u}\, \boldsymbol{W}_u^* \hat{\boldsymbol{H}}_u s_u}_{\text{Desired signal}} + \underbrace{\sqrt{G_u E_u}\, \boldsymbol{W}_u^* \tilde{\boldsymbol{H}}_u s_u}_{\text{Estimation error term}} \\ & + \underbrace{\sum_{\mathrm{u} \neq u} \sqrt{G_{\mathrm{u}} E_{\mathrm{u}}}\, \boldsymbol{W}_u^* (\hat{\boldsymbol{H}}_{\mathrm{u}} + \tilde{\boldsymbol{H}}_{\mathrm{u}}) s_{\mathrm{u}}}_{\text{Same-cell interference}} \\ & + \underbrace{\sum_{\ell \neq 0} \sum_{\mathrm{u}=0}^{U-1} \sqrt{G_{\ell,\mathrm{u}} E_{\ell,\mathrm{u}}}\, \boldsymbol{W}_u^* \boldsymbol{H}_{\ell,\mathrm{u}} s_{\ell,\mathrm{u}}}_{\text{Other-cell interference}} + \underbrace{\boldsymbol{W}_u^* \boldsymbol{v}}_{\text{Filtered noise}}, \end{aligned} \tag{10.31}$$

where the desired signal is the component projected on what the receiver regards as the true channel for this user, $\hat{\boldsymbol{H}}_u$, while the projection on the estimation error $\tilde{\boldsymbol{H}}_u$ is treated as additional Gaussian noise (filtered by $\boldsymbol{W}_u$). The other-cell interference, subsumed within the noise in most of this book, is made explicit in this formulation, but it is also treated as Gaussian noise (filtered by $\boldsymbol{W}_u$).

Conditioned on $\hat{\boldsymbol{H}}_0, \ldots, \hat{\boldsymbol{H}}_{U-1}$ and $\boldsymbol{W}_u$, which the base station of interest is privy to, the output SINR for user u equals

$$\mathsf{sinr}_u = \frac{\frac{E_u}{E_s^r} \mathsf{SNR}_u^r \left| \boldsymbol{W}_u^* \hat{\boldsymbol{H}}_u \right|^2}{\text{Den}}, \tag{10.32}$$

with

$$\begin{aligned} \text{Den} = & \underbrace{\sum_{\mathrm{u} \neq u} \frac{E_{\mathrm{u}}}{E_s^r} \mathsf{SNR}_{\mathrm{u}}^r \left| \boldsymbol{W}_u^* \hat{\boldsymbol{H}}_{\mathrm{u}} \right|^2}_{\text{Same-cell interference (estimated channels)}} + \underbrace{\left(1 + \sum_{\mathrm{u}=0}^{U-1} \frac{E_{\mathrm{u}}}{E_s^r} \mathsf{SNR}_{\mathrm{u}}^r \, \mathsf{MMSE}_{\mathrm{u}} \right) \| \boldsymbol{W}_u \|^2}_{\text{Noise plus same-cell estimation error terms}} \\ & + \underbrace{\sum_{\ell \neq 0} \sum_{\mathrm{u}=0}^{U-1} \frac{E_{\ell,\mathrm{u}}}{E_s^r} \mathsf{SNR}_{\ell,\mathrm{u}}^r \, \mathbb{E}\left[\left| \boldsymbol{W}_u^* \boldsymbol{H}_{\ell,\mathrm{u}} \right|^2 \middle| \boldsymbol{W}_u \right]}_{\text{Other-cell interference}}, \end{aligned} \tag{10.33}$$

where we have capitalized on the independence between $\boldsymbol{W}_u$ and $\tilde{\boldsymbol{H}}_{\mathrm{u}}$, for $\mathrm{u} = 0, \ldots, U-1$, to develop

$$\mathbb{E}\left[\left| \boldsymbol{W}_u^* \tilde{\boldsymbol{H}}_{\mathrm{u}} \right|^2 \middle| \boldsymbol{W}_u \right] = \mathbb{E}\left[\boldsymbol{W}_u^* \tilde{\boldsymbol{H}}_{\mathrm{u}} \tilde{\boldsymbol{H}}_{\mathrm{u}}^* \boldsymbol{W}_u \middle| \boldsymbol{W}_u \right] \tag{10.34}$$

$$= \boldsymbol{W}_u^* \, \mathbb{E}\left[\tilde{\boldsymbol{H}}_{\mathrm{u}} \tilde{\boldsymbol{H}}_{\mathrm{u}}^* \right] \boldsymbol{W}_u \tag{10.35}$$

$$= \boldsymbol{W}_u^* \left(\mathsf{MMSE}_{\mathrm{u}} \cdot \boldsymbol{I}\right) \boldsymbol{W}_u \tag{10.36}$$

$$= \mathsf{MMSE}_{\mathrm{u}} \, \|\boldsymbol{W}_u\|^2 . \tag{10.37}$$

As far as the other-cell interference, $\mathbb{E}\left[|\boldsymbol{W}_u^* \boldsymbol{H}_{\ell,\mathrm{u}}|^2 \,|\boldsymbol{W}_u\right] = \boldsymbol{W}_u^* \, \mathbb{E}\left[\boldsymbol{H}_{\ell,\mathrm{u}} \boldsymbol{H}_{\ell,\mathrm{u}}^* \,|\boldsymbol{W}_u\right] \boldsymbol{W}_u$ cannot be further elaborated in complete generality for copilot interferers, i.e., for $\ell \in \mathcal{C}$ and $\mathrm{u} = u$, because $\boldsymbol{W}_u$ is contaminated by, and therefore it is not independent of, $\boldsymbol{H}_{\ell,u}$ whenever $\ell \in \mathcal{C}$. For noncopilot interferers, though, $\boldsymbol{W}_u$ and $\boldsymbol{H}_{\ell,\mathrm{u}}$ do exhibit independence and thus

$$\boldsymbol{W}_u^* \, \mathbb{E}\left[\boldsymbol{H}_{\ell,\mathrm{u}} \boldsymbol{H}_{\ell,\mathrm{u}}^* \,|\boldsymbol{W}_u\right] \boldsymbol{W}_u = \boldsymbol{W}_u^* \, \mathbb{E}\left[\boldsymbol{H}_{\ell,\mathrm{u}} \boldsymbol{H}_{\ell,\mathrm{u}}^*\right] \boldsymbol{W}_u \tag{10.38}$$

$$= \|\boldsymbol{W}_u\|^2 , \tag{10.39}$$

allowing for (10.33) to be written as

$$\begin{aligned} \mathrm{Den} = {} & \sum_{\mathrm{u} \neq u} \frac{E_{\mathrm{u}}}{E_{\mathrm{s}}^{\mathrm{r}}} \mathsf{SNR}_{\mathrm{u}}^{\mathrm{r}} \left|\boldsymbol{W}_u^* \hat{\boldsymbol{H}}_{\mathrm{u}}\right|^2 \\ & + \left(1 + \sum_{\mathrm{u}=0}^{U-1} \frac{E_{\mathrm{u}}}{E_{\mathrm{s}}^{\mathrm{r}}} \mathsf{SNR}_{\mathrm{u}}^{\mathrm{r}} \, \mathsf{MMSE}_{\mathrm{u}} + \sum_{\substack{\ell \neq 0 \\ \ell \notin \mathcal{C}}} \sum_{\mathrm{u}=0}^{U-1} \frac{E_{\ell,\mathrm{u}}}{E_{\mathrm{s}}^{\mathrm{r}}} \mathsf{SNR}_{\ell,\mathrm{u}}^{\mathrm{r}} + \sum_{\ell \in \mathcal{C}} \sum_{\mathrm{u} \neq u} \frac{E_{\ell,\mathrm{u}}}{E_{\mathrm{s}}^{\mathrm{r}}} \mathsf{SNR}_{\ell,\mathrm{u}}^{\mathrm{r}} \right) \|\boldsymbol{W}_u\|^2 \\ & + \sum_{\ell \in \mathcal{C}} \frac{E_{\ell,u}}{E_{\mathrm{s}}^{\mathrm{r}}} \mathsf{SNR}_{\ell,u}^{\mathrm{r}} \, \boldsymbol{W}_u^* \, \mathbb{E}\left[\boldsymbol{H}_{\ell,u} \boldsymbol{H}_{\ell,u}^* \,|\boldsymbol{W}_u\right] \boldsymbol{W}_u , \end{aligned} \tag{10.40}$$

where the final term contains the copilot interference. Plugging (10.40) into (10.32), we obtain sinr_u for specific $\hat{\boldsymbol{H}}_0, \ldots, \hat{\boldsymbol{H}}_{U-1}$ and $\boldsymbol{W}_u$. If the channel estimation were perfect, the pilot contamination were negligible, and the other-cell interference were subsumed within the noise, we would recover from sinr_u the corresponding MU-SIMO MAC expression in (9.6). All these effects, essential for the derivations at hand, are herein explicit.

Let us distinguish as *gross* spectral efficiencies the values measured over the data symbols only, before discounting the pilot overheads. At the cell of interest, with the other-cell interference—whose fading is unknown to the base of interest—and the channel estimation errors all treated as filtered Gaussian noise, the gross reverse-link user spectral efficiencies

$$\frac{R_u}{B} = \mathbb{E}\left[\log_2(1 + \mathsf{sinr}_u)\right] \qquad u = 0, \ldots, U-1 \tag{10.41}$$

can be achieved, with expectation over $\hat{\boldsymbol{H}}_u$ and consequently also over $\boldsymbol{W}_u$. Such spectral efficiencies should be optimized over the choice of U. Subsequently, the combined gross spectral efficiencies over the reverse and forward data symbols need to be adjusted down by the pilot overhead, $N_{\mathrm{p}}/N_{\mathrm{c}}$, and optimized also over such overhead, and possibly also over the pilot power boosting.

10.4.1 Channel hardening

The foregoing derivations hinge on that, having gathered the estimate $\hat{\boldsymbol{H}}_u$, the decoder for user u at the base station of interest regards $\boldsymbol{W}_u^* \hat{\boldsymbol{H}}_u$ as the filtered channel bearing the desired signal.

As mentioned early in the chapter, one of the benefits of having excess antennas is the hardening of the filtered signals. To see how a receiver can capitalize on this phenomenon, let us decompose the filtered channel of user u at the base station of interest as

$$\boldsymbol{W}_u^* \boldsymbol{H}_u = \mathbb{E}[\boldsymbol{W}_u^* \boldsymbol{H}_u] + \left(\boldsymbol{W}_u^* \boldsymbol{H}_u - \mathbb{E}[\boldsymbol{W}_u^* \boldsymbol{H}_u] \right). \tag{10.42}$$

Suppose that, rather than $\boldsymbol{W}_u^* \hat{\boldsymbol{H}}_u$, the decoder regards $\mathbb{E}[\boldsymbol{W}_u^* \boldsymbol{H}_u]$ as the filtered channel—the receiver can compute this value from the channel statistics—that bears the desired signal and regards the signal borne by the bracketed term in (10.42) as self-interference, treated as additional noise. Since, for $N_{\mathrm{a}} \to \infty$, we have that $\frac{1}{N_{\mathrm{a}}} \boldsymbol{W}_u^* \boldsymbol{H}_u \stackrel{\text{a.s.}}{\to} \frac{1}{N_{\mathrm{a}}} \mathbb{E}[\boldsymbol{W}_u^* \boldsymbol{H}_u]$, it follows that, for growing N_{a}/U, the term $\mathbb{E}[\boldsymbol{W}_u^* \boldsymbol{H}_u]$ will come to accurately approximate the true filtered channel and the bracketed term in (10.42) will be small. (In non-massive MIMO, the same approach could be taken, but then $\mathbb{E}[\boldsymbol{W}_u^* \boldsymbol{H}_u]$ would be a lousy approximation to $\boldsymbol{W}_u^* \boldsymbol{H}_u$ and the self-interference would be correspondingly strong, altogether leading to poor performance.)

For a receiver regarding $\mathbb{E}[\boldsymbol{W}_u^* \boldsymbol{H}_u]$ as the filtered channel, the counterpart to (10.31) is

$$\begin{aligned} \boldsymbol{W}_u^* \boldsymbol{y} = {} & \underbrace{\sqrt{G_u E_u}\, \mathbb{E}[\boldsymbol{W}_u^* \boldsymbol{H}_u]\, s_u}_{\text{Desired signal}} + \underbrace{\sqrt{G_u E_u}\, (\boldsymbol{W}_u^* \boldsymbol{H}_u - \mathbb{E}[\boldsymbol{W}_u^* \boldsymbol{H}_u])\, s_u}_{\text{Self-interference}} \\ & + \underbrace{\sum_{\mathrm{u} \neq u} \sqrt{G_{\mathrm{u}} E_{\mathrm{u}}}\, \boldsymbol{W}_u^* \boldsymbol{H}_{\mathrm{u}}\, s_{\mathrm{u}}}_{\text{Same-cell interference}} + \underbrace{\sum_{\ell \neq 0} \sum_{\mathrm{u}=0}^{U-1} \sqrt{G_{\ell,\mathrm{u}} E_{\ell,\mathrm{u}}}\, \boldsymbol{W}_u^* \boldsymbol{H}_{\ell,\mathrm{u}}\, s_{\ell,\mathrm{u}}}_{\text{Other-cell interference}} + \underbrace{\boldsymbol{W}_u^* \boldsymbol{v}}_{\text{Filtered noise}} \end{aligned} \tag{10.43}$$

and the output SINR for user u at the base station of interest is

$$\mathsf{sinr}_u = \frac{\frac{E_u}{E_{\mathrm{s}}^{\mathrm{r}}} \mathsf{SNR}_u^{\mathrm{r}} \left| \mathbb{E}[\boldsymbol{W}_u^* \boldsymbol{H}_u] \right|^2}{\overline{\mathrm{Den}}}, \tag{10.44}$$

where

$$\begin{aligned} \overline{\mathrm{Den}} = {} & \frac{E_u}{E_{\mathrm{s}}^{\mathrm{r}}} \mathsf{SNR}_u^{\mathrm{r}}\, \mathrm{var}[\boldsymbol{W}_u^* \boldsymbol{H}_u] + \sum_{\mathrm{u} \neq u} \frac{E_{\mathrm{u}}}{E_{\mathrm{s}}^{\mathrm{r}}} \mathsf{SNR}_{\mathrm{u}}^{\mathrm{r}}\, \mathbb{E}\left[\left| \boldsymbol{W}_u^* \boldsymbol{H}_{\mathrm{u}} \right|^2 \right] \\ & + \sum_{\ell \neq 0} \sum_{\mathrm{u}=0}^{U-1} \frac{E_{\ell,\mathrm{u}}}{E_{\mathrm{s}}^{\mathrm{r}}} \mathsf{SNR}_{\ell,\mathrm{u}}^{\mathrm{r}}\, \mathbb{E}\left[\left| \boldsymbol{W}_u^* \boldsymbol{H}_{\ell,\mathrm{u}} \right|^2 \right] + \mathbb{E}\left[\| \boldsymbol{W}_u \|^2 \right]. \end{aligned} \tag{10.45}$$

Recalling the considerations made for the other-cell interference in the case of a receiver reliant on channel estimates, namely that $\boldsymbol{W}_u$ is independent of $\boldsymbol{H}_{\ell,\mathrm{u}}$ except for $\ell \in \mathcal{C}$ and $\mathrm{u} = u$ we can further elaborate $\overline{\mathrm{Den}}$ into

$$\begin{aligned} \overline{\mathrm{Den}} = {} & \frac{E_u}{E_{\mathrm{s}}^{\mathrm{r}}} \mathsf{SNR}_u^{\mathrm{r}}\, \mathrm{var}[\boldsymbol{W}_u^* \boldsymbol{H}_u] + \sum_{\mathrm{u} \neq u} \frac{E_{\mathrm{u}}}{E_{\mathrm{s}}^{\mathrm{r}}} \mathsf{SNR}_{\mathrm{u}}^{\mathrm{r}}\, \mathbb{E}\left[\left| \boldsymbol{W}_u^* \boldsymbol{H}_{\mathrm{u}} \right|^2 \right] \\ & + \left(1 + \sum_{\substack{\ell \neq 0 \\ \ell \notin \mathcal{C}}} \sum_{\mathrm{u}=0}^{U-1} \frac{E_{\ell,\mathrm{u}}}{E_{\mathrm{s}}^{\mathrm{r}}} \mathsf{SNR}_{\ell,\mathrm{u}}^{\mathrm{r}} + \sum_{\ell \in \mathcal{C}} \sum_{\mathrm{u} \neq u} \frac{E_{\ell,\mathrm{u}}}{E_{\mathrm{s}}^{\mathrm{r}}} \mathsf{SNR}_{\ell,\mathrm{u}}^{\mathrm{r}} \right) \mathbb{E}\left[\| \boldsymbol{W}_u \|^2 \right] \end{aligned}$$

$$+\sum_{\ell\in\mathcal{C}} \frac{E_{\ell,u}}{E_{\rm s}^{\rm r}} \mathsf{SNR}^{\rm r}_{\ell,u}\, \mathbb{E}\Big[\big|\boldsymbol{W}_u^* \boldsymbol{H}_{\ell,u}\big|^2\Big]. \tag{10.46}$$

Note that, in $\overline{\mathsf{sinr}}_u$ and $\overline{\text{Den}}$, we have introduced an overline, which is how we distinguish these hardening-based quantities from their channel-estimation-based brethren. Undoubtedly, $\overline{\mathsf{sinr}}_u \leq \mathbb{E}\big[\mathsf{sinr}_u\big]$ for $u = 0, \ldots, U-1$ because, to compute $\overline{\mathsf{sinr}}_0, \ldots, \overline{\mathsf{sinr}}_{U-1}$, we have unconditioned on $\hat{\boldsymbol{H}}_0, \ldots, \hat{\boldsymbol{H}}_{U-1}$ and $\boldsymbol{W}_u$, depriving the receiver of information. Comparisons between $\overline{\mathsf{sinr}}_u$ and $\mathbb{E}\big[\mathsf{sinr}_u\big]$ are provided in the section that begins hereafter.

With $\overline{\mathsf{sinr}}_0, \ldots, \overline{\mathsf{sinr}}_{U-1}$ stable over the respective local neighborhoods, the ensuing gross spectral efficiencies would not require expectations over the fading, but rather they would directly be

$$\frac{\bar{R}_u}{B} = \log_2\big(1 + \overline{\mathsf{sinr}}_u\big) \qquad u = 0, \ldots, U-1. \tag{10.47}$$

Since the base station does need to gather channel estimates to compute the receive filters, there is no operational advantage in exploiting channel hardening in the reverse link. Moreover, as argued, $\bar{R}_u/B \leq R_u/B$. However, the performance of a hardening-based receiver is easier to evaluate and, deep into the massive MIMO regime, it is hardly inferior, thereby serving as a tight lower bound and a convenient analytical instrument. And, by introducing hardening-based receivers at this point, we set the stage for the forward link, where they do have an operational advantage.

10.4.2 Matched-filter receiver

For nonmassive settings, previous chapters focus on the ZF and LMMSE receivers, which are crafted to reject, completely or partially, the interference from same-cell users. The simpler matched-filter receiver is disregarded because its performance in those settings is poor. For $N_{\rm a}/U \gg 1$, this is no longer the case, and a matched filter that ignores interference and merely seeks to capture as much desired power as possible can perform quite well. In fact, because of the asymptotic orthogonality of any two distinct channel vectors, the ZF and LMMSE receivers revert to a matched filter for $N_{\rm a}/U \to \infty$ and thus the appeal of the latter in this regime is based on a sound argument. Invoking the parallel beamforming interpretation of MIMO, what occurs for growing $N_{\rm a}/U$ is that the beam through which each user's signal is received becomes progressively sharper and the interference is naturally rejected. Indeed, this is one of the stimuli for having $N_{\rm a}/U \gg 1$.

A matched filter or maximum-ratio combiner for user u satisfies $\boldsymbol{W}_u^{\mathsf{MF}} \propto \hat{\boldsymbol{H}}_u$; the scaling factor is important to operate the decoder, but immaterial (because it equally affects signal, interference, and noise) to the output SINR and thus to the spectral efficiency. Recalling from (10.20) the estimate $\hat{\boldsymbol{H}}_u$ gathered by the base station, the matched filter for user u at the cell of interest can therefore be taken to be

$$\boldsymbol{W}_u^{\mathsf{MF}} = \sqrt{\frac{\frac{E_u}{E_{\rm s}^{\rm r}}\mathsf{SNR}^{\rm r}_u}{1 + \frac{E_u}{E_{\rm s}^{\rm r}}\mathsf{SNR}^{\rm r}_u + \sum_{\ell\in\mathcal{C}} \frac{E_{\ell,u}}{E_{\rm s}^{\rm r}}\mathsf{SNR}^{\rm r}_{\ell,u}}} \left(\boldsymbol{H}_u + \sum_{\ell\in\mathcal{C}} \sqrt{\frac{\frac{E_{\ell,u}}{E_{\rm s}^{\rm r}}\mathsf{SNR}^{\rm r}_{\ell,u}}{\frac{E_u}{E_{\rm s}^{\rm r}}\mathsf{SNR}^{\rm r}_u}}\, \boldsymbol{H}_{\ell,u} + \boldsymbol{v}_u' \right) \tag{10.48}$$

with scaling such that $E\big[\|\boldsymbol{W}_u^{\mathsf{MF}}\|^2\big] = N_\mathrm{a}$ and with the entries of $\boldsymbol{v}_u'$, recall, having power $1/(\frac{E_u}{E_\mathrm{s}^\mathrm{r}}\mathsf{SNR}_u^\mathrm{r})$. The pilots are presumed regular and aligned at every cell.

Output SINR

For $\boldsymbol{W}_u^{\mathsf{MF}}$, the conditional covariance of the copilot interference can be evaluated (refer to Problem 10.7) as

$$\mathbb{E}\big[\boldsymbol{H}_{\ell,u}\boldsymbol{H}_{\ell,u}^*\,|\boldsymbol{W}_u^{\mathsf{MF}}\big] = \boldsymbol{I} + \frac{\frac{E_{\ell,u}}{E_\mathrm{s}^\mathrm{r}}\mathsf{SNR}_{\ell,u}^\mathrm{r}}{1+\frac{E_u}{E_\mathrm{s}^\mathrm{r}}\mathsf{SNR}_u^\mathrm{r} + \sum_{l\in\mathcal{C}}\frac{E_{l,u}}{E_\mathrm{s}^\mathrm{r}}\mathsf{SNR}_{l,u}^\mathrm{r}}\big(\boldsymbol{W}_u^{\mathsf{MF}}\boldsymbol{W}_u^{\mathsf{MF}*} - \boldsymbol{I}\big) \tag{10.49}$$

from which, regrouping some terms, (10.32) and (10.40) become

$$\mathsf{sinr}_u^{\mathsf{MF}} = \frac{\frac{E_u}{E_\mathrm{s}^\mathrm{r}}\mathsf{SNR}_u^\mathrm{r}\,\big|\boldsymbol{W}_u^{\mathsf{MF}*}\hat{\boldsymbol{H}}_u\big|^2}{\mathrm{Den}^{\mathsf{MF}}} \tag{10.50}$$

and

$$\begin{aligned}\mathrm{Den}^{\mathsf{MF}} &= \sum_{\mathrm{u}\neq u}\frac{E_\mathrm{u}}{E_\mathrm{s}^\mathrm{r}}\mathsf{SNR}_\mathrm{u}^\mathrm{r}\,\big|\boldsymbol{W}_u^{\mathsf{MF}*}\hat{\boldsymbol{H}}_\mathrm{u}\big|^2 \\ &+\left(1+\sum_{\mathrm{u}=0}^{U-1}\frac{E_\mathrm{u}}{E_\mathrm{s}^\mathrm{r}}\mathsf{SNR}_\mathrm{u}^\mathrm{r}\,\mathsf{MMSE}_\mathrm{u} + \sum_{\ell\neq 0}\sum_{\mathrm{u}=0}^{U-1}\frac{E_{\ell,\mathrm{u}}}{E_\mathrm{s}^\mathrm{r}}\mathsf{SNR}_{\ell,\mathrm{u}}^\mathrm{r}\right)\|\boldsymbol{W}_u^{\mathsf{MF}}\|^2 \\ &+\frac{\sum_{\ell\in\mathcal{C}}\big(\frac{E_{\ell,u}}{E_\mathrm{s}^\mathrm{r}}\mathsf{SNR}_{\ell,u}^\mathrm{r}\big)^2}{1+\frac{E_u}{E_\mathrm{s}^\mathrm{r}}\mathsf{SNR}_u^\mathrm{r}+\sum_{\ell\in\mathcal{C}}\frac{E_{\ell,u}}{E_\mathrm{s}^\mathrm{r}}\mathsf{SNR}_{\ell,u}^\mathrm{r}}\big(\|\boldsymbol{W}_u^{\mathsf{MF}}\|^4 - \|\boldsymbol{W}_u^{\mathsf{MF}}\|^2\big)\end{aligned} \tag{10.51}$$

where the final term is the interference that other-cell copilot users provoke *in excess of* the interference they would cause if they did not reuse the pilots of the cell of interest. This term features $\|\boldsymbol{W}_u^{\mathsf{MF}}\|^4$, which behaves as $\mathcal{O}(N_\mathrm{a}^2)$ thereby reflecting the partial pointing of the receive beam formed by $\boldsymbol{W}_u^{\mathsf{MF}}$ to those other-cell copilot users.

Turning to a receiver reliant on channel hardening, (10.44) and (10.45) specialize, with a matched filter satisfying $\mathbb{E}\big[\|\boldsymbol{W}_u^{\mathsf{MF}}\|^2\big] = N_\mathrm{a}$, to

$$\overline{\mathsf{sinr}}_u^{\mathsf{MF}} = \frac{\frac{E_u}{E_\mathrm{s}^\mathrm{r}}\mathsf{SNR}_u^\mathrm{r}\,\big|\mathbb{E}[\boldsymbol{W}_u^{\mathsf{MF}*}\boldsymbol{H}_u]\big|^2}{\overline{\mathrm{Den}}^{\mathsf{MF}}} \tag{10.52}$$

$$= \frac{\big(\frac{E_u}{E_\mathrm{s}^\mathrm{r}}\mathsf{SNR}_u^\mathrm{r}\big)^2 N_\mathrm{a}^2}{\Big(1+\frac{E_u}{E_\mathrm{s}^\mathrm{r}}\mathsf{SNR}_u^\mathrm{r}+\sum_{\ell\in\mathcal{C}}\frac{E_{\ell,u}}{E_\mathrm{s}^\mathrm{r}}\mathsf{SNR}_{\ell,u}^\mathrm{r}\Big)\,\overline{\mathrm{Den}}^{\mathsf{MF}}}, \tag{10.53}$$

with

$$\begin{aligned}\overline{\mathrm{Den}}^{\mathsf{MF}} &= \frac{E_u}{E_\mathrm{s}^\mathrm{r}}\mathsf{SNR}_u^\mathrm{r}\,\mathrm{var}\big[\boldsymbol{W}_u^{\mathsf{MF}*}\boldsymbol{H}_u\big] + \sum_{\mathrm{u}\neq u}\frac{E_\mathrm{u}}{E_\mathrm{s}^\mathrm{r}}\mathsf{SNR}_\mathrm{u}^\mathrm{r}\,\mathbb{E}\Big[\big|\boldsymbol{W}_u^{\mathsf{MF}*}\boldsymbol{H}_\mathrm{u}\big|^2\Big] \\ &+\left(1+\sum_{\substack{\ell\neq 0\\ \ell\notin\mathcal{C}}}\sum_{\mathrm{u}=0}^{U-1}\frac{E_{\ell,\mathrm{u}}}{E_\mathrm{s}^\mathrm{r}}\mathsf{SNR}_{\ell,\mathrm{u}}^\mathrm{r} + \sum_{\ell\in\mathcal{C}}\sum_{\mathrm{u}\neq u}\frac{E_{\ell,\mathrm{u}}}{E_\mathrm{s}^\mathrm{r}}\mathsf{SNR}_{\ell,\mathrm{u}}^\mathrm{r}\right)N_\mathrm{a}\end{aligned}$$

$$+\sum_{\ell \in \mathcal{C}} \frac{E_{\ell,u}}{E_s^r} \mathsf{SNR}^r_{\ell,u} \, \mathbb{E}\left[\left|\boldsymbol{W}_u^{\mathsf{MF}*} \boldsymbol{H}_{\ell,u}\right|^2\right], \tag{10.54}$$

whose various terms are elaborated next. First of all (refer to Problem 10.11),

$$\mathrm{var}\left[\boldsymbol{W}_u^{\mathsf{MF}*} \boldsymbol{H}_u\right] = \mathbb{E}\left[\left|\boldsymbol{W}_u^{\mathsf{MF}*} \boldsymbol{H}_u\right|^2\right] - \frac{\frac{E_u}{E_s^r}\mathsf{SNR}^r_u}{1+\frac{E_u}{E_s^r}\mathsf{SNR}^r_u + \sum_{\ell \in \mathcal{C}} \frac{E_{\ell,u}}{E_s^r}\mathsf{SNR}^r_{\ell,u}} N_a^2 \tag{10.55}$$

$$= N_a. \tag{10.56}$$

Then, since $\boldsymbol{W}_u^{\mathsf{MF}}$ and $\boldsymbol{H}_\mathrm{u}$ are independent for $\mathrm{u} \neq u$,

$$\mathbb{E}\left[\left|\boldsymbol{W}_u^{\mathsf{MF}*} \boldsymbol{H}_\mathrm{u}\right|^2\right] = \mathbb{E}\left[\boldsymbol{W}_u^{\mathsf{MF}*} \, \mathbb{E}\left[\boldsymbol{H}_\mathrm{u} \boldsymbol{H}_\mathrm{u}^*\right] \boldsymbol{W}_u^{\mathsf{MF}}\right] \tag{10.57}$$

$$= \mathbb{E}\left[\|\boldsymbol{W}_u^{\mathsf{MF}}\|^2\right] \tag{10.58}$$

$$= N_a \tag{10.59}$$

whereas the unconditional copilot interference power (refer to Problem 10.8) equals

$$\mathbb{E}\left[\left|\boldsymbol{W}_u^{\mathsf{MF}*} \boldsymbol{H}_{\ell,u}\right|^2\right] = N_a + \frac{\frac{E_{\ell,u}}{E_s^r}\mathsf{SNR}^r_{\ell,u}}{1+\frac{E_u}{E_s^r}\mathsf{SNR}^r_u + \sum_{l \in \mathcal{C}} \frac{E_{l,u}}{E_s^r}\mathsf{SNR}^r_{l,u}} N_a^2 \qquad \ell \in \mathcal{C}. \tag{10.60}$$

Altogether,

$$\overline{\mathrm{Den}}^{\mathsf{MF}} = \left(1 + \sum_{\ell}\sum_{\mathrm{u}=0}^{U-1} \frac{E_{\ell,\mathrm{u}}}{E_s^r}\mathsf{SNR}^r_{\ell,\mathrm{u}}\right) N_a + \frac{\sum_{\ell \in \mathcal{C}} \left(\frac{E_{\ell,u}}{E_s^r}\mathsf{SNR}^r_{\ell,u}\right)^2}{1+\frac{E_u}{E_s^r}\mathsf{SNR}^r_u + \sum_{\ell \in \mathcal{C}} \frac{E_{\ell,u}}{E_s^r}\mathsf{SNR}^r_{\ell,u}} N_a^2 \tag{10.61}$$

from which, finally,

$$\overline{\mathsf{sinr}}_u^{\mathsf{MF}} = \frac{\left(\frac{E_u}{E_s^r}\mathsf{SNR}^r_u\right)^2}{\frac{1+\frac{E_u}{E_s^r}\mathsf{SNR}^r_u + \sum_{\ell \in \mathcal{C}} \frac{E_{\ell,u}}{E_s^r}\mathsf{SNR}^r_{\ell,u}}{N_a}\left(1 + \sum_{\ell}\sum_{\mathrm{u}=0}^{U-1} \frac{E_{\ell,\mathrm{u}}}{E_s^r}\mathsf{SNR}^r_{\ell,\mathrm{u}}\right) + \sum_{\ell \in \mathcal{C}}\left(\frac{E_{\ell,u}}{E_s^r}\mathsf{SNR}^r_{\ell,u}\right)^2}. \tag{10.62}$$

For the purpose of mirroring this expression with its forward-link counterpart later in the chapter, we also present it as

$$\overline{\mathsf{sinr}}_u^{\mathsf{MF}} = \frac{\frac{N_a}{1+\frac{E_u}{E_s^r}\mathsf{SNR}^r_u + \sum_{\ell \in \mathcal{C}} \frac{E_{\ell,u}}{E_s^r}\mathsf{SNR}^r_{\ell,u}} \left(\frac{E_u}{E_s^r}\mathsf{SNR}^r_u\right)^2}{1 + \sum_{\ell}\sum_{\mathrm{u}=0}^{U-1} \frac{E_{\ell,\mathrm{u}}}{E_s^r}\mathsf{SNR}^r_{\ell,\mathrm{u}} + \frac{N_a}{1+\frac{E_u}{E_s^r}\mathsf{SNR}^r_u + \sum_{\ell \in \mathcal{C}} \frac{E_{\ell,u}}{E_s^r}\mathsf{SNR}^r_{\ell,u}} \sum_{\ell \in \mathcal{C}}\left(\frac{E_{\ell,u}}{E_s^r}\mathsf{SNR}^r_{\ell,u}\right)^2}. \tag{10.63}$$

A first observation that can be made, by mere inspection, is that $\overline{\mathsf{sinr}}_u^{\mathsf{MF}} \to N_a$ for $\mathsf{SNR}^r_u \to \infty$. This SINR ceiling, caused by self-interference, is approached by users with the most favorable combinations of short distance and shadow fading to the serving base station, but cannot be exceeded by a matched-filter receiver reliant on channel hardening.

Beyond this observation, let us see how $\mathsf{sinr}_u^{\mathsf{MF}}$ and $\overline{\mathsf{sinr}}_u^{\mathsf{MF}}$ behave for relevant values of N_{a}/U, initially without power control.

Example 10.2

Consider the reverse link of a hexagonal lattice network where $\eta = 4$ and the shadow fading is log-normal with $\sigma_{\mathsf{dB}} = 8$ dB. The thermal noise is neglected, meaning that the system is interference-limited. Let $\frac{E_{\ell,u}}{E_{\mathrm{s}}^{\mathrm{r}}} = 1\ \forall \ell, u$. Further let $N_{\mathrm{p}} \to \infty$ such that there is no pilot contamination. Set $U = 10$. For $N_{\mathrm{a}}/U = 10$ and $N_{\mathrm{a}}/U = 100$, plot the CDFs of $\overline{\mathsf{sir}}_u^{\mathsf{MF}}$ and of $\mathbb{E}\big[\mathsf{sir}_u^{\mathsf{MF}}\big]$.

Solution

These distributions, and subsequent ones throughout the chapter, are produced by a Monte-Carlo simulator (versions of which are provided in the book's companion webpage). It encompasses a hexagonal network subset featuring a central cell of interest, where the performance statistics are computed, surrounded by two tiers of interfering cells; these 18 surrounding cells (6 on the first tier and 12 on the second) contribute a vast majority of the other-cell interference experienced by the central cell. In examples featuring pilot contamination, if the pilot reuse factor happens to be large enough for no copilot cells to be included within these two tiers, then an additional tier of six copilot cells is incorporated as in Fig. 10.1. All antennas have uniform patterns in azimuth. Users associate with the cell from whose base they have the strongest large-scale channel gain. Large-scale statistics are obtained from drops, each involving the uniformly random positioning with independent shadow fading instantiation of as many users as required for U users to associate with each cell. There is a tiny exclusion region around each base station—0.1% of the cell's area—to account for the base station height and avoid distance singularities. The number of drops is adjusted to push the 95% confidence interval below 0.25 dB in SIR. On each drop, the local-average expectation of sir_u and/or the corresponding spectral efficiency are computed by means of an inner Monte-Carlo drawing realizations of the IID Rayleigh fading.

The CDFs requested in the present example are displayed in Fig. 10.3. We observe that $\overline{\mathsf{sir}}_u^{\mathsf{MF}}$ tracks $\mathbb{E}\big[\mathsf{sir}_u^{\mathsf{MF}}\big]$ most of the way with a gap of about 2 dB that, in the absence of pilot contamination, remains roughly unchanged in absolute value and therefore shrinks relatively as N_{a}/U grows. In its upper tail, $\overline{\mathsf{sir}}_u^{\mathsf{MF}} \leq N_{\mathrm{a}}$ as anticipated.

For interested readers, the validity of the interference-limited premise in typical macrocellular conditions is verified in Problems 10.26 and 10.37.

A conclusion of broad significance can be drawn from the foregoing example: with matched-filter receivers and no power control, N_{a}/U does need to be truly large lest a substantial share of the users be starved of service. With $N_{\mathrm{a}}/U = 10$, as many as 25% of the users would have a local-average SIR below -5 dB, a value that we can regard as a reasonable threshold for service. (Equivalently, a service threshold could be imposed in terms of user spectral efficiency rather than SIR.) It takes $N_{\mathrm{a}}/U = 100$ for the percentage of users not reaching -5 dB to shrink to a more palatable, if still considerable, 8% of users. This testifies to the shortcomings of a matched-filter receiver for finite N_{a}/U, and

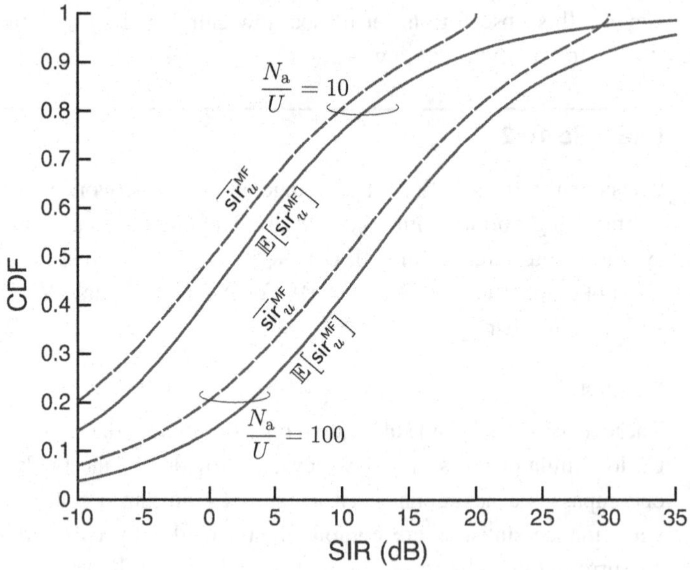

Fig. 10.3 Large-scale SIR distributions in the reverse link of a hexagonal network with no power control, $\eta = 4$, $\sigma_{\text{dB}} = 8$ dB, $N_\text{p} \to \infty$, and $U = 10$, for $N_\text{a}/U = 10$ and 100. The dashed lines depict $\overline{\text{sir}}_u^{\text{MF}}$ whereas the solid lines depict $\mathbb{E}\big[\text{sir}_u^{\text{MF}}\big]$.

the consequence is a highly spread distribution for the local-average SIR without power control. The lower tail, which essentially maps to users in cell-edge locations and/or with adverse shadow fading, is heavy: too many users are basically in large-scale outage, deprived of service. Lifting this lower tail requires an exceedingly large N_a/U and, for some given N_a, the corollary is a reduction in U and, ultimately, in the sum spectral efficiency. Referring back to Example 10.2, even with $N_\text{a} = 1000$, a ratio $N_\text{a}/U = 100$ allows for only $U = 10$.

Fractional power control

Enter power control. Formally, $\frac{E_{\ell,0}}{E_\text{s}^\text{r}}, \ldots, \frac{E_{\ell,U-1}}{E_\text{s}^\text{r}}$ should be optimized, at each cell ℓ, on the basis of that cell's weighted sum spectral efficiency, further with a view to causing the least amount of interference to other cells. And, in the absence of CSIT at the users, such optimization would have to be based on channel statistics only. This is a generalization of the power control problem encountered in Section 9.4, with the nonconvexity aggravated by the interactions among cells: a user's power increase worsens the interference, not only for same-cell users but, furthermore, for other-cell users. Maxima can be found, but with no assurance of global optimality [901, 902]. It is an imposing optimization, which relaxes somewhat only if all users are simultaneously at high SINR or for certain metrics that are not the weighted sum spectral efficiency [822, 823].

A pragmatic but very effective alternative to the formal optimization of the weighted sum spectral efficiency is the fractional power control policy, featured in LTE and NR, whereby [903–906]

$$\frac{E_{\ell,u}}{E_{\mathrm{s}}^{\mathrm{r}}} \propto \frac{1}{G_{\ell;\ell,u}^{\vartheta}} \qquad u = 0, \ldots, U-1, \tag{10.64}$$

where the proportionality factor should ensure that $\frac{E_{\ell,u}}{E_{\mathrm{s}}^{\mathrm{r}}} \leq 1$ for $u = 0, \ldots, U-1$, meaning that transmit powers do not exceed their maximum value. Besides such proportionality factor, advantageously, this policy features a single knob: the exponent $\vartheta \in [0,1]$. For $\vartheta = 0$, the fractional power control policy reverts to a fixed-power transmission. For $\vartheta > 0$, conversely, it compresses the dynamic range of the received powers in dB by a factor ϑ, lifting the lower tail of the SIR distribution at the expense of a cutback in the upper tail and, ultimately, in the cell's sum spectral efficiency. Exponent values on the order of $\vartheta = 0.5$–0.7 have been identified as providing a satisfactory balance between cell-edge and aggregate performance.

Example 10.3

Repeat Example 10.2 with fractional power control ($\vartheta = 0.7$), incorporating also the curves for $N_{\mathrm{a}}/U = 4$.

Solution

See Fig. 10.4, which confirms that the fractional power control has brought the percentage of users below -5 dB from 25% and 8%, respectively for $N_{\mathrm{a}}/U = 10$ and $N_{\mathrm{a}}/U = 100$, down to essentially zero. Even for $N_{\mathrm{a}}/U = 4$, the percentage with fractional power control is still essentially zero, indicating that N_{a}/U can be pushed even further downward.

Power control appears instrumental in massive MIMO with matched-filter receivers, and a fractional control policy can be an effective means to incorporate it. We therefore consider it in the sequel.

Behavior for $N_{\mathrm{a}}/U \to \infty$

To gauge the limits of massive MIMO, let us examine the behavior of the user SINRs for $N_{\mathrm{a}}/U \to \infty$.

Suppose first of all that we could afford having $N_{\mathrm{p}} \to \infty$ and hence no pilot contamination, such that $\hat{\boldsymbol{H}}_u = \boldsymbol{H}_u + \boldsymbol{v}_u'$ and $\boldsymbol{W}_u^{\mathsf{MF}} \propto \boldsymbol{H}_u + \boldsymbol{v}_u'$. Referring back to the expression for $\mathsf{sinr}_u^{\mathsf{MF}}$ in (10.50), since $\frac{1}{N_{\mathrm{a}}}\|\boldsymbol{H}_u\|^2 \overset{\text{a.s.}}{\to} 1$, $\frac{1}{N_{\mathrm{a}}}\boldsymbol{H}_u^*\boldsymbol{v}_u' \overset{\text{a.s.}}{\to} 0$ and $\frac{1}{N_{\mathrm{a}}}\|\boldsymbol{v}_u'\|^2 \overset{\text{a.s.}}{\to} 1/\mathsf{SNR}_u^{\mathrm{r}}$, the numerator of $\mathsf{sinr}_u^{\mathsf{MF}}$ would behave as

$$\frac{E_u}{E_{\mathrm{s}}^{\mathrm{r}}}\,\mathsf{SNR}_u^{\mathrm{r}}\left|\boldsymbol{W}_u^{\mathsf{MF}*}\hat{\boldsymbol{H}}_u\right|^2 \propto \frac{E_u}{E_{\mathrm{s}}^{\mathrm{r}}}\,\mathsf{SNR}_u^{\mathrm{r}}\left|\|\boldsymbol{H}_u\|^2 + \|\boldsymbol{v}_u'\|^2 + 2\,\Re[\boldsymbol{H}_u^*\boldsymbol{v}_u']\right|^2 \tag{10.65}$$

$$= \mathcal{O}(N_{\mathrm{a}}^2). \tag{10.66}$$

In turn, the denominator would behave as

$$\mathrm{Den}^{\mathsf{MF}} = \mathcal{O}(N_{\mathrm{a}}), \tag{10.67}$$

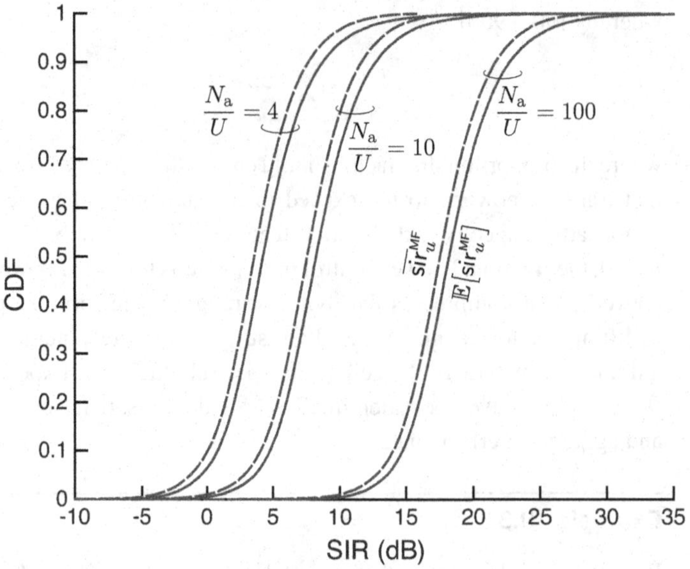

Fig. 10.4 Large-scale SIR distributions in the reverse link of a hexagonal network with fractional power control ($\vartheta = 0.7$), $\eta = 4$, $\sigma_{\text{dB}} = 8$ dB, $N_{\text{p}} \to \infty$, and $U = 10$, for $N_{\text{a}}/U = 4, 10,$ and 100. In dashed and in solid lines, respectively, $\overline{\text{sir}}_u^{\text{MF}}$ and $\mathbb{E}\big[\text{sir}_u^{\text{MF}}\big]$.

meaning that $\text{sinr}_u^{\text{MF}}$ would grow linearly and unboundedly with N_{a}. The same verdict would be reached for $\overline{\text{sir}}_u^{\text{MF}}$.

This powerful result corroborates that the effects of fading, interference from same- and other-cell transmissions, and noise could all be eradicated by having sufficiently many antennas per base station. Moreover, this could be accomplished with very simple linear receivers based on pilot-assisted channel estimates or even on channel hardening.

Now, let us see how the limiting behavior changes with pilot contamination taken into account, i.e., with $\boldsymbol{W}_u^{\text{MF}}$ as in (10.48). The numerator of $\text{sinr}_u^{\text{MF}}$ continues to exhibit the behavior in (10.66), precisely

$$\frac{E_u}{E_{\text{s}}^{\text{r}}}\,\text{SNR}_u^{\text{r}}\,\big|\boldsymbol{W}_u^{\text{MF}*}\hat{\boldsymbol{H}}_u\big|^2 = \frac{\left(\frac{E_u}{E_{\text{s}}^{\text{r}}}\text{SNR}_u^{\text{r}}\right)^2}{1+\frac{E_u}{E_{\text{s}}^{\text{r}}}\text{SNR}_u^{\text{r}}+\sum_{\ell\in\mathcal{C}}\frac{E_{\ell,u}}{E_{\text{s}}^{\text{r}}}\text{SNR}_{\ell,u}^{\text{r}}}\,N_{\text{a}}^2+\mathcal{O}(N_{\text{a}}), \tag{10.68}$$

but the denominator now satisfies

$$\text{Den}^{\text{MF}} = \frac{\sum_{\ell\in\mathcal{C}}\left(\frac{E_{\ell,u}}{E_{\text{s}}^{\text{r}}}\text{SNR}_{\ell,u}^{\text{r}}\right)^2}{1+\frac{E_u}{E_{\text{s}}^{\text{r}}}\text{SNR}_u^{\text{r}}+\sum_{\ell\in\mathcal{C}}\frac{E_{\ell,u}}{E_{\text{s}}^{\text{r}}}\text{SNR}_{\ell,u}^{\text{r}}}\,N_{\text{a}}^2+\mathcal{O}(N_{\text{a}}) \tag{10.69}$$

meaning that

$$\text{sinr}_u^{\text{MF}} = \frac{\left(\frac{E_u}{E_{\text{s}}^{\text{r}}}\text{SNR}_u^{\text{r}}\right)^2}{\sum_{\ell\in\mathcal{C}}\left(\frac{E_{\ell,u}}{E_{\text{s}}^{\text{r}}}\text{SNR}_{\ell,u}^{\text{r}}\right)^2}+\mathcal{O}\!\left(\frac{1}{N_{\text{a}}}\right) \tag{10.70}$$

and confirming that pilot contamination does impose a limit on the SINRs. By inspection, it can be verified that $\overline{\mathsf{sinr}}_u^{\mathsf{MF}}$ in (10.62) also abides by (10.70) or, equivalently, by

$$\overline{\mathsf{sinr}}_u^{\mathsf{MF}} \to \frac{\left(\frac{E_u}{E_s^r} G_u\right)^2}{\sum_{\ell \in \mathcal{C}} \left(\frac{E_{\ell,u}}{E_s^r} G_{\ell,u}\right)^2}, \tag{10.71}$$

which, minus the power control, is how it was originally expressed by Marzetta [883] and how it is typically found in the literature.

The bottom line is that, as N_a grows unboundedly and the user beams become exceedingly sharp, the desired signal surges over the noise and the interference, but, along with it, the pilot contamination—a very minor term for small N_a—also surges. Eventually, both the signal and the pilot contamination come to dominate and, since they scale similarly, the SINR is curbed no matter how many more antennas are added. But how much of an operational limitation does this represent for large but finite values of N_a?

Impact of pilot contamination

To answer this question, observe that, without pilot contamination, the set $\mathcal{C}$ would be empty and thus (10.62) would reduce to

$$\overline{\mathsf{sinr}}_u^{\mathsf{MF}} = \frac{\left(\frac{E_u}{E_s^r} \mathsf{SNR}_u^r\right)^2}{\frac{1+\frac{E_u}{E_s^r}\mathsf{SNR}_u^r}{N_a}\left(1+\sum_\ell \sum_{u=1}^{U} \frac{E_{\ell,u}}{E_s^r} \mathsf{SNR}_{\ell,u}^r\right)}. \tag{10.72}$$

Pilot contamination would have negligible impact provided that the difference between the denominators of (10.62) and (10.72) is itself negligible, i.e., if

$$\frac{\sum_{\ell \in \mathcal{C}} \frac{E_{\ell,u}}{E_s^r} \mathsf{SNR}_{\ell,u}^r}{N_a} \left(1 + \sum_\ell \sum_{u=0}^{U-1} \frac{E_{\ell,u}}{E_s^r} \mathsf{SNR}_{\ell,u}^r\right) + \sum_{\ell \in \mathcal{C}} \left(\frac{E_{\ell,u}}{E_s^r} \mathsf{SNR}_{\ell,u}^r\right)^2 \ll \frac{1 + \frac{E_u}{E_s^r}\mathsf{SNR}_u^r}{N_a} \left(1 + \sum_\ell \sum_{u=0}^{U-1} \frac{E_{\ell,u}}{E_s^r} \mathsf{SNR}_{\ell,u}^r\right). \tag{10.73}$$

Since this is tantamount to requiring that the contamination-induced interference from copilot users be negligible relative to the contamination-unrelated interference plus the noise, an even stronger condition can be derived with the noise disregarded, namely

$$\frac{\sum_{\ell \in \mathcal{C}} \frac{E_{\ell,u}}{E_s^r} G_{\ell,u}}{N_a} \sum_\ell \sum_{u=0}^{U-1} \frac{E_{\ell,u}}{E_s^r} G_{\ell,u} + \sum_{\ell \in \mathcal{C}} \left(\frac{E_{\ell,u}}{E_s^r} G_{\ell,u}\right)^2 \ll \frac{\frac{E_u}{E_s^r} G_u}{N_a} \sum_\ell \sum_{u=0}^{U-1} \frac{E_{\ell,u}}{E_s^r} G_{\ell,u}$$

or, equivalently,

$$\sum_{\ell \in \mathcal{C}} \left(\frac{E_{\ell,u}}{E_s^r} G_{\ell,u}\right)^2 \ll \frac{\frac{E_u}{E_s^r} G_u - \sum_{\ell \in \mathcal{C}} \frac{E_{\ell,u}}{E_s^r} G_{\ell,u}}{N_a} \sum_\ell \sum_{u=0}^{U-1} \frac{E_{\ell,u}}{E_s^r} G_{\ell,u}. \tag{10.74}$$

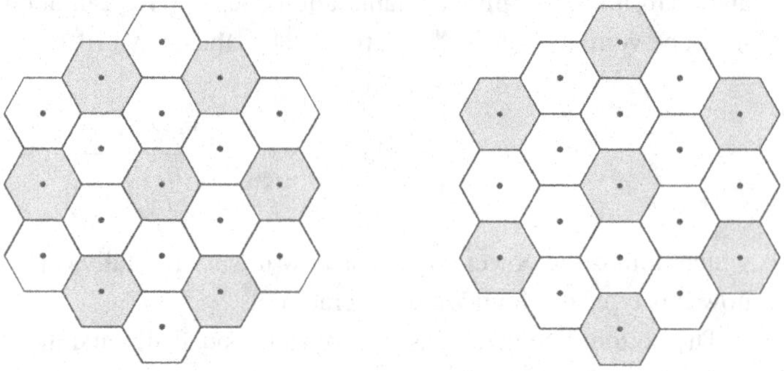

Fig. 10.5 Pilot reuse patterns: left, for $L_{\text{cluster}} = 3$; right, for $L_{\text{cluster}} = 4$.

Scaling both sides by $\left(\frac{E_u}{E_s^r} G_u\right)^2$, we obtain

$$\sum_{\ell \in \mathcal{C}} \left(\frac{\frac{E_{\ell,u}}{E_s^r} G_{\ell,u}}{\frac{E_u}{E_s^r} G_u} \right)^2 \ll \left(1 - \sum_{\ell \in \mathcal{C}} \frac{\frac{E_{\ell,u}}{E_s^r} G_{\ell,u}}{\frac{E_u}{E_s^r} G_u} \right) \frac{1}{N_a} \sum_{\ell} \sum_{u=0}^{U-1} \frac{\frac{E_{\ell,u}}{E_s^r} G_{\ell,u}}{\frac{E_u}{E_s^r} G_u}, \tag{10.75}$$

from which, by considering worst-case situations in terms of strength of the contamination-induced interference and weakness of the contamination-unrelated interference, sufficient conditions for pilot contamination negligibleness can be derived. This, however, requires positing specific network topologies.

Example 10.4 (Impact of pilot contamination in a hexagonal network)

Consider a hexagonal lattice network like the one in Fig. 10.1, with a pilot reuse pattern whereby each pilot is used up in only one cell within every cluster of L_{cluster} adjacent cells. Only certain values of L_{cluster} lead to regular reuse patterns on a hexagonal network, precisely $L_{\text{cluster}} = (k + \ell)^2 - k\ell$ for integer k and ℓ. Such values $L_{\text{cluster}} = 1, 3, 4, 7, \ldots$ are the ones considered in this text, with the pilot reuse patterns for two of them depicted in Fig. 10.5. (Other values for L_{cluster}, even noninteger ones, could be implemented via pilot hopping [907, 908] or by adopting fractional reuse ideas [909–911] whereby cells would be divided in concentric parts with different pilot reuse factors on each [912, 913].) Given L_{cluster}, the number of required pilot symbols readily equals $N_p = L_{\text{cluster}} U$.

Let us carry off a coarse assessment based only on pathloss, with shadow fading disregarded. The worst-case situation in terms of contamination-induced versus contamination-unrelated interference corresponds to having every copilot user placed at its cell corner closest to the base station of interest while simultaneously having every noncopilot user on its farthest cell corner, as far as possible from the base of interest. In this situation, illustrated in Fig. 10.6, all users are at cell corners and thus their power control coefficients coincide, factoring out of (10.75).

The distance from the base of interest to the bases of its first-tier copilot cells is [212]

$$D_{\text{copilot}} = \sqrt{3 L_{\text{cluster}}}\, D_{\text{cell}}, \tag{10.76}$$

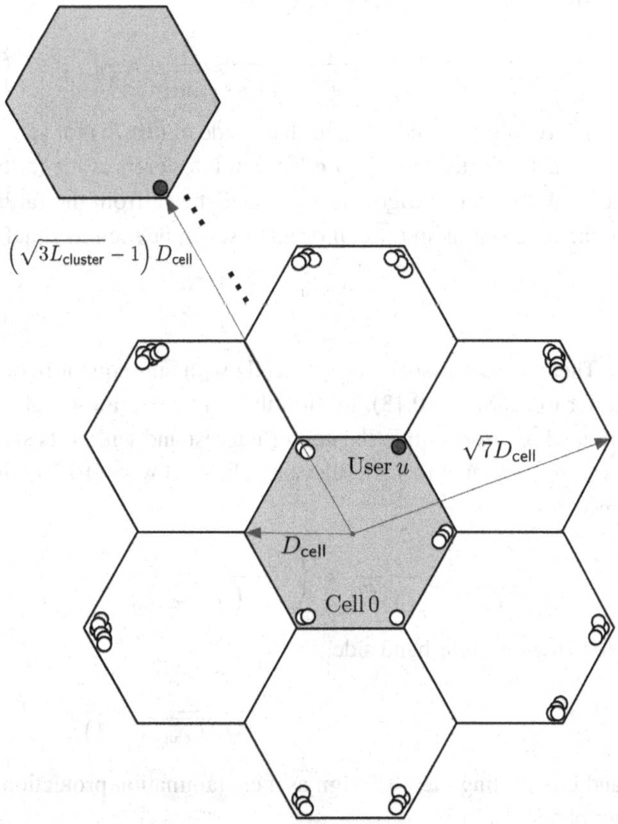

Fig. 10.6 Cell of interest and first tier of neighboring cells, with all the users therein (indicated by circles) at their farthest distances from the base station of interest; this corresponds to the weakest possible contamination-unrelated interference (i.e., from noncopilot users). In turn, the first tier of copilot users are at their minimum distance from the base of interest; this corresponds to the strongest possible copilot interference (i.e., from copilot users).

where D_{cell} is the cell radius. Hence, the minimum possible distance from the base of interest to the closest corners of the first-tier copilot cells is

$$D_{\text{copilot}} - D_{\text{cell}} = \left(\sqrt{3L_{\text{cluster}}}-1\right)D_{\text{cell}}. \tag{10.77}$$

Since the pathloss decays with $D^{-\eta}$, if cell ℓ is a first-tier copilot neighbor, then, as far as pathloss is concerned

$$G_{\ell,u} = \frac{G_u|_{D_u=D_{\text{cell}}}}{\left(\sqrt{3L_{\text{cluster}}}-1\right)^{\eta}}, \tag{10.78}$$

where the numerator equals G_u evaluated at $D_u = D_{\text{cell}}$. Thus, we can upper-bound the left-hand side of (10.75) by means of

$$\frac{G_{\ell,u}}{G_u} = \frac{1}{\left(\sqrt{3L_{\text{cluster}}} - 1\right)^{\eta}} \qquad \ell \in \mathcal{C}. \tag{10.79}$$

Next, we lower-bound the right-hand side of (10.75) for same-cell interferers at the corners of the cell of interest and for other-cell interferers at the farthest corners of their respective cells. With a bit of trigonometry, the distance from the farthest corner of an adjacent cell to the base station in the cell of interest can be seen to equal $\sqrt{7}D_{\text{cell}}$ and therefore

$$\frac{G_{\ell,\text{u}}}{G_u} = \begin{cases} 1 & \ell = 0 \\ 1/7^{\eta/2} & \ell \neq 0. \end{cases} \tag{10.80}$$

There are six first-tier copilot cells with subsequent tiers having a rather minor impact (refer to Problem 10.13). In turn, the contamination-unrelated interference is mostly contributed by users within the cell of interest and within its six adjacent cells. With the summations restricted to such relevant cells, and with (10.79) and (10.80) plugged in, (10.75) becomes

$$\frac{6}{\left(\sqrt{3L_{\text{cluster}}} - 1\right)^{2\eta}} \ll \left(1 - \frac{6}{\left(\sqrt{3L_{\text{cluster}}} - 1\right)^{\eta}}\right) \frac{U}{N_{\text{a}}} \left(1 + \frac{6}{7^{\eta/2}}\right). \tag{10.81}$$

Relaxing the right-hand side via

$$1 - \frac{6}{\left(\sqrt{3L_{\text{cluster}}} - 1\right)^{\eta}} \approx 1 \tag{10.82}$$

and interpreting the "$\ll$" sign as a contamination protection factor ε (say $\varepsilon = \frac{1}{10}$ or $\frac{1}{100}$), we obtain

$$\frac{6}{\left(\sqrt{3L_{\text{cluster}}} - 1\right)^{2\eta}} \lesssim \varepsilon \frac{U}{N_{\text{a}}} \left(1 + \frac{6}{7^{\eta/2}}\right) \tag{10.83}$$

and, recalling that $N_{\text{p}} = L_{\text{cluster}} U$, a bit of algebra finally leads to

$$\frac{N_{\text{p}}}{U} \gtrsim \frac{1}{3}\left[1 + \left(\frac{6/\varepsilon \cdot N_{\text{a}}/U}{1 + 6/7^{\eta/2}}\right)^{1/2\eta}\right]^2. \tag{10.84}$$

In Problem 10.14, the reader is invited to verify that the relaxation in (10.82) has a minor effect on (10.84). This final condition, which does not account for shadow fading but is otherwise highly conservative—it corresponds to an interference-limited network with the *simultaneous* worst-case positions for all users involved—depends only on the pathloss exponent η, on N_{a}/U, and on the contamination protection factor ε. Let us proceed to test this condition, with and without power control.

Example 10.5

For $\eta = 4$ and $N_{\text{a}}/U \leq 10$, and with the contamination protection factor set to $\varepsilon = \frac{1}{10}$, (10.84) gives

$$\frac{N_{\text{p}}}{U} \gtrsim 3.4, \tag{10.85}$$

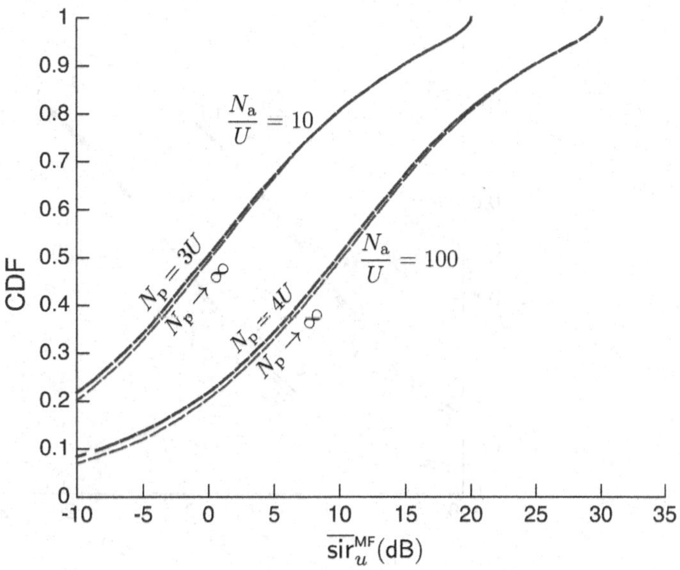

Fig. 10.7 Large-scale distributions of $\overline{\text{sir}}_u^{\text{MF}}$ in the reverse link of an interference-limited hexagonal network with no power control, $\eta = 4$, $\sigma_{\text{dB}} = 8$ dB, and $U = 10$. For $N_a/U = 10$, comparison between $N_p \to \infty$ and $N_p = 3U$. For $N_a/U = 100$, comparison between $N_p \to \infty$ and $N_p = 4U$.

which we can conveniently round to $N_p/U = 3$, whereby the pilot reuse clusters feature three cells. This value, which we found analytically, agrees with the simulation-based result reported in [914].

Consider an interference-limited hexagonal network with no power control, $\eta = 4$, $\sigma_{\text{dB}} = 8$ dB, and $U = 10$. For $N_a/U = 10$, plot the CDF of $\overline{\text{sir}}_u^{\text{MF}}$ both for $N_p \to \infty$ (no pilot contamination) and for $N_p = 3U = 30$.

Solution

The CDFs, presented in Fig. 10.7, corroborate that $N_p = 3U$ suffices for the effect of the contamination to be shy of 1 dB at low SIRs, and outright imperceptible at high SIRs.

Example 10.6

For $\eta = 4$ and $N_a/U \leq 100$, and with the interference protection factor set to $\varepsilon = \frac{1}{10}$, (10.84) gives $\frac{N_p}{U} \gtrsim 5.1$, which we aggressively round to $N_p/U = 4$. Repeat Example 10.5 to test such condition with $N_a/U = 100$.

Solution

The CDFs, also presented in Fig. 10.7, again confirm the efficacy of (10.84) with $\varepsilon = \frac{1}{10}$, even with N_a/U aggressively rounded.

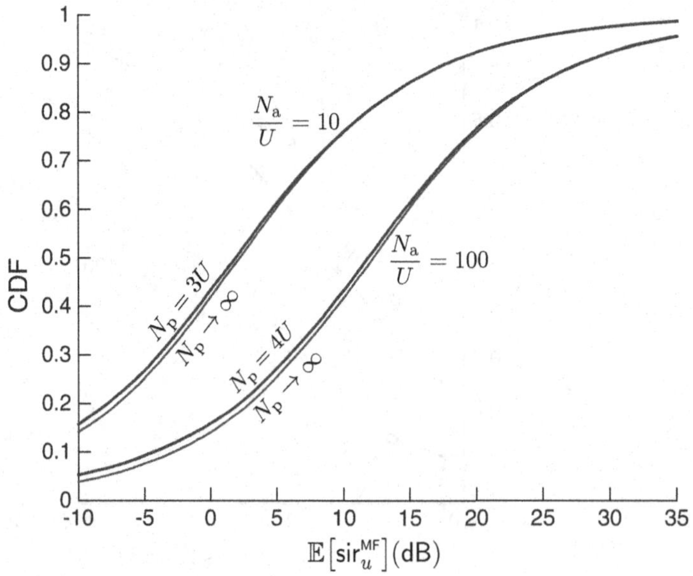

Fig. 10.8 Large-scale distributions of $\mathbb{E}\left[\mathsf{sir}_u^{\mathsf{MF}}\right]$ in the reverse link of an interference-limited hexagonal network with no power control, $\eta = 4$, $\sigma_{\mathsf{dB}} = 8$ dB, and $U = 10$. For $N_a/U = 10$, comparison between $N_p \to \infty$ and $N_p = 3U$. For $N_a/U = 100$, comparison between $N_p \to \infty$ and $N_p = 4U$.

Example 10.7

Repeat Examples 10.5 and 10.6 for $\mathbb{E}\left[\mathsf{sir}_u^{\mathsf{MF}}\right]$ instead of $\overline{\mathsf{sir}}_u^{\mathsf{MF}}$.

Solution

See Fig. 10.8, buttressing the validity of (10.84) with $\varepsilon = \frac{1}{10}$ also for a receiver reliant on channel estimates rather than on channel hardening.

Example 10.8

Repeat Example 10.7 with fractional power control ($\vartheta = 0.7$), further testing the case $N_a/U = 4$ for which (10.84) with $\varepsilon = \frac{1}{10}$ gives $N_p/U \gtrsim 2.7$, rounded to $N_p/U = 3$.

Solution

See Fig. 10.9. Except perhaps for the need of a slight tightening in the case $N_a/U = 100$, the pilot reuse factors derived analytically again prove satisfactory.

With the condition in (10.84) thoroughly verified we can, even without a formal optimization of N_a/U, begin to gauge the range of channel coherences where pilot contamination becomes noticeable.

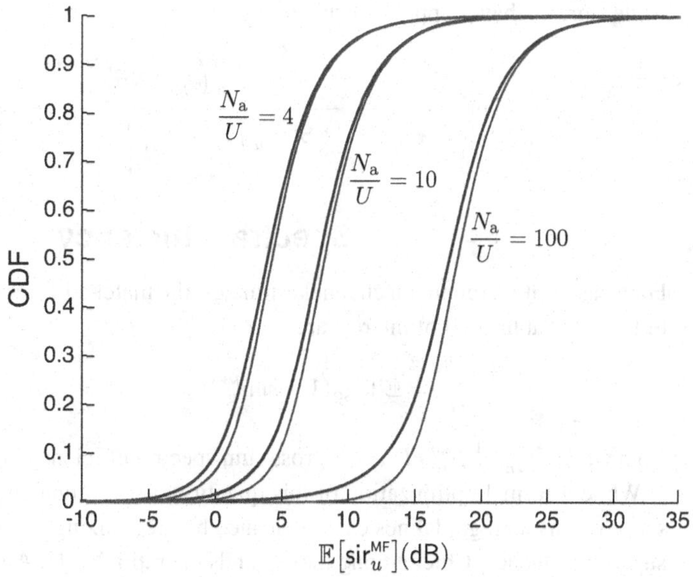

Fig. 10.9 Large-scale distributions of $\mathbb{E}[\mathsf{sir}_u^{\mathsf{MF}}]$ in the reverse link of an interference-limited hexagonal network with fractional power control ($\vartheta = 0.7$), $\eta = 4$, $\sigma_{\mathrm{dB}} = 8$ dB and $U = 10$. For both $N_a/U = 4$ and $N_a/U = 10$, comparison between $N_p \to \infty$ and $N_p = 3U$. For $N_a/U = 100$, comparison between $N_p \to \infty$ and $N_p = 4U$.

Example 10.9

The coherence range $N_c \geq 1000$ considered in most of this book, with a reasonable overhead of $N_p/N_c = 0.15$, gives $N_p \geq 150$. For $N_a/U = 4$, 10, and 100, establish the values of N_a where pilot contamination would begin to be noticeable.

Solution

From (10.84) with $\varepsilon = \frac{1}{10}$, $N_p/U \gtrsim 3$ for $N_a/U = 4$ and $N_a/U = 10$, which gives $U \lesssim 50$; this marks $N_a = 200$ and $N_a = 500$ antennas as the respective points where pilot contamination would begin to be noticeable. For $N_a/U = 100$, that number goes up to a whopping $N_a = 3000$ antennas.

This example, to be sharpened once we optimize N_a/U, suggests that pilot contamination would only impair matched-filter receivers in truly massive MIMO settings and/or in situations of unusually low fading coherence, say simultaneous high carrier frequencies, long delay spreads, and extreme velocities. Barring this, the contamination can be rendered negligible with acceptable overheads leading $\mathsf{sinr}_u^{\mathsf{MF}}$ to behave approximately as

$$\frac{\frac{E_u}{E_s^r}\mathsf{SNR}_u^r \left|\boldsymbol{W}_u^{\mathsf{MF}*}\hat{\boldsymbol{H}}_u\right|^2}{\sum_{u\neq u}\frac{E_u}{E_s^r}\mathsf{SNR}_u^r \left|\boldsymbol{W}_u^{\mathsf{MF}*}\hat{\boldsymbol{H}}_u\right|^2 + \left(1+\sum_u \frac{\frac{E_u}{E_s^r}\mathsf{SNR}_u^r}{1+\frac{E_u}{E_s^r}\mathsf{SNR}_u^r} + \sum_{\ell\neq 0}\sum_u \frac{E_{\ell,u}}{E_s^r}\mathsf{SNR}_{\ell,u}^r\right)\|\boldsymbol{W}_u^{\mathsf{MF}}\|^2}$$

while $\overline{\mathsf{sinr}}_u^{\mathsf{MF}}$ behaves approximately as

$$\overline{\mathsf{sinr}}_u^{\mathsf{MF}} \approx \frac{N_{\mathrm{a}} \left(\frac{E_u}{E_{\mathrm{s}}^{\mathrm{r}}} \mathsf{SNR}_u^{\mathrm{r}} \right)^2}{\left(1 + \frac{E_u}{E_{\mathrm{s}}^{\mathrm{r}}} \mathsf{SNR}_u^{\mathrm{r}}\right) \left(1 + \sum_\ell \sum_{\mathrm{u}=0}^{U-1} \frac{E_{\ell,\mathrm{u}}}{E_{\mathrm{s}}^{\mathrm{r}}} \mathsf{SNR}_{\ell,\mathrm{u}}^{\mathrm{r}}\right)}. \tag{10.86}$$

Spectral efficiency

For a base station reliant on channel estimates, the matched-filter gross spectral efficiencies of the users at the cell of interest are

$$\frac{R_u^{\mathsf{MF}}}{B} = \mathbb{E}[\log_2(1 + \mathsf{sinr}_u^{\mathsf{MF}})] \qquad u = 0, \ldots, U-1, \tag{10.87}$$

from which $\sum_{u=0}^{U-1} R_u^{\mathsf{MF}}/B$ is the gross sum spectral efficiency at that cell.

While a formal optimization of this quantity over all tunable parameters is beyond the scope of this text, guidelines can be gleaned by recognizing that certain parameters have a superior influence. Chief among these stands the ratio N_{a}/U. And, as long as its optimization falls within the range established in Example 10.9, it can advantageously be conducted directly upon (10.87), with pilot contamination neglected.

Example 10.10

Let $N_{\mathrm{a}} = 100$ in an interference-limited hexagonal network with matched-filter receivers reliant on channel estimates, fractional power control ($\vartheta = 0.7$), $\eta = 4$ and $\sigma_{\mathrm{dB}} = 8$ dB. Which N_{a}/U maximizes the gross sum spectral efficiency in the absence of pilot contamination, subject to less than 3% of the users having a local-average SIR below -5 dB?

Solution

Consider the distribution, over the locations and shadow fadings of the U users, of the gross sum spectral efficiency per cell. Shown in the main plot of Fig. 10.10 is the evolution of such distribution as N_{a}/U is swept from $100/25$ down to $100/67$. The inset, meanwhile, displays the corresponding distributions of $\mathbb{E}[\mathsf{sinr}_u^{\mathsf{MF}}]$. The optimum ratio is $N_{\mathrm{a}}/U = 100/50 = 2$, for which the average gross sum spectral efficiency is 60.2 b/s/Hz per cell with 2.8% of users below -5 dB.

Altogether, for typical propagation conditions and with pilot overhead and pilot contamination ignored, ratios on the order of $N_{\mathrm{a}}/U \approx 2$ yield interesting operating points in terms of sum spectral efficiency versus large-scale outage with matched-filter receivers. (Loosening the large-scale outage constraint would allow for lower ratios N_{a}/U and higher sum spectral efficiencies, and vice versa. A more aggressive power control, $\vartheta > 0.7$, would also allow for lower ratios N_{a}/U, although curtailing the spectral efficiency of users in favorable locations.) With user selection, it would be desirable that the ratio N_{a}/U be rendered dynamic rather than fixed: higher when low-SIR users are selected, to increase their interference protection, and vice versa.

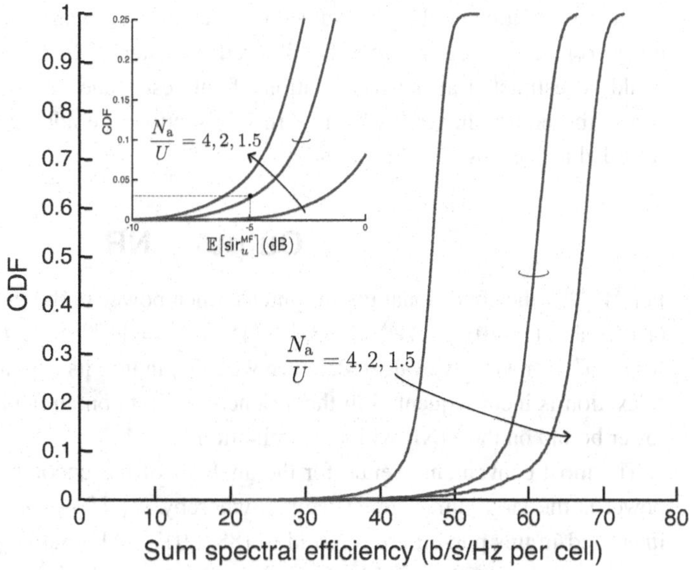

Fig. 10.10 On the main plot, large-scale distribution of the gross sum spectral efficiency per cell in the reverse link of an interference-limited hexagonal network with no pilot contamination, matched-filter receivers reliant on channel estimates, fractional power control ($\vartheta = 0.7$), $\eta = 4$, $\sigma_{\text{dB}} = 8$ dB, and $N_a = 100$, for $N_a/U = 4$, 2, and 1.5. On the inset, respective distributions of $\mathbb{E}[\text{sinr}_u^{\text{MF}}]$.

A further elaboration of the spectral efficiency requires accounting for the pilot contamination and factoring the pilot overhead into (10.87), something that we defer to the next section, in the context of the LMMSE receiver.

10.4.3 LMMSE receiver

The deficiency of the matched-filter receiver is its obliviousness to interference, and the remedy for this deficiency is an interference-aware receiver structure. Since, as learned in previous chapters, the SINR-maximizing structure is the LMMSE receiver, we proceed directly to this structure. Retrieving and specializing to single-antenna users and channel estimates the solution derived in (9.282), we obtain

$$W_u^{\text{MMSE}} \propto \left[\sum_{\text{u}=0}^{U-1} \frac{E_\text{u}}{E_\text{s}^\text{r}} \text{SNR}_\text{u}^\text{r} \hat{H}_\text{u} \hat{H}_\text{u}^* + \left(1 + \sum_{\text{u}=0}^{U-1} \frac{E_\text{u}}{E_\text{s}^\text{r}} \text{SNR}_\text{u}^\text{r} \, \text{MMSE}_\text{u} + \sum_{\ell \neq 0} \sum_{\text{u}=0}^{U-1} \frac{E_{\ell,\text{u}}}{E_\text{s}^\text{r}} \text{SNR}_{\ell,\text{u}}^\text{r} \right) I \right]^{-1} \hat{H}_u, \tag{10.88}$$

which is a function of $\hat{H}_0, \ldots, \hat{H}_{U-1}$, meaning that it explicitly accounts for the same-cell interference, while treating other-cell interference and same-cell channel estimation errors as noise.

The formulation could be extended to actively encompass the channels for users in the other $(L_{\text{cluster}} - 1)$ cells within the pilot reuse cluster of the cell of interest; those channels could be estimated at the base station of interest since orthogonal pilots are employed across the entire cluster [914–916]. In this section, we adhere to (10.88), which can be regarded as the single-cell baseline.

Output SINR

For $\boldsymbol{W}_u^{\text{MMSE}}$, the conditional copilot interference power in the general SINR expression of (10.32) and (10.40), $\mathbb{E}\big[|\boldsymbol{W}_u^{\text{MMSE}*}\boldsymbol{H}_{\ell,u}|^2\,|\boldsymbol{W}_u^{\text{MMSE}}\big] = \boldsymbol{W}_u^{\text{MMSE}*}\,\mathbb{E}\big[\boldsymbol{H}_{\ell,u}\boldsymbol{H}_{\ell,u}^*\,|\boldsymbol{W}_u^{\text{MMSE}}\big]\boldsymbol{W}_u^{\text{MMSE}}$ for $\ell \in \mathcal{C}$, is not easily evaluated, hence we relax in into its unconditional counterpart. This relaxation is inconsequential in the absence of pilot contamination, while it yields a tight lower bound on the SINR with contamination.

The most convenient avenue for the analysis of the unconditional copilot interference power is the large-dimensional regime, whereby $N_{\text{a}}, U \to \infty$ with fixed N_{a}/U; readers interested in this pursuit are referred to [884, 917]. Alternatively, $\mathbb{E}\big[|\boldsymbol{W}_u^{\text{MMSE}*}\boldsymbol{H}_{\ell,u}|^2\big]$ for $\ell \in \mathcal{C}$ can be computed via Monte-Carlo and fed into (10.40) to obtain $\text{sir}_u^{\text{MMSE}}$; this is how the results that follow are produced.

Gone the analytical edge offered by channel hardening in the case of matched-filter receivers, here we concentrate on receivers reliant on channel estimates. Also, since the presence of noise blurs the difference between the matched-filter and the LMMSE receivers, all the examples are again for interference-limited networks.

Example 10.11

Consider a hexagonal network where $\eta = 4$, the shadow fading is log-normal with $\sigma_{\text{dB}} = 8$ dB, the thermal noise is neglected, and there is no power control. Let $N_{\text{a}} = 100$ and let $N_{\text{p}} \to \infty$ such that there is no pilot contamination. For $N_{\text{a}}/U = 4$ and 10, plot the CDFs of $\mathbb{E}\big[\text{sir}_u^{\text{MMSE}}\big]$ and compare them with those of $\mathbb{E}\big[\text{sir}_u^{\text{MF}}\big]$.

Solution

The requested distributions are displayed in Fig. 10.11. Despite the substantial level of excess antennas for $N_{\text{a}}/U = 10$, even in that case, and let alone for $N_{\text{a}}/U = 4$, the LMMSE receiver markedly improves the SINRs over the entire range. The lower tails, in particular, become much better behaved than with matched filters, rendering power control less critical.

Example 10.12

Repeat Example 10.11 with fractional power control ($\vartheta = 0.7$).

Solution

See Fig. 10.12. With this rather stringent power control, the advantage of the LMMSE receiver relative to the matched filter subsides.

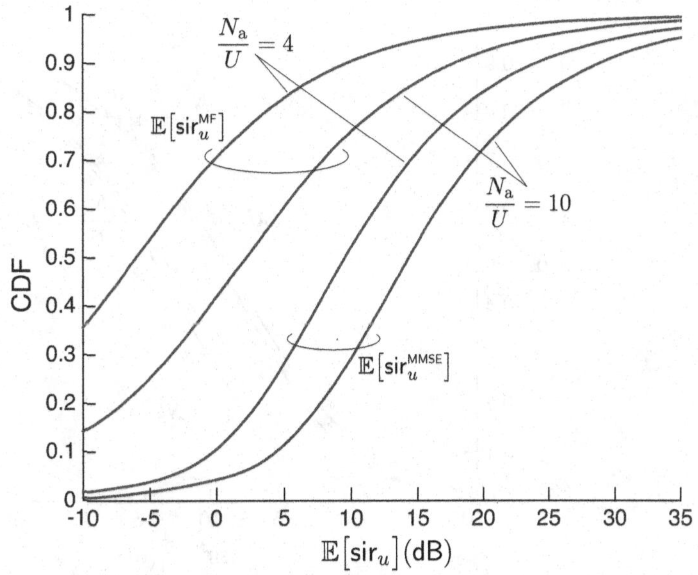

Fig. 10.11 Large-scale distributions of $\mathbb{E}[\mathsf{sir}_u^{\mathsf{MMSE}}]$ and $\mathbb{E}[\mathsf{sir}_u^{\mathsf{MF}}]$ in the reverse link of an interference-limited hexagonal network with no power control, $\eta = 4$, $\sigma_{\mathsf{dB}} = 8$ dB, $N_a = 100$, and $N_p \to \infty$ for $N_a/U = 4$ and 10.

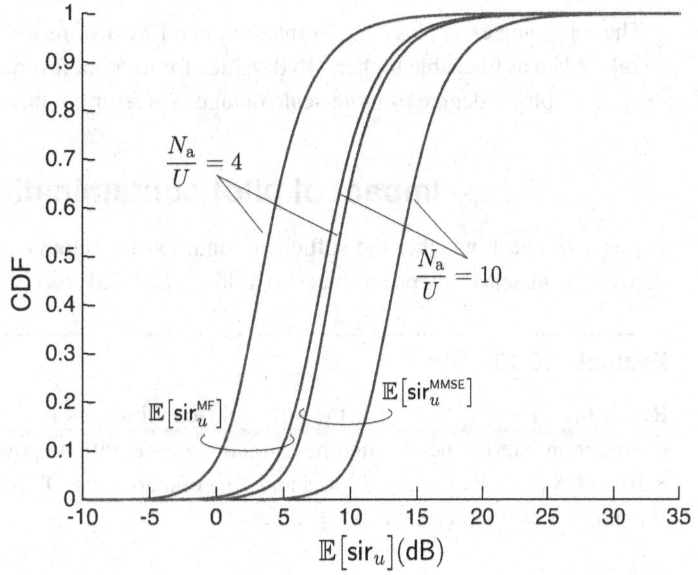

Fig. 10.12 Large-scale distributions of $\mathbb{E}[\mathsf{sir}_u^{\mathsf{MMSE}}]$ and $\mathbb{E}[\mathsf{sir}_u^{\mathsf{MF}}]$ in the reverse link of an interference-limited hexagonal network with fractional power control ($\vartheta = 0.7$), $\eta = 4$, $\sigma_{\mathsf{dB}} = 8$ dB, $N_a = 100$, and $N_p \to \infty$ for $N_a/U = 4$ and 10.

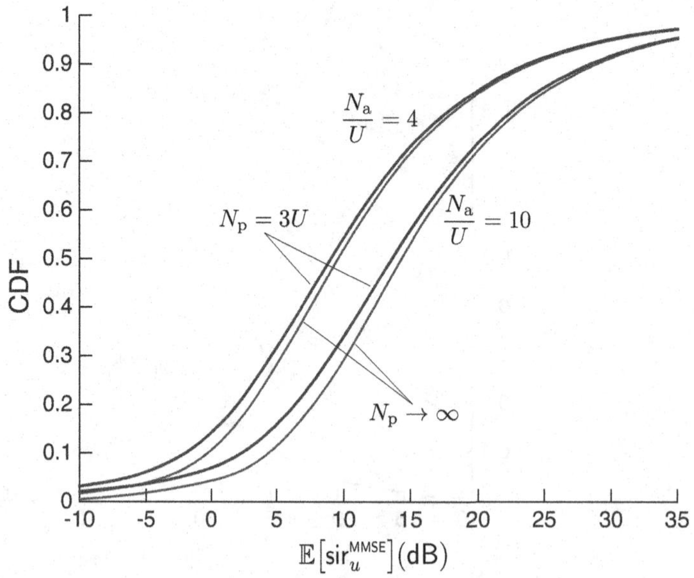

Fig. 10.13 Large-scale distributions of $\mathbb{E}\left[\mathsf{sir}_u^{\mathsf{MMSE}}\right]$ in the reverse link of an interference-limited hexagonal network with no power control, $\eta = 4$, $\sigma_{\mathrm{dB}} = 8$ dB, and $N_\mathrm{a} = 100$. For $N_\mathrm{a}/U = 4$ and 10, comparison between $N_\mathrm{p} \to \infty$ and $N_\mathrm{p} = 3U$.

The foregoing examples suggest that, with an LMMSE receiver, the power control may be relaxed so as to enable higher SINR values for users in favorable situations while keeping an acceptable degree of large-scale outage. We return to this at the end of the section.

Impact of pilot contamination

Let us now check whether the sufficient condition for pilot contamination negligibleness derived for matched filters continues to hold for LMMSE receivers.

Example 10.13

Recall that, for $N_\mathrm{a}/U \leq 10$, the sufficient condition with $\varepsilon = \frac{1}{10}$ gives $N_\mathrm{p}/U \gtrsim 3$. Consider an interference-limited hexagonal network with no power control, $\eta = 4$, $\sigma_{\mathrm{dB}} = 8$ dB and $N_\mathrm{a} = 100$. For $N_\mathrm{a}/U = 4$ and 10, compare the CDFs of $\mathbb{E}\left[\mathsf{sir}_u^{\mathsf{MMSE}}\right]$ for $N_\mathrm{p} \to \infty$ (no pilot contamination) and $N_\mathrm{p} = 3U$.

Solution

See Fig. 10.13.

The sufficient condition remains applicable, even if it is slightly looser than with a matched filter (recall Fig. 10.8). This can undoubtedly be attributed to the enhanced ability of the LMMSE receiver to suppress contamination-unrelated interference, making the contamination-induced interference relatively more significant. But the condition holds up and, possibly with a slight tightening of the factor ε, it can continue to ensure the negligibleness of pilot contamination.

Spectral efficiency

For a base station reliant on channel estimates, the LMMSE reverse-link gross spectral efficiencies at the cell of interest are

$$\frac{R_u^{\mathsf{MMSE}}}{B} = \mathbb{E}\big[\log_2(1 + \mathsf{sinr}_u^{\mathsf{MMSE}})\big] \qquad u = 0, \ldots, U-1, \tag{10.89}$$

from which $\sum_{u=0}^{U-1} R_u^{\mathsf{MMSE}}/B$ is the gross sum spectral efficiency at that cell. Within the range of N_a where pilot contamination can be rendered negligible with acceptable overhead, the optimization of this quantity directly yields suitable values for N_a/U and ϑ.

Example 10.14

Let $N_\mathrm{a} = 100$ in an interference-limited hexagonal network with LMMSE receivers, fractional power control, $\eta = 4$ and $\sigma_{\mathsf{dB}} = 8$ dB. Which ratio N_a/U and which power control parameter ϑ maximize the reverse-link gross sum spectral efficiency in the absence of pilot contamination, subject to less than 3% of users having a local-average SIR below -5 dB?

Solution

An optimization over the power control parameter assumed to change in steps of 0.1 returns $\vartheta = 0.4$, which, as anticipated, is smaller than with matched-filter receivers, thereby allowing for higher-power transmissions and higher SINRs. Shown in the main plot of Fig. 10.14 is the distribution of the gross sum spectral efficiency as N_a/U is swept from $100/33$ down to $100/83$. The inset, meanwhile, displays the corresponding distributions of $\mathbb{E}[\mathsf{sinr}_u^{\mathsf{MMSE}}]$. The optimum ratio is $N_\mathrm{a}/U = 100/62 = 1.6$, for which the average gross sum spectral efficiency is 146 b/s/Hz per cell with 3% of users below -5 dB.

Combining Examples 10.10 and 10.14 we surmise that, for typical propagation conditions and with pilot overhead and pilot contamination ignored, ratios on the order of $N_\mathrm{a}/U \approx 1.6$–2 yield interesting operating points in terms of sum spectral efficiency versus large-scale outage. The difference in the optimum value of this ratio is only marginal when moving from matched-filter to LMMSE receivers, but the gross sum spectral efficiency more than doubles. While, without pilot contamination, the LMMSE receiver is no better than a matched filter for $N_\mathrm{a}/U \to \infty$, for large but finite numbers of antennas the former is considerably better.

Example 10.15

From the result for $N_\mathrm{a} = 100$ antennas in Example 10.14, extrapolate the average gross sum spectral efficiency as a function of N_a assuming that its value per antenna remains stable.

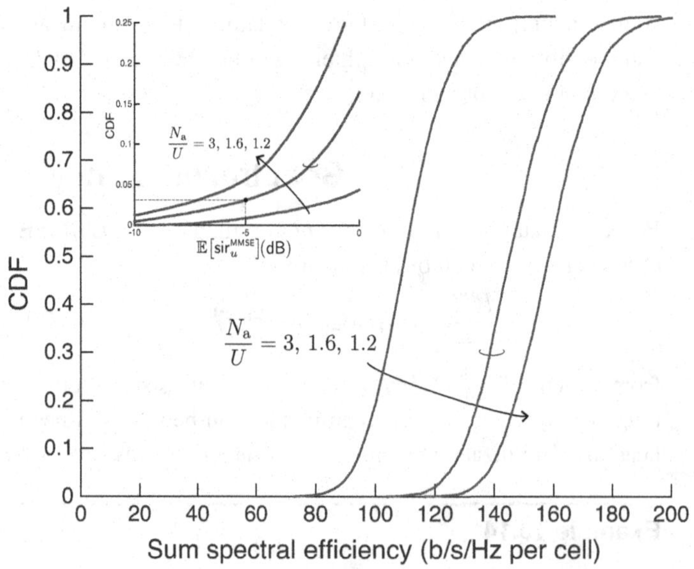

Fig. 10.14 On the main plot, large-scale distribution of the gross sum spectral efficiency per cell in the reverse link of an interference-limited cellular network with no pilot contamination, LMMSE receivers, fractional power control ($\vartheta = 0.4$), $\eta = 4$, $\sigma_{\text{dB}} = 8$ dB, and $N_a = 100$, for $N_a/U = 3$, 1.6, and 1.2. On the inset, respective distributions of $\mathbb{E}[\mathsf{sinr}_u^{\mathsf{MMSE}}]$.

Solution

From 146 b/s/Hz per cell at $N_a = 100$,

$$\frac{\mathbb{E}\left[R^{\mathsf{MMSE}}\right]}{B} \approx 1.46\, N_a, \tag{10.90}$$

where the expectation is over the large-scale quantities (user locations and shadow fadings). This extrapolation is depicted in Fig. 10.15.

Next, incorporating pilot contamination and pilot overhead calls for an aggregation of the reverse and forward spectral efficiencies or, in its place, for a partition of the overhead between the reverse and the forward links. As an exercise, we can ascribe half of the overhead to the reverse-link efficiency and rewrite the gross quantity in (10.89) into its net counterpart

$$\frac{R_u^{\mathsf{MMSE}}}{B} = \left(1 - \frac{N_p}{2N_c}\right) \mathbb{E}\left[\log_2(1 + \mathsf{sinr}_u^{\mathsf{MMSE}})\right] \qquad u = 0, \ldots, U-1, \tag{10.91}$$

where the SINRs are now computed with the inclusion of pilot contamination.

For the range we have identified earlier, $N_a/U \approx 1.6$–2, condition (10.84) with $\varepsilon = \frac{1}{10}$ leads to $N_p \gtrsim 1.5 N_a$. The transition from gross to net spectral efficiencies then depends crucially on the fading coherence.

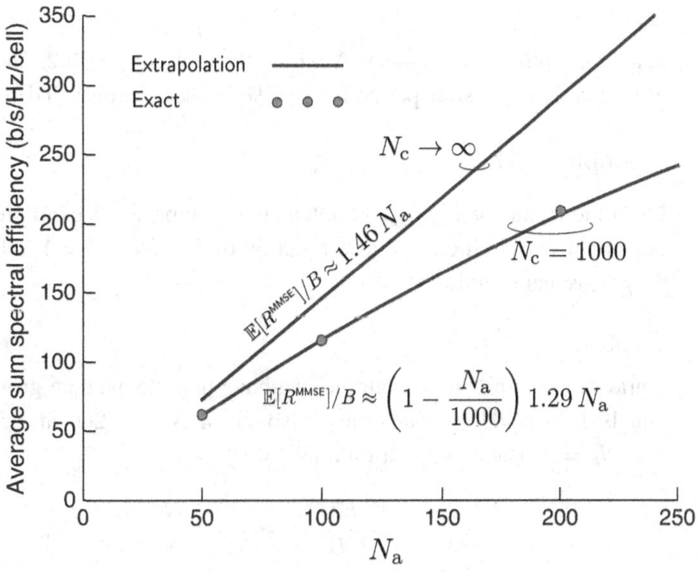

Fig. 10.15 As a function of N_a, reverse-link average sum spectral efficiency with LMMSE receivers, optimized fractional power control ($\vartheta = 0.4$), and optimized N_a/U, with no more than 3% of users below -5 dB. Top curve: gross-value extrapolation, which approximates the net value for $N_c \to \infty$. Bottom curve: net-value extrapolation for $N_c = 1000$. In circles, exact overhead-optimized values for $N_c = 1000$. All the results are for an interference-limited hexagonal network with $\eta = 4$ and $\sigma_{\text{dB}} = 8$ dB.

- For pedestrian users ($N_c = 20\,000$) and almost any conceivable number of antennas,

$$\left(1 - \frac{N_p}{2N_c}\right) \approx 1 \tag{10.92}$$

even with N_p large enough to keep the contamination at a negligible level. The net spectral efficiency then essentially equals the gross spectral efficiency, meaning that the top curve in Fig. 10.15 is directly representative of pedestrian settings.

- For vehicular users ($N_c = 1000$), in contrast,

$$\left(1 - \frac{N_p}{2N_c}\right) < 1, \tag{10.93}$$

and thus the net and gross spectral efficiencies do differ. We next elaborate on this case through a string of examples.

Example 10.16

For $N_c = 1000$, and keeping the power control parameter at $\vartheta = 0.4$, redo Example 10.14 with pilot contamination accounted for and with the net spectral efficiency as per (10.91).

Solution

A tedious optimization returns $N_a/U = 2$ and $N_p/N_c = 0.2$. The average net sum spectral efficiency is 116 b/s/Hz per cell, with 3% of users below -5 dB.

Example 10.17

From the result for $N_a = 100$ antennas in the previous example, extrapolate the average net sum spectral efficiency, as a function of N_a, for $N_c = 1000$. Assume that N_a/U and N_p/N_a remain stable.

Solution

Unraveling the previous example, we have that the average gross sum spectral efficiency equals 1.29 b/s/Hz per antenna. Also, from $N_p = 200$ and $N_a = 100$ we have that $N_p/N_a = 2$. The extrapolation thus gives

$$\frac{\mathbb{E}\left[R^{\text{MMSE}}\right]}{B} \approx \left(1 - \frac{N_a}{N_c}\right) 1.29\, N_a \tag{10.94}$$

$$= \left(1 - \frac{N_a}{1000}\right) 1.29\, N_a, \tag{10.95}$$

which indeed returns 116 b/s/Hz per cell for $N_a = 100$. This extrapolation, also depicted in Fig. 10.15, evidences how, as N_a becomes comparable to N_c, the fading coherence becomes a fundamental limitation.

Example 10.18

To test the average net sum spectral efficiency extrapolation derived above for $N_c = 1000$, find the exact values for $N_a = 50$ and $N_a = 200$.

Solution

The exact solutions obtained via Monte-Carlo are as follows.

- For $N_a = 50$, the optimization returns $N_a/U = 1.85$ and $N_p/N_c = 0.19$. The average net sum spectral efficiency is 61.8 b/s/Hz per cell, with 2.9% of users below -5 dB.
- For $N_a = 200$, the optimization returns $N_a/U = 2.22$ and $N_p/N_c = 0.27$. The average net sum spectral efficiency is 208.7 b/s/Hz per cell, with 3% of users below -5 dB.

These exact average net sum spectral efficiencies, shown in Fig. 10.15 alongside the extrapolation, confirm the validity of the latter up to hundreds of antennas.

As a final elaboration of the LMMSE receiver, the spectral efficiencies could be further optimized by releasing our premise that, on every coherence tile, every user transmits a single pilot symbol [918, 919]. The tradeoff between devoting additional symbols to overhead in exchange for more precise channel estimates could be tackled as done in Section 4.8 for SU-SISO. Alternatively, it is possible to keep the overhead as is and optimize the pilot power boosting in its stead [920]; this is simpler, as it can be effected directly over the SINRs. Again, the reader is referred to Section 4.8 for an SU-SISO version of this optimization.

In order to push N_{a}/U into even lower values, nonlinear receivers could be entertained, for instance by incorporating SIC at the base stations. The incorporation of SIC, however, is not without complications, e.g., higher complexity and an enhanced exposure to error propagation. Parallel interference cancelation, originally developed for CDMA [921], is an alternative for nonlinear reception [922].

10.5 Forward-link data transmission

Upon data transmission from the base stations, the users at the cell of interest (cell 0) observe

$$y_{0,u} = \sum_{\ell} \sqrt{G_{\ell;0,u}}\, \boldsymbol{H}_{\ell;0,u}^{*} \boldsymbol{x}_{\ell} + v_{0,u} \qquad u = 0, \ldots, U-1, \tag{10.96}$$

where, by virtue of reciprocity, $\sqrt{G_{\ell;0,u}}\, \boldsymbol{H}_{\ell;0,u}^{*}$ is the channel connecting the ℓth base with the uth user at cell 0. The signal emitted by the ℓth base is

$$\boldsymbol{x}_{\ell} = \sum_{\mathrm{u}=0}^{U-1} \sqrt{\frac{E_{\ell,\mathrm{u}}}{N_{\mathrm{a}}}}\, \boldsymbol{F}_{\ell,\mathrm{u}}\, s_{\ell,\mathrm{u}}, \tag{10.97}$$

where $\|\boldsymbol{F}_{\ell,\mathrm{u}}\|^2 = N_{\mathrm{a}}$ (under a per-symbol power constraint) or $\mathbb{E}\big[\|\boldsymbol{F}_{\ell,\mathrm{u}}\|^2\big] = N_{\mathrm{a}}$ (under a per-codeword power constraint). In turn, and provided that the ℓth base station radiates its full power, $\sum_{\mathrm{u}=0}^{U-1} E_{\ell,\mathrm{u}} = E_{\mathrm{s}}$. While, in the single-cell setups of previous chapters, it is pointless—from a spectral efficiency vantage—to power down a base station, in a multicell context and with linear transceivers it is conceivable that, in times or places of low traffic, $\sum_{\mathrm{u}=0}^{U-1} E_{\ell,\mathrm{u}} < E_{\mathrm{s}}$ so as to reduce other-cell interference and power consumption. However, for the purpose of exploring the limits of massive MIMO in a highly loaded network, full-power transmission at each base is fitting.

Altogether, and with the index of the cell of interest dropped,

$$y_u = \sum_{\mathrm{u}=0}^{U-1} \sqrt{\frac{G_u E_{\mathrm{u}}}{N_{\mathrm{a}}}}\, \boldsymbol{H}_u^{*} \boldsymbol{F}_{\mathrm{u}} s_{\mathrm{u}} + \sum_{\ell \neq 0} \sum_{\mathrm{u}=0}^{U-1} \sqrt{\frac{G_{\ell;u} E_{\ell,\mathrm{u}}}{N_{\mathrm{a}}}}\, \boldsymbol{H}_{\ell;u}^{*} \boldsymbol{F}_{\ell,\mathrm{u}} s_{\ell,\mathrm{u}} + v_u \qquad u = 0, \ldots, U-1, \tag{10.98}$$

where, with the compacted indexing, $G_{\ell;u} \boldsymbol{H}_{\ell;u}^{*}$ is the channel between the ℓth base station and user u at the cell of interest.

While, in the reverse link, channel estimates are readily available to the base stations from the corresponding pilot observations, users are in principle not cognizant of either their forward-link fading realizations or of the precoders selected by their serving base. To enable the estimation of these quantities at the user receivers, precoded forward pilots (at least one per user and coherence tile) would have to be sent. Alternatively, the user receivers can rely on hardening, possibly reinforced by blind methods operating on payload data observations [923].

In light of the closeness, observed for the reverse link, between the performance of channel-estimate-reliant receivers and hardening-reliant receivers, for our forward-link analysis we consider only the latter. This is not to advocate that the forward link should be devoid of pilots, but rather it is an analytical shortcut whose fruits can be regarded as tight lower bounds on the performance with sufficiently many antennas. In all likelihood, massive MIMO is to be implemented with forward pilots, and the corresponding analysis with pilot-assisted channel estimation at the users could be effected building on the teachings of Section 9.7.6. Readers interested in this extension are referred to [924–926].

Focusing then on receivers reliant on channel hardening, user u regards $\mathbb{E}[\boldsymbol{H}_u^* \boldsymbol{F}_u]$ as its precoded channel. With this in mind, (10.98) can be rewritten as

$$y_u = \underbrace{\sqrt{\frac{G_u E_u}{N_\text{a}}}\,\mathbb{E}\big[\boldsymbol{H}_u^* \boldsymbol{F}_u\big] s_u}_{\text{Desired signal}} + \underbrace{\sqrt{\frac{G_u E_u}{N_\text{a}}}\Big(\boldsymbol{H}_u^* \boldsymbol{F}_u - \mathbb{E}\big[\boldsymbol{H}_u^* \boldsymbol{F}_u\big]\Big) s_u}_{\text{Self-interference}} \tag{10.99}$$
$$+ \underbrace{\sum_{\text{u}\neq u} \sqrt{\frac{G_u E_\text{u}}{N_\text{a}}}\,\boldsymbol{H}_u^* \boldsymbol{F}_\text{u}\, s_\text{u}}_{\text{Same-cell interference}} + \underbrace{\sum_{\ell\neq 0}\sum_{\text{u}=0}^{U-1} \sqrt{\frac{G_{\ell;u} E_{\ell,\text{u}}}{N_\text{a}}}\,\boldsymbol{H}_{\ell;u}^* \boldsymbol{F}_{\ell,\text{u}}\, s_{\ell,\text{u}}}_{\text{Other-cell interference}} + \underbrace{v_u}_{\text{Noise}}$$

such that the SINR at user u in the cell of interest is

$$\overline{\mathsf{sinr}}_u = \frac{\frac{E_u}{E_\text{s}}\mathsf{SNR}_u\,\big|\mathbb{E}[\boldsymbol{H}_u^* \boldsymbol{F}_u]\big|^2}{\overline{\text{Den}}}, \tag{10.100}$$

with

$$\overline{\text{Den}} = \frac{E_u}{E_\text{s}}\mathsf{SNR}_u \,\text{var}[\boldsymbol{H}_u^* \boldsymbol{F}_u] + \sum_{\text{u}\neq u} \frac{E_\text{u}}{E_\text{s}}\mathsf{SNR}_u\,\mathbb{E}\Big[\big|\boldsymbol{H}_u^* \boldsymbol{F}_\text{u}\big|^2\Big]$$
$$+ \sum_{\ell\neq 0}\sum_{\text{u}=0}^{U-1} \frac{E_{\ell,\text{u}}}{E_\text{s}}\mathsf{SNR}_{\ell;u}\,\mathbb{E}\Big[\big|\boldsymbol{H}_{\ell;u}^* \boldsymbol{F}_{\ell,\text{u}}\big|^2\Big] + N_\text{a}. \tag{10.101}$$

10.5.1 Matched-filter transmitter

With matched-filter or maximum-ratio transmissions, sometimes also termed *conjugate beamforming*, the precoders at cell ℓ are given by

$$\boldsymbol{F}_{\ell,u}^{\text{MF}} = \sqrt{N_\text{a}}\frac{\hat{\boldsymbol{H}}_{\ell;\ell,u}}{\sqrt{\mathbb{E}\big[\|\hat{\boldsymbol{H}}_{\ell;\ell,u}\|^2\big]}} \qquad u = 0,\ldots,U-1, \tag{10.102}$$

where $\hat{\boldsymbol{H}}_{\ell;\ell,0},\ldots,\hat{\boldsymbol{H}}_{\ell;\ell,U-1}$ are the channel estimates gathered by base ℓ from the reverse-link pilots and where a per-codeword power constraint has been applied; this constraint is more amenable to the analysis that follows and, as argued in earlier chapters, appropriate for wideband systems featuring many OFDM subcarriers. Readers interested in the slightly more involved analysis under a per-symbol power constraint are directed to Problem 10.39.

Recalling the expression for the reverse-link channel estimates in (10.20), at the cell of interest specifically, for $u = 0, \ldots, U-1$,

$$\boldsymbol{F}_u^{\mathsf{MF}} = \frac{\boldsymbol{H}_u + \sum_{\ell \in \mathcal{C}} \sqrt{\frac{\frac{E_{\ell,u}}{E_s^r}\mathsf{SNR}_{\ell,u}^r}{\frac{E_u}{E_s^r}\mathsf{SNR}_u^r}}\boldsymbol{H}_{\ell,u} + \boldsymbol{v}_u'}{\sqrt{\frac{1}{N_a}\mathbb{E}\left[\left\|\boldsymbol{H}_u + \sum_{\ell \in \mathcal{C}} \sqrt{\frac{\frac{E_{\ell,u}}{E_s^r}\mathsf{SNR}_{\ell,u}^r}{\frac{E_u}{E_s^r}\mathsf{SNR}_u^r}}\boldsymbol{H}_{\ell,u} + \boldsymbol{v}_u'\right\|^2\right]}} \tag{10.103}$$

$$= \sqrt{\frac{\frac{E_u}{E_s^r}\mathsf{SNR}_u^r}{1 + \frac{E_u}{E_s^r}\mathsf{SNR}_u^r + \sum_{\ell \in \mathcal{C}}\frac{E_{\ell,u}}{E_s^r}\mathsf{SNR}_{\ell,u}^r}}\left(\boldsymbol{H}_u + \sum_{\ell \in \mathcal{C}}\sqrt{\frac{\frac{E_{\ell,u}}{E_s^r}\mathsf{SNR}_{\ell,u}^r}{\frac{E_u}{E_s^r}\mathsf{SNR}_u^r}}\boldsymbol{H}_{\ell,u} + \boldsymbol{v}_u'\right), \tag{10.104}$$

which coincides with the reverse-link matched-filter receiver in (10.48).

Output SINR

With matched-filter transmissions, (10.100) and (10.101) specialize—the reader is invited to verify the derivation in Problem 10.36—to

$$\overline{\mathsf{sinr}}_u^{\mathsf{MF}} = \frac{N_a \frac{\frac{E_u}{E_s^r}\mathsf{SNR}_u^r}{1 + \frac{E_u}{E_s^r}\mathsf{SNR}_u^r + \sum_{\ell \in \mathcal{C}}\frac{E_{\ell,u}}{E_s^r}\mathsf{SNR}_{\ell,u}^r}\frac{E_u}{E_s}\mathsf{SNR}_u}{1 + \sum_{\ell}\mathsf{SNR}_{\ell;u} + N_a\sum_{\ell \in \mathcal{C}}\frac{\frac{E_u}{E_s^r}\mathsf{SNR}_{\ell;u}^r}{1 + \frac{E_u}{E_s^r}\mathsf{SNR}_{\ell;u}^r + \sum_{l \in \mathcal{C}}\frac{E_{l,u}}{E_s^r}\mathsf{SNR}_{\ell;l,u}^r}\frac{E_{\ell,u}}{E_s}\mathsf{SNR}_{\ell;u}}, \tag{10.105}$$

which, invoking the forward–reverse relationship $\mathsf{SNR}_{\ell;u}^r = \mathsf{SNR}_{\ell;u}/\rho$, can be rewritten as

$$\overline{\mathsf{sinr}}_u^{\mathsf{MF}} = \frac{\frac{N_a}{\rho + \frac{E_u}{E_s^r}\mathsf{SNR}_u + \sum_{\ell \in \mathcal{C}}\frac{E_{\ell,u}}{E_s^r}\mathsf{SNR}_{\ell,u}}\frac{E_u}{E_s^r}\frac{E_u}{E_s}\mathsf{SNR}_u^2}{1 + \sum_{\ell}\mathsf{SNR}_{\ell;u} + \sum_{\ell \in \mathcal{C}}\frac{N_a}{\rho + \frac{E_u}{E_s^r}\mathsf{SNR}_{\ell;u} + \sum_{l \in \mathcal{C}}\frac{E_{l,u}}{E_s^r}\mathsf{SNR}_{\ell;l,u}}\frac{E_u}{E_s^r}\frac{E_{\ell,u}}{E_s}\mathsf{SNR}_{\ell;u}^2}. \tag{10.106}$$

Readers should watch out for the subtleties in the notation: $\mathsf{SNR}_{\ell,u} = \mathsf{SNR}_{0;\ell,u}$ relates the base station of interest with user u at cell ℓ whereas $\mathsf{SNR}_{\ell;u} = \mathsf{SNR}_{\ell;0,u}$ relates the ℓth base with user u at the cell of interest. The power control coefficient applied to the reverse-link pilot is $\frac{E_u}{E_s^r}$, whereas $\frac{E_u}{E_s}$ is the forward-link power allocation coefficient.

The form in (10.106) has the advantage of being solely a function of the forward-link SNRs, facilitating a contrast with the reverse-link SINR expression in (10.63). As noted in [883], these expressions are not dual in the sense of the MAC–BC duality expounded in Chapters 8 and 9. However, if pilot contamination is negligible, then (10.106) reduces to

$$\overline{\mathsf{sinr}}_u^{\mathsf{MF}} \approx \frac{\frac{E_u}{E_s^r}\frac{E_u}{E_s}\mathsf{SNR}_u^2\, N_a}{\left(\rho + \frac{E_u}{E_s^r}\mathsf{SNR}_u\right)\left(1 + \sum_{\ell}\mathsf{SNR}_{\ell;u}\right)}, \tag{10.107}$$

which does admit a duality with its reverse-link counterpart in (10.86). Under the premise that $\rho + \frac{E_u}{E_s^r}\mathsf{SNR}_u \approx \frac{E_u}{E_s^r}\mathsf{SNR}_u$, meaning perfect channel estimation, any combination of SINRs achievable in the reverse link can also be achieved in the forward link if the same total power is transmitted. This is consistent with results that, under the premise of CSI, extend to multicell networks the linear-transceiver MAC–BC duality [927–929].

Power allocation

Unlike the reverse-link transmitters, which might be deprived of CSIT, the forward-link transmitters are always privy to channel estimates for all the users in their respective cells and could therefore perform CSIT-based power allocation. However, the channel hardening makes CSIT-based power allocations hardly better than simpler counterparts based on statistical information, hence only the latter are considered here.

As in the reverse link, the forward power allocation embodies a demanding nonconvex optimization: a user's power rise comes at the expense of all other same-cell users, and it increases the interference to other-cell users. In the forward-link version of the problem, maxima found through techniques such as those in [901, 902] can be appraised through the iterative procedure in [930], which can identify the global optimum with a precision contingent on the number of iterations (a vanishing error requires a diverging number of iterations). However, this procedure requires centralized processing, hence it serves to benchmark distributed power allocation schemes rather than being itself an allocation scheme.

The principle of fractional power control that, in the reverse link, conveniently effects a partial SINR equalization, turns out not to be effective in the forward link. Indeed, reducing the power transmitted to nearby users could cause them to drown in interference from stronger transmissions to more distant users. Other guiding principles are therefore necessary to effect power allocation. As suggested in [914], the duality observed in the absence of pilot contamination and of channel estimation errors could be leveraged to formulate a power allocation policy whereby $\frac{E_0}{E_s}, \ldots, \frac{E_{U-1}}{E_s}$ are computed on the basis of the reverse-link power control coefficients obtained, say, through fractional power control.

Alternatively, and observing that, in (10.107), the power allocation coefficient $\frac{E_u}{E_s}$ appears only in the numerator, the SINRs in the absence of pilot contamination can be fully equalized for the U users by setting [931]

$$\frac{E_u}{E_s} \propto \frac{\left(\rho + \frac{E_u}{E_s^r}\mathsf{SNR}_u\right)\left(1 + \sum_\ell \mathsf{SNR}_{\ell;u}\right)}{\frac{E_u}{E_s^r}\mathsf{SNR}_u^2} \qquad u = 0, \ldots, U-1, \tag{10.108}$$

with the proportionality constant ensuring that $\sum_{u=0}^{U-1} \frac{E_u}{E_s} = 1$. Generalizing this idea, the SINRs can be partially equalized by setting

$$\frac{E_u}{E_s} \propto \left[\frac{\left(\rho + \frac{E_u}{E_s^r}\mathsf{SNR}_u\right)\left(1 + \sum_\ell \mathsf{SNR}_{\ell;u}\right)}{\frac{E_u}{E_s^r}\mathsf{SNR}_u^2}\right]^{\vartheta} \qquad u = 0, \ldots, U-1, \tag{10.109}$$

where $\vartheta \in [0, 1]$. By adjusting ϑ, the tradeoff between cell-edge performance and sum spectral efficiency can be regulated.

Example 10.19

Specialize (10.109) to interference-limited conditions.

Solution

In interference-limited conditions,

$$\frac{E_u}{E_\mathrm{s}} = \frac{\left(\frac{\sum_\ell G_{\ell;u}}{G_u}\right)^\vartheta}{\sum_{\mathrm{u}=0}^{U-1}\left(\frac{\sum_\ell G_{\ell;\mathrm{u}}}{G_\mathrm{u}}\right)^\vartheta}, \tag{10.110}$$

which does not depend on the reverse-link power control.

Example 10.20

Consider an interference-limited hexagonal network with $\eta = 4$ and $\sigma_{\mathrm{dB}} = 8$ dB. Let $N_\mathrm{p} \to \infty$ such that there is no pilot contamination and set $N_\mathrm{a} = 100$. For $N_\mathrm{a}/U = 4$ and 10, plot the CDFs of $\overline{\mathsf{sir}}_u^{\mathsf{MF}}$, both with a uniform power allocation ($\vartheta = 0$) and with a partially equalizing one ($\vartheta = 0.5$).

Solution

See Fig. 10.16, which confirms the effectiveness of the power allocation in (10.109) as a mechanism to regulate fairness: as ϑ is driven away from zero, the lower tail improves at the expense of the upper part of the distribution.

Behavior for $N_\mathrm{a}/U \to \infty$

If U is held fixed while $N_\mathrm{a} \to \infty$, pilot contamination eventually comes to dominate the performance and (10.106) behaves as [932]

$$\overline{\mathsf{sinr}}_u^{\mathsf{MF}} = \frac{\frac{1}{\rho + \frac{E_u}{E_\mathrm{s}^\mathrm{r}}\mathsf{SNR}_u + \sum_{\ell\in\mathcal{C}}\frac{E_{\ell,u}}{E_\mathrm{s}^\mathrm{r}}\mathsf{SNR}_{\ell,u}}\frac{E_u}{E_\mathrm{s}^\mathrm{r}}\frac{E_u}{E_\mathrm{s}}\mathsf{SNR}_u^2}{\sum_{\ell\in\mathcal{C}}\frac{1}{\rho + \frac{E_u}{E_\mathrm{s}^\mathrm{r}}\mathsf{SNR}_{\ell;u} + \sum_{l\in\mathcal{C}}\frac{E_{l,u}}{E_\mathrm{s}^\mathrm{r}}\mathsf{SNR}_{\ell;l,u}}\frac{E_u}{E_\mathrm{s}^\mathrm{r}}\frac{E_{\ell,u}}{E_\mathrm{s}}\mathsf{SNR}_{\ell;u}^2}, \tag{10.111}$$

which is the forward-link counterpart to (10.71). As anticipated, these expressions do not exhibit the duality relationship that is encountered when pilot contamination is negligible.

Impact of pilot contamination

The derivation of a sufficient condition for pilot contamination negligibleness is not straightforward in the forward link. Unlike in the reverse link, where there is an easy-to-identify worst-case combination of user locations, in the forward link the worst-case user locations in terms of exposure to pilot contamination are not evident (refer to Problem 10.45). However, since the value of N_p would ultimately be determined by the most stringent condition

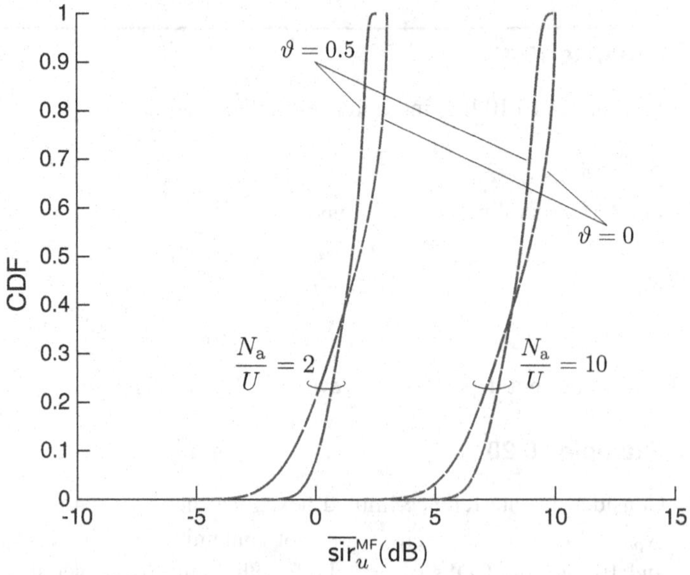

Fig. 10.16 Large-scale distributions of $\overline{\mathsf{sir}}_u^{\mathsf{MF}}$ in the forward link of an interference-limited hexagonal network with uniform ($\vartheta = 0$) and nonuniform ($\vartheta = 0.5$) power allocations, $\eta = 4$, $\sigma_{\mathsf{dB}} = 8$ dB, $N_{\mathrm{p}} \to \infty$, and $N_{\mathrm{a}} = 100$ for both $N_{\mathrm{a}}/U = 2$ and 10.

on either direction, the reverse-link condition in (10.84) can serve to establish this value provided that it guarantees a negligible degree of contamination in the forward link as well. Let us see that this is indeed the case, with slightly more conservative pilot reuse factors.

Example 10.21

Consider an interference-limited hexagonal network with uniform power allocation, $\eta = 4$, $\sigma_{\mathsf{dB}} = 8$ dB, and $N_{\mathrm{a}} = 100$. For $N_{\mathrm{a}}/U = 2$ and 10, plot the CDF of $\overline{\mathsf{sir}}_u^{\mathsf{MF}}$ for both $N_{\mathrm{p}} \to \infty$ and $N_{\mathrm{p}} = 4U$.

Solution

See Fig. 10.17.

The foregoing example illustrates how, with a reasonable ratio N_{p}/U, pilot contamination can be caused to be negligible on both the reverse and forward links. At this point, and rather than optimize the forward-link spectral efficiency for matched-filter transmitters, we defer this step to the next section, in the context of the superior regularized ZF transmitters.

10.5.2 Regularized ZF transmitter

Refashioning for channel estimates and a per-codeword power constraint the MU-MISO regularized ZF solution derived in Section 9.4, we obtain, for $u = 0, \ldots, U-1$ at the cell

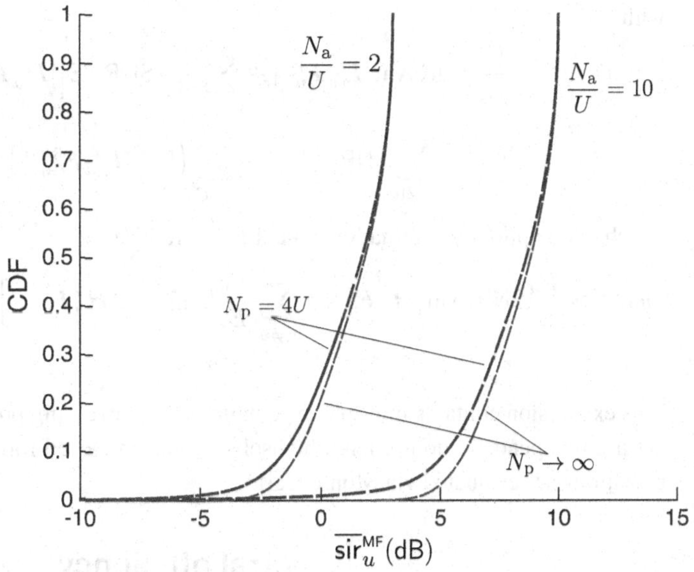

Fig. 10.17 Large-scale distributions of $\overline{\mathsf{sir}}_u^{\mathsf{MF}}$ in the forward link of an interference-limited hexagonal network with a uniform power allocation, $\eta = 4$, $\sigma_{\mathsf{dB}} = 8$ dB, and $N_a = 100$ for both $N_a/U = 2$ and $N_a/U = 10$. Comparison between $N_p \to \infty$ and $N_p = 4U$.

of interest,

$$\boldsymbol{F}_u^{\mathsf{Reg}} = \sqrt{N_a} \frac{\left(\sum_{u=1}^{U} \mathsf{SNR}_u \hat{\boldsymbol{H}}_u \hat{\boldsymbol{H}}_u^* + N_a U \left(1 + \sum_{\ell \neq 0} \mathsf{SNR}_{\ell;u}\right) \boldsymbol{I}\right)^{-1} \hat{\boldsymbol{H}}_u}{\sqrt{\mathbb{E}\left[\left\| \left(\sum_{u=0}^{U-1} \mathsf{SNR}_u \hat{\boldsymbol{H}}_u \hat{\boldsymbol{H}}_u^* + N_a U \left(1 + \sum_{\ell \neq 0} \mathsf{SNR}_{\ell;u}\right) \boldsymbol{I}\right)^{-1} \hat{\boldsymbol{H}}_u \right\|^2\right]}}, \tag{10.112}$$

which is a function of $\hat{\boldsymbol{H}}_0, \ldots, \hat{\boldsymbol{H}}_{U-1}$, meaning that it explicitly accounts for the same-cell interference while treating other-cell interference as noise.

As for the reverse-link LMMSE receiver, one might argue that the regularized ZF precoder could be broadened so as to explicitly account, not only for the interference inflicted upon same-cell users, but rather upon the entire pilot reuse cluster of the cell of interest [932, 933]. In the sequel though, we abide by (10.112), which can be regarded as the single-cell baseline. Also note that, following the argumentation in Section 9.9, the regularization term is set so as to maximize the SLNR.

Output SINR

Plugging (10.112) into (10.100) and (10.101), and taking advantage of the independence between the precoders at the cell of interest and the channels to other-cell users, we obtain

$$\mathsf{sinr}_u^{\mathsf{Reg}} = \frac{\left|\mathbb{E}[\boldsymbol{H}_u^* \boldsymbol{F}_u^{\mathsf{Reg}}]\right|^2 \frac{E_u}{E_s} \mathsf{SNR}_u}{\mathsf{Den}^{\mathsf{Reg}}} \tag{10.113}$$

with

$$\overline{\mathrm{Den}}^{\mathrm{Reg}} = \frac{E_u}{E_\mathrm{s}} \mathsf{SNR}_u \mathrm{var}[\boldsymbol{H}_u^* \boldsymbol{F}_u^{\mathrm{Reg}}] + \sum_{\mathrm{u} \neq u} \frac{E_\mathrm{u}}{E_\mathrm{s}} \mathsf{SNR}_u \, \mathbb{E}\Big[|\boldsymbol{H}_u^* \boldsymbol{F}_\mathrm{u}^{\mathrm{Reg}}|^2\Big] \tag{10.114}$$

$$+ \Big(1 + \sum_{\ell \neq 0} \mathsf{SNR}_{\ell;u}\Big) N_\mathrm{a} + \sum_{\ell \in \mathcal{C}} \Big(\mathbb{E}\Big[|\boldsymbol{H}_{\ell;u}^* \boldsymbol{F}_{\ell,u}^{\mathrm{Reg}}|^2\Big] - N_\mathrm{a}\Big) \frac{E_{\ell,u}}{E_\mathrm{s}} \mathsf{SNR}_{\ell;u}$$

which, when pilot contamination is negligible, reduces to

$$\overline{\mathrm{Den}}^{\mathrm{Reg}} \approx \frac{E_u}{E_\mathrm{s}} \mathsf{SNR}_u \, \mathrm{var}[\boldsymbol{H}_u^* \boldsymbol{F}_u^{\mathrm{Reg}}] + \sum_{\mathrm{u} \neq u} \frac{E_\mathrm{u}}{E_\mathrm{s}} \mathsf{SNR}_u \, \mathbb{E}\Big[|\boldsymbol{H}_u^* \boldsymbol{F}_\mathrm{u}^{\mathrm{Reg}}|^2\Big] + \Big(1 + \sum_{\ell \neq 0} \mathsf{SNR}_{\ell;u}\Big) N_\mathrm{a}. \tag{10.115}$$

This expression contains multiple expectations that, involving precoders and channels that are not independent, do not lend themselves to clean closed forms. Henceforth, these expectations are evaluated via Monte-Carlo.

Spectral efficiency

For receivers reliant on channel hardening, without precoded forward pilots, the regularized ZF forward-link gross spectral efficiencies at the cell of interest are

$$\frac{\bar{R}_u^{\mathrm{Reg}}}{B} = \log_2\big(1 + \overline{\mathsf{sinr}}_u^{\mathrm{Reg}}\big) \qquad u = 0, \ldots, U-1, \tag{10.116}$$

from which $\sum_{u=0}^{U-1} \bar{R}_u^{\mathrm{Reg}}/B$ is the gross sum spectral efficiency at that cell. Within the range where pilot contamination can be rendered negligible, the optimization of this quantity directly yields suitable values for N_a/U.

Example 10.22

Let $N_\mathrm{a} = 100$ in an interference-limited hexagonal network with regularized ZF transmitters, uniform power allocation, $\eta = 4$, and $\sigma_{\mathrm{dB}} = 8$ dB. Which ratio N_a/U maximizes the forward-link gross sum spectral efficiency in the absence of pilot contamination, subject to no more than 3% of users having an SIR below -5 dB?

Solution

The ratio can be as low as $N_\mathrm{a}/U = 1$ without exceeding the 3% large-scale outage. The corresponding distributions of gross sum spectral efficiency and user SIR are presented in Fig. 10.18. The average gross sum spectral efficiency is 291.2 b/s/Hz per cell with 1% of users below -5 dB.

The forward-link performance could be improved further with nonuniform power allocations, albeit again with the obstacle of the lack of convexity. To skirt this hurdle, the simple power allocation in (10.109) could be considered. Alternatively, strict ZF precoders could be applied; this would open the door to a power allocation based on waterfilling [934], which, recall from Section 9.7, maximizes the sum spectral efficiency with ZF transmission

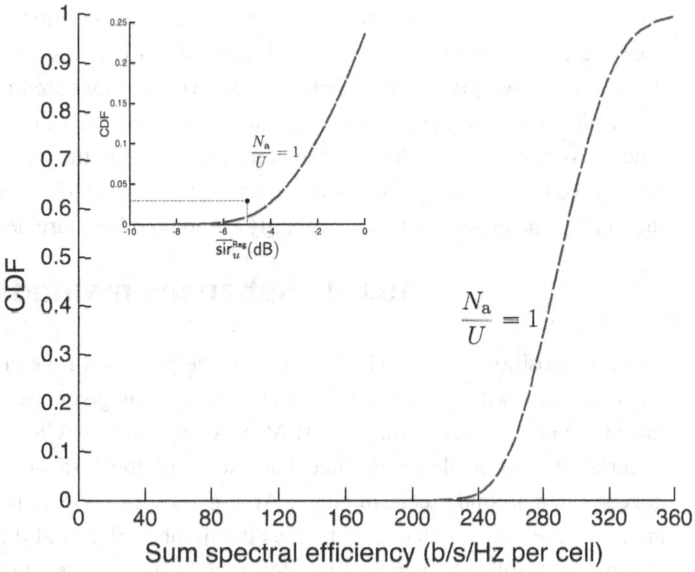

Fig. 10.18 In the main plot, large-scale distribution of the gross sum spectral efficiency per cell in the forward link of an interference-limited hexagonal network with no pilot contamination, regularized ZF transmitters, a uniform power allocation, $\eta = 4$, $\sigma_{\text{dB}} = 8$ dB, and $N_a = 100$, for $N_a/U = 1$. In the inset, the corresponding distribution of $\overline{\text{sir}}_u^{\text{Reg}}$.

and CSIT. The variations of waterfilling proposed in Problem 9.16, which maximize the weighted sum spectral efficiency, could allow giving higher weights to users in detrimental locations so as to avoid starving those users while maximizing the aggregate performance.

The final steps in the optimization of the spectral efficiency, moving from gross to net, would be to explicitly account for the pilot overhead (whichever share is ascribed to the forward link) and for the effect (minor provided the sufficient condition is respected) of pilot contamination, as well as for the possibility of increasing the number of pilot symbols per user or adjusting the pilot power boost. We also hasten to recall that the forward-link analysis and results throughout this section are for hardening-reliant receivers; with additional precoded pilots and pilot-observation-reliant receivers, the performance could further improve in many settings, e.g., pedestrian.

10.6 Mitigation of pilot contamination

We have seen throughout the chapter that, other than for extreme—in excess of several hundred—numbers of antennas, contamination can be rendered minor with reasonable overheads thanks to the many pilot opportunities allowed by underspread fading as well as the fast decay of power over distance. Furthermore, this conclusion emanates from interference-limited evaluations, and the presence of noise would further conceal the contamination. Additional mechanisms, not included in the models invoked throughout the book because they only become significant over distances exceeding the size of a cell, would tone down pilot contamination to an even greater extent; these include vertical antenna tilts at the base stations and pathloss exponents that increase with distance, among others. Notwithstanding these considerations, and for the sake of completeness, in this section we briefly survey some techniques that can serve to actively mitigate pilot contamination. Yet other schemes, not explicitly elaborated here, are developed in [935–940].

10.6.1 Subspace methods

These propositions [854, 941–944] rely on the premise of the channel being sparse in the angle domain, with the ensuing antenna correlations given the reduced antenna spacings that are to be expected in massive MIMO. As discussed in Chapter 3, this sparsity premise is certainly reasonable in elevated base stations: the PAS at an elevated base is highly concentrated around a certain angle. And, even if the PAS is not particularly compact in angle, it is reasonable to expect that, as the number of spatial dimensions grows large and the angular resolution sharpens, the PAS may become sparse: most power from each given user may be received over a limited number of angular directions.

Consider the transmission of a pilot from an intended user within the cell of interest. The contamination impinging on the corresponding base station of interest outside the angular directions occupied by that intended pilot can be filtered out. In fact, a channel estimator equipped with antenna correlation information for the intended pilots would automatically reject the nonoverlapping contamination, in a sense by forming one or multiple beams that avoid it while targeting the intended pilot. Alternatively, machine learning techniques could be applied to isolate the channel estimate of the intended user from the nonoverlapping contamination. Interestingly then, antenna correlations, often detrimental, are advantageous when it comes to channel estimation—not only because of the reduced uncertainty in the channel vectors, but, more subtly, because of the possibility of rejecting the contamination in the estimates.

Besides the angle domain, sparsity may also exist in the delay domain: the delay spread may be short enough that the power delay profile of the intended user is essentially nonoverlapping with those of the more distant (and thus later-arriving) contaminating users [945, 946]. Again, a channel estimator equipped with the delay information for the intended user, or a suitable machine learning algorithm, could strip the nonoverlapping contamination.

An alternative perspective on these ideas, referring back to Section 5.7, would be that pilot contamination is less harmful if it exhibits color in the space or frequency domains, as some dimensions are then less contaminated and the receiver can project the intended signal onto the subspace defined by those dimensions. Conversely, in the absence of color, all dimensions are equally contaminated and projections have no effect.

10.6.2 Coordinated pilot assignment

The rejection of contamination based on subspace projections can be fostered by assigning pilots to users in such a way that the contamination is as disjoint as possible, in angle and/or delay, to the intended pilot transmissions at nearby copilot base stations. This requires integrating into the user selection procedures a certain degree of coordination among cells, and is subject to the possibilities offered by the propagation environment and the user locations.

This relates to the idea, discussed in Chapter 9, of exploiting correlations so as to reuse each pilot multiple times *within* each cell. In fact, what we are pondering here may be interpreted as a multicell version of that idea. Ironically, this multicell version, seemingly less aggressive, is harder to coordinate, as the copilot users in question are connected to distinct base stations rather than to the same one.

10.6.3 Reception and precoding with other-cell awareness

Suppose that, because it does not affect the formulation that follows, the power control is inactive. Recall from (10.20) that the MMSE channel estimate gathered by the base of interest for user u, in the face of pilot contamination and without antenna correlations, is

$$\hat{\boldsymbol{H}}_u = \frac{\sqrt{\mathsf{SNR}^{\mathsf{r}}_u}}{1+\mathsf{SNR}^{\mathsf{r}}_u + \sum_{\ell \in \mathcal{C}} \mathsf{SNR}^{\mathsf{r}}_{\ell,u}} \left(\sqrt{\mathsf{SNR}^{\mathsf{r}}_u}\, \boldsymbol{H}_u + \sum_{\ell \in \mathcal{C}} \sqrt{\mathsf{SNR}^{\mathsf{r}}_{\ell,u}}\, \boldsymbol{H}_{\ell,u} + \boldsymbol{v}_u \right). \tag{10.117}$$

If the same base attempted to estimate the channel for user u in copilot cell $\ell \in \mathcal{C}$, that estimate would be

$$\hat{\boldsymbol{H}}_{\ell,u} = \frac{\sqrt{\mathsf{SNR}^{\mathsf{r}}_{\ell,u}}}{1+\mathsf{SNR}^{\mathsf{r}}_u + \sum_{l \in \mathcal{C}} \mathsf{SNR}^{\mathsf{r}}_{l,u}} \left(\sqrt{\mathsf{SNR}^{\mathsf{r}}_u}\, \boldsymbol{H}_u + \sum_{l \in \mathcal{C}} \sqrt{\mathsf{SNR}^{\mathsf{r}}_{l,u}}\, \boldsymbol{H}_{l,u} + \boldsymbol{v}_u \right), \tag{10.118}$$

which is a scaled version of $\boldsymbol{H}_u$. These two estimates, and those of any other copilot users, are hence colinear and there is no hope of disentangling the corresponding signals through the projections effected by linear receivers. Hence the contamination.

Enter antenna correlations, with $\boldsymbol{R}_{\ell,u}$ denoting the correlation matrix experienced at the base of interest by user u in cell ℓ. The above MMSE channel estimates then generalize (for both $\ell = 0$ and $\ell \in \mathcal{C}$) to

$$\begin{aligned}\hat{\boldsymbol{H}}_{\ell,u} = {} & \sqrt{\mathsf{SNR}^{\mathsf{r}}_{\ell,u}} \left(\boldsymbol{I} + \mathsf{SNR}^{\mathsf{r}}_u \boldsymbol{R}_u + \sum_{l \in \mathcal{C}} \mathsf{SNR}^{\mathsf{r}}_{l,u} \boldsymbol{R}_{l,u} \right)^{-1} \\ & \cdot \boldsymbol{R}_{\ell,u} \left(\sqrt{\mathsf{SNR}^{\mathsf{r}}_u}\, \boldsymbol{H}_u + \sum_{l \in \mathcal{C}} \sqrt{\mathsf{SNR}^{\mathsf{r}}_{l,u}}\, \boldsymbol{H}_{l,u} + \boldsymbol{v}_u \right),\end{aligned} \tag{10.119}$$

satisfying $\boldsymbol{H}_{\ell,u} = \hat{\boldsymbol{H}}_{\ell,u} + \tilde{\boldsymbol{H}}_{\ell,u}$ where $\hat{\boldsymbol{H}}_{\ell,u} \sim \mathcal{N}_{\mathbb{C}}(\boldsymbol{0}, \boldsymbol{R}_{\ell,u} - \boldsymbol{E}_{\ell,u})$ while the estimation error, uncorrelated with the estimates, is $\tilde{\boldsymbol{H}}_{\ell,u} \sim \mathcal{N}_{\mathbb{C}}(\boldsymbol{0}, \boldsymbol{E}_{\ell,u})$. The MMSE matrix equals

$$\boldsymbol{E}_{\ell,u} = \boldsymbol{R}_{\ell,u} - \mathsf{SNR}^{\mathsf{r}}_{\ell,u} \boldsymbol{R}_{\ell,u} \left(\boldsymbol{I} + \mathsf{SNR}^{\mathsf{r}}_u \boldsymbol{R}_u + \sum_{l \in \mathcal{C}} \mathsf{SNR}^{\mathsf{r}}_{l,u} \boldsymbol{R}_{l,u} \right)^{-1} \boldsymbol{R}_{\ell,u}. \tag{10.120}$$

The key point is that, in the face of antenna correlations, $\hat{\boldsymbol{H}}_u$ and $\hat{\boldsymbol{H}}_{\ell,u}$ need no longer be colinear if $\boldsymbol{R}_u$ and $\boldsymbol{R}_{\ell,u}$ are different. In fact, if $\boldsymbol{R}_u \boldsymbol{R}_{\ell,u} = \boldsymbol{0}$, the estimates $\hat{\boldsymbol{H}}_u$ and $\hat{\boldsymbol{H}}_{\ell,u}$ are uncorrelated and there is no contamination between them, which is what the coordinated pilot allocation in the previous section seeks to foster. Even without this strong condition though, if simply the estimates are not colinear, the door is opened to rejecting the contamination from user u at cell ℓ without simultaneously rejecting all the wanted signal from user u at the cell of interest. As claimed in [947], this could be achieved by an LMMSE receiver that explicitly accounted for the channels of copilot users in other cells, rather than treat them as noise. In its most ambitious form, with all other-cell users explicitly incorporated, this would entail changing (10.88) into

$$\boldsymbol{W}_u^{\mathsf{MMSE}} \propto \left[\sum_{\ell} \sum_{\mathrm{u}=0}^{U-1} \frac{E_{\ell,\mathrm{u}}}{E_{\mathrm{s}}^{\mathrm{r}}} \mathsf{SNR}_{\ell,\mathrm{u}}^{\mathrm{r}} \hat{\boldsymbol{H}}_{\ell,\mathrm{u}} \hat{\boldsymbol{H}}_{\ell,\mathrm{u}}^{*} + \left(\boldsymbol{I} + \sum_{\ell} \sum_{\mathrm{u}=0}^{U-1} \frac{E_{\ell,\mathrm{u}}}{E_{\mathrm{s}}^{\mathrm{r}}} \mathsf{SNR}_{\ell,\mathrm{u}}^{\mathrm{r}} \boldsymbol{E}_{\ell,\mathrm{u}} \right) \right]^{-1} \hat{\boldsymbol{H}}_u$$

$$u = 0, \ldots, U-1. \qquad (10.121)$$

Provided that, at the base of interest, copilot users exhibit correlations that are "sufficiently different" from those of the same-cell users, the above LMMSE receivers would automatically reject the contamination and enable an unbounded growth of the SINRs for $N_{\mathrm{a}}/U \to \infty$. Technical conditions given in [947] make precise the notion of "sufficiently different," which for a certain user u essentially amounts to

$$\lim_{N_{\mathrm{a}} \to \infty} \frac{1}{N_{\mathrm{a}}} \left\| \boldsymbol{R}_u - \xi \, \boldsymbol{R}_{\ell,u} \right\|_{\mathrm{F}}^2 > 0 \qquad (10.122)$$

for every ξ. Intuitively, this ensures that $\boldsymbol{R}_u$ and $\boldsymbol{R}_{\ell,u}$ do not become progressively similar as they grow large, which in turn ensures that the respective MMSE channel estimates remain noncolinear, rendering $\boldsymbol{W}_u^{\mathsf{MMSE}}$ effective at rejecting the contamination from cell ℓ as $N_{\mathrm{a}} \to \infty$.

The same mitigation of pilot contamination could be achieved in the forward link with a regularized ZF transmitter that explicitly accounted for copilot users in other cells [947].

The implementation of receivers and transmitters that explicitly account for other-cell channels requires the estimation of those channels with knowledge of the corresponding individual antenna correlation matrices. While antenna correlations can in principle be computed empirically from either pilot or data observations, if obtaining individual correlation matrices for each user demanded silence from all other neighboring users during the necessary number of symbols, that would certainly be a complication. Methods to disentangle individual correlation matrices would be a welcome support to this method.

10.6.4 Large-scale multicell processing

In contrast with the preceding techniques, which hinge upon properties of the antenna correlations, this idea does not involve correlations. Rather, it entails transcending the framework, firmly held throughout the chapter, of single-cell processing. Precisely, the idea involves augmenting the single-cell reception and precoding procedures studied throughout the chapter with an additional stage of multicell processing [948].

For the reverse link, let us recall from (10.31) what the base station of interest observes.

As established, for $N_\mathrm{a}/U \to \infty$ the intended signal and those from copilot users surge above everything else (channel estimation errors, same-cell and other-cell interference, and noise) such that, scaled by $\sqrt{G_u}/N_\mathrm{a}^2$,

$$\frac{\sqrt{G_u}}{N_\mathrm{a}^2} \boldsymbol{W}_u^* \boldsymbol{y} \to G_u \sqrt{E_u} s_u + \sum_{\ell \in \mathcal{C}} G_{\ell,u} \sqrt{E_{\ell,u}} s_{\ell,u} \qquad u = 0, \ldots, U-1, \tag{10.123}$$

which readily leads to the limiting SINR expression in (10.71). Based on its individual observation for user u, the base cannot untangle the intended and contaminating symbols, namely s_u and $s_{\ell,u}$ for $\ell \in \mathcal{C}$. Even if the transmit energies (E_0 and $E_{\ell,u}$ for $\ell \in \mathcal{C}$) and the large-scale gains (G_u and $G_{\ell,u}$ for $\ell \in \mathcal{C}$) are known, the base is faced with many unknowns and a single equation. However, combining the observations from the $\bar{L} = L/L_{\mathsf{cluster}}$ copilot bases, all of a sudden we have for each user index u a system with as many unknowns as equations. Indeed, we can write

$$\bar{\boldsymbol{y}}_u = \boldsymbol{\Xi}_u \bar{\boldsymbol{s}}_u \qquad u = 0, \ldots, U-1, \tag{10.124}$$

where $\bar{\boldsymbol{s}}_u = [s_u \; s_{1,u} \; \cdots \; s_{\bar{L}-1,u}]^\mathrm{T}$ contains the sought information symbols,

$$\bar{\boldsymbol{y}}_u = \left[\frac{\sqrt{G_u}}{N_\mathrm{a}^2} \boldsymbol{W}_u^* \boldsymbol{y} \quad \frac{\sqrt{G_{1;1,u}}}{N_\mathrm{a}^2} \boldsymbol{W}_{1,u}^* \boldsymbol{y}_1 \quad \cdots \quad \frac{\sqrt{G_{\bar{L}-1;\bar{L}-1,u}}}{N_\mathrm{a}^2} \boldsymbol{W}_{\bar{L}-1,u}^* \boldsymbol{y}_{\bar{L}-1} \right]^\mathrm{T} \tag{10.125}$$

contains scaled versions of the base station observations, and

$$\boldsymbol{\Xi}_u = \begin{bmatrix} G_u \sqrt{E_u} & G_{1,u} \sqrt{E_{1,u}} & \cdots & G_{\bar{L}-1,u} \sqrt{E_{\bar{L}-1,u}} \\ G_{1;u} \sqrt{E_u} & G_{1;1,u} \sqrt{E_{1,u}} & \cdots & G_{1;\bar{L}-1,u} \sqrt{E_{\bar{L}-1,u}} \\ \vdots & \ddots & \ddots & \vdots \\ G_{\bar{L}-1;u} \sqrt{E_u} & G_{\bar{L}-1;1,u} \sqrt{E_{1,u}} & \cdots & G_{\bar{L}-1;\bar{L}-1,u} \sqrt{E_{\bar{L}-1,u}} \end{bmatrix}. \tag{10.126}$$

The information symbols can be ridden of the pilot contamination, the only lingering impairment for $N_\mathrm{a}/U \to \infty$, via

$$\bar{\boldsymbol{s}}_u = \boldsymbol{\Xi}_u^{-1} \bar{\boldsymbol{y}}_u, \tag{10.127}$$

which can be interpreted as an added multicell ZF operation that, by virtue of channel hardening, entails only large-scale quantities.

In the forward link, the process is reversed: there is multicell ZF precoding based on large-scale quantities, whereby the information symbols for every set of copilot users are converted into a vector whose entries are then distributed to specific base stations. At each, the respective entry is scaled and further precoded, this time on the basis of the channel estimates therein.

While, in theory, the additional multicell processing stage would ensure an unlimited increase of the spectral efficiency for $N_\mathrm{a} \to \infty$, for finite N_a it suffers the usual issues of ZF: the noise and the contamination-unrelated interference, which for finite N_a are significant and possibly even dominant, get enhanced, to the point that the performance may ultimately deteriorate [949]. By regularizing the large-scale-based multicell ZF, this enhancement can be balanced with the sought mitigation of pilot contamination.

Implementationally, the multicell processing stage should take place at some central hub, which could be a separate entity or a designated base station playing that role. Although only slowly changing large-scale quantities are required for such centralized processing, and not small-scale fading coefficients, what dominates the amount of data to be exchanged between the bases and the hub are the complex data symbols, and these dictate a very high degree of cooperation across the infrastructure.

10.7 Practical considerations

There are a number of aspects that are not of particular concern for small numbers of antennas, but which have the potential to result in blockades as N_{a} grows seriously large. All these aspects pertain to the base stations, which is where the massification takes place.

Hardware nonidealities

A first aspect relates to hardware nonidealities (e.g., multiplicative phase noise due to drifts in the oscillators, distortion due to amplifier nonlinearities and filter imperfections, and quantization noise due to a finite number of bits in the digital representations) and as to whether the effects of these nonidealities compound with catastrophic consequences as N_{a} grows large. Satisfyingly, that is not the case [897, 950–952]. Because additive impairments are independent of the intended signals, and usually also independent from antenna to antenna, their effect—just like those of noise, channel estimation errors, and interference—vanishes for $N_{\mathrm{a}}/U \to \infty$. In turn, the multiplicative effect of phase noise does not vanish, but it does not worsen with N_{a} either [953].

Complexity

Also related to hardware is the issue of complexity, and specifically its increase with N_{a} and U. With the preferred LMMSE receivers and regularized ZF transmitters, the obtention of the filters and precoders entails inverting a large matrix per coherence tile.

- For the dimensionalities considered in this chapter, the computational cost is still within the practical realm [951, 954]. Moreover, polynomial approximations to the matrix inverse [887, 955–957] and certain iterative methods [958–960] are highly effective in the presence of excess antennas.
- For extreme numbers of antennas and users, the computational burden could become overwhelming. However, in that extreme, matched-filter transceivers would become effective, and their computational cost scales far more gracefully.

Power consumption

The power consumed by the baseband processors and by the analog-to-digital and digital-to-analog converters does grow steadily with N_{a}, but the radiated power need not. Keeping that power (typically tens of watts per base station) constant, with massive MIMO the power radiated by each antenna drops to hundreds or even tens of milliwatts. Put differently, massive MIMO pushes the power radiated by each antenna from the level of regular base stations down roughly to the level of user devices. This may enable, at least for part of the transmission chains, low-power consumer-grade parts rather than high-power industry-grade equipment [961].

To further restrain the power consumption and the costs associated with the digitalization, hybrid precoders and hybrid receive filters can be employed. These consist of a U-dimensional digital stage cascaded with an analog stage that brings the overall dimensionality to N_a [962, 963].

Form factors

An evident concern in massive MIMO is the size and shape of the base stations. It is simply not feasible to endlessly pack additional antennas within a given space, not only because of mutual coupling, but because the channel dimensionality over a finite volume is fundamentally limited [964]. On a volumetric array, only outer antennas contribute to the communication because the electromagnetic fields inside a volume are not independent of, but rather fully described by those on the surface [965–967]. On a cylindrical layout, for instance, no antennas should be installed inside the cylinder. At some point, therefore, additional antennas require increasing the footprint of the base station. Importantly though, that footprint refers to the electrical size, i.e., relative to the wavelength.

- If the carrier frequencies are unchanged, then the physical size of the base stations does eventually need to grow.
- If progressively higher frequencies come into use, then dense arrays could be retrofitted into existing base stations, and into possibly even smaller ones.

Another family of issues that affect the base station form factor are cell sectorization and fixed vertical beamforming, traditional complementary approaches that split each cell into multiple subcells (the sectors) whose serving antennas are collocated. While it may be tempting to combine these approaches with massive MIMO, the creation of fixed sectors and the formation of vertical beams require either bulkier directional antennas or else a sacrifice—various nondirectional antennas need to be combined to achieve directionality—in the number of antennas that remain usable for MIMO, i.e., in the value of N_a in our derivations.

Example 10.23

Consider a typical macrocellular base station, where every "antenna" is in actuality a vertical array of eight tightly spaced antennas that effect vertical beamforming. Suppose that there are three sectors, each featuring four such "antennas" for MIMO or diversity. Altogether, this base station is equipped with $3 \cdot 4 \cdot 8 = 96$ actual antennas. If these antennas, possibly rearranged, were allowed to operate individually, a 96-antenna massive MIMO base station would be obtained.

Conversely, if we introduced vertical beamforming and sectorization in a 96-antenna massive MIMO array keeping the same physical antennas, we would end up with a classic structure having four "antennas" per sector.

Note that the above equivalence refers to the antenna count for a given form factor. Operating in a massive MIMO fashion does require many additional radio front-ends and much additional baseband processing.

Massive MIMO can therefore be seen as a deconstruction of existing base station architectures that maximizes signaling dimensionality and flexibility as fixed interconnections, fixed sectors, and beams, are released and rendered adaptive to the channel and user conditions via pilot observations. This deconstruction, though, tends to require increased physical space, especially because the vertical dimension tends to be less useful for signaling than the horizontal one: as anticipated in Chapter 3, the correlation distance is longer vertically because the angle spread is smaller in elevation than in azimuth. Hence, beyond pushing to their limits existing base station arrangements, alternative structures, ideally camouflaged with the environment, ought to be considered for massive MIMO. This may include arrays deployed along rooftops, building facades, billboards, or other urban elements [968], and even contiguous surfaces of electromagnetically active material [969].

Channel reciprocity

Instrumental for massive MIMO is the reciprocity between the reverse and forward channels. Calibration algorithms specific to massive MIMO have been proposed and validated, e.g., in [970–974], and the impact of residual nonreciprocities is studied in [975].

Besides the nuisance of recurring calibration procedures, TDD requires tight synchronization across cells so as to avoid potentially damaging base-to-base and user-to-user interference [976]. This danger materializes with full duplexing, and a full-duplex network would have to be designed with these new types of interference in mind [977].

Turning now to FDD, could it be compatible with massive MIMO? What is onerous in FDD, recall, is that the pilot overhead scales, besides U, also with N_{a}. Two avenues have been proposed to avoid the ensuing overhead explosion whenever $N_{\mathrm{a}} \gg U$ [978].

- Exploit sparsity in the channel, whenever present. If a domain could be identified, say the angle domain, where the channel were markedly sparse, then $N_{\mathrm{p}} \ll N_{\mathrm{a}}$ pilot symbols per coherence interval would suffice [979–983].
- Exploit antenna correlations, e.g., by parsing the N_{a} antennas into groups and estimating the channels at only one antenna per group [984]. Alternatively, schemes can be applied that parse the U users into groups having similar correlations and then isolate the groups through correlation-based beams; each beam serves a fraction of U with a number of equivalent base antennas that is a fraction of N_{a}. Practical embodiments of this latter idea range from the JSDM scheme described in Chapter 9, which fully adapts to the correlations via two-stage precoding [985–990], to traditional sectorization, which for the purpose of FDD implementation could be a necessary recourse.

Channel conditions

The IID fading postulated throughout the chapter exhibits the so-called favorable propagation property whereby the channel vectors of different users become orthogonal for $N_{\text{a}}/U \to \infty$. While IID fading indeed suffices for this property to hold, it is not required. For single-antenna users in particular, the favorable propagation property is perfectly feasible in the face of antenna correlations at the base. As long as the channels corresponding to different users are not colinear, they may exhibit this property and be discriminated by linear transceivers [883, 991]. Even in the extreme case of LOS propagation, with full correlation across the base station antennas, the favorable propagation property holds provided there is a modicum of angular separation—the user selection process would have to ensure this within each cell—among users [951, 992, 993]. Importantly though, and consistently with the earlier comments about the limited channel dimensionality over a finite space, this holds only if the base station becomes large as $N_{\text{a}} \to \infty$. If its size is constrained, conversely, favorable propagation cannot be enabled through angular separation [994].

Turning to multiantenna users, they can always be viewed as multiple single-antenna users that happen to be collocated, with the further possibility of joint processing, say through block-diagonalization (recall Section 9.8). In that sense, multiantenna users are subsumed by the analysis in the chapter provided the fading is IID. However, antenna correlations do have an effect here because angular separation of same-user antennas is unfeasible. The impact of such correlations can be determined by applying and extending the MU-MIMO results in Chapter 9.

Fading stationarity along the arrays, a cornerstone of classic models and a consistent assumption throughout this book, may cease to hold if the arrays become comparable in size to the correlation distance of the shadow fading [202, 259, 995, 996]. This need not be detrimental, and it could for instance facilitate the technique described in Section 10.6.3 to fight pilot contamination, but if the nonstationarity becomes pronounced it might require modifying long-standing models and alter some of the expressions in this chapter [997].

Besides antenna correlations and nonstationarity, other channel features not explicitly included in the analysis and results in the chapter are Rice factors [998] and correlation among the shadow fading of links having a common base station or a common user. Some problems are proposed at the end to exercise the impact of these features.

Differences in the models and results notwithstanding, simulations conducted on experimentally measured channels welcomely indicate that analytical predictions made on the basis of IID fading mostly hold up [999–1003].

10.8 Summary and outlook

Massive MIMO culminates the idea of space-domain signaling, pushing it toward the limits imposed by the real estate at the base stations and motivating the interest in expanding that real estate. With the number of active users properly set, and with power control and power allocation ensuring that over 97% of users perform acceptably, well over 1 b/s/Hz per base antenna is achievable on average with large- and small-scale propagation, same- and other-cell interference, channel estimation, pilot contamination, and pilot overheads all accounted for. This figure could be further increased with more involved power control and power allocation policies, and with further optimizations of the pilot overheads and/or power boostings. Yet further improvements could be attained with user selection as well as with user-cell associations based, not only on channel gains, but also on cell loads [1004].

With the caveat of a diminished ability to spatially multiplex data streams to multi-antenna users, the performance is relatively robust to antenna correlations at the base. In fact, certain correlations may be advantageous if pilot contamination needs to be mitigated, or the pilot overhead otherwise reduced.

The performance may degrade if interfering users are ever in LOS to the base station of interest and, to a lesser extent, if interfering bases are in LOS to users in the cell of interest. In both cases, the interference surges because of its lower pathloss exponent, requiring a compensating increase in N_{a}/U and/or ϑ. The impact of the pathloss exponent is the subject of some of the problems at the end of the chapter.

All in all, the book begins with SISO communication and link spectral efficiencies in the low single digits, which, transplanted onto a cellular network, map to average values on the order of 1 b/s/Hz per cell [444, 1005]. We finish the book with a similar value *per antenna*, meaning average spectral efficiencies of tens or even hundreds of b/s/Hz per cell, a journey of two orders of magnitude [1006–1008]. Massive MIMO, sometimes under the moniker *full-dimension* MIMO and beginning at $N_{\mathrm{a}} = 64$, is an integral part of the LTE and NR roadmaps [1009–1012].

It is interesting to contrast the performance of the forward and reverse links, which this chapter brings together. Recalling the duality results in Chapters 8 and 9, we can state that, in a single-cell setup, the capacity would be the same in both directions if the same total power were transmitted and CSI were available at both ends. Although the power budget of a base station tends to be a couple of orders of magnitude higher than that of an individual user, the number of active users gets to be considerable in massive MIMO. Moreover, once we zoom out to an entire network, other-cell interference comes into play and, in interference-limited conditions, the absolute power values become immaterial and things even out in the forward and reverse directions. This tie is broken by the forward-link interference being localized at a single spot per cell (the corresponding base, by definition far from the cell boundary) while the reverse-link interference is generated at many geographically distributed spots (the users, some of which may be near the boundary with other cells). The forward-link interference distribution is altogether more benign [1013], causing a performance asymmetry in favor of the forward link. This is reflected by the mirror examples in this chapter, e.g., Examples 10.10 and 10.18.

In relation to the large-scale distributions in the chapter, we must recall that they are produced through Monte-Carlo on hexagonal layouts. Indeed, Monte-Carlo simulations have long been the workhorse of wireless network design. While that is still the case, an increasingly popular alternative to produce large-scale distributions of quantities such as the SINR or the spectral efficiency is stochastic geometry, whereby, in lieu of shadow fading, the base station locations are randomized [1005, 1014, 1015]. Then, with both users and base station positions drawn from appropriate stochastic point processes, a powerful and

Take-away points

1. The massive MIMO regime can be defined by the relationships $N_a \geq U \gg 1$, where the former inequality should be in the range $N_a/U \approx$ 1–2 depending on type of transceivers and the channel conditions, whereas the latter should be as pronounced as the fading coherence allows. The number of spatial DOF per cell equals U.
2. TDD or full duplex enable the CSI-related overhead to scale only with U, and not with N_a.
3. The communication procedure, tailored to the fading coherence, entails: reverse-link pilots from the users, channel estimation at the bases, receive filter and precoder computation, (optionally) precoded forward pilot transmission, and successive data transmission in both directions.
4. Pilot dimensions are finite, hence they must be reused across cells. This causes contamination in the channel estimates; unlike fading, noise, interference, and channel estimation errors, the effects of this contamination do not abate for $N_a/U \to \infty$.
5. Sufficiently many excess antennas leads to the hardening of the filtered signals, i.e., the vanishing of small-scale randomness, and renders simple matched filters feasible. Furthermore, precoded forward pilots then become dispensable as the user receivers can track the hardened precoded channels.
6. While matched-filter receivers are feasible, without power control N_a/U must be exceedingly large for the share of users at very low SINR to be acceptable. For given N_a, this curtails the number of users and thus the sum spectral efficiency.
7. Although the optimization of the reverse-link power control is nonconvex, simple suboptimum policies operating on large-scale quantities allow matched filters to operate at much smaller N_a/U while keeping the share of low-SINR users in check.
8. Despite the limiting ($N_a/U \to \infty$) optimality of matched filters in the absence of pilot contamination, up to hundreds of antennas the LMMSE receiver remains markedly superior. Power control is then less critical, yet properly adjusted it continues to be beneficial.
9. Likewise in the forward link, matched-filter transmitters are feasible but regularized ZF transmitters remain decidedly superior.
10. On average, and up to hundreds of antennas per base, well over 1 b/s/Hz per antenna is achievable with 97% of users exceeding −5 dB of SINR.
11. For $N_a/U \to \infty$, pilot contamination becomes a limiting impairment, but up to hundreds of antennas and for the desirable ratios $N_a/U \approx$ 1–2 its effects can be rendered negligible with acceptable pilot overheads. Moreover, techniques exist to mitigate the contamination by exploiting antenna correlations and multicell coordination.

Table 10.1 Parameter s for typical pathloss exponents

η	s
3.5	−0.672
3.6	−0.71
3.7	−0.747
3.8	−0.783
3.9	−0.819
4.0	−0.854

expanding toolkit of mathematical results can be applied to generate certain large-scale distributions analytically. In fact, under a mild homogeneity condition and the premise that the base station locations are agnostic to the radio propagation, a simple Poisson point process (PPP) represents the limit to which actual behaviors converge as the shadowing strengthens [1016]. Precisely, an increasing shadowing standard deviation makes the powers that a user receives from any population of base stations look as if they originated from PPP-distributed bases. Ironically then, shadow fading, a nuisance in the study of lattice networks, simplifies the stochastic modeling of networks by making them all look alike propagation-wise. And, although the convergence is asymptotic in the shadowing standard deviation, values of interest for σ_{dB} suffice for networks to look essentially Poisson.

Example 10.24 (Forward-link SIR distribution with matched-filter transmission and a uniform power allocation)

Let $\mathsf{s} < 0$ be the solution, common values for which are listed in Table 10.1, to

$$\mathsf{s}^{2/\eta}\,\gamma(-2/\eta, \mathsf{s}) = 0, \tag{10.128}$$

where $\gamma(\cdot,\cdot)$ is the lower incomplete gamma function. In the absence of pilot contamination, and with a uniform power allocation, the distribution of $\overline{\mathsf{sir}}_u^{\mathsf{MF}}$ with PPP-distributed BSs satisfies [1017]

$$\begin{cases} F_{\overline{\mathsf{sir}}_u^{\mathsf{MF}}}(\xi) \simeq \exp\!\left(\mathsf{s}\left[\frac{N_{\mathrm{a}}/U}{\xi} - 1\right]\right) & 0 \le \xi < \frac{N_{\mathrm{a}}/U}{3+\epsilon} \\ F_{\overline{\mathsf{sir}}_u^{\mathsf{MF}}}(\xi) = 1 - \left[\frac{N_{\mathrm{a}}/U}{\xi} - 1\right]^{2/\eta} \operatorname{sinc}\frac{2}{\eta} + \mathsf{B}\!\left(\frac{\xi}{N_{\mathrm{a}}/U - 2\,\xi}\right) & \frac{N_{\mathrm{a}}/U}{3} \le \xi < \frac{N_{\mathrm{a}}/U}{2} \\ F_{\overline{\mathsf{sir}}_u^{\mathsf{MF}}}(\xi) = 1 - \left[\frac{N_{\mathrm{a}}/U}{\xi} - 1\right]^{2/\eta} \operatorname{sinc}\frac{2}{\eta} & \frac{N_{\mathrm{a}}/U}{2} \le \xi < \frac{N_{\mathrm{a}}}{U}, \end{cases} \tag{10.129}$$

where $\simeq$ denotes asymptotic equality for $\xi \to 0$ and

$$\mathsf{B}(z) = \frac{2}{\eta}\,\frac{{}_2F_1\!\left(1, 1+\frac{2}{\eta}; 2+\frac{4}{\eta}; -\frac{1}{z}\right)}{z^{1+4/\eta}\,\Gamma\!\left(2+\frac{4}{\eta}\right)\Gamma^2\!\left(1-\frac{2}{\eta}\right)}, \tag{10.130}$$

with ${}_2F_1(\cdot)$ the Gauss hypergeometric function (see Appendix E.6). Then, ϵ can be set such that $F_{\overline{\mathsf{sir}}_u^{\mathsf{MF}}}\!\left(\frac{N_{\mathrm{a}}/U}{3+\epsilon}\right) = F_{\overline{\mathsf{sir}}_u^{\mathsf{MF}}}\!\left(\frac{N_{\mathrm{a}}/U}{3}\right)$ and the CDF taken as constant therewithin.

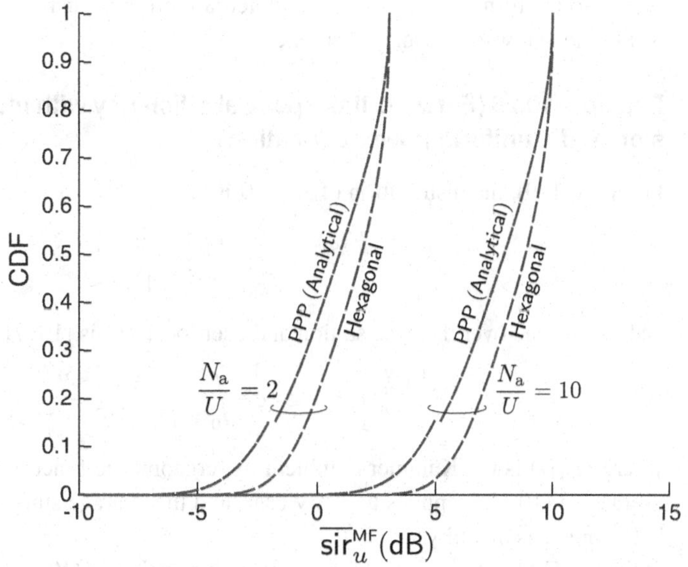

Fig. 10.19 Large-scale distributions of $\overline{\mathsf{sir}}_u^{\mathsf{MF}}$ in the forward link of an interference-limited network with uniform power allocation, $\eta = 4$, $N_p \to \infty$, and $N_a = 100$ for both $N_a/U = 2$ and 10. Comparison between the analytical solution in (10.131) and the simulation-based results with $\sigma_{\mathsf{dB}} = 8$ dB on a hexagonal lattice network.

Example 10.25

Specialize (10.129) to $\eta = 4$ and compare it against the simulation-based distribution for a hexagonal lattice network with $\sigma_{\mathsf{dB}} = 8$ dB, $N_a = 100$, and $N_a/U = 10$.

Solution

For $\eta = 4$, s $= -0.854$ and (10.129) simplifies into

$$\begin{cases} F_{\overline{\mathsf{sir}}_u^{\mathsf{MF}}}(\xi) \simeq \exp\left(-0.854\left[\frac{N_a/U}{\xi} - 1\right]\right) & 0 \le \xi < \frac{N_a/U}{3.19} \\ F_{\overline{\mathsf{sir}}_u^{\mathsf{MF}}}(\xi) = 1 - \frac{4}{\pi}\sqrt{\frac{N_a/U}{\xi} - 1} + \frac{N_a/U}{\pi\xi} & \frac{N_a/U}{3} \le \xi < \frac{N_a/U}{2} \\ F_{\overline{\mathsf{sir}}_u^{\mathsf{MF}}}(\xi) = 1 - \frac{2}{\pi}\sqrt{\frac{N_a/U}{\xi} - 1} & \frac{N_a/U}{2} \le \xi < \frac{N_a}{U}. \end{cases} \tag{10.131}$$

The comparison with the simulation-based results, recovered from Fig. 10.16, is presented in Fig. 10.19. The gap between the analytical results (which are asymptotic in the number of cells and the shadowing standard deviation) and the simulation results (with only 19 cells and with $\sigma_{\mathsf{dB}} = 8$ dB) is always below 2 dB. If the strength of the shadow fading and the number of cells in the simulator were to grow, the gap would shrink as the simulation curves converge to their analytical counterparts. Likewise, if the base station locations ceased to conform to a lattice and adopted a more irregular layout, the simulation curves would move toward the analytical ones.

Example 10.26 (Forward-link spectral efficiency with matched-filter transmission and a uniform power allocation)

From (10.129), the distribution of R_u^{MF}/B is

$$F(\xi) = \mathbb{P}\big[\log_2(1 + \overline{\mathsf{sir}}_u^{\mathsf{MF}}) < \xi\big] \tag{10.132}$$

$$= F_{\overline{\mathsf{sir}}_u^{\mathsf{MF}}}\big(2^{\xi} - 1\big) \tag{10.133}$$

and its average over the base station and user locations is [1017]

$$\mathbb{E}\left[\frac{R_u^{\mathsf{MF}}}{B}\right] = \log_2(e) \int_0^{\infty} \frac{1 - e^{-zN_a/U}}{{}_1F_1\big(1; 1 - \frac{2}{\eta}; z\big)} \frac{\mathrm{d}z}{z}, \tag{10.134}$$

where ${}_1F_1(\cdot)$ is the Kummer confluent hypergeometric function (see Appendix E.6). The average in (10.134) applies to every user, and the average sum spectral efficiency per cell is U times this quantity.

Since ${}_1F_1\big(1; 1 - \frac{2}{\eta}; z\big) > 0$ for $z \geq 0$, an inspection of (10.134) indicates that $E[R_u^{\mathsf{MF}}/B]$ decreases as N_a/U shrinks, i.e., as we add users with N_a fixed. However, this decrease is sublinear in U, hence the average sum spectral efficiency grows as users are added.

Example 10.27

Compare the analytical solution for the average sum spectral efficiency in the previous example against simulation-based results for $\eta = 4$, $N_a = 100$, and $N_a/U = 10$.

Solution

For $N_a/U = 10$, (10.134) evaluates into 27.6 b/s/Hz per cell while the simulation returns 29.8 b/s/Hz per cell.

The analytical solutions in the foregoing string of examples hardly convey much intuition, but they do eliminate the need for tedious and time-consuming simulations in the situations they cover. Also, they allow calibrating the simulators, e.g., determining how many cells are needed—too few cells lead to optimistic results because of insufficient interference—for some desired accuracy. Given η and N_a/U, the gap between these analytical solutions and case-specific simulations captures the variability associated with the network layout, the shadow fading, and the number of cells. This gap is remarkably small, rendering the analysis broadly applicable in those respects and making it desirable that other large-scale distributions and averages could similarly be produced analytically.

Problems

10.1 Replicate Example 10.1 with FDD, rather than TDD/full-duplex.

10.2 Express the variances of the left-hand sides of (10.9) and (10.10) and then plot those variances for $N_{\text{a}} = 1, \ldots, 100$.

10.3 How would you generalize (10.9) and (10.10) to channel vectors with non-IID entries?

10.4 Verify (10.20)–(10.22).

10.5 Verify (10.26)–(10.28).

10.6 Formulate a matched-filter receiver for the reverse link, scaled to yield a unit-power output signal, by letting $N_{\text{a}}/U \to \infty$ in the corresponding LMMSE MAC receiver.

10.7 Verify (10.49).

Hint: Use (C.19)–(C.21).

10.8 Verify (10.60).

10.9 Consider a staggered pilot disposition, such that $\hat{\boldsymbol{H}}_u$ and MMSE_u are as given in Discussion 10.2.

(a) Derive $\mathsf{sinr}_u^{\mathsf{MF}}$, the staggered-pilot counterpart to (10.50).

(b) Derive $\overline{\mathsf{sinr}}_u^{\mathsf{MF}}$, the staggered-pilot counterpart to (10.62).

(c) From part (b), identify the limiting behavior of $\overline{\mathsf{sinr}}_u^{\mathsf{MF}}$ for $N_{\text{a}}/U \to \infty$.

10.10 Consider a staggered pilot disposition such as the one in Fig. 10.2, but explicitly account for the fact that only $N_{\text{c}}/N_{\text{p}}$ pilot sequence sets can be staggered into nonoverlapping positions. Express $\hat{\boldsymbol{H}}_u$, generalizing both (10.20) and (10.26).

10.11 Verify (10.56).

10.12 Verify (10.66), (10.67), and (10.69).

10.13 Consider a hexagonal lattice network with a pilot reuse pattern based on clusters of L_{cluster} cells. User u is located at a corner of the cell of interest. Suppose that, around that cell, there are an infinite number of concentric tiers of copilot cells, with the kth tier containing $6k$ cells and being at distance kD_{copilot} from the cell of interest. Further suppose that every copilot user is at the center of its own cell.

(a) Express, as a function of the pathloss exponent and the shadow fading coefficients, the share of the term $\sum_{\ell \in \mathcal{C}} G_{\ell,u}$ that corresponds to the first tier.

(b) Quantify part (a) for $\eta = 4$, with no shadow fading.

10.14 For $\eta = 4$ and $\varepsilon = \frac{1}{10}$, quantify the impact that the relaxation in (10.82) has on the condition in (10.85) in the following cases:

(a) $N_{\text{a}}/U = 10$.

(b) $N_{\text{a}}/U = 4$.

10.15 Recompute the condition in (10.85) with $\varepsilon = \frac{1}{100}$ rather than $\epsilon = \frac{1}{10}$.

10.16 Quantify the change in the condition in (10.85) as the pathloss exponent shrinks from $\eta = 4$ to $\eta = 3.5$.

10.17 Consider the reverse link of a hexagonal lattice network with a pilot reuse pattern based on clusters of L_{cluster} cells. Neglect shadow fading and account for only the first tier of copilot cells around the cell of interest.

(a) Derive, as a function of N_p, U, and η, the limiting ($N_\mathrm{a}/U \to \infty$) cell-edge SINR with the copilot users at the corner of their respective cells that is closest to the cell of interest.

(b) Let $\eta = 4$ and suppose that $N_\mathrm{p}/N_\mathrm{c} = 0.1$, a conservatively low overhead. Express, for vehicular users, the worst-case limiting SINR as a function of U.

(c) Establish the range of U where the SINR derived in part (b) is above 20 dB, which, with fading quelled by the unbounded spatial diversity, suffices to support 64-QAM (recall Fig. 4.3).

10.18 Repeat Problem 10.17 with a second tier of copilot cells, having twice as many cells as the first tier and being at twice the distance, also accounted for. Does the result of part (c) change significantly?

10.19 Repeat Problem 10.17, but with the users in the copilot cells located at the corner of their respective cells that is farthest from the cell of interest.

10.20 Repeat Problem 10.17 for the following cases in terms of the cell of interest.

(a) A user halfway between the center and the edge of the cell.

(b) A pedestrian cell-edge user.

(c) A vehicular cell-edge user requiring an SINR of only 10 dB.

10.21 In the context of Problem 10.17, establish the pilot overhead required to support $U = 30$ vehicular users with a worst-case SINR of 25 dB.

10.22 Consider the reverse link of a hexagonal network with $\eta = 4$, $\sigma_{\mathrm{dB}} = 8$ dB, and no power control. Plot the CDF of $\mathsf{sir}_u^{\mathsf{MF}}$ for $N_\mathrm{a}/U \to \infty$ in the following cases:

(a) $N_\mathrm{p}/U = 1$.

(b) $N_\mathrm{p}/U = 3$.

(c) $N_\mathrm{p}/U = 7$.

10.23 Repeat Problem 10.22 with fractional power control ($\vartheta = 0.7$).

10.24 Generalize (10.62) to an arbitrary fading distribution, not necessarily Rayleigh.
Note: Leave the solution in terms of the kurtosis of the fading magnitude.

10.25 Reproduce Example 10.2.

10.26 Consider a macrocellular network with $D_{\mathsf{cell}} = 1$ km (such that the distance between adjacent base stations is $\sqrt{3}$ km), a transmit power of $P_\mathrm{t} = 200$ mW per user, a noise figure of 3 dB at the base station receivers, a pathloss intercept of -128 dB at 1 km, and a bandwidth of $B = 20$ MHz. Let $N_\mathrm{a} = 100$, $U = 10$, and $N_\mathrm{p} = 3U$. Further let the user antennas be omnidirectional and, for the gain of the base station antennas, consider the following cases.

(a) A 12-dB gain, corresponding to antennas that focus power in the elevation domain.

(a) Omnidirectional.

Contrasting the reverse-link distribution of $\mathbb{E}\left[\mathsf{sir}_u^{\mathsf{MF}}\right]$ with its interference-limited brethren in Fig. 10.8, how significant is the effect of noise?
Note: This problem allows gauging the validity of the interference-limited premise in the reverse link for typical macrocellular conditions.

10.27 In reference to Example 10.3.

(a) Reproduce the example verbatim, with $\vartheta = 0.7$.
(b) Repeat for $\vartheta = 0.5$.
Comment on the impact of varying the fractional power control exponent.

10.28 Repeat Example 10.3 with the pathloss exponent modified as follows:
(a) $\eta = 3.5$.
(a) $\eta = 4.5$.
Comment on the impact of the pathloss exponent.

10.29 Repeat Example 10.3 with the shadow fading strengthened to $\sigma_{\text{dB}} = 10$ dB. Comment on the impact of this change.

10.30 Repeat Problem 10.8 with $\varepsilon = \frac{1}{100}$ and with the ensuing N_p/U rounded to the closest feasible value.

10.31 Repeat Example 10.9 for $N_\text{c} = 500$, corresponding to an extreme vehicular setting, and for $N_\text{c} = 2000$, corresponding to a relaxed one.

10.32 In reference to Example 10.10.
(a) Reproduce the example verbatim, with $\vartheta = 0.7$.
(b) Repeat for $\vartheta = 0.5$.
Verify that this reduction in the fractional power control exponent increases the optimum ratio N_a/U and decreases the average sum spectral efficiency.

10.33 Redo Example 10.10 with the allowance of users below -5 dB loosened to 5%. By how much does the sum spectral efficiency increase?

10.34 For a reverse link with matched filter reception, rederive $\text{sinr}_u^{\text{MF}}$ and $\overline{\text{sinr}}_u^{\text{MF}}$ under the assumption that the payload data are subject to power control, but the pilot symbols are not. What would the advantages and disadvantages be? Is there a regime in which the performance is irrespective of whether the pilots are power-controlled?
Note: This derivation can serve as a stepping stone toward a matched-filter formulation with an arbitrary pilot power boosting.

10.35 Reformulate the LMMSE receiver in (10.88) with further conditioning on the channel estimates from users in the L_{cluster} cells within the pilot reuse cluster of the cell of interest.

10.36 Verify (10.105).

10.37 Consider a macrocellular network with $D_{\text{cell}} = 1$ km (such that the distance between adjacent base stations is $\sqrt{3}$ km), a transmit power of $P_\text{t} = 60$ W per base, a user noise figure of 7 dB, a pathloss intercept of -128 dB at 1 km, and a bandwidth of $B = 20$ MHz. The forward–reverse SNR ratio is $\rho = 20$ dB. Let $N_\text{a} = 100$, $U = 25$, and $N_\text{p} = 4U$. Further let the user antennas be omnidirectional and, for the gain of the base station antennas, consider the following extreme cases.
(a) A 12-dB gain, corresponding to antennas that focus power in elevation.
(a) Omnidirectional.
Contrasting the forward-link distribution of $\mathbb{E}\left[\text{sir}_u^{\text{MF}}\right]$ with its interference-limited brethren in Fig. 10.17, how significant is the effect of noise?
Note: This problem allows gauging the validity of the interference-limited premise in the forward link for typical macrocellular conditions.

10.38 For a forward link with conjugate beamforming, rederive $\overline{\mathsf{sinr}}_u^{\mathsf{MF}}$ under the premise that the payload data are subject to power control, but the pilot symbols are not.

10.39 With a per-symbol power constraint, the matched-filter precoders in (10.102) morph into

$$\boldsymbol{F}_{\ell,u}^{\mathsf{MF}} = \sqrt{N_{\mathrm{a}}} \frac{\hat{\boldsymbol{H}}_{\ell;\ell,u}}{\|\hat{\boldsymbol{H}}_{\ell;\ell,u}\|} \qquad u = 0, \ldots, U-1. \tag{10.135}$$

Let us consider such precoders.

(a) Compute $\mathbb{E}[\boldsymbol{H}_u^* \boldsymbol{F}_u^{\mathsf{MF}}]$ and, from this expectation, rewrite the numerator of the SINR expression in (10.105).

(b) Express the ratio between the numerator in part (a) and that in (10.105).

(c) From the ratio in part (b), calculate the dB-loss in intended signal power when going from a per-codeword to a per-symbol power constraint with $N_{\mathrm{a}} = 100$.

(d) Show that, for $N_{\mathrm{a}} \to \infty$, the loss vanishes.

Hint: Useful to this problem may be the chi distribution (see Appendix C.1.9), the relationship $\Gamma(M + 1/2) = \sqrt{\pi}\, 2^{1-2M} \Gamma(2M)/\Gamma(M)$, *and Stirling's formula*

$$\lim_{M \to \infty} \frac{M!}{\sqrt{2\pi M} M^M e^{-M}} = 1. \tag{10.136}$$

10.40 Reproduce Example 10.20.

10.41 Derive the counterpart to (10.107) with additional side information whereby each user is privy to its own precoded channel. Then, plot the CDF of such $\mathsf{sinr}_u^{\mathsf{MF}}$ against that of $\overline{\mathsf{sinr}}_u^{\mathsf{MF}}$ in (10.107). Apply the setting of Example 10.20 with a uniform power allocation. What do you observe?

10.42 Consider the forward link of a hexagonal network with $\eta = 4$, $\sigma_{\mathrm{dB}} = 8$ dB, and a uniform power allocation. Plot the CDF of $\overline{\mathsf{sir}}_u^{\mathsf{MF}}$ for $N_{\mathrm{a}}/U \to \infty$ in these cases:

(a) $N_{\mathrm{p}}/U = 1$.

(b) $N_{\mathrm{p}}/U = 3$.

(c) $N_{\mathrm{p}}/U = 7$.

10.43 Letting $E_{\mathrm{s}}/N_0 \to \infty$ in the reverse- and forward-link SINR expressions in (10.86) and (10.107), express the forward-link power allocation $\frac{E_0}{E_{\mathrm{s}}}, \ldots, \frac{E_{U-1}}{E_{\mathrm{s}}}$ that achieves some specific $\overline{\mathsf{sinr}}_0, \ldots, \overline{\mathsf{sinr}}_{U-1}$.

Hint: Apply the MAC–BC duality for linear transceivers with the role of thermal noise played by the sum of thermal noise plus other-cell interference.

10.44 Show that the fully equalizing power allocation policy in (10.108), implemented in the forward link with matched-filter transmitters, can only reduce $\frac{1}{U}\sum_{u=0}^{U-1} \overline{\mathsf{sinr}}_u$ relative to a uniform power allocation.

Hint: The Cauchy–Schwarz inequality states that

$$(a_0 b_0 + \cdots + a_{N-1} b_{N-1})^2 \leq \left(a_0^2 + \cdots + a_{N-1}^2\right)\left(b_0^2 + \cdots + b_{N-1}^2\right). \tag{10.137}$$

10.45 In relation to establishing a sufficient condition for pilot contamination negligibleness in the forward link.

(a) Why is a corner location within the cell of interest not a worst case in terms of exposure to contamination?

(b) Characterize the worst-case user location in terms of same-cell and other-cell interference.

(c) Does the location identified in part (b) serve to formulate a sufficient condition?

10.46 Verify that, on the basis of channel estimates, with a per-codeword power constraint, and with other-cell interference incorporated and treated as noise, (9.282) morphs into (10.112).

10.47 Reformulate the regularized ZF precoder in (10.112) with further conditioning on the channel estimates from users in the L_{cluster} cells within the pilot reuse cluster of the cell of interest.

10.48 Verify (10.113)–(10.115).

10.49 Consider a forward link with regularized ZF transmission. What is the limiting value of the precoder $\boldsymbol{F}_u^{\text{Reg}}$ for $N_{\text{a}}/U \to \infty$?

10.50 Derive (10.121).

10.51 Reformulate (10.86) with user u at cell ℓ subject, at the base station of interest, to an antenna correlation matrix $\boldsymbol{R}_{\ell,u}$.

10.52 Reformulate (10.107) with user u at cell ℓ experiencing, at the cell of interest, a correlation matrix $\boldsymbol{R}_{\ell;u}$.

10.53 Reformulate (10.86) with Rice fading. For a hexagonal interference-limited network with $\eta = 4$, $\sigma_{\text{dB}} = 8$ dB, $N_{\text{a}} = 100$, and $N_{\text{a}}/U = 10$, plot the CDF of this quantity for $\mathsf{K} = 0$, $\mathsf{K} = 1$, and $\mathsf{K} = 10$. Assume no power control.

10.54 Reformulate (10.107) with Rice fading. For a hexagonal interference-limited network with $\eta = 4$, $\sigma_{\text{dB}} = 8$ dB, $N_{\text{a}} = 100$, and $N_{\text{a}}/U = 10$, plot the CDF of this quantity for $\mathsf{K} = 0$, $\mathsf{K} = 1$, and $\mathsf{K} = 10$. Assume a uniform power allocation.

10.55 Consider a hexagonal interference-limited network with $\eta = 4$, $\sigma_{\text{dB}} = 8$ dB, $N_{\text{a}} = 100$, and $N_{\text{a}}/U = 10$. Assume no power control. Plot the CDF of (10.86) without macroscopic diversity, meaning with users connecting to their closest base station rather than the one from which they receive the strongest local-average power. Contrast this plot with Fig. 10.3.

10.56 Consider a hexagonal interference-limited network with $\eta = 4$, $\sigma_{\text{dB}} = 8$ dB, $N_{\text{a}} = 100$ and $N_{\text{a}}/U = 10$. Assume a uniform power allocation. Plot the CDF of (10.107) without macroscopic diversity, meaning with users connecting to their closest base station rather than the one from which they receive the strongest local-average power. Contrast this plot with Fig. 10.16.

10.57 Express ϵ so as to ensure that $F_{\overline{\text{sir}}_u^{\text{MF}}}\left(\frac{N_{\text{a}}/U}{3+\epsilon}\right) = F_{\overline{\text{sir}}_u^{\text{MF}}}\left(\frac{N_{\text{a}}/U}{3}\right)$ in (10.129).

10.58 Repeat Example 10.25 for $N_{\text{a}}/U = 4$.

10.59 From the SIR distribution in (10.131), express the distribution of the corresponding user spectral efficiency.

10.60 Consider the forward link of an interference-limited network with PPP-distributed base stations, matched-filter transmission and a uniform power allocation. Pilot contamination is negligible.

(a) For $N_a/U = 4$, plot the average spectral efficiency per cell as function of $\eta \in [3.5, 4]$. Alongside, plot the simulation-based results for a hexagonal lattice network with $N_a = 100$ and $\sigma_{\text{dB}} = 8$ dB.

(b) For $\eta = 4$, plot the average spectral efficiency per cell as function of $N_a/U \in [2, 10]$. Alongside, plot the simulation-based results for a hexagonal lattice network with $N_a = 100$ and $\sigma_{\text{dB}} = 8$ dB.

10.61 Consider the forward link of an interference-limited network with PPP-distributed base stations, matched-filter transmission, and a uniform power allocation. Pilot contamination is negligible and $N_a = 128$.

(a) Plot, as a function of $\eta \in [3.5, 4]$, the largest possible U such that $\overline{\mathsf{sir}}_u^{\mathsf{MF}} \leq -5$ dB for no more than 3% of users.

(b) On the same chart as part (a), plot the average sum spectral efficiency for the found values of U.

11 Afterword

11.1 Beyond cellular

What is next in the evolution of cellular networks, and of MIMO in particular? Besides the push toward progressively higher carrier frequencies and wider bandwidths, some trends seem highly relevant.

- The delocalization of base station antennas, so as to distribute them over the cell area by extending the lines that connect them to the bases [257, 1018, 1019].
- The blurring of the cell boundaries via different intensities of base station cooperation, which, besides pilot contamination, can serve to fight other-cell interference as well. These cooperation intensities range from loosely coordinating user assignments, in order to keep other-cell interference at bay, all the way to joint reception and transmission on the part of neighboring bases, ambitiously intending to turn interference into usable signal—at the expense of having to acquire more extensive and precise channel estimates [475, 1020–1030].

Ultimately, a very aggressive degree of antenna delocalization and base station cooperation with increased processing at central hubs may naturally lead to so-called *cell-free* or *cell-less* networks [1031–1035]: fields of antennas connected directly to central hubs, where all the baseband processing takes place, and without actual base stations. The network would then look like a canopy that could offer superior coverage and a dynamic load balance. Furthermore, this would open the door to cloud-based implementations of massive MIMO, with all the baseband processing progressively software-defined on the cloud rather than hardwired at physical base stations [1036, 1037]. This, and other visions for the evolution and revolution of cellular networks, are sure to occupy the research community for years to come, with MIMO ideas at the center of it all.

Beyond cellular networks, and beyond WLANs, there are communication setups that have not been explicitly considered in this book, but to which the MIMO ideas expounded herein readily apply, for instance, device-to-device communication, relays, or ad hoc networks. The contents of this book are further a stepping stone to the following paradigms.

Millimeter-wave and terahertz communication

As mentioned in Chapter 3, the abundance of idle spectrum in the millimeter-wave band makes these high frequencies very attractive. With a longer view, and for very-short-range transmissions, even terahertz frequencies could be tapped [1038].

The primary application of MIMO in these realms is power gain, to overcome the high propagation losses, and hence the signaling strategies of interest are primarily those presented for the low-SNR regime, namely beamforming and reduced-rank transmissions. Thanks to the small wavelength, large numbers of antennas can be packed for that purpose into small form factors. Incorporating the specificities of radio propagation at these frequencies, many results in the book become immediately relevant.

Interference alignment

Interference alignment (IA) is a technique whereby multiple base stations share their CSIT and jointly design their precoders such that, from the vantage of each user, all unintended signals align on a certain subspace leaving the remaining dimensions free of interference [1039, 1040]. On those dimensions, the intended signals encounter only noise. The joint design of the precoders is most conveniently conducted on a centralized fashion, which would make IA a form of base station cooperation, but distributed iterative implementations are possible [1041, 1042].

If all the base stations could participate in the alignment and the CSIT were perfect, IA could deliver an unbounded growth of the spectral efficiency with the SNR. Unfortunately, only a limited number of bases, depending on the number of antennas, can participate in the alignment: with two antennas, for instance, at most three bases can participate. This necessarily leads to the formation of IA clusters that are inevitably exposed to other-cluster interference. Thus, even the subspaces that IA protects from in-cluster interference are bound to experience interference [475]. In addition, IA restricts the spatial dimensionality of the transmit signals; in the three-base two-antenna example, the interference-free subspaces have spatial dimension one; without IA, in contrast, a two-dimensional signal could be transmitted by each base. Altogether then, IA can create subspaces with reduced interference in exchange for a sacrifice in signal dimensions.

While this advises against a blanket use of IA in cellular networks, it is a technique to consider in cases where isolated clusters of transmitters naturally occur, say for WLANs within physically separated residences or small buildings [1043], or for in-vehicle networks.

11.2 Beyond wireless

The applicability of MIMO principles does not end in the province of wireless communication, and as closure we briefly point to other domains of application.

Fiber optics

Fiber optics constitute the backbone of communication networks. In what is called multimode fiber, specifically, multiple propagation modes are excited at once and, in certain deployments, mode mixing limits the performance. This problem can be viewed through the lens of MIMO communication, and multiple works have explored the feasibility of this approach [1044–1046].

Digital subscriber lines

Digital subscriber lines (DSL) enable high-bit-rate transmissions over legacy wireline telephone networks. A chief impairment in DSL is crosstalk: near-end crosstalk between a transmitter and receiver at the same end of the full-duplex link, and far-end crosstalk between a transmitter and receiver at opposite ends. The former is avoided by resorting to FDD, rather than full-duplex, while the latter can be handled by applying MIMO principles—termed *vectoring* in this context [1047].

Visible-light communication

Visible-light communication is a form of free space optics, suitable for WLANs. Huge unlicensed bandwidths are available, limited only by safety regulations and device capabilities, and with the distinguishing feature of in-room signal confinement and lack of obstacle penetration. This is a double-edged sword, constraining the coverage but also eliminating interference and restricting the possibility of eavesdropping.

With the advancement of solid-state technology, the light-emitting diode has evolved from dowdy indicator light to the main instrument for lighting technology and, thanks to its fast response time, for visible-light transmission. Although these devices currently have a limited modulation bandwidth [1048], often multiple light-emitting diodes are available, making MIMO a possibility [1049, 1050].

Underwater acoustic communication

Marine research, oceanography, and oil drilling are example applications of underwater acoustic communication. The carrier here is sound rather than an electromagnetic wave. This is because sound propagates best in seawater, where long-range propagation of radio waves is feasible only at very low frequency (30–300 Hz) with prohibitively large antennas. The bandwidth of underwater acoustic transmissions is very limited, which puts a premium on spectral efficiency thereby rendering MIMO attractive [1051, 1052].

APPENDICES

A Appendix A Transforms

This appendix reviews several Fourier transforms, instrumental in the transition between the time and the frequency domains, as well as the Z-transform.

A.1 Fourier transforms

A.1.1 Continuous-time Fourier transform

A well-behaved continuous-time function $x(t)$ and its Fourier transform $\mathsf{x}(f)$ are related by the analysis and synthesis equations

$$\text{Analysis} \qquad \mathsf{x}(f) = \int_{-\infty}^{\infty} x(t)\, e^{-\mathrm{j}2\pi f t}\, \mathrm{d}t \tag{A.1}$$

$$\text{Synthesis} \qquad x(t) = \int_{-\infty}^{\infty} \mathsf{x}(f)\, e^{\mathrm{j}2\pi f t}\, \mathrm{d}f. \tag{A.2}$$

A detailed account of the technical conditions that are necessary for a Fourier transform to exist is beyond the scope of this book, and interested readers are referred for instance to [1053]. It suffices to say that all physically realizable signals do have Fourier transforms. For our purposes, therefore, we can take "well-behaved" to simply mean a signal for which the integral in (A.1) exists.

Those properties of the continuous-time Fourier transform that are invoked somewhere in the text are listed in Table A.1. Likewise, relevant Fourier transform pairs are presented in Table A.2.

A.1.2 Continuous-time Fourier series

Periodic functions do not fall under the umbrella of well-behaved functions and yet they are very important in the analysis of communication signals. This obstacle can be side-stepped through the definition of the Fourier series and by invoking the Dirac delta function, $\delta(\cdot)$.

Consider a periodic signal $x(t)$ whose period, $T > 0$, is the smallest real number such that $x(t) = x(t + T)$. The continuous-time Fourier series of such signal is defined as

$$\text{Analysis} \qquad \mathsf{x}[n] = \frac{1}{T}\int_{0}^{T} x(t)\, e^{-\mathrm{j}\frac{2\pi}{T}nt}\, \mathrm{d}t \tag{A.3}$$

$$\text{Synthesis} \qquad x(t) = \sum_{n} \mathsf{x}[n]\, e^{\mathrm{j}\frac{2\pi}{T}nt}. \tag{A.4}$$

Table A.1 Continuous-time Fourier transform properties

Property	Aperiodic signal	Fourier transform
	$x(t)$	$\mathsf{x}(f)$
Linearity	$a\,x(t)+b\,y(t)$	$a\,\mathsf{x}(f)+b\,\mathsf{y}(f)$
Time shift	$x(t-t_0)$	$e^{-j2\pi f t_0}\,\mathsf{x}(f)$
Conjugation	$x^*(t)$	$\mathsf{x}^*(-f)$
Time reversal	$x(-t)$	$\mathsf{x}(-f)$
Time scaling	$x(at)$	$\frac{1}{\lvert a\rvert}\,\mathsf{x}(\frac{f}{a})$
Convolution	$x(t)*y(t)=\int x(\tau)\,y(t-\tau)\,d\tau$	$\mathsf{x}(f)\,\mathsf{y}(f)$
Autocorrelation	$x(t)*x^*(-t)$	$\lvert\mathsf{x}(f)\rvert^2$
Modulation	$x(t)\,e^{j2\pi f_0 t}$	$\mathsf{x}(f-f_0)$
Conjugate symmetry	$x(t)$ is real	$\mathsf{x}(f)=\mathsf{x}^*(-f)$
Duality	$x(t)\longleftrightarrow\mathsf{x}(f)$	$\mathsf{x}(t)\longleftrightarrow x(-f)$
Parseval's theorem	$\int x(t)\,y^*(t)\,dt=\int\mathsf{x}(f)\,\mathsf{y}^*(f)\,df$	

Table A.2 Continuous-time Fourier transform pairs

Function	Time-domain	Frequency-domain
Impulse	$\delta(t)$	1
Constant function	1	$\delta(f)$
Complex exponential	$e^{j2\pi f_0 t}$	$\delta(f-f_0)$
Cosine	$\cos(2\pi f_0 t+\theta)$	$\frac{1}{2}\left[e^{j\theta}\delta(f-f_0)+e^{-j\theta}\delta(f+f_0)\right]$
Sine	$\sin(2\pi f_0 t+\theta)$	$\frac{1}{2j}\left[e^{j\theta}\delta(f-f_0)-e^{-j\theta}\delta(f+f_0)\right]$
Impulse train	$\sum_k \delta(t-kT)$	$\frac{1}{T}\sum_n \delta(f-\frac{n}{T})$
Rectangular pulse	$\mathrm{rect}(\frac{t}{T})=\begin{cases}1 & \lvert t\rvert\leqslant\frac{T}{2}\\ 0 & \text{else}\end{cases}$	$T\,\mathrm{sinc}(fT)=\frac{\sin(\pi fT)}{\pi f}$
	$W\,\mathrm{sinc}^2(fW)$	
Sinc pulse	$\mathrm{sinc}(Wt)=\frac{\sin(\pi Wt)}{\pi Wt}$	$\frac{1}{W}\mathrm{rect}(\frac{f}{W})$

The continuous-time Fourier series creates as an output a weighting of the fundamental frequency of the signal $e^{-j2\pi/T}$ and its harmonics. From the series it is possible to express the Fourier transform of a periodic signal as

$$\mathsf{x}(f)=\sum_n \mathsf{x}[n]\,\delta\Big(f-\frac{n}{T}\Big) \tag{A.5}$$

$$x(t)=\int_{-\infty}^{\infty}\mathsf{x}(f)\,e^{j2\pi ft}\,df, \tag{A.6}$$

where the unboundedness of (A.1) has been circumvented by means of $\delta(\cdot)$.

A.1.3 Discrete-time Fourier transform

Consider now a well-behaved discrete-time signal $x[n]$. Its discrete-time Fourier transform analysis and synthesis relationships are

$$\mathrm{x}(\nu) = \sum_n x[n]\, e^{-\mathrm{j}2\pi n\nu} \tag{A.7}$$

$$x[n] = \int_{-1/2}^{1/2} \mathrm{x}(\nu)\, e^{\mathrm{j}2\pi n\nu}\, d\nu, \tag{A.8}$$

where the frequency ν is defined on any finite interval of unit length, typically $[-1/2, 1/2]$.

A.1.4 Discrete Fourier transform

While the discrete-time Fourier transform offers a sound analytical framework for discrete-time signals, it is inconvenient due to its continuous frequency. An alternative is the discrete Fourier transform (DFT) and its easy-to-implement cousin, the fast Fourier transform (FFT). The DFT is extremely important in digital signal processing and communications, and instrumental in the context of OFDM. It applies to finite-length discrete-time signals and, since any finite-length signal can be repeated to form a periodic discrete-time signal, the DFT can also be interpreted as applying to periodic signals. The DFT synthesis and analysis equations are

$$\mathrm{x}[k] = \sum_{n=0}^{N-1} x[n]\, e^{-\mathrm{j}\frac{2\pi}{N}kn} \qquad k = 0, \ldots, N-1 \tag{A.9}$$

$$x[n] = \frac{1}{N} \sum_{k=0}^{N-1} \mathrm{x}[k]\, e^{\mathrm{j}\frac{2\pi}{N}kn} \qquad n = 0, \ldots, N-1, \tag{A.10}$$

where the discrete and finite-range nature of both time and frequency are noteworthy. To compactly denote the direct (analysis) and inverse (synthesis) N-point DFT transforms, we introduce the terminology

$$\mathrm{x}[k] = \mathrm{DFT}_N\{x[n]\} \tag{A.11}$$

$$x[n] = \mathrm{IDFT}_N\{\mathrm{x}[k]\}. \tag{A.12}$$

Some relevant DFT properties are listed in Table A.3, where $((\cdot))_N$ denotes modulo-N operation.

A.2 Z-transform

The Z-transform converts a function of a discrete real variable (say the discrete time, n) to a function of a complex variable z. This converts difference equations into algebraic

Table A.3 DFT properties

Length-N sequence	N-point DFT
$x[n]$	$\mathsf{x}[k]$
$a\,x[n] + b\,y[n]$	$a\,\mathsf{x}[k] + b\,\mathsf{y}[k]$
$\mathsf{x}[n]$	$N\,x[((-k))_N]$
$x[((n-m))_N]$	$e^{j2\pi km/N}\,\mathsf{x}[k]$
$\sum_m x[m]\,y[((n-m))_N]$	$\mathsf{x}[k]\,\mathsf{y}[k]$
$x^*[n]$	$\mathsf{x}^*[((-k))_N]$
$x[n]$ is real	$\mathsf{x}[k] = \mathsf{x}^*[((-k))_N]$

equations and convolution into product. The Z-transform of a causal function $x[n]$ is

$$\mathsf{x}(z) = \sum_{n=0}^{\infty} x[n]\,z^{-n}, \tag{A.13}$$

while the inversion of $\mathsf{x}(z)$ back onto $x[n]$ requires an integration on the complex plane [133].

Example A.1

Obtain the Z-transform of $x[n] = \delta[n-\Delta]$.

Solution

$\mathsf{x}(z) = z^{-\Delta}$.

The foregoing example generalizes into a time-shifting property according to which, if the Z-transform of $x[n]$ is $\mathsf{x}(z)$, then the Z-transform of $x[n-\Delta]$ is $z^{-\Delta}\,\mathsf{x}(z)$.

B Appendix B Matrix algebra

B.1 Column space, row space, null spaces

The *column space* of an $N \times M$ matrix $\boldsymbol{A} = [\boldsymbol{a}_0 \cdots \boldsymbol{a}_{M-1}]$ is the set of all linear combinations of its column vectors $\boldsymbol{a}_0, \ldots, \boldsymbol{a}_{M-1}$. It is therefore a subspace (whose dimension is at most M) of the N-dimensional vector space. Since a linear combination with arbitrary coefficients $x_0, \ldots, x_{M-1}$ of the columns of $\boldsymbol{A}$ can be written as the product of $\boldsymbol{A}$ with the vector $\boldsymbol{x} = [x_0 \cdots x_{M-1}]^{\mathrm{T}}$, i.e.,

$$x_0\,\boldsymbol{a}_0 + \cdots + x_{M-1}\,\boldsymbol{a}_{M-1} = \boldsymbol{A}\begin{bmatrix} x_0 \\ \vdots \\ x_{M-1} \end{bmatrix}, \tag{B.1}$$

the column space of $\boldsymbol{A}$ consists of all possible vectors $\boldsymbol{y} = \boldsymbol{A}\boldsymbol{x}$.

The *row space* of $\boldsymbol{A}$, in turn, equals the column space of $\boldsymbol{A}^{\mathrm{T}}$ or, equivalently, of $\boldsymbol{A}^*$. Thus, it is a subspace (whose dimension is at most N) of the M-dimensional vector space.

Example B.1

The column space of

$$\boldsymbol{A} = \begin{bmatrix} 0 & 3 \\ 2 & 0 \\ 0 & 1 \end{bmatrix} \tag{B.2}$$

is the set of vectors $\boldsymbol{y} = [y_0\ y_1\ y_2]^{\mathrm{T}}$ having the form

$$\boldsymbol{y} = \boldsymbol{A}\boldsymbol{x} \tag{B.3}$$

$$= \begin{bmatrix} 3\,x_1 \\ 2\,x_0 \\ x_1 \end{bmatrix}. \tag{B.4}$$

These vectors satisfy $y_0 = 3\,y_2$, which defines a subspace of dimension $M = 2$ (that is, a plane) on a vector space of dimension $N = 3$.

Example B.2

The row space of $\boldsymbol{A}$ in (B.2), in turn, is the set of vectors $\boldsymbol{y}$ having the form

$$\boldsymbol{y} = \boldsymbol{A}^{\mathrm{T}}\boldsymbol{x} \tag{B.5}$$

$$= \begin{bmatrix} 2\,x_1 \\ 3\,x_0 + x_2 \end{bmatrix}, \tag{B.6}$$

which defines the entire vector space of dimension $M = 2$.

The column rank and row rank of $\boldsymbol{A}$ are the dimensions of its column space and row space, respectively. For the matrix in Examples B.1 and B.2, for instance, both equal 2. The fact that the column and row ranks coincide in this case is not a coincidence. Indeed, the row and column ranks always coincide, giving the *rank* of the matrix. A matrix is said to be *full-rank* if its rank equals the largest possible, which is $\min(N, M)$, and it is said to be *rank-deficient* otherwise.

In addition to its column and row spaces, a matrix $\boldsymbol{A}$ elicits two additional subspaces, respectively the orthogonal complements of such column and row spaces.

- The orthogonal complement of the row space, termed the *null space* of $\boldsymbol{A}$, is the collection of those vectors that are orthogonal to every row of $\boldsymbol{A}$, i.e., of those vectors $\boldsymbol{x}$ satisfying $\boldsymbol{A}\boldsymbol{x} = \boldsymbol{0}$. The null space of $\boldsymbol{A}$ has dimension $M - \text{rank}(\boldsymbol{A})$.
- The orthogonal complement of the column space of $\boldsymbol{A}$ equals the null space of $\boldsymbol{A}^{\mathrm{T}}$, with dimension $N - \text{rank}(\boldsymbol{A})$.

Example B.3

For the matrix $\boldsymbol{A}$ in Examples B.1 and B.2, the null space is empty while the null space of $\boldsymbol{A}^{\mathrm{T}}$ contains all the colinear vectors $\boldsymbol{y} = [y_0 \ y_1 \ y_2]^{\mathrm{T}}$ that are orthogonal to the plane defined by $y_0 = 3\,y_2$.

B.2 Special matrices

B.2.1 Hermitian matrices

A complex matrix $\boldsymbol{A}$ is said to be *Hermitian* if $\boldsymbol{A}^* = \boldsymbol{A}$.

B.2.2 Unitary matrices

A complex matrix $\boldsymbol{U}$ is said to be *unitary* if $\boldsymbol{U}^*\boldsymbol{U} = \boldsymbol{U}\boldsymbol{U}^* = \boldsymbol{I}$. In addition:

- $\boldsymbol{U}$ is nonsingular and $\boldsymbol{U}^* = \boldsymbol{U}^{-1}$.
- The columns of $\boldsymbol{U}$ form an orthonormal set, as do the rows of $\boldsymbol{U}$.
- For any complex vector $\boldsymbol{x}$, the vector $\boldsymbol{y} = \boldsymbol{U}\boldsymbol{x}$ satisfies $|\boldsymbol{y}| = |\boldsymbol{x}|$. Thus, $\boldsymbol{y}$ is a rotated version of $\boldsymbol{x}$ and $\boldsymbol{U}$ embodies that rotation.

B.2.3 Fourier matrices

An $N \times N$ Fourier matrix $\boldsymbol{U}$ is a unitary matrix whose (i,j)th entry equals $e^{j2\pi ij/N}$. It follows that the jth column, for $j = 0, \ldots, N-1$, is given by

$$\boldsymbol{u}_j = \frac{1}{\sqrt{N}} \begin{bmatrix} 1 \\ e^{j2\pi j/N} \\ \vdots \\ e^{j2\pi (N-1)j/N} \end{bmatrix}. \tag{B.7}$$

The DFT of a vector $\boldsymbol{x}$ is

$$\mathsf{x} = \sqrt{N}\,\boldsymbol{U}^*\boldsymbol{x} \tag{B.8}$$

whereas the IDFT is

$$\boldsymbol{x} = \frac{1}{\sqrt{N}}\boldsymbol{U}\mathsf{x}. \tag{B.9}$$

Indeed, by interpreting the entries of $\boldsymbol{x}$ and x as sequences, (B.8) and (B.9) are scaled versions of the $\mathrm{DFT}_N\{\cdot\}$ and $\mathrm{IDFT}_N\{\cdot\}$ transforms in (A.9) and (A.10).

B.2.4 Toeplitz and circulant matrices

A matrix is *Toeplitz* if it is constant along each of its diagonals. A Toeplitz matrix is further *circulant* if it is completely described by any of its rows, of which the other rows are just circular shifts with offsets equal to the row indices. Alternatively, a circulant matrix is described by any of its columns with the other columns just circular shifts thereof.

Example B.4

The real matrices $\boldsymbol{A}$ and $\boldsymbol{B}$ below are Toeplitz and circulant, respectively.

$$\boldsymbol{A} = \begin{bmatrix} 2 & 5 & 1 \\ 4 & 2 & 5 \\ 3 & 4 & 2 \end{bmatrix} \qquad \boldsymbol{B} = \begin{bmatrix} 2 & 5 & 1 \\ 1 & 2 & 5 \\ 5 & 1 & 2 \end{bmatrix}. \tag{B.10}$$

If $\boldsymbol{A}$ is an $N \times N$ circulant matrix, then the following holds:

- The eigenvectors of $\boldsymbol{A}$ equal the columns of the Fourier matrix $\boldsymbol{U}$ in (B.7).
- The eigenvalues of $\boldsymbol{A}$ equal the entries of $\boldsymbol{U}^*\boldsymbol{a}$ where $\boldsymbol{a}$ is any column of $\boldsymbol{A}$.

Hence, the eigenvalues of a circulant matrix are directly the DFT of any of the columns (or rows) of that matrix [139].

B.2.5 Hankel matrices

A Hankel matrix is an upside-down Toeplitz matrix, i.e., a matrix in which each rightward-ascending diagonal is constant.

Example B.5

The real matrix $\boldsymbol{A}$ below is a Hankel matrix.

$$\boldsymbol{A} = \begin{bmatrix} 3 & 4 & 2 \\ 4 & 2 & 5 \\ 2 & 5 & 1 \end{bmatrix} \tag{B.11}$$

B.3 Matrix decompositions

B.3.1 Eigenvalue decomposition

Any $N \times N$ Hermitian matrix $\boldsymbol{A}$ can be factored into the canonical form

$$\boldsymbol{A} = \boldsymbol{U}\boldsymbol{\Lambda}\boldsymbol{U}^* \tag{B.12}$$

$$= \sum_{i=0}^{N-1} \lambda_i(\boldsymbol{A})\, \boldsymbol{u}_i \boldsymbol{u}_i^*, \tag{B.13}$$

where $\boldsymbol{U} = [\boldsymbol{u}_0 \cdots \boldsymbol{u}_{N-1}]$ is a unitary matrix whose columns are the eigenvectors of $\boldsymbol{A}$ whereas $\boldsymbol{\Lambda} = \text{diag}(\lambda_0(\boldsymbol{A}), \ldots, \lambda_{N-1}(\boldsymbol{A}))$ is a diagonal matrix whose diagonal entries are the eigenvalues of $\boldsymbol{A}$. The eigenvectors are the vectors (normalized to have unit norm) that the linear transformation embodied by $\boldsymbol{A}$ does not rotate but merely stretches, and the eigenvalues are the factors by which they are stretched. Thus,

$$\boldsymbol{A}\, \boldsymbol{u}_i = \lambda_i(\boldsymbol{A})\, \boldsymbol{u}_i \qquad i = 0, \ldots, N-1. \tag{B.14}$$

The eigenvalues of a Hermitian matrix are always real. Furthermore, a Hermitian matrix is *positive-semidefinite* if, for every nonzero complex vector $\boldsymbol{x}$,

$$\boldsymbol{x}^* \boldsymbol{A}\, \boldsymbol{x} \geq 0. \tag{B.15}$$

If the equality in (B.15) is strict, then $\boldsymbol{A}$ is *positive-definite*. The eigenvalues of a positive-definite matrix are strictly positive whereas those of a positive-semidefinite matrix may be zero.

The rank of $\boldsymbol{A}$ equals the number of nonzero eigenvalues and $\boldsymbol{A}$ is invertible if it is full rank, i.e., if $\text{rank}(\boldsymbol{A}) = N$. Then, the eigenvectors of $\boldsymbol{A}$ and $\boldsymbol{A}^{-1}$ coincide and

$$\boldsymbol{A}^{-1} = \boldsymbol{U}\boldsymbol{\Lambda}^{-1}\boldsymbol{U}^* \tag{B.16}$$

$$= \boldsymbol{U}\, \text{diag}\left(\frac{1}{\lambda_0(\boldsymbol{A})}, \ldots, \frac{1}{\lambda_{N-1}(\boldsymbol{A})}\right) \boldsymbol{U}^*. \tag{B.17}$$

The eigenvectors are defined up to phase rotations, meaning that $e^{j\phi}\boldsymbol{u}_i$ is as valid a choice as $\boldsymbol{u}_i$ for the ith eigenvector.

Table B.1 Bases for the column, row, and null spaces of an $N \times M$ matrix $\boldsymbol{A} = \boldsymbol{U}\boldsymbol{\Sigma}\boldsymbol{V}^*$

Subspace	Dimension	Basis
Column space of $\boldsymbol{A}$	$r = \text{rank}(\boldsymbol{A})$	First r columns of $\boldsymbol{U}$
Row space of $\boldsymbol{A}$	r	First r columns of $\boldsymbol{V}$
Null space of $\boldsymbol{A}^{\mathrm{T}}$	$(N-r)$	Last $(N-r)$ columns of $\boldsymbol{U}$
Null space of $\boldsymbol{A}$	$(M-r)$	Last $(M-r)$ columns of $\boldsymbol{V}$

B.3.2 Singular-value decomposition

By means of the singular-value decomposition (SVD), any $N \times M$ matrix $\boldsymbol{A}$ can be factored as

$$\boldsymbol{A} = \boldsymbol{U}\boldsymbol{\Sigma}\boldsymbol{V}^*, \tag{B.18}$$

where $\boldsymbol{U}$ and $\boldsymbol{V}$ are unitary, respectively $N \times N$ and $M \times M$, while $\boldsymbol{\Sigma}$ is an $N \times M$ matrix with real entries on its main diagonal and zero elsewhere. The entries on the main diagonal of $\boldsymbol{\Sigma}$ are termed the *singular values* of $\boldsymbol{A}$, whereas the leading $\min(M, N)$ columns of $\boldsymbol{U}$ and of $\boldsymbol{V}$ are the left and the right singular vectors of $\boldsymbol{A}$, respectively, such that

$$\boldsymbol{A}\boldsymbol{v}_j = \sigma_j \boldsymbol{u}_j \qquad j = 0, \ldots, \min(M, N) - 1 \tag{B.19}$$

and

$$\boldsymbol{A}^*\boldsymbol{u}_j = \sigma_j \boldsymbol{v}_j \qquad j = 0, \ldots, \min(M, N) - 1 \tag{B.20}$$

where σ_j is the jth singular value whereas $\boldsymbol{u}_j = [\boldsymbol{U}]_{:,j}$ is the jth left singular vector and $\boldsymbol{v}_j = [\boldsymbol{V}]_{:,j}$ is the jth right singular vector.

The SVD of $\boldsymbol{A}$ has an intimate relationship with the eigenvalue decompositions of $\boldsymbol{A}\boldsymbol{A}^*$ and $\boldsymbol{A}^*\boldsymbol{A}$.

- The squared singular values of $\boldsymbol{A}$ are the nonzero eigenvalues of $\boldsymbol{A}\boldsymbol{A}^*$ and $\boldsymbol{A}^*\boldsymbol{A}$.
- The left singular vectors of $\boldsymbol{A}$ are the eigenvectors of $\boldsymbol{A}\boldsymbol{A}^*$. Indeed, $\boldsymbol{A}\boldsymbol{A}^* = \boldsymbol{U}\boldsymbol{\Sigma}\boldsymbol{\Sigma}^*\boldsymbol{U}^*$.
- Conversely, $\boldsymbol{A}^*\boldsymbol{A} = \boldsymbol{V}\boldsymbol{\Sigma}^*\boldsymbol{\Sigma}\boldsymbol{V}^*$ and thus the right singular vectors of $\boldsymbol{A}$ are the eigenvectors of $\boldsymbol{A}^*\boldsymbol{A}$.

The number of nonzero singular values of $\boldsymbol{A}$ equals its rank. Moreover, the columns of $\boldsymbol{U}$ and $\boldsymbol{V}$ provide bases for the column, row, and null spaces of $\boldsymbol{A}$ as detailed in Table B.1.

B.3.3 QR decomposition

Given $N \geq M$, any $N \times M$ matrix $\boldsymbol{A}$ can be factored as

$$\boldsymbol{A} = \boldsymbol{Q}\boldsymbol{R}, \tag{B.21}$$

where $\boldsymbol{Q}$ is an $N \times N$ unitary matrix while $\boldsymbol{R}$ is an $N \times M$ upper-diagonal matrix whose bottom $(N - M)$ rows are all-zero.

B.4 Trace and determinant

The trace of a square matrix equals the sum of the entries on its main diagonal and, also, the sum of its eigenvalues. It is a linear operation that is invariant to a change of basis and, therefore, to pre- or post-multiplication by unitary matrices.

The determinant of a square matrix equals the product of its eigenvalues, counted with their corresponding multiplicities. If the determinant is nonzero, the matrix is invertible and

$$\det\left(\boldsymbol{A}^{-1}\right) = \frac{1}{\det(\boldsymbol{A})}. \tag{B.22}$$

If the determinant is zero, the matrix is said to be singular and it is not invertible.

Although matrix multiplication is generally noncommutative, there are some very useful commutative properties involving the determinant and trace of products of matrices. In particular, for properly dimensioned matrices $\boldsymbol{A}$ and $\boldsymbol{B}$,

$$\det(\boldsymbol{I} + \boldsymbol{A}\boldsymbol{B}) = \det(\boldsymbol{I} + \boldsymbol{B}\boldsymbol{A}), \tag{B.23}$$

with the identity matrix sized accordingly, and

$$\det(\boldsymbol{A}\boldsymbol{B}) = \det(\boldsymbol{B}\boldsymbol{A}) \tag{B.24}$$

$$= \det(\boldsymbol{A})\det(\boldsymbol{B}). \tag{B.25}$$

Also,

$$\mathrm{tr}(\boldsymbol{A}\boldsymbol{B}) = \mathrm{tr}(\boldsymbol{B}\boldsymbol{A}), \tag{B.26}$$

which, more generally, makes the trace invariant to cyclic permutations.

B.5 Frobenius norm

The Frobenius norm of an $N \times M$ matrix $\boldsymbol{A}$ equals

$$\|\boldsymbol{A}\|_{\mathrm{F}} = \sqrt{\mathrm{tr}(\boldsymbol{A}\boldsymbol{A}^*)} \tag{B.27}$$

$$= \sqrt{\mathrm{tr}(\boldsymbol{A}^*\boldsymbol{A})}. \tag{B.28}$$

Expanding the argument of the square root, what results is

$$\|\boldsymbol{A}\|_{\mathrm{F}} = \sqrt{\sum_{i=0}^{N-1}\sum_{j=0}^{M-1} |[\boldsymbol{A}]_{i,j}|^2}, \tag{B.29}$$

which, in the special case that $\boldsymbol{A}$ is a vector, reduces to the standard Euclidean norm. Unless otherwise stated, this is the norm definition throughout the book. And, since the distance between two vectors is the norm of their difference, unless otherwise stated the distances in this book are Euclidean distances.

B.6 Moore–Penrose pseudoinverse

The pseudoinverse generalizes the concept of an inverse. In particular, the Moore–Penrose pseudoinverse of an $N \times M$ rectangular matrix $\boldsymbol{A}$ is the unique matrix $\boldsymbol{A}^\dagger$ satisfying

$$\boldsymbol{A}\boldsymbol{A}^\dagger\boldsymbol{A} = \boldsymbol{A} \tag{B.30}$$

$$\boldsymbol{A}^\dagger\boldsymbol{A}\boldsymbol{A}^\dagger = \boldsymbol{A}^\dagger \tag{B.31}$$

and such that $\boldsymbol{A}\boldsymbol{A}^\dagger$ and $\boldsymbol{A}^\dagger\boldsymbol{A}$ are both Hermitian.

- If $\boldsymbol{A}^*\boldsymbol{A}$ is invertible, then it is easily verified that

$$\boldsymbol{A}^\dagger = (\boldsymbol{A}^*\boldsymbol{A})^{-1}\boldsymbol{A}^* \tag{B.32}$$

  satisfies the above conditions and that $\boldsymbol{A}^\dagger\boldsymbol{A} = \boldsymbol{I}$. This may be the case if $N \geq M$.
- In turn, if it is $\boldsymbol{A}\boldsymbol{A}^*$ that is invertible, then

$$\boldsymbol{A}^\dagger = \boldsymbol{A}^*(\boldsymbol{A}\boldsymbol{A}^*)^{-1} \tag{B.33}$$

  and $\boldsymbol{A}\boldsymbol{A}^\dagger = \boldsymbol{I}$. This may be the case if $N \leq M$.
- If $\boldsymbol{A}$ is square and invertible, then (B.32) and (B.33) coincide and the pseudoinverse reduces to the regular inverse, $\boldsymbol{A}^\dagger = \boldsymbol{A}^{-1}$.

B.7 Matrix inversion lemma

An identity that often comes handy in many derivations is the so-called matrix inversion lemma, which states that

$$(\boldsymbol{A} + \boldsymbol{B}\boldsymbol{C}\boldsymbol{D})^{-1} = \boldsymbol{A}^{-1} - \boldsymbol{A}^{-1}\boldsymbol{B}\left(\boldsymbol{C}^{-1} + \boldsymbol{D}\boldsymbol{A}^{-1}\boldsymbol{B}\right)^{-1}\boldsymbol{D}\boldsymbol{A}^{-1}. \tag{B.34}$$

This formula, also termed the Woodbury matrix identity, allows computing the inverse of the linear operator on the left-hand side of (B.34) whenever the inverse of its two main pieces, $\boldsymbol{A}$ and $\boldsymbol{C}$, are known.

B.8 Kronecker product

The Kronecker product extends to matrices the outer product vector operator. Given an $N_{\boldsymbol{A}} \times M_{\boldsymbol{A}}$ matrix $\boldsymbol{A}$ and an $N_{\boldsymbol{B}} \times M_{\boldsymbol{B}}$ matrix $\boldsymbol{B}$, their Kronecker product yields an $N_{\boldsymbol{A}}N_{\boldsymbol{B}} \times M_{\boldsymbol{A}}M_{\boldsymbol{B}}$ matrix

$$\boldsymbol{A} \otimes \boldsymbol{B} = \begin{bmatrix} [\boldsymbol{A}]_{0,0}\boldsymbol{B} & \cdots & [\boldsymbol{A}]_{0,M_{\boldsymbol{A}}-1}\boldsymbol{B} \\ \vdots & \ddots & \vdots \\ [\boldsymbol{A}]_{N_{\boldsymbol{A}}-1,0}\boldsymbol{B} & \cdots & [\boldsymbol{A}]_{N_{\boldsymbol{A}}-1,M_{\boldsymbol{A}}-1}\boldsymbol{B} \end{bmatrix}. \tag{B.35}$$

Like the regular matrix product, the Kronecker product is nonconmutative, linear, and associative. In addition,

$$(\boldsymbol{A} \otimes \boldsymbol{B})^* = \boldsymbol{A}^* \otimes \boldsymbol{B}^* \tag{B.36}$$

and, if both $\boldsymbol{A}$ and $\boldsymbol{B}$ are invertible,

$$(\boldsymbol{A} \otimes \boldsymbol{B})^{-1} = \boldsymbol{A}^{-1} \otimes \boldsymbol{B}^{-1}. \tag{B.37}$$

Finally, if $\boldsymbol{A}$ and $\boldsymbol{B}$ have $r_{\boldsymbol{A}}$ and $r_{\boldsymbol{B}}$ nonzero singular values, respectively, then $(\boldsymbol{A} \otimes \boldsymbol{B})$ has $r_{\boldsymbol{A}} r_{\boldsymbol{B}}$ nonzero singular values given by all the cross products thereof.

Appendix C Random variables and processes

This appendix collects a host of relevant results on random variables (scalars, vectors, and matrices), presents the distributions used throughout the book, and provides a classification of the principal forms of convergence of random sequences. Special treatment is given to the behavior of certain random matrices as their size grows without bound, as this behavior is exploited to provide large-dimensional MIMO characterizations of quantities of interest. Shifting the attention from random variables to random processes, the all-important concepts of stationarity and ergodicity are then set forth.

C.1 Random variables

The random variables in this section are regarded as continuous unless otherwise indicated, yet most of the concepts extend to discrete distributions with a probability mass function (PMF) in place of the probability density function (PDF), and with a proper replacement of integrals by summations.

C.1.1 Bayes' theorem

Given two random variables x and y with joint PDF $f_{xy}(\cdot,\cdot)$ and with marginals $f_x(\cdot)$ and $f_y(\cdot)$, the respective conditional distributions are obtained as

$$f_{y|x}(\mathrm{y}|\mathrm{x}) = \frac{f_{xy}(\mathrm{x},\mathrm{y})}{f_x(\mathrm{x})} \tag{C.1}$$

$$f_{x|y}(\mathrm{x}|\mathrm{y}) = \frac{f_{xy}(\mathrm{x},\mathrm{y})}{f_y(\mathrm{y})} \tag{C.2}$$

respectively for $f_x(\mathrm{x}) > 0$ and $f_y(\mathrm{y}) > 0$. Bayes' theorem states that

$$f_{x|y}(\mathrm{x}|\mathrm{y}) = \frac{f_{y|x}(\mathrm{y}|\mathrm{x})\, f_x(\mathrm{x})}{f_y(\mathrm{y})}. \tag{C.3}$$

Bayes' theorem has been said to play a role similar to that of Pythagoras' theorem in geometry [1054], a comparison that seems certainly appropriate in light of its importance and of the triangular relationship it establishes between the joint distribution and the two conditionals.

C.1.2 Expectation

Given a real scalar x with PDF $f_x(\cdot)$, the expected value of x can be obtained directly from $f_x(\cdot)$ via

$$\mathbb{E}[x] = \int_{-\infty}^{\infty} \mathrm{x}\, f_x(\mathrm{x})\, \mathrm{dx} \tag{C.4}$$

and also from the corresponding cumulative distribution function (CDF), $F_x(\cdot)$, through the relationship

$$\mathbb{E}[x] = \int_0^{\infty} \left(1 - F_x(\mathrm{x})\right) \mathrm{dx} - \int_{-\infty}^{0} F_x(\mathrm{x})\, \mathrm{dx}. \tag{C.5}$$

If x is complex, then (C.4) applies with integration over the complex plane. Sometimes, the integration limits in the expectations are not explicitly indicated; then, they should be taken over the support of the corresponding random variables, i.e., the set of values on which their probabilities are nonzero.

C.1.3 Correlation

The covariance between two random scalars x and y, with respective means $\mu_x = \mathbb{E}[x]$ and $\mu_y = \mathbb{E}[y]$, is given by

$$R_{xy} = \mathbb{E}\big[(x - \mu_x)(y - \mu_y)^*\big], \tag{C.6}$$

which, if $x = y$, reduces to the variance $\mathrm{var}[x] = \sigma_x^2$. As is common practice in signal processing and communications, we use the term *correlation* interchangeably with *covariance*. In other disciplines, chiefly in statistics, the term correlation instead refers to $\frac{1}{\sigma_x \sigma_y} R_{xy}$, scaled so it ranges within ± 1. In this text, this scaled version is referred to as *correlation coefficient* rather than merely *correlation*.

Similarly, for two random vectors $\boldsymbol{x}$ and $\boldsymbol{y}$ we can define the covariance/correlation matrix

$$\boldsymbol{R}_{\boldsymbol{xy}} = \mathbb{E}\big[(\boldsymbol{x} - \boldsymbol{\mu}_{\boldsymbol{x}})(\boldsymbol{y} - \boldsymbol{\mu}_{\boldsymbol{y}})^*\big], \tag{C.7}$$

which, if $\boldsymbol{x} = \boldsymbol{y}$, reduces to the covariance/correlation matrix of $\boldsymbol{x}$, denoted by $\boldsymbol{R}_{\boldsymbol{x}}$. By properly scaling the entries of $\boldsymbol{R}_{\boldsymbol{xy}}$, a matrix of correlation coefficients could be obtained.

A concept that arises frequently in MIMO is that of uncorrelatedness: we say that $\boldsymbol{x}$ and $\boldsymbol{y}$ are uncorrelated if $\boldsymbol{R}_{\boldsymbol{xy}} = \boldsymbol{0}$. If two random variables are independent, then they are also uncorrelated because

$$\begin{aligned} \boldsymbol{R}_{\boldsymbol{xy}} &= \mathbb{E}\big[(\boldsymbol{x} - \boldsymbol{\mu}_{\boldsymbol{x}})(\boldsymbol{y} - \boldsymbol{\mu}_{\boldsymbol{y}})^*\big] && \text{(C.8)} \\ &= \mathbb{E}\big[\boldsymbol{x} - \boldsymbol{\mu}_{\boldsymbol{x}}\big]\, \mathbb{E}\big[\boldsymbol{y}^* - \boldsymbol{\mu}_{\boldsymbol{y}}^*\big] && \text{(C.9)} \\ &= \boldsymbol{0}. && \text{(C.10)} \end{aligned}$$

However, two uncorrelated variables need not be independent.

C.1.4 Properness

Another relevant notion, which distinguishes a relevant class of signals, is properness [1055]. A complex random scalar x is said to be *proper complex* if $\mathbb{E}[x^2] = \mathbb{E}[x]^2$. As can be verified, properness requires that the real and imaginary parts of x, respectively $\Re\{x\}$ and $\Im\{x\}$, be uncorrelated and have the same variance. The concept generalizes to vectors in a straightforward manner: $\boldsymbol{x}$ is a proper complex random vector if $\mathbb{E}[\boldsymbol{x}\boldsymbol{x}^{\mathrm{T}}] = \mathbb{E}[\boldsymbol{x}]\mathbb{E}[\boldsymbol{x}^{\mathrm{T}}]$. This is tantamount to $\Re\{\boldsymbol{x}\}$ and $\Im\{\boldsymbol{x}\}$ having identical covariance matrices while the cross-covariance of $\Re\{\boldsymbol{x}\}$ and $\Im\{\boldsymbol{x}\}$ is zero.

Any subvector of a proper complex vector is also proper complex, i.e., if $[\boldsymbol{x}_0 \; \boldsymbol{x}_1]^{\mathrm{T}}$ is proper complex then both $\boldsymbol{x}_0$ and $\boldsymbol{x}_1$ are proper complex. The converse, however, is not true: two vectors that are individually proper complex need not be jointly proper complex.

Properness is preserved under affine transformations, that is, if $\boldsymbol{x}$ is proper complex then $\boldsymbol{y} = \boldsymbol{A}\boldsymbol{x} + \boldsymbol{b}$ is also proper complex for any constant matrix $\boldsymbol{A}$ and vector $\boldsymbol{b}$.

C.1.5 Circular symmetry

A random scalar x is circularly symmetric if its distribution remains unchanged when x is rotated around its mean by an arbitrary angle, i.e., if the distribution of $(x - \mathbb{E}[x])e^{\mathrm{j}\phi}$ is identical to that of $(x - \mathbb{E}[x])$ for any arbitrary ϕ. The property generalizes directly to vectors and is preserved under affine transformations. For matrices, the notion of circular symmetry generalizes into the unitary invariance property described next.

C.1.6 Unitary invariance

A random matrix $\boldsymbol{X}$ is left unitarily invariant if its distribution equals that of $\boldsymbol{U}\boldsymbol{X}$ for any unitary matrix $\boldsymbol{U}$ independent of $\boldsymbol{X}$. Alternatively, $\boldsymbol{X}$ is right unitarily invariant if its distribution equals that of $\boldsymbol{X}\boldsymbol{V}^*$ for any unitary matrix $\boldsymbol{V}$ independent of $\boldsymbol{X}$. If $\boldsymbol{X}$ is both left and right unitarily invariant, then it is bi-unitarily invariant.

C.1.7 Linear transformations

If $\boldsymbol{x}$ is a complex random vector with PDF $f_{\boldsymbol{x}}(\cdot)$ and $\boldsymbol{A}$ is a nonsingular matrix, then the PDF of $\boldsymbol{y} = \boldsymbol{A}\boldsymbol{x}$ is

$$f_{\boldsymbol{y}}(\mathrm{y}) = \frac{f_{\boldsymbol{x}}(\boldsymbol{A}^{-1}\mathrm{y})}{|\det(\boldsymbol{A})|^2}. \tag{C.11}$$

C.1.8 Kurtosis

A descriptor of random variables that comes in handy to describe certain constraints that the hardware poses on signals as well as relevant properties of fading channels is the *kurtosis*, which informs of the shape of the PDF. A large kurtosis indicates that, for a given power, a random variable exhibits infrequent but extreme deviations—as opposed to frequent but mild ones if the kurtosis is small. This is captured by the relationship between the fourth- and second-order moments, which can be formulated in various ways. The definition we adopt for the kurtosis of a real scalar x is

$$\kappa(x) = \frac{\mathbb{E}\left[x^4\right]}{\mathbb{E}^2\left[x^2\right]}, \tag{C.12}$$

which satisfies $\kappa(x) \geq 1$ with equality if x is nonrandom.

If x is zero-mean, then the above definition coincides with the common one in statistics, where the fourth- and second-order moments are centered on the mean. That is appropriate to analyze transmit signals, which are indeed zero-mean, but for fading channels (C.12) turns out to be a more convenient definition.

As yet another variant, the kurtosis may be adjusted so as to equal zero for a Gaussian variable, a definition encountered in signal processing and that is sometimes dubbed *excess kurtosis* [1056].

C.1.9 Relevant distributions

The Gaussian distribution

The PDF of a Gaussian scalar x with mean μ_x and variance σ_x^2 is

$$f_x(\mathrm{x}) = \frac{1}{\sqrt{2\pi}\sigma_x} e^{-\frac{1}{2}(\mathrm{x}-\mu_x)^2/\sigma_x^2}, \tag{C.13}$$

denoted as $x \sim \mathcal{N}(\mu_x, \sigma_x^2)$. A Gaussian scalar is *standard* if $\mu_x = 0$ and $\sigma_x^2 = 1$.

It is frequently necessary to evaluate the tail probability of a standard Gaussian, which is quantified by the Q-function described in Section E.5.

A result that casts great importance on the Gaussian distribution is the central limit theorem (see Example C.1), which states that the normalized sum of N IID random observations having a bounded variance converges to a Gaussian random variable for $N \to \infty$. Often, approximate Gaussianity holds even for modest values of N and even if the composing variables are not IID. The central limit theorem is, for instance, the reason that the background noise encountered in communications is Gaussian, as it descends from the superposition of many spurious fluctuations of thermal origin.

The complex Gaussian distribution

The complex Gaussian distribution plays a central role in many of the derivations in this book. A complex Gaussian scalar x has Gaussian real and imaginary parts. Unless otherwise stated, properness is always assumed and hence those real and imaginary parts are independent. The magnitude and phase of x are also mutually independent. With mean μ_x and variance σ_x^2, the PDF of x is

$$f_x(\mathrm{x}) = \frac{1}{\pi\sigma_x^2} e^{-|\mathrm{x}-\mu_x|^2/\sigma_x^2} \tag{C.14}$$

and we write $x \sim \mathcal{N}_{\mathbb{C}}(\mu_x, \sigma_x^2)$.[1] The phase of x is uniform over $[0, 2\pi)$ and independent of its magnitude, which abides by the Rayleigh distribution given later in this section. As in the case of a real Gaussian, a complex Gaussian is standard if $\mu_x = 0$ and $\sigma_x^2 = 1$.

Moving on to vectors, we define a complex Gaussian random vector as one that has jointly Gaussian entries. Again, properness is assumed unless otherwise stated. Such a vector $\boldsymbol{x}$ with mean $\boldsymbol{\mu_x}$ and covariance matrix $\boldsymbol{R_x}$ has the PDF

$$f_{\boldsymbol{x}}(\mathbf{x}) = \frac{1}{\det(\pi \boldsymbol{R_x})} e^{-(\mathbf{x}-\boldsymbol{\mu_x})^* \boldsymbol{R_x}^{-1}(\mathbf{x}-\boldsymbol{\mu_x})}, \tag{C.15}$$

which we write as $\boldsymbol{x} \sim \mathcal{N}_{\mathbb{C}}(\boldsymbol{\mu_x}, \boldsymbol{R_x})$.[2] If $\boldsymbol{R_x} = \sigma^2 \boldsymbol{I}$, meaning that the entries of $\boldsymbol{x}$ are IID, then the magnitude $\|\boldsymbol{x}\|$ and the vector direction $\boldsymbol{x}/\|\boldsymbol{x}\|$ are mutually independent and the latter is uniformly distributed.

Extending the definition from vectors to matrices, the PDF of a complex Gaussian matrix $\boldsymbol{X} = [\boldsymbol{x}_0 \cdots \boldsymbol{x}_{M-1}]$ whose columns are independent with the jth column, $\boldsymbol{x}_j$, having mean $\boldsymbol{\mu}_j$ and covariance $\boldsymbol{R}_j$, equals

$$f_{\boldsymbol{X}}(\mathbf{X}) = \frac{1}{\prod_{j=0}^{M-1} \det(\pi \boldsymbol{R}_j)} \exp\left(- \sum_{j=0}^{M-1} (\mathbf{x}_j - \boldsymbol{\mu}_j)^* \boldsymbol{R}_j^{-1} (\mathbf{x}_j - \boldsymbol{\mu}_j) \right). \tag{C.16}$$

If the columns of $\boldsymbol{X}$ are not only independent but also identically distributed, i.e., $\boldsymbol{\mu}_j = \boldsymbol{\mu}$ and $\boldsymbol{R}_j = \boldsymbol{R}$ for $j = 0, \ldots, M-1$, then (C.16) can be more conveniently written as

$$f_{\boldsymbol{X}}(\mathbf{X}) = \frac{1}{(\det(\pi \boldsymbol{R}))^M} e^{-\mathrm{tr}\left((\mathbf{X}-\boldsymbol{\mu})^* \boldsymbol{R}^{-1} (\mathbf{X}-\boldsymbol{\mu})\right)} \tag{C.17}$$

which, if the entries of $\boldsymbol{X}$ are further IID and zero-mean with variance σ^2, simplifies to

$$f_{\boldsymbol{X}}(\mathbf{X}) = \frac{1}{(\pi \sigma^2)^{MN}} e^{-\mathrm{tr}(\mathbf{X}^* \mathbf{X})/\sigma^2} \tag{C.18}$$

where N and M are the dimensions of $\mathbf{X}$.

Some relevant properties of the complex Gaussian distribution are as follows.

- Uncorrelatedness and independence are equivalent attributes for Gaussian random variables.
- If a random vector is both proper complex and Gaussian, then it is also circularly symmetric. In the complex Gaussian case, therefore, circular symmetry and properness are equivalent concepts.
- Gaussianity is preserved under linear transformations. Thus, linear combinations of complex Gaussian scalars, vectors, or matrices are themselves complex Gaussian.
- The IID zero-mean complex Gaussian distribution is bi-unitarily invariant. It does not have preference for any spatial direction and it is therefore invariant to rotations and reflections.

[1] If x were complex Gaussian but not proper, then it would not be completely characterized by μ_x and σ_x^2. We would further require its pseudo-covariance $\mathbb{E}[(x - \mu_x)^2]$, which for proper complex Gaussians is zero.

[2] If $\boldsymbol{x}$ were complex Gaussian but not proper, its characterization would further require the pseudo-covariance $\mathbb{E}[(\boldsymbol{x} - \boldsymbol{\mu_x})(\boldsymbol{x} - \boldsymbol{\mu_x})^{\mathrm{T}}]$, which would no longer be zero. When referring to a complex Gaussian vector with the notation $\mathcal{N}_{\mathbb{C}}(\cdot, \cdot)$ we always mean a vector abiding by (C.15).

A further relevant result is that, if $\boldsymbol{x}$ and $\boldsymbol{y}$ are jointly Gaussian, i.e.,

$$\begin{bmatrix} \boldsymbol{y} \\ \boldsymbol{x} \end{bmatrix} \sim \mathcal{N}_{\mathbb{C}}\left(\begin{bmatrix} \boldsymbol{\mu}_y \\ \boldsymbol{\mu}_x \end{bmatrix}, \begin{bmatrix} \boldsymbol{R}_y & \boldsymbol{R}_{yx} \\ \boldsymbol{R}_{xy} & \boldsymbol{R}_x \end{bmatrix} \right), \tag{C.19}$$

then $\boldsymbol{y}|\boldsymbol{x}=\mathbf{x}$ is complex Gaussian with mean

$$\boldsymbol{\mu} = \boldsymbol{R}_{yx}\boldsymbol{R}_x^{-1}\mathbf{x} \tag{C.20}$$

and covariance

$$\boldsymbol{R} = \boldsymbol{R}_y - \boldsymbol{R}_{yx}\boldsymbol{R}_x^{-1}\boldsymbol{R}_{xy}. \tag{C.21}$$

The Wishart distribution

Also arising frequently in MIMO and closely related to the complex Gaussian distribution is the Wishart distribution [1057]. Let $\boldsymbol{X}$ be an $N \times M$ matrix distributed as per (C.17) with $\boldsymbol{\mu} = \mathbf{0}$. If $M \geq N$, then $\boldsymbol{W} = \boldsymbol{X}\boldsymbol{X}^*$ is an $N \times N$ central Wishart matrix with PDF

$$f_{\boldsymbol{W}}(\mathbf{W}) = \frac{\pi^{-N(N-1)/2}}{(\det \boldsymbol{R})^M \prod_{i=1}^{N}(M-i)!} e^{-\mathrm{tr}(\boldsymbol{R}^{-1}\mathbf{W})} (\det \mathbf{W})^{N-M} \tag{C.22}$$

indicated as $\boldsymbol{W} \sim \mathcal{W}_N(M, \boldsymbol{R})$. Alternatively, if $M \leq N$, then $\boldsymbol{W} = \boldsymbol{X}^*\boldsymbol{X} \sim \mathcal{W}_M(N, \boldsymbol{R})$. (In both cases, if $\boldsymbol{\mu} \neq \mathbf{0}$ then $\boldsymbol{W}$ conforms to the more involved noncentral Wishart distribution.)

Some results of interest pertaining to $\boldsymbol{W} \sim \mathcal{W}_N(M, \boldsymbol{I})$ are [546, 1058]

$$\mathbb{E}\big[\mathrm{tr}(\boldsymbol{W})\big] = MN \tag{C.23}$$

$$\mathbb{E}\big[\mathrm{tr}(\boldsymbol{W}^2)\big] = MN(M+N) \tag{C.24}$$

$$\mathbb{E}\big[\mathrm{tr}^2(\boldsymbol{W})\big] = MN(MN+1) \tag{C.25}$$

$$\mathbb{E}\big[\mathrm{tr}(\boldsymbol{W}^{-1})\big] = \frac{N}{M-N} \tag{C.26}$$

$$\mathbb{E}\big[(\det \boldsymbol{W})^K\big] = \prod_{n=0}^{N-1} \frac{\Gamma(M+K-n)}{\Gamma(M-n)} \tag{C.27}$$

$$\mathbb{E}\big[\log_e \det \boldsymbol{W}\big] = \sum_{n=0}^{N-1} \psi(M-n), \tag{C.28}$$

where $\Gamma(\cdot)$ and $\psi(\cdot)$ are the gamma and digamma functions introduced in Appendix E.

As important as the distribution of a Wishart matrix $\boldsymbol{W} \sim \mathcal{W}_N(M, \boldsymbol{I})$ is that of its eigenvalues, and specifically the marginal distribution of an unordered eigenvalue λ, which equals [1059]

$$f_\lambda(\xi) = \frac{1}{N} \sum_{k=0}^{N-1} \frac{k!}{(k+M-N)!} \left(L_k^{M-N}(\xi)\right)^2 \xi^{M-N} e^{-\xi}, \tag{C.29}$$

where $L_k^q(\cdot)$ is the associated Laguerre polynomial

$$L_k^q(\xi) = \frac{e^{\xi}\xi^{-q}}{k!}\frac{d^k}{d\xi^k}\left(e^{-\xi}\xi^{q+k}\right) \tag{C.30}$$

$$= \sum_{i=0}^{k}(-1)^i \frac{(k+q)!}{(k-i)!(q+i)!i!}\,\xi^i. \tag{C.31}$$

The chi-square distribution

For $N = 1$, the central Wishart distribution reduces to the chi-square distribution. Specifically, if $x_i \sim \mathcal{N}_{\mathbb{C}}(0,1)$ for $i = 0, \ldots, M-1$, then

$$w = \sum_{i=0}^{M-1} |x_i|^2 \tag{C.32}$$

is said to be a chi-square random variable with $2M$ degrees of freedom (there are M complex terms in the summation, hence $2M$ real terms), denoted $w \sim \chi^2_{2M}$ and with PDF

$$f_w(\xi) = \frac{\xi^{M-1}e^{-\xi}}{(M-1)!}. \tag{C.33}$$

The chi-square is a special case of the gamma distribution, which features two distinct parameters as opposed to only the number of degrees of freedom.

An expectation of particular interest in MIMO analysis, and which is invoked repeatedly throughout the text, is

$$\mathbb{E}\left[\log_e\left(1 + A\sum_{i=0}^{M-1}|x_i|^2\right)\right] = \int_0^{\infty} \log_e(1+A\,\xi)\,\frac{\xi^{M-1}e^{-\xi}}{(M-1)!}\,\mathrm{d}\xi \tag{C.34}$$

$$= e^{1/A}\sum_{q=1}^{M}\frac{\Gamma(q-M,1/A)}{A^{M-q}}, \tag{C.35}$$

where $\Gamma(\cdot,\cdot)$ is the incomplete gamma function introduced in Appendix E and where we have applied the identity [1060, 1061]

$$\int_0^{\infty} \log_e(1+A\,\xi)\,e^{-c\xi}\,\xi^{M-1}\mathrm{d}\xi = \Gamma(n)\,e^{c/A}\sum_{q=1}^{M}\frac{\Gamma(q-M,c/A)}{c^q\,A^{M-q}}. \tag{C.36}$$

By means of the relationship provided in (E.12), the expectation in (C.35) can be rewritten in the alternative form

$$\mathbb{E}\left[\log_e\left(1 + A\sum_{i=0}^{M-1}|x_i|^2\right)\right] = e^{1/A}\sum_{q=1}^{M}\mathcal{E}_q\left(\frac{1}{A}\right), \tag{C.37}$$

where $\mathcal{E}_q(\cdot)$ is an exponential integral function also introduced in Appendix E.

The exponential distribution

For $M = 1$, the chi-square reverts to the exponential distribution, which thereby gives the distribution of the squared magnitude of a zero-mean complex Gaussian. If $x \sim \mathcal{N}_{\mathbb{C}}(0, \sigma^2)$, then

$$f_{|x|^2}(\xi) = \frac{1}{\sigma^2} e^{-\xi/\sigma^2}, \tag{C.38}$$

which, for $\sigma^2 = 1$, reduces to $f_{|x|^2}(\xi) = e^{-\xi}$. Of particular interest to MIMO is the expectation

$$\mathbb{E}\Big[\log_e\Big(1 + A\,|x|^2\Big)\Big] = \int_0^\infty \log_e(1 + A\xi)\, e^{-\xi}\, \mathrm{d}\xi \tag{C.39}$$

$$= e^{1/A}\, \mathcal{E}_1\left(\frac{1}{A}\right), \tag{C.40}$$

which is the special case of (C.37) corresponding to $M = 1$.

The chi distribution

If $x_i \sim \mathcal{N}_{\mathbb{C}}(0, 1)$ for $i = 0, \ldots, M-1$, then

$$w = \sqrt{\sum_{i=0}^{M-1} |x_i|^2} \tag{C.41}$$

is a chi random variable with $2M$ degrees of freedom and mean

$$\mathbb{E}[w] = \frac{\Gamma(M + 1/2)}{\Gamma(M)}. \tag{C.42}$$

Squaring a chi random variable, we obtain a chi-square random variable.

The Rayleigh distribution

For $M = 1$, the chi distribution reduces to the Rayleigh distribution. Therefore, the square-root of an exponentially distributed variable gives a Rayleigh-distributed variable. Letting $x \sim \mathcal{N}_{\mathbb{C}}(0, \sigma^2)$,

$$f_{|x|}(\xi) = \frac{\xi}{\sigma^2} e^{-\frac{1}{2}\xi^2/\sigma^2}. \tag{C.43}$$

C.1.10 Convergence of sequences

Results on the convergence of sequences of random variables play a very important role in information theory. We next classify the most relevant forms of convergence, focusing on scalar random variables; similar notions apply to vectors, matrices, and functions.

Convergence in distribution

A sequence of random variables $\{x_n\}$ is said to converge *in distribution* to another random variable x if

$$\lim_{n\to\infty} F_{x_n}(\mathrm{x}) = F_x(\mathrm{x}) \tag{C.44}$$

at every x where $F_x(\mathrm{x})$ is continuous. The formulation in terms of CDFs rather than PDFs is not irrelevant, as the latter do not always converge. While relatively weak, in the sense that it is implied by all other types of convergence in this section, the notion of convergence in distribution can be extremely useful, as the next result illustrates.

Example C.1 (Central limit theorem)

Let $x_0, \ldots, x_{N-1}$ be IID random variables with mean μ and variance $\sigma^2 < \infty$ and let

$$y_N = \frac{1}{\sqrt{N}} \sum_{n=0}^{N-1} (x_n - \mu). \tag{C.45}$$

Then, for $N \to \infty$, y_N converges in distribution to $x \sim \mathcal{N}(0, \sigma^2)$. This result, indispensable in engineering and in countless other disciplines, states that the normalized sum of an increasing number of independent observations of a random event becomes progressively Gaussian regardless of the underlying distribution—provided that its variance is bounded. (If the variance is not bounded, there may be still convergence in distribution, only to a non-Gaussian function.)

Next, we turn our attention to stronger forms of convergence that go beyond the limiting distribution of a sequence.

Convergence in probability

A sequence of random variables $\{x_n\}$ is said to converge *in probability* (or weakly) to x if, for every $\epsilon > 0$,

$$\lim_{n\to\infty} \mathbb{P}\big[\, |x_n - x| > \epsilon \big] = 0. \tag{C.46}$$

This indicates that the probability of outcomes deviating from x diminishes as the sequence progresses. For every $\epsilon > 0$ and $\delta > 0$, there is an N such that $\mathbb{P}[|x_{n\geq N} - x| > \epsilon] < \delta$. However, there are no guarantees about specific realizations of $x_{n\geq N}$ being within ϵ of x; the guarantee is only that the probability thereof be smaller than δ.

Example C.2 (Weak law of large numbers)

Let $x_0, \ldots, x_{N-1}$ be IID random variables with mean $\mu < \infty$ and let

$$\bar{x}_N = \frac{1}{N} \sum_{n=0}^{N-1} x_n. \tag{C.47}$$

Then, for $N \to \infty$, $\bar{x}_N$ converges in probability to μ.

This supremely important result affirms that the sample average converges to the expected value, thereby ensuring, in probability, an increasing stability as more and more independent observations are thrown into the average. No condition is placed on the variance $\sigma^2_{x_n}$ because the convergence takes place (possibly at a slower rate) even if $\sigma^2_{x_n}$ is not bounded.

Convergence almost surely

A sequence of random variables $\{x_n\}$ is said to converge *almost surely* (or strongly) to x if it holds that

$$\mathbb{P}\Big[\lim_{n\to\infty} x_n = x\Big] = 1. \tag{C.48}$$

This form of convergence, denoted by $x_n \stackrel{\text{a.s.}}{\to} x$, signifies that all the realizations not convergent upon x have an aggregate probability of zero. For this reason, convergence almost surely is also termed convergence with probability 1. Furthermore, this strong type of convergence immediately implies the weaker convergence in probability.

Example C.3 (Strong law of large numbers)

Let $x_0, \ldots, x_{N-1}$ be IID random variables with mean $\mu < \infty$ and let

$$\bar{x}_N = \frac{1}{N} \sum_{n=0}^{N-1} x_n. \tag{C.49}$$

Then, for $N \to \infty$, we have that $\bar{x}_N \stackrel{\text{a.s.}}{\to} \mu$. Under technical conditions only slightly more stringent than those required for its weak counterpart, the strong law of large numbers indicates that, beyond a certain N that depends on ϵ, all realizations of $\bar{x}_n$ are within ϵ of μ with probability 1.

The strong law continues to apply even if $x_0, \ldots, x_{N-1}$ are independent but nonidentically distributed (IND), provided that their variances $\sigma^2_{x_n}$ are bounded and $\sum_{n=0}^{N-1} \sigma^2_{x_n}/n^2$ remains bounded for $N \to \infty$.

Convergence in the mean-square sense

Finally, and given the importance of the concept of mean-square distortion, we present a form of convergence involving a vanishing mean-square difference. A sequence of random variables $\{x_n\}$ is said to converge *in the mean-square sense* to x if

$$\lim_{n\to\infty} \mathbb{E}\Big[|x_n - x|^2\Big] = 0. \tag{C.50}$$

While weaker than convergence almost surely, convergence in the mean-square sense also implies convergence in probability.

C.2 Large random matrices

Many performance metrics in MIMO communication are functions of the eigenvalues of a random matrix embodying the channel. For a Hermitian $N \times N$ matrix $\boldsymbol{A}$, such eigenvalues can be fully characterized by the empirical cumulative distribution

$$F_{\boldsymbol{A}}^N(\xi) = \frac{1}{N} \sum_{i=0}^{N-1} 1\{\lambda_i(\boldsymbol{A}) \leq \xi\}, \tag{C.51}$$

where $1\{\cdot\}$ is the indicator function, returning 1 if its argument is true and 0 otherwise. $F_{\boldsymbol{A}}^N(\xi)$, which gives the fraction of the eigenvalues $\lambda_0(\boldsymbol{A}), \ldots, \lambda_{N-1}(\boldsymbol{A})$ that fall below ξ, is itself random. In many instances, however, $F_{\boldsymbol{A}}^N(\xi)$ converges to a nonrandom limit, $F_{\boldsymbol{A}}(\xi)$, as $N \to \infty$. This deterministic function, termed the asymptotic eigenvalue distribution, enables robust large-dimensional characterizations that are not subject to the vagaries of specific realizations of $\boldsymbol{A}$.

Driven by nuclear physics, the first results on the asymptotic eigenvalue distribution of random matrices were derived by the physicist Eugene Wigner in the 1950s [1062, 1063]. By means of these asymptotic distributions, Wigner explained the statistics of experimentally measured atomic energy levels. Since then, research on the large-dimensional behavior of random matrices has continued to draw interest in physics, probability, statistics, and, more recently, engineering. At first, Wigner considered an $N \times N$ symmetric matrix $\boldsymbol{A}$ with zeroes along the diagonal, and independent equiprobable ± 1 upper-triangular entries [1062]. For $N \to \infty$, the averaged empirical distribution of the eigenvalues of $\boldsymbol{A}/\sqrt{N}$ was shown to converge to the semicircle law

$$f_{\boldsymbol{A}/\sqrt{N}}(\xi) = \frac{1}{2\pi}\sqrt{4 - \xi^2} \qquad |\xi| \in [0, 2]. \tag{C.52}$$

Subsequently, Wigner realized that the same result is obtained if the upper-triangle entries are simply independent and zero-mean, not necessarily ± 1 [1063].

If $\boldsymbol{A}$ is not symmetric and all its entries are IID, then the eigenvalues of $\boldsymbol{A}/\sqrt{N}$ are complex and fall asymptotically uniformly on the unit circle of the complex plane, a result commonly referred to as Girko's full-circle law [1064].

A quantum leap was made in 1967, when Marčenko and Pastur [1065] derived the asymptotic eigenvalue distribution of $\boldsymbol{A} = \boldsymbol{B} + \boldsymbol{H}\boldsymbol{D}\boldsymbol{H}^*$ where $\boldsymbol{B}$ is deterministic and Hermitian, $\boldsymbol{H}$ is $N \times M$ with IID entries, and $\boldsymbol{D}$ is real, diagonal, and independent of $\boldsymbol{H}$. In its general form, the asymptotic eigenvalue distribution is not characterized explicitly, but indirectly through its Stieltjes transform, which uniquely determines it. Since then, this transform, which can be interpreted as an iterated Laplace transform, has played a fundamental role in the theory of large random matrices. In the case that $\boldsymbol{B} = \boldsymbol{0}$ and $\boldsymbol{D} = \boldsymbol{I}$, with $\boldsymbol{H}$ having unit-variance entries, the Marčenko–Pastur law for $\frac{1}{M}\boldsymbol{H}\boldsymbol{H}^*$ emerges explicitly as

$$f_{\frac{1}{M}\boldsymbol{H}\boldsymbol{H}^*}(\xi) = [1 - \beta]^+ \delta(\xi) + \beta \frac{\sqrt{(\xi - a)(b - \xi)}}{2\pi\xi} \qquad \xi \in [a, b], \tag{C.53}$$

where $\beta = M/N$, $[z]^+ = \max(0, z)$ and

$$a = \left(1 - \frac{1}{\sqrt{\beta}}\right)^2 \qquad b = \left(1 + \frac{1}{\sqrt{\beta}}\right)^2. \tag{C.54}$$

In turn,

$$f_{\frac{1}{M}\boldsymbol{H}^*\boldsymbol{H}}(\xi) = \left[1 - \frac{1}{\beta}\right]^+ \delta(\xi) + \frac{\sqrt{(\xi - a)(b - \xi)}}{2\pi\xi}. \tag{C.55}$$

Recall that the nonzero eigenvalues of $\boldsymbol{H}\boldsymbol{H}^*$ and $\boldsymbol{H}^*\boldsymbol{H}$ coincide. For $\beta \leq 1$, on the one hand, $\boldsymbol{H}\boldsymbol{H}^*$ has M nonzero and $(N - M)$ zero eigenvalues; a share $(1 - \beta)$ is thus zero, as reflected by the mass point in (C.53). For $\beta > 1$, on the other hand, all eigenvalues of $\boldsymbol{H}\boldsymbol{H}^*$ are nonzero and the mass point then moves to $f_{\frac{1}{M}\boldsymbol{H}^*\boldsymbol{H}}(\cdot)$.

The counterpart to the Marčenko–Pastur law with $\boldsymbol{B} = \boldsymbol{0}$ and with $\boldsymbol{D}$ no longer diagonal but Hermitian was reported by Silverstein, also in terms of its Stieltjes transform [1066].

Characterizations for progressively more general matrices, e.g., Gaussian matrices with certain correlation structures, have continued to appear in the literature, often in the Stieltjes domain or else through their asymptotic moments. Recently, an important advance was the realization that the noncommutative free probability theory introduced by Voiculescu in the 1980s applies to random matrices [1067]. In this theory, the traditional notion of independence of random variables is replaced by the concept of *freeness*. While, unless they share the same eigenvectors, we cannot find the eigenvalues of a sum of matrices from their individual eigenvalues, for asymptotically free random matrices [1068, 1069] the asymptotic eigenvalue distribution of the sum is obtainable from their individual ones. In free probability, the role of the Gaussian distribution in the central limit theorem of classical probability is taken by the semicircle law in the sense that the asymptotic eigenvalue distribution of the normalized sum of free random matrices converges to (C.52).

The fact that abstract mathematical tools such as free probability can now be applied to study MIMO communication is a reflection of the "unreasonable effectiveness of mathematics" that Wigner himself, and others have marveled at [1070, 1071].

C.3 Random processes

A random process is a random function of time or, alternatively, a collection of random variables (possibly vectors or matrices) indexed by time. We next illustrate some relevant notions, concentrating for the sake of brevity on continuous-time processes; with the appropriate integrals replaced by suitable summations, the same ideas apply to discrete-time processes. Of particular interest are Gaussian random processes, any finite collection of whose samples are jointly Gaussian. If a Gaussian process is input to a linear time-invariant filter, the output is sure to be also a Gaussian process.

Given the mean $\mu_x(t) = \mathbb{E}\big[x(t)\big]$, the autocovariance of a random process $x(t)$ at time t and lag τ is

$$R_x(t, \tau) = \mathbb{E}\Big[\big(x(t) - \mu_x(t)\big)\big(x(t+\tau) - \mu_x(t+\tau)\big)^*\Big], \tag{C.56}$$

which in general is a function of both t and τ. As for random variables, we use the term *autocorrelation* interchangeably with *autocovariance*.

C.3.1 Stationarity

A random process $x(t)$ is *stationary* if time shifts do not affect its distribution. Then, μ_x is not a function of time. Likewise, the autocorrelation is not a function of time, but only of lag, with $R_x(\tau)$ measuring the similarity between samples of the process separated by τ; at lag $\tau = 0$, in particular, $R_x(0)$ gives the power of the process. All higher-order moments are equally invariant to time shifts.

Having a mean and autocorrelation that are invariant to time shifts suffices for a process to be *wide-sense stationary*, even if higher-order moments are not invariant. Clearly, all stationary processes are wide-sense stationary but not vice versa and, to emphasize the difference, stationary processes are sometimes dubbed *strict-sense stationary*. In the special case of Gaussian processes, the notions of wide-sense and strict-sense stationarity are equivalent (all higher moments derive from the first two), but in general one is more lax than the other.

The power spectral density or power spectrum of a wide-sense stationary process is defined as the Fourier transform (with respect to the lag) of the autocorrelation, that is,

$$S_x(\nu) = \int_{-\infty}^{\infty} R_x(\tau)\, e^{-\mathrm{j}2\pi\nu\tau}\, \mathrm{d}\tau. \tag{C.57}$$

C.3.2 Ergodicity

A random process $x(t)$ is ergodic if its distribution can be deduced from a single, sufficiently long, realization. This requires that the time-average of $x(t)$ and of $x(t)x^*(t+\tau)$ equal (asymptotically) the mean and the autocorrelation, respectively, i.e., that

$$\lim_{T\to\infty} \frac{1}{T} \int_{-T/2}^{T/2} x(t)\, \mathrm{d}t = \mathbb{E}\big[x(t)\big] \tag{C.58}$$

$$\lim_{T\to\infty} \frac{1}{T} \int_{-T/2}^{T/2} x(t)x^*(t+\tau)\, \mathrm{d}t = \mathbb{E}\big[x(t)x^*(t+\tau)\big]. \tag{C.59}$$

The left-hand sides of the above identities do not depend on t and hence there is only hope for them to hold if $x(t)$ is wide-sense stationary. Moreover, for the ergodicity to extend to all other moments, strict-sense stationarity is required and thus an ergodic process needs to be stationary. Although the converse need not hold, i.e., a stationary process need not be ergodic, under mild conditions it does: a stationary process is ergodic if its autocorrelation decays to zero sufficiently rapidly in the lag, sufficing that

$$\int_{0}^{\infty} |R_x(\tau)|\, \mathrm{d}\tau < \infty. \tag{C.60}$$

Equivalently, a stationary process is ergodic if its power spectrum exists free of delta functions [68].

Appendix D Gradient operator

The gradient of a scalar-valued function $f(\boldsymbol{x})$ of vector argument $\boldsymbol{x}$, denoted by $\nabla_{\boldsymbol{x}} f(\boldsymbol{x})$, yields a vector-valued function that, at each point $\boldsymbol{x}$, identifies the direction of greatest increase of $f(\cdot)$, with a magnitude that equals the corresponding rate of increase. It is nothing but the generalization to multiple dimensions of the standard one-dimensional derivative.

In rectangular coordinates, where $\boldsymbol{x} = [x_0 \; x_1 \; x_2]$,

$$\nabla f(x_0, x_1, x_2) = \frac{\partial f(x_0, x_1, x_2)}{\partial x_0} \boldsymbol{e}_0 + \frac{\partial f(x_0, x_1, x_2)}{\partial x_1} \boldsymbol{e}_1 + \frac{\partial f(x_0, x_1, x_2)}{\partial x_2} \boldsymbol{e}_2, \quad \text{(D.1)}$$

where $\boldsymbol{e}_0$, $\boldsymbol{e}_1$, and $\boldsymbol{e}_2$ are unit vectors along the corresponding coordinates. More generally, for vectors $\boldsymbol{x}$ with an arbitrary number of dimensions, and with the allowance of them being complex, we can compactly write

$$[\nabla_{\boldsymbol{x}} f(\boldsymbol{x})]_j = \frac{\partial f(\boldsymbol{x})}{\partial [\boldsymbol{x}^*]_j}, \quad \text{(D.2)}$$

which can be further extended to express the gradient of a scalar-valued function $f(\boldsymbol{X})$ of a complex matrix argument $\boldsymbol{X}$ as

$$[\nabla_{\boldsymbol{X}} f(\boldsymbol{X})]_{i,j} = \frac{\partial f(\boldsymbol{X})}{\partial [\boldsymbol{X}^*]_{i,j}}. \quad \text{(D.3)}$$

Example D.1

For the linear function of complex matrix argument $f(\boldsymbol{X}) = \operatorname{tr}(\boldsymbol{R}_0 \boldsymbol{X}^* \boldsymbol{R}_1)$,

$$\nabla_{\boldsymbol{X}} \operatorname{tr}(\boldsymbol{R}_0 \boldsymbol{X}^* \boldsymbol{R}_1) = \boldsymbol{R}_1 \boldsymbol{R}_0 \quad \text{(D.4)}$$

and, as a corollary [127, chapter 2]

$$\nabla_{\boldsymbol{x}} (\boldsymbol{x}^* \boldsymbol{r}) = \boldsymbol{r}. \quad \text{(D.5)}$$

Furthermore, because a vector and its conjugate can be regarded as independent for the purpose of gradient computations,

$$\nabla_{\boldsymbol{x}} (\boldsymbol{r}^* \boldsymbol{x}) = \boldsymbol{0}. \quad \text{(D.6)}$$

Example D.2

For the quadratic form $f(\boldsymbol{x}) = \boldsymbol{x}^* \boldsymbol{R} \boldsymbol{x}$, the gradient equals [127, chapter 2]

$$\nabla_{\boldsymbol{x}} (\boldsymbol{x}^* \boldsymbol{R} \boldsymbol{x}) = \boldsymbol{R} \boldsymbol{x} \quad \text{(D.7)}$$

while, for $f(\boldsymbol{X}) = \operatorname{tr}(\boldsymbol{R}_0\boldsymbol{X}\boldsymbol{R}_1\boldsymbol{X}^*)$,

$$\nabla_{\boldsymbol{X}} \operatorname{tr}(\boldsymbol{R}_0\boldsymbol{X}\boldsymbol{R}_1\boldsymbol{X}^*) = \boldsymbol{R}_0\boldsymbol{X}\boldsymbol{R}_1 \tag{D.8}$$

Example D.3

A scalar function of matrix argument that appears often in this text is $f(\boldsymbol{X}) = \log_e \det(\boldsymbol{X})$, with gradient

$$\nabla_{\boldsymbol{X}} \log_e \det(\boldsymbol{X}) = \boldsymbol{X}^{-1}. \tag{D.9}$$

From this expression, in turn, (D.8) and the chain rule of differentiation lead to

$$\nabla_{\boldsymbol{X}} \log_e \det\left(\boldsymbol{I} + \boldsymbol{X}^*\boldsymbol{R}_0\boldsymbol{X}\boldsymbol{R}_1\right) = \boldsymbol{R}_0\boldsymbol{X}\boldsymbol{R}_1\left(\boldsymbol{I} + \boldsymbol{X}^*\boldsymbol{R}_0\boldsymbol{X}\boldsymbol{R}_1\right)^{-1}. \tag{D.10}$$

For definitions of the gradient operator in cylindrical and spherical coordinates, the reader is referred to [1072].

E Appendix E Special functions

In order to maximize the generality, elegance, and meaning of their analysis, researchers strive for closed-form results, meaning a combination and composition of elementary functions via the four basic operations. These elementary functions include algebraic functions (solutions of a polynomial equation with integer coefficients) and transcendental functions (including exponentials, logarithms, trigonometric functions, and their inverses).

There are other functions that, while not elementary, appear frequently enough and display sufficiently important properties to have their own names and to be readily tabulated or computable. It is common to relax the interpretation of "closed-form" to also encompass expressions involving these special functions. In this appendix, we survey the special functions that appear throughout this textbook [1073].

E.1 Gamma function

Perhaps the most common special function, and therefore the least special of them all, is the gamma function. For arguments whose real part is positive,

$$\Gamma(z) = \int_0^{\infty} \xi^{z-1} e^{-\xi} \, d\xi, \tag{E.1}$$

which, for positive integers arguments, reduces to $\Gamma(n) = (n-1)!$, indicating that the gamma function can be interpreted as an interpolator for the factorial function. For noninteger arguments, the best known value is $\Gamma(1/2) = \sqrt{\pi}$.

Partial integrals with the same integrand as (E.1) are referred to as incomplete gamma functions, of which two varieties exist: the upper incomplete gamma function

$$\Gamma(z, s) = \int_s^{\infty} \xi^{z-1} e^{-\xi} \, d\xi, \tag{E.2}$$

and the lower incomplete gamma function

$$\gamma(z, s) = \int_0^{s} \xi^{z-1} e^{-\xi} \, d\xi. \tag{E.3}$$

E.2 Digamma function

The digamma function is the logarithmic derivative of the gamma function, i.e.,

$$\psi(z) = \frac{\mathrm{d}}{\mathrm{d}z} \log_e \Gamma(z) \tag{E.4}$$

$$= \frac{\dot{\Gamma}(z)}{\Gamma(z)}, \tag{E.5}$$

which, for integer arguments, can be expressed as

$$\psi(N) = -\gamma_{\mathsf{EM}} + \sum_{\ell=1}^{N-1} \frac{1}{\ell} \tag{E.6}$$

given the Euler–Mascheroni constant

$$\gamma_{\mathsf{EM}} = \lim_{N\to\infty} \left(\sum_{n=1}^{N} \frac{1}{n} - \log_e N \right) \tag{E.7}$$

$$\approx 0.5772. \tag{E.8}$$

The digamma function satisfies the recursion

$$\psi(N+1) = 1 + \frac{1}{N} \sum_{n=1}^{N} \psi(n). \tag{E.9}$$

E.3 Exponential integrals

For real nonzero arguments, the exponential integral function is

$$\mathcal{E}_{\mathrm{i}}(z) = \int_{-\infty}^{z} \frac{e^{\xi}}{\xi}\, \mathrm{d}\xi, \tag{E.10}$$

from which a class of functions, parameterized by an order n, is defined as

$$\mathcal{E}_n(z) = \int_{1}^{\infty} \frac{e^{-z\xi}}{\xi^n}\, \mathrm{d}\xi. \tag{E.11}$$

These functions, which we loosely refer to as exponential integrals, are related with the gamma function via

$$\mathcal{E}_n(z) = z^{n-1}\, \Gamma(1-n, z), \tag{E.12}$$

with the most commonly encountered exponential integral being

$$\mathcal{E}_1(z) = -\mathcal{E}_{\mathrm{i}}(-z) \tag{E.13}$$

$$= \Gamma(0, z). \tag{E.14}$$

E.4 Bessel functions

Bessel functions are solutions to a famed differential equation, Bessel's equation. These solutions are parameterized by an order, the most important such orders being integers or half-integers.

Bessel functions of the first kind, denoted by $J_n(x)$, are solutions that are finite at the origin for integer n. They can be expressed as an infinite series or in the integral forms

$$J_n(x) = \frac{1}{2\pi} \int_{-\pi}^{\pi} e^{-\mathrm{j}(n\xi - x \sin \xi)} \, \mathrm{d}\xi \tag{E.15}$$

$$= \frac{1}{\pi} \int_{0}^{\pi} \cos(n\xi - x \sin \xi) \, \mathrm{d}\xi. \tag{E.16}$$

From $J_n(\cdot)$, one can further define the modified Bessel functions of the first kind as

$$I_n(x) = \mathrm{j}^{-n} J_n(\mathrm{j}x) \tag{E.17}$$

and, in due course, the modified Bessel functions of the second kind as

$$K_n(x) = \frac{\pi}{2} \frac{I_{-n}(x) - I_n(x)}{\sin(n\pi)}. \tag{E.18}$$

E.5 Q-function

The Q-function is the tail probability of a standard Gaussian distribution. If $x \sim \mathcal{N}(0,1)$, then $Q(\mathrm{x}) = \mathbb{P}[x > \mathrm{x}]$ and thus the CDF of x satisfies $F_x(\mathrm{x}) = 1 - Q(\mathrm{x})$. The usual form for $Q(\cdot)$ is

$$Q(\mathrm{x}) = \frac{1}{\sqrt{2\pi}} \int_{\mathrm{x}}^{\infty} e^{-\xi^2/2} \, \mathrm{d}\xi, \tag{E.19}$$

which can be rewritten as the finite-range integral

$$Q(\mathrm{x}) = \frac{1}{\pi} \int_{0}^{\pi/2} \exp\left(-\frac{\mathrm{x}^2}{2 \sin^2 \phi} \right) \mathrm{d}\phi. \tag{E.20}$$

For positive arguments, the Q-function can be bounded as

$$\frac{\mathrm{x}}{1+\mathrm{x}^2} \frac{e^{-\mathrm{x}^2/2}}{\sqrt{2\pi}} < Q(\mathrm{x}) < \frac{1}{\mathrm{x}} \frac{e^{-\mathrm{x}^2/2}}{\sqrt{2\pi}}, \tag{E.21}$$

where the upper bound can be relaxed into the popular Chernoff bound

$$Q(\mathrm{x}) < \frac{1}{2} e^{-\mathrm{x}^2/2}. \tag{E.22}$$

Other upper and lower bounds are given in [1074]. Alternatively, a widely valid approximation is

$$Q(\mathrm{x}) \approx \frac{1}{\frac{\pi-1}{\pi}\mathrm{x} + \frac{1}{\pi}\sqrt{\mathrm{x}^2 + 2\pi}} \frac{e^{-\mathrm{x}^2/2}}{\sqrt{2\pi}}, \tag{E.23}$$

which, like the bounds, evidences that the tail decays exponentially fast.

One can also relate the Q-function with the complementary error function via

$$Q(\mathrm{x}) = \frac{1}{2}\,\mathrm{erfc}\left(\frac{\mathrm{x}}{\sqrt{2}}\right), \tag{E.24}$$

where

$$\mathrm{erfc}(\mathrm{x}) = \frac{2}{\sqrt{\pi}} \int_{\mathrm{x}}^{\infty} e^{-\xi^2}\, d\xi. \tag{E.25}$$

E.6 Hypergeometric functions

A hypergeometric function ${}_pF_q(a_0, \ldots, a_{p-1}; b_0, \ldots, b_{q-1}; x)$ is defined by its series being hypergeometric, meaning that the ratio of consecutive terms is a rational function of the summation index. If c_n and c_{n+1} are the nth and $(n+1)$th terms in the series, then

$$\frac{c_{n+1}}{c_n} = \frac{P(n)}{Q(n)}, \tag{E.26}$$

where $P(n)$ and $Q(n)$ are polynomials. We identify two hypergeometric functions in particular.

- The Kummer confluent hypergeometric function is

$$ {}_1F_1(a; b; x) = \sum_{n=0}^{\infty} \frac{(a)_n}{(b)_n} \frac{x^n}{n!}, \tag{E.27}$$

  where $(a)_n = x \cdot (x+1) \cdots (x+n-1)$ with $(a)_0 = 1$. In integral form,

$$ {}_1F_1(a; b; x) = \frac{\Gamma(b)}{\Gamma(b-a)\,\Gamma(a)} \int_0^1 e^{xt} t^{a-1} (1-t)^{b-a-1} dt. \tag{E.28}$$

- The Gauss hypergeometric function is

$$ {}_2F_1(a_0, a_1; b; x) = \sum_{n=0}^{\infty} \frac{(a_0)_n (a_1)_n}{(b)_n} \frac{x^n}{n!}. \tag{E.29}$$

Appendix F Landau symbols

The Landau symbols $\mathcal{O}(\cdot)$ and $o(\cdot)$ allow describing, in terms of simpler functions, the limiting behavior of a function as its argument approaches a particular value or tends to infinity. We present these two symbols in the simplest possible way that suffices for their usage in this book.

A function $f(x)$ is said to be $\mathcal{O}\big(g(x)\big)$, with $g(\cdot)$ being another—ideally simpler—function, if [1075]

$$|f(x)| \leq b\,|g(x)| \tag{F.1}$$

for some constant b and all values of x.

In turn, $f(x)$ is said to be $o\big(g(x)\big)$ if

$$\frac{f(x)}{g(x)} \to 0 \tag{F.2}$$

for x approaching a certain value or tending to infinity, depending on the behavior being described.

G Appendix G Convex optimization

G.1 Convex sets

A set is convex if it contains all segments between any two of its points, meaning that it is free of indentations. An example of convex set is the one depicted in Fig. 5.4.

G.2 Convex and concave functions

A real-valued function $f(\boldsymbol{x})$ is convex if it satisfies

$$f(\theta \boldsymbol{x}_0 + (1-\theta)\boldsymbol{x}_1) \leq \theta f(\boldsymbol{x}_0) + (1-\theta) f(\boldsymbol{x}_1) \tag{G.1}$$

for any real vectors $\boldsymbol{x}_0$ and $\boldsymbol{x}_1$ on a certain domain and for every real scalar $\theta \in [0, 1]$. This implies that the graph of the function lies below any chord, i.e., below any segment connecting two points of that graph.

If the inequality in (G.1) is strict, then the function is strictly convex. Conversely, if the inequality is a strict equality, then the function is linear.

A function $f(\cdot)$ is concave (respectively strictly concave) if $-f(\cdot)$ is convex (respectively strictly convex), meaning that its graph lies above any chord. Examples of concave functions are shown in Fig. G.1.

G.3 Convex optimization problems

An optimization problem has the form

$$\begin{aligned} &\text{minimize} \;\; f(\boldsymbol{x}) \\ &\text{subject to} \;\; g_i(\boldsymbol{x}) \leq d_i \qquad i = 0, \ldots, N-1, \end{aligned} \tag{G.2}$$

where $f(\cdot)$ is the objective or cost function whereas $g_i(\cdot)$ are the constraint functions. The solution $\boldsymbol{x}^\star$ satisfies $f(\boldsymbol{x}^\star) \leq f(\boldsymbol{x})$ for every vector $\boldsymbol{x}$ meeting the N constraints.

If $f(\cdot)$ as well as $g_0(\cdot), \ldots, g_{N-1}(\cdot)$ are convex, then the optimization problem is said to be convex. This means that the set defined by the constraint functions, over which the optimum is to be found, is convex and the objective function thereon is also convex. Convex problems can be interpreted as a generalization of linear problems, which are those where $f(\cdot)$ and $g_0(\cdot), \ldots, g_{N-1}(\cdot)$ are linear functions.

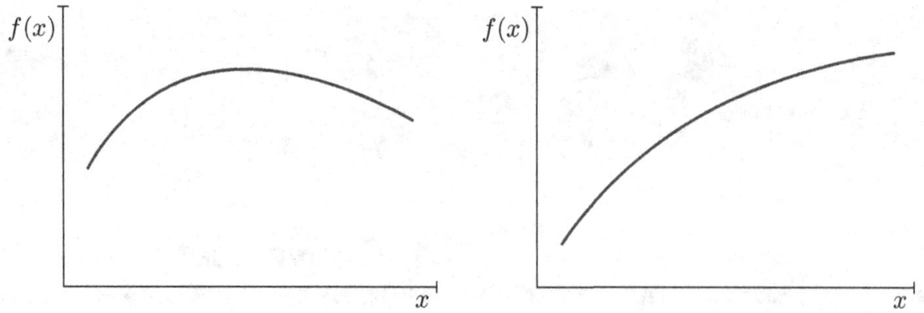

Fig. G.1 Left, nonmonotonic concave function of a real scalar argument. Right, monotonically increasing concave function.

If $f(\cdot)$ is concave, then the problem of finding its maximum over a convex set readily maps to (G.2) with the objective function given by $-f(\cdot)$.

While, in general, the solution of generic optimizations may pose considerable difficulty, convex problems can be solved reliably and efficiently even when they involve vectors with many dimensions. For an overview of convex optimization procedures and algorithms, the reader is referred to textbooks such as [1076]. Convex optimization problems arise frequently in communications and, in particular, in MIMO.

A key advantage of convex problems over nonconvex ones is that, if a local minimum exists, then it is a global minimum. As a result, any set of necessary conditions characterizing a minimum are also sufficient for global optimality.

G.4 KKT optimality conditions

The Karush–Kuhn–Tucker (KKT) conditions are necessary and sufficient conditions characterizing the solution to a convex problem. In order to present them, it is useful to rewrite (G.2) distinguishing between those constraints that are inequalities and those that are strict equalities. Also, by suitably modifying the constraint functions, we can set all the constraints to zero, obtaining

$$\begin{aligned} \min \quad & f(\boldsymbol{x}) \\ \text{s.t.} \quad & g_i(\boldsymbol{x}) \leq 0 \qquad i = 0, \ldots, M'-1 \\ & h_j(\boldsymbol{x}) = 0 \qquad j = 0, \ldots, M-1. \end{aligned} \tag{G.3}$$

Then, the KKT conditions characterizing $\boldsymbol{x}^\star$ are

$$\nabla f(\boldsymbol{x}^\star) + \sum_{i=0}^{M'-1} \lambda'_i \nabla g_i(\boldsymbol{x}^\star) + \sum_{j=0}^{M-1} \lambda_j \nabla h_j(\boldsymbol{x}^\star) = \mathbf{0} \tag{G.4}$$

$$g_i(\boldsymbol{x}^\star) \leq 0 \qquad i = 0, \ldots, M'-1$$

$$\begin{aligned} h_j(\boldsymbol{x}^\star) &= 0 \qquad & j &= 0, \ldots, M-1 \\ \lambda'_i &\geq 0 \qquad & i &= 0, \ldots, M'-1 \\ \lambda'_i \, g_i(\boldsymbol{x}^\star) &= 0 \qquad & i &= 0, \ldots, M'-1, \end{aligned}$$

where $\lambda'_0, \ldots, \lambda'_{M'-1}$ and $\lambda_0, \ldots, \lambda_{M-1}$ are the so-called KKT multipliers.

The KKT conditions play an important role in convex optimization, often serving as targets for the algorithms. Moreover, sometimes the KKT conditions can be analytically resolved, yielding expressions that characterize the solution directly.

G.5 Lagrange multipliers

If there are no inequality constraints, but only equality ones, the solution of the KKT conditions reverts to the method of Lagrange multipliers. Consider the problem

$$\begin{aligned} &\min \;\; f(\boldsymbol{x}) \\ &\text{s.t.} \;\; h_j(\boldsymbol{x}) = 0 \qquad j = 0, \ldots, M-1. \end{aligned} \tag{G.5}$$

If we augment the objective function with a weighted sum of the constraint functions we obtain the so-called *Lagrangian* function

$$L(\boldsymbol{x}, \lambda_0, \ldots, \lambda_{M-1}) = f(\boldsymbol{x}) + \sum_{j=0}^{M-1} \lambda_j h_j(\boldsymbol{x}), \tag{G.6}$$

where $\lambda_0, \ldots, \lambda_{M-1}$ are termed *Lagrange multipliers.*

Any point $\boldsymbol{x}^\star$ where all partial derivatives of $L(\cdot)$ are zero corresponds to an extreme point (minimum or maximum) that satisfies the equality constraints. If $f(\cdot)$ is convex or concave, then this extreme point is guaranteed to be the global one. The conditions satisfied by $\boldsymbol{x}^\star$ can be written as

$$\begin{aligned} \nabla_{\boldsymbol{x}} L(\boldsymbol{x}^\star) = \nabla f(\boldsymbol{x}^\star) + \sum_{j=0}^{M-1} \lambda_j \nabla h_j(\boldsymbol{x}^\star) &= \mathbf{0} \\ h_j(\boldsymbol{x}^\star) &= 0 \qquad j = 0, \ldots, M-1, \end{aligned} \tag{G.7}$$

which, indeed, can be seen to be a special case of (G.4).

Often, the objective functions encountered in communications are not only convex or concave, but further monotonic in quantities of interest (see Fig. G.1). Taking advantage of that, inequality constraints can be tightened into strict equalities, paving the way for the application of Lagrange multipliers directly, rather than the more general KKT conditions. Examples include the spectral efficiency or the error probability, both of which improve monotonically with the signal power in most settings; then, any applicable power constraints can be considered directly as equalities.

G.6 Jensen's inequality

As stated earlier, the inequality in (G.1) reflects that the graph of a convex function lies below any chord. This generalizes into the widely utilized Jensen's inequality [1077]. Given a convex function $f(\boldsymbol{x})$ and the vectors $\boldsymbol{x}_0, \ldots, \boldsymbol{x}_{N-1}$ in its domain,

$$f\left(\frac{\sum_{i=0}^{N-1} a_i \boldsymbol{x}_i}{\sum_{i=0}^{N-1} a_i}\right) \leq \frac{\sum_{i=0}^{N-1} a_i f(\boldsymbol{x}_i)}{\sum_{i=0}^{N-1} a_i} \tag{G.8}$$

for any positive coefficients $a_0, \ldots, a_{N-1}$. As a special case, if $a_i = 1$ for $0 = 1, \ldots, N$–1, the weighted sum gives the average and thus, if the vectors are random variables, we can write

$$f(\mathbb{E}[\boldsymbol{x}]) \leq \mathbb{E}\big[f(\boldsymbol{x})\big]. \tag{G.9}$$

If $f(\cdot)$ is concave, rather than convex, the above inequalities are simply reversed.

References

[1] S. Verdú, "Wireless bandwidth in the making," *IEEE Commun. Mag.*, vol. 38, no. 7, pp. 53–58, 2000.

[2] P. K. Bondyopadhyay, "The first application of array antenna," *IEEE Int'l Conf. Phased Array Systems and Techn.*, pp. 29–32, May 2000.

[3] A. C. Schell, "Antenna developments of the 1950s to the 1980s," *IEEE Int'l Symp. Antennas Propag.*, vol. 1, pp. 30–33, Jul. 2001.

[4] A. J. Paulraj and C. B. Papadias, "Space–time processing for wireless communications," *IEEE Signal Process. Mag.*, vol. 14, no. 6, pp. 49–83, 1997.

[5] L. R. Kahn, "Ratio squarer," *Proc. IRE*, vol. 42, p. 1704, 1954.

[6] A. Wittneben, "Basestation modulation diversity for digital simulcast," *IEEE Veh. Techn. Conf. (VTC'91)*, pp. 848–853, May 1991.

[7] N. Seshadri and J. Winters, "Two signaling schemes for improving the error performance of frequency-division-duplex (FDD) transmission systems using transmitted antenna diversity," *Int'l J. Wireless Inform. Networks*, vol. 1, no. 1, pp. 49–60, 1994.

[8] A. Hiroike, F. Adachi, and N. Nakajima, "Combined effects of phase sweeping transmitter diversity and channel coding," *IEEE Trans. Veh. Techn.*, vol. 41, no. 2, pp. 170–176, May 1992.

[9] S. M. Alamouti, "A simple transmit diversity technique for wireless communications," *IEEE J. Sel. Areas Commun.*, vol. 16, no. 8, pp. 1451–1458, 1998.

[10] H. Jafarkhani and V. Tarokh, "Multiple transmit antenna differential detection from generalized orthogonal designs," *IEEE Trans. Inform. Theory*, vol. 47, no. 6, pp. 2626–2631, 2001.

[11] J. Winters, "Optimum combining in digital mobile radio with cochannel interference," *IEEE J. Sel. Areas Commun.*, vol. 2, no. 4, pp. 528–539, 1984.

[12] B. S. Tsybakov, "The capacity of a memoryless Gaussian vector channel," *Prob. Inform. Transm.*, vol. 1, no. 1, pp. 18–29, 1965.

[13] W. L. Root and P. P. Varaiya, "Capacity of classes of Gaussian channels," *SIAM J. Appl. Math.*, vol. 16, no. 6, pp. 1350–1393, 1968.

[14] T. Cover and J. Thomas, *Elements of information theory*. Wiley & Sons, 1991.

[15] L. H. Brandenburg and A. D. Wyner, "Capacity of the Gaussian channel with memory: the multivariate case," *Bell Labs Tech. J.*, vol. 53, no. 5, pp. 745–778, 1974.

[16] J. Salz, "Digital transmission over cross-coupled linear channels," *Bell Labs Tech. J.*, vol. 64, no. 6, pp. 1147–1159, 1985.

[17] A. Duel-Hallen, "Equalizers for multiple input/multiple output channels and PAM systems with cyclostationary input sequences," *IEEE J. Sel. Areas Commun.*, vol. 10, no. 3, pp. 630–639, 1992.

[18] J. Yang and S. Roy, "On joint transmitter and receiver optimization for multiple-input–multiple-output (MIMO) transmission systems," *IEEE Trans. Commun.*, vol. 42, no. 12, pp. 3221–3231, 1994.

[19] S. Verdú, "Minimum probability of error for asynchronous Gaussian multiple-access channels," *IEEE Trans. Inform. Theory*, vol. 32, no. 1, pp. 85–96, 1986.

[20] ——, *Multiuser detection*. Cambridge University Press, 1998.

[21] N. Amitay and J. Salz, "Linear equalization theory in digital data transmission over dually polarized fading radio channels," *Bell Labs Tech. J.*, vol. 63, no. 10, pp. 2215–2259, 1984.

[22] J. Winters, "On the capacity of radio communication systems with diversity in a Rayleigh fading environment," *IEEE J. Sel. Areas Commun.*, vol. 5, no. 5, pp. 871–878, 1987.

[23] G. J. Foschini, "Layered space–time architecture for wireless communication in a fading environment when using multi-element antennas," *Bell Labs Tech. J.*, vol. 1, no. 2, pp. 41–59, 1996.

[24] P. W. Wolniansky, G. J. Foschini, G. D. Golden, and R. A. Valenzuela, "V-BLAST: an architecture for realizing very high data rates over the rich-scattering wireless channel," *URSI Int'l Symp. Signals, Systems, and Electronics (ISSSE'98)*, pp. 295–300, Sep. 1998.

[25] I. E. Telatar, "Capacity of multi-antenna Gaussian channels," *Eur. Trans. Telecommun.*, vol. 10, pp. 585–595, Nov. 1999.

[26] A. Paulraj and T. Kailath, "U.S. Patent 5345599: Increasing capacity in wireless broadcast systems using distributed transmission/directional reception (DTDR)," Sep. 1994.

[27] G. G. Raleigh and J. M. Cioffi, "Spatio-temporal coding for wireless communication," *IEEE Trans. Commun.*, vol. 46, no. 3, pp. 357–366, 1998.

[28] G. G. Raleigh and V. K. Jones, "Multivariate modulation and coding for wireless communication," *IEEE J. Sel. Areas Commun.*, vol. 17, no. 5, pp. 851–866, 1999.

[29] W. Isaacson, *The innovators: how a group of inventors, hackers, geniuses and geeks created the digital revolution*. Simon and Schuster, 2014.

[30] A. Paulraj, R. Nabar, and D. Gore, *Introduction to space–time wireless communications*. Cambridge University Press, 2003.

[31] B. Vucetic and J. Yuan, *Space-time coding*. John Wiley & Sons, 2003.

[32] T. M. Duman and A. Ghrayeb, *Coding for MIMO communication systems*. John Wiley & Sons, 2008.

[33] C. Oestges and B. Clerckx, *MIMO wireless communications: from real-world propagation to space–time code design*. Academic Press, 2010.

[34] H. C. Huang, C. B. Papadias, and S. Venkatesan, *MIMO communication for cellular networks*. Springer, 2012.

[35] L. Bai and J. Choi, *Low complexity MIMO detection*. Springer Science & Business Media, 2012.

[36] L. Hanzo, J. Akhtman, L. Wang, and M. Jiang, *MIMO-OFDM for LTE, WiFi and WiMAX: coherent versus non-coherent and cooperative turbo transceivers.* John Wiley & Sons, 2010.

[37] B. Clerckx and C. Oestges, *MIMO wireless networks: channels, techniques and standards for multi-antenna, multi-user and multi-cell systems.* Academic Press, 2013.

[38] J. R. Hampton, *Introduction to MIMO communications.* Cambridge University Press, 2013.

[39] A. Chockalingam and B. S. Rajan, *Large MIMO systems.* Cambridge University Press, 2014.

[40] T. L. Marzetta, E. G. Larsson, H. Yang, and H. Q. Ngo, *Fundamentals of massive MIMO.* Cambridge University Press, 2016.

[41] A. Gershman and N. Sidiropoulos, *Space–time processing for MIMO communications.* John Wiley & Sons, 2005.

[42] G. Tsoulos, *MIMO system technology for wireless communications.* CRC Press, 2006.

[43] E. Biglieri, R. Calderbank, A. Constantinides, A. Goldsmith, A. Paulraj, and H. V. Poor, *MIMO wireless communications.* Cambridge University Press, 2007.

[44] D. P. Palomar and Y. Jiang, "MIMO transceiver design via majorization theory," *Found. Trends Commun. Inform. Theory*, vol. 3, no. 4–5, pp. 331–551, 2007.

[45] M. Sellathurai and S. Haykin, *Space–time layered information processing for wireless communications.* John Wiley & Sons, 2009.

[46] A. Sibille, C. Oestges, and A. Zanella, *MIMO: from theory to implementation.* Academic Press, 2010.

[47] A. Lozano, F. R. Farrokhi, and R. A. Valenzuela, "Lifting the limits on high speed wireless data access using antenna arrays," *IEEE Commun. Mag.*, vol. 39, no. 9, pp. 156–162, 2001.

[48] A. Goldsmith, S. A. Jafar, N. Jindal, and S. Vishwanath, "Capacity limits of MIMO channels," *IEEE J. Sel. Areas Commun.*, vol. 21, no. 5, pp. 684–702, 2003.

[49] D. Gesbert, M. Shafi, D. Shiu, P. Smith, and A. Naguib, "From theory to practice: an overview of MIMO space–time coded wireless systems," *IEEE J. Sel. Areas Commun.*, vol. 21, no. 3, pp. 281–302, 2003.

[50] D. Tse and P. Viswanath, *Fundamentals of wireless communication.* Cambridge University Press, 2005.

[51] A. Goldsmith, *Wireless communications.* Cambridge University Press, 2005.

[52] J. L. Massey, "Information theory: the Copernican system of communications," *IEEE Commun. Mag.*, vol. 22, no. 12, pp. 26–28, 1984.

[53] M. Duarte and A. Sabharwal, "Full-duplex wireless communications using off-the-shelf radios: feasibility and first results," *Asilomar Conf. Signals, Systems and Computers*, pp. 1558–1562, Nov. 2010.

[54] J. I. Choi, M. Jain, K. Srinivasan, P. Levis, and S. Katti, "Achieving single channel, full duplex wireless communication," *ACM Int'l Conf. Mobile Computing and Networking*, pp. 1–12, Sep. 2010.

[55] D. Bharadia, E. McMilin, and S. Katti, "Full duplex radios," *ACM Int'l Conf. Mobile Computing and Networking*, vol. 43, no. 4, pp. 375–386, Aug. 2013.
[56] R. W. Chang, "High-speed multichannel data transmission with bandlimited orthogonal signals," *Bell System Tech. J.*, vol. 45, no. 10, pp. 1775–1796, 1966.
[57] S. B. Weinstein, "The history of orthogonal frequency-division multiplexing," *IEEE Commun. Mag.*, vol. 47, no. 11, 2009.
[58] C. E. Shannon, "A mathematical theory of communication," *Bell System Tech. J.*, vol. 27, pp. 379–423, 1948.
[59] R. Gallager, *Information theory and reliable communication.* Wiley, 1968.
[60] D. J. C. MacKay, *Information theory, inference, and learning algorithms.* Cambridge University Press, 2003.
[61] D. Guo, S. Shamai, and S. Verdú, "The interplay between information and estimation measures," *Found. Trends Signal Process.*, vol. 6, no. 4, 2012.
[62] H. L. Van Trees, *Detection, estimation, and modulation theory.* John Wiley & Sons, 2004.
[63] I. C. Abou-Faycal, M. D. Trott, and S. Shamai, "The capacity of discrete-time memoryless Rayleigh-fading channels," *IEEE Trans. Inform. Theory*, vol. 47, pp. 1290–1301, Apr. 2001.
[64] S. Verdú, "Spectral efficiency in the wideband regime," *IEEE Trans. Inform. Theory*, vol. 48, no. 6, pp. 1319–1343, 2002.
[65] M. J. E. Golay, "Note on the theoretical efficiency of information reception with PPM," *Proc. IRE*, vol. 37, p. 1031, Sep. 1949.
[66] M. Z. Win and R. A. Scholtz, "Ultra-wide bandwidth time-hopping spread-spectrum impulse radio for wireless multiple-access communications," *IEEE Trans. Commun.*, vol. 48, no. 4, pp. 679–691, 2000.
[67] S. Verdú, "Fifty years of Shannon theory," *IEEE Trans. Inform. Theory*, vol. 44, no. 6, pp. 2057–2078, 1998.
[68] J. Doob, *Stochastic processes.* Wiley, 1990.
[69] J. G. Kreer, "A question of terminology," *IEEE Trans. Inform. Theory*, vol. 3, p. 208, Sep. 1957.
[70] J. M. Geist, "Capacity and cutoff rate for dense M-ary constellation," *Proc. IEEE Military Commun. Conf. (MILCOM'90)*, pp. 768–770, Sep. 1990.
[71] A. Lozano, A. M. Tulino, and S. Verdú, "Optimum power allocation for parallel Gaussian channels with arbitrary input distributions," *IEEE Trans. Inform. Theory*, vol. 52, no. 7, pp. 3033–3051, 2006.
[72] T. J. Richardson and R. L. Urbanke, *Modern coding theory.* Cambridge University Press, 2008.
[73] P. J. Davis and P. Rabinowitz, *Methods of numerical integration.* Courier Corporation, 2007.
[74] T. Ketseoglou and E. Ayanoglu, "Linear precoding gain for large MIMO configurations with QAM and reduced complexity," *IEEE Trans. Commun.*, vol. 64, no. 10, pp. 4196–4208, 2016.
[75] V. V. Prelov and S. Verdú, "Second-order asymptotics of mutual information," *IEEE Trans. Inform. Theory*, vol. 50, no. 8, pp. 1567–1580, 2004.

[76] A. Ephremides, "The historian's column," *IEEE Inform. Theory Newsl.*, vol. 65, no. 4, p. 35, 2015.

[77] C. E. Shannon, "Communication in the presence of noise," *Proc. IRE*, vol. 37, pp. 10–21, Jan. 1949.

[78] M. S. Pinsker, *Information and information stability of random variables and processes*. Holden-Day, 1965.

[79] S. Verdú and T. S. Han, "A general formula for channel capacity," *IEEE Trans. Inform. Theory*, vol. 40, no. 4, pp. 1147–1157, 1994.

[80] T. S. Han, *Information-spectrum methods in information theory*. Springer Science & Business Media, 2003.

[81] J. Wolfowitz, *Coding theorems of information theory*. Springer-Verlag, 1978.

[82] S. G. Glisic and P. A. Leppänen, *Code division multiple access communications*. Springer Science & Business Media, 2012.

[83] E. Biglieri, *Coding for wireless channels*. Kluwer Academic, 2005.

[84] A. J. Viterbi, "Wireless digital communication: a view based on three lessons learned," *IEEE Commun. Mag.*, vol. 29, pp. 33–36, Sep. 1991.

[85] D. J. Costello Jr., J. Hagenauer, H. Imai, and S. B. Wicker, "Applications of error-control coding," *IEEE Trans. Inform. Theory*, vol. 44, no. 6, pp. 2531–2560, 1998.

[86] C. Berrou and A. Glavieux, "Near optimum error correcting coding and decoding: turbo-codes," *IEEE Trans. Commun.*, vol. 44, no. 10, pp. 1261–1271, 1996.

[87] D. Declercq, M. Fossorier, and E. Biglieri, *Channel coding: theory, algorithms, and applications*. Academic Press, 2014.

[88] G. Ungerböck, "Channel coding with multilevel/phase signals," *IEEE Trans. Inform. Theory*, vol. 28, no. 1, pp. 55–67, 1982.

[89] J. Hagenauer, E. Offer, and L. Papke, "Iterative decoding of binary block and convolutional codes," *IEEE Trans. Inform. Theory*, vol. 42, no. 2, pp. 429–445, 1996.

[90] L. Bahl, J. Cocke, F. Jelinek, and J. Raviv, "Optimal decoding of linear codes for minimizing symbol error rate," *IEEE Trans. Inform. Theory*, vol. 20, pp. 284–287, Mar. 1974.

[91] J. Boutros, G. Caire, E. Viterbo, H. Sawaya, and S. Vialle, "Turbo code at 0.03 dB from capacity limit," *IEEE Int'l Symp. Inform. Theory (ISIT'02)*, p. 56, 2002.

[92] T. J. Richardson, M. A. Shokrollahi, and R. L. Urbanke, "Design of capacity-approaching irregular low-density parity-check codes," *IEEE Trans. Inform. Theory*, vol. 47, pp. 619–637, Feb. 2001.

[93] E. Zehavi, "8-PSK trellis codes for a Rayleigh channel," *IEEE Trans. Commun.*, vol. 40, no. 5, pp. 873–884, 1992.

[94] G. Caire, G. Taricco, and E. Biglieri, "Bit-interleaved coded modulation," *IEEE Trans. Inform. Theory*, vol. 44, no. 3, pp. 927–946, 1998.

[95] E. Agrell, J. Lassing, E. G. Strom, and T. Ottosson, "On the optimality of the binary reflected Gray code," *IEEE Trans. Inform. Theory*, vol. 50, no. 12, pp. 3170–3182, 2004.

[96] A. Guillén, A. Martinez, and G. Caire, "Bit-interleaved coded modulation," *Found. Trends Commun. Inform. Theory*, vol. 5, no. 1–2, 2008.

[97] X. Li and J. A. Ritcey, "Bit-interleaved coded modulation with iterative decoding," *IEEE Commun. Letters*, vol. 1, no. 6, pp. 169–171, 1997.

[98] S. Ten Brink, J. Speidel, and R. H. Yan, "Iterative demapping and decoding for multilevel modulation," *IEEE Global Commun. Conf. (GLOBECOM'98)*, vol. 1, pp. 579–584, 1998.

[99] F. Simoens, H. Wymeersch, H. Bruneel, and M. Moeneclaey, "Multidimensional mapping for bit-interleaved coded modulation with BPSK/QPSK signaling," *IEEE Commun. Letters*, vol. 9, no. 5, pp. 453–455, 2005.

[100] H. M. Navazi and M. J. Hossain, "Efficient multi-dimensional mapping using QAM constellations for BICM-ID," *IEEE Trans. Wireless Commun.*, vol. 16, no. 12, pp. 8067–8076, 2017.

[101] Y. Polyanskiy, H. V. Poor, and S. Verdú, "Channel coding rate in the finite blocklength regime," *IEEE Trans. Inform. Theory*, vol. 56, pp. 2307–2359, May 2010.

[102] T. Erseghe, "On the evaluation of the Polyanskiy–Poor–Verdú converse bound for finite block-length coding in AWGN," *IEEE Trans. Inform. Theory*, vol. 61, pp. 6578–6590, Dec. 2015.

[103] F. Dowla, *Handbook of RF and wireless technologies*. Newnes, 2003.

[104] T. Richardson and S. Kudekar, "Design of low-density parity check codes for 5G new radio," *IEEE Commun. Mag.*, vol. 56, no. 3, pp. 28–34, 2018.

[105] J. Andrews, S. Buzzi, W. Choi, S. Hanly, A. Lozano, A. Soong, and J. Zhang, "What will 5G be?" *IEEE J. Sel. Areas Commun.*, vol. 32, pp. 1065–1082, 2014.

[106] F. Boccardi, R. W. Heath Jr., A. Lozano, T. Marzetta, and P. Popovski, "Five disruptive technology directions for 5G," *IEEE Commun. Mag.*, vol. 52, pp. 74–80, 2014.

[107] P. S. Sindhu, "Retransmission error control with memory," *IEEE Trans. Commun.*, vol. 25, no. 5, pp. 473–479, 1977.

[108] S. Lin and P. S. Yu, "A hybrid ARQ scheme with parity retransmission for error control of satellite channels," *IEEE Trans. Commun.*, vol. 30, no. 7, pp. 1701–1719, 1982.

[109] D. Chase, "Code combining—a maximum-likelihood decoding approach for combining an arbitrary number of noisy packets," *IEEE Trans. Commun.*, vol. 33, no. 5, pp. 385–393, 1985.

[110] E. Biglieri, G. Taricco, and E. Viterbo, "Bit-interleaved time-space codes for fading channels," *Proc. Conf. Inform. Science and Systems (CISS'00)*, pp. 15–17, 2000.

[111] N. Gresset, L. Brunel, and J. J. Boutros, "Space–time coding techniques with bit-interleaved coded modulations for MIMO block-fading channels," *IEEE Trans. Inform. Theory*, vol. 54, no. 5, pp. 2156–2178, 2008.

[112] A. M. Tonello, "Space–time bit-interleaved coded modulation with an iterative decoding strategy," *IEEE Veh. Techn. Conf. (VTC'00 Fall)*, vol. 1, pp. 473–478, 2000.

[113] B. M. Hochwald and S. Ten Brink, "Achieving near-capacity on a multiple-antenna channel," *IEEE Trans. Commun.*, vol. 51, no. 3, pp. 389–399, 2003.

[114] T. Kailath, A. H. Sayed, and B. Hassibi, *Linear estimation*. Prentice Hall, 2000.

[115] S. M. Kay, *Fundamentals of statistical signal processing, volume I: estimation theory*. Prentice Hall, 1993.

[116] C. F. Gauss, *Theoria motus corporum coelestium in sectionibus conicis solem ambientium.* Friedrich Perthes and I. H. Besser, 1809.

[117] A. M. Legendre, *Nouvelles méthodes pour la détermination des orbites des comètes.* Courcier, 1806.

[118] D. Guo, S. Shamai, and S. Verdú, "Estimation of non-Gaussian random variables in Gaussian noise: properties of the MMSE," *IEEE Int'l Symp. Inform. Theory (ISIT'08)*, pp. 1083–1087, 2008.

[119] D. Guo, Y. Wu, S. Shamai, and S. Verdú, "Estimation in Gaussian noise: properties of the minimum mean-square error," *IEEE Trans. Inform. Theory*, vol. 57, no. 4, pp. 2371–2385, 2011.

[120] Y. Wu and S. Verdú, "MMSE dimension," *IEEE Trans. Inform. Theory*, vol. 57, pp. 4857–4879, Aug. 2011.

[121] M. R. D. Rodrigues, "Multiple-antenna fading channels with arbitrary inputs: characterization and optimization of the information rate," *IEEE Trans. Inform. Theory*, vol. 60, pp. 569–585, Jan. 2014.

[122] A. Alvarado, F. Brannstrom, E. Agrell, and T. Koch, "High-SNR asymptotics of mutual information for discrete constellations with applications to BICM," *IEEE Trans. Inform. Theory*, vol. 60, pp. 1061–1076, Feb. 2014.

[123] D. P. Palomar, J. M. Cioffi, and M. A. Lagunas, "Joint Tx-Rx beamforming design for multicarrier MIMO channels: a unified framework for convex optimization," *IEEE Trans. Signal Process.*, vol. 51, no. 9, pp. 2381–2401, 2003.

[124] D. Guo, S. Shamai, and S. Verdú, "Mutual information and minimum mean-square error in Gaussian channels," *IEEE Trans. Inform. Theory*, vol. 51, no. 4, pp. 1261–1283, 2005.

[125] D. P. Palomar and S. Verdú, "Gradient of mutual information in linear vector Gaussian channels," *IEEE Trans. Inform. Theory*, vol. 52, no. 1, pp. 141–154, 2006.

[126] N. Wiener, *Extrapolation, interpolation, and smoothing of stationary time series.* MIT Press, 1964.

[127] M. Hayes, *Statistical digital signal processing and modeling.* Wiley, 1996.

[128] I. E. Telatar and D. N. C. Tse, "Capacity and mutual information of wideband multipath fading channels," *IEEE Trans. Inform. Theory*, vol. 46, no. 4, pp. 1384–1400, 2000.

[129] G. Durisi, U. G. Schuster, H. Bölcskei, and S. Shamai, "Noncoherent capacity of underspread fading channels," *IEEE Trans. Inform. Theory*, vol. 56, no. 1, pp. 367–395, 2010.

[130] E. Arikan, "Capacity bounds for an ultra-wideband channel model," *IEEE Inform. Theory Workshop (ITW'04)*, pp. 176–181, Oct. 2004.

[131] A. Lozano and D. Porrat, "Non-peaky signals in wideband fading channels: achievable bit rates and optimal bandwidth," *IEEE Trans. Wireless Commun.*, vol. 11, no. 1, pp. 246–257, 2012.

[132] B. Razavi, *RF microelectronics.* Prentice Hall, 1997.

[133] A. V. Oppenheim, R. W. Schafer, and J. R. Buck, *Discrete-time signal processing*, 2nd ed. Prentice Hall, 1999.

[134] U. Madhow, *Fundamentals of digital communication*. Cambridge University Press, 2008.

[135] J. Reed, *Software radio: a modern approach to radio engineering*. Prentice Hall, 2002.

[136] J. G. Proakis and M. Salehi, *Communication systems engineering*, 2nd ed. Prentice Hall, 2002.

[137] C. Cheng and K. Parhi, "Hardware efficient fast parallel FIR filter structures based on iterated short convolution," *Int'l Symp. Circuits and Syst.*, vol. 3, pp. III, 361–364, May 2004.

[138] H. J. Nussbaumer, *Fast Fourier transform and convolution algorithms*. Springer Science & Business Media, vol. 2, 2012.

[139] R. M. Gray, "Toeplitz and circulant matrices: a review," *Found. Trends Commun. Inform. Theory*, vol. 2, no. 3, 2006.

[140] W. J. Rugh, *Linear system theory*. Prentice Hall, 1995.

[141] V. Tarokh, N. Seshadri, and A. Calderbank, "Space–time codes for high data rate wireless communication: performance criterion and code construction," *IEEE Trans. Inform. Theory*, vol. 44, no. 2, pp. 744–765, 1998.

[142] V. Tarokh, A. Naguib, N. Seshadri, and A. R. Calderbank, "Space–time codes for high data rate wireless communication: performance criteria in the presence of channel estimation errors, mobility, and multiple paths," *IEEE Trans. Commun.*, vol. 47, no. 2, pp. 199–207, 1999.

[143] H. Sampath, P. Stoica, and A. Paulraj, "Generalized linear precoder and decoder design for MIMO channels using the weighted MMSE criterion," *IEEE Trans. Commun.*, vol. 49, pp. 2198–2206, Dec. 2001.

[144] H. Sampath and A. Paulraj, "Linear precoding for space–time coded systems with known fading correlations," *Asilomar Conf. Signals, Systems and Computers*, vol. 1, pp. 246–251, 2001.

[145] A. Scaglione, P. Stoica, S. Barbarossa, G. Giannakis, and H. Sampath, "Optimal designs for space–time linear precoders and decoders," *IEEE Trans. Signal Processing*, vol. 50, no. 5, pp. 1051–1064, 2002.

[146] J. Tellado, *Multicarrier modulation with low PAR: applications to DSL and wireless*. Springer, 2000.

[147] T. Jiang and Y. Wu, "An overview: peak-to-average power ratio reduction techniques for OFDM signals," *IEEE Trans. Broadcasting*, vol. 54, no. 2, pp. 257–268, 2008.

[148] "Cubic metric in 3GPP-LTE," 3GPP TSG RAN WG1 TDoc R1-060023, Tech. Rep., Jan. 2006.

[149] X. Zhang and S.-Y. Kung, "Capacity bound analysis for FIR Bézout equalizers in ISI MIMO channels," *IEEE Trans. Signal Processing*, vol. 53, no. 6, pp. 2193–2204, 2005.

[150] R. Rajagopal and L. Potter, "Multivariate MIMO FIR inverses," *IEEE Trans. Image Process.*, vol. 12, no. 4, pp. 458–465, 2003.

[151] B.-G. Song and J. A. Ritcey, "Spatial diversity equalization for MIMO ocean acoustic communication channels," *IEEE J. Ocean. Eng.*, vol. 21, no. 4, pp. 505–512, 1996.

[152] D. Falconer, S. Ariyavisitakul, A. Benyamin-Seeyar, and B. Eidson, "Frequency domain equalization for single-carrier broadband wireless systems," *IEEE Commun. Mag.*, vol. 40, no. 4, pp. 58–66, 2002.

[153] L. Deneire, B. Gyselinckx, and M. Engels, "Training sequence versus cyclic prefix—a new look on single carrier communication," *IEEE Commun. Letters*, vol. 5, no. 7, pp. 292–294, 2001.

[154] Y. Hou and T. Hase, "Improvement on the channel estimation of pilot cyclic prefixed single carrier (PCP-SC) system," *IEEE Signal Process. Letters*, vol. 16, no. 8, pp. 719–722, 2009.

[155] B. Muquet, Z. Wang, G. Giannakis, M. de Courville, and P. Duhamel, "Cyclic prefixing or zero padding for wireless multicarrier transmissions?" *IEEE Trans Commun.*, vol. 50, no. 12, pp. 2136–2148, 2002.

[156] Y. Li, H. Minn, and R. Rajatheva, "Synchronization, channel estimation, and equalization in MB-OFDM systems," *IEEE Trans. Wireless Commun.*, vol. 7, no. 11, pp. 4341–4352, 2008.

[157] X. Zhu and R. D. Murch, "Layered space–frequency equalization in a single-carrier MIMO system for frequency-selective channels," *IEEE Trans. Wireless Commun.*, vol. 3, no. 3, pp. 701–708, 2004.

[158] J. Coon, S. Armour, M. Beach, and J. McGeehan, "Adaptive frequency-domain equalization for single-carrier multiple-input multiple-output wireless transmissions," *IEEE Trans. Signal Processing*, vol. 53, pp. 3247–3256, Aug. 2005.

[159] F. Pancaldi, G. M. Vitetta, R. Kalbasi, N. Al-Dhahir, M. Uysal, and H. Mheidat, "Single-carrier frequency domain equalization," *IEEE Signal Process. Mag.*, vol. 25, no. 5, pp. 37–56, 2008.

[160] G. L. Stüber, J. R. Barry, S. W. McLaughlin, Y. Li, M. A. Ingram, and T. G. Pratt, "Broadband MIMO-OFDM wireless communications," *Proc. IEEE*, vol. 92, no. 2, pp. 271–294, 2004.

[161] R. van Nee and R. Prasad, *OFDM for wireless multimedia communications*. Artech House, 2000.

[162] L. Zhang, A. Ijaz, P. Xiao, M. M. Molu, and R. Tafazolli, "Filtered OFDM systems, algorithms, and performance analysis for 5G and beyond," *IEEE Trans. Commun.*, vol. 66, no. 3, pp. 1205–1218, 2018.

[163] H. Sari, G. Karam, and I. Jeanclaude, "Transmission techniques for digital terrestrial TV broadcasting," *IEEE Commun. Mag.*, vol. 33, no. 2, pp. 100–109, 1995.

[164] Q. Shi, "OFDM in bandpass nonlinearity," *IEEE Trans. Consum. Electronics*, vol. 42, no. 3, pp. 253–258, 1996.

[165] F. Classen and H. Meyr, "Frequency synchronization algorithms for OFDM systems suitable for communication over frequency selective fading channels," *IEEE Veh. Techn. Conf. (VTC'94)*, pp. 1655–1659, Jun. 1994.

[166] R. Parot and F. Harris, "Resolving and correcting gain and phase mismatch in

transmitters and receivers for wideband OFDM systems," *Asilomar Conf. Signals, Systems and Computers*, vol. 2, pp. 1005–1009, Nov. 2002.

[167] T. Pollet, M. Van Bladel, and M. Moeneclaey, "BER sensitivity of OFDM systems to carrier frequency offset and Wiener phase noise," *IEEE Trans. Commun.*, vol. 43, no. 2–4, pp. 191–193, 1995.

[168] J. Coon, J. Siew, M. Beach, A. Nix, S. Armour, and J. McGeehan, "A comparison of MIMO-OFDM and MIMO-SCFDE in WLAN environments," *IEEE Global Commun. Conf. (GLOBECOM'03)*, vol. 6, pp. 3296–3301, Dec. 2003.

[169] H. Bölcskei, "MIMO-OFDM wireless systems: basics, perspectives, and challenges," *IEEE Wireless Communications*, vol. 13, no. 4, pp. 31–37, 2006.

[170] H. Yang, "A road to future broadband wireless access: MIMO-OFDM-based air interface," *IEEE Commun. Mag.*, vol. 43, no. 1, pp. 53–60, 2005.

[171] Z. Ding and L. Qiu, "Blind MIMO channel identification from second order statistics using rank deficient channel convolution matrix," *IEEE Trans. Signal Process.*, vol. 51, no. 2, pp. 535–544, 2003.

[172] H. Bölcskei, R. W. Heath Jr., and A. J. Paulraj, "Blind channel identification and equalization in OFDM-based multiantenna systems," *IEEE Trans. Signal Process.*, vol. 50, no. 1, pp. 96–109, 2002.

[173] A. Chevreuil and P. Loubaton, "MIMO blind second-order equalization method and conjugate cyclostationarity," *IEEE Trans. Signal Process.*, vol. 47, no. 2, pp. 572–578, 1999.

[174] C. Shin, R. W. Heath Jr., and E. J. Powers, "Blind channel estimation for MIMO-OFDM systems," *IEEE Trans. Veh. Technol.*, vol. 56, no. 2, pp. 670–685, 2007.

[175] J. K. Cavers, "An analysis of pilot symbol assisted modulation for Rayleigh fading channels," *IEEE Trans. Veh. Techn.*, vol. 40, pp. 686–693, Nov 1991.

[176] R. Negi and J. Cioffi, "Pilot tone selection for channel estimation in a mobile OFDM system," *IEEE Trans. Consumer Electronics*, vol. 44, no. 3, pp. 1122–1128, 1998.

[177] L. Tong, B. M. Sadler, and M. Dong, "Pilot-assisted wireless transmissions: general model, design criteria, and signal processing," *IEEE Signal Proc. Mag.*, vol. 21, no. 6, pp. 12–25, 2004.

[178] G. H. Golub and C. F. V. Loan, *Matrix computations*, 3rd ed. Johns Hopkins University Press, 1996.

[179] R. Frank and S. Zadoff, "Phase shift pulse codes with good periodic correlation properties," *IRE Trans. Inform. Theory*, vol. 8, no. 6, pp. 381–382, 1962.

[180] D. Chu, "Polyphase codes with good periodic correlation properties," *IEEE Trans. Inform. Theory*, vol. 18, no. 4, pp. 531–532, 1972.

[181] B. Popovic, "Generalized chirp-like polyphase sequences with optimum correlation properties," *IEEE Trans. Inform. Theory*, vol. 38, no. 4, pp. 1406–1409, 1992.

[182] I. Barhumi, G. Leus, and M. Moonen, "Optimal training design for MIMO OFDM systems in mobile wireless channels," *IEEE Trans. Signal Process.*, vol. 51, no. 6, pp. 1615–1624, 2003.

[183] M. Biguesh and A. B. Gershman, "Training-based MIMO channel estimation: a study of estimator tradeoffs and optimal training signals," *IEEE Trans. Signal Process.*, vol. 54, no. 3, pp. 884–893, 2006.

[184] T. Kashima, K. Fukawa, and H. Suzuki, "Adaptive MAP receiver via the EM algorithm and message passings for MIMO-OFDM mobile communications," *IEEE J. Sel. Areas Commun.*, vol. 24, no. 3, pp. 437–447, 2006.

[185] A. H. Sayed, *Fundamentals of adaptive filtering*. Wiley-IEEE Press, 2003.

[186] Y. Li, L. J. Cimini, and N. R. Sollenberger, "Robust channel estimation for OFDM systems with rapid dispersive fading channels," *IEEE Trans. Commun.*, vol. 46, no. 7, pp. 902–915, 1998.

[187] Y. G. Li, J. H. Winters, and N. R. Sollenberger, "MIMO-OFDM for wireless communications: signal detection with enhanced channel estimation," *IEEE Trans. Commun.*, vol. 50, no. 9, pp. 1471–1477, 2002.

[188] Y. Li, N. Seshadri, and S. Ariyavisitakul, "Channel estimation for OFDM systems with transmitter diversity in mobile wireless channels," *IEEE J. Sel. Areas Commun.*, vol. 17, no. 3, pp. 461–471, 1999.

[189] M. Larsen, A. L. Swindlehurst, and T. Svantesson, "A performance bound for interpolation of MIMO-OFDM channels," *Asilomar Conf. Signals, Systems and Computers*, pp. 1801–1805, Oct./Nov. 2006.

[190] W. U. Bajwa, J. Haupt, A. Sayeed, and R. Nowak, "Compressed channel sensing: a new approach to estimating sparse multipath channels," *Proc. IEEE*, vol. 98, no. 6, pp. 1058–1076, 2010.

[191] A. M. Sayeed and V. Raghavan, "Maximizing MIMO capacity in sparse multipath with reconfigurable antenna arrays," *IEEE J. Sel. Topics Signal Process.*, vol. 1, no. 1, pp. 156–166, 2007.

[192] A. Alkhateeb, O. El Ayach, G. Leus, and R. W. Heath Jr., "Channel estimation and hybrid precoding for millimeter wave cellular systems," *IEEE J. Sel. Topics Signal Process.*, vol. 8, no. 5, pp. 831–846, 2014.

[193] R. W. Heath Jr., N. Gonzalez-Prelcic, S. Rangan, W. Roh, and A. M. Sayeed, "An overview of signal processing techniques for millimeter wave MIMO systems," *IEEE J. Sel. Topics Signal Process.*, vol. 10, no. 3, pp. 436–453, 2016.

[194] M. Dohler, R. W. Heath Jr., A. Lozano, C. B. Papadias, and R. A. Valenzuela, "Is the PHY layer dead?" *IEEE Commun. Mag.*, vol. 49, no. 4, pp. 159–165, 2011.

[195] A. Lozano and N. Jindal, "Are yesterday's information-theoretic fading models and performance metrics adequate for the analysis of today's wireless systems?" *IEEE Commun. Mag.*, vol. 50, no. 11, pp. 210–217, 2012.

[196] H. Hashemi, "The indoor radio propagation channel," *Proc. IEEE*, vol. 81, no. 7, pp. 943–968, 1993.

[197] T. S. Rappaport, *Wireless communications: principles and practice*, 2nd ed. Prentice Hall, 2004.

[198] A. F. Molisch, *Wireless communications*. John Wiley & Sons, 2007.

[199] C.-F. Yang, B.-C. Wu, and C.-J. Ko, "A ray-tracing method for modeling indoor wave propagation and penetration," *IEEE Trans. Antennas Propag.*, vol. 46, no. 6, pp. 907–919, 1998.

[200] J. Tarng, W.-S. Liu, Y.-F. Huang, and J.-M. Huang, "A novel and efficient hybrid model of radio multipath-fading channels in indoor environments," *IEEE Trans. Antennas Propag.*, vol. 51, no. 3, pp. 585–594, 2003.

[201] W. Zhang, "Fast two-dimensional diffraction modeling for site-specific propagation prediction in urban microcellular environments," *IEEE Trans. Veh. Techn.*, vol. 49, no. 2, pp. 428–436, 2000.

[202] P. Ferrand, M. Amara, S. Valentin, and M. Guillaud, "Trends and challenges in wireless channel modeling for evolving radio access," *IEEE Commun. Mag.*, vol. 54, no. 7, pp. 93–99, 2016.

[203] M. Franceschetti, J. Bruck, and L. J. Schulman, "A random walk model of wave propagation," *IEEE Trans. Antennas Propag.*, vol. 52, no. 5, pp. 1304–1317, 2004.

[204] D. Chizhik, J. Ling, and R. A. Valenzuela, "Radio wave diffusion indoors and throughput scaling with cell density," *IEEE Trans. Wireless Commun.*, vol. 11, no. 9, pp. 3284–3291, 2012.

[205] G. L. Turin, F. D. Clapp, T. L. Johnston, S. B. Fine, and D. Lavry, "A statistical model of urban multipath propagation," *IEEE Trans. Veh. Techn.*, vol. 21, no. 1, pp. 1–9, 1972.

[206] G. D. Ott and A. Plitkins, "Urban path-loss characteristics at 820 MHz," *IEEE Trans. Veh. Techn.*, vol. 27, no. 4, pp. 189–197, 1978.

[207] D. C. Cox, "Universal digital portable radio communications," *Proc. IEEE*, vol. 75, no. 4, pp. 436–477, 1987.

[208] M. Gudmunson, "Correlation model for shadow fading in mobile radio systems," *Electronics Letters*, vol. 27, no. 23, pp. 2145–2146, 1991.

[209] N. Jaldén, P. Zetterberg, B. Ottersten, A. Hong, and R. Thomä, "Correlation properties of large scale fading based on indoor measurements," *IEEE Wireless Commun. and Networking Conf. (WCNC'07)*, pp. 1894–1899, 2007.

[210] F. Graziosi and F. Santucci, "A general correlation model for shadow fading in mobile radio systems," *IEEE Commun. Letters*, vol. 6, no. 3, pp. 102–104, 2002.

[211] S. S. Szyszkowicz, H. Yanikomeroglu, and J. S. Thompson, "On the feasibility of wireless shadowing correlation models," *IEEE Trans. Veh. Techn.*, vol. 59, no. 9, pp. 4222–4236, 2010.

[212] W. C. Jakes, *Microwave mobile communications*, 2nd ed. Wiley-IEEE Press, 1994.

[213] V. Erceg, L. J. Greenstein, S. Y. Tjandra, S. R. Parkoff, A. Gupta, B. Kulic, A. A. Julius, and R. Bianchi, "An empirically based path loss model for wireless channels in suburban environments," *IEEE J. Sel. Areas Commun.*, vol. 17, no. 7, pp. 1205–1211, 1999.

[214] T. S. Chu and L. J. Greenstein, "A semi-empirical representation of antenna diversity gain at cellular and PCS base stations," *IEEE Trans. Commun.*, vol. 45, no. 6, pp. 644–646, 1997.

[215] L. J. Greenstein, V. Erceg, Y. S. Yeh, and M. V. Clark, "A new path-gain/delay-spread propagation model for digital cellular channels," *IEEE Trans. Veh. Techn.*, vol. 46, no. 2, pp. 477–485, 1997.

[216] H. T. Friis, "A note on a simple transmission formula," *Proc. IRE*, vol. 34, no. 5, pp. 254–256, 1946.
[217] M. Hata, "Empirical formula for propagation loss in land mobile radio services," *IEEE Trans. Veh. Techn.*, vol. 29, no. 3, pp. 317–325, 1980.
[218] "Urban transmission loss models for mobile radio in the 900 and 1800 MHz bands," European Cooperation in the Field of Scientific and Technical Research EURO-COST 231, Tech. Rep. 2, Sep. 1991.
[219] V. Erceg *et al.*, "IEEE 802.16 broadband wireless access working group—IEEE 802.16.3c-01/29r4," Tech. Rep., Jul. 2001.
[220] C. Phillips, D. Sicker, and D. Grunwald, "Bounding the error of path loss models," *IEEE Int'l Symp. Dynamic Spectrum Access Networks (DySPAN'11)*, pp. 71–82, May 2011.
[221] G. Taricco, "On the convergence of multipath fading channel gains to the Rayleigh distribution," *IEEE Wireless Commun. Letters*, vol. 4, no. 5, pp. 549–552, 2015.
[222] S. O. Rice, "Mathematical analysis of random noise," *Bell System Tech. J.*, vol. 23, pp. 282–332, 1944.
[223] M. D. Yacoub, "The κ-μ distribution and the η-μ distribution," *IEEE Antennas Propag. Mag.*, vol. 49, no. 1, pp. 68–81, 2007.
[224] L. M. Correia, *Wireless flexible personalised communications.* John Wiley and Sons, 2001.
[225] R. H. Clarke, "A statistical theory of mobile radio reception," *Bell System Tech. J.*, vol. 47, no. 6, pp. 957–1000, 1968.
[226] W. C. Y. Lee, "Correlation between two mobile radio base-station antennas," *IEEE Trans. Commun.*, vol. 21, no. 11, pp. 1214–1224, Nov. 1973.
[227] J. Salz and J. H. Winters, "Effect of fading correlation on adaptive arrays in digital mobile radio," *IEEE Trans. Veh. Techn.*, vol. 43, no. 4, pp. 1049–1057, 1994.
[228] R. Vaughan, "Spaced directive antennas for mobile communications by the Fourier transform method," *IEEE Trans. Antennas Propag.*, vol. 48, no. 7, pp. 1025–1032, 2000.
[229] F. Adachi, M. Feeny, A. Williamson, and J. Parsons, "Cross-correlation between the envelopes of 900 MHz signals received at a mobile radio base station site," *IEE Proc. Radar and Signal Process.*, vol. 133, pp. 506–512, 1986.
[230] A. Saleh and R. Valenzuela, "A statistical model for indoor multipath propagation," *IEEE J. Sel. Areas Commun.*, vol. 5, no. 2, pp. 128–137, 1987.
[231] Q. Spencer, M. Rice, B. Jeffs, and M. Jensen, "A statistical model for angle of arrival in indoor multipath propagation," *IEEE Veh. Techn. Conf. (VTC'97)*, vol. 3, pp. 1415–1419, May 1997.
[232] Q. H. Spencer, B. D. Jeffs, M. A. Jensen, and A. L. Swindlehurst, "Modeling the statistical time and angle of arrival characteristics of an indoor multipath channel," *IEEE J. Sel. Areas Commun.*, vol. 18, no. 3, pp. 347–360, Mar. 2000.
[233] G. German, Q. Spencer, L. Swindlehust, and R. A. Valenzuela, "Wireless indoor channel modeling: statistical agreement of ray tracing simulations and channel sounding measurements," *IEEE Int'l Conf. Acoustics, Speech and Signal Process. (ICASSP'01)*, vol. 4, pp. 2501–2504, May 2001.

[234] A. S. Y. Poon and M. Ho, "Indoor multiple-antenna channel characterization from 2 to 8 GHz," *IEEE Int'l Conf. Commun. (ICC'03)*, vol. 5, pp. 3519–3523, May 2003.

[235] K. I. Pedersen, P. E. Mogensen, and B. H. Fleury, "Power azimuth spectrum in outdoor environments," *Electronics Letters*, vol. 33, no. 18, pp. 1583–1584, 1997.

[236] ——, "Spatial channel characteristics in outdoor environments and their impact on BS antenna system performance," *IEEE Veh. Techn. Conf. (VTC'98)*, vol. 2, pp. 719–723, 1998.

[237] ——, "A stochastic model of the temporal and azimuthal dispersion seen at the base station in outdoor propagation environments," *IEEE Trans. Veh. Techn.*, vol. 49, no. 2, pp. 437–447, 2000.

[238] L. Schumacher, K. I. Pedersen, and P. Mogensen, "From antenna spacings to theoretical capacities—guidelines for simulating MIMO systems," *IEEE Int'l Symp. Personal, Indoor and Mobile Radio Commun. (PIMRC'02)*, vol. 2, pp. 587–592, Sep. 2002.

[239] D. Chizhik, "Slowing the time-fluctuating MIMO channel by beam forming," *IEEE Trans. Wireless Commun.*, vol. 3, no. 5, pp. 1554–1565, 2004.

[240] V. Va and R. W. Heath Jr., "Basic relationship between channel coherence time and beamwidth in vehicular channels," *IEEE Veh. Techn. Conf. (VTC'15 Fall)*, Sep. 2015.

[241] P. D. Teal, T. D. Abhayapala, and R. A. Kennedy, "Spatial correlation for general distributions of scatterers," *IEEE Signal Process. Letters*, vol. 9, no. 10, pp. 305–308, 2002.

[242] R. K. Cook, R. V. Waterhouse, R. D. Berendt, S. Edelman, and M. C. Thompson, "Measurement of correlation coefficients in reverberant sound fields," *J. Acoust. Soc. Am.*, vol. 27, no. 6, pp. 1072–1077, 1955.

[243] R. Narasimhan and D. C. Cox, "A generalized Doppler power spectrum for wireless environments," *IEEE Commun. Letters*, vol. 3, no. 6, pp. 164–165, 1999.

[244] W. Yamada, K. Nishimori, Y. Takatori, and Y. Asai, "Statistical analysis and characterization of Doppler spectrum in large office environment," *Int'l Symp. Antennas Propag. (ISAP'09)*, pp. 564–567, 2009.

[245] Y. Liang and V. V. Veeravalli, "Capacity of noncoherent time-selective Rayleigh-fading channels," *IEEE Trans. Inform. Theory*, vol. 50, no. 12, pp. 3095–3110, 2004.

[246] V. I. Morgenshtern, E. Riegler, W. Yang, G. Durisi, S. Lin, B. Sturmfels, and H. Bölcskei, "Capacity pre-log of noncoherent SIMO channels via Hironaka's Theorem," *IEEE Trans. Inform. Theory*, vol. 59, pp. 4213–4229, Jul. 2013.

[247] G. Fettweis, "The development of GSM," *IEEE Int'l Conf. Commun. (ICC'13)*, 2013.

[248] G. L. Stüber, *Principles of mobile communication.* Springer, 2011.

[249] "Guidelines for the evaluation of radio-transmission technologies for IMT-2000," Recommendation ITU-R M.1225, Tech. Rep., 1997.

[250] P. Bello, "Characterization of randomly time-variant linear channels," *IEEE Trans. Commun.*, vol. 11, no. 4, pp. 360–393, 1963.

[251] R. S. Kennedy, *Fading dispersive communication channels.* Wiley, 1969.

[252] T. Kailath, "Measurements on time-variant communication channels," *IRE Trans. Inform. Theory*, vol. 8, pp. 229–236, Sep. 1962.

[253] W. Kozek and A. F. Molisch, "Nonorthogonal pulseshapes for multicarrier communications in doubly dispersive channels," *IEEE J. Sel. Areas Commun.*, vol. 16, no. 8, pp. 1579–1589, 1998.

[254] D. H. Johnson and D. E. Dudgeon, *Array signal process.* Prentice Hall, 1993.

[255] M. Stojanovic, "Recent advances in high-speed underwater acoustic communications," *IEEE J. Ocean. Eng.*, vol. 21, no. 2, pp. 125–136, 1996.

[256] B. Li, J. Huang, S. Zhou, K. Ball, M. Stojanovic, L. Freitag, and P. Willett, "MIMO-OFDM for high-rate underwater acoustic communications," *IEEE J. Ocean. Eng.*, vol. 34, no. 4, pp. 634–644, 2009.

[257] R. W. Heath Jr., S. Peters, Y. Wang, and J. Zhang, "A current perspective on distributed antenna systems for the downlink of cellular systems," *IEEE Commun. Mag.*, vol. 51, no. 4, pp. 161–167, 2013.

[258] J.-S. Jiang and M. Ingram, "Spherical-wave model for short-range MIMO," *IEEE Trans. Commun.*, vol. 53, no. 9, pp. 1534–1541, 2005.

[259] S. Wu, C. X. Wang, H. Haas, e.-H. M. Aggoune, M. M. Alwakeel, and B. Ai, "A non-stationary wideband channel model for massive MIMO communication systems," *IEEE Trans. Wireless Commun.*, vol. 14, no. 3, pp. 1434–1446, 2015.

[260] A. Forenza and R. W. Heath Jr., "Impact of antenna geometry on MIMO communication in indoor clustered channels," *IEEE Antennas Propag. Society Int'l Symp.*, vol. 2, pp. 1700–1703, 2004.

[261] S. J. Orfanidis, *Electromagnetic waves and antennas.* S. J. Orfanidis, 2014.

[262] A. Forenza, S. Perlman, F. Saibi, M. Di Dio, R. van der Laan, and G. Caire, "Achieving large multiplexing gain in distributed antenna systems via cooperation with pCell technology," *Asilomar Conf. Signals, Systems and Computers*, pp. 286–293, 2015.

[263] M. Stege, J. Jelitto, M. Bronzel, and G. Fettweis, "A multiple input–multiple output channel model for simulation of Tx- and Rx-diversity wireless systems," *IEEE Veh. Techn. Conf. (VTC'00)*, vol. 2, pp. 833–839, 2000.

[264] K. I. Pedersen, J. B. Andersen, J. P. Kermoal, and P. Mogensen, "A stochastic multiple-input–multiple-output radio channel model for evaluation of space–time coding algorithms," *IEEE Veh. Techn. Conf. (VTC'00)*, vol. 2, pp. 893–897, 2000.

[265] T. Svantesson, "A physical MIMO radio channel model for multi-element multi-polarized antenna systems," *IEEE Veh. Techn. Conf. (VTC'01)*, vol. 2, pp. 1083–1087, 2001.

[266] K. Yu and B. Ottersten, "Models for MIMO propagation channels: a review," *Wiley J. Wireless Commun. Mobile Comput.*, vol. 2, no. 7, pp. 653–666, Nov. 2002.

[267] A. Taparugssanagorn, T. Jasma, and J. Ylitalo, "Spatial correlation and eigenvalue statistics investigation of wideband MIMO channel measurements," *Int'l Symp. Personal, Indoor and Mobile Radio Commun. (PIMRC'06)*, pp. 1–5, 2006.

[268] V. Erceg, H. Sampath, and S. Catreux-Erceg, "An experimental investigation of wideband MIMO channel characteristics based on outdoor non-LOS measurements at 1.8 GHz," *IEEE Trans. Wireless Commun.*, vol. 5, no. 1, pp. 24–33, 2006.

[269] P. Almers, E. Bonek, A. Burr, N. Czink, M. Debbah, V. Degli-Esposti, H. Hofstetter, P. Kyosti, D. Laurenson, G. Matz, A. Molisch, C. Oestges, and H. Ozcelik, "Survey of channel and radio propagation models for wireless MIMO systems," *EURASIP J. Wireless Commun. Netw.*, vol. 2007, 2007.

[270] P. F. Driessen and G. J. Foschini, "On the capacity formula for multiple input-multiple output wireless channels: a geometric interpretation," *IEEE Trans. Commun.*, vol. 47, no. 2, pp. 173–176, 1999.

[271] F. Bohagen, P. Orten, and G. E. Oien, "Design of optimal high-rank line-of-sight MIMO channels," *IEEE Trans. Wireless Commun.*, vol. 6, no. 4, pp. 1420–1425, 2007.

[272] C.-N. Chuah, J. Kahn, and D. Tse, "Capacity of multi-antenna array systems in indoor wireless environment," *IEEE Global Commun. Conf. (GLOBECOM'98)*, vol. 4, pp. 1894–1899, 1998.

[273] D. Chizhik, F. Rashid-Farrokhi, J. Ling, and A. Lozano, "Effect of antenna separation on the capacity of BLAST in correlated channels," *IEEE Commun. Letters*, vol. 4, no. 11, pp. 337–339, 2000.

[274] D. S. Shiu, G. J. Foschini, M. J. Gans, and J. M. Kahn, "Fading correlation and its effect on the capacity of multielement antenna systems," *IEEE Trans. Commun.*, vol. 48, no. 3, pp. 502–513, 2000.

[275] A. Forenza and R. W. Heath Jr., "Benefit of pattern diversity via 2-element array of circular patch antennas in indoor clustered MIMO channels," *IEEE Trans. Commun.*, vol. 54, pp. 943–954, 2006.

[276] J. Brewer, "Kronecker products and matrix calculus in system theory," *IEEE Trans. Circuits and Syst.*, vol. 25, no. 9, pp. 772–781, 1978.

[277] E. Bonek, "Experimental validation of analytical MIMO channel models," *Elektrotechnik und Informationstechnik*, pp. 196–205, 2005.

[278] M. Bengtsson and B. Ottersten, "Low-complexity estimators for distributed sources," *IEEE Trans. Signal Process.*, vol. 48, no. 8, pp. 2185–2194, 2000.

[279] A. Forenza, D. J. Love, and R. W. Heath Jr., "Simplified spatial correlation models for clustered MIMO channels with different array configurations," *IEEE Trans. Veh. Techn.*, vol. 56, no. 4, pp. 1924–1934, 2007.

[280] S. Loyka, "Channel capacity of MIMO architecture using the exponential correlation model," *IEEE Commun. Letters*, vol. 5, no. 9, pp. 369–371, 2001.

[281] H. Lim, Y. Jang, and D. Yoon, "Bounds for eigenvalues of spatial correlation matrices with the exponential model in MIMO systems," *IEEE Trans. Wireless Commun.*, vol. 16, no. 2, pp. 1196–1204, 2017.

[282] J. Kermoal, L. Schumacher, K. Pedersen, P. Mogensen, and F. Frederiksen, "A stochastic MIMO radio channel model with experimental validation," *IEEE J. Sel. Areas Commun.*, vol. 20, no. 6, pp. 1211–1226, 2002.

[283] K. Yu, M. Bengtsson, B. Ottersten, D. McNamara, P. Karlsson, and M. Beach, "A

wideband statistical model for NLOS indoor MIMO channels," *IEEE Veh. Techn. Conf. (VTC'02)*, vol. 1, pp. 370–374, 2002.

[284] ——, "Modeling of wide-band MIMO radio channels based on NLOS indoor measurements," *IEEE Trans. Veh. Techn.*, vol. 53, no. 3, pp. 655–665, 2004.

[285] D. Chizhik, J. Ling, P. W. Wolniansky, R. A. Valenzuela, N. Costa, and K. Huber, "Multiple-input–multiple-output measurements and modeling in Manhattan," *IEEE J. Sel. Areas Commun.*, vol. 21, no. 3, pp. 321–331, 2003.

[286] A. Abdi and M. Kaveh, "A space–time correlation model for multielement antenna systems in mobile fading channels," *IEEE J. Sel. Areas Commun.*, vol. 20, no. 3, pp. 550–560, 2002.

[287] H. Ozcelik, M. Herdin, W. Weichselberger, J. Wallace, and E. Bonek, "Deficiencies of 'Kronecker' MIMO radio channel model," *Electronics Letters*, vol. 39, no. 16, pp. 1209–1210, 2003.

[288] C. Oestges and A. Paulraj, "Beneficial impact of channel correlations on MIMO capacity," *Electronics Letters*, vol. 40, no. 10, pp. 606–608, 2004.

[289] C. Oestges, "Validity of the Kronecker model for MIMO correlated channels," *IEEE Veh. Techn. Conf. (VTC'06 Spring)*, vol. 6, pp. 2818–2822, 2006.

[290] V. Raghavan, J. H. Kotecha, and A. M. Sayeed, "Why does the Kronecker model result in misleading capacity estimates?" *IEEE Trans. Inform. Theory*, vol. 56, no. 10, pp. 4843–4864, 2010.

[291] A. M. Tulino, A. Lozano, and S. Verdú, "Capacity-achieving input covariance for single-user multi-antenna channels," *IEEE Trans. Wireless Commun.*, vol. 5, no. 3, pp. 662–671, 2006.

[292] J. H. Kotecha and A. M. Sayeed, "Optimal signal design for estimation of correlated MIMO channels," *IEEE Int'l Conf. Commun. (ICC'03)*, vol. 5, pp. 3170–3174, 2003.

[293] W. Weichselberger, M. Herdin, H. Ozcelik, and E. Bonek, "A stochastic MIMO channel model with joint correlation of both link ends," *IEEE Trans. Wireless Commun.*, vol. 5, no. 1, pp. 90–100, 2006.

[294] R. Vaughan, "Switched parasitic elements for antenna diversity," *IEEE Trans. Antennas Propag.*, vol. 47, pp. 399–405, 1999.

[295] P. Mattheijssen, M. H. A. J. Herben, G. Dolmans, and L. Leyten, "Antenna-pattern diversity versus space diversity for use at handhelds," *IEEE Trans. Veh. Techn.*, vol. 53, pp. 1035–1042, 2004.

[296] L. Dong, H. Ling, and R. W. Heath Jr., "Multiple-input multiple-output wireless communication systems using antenna pattern diversity," *IEEE Global Commun. Conf. (GLOBECOM'02)*, vol. 1, pp. 997–1001, Nov. 2002.

[297] C. B. Dietrich Jr., K. Dietze, J. R. Nealy, and W. L. Stutzman, "Spatial, polarization, and pattern diversity for wireless handheld terminals," *IEEE Antennas Prop. Symp.*, vol. 49, pp. 1271–1281, Sep. 2001.

[298] R. G. Vaughan, "Polarization diversity in mobile communications," *IEEE Trans. Veh. Techn.*, vol. 39, no. 3, pp. 177–186, 1990.

[299] J. Hamalainen, R. Wichman, J. P. Nuutinen, J. Ylitalo, and T. Jamsa, "Analysis and measurements for indoor polarization MIMO in 5.25 GHz band," *IEEE Vehic. Techn. Conf. (VTC'05 Spring)*, vol. 1, pp. 252–256, 2005.

[300] M. Shafi, M. Zhang, A. L. Moustakas, P. J. Smith, A. F. Molisch, F. Tufvesson, and S. H. Simon, "Polarized MIMO channels in 3-D: models, measurements and mutual information," *IEEE J. Sel. Areas Commun.*, vol. 24, no. 3, pp. 514–527, 2006.

[301] V. R. Anreddy and M. A. Ingram, "Capacity of measured Ricean and Rayleigh indoor MIMO channels at 2.4 GHz with polarization and spatial diversity," *IEEE Wireless Commun. and Networking Conf. (WCNC'06)*, vol. 2, pp. 946–951, 2006.

[302] V. Erceg, H. Sampath, and S. Catreux-Erceg, "Dual-polarization versus single-polarization MIMO channel measurement results and modeling," *IEEE Trans. Wireless Commun.*, vol. 5, no. 1, pp. 24–33, 2006.

[303] R. U. Nabar, H. Bölcskei, V. Erceg, D. Gesbert, and A. J. Paulraj, "Performance of spatial-multiplexing in the presence of polarization diversity," *IEEE Int'l Conf. Acoustics, Speech and Signal Proc. (ICASSP'01)*, vol. 4, pp. 2437–2440, May 2001.

[304] C. C. Martin, J. H. Winters, and N. R. Sollenberger, "MIMO radio channel measurements: performance comparison of antenna configurations," *IEEE Veh. Techn. Conf. (VTC'01)*, vol. 2, pp. 1225–1229, Oct. 2001.

[305] H. J. Li and C. H. Yu, "Correlation properties and capacity of antenna polarization combinations for MIMO radio channel," *IEEE Antennas Prop. Symp.*, vol. 2, pp. 503–506, Jun. 2003.

[306] V. Jungnickel, V. Pohl, H. Nguyen, U. Kruger, T. Haustein, and C. V. Helmolt, "High capacity antennas for MIMO radio systems," *Int'l Symp. Wireless Personal Commun.*, vol. 2, pp. 407–411, Oct. 2002.

[307] A. M. Sayeed, "Deconstructing multiantenna fading channels," *IEEE Trans. Signal Process.*, vol. 50, no. 10, pp. 2563–2579, 2002.

[308] D. Chizhik, G. J. Foschini, and R. A. Valenzuela, "Capacities of multi-element transmit and receive antennas: correlations and keyholes," *Electronics Letters*, vol. 36, no. 13, pp. 1099–1100, 2000.

[309] D. Chizhik, G. J. Foschini, M. J. Gans, and R. A. Valenzuela, "Keyholes, correlations, and capacities of multielement transmit and receive antennas," *IEEE Trans. Wireless Commun.*, vol. 1, no. 2, pp. 361–368, 2002.

[310] D. Gesbert, H. Bölcskei, D. A. Gore, and A. J. Paulraj, "Outdoor MIMO wireless channels: models and performance prediction," *IEEE Trans. Commun.*, vol. 50, no. 12, pp. 1926–1934, 2002.

[311] S. Loyka and A. Kouki, "On MIMO channel capacity, correlations, and keyholes: analysis of degenerate channels," *IEEE Trans. Commun.*, vol. 50, no. 12, pp. 1886–1888, 2002.

[312] V. Erceg, S. J. Fortune, J. Ling, A. Rustako Jr., and R. A. Valenzuela, "Comparisons of a computer-based propagation prediction tool with experimental data collected in urban microcellular environments," *IEEE J. Sel. Areas Commun.*, vol. 15, no. 4, pp. 677–684, 1997.

[313] P. Almers, F. Tufvesson, and A. F. Molisch, "Measurement of keyhole effect in a wireless multiple-input multiple-output (MIMO) channel," *IEEE Commun. Letters*, vol. 7, no. 8, pp. 373–375, 2003.

[314] ——, "Keyhole effect in MIMO wireless channels: measurements and theory," *IEEE Trans. Wireless Commun.*, vol. 5, no. 12, pp. 3596–3604, 2006.

[315] P. Petrus, J. H. Reed, and T. S. Rappaport, "Geometrical-based statistical macrocell channel model for mobile environments," *IEEE Trans. Commun.*, vol. 50, no. 3, pp. 495–502, 2002.

[316] K. V. Mardia and P. E. Jupp, *Directional statistics*. Wiley, 1999.

[317] G. J. Byers and F. Takawira, "The influence of spatial and temporal correlation on the capacity of MIMO channels," *IEEE Wireless Commun. and Networking (WCNC'03)*, pp. 359–364, 2003.

[318] J. W. Wallace and M. A. Jensen, "Statistical characteristics of measured MIMO wireless channel data and comparison to conventional models," *IEEE Veh. Techn. Conf. (VTC'01)*, vol. 2, pp. 1078–1082, 2001.

[319] K.-H. Li, M. Ingram, and A. Van Nguyen, "Impact of clustering in statistical indoor propagation models on link capacity," *IEEE Trans. Commun.*, vol. 50, no. 4, pp. 521–523, 2002.

[320] E. Biglieri, E. Grossi, and M. Lops, "Random-set theory and wireless communications," *Found. Trends Commun. Inform. Theory*, vol. 7, no. 4, pp. 317–462, 2012.

[321] S. Ferson, V. Kreinovich, L. Ginzburg, D. Myers, and K. Sentz, *Constructing probability boxes and Dempster–Shafer structures*. Sandia National Laboratories Report SAND 20024015, 2002.

[322] E. Biglieri, "Dealing with uncertain models in wireless communications," *IEEE Int'l Conf. Acoustics, Speech and Signal Process. (ICASSP'16)*, 2016.

[323] M. Mohammadkarimi, E. Karami, O. A. Dobre, and M. Z. Win, "Doppler spread estimation in MIMO frequency-selective fading channels," *IEEE Trans. Wireless Commun.*, vol. 17, no. 3, pp. 1951–1965, 2018.

[324] A. Lozano, "Interplay of spectral efficiency, power and Doppler spectrum for reference-signal-assisted wireless communication," *IEEE Trans. Commun.*, vol. 56, no. 12, pp. 5020–5029, 2008.

[325] A. F. Molisch, H. Asplund, R. Heddergot, M. Steinbauer, and T. Zwick, "The COST 259 directional channel model—part I: overview and methodology," *IEEE Trans. Wireless Commun.*, vol. 5, no. 12, pp. 3421–3433, 2006.

[326] H. Asplund, A. A. Glazunov, A. F. Molisch, K. Pedersen, and M. Steinbauer, "The COST 259 directional channel model—part II: Macrocells," *IEEE Trans. Wireless Commun.*, vol. 5, no. 12, pp. 3434–3450, 2006.

[327] L. Correia, *COST 273 final report*. Springer, 2006.

[328] D. S. Baum, "Final report on link level and system level channel models," Document IST-2003-507581, Tech. Rep., Nov. 2006.

[329] P. Kyösti, "WINNER II channel models," Document IST-4-027756, Tech. Rep., Apr. 2008.

[330] 3GPP Technical Specification Group, "Spatial channel model, SCM-134 text V6.0," Spatial Channel Model AHG (Combined ad-hoc from 3GPP and 3GPP2), Tech. Rep., Apr. 2003.

[331] C. Oestges, V. Erceg, and A. J. Paulraj, "A physical scattering model for MIMO macrocellular broadband wireless channels," *IEEE J. Sel. Areas Commun.*, vol. 21, no. 5, pp. 721–729, 2003.

[332] M. Lu, T. Lo, and J. Litva, "A physical spatio-temporal model of multipath propagation channels," *IEEE Veh. Techn. Conf. (VTC'97)*, vol. 2, pp. 810–814, 1997.

[333] O. Norklit and J. B. Andersen, "Diffuse channel model and experimental results for array antennas in mobile environments," *IEEE Trans. Antennas Propag.*, vol. 46, no. 6, pp. 834–840, 1998.

[334] V. Erceg *et al.*, "TGn channel models IEEE 802.11-03/940r4," Tech. Rep., May 2004.

[335] T. S. Rappaport, S. Shu, R. Mayzus, Z. Hang, Y. Azar, K. Wang, G. N. Wong, J. K. Schulz, M. Samimi, and F. Gutierrez, "Millimeter wave mobile communications for 5G cellular: it will work!" *IEEE Access*, vol. 1, pp. 335–349, 2013.

[336] X. Cheng, B. Yu, L. Yang, J. Zhang, G. Liu, Y. Wu, and L. Wan, "Communicating in the real world: 3D MIMO," *IEEE Commun. Mag.*, vol. 52, no. 8, pp. 136–144, 2014.

[337] J. Zhang, C. Pan, F. Pei, G. Liu, and X. Cheng, "Three-dimensional fading channel models: a survey of elevation angle research," *IEEE Commun. Mag.*, vol. 52, no. 6, pp. 218–226, 2014.

[338] R. N. Almesaeed, A. S. Ameen, E. Mellios, A. Doufexi, and A. Nix, "3D channel models: principles, characteristics, and system implications," *IEEE Commun. Mag.*, vol. 55, no. 4, pp. 152–159, 2017.

[339] J. Zhang, Y. Zhang, Y. Yu, R. Xu, Q. Zheng, and P. Zhang, "3-D MIMO: how much does it meet our expectations observed from channel measurements?" *IEEE J. Sel. Areas Commun.*, vol. 35, no. 8, pp. 1887–1903, 2017.

[340] B. Mondal, T. Thomas, E. Visotsky, F. Vook, A. Ghosh, Y.-H. Nam, Y. Li, J. Zhang, M. Zhang, Q. Luo, Y. Kakishima, and K. Kitao, "3D channel model in 3GPP," *IEEE Commun. Mag.*, vol. 53, no. 3, pp. 16–23, 2015.

[341] J. Meinila, P. Kyösti, L. Hentila, T. Jamsa, E. K. E. Suikkanen, and M. Narandzia, "WINNER+ final channel models," Document CELTIC/CP5-026 D5.3, Tech. Rep., Jun. 2010.

[342] L. Thiele, T. Wirth, K. Brner, M. Olbrich, V. Jungnickel, J. Rumold, and S. Fritze, "Modeling of 3D field patterns of downtilted antennas and their impact on cellular systems," *Int'l ITG Workshop on Smart Antennas*, Feb. 2009.

[343] X. Li, R. W. Heath Jr., K. Linehan, and R. Butler, "Metrocell antennas: the positive impact of a narrow vertical beamwidth and electrical downtilt," *IEEE Veh. Techn. Mag.*, vol. 10, no. 3, pp. 51–59, 2015.

[344] X. Lin, J. Andrews, A. Ghosh, and R. Ratasuk, "An overview of 3GPP device-to-device proximity services," *IEEE Commun. Mag.*, vol. 52, pp. 40–48, Apr. 2014.

[345] G. George, R. K. Mungara, and A. Lozano, "An analytical framework for device-to-device communication in cellular networks," *IEEE Trans. Wireless Commun.*, vol. 14, pp. 6297–6310, Nov. 2015.

[346] K. E. Stocker, B. E. Gschwendtner, and F. M. Landstorfer, "Neural network approach to prediction of terrestrial wave propagation for mobile radio," *IEE Proc. H (Microwaves, Antennas and Prop.)*, vol. 140, no. 4, pp. 315–320, 1993.

[347] E. Ostlin, H.-J. Zepernick, and H. Suzuki, "Macrocell path-loss prediction using artificial neural networks," *IEEE Trans. Veh. Techn.*, vol. 59, no. 6, pp. 2735–2747, 2010.

[348] E. Dall'Anese, S. J. Kim, and G. B. Giannakis, "Channel gain map tracking via distributed Kriging," *IEEE Trans. Veh. Techn.*, vol. 60, no. 3, pp. 1205–1211, 2011.

[349] L. S. Muppirisetty, T. Svensson, and H. Wymeersch, "Spatial wireless channel prediction under location uncertainty," *IEEE Trans. Wireless Commun.*, vol. 15, no. 2, pp. 1031–1044, 2016.

[350] J. Chen, U. Yatnalli, and D. Gesbert, "Learning radio maps for UAV-aided wireless networks: a segmented regression approach," *IEEE Int'l Conf. Commun. (ICC'17)*, pp. 1–6, May 2017.

[351] M. Kasparick, R. L. G. Cavalcante, S. Valentin, S. Stanczak, and M. Yukawa, "Kernel-based adaptive online reconstruction of coverage maps with side information," *IEEE Trans. Veh. Technol.*, vol. 65, no. 7, pp. 5461–5473, 2016.

[352] D. Romero, S.-J. Kim, G. B. Giannakis, and R. Lopez-Valcarce, "Learning power spectrum maps from quantized power measurements," *IEEE Trans. Signal Process.*, vol. 65, no. 10, pp. 2547–2560, 2017.

[353] R. Nikbakht, A. Jonsson, and A. Lozano, "Dual-kernel online reconstruction of power maps," *IEEE Global Commun. Conf. (GLOBECOM'18)*, 2018.

[354] S. Verdú, "On channel capacity per unit cost," *IEEE Trans. Inform. Theory*, vol. 36, no. 5, pp. 1019–1030, 1990.

[355] A. Lozano, A. Tulino, and S. Verdú, "High-SNR power offset in multiantenna communication," *IEEE Trans. Inform. Theory*, vol. 51, no. 12, pp. 4134–4151, 2005.

[356] J. L. Massey, "All signal sets centered about the origin are optimal at low energy-to-noise ratios on the AWGN channel," *IEEE Int'l Symp. Inform. Theory (ISIT'76)*, pp. 80–81, 1976.

[357] G. D. Forney Jr. and L. F. Wei, "Multidimensional constellations. I. Introduction, figures of merit, and generalized cross constellations," *IEEE J. Sel. Areas Commun.*, vol. 7, no. 6, pp. 877–892, 1989.

[358] A. R. Calderbank and L. H. Ozarow, "Nonequiprobable signaling on the Gaussian channel," *IEEE Trans. Inform. Theory*, vol. 36, no. 4, pp. 726–740, 1990.

[359] D. S. Shiu and J. M. Kahn, "Shaping and nonequiprobable signaling for intensity-modulated signals," *IEEE Trans. Inform. Theory*, vol. 45, no. 7, pp. 2661–2668, 1999.

[360] M. Abdelaziz and T. A. Gulliver, "Triangular constellations for adaptive modulation," *IEEE Trans. Commun.*, vol. 66, no. 2, pp. 756–766, 2018.

[361] P. Wu and N. Jindal, "Coding versus ARQ in fading channels: how reliable should the PHY be?" *IEEE Trans. Commun.*, vol. 59, no. 12, pp. 3363–3374, 2011.

[362] "Coded modulation library," http://www.iterativesolutions.com/Matlab.htm.

[363] W. Hirt and J. L. Massey, "Capacity of the discrete-time Gaussian channel with intersymbol interference," *IEEE Trans. Inform. Theory*, vol. 34, pp. 380–388, May 1988.

[364] S. Kasturia, J. T. Aslanis, and J. M. Cioffi, "Vector coding for partial response channels," *IEEE Trans. Inform. Theory*, vol. 36, no. 4, pp. 741–762, 1990.

[365] J. L. Holsinger, "Digital communications over fixed time-continuous channels with memory, with special application to telephone channel," *MIT Res. Lab. Electron. Rep.*, vol. 430, p. 460, 1964.

[366] B. S. Tsybakov, "Capacity of a discrete-time Gaussian channel with a filter," *Prob. Inform. Transm.*, vol. 6, pp. 253–256, Jul.–Sep. 1970.

[367] R. Price, "A conversation with Claude Shannon," *IEEE Commun. Mag.*, vol. 22, pp. 123–126, May 1984.

[368] J. M. Cioffi, G. P. Dudevoir, M. V. Eyuboglu, and J. G. D. Forney, "MMSE decision-feedback equalizers and coding. I. equalization results," *IEEE Trans. Commun.*, vol. 43, no. 10, pp. 2582–2594, 1995.

[369] A. Lozano, A. Tulino, and S. Verdú, "Mercury/waterfilling: optimum power allocation with arbitrary input constellations," *Int'l Symp. Inform. Theory (ISIT'05)*, pp. 1773–1777, 2005.

[370] G. D. Forney and G. Ungerboeck, "Modulation and coding for linear Gaussian channels," *IEEE Trans. Inform. Theory*, vol. 44, no. 6, pp. 2384–2415, 1998.

[371] P. S. Chow, J. M. Cioffi, and J. A. C. Bingham, "A practical discrete multitone transceiver loading algorithm for data transmission over spectrally shaped channels," *IEEE Trans. Commun.*, vol. 43, no. 2–4, pp. 773–775, 1995.

[372] P. S. Chow, "Bandwidth optimized digital transmission techniques for spectrally shaped channels with impulse noise," Ph.D. dissertation, Stanford University, May 1993.

[373] D. Hughes-Hartogs, "Ensemble modem structure for imperfect transmission media," *U.S. Patent 4 679 227*, Jul. 1987.

[374] S. Shamai and R. Laroia, "The intersymbol interference channel: lower bounds on capacity and channel precoding loss," *IEEE Trans. Inform. Theory*, vol. 42, no. 5, pp. 1388–1404, 1996.

[375] Y. Carmon, S. Shamai, and T. Weissman, "Comparison of the achievable rates in OFDM and single carrier modulation with I.I.D. inputs," *IEEE Trans. Inform. Theory*, vol. 61, no. 4, pp. 1795–1818, 2015.

[376] Y. Polyanskiy, H. V. Poor, and S. Verdú, "Dispersion of Gaussian channels," *IEEE Int'l Symp. Inform. Theory (ISIT'09)*, pp. 2204–2208, 2009.

[377] K. Brueninghaus, D. Astely, T. Salzer, S. Visuri, A. Alexiou, S. Karger, and G.-A. Seraji, "Link performance models for system level simulations of broadband radio access systems," *IEEE Int'l Symp. Personal, Indoor and Mobile Radio Commun. (PIMRC'05)*, vol. 4, pp. 2306–2311, 2005.

[378] S. S. Tsai and A. Soong, "Effective-SNR mapping for modeling frame error rates in multiple-state channels," 3GPP2-C30-20030429-010, Tech. Rep., 2003.

[379] Ericsson, "System-level evaluation of OFDM—further considerations," TSG-RAN WG1 meeting 35, Tech. Rep., Nov. 2003.

[380] Y. Blankenship, P. Sartori, B. Classon, V. Desai, and K. Baum, "Link error prediction methods for multicarrier systems," *IEEE Veh. Techn. Conf. (VTC'04 Fall)*, vol. 6, pp. 4175–4179, 2004.

[381] S. Schwarz, "Limited feedback transceiver design for downlink MIMO OFDM cellular networks," Ph.D. dissertation, Vienna University of Technology, 2013.

[382] A. M. Cipriano, R. Visoz, and T. Sälzer, "Calibration issues of PHY layer abstractions for wireless broadband systems," *IEEE Veh. Techn. Conf. (VTC'08 Fall)*, pp. 1–5, 2008.

[383] R. C. Daniels and R. W. Heath Jr., "Modeling ordered subcarrier SNR in MIMO-OFDM wireless links," *Physical Commun.*, vol. 4, no. 4, pp. 275–285, 2011.

[384] R. C. Daniels, C. M. Caramanis, and R. W. Heath Jr., "Adaptation in convolutionally coded MIMO-OFDM wireless systems through supervised learning and SNR ordering," *IEEE Trans. Veh. Techn.*, vol. 59, no. 1, pp. 114–126, 2010.

[385] L. H. Ozarow, S. Shamai, and A. D. Wyner, "Information theoretic considerations for cellular mobile radio," *IEEE Trans. Veh. Techn.*, vol. 43, no. 2, pp. 359–378, 1994.

[386] E. Biglieri, J. Proakis, and S. Shamai, "Fading channels: information-theoretic and communication aspects," *IEEE Trans. Inform. Theory*, vol. 44, no. 6, pp. 2619–2692, 1998.

[387] F. Hlawatsch and G. Matz, *Wireless communications over rapidly time-varying channels*. Academic Press, 2011.

[388] J. Cavers, "Variable-rate transmission for Rayleigh fading channels," *IEEE Trans. Commun.*, vol. 20, no. 1, pp. 15–22, 1972.

[389] G. Caire, G. Taricco, and E. Biglieri, "Optimum power control over fading channels," *IEEE Trans. Inform. Theory*, vol. 45, no. 5, pp. 1468–1489, 1999.

[390] R. Negi and J. M. Cioffi, "Delay-constrained capacity with causal feedback," *IEEE Trans. Inform. Theory*, vol. 48, no. 9, pp. 2478–2494, 2002.

[391] M. S. Alouini and A. J. Goldsmith, "Capacity of Rayleigh fading channels under different adaptive transmission and diversity-combining techniques," *IEEE Trans. Veh. Techn.*, vol. 48, no. 4, pp. 1165–1181, 1999.

[392] S. Borade and L. Zheng, "Wideband fading channels with feedback," *IEEE Trans. Inform. Theory*, vol. 56, no. 12, pp. 6058–6065, 2010.

[393] M. Khoshnevisan and J. N. Laneman, "Power allocation in multi-antenna wireless systems subject to simultaneous power constraints," *IEEE Trans. Commun.*, vol. 60, no. 12, pp. 3855–3864, 2012.

[394] Z. Wang and G. B. Giannakis, "A simple and general parameterization quantifying performance in fading channels," *IEEE Trans. Commun.*, vol. 51, no. 8, pp. 1389–1398, 2003.

[395] L. Zheng and D. N. C. Tse, "Diversity and multiplexing: a fundamental tradeoff in multiple-antenna channels," *IEEE Trans. Inform. Theory*, vol. 49, no. 5, pp. 1073–1096, 2003.

[396] W. Y. Shin, S. Y. Chung, and Y. H. Lee, "Diversity-multiplexing tradeoff and outage performance for Rician MIMO channels," *IEEE Trans. Inform. Theory*, vol. 54, no. 3, pp. 1186–1196, 2008.

[397] L. Zhao, W. Mo, Y. Ma, and Z. Wang, "Diversity and multiplexing tradeoff in general fading channels," *IEEE Trans. Inform. Theory*, vol. 53, no. 4, pp. 1549–1557, 2007.

[398] A. Lozano and N. Jindal, "Transmit diversity vs. spatial multiplexing in modern MIMO systems," *IEEE Trans. Wireless Commun.*, vol. 9, no. 1, pp. 186–197, 2010.

[399] R. Narasimhan, "Finite-SNR diversity-multiplexing tradeoff for correlated Rayleigh and Rician MIMO channels," *IEEE Trans. Inform. Theory*, vol. 52, no. 9, pp. 3956–3979, 2006.

[400] K. Azarian and H. El Gamal, "The throughput–reliability tradeoff in block-fading MIMO channels," *IEEE Trans. Inform. Theory*, vol. 53, no. 2, pp. 488–501, 2007.

[401] W. C. Y. Lee, "Estimate of channel capacity in Rayleigh fading environment," *IEEE Trans. Veh. Techn.*, vol. 39, no. 3, pp. 187–189, 1990.

[402] M. S. Alouini and A. Goldsmith, "Capacity of Nakagami multipath fading channels," *IEEE Veh. Techn. Conf. (VTC'97)*, vol. 1, pp. 358–362, May 1997.

[403] T. F. Wong, "Numerical calculation of symmetric capacity of Rayleigh fading channel with BPSK/QPSK," *IEEE Commun. Letters*, vol. 5, no. 8, pp. 328–330, 2001.

[404] A. Lapidoth and S. M. Moser, "Capacity bounds via duality with applications to multiple-antenna systems on flat-fading channels," *IEEE Trans. Inform. Theory*, vol. 49, no. 10, pp. 2426–2467, 2003.

[405] J. S. Richters, "Communication over fading dispersive channels," *Tech. Rep. 464, MIT Research Laboratory Electronics*, Nov. 1967.

[406] S. de la Kethulle de Ryhove, N. Marina, and G. E. Oien, "On the mutual information and low-SNR capacity of memoryless noncoherent Rayleigh-fading channels," *IEEE Trans. Inform. Theory*, vol. 54, no. 7, pp. 3221–3231, 2008.

[407] L. Zheng, D. N. C. Tse, and M. Médard, "Channel coherence in the low-SNR regime," *IEEE Trans. Inform. Theory*, vol. 53, no. 3, pp. 976–997, 2007.

[408] G. Taricco and M. Elia, "Capacity of fading channel with no side information," *Electronics Letters*, vol. 33, no. 16, pp. 1368–1370, 1997.

[409] T. L. Marzetta and B. M. Hochwald, "Capacity of a mobile multiple-antenna communication link in Rayleigh flat fading," *IEEE Trans. Inform. Theory*, vol. 45, no. 1, pp. 139–157, 1999.

[410] I. Bergel and S. Benedetto, "Bounds on the capacity of OFDM underspread frequency selective fading channels," *IEEE Trans. Inform. Theory*, vol. 58, no. 10, pp. 6446–6470, 2012.

[411] M. Médard and R. G. Gallager, "Bandwidth scaling for fading multipath channels," *IEEE Trans. Inform. Theory*, vol. 48, no. 4, pp. 840–852, 2002.

[412] C. Rao and B. Hassibi, "Analysis of multiple-antenna wireless links at low SNR," *IEEE Trans. Inform. Theory*, vol. 50, no. 9, pp. 2123–2130, 2004.

[413] S. G. Srinivasan and M. K. Varanasi, "Optimal constellations for the low-SNR noncoherent MIMO block Rayleigh-fading channel," *IEEE Trans. Inform. Theory*, vol. 55, no. 2, pp. 776–796, 2009.

[414] V. Sethuraman, L. Wang, B. Hajek, and A. Lapidoth, "Low-SNR capacity of noncoherent fading channels," *IEEE Trans. Inform. Theory*, vol. 55, no. 4, pp. 1555–1574, 2009.

[415] B. M. Hochwald and T. L. Marzetta, "Unitary space–time modulation for multiple-antenna communications in Rayleigh flat fading," *IEEE Trans. Inform. Theory*, vol. 46, no. 2, pp. 543–564, 2000.

[416] L. Zheng and D. N. C. Tse, "Communication on the Grassmann manifold: a geometric approach to the noncoherent multiple-antenna channel," *IEEE Trans. Inform. Theory*, vol. 48, no. 2, pp. 359–383, 2002.

[417] A. Lapidoth, "On the asymptotic capacity of stationary Gaussian fading channels," *IEEE Trans. Inform. Theory*, vol. 51, no. 2, pp. 437–446, 2005.

[418] G. Durisi, V. I. Morgenshtern, and H. Bölcskei, "On the sensitivity of continuous-time noncoherent fading channel capacity," *IEEE Trans. Inform. Theory*, vol. 58, no. 10, pp. 6372–6391, 2012.

[419] W. Zhang and J. N. Laneman, "How good is PSK for peak-limited fading channels in the low-SNR regime?" *IEEE Trans. Inform. Theory*, vol. 53, no. 1, pp. 236–251, 2007.

[420] F. Rusek, A. Lozano, and N. Jindal, "Mutual information of IID complex gaussian signals on block Rayleigh-faded channels," *IEEE Trans. Inform. Theory*, vol. 58, no. 1, pp. 331–340, 2012.

[421] M. Dörpinghaus, H. Meyr, and R. Mathar, "On the achievable rate of stationary Rayleigh flat-fading channels with Gaussian inputs," *IEEE Trans. Inform. Theory*, vol. 59, no. 4, pp. 2208–2220, 2013.

[422] G. Caire, "On the ergodic rate lower bounds with applications to massive MIMO," *IEEE Trans. Wireless Commun.*, vol. 17, no. 5, pp. 3258–3268, 2018.

[423] S. N. Diggavi, "On achievable performance of spatial diversity fading channels," *IEEE Trans. Inform. Theory*, vol. 47, no. 1, pp. 308–325, 2001.

[424] H. Bölcskei, D. Gesbert, and A. J. Paulraj, "On the capacity of OFDM-based spatial multiplexing systems," *IEEE Trans. Commun.*, vol. 50, no. 2, pp. 225–234, 2002.

[425] M. Médard, "The effect upon channel capacity in wireless communications of perfect and imperfect knowledge of the channel," *IEEE Trans. Inform. Theory*, vol. 46, no. 3, pp. 933–946, 2000.

[426] B. Hassibi and B. M. Hochwald, "How much training is needed in multiple-antenna wireless links?" *IEEE Trans. Inform. Theory*, vol. 49, no. 4, pp. 951–963, 2003.

[427] X. Ma, L. Yang, and G. B. Giannakis, "Optimal training for MIMO frequency-selective fading channels," *IEEE Trans. Wireless Commun.*, vol. 4, no. 2, pp. 453–466, 2005.

[428] S. Furrer and D. Dahlhaus, "Multiple-antenna signaling over fading channels with estimated channel state information: capacity analysis," *IEEE Trans. Inform. Theory*, vol. 53, no. 6, pp. 2028–2043, 2007.

[429] J. Baltersee, G. Fock, and H. Meyr, "An information theoretic foundation of synchronized detection," *IEEE Trans. Commun.*, vol. 49, no. 12, pp. 2115–2123, 2001.

[430] S. Ohno and G. B. Giannakis, "Average-rate optimal PSAM transmissions over time-selective fading channels," *IEEE Trans. Wireless Commun.*, vol. 1, no. 4, pp. 712–720, 2002.

[431] X. Deng and A. M. Haimovich, "Achievable rates over time-varying Rayleigh fading channels," *IEEE Trans. Commun.*, vol. 55, no. 7, pp. 1397–1406, 2007.

[432] C. B. Peel and A. L. Swindlehurst, "Throughput-optimal training for a time-varying multi-antenna channel," *IEEE Trans. Wireless Commun.*, vol. 6, no. 9, pp. 3364–3373, 2007.

[433] M. C. Gursoy, "On the capacity and energy efficiency of training-based transmissions over fading channels," *IEEE Trans. Inform. Theory*, vol. 55, no. 10, pp. 4543–4567, 2009.

[434] S. Savazzi and U. Spagnolini, "Optimizing training lengths and training intervals in time-varying fading channels," *IEEE Trans. Signal Process.*, vol. 57, no. 3, pp. 1098–1112, 2009.

[435] T. Li and O. Collins, "A successive decoding strategy for channels with memory," *IEEE Int'l Symp. Inform. Theory (ISIT'05)*, Sep. 2005.

[436] M. Coldrey and P. Bohlin, "Training-based MIMO systems—part II: improvements using detected symbol information," *IEEE Trans. Signal Process.*, vol. 56, no. 1, pp. 296–303, 2008.

[437] T. L. Marzetta, N. Jindal, and A. Lozano, "What is the value of joint processing of pilots and data in block-fading channels?" *IEEE Int'l Symp. Inform. Theory (ISIT'09)*, pp. 2189–2193, 2009.

[438] M. Dörpinghaus, A. Ispas, and H. Meyr, "On the gain of joint processing of pilot and data symbols in stationary Rayleigh fading channels," *IEEE Trans. Inform. Theory*, vol. 58, no. 5, pp. 2963–2982, 2011.

[439] S. N. Diggavi and T. M. Cover, "The worst additive noise under a covariance constraint," *IEEE Trans. Inform. Theory*, vol. 47, no. 7, pp. 3072–3081, 2001.

[440] A. Lapidoth and S. Shamai, "Fading channels: how perfect need 'perfect side information' be?" *IEEE Trans. Inform. Theory*, vol. 48, no. 5, pp. 1118–1134, 2002.

[441] N. Jindal and A. Lozano, "A unified treatment of optimum pilot overhead in multipath fading channels," *IEEE Trans. Commun.*, vol. 58, no. 10, pp. 2939–2948, 2010.

[442] S. Adireddy, L. Tong, and H. Viswanathan, "Optimal placement of training for frequency-selective block-fading channels," *IEEE Trans. Inform. Theory*, vol. 48, no. 8, pp. 2338–2353, 2002.

[443] M. Simko, P. S. R. Diniz, Q. Wang, and M. Rupp, "Adaptive pilot-symbol patterns for MIMO OFDM systems," *IEEE Trans. Wireless Commun.*, vol. 12, pp. 4705–4715, Sep. 2013.

[444] G. George, R. K. Mungara, A. Lozano, and M. Haenggi, "Ergodic spectral efficiency in MIMO cellular networks," *IEEE Trans. Wireless Commun.*, vol. 16, no. 5, pp. 2835–2849, 2017.

[445] J. Wu, N. Mehta, A. Molisch, and J. Zhang, "Unified spectral efficiency analysis of cellular systems with channel-aware schedulers," *IEEE Trans. Commun.*, no. 99, pp. 1–12, 2011.

[446] M. Chiani, M. Z. Win, and H. Shin, "MIMO networks: the effects of interference," *IEEE Trans. Inform. Theory*, vol. 56, no. 1, pp. 336–349, 2010.

[447] A. Lozano, A. M. Tulino, and S. Verdú, "Multiple-antenna capacity in the low-power regime," *IEEE Trans. Inform. Theory*, vol. 49, no. 10, pp. 2527–2544, 2003.

[448] P. Mogensen, W. Na, I. Z. Kovács, F. Frederiksen, A. Pokhariyal, K. I. Pedersen, T. Kolding, K. Hugl, and M. Kuusela, "LTE capacity compared to the Shannon bound," *IEEE Veh. Techn. Conf. (VTC'07)*, pp. 1234–1238, 2007.

[449] C. Mehlführer, S. Caban, and M. Rupp, "Cellular system physical layer throughput: how far off are we from the Shannon bound?" *IEEE Wireless Commun.*, vol. 18, no. 6, pp. 54–63, 2011.

[450] S. Caban, M. Rupp, C. Mehlführer, and M. Wrulich, *Evaluation of HSDPA and LTE: from testbed measurements to system level performance.* John Wiley & Sons, 2011.

[451] M. Lerch and M. Rupp, "Measurement-based evaluation of the LTE MIMO downlink at different antenna configurations," *Int'l ITG Workshop Smart Antennas (WSA'13)*, pp. 1–6, 2013.

[452] G. Caire and D. Tuninetti, "The throughput of hybrid-ARQ protocols for the Gaussian collision channel," *IEEE Trans. Inform. Theory*, vol. 47, pp. 1971–1988, 2001.

[453] P. Larsson, L. K. Rasmussen, and M. Skoglund, "Throughput analysis of ARQ schemes in Gaussian block fading channels," *IEEE Trans. Commun.*, vol. 62, pp. 2569–2588, Jul. 2014.

[454] Z. D. Bai, "Convergence rate of expected spectral distributions of large random matrices. Part I. Wigner matrices," *Ann. Probab.*, vol. 21, pp. 625–648, 1993.

[455] E. Biglieri, G. Taricco, and A. Tulino, "How far away is infinity? Using asymptotic analyses in multiple-antenna systems," *IEEE Int'l Symp. Spread Spectrum Techniques and Applications (ISSSTA'02)*, vol. 1, pp. 1–6, 2002.

[456] D. N. C. Tse and S. V. Hanly, "Linear multiuser receivers: effective interference, effective bandwidth and user capacity," *IEEE Trans. Inform. Theory*, vol. 45, no. 2, pp. 641–657, 1999.

[457] S. Verdú and S. Shamai, "Spectral efficiency of CDMA with random spreading," *IEEE Trans. Inform. Theory*, vol. 45, no. 2, pp. 622–640, 1999.

[458] P. B. Rapajic and D. Popescu, "Information capacity of a random signature multiple-input multiple-output channel," *IEEE Trans. Commun.*, vol. 48, no. 8, pp. 1245–1248, 2000.

[459] J. W. Wallace and M. A. Jensen, "Mutual coupling in MIMO wireless systems: a rigorous network theory analysis," *IEEE Trans. Wireless Commun.*, vol. 3, no. 4, pp. 1317–1325, 2004.

[460] D. Gesbert, T. Ekman, and N. Christophersen, "Capacity limits of dense palm-sized MIMO arrays," *IEEE Global Commun. Conf. (GLOBECOM'02)*, vol. 2, pp. 1187–1191, 2002.

[461] N. Chiurtu, B. Rimoldi, E. Telatar, and V. Pauli, "Impact of correlation and coupling on the capacity of MIMO systems," *IEEE Int'l Symp. Signal Process. and Inform. Techn. (ISSPIT'03)*, pp. 154–157, 2003.

[462] T. S. Pollock, T. D. Abhayapala, and R. A. Kennedy, "Antenna saturation effects on dense array MIMO capacity," *IEEE Int'l Conf. Acoustics, Speech, and Signal Process. (ICASSP'03)*, vol. 4, pp. IV–361, 2003.

[463] A. Sayeed, V. Raghavan, and J. Kotecha, "Capacity of space–time wireless channels: a physical perspective," *IEEE Inform. Theory Workshop (ITW'04)*, pp. 434–439, 2004.

[464] T. Muharemovic, A. Sabharwal, and B. Aazhang, "Antenna packing in low-power systems: communication limits and array design," *IEEE Trans. Inform. Theory*, vol. 54, no. 1, pp. 429–440, 2008.

[465] C. Masouros, M. Sellathurai, and T. Ratnarajah, "Large-scale MIMO transmitters in fixed physical spaces: the effect of transmit correlation and mutual coupling," *IEEE Trans. Commun.*, vol. 61, no. 7, pp. 2794–2804, 2013.

[466] M. Payaró and D. P. Palomar, "On optimal precoding in linear vector Gaussian channels with arbitrary input distribution," *IEEE Int'l Symp. Inform. Theory (ISIT'09)*, pp. 1085–1089, 2009.

[467] F. Pérez-Cruz, M. R. D. Rodrigues, and S. Verdú, "MIMO Gaussian channels with arbitrary inputs: optimal precoding and power allocation," *IEEE Trans. Inform. Theory*, vol. 56, no. 3, pp. 1070–1084, 2010.

[468] M. Lamarca, "Linear precoding for mutual information maximization in MIMO systems," *Int'l Symp. Wireless Commun. Systems (ISWCS'09)*, pp. 26–30, 2009.

[469] C. Xiao, Y. R. Zheng, and Z. Ding, "Globally optimal linear precoders for finite alphabet signals over complex vector Gaussian channels," *IEEE Trans. Signal Process.*, vol. 59, no. 7, pp. 3301–3314, 2011.

[470] N.-Q. Nhan, P. Rostaing, K. Amis, L. Collin, and E. Radoi, "Complexity reduction for the optimization of linear precoders over random MIMO channels," *IEEE Trans. Commun.*, vol. 65, no. 10, pp. 4205–4217, 2017.

[471] S. K. Mohammed, E. Viterbo, Y. Hong, and A. Chockalingam, "Precoding by pairing subchannels to increase MIMO capacity with discrete input alphabets," *IEEE Trans. Inform. Theory*, vol. 57, no. 7, pp. 4156–4169, 2011.

[472] Y. Wu, C.-K. Wen, D. W. K. Ng, R. Schober, and A. Lozano, "Low-complexity MIMO precoding with discrete signals and statistical CSI," *IEEE Int'l Conf. Commun. (ICC'16)*, May 2016.

[473] Y. Wu, D. W. K. Ng, C.-K. Wen, R. Schober, and A. Lozano, "Low-complexity MIMO precoding for finite-alphabet signals," *IEEE Trans. Wireless Commun.*, vol. 16, no. 7, pp. 4571–4584, 2017.

[474] A. Lozano, J. G. Andrews, and R. W. Heath Jr., "On the limitations of cooperation in wireless networks," *Inform. Theory and Applications Workshop (ITA'12)*, pp. 123–130, 2012.

[475] A. Lozano, R. W. Heath Jr., and J. G. Andrews, "Fundamental limits of cooperation," *IEEE Trans. Inform. Theory*, vol. 59, no. 9, pp. 5213–5226, 2013.

[476] S. K. Jayaweera and H. V. Poor, "Capacity of multiple-antenna systems with both receiver and transmitter channel state information," *IEEE Trans. Inform. Theory*, vol. 49, no. 10, pp. 2697–2709, 2003.

[477] S. Wang and A. Abdi, "Instantaneous mutual information and eigen-channels in MIMO mobile Rayleigh fading," *IEEE Trans. Inform. Theory*, vol. 58, no. 1, pp. 353–368, 2012.

[478] C. N. Chuah, D. N. C. Tse, J. M. Kahn, and R. A. Valenzuela, "Capacity scaling in MIMO wireless systems under correlated fading," *IEEE Trans. Inform. Theory*, vol. 48, no. 3, pp. 637–650, 2002.

[479] A. Grant, "Rayleigh fading multi-antenna channels," *EURASIP J. Appl. Signal Process.*, vol. 2002, no. 1, pp. 316–329, 2002.

[480] A. Tulino, A. Lozano, and S. Verdú, "MIMO capacity with channel state information at the transmitter," *IEEE Int'l Symp. Spread Spectrum Techniques and Applications (ISSSTA'04)*, pp. 22–26, 2004.

[481] J. B. Andersen, "Array gain and capacity for known random channels with multiple element arrays at both ends," *IEEE J. Sel. Areas Commun.*, vol. 18, no. 11, pp. 2172–2178, 2000.

[482] J. W. Silverstein and Z. D. Bai, "On the empirical distribution of eigenvalues of a class of large dimensional random matrices," *J. Multivar. Anal.*, vol. 54, no. 2, pp. 175–192, 1995.

[483] A. M. Sengupta and P. P. Mitra, "Capacity of multivariate channels with multiplicative noise: I. Random matrix techniques and large-N expansions for full transfer matrices," *arXiv:physics/0010081*, 2000.

[484] O. Oyman, R. U. Nabar, H. Bölcskei, and A. J. Paulraj, "Characterizing the statistical properties of mutual information in MIMO channels," *IEEE Trans. Signal Process.*, vol. 51, no. 11, pp. 2784–2795, 2003.

[485] L. W. Hanlen and A. J. Grant, "Capacity analysis of correlated MIMO channels," *IEEE Trans. Inform. Theory*, vol. 58, 2012.

[486] E. Abbe, S. L. Huang, and E. Telatar, "Proof of the outage probability conjecture for MISO channels," *IEEE Trans. Inform. Theory*, vol. 59, no. 5, pp. 2596–2602, 2013.

[487] S. H. Simon and A. L. Moustakas, "Optimizing MIMO antenna systems with channel covariance feedback," *IEEE J. Sel. Areas Commun.*, vol. 21, no. 3, pp. 406–417, 2003.

[488] M. A. Kamath, B. L. Hughes, and X. Yu, "Gaussian approximations for the capacity of MIMO Rayleigh fading channels," *Asilomar Conf. Signals, Systems and Computers*, vol. 1, pp. 614–618, 2002.

[489] A. L. Moustakas, S. H. Simon, and A. M. Sengupta, "MIMO capacity through correlated channels in the presence of correlated interferers and noise: a (not so) large N analysis," *IEEE Trans. Inform. Theory*, vol. 49, no. 10, pp. 2545–2561, 2003.

[490] B. M. Hochwald, T. L. Marzetta, and V. Tarokh, "Multiple-antenna channel hardening and its implications for rate feedback and scheduling," *IEEE Trans. Inform. Theory*, vol. 50, no. 9, pp. 1893–1909, 2004.

[491] A. M. Tulino and S. Verdú, "Asymptotic outage capacity of multiantenna channels," *IEEE Int'l Conf. Acoustics, Speech, and Signal Process. (ICASSP'05)*, vol. 5, pp. 825–828, 2005.

[492] A. L. Moustakas and S. H. Simon, "On the outage capacity of correlated multiple-path MIMO channels," *IEEE Trans. Inform. Theory*, vol. 53, no. 11, pp. 3887–3903, 2007.

[493] G. Taricco, "Asymptotic mutual information statistics of separately correlated Rician fading MIMO channels," *IEEE Trans. Inform. Theory*, vol. 54, no. 8, pp. 3490–3504, 2008.

[494] H. Shin, M. Z. Win, and M. Chiani, "Asymptotic statistics of mutual information for doubly correlated MIMO channels," *IEEE Trans. Wireless Commun.*, vol. 7, no. 2, pp. 562–573, 2008.

[495] W. Hachem, O. Khorunzhiy, P. Loubaton, J. Najim, and L. Pastur, "A new approach for mutual information analysis of large dimensional multi-antenna channels," *IEEE Trans. Inform. Theory*, vol. 54, no. 9, pp. 3987–4004, 2008.

[496] P. Kazakopoulos, P. Mertikopoulos, A. L. Moustakas, and G. Caire, "Living at the edge: a large deviations approach to the outage MIMO capacity," *IEEE Trans. Inform. Theory*, vol. 57, no. 4, pp. 1984–2007, 2011.

[497] Z. Bao, G. Pan, and W. Zhou, "Asymptotic mutual information statistics of MIMO channels and CLT of sample covariance matrices," *IEEE Trans. Inform. Theory*, vol. 61, no. 6, pp. 3413–3426, 2015.

[498] B. M. Hochwald, T. L. Marzetta, and B. Hassibi, "Space–time autocoding," *IEEE Trans. Inform. Theory*, vol. 47, no. 7, pp. 2761–2781, 2001.

[499] M. Chowdhury and A. Goldsmith, "Reliable uncoded communication in the SIMO MAC," *IEEE Trans. Inform. Theory*, vol. 61, no. 1, pp. 388–403, 2015.

[500] M. Vu and A. Paulraj, "Capacity optimization for Rician correlated MIMO wireless channels," *Asilomar Conf. Signals, Systems and Computers*, pp. 133–138, 2005.

[501] W. Rhee and G. Taricco, "On the ergodic capacity-achieving covariance matrix of certain classes of MIMO channels," *IEEE Trans. Inform. Theory*, vol. 52, no. 8, pp. 3810–3817, 2006.

[502] E. Visotsky and U. Madhow, "Space–time transmit precoding with imperfect feedback," *IEEE Trans. Inform. Theory*, vol. 47, no. 6, pp. 2632–2639, 2001.

[503] J. H. Kotecha and A. M. Sayeed, "On the capacity of correlated MIMO channels," *IEEE Int'l Symp. Inform. Theory (ISIT'03)*, pp. 355–355, 2003.

[504] E. A. Jorswieck and H. Boche, "Optimal transmission with imperfect channel state information at the transmit antenna array," *Wireless Personal Commun.*, vol. 27, no. 1, pp. 33–56, 2003.

[505] S. Venkatesan, S. H. Simon, and R. A. Valenzuela, "Capacity of a Gaussian MIMO channel with nonzero mean," *IEEE Veh. Techn. Conf. (VTC'03)*, vol. 3, pp. 1767–1771, 2003.

[506] D. Hosli and A. Lapidoth, "The capacity of a MIMO Ricean channel is monotonic in the singular values of the mean," *5th Int'l ITG Conf. Source and Channel Coding*, pp. 381–386, 2004.

[507] J. Li and Q. Zhang, "Transmitter optimization for correlated MISO fading channels with generic mean and covariance feedback," *IEEE Trans. Wireless Commun.*, vol. 7, no. 9, pp. 3312–3317, 2008.

[508] A. Lozano, A. M. Tulino, and S. Verdú, "Multiantenna capacity: myths and realities," *Space–time wireless systems: from array processing to MIMO communications*. Cambridge University Press, 2005.

[509] L. W. Hanlen and A. J. Grant, "Optimal transmit covariance for MIMO channels with statistical transmitter side information," *IEEE Int'l Symp. Inform. Theory (ISIT'05)*, pp. 1818–1822, 2005.

[510] A. M. Tulino, A. Lozano, and S. Verdu, "Power allocation in multiantenna communication with statistical channel information at the transmitter," *IEEE Int'l Symp. Personal, Indoor and Mobile Radio Commun. (PIMRC'04)*, vol. 3, pp. 2003–2007, Sep. 2004.

[511] X. Li, S. Jin, X. Gao, and K. K. Wong, "Near-optimal power allocation for MIMO channels with mean or covariance feedback," *IEEE Trans. Commun.*, vol. 58, no. 1, pp. 289–300, 2010.

[512] A. Soysal and S. Ulukus, "Optimum power allocation for single-user MIMO and multi-user MIMO-MAC with partial CSI," *IEEE J. Sel. Areas Commun.*, vol. 25, no. 7, pp. 1402–1412, 2007.

[513] E. A. Jorswieck and H. Boche, "Optimal transmission strategies and impact of correlation in multiantenna systems with different types of channel state information," *IEEE Trans. Signal Process.*, vol. 52, no. 12, pp. 3440–3453, 2004.

[514] J. S. Kwak, J. G. Andrews, and A. Lozano, "MIMO capacity in correlated interference-limited channels," *IEEE Int'l Symp. Inform. Theory (ISIT'07)*, pp. 106–110, 2007.

[515] V. V. Mai, J. S. Kwak, Y. Jeong, and H. Shin, "Optimal power allocation in MIMO interference networks," *IEEE Trans. Wireless Commun. (to appear)*, 2018.

[516] X. Gao, B. Jiang, X. Li, A. B. Gershman, and M. R. McKay, "Statistical eigenmode transmission over jointly correlated MIMO channels," *IEEE Trans. Inform. Theory*, vol. 55, no. 8, pp. 3735–3750, 2009.

[517] V. Raghavan, A. M. Sayeed, and V. V. Veeravalli, "Semiunitary precoding for spatially correlated MIMO channels," *IEEE Trans. Inform. Theory*, vol. 57, no. 3, pp. 1284–1298, 2011.

[518] W. Rhee and J. M. Cioffi, "On the capacity of multiuser wireless channels with multiple antennas," *IEEE Trans. Inform. Theory*, vol. 49, no. 10, pp. 2580–2595, 2003.

[519] E. Abbe, E. Telatar, and L. Zheng, "The algebra of MIMO channels," *Allerton Annual Conf. Commun., Control and Computing*, 2005.

[520] H. Shin and J. H. Lee, "Capacity of multiple-antenna fading channels: spatial fading correlation, double scattering, and keyhole," *IEEE Trans. Inform. Theory*, vol. 49, no. 10, pp. 2636–2647, 2003.

[521] M. Dohler and H. Aghvami, "A closed form expression of MIMO capacity over ergodic narrowband channels," Unpublished manuscript, 2003.

[522] R. Janaswamy, "Analytical expressions for the ergodic capacities of certain MIMO systems by the Mellin transform," *IEEE Global Commun. Conf. (GLOBE-COM'03)*, vol. 1, pp. 287–291, 2003.

[523] P. J. Smith, S. Roy, and M. Shafi, "Capacity of MIMO systems with semicorrelated flat fading," *IEEE Trans. Inform. Theory*, vol. 49, no. 10, pp. 2781–2788, 2003.

[524] G. Alfano, A. M. Tulino, A. Lozano, and S. Verdú, "Capacity of MIMO channels with one-sided correlation," *IEEE Int'l Symp. Spread Spectrum Techniques and Applications (ISSSTA'04)*, pp. 515–519, 2004.

[525] M. Kang and M. S. Alouini, "Capacity of correlated MIMO Rayleigh channels," *IEEE Trans. Wireless Commun.*, vol. 5, no. 1, pp. 143–155, 2006.

[526] P. M. Marques and S. A. Abrantes, "On the derivation of the exact, closed-form capacity formulas for receiver-sided correlated MIMO channels," *IEEE Trans. Inform. Theory*, vol. 54, no. 3, pp. 1139–1161, 2008.

[527] G. Alfano, A. Lozano, A. M. Tulino, and S. Verdú, "Mutual information and eigenvalue distribution of MIMO Ricean channels," *Int'l Symp. Inform. Theory and Applications (ISITA'04)*, vol. 4, pp. 1040–1045, 2004.

[528] M. Kiessling and J. Speidel, "Mutual information of MIMO channels in correlated Rayleigh fading environments—a general solution," *IEEE Int'l Conf. Commun. (ICC'04)*, vol. 2, pp. 814–818, 2004.

[529] S. K. Jayaweera and H. V. Poor, "On the capacity of multiple-antenna systems in Rician fading," *IEEE Trans. Wireless Commun.*, vol. 4, no. 3, pp. 1102–1111, 2005.

[530] M. R. McKay and I. B. Collings, "General capacity bounds for spatially correlated Rician MIMO channels," *IEEE Trans. Inform. Theory*, vol. 51, no. 9, pp. 3121–3145, 2005.

[531] H. Shin, M. Z. Win, J. H. Lee, and M. Chiani, "On the capacity of doubly correlated MIMO channels," *IEEE Trans. Wireless Commun.*, vol. 5, no. 8, pp. 2253–2265, 2006.

[532] M. Kang and M. S. Alouini, "Capacity of MIMO Rician channels," *IEEE Trans. Wireless Commun.*, vol. 5, no. 1, pp. 112–122, 2006.

[533] M. R. McKay and I. B. Collings, "Improved general lower bound for spatially-correlated Rician MIMO capacity," *IEEE Commun. Letters*, vol. 10, no. 3, pp. 162–164, 2006.

[534] S. Jin, X. Gao, and X. You, "On the ergodic capacity of rank-1 Ricean-fading MIMO channels," *IEEE Trans. Inform. Theory*, vol. 53, no. 2, pp. 502–517, 2007.

[535] C. Zhong, K. K. Wong, and S. Jin, "Capacity bounds for MIMO Nakagami-m fading channels," *IEEE Trans. Signal Process.*, vol. 57, no. 9, pp. 3613–3623, 2009.

[536] A. Ghaderipoor, C. Tellambura, and A. Paulraj, "On the application of character expansions for MIMO capacity analysis," *IEEE Trans. Inform. Theory*, vol. 58, no. 5, pp. 2950–2962, 2011.

[537] E. A. Jorswieck and H. Boche, "Average mutual information in spatial correlated MIMO systems with uninformed transmitter," *Conf. Inform. Science and Systems (CISS'04)*, 2004.

[538] W. Zeng, C. Xiao, M. Wang, and J. Lu, "On the linear precoder design for MIMO channels with finite-alphabet inputs and statistical CSI," *IEEE Global Commun. Conf. (GLOBECOM 2011)*, 2011.

[539] W. He and C. N. Georghiades, "Computing the capacity of a MIMO fading channel under PSK signaling," *IEEE Trans. Inform. Theory*, vol. 51, no. 5, pp. 1794–1803, 2005.

[540] R. H. Gohary and T. N. Davidson, "On rate-optimal MIMO signalling with mean and covariance feedback," *IEEE Trans. Wireless Commun.*, vol. 8, no. 2, pp. 912–921, 2009.

[541] E. Vagenas, G. S. Paschos, and S. A. Kotsopoulos, "Beamforming capacity optimization for MISO systems with both mean and covariance feedback," *IEEE Trans. Wireless Commun.*, vol. 10, no. 9, p. 2994, 2011.

[542] S. A. Jafar and A. Goldsmith, "Transmitter optimization and optimality of beamforming for multiple antenna systems," *IEEE Trans. Wireless Commun.*, vol. 3, no. 4, pp. 1165–1175, 2004.

[543] E. A. Jorswieck and H. Boche, "Channel capacity and capacity-range of beamforming in MIMO wireless systems under correlated fading with covariance feedback," *IEEE Trans. Wireless Commun.*, vol. 3, no. 5, pp. 1543–1553, 2004.

[544] S. Srinivasa and S. A. Jafar, "The optimality of transmit beamforming: a unified view," *IEEE Trans. Inform. Theory*, vol. 53, no. 4, pp. 1558–1564, 2007.

[545] J. Hansen and H. Bölcskei, "A geometrical investigation of the rank-1 Ricean MIMO channel at high SNR," *IEEE Int'l Symp. Inform. Theory (ISIT'04)*, p. 64, 2004.

[546] A. M. Tulino and S. Verdú, "Random matrix theory and wireless communications," *Found. and Trends Commun. Inform. Theory*, vol. 1, 2004.

[547] R. Couillet and M. Debbah, *Random matrix methods for wireless communications*. Cambridge University Press, 2011.

[548] A. M. Tulino, A. Lozano, and S. Verdú, "Impact of antenna correlation on the capacity of multiantenna channels," *IEEE Trans. Inform. Theory*, vol. 51, no. 7, pp. 2491–2509, 2005.

[549] A. M. Tulino, S. Verdú, and A. Lozano, "Capacity of antenna arrays with space, polarization and pattern diversity," *IEEE Inform. Theory Workshop (ITW'03)*, pp. 324–327, 2003.

[550] X. Mestre, J. R. Fonollosa, and A. Pages-Zamora, "Capacity of MIMO channels: asymptotic evaluation under correlated fading," *IEEE J. Sel. Areas Commun.*, vol. 21, no. 5, pp. 829–838, 2003.

[551] V. V. Veeravalli, Y. Liang, and A. M. Sayeed, "Correlated MIMO wireless channels: capacity, optimal signaling, and asymptotics," *IEEE Trans. Inform. Theory*, vol. 51, no. 6, pp. 2058–2072, 2005.

[552] J. Dumont, W. Hachem, S. Lasaulce, P. Loubaton, and J. Najim, "On the capacity achieving covariance matrix for Rician MIMO channels: an asymptotic approach," *IEEE Trans. Inform. Theory*, vol. 56, no. 3, pp. 1048–1069, 2010.

[553] Z. D. Bai, "Methodologies in spectral analysis of large-dimensional random matrices, A review," *Statist. Sinica*, vol. 9, no. 3, pp. 611–677, 1999.

[554] W. Yang, G. Durisi, and E. Riegler, "On the capacity of large-MIMO block-fading channels," *IEEE J. Sel. Areas Commun.*, vol. 31, no. 2, pp. 117–132, 2013.

[555] S. Ray, M. Médard, and L. Zheng, "On noncoherent MIMO channels in the wideband regime: capacity and reliability," *IEEE Trans. Inform. Theory*, vol. 53, no. 6, pp. 1983–2009, 2007.

[556] B. Hochwald, T. Marzetta, T. Richardson, W. Sweldens, and R. Urbanke, "Systematic design of unitary space–time constellations," *IEEE Trans. Inform. Theory*, vol. 46, no. 6, pp. 1962–1973, 2000.

[557] B. Hassibi and T. L. Marzetta, "Multiple-antennas and isotropically random unitary inputs: the received signal density in closed form," *IEEE Trans. Inform. Theory*, vol. 48, no. 6, pp. 1473–1484, 2002.

[558] S. Moser, "The fading number of multiple-input multiple-output fading channels with memory," *IEEE Trans. Inform. Theory*, vol. 55, no. 6, pp. 2716–2755, 2009.

[559] J. C. Guey, M. P. Fitz, M. R. Bell, and W. Y. Kuo, "Signal design for transmitter diversity wireless communication systems over Rayleigh fading channels," *IEEE Trans. Commun.*, vol. 47, no. 4, pp. 527–537, 1999.

[560] T. L. Marzetta, "BLAST training: estimating channel characteristics for high capacity space–time wireless," *Allerton Conf. Commun., Control and Computing*, vol. 37, pp. 958–966, 1999.

[561] Q. Sun, D. C. Cox, H. C. Huang, and A. Lozano, "Estimation of continuous flat fading MIMO channels," *IEEE Trans. Wireless Commun.*, vol. 1, no. 4, pp. 549–553, 2002.

[562] Q. Sun, D. C. Cox, A. Lozano, and H. C. Huang, "Training-based channel estimation for continuous flat fading BLAST," *IEEE Int'l Conf. Commun. (ICC'02)*, vol. 1, pp. 325–329, 2002.

[563] Q. Sun, D. C. Cox, H. C. Huang, and A. Lozano, "Estimation of continuous flat fading MIMO channels," *IEEE Wireless Commun. and Networking Conf. (WCNC'02)*, vol. 1, pp. 189–193, Mar. 2002.

[564] M. Coldrey and P. Bohlin, "Training-based MIMO systems—part I: performance comparison," *IEEE Trans. Signal Process.*, vol. 55, no. 11, pp. 5464–5476, 2007.

[565] A. T. Asyhari and S. ten Brink, "Orthogonal or superimposed pilots? A rate-efficient channel estimation strategy for stationary MIMO fading channels," *IEEE Trans. Wireless Commun.*, vol. 16, no. 5, pp. 2776–2789, 2017.

[566] A. Pastore, M. Joham, and J. R. Fonollosa, "A framework for joint design of pilot sequence and linear precoder," *IEEE Trans. Inform. Theory*, vol. 62, no. 9, pp. 5059–5079, 2016.

[567] T. Yoo and A. Goldsmith, "Capacity and power allocation for fading MIMO channels with channel estimation error," *IEEE Trans. Inform. Theory*, vol. 52, no. 5, pp. 2203–2214, 2006.

[568] L. Musavian, M. R. Nakhai, M. Dohler, and A. H. Aghvami, "Effect of channel uncertainty on the mutual information of MIMO fading channels," *IEEE Trans. Veh. Techn.*, vol. 56, no. 5, pp. 2798–2806, 2007.

[569] M. Ding and S. D. Blostein, "Maximum mutual information design for MIMO

systems with imperfect channel knowledge," *IEEE Trans. Inform. Theory*, vol. 56, no. 10, pp. 4793–4801, 2010.

[570] F. R. Farrokhi, A. Lozano, G. J. Foschini, and R. A. Valenzuela, "Spectral efficiency of wireless systems with multiple transmit and receive antennas," *IEEE Int'l Symp. Personal, Indoor and Mobile Radio Commun. (PIMRC'00)*, vol. 1, pp. 373–377, 2000.

[571] F. R. Farrokhi, G. J. Foschini, A. Lozano, and R. A. Valenzuela, "Link-optimal BLAST processing with multiple-access interference," *IEEE Veh. Techn. Conf. (VTC'00)*, vol. 1, pp. 87–91, 2000.

[572] ——, "Link-optimal space–time processing with multiple transmit and receive antennas," *IEEE Commun. Letters*, vol. 5, no. 3, pp. 85–87, 2001.

[573] F. R. Farrokhi, A. Lozano, G. J. Foschini, and R. A. Valenzuela, "Spectral efficiency of FDMA/TDMA wireless systems with transmit and receive antenna arrays," *IEEE Trans. Wireless Commun.*, vol. 1, no. 4, pp. 591–599, 2002.

[574] S. Ye and R. S. Blum, "Optimized signaling for MIMO interference systems with feedback," *IEEE Trans. Signal Process.*, vol. 51, no. 11, pp. 2839–2848, 2003.

[575] E. A. Jorswieck and H. Boche, "Performance analysis of capacity of MIMO systems under multiuser interference based on worst-case noise behavior," *EURASIP J. Wireless Commun. Netw.*, vol. 2004, no. 2, pp. 273–285, 2004.

[576] E. A. Jorswieck, E. G. Larsson, and D. Danev, "Complete characterization of the Pareto boundary for the MISO interference channel," *IEEE Trans. Signal Process.*, vol. 56, no. 10, pp. 5292–5296, 2008.

[577] G. Scutari, D. Palomar, and S. Barbarossa, "Competitive design of multiuser MIMO systems based on game theory: a unified view," *IEEE J. Sel. Areas Commun.*, vol. 26, no. 7, pp. 1089–1103, 2008.

[578] G. Scutari, D. P. Palomar, and S. Barbarossa, "Optimal linear precoding strategies for wideband noncooperative systems based on game theory—part I: Nash equilibria," *IEEE Trans. Signal Process.*, vol. 56, no. 3, pp. 1230–1249, 2008.

[579] E. G. Larsson, E. A. Jorswieck, J. Lindblom, and R. Mochaourab, "Game theory and the flat-fading Gaussian interference channel," *IEEE Signal Process. Mag.*, vol. 26, no. 5, pp. 18–27, 2009.

[580] R. S. Blum, "MIMO capacity with interference," *IEEE J. Sel. Areas Commun.*, vol. 21, no. 5, pp. 793–801, 2003.

[581] G. Alfano, A. M. Tulino, A. Lozano, and S. Verdu, "Eigenvalue statistics of finite-dimensional random matrices for MIMO wireless communications," *IEEE Int'l Conf. Commun. (ICC'06)*, vol. 9, pp. 4125–4129, Jun. 2006.

[582] A. Lozano and A. M. Tulino, "Capacity of multiple-transmit multiple-receive antenna architectures," *IEEE Trans. Inform. Theory*, vol. 48, no. 12, pp. 3117–3128, 2002.

[583] G. Taricco and E. Riegler, "On the ergodic capacity of correlated Rician fading MIMO channels with interference," *IEEE Trans. Inform. Theory*, vol. 57, no. 7, pp. 4123–4137, 2011.

[584] S. L. Ariyavisitakul, "Turbo space–time processing to improve wireless channel capacity," *IEEE Trans. Commun.*, vol. 48, no. 8, pp. 1347–1359, 2000.

[585] S. T. Chung, A. Lozano, and H. C. Huang, "Approaching eigenmode BLAST channel capacity using V-BLAST with rate and power feedback," *IEEE Veh. Techn. Conf. (VTC'01 Fall)*, vol. 2, pp. 915–919, 2001.

[586] ——, "Low complexity algorithm for rate and power quantization in extended V-BLAST," *IEEE Veh. Techn. Conf. (VTC'01 Fall)*, vol. 2, pp. 910–914, 2001.

[587] S. T. Chung, A. Lozano, H. C. Huang, A. Sutivong, and J. M. Cioffi, "Approaching the MIMO capacity with V-BLAST: theory and practice," *EURASIP J. Appl. Signal Process.*, vol. 2002, pp. 762–771, 2004.

[588] A. Lozano, "Capacity-approaching rate function for layered multiantenna architectures," *IEEE Trans. Wireless Commun.*, vol. 2, no. 4, pp. 616–620, 2003.

[589] ——, "Per-antenna rate and power control for MIMO layered architectures in the low-and high-power regimes," *IEEE Trans. Commun.*, vol. 58, no. 2, pp. 652–659, 2010.

[590] B. Hassibi, "An efficient square-root algorithm for BLAST," *IEEE Int'l Conf. Acous., Speech, and Signal Proc. (ICASSP'00)*, vol. 2, Jun. 2000.

[591] M. K. Varanasi and T. Guess, "Optimum decision feedback multiuser equalization with successive decoding achieves the total capacity of the Gaussian multiple-access channel," *Asilomar Conf. Signals, Systems & Computers*, vol. 2, pp. 1405–1409, 1997.

[592] G. Ginis and J. M. Cioffi, "On the relation between V-BLAST and the GDFE," *IEEE Commun. Letters*, vol. 5, no. 9, pp. 364–366, 2001.

[593] J. G. Andrews, "Interference cancellation for cellular systems: a contemporary overview," *IEEE Wireless Commun.*, vol. 12, no. 2, pp. 19–29, 2005.

[594] G. Bauch and N. Al-Dhahir, "Reduced-complexity space–time turbo-equalization for frequency-selective MIMO channels," *IEEE Trans. Wireless Commun.*, vol. 1, no. 4, pp. 819–828, 2002.

[595] G. J. Foschini, G. D. Golden, R. A. Valenzuela, and P. W. Wolniansky, "Simplified processing for high spectral efficiency wireless communication employing multi-element arrays," *IEEE J. Sel. Areas Commun.*, vol. 17, no. 11, pp. 1841–1852, 1999.

[596] A. Lozano and C. B. Papadias, "Space–time receiver for wideband BLAST in rich-scattering wireless channels," *IEEE Veh. Techn. Conf. (VTC'00)*, vol. 1, pp. 186–190, 2000.

[597] ——, "Layered space–time receivers for frequency-selective wireless channels," *IEEE Trans. Commun.*, vol. 50, pp. 65–73, Jan. 2002.

[598] P. Liu and I. Kim, "Exact and closed-form error performance analysis for hard MMSE-SIC detection in MIMO systems," *IEEE Trans. Commun.*, vol. 59, no. 9, pp. 2463–2477, 2011.

[599] G. J. Foschini, D. Chizhik, M. J. Gans, C. Papadias, and R. A. Valenzuela, "Analysis and performance of some basic space–time architectures," *IEEE J. Sel. Areas Commun.*, vol. 21, no. 3, pp. 303–320, 2003.

[600] H. El Gamal and A. R. Hammons, "A new approach to layered space–time coding and signal processing," *IEEE Trans. Inform. Theory*, vol. 47, no. 6, pp. 2321–2334, 2001.

[601] M. Sellathurai and S. Haykin, "Turbo-BLAST for wireless communications: theory and experiments," *IEEE Trans. Signal Process.*, vol. 50, no. 10, pp. 2538–2546, 2002.

[602] D. Zhang and H. Meyr, "On the performance gap between ML and iterative decoding of finite-length turbo-coded BICM in MIMO systems," *IEEE Trans. Commun.*, vol. 65, no. 8, pp. 3201–3213, 2017.

[603] P. Robertson, E. Villebrun, and P. Hoeher, "A comparison of optimal and suboptimal MAP decoding algorithms operating in the log domain," *IEEE Int'l Conf. Commun. (ICC'95)*, vol. 2, pp. 1009–1013, 1995.

[604] M. Pohst, "On the computation of lattice vectors of minimal length, successive minima, and reduced bases with applications," *SIGSAM Bull.*, vol. 15, no. 1, pp. 37–44, Feb. 1981.

[605] U. Fincke and M. Pohst, "Improved methods for calculating vectors of short length in a lattice, including a complexity analysis," *Math. Comput.*, vol. 44, no. 170, pp. 463–471, 1985.

[606] E. Viterbo and J. Boutros, "A universal lattice code decoder for fading channels," *IEEE Trans. Inform. Theory*, vol. 45, no. 5, pp. 1639–1642, 1999.

[607] B. Hassibi and B. M. Hochwald, "High-rate codes that are linear in space and time," *IEEE Trans. Inform. Theory*, vol. 48, no. 7, pp. 1804–1824, 2002.

[608] M. O. Damen, H. E. Gamal, and G. Caire, "On maximum-likelihood detection and the search for the closest lattice point," *IEEE Trans. Inform. Theory*, vol. 49, no. 10, pp. 2389–2402, 2003.

[609] A. Burg, M. Borgmann, M. Wenk, M. Zellweger, W. Fichtner, and H. Bölcskei, "VLSI implementation of MIMO detection using the sphere decoding algorithm," *IEEE J. Solid-State Circuits*, vol. 40, no. 7, pp. 1566–1577, 2005.

[610] L. G. Barbero and J. S. Thompson, "Fixing the complexity of the sphere decoder for MIMO detection," *IEEE Trans. Wireless Commun.*, vol. 7, no. 6, pp. 2131–2142, 2008.

[611] S. Chen, T. Zhang, and Y. Xin, "Relaxed K-best MIMO signal detector design and VLSI implementation," *IEEE Trans. VLSI Syst.*, vol. 15, no. 3, pp. 328–337, 2007.

[612] X. Wang and H. V. Poor, "Iterative (turbo) soft interference cancellation and decoding for coded CDMA," *IEEE Trans. Commun.*, vol. 47, no. 7, pp. 1046–1061, 1999.

[613] T. Abe and T. Matsumoto, "Space–time turbo equalization in frequency-selective MIMO channels," *IEEE Trans. Veh. Technol.*, vol. 52, no. 3, pp. 469–475, 2003.

[614] W.-J. Choi, K.-W. Cheong, and J. M. Cioffi, "Iterative soft interference cancellation for multiple antenna systems," *IEEE Wireless Commun. Netw. Conf. (WCNC'00)*, vol. 1, pp. 304–309, 2000.

[615] X. Li, H. C. Huang, A. Lozano, and G. J. Foschini, "Reduced-complexity detection algorithms for systems using multi-element arrays," *IEEE Global Commun. Conf. (GLOBECOM'00)*, vol. 2, pp. 1072–1076, 2000.

[616] S. ten Brink, "Convergence behavior of iteratively decoded parallel concatenated codes," *IEEE Trans. Commun.*, vol. 49, pp. 1727–1737, Oct. 2001.

[617] C. Hermosilla and L. Szczeciński, "Performance evaluation of linear turbo receivers using analytical extrinsic information transfer functions," *EURASIP J. Advances Signal Process.*, vol. 2005, no. 6, pp. 1–14, 2005.

[618] H. Zheng, A. Lozano, and M. Haleem, "Multiple ARQ processes for MIMO systems," *IEEE Int'l Symp. Personal, Indoor and Mobile Radio Commun. (PIMRC'02)*, vol. 3, pp. 1023–1026, 2002.

[619] ——, "Multiple ARQ processes for MIMO systems," *EURASIP J. Appl. Signal Process.*, vol. 2004, pp. 772–782, 2004.

[620] C. A. Balanis, *Antenna theory: analysis and design.* John Wiley & Sons, 2012.

[621] C. K. Au-Yeung and D. J. Love, "Optimization and tradeoff analysis of two-way limited feedback beamforming systems," *IEEE Trans. Wireless Commun.*, vol. 8, no. 5, pp. 2570–2579, 2009.

[622] N. E. Buris, "Reciprocity calibration of TDD smart antenna systems," *IEEE Antennas Propag. Society Int'l Symp.*, pp. 1–4, Jul. 2010.

[623] J. Liu, A. Bourdoux, J. Craninckx, P. Wambacq, B. Come, S. Donnay, and A. Barel, "OFDM-MIMO WLAN AP front-end gain and phase mismatch calibration," *IEEE Radio and Wireless Conf.*, pp. 151–154, Sep. 2004.

[624] T. Schenk, *RF imperfections in high-rate wireless systems: impact and digital compensation.* Springer-Verlag, 2008.

[625] M. Petermann, M. Stefer, F. Ludwig, D. Wubben, M. Schneider, S. Paul, and K. Kammeyer, "Multi-user pre-processing in multi-antenna OFDM TDD systems with non-reciprocal transceivers," *IEEE Trans. Commun.*, vol. 61, pp. 3781–3793, Sep. 2013.

[626] M. Guillaud, D. T. M. Slock, and R. Knopp, "A practical method for wireless channel reciprocity exploitation through relative calibration," *ISSPA*, pp. 403–406, 2005.

[627] J. Liu, G. Vandersteen, J. Craninckx, M. Libois, M. Wouters, F. Petre, and A. Barel, "A novel and low-cost analog front-end mismatch calibration scheme for MIMO-OFDM WLANs," *IEEE Radio and Wireless Symp.*, pp. 219–222, Jan. 2006.

[628] J. Shi, Q. Luo, and M. You, "An efficient method for enhancing TDD over the air reciprocity calibration," *IEEE Wireless Commun. Netw. Conf. (WCNC'11)*, pp. 339–344, Mar. 2011.

[629] K.-H. Lee and D. Petersen, "Optimal linear coding for vector channels," *IEEE Trans. Commun.*, vol. 24, no. 12, pp. 1283–1290, 1976.

[630] T. L. Marzetta and B. M. Hochwald, "Fast transfer of channel state information in wireless systems," *IEEE Trans. Signal Process.*, vol. 54, no. 4, pp. 1268–1278, 2006.

[631] D. Samardzija and N. Mandayam, "Unquantized and uncoded channel state information feedback in multiple-antenna multiuser systems," *IEEE Trans. Commun.*, vol. 54, no. 7, pp. 1335–1345, 2006.

[632] M. M. Shanechi, R. Porat, and U. Erez, "Comparison of practical feedback algorithms for multiuser MIMO," *IEEE Trans. Commun.*, vol. 58, no. 8, pp. 2436–2446, 2010.

[633] E. Akyol, K. Rose, and T. Ramstad, "Optimal mappings for joint source channel coding," *IEEE Inform. Theory Workshop (ITW'10)*, pp. 1–5, 2010.

[634] O. E. Ayach, A. Lozano, and R. W. Heath Jr., "On the overhead of interference alignment: training, feedback, and cooperation," *IEEE Trans. Wireless Commun.*, vol. 11, no. 11, pp. 4192–4203, 2012.

[635] D. J. Love, R. W. Heath Jr., W. Santipach, and M. Honig, "What is the value of limited feedback for MIMO channels?" *IEEE Commun. Mag.*, vol. 42, no. 10, pp. 54–59, 2004.

[636] B. Mondal and R. W. Heath Jr., "Algorithms for quantized precoded MIMO-OFDM systems," *Asilomar Conf. Signals, Systems and Computers*, pp. 381–385, Oct.-Nov. 2005.

[637] J. Choi and R. W. Heath Jr., "Interpolation based transmit beamforming for MIMO-OFDM with limited feedback," *IEEE Trans. Signal Processing*, vol. 53, no. 11, pp. 4125–4135, 2005.

[638] J. C. Roh and B. D. Rao, "An efficient feedback method for MIMO systems with slowly time-varying channels," *IEEE Wireless Commun. Netw. Conf. (WCNC'04)*, vol. 2, pp. 760–764, Mar. 2004.

[639] K.-B. Huang, B. Mondal, R. W. Heath Jr., and J. G. Andrews, "Multi-antenna limited feedback for temporally correlated channel: feedback compression," *IEEE Global Commun. Conf. (GLOBECOM'06)*, Nov. 2006.

[640] C. Simon and G. Leus, "Adaptive feedback reduction for precoded spatial multiplexing MIMO systems," *Int'l ITG/IEEE Workshop on Smart Antennas (WSA'07)*, 2007.

[641] B. Mondal and R. W. Heath Jr., "Channel adaptive quantization for limited feedback MIMO beamforming systems," *IEEE Trans. Signal Process.*, vol. 54, pp. 4741–4740, 2006.

[642] V. Raghavan, R. W. Heath Jr., and A. M. Sayeed, "Systematic codebook designs for quantized beamforming in correlated MIMO channels," *IEEE J. Sel. Areas Commun.*, vol. 25, no. 7, pp. 1298–1310, Sep. 2007.

[643] W. Santipach and M. L. Honig, "Asymptotic performance of MIMO wireless channels with limited feedback," *IEEE Military Commun. Conf. (MILCOM'03)*, vol. 1, pp. 141–146, Oct. 2003.

[644] ——, "Capacity of a multiple-antenna fading channel with a quantized precoding matrix," *IEEE Trans. Inform. Theory*, vol. 55, no. 3, pp. 1218–1234, 2009.

[645] C. K. Au-Yeung and D. J. Love, "On the performance of random vector quantization limited feedback beamforming in a MISO system," *IEEE Trans. Wireless Commun.*, vol. 6, no. 2, pp. 458–462, 2007.

[646] A. T. James, "Distributions of matrix variates and latent roots derived from normal samples," *Ann. Math. Stat.*, vol. 35, pp. 475–501, 1964.

[647] L. Zheng and D. N. C. Tse, "Communication on the Grassmann manifold: a geometric approach to the noncoherent multiple-antenna channel," *IEEE Trans. Inform. Theory*, vol. 48, no. 2, pp. 359–383, 2002.

[648] D. J. Love, R. W. Heath Jr., and T. Strohmer, "Grassmannian beamforming for

multiple-input multiple-output wireless systems," *IEEE Trans. Inform. Theory*, vol. 49, no. 10, pp. 2735–2747, 2003.

[649] J. H. Conway, R. H. Hardin, and N. J. A. Sloane, "Packing lines, planes, etc.: packings in Grassmannian spaces," *Experiment. Math.*, vol. 5, no. 2, pp. 139–159, 1996.

[650] I. S. Dhillon, R. W. Heath Jr., T. Strohmer, and J. A. Tropp, "Constructing packings in Grassmannian manifolds via alternating projection," *Experiment. Math.*, 2007.

[651] P. Xia, S. Zhou, and G. B. Giannakis, "Achieving the Welch bound with difference sets," *IEEE Trans. Inform. Theory*, vol. 51, no. 5, pp. 1900–1907, 2005.

[652] D. J. Love and R. W. Heath Jr., "Necessary and sufficient conditions for full diversity order in correlated Rayleigh fading beamforming and combining systems," *IEEE Trans. Wireless Commun.*, vol. 4, no. 1, pp. 20–23, 2005.

[653] L. Liu and H. Jafarkhani, "Novel transmit beamforming schemes for time-selective fading multiantenna systems," *IEEE Trans. Signal Process.*, vol. 54, no. 12, pp. 4767–4781, 2006.

[654] B. Mondal, S. Dutta, and R. W. Heath Jr., "Quantization on the Grassmann manifold," *IEEE Trans. Signal Process.*, vol. 55, no. 8, pp. 4208–4216, 2007.

[655] W. Dai, Y. Liu, and B. Rider, "Quantization bounds on Grassmann manifolds and applications to MIMO communications," *IEEE Trans. Inform. Theory*, vol. 54, no. 3, pp. 1108–1123, 2008.

[656] R. A. Pitaval and O. Tirkkonen, "Joint Grassmann–Stiefel quantization for MIMO product codebooks," *IEEE Trans. Wireless Commun.*, vol. 13, no. 1, pp. 210–222, 2014.

[657] P. Zador, "Asymptotic quantization error of continuous signals and the quantization dimension," *IEEE Trans. Inform. Theory*, vol. 28, no. 2, pp. 139–149, 1982.

[658] R. M. Gray and D. L. Neuhoff, "Quantization," *IEEE Trans. Inform. Theory*, vol. 44, no. 6, pp. 2325–2383, 1998.

[659] W. Dai, Y. Liu, B. Rider, and V. K. N. Lau, "On the information rate of MIMO systems with finite rate channel state feedback using beamforming and power on/off strategy," *IEEE Trans. Inform. Theory*, vol. 55, no. 11, pp. 5032–5047, 2009.

[660] D. J. Love and R. W. Heath Jr., "Limited feedback unitary precoding for spatial multiplexing systems," *IEEE Trans. Inform. Theory*, vol. 51, no. 8, pp. 2967–2976, 2005.

[661] R. W. Heath Jr. and A. J. Paulraj, "Switching between multiplexing and diversity based on constellation distance," *Allerton Conf. Commun., Control and Computing*, pp. 212–221, 2000.

[662] R. W. Heath Jr. and A. Paulraj, "Switching between diversity and multiplexing in MIMO systems," *IEEE Trans. Commun.*, vol. 53, no. 6, pp. 962–968, 2005.

[663] R. W. Heath Jr. and D. J. Love, "Multimode antenna selection for spatial multiplexing systems with linear receivers," *IEEE Trans. Signal Processing*, vol. 53, pp. 3042–3056, 2005.

[664] D. Love and R. W. Heath Jr., "Multimode precoding for MIMO wireless systems," *IEEE Trans. Signal Processing*, vol. 53, pp. 3674–3687, 2005.

[665] J. C. Roh and B. D. Rao, "Design and analysis of MIMO spatial multiplexing systems with quantized feedback," *IEEE Trans. Signal Process.*, vol. 54, no. 8, pp. 2874–2886, 2006.

[666] A. Gersho and R. M. Gray, *Vector quantization and signal compression.* Springer, 1991.

[667] J. C. Roh and B. D. Rao, "Transmit beamforming in multiple-antenna systems with finite rate feedback: a VQ-based approach," *IEEE Trans. Inform. Theory*, vol. 52, no. 3, pp. 1101–1112, 2006.

[668] V. Lau, Y. Liu, and T. A. Chen, "On the design of MIMO block-fading channels with feedback-link capacity constraint," *IEEE Trans. Commun.*, vol. 52, no. 1, pp. 62–70, 2004.

[669] Y. Linde, A. Buzo, and R. Gray, "An algorithm for vector quantizer design," *IEEE Trans. Commun.*, vol. 28, no. 1, pp. 84–95, 1980.

[670] S. Lloyd, "Least squares quantization in PCM," *IEEE Trans. Inform. Theory*, vol. 28, no. 2, pp. 129–137, 1982.

[671] P. G. Szabo, M. C. Markot, T. Csendes, E. Specht, L. G. Casado, and I. Garcia, *New approaches to circle packing in a square: with program codes.* Springer Science & Business Media, 2007.

[672] E. Specht, "The best known packings of equal circles in a square (up to n = 10000)," http://hydra.nat.uni-magdeburg.de/packing/csq/csq.html.

[673] N. J. A. Sloane, "Packings in Grassmannian spaces," http://neilsloane.com/grass/.

[674] T. Strohmer and R. W. Heath Jr., "Grassmannian frames with applications to coding and communications," *Appl. and Computational Harmonic Analysis*, vol. 14, no. 3, pp. 257–275, 2003.

[675] B. M. Hochwald, T. L. Marzetta, T. J. Richardson, W. Sweldens, and R. Urbanke, "Systematic design of unitary space–time constellations," *IEEE Trans. Inform. Theory*, vol. 46, no. 6, pp. 1962–1973, 2000.

[676] D. Love and R. W. Heath Jr., "Limited feedback unitary precoding for orthogonal space–time block codes," *IEEE Trans. Signal Processing*, vol. 53, no. 1, pp. 64–73, 2005.

[677] A. Kerdock, "Studies of low-rate binary codes," *IEEE Trans. Inform. Theory*, vol. 18, no. 2, pp. 316–316, 1972.

[678] A. R. Calderbank, P. J. Cameron, W. M. Kantor, and J. J. Seidel, "Z_4-Kerdock codes, orthogonal spreads, and extremal Euclidean line-sets," *Proc. London Math. Soc. (3)*, vol. 75, no. 2, pp. 436–480, 1997.

[679] A. Klappenecker and M. Roetteler, "Constructions of mutually unbiased bases," *Int'l Conf. Finite Fields and Applications*, pp. 137–144, 2004.

[680] T. Inoue and R. W. Heath Jr., "Kerdock codes for limited feedback precoded MIMO systems," *IEEE Trans. Signal Process.*, vol. 57, no. 9, pp. 3711–3716, 2009.

[681] R. W. Heath Jr., T. Strohmer, and A. J. Paulraj, "On quasi-orthogonal signatures for CDMA systems," *IEEE Trans. Inform. Theory*, vol. 52, no. 3, pp. 1217–1226, 2006.

[682] R. Gow, "Generation of mutually unbiased bases as powers of a unitary matrix in 2-power dimensions," *arXiv:math/0703333*, 2007.

[683] T. Durt, B.-G. Englert, I. Bengtsson, and K. Zyczkowski, "On mutually unbiased bases," *Int'l J. Quantum Inform.*, vol. 8, pp. 535–640, 2010.

[684] B. Mondal, T. A. Thomas, and M. Harrison, "Rank-independent codebook design from a quaternary alphabet," *Asilomar Conf. Signals, Systems, and Comp.*, pp. 297–301, Nov. 2007.

[685] T. Inoue and R. W. Heath Jr., "Kerdock codes for limited feedback MIMO systems," *IEEE Int'l Conf. Acoustics, Speech and Signal Process. (ICASSP'08)*, pp. 3113–3116, Mar. 2008.

[686] A. Klappenecker and M. Roetteler, "Mutually unbiased bases, spherical designs, and frames," *Proc. of SPIE*, vol. 5914, no. 1, 2005.

[687] J. Wang, M. Wu, and F. Zheng, "The codebook design for MIMO precoding systems in LTE and LTE-A," *Int'l Conf. Wireless Commun. Networking and Mobile Computing (WiCOM'10)*, pp. 1–4, Sep. 2010.

[688] M. Vu, "MISO capacity with per-antenna power constraint," *IEEE Trans. Commun.*, vol. 59, no. 5, pp. 1268–1274, 2011.

[689] ——, "MIMO capacity with per-antenna power constraint," *IEEE Global Commun. Conf. (GLOBECOM '11)*, Dec. 2011.

[690] S. Loyka, "The capacity of Gaussian MIMO channels under total and per-antenna power constraints," *IEEE Trans. Commun.*, vol. 65, no. 3, pp. 1035–1042, 2017.

[691] J. Dai, C. Chang, W. Xu, and Z. Ye, "Linear precoder optimization for MIMO systems with joint power constraints," *IEEE Trans. Commun.*, vol. 60, no. 8, pp. 2240–2254, 2012.

[692] D. Tuninetti, "On the capacity of the AWGN MIMO channel under per-antenna power constraints," *IEEE Int'l Conf. Commun. (ICC'14)*, pp. 2153–2157, Jun. 2014.

[693] S. K. Mohammed and E. G. Larsson, "Single-user beamforming in large-scale MISO systems with per-antenna constant-envelope constraints: the doughnut channel," *IEEE Trans. Wireless Commun.*, vol. 11, no. 11, pp. 3992–4005, 2012.

[694] J. Pan and W.-K. Ma, "Constant envelope precoding for single-user large-scale MISO channels: efficient precoding and optimal designs," *IEEE J. Sel. Topics Signal Process.*, vol. 8, no. 5, pp. 982–995, 2014.

[695] S. Zhang, R. Zhang, and T. J. Lim, "Constant envelope precoding for MIMO systems," *IEEE Trans. Commun.*, vol. 66, no. 1, pp. 149–162, 2018.

[696] C. Potter, K. Kosbar, and A. Panagos, "On achievable rates for MIMO systems with imperfect channel state information in the finite length regime," *IEEE Trans. Commun.*, vol. 61, no. 7, pp. 2772–2781, 2013.

[697] D. Wu and R. Negi, "Effective capacity: a wireless link model for support of quality of service," *IEEE Trans. Wireless Commun.*, vol. 2, no. 4, pp. 630–643, 2003.

[698] Y. Chau and S.-H. Yu, "Space modulation on wireless fading channels," *IEEE Veh. Techn. Conf. (VTC'01)*, vol. 3, pp. 1668–1671, 2001.

[699] H. Haas, E. Costa, and E. Schulz, "Increasing spectral efficiency by data multiplexing using antenna arrays," *IEEE Int'l Symp. Personal, Indoor and Mobile Radio Commun. (PIMRC'02)*, vol. 2, pp. 610–613, Sep. 2002.

[700] R. Y. Mesleh, H. Haas, S. Sinanović, C. W. Ahn, and S. Yun, "Spatial modulation," *IEEE Trans. Veh. Techn.*, vol. 57, no. 4, pp. 2228–2241, 2008.

[701] J. Jeganathan, A. Ghrayeb, L. Szczecinski, and A. Ceron, "Space shift keying modulation for MIMO channels," *IEEE Trans. Wireless Commun.*, vol. 8, no. 7, pp. 3692–3703, 2009.

[702] M. Di Renzo, H. Haas, and P. M. Grant, "Spatial modulation for multiple-antenna wireless systems: a survey," *IEEE Commun. Mag.*, vol. 49, no. 12, pp. 182–191, 2011.

[703] M. Di Renzo, H. Haas, A. Ghrayeb, S. Sugiura, and L. Hanzo, "Spatial modulation for generalized MIMO: challenges, opportunities, and implementation," *Proc. IEEE*, vol. 102, no. 1, pp. 56–103, 2014.

[704] Z. Bouida, H. El-Sallabi, A. Ghrayeb, and K. A. Qaraqe, "Reconfigurable antenna-based space-shift keying (SSK) for MIMO Rician channels," *IEEE Trans. Wireless Commun.*, vol. 15, pp. 446–457, Jan. 2016.

[705] A. A. I. Ibrahim, T. Kim, and D. J. Love, "On the achievable rate of generalized spatial modulation using multiplexing under a Gaussian mixture model," *IEEE Trans. Commun.*, vol. 64, pp. 1588–1599, Apr. 2016.

[706] A. Younis, N. Serafimovski, R. Mesleh, and H. Haas, "Generalised spatial modulation," *Asilomar Conf. Signals, Systems and Computers*, pp. 1498–1502, 2010.

[707] Z. An, J. Wang, J. Wang, and J. Song, "Mutual information and error probability analysis on generalized spatial modulation system," *IEEE Trans. Commun.*, vol. 65, no. 3, pp. 1044–1060, 2017.

[708] C. Sun, A. Hirata, T. Ohira, and N. C. Karmakar, "Fast beamforming of electronically steerable parasitic array radiator antennas: theory and experiment," *IEEE Trans. Antennas Propag.*, vol. 52, no. 7, pp. 1819–1832, 2004.

[709] O. N. Alrabadi, C. B. Papadias, A. Kalis, and R. Prasad, "A universal encoding scheme for MIMO transmission using a single active element for PSK modulation schemes," *IEEE Trans. Wireless Commun.*, vol. 8, no. 10, pp. 5133–5142, 2009.

[710] O. N. Alrabadi, C. Divarathne, P. Tragas, A. Kalis, N. Marchetti, C. B. Papadias, and R. Prasad, "Spatial multiplexing with a single radio: proof-of-concept experiments in an indoor environment with a 2.6-GHz prototype," *IEEE Commun. Letters*, vol. 15, no. 2, pp. 178–180, 2011.

[711] A. Kalis, A. G. Kanatas, and C. B. Papadias, *Parasitic antenna arrays for wireless MIMO systems*. Springer, 2014.

[712] A. K. Khandani, "Media-based modulation: a new approach to wireless transmission," *IEEE Int'l Symp. Inform. Theory (ISIT'13)*, pp. 3050–3054, Jul. 2013.

[713] ——, "Media-based modulation: converting static Rayleigh fading to AWGN," *IEEE Int'l Symp. Inform. Theory*, pp. 1549–1553, Jun. 2014.

[714] A. Alkhateeb, J. Mo, N. González-Prelcic, and R. W. Heath Jr., "MIMO precoding and combining solutions for millimeter-wave systems," *IEEE Commun. Mag.*, vol. 52, no. 12, pp. 122–131, 2014.

[715] F. Sohrabi and W. Yu, "Hybrid analog and digital beamforming for mmWave OFDM large-scale antenna arrays," *IEEE J. Sel. Areas Commun.*, vol. 35, no. 7, pp. 1432–1443, 2017.

[716] C. Artigue and P. Loubaton, "On the precoder design of flat fading MIMO systems equipped with MMSE receivers: a large-system approach," *IEEE Trans. Inform. Theory*, vol. 57, no. 7, pp. 4138–4155, 2011.

[717] M. Kiessling and J. Speidel, "Analytical performance of MIMO zero-forcing receivers in correlated Rayleigh fading environments," *IEEE Workshop on Signal Process. Advances in Wireless Commun. (SPAWC'03)*, pp. 383–387, 2003.

[718] D. Gore, R. W. Heath Jr., and A. Paulraj, "On performance of the zero forcing receiver in presence of transmit correlation," *IEEE Int'l Symp. Inform. Theory (ISIT'02)*, p. 159, 2002.

[719] P. Li, D. Paul, R. Narasimhan, and J. Cioffi, "On the distribution of SINR for the MMSE MIMO receiver and performance analysis," *IEEE Trans. Inform. Theory*, vol. 52, no. 1, pp. 271–286, 2006.

[720] C. Siriteanu, S. D. Blostein, A. Takemura, H. Shin, S. Yousefi, and S. Kuriki, "Exact MIMO zero-forcing detection analysis for transmit-correlated Rician fading," *IEEE Trans. Wireless Commun.*, vol. 13, no. 3, pp. 1514–1527, 2014.

[721] C. Siriteanu, A. Takemura, S. Kuriki, H. Shin, and C. Koutschan, "MIMO zero-forcing performance evaluation using the holonomic gradient method," *IEEE Trans. Wireless Commun.*, vol. 14, no. 4, pp. 2322–2335, 2015.

[722] K. R. Kumar, G. Caire, and A. L. Moustakas, "The diversity-multiplexing tradeoff of linear MIMO receivers," *IEEE Inform. Theory Workshop (ITW'07)*, pp. 487–492, 2007.

[723] H. Gao, P. J. Smith, and M. V. Clark, "Theoretical reliability of MMSE linear diversity combining in Rayleigh-fading additive interference channels," *IEEE Trans. Commun.*, vol. 46, no. 5, pp. 666–672, 1998.

[724] P. J. Smith, "Exact performance analysis of optimum combining with multiple interferers in flat Rayleigh fading," *IEEE Trans. Commun.*, vol. 55, no. 9, pp. 1674–1677, 2007.

[725] Y. Jiang, M. K. Varanasi, and J. Li, "Performance analysis of ZF and MMSE equalizers for MIMO systems: an in-depth study of the high SNR regime," *IEEE Trans. Inform. Theory*, vol. 57, no. 4, pp. 2008–2026, 2011.

[726] M. R. McKay, A. Zanella, I. B. Collings, and M. Chiani, "Error probability and SINR analysis of optimum combining in Rician fading," *IEEE Trans. Commun.*, vol. 57, no. 3, pp. 676–687, 2009.

[727] R. Louie, M. R. McKay, and I. B. Collings, "New performance results for multiuser optimum combining in the presence of Rician fading," *IEEE Trans. Commun.*, vol. 57, no. 8, pp. 2348–2358, 2009.

[728] H. V. Poor and S. Verdú, "Probability of error in MMSE multiuser detection," *IEEE Trans. Inform. Theory*, vol. 43, no. 3, pp. 858–871, 1997.

[729] R. H. Y. Louie, M. R. McKay, and I. B. Collings, "Maximum sum-rate of MIMO multiuser scheduling with linear receivers," *IEEE Trans. Commun.*, vol. 57, no. 11, pp. 3500–3510, 2009.

[730] M. R. McKay, I. B. Collings, and A. M. Tulino, "Achievable sum rate of MIMO MMSE receivers: a general analytic framework," *IEEE Trans. Inform. Theory*, vol. 56, no. 1, pp. 396–410, 2010.

[731] P. H. Tan, Y. Wu, and S. Sun, "Link adaptation based on adaptive modulation and coding for multiple-antenna OFDM system," *IEEE J. Sel. Areas Commun.*, vol. 26, no. 8, pp. 1599–1606, 2008.

[732] T. L. Jensen, S. Kant, J. Wehinger, and B. H. Fleury, "Fast link adaptation for MIMO OFDM," *IEEE Trans. Veh. Techn.*, vol. 59, no. 8, pp. 3766–3778, 2010.

[733] P. Bergmans and T. M. Cover, "Cooperative broadcasting," *IEEE Trans. Inform. Theory*, vol. 20, no. 3, pp. 317–324, 1974.

[734] Y. Saito, Y. Kishiyama, A. Benjebbour, T. Nakamura, A. Li, and K. Higuchi, "Non-orthogonal multiple access (NOMA) for cellular future radio access," *IEEE Veh. Techn. Conf. (VTC'03 Spring)*, pp. 1–5, 2013.

[735] P.-H. Kuo, "New physical layer features of 3GPP LTE Release-13," *IEEE Wireless Commun.*, vol. 22, pp. 4–5, Aug. 2015.

[736] Z. Ding, Z. Yang, P. Fan, and H. V. Poor, "On the performance of non-orthogonal multiple access in 5G systems with randomly deployed users," *IEEE Signal Process. Letters*, vol. 21, no. 12, pp. 1501–1505, 2014.

[737] Q. Wang, R. Zhang, L. L. Yang, and L. Hanzo, "Non-orthogonal multiple access: a unified perspective," *IEEE Wireless Commun.*, vol. 25, no. 2, pp. 10–16, 2018.

[738] L. Liu, R. Chen, S. Geirhofer, K. Sayana, Z. Shi, and Y. Zhou, "Downlink MIMO in LTE-advanced: SU-MIMO vs. MU-MIMO," *IEEE Commun. Mag.*, vol. 50, no. 2, pp. 140–147, 2012.

[739] A. Wiesel, Y. C. Eldar, and S. Shamai, "Linear precoding via conic optimization for fixed MIMO receivers," *IEEE Trans. Signal Process.*, vol. 54, no. 1, pp. 161–176, 2006.

[740] F. Kelly, "Charging and rate control for elastic traffic," *Eur. Trans. Telecommun.*, vol. 8, no. 1, pp. 33–37, 1997.

[741] J. Mo and J. Walrand, "Fair end-to-end window-based congestion control," *IEEE/ACM Trans. Netw.*, vol. 8, no. 5, pp. 556–567, 2000.

[742] P. Viswanath, D. N. C. Tse, and R. Laroia, "Opportunistic beamforming using dumb antennas," *IEEE Trans. Inform. Theory*, vol. 48, no. 6, pp. 1277–1294, 2002.

[743] T. Yoo, N. Jindal, and A. Goldsmith, "Multi-antenna downlink channels with limited feedback and user selection," *IEEE J. Sel. Areas Commun.*, vol. 25, no. 7, pp. 1478–1491, 2007.

[744] Z. Tu and R. S. Blum, "Multiuser diversity for a dirty paper approach," *IEEE Commun. Letters*, vol. 7, no. 8, pp. 370–372, 2003.

[745] G. Dimic and N. D. Sidiropoulos, "On downlink beamforming with greedy user selection: performance analysis and a simple new algorithm," *IEEE Trans. Signal Process.*, vol. 53, no. 10, pp. 3857–3868, 2005.

[746] M. Sharif and B. Hassibi, "On the capacity of MIMO broadcast channels with partial side information," *IEEE Trans. Inform. Theory*, vol. 51, no. 2, pp. 506–522, 2005.

[747] M. Kang, Y. J. Sang, H. G. Hwang, H. Y. Lee, and K. S. Kim, "Performance analysis of proportional fair scheduling with partial feedback information for multiuser multicarrier systems," *IEEE Veh. Techn. Conf. (VTC'09 Spring)*, pp. 1–5, 2009.

[748] D. Morales-Jimenez and A. Lozano, "Ergodic sum-rate of proportional fair scheduling with multiple antennas," *IEEE Int'l Symp. Inform. Theory (ISIT'13)*, pp. 2124–2128, 2013.

[749] F. P. Kelly, A. K. Maulloo, and D. K. H. Tan, "Rate control for communication networks: shadow prices, proportional fairness and stability," *J. Oper. Res. Soc.*, vol. 49, no. 3, pp. 237–252, 1998.

[750] A. L. Stolyar, "On the asymptotic optimality of the gradient scheduling algorithm for multiuser throughput allocation," *Oper. Res.*, vol. 53, no. 1, pp. 12–25, 2005.

[751] A. Jalali, R. Padovani, and R. Pankaj, "Data throughput of CDMA-HDR a high efficiency-high data rate personal communication wireless system," *IEEE Veh. Techn. Conf. (VTC'00 Spring)*, vol. 3, pp. 1854–1858, 2000.

[752] J.-G. Choi and S. Bahk, "Cell-throughput analysis of the proportional fair scheduler in the single-cell environment," *IEEE Trans. Veh. Techn.*, vol. 56, no. 2, pp. 766–778, 2007.

[753] A. L. Stolyar, "Greedy primal-dual algorithm for dynamic resource allocation in complex networks," *Queue. Syst.*, vol. 54, no. 3, pp. 203–220, 2006.

[754] G. Caire, R. R. Muller, and R. Knopp, "Hard fairness versus proportional fairness in wireless communications: the single-cell case," *IEEE Trans. Inform. Theory*, vol. 53, no. 4, pp. 1366–1385, 2007.

[755] G. Caire, D. Tuninetti, and S. Verdú, "Suboptimality of TDMA in the low-power regime," *IEEE Trans. Inform. Theory*, vol. 50, no. 4, pp. 608–620, 2004.

[756] T. Philosof and R. Zamir, "The cost of uncorrelation and noncooperation in MIMO channels," *IEEE Trans. Inform. Theory*, vol. 53, no. 11, pp. 3904–3920, 2007.

[757] R. Ahlswede, "Multi-way communication channels," *IEEE Int'l Symp. Inform. Theory (ISIT'71)*, pp. 23–52, 1971.

[758] H. Liao, "Multiple access channels," Ph.D. dissertation, University of Hawaii, 1972.

[759] R. S. Cheng and S. Verdú, "Gaussian multiaccess channels with ISI: capacity region and multiuser water-filling," *IEEE Trans. Inform. Theory*, vol. 39, no. 3, pp. 773–785, 1993.

[760] H.-F. Chong and M. Motani, "Capacity region of the asynchronous Gaussian vector multiple-access channel," *IEEE Trans. Inform. Theory*, vol. 59, no. 9, pp. 5398–5420, 2013.

[761] R. Knopp and P. A. Humblet, "Information capacity and power control in single-cell multiuser communications," *IEEE Int'l Conf. Commun. (ICC'95)*, vol. 1, pp. 331–335, Jun. 1995.

[762] D. N. C. Tse and S. V. Hanly, "Multiaccess fading channels. I. Polymatroid structure, optimal resource allocation and throughput capacities," *IEEE Trans. Inform. Theory*, vol. 44, no. 7, pp. 2796–2815, 1998.

[763] P. Viswanath, D. N. C. Tse, and V. Anantharam, "Asymptotically optimal water-filling in vector multiple-access channels," *IEEE Trans. Inform. Theory*, vol. 47, no. 1, pp. 241–267, 2001.

[764] W. Yu, W. Rhee, and J. M. Cioffi, "Optimal power control in multiple access fading channels with multiple antennas," *IEEE Int'l Conf. Commun. (ICC'01)*, vol. 2, pp. 575–579, Jun. 2001.

[765] W. Yu and W. Rhee, "Degrees of freedom in wireless multiuser spatial multiplex systems with multiple antennas," *IEEE Trans. Commun.*, vol. 54, no. 10, pp. 1747–1753, 2006.

[766] M. A. Maddah-Ali, A. Mobasher, and A. K. Khandani, "Fairness in multiuser systems with polymatroid capacity region," *IEEE Trans. Inform. Theory*, vol. 55, no. 5, pp. 2128–2138, 2009.

[767] D. Calabuig, R. H. Gohary, and H. Yanikomeroglu, "Optimum transmission through the multiple-antenna Gaussian multiple access channel," *IEEE Trans. Inform. Theory*, vol. 62, no. 1, pp. 230–243, 2016.

[768] W. Yu, W. Rhee, S. Boyd, and J. M. Cioffi, "Iterative water-filling for Gaussian vector multiple-access channels," *IEEE Trans. Inform. Theory*, vol. 50, no. 1, pp. 145–152, 2004.

[769] M. Wang, W. Zeng, and C. Xiao, "Linear precoding for MIMO multiple access channels with finite discrete inputs," *IEEE Trans. Wireless Commun.*, vol. 10, no. 11, pp. 3934–3942, 2011.

[770] J. Harshan and B. S. Rajan, "On two-user Gaussian multiple access channels with finite input constellations," *IEEE Trans. Inform. Theory*, vol. 57, no. 3, pp. 1299–1327, 2011.

[771] S. Bellofiore, C. A. Balanis, J. Foutz, and A. S. Spanias, "Smart-antenna systems for mobile communication networks. Part 1. Overview and antenna design," *IEEE Antennas Propag. Mag.*, vol. 44, no. 3, pp. 145–154, 2002.

[772] P. Zetterberg and B. Ottersten, "The spectrum efficiency of a base station antenna array system for spatially selective transmission," *IEEE Trans. Veh. Techn.*, vol. 44, no. 3, pp. 651–660, 1995.

[773] B. Suard, G. Xu, H. Liu, and T. Kailath, "Uplink channel capacity of space-division-multiple-access schemes," *IEEE Trans. Inform. Theory*, vol. 44, no. 4, pp. 1468–1476, 1998.

[774] J. Lee and N. Jindal, "High SNR analysis for MIMO broadcast channels: dirty paper coding versus linear precoding," *IEEE Trans. Inform. Theory*, vol. 53, no. 12, pp. 4787–4792, 2007.

[775] D. N. C. Tse, P. Viswanath, and L. Zheng, "Diversity-multiplexing tradeoff in multiple-access channels," *IEEE Trans. Inform. Theory*, vol. 50, no. 9, pp. 1859–1874, 2004.

[776] S. Shamai and A. D. Wyner, "Information-theoretic considerations for symmetric, cellular, multiple-access fading channels," *IEEE Trans. Inform. Theory*, vol. 43, no. 6, pp. 1877–1894, 1997.

[777] S. A. Jafar, S. Vishwanath, and A. Goldsmith, "Vector MAC capacity region with covariance feedback," *IEEE Int'l Symp. Inform. Theory (ISIT'01)*, vol. 1, p. 54, Jun. 2001.

[778] A. Soysal and S. Ulukus, "Optimality of beamforming in fading MIMO multiple access channels," *IEEE Trans. Commun.*, vol. 57, no. 4, pp. 1171–1183, 2009.

[779] Y. Wu, C.-K. Wen, C. Xiao, X. Gao, and R. Schober, "Linear precoding for the MIMO multiple access channel with finite alphabet inputs and statistical CSI," *IEEE Trans. Wireless Commun.*, vol. 14, no. 2, pp. 983–997, 2015.

[780] M. J. M. Peacock, I. B. Collings, and M. L. Honig, "Eigenvalue distributions of sums and products of large random matrices via incremental matrix expansions," *IEEE Trans. Inform. Theory*, vol. 54, no. 5, pp. 2123–2138, 2008.

[781] R. Couillet, M. Debbah, and J. W. Silverstein, "A deterministic equivalent for the analysis of correlated MIMO multiple access channels," *IEEE Trans. Inform. Theory*, vol. 57, no. 6, pp. 3493–3514, 2011.

[782] D. Aktas, M. N. Bacha, J. S. Evans, and S. V. Hanly, "Scaling results on the sum capacity of cellular networks with MIMO links," *IEEE Trans. Inform. Theory*, vol. 52, no. 7, pp. 3264–3274, 2006.

[783] H. Huh, S.-H. Moon, Y.-T. Kim, I. Lee, and G. Caire, "Multi-cell MIMO downlink with cell cooperation and fair scheduling: a large-system limit analysis," *IEEE Trans. Inform. Theory*, vol. 57, no. 12, pp. 7771–7786, 2011.

[784] A. Aubry, I. Esnaola, A. M. Tulino, and S. Venkatesan, "Achievable rate region for Gaussian MIMO MAC with partial CSI," *IEEE Trans. Inform. Theory*, vol. 59, no. 7, pp. 4139–4170, 2013.

[785] A. Lozano, J. G. Andrews, and R. W. Heath Jr., "Spectral efficiency limits in pilot-assisted cooperative communications," *IEEE Int'l Symp. Inform. Theory (ISIT'12)*, pp. 1132–1136, 2012.

[786] P. Zhao, G. Fodor, G. Dan, and M. Telek, "A game theoretic approach to setting the pilot power ratio in multi-user MIMO systems," *IEEE Trans. Commun.*, vol. 66, no. 3, pp. 999–1012, 2018.

[787] F. Rashid-Farrokhi, K. J. R. Liu, and L. Tassiulas, "Transmit beamforming and power control for cellular wireless systems," *IEEE J. Sel. Areas Commun.*, vol. 16, no. 8, pp. 1437–1450, 1998.

[788] F. Rashid-Farrokhi, L. Tassiulas, and K. J. R. Liu, "Joint optimal power control and beamforming in wireless networks using antenna arrays," *IEEE Trans. Commun.*, vol. 46, no. 10, pp. 1313–1324, 1998.

[789] S. Vishwanath, N. Jindal, and A. Goldsmith, "Duality, achievable rates, and sum-rate capacity of Gaussian MIMO broadcast channels," *IEEE Trans. Inform. Theory*, vol. 49, no. 10, pp. 2658–2668, 2003.

[790] P. Viswanath and D. N. C. Tse, "Sum capacity of the vector Gaussian broadcast channel and uplink–downlink duality," *IEEE Trans. Inform. Theory*, vol. 49, no. 8, pp. 1912–1921, 2003.

[791] N. Jindal, S. Vishwanath, and A. Goldsmith, "On the duality of Gaussian multiple-access and broadcast channels," *IEEE Trans. Inform. Theory*, vol. 50, no. 5, pp. 768–783, 2004.

[792] W. Yu and T. Lan, "Transmitter optimization for the multi-antenna downlink with per-antenna power constraints," *IEEE Trans. Signal Process.*, vol. 55, no. 6, pp. 2646–2660, 2007.

[793] P. Bergmans, "A simple converse for broadcast channels with additive white Gaussian noise," *IEEE Trans. Inform. Theory*, vol. 20, no. 2, pp. 279–280, 1974.

[794] J. Korner and K. Marton, "General broadcast channels with degraded message sets," *IEEE Trans. Inform. Theory*, vol. 23, no. 1, pp. 60–64, 1977.

[795] A. El Gamal, "The capacity of a class of broadcast channels," *IEEE Trans. Inform. Theory*, vol. 25, no. 2, pp. 166–169, 1979.

[796] G. Caire and S. Shamai, "On the achievable throughput of a multiantenna Gaussian broadcast channel," *IEEE Trans. Inform. Theory*, vol. 49, no. 7, pp. 1691–1706, 2003.

[797] H. Weingarten, Y. Steinberg, and S. Shamai, "The capacity region of the Gaussian multiple-input multiple-output broadcast channel," *IEEE Trans. Inform. Theory*, vol. 52, no. 9, pp. 3936–3964, 2006.

[798] M. Tomlinson, "New automatic equaliser employing modulo arithmetic," *Electronics Letters*, vol. 7, no. 5, pp. 138–139, 1971.

[799] H. Harashima and H. Miyakawa, "Matched-transmission technique for channels with intersymbol interference," *IEEE Trans. Commun.*, vol. 20, no. 4, pp. 774–780, 1972.

[800] W. Yu and J. M. Cioffi, "Trellis precoding for the broadcast channel," *IEEE Global Commun. Conf. (GLOBECOM'01)*, vol. 2, pp. 1344–1348, 2001.

[801] C. Windpassinger, R. F. H. Fischer, T. Vencel, and J. B. Huber, "Precoding in multiantenna and multiuser communications," *IEEE Trans. Wireless Commun.*, vol. 3, no. 4, pp. 1305–1316, 2004.

[802] C. Windpassinger, R. F. H. Fischer, and J. B. Huber, "Lattice-reduction-aided broadcast precoding," *IEEE Trans. Commun.*, vol. 52, no. 12, pp. 2057–2060, 2004.

[803] M. Joham, J. Brehmer, and W. Utschick, "MMSE approaches to multiuser spatio-temporal Tomlinson–Harashima precoding," *ITG Conf. Source Channel Coding*, pp. 387–394, 2004.

[804] B. M. Hochwald, C. B. Peel, and A. L. Swindlehurst, "A vector-perturbation technique for near-capacity multiantenna multiuser communication—part II: perturbation," *IEEE Trans. Commun.*, vol. 53, no. 3, pp. 537–544, 2005.

[805] M. Barrenechea, M. Joham, M. Mendicute, and W. Utschick, "Analysis of vector precoding at high SNR: rate bounds and ergodic results," *IEEE Global Commun. Conf. (GLOBECOM'10)*, pp. 1–5, 2010.

[806] M. H. M. Costa, "Writing on dirty paper," *IEEE Trans. Inform. Theory*, vol. 29, no. 3, pp. 439–441, 1983.

[807] U. Erez, S. Shamai, and R. Zamir, "Capacity and lattice strategies for canceling known interference," *IEEE Trans. Inform. Theory*, vol. 51, no. 11, pp. 3820–3833, 2005.

[808] U. Erez and S. ten Brink, "A close-to-capacity dirty paper coding scheme," *IEEE Trans. Inform. Theory*, vol. 51, no. 10, pp. 3417–3432, 2005.

[809] A. Khina and U. Erez, "On the robustness of dirty paper coding," *IEEE Trans. Commun.*, vol. 58, no. 5, pp. 1437–1446, 2010.

[810] N. Jindal, W. Rhee, S. Vishwanath, S. A. Jafar, and A. Goldsmith, "Sum power iterative water-filling for multi-antenna Gaussian broadcast channels," *IEEE Trans. Inform. Theory*, vol. 51, no. 4, pp. 1570–1580, 2005.

[811] M. Kobayashi and G. Caire, "An iterative water-filling algorithm for maximum weighted sum-rate of Gaussian MIMO-BC," *IEEE J. Sel. Areas Commun.*, vol. 24, no. 8, pp. 1640–1646, 2006.

[812] Y. Wu, M. Wang, C. Xiao, Z. Ding, and X. Gao, "Linear precoding for MIMO broadcast channels with finite-alphabet constraints," *IEEE Trans. Wireless Commun.*, vol. 11, no. 8, pp. 2906–2920, 2012.

[813] S. A. Jafar and A. Goldsmith, "Isotropic fading vector broadcast channels: the scalar upper bound and loss in degrees of freedom," *IEEE Trans. Inform. Theory*, vol. 51, no. 3, pp. 848–857, 2005.

[814] A. G. Davoodi and S. A. Jafar, "Aligned image sets under channel uncertainty: settling conjectures on the collapse of degrees of freedom under finite precision CSIT," *IEEE Trans. Inform. Theory*, vol. 62, pp. 5603–5618, 2016.

[815] A. Lapidoth, S. Shamai, and M. Wigger, "On the capacity of a MIMO fading broadcast channel with imperfect transmitter side-information," *Allerton Conf. Commun., Control and Computing*, 2005.

[816] G. Caire, N. Jindal, and S. Shamai, "On the required accuracy of transmitter channel state information in multiple antenna broadcast channels," *Asilomar Conf. Signals, Systems and Computers*, pp. 287–291, 2007.

[817] D. Hammarwall, M. Bengtsson, and B. Ottersten, "Utilizing the spatial information provided by channel norm feedback in SDMA systems," *IEEE Trans. Signal Process.*, vol. 56, no. 7, pp. 3278–3293, 2008.

[818] V. Raghavan, S. V. Hanly, and V. V. Veeravalli, "Statistical beamforming on the Grassmann manifold for the two-user broadcast channel," *IEEE Trans. Inform. Theory*, vol. 59, no. 10, pp. 6464–6489, 2013.

[819] J. Wang, S. Jin, X. Gao, K.-K. Wong, and E. Au, "Statistical eigenmode-based SDMA for two-user downlink," *IEEE Trans. Signal Process.*, vol. 60, no. 10, pp. 5371–5383, 2012.

[820] S. Serbetli and A. Yener, "Transceiver optimization for multiuser MIMO systems," *IEEE Trans. Signal Process.*, vol. 52, no. 1, pp. 214–226, 2004.

[821] E. Visotsky and U. Madhow, "Optimum beamforming using transmit antenna arrays," *IEEE Veh. Techn. Conf. (VTC'99)*, vol. 1, pp. 851–856, 1999.

[822] M. Chiang, C. W. Tan, D. P. Palomar, D. O'Neill, and D. Julian, "Power control by geometric programming," *IEEE Trans. Wireless Commun.*, vol. 6, no. 7, pp. 2640–2651, 2007.

[823] P. Hande, S. Rangan, M. Chiang, and X. Wu, "Distributed uplink power control for optimal SIR assignment in cellular data networks," *IEEE/ACM Trans. Netw.*, vol. 16, no. 6, pp. 1420–1433, 2008.

[824] A. Wiesel, Y. C. Eldar, and S. Shamai, "Zero-forcing precoding and generalized inverses," *IEEE Trans. Signal Process.*, vol. 56, no. 9, pp. 4409–4418, 2008.

[825] A. Lozano, A. M. Tulino, and S. Verdú, "Optimum power allocation for multiuser OFDM with arbitrary signal constellations," *IEEE Trans. Commun.*, vol. 56, pp. 828–837, 2008.

[826] ——, "Multiuser mercury/waterfilling for downlink OFDM with arbitrary signal constellations," *Proc. Int'l Symp. Spread Spectrum Tech. and Applications (ISSSTA'06)*, 2006.

[827] B. Hochwald and S. Vishwanath, "Space–time multiple access: linear growth in the sum rate," *Allerton Conf. Commun., Control and Computing*, 2002.

[828] T. Yoo and A. Goldsmith, "On the optimality of multiantenna broadcast scheduling using zero-forcing beamforming," *IEEE J. Sel. Areas Commun.*, vol. 24, no. 3, pp. 528–541, 2006.

[829] F. Boccardi and H. Huang, "Zero-forcing precoding for the MIMO broadcast channel under per-antenna power constraints," *IEEE Workshop on Signal Process. Adv. in Wireless Commun. (SPAWC'06)*, pp. 1–5, 2006.

[830] G. Caire, N. Jindal, M. Kobayashi, and N. Ravindran, "Multiuser MIMO achievable rates with downlink training and channel state feedback," *IEEE Trans. Inform. Theory*, vol. 56, no. 6, pp. 2845–2866, 2010.

[831] T. Tang, R. W. Heath Jr., S. Cho, and S. Yun, "Opportunistic feedback for multiuser MIMO systems with linear receivers," *IEEE Trans. Commun.*, vol. 55, no. 5, pp. 1020–1032, 2007.

[832] D. Gesbert and M.-S. Alouini, "How much feedback is multi-user diversity really worth?" *IEEE Int'l Conf. Commun. (ICC'04)*, vol. 1, pp. 234–238, June 2004.

[833] S. Sanayei and A. Nosratinia, "Exploiting multiuser diversity with only 1-bit feedback," *IEEE Wireless Commun. Netw. Conf. (WCNC'05)*, vol. 2, pp. 978–983, 2005.

[834] C. Swannack, E. Uysal-Biyikoglu, and G. W. Wornell, "Finding NEMO: near mutually orthogonal sets and applications to MIMO broadcast scheduling," *Int'l Conf. Wireless Networks, Commun., Mobile Computing*, June 2005.

[835] ——, "MIMO broadcast scheduling with limited channel state information," *Allerton Conf. on Comm. Control and Comp.*, Sep. 2005.

[836] T. Yoo, N. Jindal, and A. Goldsmith, "Multi-antenna downlink channels with limited feedback and user selection," *IEEE J. Sel. Areas Commun.*, vol. 25, no. 7, pp. 1478–1491, 2007.

[837] N. Jindal, "MIMO broadcast channels with finite-rate feedback," *IEEE Trans. Inform. Theory*, vol. 52, no. 11, pp. 5045–5060, 2006.

[838] P. Ding, D. J. Love, and M. D. Zoltowski, "Multiple antenna broadcast channels with shape feedback and limited feedback," *IEEE Trans. Signal Process.*, vol. 55, pp. 3417–3428, July 2007.

[839] W. Choi, A. Forenza, J. G. Andrews, and R. W. Heath Jr., "Opportunistic space-division multiple access with beam selection," *IEEE Trans. Commun.*, vol. 55, no. 12, pp. 2371–2380, 2007.

[840] K. Huang, J. G. Andrews, and R. W. Heath Jr., "Performance of orthogonal beamforming for SDMA with limited feedback," *IEEE Trans. Veh. Techn.*, vol. 58, no. 1, pp. 152–164, 2009.

[841] N. Jindal, "Antenna combining for the MIMO downlink channel," *IEEE Trans. Wireless Commun.*, vol. 7, no. 10, pp. 3834–3844, 2008.

[842] K.-B. Huang, R. W. Heath Jr., and J. G. Andrews, "SDMA with a sum feedback rate constraint," *IEEE Trans. Signal Process.*, vol. 55, no. 7, pp. 3879–3891, 2007.

[843] C. Swannack, G. W. Wornell, and E. Uysal-Biyikoglu, "MIMO broadcast scheduling with quantized channel state information," *IEEE Int'l Symp. Inform. Theory (ISIT'06)*, pp. 1788–1792, July 2006.

[844] "Downlink MIMO for EUTRA," 3GPP TSG RAN WG1 # 44/R1-060335, Tech. Rep., Feb. 2006.

[845] C. B. Chae, D. Mazzarese, N. Jindal, and R. W. Heath Jr., "Coordinated beamforming with limited feedback in the MIMO broadcast channel," *IEEE J. Sel. Areas Commun.*, vol. 26, no. 8, pp. 1505–1515, 2008.

[846] L. Thiele, M. Schellmann, W. Zirwas, and V. Jungnickel, "Capacity scaling of multiuser MIMO with limited feedback in a multicell environment," *Asilomar Conf. Signals, Systems and Computers*, Nov. 2007.

[847] N. Ravindran and N. Jindal, "Limited feedback-based block diagonalization for the MIMO broadcast channel," *IEEE J. Sel. Areas Commun.*, vol. 26, no. 8, pp. 1473–1482, 2008.

[848] R. Bhagavatula and R. W. Heath Jr., "Adaptive bit partitioning for multicell intercell interference nulling with delayed limited feedback," *IEEE Trans. Signal Process.*, vol. 59, no. 8, pp. 3824–3836, 2011.

[849] ——, "Adaptive limited feedback for sum-rate maximizing beamforming in cooperative multicell systems," *IEEE Trans. Signal Process.*, vol. 59, no. 2, pp. 800–811, 2011.

[850] M. Kobayashi, N. Jindal, and G. Caire, "Training and feedback optimization for multiuser MIMO downlink," *IEEE Trans. Commun.*, vol. 59, no. 8, pp. 2228–2240, 2011.

[851] X. Du, J. Tadrous, and A. Sabharwal, "Sequential beamforming for multiuser MIMO with full-duplex training," *IEEE Trans. Wireless Commun.*, vol. 15, no. 12, pp. 8551–8564, 2016.

[852] J. Nam and J.-Y. Ahn, "Joint spatial division and multiplexing—benefits of antenna correlation in multi-user MIMO," *IEEE Int'l Symp. Inform. Theory (ISIT'13)*, pp. 619–623, Jul. 2013.

[853] J. Nam, G. Caire, and J. Ha, "On the role of transmit correlation diversity in multiuser MIMO systems," *IEEE Trans. Inform. Theory*, vol. 63, no. 1, pp. 336–354, 2017.

[854] H. Yin, D. Gesbert, M. Filippou, and Y. Liu, "A coordinated approach to channel estimation in large-scale multiple-antenna systems," *IEEE J. Sel. Areas Commun.*, vol. 31, no. 2, pp. 264–273, 2013.

[855] Y. Jeon, C. Song, S. R. Lee, S. Maeng, J. Jung, and I. Lee, "New beamforming designs for joint spatial division and multiplexing in large-scale MISO multi-user systems," *IEEE Trans. Wireless Commun.*, vol. 16, no. 5, pp. 3029–3041, 2017.

[856] L. You, X. Gao, X.-G. Xia, N. Ma, and Y. Peng, "Pilot reuse for massive MIMO transmission over spatially correlated Rayleigh fading channels," *IEEE Trans. Wireless Commun.*, vol. 14, no. 6, pp. 3352–3366, 2015.

[857] K. K. Wong, "Maximizing the sum-rate and minimizing the sum-power of a broadcast 2-user 2-input multiple-output antenna system using a generalized zeroforcing approach," *IEEE Trans. Wireless Commun.*, vol. 5, no. 12, pp. 3406–3412, 2006.

[858] H. Viswanathan, S. Venkatesan, and H. Huang, "Downlink capacity evaluation of cellular networks with known-interference cancellation," *IEEE J. Sel. Areas Commun.*, vol. 21, no. 5, pp. 802–811, 2003.

[859] L.-U. Choi and R. D. Murch, "A transmit preprocessing technique for multiuser MIMO systems using a decomposition approach," *IEEE Trans. Wireless Commun.*, vol. 3, no. 1, pp. 20–24, 2004.

[860] Q. H. Spencer, A. L. Swindlehurst, and M. Haardt, "Zero-forcing methods for downlink spatial multiplexing in multiuser MIMO channels," *IEEE Trans. Signal Process.*, vol. 52, no. 2, pp. 461–471, 2004.

[861] F. Boccardi and H. C. Huang, "A near-optimum technique using linear precoding for the MIMO broadcast channel," *IEEE Int'l Conf. Acoustics, Speech and Signal Process. (ICASSP'07)*, vol. 3, pp. III–17, 2007.

[862] R. Zhang, "Cooperative multi-cell block diagonalization with per-base-station power constraints," *IEEE J. Sel. Areas Commun.*, vol. 28, no. 9, pp. 1435–1445, 2010.

[863] Z. Shen, R. Chen, J. G. Andrews, R. W. Heath Jr., and B. L. Evans, "Low complexity user selection algorithms for multiuser MIMO systems with block diagonalization," *IEEE Trans. Signal Process.*, vol. 54, no. 9, pp. 3658–3663, 2006.

[864] K. Ko and J. Lee, "Multiuser MIMO user selection based on chordal distance," *IEEE Trans. Commun.*, vol. 60, no. 3, pp. 649–654, 2012.

[865] M. Sadek, A. Tarighat, and A. H. Sayed, "A leakage-based precoding scheme for downlink multi-user MIMO channels," *IEEE Trans. Wireless Commun.*, vol. 6, no. 5, pp. 1711–1721, 2007.

[866] C. Geng, N. Naderializadeh, A. S. Avestimehr, and S. A. Jafar, "On the optimality of treating interference as noise," *IEEE Trans. Inform. Theory*, vol. 61, no. 4, pp. 1753–1767, 2015.

[867] E. Björnson, M. Bengtsson, and B. Ottersten, "Optimal multiuser transmit beamforming: a difficult problem with a simple solution structure," *IEEE Signal Process. Mag.*, vol. 31, no. 4, pp. 142–148, 2014.

[868] Y.-F. Liu, Y.-H. Dai, and Z.-Q. Luo, "Coordinated beamforming for MISO interference channel: complexity analysis and efficient algorithms," *IEEE Trans. Signal Process.*, vol. 59, no. 3, pp. 1142–1157, 2011.

[869] E. Björnson and E. Jorswieck, "Optimal resource allocation in coordinated multicell systems," *Found. Trends Commun. Inform. Theory*, vol. 9, no. 2–3, pp. 113–381, 2013.

[870] C. B. Peel, B. M. Hochwald, and A. L. Swindlehurst, "A vector-perturbation technique for near-capacity multiantenna multiuser communication—part I: channel inversion and regularization," *IEEE Trans. Commun.*, vol. 53, no. 1, pp. 195–202, 2005.

[871] S. Park, J. Park, A. Yazdan, and R. W. Heath Jr., "Optimal user loading in massive

MIMO systems with regularized zero forcing precoding," *IEEE Wireless Commun. Letters*, vol. 6, no. 1, pp. 118–121, 2017.

[872] B. R. Vojcic and W. M. Jang, "Transmitter precoding in synchronous multiuser communications," *IEEE Trans. Commun.*, vol. 46, no. 10, pp. 1346–1355, 1998.

[873] M. Joham, W. Utschick, and J. A. Nossek, "Linear transmit processing in MIMO communications systems," *IEEE Trans. Signal Process.*, vol. 53, no. 8, pp. 2700–2712, 2005.

[874] M. Stojnic, H. Vikalo, and B. Hassibi, "Rate maximization in multi-antenna broadcast channels with linear preprocessing," *IEEE Trans. Wireless Commun.*, vol. 5, no. 9, pp. 2338–2342, 2006.

[875] C. Jie and A. L. Swindlehurst, "Applying bargaining solutions to resource allocation in multiuser MIMO-OFDMA broadcast systems," *IEEE J. Sel. Topics Signal Process.*, vol. 6, no. 2, pp. 127–139, 2012.

[876] P. C. Weeraddana, M. Codreanu, M. Latva-aho, and A. Ephremides, "Weighted sum-rate maximization for a set of interfering links via branch and bound," *IEEE Trans. Signal Process.*, vol. 59, no. 8, pp. 3977–3996, 2011.

[877] S. Shi, M. Schubert, and H. Boche, "Rate optimization for multiuser MIMO systems with linear processing," *IEEE Trans. Signal Process.*, vol. 56, no. 8, pp. 4020–4030, 2008.

[878] V. Stankovic and M. Haardt, "Generalized design of multi-user MIMO precoding matrices," *IEEE Trans. Wireless Commun.*, vol. 7, no. 3, pp. 953–961, 2008.

[879] S. S. Christensen, R. Agarwal, E. de Carvalho, and J. M. Cioffi, "Weighted sum-rate maximization using weighted MMSE for MIMO-BC beamforming design," *IEEE Trans. Wireless Commun.*, vol. 7, no. 12, pp. 4792–4799, 2008.

[880] Q. Shi, M. Razaviyayn, Z.-Q. Luo, and C. He, "An iteratively weighted MMSE approach to distributed sum-utility maximization for a MIMO interfering broadcast channel," *IEEE Int'l Conf. Acoustics, Speech and Signal Process. (ICASSP'11)*, pp. 3060–3063, 2011.

[881] M. A. Girnyk, A. Müller, M. Vehkaperä, L. K. Rasmussen, and M. Debbah, "On the asymptotic sum rate of downlink cellular systems with random user locations," *IEEE Wireless Commun. Letters*, vol. 4, no. 3, pp. 333–336, 2015.

[882] T. L. Marzetta, "How much training is required for multiuser MIMO?" *Asilomar Conf. Signals, Systems and Computers*, pp. 359–363, 2006.

[883] ——, "Noncooperative cellular wireless with unlimited numbers of base station antennas," *IEEE Trans. Wireless Commun.*, vol. 9, no. 11, pp. 3590–3600, 2010.

[884] J. Hoydis, S. ten Brink, and M. Debbah, "Massive MIMO in the UL/DL of cellular networks: how many antennas do we need?" *IEEE J. Sel. Areas Commun.*, vol. 31, no. 2, pp. 160–171, 2013.

[885] H. Yang and T. L. Marzetta, "Performance of conjugate and zero-forcing beamforming in large-scale antenna systems," *IEEE J. Sel. Areas Commun.*, vol. 31, no. 2, pp. 172–179, 2013.

[886] H. Q. Ngo, E. G. Larsson, and T. L. Marzetta, "Energy and spectral efficiency of very large multiuser MIMO systems," *IEEE Trans. Commun.*, vol. 61, no. 4, pp. 1436–1449, 2013.

[887] F. Rusek, D. Persson, B. K. Lau, E. G. Larsson, T. L. Marzetta, O. Edfors, and F. Tufvesson, "Scaling up MIMO: opportunities and challenges with very large arrays," *IEEE Signal Process. Mag.*, vol. 30, no. 1, pp. 40–60, 2013.

[888] L. Lu, G. Y. Li, A. L. Swindlehurst, A. Ashikhmin, and R. Zhang, "An overview of massive MIMO: benefits and challenges," *IEEE J. Sel. Topics Signal Process.*, vol. 8, no. 5, pp. 742–758, 2014.

[889] K. T. Truong, A. Lozano, and R. W. Heath Jr., "Optimal training in continuous flat-fading massive MIMO systems," *Eur. Wireless Conf.*, pp. 1–6, 2014.

[890] E. G. Larsson, O. Edfors, F. Tufvesson, and T. L. Marzetta, "Massive MIMO for next generation wireless systems," *IEEE Commun. Mag.*, vol. 52, no. 2, pp. 186–195, 2014.

[891] K. Zheng, L. Zhao, J. Mei, B. Shao, W. Xiang, and L. Hanzo, "Survey of large-scale MIMO systems," *IEEE Commun. Surv. Tutor.*, vol. 17, no. 3, pp. 1738–1760, 2015.

[892] X. Guo, S. Chen, J. Zhang, X. Mu, and L. Hanzo, "Optimal pilot design for pilot contamination elimination/reduction in large-scale multiple-antenna aided OFDM systems," *IEEE Trans. Wireless Commun.*, vol. 15, no. 11, pp. 7229–7243, 2016.

[893] A. Garcia-Rodriguez and C. Masouros, "Exploiting the increasing correlation of space constrained massive MIMO for CSI relaxation," *IEEE Trans. Commun.*, vol. 64, no. 4, pp. 1572–1587, 2016.

[894] C. Mollén, E. G. Larsson, and T. Eriksson, "Waveforms for the massive MIMO downlink: amplifier efficiency, distortion, and performance," *IEEE Trans. Commun.*, vol. 64, no. 12, pp. 5050–5063, 2016.

[895] Y. Nan, L. Zhang, and X. Sun, "Efficient downlink channel estimation scheme based on block-structured compressive sensing for TDD massive MU-MIMO systems," *IEEE Wireless Commun. Letters*, vol. 4, no. 4, pp. 345–348, 2015.

[896] S. Jacobsson, G. Durisi, M. Coldrey, T. Goldstein, and C. Studer, "Quantized precoding for massive MU-MIMO," *IEEE Trans. Commun.*, vol. 65, no. 11, pp. 4670–4684, 2017.

[897] E. Björnson, J. Hoydis, and L. Sanguinetti, "Massive MIMO networks: spectral, energy, and hardware efficiency," *Found. Trends Signal Process.*, vol. 11, no. 3–4, pp. 154–655, 2017.

[898] K. Appaiah, A. Ashikhmin, and T. L. Marzetta, "Pilot contamination reduction in multi-user TDD systems," *IEEE Int'l Conf. Commun. (ICC'10)*, pp. 1–5, 2010.

[899] F. Fernandes, A. Ashikhmin, and T. L. Marzetta, "Inter-cell interference in noncooperative TDD large scale antenna systems," *IEEE J. Sel. Areas Commun.*, vol. 31, no. 2, pp. 192–201, 2013.

[900] K. Upadhya, S. A. Vorobyov, and M. Vehkaper a, "Downlink performance of superimposed pilots in massive MIMO systems," *IEEE Trans. Wireless Commun.*, vol. 17, no. 10, pp. 6630–6644, 2018.

[901] J. Huang, R. A. Berry, and M. L. Honig, "Distributed interference compensation for wireless networks," *IEEE J. Sel. Areas Commun.*, vol. 24, no. 5, pp. 1074–1084, 2006.

[902] K. Shen and W. Yu, "Fractional programming for communication systems—part I: power control and beamforming," *IEEE Trans. Signal Process.*, vol. 66, no. 10, pp. 2616–2630, 2018.

[903] J. F. Whitehead, "Signal-level-based dynamic power control for co-channel interference management," *IEEE Veh. Techn. Conf. (VTC'93)*, pp. 499–502, 1993.

[904] R. D. Yates, "A framework for uplink power control in cellular radio systems," *IEEE J. Sel. Areas Commun.*, vol. 13, no. 7, pp. 1341–1347, 1995.

[905] A. Simonsson and A. Furuskar, "Uplink power control in LTE—overview and performance," *IEEE Veh. Techn. Conf. (VTC'08 Fall)*, pp. 1–5, 2008.

[906] C. U. Castellanos, D. L. Villa, C. Rosa, K. I. Pedersen, F. D. Calabrese, P.-H. Michaelsen, and J. Michel, "Performance of uplink fractional power control in UTRAN LTE," *IEEE Veh. Techn. Conf. (VTC'08 Spring)*, pp. 2517–2521, 2008.

[907] J. H. Sørensen and E. De Carvalho, "Pilot decontamination through pilot sequence hopping in massive MIMO systems," *IEEE Global Commun. Conf. (GLOBECOM'14)*, pp. 3285–3290, 2014.

[908] A. S. Alwakeel and A. M. H. Mehana, "Achievable rates in uplink massive MIMO systems with pilot hopping," *IEEE Trans. Commun.*, vol. 65, no. 10, pp. 4232–4246, 2017.

[909] H. Lei, L. Zhang, X. Zhang, and D. Yang, "A novel multi-cell OFDMA system structure using fractional frequency reuse," *IEEE Int'l Symp. Personal, Indoor and Mobile Radio Commun. (PIMRC'07)*, pp. 1–5, 2007.

[910] I. Atzeni, J. Arnau, and M. Debbah, "Fractional pilot reuse in massive MIMO systems," *IEEE Int'l Conf. Commun. Workshop (ICCW'15)*, pp. 1030–1035, 2015.

[911] X. Zhu, Z. Wang, C. Qian, L. Dai, J. Chen, S. Chen, and L. Hanzo, "Soft pilot reuse and multicell block diagonalization precoding for massive MIMO systems," *IEEE Trans. Veh. Techn.*, vol. 65, no. 5, pp. 3285–3298, 2016.

[912] T. Lee, H. S. Kim, S. Park, and S. Bahk, "Mitigation of sounding pilot contamination in massive MIMO systems," *IEEE Int'l Conf. Commun. (ICC'14)*, pp. 1191–1196, Jun. 2014.

[913] Y. Li, R. Wang, and Z. Zhang, "Massive MIMO downlink goodput analysis with soft pilot or frequency reuse," *IEEE Wireless Commun. Letters*, vol. 7, no. 3, pp. 448–451, 2018.

[914] E. Björnson, E. G. Larsson, and M. Debbah, "Massive MIMO for maximal spectral efficiency: how many users and pilots should be allocated?" *IEEE Trans. Wireless Commun.*, vol. 15, no. 2, pp. 1293–1308, 2016.

[915] K. Guo and G. Ascheid, "Performance analysis of multi-cell MMSE based receivers in MU-MIMO systems with very large antenna arrays," *IEEE Wireless Commun. Netw. Conf. (WCNC'13)*, pp. 3175–3179, Apr. 2013.

[916] X. Li, E. Björnson, E. G. Larsson, S. Zhou, and J. Wang, "Massive MIMO with multi-cell MMSE processing: exploiting all pilots for interference suppression," *EURASIP J. Wireless Commun. Netw.*, vol. 2017, no. 1, p. 117, 2017.

[917] N. Krishnan, R. D. Yates, and N. B. Mandayam, "Uplink linear receivers for multi-cell multiuser MIMO with pilot contamination: large system analysis," *IEEE Trans. Wireless Commun.*, vol. 13, no. 8, pp. 4360–4373, 2014.

[918] Q. Zhangi, S. Jin, D. Morales, M. McKay, and H. Zhu, "Optimal pilot length for uplink massive MIMO systems with pilot reuse," *IEEE Int'l Conf. Acoustics, Speech and Signal Process. (ICASSP'16)*, pp. 3536–3540, 2016.

[919] T. V. Chien, E. Björnson, and E. G. Larsson, "Joint pilot design and uplink power allocation in multi-cell massive MIMO systems," *IEEE Trans. Wireless Commun.*, vol. 17, no. 3, pp. 2000–2015, 2018.

[920] H. Q. Ngo, M. Matthaiou, and E. G. Larsson, "Massive MIMO with optimal power and training duration allocation," *IEEE Wireless Commun. Letters*, vol. 3, no. 6, pp. 605–608, 2014.

[921] M. K. Varanasi and B. Aazhang, "Multistage detection in asynchronous code-division multiple-access communications," *IEEE Trans. Commun.*, vol. 38, no. 4, pp. 509–519, 1990.

[922] O. Shental, S. Venkatesan, A. Ashikhmin, and R. A. Valenzuela, "Massive BLAST: an architecture for realizing ultra-high data rates for large-scale MIMO," *IEEE Wireless Commun. Letters*, vol. 7, no. 3, pp. 404–407, 2018.

[923] H. Q. Ngo and E. G. Larsson, "No downlink pilots are needed in TDD massive MIMO," *IEEE Trans. Wireless Commun.*, vol. 16, no. 5, pp. 2921–2935, 2017.

[924] H. Q. Ngo, E. G. Larsson, and T. L. Marzetta, "Massive MU-MIMO downlink TDD systems with linear precoding and downlink pilots," *Allerton Conf. Commun., Control, and Computing*, pp. 293–298, 2013.

[925] A. Khansefid and H. Minn, "Achievable downlink rates of MRC and ZF precoders in massive MIMO with uplink and downlink pilot contamination," *IEEE Trans. Commun.*, vol. 63, no. 12, pp. 4849–4864, 2015.

[926] J. Zuo, J. Zhang, C. Yuen, W. Jiang, and W. Luo, "Multicell multiuser massive MIMO transmission with downlink training and pilot contamination precoding," *IEEE Trans. Veh. Techn.*, vol. 65, no. 8, pp. 6301–6314, 2016.

[927] B. Song, R. L. Cruz, and B. D. Rao, "Network duality for multiuser MIMO beamforming networks and applications," *IEEE Trans. Commun.*, vol. 55, no. 3, pp. 618–630, 2007.

[928] H. Dahrouj and W. Yu, "Coordinated beamforming for the multicell multi-antenna wireless system," *IEEE Trans. Wireless Commun.*, vol. 9, no. 5, pp. 1748–1759, 2010.

[929] Y. Huang, C. W. Tan, and B. D. Rao, "Joint beamforming and power control in coordinated multicell: max–min duality, effective network and large system transition," *IEEE Trans. Wireless Commun.*, vol. 12, no. 6, pp. 2730–2742, 2013.

[930] L. P. Qian, Y. J. Zhang, and J. Huang, "MAPEL: achieving global optimality for a non-convex wireless power control problem," *IEEE Trans. Wireless Commun.*, vol. 8, no. 3, pp. 1553–1563, 2009.

[931] H. Yang and T. L. Marzetta, "A macro cellular wireless network with uniformly high user throughputs," *IEEE Veh. Techn. Conf. (VTC'14 Fall)*, pp. 1–5, 2014.

[932] J. Jose, A. Ashikhmin, T. L. Marzetta, and S. Vishwanath, "Pilot contamination and precoding in multi-cell TDD systems," *IEEE Trans. Wireless Commun.*, vol. 10, no. 8, pp. 2640–2651, 2011.

[933] X. Li, E. Björnson, E. G. Larsson, S. Zhou, and J. Wang, "A multi-cell MMSE precoder for massive MIMO systems and new large system analysis," *IEEE Global Commun. Conf. (GLOBECOM'15)*, pp. 1–6, Dec. 2015.

[934] Q. Zhang, S. Jin, M. McKay, D. Morales-Jimenez, and H. Zhu, "Power allocation schemes for multicell massive MIMO systems," *IEEE Trans. Wireless Commun.*, vol. 14, no. 11, pp. 5941–5955, 2015.

[935] H. Q. Ngo and E. G. Larsson, "EVD-based channel estimation in multicell multiuser MIMO systems with very large antenna arrays," *IEEE Int'l Conf. Acoustics, Speech and Signal Process. (ICASSP'12)*, pp. 3249–3252, 2012.

[936] R. R. Müller, L. Cottatellucci, and M. Vehkaperä, "Blind pilot decontamination," *IEEE J. Sel. Topics Signal Process.*, vol. 8, no. 5, pp. 773–786, 2014.

[937] X. Zhu, Z. Wang, L. Dai, and C. Qian, "Smart pilot assignment for massive MIMO," *IEEE Commun. Letters*, vol. 19, no. 9, pp. 1644–1647, 2015.

[938] D. Hu, L. He, and X. Wang, "Semi-blind pilot decontamination for massive MIMO systems," *IEEE Trans. Wireless Commun.*, vol. 15, no. 1, pp. 525–536, 2016.

[939] J. Y. Sohn, S. W. Yoon, and J. Moon, "On reusing pilots among interfering cells in massive MIMO," *IEEE Trans. Wireless Commun.*, vol. 16, no. 12, pp. 8092–8104, 2017.

[940] O. Elijah, C. Y. Leow, T. A. Rahman, S. Nunoo, and S. Z. Iliya, "A comprehensive survey of pilot contamination in massive MIMO—5G system," *IEEE Commun. Surv. Tutor.*, vol. 18, no. 2, pp. 905–923, 2016.

[941] M. Li, S. Jin, and X. Gao, "Spatial orthogonality-based pilot reuse for multi-cell massive MIMO transmission," *Int'l Conf. Wireless Commun. and Signal Process.*, pp. 1–6, Oct. 2013.

[942] H. Yin, D. Gesbert, and L. Cottatellucci, "Dealing with interference in distributed large-scale MIMO systems: a statistical approach," *IEEE J. Sel. Topics Signal Process.*, vol. 8, no. 5, pp. 942–953, 2014.

[943] Z. Wang, C. Qian, L. Dai, J. Chen, C. Sun, and S. Chen, "Location-based channel estimation and pilot assignment for massive MIMO systems," *IEEE Int'l Conf. Commun. Workshop (ICCW'15)*, pp. 1264–1268, Jun. 2015.

[944] H. Yin, L. Cottatellucci, D. Gesbert, R. R. Müller, and G. He, "Robust pilot decontamination based on joint angle and power domain discrimination," *IEEE Trans. Signal Process.*, vol. 64, no. 11, pp. 2990–3003, 2016.

[945] Z. Chen and C. Yang, "Pilot decontamination in wideband massive MIMO systems by exploiting channel sparsity," *IEEE Trans. Wireless Commun.*, vol. 15, no. 7, pp. 5087–5100, 2016.

[946] S. Haghighatshoar and G. Caire, "Massive MIMO pilot decontamination and channel interpolation via wideband sparse channel estimation," *IEEE Trans. Wireless Commun.*, vol. 16, no. 12, pp. 8316–8332, 2017.

[947] E. Björnson, J. Hoydis, and L. Sanguinetti, "Massive MIMO has unlimited capacity," *IEEE Trans. Wireless Commun.*, vol. 17, no. 1, pp. 574–590, 2018.

[948] A. Ashikhmin, L. Li, and T. L. Marzetta, "Interference reduction in multi-cell massive MIMO systems with large-scale fading precoding," *IEEE Trans. Inform. Theory*, vol. 64, no. 9, pp. 6340–6361, 2018.

[949] A. Adhikary, A. Ashikhmin, and T. L. Marzetta, "Uplink interference reduction in large-scale antenna systems," *IEEE Trans. Commun.*, vol. 65, no. 5, pp. 2194–2206, 2017.

[950] E. Björnson, J. Hoydis, M. Kountouris, and M. Debbah, "Massive MIMO systems with non-ideal hardware: energy efficiency, estimation, and capacity limits," *IEEE Trans. Inform. Theory*, vol. 60, no. 11, pp. 7112–7139, 2014.

[951] E. Björnson, E. G. Larsson, and T. L. Marzetta, "Massive MIMO: ten myths and one critical question," *IEEE Commun. Mag.*, vol. 54, no. 2, pp. 114–123, 2016.

[952] C. Mollén, J. Choi, E. G. Larsson, and R. W. Heath Jr., "Uplink performance of wideband massive MIMO with one-bit ADCs," *IEEE Trans. Wireless Commun.*, vol. 16, no. 1, pp. 87–100, 2017.

[953] E. Björnson, M. Matthaiou, and M. Debbah, "Massive MIMO with non-ideal arbitrary arrays: hardware scaling laws and circuit-aware design," *IEEE Trans. Wireless Commun.*, vol. 14, no. 8, pp. 4353–4368, 2015.

[954] H. Prabhu, J. N. Rodrigues, L. Liu, and O. Edfors, "A 60pJ/b 300Mb/s 1288 massive MIMO precoder-detector in 28nm FD-SOI," in *IEEE Int'l Solid-State Circuits Conf. (ISSCC'17)*, 2017, pp. 60–61.

[955] H. Prabhu, J. Rodrigues, O. Edfors, and F. Rusek, "Approximative matrix inverse computations for very-large MIMO and applications to linear pre-coding systems," *IEEE Wireless Commun. Netw. Conf. (WCNC'13)*, pp. 2710–2715, 2013.

[956] A. Kammoun, A. Müller, E. Björnson, and M. Debbah, "Linear precoding based on polynomial expansion: large-scale multi-cell MIMO systems," *IEEE J. Sel. Topics Signal Process.*, vol. 8, no. 5, pp. 861–875, 2014.

[957] B. Nagy, M. Elsabrouty, and S. Elramly, "Fast converging weighted Neumann series precoding for massive MIMO systems," *IEEE Wireless Commun. Letters*, vol. 7, no. 2, pp. 154–157, 2018.

[958] X. Qin, Z. Yan, and G. He, "A near-optimal detection scheme based on joint steepest descent and Jacobi method for uplink massive MIMO systems," *IEEE Commun. Letters*, vol. 20, no. 2, pp. 276–279, 2016.

[959] L. Dai, X. Gao, X. Su, S. Han, C.-L. I, and Z. Wang, "Low-complexity soft-output signal detection based on Gauss–Seidel method for uplink multiuser large-scale MIMO systems," *IEEE Trans. Veh. Techn.*, vol. 64, no. 10, pp. 4839–4845, 2015.

[960] M. N. Boroujerdi, S. Haghighatshoar, and G. Caire, "Low-complexity statistically robust precoder/detector computation for massive MIMO systems," *IEEE Trans. Wireless Commun.*, vol. 17, no. 10, pp. 6516–6530, 2018.

[961] E. Björnson, L. Sanguinetti, J. Hoydis, and M. Debbah, "Optimal design of energy-efficient multi-user MIMO systems: is massive MIMO the answer?" *IEEE Trans. Wireless Commun.*, vol. 14, no. 6, pp. 3059–3075, 2015.

[962] A. F. Molisch, V. V. Ratnam, S. Han, Z. Li, S. L. H. Nguyen, L. Li, and K. Haneda, "Hybrid beamforming for massive MIMO: a survey," *IEEE Commun. Mag.*, vol. 55, no. 9, pp. 134–141, 2017.

[963] D. Kudathanthirige and G. Amarasuriya, "Sum rate analysis of massive MIMO downlink with hybrid beamforming," *IEEE Global Commun. Conf. (GLOBECOM'17)*, 2017.

[964] R. A. Kennedy, P. Sadeghi, T. D. Abhayapala, and H. M. Jones, "Intrinsic limits of dimensionality and richness in random multipath fields," *IEEE Trans. Signal Process.*, vol. 55, no. 6, pp. 2542–2556, 2007.

[965] A. L. Moustakas, H. U. Baranger, L. Balents, A. M. Sengupta, and S. H. Simon, "Communication through a diffusive medium: coherence and capacity," *Science*, vol. 287, no. 5451, pp. 287–290, 2000.

[966] A. S. Y. Poon, R. W. Brodersen, and D. N. C. Tse, "Degrees of freedom in multiple-antenna channels: a signal space approach," *IEEE Trans. Inform. Theory*, vol. 51, no. 2, pp. 523–536, 2005.

[967] T. L. Marzetta, "Spatially-stationary propagating random field model for massive MIMO small-scale fading," *IEEE Int'l Symp. Inform. Theory (ISIT'18)*, pp. 391–395, 2018.

[968] Y. Huang, Y. Li, H. Ren, J. Lu, and W. Zhang, "Multi-panel MIMO in 5G," *IEEE Commun. Mag.*, vol. 56, no. 3, pp. 56–61, 2018.

[969] S. Hu, F. Rusek, and O. Edfors, "Beyond massive MIMO: the potential of data transmission with large intelligent surfaces," *IEEE Trans. Signal Process.*, vol. 66, no. 10, pp. 2746–2758, 2018.

[970] D. Mi, M. Dianati, L. Zhang, S. Muhaidat, and R. Tafazolli, "Massive MIMO performance with imperfect channel reciprocity and channel estimation error," *IEEE Trans. Commun.*, vol. 65, no. 9, pp. 3734–3749, 2017.

[971] R. Rogalin, O. Y. Bursalioglu, H. Papadopoulos, G. Caire, A. F. Molisch, A. Michaloliakos, V. Balan, and K. Psounis, "Scalable synchronization and reciprocity calibration for distributed multiuser MIMO," *IEEE Trans. Wireless Commun.*, vol. 13, no. 4, pp. 1815–1831, 2014.

[972] H. Wei, D. Wang, H. Zhu, J. Wang, S. Sun, and X. You, "Mutual coupling calibration for multiuser massive MIMO systems," *IEEE Trans. Wireless Commun.*, vol. 15, no. 1, pp. 606–619, 2016.

[973] J. Vieira, F. Rusek, O. Edfors, S. Malkowsky, L. Liu, and F. Tufvesson, "Reciprocity calibration for massive MIMO: proposal, modeling, and validation," *IEEE Trans. Wireless Commun.*, vol. 16, no. 5, pp. 3042–3056, 2017.

[974] X. Jiang, A. Decurninge, K. Gopala, F. Kaltenberger, M. Guillaud, D. Slock, and L. Deneire, "A framework for over-the-air reciprocity calibration for TDD massive MIMO systems," *IEEE Trans. Wireless Commun.*, vol. 17, no. 9, pp. 5975–5990, 2018.

[975] O. Raeesi, A. Gokceoglu, Y. Zou, E. Björnson, and M. Valkama, "Performance analysis of multi-user massive MIMO downlink under channel non-reciprocity and imperfect CSI," *IEEE Trans. Commun.*, vol. 66, no. 6, pp. 2456–2471, 2018.

[976] A. Alexiou *et al.*, "Duplexing, resource allocation and inter-cell coordination: design recommendations for next generation wireless systems," *Wireless Commun. Mob. Comput.*, vol. 5, no. 1, pp. 77–93, 2005.

[977] R. K. Mungara, I. Thibault, and A. Lozano, "Full-duplex MIMO in cellular networks: system-level performance," *IEEE Trans. Wireless Commun.*, vol. 16, no. 5, pp. 3124–3137, 2017.

[978] Z. Jiang, A. F. Molisch, G. Caire, and Z. Niu, "Achievable rates of FDD massive MIMO systems with spatial channel correlation," *IEEE Trans. Wireless Commun.*, vol. 14, no. 5, pp. 2868–2882, 2015.

[979] X. Rao and V. K. N. Lau, "Distributed compressive CSIT estimation and feedback for FDD multi-user massive MIMO systems," *IEEE Trans. Signal Process.*, vol. 62, no. 12, pp. 3261–3271, 2014.

[980] L. You, X. Gao, A. L. Swindlehurst, and W. Zhong, "Channel acquisition for massive MIMO-OFDM with adjustable phase shift pilots," *IEEE Trans. Signal Process.*, vol. 64, no. 6, pp. 1461–1476, 2016.

[981] M. Masood, L. H. Afify, and T. Y. Al-Naffouri, "Efficient coordinated recovery of sparse channels in massive MIMO," *IEEE Trans. Signal Process.*, vol. 63, no. 1, pp. 104–118, 2015.

[982] X. Zhang, L. Zhong, and A. Sabharwal, "Directional training for FDD massive MIMO," *IEEE Trans. Wireless Commun.*, vol. 17, no. 8, pp. 5183–5197, 2018.

[983] Y. Ding and B. D. Rao, "Dictionary learning-based sparse channel representation and estimation for FDD massive MIMO systems," *IEEE Trans. Wireless Commun.*, vol. 17, no. 8, pp. 5437–5451, 2018.

[984] B. Lee, J. Choi, J.-Y. Seol, D. J. Love, and B. Shim, "Antenna grouping based feedback compression for FDD-based massive MIMO systems," *IEEE Trans. Commun.*, vol. 63, no. 9, pp. 3261–3274, 2015.

[985] A. Adhikary, J. Nam, J.-Y. Ahn, and G. Caire, "Joint spatial division and multiplexing—the large-scale array regime," *IEEE Trans. Inform. Theory*, vol. 59, no. 10, pp. 6441–6463, 2013.

[986] Y. Xu, G. Yue, and S. Mao, "User grouping for massive MIMO in FDD systems: new design methods and analysis," *IEEE Access*, vol. 2, pp. 947–959, 2014.

[987] J. Ma, S. Zhang, H. Li, N. Zhao, and V. C. M. Leung, "Base station selection for massive MIMO networks with two-stage precoding," *IEEE Wireless Commun. Letters*, vol. 6, no. 5, pp. 598–601, 2017.

[988] J. Chen and V. K. N. Lau, "Two-tier precoding for FDD multi-cell massive MIMO time-varying interference networks," *IEEE J. Sel. Areas Commun.*, vol. 32, no. 6, pp. 1230–1238, 2014.

[989] Z. Li, S. Han, S. Sangodoyin, R. Wang, and A. F. Molisch, "Joint optimization of hybrid beamforming for multi-user massive MIMO downlink," *IEEE Trans. Wireless Commun.*, vol. 17, no. 6, pp. 3600–3614, 2018.

[990] S. Haghighatshoar and G. Caire, "Massive MIMO channel subspace estimation from low-dimensional projections," *IEEE Trans. Signal Process.*, vol. 65, no. 2, pp. 303–318, 2017.

[991] H. Q. Ngo, T. L. Marzetta, and E. G. Larsson, "Analysis of the pilot contamination effect in very large multicell multiuser MIMO systems for physical channel models," *IEEE Int'l Conf. Acoustics, Speech and Signal Process. (ICASSP'11)*, pp. 3464–3467, 2011.

[992] C. Sun, X. Gao, S. Jin, M. Matthaiou, Z. Ding, and C. Xiao, "Beam division multiple access transmission for massive MIMO communications," *IEEE Trans. Commun.*, vol. 63, no. 6, pp. 2170–2184, 2015.

[993] H. Yang and T. L. Marzetta, "Massive MIMO with max–min power control in line-of-sight propagation environment," *IEEE Trans. Commun.*, vol. 65, no. 11, pp. 4685–4693, 2017.

[994] C. Masouros and M. Matthaiou, "Space-constrained massive MIMO: hitting the wall of favorable propagation," *IEEE Commun. Letters*, vol. 19, no. 5, pp. 771–774, 2015.

[995] S. Payami and F. Tufvesson, "Channel measurements and analysis for very large array systems at 2.6 GHz," *Eur. Conf. Antennas and Propagation (EUCAP'12)*, pp. 433–437, 2012.

[996] A. Amiri, M. Angjelichinoski, E. de Carvalho, and R. W. Heath Jr., "Extremely large aperture massive MIMO: low complexity receiver architectures," *IEEE Global Commun. Conf. Workshop (GLOBECOM Workshop'18)*, Dec. 2018.

[997] X. Li, S. Zhou, E. Björnson, and J. Wang, "Capacity analysis for spatially non-wide sense stationary uplink massive MIMO systems," *IEEE Trans. Wireless Commun.*, vol. 14, no. 12, pp. 7044–7056, 2015.

[998] L. Sanguinetti, A. Kammoun, and M. Debbah, "Asymptotic analysis of multicell massive MIMO over Rician fading channels," *IEEE Int'l Conf. Acoustics, Speech and Signal Process. (ICASSP'17)*, pp. 3539–3543, 2017.

[999] X. Gao, O. Edfors, F. Rusek, and F. Tufvesson, "Linear pre-coding performance in measured very-large MIMO channels," *IEEE Veh. Techn. Conf. (VTC'11 Fall)*, pp. 1–5, 2011.

[1000] J. Hoydis, C. Hoek, T. Wild, and S. ten Brink, "Channel measurements for large antenna arrays," *Int'l Symp. Wireless Commun. Systems (ISWCS'12)*, pp. 811–815, 2012.

[1001] X. Gao, O. Edfors, F. Rusek, and F. Tufvesson, "Massive MIMO performance evaluation based on measured propagation data," *IEEE Trans. Wireless Commun.*, vol. 14, no. 7, pp. 3899–3911, 2015.

[1002] P. Harris, S. Malkowsky, J. Vieira, E. Bengtsson, F. Tufvesson, W. B. Hasan, L. Liu, M. Beach, S. Armour, and O. Edfors, "Performance characterization of a real-time massive MIMO system with LOS mobile channels," *IEEE J. Sel. Areas Commun.*, vol. 35, no. 6, pp. 1244–1253, 2017.

[1003] J. Zhang, Z. Zheng, Y. Zhang, J. Xi, X. Zhao, and G. Gui, "3D MIMO for 5G NR: several observations from 32 to massive 256 antennas based on channel measurement," *IEEE Commun. Mag.*, vol. 56, no. 3, pp. 62–70, 2018.

[1004] D. Bethanabhotla, O. Y. Bursalioglu, H. C. Papadopoulos, and G. Caire, "Optimal user-cell association for massive MIMO wireless networks," *IEEE Trans. Wireless Commun.*, vol. 15, no. 3, pp. 1835–1850, 2016.

[1005] J. G. Andrews, F. Baccelli, and R. K. Ganti, "A tractable approach to coverage and rate in cellular networks," *IEEE Trans. Commun.*, vol. 59, no. 11, pp. 3122–3134, 2011.

[1006] C. Shepard, H. Yu, N. Anand, E. Li, T. Marzetta, R. Yang, and L. Zhong, "Argos: practical many-antenna base stations," *Int'l Conf. Mobile Computing and Networking*, pp. 53–64, 2012.

[1007] "Joint massive MIMO test between British Telecom and the University of Bristol," https://spectrum.ieee.org/tech-talk/telecom/wireless/5g-researchers-achieve-new-spectrum-efficiency-record.

[1008] "The MAMMOET european project," https://mammoet-project.eu/.

[1009] Y.-H. Nam, B. L. Ng, K. Sayana, Y. Li, J. Zhang, Y. Kim, and J. Lee, "Full-dimension MIMO (FD-MIMO) for next generation cellular technology," *IEEE Commun. Mag.*, vol. 51, no. 6, pp. 172–179, 2013.

[1010] K. A. Alnajjar, P. J. Smith, P. Whiting, and G. K. Woodward, "Size and array shape for massive MIMO," *IEEE Wireless Commun. Letters*, vol. 4, no. 6, pp. 653–656, 2015.

[1011] H. Ji, Y. Kim, J. Lee, E. Onggosanusi, Y. Nam, J. Zhang, B. Lee, and B. Shim, "Overview of full-dimension MIMO in LTE-Advanced Pro," *IEEE Commun. Mag.*, vol. 55, no. 2, pp. 176–184, 2017.

[1012] Q. U. A. Nadeem, A. Kammoun, M. Debbah, and M. S. Alouini, "Design of 5G full dimension massive MIMO systems," *IEEE Trans. Commun.*, vol. 66, no. 2, pp. 726–740, 2018.

[1013] A. Lozano, D. C. Cox, and T. R. Bourk, "Uplink–downlink imbalance in TDMA personal communication systems," *IEEE Int'l Conf. Universal Personal Commun. (ICUPC'98)*, vol. 1, pp. 151–155, 1998.

[1014] M. Haenggi, *Stochastic geometry for wireless networks.* Cambridge University Press, 2012.

[1015] G. George, A. Lozano, and M. Haenggi, "Distribution of the number of users per base station in cellular networks," *IEEE Wireless Commun. Letters*, vol. 8, 2019.

[1016] B. Blaszczyszyn, M. K. Karray, and H. P. Keeler, "Wireless networks appear Poissonian due to strong shadowing," *IEEE Trans. Wireless Commun.*, vol. 14, no. 8, pp. 4379–4390, 2015.

[1017] G. George, A. Lozano, and M. Haenggi, "Massive MIMO forward link analysis for cellular networks," Preprint, 2018.

[1018] A. A. M. Saleh, A. J. Rustako, and R. Roman, "Distributed antennas for indoor radio communications," *IEEE Trans. Commun.*, vol. 35, no. 12, pp. 1245–1251, 1987.

[1019] R. W. Heath Jr., T. Wu, Y. H. Kwon, and A. C. K. Soong, "Multiuser MIMO in distributed antenna systems with out-of-cell interference," *IEEE Trans. Signal Process.*, vol. 59, no. 10, pp. 4885–4899, 2011.

[1020] G. J. Foschini, K. Karakayali, and R. A. Valenzuela, "Coordinating multiple antenna cellular networks to achieve enormous spectral efficiency," *IEE Proc.*, vol. 153, no. 4, pp. 548–555, 2006.

[1021] S. Venkatesan, A. Lozano, and R. Valenzuela, "Network MIMO: overcoming intercell interference in indoor wireless systems," *Asilomar Conf. Signals, Systems and Computers*, pp. 83–87, 2007.

[1022] S. Venkatesan, H. Huang, A. Lozano, and R. Valenzuela, "A WiMAX-based implementation of network MIMO for indoor wireless systems," *EURASIP J. Adv. Signal Process.*, vol. 2009, p. 9, Oct. 2009.

[1023] H. Huang, M. Trivellato, A. Hottinen, M. Shafi, P. Smith, and R. Valenzuela, "Increasing downlink cellular throughput with limited network MIMO coordination," *IEEE Trans. Wireless Commun.*, vol. 8, no. 6, pp. 2983–2989, 2009.

[1024] D. Gesbert, S. Hanly, H. Huang, S. Shamai, O. Simeone, and W. Yu, "Multi-cell MIMO cooperative networks: a new look at interference," *IEEE J. Sel. Areas Commun.*, vol. 28, no. 9, pp. 1380–1408, 2010.

[1025] P. Wang, H. Wang, L. Ping, and X. Lin, "On the capacity of MIMO cellular systems with base station cooperation," *IEEE Trans. Wireless Commun.*, vol. 10, no. 11, pp. 3720–3731, 2011.

[1026] O. Simeone, N. Levy, A. Sanderovich, O. Somekh, B. M. Zaidel, H. V. Poor, and S. Shamai, "Information theoretic considerations for wireless cellular systems: the impact of cooperation," *Found. Trends Commun. Inform. Theory*, vol. 7, 2012.

[1027] H. Huh, A. M. Tulino, and G. Caire, "Network MIMO with linear zero-forcing beamforming: large system analysis, impact of channel estimation, and reduced-complexity scheduling," *IEEE Trans. Inform. Theory*, vol. 58, no. 5, pp. 2911–2934, 2012.

[1028] H. Huh, G. Caire, H. C. Papadopoulos, and S. A. Ramprashad, "Achieving 'massive MIMO' spectral efficiency with a not-so-large number of antennas," *IEEE Trans. Wireless Commun.*, vol. 11, no. 9, pp. 3226–3239, 2012.

[1029] N. Lee, R. W. Heath Jr., D. Morales-Jimenez, and A. Lozano, "Base station cooperation with dynamic clustering in super-dense cloud-RAN," *IEEE Global Commun. Conf. Workshops (GLOBECOM'13)*, pp. 784–788, Dec. 2013.

[1030] N. Lee, D. Morales-Jimenez, A. Lozano, and R. W. Heath Jr., "Spectral efficiency of dynamic coordinated beamforming: a stochastic geometry approach," *IEEE Trans. Wireless Commun.*, vol. 14, no. 1, pp. 230–241, 2015.

[1031] H. Q. Ngo, A. Ashikhmin, H. Yang, E. G. Larsson, and T. L. Marzetta, "Cell-free massive MIMO versus small cells," *IEEE Trans. Wireless Commun.*, vol. 16, no. 3, pp. 1834–1850, 2017.

[1032] L. D. Nguyen, T. Q. Duong, H. Q. Ngo, and K. Tourki, "Energy efficiency in cell-free massive MIMO with zero-forcing precoding design," *IEEE Commun. Letters*, vol. 21, no. 8, pp. 1871–1874, 2017.

[1033] E. Nayebi, A. Ashikhmin, T. L. Marzetta, H. Yang, and B. D. Rao, "Precoding and power optimization in cell-free massive MIMO systems," *IEEE Trans. Wireless Commun.*, vol. 16, no. 7, pp. 4445–4459, 2017.

[1034] H. Q. Ngo, L. N. Tran, T. Q. Duong, M. Matthaiou, and E. G. Larsson, "On the total energy efficiency of cell-free massive MIMO," *IEEE Trans. Green Commun. Netw.*, 2017.

[1035] M. Attarifar, A. Abbasfar, and A. Lozano, "Random vs structured pilot assignment in cell-free massive MIMO wireless networks," *IEEE Int'l Conf. Commun. Workshops (ICCW'18)*, Jun. 2018.

[1036] "C-RAN the road towards green RAN," China Mobile, Tech. Rep., Oct. 2011.

[1037] A. Checko, H. L. Christiansen, Y. Yan, L. Scolari, G. Kardaras, M. S. Berger, and L. Dittmann, "Cloud RAN for mobile networks—a technology overview," *IEEE Commun. Surv. Tutor.*, vol. 17, no. 1, pp. 405–426, 2015.

[1038] I. F. Akyildiz, J. M. Jornet, and C. Han, "Terahertz band: next frontier for wireless communications," *Physical Commun.*, vol. 12, pp. 16–32, 2014.

[1039] V. Cadambe and S. A. Jafar, "Interference alignment and the degrees of freedom of the K-user interference channel," *IEEE Trans. Inform. Theory*, vol. 54, no. 8, pp. 3425–3441, 2008.

[1040] M. Maddah-Ali, A. Motahari, and A. Khandani, "Communication over MIMO X channels: interference alignment, decomposition, and performance analysis," *IEEE Trans. Inform. Theory*, vol. 54, no. 8, pp. 3457–3470, 2008.

[1041] K. Gomadam, V. Cadambe, and S. Jafar, "A distributed numerical approach to interference alignment and applications to wireless interference networks," *IEEE Trans. Inform. Theory*, vol. 57, no. 6, pp. 3309–3322, 2011.

[1042] S. W. Peters and R. W. Heath Jr., "Interference alignment via alternating minimization," *IEEE Int'l Conf. Acoustics, Speech, and Signal Process. (ICASSP'09)*, pp. 2445–2448, Apr. 2009.

[1043] R. K. Mungara, D. Morales-Jimenez, and A. Lozano, "System-level performance of interference alignment," *IEEE Trans. Wireless Commun.*, vol. 14, no. 2, pp. 1060–1070, 2015.

[1044] H. R. Stuart, "Dispersive multiplexing in multimode optical fiber," *Science*, vol. 289, no. 5477, pp. 281–283, 2000.

[1045] K. Appaiah, S. Vishwanath, and S. R. Bank, "Advanced modulation and multiple-input multiple-output for multimode fiber links," *IEEE Photonics Techn. Letters*, vol. 23, no. 20, pp. 1424–1426, 2011.

[1046] ——, "Vector intensity-modulation and channel state feedback for multimode fiber optic links," *IEEE Trans. Commun.*, vol. 61, no. 7, pp. 2958–2969, 2013.

[1047] M. Timmers, M. Guenach, C. Nuzman, and J. Maes, "G.fast: evolving the copper access network," *IEEE Commun. Mag.*, vol. 51, no. 8, pp. 74–79, 2013.

[1048] H. Elgala, R. Mesleh, and H. Haas, "Indoor optical wireless communication: potential and state-of-the-art," *IEEE Commun. Mag.*, vol. 49, no. 9, pp. 56–62, 2011.

[1049] L. Zeng, D. C. O'Brien, H. L. Minh, G. E. Faulkner, K. Lee, D. Jung, Y. Oh, and E. T. Won, "High data rate multiple input multiple output (MIMO) optical wireless communications using white LED lighting," *IEEE J. Sel. Areas Commun.*, vol. 27, no. 9, pp. 1654–1662, 2009.

[1050] A. H. Azhar, T. Tran, and D. O'Brien, "A gigabit/s indoor wireless transmission using MIMO-OFDM visible-light communications," *IEEE Photonics Techn. Letters*, vol. 25, no. 2, pp. 171–174, 2013.

[1051] M. Stojanovic and J. Preisig, "Underwater acoustic communication channels: propagation models and statistical characterization," *IEEE Commun. Mag.*, vol. 47, no. 1, pp. 84–89, 2009.

[1052] S. Roy, T. M. Duman, V. McDonald, and J. G. Proakis, "High-rate communication for underwater acoustic channels using multiple transmitters and space–time coding: receiver structures and experimental results," *IEEE J. Ocean. Eng.*, vol. 32, no. 3, pp. 663–688, 2007.

[1053] R. N. Bracewell, *The Fourier transform and its applications*. McGraw Hill, 1978.

[1054] E. T. Jaynes, *Probability theory*. Cambridge University Press, 2003.

[1055] F. Neeser and J. L. Massey, "Proper complex random processes with applications to information theory," *IEEE Trans. Inform. Theory*, vol. 39, no. 4, pp. 1293–1302, 1993.

[1056] O. Shalvi and E. Weinstein, "New criteria for blind deconvolution of nonminimum phase systems (channels)," *IEEE Trans. Inform. Theory*, vol. 36, pp. 312–321, 1990.

[1057] J. Wishart, "The generalised product moment distribution in samples from a normal multivariate population," *Biometrika*, vol. 20A, no. 1–2, pp. 32–52, 1928.

[1058] R. J. Muirhead, *Aspects of multivariate statistical theory*. Wiley, 1982.

[1059] B. V. Bronk, "Exponential ensemble for random matrices," *J. Math. Phys.*, vol. 6, p. 228, 1965.

[1060] M. Kang and M. S. Alouini, "Water-filling capacity and beamforming performance of MIMO systems with covariance feedback," *IEEE Workshop Signal Process. Adv. in Wireless Commun. (SPAWC'03)*, pp. 556–560, 2003.

[1061] M. Park, C. B. Chae, and R. W. Heath Jr., "Ergodic capacity of spatial multiplexing MIMO systems with ZF receivers for log-normal shadowing and Rayleigh fading channels," *IEEE Int'l Symp. Personal, Indoor and Mobile Radio Commun. (PIMRC'07)*, pp. 1–5, 2007.

[1062] E. Wigner, "Characteristic vectors of bordered matrices with infinite dimensions," *Ann. Math.*, vol. 62, pp. 546–564, 1955.

[1063] ——, "Statistical properties of real symmetric matrices with many dimensions," *4th Canadian Math. Congress*, pp. 174–176, 1959.

[1064] V. L. Girko, "Circular law," *Theory Prob. Appl.*, vol. 29, pp. 694–706, 1984.

[1065] V. A. Marčenko and L. A. Pastur, "Distribution of eigenvalues for some sets of random matrices," *Math. USSR-Sbornik*, vol. 1, p. 457, 1967.

[1066] J. W. Silverstein, "Strong convergence of the empirical distribution of eigenvalues of large dimensional random matrices," *J. Multivar. Anal.*, vol. 55, pp. 331–339, 1995.

[1067] D. Voiculescu, "Asymptotically commuting finite rank unitary operators without commuting approximants," *Acta Sci. Math.*, vol. 45, pp. 429–431, 1983.

[1068] ——, "Addition of certain non-commuting random variables," *Funct. Anal.*, vol. 66, pp. 323–346, 1986.

[1069] ——, "Limit laws for random matrices and free products," *Inventiones Mathematicae*, vol. 104, no. 1, pp. 201–220, 1991.

[1070] E. P. Wigner, "The unreasonable effectiveness of mathematics in the natural sciences. Richard Courant lecture in mathematical sciences delivered at New York University, May 11, 1959," *Comm. Pure Appl. Math.*, vol. 13, no. 1, pp. 1–14, 1960.

[1071] M. Livio, *Is God a mathematician?* Simon and Schuster, 2009.

[1072] H. M. Schey, *Div, grad, curl, and all that: an informal text on vector calculus*. W. W. Norton and Company, 2005.

[1073] M. Abramowitz and I. A. Stegun, *Handbook of mathematical functions with formulas, graphs, and mathematical tables*. Dover Publications, 1972.

[1074] G. Abreu, "Very simple tight bounds on the Q-function," *IEEE Trans. Commun.*, vol. 60, no. 9, pp. 2415–2420, 2012.

[1075] G. H. Hardy and E. M. Wright, *An introduction to the theory of numbers*. Oxford University Press, 1979.

[1076] S. Boyd and L. Vandenberghe, *Convex optimization*. Cambridge University Press, 2004.

[1077] J. L. W. V. Jensen, "Sur les fonctions convexes et les inégalités entre les valeurs moyennes," *Acta Mathematica*, vol. 30, no. 1, pp. 175–193, 1906.

Index

2G, xv
3G, xxi
3GPP, 142, 201
4G, xv, 36
5G, xv, xxi, 36

a-posteriori probability, 30
a-posteriori probability decoder, 29
acoustic communication, 649
ad hoc networks, 647
adaptive array, xix, 176
adaptive modulation and coding, 228
additive noise, 61
additive white Gaussian noise, *see* AWGN
affine transformation, 667
Amitay, xx
amplitude constraint, 86
analog feedback, 248, 364, 365, 538
analog-to-digital conversion, 66, 71, 380, 633
angle diversity, 186
angle spread, 147, 182, 183, 191, 203, 452, 492, 633
antenna, 59, 72, 85
antenna array, 136, 171, 173, 185, 302, 306, 380, 633, 634
antenna correlation, 408, 461, 583, 628
  base station, 629, 630, 635
  beamforming, 310
  capacity, 324, 334
  channel estimation, 200, 347
  channel normalization, 302
  Kronecker model, 120, 180, 312, 330, 334, 340
  link adaptation, 361
  polarization diversity, 186
  receiver, 336
  steering vectors, 173
  transmitter, 313, 315, 390, 394, 405, 409, 461, 492
  users, 636
  virtual channel model, 185, 188
antenna height
  base station, 140, 141, 143, 598
  user, 140, 142, 143
antenna pattern, 150, 151, 154, 173
antenna tilt, 204, 628
APP decoder, 29, 34
array factor, 174
array manifold, 173
array response, 306
array steering vector, 171, 191
array topology, 173, 179, 191
asymptotic eigenvalue distribution, 336, 675
autocorrelation, 138, 154, 677
autocovariance, 72, 677
autoregressive process, 158, 167
AWGN, 72
AWGN channel, 212, 217, 220, 222, 245, 305, 360
azimuth, 147, 151, 173, 182

Bahl–Cocke–Jelinek–Raviv, *see* BCJR
ban, 7
bandwidth, 196, 329, 430, 647
  bandlimited signal, 58
  bandpass filter, 63
  capacity, 22, 216, 219, 220, 226
  Doppler spread, 200
  FDMA, 421
  frequency selectivity, 159, 160, 194
  frequency-division multiplexing, 432
  in-phase and quadrature signals, 59
  latency, 36
  noise power, 440
  OFDM, 108, 109, 236, 428
  sampling theorem, 64
  SNR, 72, 74, 144
  spectral efficiency, 211, 214
  TDMA, 422
bandwidth-limited regime, 217
base station, 452, 466, 469, 527, 539, 547, 549–551, 557, 582, 585, 586, 598, 603, 605, 629, 631
  antenna correlation, 191, 192
  antenna gain, 144
  antenna tilt, 628
  channel estimation, 585, 586, 588
  correlation matrices, 182, 628
  energy, 73, 584, 633
  form factor, 633
  massive MIMO, 601, 632, 634, 635
  power angle spectrum, 148
  receiver, 182, 439, 459, 590, 593–595, 611, 614
  transmit power, 143
  transmitter, 182, 481, 487, 489, 490, 492, 619, 621
Bayes' theorem, 23, 27, 40, 41, 665
BCJR algorithm, 29
beam steering, 176

beamforming, 174, 176, 306, 310, 311, 326, 327, 331, 367, 372, 380, 433, 440, 450, 557, 595, 629
Bell Laboratories, xx
Bell Labs layered space–time, *see* BLAST
bell-shape Doppler spectrum, 155
Bello, 166
Bernoulli distribution, 7
Bessel function, 151, 682
bi-unitary invariance, 667, 669
bias, 40, 43, 48, 114, 117, 119, 120
BICM, 32, 38, 229, 355, 358, 361, 362, 418
binary codeword, 25
binary entropy function, 7
binary phase-shift keying, *see* BPSK
bit, 6, 22
bit error probability, 30
bit rate, 22, 211, 221, 428
bit-interleaved coded modulation, *see* BICM
BLAST, 354
blind channel estimation, 619
block error probability, 21
block fading, 158, 167, 168, 197, 274, 341–344, 346, 531, 583
block Toeplitz matrix, 80, 91
block-diagonal, 440, 446, 492, 501
block-diagonalization, 508, 558, 574, 635
Boltzmann, 7
BPSK
  MMSE, 43, 45
  mutual information, 14, 15
  signal, 5, 6, 35, 38, 43, 70
  spectral efficiency, 33, 227, 260, 261, 330
branch-and-bound, 572
broadcast channel, 417, 418
broadside, 173
building height, 143
building penetration losses, 138
building-to-building distance, 143
byte, 7

cable losses, 138
calibration, 244, 248, 365, 490, 634
canonical channel, 178
capacity
  AWGN channel, 222, 229
  codeword length, 36
  CSI, 246, 305
  ergodic, 258, 271, 309, 316
  fading channel, 260, 263, 309, 322, 324
  frequency-selective channel, 232, 239, 240
  interference, 282, 348
  number of codewords, 351
  outage, 253
  per-symbol vs per-bit, 217
  Shannon, 3, 21, 25, 253, 419
  spectral efficiency, 212, 213
capacity boundary, 419, 424, 429, 435, 442, 443, 445, 464, 472, 475, 482, 483, 486, 490, 492
capacity per unit cost, 212
capacity region, 419
capacity-effective SNR mapping, *see* CESM method
carrier frequency, 58–60, 140, 142, 154, 156, 170, 380, 633
causal MMSE, 51
causality, 76, 79
CDF, 666
CDMA, xx, 300, 343, 422
cell-free, 647
cell-less, 647
central limit theorem, 137, 145, 146, 668, 673, 676
centroid, 370
CESM method, 243, 245, 270, 360, 362
chain rule
  differential entropy, 10
  entropy, 8
  mutual information, 17, 268, 275, 353, 534
channel capacity, 222
channel diagonalization, 305, 307, 308, 326, 361, 362, 449, 450, 461, 521
channel dispersion, 109
channel estimation, 111–113, 117, 121, 123, 195, 275, 364, 410, 532, 534, 583, 585, 628, 630
channel hardening, 336, 583, 593, 619, 622, 631
channel inversion power control, 251
channel law, 18, 19, 22, 31, 38, 39, 222, 252, 357
channel order, 76, 79
channel-state information, *see* CSI
channel-state information at the receiver, *see* CSIR
channel-state information at the transmitter, *see* CSIT
Chase combining, 37
Chernoff bound, 682
chi distribution, 672
chi-square distribution, 313, 322, 393, 504, 521, 524, 671
chip period, 107, 108
Cioffi, xx
circuit power consumption, 266, 380, 633
circulant matrix, 239, 659
circular array, 173
circular convolution, 98–100, 105, 106, 655
circular polarization, 186
circular symmetry, 667, 669
Clarke, 148, 151, 153, 155, 182, 191, 198, 201, 273, 277, 279, 344
clipped Gaussian distribution, 87
clipping, 87
cloud, 647
code-division multiple access, *see* CDMA
codebook, 18, 25, 38, 74, 366, 442, 471, 475
coded bit, 25
coded modulation, 25
coded modulation library, 229

codeword, 17, 69, 76, 104, 446
BICM, 30, 355, 358
complex Gaussian, 405
dirty-paper coding, 474, 482
error probability, 20
latency, 36
layered architecture, 355
length, 35, 211
memoryless channel, 19
MIMO, 38
minimum-distance decoding, 24
OFDM, 243
power constraint, 81, 85
random coding, 25
single vs multiple, 350, 351, 389, 447, 501
coding rate, 25, 355, 420
coherence bandwidth, 160, 168, 169, 196, 197, 267, 269, 271, 275, 281, 282, 343, 344, 367, 463, 541
coherence distance, 151, 173, 176, 196
coherence time, 153, 157, 634
block fading, 168, 196, 282, 343, 344, 463, 541
feedback, 248, 365, 367
OFDM, 236, 269
pilot overhead, 281, 535
power control, 516
precoding, 516, 574
collaborative upper bound, 454, 469, 484, 492
column rank, 90, 658
column space, 657, 661
complex baseband equivalent, 59, 77
complex baseband equivalent channel, 62
complex envelope, 59
complex Gaussian
differential entropy, 10
dirty-paper coding, 475
distribution, 668
interference, 282
ML estimation, 113
MMSE estimation, 41
signal distribution, 4, 13, 84, 223, 282, 308, 313, 343, 394, 405, 442, 475
Wishart distribution, 670
complex Gaussian matrix, 669
complex Gaussian vector, 669
complex pseudo-baseband equivalent, 63, 77, 144
complexity, 36, 99, 100, 104, 109, 211, 321, 332, 351, 358, 364, 389, 574, 618, 633
compressed sensing, 123
concave function, 685, 688
capacity, 219, 226
mutual information, 241, 308
power allocation, 523
power control, 515
precoding, 449, 469, 479, 480, 515
spectral efficiency, 265, 277, 279, 344
xESM methods, 244
conditional Gaussian distribution, 670
conditional-mean estimator, 40, 41, 44, 50, 276, 277
confidence interval, 598, 612
conjugate beamforming, 620
constant envelope, 86
constellation-constrained capacity, 226
constrained capacity, 226
contours of constant weighted sum spectral efficiency, 424
convergence
almost surely, 250, 259, 265, 310, 336, 408, 674
in distribution, 406, 673
in probability, 113, 673
mean-square sense, 674
convex function, 49, 265, 279, 284, 685, 688
convex hull, 478, 479
convex optimization, 685
BC power allocation, 523, 563, 574
dual-MAC precoding, 466, 480, 487, 526, 561
MAC power control, 517
MAC precoding, 445, 449, 460, 464
pilot overhead, 541, 555
single-user power allocation, 318
single-user precoding, 316
waterfilling, 237
convex relaxation, 572
convex set, 478, 685
convolution, 62, 65, 66, 77, 89, 98–100, 105, 106, 160, 231, 655
convolution matrix, 77, 231
cooperation, 632, 647
Cooperation in Science and Technology, *see* COST
coordination, 629
coprimeness, 90
corner point, 443, 444, 447, 448, 452, 456, 457, 472
correlation, 368, 666
correlation coefficient, 666
correlation distance, 138
correlation matrix, 120, 180, 334, 461, 629, 630, 666
COST-231, 201
Hata, 141
Walfisch–Ikegami, 142
COST-273, 201
Costa, 475
covariance matrix, 666
crest factor, 86
cross-polar discrimination, 185, 186, 339
CSI, 81, 85, 247, 263, 305, 321, 340, 389, 462, 463, 490, 531, 543, 550, 581, 585, 599
CSIR, 247, 248
BC with CSIT, 471, 521, 557
duality, 305, 466, 467, 471, 517
MAC with CSIT, 442, 444, 448
MAC without CSIT, 462, 463
pilot-assisted, 273, 275, 279, 283
SDMA, 452

side information, 267, 268, 270, 490
single-user with CSIT, 247, 305, 431
single-user without CSIT, 252, 260, 279, 321, 389, 390, 431
CSIT, 248
BC, 471, 485, 491, 492, 523, 531, 554, 557, 574, 621
duality, 305, 466, 467, 469, 471, 517, 521
MAC, 442, 444, 513
single-user, 247, 305, 311, 321, 364, 389, 431
cubic metric, 86
cumulative distribution function, *see* CDF
cyclic permutation, 662
cyclic prefix, 77, 98–100, 102, 104, 105, 107, 110, 116, 162, 168, 171, 233, 235, 236, 239, 269, 287
cyclostationarity, 158

D-BLAST, 355
data stream, 81
decimation, 197
decision region, 23
decoder, 18, 23, 31, 35, 38, 84, 472
decoding order, 353, 446–448, 454, 466, 479
decorrelator, 390, 503
degradedness, 477
degrees of freedom, *see* DOF
delay, 63, 64, 66, 76, 88, 91, 93, 95, 171, 193, 365
delay spread, 164, 166, 170, 203, 236, 269, 272, 441, 628
delay-limited capacity, 249
delta function, 63, 145, 147, 155, 160, 163, 335, 653, 677
demodulation, 24, 26, 66
determinant, 662
deterministic equivalent, 335
device-to-device, 204, 647
DFT, 98, 100, 102, 104, 107, 234, 239, 655
diagonal BLAST, *see* D-BLAST
differential entropy, 9, 11, 442
diffraction, 134, 144
digamma function, 312, 670, 681
digital feedback, 248, 364, 366
digital subscriber lines, 649
digital-to-analog conversion, 66, 71, 380, 633
dimensional overloading, 321, 327, 333, 423, 444, 452
Dirac delta function, *see* delta function
direction of arrival, xix
direction of departure, xix
dirty-paper coding, *see* DPC
discrete constellations, 4
BICM, 30, 358
coded modulation, 25
entropy, 8
extended constellation, 472
link adaptation, 228, 263
MMSE estimation, 43
mutual information, 15, 396
precoding, 307, 309, 325, 445, 448, 450, 462, 482
signal-space coding, 25
spatial modulation, 379
discrete Fourier transform, *see* DFT, 185, 191, 655
discrete-time Fourier transform, 239, 655
dispersive channel, 61, 66, 67, 70, 160
diversity, xix, 81, 173, 183, 186, 255, 313, 315, 398, 634
diversity–multiplexing tradeoff, *see* DMT
DMT, 255, 315, 355, 398, 407
DOF
BC, 484, 487, 489–491, 529, 538, 549, 554
MAC, 451, 452, 457, 459, 462–464, 508, 516
multiuser, 433, 492, 519, 583
single-user, 221, 266, 280, 307, 311, 313, 315, 332, 333, 342, 347, 350, 398, 406
Doppler spectrum, 153, 158, 168, 197, 198, 200, 246, 267, 269, 275, 277, 279, 343, 344, 346
Doppler-delay spreading function, 167
downconversion, 59, 61, 63, 75, 77
DPC, 475–477, 480, 482, 490, 493
drop, 201, 598
duality, 305, 417, 464, 471, 472, 479–482, 485, 486, 490, 517, 521, 526, 561, 567, 621, 636
duplexing, 490, 531, 550, 582, 583, 634
dynamic programming, 250

EESM method, 243, 245, 270, 360, 362
eigenfunction, 184
eigenvalue, 47, 403, 659, 660, 662, 675, 676
channel, 305, 306, 308, 309, 323, 337, 468
correlation, 184, 325–327, 330, 331, 394, 492
LOS component, 317
precoded channel, 335, 340, 361, 393, 559
precoder, 376
eigenvalue decomposition, 184, 317, 659, 660
eigenvector, 184, 189, 306, 309, 315, 317, 320, 326, 327, 376, 393, 394, 403, 492, 659, 660
electrically-steerable parasitic array radiator, *see* ESPAR
elevation, 147, 151, 173, 204
empirical eigenvalue distribution, 335, 337, 340
encoder, 17, 25, 38, 84, 472
encoding order, 466, 474, 477, 478, 480
endfire, 173, 182, 185
energy, 73
energy per bit, 212, 218, 226, 227, 251, 261, 265, 310, 326, 330, 342, 350, 397
energy per symbol, 74, 85, 282, 439
entropy, 7, 9
entropy rate, 10
equal spectral efficiencies, 426
equalization, 70, 88, 161, 241
equalizer, 240
ergodic capacity, 259–261, 263, 270, 271, 284, 309, 316, 322, 337, 339, 348, 360, 362, 457, 460

ergodic setting, 247, 249, 258, 271, 309, 316, 439, 456, 460, 486
ergodicity, 21, 72, 85, 123, 155, 677
error function, 683
error probability, 20, 22, 35, 360, 361, 418
ESPAR, 380
estimation, 113
estimation bias, 40, 117, 120
estimation theory, 39
Euclidean distance, 662
Euclidean norm, 662
Euler, 59
Euler–Mascheroni constant, 262, 280, 681
excess antennas, 584, 589, 599, 612, 627, 633
excess bandwidth, 22, 70, 74, 75, 108, 109, 287
excess kurtosis, 668
expectation, 666
exponential correlation model, 183
exponential distribution, 10, 146, 393, 504, 522, 672
exponential integral, 251, 259, 260, 263, 322, 323, 681
exponential-effective SNR mapping, *see* EESM method
extended constellation, 473
extrinsic information, 27, 34

factorial function, 680
fading, 144
fairness, 424, 426–429, 444, 508
Fano, 237
far field, 61, 139, 172
fast-Fourier transform, *see* FFT, 655
favorable propagation, 584, 635
FDD, xxiv, 246, 248, 365, 490, 538, 549, 582, 583, 634
FDMA, 421, 424, 431, 451, 452
feedback, 532
feedback symbols, 490, 491, 549, 557
FFT, 98, 100, 105, 110, 235, 655
filtered noise, 391, 592
filtering, 51
finite impulse response, *see* FIR
finite-length coding, 36, 243, 351, 360, 418
FIR, 76, 79, 88, 93, 95, 99
forward link, 182, 191, 417, 465, 582, 583, 618, 621, 630, 631
forward–reverse SNR ratio, 544, 582, 584
Foschini, xx, 300, 354, 355
Fourier, 653
Fourier codebooks, 373
Fourier matrix, 100, 185, 189, 375, 659
Fourier transform, 58, 59, 62, 63, 98, 154–156, 166, 168, 187, 653, 677
fractional power control, 599, 609, 612, 622
fractional reuse, 603
frame error probability, 21
free probability, 676
free space, 139
frequency band, 421, 428
frequency correlation, 159
frequency offset, 109
frequency selectivity, 159
frequency-division, 421, 424, 433, 451, 452
frequency-division duplexing, *see* FDD
frequency-division multiple access, *see* FDMA
frequency-domain equalization, 98
frequency-flat channel, 76, 79, 97, 107, 110, 115, 118, 172
frequency-flat fading, 159
  BC, 470
  channel estimation, 197, 199, 274
  MAC, 439
  single-user, 246, 248, 253, 263, 264, 277, 313, 341, 343, 346, 362, 389, 390
frequency-selective channel, 76, 79, 230
frequency-selective fading, 160, 190, 269, 271, 360, 389
Friis, 139
Frobenius norm, 79, 662
front-end, 72, 73, 77
full duplexing, xxiv, 246, 248, 364, 490, 550, 634
full-dimension MIMO, 636
full-rank, 92, 98, 113, 190, 305, 307, 311, 331, 376, 452, 484, 485, 658, 660

Gallager, 25
game theory, 572
gamma distribution, 405
gamma function, 146, 260, 286, 314, 369, 670, 672, 680, 681
Gans, xx
Gauss, 40
Gauss–Markov, 158, 167, 274
Gaussian, 223, 240, 282, 668
Gaussian channel, 212, 217, 220, 222
Gaussian noise, *see* noise
Gaussian random process, 676, 677
generalized decision-feedback equalizer, 353
generalized proportional fairness, 427
generalized spatial modulation, 380
geometric programming, 517, 574
Girko, 675
Gold codes, 115
Golden, 354
gradient, 46, 49, 92, 678
Grassmann manifold, 368, 370
Grassmannian codebook, 371, 376
Grassmannian packing, 369
Gray mapping, 8, 28, 32, 34, 357
gross spectral efficiency, 593, 595, 609, 614, 616, 626
group detection, 360
GSM, xv, 161

Hadamard product, 185
hallway, 189

Hankel matrix, 112, 115, 659
hard decision, 24
hardware, 633
Hata, 140
Hermitian matrix, 658, 660, 663, 675
hexagonal network, 586, 598, 603, 605, 612, 623
high-SNR regime, 217
  capacity, 220, 225, 240, 306, 311, 331, 337, 342, 440, 451, 454, 461, 482, 484, 487
  MMSE, 43
  mutual information, 13, 14, 16
  spectral efficiency, 240, 280, 281, 286, 346, 396, 406, 433, 484, 528, 543, 562, 563, 581
Householder codebook, 376
Householder matrix, 376
hybrid precoding, 380, 633
hybrid-ARQ, 37, 271, 273, 289, 361, 364
hypergeometric function
  Gauss, 638, 683
  Kummer confluent, 639, 683
hyperplane, 421, 425, 444, 447

I-MMSE relationship, 45, 241, 308, 450
IDFT, 100, 102, 104, 107, 234, 655
IEEE 802.11ac, xxi
IEEE 802.11ad, xxi, 204
IEEE 802.11ax, xxi
IEEE 802.11ay, xxi, 204
IEEE 802.11n, xxi
IEEE 802.16, xxi, 201
IIR, 88
ill-conditioned, 393
impulse response, 62, 108
in-phase, 4, 5, 31, 32, 43, 59
incomplete gamma function, 260, 286, 314, 680
incremental redundancy, 37
independence, 533
  additive impairments, 632
  channel entries, 120, 184, 188, 189, 200, 321, 407, 508, 675
  channel matrices, 189
  coded bits, 27, 38
  codeword symbols, 19, 22, 30, 73, 264, 341, 342, 475
  fading, 285, 289, 336
  fading blocks, 159
  interference, 283
  MMSE estimation, 532, 585, 592, 597
  multipath components, 145, 150
  noise, 78, 539
  pilot and data symbols, 275
  precoders, 522, 540, 626
  random variables, 8, 10, 12, 137, 666, 669, 674
  ray delays, 164
  real and imaginary parts, 5, 32
  reverse and forward fading, 248
  signal and noise, 13, 16, 44, 45, 222
  signal streams, 240, 300, 305
  signals, 463, 470, 511
  subcarrier fading, 270
  WSSUS, 166
indicator function, 675
indoor environment, 143, 147, 149, 163, 164, 166, 170, 189
infinite impulse response, *see* IIR
infinite-PSK, 5, 13
infinite-QAM, 5, 14
information, 6, 11, 18, 21
information density, 35
information divergence, 11, 12
information stability, 21, 231, 255, 271, 273
INR, 286, 348
intercarrier interference, 171
intercept, 139, 142
interference, 519, 611
  colored, 410
  data upon pilot symbols, 590
  multiantenna, 88, 93, 94, 328, 585
  other-beam, 492
  other-cell, 418, 592, 599, 602, 605, 606, 619, 625, 628, 631
  other-stream, 406, 409, 491, 512
  other-user, 71, 282, 347, 431, 433, 446–449, 453, 474–476, 480, 511, 595, 602, 625, 631
  pilot contamination, 586
interference alignment, 648
interference-to-noise ratio, *see* INR
interleaving, 25, 30, 32, 34
interpolation, 51, 123
intersymbol interference, *see* ISI
intrinsic information, 27
inverse discrete Fourier transform, *see* IDFT
Iospan Wireless, xx
irregular pilot sequences, 589
ISI, 66, 69, 70, 88, 107, 160, 171, 241, 353, 472
isotropic matrix, 341
isotropic signaling, 82
isotropic vector, 265, 341
iterative decoding, 27, 35
iterative waterfilling, 449, 450, 457, 480

Jakes, 148, 151, 153, 155, 182, 191, 198, 201, 273, 277, 279, 344
Jensen's inequality, 260, 268, 284, 313, 321, 331, 537, 538, 688
joint spatial division and multiplexing, 557

k-means clustering, 370
Kailath, xx
Kalman filter, 51
Karhunen–Loève expansion, 184
Karush–Kuhn–Tucker conditions, *see* KKT conditions
Kerdock codebooks, 375
keyhole channel, 189, 317
KKT conditions, 686

Kronecker model, 120, 180, 184, 187, 309, 312, 315, 317, 330, 334–336, 340, 409, 461
Kronecker product, 112, 114, 181, 663
Kullback–Leibler divergence, 11, 12
kurtosis, 86, 261, 265, 310, 327, 328, 667

L-value, 26, 29, 34, 357, 358, 360, 361
Lagrange multipliers, 235, 240, 484, 687
Lagrangian, 237, 483, 527, 570
Landau symbols, 13, 217, 684
Laplace transform, 675
Laplacian, 149, 183, 192, 193
large city, 141
large-dimensional regime, 300, 320, 335, 340, 349, 407, 419, 462, 493, 581, 585, 611, 675
large-scale channel gain, 138, 145, 150, 173, 186, 195, 212, 429, 439, 539
large-scale phenomena, 135, 177, 195, 302
latency, 25, 34, 36, 429, 434, 508
law of large numbers, 276, 336, 585, 673, 674
layered architecture, xx, 354
LDPC code, 25, 30, 36
least mean-squares, 117
least-squares error, 114
least-squares estimation, 114, 122
left singular vector, 661
left unitary invariance, 370, 667
Legendre, 40
likelihood function, 24, 27, 39, 113
likelihood ratio, 26, 29
limited feedback, 364, 366
Linde–Buzo–Gray algorithm, 370
line-of-sight, 134, 143, 146, 155, 172, 174, 176, 179, 202, 635
linear array, 173, 176
linear convolution, 62, 65, 66, 77, 89, 98, 99, 106, 160, 231, 655
linear estimation, 42
linear estimator, 48
linear minimum mean-square error, *see* LMMSE
linear polarization, 186
link adaptation, 26, 243, 245, 248, 270, 271, 306, 308, 355, 360, 370, 410, 420
link budget, 138
list sphere decoder, 357
Lloyd algorithm, 370
LMMSE
  BC, 482
  channel estimation, 196
  dual-MAC, 464, 466
  equalization, 94, 390
  estimation, 48, 49
  MAC, 446–448, 453, 456
  single-user receiver, 352–355, 358, 362, 389, 398, 400, 402, 404–407, 409
local neighborhood, 135
local oscillator, 59, 75, 108
local stationarity, 135
log-likelihood ratio, 26, 27, 29
log-normal, 137, 143
long-term evolution, *see* LTE
low-density parity check code, *see* LDPC code
low-noise amplifier, 93
low-SNR regime, 217
  capacity, 217, 225, 240, 260, 306, 309, 326, 337, 341, 431, 461
  MMSE, 43, 46
  mutual information, 13–16
  spectral efficiency, 240, 278, 281, 285, 346, 396, 406, 431, 433
low-SNR slope, 218, 226, 261, 265, 309, 326, 342, 406, 432, 433
lower incomplete gamma function, 680
lowpass filter, 60, 63, 64, 69
LTE, xv, xxi, 36, 109, 161, 228, 229, 236, 244, 248, 272, 273, 278, 360, 362, 375, 376, 428, 599, 636

machine learning, 204, 628
macrocell, 140, 142, 166, 201
magnitude constraint, 86, 115
Marconi, xix
Marzetta, 583, 602
Marčenko–Pastur, 338, 408, 675
Massey, xxii
massive MIMO, xxi, 123, 136, 173, 177, 196, 341, 343, 389, 419, 462, 490, 502, 510, 551, 581, 583
matched filter, 176, 389, 399, 502, 511, 565, 582, 583, 585, 595, 620
matrix inversion lemma, 50, 399, 400, 511, 567, 663
max-log approximation, 357
maximum a-posteriori, 23, 30
maximum a-posteriori estimation, 39, 42
maximum-eigenvalue eigenvector, 306, 309, 326, 327, 367, 393, 403, 492
maximum likelihood, 23, 113, 358
maximum-likelihood estimation, 40, 42, 113, 358
maximum-ratio combining, 310, 595
maximum-ratio transmission, 310, 565, 620
MCS, 228, 245, 271, 360, 362
mean doubly regular, 339
media-based communication, 380
memoryless channel, 19, 21, 24, 27, 30, 31, 38, 66, 222, 231, 233, 247, 252, 264
mercury/waterfilling, 241, 251, 308, 523
message, 23, 25, 34
microcell, 142, 192, 201
MIESM method, 243, 245, 270, 360, 362
millimeter wave, 204, 380, 633, 647
minimum distance, 5, 16
minimum-distance decoding, 24, 276, 278, 283, 535
minimum energy per bit, 218, 226, 227, 251, 261, 265, 309, 326, 330, 341, 342, 350, 397, 406, 433
minimum mean-square error, *see* MMSE
minimum-norm equalizer, 92, 97

minimum spectral efficiency, 426
MISO, 310, 314, 369, 480, 481
mixing matrix, 82–84
MMSE, 40, 45, 48, 94, 117, 122, 197, 200, 240, 276, 305, 318, 344, 353, 398, 532, 573, 582, 611, 630
MMSE beamformer, 569
MMSE matrix, 44, 46, 49, 118, 318, 399, 630
modified Bessel function, 146, 189, 192, 682
modulation, 25, 66
modulation and coding scheme, *see* MCS
modulo operation, 655
Monte-Carlo, 248, 311, 324, 334, 395, 405, 598, 611, 612, 626, 638
Moore–Penrose pseudoinverse, 92, 94, 391, 503, 521, 565, 663
MU-MISO, 423, 490, 517, 520, 521, 523, 524, 624
MU-SIMO, 423, 441, 502, 504, 505, 517, 593
MU-SISO, 423
multicarrier, 98, 105, 108, 168, 171
multicell, 586, 619, 621, 629
multidimensional symbol mapping, 35
multipath, 61, 380
  fading, 145, 154
  propagation, 144
multiple codewords, 358
multiple-access channel, 353, 417, 418, 431, 439, 464, 501, 517
multiple-input single-output, *see* MISO
multiplexing gain, 255, 315, 398, 460
multiplicative noise, 246, 632
multitone, 105
multiuser detection, xx, 300, 409
multiuser diversity, 429, 439
multiuser eigenmode transmission, 562
multiuser MISO, *see* MU-MISO
multiuser SIMO, *see* MU-SIMO
multiuser SISO, *see* MU-SISO
multivariate impulse response, 78
mutual coupling, 302, 380, 633
mutual information, 12, 25, 313, 343, 348, 351, 442, 460, 534
  BICM, 31, 35
  BPSK, 14
  complex Gaussian signal, 13, 282
  discrete constellations, 15, 16, 226, 229, 308, 325, 450
  Gaussian noise, 13, 282
  I-MMSE relationship, 45
  infinite PSK, 13
  infinite QAM, 14
  information stability, 21
  large-dimensional regime, 315
  link adaptation, 361
  nonsingle-letter formulation, 231, 232
  precoding, 84, 304
  QPSK, 15
  single-letter formulation, 222
mutual-information-effective SNR mapping, *see* MIESM method
mutually unbiased bases, 375
Nakagami fading, 146, 178, 260, 263, 429
narrowband, 58, 159
nat, 7
near–far effect, 448
nearest-neighbor decoding, 24, 276, 278, 283
new radio, *see* NR
noise, 13, 71, 75, 80, 300, 328, 341, 342, 348, 350, 397, 405, 410, 421, 422, 431, 440, 442, 446
  bandwidth, 72, 74
  BICM, 31, 33
  channel law, 19, 20
  enhancement, 102, 398, 525, 632
  equalization, 93, 94, 100, 104
  estimation, 40
  figure, 72, 73, 547
  filtered, 390, 391, 592
  Gaussian, 71, 276, 282, 308, 668
  I-MMSE relationship, 45, 46
  log-likelihood ratio, 29
  ML decoding, 24
  MMSE estimation, 39, 41, 43, 50, 120, 588
  OFDM, 107
  power, 144, 398, 466, 473, 475, 491, 589
  spectral density, 72
NOMA, 423
noncausal MMSE, 51, 123
non-line-of-sight, 134, 143
nonmemoryless, 264, 341
non-orthogonal multiple access, *see* NOMA
nonregular fading, 11, 157, 266
nonregular random process, 11, 51
nonsingle-letter formulation, 20, 21, 231, 264, 341
nonsquare QAM, 5
nonuniform linear array, 173
normalization
  antenna pattern, 173
  capacity, 335
  channel, 75, 79, 139, 145, 177, 179, 185, 230, 300, 302, 439
  power angle spectrum, 147, 149
  power delay profile, 163, 164
  precoder, 81, 467, 470, 479, 567
  receiver, 517
  signal, 341
  SNR, 212
NR, xv, 36, 109, 230, 236, 273, 278, 428, 599
null space, 181, 558, 657, 658, 661
Nyquist, 69
Nyquist criterion, 69, 70, 72
OFDM, 20, 37, 70, 79, 104, 269, 390, 422, 441, 655
  channel estimation, 121, 199
  chip, 107

cyclic prefix, 233
equalization, 107, 241
link adaptation, 361
LTE, 161, 236
multicarrier, 105, 171
multitone, 105, 171
peakedness, 87
power constraint, 85, 620
resource element, 110, 168, 171, 196, 197, 200, 211, 212, 222, 246, 271, 282, 428, 583, 589
symbol, 107, 109, 110, 121, 234–236, 239, 240, 243, 244, 270, 287, 428, 589
symbol period, 107
xESM methods, 361, 362
OFDMA, 422, 423
Okumura–Hata, 140
on–off keying, 6, 264
one-ring model, 191
one-shot BICM, 29, 34, 38, 357
one-step prediction error, 158
opportunistic transmissions, 429
orthogonal frequency-division multiple access, *see* OFDMA
orthogonal frequency-division multiplexing, *see* OFDM
orthogonality principle, 40, 49, 55
outage, 253, 439
outage capacity, 254, 273, 313, 355, 460
outage probability, 253, 273, 313, 460
output SINR, 400, 512, 596, 611, 626
output SNR, 392, 504
over-the-air calibration, 365
oversampling, 72

PAM signal, 472
PAPR, 86, 109, 223
parallel interference cancelation, 618
parallel subchannels, 234, 240, 305, 309, 469, 521, 559
parasitic antenna arrays, 380
PARC, 353, 362
passband, 58
pathloss, 137, 195, 203, 212, 380, 598, 603, 605, 612, 623
pathloss exponent, 139, 142, 638
pathloss intercept, 139, 142
pattern diversity, 186
Paulraj, xx
PDF, 665
peak-to-average power ratio, *see* PAPR
peakedness, 5, 86, 264, 281, 342
pedestrian setting, 153, 163, 170, 278, 279, 344, 345, 541, 542, 548, 549, 555, 582, 616, 627
pentagonal region, 443, 445, 450, 476
per-antenna rate control, *see* PARC
perfect reconstruction, 90
phase, 59
phase invariance, 368
phase noise, 61, 71, 109, 632
phased array, xix, 176
picocell, 143, 147, 201
pilot clustering, 603
pilot contamination, 586, 590, 603, 611, 613, 614, 621, 624
pilot hopping, 603
pilot pollution, 586
pilot power boosting, 280, 282, 347, 464, 550, 585, 589, 618, 627, 636
pilot reuse, 557, 586, 611, 629
pilot reuse cluster, 611, 625, 629
pilot reuse pattern, 603
pilot symbols, 111, 197
common, 196, 246, 273, 282, 343, 410, 463, 531, 536, 543, 551, 557, 585, 618
dedicated, 533, 536, 543, 551, 557, 582, 619, 627
precoded, 347, 533, 536, 543, 551, 557, 582, 619, 627
unprecoded, 531, 536, 543, 551, 557, 585
pilot word, 102
pinhole channel, 189, 317
planar wavefront, 171
PMF, 665
Poincaré, 135
Poisson distribution, 164
Poisson point process, 638
polarization, 185, 339
diversity, 186
mismatch, 138
polyhedron, 444
polytope, 444
positive-definite, 660
positive-semidefinite, 660
post-processing SNR, 392
posterior probability, 39, 41
power, 73, 85, 137
power allocation
BC, 470, 483, 484
frequency-division multiplexing, 422, 432
hardening-based, 622
matrix, 82, 369, 461
outage capacity, 314
parallel subchannels, 237, 469, 521
single-user without CSIT, 318
uniform, 327, 433, 492
waterfilling, 237, 523, 559, 627
power amplifier, 74, 85
power angle spectrum, 147, 183, 380, 557, 628
power azimuth spectrum, 147
power constraint, 466, 480
per-antenna, 85, 302, 440, 469–471, 519, 521, 527, 562
per-codeword, 74, 81, 84, 222, 231, 270, 280, 302, 313, 440, 442, 460, 470, 489, 619, 625

per-symbol, 85, 222, 231, 280, 302, 313, 410, 440, 442, 460, 470, 489, 521, 527, 567, 589, 619
power control
ergodic setting, 250
fractional, 599, 610, 612, 622
MAC, 440, 442, 470, 501, 515–517, 577, 584, 585
near–far effect, 448
single-user, 74, 81, 138, 212, 260, 301, 303, 422
waterfilling, 251
power cosine, 148
power delay profile, 163, 168, 628
power efficiency, 212, 218, 226, 241, 266, 289, 309, 633
power-limited regime, 217
power offset
BC, 485, 486, 488, 529, 538, 563, 574
MAC, 452, 508, 517
single-user, 221, 262, 286, 307, 311, 312, 315, 331–334, 343, 398, 406, 458
power spectral density, 72, 154, 677
power spectrum, 72, 154, 677
pre-log factor, 220
precoding, 37, 81, 98, 335, 362, 440, 444, 492
block-diagonalization, 508, 558, 559
capacity, 305, 316, 321, 326, 331, 390, 449, 461, 479, 480
collaborative upper bound, 454, 484
discrete constellations, 308, 325
duality, 466, 467, 469, 479, 480
LMMSE receiver, 408
matched filter, 585, 619
outage probability, 314
pilot symbols, 347
regularized ZF, 513, 565, 625
semiunitary, 369
statistical beamforming, 492
total channel gain, 302
unitary, 406
ZF, 520, 557
ZF receiver, 394
prediction, 51
prior probability, 39
priority, 424
probability boxes, 195
probability density function, *see* PDF
probability mass function, *see* PMF
product correlation model, *see* Kronecker model
propagation clutter, 134, 139, 148, 182, 191, 492
properness, 6, 15, 16, 43, 227, 241, 261, 310, 330, 667, 669
proportional fairness, 427, 428, 430, 435, 482
PSK
mutual information, 15
signal, 5, 86
spectral efficiency, 226, 240, 325
pulse shaping, 22, 66, 72, 76, 87, 107, 108
pulse-amplitude modulation, *see* PAM
pulse-position modulation, 6
Q-function, 14, 35, 149, 223, 668, 682
QAM
mutual information, 15
signal, 5, 28, 34, 87
spectral efficiency, 226, 240, 263
QPSK
MMSE, 43, 45, 46
mutual information, 15, 450
signal, 5, 35, 38, 43, 82
spectral efficiency, 33, 227, 260, 450
QR decomposition, 661
quadrature, 4, 5, 15, 31–33, 43, 59
quantization, 366, 367
noise, 632
scalar, 475
vector, 370, 475
quasi-static setting, 247, 248, 252, 271, 303, 313, 360, 362, 398, 442, 454, 460, 475
radio front-end, 61
radio propagation, 133
raised-cosine filter, 70, 76, 109
Raleigh, xx
random coding, 25
random matrix theory, 336
random process, 10, 50, 72, 676
random set theory, 195
rank, 92, 179, 189, 300, 305, 307, 311, 315, 321, 331, 333, 367, 375, 403, 452, 658, 660, 661
rank adaptation, 370, 375
rank-deficient, 658
rate control, 228, 360
rate region, 419
ray tracing, 135, 204
Rayleigh distribution, 669, 672
Rayleigh fading
BC block-diagonalization power offset, 563, 564
BC capacity, 487, 488
BC pilot-assisted transmission, 555
BC ZF power offset, 529, 530
BC ZF spectral efficiency, 524, 530
channel matrix, 178, 309
distribution, 145
DMT, 315
interference, 284, 286, 350
large-dimensional capacity, 338
LMMSE spectral efficiency, 404–406, 410
MAC capacity, 456, 459, 461
MAC ZF spectral efficiency, 504, 505, 508, 510
Monte-Carlo simulation, 598
multiuser diversity, 429, 439
outage probability, 313
pilot-assisted transmission, 277, 344
power offset, 311
precoding, 314

ray, 164
single-user capacity, 251, 253, 254, 259, 261, 265, 266, 268, 322, 323, 328, 332, 334, 341
throughput, 249, 272, 362
ZF spectral efficiency, 393, 395
reciprocity, 248, 364, 466, 532, 550, 619, 634
recursive least-squares, 117
reference distance, 139, 142
reference symbols, 111
reflection, 134, 144
regular fading, 11, 157, 158, 267, 343
regular pilot sequences, 589
regular random process, 11, 51
regularization, 98, 102, 120, 565, 624, 627
regularized zero-forcing, 565, 582, 625
regulatory agencies, 85
relative entropy, 11, 12
relay, 647
resource allocation, 428, 429, 439
resource block, 272, 428, 430, 439
resource element, 110
reverse link, 182, 417, 439, 465, 469, 539, 543, 549, 550, 583, 585, 590, 631
Rice fading, 146, 179, 186, 256, 263, 317, 324, 330, 334–336, 340, 394, 405, 409, 429, 439, 461
rich scattering, 179
right singular vector, 661
right unitary invariance, 667
rolloff factor, 70
roof-edge diffraction, 189
root raised-cosine filter, 70
row rank, 658
row space, 657, 661
rural area, 141, 183

saddle point, 276, 282, 308, 535, 565
Saleh–Valenzuela, 164, 193
Salz, xx
sampling, 18, 64–67, 69, 70, 75, 78, 108, 144, 171
sampling offset, 66
sampling period, 108, 157, 197
sampling rate, 70
scalar metric, 424, 435
scattering, 134, 144
scattering function, 168
scheduling, 430, 435
SDMA, 452, 492, 557
sectorization, 492, 557, 634, 636
self-calibration, 365
self-interference, 594, 597
semicircle law, 675, 676
semiunitary matrix, 341, 559
semiunitary precoding, 370
separable correlation model, *see* Kronecker model
shadow fading, 137, 143, 195, 203, 212, 440, 583, 598, 599, 603, 605, 607, 609, 612, 623, 635, 638
shadow fading correlation distance, 138, 196, 635
shadow fading correlation process, 583
Shannon, 3, 4, 6, 11, 12, 17, 18, 22, 25, 218, 237, 253, 254
Shannon bandwidth, 22
shaping gain, 228
SIC, 353, 354, 358, 362, 389, 409, 446–448, 453, 456, 464, 466, 472–477, 480, 482, 618
side information, 246, 252, 267, 274, 342, 389, 490, 491, 534, 582
signal streams, 82
signal-space coding, 25, 30, 33, 34, 38
signal-to-interference ratio, *see* SIR
signal-to-interference-plus-noise ratio, *see* SINR
signal-to-leakage-plus-noise ratio, *see* SLNR
signal-to-noise ratio, *see* SNR
Silverstein, 676
SIMO, 310, 313, 322, 324, 328, 333, 341, 393, 395
single carrier, 100, 108
single-carrier frequency-domain equalization, 98
single-carrier transmission, 240, 241
single codeword, 356
single-input multiple-output, *see* SIMO
single-letter formulation, 20, 22, 27, 36, 38, 222, 223, 252, 264, 276, 282, 300, 439, 470
singular matrix, 662
singular value, 506, 525, 527, 558, 661
singular-value decomposition, *see* SVD
SINR, 197
local-average, 283, 285, 349
per-stream, 352, 400, 402, 404, 408, 512, 517, 566–568, 570, 585, 594, 596, 600, 611, 620, 626
per-subcarrier, 362
SIR, 283, 598, 600, 606, 609, 612, 614, 623, 626
SLNR, 568, 625
SLNR-maximizing precoder, 569
slope, 218, 226, 261, 265, 309, 326, 332, 342, 406, 432
small-scale fading, 144, 429
small-scale phenomena, 135, 177, 195, 302
small-to-medium cities, 141
smart antenna, xix
smoothing, 51
snapshot, 201
SNR, 13
SNR gap, 241
soft-input soft-output, 30
space correlation, 150
space diversity, 173
space selectivity, 146
space-division multiple access, *see* SDMA
space-time coding, 81
sparsity, 123, 628
spatial aliasing, 176
spatial channel model, 201, 245
spatial correlation, 176

spatial DOF, *see* DOF
spatial modulation, 379
spatial multiplexing, xx
special functions, 680
spectral broadening, 157
spectral efficiency, 22, 25, 212, 213
  cyclic prefix, 239
  discrete constellations, 226, 260, 263, 308, 325
  energy efficiency, 217
  excess bandwidth, 75, 109
  finite blocklength, 225
  link adaptation, 228, 245, 362
  per-stream, 353, 362, 394, 404
  power control, 251
  region, 418, 420, 424, 432
  single-user, 240, 241, 265, 266, 275, 278, 280, 281, 318, 324, 335, 340, 343, 409
speed of light, 150, 160, 169
speed of sound, 169
sphere decoder, 357
spherical wavefront, 171
square QAM, 5
square-root raised-cosine filter, 70
staggered pilots, 590
standard Gaussian, 13, 223, 668, 669
Stanford University, xx
stationarity, 11, 20, 21, 30, 38, 51, 72, 135, 145, 155, 158, 166, 168, 223, 250, 677
statistical beamforming, 326, 330, 331, 433, 461, 492
statistical CSI, 246
statistical waterfilling, 320
steering matrix, 82, 300, 304, 308, 314, 316–318, 347, 369, 408, 434, 461, 559
Stieltjes transform, 675, 676
Stirling's formula, 266, 644
stochastic geometry, 638
stochastic modeling, 135
street width, 143
strict-sense stationarity, 11, 20, 21, 30, 51, 135, 145, 155, 168, 223, 250, 677
strong law of large numbers, 674
subcarrier, 104, 105, 110, 235, 243
  channel estimation, 121, 122
  fading, 107, 269, 271, 389
  link adaptation, 243, 244, 271, 361
  number of subcarriers, 85, 168, 270, 428
  parallel subchannels, 234, 235, 240
  power constraint, 281
  resource allocation, 422
  shape, 107, 171
  spacing, 107–109, 162, 236, 272, 428
subspace distance, 368, 372, 373
suburban area, 141, 182, 183
successive interference cancelation, *see* SIC
sum spectral efficiency, 424, 428, 429
  BC, 523, 525, 528, 530, 555, 560–563, 565, 570, 571, 574
  forward link, 622, 626
  MAC, 441, 444, 456, 458, 505, 508, 509, 513
  reverse link, 600, 609, 614
sum-capacity, 424
  BC, 481, 482, 484, 487, 489, 527, 555, 561, 571
  MAC, 443, 446–451, 453, 454, 456–458, 460, 462
superposition, 423, 475–477
superposition coding, 423
support of a distribution, 7, 11, 666
SVD, 82, 232, 304, 389, 449, 450, 467, 558, 661
symbol period, 22, 66, 74, 76, 107, 157, 161, 190, 197, 200
symbol rate, 69, 72, 74
synchronization, 22, 66, 70, 72, 100, 111, 441, 471
TDD, xxiv, 246, 248, 364, 490, 550, 582, 583, 633
TDMA, 420–422, 424, 426, 427, 431, 444, 451, 452, 476
Telatar, xx, 314
telephony, 428
tensor, 180, 184
terrain types, 142
thermal noise, 71
throughput, 36, 229, 244, 245, 249, 287, 360–362, 430
time correlation, 152, 197
time-division, 420, 422, 424, 426, 433, 444, 447, 451, 452, 476
time-division duplexing, *see* TDD
time-division multiple access, *see* TDMA
time–frequency correlation, 167, 168
time–frequency double selectivity, 167
time selectivity, 152, 177
time slot, 420, 422, 423, 428, 430
time-varying transfer function, 171
Toeplitz matrix, 77, 80, 96, 232, 239, 659
Tomlinson–Harashima, 472, 474
total channel gain, 212, 251, 301, 341
trace, 662
training symbols, 111
transmit Wiener filter, 569
treat interference as noise, 276, 283, 344, 418, 535, 592, 593, 611, 620, 625, 630
tree decoder, 358
trellis coded modulation, 25
truncated channel inversion power control, 251
truncated Gaussian, 148
truncated Laplacian, 149, 183, 192, 193, 201
tunnel, 189
turbo BLAST, 360
turbo code, 25, 30, 36, 229
two-ring model, 192
two-slope model, 139
typical urban, 163, 272
UIU model, 183, 339
ultrawideband, 6, 59

uncoded, 25, 354
uncorrelated scattering, 166, 188
uncorrelatedness, 120, 166, 188, 189, 276, 317, 344, 462, 535, 585, 588, 630, 666, 669
underspread fading, 169, 248, 252, 264–270, 273, 278, 321, 327, 333, 340–343, 463, 490, 628
underwater channel, 169, 172
underwater communication, 649
Ungerboeck, 25
uniform circular array, 173
uniform linear array, 173, 176
unit-rank, 189, 367
unitary invariance, 107, 314, 317, 406, 408, 559, 667
unitary matrix, 82, 85, 116, 120, 183, 184, 300, 304, 317, 376, 406, 408, 559, 563, 658, 661, 662
unitary rotation, 658
unitary–independent–unitary model, *see* UIU model
upconversion, 59, 61, 75, 77
upper incomplete gamma function, 680
urban area, 142, 147, 166
user association, 636
user selection, 428, 435, 492, 502, 508, 527, 562, 610, 629, 635, 636

V-BLAST, 354
Valenti, 229
Valenzuela, xx, 164, 193, 354
Varanasi, 353
vector coding, 234, 235, 239, 240, 269
vector perturbation, 474
vector quantization codebook, 370
vectorization, 112, 520
vehicular setting, 163, 170, 196, 268, 277, 279, 344, 345, 463, 541–555, 582, 617
velocity, 152, 156
vertical BLAST, *see* V-BLAST
virtual channel model, 185, 191
visible-light communication, 649
Voiculescu, 676
von-Mises, 149, 192

Walfisch–Ikegami, 141, 142
waterfilling, 237, 241, 250, 251, 305, 306, 320, 442, 449–451, 468, 523–529, 545, 559–562, 627
waveform, 58, 87, 107
waveguide, 189
wavelength, 134, 135, 139, 140, 145, 150, 151, 380, 633
weak law of large numbers, 673
Weichselberger model, 183
weighted sum-capacity, 426
  BC, 477, 479, 480, 486
  MAC, 445, 449, 450, 456, 461
weighted sum spectral efficiency, 424, 430, 435
  BC, 478, 523, 527, 535, 540, 560, 570, 574
  forward link, 627, 628
  MAC, 445, 447, 449, 450, 505, 513
  reverse link, 599
whitening filter, 72
wide-sense stationarity, 135, 145, 155, 166, 168, 677
wide-sense stationary uncorrelated scattering, *see* WSSUS
wideband, 160
Wiener, 51, 94
Wiener filter, 51
Wigner, 675, 676
windowing, 109
WINNER models, 201, 204
Winters, xx
wireless local-area network, *see* WLAN
Wishart, 309, 312, 323, 333, 350, 459, 488, 564, 670
Wishart distribution, 178, 309, 312, 323, 333, 350, 564, 670
Wishart matrix, 309, 312, 323, 333, 350, 459, 488, 564, 670
WLAN, xxi, 143, 201
Woodbury matrix identity, 663
word error probability, 21
WSSUS, 166, 168, 188

X-codes, 308

Z-transform, 88, 89, 655
Zadoff–Chu sequences, 115, 343, 589
zero-forcing, *see* ZF
ZF
  equalizer, 88, 89, 92, 98
  layered architecture, 355
  OFDM, 107
  receiver, 390, 399, 405, 406, 503–505, 508, 509, 513, 516
  single-carrier frequency-domain equalizer, 99, 100, 102
  transmitter, 520, 521, 524, 525, 528, 529, 531, 536, 540, 543, 546, 550, 553, 554, 565, 627